Symbol	Meaning
$\mathbb{R}$	set of real numbers
$\mathbb{R}^+$	set of positive real numbers
$\mathbb{C}$	set of complex numbers
$\overline{z}$	complex conjugate of z
$\mathbb{R}^n$	set of ordered n-tuples of real numbers
$\mathbf{e}_i$	ith standard basis vector in $\mathbb{R}^n$
$\mathbb{C}^n$	set of ordered n-tuples of complex numbers
P_n	set of polynomials with real coefficients of degree $\leq n$
$M_n(\mathbb{R})$	set of $n \times n$ matrices with real elements
$M_{m \times n}(\mathbb{R})$	set of $m \times n$ matrices with real elements
$C[a, b]$	set of continuous functions on the interval $[a, b]$
$C^k(I)$	set of continuous functions with at least k continuous derivatives on I
$V_n(I)$	set of column n-vector functions on I
$\text{span}\{\mathbf{v}_1, \mathbf{v}_2, \ldots, \mathbf{v}_k\}$	linear span of $\{\mathbf{v}_1, \mathbf{v}_2, \ldots, \mathbf{v}_k\}$
$W[f_1, f_2, \ldots, f_k]$	Wronskian of functions $f_1, f_2, \ldots, f_k$
$W[\mathbf{x}_1, \mathbf{x}_2, \ldots, \mathbf{x}_k]$	Wronskian of vector functions $\mathbf{x}_1, \mathbf{x}_2, \ldots, \mathbf{x}_k$
$\dim[V]$	dimension of the vector space V
$[\mathbf{v}]_B$	components of $\mathbf{v}$ relative to ordered B
$P_{C \leftarrow B}$	change of basis matrix from B to C
$\langle , \rangle$	inner product
$\mathbf{a} \cdot \mathbf{b}$	dot product of vectors $\mathbf{a}$ and $\mathbf{b}$
$\mathbf{a} \times \mathbf{b}$	cross product of vectors $\mathbf{a}$ and $\mathbf{b}$
$\mathbf{i}, \mathbf{j}, \mathbf{k}$	unit vectors pointing along coordinate axes in 3-space
$\mathbf{0}$	zero vector
$-\mathbf{v}$	additive inverse of $\mathbf{v}$
$||\mathbf{v}||$	norm of the vector $\mathbf{v}$
$\mathbf{P}(\mathbf{w}, \mathbf{v})$	orthogonal projection of $\mathbf{w}$ on $\mathbf{v}$
$W^{\perp}$	orthogonal complement of the subspace W
e^{At}	matrix exponential function
$D : C^1(I) \to C^0(I)$	derivative operator
$P(D)$	polynomial differential operator
$K(x, t)$	Green's function
$X(t)$	fundamental matrix
$X_0(t)$	transition matrix
$J(x, y)$	Jacobian matrix
$L[f]$	Laplace transform of f
$L^{-1}[F]$	inverse Laplace transform of F
$u_a(t)$	Heaviside (unit) step function
$\delta(t - a)$	Dirac-delta function
$f * g$	convolution product of f and g
$\Gamma(p)$	gamma function
$J_p(x)$	Bessel function of the first kind of order p

BASIC INTEGRALS

Function $F(x)$	**Integral** $\int F(x)\,dx$
$x^n \quad (n \neq -1)$	$\dfrac{1}{n+1}x^{n+1} + c$
x^{-1}	$\ln|x| + c$
$e^{ax} \quad (a \neq 0)$	$\dfrac{1}{a}e^{ax} + c$
$\sin x$	$-\cos x + c$
$\cos x$	$\sin x + c$
$\tan x$	$-\ln|\cos x| + c$
$\cot x$	$\ln|\sin x| + c$
$\sec x$	$\ln|\sec x + \tan x| + c$
$\csc x$	$\ln|\csc x - \cot x| + c$
$e^{ax}\sin bx$	$\dfrac{1}{a^2+b^2}e^{ax}(a\sin bx - b\cos bx) + c$
$e^{ax}\cos bx$	$\dfrac{1}{a^2+b^2}e^{ax}(a\cos bx + b\sin bx) + c$
$\ln x$	$x\ln x - x + c$
$\dfrac{1}{a^2+x^2}$	$\dfrac{1}{a}\tan^{-1}(x/a) + c$
$\dfrac{1}{\sqrt{a^2-x^2}} \quad (a > 0)$	$\sin^{-1}(x/a) + c$
$\dfrac{1}{\sqrt{a^2+x^2}}$	$\ln|x + \sqrt{a^2+x^2}| + c$
$\dfrac{f'(x)}{f(x)}$	$\ln|f(x)| + c$
$e^{u(x)}\dfrac{du}{dx}$	$e^{u(x)} + c$

SOME SOLUTION TECHNIQUES FOR $y' = f(x, y)$

Type	**Standard Form**	**Technique**
Separable Equation (Section 1.4)	$p(y)y' = q(x)$	Separate the variables and integrate directly: $$\int p(y)dy = \int q(x)dx$$
First-Order Linear Equation (Section 1.6)	$y' + p(x)y = q(x)$	Rewrite as $\dfrac{d}{dx}(I \cdot y) = qI$, where $I = e^{\int p(x)dx}$, and integrate with respect to x
First-Order Homogeneous (Section 1.8)	$y' = f(x, y)$ with f homogeneous of degree zero: $f(tx, ty) = f(x, y)$	Change variables: $y = xV(x)$ and reduce to a separable equation
Bernoulli Equation (Section 1.8)	$y' + p(x)y = q(x)y^n$	Divide by y^n and make the change of variables $u = y^{1-n}$ and reduce to a linear equation
Exact Equation (Section 1.9)	$M(x, y)dx + N(x, y)dy = 0$, with $M_y = N_x$	Solution is $\phi(x, y) = c$, where ϕ is determined by integrating $\phi_x = M$, $\phi_y = N$

THIRD EDITION

Differential Equations and Linear Algebra

Stephen W. Goode

and

Scott A. Annin

California State University, Fullerton

PEARSON

Prentice Hall

Upper Saddle River, NJ 07458

Library of Congress Cataloging-in-Publication Data

Goode, Stephen W., 1957–
  Differential equations and linear algebra. –3rd. ed. / Stephen W. Goode, Scott A. Annin.
      p. cm.
    Includes index
    ISBN 0–13–045794–9
    1. Differential equations. 2. Algebras, Linear. I. Annin. II. Title.
QA371.G644   2007
515'.35—dc22   2006051573

Vice President and Editorial Director: *Marcia J. Horton*
Senior Editor: *Holly Stark*
Editorial Assistant: *Jennifer Lonschein*
Senior Managing Editor: *David A. George*
Production Editor: *Wendy Kopf*
Creative Director: *Jayne Conte*
Cover Designer: *Bruce Kenselaar*
Art Editor: *Gregory Dulles*
Manufacturing Manager: *Alexis Heydt-Long*
Manufacturing Buyer: *Lisa McDowell*
Executive Marketing Manager: *Tim Galligan*

© 2007 Pearson Education, Inc.
Pearson Prentice Hall
Pearson Education, Inc.
Upper Saddle River, NJ 07458

The author and publisher of this book have used their best efforts in preparing this book. These efforts include the development, research, and testing of the theories and programs to determine their effectiveness. The author and publisher make no warranty of any kind, expressed or implied, with regard to these programs or the documentation contained in this book. The author and publisher shall not be liable in any event for incidental or consequential damages in connection with, or arising out of, the furnishing, performance, or use of these programs.

Printed in the United States of America

10 9 8 7 6 5 4 3 2 1

ISBN 0-13-045794-9

Pearson Education, Inc., *Upper Saddle River, New Jersey*
Pearson Education Ltd., *London*
Pearson Education Australia Pty. Ltd., *Sydney*
Pearson Education Singapore, Pte. Ltd.
Pearson Education North Asia Ltd., *Hong Kong*
Pearson Education Canada, Inc., *Toronto*
Pearson Educación de Mexico, S.A. de C.V.
Pearson Education—Japan, *Tokyo*
Pearson Education—Malaysia, Pte. Ltd.

S. W. Goode dedicates this book to Megan and Tobi

S. A. Annin dedicates this book to Arthur and Juliann, the best parents anyone could ask for

Contents

Preface

In *Differential Equations and Linear Algebra*, Third Edition, the material on differential equations and linear algebra required in many sophomore courses for mathematics, science, and engineering majors is introduced. In writing this text we have endeavored to develop an appreciation for the power of the general vector space framework in formulating and solving linear problems. We have made every effort to lay forth the subject as we ourselves would naturally teach it, using an abundance of examples and illustrations throughout the exposition, but not at the expense of a deliberate, rigorous treatment. Our aim has been to present the material so that it is accessible to the student who has successfully completed three semesters of calculus, and it is definitely the intention that the student read the text, not just the examples. Almost all results are proved in detail. However, it is certainly possible to by-pass many of the proofs and use the text in a more problem solving based setting. Such an approach to the course could potentially incorporate some form of technology (computer algebra system (CAS) or graphing calculator) and there are many instances in the text where the power of technology is illustrated using the CAS Maple. Furthermore, a large majority of the exercise sets have problems that require some form of technology for their solution. These problems are designated with a ◇.

As with the previous editions of this text, in developing the third edition we have kept maximum flexibility of the material in mind. In so doing, the text can effectively accommodate the different emphases that can be placed in a combined differential equations and linear algebra course, the varying backgrounds of students who enroll in this type of course, and the fact that different institutions have different credit values for such a course. The whole text can be covered in a five credit-hour course. For courses with a lower credit-hour value, some selectivity will have to be exercised. For example, in the differential equations part of the text, much (or all) of Chapter 1 may be omitted since most students will have seen many of these topics in an earlier calculus course, and the remainder of the text does not depend on the techniques introduced in this chapter. In addition, Sections 6.4, 6.8, and 6.9 could be omitted, and, depending on the goals of the course, Sections 6.5 and 6.6 could either be de-emphasized or omitted completely. Similar remarks apply to Sections 7.7–7.10. The core material in linear algebra is given in Sections 2.1–2.6, 3.4 (for instructors who wish to de-emphasize the determinant), 4.1–4.6, 4.8, 4.11, 4.12, 5.1, 5.3, 5.4, and 5.6–5.8. At California State University, Fullerton we have a four credit-hour course for sophomores that is based around the material in Chapters 1-7.

Major Changes in the Third Edition

Almost all sections of the text have been altered to improve the clarity of presentation. Other significant changes within the text are listed below.

1. At the end of each section in the text a summary of key terms, expected skills, and a true/false review are given. Nearly 600 true/false review items are included in the text.

2. Most exercise sets have been enlarged. Over 2600 problems are now contained within the text.

3. Every chapter concludes with a review section that includes a set of review problems. Some chapters also contain project ideas for students interested in deeper applications of the material.

4. The following reorganization of material has been incorporated:

    (a) The material on second-order linear differential equations has been moved into the chapter on general nth order differential equations, Chapter 6.

    (b) Matrix functions are now introduced in Chapter 2.

    (c) The matrix exponential function is now introduced in the linear transformations chapter, Chapter 5.

5. The following new sections have been included:

    (a) Sections 2.8 and 4.10 keep track of the many characterizations of the invertibility of an $n \times n$ matrix.

    (b) Section 4.7 - Change of Basis - this section introduces the idea of the change of basis matrix, and how components of vectors relative to different bases are related.

    (c) Section 5.5 - The Matrix of a Linear Transformation - this section illustrates how an arbitrary linear transformation between finite-dimensional vector spaces can be represented by a matrix, once a basis for each vector space has been specified, and shows how linear transformation concepts can be described in terms of matrix algebra.

    (d) Section 5.11 - Jordan Canonical Forms - this gives an elementary introduction to the Jordan canonical form of an $n \times n$ matrix.

Acknowledgments

We would like to acknowledge the thoughtful input from the following reviewers of the third edition: Paul M. Yun, El Camino College; Jianzhong Su, University of Texas at Arlington; Lance Littlejohn, Utah State University; Donatella Danielli Garofalo, Purdue University; and Barbara Shipman, University of Texas at Arlington.

All of their comments were considered carefully in the preparation of the text. Bill Campbell, III and Reginald Jackman have also provided us with several suggestions that have been incorporated into this edition.

S.W. Goode: I would like to thank my wife, Christina, and daughters, Megan and Tobi, for their continued support, encouragement, and understanding throughout the development of this project.

S.A. Annin: I would especially like to thank my parents, Arthur and Juliann Annin, for their endless love and encouragement, as well as the many students over the years who have enriched my professional life and inspired me to be the best instructor I can possibly be.

First-Order Differential Equations

Among all of the mathematical disciplines the theory of differential equations is the most important. It furnishes the explanation of all those elementary manifestations of nature which involve time. — Sophus Lie

1.1 How Differential Equations Arise

In this section we will introduce the idea of a differential equation through the mathematical formulation of a variety of problems. We then use these problems throughout the chapter to illustrate the applicability of the techniques introduced.

Newton's Second Law of Motion

Newton's second law of motion states that, for an object of constant mass m, the sum of the applied forces acting on the object is equal to the mass of the object multiplied by the acceleration of the object. If the object is moving in one dimension under the influence of a force F, then the mathematical statement of this law is

$$m\frac{dv}{dt} = F, \tag{1.1.1}$$

where $v(t)$ denotes the velocity of the object at time t. We let $y(t)$ denote the displacement of the object at time t. Then, using the fact that velocity and displacement are related via

$$v = \frac{dy}{dt},$$

we can write (1.1.1) as

$$m\frac{d^2y}{dt^2} = F. \tag{1.1.2}$$

This is an example of a **differential equation**, so called because it involves *derivatives* of the unknown function $y(t)$.

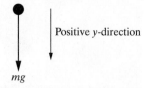

Positive y-direction

mg

Figure 1.1.1: Object falling under the influence of gravity.

Gravitational Force: As a specific example, consider the case of an object falling freely under the influence of gravity (see Figure 1.1.1). In this case the only force acting on the object is $F = mg$, where g denotes the (constant) acceleration due to gravity. Choosing the positive y-direction as downward, it follows from Equation (1.1.2) that the motion of the object is governed by the differential equation

$$m\frac{d^2y}{dt^2} = mg, \qquad (1.1.3)$$

or equivalently,

$$\frac{d^2y}{dt^2} = g.$$

Since g is a (positive) constant, we can integrate this equation to determine $y(t)$. Performing one integration yields

$$\frac{dy}{dt} = gt + c_1,$$

where c_1 is an arbitrary integration constant. Integrating once more with respect to t, we obtain

$$y(t) = \frac{1}{2}gt^2 + c_1t + c_2, \qquad (1.1.4)$$

where c_2 is a second integration constant. We see that the differential equation has an infinite number of solutions parameterized by the constants c_1 and c_2. In order to uniquely specify the motion, we must augment the differential equation with initial conditions that specify the initial position and initial velocity of the object. For example, if the object is released at $t = 0$ from $y = y_0$ with a velocity v_0, then, in addition to the differential equation, we have the initial conditions

$$y(0) = y_0, \qquad \frac{dy}{dt}(0) = v_0. \qquad (1.1.5)$$

These conditions must be imposed on the solution (1.1.4) in order to determine the values of c_1 and c_2 that correspond to the particular problem under investigation. Setting $t = 0$ in (1.1.4) and using the first initial condition from (1.1.5), we find that

$$y_0 = c_2.$$

Substituting this into Equation (1.1.4), we get

$$y(t) = \frac{1}{2}gt^2 + c_1t + y_0. \qquad (1.1.6)$$

In order to impose the second initial condition from (1.1.5), we first differentiate Equation (1.1.6) to obtain

$$\frac{dy}{dt} = gt + c_1.$$

Consequently the second initial condition in (1.1.5) requires

$$c_1 = v_0.$$

From (1.1.6), it follows that the position of the object at time t is

$$y(t) = \frac{1}{2}gt^2 + v_0 t + y_0.$$

The differential equation (1.1.3) together with the initial conditions (1.1.5) is an example of an **initial-value problem**.

Spring Force: As a second application of Newton's law of motion, consider the spring–mass system depicted in Figure 1.1.2, where, for simplicity, we are neglecting frictional and external forces. In this case, the only force acting on the mass is the restoring force (or spring force), F_s, due to the displacement of the spring from its equilibrium (unstretched) position. We use Hooke's law to model this force:

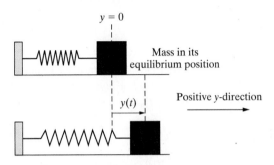

Figure 1.1.2: A simple harmonic oscillator.

Hooke's Law: The restoring force of a spring is directly proportional to the displacement of the spring from its equilibrium position and is directed toward the equilibrium position.

If $y(t)$ denotes the displacement of the spring from its equilibrium position at time t (see Figure 1.1.2), then according to Hooke's law, the restoring force is

$$F_s = -ky,$$

where k is a positive constant called the **spring constant**. Consequently, Newton's second law of motion implies that the motion of the spring–mass system is governed by the differential equation

$$m\frac{d^2y}{dt^2} = -ky,$$

which we write in the equivalent form

$$\frac{d^2y}{dt^2} + \omega^2 y = 0, \tag{1.1.7}$$

where $\omega = \sqrt{k/m}$. At present we cannot solve this differential equation. However, we leave it as an exercise (Problem 7) to verify by direct substitution that

$$y(t) = A\cos(\omega t - \phi)$$

is a solution to the differential equation (1.1.7), where A and ϕ are constants (determined from the initial conditions for the problem). We see that the resulting motion is periodic with amplitude A. This is consistent with what we might expect physically, since no frictional forces or external forces are acting on the system. This type of motion is referred to as **simple harmonic motion**, and the physical system is called a **simple harmonic oscillator**.

Newton's Law of Cooling

We now build a mathematical model describing the cooling (or heating) of an object. Suppose that we bring an object into a room. If the temperature of the object is hotter than that of the room, then the object will begin to cool. Further, we might expect that the major factor governing the rate at which the object cools is the temperature difference between it and the room.

Newton's Law of Cooling: The rate of change of temperature of an object is proportional to the temperature difference between the object and its surrounding medium.

To formulate this law mathematically, we let $T(t)$ denote the temperature of the object at time t, and let $T_m(t)$ denote the temperature of the surrounding medium. Newton's law of cooling can then be expressed as the differential equation

$$\frac{dT}{dt} = -k(T - T_m), \tag{1.1.8}$$

where k is a constant. The minus sign in front of the constant k is traditional. It ensures that k will always be positive.[1] After we study Section 1.4, it will be easy to show that, when T_m is constant, the solution to this differential equation is

$$T(t) = T_m + ce^{-kt}, \tag{1.1.9}$$

where c is a constant (see also Problem 12). Newton's law of cooling therefore predicts that as t approaches infinity ($t \to \infty$), the temperature of the object approaches that of the surrounding medium ($T \to T_m$). This is certainly consistent with our everyday experience (see Figure 1.1.3).

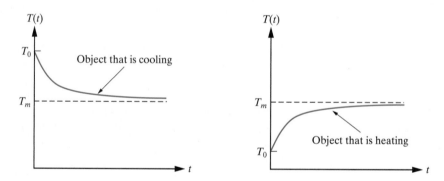

Figure 1.1.3: According to Newton's law of cooling, the temperature of an object approaches room temperature exponentially.

The Orthogonal Trajectory Problem

Next we consider a geometric problem that has many interesting and important applications. Suppose

$$F(x, y, c) = 0 \tag{1.1.10}$$

---

[1] If $T > T_m$, then the object will cool, so that $dT/dt < 0$. Hence, from Equation (1.1.8), k must be positive. Similarly, if $T < T_m$, then $dT/dt > 0$, and once more Equation (1.1.8) implies that k must be positive.

defines a family of curves in the xy-plane, where the constant c labels the different curves. For instance, the equation

$$x^2 + y^2 - c = 0$$

describes a family of concentric circles with center at the origin, whereas

$$-x^2 + y - c = 0$$

describes a family of parabolas that are vertical shifts of the standard parabola $y = x^2$.

We assume that every curve in the family $F(x, y, c) = 0$ has a well-defined tangent line at each point. Associated with this family is a second family of curves, say,

$$G(x, y, k) = 0, \tag{1.1.11}$$

with the property that whenever a curve from the family (1.1.10) intersects a curve from the family (1.1.11), it does so at right angles.[2] We say that the curves in the family (1.1.11) are **orthogonal trajectories** of the family (1.1.10), and vice versa. For example, from elementary geometry, it follows that the lines $y = kx$ in the family $G(x, y, k) = y - kx = 0$ are orthogonal trajectories of the family of concentric circles $x^2 + y^2 = c^2$. (See Figure 1.1.4.)

Orthogonal trajectories arise in various applications. For example, a family of curves and its orthogonal trajectories can be used to define an orthogonal coordinate system in the xy-plane. In Figure 1.1.4 the families $x^2 + y^2 = c^2$ and $y = kx$ are the coordinate curves of a polar coordinate system (that is, the curves $r = $ constant and $\theta = $ constant, respectively). In physics, the lines of electric force of a static configuration are the orthogonal trajectories of the family of equipotential curves. As a final example, if we consider a two-dimensional heated plate, then the heat energy flows along the orthogonal trajectories to the constant-temperature curves (isotherms).

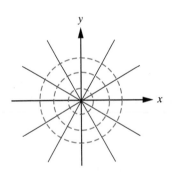

Figure 1.1.4: The family of curves $x^2 + y^2 = c^2$ and the orthogonal trajectories $y = kx$.

Statement of the Problem: Given the equation of a family of curves, find the equation of the family of orthogonal trajectories.

Mathematical Formulation: We recall that curves that intersect at right angles satisfy the following:

The product of the slopes[3] at the point of intersection is -1.

Thus if the given family $F(x, y, c) = 0$ has slope $m_1 = f(x, y)$ at the point (x, y), then the slope of the family of orthogonal trajectories $G(x, y, k) = 0$ is $m_2 = -1/f(x, y)$, and therefore the differential equation that determines the orthogonal trajectories is

$$\frac{dy}{dx} = -\frac{1}{f(x, y)}.$$

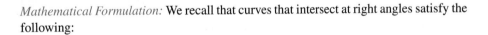

---

[2]That is, the tangent lines to each curve are perpendicular at any point of intersection.

[3]By the slope of a curve at a given point, we mean the slope of the tangent line to the curve at that point.

Example 1.1.1   Determine the equation of the family of orthogonal trajectories to the curves with equation

$$y^2 = cx. \tag{1.1.12}$$

Solution: According to the preceding discussion, the differential equation determining the orthogonal trajectories is

$$\frac{dy}{dx} = -\frac{1}{f(x, y)},$$

where $f(x, y)$ denotes the slope of the given family at the point (x, y). To determine $f(x, y)$, we differentiate Equation (1.1.12) implicitly with respect to x to obtain

$$2y\frac{dy}{dx} = c. \tag{1.1.13}$$

We must now eliminate c from the previous equation to obtain an expression that gives the slope at the point (x, y). From Equation (1.1.12) we have

$$c = \frac{y^2}{x},$$

which, when substituted into Equation (1.1.13), yields

$$\frac{dy}{dx} = \frac{y}{2x}.$$

Consequently, the slope of the given family at the point (x, y) is

$$f(x, y) = \frac{y}{2x},$$

so that the orthogonal trajectories are obtained by solving the differential equation

$$\frac{dy}{dx} = -\frac{2x}{y}.$$

A key point to notice is that we cannot solve this differential equation by simply integrating with respect to x, since the function on the right-hand side of the differential equation depends on both x and y. However, multiplying by y, we see that

$$y\frac{dy}{dx} = -2x,$$

or equivalently,

$$\frac{d}{dx}\left(\frac{1}{2}y^2\right) = -2x.$$

Since the right-hand side of this equation depends only on x, whereas the term on the left-hand side is a derivative with respect to x, we can integrate both sides of the equation with respect to x to obtain

$$\frac{1}{2}y^2 = -x^2 + c_1,$$

which we write as

$$2x^2 + y^2 = k, \tag{1.1.14}$$

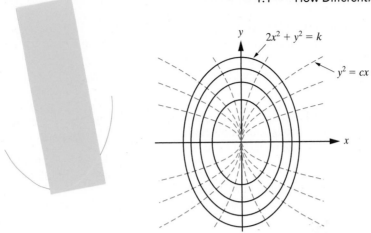

Figure 1.1.5: The family of curves $y^2 = cx$ and its orthogonal trajectories $2x^2 + y^2 = k$.

where $k = 2c_1$. We see that the curves in the given family (1.1.12) are parabolas, and the orthogonal trajectories (1.1.14) are a family of ellipses. This is illustrated in Figure 1.1.5.

$\square$

Exercises for 1.1

Key Terms

Differential equation, Initial conditions, Initial-value problem, Newton's second law of motion, Hooke's law, Spring constant, Simple harmonic motion, Simple harmonic oscillator, Newton's law of cooling, Orthogonal trajectories.

Skills

- Given a differential equation, be able to check whether or not a given function $y = f(x)$ is indeed a solution to the differential equation.

- Be able to find the distance, velocity, and acceleration functions for an object moving freely under the influence of gravity.

- Be able to determine the motion of an object in a spring–mass system with no frictional or external forces.

- Be able to describe qualitatively how the temperature of an object changes as a function of time according to Newton's law of cooling.

- Be able to find the equation of the orthogonal trajectories to a given family of curves. In simple geometric cases, be prepared to provide rough sketches of some representative orthogonal trajectories.

True-False Review

For Questions 1–11, decide if the given statement is **true** or **false**, and give a brief justification for your answer. If true, you can quote a relevant definition or theorem from the text. If false, provide an example, illustration, or brief explanation of why the statement is false.

1. A differential equation for a function $y = f(x)$ must contain the first derivative $y' = f'(x)$.

2. The numerical values $y(0)$ and $y'(0)$ accompanying a differential equation for a function $y = f(x)$ are called initial conditions of the differential equation.

3. The relationship between the velocity and the acceleration of an object falling under the influence of gravity can be expressed mathematically as a differential equation.

4. A sketch of the height of an object falling freely under the influence of gravity as a function of time takes the shape of a parabola.

5. Hooke's law states that the restoring force of a spring is directly proportional to the displacement of the spring from its equilibrium position and is directed in the direction of the displacement from the equilibrium position.

6. If room temperature is 70°F, then an object whose temperature is 100°F at a particular time cools faster at that time than an object whose temperature at that time is 90°F.

7. According to Newton's law of cooling, the temperature of an object eventually becomes the same as the temperature of the surrounding medium.

8. A hot cup of coffee that is put into a cold room cools more in the first hour than the second hour.

9. At a point of intersection of a curve and one of its orthogonal trajectories, the slopes of the two curves are reciprocals of one another.

10. The family of orthogonal trajectories for a family of parallel lines is another family of parallel lines.

11. The family of orthogonal trajectories for a family of circles that are centered at the origin is another family of circles centered at the origin.

Problems

1. An object is released from rest at a height of 100 meters above the ground. Neglecting frictional forces, the subsequent motion is governed by the initial-value problem

$$\frac{d^2y}{dt^2} = g, \qquad y(0) = 0, \qquad \frac{dy}{dt}(0) = 0,$$

where $y(t)$ denotes the displacement of the object from its initial position at time t. Solve this initial-value problem and use your solution to determine the time when the object hits the ground.

2. A five-foot-tall boy tosses a tennis ball straight up from the level of the top of his head. Neglecting frictional forces, the subsequent motion is governed by the differential equation

$$\frac{d^2y}{dt^2} = g.$$

If the object hits the ground 8 seconds after the boy releases it, find

(a) the time when the tennis ball reaches its maximum height.

(b) the maximum height of the tennis ball.

3. A pyrotechnic rocket is to be launched vertically upward from the ground. For optimal viewing, the rocket should reach a maximum height of 90 meters above the ground. Ignore frictional forces.

(a) How fast must the rocket be launched in order to achieve optimal viewing?

(b) Assuming the rocket is launched with the speed determined in part (a), how long after it is launched will it reach its maximum height?

4. Repeat Problem 3 under the assumption that the rocket is launched from a platform 5 meters above the ground.

5. An object thrown vertically upward with a speed of 2 m/s from a height of h meters takes 10 seconds to reach the ground. Set up and solve the initial-value problem that governs the motion of the object, and determine h.

6. An object released from a height h meters above the ground with a vertical velocity of v_0 m/s hits the ground after t_0 seconds. Neglecting frictional forces, set up and solve the initial-value problem governing the motion, and use your solution to show that

$$v_0 = \frac{1}{2t_0}(2h - gt_0^2).$$

7. Verify that $y(t) = A\cos(\omega t - \phi)$ is a solution to the differential equation (1.1.7), where A, ω, and ϕ are constants with A and ω nonzero. Determine the constants A and ϕ (with $|\phi| < \pi$ radians) in the particular case when the initial conditions are

$$y(0) = a, \qquad \frac{dy}{dt}(0) = 0.$$

8. Verify that

$$y(t) = c_1 \cos \omega t + c_2 \sin \omega t$$

is a solution to the differential equation (1.1.7). Show that the amplitude of the motion is

$$A = \sqrt{c_1^2 + c_2^2}.$$

9. Verify that, for $t > 0$, $y(t) = \ln t$ is a solution to the differential equation

$$2\left(\frac{dy}{dt}\right)^3 = \frac{d^3y}{dt^3}.$$

10. Verify that $y(x) = x/(x+1)$ is a solution to the differential equation

$$y + \frac{d^2 y}{dx^2} = \frac{dy}{dx} + \frac{x^3 + 2x^2 - 3}{(1+x)^3}.$$

11. Verify that $y(x) = e^x \sin x$ is a solution to the differential equation

$$2y \cot x - \frac{d^2 y}{dx^2} = 0.$$

12. By writing Equation (1.1.8) in the form

$$\frac{1}{T - T_m} \frac{dT}{dt} = -k$$

and using $u^{-1} \dfrac{du}{dt} = \dfrac{d}{dt} (\ln u)$, derive (1.1.9).

13. A glass of water whose temperature is 50°F is taken outside at noon on a day whose temperature is constant at 70°F. If the water's temperature is 55°F at 2 p.m., do you expect the water's temperature to reach 60°F before 4 p.m. or after 4 p.m.? Use Newton's law of cooling to explain your answer.

14. On a cold winter day (10°F), an object is brought outside from a 70°F room. If it takes 40 minutes for the object to cool from 70°F to 30°F, did it take more or less than 20 minutes for the object to reach 50°F? Use Newton's law of cooling to explain your answer.

For Problems 15–20, find the equation of the orthogonal trajectories to the given family of curves. In each case, sketch some curves from each family.

15. $x^2 + 4y^2 = c$.

16. $y = c/x$.

17. $y = cx^2$.

18. $y = cx^4$.

19. $y^2 = 2x + c$.

20. $y = ce^x$.

For Problems 21–24, m denotes a fixed nonzero constant, and c is the constant distinguishing the different curves in the given family. In each case, find the equation of the orthogonal trajectories.

21. $y = mx + c$.

22. $y = cx^m$.

23. $y^2 + mx^2 = c$.

24. $y^2 = mx + c$.

25. We call a coordinate system (u, v) **orthogonal** if its coordinate curves (the two families of curves $u = $ constant and $v = $ constant) are orthogonal trajectories (for example, a Cartesian coordinate system or a polar coordinate system). Let (u, v) be orthogonal coordinates, where $u = x^2 + 2y^2$, and x and y are Cartesian coordinates. Find the Cartesian equation of the v-coordinate curves, and sketch the (u, v) coordinate system.

26. Any curve with the property that whenever it intersects a curve of a given family it does so at an angle $a \neq \pi/2$ is called an **oblique trajectory** of the given family. (See Figure 1.1.6.) Let m_1 (equal to $\tan a_1$) denote the slope of the required family at the point (x, y), and let m_2 (equal to $\tan a_2$) denote the slope of the given family. Show that

$$m_1 = \frac{m_2 - \tan a}{1 + m_2 \tan a}.$$

[**Hint**: From Figure 1.1.6, $\tan a_1 = \tan(a_2 - a)$. Thus, the equation of the family of oblique trajectories is obtained by solving

$$\frac{dy}{dx} = \frac{m_2 - \tan a}{1 + m_2 \tan a}.]$$

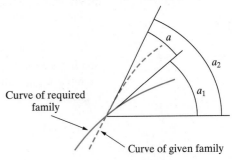

$m_1 = \tan a_1 = $ slope of required family
$m_2 = \tan a_2 = $ slope of given family

Curve of required family

Curve of given family

Figure 1.1.6: Oblique trajectories intersecting at an angle a.

1.2   Basic Ideas and Terminology

In the preceding section we have used some applied problems to illustrate how differential equations arise. We now undertake to formalize mathematically several ideas introduced through these examples. We begin with a very general definition of a differential equation.

DEFINITION   1.2.1

A **differential equation** is an equation involving one or more derivatives of an unknown function.

Example 1.2.2   The following are all differential equations:

$$\text{(a)} \ \frac{dy}{dx} + y = x^2, \qquad \text{(b)} \ \frac{d^2 y}{dx^2} = -k^2 y, \qquad \text{(c)} \ \frac{d^3 y}{dx^3} + \left(\frac{d^2 y}{dx^2}\right)^5 + \cos x = 0,$$

$$\text{(d)} \ \sin\left(\frac{dy}{dx}\right) + \tan^{-1} y = 1, \qquad \text{(e)} \ \phi_{xx} + \phi_{yy} - \phi_x = e^x + x \sin y.$$

The differential equations occurring in (a) through (d) are called **ordinary differential equations**, since the unknown function $y(x)$ depends only on one variable, x. In (e), the unknown function $\phi(x, y)$ depends on more than one variable; hence the equation involves partial derivatives. Such a differential equation is called a **partial differential equation**. In this text we consider only ordinary differential equations. □

We now introduce some more definitions and terminology.

DEFINITION   1.2.3

The order of the highest derivative occurring in a differential equation is called the **order** of the differential equation.

In Example 1.2.2, (a) has order 1, (b) has order 2, (c) has order 3, and (d) has order 1. If we look back at the examples from the previous section, we see that problems formulated using Newton's second law of motion will always be governed by a second-order differential equation (for the position of the object). Indeed, second-order differential equations play a very fundamental role in applied problems, although differential equations of other orders also arise. For example, the differential equation obtained from Newton's law of cooling is a first-order differential equation, as is the differential equation for determining the orthogonal trajectories to a given family of curves. As another example, we note that under certain conditions, the deflection, $y(x)$, of a horizontal beam is governed by the fourth-order differential equation

$$\frac{d^4 y}{dx^4} = F(x)$$

for an appropriate function $F(x)$.

Any differential equation of order n can be written in the form

$$G(x, y, y', y'', \ldots, y^{(n)}) = 0, \tag{1.2.1}$$

where we have introduced the prime notation to denote derivatives, and $y^{(n)}$ denotes the nth derivative of y with respect to x (not y to the power of n). Of particular interest to us throughout the text will be *linear* differential equations. These arise as the special case of Equation (1.2.1), when $y, y', \ldots, y^{(n)}$ occur to the first degree only, and not as products or arguments of other functions. The general form for such a differential equation is given in the next definition.

DEFINITION   1.2.4

A differential equation that can be written in the form

$$a_0(x)y^{(n)} + a_1(x)y^{(n-1)} + \cdots + a_n(x)y = F(x),$$

where $a_0, a_1, \ldots, a_n$ and F are functions of x only, is called a **linear** differential equation of order n. Such a differential equation is linear in $y, y', y'', \ldots, y^{(n)}$.

A differential equation that does not satisfy this definition is called a **nonlinear** differential equation.

Example 1.2.5   The equations

$$y'' + x^2 y' + (\sin x)y = e^x \qquad \text{and} \qquad xy''' + 4x^2 y' - \frac{2}{1+x^2}y = 0$$

are linear differential equations of order 2 and order 3, respectively, whereas the differential equations

$$y'' + x\sin(y') - xy = x^2 \qquad \text{and} \qquad y'' - x^2 y' + y^2 = 0$$

are nonlinear. In the first case the nonlinearity arises from the $\sin(y')$ term, whereas in the second, the nonlinearity is due to the y^2 term.   □

Example 1.2.6   The general forms for first- and second-order linear differential equations are

$$a_0(x)\frac{dy}{dx} + a_1(x)y = F(x)$$

and

$$a_0(x)\frac{d^2 y}{dx^2} + a_1(x)\frac{dy}{dx} + a_2(x)y = F(x),$$

respectively.   □

If we consider the examples from the previous section, we see that the differential equation governing the simple harmonic oscillator is a second-order linear differential equation. In this case the linearity was imposed in the modeling process when we assumed that the restoring force was directly proportional to the displacement from equilibrium (Hooke's law). Not all springs satisfy this relationship. For example, Duffing's equation

$$m\frac{d^2 y}{dx^2} + k_1 y + k_2 y^3 = 0$$

gives a mathematical model of a nonlinear spring–mass system. If $k_2 = 0$, this reduces to the simple harmonic oscillator equation. Newton's law of cooling assumes a linear relationship between the rate of change of the temperature of an object and the temperature difference between the object and that of the surrounding medium. Hence, the resulting differential equation is linear. This can be seen explicitly by writing Equation (1.1.8) as

$$\frac{dT}{dt} + kT = kT_m,$$

which is a first-order linear differential equation. Finally, the differential equation for determining the orthogonal trajectories of a given family of curves will in general be nonlinear, as seen in Example 1.1.1.

Solutions of Differential Equations

We now define precisely what is meant by a solution to a differential equation.

DEFINITION   1.2.7

A function $y = f(x)$ that is (at least) n times differentiable on an interval I is called a **solution** to the differential equation (1.2.1) on I if the substitution $y = f(x)$, $y' = f'(x)$, $\ldots$, $y^{(n)} = f^{(n)}(x)$ reduces the differential equation (1.2.1) to an identity valid for all x in I. In this case we say that $y = f(x)$ **satisfies** the differential equation.

Example 1.2.8   Verify that for all constants c_1 and c_2, $y(x) = c_1 \sin x + c_2 \cos x$ is a solution to the linear differential equation $y'' + y = 0$ for x in the interval $(-\infty, \infty)$.

Solution: The function $y(x)$ is certainly twice differentiable for all real x. Furthermore,

$$y'(x) = c_1 \cos x - c_2 \sin x$$

and

$$y''(x) = -(c_1 \sin x + c_2 \cos x).$$

Consequently,

$$y'' + y = -(c_1 \sin x + c_2 \cos x) + c_1 \sin x + c_2 \cos x = 0,$$

so that $y'' + y = 0$ for every x in $(-\infty, \infty)$. It follows from the preceding definition that the given function is a solution to the differential equation on $(-\infty, \infty)$.   □

In the preceding example, x could assume all real values. Often, however, the independent variable will be restricted in some manner. For example, the differential equation

$$\frac{dy}{dx} = \frac{1}{2\sqrt{x}}(y - 1)$$

is undefined when $x \leq 0$, and so any solution would be defined only for $x > 0$. In fact this linear differential equation has solution

$$y(x) = ce^{\sqrt{x}} + 1, \qquad x > 0,$$

where c is a constant. (The reader can check this by plugging in to the given differential equation, as was done in Example 1.2.8. In Section 1.4 we will introduce a technique that will enable us to derive this solution.) We now distinguish two ways in which solutions to a differential equation can be expressed. Often, as in Example 1.2.8, we will be able to obtain a solution to a differential equation in the explicit form $y = f(x)$, for some function f. However, when dealing with nonlinear differential equations, we usually have to be content with a solution written in implicit form

$$F(x, y) = 0,$$

where the function F defines the solution, $y(x)$, implicitly as a function of x. This is illustrated in Example 1.2.9.

Example 1.2.9   Verify that the relation $x^2 + y^2 - 4 = 0$ defines an implicit solution to the nonlinear differential equation

$$\frac{dy}{dx} = -\frac{x}{y}.$$

Solution: We regard the given relation as defining y as a function of x. Differentiating this relation with respect to x yields[4]

$$2x + 2y\frac{dy}{dx} = 0.$$

That is,

$$\frac{dy}{dx} = -\frac{x}{y},$$

as required. In this example we can obtain y explicitly in terms of x, since $x^2 + y^2 - 4 = 0$ implies that

$$y = \pm\sqrt{4 - x^2}.$$

The implicit relation therefore contains the two explicit solutions

$$y(x) = \sqrt{4 - x^2}, \qquad y(x) = -\sqrt{4 - x^2},$$

which correspond graphically to the two semi-circles sketched in Figure 1.2.1.

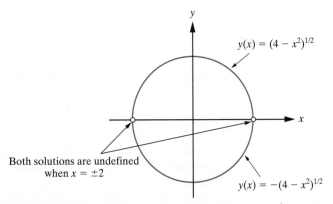

Figure 1.2.1: Two solutions to the differential equation $y' = -x/y$.

---

[4]Note that we have used implicit differentiation in obtaining $d(y^2)/dx = 2y \cdot (dy/dx)$.

Since $x = \pm 2$ corresponds to $y = 0$ in both of these equations, whereas the differential equation is defined only for $y \neq 0$, we must omit $x = \pm 2$ from the domains of the solutions. Consequently, both of the foregoing solutions to the differential equation are valid for $-2 < x < 2$.

$\square$

In the preceding example the solutions to the differential equation are more simply expressed in implicit form, although, as we have shown, it is quite easy to obtain the corresponding explicit solutions. In the following example the solution must be expressed in implicit form, since it is impossible to solve the implicit relation (analytically) for y as a function of x.

Example 1.2.10   Show that the relation $\sin(xy) + y^2 - x = 0$ defines a solution to

$$\frac{dy}{dx} = \frac{1 - y\cos(xy)}{x\cos(xy) + 2y}.$$

Solution: Differentiating the given relationship implicitly with respect to x yields

$$\cos(xy)\left(y + x\frac{dy}{dx}\right) + 2y\frac{dy}{dx} - 1 = 0.$$

That is,

$$\frac{dy}{dx}[x\cos(xy) + 2y] = 1 - y\cos(xy),$$

which implies that

$$\frac{dy}{dx} = \frac{1 - y\cos(xy)}{x\cos(xy) + 2y}$$

as required. $\square$

Now consider the simple differential equation

$$\frac{d^2y}{dx^2} = 12x.$$

From elementary calculus we know that all functions whose second derivative is $12x$ can be obtained by performing two integrations. Integrating the given differential equation once yields

$$\frac{dy}{dx} = 6x^2 + c_1,$$

where c_1 is an arbitrary constant. Integrating again, we obtain

$$y(x) = 2x^3 + c_1 x + c_2, \tag{1.2.2}$$

where c_2 is another arbitrary constant. The point to notice about this solution is that it contains two arbitrary constants. Further, by assigning appropriate values to these constants, we can determine all solutions to the differential equation. We call (1.2.2) the general solution to the differential equation. In this example the given differential equation was of second-order, and the general solution contained two arbitrary constants, which arose because two integrations were required to solve the differential equation. In the case of an nth-order differential equation we might suspect that the most general form of solution that can arise would contain n arbitrary constants. This is indeed the case and motivates the following definition.

DEFINITION   1.2.11

A solution to an nth-order differential equation on an interval I is called the **general solution on** I if it satisfies the following conditions:

1. The solution contains n constants $c_1, c_2, \ldots, c_n$.
2. All solutions to the differential equation can be obtained by assigning appropriate values to the constants.

Remark   Not all differential equations have a general solution. For example, consider

$$(y')^2 + (y - 1)^2 = 0.$$

The only solution to this differential equation is $y(x) = 1$, and hence the differential equation does not have a solution containing an arbitrary constant.

Example 1.2.12   Find the general solution to the differential equation $y'' = e^{-x}$.

Solution: Integrating the given differential equation with respect to x yields

$$y' = -e^{-x} + c_1,$$

where c_1 is an integration constant. Integrating this equation, we obtain

$$y(x) = e^{-x} + c_1 x + c_2 \qquad (1.2.3)$$

where c_2 is another integration constant. Consequently, all solutions to $y'' = e^{-x}$ are of the form (1.2.3), and therefore, according to Definition 1.2.11, this is the general solution to $y'' = e^{-x}$ on any interval.   □

As the preceding example illustrates, we can, in principle, always find the general solution to a differential equation of the form

$$\frac{d^n y}{dx^n} = f(x) \qquad (1.2.4)$$

by performing n integrations. However, if the function on the right-hand side of the differential equation is not a function of x only, this procedure cannot be used. Indeed, one of the major aims of this text is to determine solution techniques for differential equations that are more complicated than Equation (1.2.4). A solution to a differential equation is called a **particular solution** if it does not contain any arbitrary constants not present in the differential equation itself. One way in which particular solutions arise is by our assigning specific values to the arbitrary constants occurring in the general solution to a differential equation. For example, from (1.2.3),

$$y(x) = e^{-x} + x$$

is a particular solution to the differential equation $d^2 y/dx^2 = e^{-x}$ (the solution corresponding to $c_1 = 1, c_2 = 0$).

Initial-Value Problems

As discussed in the preceding section, the unique specification of an applied problem requires more than just a differential equation. We must also give appropriate auxiliary conditions that characterize the problem under investigation. Of particular interest to us is the case of the initial-value problem defined for an nth-order differential equation as follows.

> **DEFINITION   1.2.13**
>
> An nth-order differential equation together with n auxiliary conditions of the form
>
> $$y(x_0) = y_0, \quad y'(x_0) = y_1, \quad \ldots, \quad y^{(n-1)}(x_0) = y_{n-1},$$
>
> where $y_0, y_1, \ldots, y_{n-1}$ are constants, is called an **initial-value problem**.

Example 1.2.14   Solve the initial-value problem

$$y'' = e^{-x}, \tag{1.2.5}$$
$$y(0) = 1, \qquad y'(0) = 4. \tag{1.2.6}$$

Solution: From Example 1.2.12, the general solution to Equation (1.2.5) is

$$y(x) = e^{-x} + c_1 x + c_2. \tag{1.2.7}$$

We now impose the auxiliary conditions (1.2.6). Setting $x = 0$ in (1.2.7), we see that

$$y(0) = 1 \quad \text{if and only if} \quad 1 = 1 + c_2.$$

So $c_2 = 0$. Using this value for c_2 in (1.2.7) and differentiating the result yields

$$y'(x) = -e^{-x} + c_1.$$

Consequently

$$y'(0) = 4 \quad \text{if and only if} \quad 4 = -1 + c_1,$$

and hence $c_1 = 5$. Thus the given auxiliary conditions pick out the particular solution to the differential equation (1.2.5) with $c_1 = 5$ and $c_2 = 0$, so that the initial-value problem has the unique solution

$$y(x) = e^{-x} + 5x.$$

$\square$

Initial-value problems play a fundamental role in the theory and applications of differential equations. In the previous example, the initial-value problem had a unique solution. More generally, suppose we have a differential equation that can be written in the **normal** form

$$y^{(n)} = f(x, y, y', \ldots, y^{(n-1)}).$$

According to Definition 1.2.13, the initial-value problem for such an nth-order differential equation is the following:

Statement of the initial-value problem: Solve

$$y^{(n)} = f(x, y, y', \ldots, y^{(n-1)})$$

subject to

$$y(x_0) = y_0, \quad y'(x_0) = y_1, \quad \ldots, \quad y^{(n-1)}(x_0) = y_{n-1},$$

where $y_0, y_1, \ldots, y_{n-1}$ are constants.

It can be shown that this initial-value problem always has a unique solution, provided that f and its partial derivatives with respect to $y, y', \ldots, y^{(n-1)}$ are continuous in an appropriate region. This is a fundamental result in the theory of differential equations. In Chapter 6 we will show how the following special case can be used to develop the theory for *linear* differential equations.

Theorem 1.2.15   Let $a_1, a_2, \ldots, a_n, F$ be functions that are continuous on an interval I. Then, for any x_0 in I, the initial-value problem

$$y^{(n)} + a_1(x)y^{(n-1)} + \cdots + a_{n-1}(x)y' + a_n(x)y = F(x),$$

$$y(x_0) = y_0, \quad y'(x_0) = y_1, \quad \ldots, \quad y^{(n-1)}(x_0) = y_{n-1}$$

has a unique solution on I.

The next example, which we will refer back to on many occasions throughout the remainder of the text, illustrates the power of the preceding theorem.

Example 1.2.16   Prove that the general solution to the differential equation

$$y'' + \omega^2 y = 0, \quad -\infty < x < \infty \tag{1.2.8}$$

where ω is a nonzero constant, is

$$y(x) = c_1 \cos \omega x + c_2 \sin \omega x, \tag{1.2.9}$$

where c_1, c_2 are arbitrary constants.

Solution: It is a routine computation to verify that $y(x) = c_1 \cos \omega x + c_2 \sin \omega x$ is a solution to the differential equation (1.2.8) on $(-\infty, \infty)$. According to Definition 1.2.11 we must now establish that *every* solution to (1.2.8) is of the form (1.2.9). To that end, suppose that $y = f(x)$ is any solution to (1.2.8). Then according to the preceding theorem, $y = f(x)$ is the *unique* solution to the initial-value problem

$$y'' + \omega^2 y = 0, \quad y(0) = f(0), \quad y'(0) = f'(0). \tag{1.2.10}$$

However, consider the function

$$y(x) = f(0) \cos \omega x + \frac{f'(0)}{\omega} \sin \omega x \tag{1.2.11}$$

This is of the form $y(x) = c_1 \cos \omega x + c_2 \sin \omega x$, where $c_1 = f(0)$ and $c_2 = f'(0)/\omega$, and therefore solves the differential equation (1.2.8). Further, evaluating (1.2.11) at $x = 0$ yields

$$y(0) = f(0) \quad \text{and} \quad y'(0) = f'(0).$$

Consequently, (1.2.11) solves the initial-value problem (1.2.10). But, by assumption, $y(x) = f(x)$ solves the same initial-value problem. Owing to the uniqueness of the solution to this initial-value problem, it follows that these two solutions must coincide. Therefore,

$$f(x) = f(0) \cos \omega x + \frac{f'(0)}{\omega} \sin \omega x = c_1 \cos \omega x + c_2 \sin \omega x.$$

Since $f(x)$ was an arbitrary solution to the differential equation (1.2.8), we can conclude that every solution to (1.2.8) is of the form

$$y(x) = c_1 \cos \omega x + c_2 \sin \omega x$$

and therefore this is the general solution on $(-\infty, \infty)$.   □

In the remainder of this chapter we will focus primarily on first-order differential equations and some of their elementary applications. We will investigate such differential equations qualitatively, analytically, and numerically.

Exercises for 1.2

Key Terms

Differential equation, Order of a differential equation, Linear differential equation, Nonlinear differential equation, General solution to a differential equation, Particular solution to a differential equation, Initial-value problem.

Skills

- Be able to determine the order of a differential equation.

- Be able to determine whether a given differential equation is linear or nonlinear.

- Be able to determine whether or not a given function $y(x)$ is a particular solution to a given differential equation.

- Be able to determine whether or not a given implicit relation defines a particular solution to a given differential equation.

- Be able to find the general solution to differential equations of the form $y^{(n)} = f(x)$ via n integrations.

- Be able to use initial conditions to find the solution to an initial-value problem.

True-False Review

For Questions 1–6, decide if the given statement is **true** or **false**, and give a brief justification for your answer. If true,

you can quote a relevant definition or theorem from the text. If false, provide an example, illustration, or brief explanation of why the statement is false.

1. The order of a differential equation is the order of the lowest derivative appearing in the differential equation.

2. The general solution to a third-order differential equation must contain three constants.

3. An initial-value problem always has a unique solution if the functions and partial derivatives involved are continuous.

4. The general solution to $y'' + y = 0$ is $y(x) = c_1 \cos x + 5c_2 \cos x$.

5. The general solution to $y'' + y = 0$ is $y(x) = c_1 \cos x + 5c_1 \sin x$.

6. The general solution to a differential equation of the form $y^{(n)} = F(x)$ can be obtained by n consecutive integrations of the function $F(x)$.

Problems

For Problems 1–6, determine the order of the given differential equation and state whether it is linear or nonlinear.

1. $\dfrac{d^2 y}{dx^2} + e^{xy} \dfrac{dy}{dx} = x^2.$

2. $\dfrac{d^3y}{dx^3} + 4\dfrac{d^2y}{dx^2} + \sin x \dfrac{dy}{dx} = xy + \tan x.$

3. $y'' + 3x(y')^3 - y = 1 + 3x.$

4. $\sin x \cdot e^{y''} + y' - \tan y = \cos x.$

5. $\dfrac{d^4y}{dx^4} + 3\dfrac{d^2y}{dx^2} = x.$

6. $\sqrt{x}\,y'' + \dfrac{\ln x}{y'''} = 3x^3.$

For Problems 7–18, verify that the given function is a solution to the given differential equation (c_1 and c_2 are arbitrary constants), and state the maximum interval over which the solution is valid.

7. $y(x) = c_1 e^x \cos 2x + c_2 e^x \sin 2x, \quad y'' - 2y' + 5y = 0.$

8. $y(x) = c_1 e^x + c_2 e^{-2x}, \quad y'' + y' - 2y = 0.$

9. $y(x) = \dfrac{1}{x+4}, \quad y' = -y^2.$

10. $y(x) = c_1 x^{1/2}, \quad y' = \dfrac{y}{2x}.$

11. $y(x) = e^{-x} \sin 2x, \quad y'' + 2y' + 5y = 0.$

12. $y(x) = c_1 \cosh 3x + c_2 \sinh 3x, \quad y'' - 9y = 0.$

13. $y(x) = c_1 x^{-3} + c_2 x^{-1}, \quad x^2 y'' + 5xy' + 3y = 0.$

14. $y(x) = c_1 x^{1/2} + 3x^2, \quad 2x^2 y'' - xy' + y = 9x^2.$

15. $y(x) = c_1 x^2 + c_2 x^3 - x^2 \sin x,$
$x^2 y'' - 4xy' + 6y = x^4 \sin x.$

16. $y(x) = c_1 e^{ax} + c_2 e^{bx}, \quad y'' - (a+b)y' + aby = 0,$
where a and b are constants and $a \neq b.$

17. $y(x) = e^{ax}(c_1 + c_2 x), \quad y'' - 2ay' + a^2 y = 0,$ where a is a constant.

18. $y(x) = e^{ax}(c_1 \cos bx + c_2 \sin bx),$
$y'' - 2ay' + (a^2 + b^2)y = 0,$ where a and b are constants.

For Problems 19–22, determine all values of the constant r such that the given function solves the given differential equation.

19. $y(x) = e^{rx}, \quad y'' + 2y' - 3y = 0.$

20. $y(x) = e^{rx}, \quad y'' - 8y' + 16y = 0.$

21. $y(x) = x^r, \quad x^2 y'' + xy' - y = 0.$

22. $y(x) = x^r, \quad x^2 y'' + 5xy' + 4y = 0.$

23. When N is a positive integer, the **Legendre equation**

$$(1 - x^2)y'' - 2xy' + N(N+1)y = 0,$$

with $-1 < x < 1$, has a solution that is a polynomial of degree N. Show by substitution into the differential equation that in the case $N = 3$ such a solution is

$$y(x) = \frac{1}{2}x(5x^2 - 3).$$

24. Determine a solution to the differential equation

$$(1 - x^2)y'' - xy' + 4y = 0$$

of the form $y(x) = a_0 + a_1 x + a_2 x^2$ satisfying the normalization condition $y(1) = 1.$

For Problems 25–29, show that the given relation defines an implicit solution to the given differential equation, where c is an arbitrary constant.

25. $x \sin y - e^x = c, \quad y' = \dfrac{e^x - \sin y}{x \cos y}.$

26. $xy^2 + 2y - x = c, \quad y' = \dfrac{1 - y^2}{2(1 + xy)}.$

27. $e^{xy} - x = c, \quad y' = \dfrac{1 - ye^{xy}}{xe^{xy}}.$
Determine the solution with $y(1) = 0.$

28. $e^{y/x} + xy^2 - x = c, \quad y' = \dfrac{x^2(1 - y^2) + ye^{y/x}}{x(e^{y/x} + 2x^2 y)}.$

29. $x^2 y^2 - \sin x = c, \quad y' = \dfrac{\cos x - 2xy^2}{2x^2 y}.$
Determine the explicit solution that satisfies $y(\pi) = 1/\pi.$

For Problems 30–33, find the general solution to the given differential equation and the maximum interval on which the solution is valid.

30. $y' = \sin x.$

31. $y' = x^{-1/2}.$

32. $y'' = xe^x.$

33. $y'' = x^n, n$ an integer.

For Problems 34–38, solve the given initial-value problem.

34. $y' = \ln x, \quad y(1) = 2.$

35. $y'' = \cos x$, $y(0) = 2$, $y'(0) = 1$.

36. $y''' = 6x$, $y(0) = 1$, $y'(0) = -1$, $y''(0) = 4$.

37. $y'' = xe^x$, $y(0) = 3$, $y'(0) = 4$.

38. Prove that the general solution to $y'' - y = 0$ on any interval I is $y(x) = c_1 e^x + c_2 e^{-x}$.

A second-order differential equation together with two auxiliary conditions imposed at different values of the independent variable is called a **boundary-value problem**. For Problems 39–40, solve the given boundary-value problem.

39. $y'' = e^{-x}$, $y(0) = 1$, $y(1) = 0$.

40. $y'' = -2(3 + 2\ln x)$, $y(1) = y(e) = 0$.

41. The differential equation $y'' + y = 0$ has the general solution $y(x) = c_1 \cos x + c_2 \sin x$.

   (a) Show that the boundary-value problem $y'' + y = 0$, $y(0) = 0$, $y(\pi) = 1$ has no solutions.

   (b) Show that the boundary-value problem $y'' + y = 0$, $y(0) = 0$, $y(\pi) = 0$, has an infinite number of solutions.

For Problems 42–47, verify that the given function is a solution to the given differential equation. In these problems, c_1 and c_2 are arbitrary constants. Throughout the text, the symbol $\diamond$ refers to exercises for which some form of technology, such as a graphing calculator or computer algebra system (CAS), is recommended.

42. $\diamond$ $y(x) = c_1 e^{2x} + c_2 e^{-3x}$, $y'' + y' - 6y = 0$.

43. $\diamond$ $y(x) = c_1 x^4 + c_2 x^{-2}$, $x^2 y'' - xy' - 8y = 0$, $x > 0$.

44. $\diamond$ $y(x) = c_1 x^2 + c_2 x^2 \ln x + \frac{1}{6}x^2 (\ln x)^3$,
   $x^2 y'' - 3xy' + 4y = x^2 \ln x$, $x > 0$.

45. $\diamond$ $y(x) = x^a [c_1 \cos(b \ln x) + c_2 \sin(b \ln x)]$,
   $x^2 y'' + (1 - 2a)xy' + (a^2 + b^2)y = 0$, $x > 0$, where a and b are arbitrary constants.

46. $\diamond$ $y(x) = c_1 e^x + c_2 e^{-x}(1 + 2x + 2x^2)$,
   $xy'' - 2y' + (2 - x)y = 0$, $x > 0$.

47. $\diamond$ $y(x) = \sum_{k=0}^{10} \frac{1}{k!} x^k$, $xy'' - (x + 10)y' + 10y = 0$,
   $x > 0$.

48. $\diamond$

   (a) Derive the polynomial of degree five that satisfies both the Legendre equation

   $$(1 - x^2)y'' - 2xy' + 30y = 0$$

   and the normalization condition $y(1) = 1$.

   (b) $\diamond$ Sketch your solution from (a) and determine approximations to all zeros and local maxima and local minima on the interval $(-1, 1)$.

49. $\diamond$ One solution to the Bessel equation of (nonnegative) integer order N

   $$x^2 y'' + xy' + (x^2 - N^2)y = 0$$

   is

   $$y(x) = J_N(x) = \sum_{k=0}^{\infty} \frac{(-1)^k}{k!(N+k)!} \left(\frac{x}{2}\right)^{2k+N}.$$

   (a) Write the first three terms of $J_0(x)$.

   (b) Let $J(0, x, m)$ denote the mth partial sum

   $$J(0, x, m) = \sum_{k=0}^{m} \frac{(-1)^k}{(k!)^2} \left(\frac{x}{2}\right)^{2k}.$$

   Plot $J(0, x, 4)$ and use your plot to approximate the first positive zero of $J_0(x)$. Compare your value against a tabulated value or one generated by a computer algebra system.

   (c) Plot $J_0(x)$ and $J(0, x, 4)$ on the same axes over the interval $[0, 2]$. How well do they compare?

   (d) If your system has built-in Bessel functions, plot $J_0(x)$ and $J(0, x, m)$ on the same axes over the interval $[0, 10]$ for various values of m. What is the smallest value of m that gives an accurate approximation to the first *three* positive zeros of $J_0(x)$?

1.3   The Geometry of First-Order Differential Equations

The primary aim of this chapter is to study the first-order differential equation

$$\frac{dy}{dx} = f(x, y), \tag{1.3.1}$$

where $f(x, y)$ is a given function of x and y. In this section we focus our attention mainly on the geometric aspects of the differential equation and its solutions. The graph of any solution to the differential equation (1.3.1) is called a **solution curve**. If we recall the geometric interpretation of the derivative dy/dx as giving the slope of the tangent line at any point on the curve with equation $y = y(x)$, we see that the function $f(x, y)$ in (1.3.1) gives the slope of the tangent line to the solution curve passing through the point (x, y). Consequently, when we solve Equation (1.3.1), we are finding all curves whose slope at the point (x, y) is given by the function $f(x, y)$. According to our definition in the previous section, the general solution to the differential equation (1.3.1) will involve one arbitrary constant, and therefore, geometrically, the general solution gives a family of solution curves in the xy-plane, one solution curve corresponding to each value of the arbitrary constant.

Example 1.3.1   Find the general solution to the differential equation $dy/dx = 2x$, and sketch the corresponding solution curves.

Solution: The differential equation can be integrated directly to obtain $y(x) = x^2 + c$. Consequently the solution curves are a family of parabolas in the xy-plane. This is illustrated in Figure 1.3.1.   □

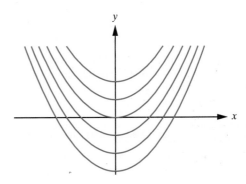

Figure 1.3.1: Some solution curves for the differential equation $dy/dx = 2x$.

Figure 1.3.2 gives a Mathematica plot of some solution curves to the differential equation

$$\frac{dy}{dx} = y - x^2.$$

This illustrates that generally the solution curves of a differential equation are quite complicated. Upon completion of the material in this section, the reader will be able to obtain Figure 1.3.2 without needing a computer algebra system.

Existence and Uniqueness of Solutions

It is useful for the further analysis of the differential equation (1.3.1) to give at this point a brief discussion of the existence and uniqueness of solutions to the corresponding initial-value problem

$$\frac{dy}{dx} = f(x, y), \qquad y(x_0) = y_0. \tag{1.3.2}$$

Geometrically, we are interested in finding the particular solution curve to the differential equation that passes through the point in the xy-plane with coordinates (x_0, y_0). The following questions arise regarding the initial-value problem:

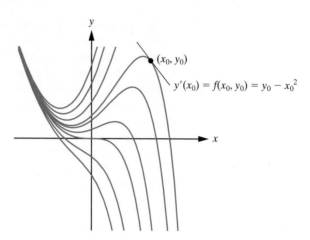

Figure 1.3.2: Some solution curves for the differential equation $dy/dx = y - x^2$.

1. Existence: Does the initial-value problem have any solutions?

2. Uniqueness: If the answer to question 1 is yes, does the initial-value problem have only one solution?

Certainly in the case of an applied problem we would be interested only in initial-value problems that have precisely one solution. The following theorem establishes conditions on f that guarantee the existence and uniqueness of a solution to the initial-value problem (1.3.2).

Theorem 1.3.2   **(Existence and Uniqueness Theorem)**

Let $f(x, y)$ be a function that is continuous on the rectangle

$$R = \{(x, y) : a \leq x \leq b, c \leq y \leq d\}.$$

Suppose further that $\partial f/\partial y$ is continuous in R. Then for any interior point (x_0, y_0) in the rectangle R, there exists an interval I containing x_0 such that the initial-value problem (1.3.2) has a unique solution for x in I.

Proof A complete proof of this theorem can be found, for example, in G. F. Simmons, *Differential Equations* (New York: McGraw-Hill, 1972). Figure 1.3.3 gives a geometric illustration of the result.   ∎

Remark   From a geometric viewpoint, if $f(x, y)$ satisfies the hypotheses of the existence and uniqueness theorem in a region R of the xy-plane, then throughout that region the solution curves of the differential equation $dy/dx = f(x, y)$ cannot intersect. For if two solution curves did intersect at (x_0, y_0) in R, then that would imply there was more than one solution to the initial-value problem

$$\frac{dy}{dx} = f(x, y), \qquad y(x_0) = y_0,$$

which would contradict the existence and uniqueness theorem.

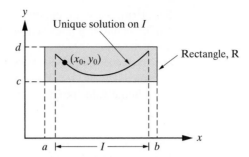

Figure 1.3.3: Illustration of the existence and uniqueness theorem for first-order differential equations.

The following example illustrates how the preceding theorem can be used to establish the existence of a unique solution to a differential equation, even though at present we do not know how to determine the solution.

Example 1.3.3   Prove that the initial-value problem

$$\frac{dy}{dx} = 3xy^{1/3}, \qquad y(0) = a$$

has a unique solution whenever $a \neq 0$.

Solution: In this case the initial point is $x_0 = 0$, $y_0 = a$, and $f(x, y) = 3xy^{1/3}$. Hence, $\partial f/\partial y = xy^{-2/3}$. Consequently, f is continuous at all points in the xy-plane, whereas $\partial f/\partial y$ is continuous at all points not lying on the x-axis ($y \neq 0$). Provided $a \neq 0$, we can certainly draw a rectangle containing $(0, a)$ that does not intersect the x-axis. (See Figure 1.3.4.) In any such rectangle the hypotheses of the existence and uniqueness theorem are satisfied, and therefore the initial-value problem does indeed have a unique solution. $\square$

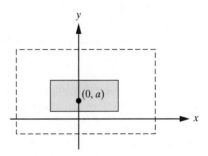

Figure 1.3.4: The initial-value problem in Example 1.3.3 satisfies the hypotheses of the existence and uniqueness theorem in the small rectangle, but not in the large rectangle.

Example 1.3.4   Discuss the existence and uniqueness of solutions to the initial-value problem

$$\frac{dy}{dx} = 3xy^{1/3}, \qquad y(0) = 0.$$

Solution: The differential equation is the same as in the previous example, but the initial condition is imposed on the x-axis. Since $\partial f/\partial y = xy^{-2/3}$ is not continuous along the x-axis, there is no rectangle containing $(0, 0)$ in which the hypotheses of the existence and uniqueness theorem are satisfied. *We can therefore draw no conclusion from the theorem itself.* We leave it as an exercise to verify by direct substitution that the given initial-value problem does in fact have the following two solutions:

$$y(x) = 0 \qquad \text{and} \qquad y(x) = x^3.$$

Consequently in this case the initial-value problem does not have a unique solution.   □

Slope Fields

We now return to our discussion of the geometry of solutions to the differential equation

$$\frac{dy}{dx} = f(x, y).$$

The fact that the function $f(x, y)$ gives the slope of the tangent line to the solution curves of this differential equation leads to a simple and important idea for determining the overall shape of the solution curves. We compute the value of $f(x, y)$ at several points and draw through each of the corresponding points in the xy-plane small line segments having $f(x, y)$ as their slopes. The resulting sketch is called the **slope field** for the differential equation. The key point is that each solution curve must be tangent to the line segments that we have drawn, and therefore by studying the slope field we can obtain the general shape of the solution curves.

Example 1.3.5   Sketch the slope field for the differential equation $dy/dx = 2x^2$.

Solution: The slope of the solution curves to the differential equation at each point in the xy-plane depends on x only. Consequently, the slopes of the solution curves will be the same at every point on any line parallel to the y-axis (on such a line, x is constant). Table 1.3.1 contains the values of the slope of the solution curves at various points in the interval $[-1, 1]$.

Using this information, we obtain the slope field shown in Figure 1.3.5. In this example, we can integrate the differential equation to obtain the general solution

x	Slope $= 2x^2$
0	0
± 0.2	0.08
± 0.4	0.32
± 0.6	0.72
± 0.8	1.28
± 1.0	2

Table 1.3.1: Values of the slope for the differential equation in Example 1.3.5.

$$y(x) = \frac{2}{3}x^3 + c.$$

Some solution curves and their relation to the slope field are also shown in Figure 1.3.5.   □

In the preceding example, the slope field could be obtained fairly easily because the slopes of the solution curves to the differential equation were constant on lines parallel to the y-axis. For more complicated differential equations, further analysis is generally required if we wish to obtain an accurate plot of the slope field and the behavior of the corresponding solution curves. Below we have listed three useful procedures.

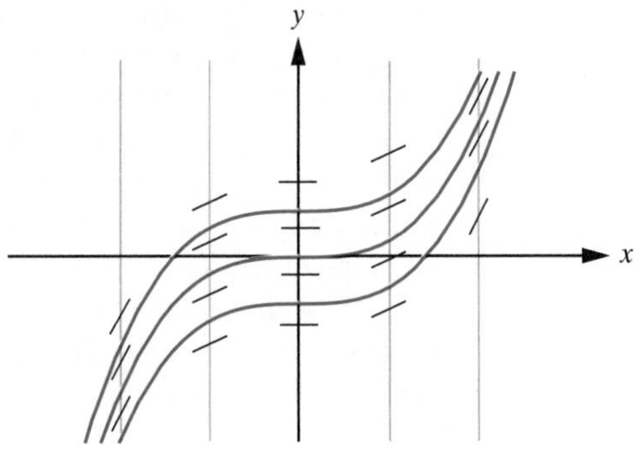

Figure 1.3.5: Slope field and some representative solution curves for the differential equation $dy/dx = 2x^2$.

1. *Isoclines*: For the differential equation

$$\frac{dy}{dx} = f(x, y), \tag{1.3.3}$$

the function $f(x, y)$ determines the regions in the xy-plane where the slope of the solution curves is positive, as well as those where it is negative. Furthermore, each solution curve will have the same slope k along the family of curves

$$f(x, y) = k.$$

These curves are called the **isoclines** of the differential equation, and they can be very useful in determining slope fields. When sketching a slope field, we often start by drawing several isoclines and the corresponding line segments with slope k at various points along them.

2. *Equilibrium Solutions*: Any solution to the differential equation (1.3.3) of the form $y(x) = y_0$, where y_0 is a constant, is called an **equilibrium solution** to the differential equation. The corresponding solution curve is a line parallel to the x-axis. From Equation (1.3.3), equilibrium solutions are given by any constant values of y for which $f(x, y) = 0$, and therefore can often be obtained by inspection. For example, the differential equation

$$\frac{dy}{dx} = (y - x)(y + 1)$$

has the equilibrium solution $y(x) = -1$. One reason that equilibrium solutions are useful in sketching slope fields and determining the general behavior of the full family of solution curves is that, from the existence and uniqueness theorem, we know that no other solution curves can intersect the solution curve corresponding to an equilibrium solution. Consequently, equilibrium solutions serve to divide the xy-plane into different regions.

3. *Concavity Changes*: By differentiating Equation (1.3.3) (implicitly) with respect to x we can obtain an expression for d^2y/dx^2 in terms of x and y. This can be useful in determining the behavior of the concavity of the solution curves to the differential equation (1.3.3). The remaining examples illustrate the application of the foregoing procedures.

Example 1.3.6   Sketch the slope field for the differential equation

$$\frac{dy}{dx} = y - x. \tag{1.3.4}$$

Solution: By inspection we see that the differential equation has no equilibrium solutions. The isoclines of the differential equation are the family of straight lines $y - x = k$. Thus each solution curve of the differential equation has slope k at all points along the line $y - x = k$. Table 1.3.2 contains several values for the slopes of the solution curves, and the equations of the corresponding isoclines. We note that the slope at all points along the isocline $y = x + 1$ is unity, which, from Table 1.3.2, coincides with the slope of any solution curve that meets it. This implies that the isocline must in fact coincide with a solution curve. Hence, one solution to the differential equation (1.3.4) is $y(x) = x + 1$, and, by the existence and uniqueness theorem, no other solution curve can intersect this one.

Slope of Solution Curves	Equation of Isocline
$k = -2$	$y = x - 2$
$k = -1$	$y = x - 1$
$k = 0$	$y = x$
$k = 1$	$y = x + 1$
$k = 2$	$y = x + 2$

Table 1.3.2: Slope and isocline information for the differential equation in Example 1.3.6.

In order to determine the behavior of the concavity of the solution curves, we differentiate the given differential equation implicitly with respect to x to obtain

$$\frac{d^2y}{dx^2} = \frac{dy}{dx} - 1 = y - x - 1,$$

where we have used (1.3.4) to substitute for dy/dx in the second step. We see that the solution curves are concave up ($y'' > 0$) at all points above the line

$$y = x + 1 \tag{1.3.5}$$

and concave down ($y'' < 0$) at all points beneath this line. We also note that Equation (1.3.5) coincides with the particular solution already identified. Putting all of this information together, we obtain the slope field sketched in Figure 1.3.6.

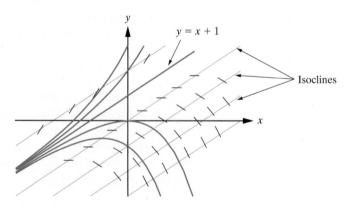

Figure 1.3.6: Hand-drawn slope field, isoclines, and some approximate solution curves for the differential equation in Example 1.3.6.

Generating Slope Fields Using Technology

Many computer algebra systems (CAS) and graphing calculators have built-in programs to generate slope fields. As an example, in the CAS Maple the command

$$\text{diffeq} := \text{diff}(y(x), x) = y(x) - x;$$

assigns the name diffeq to the differential equation considered in the previous example. The further command

$$\text{DEplot(diffeq, } y(x), x = -3..3, y = -3..3, \text{arrows=line)};$$

then produces a sketch of the slope field for the differential equation on the square $-3 \le x \le 3, -3 \le y \le 3$. Initial conditions such as $y(0) = 0, y(0) = 1, y(0) = 2, y(0) = -1$ can be specified using the command

$$\text{IC} := \{[0, 0], [0, 1], [0, 2], [0, -1]\};$$

Then the command

$$\text{DEplot(diffeq, } y(x), x = -3..3, \text{IC}, y = -3..3, \text{arrows=line)};$$

not only plots the slope field, but also gives a numerical approximation to each of the solution curves satisfying the specified initial conditions. Some of the methods that can be used to generate such numerical approximations will be discussed in Section 1.10. The preceding sequence of Maple commands was used to generate the Maple plot given in Figure 1.3.7. Clearly the generation of slope fields and approximate solution curves is one area where technology can be extremely helpful.

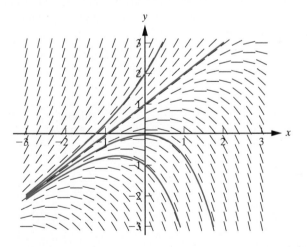

Figure 1.3.7: Maple plot of the slope field and some approximate solution curves for the differential equation in Example 1.3.6.

Example 1.3.7   Sketch the slope field and some approximate solution curves for the differential equation

$$\frac{dy}{dx} = y(2 - y). \tag{1.3.6}$$

Solution: We first note that the given differential equation has the two equilibrium solutions

$$y(x) = 0 \qquad \text{and} \qquad y(x) = 2.$$

Consequently, from Theorem 1.3.2, the xy-plane can be divided into the three distinct regions $y < 0, 0 < y < 2$, and $y > 2$. From Equation (1.3.6) the behavior of the sign of the slope of the solution curves in each of these regions is given in the following schematic.

$$\begin{array}{ccc} \textbf{sign of slope:} & - - -- \;|+ + ++\;|- - -- \\ \textbf{\textit{y}-interval:} & \quad 0 \qquad\quad 2 \end{array}$$

The isoclines are determined from

$$y(2 - y) = k.$$

That is,

$$y^2 - 2y + k = 0,$$

so that the solution curves have slope k at all points of intersection with the horizontal lines

$$y = 1 \pm \sqrt{1 - k}. \tag{1.3.7}$$

Table 1.3.3 contains some of the isocline equations. Note from Equation (1.3.7) that the largest possible positive slope is $k = 1$. We see that the slopes of the solution curves quickly become very large and negative for y outside the interval $[0, 2]$. Finally, differentiating Equation (1.3.6) implicitly with respect to x yields

$$\frac{d^2 y}{dx^2} = 2\frac{dy}{dx} - 2y\frac{dy}{dx} = 2(1 - y)\frac{dy}{dx} = 2y(1 - y)(2 - y).$$

Slope of Solution Curves	Equation of Isocline
$k = 1$	$y = 1$
$k = 0$	$y = 2$ and $y = 0$
$k = -1$	$y = 1 \pm \sqrt{2}$
$k = -2$	$y = 1 \pm \sqrt{3}$
$k = -3$	$y = 3$ and $y = -1$
$k = -n, n \geq 1$	$y = 1 \pm \sqrt{n + 1}$

Table 1.3.3: Slope and isocline information for the differential equation in Example 1.3.7.

The sign of $d^2 y/dx^2$ is given in the following schematic.

$$\begin{array}{ccc} \textbf{sign of } y'': & - - -- \;|+ + ++\;|- - -- \;|+ + + + \\ \textbf{\textit{y}-interval:} & \quad 0 \qquad\quad 1 \qquad\quad 2 \end{array}$$

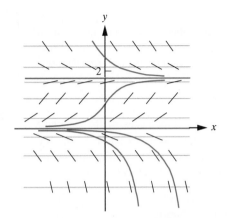

Figure 1.3.8: Hand-drawn slope field, isoclines, and some solution curves for the differential equation $dy/dx = y(2 - y)$.

Using this information leads to the slope field sketched in Figure 1.3.8. We have also included some approximate solution curves. We see from the slope field that for any initial condition $y(x_0) = y_0$, with $0 \leq y_0 \leq 2$, the corresponding unique solution to the differential equation will be bounded. In contrast, if $y_0 > 2$, the slope field suggests that all corresponding solutions approach $y = 2$ as $x \to \infty$, whereas if $y_0 < 0$, then all corresponding solutions approach $y = 0$ as $x \to -\infty$. Furthermore, the behavior of the slope field also suggests that the solution curves that do not lie in the region $0 < y < 2$ may diverge at finite values of x. We leave it as an exercise to verify (by substitution into Equation (1.3.6)) that for all values of the constant c,

$$y(x) = \frac{2ce^{2x}}{ce^{2x} - 1}$$

is a solution to the given differential equation. We see that any initial condition that yields a positive value for c will indeed lead to a solution that has a vertical asymptote at $x = \frac{1}{2} \ln(1/c)$. ☐

    The tools that we have introduced in this section enable us to analyze the solution behavior of many first-order differential equations. However, for complicated functions $f(x, y)$ in Equation (1.3.3), performing these computations by hand can be a tedious task. Fortunately, as we have illustrated, there are many computer programs available for drawing slope fields and generating solution curves (numerically). Furthermore, several graphing calculators also have these capabilities.

Exercises for 1.3

Key Terms

Solution curve, Existence and uniqueness theorem, Slope field, Isocline, Equilibrium solution.

Skills

- Be able to find isoclines for a differential equation $dy/dx = f(x, y)$.

- Be able to determine equilibrium solutions for a differential equation $dy/dx = f(x, y)$.

- Be able to sketch the slope field for a differential equation, using isoclines, equilibrium solutions, and concavity changes.

- Be able to sketch solution curves to a differential equation.

- Be able to apply the existence and uniqueness theorem to find unique solutions to initial-value problems.

True-False Review

For Questions 1–7, decide if the given statement is **true** or **false**, and give a brief justification for your answer. If true, you can quote a relevant definition or theorem from the text. If false, provide an example, illustration, or brief explanation of why the statement is false.

1. If $f(x, y)$ satisfies the hypotheses of the existence and uniqueness theorem in a region R of the xy-plane, then the solution curves to a differential equation $dy/dx = f(x, y)$ cannot intersect in R.

2. Every differential equation $dy/dx = f(x, y)$ has at least one equilibrium solution.

3. The differential equation $dy/dx = x(y^2 - 4)$ has no equilibrium solutions.

4. The circle $x^2 + y^2 = 4$ is an isocline for the differential equation $dy/dx = x^2 + y^2$.

5. The equilibrium solutions of a differential equation are always parallel to one another.

6. The isoclines for the differential equation

$$\frac{dy}{dx} = \frac{x^2 + y^2}{2y}$$

are the family of circles $x^2 + (y - k)^2 = k^2$.

7. No solution to the differential equation $dy/dx = f(x, y)$ can intersect with equilibrium solutions of the differential equation.

Problems

For Problems 1–7, determine the differential equation giving the slope of the tangent line at the point (x, y) for the given family of curves.

1. $y = c/x$.

2. $y = cx^2$.

3. $x^2 + y^2 = 2cx$.

4. $y^2 = cx$.

5. $2cy = x^2 - c^2$.

6. $y^2 - x^2 = c$.

7. $(x - c)^2 + (y - c)^2 = 2c^2$.

For Problems 8–11, verify that the given function (or relation) defines a solution to the given differential equation and sketch some of the solution curves. If an initial condition is given, label the solution curve corresponding to the resulting unique solution. (In these problems, c denotes an arbitrary constant.)

8. $x^2 + y^2 = c$, $y' = -x/y$.

9. $y = cx^3$, $y' = 3y/x$, $y(2) = 8$.

10. $y^2 = cx$, $2x\,dy - y\,dx = 0$, $y(1) = 2$.

11. $(x - c)^2 + y^2 = c^2$, $y' = \dfrac{y^2 - x^2}{2xy}$, $y(2) = 2$.

12. Prove that the initial-value problem

$$y' = x\sin(x + y), \quad y(0) = 1$$

has a unique solution.

13. Use the existence and uniqueness theorem to prove that $y(x) = 3$ is the only solution to the initial-value problem

$$y' = \frac{x}{x^2 + 1}(y^2 - 9), \quad y(0) = 3.$$

14. Do you think that the initial-value problem

$$y' = xy^{1/2}, \quad y(0) = 0$$

has a unique solution? Justify your answer.

15. Even simple-looking differential equations can have complicated solution curves. In this problem, we study the solution curves of the differential equation

$$y' = -2xy^2. \tag{1.3.8}$$

   (a) Verify that the hypotheses of the existence and uniqueness theorem (Theorem 1.3.2) are satisfied for the initial-value problem

$$y' = -2xy^2, \quad y(x_0) = y_0$$

for every (x_0, y_0). This establishes that the initial-value problem always has a unique solution on some interval containing x_0.

   (b) Verify that for all values of the constant c, $y(x) = 1/(x^2 + c)$ is a solution to (1.3.8).

(c) Use the solution to (1.3.8) given in (b) to solve the following initial-value problems. For each case, sketch the corresponding solution curve, and state the maximum interval on which your solution is valid.

    (i) $y' = -2xy^2$, $y(0) = 1$.

    (ii) $y' = -2xy^2$, $y(1) = 1$.

    (iii) $y' = -2xy^2$, $y(0) = -1$.

(d) What is the unique solution to the following initial-value problem?

$$y' = -2xy^2, \quad y(0) = 0.$$

16. Consider the initial-value problem:

$$y' = y(y - 1), \quad y(x_0) = y_0.$$

(a) Verify that the hypotheses of the existence and uniqueness theorem are satisfied for this initial-value problem for any x_0, y_0. This establishes that the initial-value problem always has a unique solution on some interval containing x_0.

(b) By inspection, determine all equilibrium solutions to the differential equation.

(c) Determine the regions in the xy-plane where the solution curves are concave up, and determine those regions where they are concave down.

(d) Sketch the slope field for the differential equation, and determine all values of y_0 for which the initial-value problem has bounded solutions. On your slope field, sketch representative solution curves in the three cases $y_0 < 0, 0 < y_0 < 1$, and $y_0 > 1$.

For Problems 17–24, sketch the slope field and some representative solution curves for the given differential equation.

17. $y' = 4x$.

18. $y' = 1/x$.

19. $y' = x + y$.

20. $y' = x/y$.

21. $y' = -4x/y$.

22. $y' = x^2 y$.

23. $y' = x^2 \cos y$.

24. $y' = x^2 + y^2$.

25. According to Newton's law of cooling (see Section 1.1), the temperature of an object at time t is governed by the differential equation

$$\frac{dT}{dt} = -k(T - T_m),$$

where T_m is the temperature of the surrounding medium, and k is a constant. Consider the case when $T_m = 70$ and $k = 1/80$. Sketch the corresponding slope field and some representative solution curves. What happens to the temperature of the object as $t \to \infty$? Note that this result is independent of the initial temperature of the object.

For Problems 26–31, determine the slope field and some representative solution curves for the given differential equation.

26. $\diamond$ $y' = -2xy$.

27. $\diamond$ $y' = \dfrac{x \sin x}{1 + y^2}$.

28. $\diamond$ $y' = 3x - y$.

29. $\diamond$ $y' = 2x^2 \sin y$.

30. $\diamond$ $y' = \dfrac{2 + y^2}{3 + 0.5x^2}$.

31. $\diamond$ $y' = \dfrac{1 - y^2}{2 + 0.5x^2}$.

32. $\diamond$

(a) Determine the slope field for the differential equation

$$y' = x^{-1}(3 \sin x - y)$$

on the interval (0, 10].

(b) Plot the solution curves corresponding to each of the following initial conditions:

$$y(0.5) = 0, \quad y(1) = -1,$$

$$y(1) = 2, \quad y(3) = 0.$$

What do you conclude about the behavior as $x \to 0^+$ of solutions to the differential equation?

(c) Plot the solution curve corresponding to the initial condition $y(\pi/2) = 6/\pi$. How does this fit in with your answer to part (b)?

(d) Describe the behavior of the solution curves for large positive x.

33. ◇ Consider the family of curves $y = kx^2$, where k is a constant.

(a) Show that the differential equation of the family of orthogonal trajectories is

$$\frac{dy}{dx} = -\frac{x}{2y}.$$

(b) On the same axes sketch the slope field for the preceding differential equation and several members of the given family of curves. Describe the family of orthogonal trajectories.

34. ◇ Consider the differential equation

$$\frac{di}{dt} + ai = b,$$

where a and b are constants. By drawing the slope fields corresponding to various values of a and b, formulate a conjecture regarding the value of

$$\lim_{t \to \infty} i(t).$$

1.4   Separable Differential Equations

In the previous section we analyzed first-order differential equations using qualitative techniques. We now begin an analytical study of these differential equations by developing some solution techniques that enable us to determine the exact solution to certain types of differential equations. The simplest differential equations for which a solution technique can be obtained are the so-called separable equations, which are defined as follows:

> **DEFINITION   1.4.1**
>
> A first-order differential equation is called **separable** if it can be written in the form
>
> $$p(y)\frac{dy}{dx} = q(x). \tag{1.4.1}$$

The solution technique for a separable differential equation is given in Theorem 1.4.2.

Theorem 1.4.2   If $p(y)$ and $q(x)$ are continuous, then Equation (1.4.1) has the general solution

$$\int p(y)\,dy = \int q(x)\,dx + c, \tag{1.4.2}$$

where c is an arbitrary constant.

Proof   We use the chain rule for derivatives to rewrite Equation (1.4.1) in the equivalent form

$$\frac{d}{dx}\left(\int p(y)\,dy\right) = q(x).$$

Integrating both sides of this equation with respect to x yields Equation (1.4.2).   ∎

Remark   In differential form, Equation (1.4.1) can be written as

$$p(y)\,dy = q(x)\,dx,$$

and the general solution (1.4.2) is obtained by integrating the left-hand side with respect to y and the right-hand side with respect to x. This is the general procedure for solving separable equations.

Example 1.4.3   Solve $(1 + y^2)\dfrac{dy}{dx} = x\cos x$.

Solution: By inspection we see that the differential equation is separable. Integrating both sides of the differential equation yields

$$\int (1 + y^2)\,dy = \int x\cos x\,dx + c.$$

Using integration by parts to evaluate the integral on the right-hand side, we obtain

$$y + \tfrac{1}{3}y^3 = x\sin x + \cos x + c,$$

or equivalently

$$y^3 + 3y = 3(x\sin x + \cos x) + c_1,$$

where $c_1 = 3c$. As often happens with separable differential equations, the solution is given in implicit form. $\qquad\square$

In general, the differential equation $dy/dx = f(x)g(y)$ is separable, since it can be written as

$$\frac{1}{g(y)}\frac{dy}{dx} = f(x),$$

which is of the form of Equation (1.4.1) with $p(y) = 1/g(y)$. It is important to note, however, that in writing the given differential equation in this way, we have assumed that $g(y) \neq 0$. Thus the general solution to the resulting differential equation may not include solutions of the original equation corresponding to any values of y for which $g(y) = 0$. (These are the equilibrium solutions for the original differential equation.) We will illustrate with an example.

Example 1.4.4   Find all solutions to

$$y' = -2y^2 x. \tag{1.4.3}$$

Solution: Separating the variables yields

$$y^{-2}dy = -2x\,dx. \tag{1.4.4}$$

Integrating both sides, we obtain

$$-y^{-1} = -x^2 + c$$

so that

$$y(x) = \frac{1}{x^2 - c}. \tag{1.4.5}$$

This is the general solution to Equation (1.4.4). It is not the general solution to Equation (1.4.3), since there is no value of the constant c for which $y(x) = 0$, whereas by inspection we see $y(x) = 0$ is a solution to Equation (1.4.3). This solution is not contained in (1.4.5), since in separating the variables, we divided by y and hence assumed implicitly that $y \neq 0$. Thus the solutions to Equation (1.4.3) are

$$y(x) = \frac{1}{x^2 - c} \qquad \text{and} \qquad y(x) = 0.$$

The slope field for the given differential equation is depicted in Figure 1.4.1, together with some representative solution curves.   □

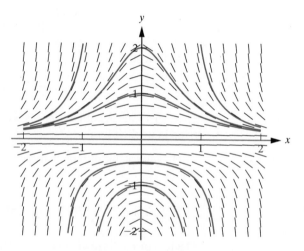

Figure 1.4.1: The slope field and some solution curves for the differential equation $dy/dx = -2xy^2$.

Many difficulties that students encounter with first-order differential equations arise not from the solution techniques themselves, but in the algebraic simplifications that are used to obtain a simple form for the resulting solution. We will explicitly illustrate some of the standard simplifications using the differential equation

$$\frac{dy}{dx} = -2xy.$$

First notice that $y(x) = 0$ is an equilibrium solution to the differential equation. Consequently, no other solution curves can cross the x-axis. For $y \neq 0$ we can separate the variables to obtain

$$\frac{1}{y}dy = -2x\,dx. \tag{1.4.6}$$

Integrating this equation yields

$$\ln |y| = -x^2 + c.$$

Exponentiating both sides of this solution gives

$$|y| = e^{-x^2 + c},$$

or equivalently,

$$|y| = e^c e^{-x^2}.$$

We now introduce a new constant c_1 defined by $c_1 = e^c$. Then the preceding expression for $|y|$ reduces to

$$|y| = c_1 e^{-x^2}. \tag{1.4.7}$$

Notice that c_1 is a positive constant. This is a perfectly acceptable form for the solution. However, a redefinition of the integration constant can be used to eliminate the absolute-value bars as follows. According to (1.4.7), the solution to the differential equation is

$$y(x) = \begin{cases} c_1 e^{-x^2}, & \text{if } y > 0, \\ -c_1 e^{-x^2}, & \text{if } y < 0. \end{cases} \tag{1.4.8}$$

We can now define a new constant c_2, by

$$c_2 = \begin{cases} c_1, & \text{if } y > 0, \\ -c_1, & \text{if } y < 0, \end{cases}$$

in terms of which the solutions given in (1.4.8) can be combined into the single formula

$$y(x) = c_2 e^{-x^2}. \tag{1.4.9}$$

The appropriate sign for c_2 will be determined from the initial conditions. For example, the initial condition $y(0) = 1$ would require that $c_2 = 1$, with corresponding unique solution

$$y(x) = e^{-x^2}.$$

Similarly the initial condition $y(0) = -1$ leads to $c_2 = -1$, so that

$$y(x) = -e^{-x^2}.$$

We make one further point about the solution (1.4.9). In obtaining the separable form (1.4.6), we divided the given differential equation by y, and so the derivation of the solution obtained assumes that $y \neq 0$. However, as we have already noted, $y(x) = 0$ is indeed a solution to this differential equation. Formally this solution is the special case $c_2 = 0$ in (1.4.9) and corresponds to the initial condition $y(0) = 0$. Thus (1.4.9) does give the general solution to the differential equation, provided we allow c_2 to assume the value zero. The slope field for the differential equation, together with some particular solution curves, is shown in Figure 1.4.2.

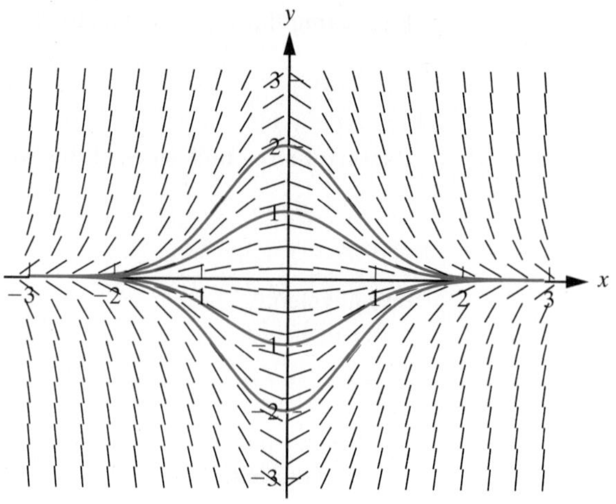

Figure 1.4.2: Slope field and some solution curves for the differential equation $dy/dx = -2xy$.

Example 1.4.5   An object of mass m falls from rest, starting at a point near the earth's surface. Assuming that the air resistance is proportional to the velocity of the object, determine the subsequent motion.

Solution: Let $y(t)$ be the distance traveled by the object at time t from the point it was released, and let the positive y-direction be downward. Then, $y(0) = 0$, and the velocity of the object is $v(t) = dy/dt$. Since the object was dropped from rest, we have $v(0) = 0$. The forces acting on the object are those due to gravity, $F_g = mg$, and the force due to air resistance, $F_r = -kv$, where k is a positive constant (see Figure 1.4.3). According to Newton's second law, the differential equation describing the motion of the object is

$$m\frac{dv}{dt} = F_g + F_r = mg - kv.$$

We are also given the initial condition $v(0) = 0$. Thus the initial-value problem governing the behavior of v is

$$\begin{cases} m\dfrac{dv}{dt} = mg - kv, \\ v(0) = 0. \end{cases} \tag{1.4.10}$$

Separating the variables in Equation (1.4.10) yields

$$\frac{m}{mg - kv}\, dv = dt,$$

which can be integrated directly to obtain

$$-\frac{m}{k}\ln|mg - kv| = t + c.$$

Multiplying both sides of this equation by $-k/m$ and exponentiating the result yields

$$|mg - kv| = c_1 e^{-(k/m)t},$$

where $c_1 = e^{-ck/m}$. By redefining the constant c_1, we can write this in the equivalent form

$$mg - kv = c_2 e^{-(k/m)t}.$$

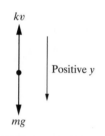

Figure 1.4.3: Particle falling under the influence of gravity and air resistance.

Hence,

$$v(t) = \frac{mg}{k} - c_3 e^{-(k/m)t}, \qquad (1.4.11)$$

where $c_3 = c_2/k$. Imposing the initial condition $v(0) = 0$ yields

$$c_3 = \frac{mg}{k}.$$

So the solution to the initial-value problem (1.4.10) is

$$v(t) = \frac{mg}{k}\left[1 - e^{-(k/m)t}\right]. \qquad (1.4.12)$$

Notice that the velocity does not increase indefinitely, but approaches a so-called limiting velocity v_L defined by

$$v_L = \lim_{t\to\infty} v(t) = \lim_{t\to\infty} \frac{mg}{k}\left[1 - e^{-(k/m)t}\right] = \frac{mg}{k}.$$

The behavior of the velocity as a function of time is shown in Figure 1.4.4. Owing to the negative exponent in (1.4.11), we see that this result is independent of the value of the initial velocity.

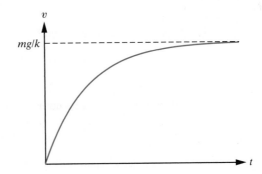

Figure 1.4.4: The behavior of the velocity of the object in Example 1.4.5.

Since $dy/dt = v$, it follows from (1.4.12) that the position of the object at time t can be determined by solving the initial-value problem

$$\frac{dy}{dt} = \frac{mg}{k}\left[1 - e^{-(k/m)t}\right], \qquad y(0) = 0.$$

The differential equation can be integrated directly to obtain

$$y(t) = \frac{mg}{k}\left[t + \frac{m}{k}e^{-(k/m)t}\right] + c.$$

Imposing the initial condition $y(0) = 0$ yields

$$c = -\frac{m^2 g}{k^2},$$

so that

$$y(t) = \frac{mg}{k}\left\{t + \frac{m}{k}\left[e^{-(k/m)t} - 1\right]\right\}.$$

□

Example 1.4.6   A hot metal bar whose temperature is 350°F is placed in a room whose temperature is constant at 70°F. After two minutes, the temperature of the bar is 210°F. Using Newton's law of cooling, determine

1. the temperature of the bar after four minutes.

2. the time required for the bar to cool to 100°F.

Solution: According to Newton's law of cooling (see Section 1.1), the temperature of the object at time t is governed by the differential equation

$$\frac{dT}{dt} = -k(T - T_m), \tag{1.4.13}$$

where, from the statement of the problem,

$$T_m = 70°F, \qquad T(0) = 350°F, \qquad T(2) = 210°F.$$

Substituting for T_m in Equation (1.4.13), we have the separable equation

$$\frac{dT}{dt} = -k(T - 70).$$

Separating the variables yields

$$\frac{1}{T - 70} \, dT = -k \, dt,$$

which we can integrate immediately to obtain

$$\ln |T - 70| = -kt + c.$$

Exponentiating both sides and solving for T yields

$$T(t) = 70 + c_1 e^{-kt}, \tag{1.4.14}$$

where we have redefined the integration constant. The two constants c_1 and k can be determined from the given auxiliary conditions as follows. The condition $T(0) = 350°F$ requires that $350 = 70 + c_1$. Hence, $c_1 = 280$. Substituting this value for c_1 into (1.4.14) yields

$$T(t) = 70(1 + 4e^{-kt}). \tag{1.4.15}$$

Consequently, $T(2) = 210°F$ if and only if

$$210 = 70(1 + 4e^{-2k}),$$

so that $e^{-2k} = \frac{1}{2}$. Hence, $k = \frac{1}{2} \ln 2$, and so, from (1.4.15),

$$T(t) = 70\left[1 + 4e^{-(t/2)\ln 2}\right]. \tag{1.4.16}$$

We can now determine the quantities requested.

1. We have $T(4) = 70(1 + 4e^{-2\ln 2}) = 70\left(1 + 4 \cdot \frac{1}{2^2}\right) = 140°F.$

2. From (1.4.16), $T(t) = 100°F$ when

$$100 = 70\left[1 + 4e^{-(t/2)\ln 2}\right]$$

—that is, when

$$e^{-(t/2)\ln 2} = \frac{3}{28}.$$

Taking the natural logarithm of both sides and solving for t yields

$$t = \frac{2\ln(28/3)}{\ln 2} \approx 6.4 \text{ minutes.} \qquad \square$$

Exercises for 1.4

Skills

- Be able to recognize whether or not a given differential equation is separable.

- Be able to solve separable differential equations.

True-False Review

For Questions 1–9, decide if the given statement is **true** or **false**, and give a brief justification for your answer. If true, you can quote a relevant definition or theorem from the text. If false, provide an example, illustration, or brief explanation of why the statement is false.

1. Every differential equation of the form $dy/dx = f(x)g(y)$ is separable.

2. The general solution to a separable differential equation contains one constant whose value can be determined from an initial condition for the differential equation.

3. Newton's law of cooling is a separable differential equation.

4. The differential equation $dy/dx = x^2 + y^2$ is separable.

5. The differential equation $dy/dx = x\sin(xy)$ is separable.

6. The differential equation $\dfrac{dy}{dx} = e^{x+y}$ is separable.

7. The differential equation

$$\frac{dy}{dx} = \frac{1}{x^2(1+y^2)}$$

is separable.

8. The differential equation

$$\frac{dy}{dx} = \frac{x+4y}{4x+y}$$

is separable.

9. The differential equation

$$\frac{dy}{dx} = \frac{x^3 y + x^2 y^2}{x^2 + xy}$$

is separable.

Problems

For Problems 1–11, solve the given differential equation.

1. $\dfrac{dy}{dx} = 2xy.$

2. $\dfrac{dy}{dx} = \dfrac{y^2}{x^2 + 1}.$

3. $e^{x+y}dy - dx = 0.$

4. $\dfrac{dy}{dx} = \dfrac{y}{x\ln x}.$

5. $ydx - (x-2)dy = 0.$

6. $\dfrac{dy}{dx} = \dfrac{2x(y-1)}{x^2+3}.$

7. $y - x\dfrac{dy}{dx} = 3 - 2x^2\dfrac{dy}{dx}$.

8. $\dfrac{dy}{dx} = \dfrac{\cos(x - y)}{\sin x \sin y} - 1$.

9. $\dfrac{dy}{dx} = \dfrac{x(y^2 - 1)}{2(x - 2)(x - 1)}$.

10. $\dfrac{dy}{dx} = \dfrac{x^2 y - 32}{16 - x^2} + 2$.

11. $(x - a)(x - b)y' - (y - c) = 0$, where a, b, c are constants.

In Problems 12–15, solve the given initial-value problem.

12. $(x^2 + 1)y' + y^2 = -1$, $y(0) = 1$.

13. $(1 - x^2)y' + xy = ax$, $y(0) = 2a$, where a is a constant.

14. $\dfrac{dy}{dx} = 1 - \dfrac{\sin(x + y)}{\sin y \cos x}$, $y(\pi/4) = \pi/4$.

15. $y' = y^3 \sin x$, $y(0) = 0$.

16. One solution to the initial-value problem

$$\frac{dy}{dx} = \frac{2}{3}(y - 1)^{1/2}, \quad y(1) = 1$$

is $y(x) = 1$. Determine another solution. Does this contradict the existence and uniqueness theorem (Theorem 1.3.2)? Explain.

17. An object of mass m falls from rest, starting at a point near the earth's surface. Assuming that the air resistance varies as the square of the velocity of the object, a simple application of Newton's second law yields the initial-value problem for the velocity, $v(t)$, of the object at time t:

$$m\frac{dv}{dt} = mg - kv^2, \quad v(0) = 0,$$

where k, m, g are positive constants.

(a) Solve the foregoing initial-value problem for v in terms of t.

(b) Does the velocity of the object increase indefinitely? Justify.

(c) Determine the position of the object at time t.

18. Find the equation of the curve that passes through the point $(0, \frac{1}{2})$ and whose slope at each point (x, y) is $-x/4y$.

19. Find the equation of the curve that passes through the point $(3, 1)$ and whose slope at each point (x, y) is e^{x-y}.

20. Find the equation of the curve that passes through the point $(-1, 1)$ and whose slope at each point (x, y) is $x^2 y^2$.

21. At time t, the velocity $v(t)$ of an object moving in a straight line satisfies

$$\frac{dv}{dt} = -(1 + v^2). \quad (1.4.17)$$

(a) Show that

$$\tan^{-1}(v) = \tan^{-1}(v_0) - t,$$

where v_0 denotes the velocity of the object at time $t = 0$ (and we assume $v_0 > 0$). Hence prove that the object comes to rest after a finite time $\tan^{-1}(v_0)$. Does the object remain at rest?

(b) Use the chain rule to show that (1.4.17) can be written as

$$v\frac{dv}{dx} = -(1 + v^2),$$

where $x(t)$ denotes the distance traveled by the object at time t, from its position at $t = 0$. Determine the distance traveled by the object when it first comes to rest.

22. The differential equation governing the velocity of an object is

$$\frac{dv}{dt} = -kv^n,$$

where $k > 0$ and n are constants. At $t = 0$, the object is set in motion with velocity v_0.

(a) Show that the object comes to rest in a finite time if and only if $n < 1$, and determine the maximum distance traveled by the object in this case.

(b) If $1 \le n < 2$, show that the maximum distance traveled by the object in a finite time is less than

$$\frac{v_0^{2-n}}{(2 - n)k}.$$

(c) If $n \ge 2$, show that there is no limit to the distance that the object can travel.

23. The pressure p, and density, ρ, of the atmosphere at a height y above the earth's surface are related by

$$dp = -g\rho \, dy.$$

Assuming that p and ρ satisfy the adiabatic equation of state $p = p_0 \left(\dfrac{\rho}{\rho_0}\right)^{\gamma}$, where $\gamma \neq 1$ is a constant and p_0 and ρ_0 denote the pressure and density at the earth's surface, respectively, show that

$$p = p_0 \left[1 - \frac{(\gamma - 1)}{\gamma} \cdot \frac{\rho_0 g y}{p_0}\right]^{\gamma/(\gamma-1)}.$$

24. An object whose temperature is 615°F is placed in a room whose temperature is 75°F. At 4 p.m. the temperature of the object is 135°F, and an hour later its temperature is 95°F. At what time was the object placed in the room?

25. A flammable substance whose initial temperature is 50°F is inadvertently placed in a hot oven whose temperature is 450°F. After 20 minutes, the substance's temperature is 150°F. Find the temperature of the substance after 40 minutes. Assuming that the substance ignites when its temperature reaches 350°F, find the time of combustion.

26. At 2 p.m. on a cool (34°F) afternoon in March, Sherlock Holmes measured the temperature of a dead body to be 38°F. One hour later, the temperature was 36°F. After a quick calculation using Newton's law of cooling, and taking the normal temperature of a living body to be 98°F, Holmes concluded that the time of death was 10 a.m. Was Holmes right?

27. At 4 p.m., a hot coal was pulled out of a furnace and allowed to cool at room temperature (75°F). If, after 10 minutes, the temperature of the coal was 415°F, and after 20 minutes, its temperature was 347°F, find the following:

(a) The temperature of the furnace.

(b) The time when the temperature of the coal was 100°F.

28. A hot object is placed in a room whose temperature is 72°F. After one minute the temperature of the object is 150°F and its rate of change of temperature is 20°F per minute. Find the initial temperature of the object and the rate at which its temperature is changing after 10 minutes.

1.5   Some Simple Population Models

In this section we consider two important models of population growth whose mathematical formulation leads to separable differential equations.

Malthusian Growth

The simplest mathematical model of population growth is obtained by assuming that the rate of increase of the population at any time is proportional to the size of the population at that time. If we let $P(t)$ denote the population at time t, then

$$\frac{dP}{dt} = kP,$$

where k is a positive constant. Separating the variables and integrating yields

$$P(t) = P_0 e^{kt}, \tag{1.5.1}$$

where P_0 denotes the population at $t = 0$. This law predicts an exponential increase in the population with time, which gives a reasonably accurate description of the growth of certain algae, bacteria, and cell cultures. It is called the **Malthusian growth model**. The time taken for such a culture to double in size is called the **doubling time**. This is the time, t_d, when $P(t_d) = 2P_0$. Substituting into (1.5.1) yields

$$2P_0 = P_0 e^{kt_d}.$$

Dividing both sides by P_0 and taking logarithms, we find

$$kt_d = \ln 2,$$

so that the doubling time is

$$t_d = \frac{1}{k} \ln 2.$$

Example 1.5.1   The number of bacteria in a certain culture grows at a rate that is proportional to the number present. If the number increased from 500 to 2000 in 2 hours, determine

1. the number present after 12 hours.
2. the doubling time.

Solution: The behavior of the system is governed by the differential equation

$$\frac{dP}{dt} = kP,$$

so that

$$P(t) = P_0 e^{kt},$$

where the time t is measured in hours. Taking $t = 0$ as the time when the population was 500, we have $P_0 = 500$. Thus,

$$P(t) = 500 e^{kt}.$$

Further, $P(2) = 2000$ implies that

$$2000 = 500 e^{2k},$$

so that

$$k = \frac{1}{2} \ln 4 = \ln 2.$$

Consequently,

$$P(t) = 500 e^{t \ln 2}.$$

1. The number of bacteria present after 12 hours is therefore

$$P(12) = 500 e^{12 \ln 2} = 500(2^{12}) = 2,048,000.$$

2. The doubling time of the system is

$$t_d = \frac{1}{k} \ln 2 = 1 \text{ hour.} \qquad \square$$

Logistic Population Model

The Malthusian growth law (1.5.1) does not provide an accurate model for the growth of a population over a long time period. To obtain a more realistic model we need to take account of the fact that as the population increases, several factors will begin to affect the growth rate. For example, there will be increased competition for the limited resources that are available, increases in disease, and overcrowding of the limited available space, all of which would serve to slow the growth rate. In order to model this situation mathematically, we modify the differential equation leading to the simple exponential growth law by adding in a term that slows the growth down as the population increases. If we consider a closed environment (neglecting factors such as immigration and emigration), then the rate of change of population can be modeled by the differential equation

$$\frac{dP}{dt} = [B(t) - D(t)]P,$$

where $B(t)$ and $D(t)$ denote the birth rate and death rate per individual, respectively. The simple exponential law corresponds to the case when $B(t) = k$ and $D(t) = 0$. In the more general situation of interest now, the increased competition as the population grows will result in a corresponding increase in the death rate per individual. Perhaps the simplest way to take account of this is to assume that the death rate per individual is directly proportional to the instantaneous population, and that the birth rate per individual remains constant. The resulting initial-value problem governing the population growth can then be written as

$$\frac{dP}{dt} = (B_0 - D_0 P)P, \qquad P(0) = P_0,$$

where B_0 and D_0 are *positive* constants. It is useful to write the differential equation in the equivalent form

$$\frac{dP}{dt} = r\left(1 - \frac{P}{C}\right)P, \tag{1.5.2}$$

where $r = B_0$, and $C = B_0/D_0$. Equation (1.5.2) is called the **logistic equation**, and the corresponding population model is called the **logistic model**. The differential equation (1.5.2) is separable and can be solved without difficulty. Before doing that, however, we give a qualitative analysis of the differential equation.

The constant C in Equation (1.5.2) is called the **carrying capacity** of the population. We see from Equation (1.5.2) that if $P < C$, then $dP/dt > 0$ and the population increases, whereas if $P > C$, then $dP/dt < 0$ and the population decreases. We can therefore interpret C as representing the maximum population that the environment can sustain. We note that $P(t) = C$ is an equilibrium solution to the differential equation, as is $P(t) = 0$. The isoclines for Equation (1.5.2) are determined from

$$r\left(1 - \frac{P}{C}\right)P = k,$$

where k is a constant. This can be written as

$$P^2 - CP + \frac{kC}{r} = 0,$$

so that the isoclines are the lines

$$P = \frac{1}{2}\left(C \pm \sqrt{C^2 - \frac{4kC}{r}}\right).$$

This tells us that the slopes of the solution curves satisfy

$$C^2 - \frac{4kC}{r} \geq 0,$$

so that

$$k \leq rC/4.$$

Furthermore, the largest value that the slope can assume is $k = rC/4$, which corresponds to $P = C/2$. We also note that the slope approaches zero as the solution curves approach the equilibrium solutions $P(t) = 0$ and $P(t) = C$. Differentiating Equation (1.5.2) yields

$$\frac{d^2 P}{dt^2} = r\left[\left(1 - \frac{P}{C}\right)\frac{dP}{dt} - \frac{P}{C}\frac{dP}{dt}\right] = r\left(1 - 2\frac{P}{C}\right)\frac{dP}{dt} = \frac{r^2}{C^2}(C - 2P)(C - P)P,$$

where we have substituted for dP/dt from (1.5.2) and simplified the result. Since $P = C$ and $P = 0$ are solutions to the differential equation (1.5.2), the only points of inflection occur along the line $P = C/2$. The behavior of the concavity is therefore given by the following schematic:

$$\text{sign of } P'' : \quad | + + + + | - - - - | + + + +$$
$$P\text{-interval:} \quad 0 \qquad C/2 \qquad C$$

This information determines the general behavior of the solution curves to the differential equation (1.5.2). Figure 1.5.1 gives a Maple plot of the slope field and some representative solution curves. Of course, such a figure could have been constructed by hand, using the information we have obtained. From Figure 1.5.1, we see that if the initial population is less than the carrying capacity, then the population increases monotonically toward the carrying capacity. Similarly, if the initial population is bigger than the carrying capacity, then the population monotonically decreases toward the carrying capacity. Once more this illustrates the power of the qualitative techniques that have been introduced for analyzing first-order differential equations.

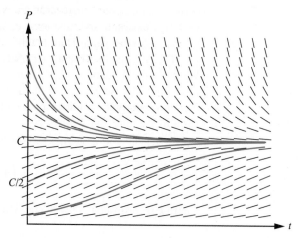

Figure 1.5.1: Representative slope field and some approximate solution curves for the logistic equation.

We turn now to obtaining an analytical solution to the differential equation (1.5.2). Separating the variables in Equation (1.5.2) and integrating yields

$$\int \frac{C}{P(C - P)} \, dP = rt + c_1,$$

where c_1 is an integration constant. Using a partial-fraction decomposition on the left-hand side, we find

$$\int \left(\frac{1}{P} + \frac{1}{C - P}\right) dP = rt + c_1,$$

which upon integration gives

$$\ln\left|\frac{P}{C-P}\right| = rt + c_1.$$

Exponentiating, and redefining the integration constant, yields

$$\frac{P}{C-P} = c_2 e^{rt},$$

which can be solved algebraically for P to obtain

$$P(t) = \frac{c_2 C e^{rt}}{1 + c_2 e^{rt}},$$

or equivalently,

$$P(t) = \frac{c_2 C}{c_2 + e^{-rt}}.$$

Imposing the initial condition $P(0) = P_0$, we find that $c_2 = P_0/(C - P_0)$. Inserting this value of c_2 into the preceding expression for $P(t)$ yields

$$P(t) = \frac{C P_0}{P_0 + (C - P_0)e^{-rt}}. \tag{1.5.3}$$

We make two comments regarding this formula. First, we see that, owing to the negative exponent of the exponential term in the denominator, as $t \to \infty$ the population does indeed tend to the carrying capacity C independently of the initial population P_0. Second, by writing (1.5.3) in the equivalent form

$$P(t) = \frac{P_0}{P_0/C + (1 - P_0/C)e^{-rt}},$$

it follows that if P_0 is very small compared to the carrying capacity, then for small t the terms involving P_0 in the denominator can be neglected, leading to the approximation

$$P(t) \approx P_0 e^{rt}.$$

Consequently, in this case, the Malthusian population model does approximate the logistic model for small time intervals.

Although we now have a formula for the solution to the logistic population model, the qualitative analysis is certainly enlightening with regard to the general overall properties of the solution. Of course if we want to investigate specific details of a particular model, then we use the corresponding exact solution (1.5.3).

Example 1.5.2 The initial population (measured in thousands) of a city is 20. After 10 years this has increased to 50.87, and after 15 years to 78.68. Use the logistic model to predict the population after 30 years.

Solution: In this problem we have $P_0 = P(0) = 20$, $P(10) = 50.87$, $P(15) = 78.68$, and we wish to find $P(30)$. Substituting for P_0 into Equation (1.5.3) yields

$$P(t) = \frac{20C}{20 + (C - 20)e^{-rt}}. \tag{1.5.4}$$

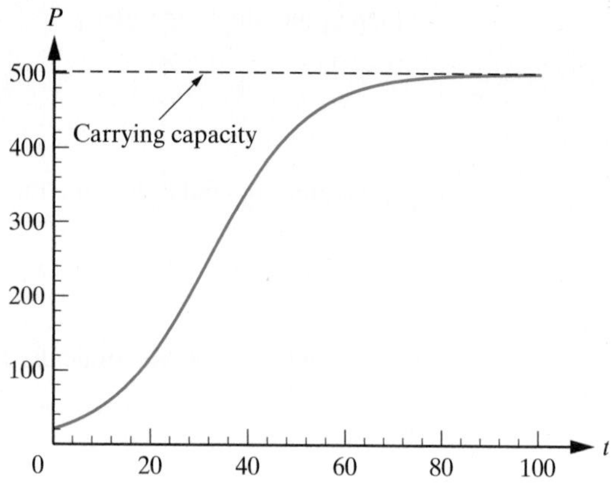

Figure 1.5.2: Solution curve corresponding to the population model in Example 1.5.2. The population is measured in thousands of people.

Imposing the two remaining auxiliary conditions leads to the following pair of equations for determining r and C:

$$50.87 = \frac{20C}{20 + (C - 20)e^{-10r}},$$

$$78.68 = \frac{20C}{20 + (C - 20)e^{-15r}}.$$

This is a pair of nonlinear equations that are tedious to solve by hand. We therefore turn to technology. Using the algebraic capabilities of Maple, we find that

$$r \approx 0.1, \qquad C \approx 500.37.$$

Substituting these values of r and C in Equation (1.5.4) yields

$$P(t) = \frac{10007.4}{20 + 480.37e^{-0.1t}}.$$

Accordingly, the predicted value of the population after 30 years is

$$P(30) = \frac{10007.4}{20 + 480.37e^{-3}} = 227.87.$$

A sketch of $P(t)$ is given in Figure 1.5.2.   □

Exercises for 1.5

Key Terms

Malthusian growth model, Doubling time, Logistic growth model, Carrying capacity.

Skills

- Be able to solve the basic differential equations describing the Malthusian and logistic population growth models.

- Be able to solve word problems involving initial conditions, doubling time, etc., for the Malthusian and logistic population growth models.

- Be able to compute the carrying capacity for a logistic population model.

- Be able to discuss the qualitative behavior of a population governed by a Malthusian or logistic model, based on initial values, doubling time, and so on as a function of time.

True-False Review

For Questions 1–10, decide if the given statement is **true** or **false**, and give a brief justification for your answer. If true, you can quote a relevant definition or theorem from the text. If false, provide an example, illustration, or brief explanation of why the statement is false.

1. A population whose growth rate at any given time is proportional to its size at that time obeys the Malthusian growth model.

2. If a population obeys the logistic growth model, then its size can never exceed the carrying capacity of the population.

3. The differential equations which describe population growth according to the Malthusian model and the logistic model are both separable.

4. The rate of change of a population whose growth is described with the logistic model eventually tends toward zero, regardless of the initial population.

5. If the doubling time of a population governed by the Malthusian growth model is five minutes, then the initial population increases 64-fold in a half-hour.

6. If a population whose growth is based on the Malthusian growth model has a doubling time of 10 years, then it takes approximately 30–40 years in order for the initial population size to increase tenfold.

7. The population growth rate according to the Malthusian growth model is always constant.

8. The logistic population model always has exactly two equilibrium solutions.

9. The concavity of the graph of population governed by the logistic model changes if and only if the initial population is less than the carrying capacity.

10. The concavity of the graph of a population governed by the Malthusian growth model never changes, regardless of the initial population.

Problems

1. The number of bacteria in a culture grows at a rate proportional to the number present. Initially there were 10 bacteria in the culture. If the doubling time of the culture is 3 hours, find the number of bacteria that were present after 24 hours.

2. The number of bacteria in a culture grows at a rate proportional to the number present. After 10 hours, there were 5000 bacteria present, and after 12 hours, 6000 bacteria. Determine the initial size of the culture and the doubling time of the population.

3. A certain cell culture has a doubling time of 4 hours. Initially there were 2000 cells present. Assuming an exponential growth law, determine the time it takes for the culture to contain 10^6 cells.

4. At time t, the population $P(t)$ of a certain city is increasing at a rate proportional to the number of residents in the city at that time. In January 1990 the population of the city was 10,000, and by 1995 it had risen to 20,000.

   (a) What will the population of the city be at the beginning of the year 2010?

   (b) In what year will the population reach one million?

In the logistic population model (1.5.3), if $P(t_1) = P_1$ and $P(2t_1) = P_2$, then it can be shown (through some algebra performed tediously by hand, or easily on a computer algebra system) that

$$r = \frac{1}{t_1} \ln \left[\frac{P_2(P_1 - P_0)}{P_0(P_2 - P_1)} \right], \qquad (1.5.5)$$

$$C = \frac{P_1[P_1(P_0 + P_2) - 2P_0P_2]}{P_1^2 - P_0P_2}. \qquad (1.5.6)$$

These formulas will be used in Problems 5–7.

5. The initial population in a small village is 500. After 5 years this has grown to 800 and after 10 years to 1000. Using the logistic population model, determine the population after 15 years.

6. An animal sanctuary had an initial population of 50 animals. After two years the population was 62 and after four years 76. Using the logistic population model, determine the carrying capacity and the number of animals in the sanctuary after 20 years.

7. (a) Using Equations (1.5.5) and (1.5.6), and the fact that r and C are positive, derive two inequalities that P_0, P_1, P_2 must satisfy in order for there to be a solution to the logistic equation satisfying the conditions

   $$P(0) = P_0, \quad P(t_1) = P_1, \quad P(2t_1) = P_2.$$

   (b) The initial population in a town is 10,000. After 5 years this has grown to 12,000, and after 10 years to 18,000. Is there a solution to the logistic equation that fits this data?

8. Of the 1500 passengers, crew, and staff that board a cruise ship, 5 have the flu. After one day of sailing, the number of infected people has risen to 10. Assuming that the rate at which the flu virus spreads is proportional to the product of the number of infected individuals and the number not yet infected, determine how many people will have the flu at the end of the 14-day cruise. Would you like to be a member of the customer relations department for the cruise line the day after the ship docks?

9. Consider the population model

$$\frac{dP}{dt} = r(P - T)P, \quad P(0) = P_0, \qquad (1.5.7)$$

where r, T, and P_0 are positive constants.

(a) Perform a qualitative analysis of the differential equation in the initial-value problem (1.5.7), following the steps used in the text for the logistic equation. Identify the equilibrium solutions, the isoclines, and the behavior of the slope and concavity of the solution curves.

(b) Using the information obtained in (a), sketch the slope field for the differential equation and include representative solution curves.

(c) What predictions can you make regarding the behavior of the population? Consider the cases $P_0 < T$ and $P_0 > T$. The constant T is called the **threshold level**. Based on your predictions, why is this an appropriate term to use for T?

10. In the preceding problem, a qualitative analysis of the differential equation in (1.5.7) was carried out. In this problem, we determine the exact solution to the differential equation and verify the predictions from the qualitative analysis.

(a) Solve the initial-value problem (1.5.7).

(b) Using your solution from (a), verify that if $P_0 < T$, then $\lim\limits_{t \to \infty} P(t) = 0$. What does this mean for the population?

(c) Using your solution from (a), verify that if $P_0 > T$, then each solution curve has a vertical asymptote at $t = t_e$, where

$$t_e = \frac{1}{rT} \ln\left(\frac{P_0}{P_0 - T}\right).$$

How do you interpret this result in terms of population growth? Note that this was not obvious from the qualitative analysis performed in the previous problem.

11. As a modification to the population model considered in the previous two problems, suppose that $P(t)$ satisfies the initial-value problem

$$\frac{dP}{dt} = r(C - P)(P - T)P, \quad P(0) = P_0,$$

where r, C, T, P_0 are positive constants, and $0 < T < C$. Perform a qualitative analysis of this model. Sketch the slope field and some representative solution curves in the three cases $0 < P_0 < T$, $T < P_0 < C$, and $P_0 > C$. Describe the behavior of the corresponding solutions.

The next two problems consider the **Gompertz** population model, which is governed by the initial-value problem

$$\frac{dP}{dt} = rP(\ln C - \ln P), \quad P(0) = P_0, \qquad (1.5.8)$$

where r, C, and P_0 are positive constants.

12. Determine all equilibrium solutions for the differential equation in (1.5.8), and the behavior of the slope and concavity of the solution curves. Use this information to sketch the slope field and some representative solution curves.

13. Solve the initial-value problem (1.5.8) and verify that all solutions satisfy $\lim\limits_{t \to \infty} P(t) = C$.

Problems 14–16 consider the phenomenon of exponential decay. This occurs when a population $P(t)$ is governed by the differential equation

$$\frac{dP}{dt} = kP,$$

where k is a *negative* constant.

14. A population of swans in a wildlife sanctuary is declining due to the presence of dangerous chemicals in the water. If the population of swans is experiencing exponential decay, and if there were 400 swans in the park at the beginning of the summer and 340 swans 30 days later,

(a) How many swans are in the park 60 days after the start of summer? 100 days after the start of summer?

(b) How long does it take for the population of swans to be cut in half? (This is known as the **half-life** of the population.)

15. At the conclusion of the Super Bowl, the number of fans remaining in the stadium decreases at a rate proportional to the number of fans in the stadium. Assume that there are 100,000 fans in the stadium at the end of the Super Bowl and ten minutes later there are 80,000 fans in the stadium.

(a) Thirty minutes after the Super Bowl will there be more or less than 40,000 fans? How do you know this without doing any calculations?

(b) What is the half-life (see the previous problem) for the fan population in the stadium?

(c) When will there be only 15,000 fans left in the stadium?

(d) Explain why the exponential decay model for the population of fans in the stadium is not realistic from a qualitative perspective.

16. Cobalt-60, an isotope used in cancer therapy, decays exponentially with a half-life of 5.2 years (i.e., half the original sample remains after 5.2 years). How long does it take for a sample of cobalt-60 to disintegrate to the extent that only 4% of the original amount remains?

17. ◇ Use some form of technology to solve the pair of equations

$$P_1 = \frac{C P_0}{P_0 + (C - P_0)e^{-rt_1}},$$

$$P_2 = \frac{C P_0}{P_0 + (C - P_0)e^{-2rt_1}},$$

for r and C, and thereby derive the expressions given in Equations (1.5.5) and (1.5.6).

18. ◇ According to data from the U.S. Bureau of the Census, the population (measured in millions of people) of the United States in 1950, 1960, and 1970 was, respectively, 151.3, 179.4, and 203.3.

(a) Using the 1950 and 1960 population figures, solve the corresponding Malthusian population model.

(b) Determine the logistic model corresponding to the given data.

(c) On the same set of axes, plot the solution curves obtained in (a) and (b). From your plots, determine the values the different models would have predicted for the population in 1980 and 1990, and compare these predictions to the actual values of 226.54 and 248.71, respectively.

19. ◇ In a period of five years, the population of a city doubles from its initial size of 50 (measured in thousands of people). After ten more years, the population has reached 250. Determine the logistic model corresponding to this data. Sketch the solution curve and use your plot to estimate the time it will take for the population to reach 95% of the carrying capacity.

1.6   First-Order Linear Differential Equations

In this section we derive a technique for determining the general solution to any first-order linear differential equation. This is the most important technique in the chapter.

DEFINITION   1.6.1

A differential equation that can be written in the form

$$a(x)\frac{dy}{dx} + b(x)y = r(x) \tag{1.6.1}$$

where $a(x)$, $b(x)$, and $r(x)$ are functions defined on an interval (a, b), is called a **first-order linear differential equation**.

We assume that $a(x) \neq 0$ on (a, b) and divide both sides of (1.6.1) by $a(x)$ to obtain the **standard form**

$$\frac{dy}{dx} + p(x)y = q(x), \tag{1.6.2}$$

where $p(x) = b(x)/a(x)$ and $q(x) = r(x)/a(x)$. The idea behind the solution technique for (1.6.2) is to rewrite the differential equation in the form

$$\frac{d}{dx}[g(x, y)] = F(x)$$

for an appropriate function $g(x, y)$. The general solution to the differential equation can then be obtained by an integration with respect to x. First consider an example.

Example 1.6.2   Solve the differential equation

$$\frac{dy}{dx} + \frac{1}{x}y = e^x, \qquad x > 0. \tag{1.6.3}$$

Solution: If we multiply (1.6.3) by x, we obtain

$$x\frac{dy}{dx} + y = xe^x.$$

But, from the product rule for differentiation, the left-hand side of this equation is just the expanded form of $\frac{d}{dx}(xy)$. Thus (1.6.3) can be written in the equivalent form

$$\frac{d}{dx}(xy) = xe^x.$$

Integrating both sides of this equation with respect to x, we obtain

$$xy = xe^x - e^x + c.$$

Dividing by x yields the general solution to (1.6.3) as

$$y(x) = x^{-1}[e^x(x - 1) + c],$$

where c is an arbitrary constant. $\square$

In the preceding example we multiplied the given differential equation by the function $I(x) = x$. This had the effect of reducing the left-hand side of the resulting differential equation to the integrable form

$$\frac{d}{dx}(xy).$$

Motivated by this example, we now consider the possibility of multiplying the general linear differential equation

$$\frac{dy}{dx} + p(x)y = q(x) \tag{1.6.4}$$

by a nonzero function $I(x)$, chosen in such a way that the left-hand side of the resulting differential equation is

$$\frac{d}{dx}[I(x)y].$$

Henceforth we will assume that the functions p and q are continuous on (a, b). Multiplying the differential equation (1.6.4) by $I(x)$ yields

$$I\frac{dy}{dx} + p(x)Iy = Iq(x). \tag{1.6.5}$$

Furthermore, from the product rule for derivatives, we know that

$$\frac{d}{dx}(Iy) = I\frac{dy}{dx} + \frac{dI}{dx}y. \tag{1.6.6}$$

Comparing Equations (1.6.5) and (1.6.6), we see that Equation (1.6.5) can indeed be written in the integrable form

$$\frac{d}{dx}(Iy) = Iq(x),$$

provided the function $I(x)$ is a solution to[5]

$$I\frac{dy}{dx} + p(x)Iy = I\frac{dy}{dx} + \frac{dI}{dx}y.$$

This will hold whenever $I(x)$ satisfies the separable differential equation

$$\frac{dI}{dx} = p(x)I. \tag{1.6.7}$$

Separating the variables and integrating yields

$$\ln|I| = \int p(x)\,dx + c,$$

so that

$$I(x) = c_1 e^{\int p(x)dx},$$

where c_1 is an arbitrary constant. Since we require only one solution to Equation (1.6.7), we set $c_1 = 1$, in which case

$$I(x) = e^{\int p(x)dx}.$$

We can therefore draw the following conclusion.
  Multiplying the linear differential equation

$$\frac{dy}{dx} + p(x)y = q(x) \tag{1.6.8}$$

by $I(x) = e^{\int p(x)dx}$ reduces it to the integrable form

$$\frac{d}{dx}\left[e^{\int p(x)dx}y\right] = q(x)e^{\int p(x)dx}. \tag{1.6.9}$$

  The general solution to (1.6.8) can now be obtained from (1.6.9) by integration. Formally we have

$$y(x) = e^{-\int p(x)\,dx}\left[\int q(x)e^{\int p(x)dx}\,dx + c\right]. \tag{1.6.10}$$

---

[5] This is obtained by equating the left-hand side of Equation (1.6.5) to the right-hand side of Equation (1.6.6).

Remarks

1. The function $I(x) = e^{\int p(x)dx}$ is called an **integrating factor** for the differential equation (1.6.8), since it enables us to reduce the differential equation to a form that is directly integrable.

2. It is not necessary to memorize (1.6.10). In a specific problem, we first evaluate the integrating factor $e^{\int p(x)dx}$ and then use (1.6.9).

Example 1.6.3   Solve the initial-value problem

$$\frac{dy}{dx} + xy = xe^{x^2/2}, \qquad y(0) = 1.$$

Solution: An appropriate integrating factor in this case is

$$I(x) = e^{\int x\,dx} = e^{x^2/2}.$$

Multiplying the given differential equation by I and using (1.6.9) yields

$$\frac{d}{dx}(e^{x^2/2}y) = xe^{x^2}.$$

Integrating both sides with respect to x, we obtain

$$e^{x^2/2}y = \tfrac{1}{2}e^{x^2} + c.$$

Hence,

$$y(x) = e^{-x^2/2}(\tfrac{1}{2}e^{x^2} + c).$$

Imposing the initial condition $y(0) = 1$ yields

$$1 = \tfrac{1}{2} + c,$$

so that $c = \tfrac{1}{2}$. Thus the required particular solution is

$$y(x) = \tfrac{1}{2}e^{-x^2/2}(e^{x^2} + 1) = \tfrac{1}{2}(e^{x^2/2} + e^{-x^2/2}) = \cosh(x^2/2). \qquad \square$$

Example 1.6.4   Solve $x\dfrac{dy}{dx} + 2y = \cos x, \; x > 0.$

Solution: We first write the given differential equation in standard form. Dividing by x yields

$$\frac{dy}{dx} + 2x^{-1}y = x^{-1}\cos x. \tag{1.6.11}$$

An integrating factor is

$$I(x) = e^{\int 2x^{-1}dx} = e^{2\ln x} = x^2,$$

so that upon multiplying Equation (1.6.11) by I, we obtain

$$\frac{d}{dx}(x^2 y) = x \cos x.$$

Integrating and rearranging gives

$$y(x) = x^{-2}(x \sin x + \cos x + c),$$

where we have used integration by parts on the right-hand side.   □

| Example 1.6.5 |

Solve the initial-value problem

$$y' - y = f(x), \qquad y(0) = 0,$$

where $f(x) = \begin{cases} 1, & \text{if } x < 1, \\ 2 - x, & \text{if } x \geq 1. \end{cases}$

Solution: We have sketched $f(x)$ in Figure 1.6.1. An integrating factor for the differential equation is $I(x) = e^{-x}$.

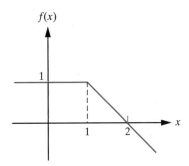

Figure 1.6.1: A sketch of the function $f(x)$ from Example 1.6.5.

Upon multiplication by the integrating factor, the differential equation reduces to

$$\frac{d}{dx}(e^{-x} y) = e^{-x} f(x).$$

We now integrate this differential equation over the interval $[0, x]$. To do so we need to use a dummy integration variable, which we denote by w. We therefore obtain

$$\left[e^{-w} y(w) \right]_0^x = \int_0^x e^{-w} f(w)\, dw,$$

or equivalently,

$$e^{-x} y(x) - y(0) = \int_0^x e^{-w} f(w)\, dw.$$

Multiplying by e^x and substituting for $y(0) = 0$ yields

$$y(x) = e^x \int_0^x e^{-w} f(w)\, dw. \tag{1.6.12}$$

Owing to the form of $f(x)$, the value of the integral on the right-hand side will depend on whether $x < 1$ or $x \geq 1$. If $x < 1$, then $f(w) = 1$, and so (1.6.12) can be written as

$$y(x) = e^x \int_0^x e^{-w} \, dw = e^x(1 - e^{-x}),$$

so that

$$y(x) = e^x - 1, \qquad x < 1.$$

If $x \geq 1$, then the interval of integration $[0, x]$ must be split into two parts. From (1.6.12) we have

$$y(x) = e^x \left[\int_0^1 e^{-w} \, dw + \int_1^x (2 - w)e^{-w} \, dw \right].$$

A straightforward integration leads to

$$y(x) = e^x \left\{ (1 - e^{-1}) + \left[-2e^{-w} + we^{-w} + e^{-w} \right]_1^x \right\},$$

which simplifies to

$$y(x) = e^x(1 - e^{-1}) + x - 1.$$

The solution to the initial-value problem can therefore be written as

$$y(x) = \begin{cases} e^x - 1, & \text{if } x < 1, \\ e^x(1 - e^{-1}) + x - 1, & \text{if } x \geq 1. \end{cases}$$

A sketch of the corresponding solution curve is given in Figure 1.6.2.

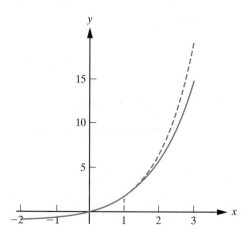

Figure 1.6.2: The solution curve for the initial-value problem in Example 1.6.5. The dashed curve is the continuation of $y(x) = e^x - 1$ for $x > 1$.

Differentiating both branches of this function, we find

$$y'(x) = \begin{cases} e^x, & \text{if } x < 1, \\ e^x(1 - e^{-1}) + 1, & \text{if } x \geq 1. \end{cases} \qquad y''(x) = \begin{cases} e^x, & \text{if } x < 1, \\ e^x(1 - e^{-1}), & \text{if } x \geq 1. \end{cases}$$

We see that even though the function f in the original differential equation was not differentiable at $x = 1$, the solution to the initial-value problem has a continuous derivative at that point. The discontinuity in the derivative of the driving term does show up in the second derivative of the solution, as indeed it must. □

Exercises for 1.6

Key Terms

First-order linear differential equation, Integrating factor.

Skills

- Be able to recognize a first-order linear differential equation.

- Be able to find an integrating factor for a given first-order linear differential equation.

- Be able to solve a first-order linear differential equation.

True-False Review

For Questions 1–5, decide if the given statement is **true** or **false**, and give a brief justification for your answer. If true, you can quote a relevant definition or theorem from the text. If false, provide an example, illustration, or brief explanation of why the statement is false.

1. There is a unique integrating factor for a differential equation of the form $y' + p(x)y = q(x)$.

2. An integrating factor for the differential equation $y' + p(x)y = q(x)$ is $e^{\int p(x)dx}$.

3. Upon multiplying the differential equation $y' + p(x)y = q(x)$ by an integrating factor $I(x)$, the differential equation becomes $(I(x) \cdot y)' = q(x)I$.

4. An integrating factor for the differential equation

$$\frac{dy}{dx} = x^2 y + \sin x$$

is $I(x) = e^{\int x^2 dx}$.

5. An integrating factor for the differential equation

$$\frac{dy}{dx} = x - \frac{y}{x}$$

is $I(x) = 5x$.

Problems

For Problems 1–14, solve the given differential equation.

1. $\dfrac{dy}{dx} - y = e^{2x}$.

2. $x^2 y' - 4xy = x^7 \sin x, \quad x > 0$.

3. $y' + 2xy = 2x^3$.

4. $\dfrac{dy}{dx} + \dfrac{2x}{1-x^2} y = 4x, \quad -1 < x < 1$.

5. $\dfrac{dy}{dx} + \dfrac{2x}{1+x^2} y = \dfrac{4}{(1+x^2)^2}$.

6. $2(\cos^2 x)y' + y \sin 2x = 4\cos^4 x$, $\quad 0 \le x < \pi/2$.

7. $y' + \dfrac{1}{x \ln x} y = 9x^2$.

8. $y' - y \tan x = 8 \sin^3 x$.

9. $t\dfrac{dx}{dt} + 2x = 4e^t, \quad t > 0$.

10. $y' = \sin x (y \sec x - 2)$.

11. $(1 - y \sin x)\, dx - (\cos x)\, dy = 0$.

12. $y' - x^{-1}y = 2x^2 \ln x$.

13. $y' + \alpha y = e^{\beta x}$, where α, β are constants.

14. $y' + mx^{-1}y = \ln x$, where m is constant.

In Problems 15–20, solve the given initial-value problem.

15. $y' + 2x^{-1}y = 4x, \quad y(1) = 2$.

16. $(\sin x)y' - y \cos x = \sin 2x, \quad y(\pi/2) = 2$.

17. $\dfrac{dx}{dt} + \dfrac{2}{4-t}x = 5, \quad x(0) = 4$.

18. $(y - e^x)\, dx + dy = 0, \quad y(0) = 1$.

19. $y' + y = f(x), \quad y(0) = 3$, where

$$f(x) = \begin{cases} 1, & \text{if } x \le 1, \\ 0, & \text{if } x > 1. \end{cases}$$

20. $y' - 2y = f(x), \quad y(0) = 1$, where

$$f(x) = \begin{cases} 1 - x, & \text{if } x < 1, \\ 0, & \text{if } x \ge 1. \end{cases}$$

21. Solve the initial-value problem in Example 1.6.5 as follows. First determine the general solution to the differential equation on each interval separately. Then use the given initial condition to find the appropriate integration constant for the interval $(-\infty, 1)$. To determine the integration constant on the interval $[1, \infty)$, use the fact that the solution must be continuous at $x = 1$.

22. Find the general solution to the second-order differential equation

$$\frac{d^2y}{dx^2} + \frac{1}{x}\frac{dy}{dx} = 9x, \quad x > 0.$$

[**Hint:** Let $u = dy/dx$.]

23. Solve the differential equation for Newton's law of cooling by viewing it as a first-order linear differential equation.

24. Suppose that an object is placed in a medium whose temperature is increasing at a constant rate of $\alpha°$F per minute. Show that, according to Newton's law of cooling, the temperature of the object at time t is given by

$$T(t) = \alpha(t - k^{-1}) + c_1 + c_2e^{-kt},$$

where c_1 and c_2 are constants.

25. Between 8 a.m. and 12 p.m. on a hot summer day, the temperature rose at a rate of $10°$F per hour from an initial temperature of $65°$F. At 9 a.m. the temperature of an object was measured to be $35°$F and was, at that time, increasing at a rate of $5°$F per hour. Show that the temperature of the object at time t was

$$T(t) = 10t - 15 + 40e^{(1-t)/8}, \quad 0 \leq t \leq 4.$$

26. It is known that a certain object has constant of proportionality $k = 1/40$ in Newton's law of cooling. When the temperature of this object is $0°$F, it is placed in a medium whose temperature is changing in time according to

$$T_m(t) = 80e^{-t/20}.$$

(a) Using Newton's law of cooling, show that the temperature of the object at time t is

$$T(t) = 80(e^{-t/40} - e^{-t/20}).$$

(b) What happens to the temperature of the object as $t \to +\infty$? Is this reasonable?

(c) Determine the time, t_{max}, when the temperature of the object is a maximum. Find $T(t_{max})$ and $T_m(t_{max})$.

(d) Make a sketch to depict the behavior of $T(t)$ and $T_m(t)$.

27. The differential equation

$$\frac{dT}{dt} = -k_1[T - T_m(t)] + A_0, \quad (1.6.13)$$

where k_1 and A_0 are positive constants, can be used to model the temperature variation $T(t)$ in a building. In this equation, the first term on the right-hand side gives the contribution due to the variation in the outside temperature, and the second term on the right-hand side gives the contribution due to the heating effect from internal sources such as machinery, lighting, people, and so on. Consider the case when

$$T_m(t) = A - B\cos\omega t, \quad \omega = \pi/12, \quad (1.6.14)$$

where A and B are constants, and t is measured in hours.

(a) Make a sketch of $T_m(t)$. Taking $t = 0$ to correspond to midnight, describe the variation of the external temperature over a 24-hour period.

(b) With T_m given in (1.6.14), solve (1.6.13) subject to the initial condition $T(0) = T_0$.

28. This problem demonstrates the **variation-of-parameters** method for first-order linear differential equations. Consider the first-order linear differential equation

$$y' + p(x)y = q(x). \quad (1.6.15)$$

(a) Show that the general solution to the associated *homogeneous* equation

$$y' + p(x)y = 0$$

is

$$y_H(x) = c_1e^{-\int p(x)dx}.$$

(b) Determine the function $u(x)$ such that

$$y(x) = u(x)e^{-\int p(x)dx}$$

is a solution to (1.6.15), and hence derive the general solution to (1.6.15).

For Problems 29–32, use the technique derived in the previous problem to solve the given differential equation.

29. $y' + x^{-1}y = \cos x, \quad x > 0.$

30. $y' + y = e^{-2x}.$

31. $y' + y\cot x = 2\cos x, \quad 0 < x < \pi.$

32. $xy' - y = x^2\ln x.$

For Problems 33–38, use a differential equation solver to determine the solution to each of the initial-value problems and sketch the corresponding solution curve.

33. ◇ The initial-value problem in Problem 15.

34. ◇ The initial-value problem in Problem 16.

35. ◇ The initial-value problem in Problem 17.

36. ◇ The initial-value problem in Problem 18.

37. ◇ The initial-value problem in Problem 19.

38. ◇ The initial-value problem in Problem 20.

1.7 Modeling Problems Using First-Order Linear Differential Equations

There are many examples of applied problems whose mathematical formulation leads to a first-order linear differential equation. In this section we analyze two in detail.

Mixing Problems

Statement of the Problem: Consider the situation depicted in Figure 1.7.1. A tank initially contains V_0 liters of a solution in which is dissolved A_0 grams of a certain chemical. A solution containing c_1 grams/liter of the same chemical flows into the tank at a constant rate of r_1 liters/minute, and the mixture flows out at a constant rate of r_2 liters/minute. We assume that the mixture is kept uniform by stirring. Then at any time t the concentration of chemical in the tank, $c_2(t)$, is the same throughout the tank and is given by

$$c_2 = \frac{A(t)}{V(t)}, \tag{1.7.1}$$

where $V(t)$ denotes the volume of solution in the tank at time t and $A(t)$ denotes the amount of chemical in the tank at time t.

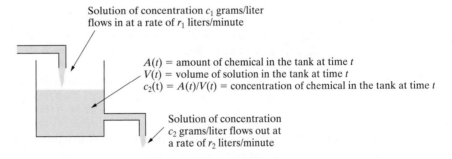

Solution of concentration c_1 grams/liter
flows in at a rate of r_1 liters/minute

$A(t)$ = amount of chemical in the tank at time t
$V(t)$ = volume of solution in the tank at time t
$c_2(t)$ = $A(t)/V(t)$ = concentration of chemical in the tank at time t

Solution of concentration
c_2 grams/liter flows out at
a rate of r_2 liters/minute

Figure 1.7.1: A mixing problem.

Mathematical Formulation: The two functions in the problem are $V(t)$ and $A(t)$. In order to determine how they change with time, we first consider their change during a short time interval, Δt minutes. In time Δt, $r_1 \Delta t$ liters of solution flow into the tank, whereas $r_2 \Delta t$ liters flow out. Thus during the time interval Δt, the *change* in the volume of solution in the tank is

$$\Delta V = r_1 \Delta t - r_2 \Delta t = (r_1 - r_2) \Delta t. \tag{1.7.2}$$

Since the concentration of chemical in the inflow is c_1 grams/liter (assumed constant), it follows that in the time interval Δt the amount of chemical that flows into the tank is $c_1 r_1 \Delta t$. Similarly, the amount of chemical that flows out in this same time interval is approximately[6] $c_2 r_2 \Delta t$. Thus, the total change in the amount of chemical in the tank

---

[6] This is only an approximation, since c_2 is *not* constant over the time interval Δt. The approximation will become more accurate as $\Delta t \to 0$.

during the time interval Δt, denoted by ΔA, is approximately

$$\Delta A \approx c_1 r_1 \, \Delta t - c_2 r_2 \, \Delta t = (c_1 r_1 - c_2 r_2) \, \Delta t. \tag{1.7.3}$$

Dividing Equations (1.7.2) and (1.7.3) by Δt yields

$$\frac{\Delta V}{\Delta t} = r_1 - r_2 \quad \text{and} \quad \frac{\Delta A}{\Delta t} \approx c_1 r_1 - c_2 r_2,$$

respectively. These equations describe the rates of change of V and A over the short, but finite, time interval Δt. In order to determine the instantaneous rates of change of V and A, we take the limit as $\Delta t \to 0$ to obtain

$$\frac{dV}{dt} = r_1 - r_2 \tag{1.7.4}$$

and

$$\frac{dA}{dt} = c_1 r_1 - \frac{A}{V} r_2, \tag{1.7.5}$$

where we have substituted for c_2 from Equation (1.7.1). Since r_1 and r_2 are constants, we can integrate Equation (1.7.4) directly, obtaining

$$V(t) = (r_1 - r_2)t + V_0,$$

where V_0 is an integration constant. Substituting for V into Equation (1.7.5) and rearranging terms yields the linear equation for $A(t)$:

$$\frac{dA}{dt} + \frac{r_2}{(r_1 - r_2)t + V_0} A = c_1 r_1. \tag{1.7.6}$$

This differential equation can be solved, subject to the initial condition $A(0) = A_0$, to determine the behavior of $A(t)$.

Remark　The reader need not memorize Equation (1.7.6), since it is better to derive it for each specific example.

Example 1.7.1　A tank contains 8 L (liters) of water in which is dissolved 32 g (grams) of chemical. A solution containing 2 g/L of the chemical flows into the tank at a rate of 4 L/min, and the well-stirred mixture flows out at a rate of 2 L/min.

1. Determine the amount of chemical in the tank after 20 minutes.
2. What is the concentration of chemical in the tank at that time?

Solution:　We are given

$$r_1 = 4 \text{ L/min}, \quad r_2 = 2 \text{ L/min}, \quad c_1 = 2 \text{ g/L}, \quad V(0) = 8 \text{ L}, \quad \text{and} \quad A(0) = 32 \text{ g}.$$

For parts 1 and 2, we must find $A(20)$ and $A(20)/V(20)$, respectively. Now,

$$\Delta V = r_1 \, \Delta t - r_2 \, \Delta t$$

implies that

$$\frac{dV}{dt} = 2.$$

Integrating this equation and imposing the initial condition that $V(0) = 8$ yields

$$V(t) = 2(t + 4). \tag{1.7.7}$$

Further,

$$\Delta A \approx c_1 r_1 \, \Delta t - c_2 r_2 \, \Delta t$$

implies that

$$\frac{dA}{dt} = 8 - 2c_2.$$

That is, since $c_2 = A/V$,

$$\frac{dA}{dt} = 8 - 2\frac{A}{V}.$$

Substituting for V from (1.7.7), we must solve

$$\frac{dA}{dt} + \frac{1}{t+4}A = 8. \tag{1.7.8}$$

This first-order linear equation has integrating factor

$$I = e^{\int 1/(t+4)dt} = t + 4.$$

Consequently (1.7.8) can be written in the equivalent form

$$\frac{d}{dt}[(t+4)A] = 8(t+4),$$

which can be integrated directly to obtain

$$(t+4)A = 4(t+4)^2 + c.$$

Hence

$$A(t) = \frac{1}{t+4}[4(t+4)^2 + c].$$

Imposing the given initial condition $A(0) = 32$ g implies that $c = 64$. Consequently

$$A(t) = \frac{4}{t+4}[(t+4)^2 + 16].$$

Setting $t = 20$ gives us the values for parts 1 and 2:

1. We have

$$A(20) = \frac{1}{6}[(24)^2 + 16] = \frac{296}{3} \text{ g}.$$

2. Furthermore, using (1.7.7),

$$\frac{A(20)}{V(20)} = \frac{1}{48} \cdot \frac{296}{3} = \frac{37}{18} \text{ g/L}. \qquad \square$$

Electric Circuits

An important application of differential equations arises from the analysis of simple electric circuits. The most basic electric circuit is obtained by connecting the ends of a wire to the terminals of a battery or generator. This causes a flow of charge, $q(t)$, measured in coulombs (C), through the wire, thereby producing a current, $i(t)$, measured in amperes (A), defined to be the rate of change of charge. Thus,

$$i(t) = \frac{dq}{dt}. \tag{1.7.9}$$

In practice a circuit will contain several components that oppose the flow of charge. As current passes through these components, work has to be done, and the loss of energy is described by the resulting voltage drop across each component. For the circuits that we will consider, the behavior of the current in the circuit is governed by Kirchoff's second law, which can be stated as follows.

Kirchoff's Second Law: The sum of the voltage drops around a closed circuit is zero.

In order to apply this law we need to know the relationship between the current passing through each component in the circuit and the resulting voltage drop. The components of interest to us are resistors, capacitors, and inductors. We briefly describe each of these next.

1. *Resistors*: A resistor is a component that, owing to its constituency, directly resists the flow of charge through it. According to *Ohm's law*, the voltage drop, ΔV_R, between the ends of a resistor is directly proportional to the current that is passing through it. This is expressed mathematically as

$$\Delta V_R = i R \tag{1.7.10}$$

   where the constant of proportionality, R, is called the **resistance** of the resistor. The units of resistance are ohms (Ω).

2. *Capacitors*: A capacitor can be thought of as a component that stores charge and thereby opposes the passage of current. If $q(t)$ denotes the charge on the capacitor at time t, then the drop in voltage, ΔV_C, as current passes through it is directly proportional to $q(t)$. It is usual to express this law in the form

$$\Delta V_C = \frac{1}{C} q, \tag{1.7.11}$$

   where the constant C is called the **capacitance** of the capacitor. The units of capacitance are farads (F).

3. *Inductors*: The third component that is of interest to us is an inductor. This can be considered as a component that opposes any change in the current flowing through it. The drop in voltage as current passes through an inductor is directly proportional to the rate at which the current is changing. We write this as

$$\Delta V_L = L \frac{di}{dt}, \tag{1.7.12}$$

   where the constant L is called the **inductance** of the inductor, measured in units of henrys (H).

4. *EMF*: The final component in our circuits will be a source of voltage that produces an electromotive force (EMF), driving the charge through the circuit. As current passes through the voltage source, there is a voltage gain, which we denote by $E(t)$ volts (that is, a voltage drop of $-E(t)$ volts).

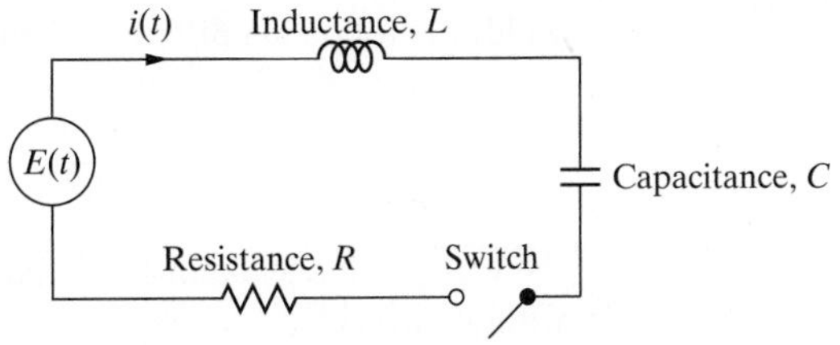

Figure 1.7.2: A simple RLC circuit.

A circuit containing all of these components is shown in Figure 1.7.2. Such a circuit is called an **RLC circuit**. According to Kirchoff's second law, the sum of the voltage drops at any instant must be zero. Applying this to the RLC circuit in Figure 1.7.2, we obtain

$$\Delta V_R + \Delta V_C + \Delta V_L - E(t) = 0. \tag{1.7.13}$$

Substituting into Equation (1.7.13) from (1.7.10)–(1.7.12) and rearranging yields the basic differential equation for an RLC circuit—namely,

$$L\frac{di}{dt} + Ri + \frac{q}{C} = E(t). \tag{1.7.14}$$

Three cases are important in applications, two of which are governed by first-order linear differential equations.

Case 1: An RL CIRCUIT.   In the case when no capacitor is present, we have what is referred to as an RL circuit. The differential equation (1.7.14) then reduces to

$$\frac{di}{dt} + \frac{R}{L}i = \frac{1}{L}E(t). \tag{1.7.15}$$

This is a first-order linear differential equation for the current in the circuit at any time t.

Case 2: An RC CIRCUIT.   Now consider the case when no inductor is present in the circuit. Setting $L = 0$ in Equation (1.7.14) yields

$$i + \frac{1}{RC}q = \frac{E}{R}.$$

In this equation we have two unknowns, $q(t)$ and $i(t)$. Substituting from (1.7.9) for $i(t) = dq/dt$, we obtain the following differential equation for $q(t)$:

$$\frac{dq}{dt} + \frac{1}{RC}q = \frac{E}{R}. \tag{1.7.16}$$

In this case, the first-order linear differential equation (1.7.16) can be solved for the charge $q(t)$ on the plates of the capacitor. The current in the circuit can then be obtained from

$$i(t) = \frac{dq}{dt}$$

by differentiation.

Case 3: An RLC CIRCUIT.   In the general case, we must consider all three components to be present in the circuit. Substituting from Equation (1.7.9) into Equation (1.7.14) yields the following differential equation for determining the charge on the capacitor:

$$\frac{d^2q}{dt^2} + \frac{R}{L}\frac{dq}{dt} + \frac{1}{LC}q = \frac{1}{L}E(t).$$

We will develop techniques in Chapter 6 that enable us to solve this differential equation without difficulty.

For the remainder of this section we restrict our attention to RL and RC circuits. Since these are both first-order linear differential equations, we can solve them using the technique derived in the previous section, once the applied EMF, $E(t)$, has been specified. The two most important forms for $E(t)$ are

$$E(t) = E_0 \quad \text{and} \quad E(t) = E_0 \cos \omega t,$$

where E_0 and ω are constants. The first of these corresponds to a source of EMF such as a battery. The resulting current is called a **direct current** (DC). The second form of EMF oscillates between $\pm E_0$ and is called an **alternating current** (AC).

Example 1.7.2   Determine the current in an RL circuit if the applied EMF is $E(t) = E_0 \cos \omega t$, where E_0 and ω are constants, and the initial current is zero.

Solution: Substituting into Equation (1.7.15) for $E(t)$ yields the differential equation

$$\frac{di}{dt} + \frac{R}{L}i = \frac{E_0}{L}\cos \omega t,$$

which we write as

$$\frac{di}{dt} + ai = \frac{E_0}{L}\cos \omega t, \tag{1.7.17}$$

where $a = R/L$. An integrating factor for (1.7.17) is $I(t) = e^{at}$, so that the equation can be written in the equivalent form

$$\frac{d}{dt}(e^{at}i) = \frac{E_0}{L}e^{at}\cos \omega t.$$

Integrating this equation using the standard integral

$$\int e^{at}\cos \omega t \, dt = \frac{1}{a^2 + \omega^2}e^{at}(a\cos \omega t + \omega \sin \omega t) + c,$$

we obtain

$$e^{at}i = \frac{E_0}{L(a^2 + \omega^2)}e^{at}(a\cos \omega t + \omega \sin \omega t) + c,$$

where c is an integration constant. Consequently,

$$i(t) = \frac{E_0}{L(a^2 + \omega^2)}(a\cos \omega t + \omega \sin \omega t) + ce^{-at}.$$

Imposing the initial condition $i(0) = 0$, we find

$$c = -\frac{E_0 a}{L(a^2 + \omega^2)},$$

so that

$$i(t) = \frac{E_0}{L(a^2 + \omega^2)}(a \cos \omega t + \omega \sin \omega t - a e^{-at}). \qquad (1.7.18)$$

This solution can be written in the form

$$i(t) = i_S(t) + i_T(t),$$

where

$$i_S(t) = \frac{E_0}{L(a^2 + \omega^2)}(a \cos \omega t + \omega \sin \omega t), \qquad i_T(t) = -\frac{aE_0}{L(a^2 + \omega^2)}e^{-at}.$$

The term $i_T(t)$ decays exponentially with time and is referred to as the **transient part** of the solution. As $t \to \infty$, the solution (1.7.18) approaches the **steady-state solution**, $i_S(t)$. The steady-state solution can be written in a more illuminating form as follows. If we construct the right-angled triangle (see Figure 1.7.3) with sides a and ω, then the hypotenuse of the triangle is $\sqrt{a^2 + \omega^2}$. Consequently, there exists a unique angle ϕ in $(0, \pi/2)$, such that

$$\cos \phi = \frac{a}{\sqrt{a^2 + \omega^2}}, \qquad \sin \phi = \frac{\omega}{\sqrt{a^2 + \omega^2}}.$$

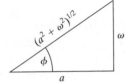

Equivalently,

$$a = \sqrt{a^2 + \omega^2}\cos \phi, \qquad \omega = \sqrt{a^2 + \omega^2}\sin \phi.$$

Figure 1.7.3: Defining the phase angle for an RL circuit.

Substituting for a and ω into the expression for i_S yields

$$i_S(t) = \frac{E_0}{L\sqrt{a^2 + \omega^2}}(\cos \omega t \cos \phi + \sin \omega t \sin \phi),$$

which can be written, using an appropriate trigonometric identity, as

$$i_S(t) = \frac{E_0}{L\sqrt{a^2 + \omega^2}}\cos(\omega t - \phi).$$

This is referred to as the phase-amplitude form of the solution. Comparing this with the original driving term, $E_0 \cos \omega t$, we see that the system has responded with a steady-state solution having the same periodic behavior, but with a phase shift of ϕ radians. Furthermore the amplitude of the response is

$$A = \frac{E_0}{L\sqrt{a^2 + \omega^2}} = \frac{E_0}{\sqrt{R^2 + \omega^2 L^2}}, \qquad (1.7.19)$$

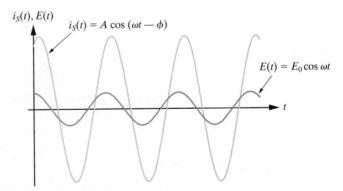

Figure 1.7.4: The response of an RL circuit to the driving term $E(t) = E_0 \cos \omega t$.

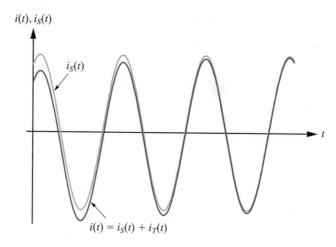

Figure 1.7.5: The transient part of the solution for an RL circuit dies out as t increases.

where we have substituted for $a = R/L$. This is illustrated in Figure 1.7.4. The general picture that we have, therefore, is that the transient part of the solution affects $i(t)$ for a short period of time, after which the current settles into a steady-state. In the case when the driving EMF has the form $E(t) = E_0 \cos \omega t$, the steady-state is a phase shift of this driving EMF with an amplitude given in Equation (1.7.19). This general behavior is illustrated in Figure 1.7.5. ☐

Our next example illustrates the procedure for solving the differential equation (1.7.16) governing the behavior of an RC circuit.

Example 1.7.3   Consider the RC circuit in which $R = 0.5\ \Omega$, $C = 0.1$ F, and $E_0 = 20$ V. Given that the capacitor has zero initial charge, determine the current in the circuit after 0.25 seconds.

Solution: In this case we first solve Equation (1.7.16) for $q(t)$ and then determine the current in the circuit by differentiating the result. Substituting for R, C and E into Equation (1.7.16) yields

$$\frac{dq}{dt} + 20q = 40,$$

which has general solution

$$q(t) = 2 + ce^{-20t},$$

where c is an integration constant. Imposing the initial condition $q(0) = 0$ yields $c = -2$, so that

$$q(t) = 2(1 - e^{-20t}).$$

Differentiating this expression for q gives the current in the circuit

$$i(t) = \frac{dq}{dt} = 40e^{-20t}.$$

Consequently,

$$i(0.25) = 40e^{-5} \approx 0.27 \text{ A.}$$

☐

Key Terms

Mixing problem, Concentration, Electric circuit, Kirchoff's second law, Resistor, Capacitor, Inductor, Electromotive force (EMF), RL circuit, RC circuit, RLC circuit, Direct current, Alternating current, Transient solution, Steady-state solution, Phase, Amplitude.

Skills

- Be able to use information about a mixing problem to provide the correct mathematical formulation of the problem.

- Be able to solve mixing problems by deriving and solving the differential equation (1.7.6) for a specific mixing problem and using initial conditions.

- Know the relationship between the charge and the current in an electric circuit.

- Be familiar with the basic components of an electric circuit, such as electromotive force, resistors, capacitors, and inductors.

- Be able to write down and solve the differential equation for the current in an RL circuit and for the charge in an RC circuit, for either a direct current or an alternating current.

- Be able to identify the transient and steady-state components of current in an electric circuit with an alternating current.

- Be able to put the steady-state component of the current in an RL circuit in phase-amplitude form, and identify the phase shift and the amplitude.

True-False Review

For Questions 1–8, decide if the given statement is **true** or **false**, and give a brief justification for your answer. If true, you can quote a relevant definition or theorem from the text. If false, provide an example, illustration, or brief explanation of why the statement is false.

1. The amount of chemical $A(t)$ in a tank at time t is obtained by multiplying the concentration of chemical $c(t)$ in the tank at time t by the volume of the solution, $V(t)$, at time t.

2. If r_1 and r_2 denote the rates at which fluid is flowing into a tank and out of the tank, respectively, then the rate of change of the volume of the tank is $r_2 - r_1$.

3. For the mixing problems described in this section, we assume that the concentration of the chemical entering the tank is independent of time.

4. For the mixing problems described in this section, we assume that the concentration of the chemical leaving the tank is independent of time.

5. Kirchoff's second law states the sum of the voltage drops around a closed circuit is independent of time.

6. The larger the resistance in a resistor, the greater the voltage drop between the ends of the resistor.

7. Given an alternating current in an RL circuit, the transient part of the current decays to zero with time, while the steady-state part of the current oscillates with the same frequency as the applied EMF.

8. The higher the frequency of an applied EMF in an RL circuit, the lower the amplitude of the steady-state current.

Problems

1. A container initially contains 10 L of water in which there is 20 g of salt dissolved. A solution containing 4 g/L of salt is pumped into the container at a rate of 2 L/min, and the well-stirred mixture runs out at a rate of 1 L/min. How much salt is in the tank after 40 minutes?

2. A tank initially contains 600 L of solution in which there is dissolved 1500 g of chemical. A solution containing 5 g/L of the chemical flows into the tank at a rate of 6 L/min, and the well-stirred mixture flows out at a rate of 3 L/min. Determine the concentration of chemical in the tank after one hour.

3. A tank whose volume is 40 L initially contains 20 L of water. A solution containing 10 g/L of salt is pumped into the tank at a rate of 4 L/min, and the well-stirred mixture flows out at a rate of 2 L/min. How much salt is in the tank just before the solution overflows?

4. A tank whose volume is 200 L is initially half full of a solution that contains 100 g of chemical. A solution containing 0.5 g/L of the same chemical flows into the tank at a rate of 6 L/min, and the well-stirred mixture flows out at a rate of 4 L/min. Determine the concentration of chemical in the tank just before the solution overflows.

5. A tank initially contains 10 L of a salt solution. Water flows into the tank at a rate of 3 L/min, and the well-stirred mixture flows out at a rate of 2 L/min. After 5 min, the concentration of salt in the tank is 0.2 g/L. Find:

   (a) The amount of salt in the tank initially.

   (b) The volume of solution in the tank when the concentration of salt is 0.1 g/L.

6. A tank initially contains 20 L of water. A solution containing 1 g/L of chemical flows into the tank at a rate of 3 L/min, and the mixture flows out at a rate of 2 L/min.

   (a) Set up and solve the initial-value problem for $A(t)$, the amount of chemical in the tank at time t.

   (b) When does the concentration of chemical in the tank reach 0.5 g/L?

7. A tank initially contains w liters of a solution in which is dissolved A_0 grams of chemical. A solution containing k g/L of this chemical flows into the tank at a rate of r L/min, and the mixture flows out at the same rate.

   (a) Show that the amount of chemical, $A(t)$, in the tank at time t is

   $$A(t) = e^{-(rt)/w}[kw(e^{(rt)/w} - 1) + A_0].$$

   (b) Show that as $t \to \infty$, the concentration of chemical in the tank approaches k g/L. Is this result reasonable? Explain.

8. Consider the double mixing problem depicted in Figure 1.7.6.

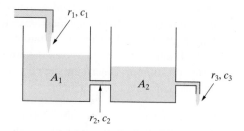

Figure 1.7.6: Double mixing problem

   (a) Show that the following are differential equations for $A_1(t)$ and $A_2(t)$:

   $$\frac{dA_1}{dt} + \frac{r_2}{(r_1 - r_2)t + V_1} A_1 = c_1 r_1,$$

   $$\frac{dA_2}{dt} + \frac{r_3}{(r_2 - r_3)t + V_2} A_2 = \frac{r_2 A_1}{(r_1 - r_2)t + V_1},$$

where V_1 and V_2 are constants.

   (b) Let $r_1 = 6$ L/min, $r_2 = 4$ L/min, $r_3 = 3$ L/min, and $c_1 = 0.5$ g/L. If the first tank initially holds 40 L of water in which 4 grams of chemical is dissolved, whereas the second tank initially contains 20 g of chemical dissolved in 20 L of water, determine the amount of chemical in the second tank after 10 min.

9. Consider the RL circuit in which $R = 4\ \Omega$, $L = 0.1$ H, and $E(t) = 20$ V. If no current is flowing initially, determine the current in the circuit for $t \geq 0$.

10. Consider the RC circuit which has $R = 5\ \Omega$, $C = \frac{1}{50}$ F, and $E(t) = 100$ V. If the capacitor is uncharged initially, determine the current in the circuit for $t \geq 0$.

11. An RL circuit has EMF $E(t) = 10 \sin 4t$ V. If $R = 2\ \Omega$, $L = \frac{2}{3}$ H, and there is no current flowing initially, determine the current for $t \geq 0$.

12. Consider the RC circuit with $R = 2\ \Omega$, $C = \frac{1}{8}$ F, and $E(t) = 10 \cos 3t$ V. If $q(0) = 1$ C, determine the current in the circuit for $t \geq 0$.

13. Consider the general RC circuit with $E(t) = 0$. Suppose that $q(0) = 5$ C. Determine the charge on the capacitor for $t > 0$. What happens as $t \to \infty$? Is this reasonable? Explain.

14. Determine the current in an RC circuit if the capacitor has zero charge initially and the driving EMF is $E = E_0$, where E_0 is a constant. Make a sketch showing the change in the charge $q(t)$ on the capacitor with time and show that $q(t)$ approaches a constant value as $t \to \infty$. What happens to the current in the circuit as $t \to \infty$?

15. Determine the current flowing in an RL circuit if the applied EMF is $E(t) = E_0 \sin \omega t$, where E_0 and ω are constants. Identify the transient part of the solution and the steady-state solution.

16. Determine the current flowing in an RL circuit if the applied EMF is constant and the initial current is zero.

17. Determine the current flowing in an RC circuit if the capacitor is initially uncharged and the driving EMF is given by $E(t) = E_0 e^{-at}$, where E_0 and a are constants.

18. Consider the special case of the RLC circuit in which the resistance is negligible and the driving EMF is zero. The differential equation governing the charge on the capacitor in this case is

$$\frac{d^2q}{dt^2} + \frac{1}{LC}q = 0.$$

If the capacitor has an initial charge of q_0 coulombs,

and no current is flowing initially, determine the charge on the capacitor for $t > 0$, and the corresponding current in the circuit. [**Hint:** Let $u = dq/dt$ and use the chain rule to show that this implies $du/dt = u(du/dq)$.]

19. Repeat the previous problem for the case in which the driving EMF is $E(t) = E_0$, a constant.

1.8   Change of Variables

So far we have introduced techniques for solving separable and first-order linear differential equations. Clearly, most first-order differential equations are not of these two types. In this section, we consider two further types of differential equations that can be solved by using a change of variables to reduce them to one of the types we know how to solve. The key point to grasp, however, is not the specific changes of variables that we discuss, but the general idea of changing variables in a differential equation. Further examples are considered in the exercises. We first require a preliminary definition.

> **DEFINITION   1.8.1**
>
> A function $f(x, y)$ is said to be **homogeneous of degree zero**[7] if
>
> $$f(tx, ty) = f(x, y)$$
>
> for all positive values of t for which (tx, ty) is in the domain of f.

Remark   Equivalently, we can say that f is homogeneous of degree zero if it is invariant under a rescaling of the variables x and y.

The simplest nonconstant functions that are homogeneous of degree zero are $f(x, y) = y/x$, and $f(x, y) = x/y$.

Example 1.8.2   If $f(x, y) = \dfrac{x^2 - y^2}{2xy + y^2}$, then

$$f(tx, ty) = \frac{t^2(x^2 - y^2)}{t^2(2xy + y^2)} = f(x, y),$$

so that f is homogeneous of degree zero.   □

In the previous example, if we factor an x^2 term from the numerator and denominator, then the function f can be written in the form

$$f(x, y) = \frac{x^2[1 - (y/x)^2]}{x^2[2(y/x) + (y/x)^2]}.$$

That is,

$$f(x, y) = \frac{1 - (y/x)^2}{2(y/x) + (y/x)^2}.$$

---

[7]More generally, $f(x, y)$ is said to be **homogeneous of degree** m if $f(tx, ty) = t^m f(x, y)$.

Thus f can be considered to depend on the single variable $V = y/x$. The following theorem establishes that this is a basic property of all functions that are homogeneous of degree zero.

Theorem 1.8.3   A function $f(x, y)$ is homogeneous of degree zero if and only if it depends on y/x only.

Proof Suppose that f is homogeneous of degree zero. We must consider two cases separately.

(a) If $x > 0$, we can take $t = 1/x$ in Definition 1.8.1 to obtain

$$f(x, y) = f(1, y/x),$$

which is a function of $V = y/x$ only.

(b) If $x < 0$, then we can take $t = -1/x$ in Definition 1.8.1. In this case we obtain

$$f(x, y) = f(-1, -y/x),$$

which once more depends on y/x only.

Conversely, suppose that $f(x, y)$ depends only on y/x. If we replace x by tx and y by ty, then f is unaltered, since $y/x = (ty)/(tx)$, and hence is homogeneous of degree zero.   ∎

Remark Do not memorize the formulas in the preceding theorem. Just remember that a function $f(x, y)$ that is homogeneous of degree zero depends only on the combination y/x and hence can be considered as a function of a single variable, say, V, where $V = y/x$.

We now consider solving differential equations that satisfy the following definition.

DEFINITION   1.8.4

If $f(x, y)$ is homogeneous of degree zero, then the differential equation

$$\frac{dy}{dx} = f(x, y)$$

is called a **homogeneous first-order differential equation**.

In general, if

$$\frac{dy}{dx} = f(x, y)$$

is a homogeneous first-order differential equation, then we cannot solve it directly. However, our preceding discussion implies that such a differential equation can be written in the equivalent form

$$\frac{dy}{dx} = F(y/x), \tag{1.8.1}$$

for an appropriate function F. This suggests that, instead of using the variables x and y, we should use the variables x and V, where $V = y/x$, or equivalently,

$$y = xV(x). \tag{1.8.2}$$

Substitution of (1.8.2) into the right-hand side of Equation (1.8.1) has the effect of reducing it to a function of V only. We must also determine how the derivative term dy/dx transforms. Differentiating (1.8.2) with respect to x using the product rule yields the following relationship between dy/dx and dV/dx:

$$\frac{dy}{dx} = x\frac{dV}{dx} + V.$$

Substituting into Equation (1.8.1), we therefore obtain

$$x\frac{dV}{dx} + V = F(V),$$

or equivalently,

$$x\frac{dV}{dx} = F(V) - V.$$

The variables can now be separated to yield

$$\frac{1}{F(V) - V}\, dV = \frac{1}{x}\, dx,$$

which can be solved directly by integration. We have therefore established the next theorem.

Theorem 1.8.5   The change of variables $y = xV(x)$ reduces a homogeneous first-order differential equation $dy/dx = f(x, y)$ to the separable equation

$$\frac{1}{F(V) - V}\, dV = \frac{1}{x}\, dx.$$

Remark   The separable equation that results in the previous technique can be integrated to obtain a relationship between V and x. We then obtain the solution to the given differential equation by substituting y/x for V in this relationship.

Example 1.8.6   Find the general solution to

$$\frac{dy}{dx} = \frac{4x + y}{x - 4y}. \tag{1.8.3}$$

Solution: The function on the right-hand side of Equation (1.8.3) is homogeneous of degree zero, so that we have a first-order homogeneous differential equation. Substituting $y = xV$ into the equation yields

$$\frac{d}{dx}(xV) = \frac{4 + V}{1 - 4V}.$$

That is,

$$x\frac{dV}{dx} + V = \frac{4 + V}{1 - 4V},$$

or equivalently,

$$x\frac{dV}{dx} = \frac{4(1 + V^2)}{1 - 4V}.$$

Separating the variables gives

$$\frac{1 - 4V}{4(1 + V^2)} dV = \frac{1}{x} dx.$$

We write this as

$$\left[\frac{1}{4(1 + V^2)} - \frac{V}{1 + V^2}\right] dV = \frac{1}{x} dx,$$

which can be integrated directly to obtain

$$\frac{1}{4} \arctan V - \frac{1}{2} \ln (1 + V^2) = \ln |x| + c.$$

Substituting $V = y/x$ and multiplying through by 2 yields

$$\frac{1}{2} \arctan \left(\frac{y}{x}\right) - \ln \left(\frac{x^2 + y^2}{x^2}\right) = \ln (x^2) + c_1,$$

which simplifies to

$$\frac{1}{2} \arctan \left(\frac{y}{x}\right) - \ln (x^2 + y^2) = c_1. \tag{1.8.4}$$

Although this technically gives the answer, the solution is more easily expressed in terms of polar coordinates:

$$x = r \cos \theta \quad \text{and} \quad y = r \sin \theta \quad \Longleftrightarrow \quad r = \sqrt{x^2 + y^2} \quad \text{and} \quad \theta = \arctan \left(\frac{y}{x}\right).$$

Substituting into Equation (1.8.4) yields

$$\frac{1}{2}\theta - \ln (r^2) = c_1,$$

or equivalently,

$$\ln r = \frac{1}{4}\theta + c_2.$$

Exponentiating both sides of this equation gives

$$r = c_3 e^{\theta/4}.$$

For each value of c_3, this is the equation of a logarithmic spiral. The particular spiral with equation $r = \frac{1}{2}e^{\theta/4}$ is shown in Figure 1.8.1.

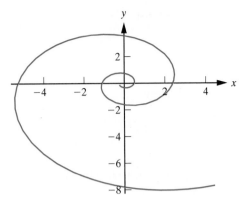

Figure 1.8.1: Graph of the logarithmic spiral with polar equation $r = \frac{1}{2}e^{\theta/4}$, $-5\pi/6 \le \theta \le 22\pi/6$.

☐

Example 1.8.7   Find the equation of the orthogonal trajectories to the family

$$x^2 + y^2 - 2cx = 0. \tag{1.8.5}$$

(Completing the square in x, we obtain $(x-c)^2 + y^2 = c^2$, which represents the family of circles centered at $(c, 0)$, with radius c.)

Solution: First we need an expression for the slope of the given family at the point (x, y). Differentiating Equation (1.8.5) implicitly with respect to x yields

$$2x + 2y\frac{dy}{dx} - 2c = 0,$$

which simplifies to

$$\frac{dy}{dx} = \frac{c - x}{y}. \tag{1.8.6}$$

This is not the differential equation of the given family, since it still contains the constant c and hence is dependent on the individual curves in the family. Therefore, we must eliminate c to obtain an expression for the slope of the family that is independent of any particular curve in the family. From Equation (1.8.5) we have

$$c = \frac{x^2 + y^2}{2x}.$$

Substituting this expression for c into Equation (1.8.6) and simplifying gives

$$\frac{dy}{dx} = \frac{y^2 - x^2}{2xy}.$$

Therefore, the differential equation for the family of orthogonal trajectories is

$$\frac{dy}{dx} = -\frac{2xy}{y^2 - x^2}. \tag{1.8.7}$$

This differential equation is first-order homogeneous. Substituting $y = xV(x)$ into Equation (1.8.7) yields

$$\frac{d}{dx}(xV) = \frac{2V}{1 - V^2},$$

so that

$$x\frac{dV}{dx} + V = \frac{2V}{1 - V^2}.$$

Hence

$$x\frac{dV}{dx} = \frac{V + V^3}{1 - V^2},$$

or in separated form,

$$\frac{1 - V^2}{V(1 + V^2)}\, dV = \frac{1}{x}\, dx.$$

Decomposing the left-hand side into partial fractions yields

$$\left(\frac{1}{V} - \frac{2V}{1 + V^2}\right) dV = \frac{1}{x}\, dx,$$

which can be integrated directly to obtain

$$\ln |V| - \ln (1 + V^2) = \ln |x| + c,$$

or equivalently,

$$\ln\left(\frac{|V|}{1 + V^2}\right) = \ln |x| + c.$$

Exponentiating both sides and redefining the constant yields

$$\frac{V}{1 + V^2} = c_1 x.$$

Substituting back for $V = y/x$, we obtain

$$\frac{xy}{x^2 + y^2} = c_1 x.$$

That is,

$$x^2 + y^2 = c_2 y,$$

where $c_2 = 1/c_1$. Completing the square in y yields

$$x^2 + (y - k)^2 = k^2, \tag{1.8.8}$$

where $k = c_2/2$. Equation (1.8.8) is the equation of the family of orthogonal trajectories. This is the family of circles centered at $(0, k)$ with radius k (circles along the y-axis). (See Figure 1.8.2.)

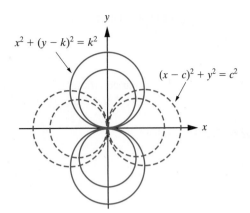

Figure 1.8.2: The family $(x - c)^2 + y^2 = c^2$ and its orthogonal trajectories $x^2 + (y - k)^2 = k^2$.

□

Bernoulli Equations

We now consider a special type of nonlinear differential equation that can be reduced to a linear equation by a change of variables.

DEFINITION   1.8.8

A differential equation that can be written in the form

$$\frac{dy}{dx} + p(x)y = q(x)y^n, \tag{1.8.9}$$

where n is a real constant, is called a **Bernoulli equation**.

If $n = 0$ or $n = 1$, Equation (1.8.9) is linear, but otherwise it is nonlinear. We can reduce it to a linear equation as follows. We first divide Equation (1.8.9) by y^n to obtain

$$y^{-n}\frac{dy}{dx} + y^{1-n}p(x) = q(x). \tag{1.8.10}$$

We now make the change of variables

$$u(x) = y^{1-n}, \tag{1.8.11}$$

which implies that

$$\frac{du}{dx} = (1 - n)y^{-n}\frac{dy}{dx}.$$

That is,

$$y^{-n}\frac{dy}{dx} = \frac{1}{1 - n}\frac{du}{dx}.$$

Substituting into Equation (1.8.10) for y^{1-n} and $y^{-n}dy/dx$ yields the linear differential equation

$$\frac{1}{1 - n}\frac{du}{dx} + p(x)u = q(x),$$

or in standard form,

$$\frac{du}{dx} + (1-n)p(x)u = (1-n)q(x).$$ (1.8.12)

The linear equation (1.8.12) can now be solved for u as a function of x. The solution to the original equation is then obtained from (1.8.11).

Example 1.8.9   Solve

$$\frac{dy}{dx} + \frac{3}{x}y = \frac{12y^{2/3}}{\sqrt{1+x^2}}, \qquad x > 0.$$

Solution: The differential equation is a Bernoulli equation. Dividing both sides of the differential equation by $y^{2/3}$ yields

$$y^{-2/3}\frac{dy}{dx} + \frac{3}{x}y^{1/3} = \frac{12}{\sqrt{1+x^2}}.$$ (1.8.13)

We make the change of variables

$$u = y^{1/3},$$ (1.8.14)

which implies that

$$\frac{du}{dx} = \frac{1}{3}y^{-2/3}\frac{dy}{dx}.$$

Substituting into Equation (1.8.13) yields

$$3\frac{du}{dx} + \frac{3}{x}u = \frac{12}{\sqrt{1+x^2}},$$

or in standard form,

$$\frac{du}{dx} + \frac{1}{x}u = \frac{4}{\sqrt{1+x^2}}.$$ (1.8.15)

An integrating factor for this linear equation is

$$I(x) = e^{\int (1/x)\,dx} = e^{\ln x} = x,$$

so that Equation (1.8.15) can be written as

$$\frac{d}{dx}(xu) = \frac{4x}{\sqrt{1+x^2}}.$$

Integrating, we obtain

$$u(x) = x^{-1}\left(4\sqrt{1+x^2} + c\right),$$

and so, from (1.8.14), the solution to the original differential equation is

$$y^{1/3} = x^{-1}\left(4\sqrt{1+x^2} + c\right). \qquad \square$$

Exercises for 1.8

Key Terms

Homogeneous of degree zero, Homogeneous first-order differential equation, Bernoulli equation.

Skills

- Be able to recognize whether or not a function $f(x, y)$ is homogeneous of degree zero, and whether or not a given differential equation is a homogeneous first-order differential equation.

- Know how to change the variables in a homogeneous first-order differential equation in order to get a differential equation that is separable and thus can be solved.

- Be able to recognize whether or not a given first-order differential equation is a Bernoulli equation.

- Know how to change the variables in a Bernoulli equation in order to get a differential equation that is first-order linear and thus can be solved.

- Be able to make other changes of variables to differential equations in order to turn them into differential equations that can be solved by methods from earlier in this chapter.

True-False Review

For Questions 1–9, decide if the given statement is **true** or **false**, and give a brief justification for your answer. If true, you can quote a relevant definition or theorem from the text. If false, provide an example, illustration, or brief explanation of why the statement is false.

1. The function

$$f(x, y) = \frac{2xy - x^2}{2xy + y^2}$$

is homogeneous of degree zero.

2. The function

$$f(x, y) = \frac{y^2}{x + y^2}$$

is homogeneous of degree zero.

3. The differential equation

$$\frac{dy}{dx} = \frac{1 + xy^2}{1 + x^2 y}$$

is a first-order homogeneous differential equation.

4. The differential equation

$$\frac{dy}{dx} = \frac{x^2 y^2}{x^4 + y^4}$$

is a first-order homogeneous differential equation.

5. The change of variables $y = xV(x)$ always turns a first-order homogeneous differential equation into a separable differential equation for V as a function of x.

6. The change of variables $u = y^{-n}$ always turns a Bernoulli differential equation into a first-order linear differential equation for u as a function of x.

7. The differential equation

$$\frac{dy}{dx} = \sqrt{xy} + \sqrt{xy}$$

is a Bernoulli differential equation.

8. The differential equation

$$\frac{dy}{dx} - e^{xy} y = 5x\sqrt{y}$$

is a Bernoulli differential equation.

9. The differential equation

$$\frac{dy}{dx} + xy = x^2 y^{2/3}$$

is a Bernoulli differential equation.

Problems

For Problems 1–8, determine whether the given function is homogeneous of degree zero. Rewrite those that are as functions of the single variable $V = y/x$.

1. $f(x, y) = \dfrac{x^2 - y^2}{xy}$.

2. $f(x, y) = x - y$.

3. $f(x, y) = \dfrac{x \sin(x/y) - y \cos(y/x)}{y}$.

4. $f(x, y) = \dfrac{\sqrt{x^2 + y^2}}{x - y}$, $x > 0$.

5. $f(x, y) = \dfrac{y}{x - 1}$.

6. $f(x, y) = \dfrac{x - 3}{y} + \dfrac{5y + 9}{3y}$.

7. $f(x, y) = \dfrac{\sqrt{x^2 + y^2}}{x}$, $\quad x < 0$.

8. $f(x, y) = \dfrac{\sqrt{x^2 + 4y^2} - x + y}{x + 3y}$, $\quad x, y \neq 0$.

For Problems 9–22, solve the given differential equation.

9. $(3x - 2y)\dfrac{dy}{dx} = 3y$.

10. $y' = \dfrac{(x + y)^2}{2x^2}$.

11. $\sin\left(\dfrac{y}{x}\right)(xy' - y) = x \cos\left(\dfrac{y}{x}\right)$.

12. $xy' = \sqrt{16x^2 - y^2} + y$, $\quad x > 0$.

13. $xy' - y = \sqrt{9x^2 + y^2}$, $\quad tx > 0$.

14. $y(x^2 - y^2)dx - x(x^2 + y^2)\, dy = 0$.

15. $xy' + y \ln x = y \ln y$.

16. $\dfrac{dy}{dx} = \dfrac{y^2 + 2xy - 2x^2}{x^2 - xy + y^2}$.

17. $2xy\, dy - (x^2 e^{-y^2/x^2} + 2y^2)\, dx = 0$.

18. $x^2 \dfrac{dy}{dx} = y^2 + 3xy + x^2$.

19. $yy' = \sqrt{x^2 + y^2} - x$, $\quad x > 0$.

20. $2x(y + 2x)y' = y(4x - y)$.

21. $x\dfrac{dy}{dx} = x \tan(y/x) + y$.

22. $\dfrac{dy}{dx} = \dfrac{x\sqrt{x^2 + y^2} + y^2}{xy}$, $\quad x > 0$.

23. Solve the differential equation in Example 1.8.6 by first transforming it into polar coordinates. [**Hint:** Write the differential equation in differential form and then express dx and dy in terms of r and θ.]

For Problems 24–26, solve the given initial-value problem.

24. $\dfrac{dy}{dx} = \dfrac{2(2y - x)}{x + y}$, $\quad y(1) = 2$.

25. $\dfrac{dy}{dx} = \dfrac{2x - y}{x + 4y}$, $\quad y(1) = 1$.

26. $\dfrac{dy}{dx} = \dfrac{y - \sqrt{x^2 + y^2}}{x}$, $\quad y(3) = 4$.

27. Find *all* solutions to
$$x\dfrac{dy}{dx} - y = \sqrt{4x^2 - y^2}, \quad x > 0.$$

28. (a) Show that the general solution to the differential equation
$$\dfrac{dy}{dx} = \dfrac{x + ay}{ax - y}$$
can be written in polar form as $r = ke^{a\theta}$.

(b) For the particular case when $a = 1/2$, determine the solution satisfying the initial condition $y(1) = 1$, and find the maximum x-interval on which this solution is valid. [**Hint:** When does the solution curve have a vertical tangent?]

(c) ◇ On the same set of axes, sketch the spiral corresponding to your solution in (b), and the line $y = x/2$. Thus verify the x-interval obtained in (b) with the graph.

For Problems 29–30, determine the orthogonal trajectories to the given family of curves. Sketch some curves from each family.

29. $x^2 + y^2 = 2cy$.

30. $(x - c)^2 + (y - c)^2 = 2c^2$.

31. Fix a real number m. Let S_1 denote the family of circles, centered on the line $y = mx$, each member of which passes through the origin.

(a) Show that the equation of S_1 can be written in the form
$$(x - a)^2 + (y - ma)^2 = a^2(m^2 + 1),$$
where a is a constant that labels particular members of the family.

(b) Determine the equation of the family of orthogonal trajectories to S_1, and show that it consists of the family of circles centered on the line $x = -my$ that pass through the origin.

(c) ◇ Sketch some curves from both families when $m = \sqrt{3}/3$.

Let F_1 and F_2 be two families of curves with the property that whenever a curve from the family F_1 intersects one from the family F_2, it does so at an angle $\alpha \neq \pi/2$. If we know the equation of F_2, then it can be shown (see Problem 26 in Section 1.1) that the differential equation for determining F_1 is

$$\frac{dy}{dx} = \frac{m_2 - \tan\alpha}{1 + m_2 \tan\alpha}, \qquad (1.8.16)$$

where m_2 denotes the slope of the family F_2 at the point (x, y).

For Problems 32–34, use Equation (1.8.16) to determine the equation of the family of curves that cuts the given family at an angle $\alpha = \pi/4$.

32. $x^2 + y^2 = c$.

33. $y = cx^6$.

34. $x^2 + y^2 = 2cx$.

35. (a) Use Equation (1.8.16) to find the equation of the family of curves that intersects the family of hyperbolas $y = c/x$ at an angle $\alpha = \alpha_0$.

   (b) ◇ When $\alpha_0 = \pi/4$, sketch several curves from each family.

36. (a) Use Equation (1.8.16) to show that the family of curves that intersects the family of concentric circles $x^2 + y^2 = c$ at an angle $\alpha = \tan^{-1} m$ has polar equation $r = ke^{m\theta}$.

   (b) ◇ When $\alpha = \pi/6$, sketch several curves from each family.

For Problems 37–49, solve the given differential equation.

37. $y' - x^{-1}y = 4x^2 y^{-1}\cos x, \quad x > 0$.

38. $\dfrac{dy}{dx} + \dfrac{1}{2}(\tan x)y = 2y^3 \sin x$.

39. $\dfrac{dy}{dx} - \dfrac{3}{2x}y = 6y^{1/3}x^2 \ln x$.

40. $y' + 2x^{-1}y = 6\sqrt{1 + x^2}\sqrt{y}, \quad x > 0$.

41. $y' + 2x^{-1}y = 6y^2 x^4$.

42. $2x(y' + y^3 x^2) + y = 0$.

43. $(x - a)(x - b)(y' - \sqrt{y}) = 2(b - a)y$, where a, b are constants.

44. $y' + 6x^{-1}y = 3x^{-1}y^{2/3}\cos x, \quad x > 0$.

45. $y' + 4xy = 4x^3 y^{1/2}$.

46. $\dfrac{dy}{dx} - \dfrac{1}{2x\ln x}y = 2xy^3$.

47. $\dfrac{dy}{dx} - \dfrac{1}{(\pi - 1)x}y = \dfrac{3}{1 - \pi}xy^\pi$.

48. $2y' + y\cot x = 8y^{-1}\cos^3 x$.

49. $(1 - \sqrt{3})y' + y\sec x = y^{\sqrt{3}}\sec x$.

For Problems 50–51, solve the given initial-value problem.

50. $\dfrac{dy}{dx} + \dfrac{2x}{1 + x^2}y = xy^2, \quad y(0) = 1$.

51. $y' + y\cot x = y^3 \sin^3 x, \quad y(\pi/2) = 1$.

52. Consider the differential equation

$$y' = F(ax + by + c), \qquad (1.8.17)$$

where $a, b \neq 0$, and c are constants. Show that the change of variables from x, y to x, V, where

$$V = ax + by + c$$

reduces Equation (1.8.17) to the separable form

$$\frac{1}{bF(V) + a}\,dV = dx.$$

For Problems 53–55, use the result from the previous problem to solve the given differential equation. For Problem 53, impose the given initial condition as well.

53. $y' = (9x - y)^2, \quad y(0) = 0$.

54. $y' = (4x + y + 2)^2$.

55. $y' = \sin^2(3x - 3y + 1)$.

56. Show that the change of variables $V = xy$ transforms the differential equation

$$\frac{dy}{dx} = \frac{y}{x}F(xy)$$

into the separable differential equation

$$\frac{1}{V[F(V) + 1]}\frac{dV}{dx} = \frac{1}{x}.$$

57. Use the result from the previous problem to solve

$$\frac{dy}{dx} = \frac{y}{x}[\ln(xy) - 1].$$

58. Consider the differential equation

$$\frac{dy}{dx} = \frac{x + 2y - 1}{2x - y + 3}. \qquad (1.8.18)$$

(a) Show that the change of variables defined by

$$x = u - 1, \quad y = v + 1$$

transforms Equation (1.8.18) into the homogeneous equation

$$\frac{dv}{du} = \frac{u + 2v}{2u - v}. \qquad (1.8.19)$$

(b) Find the general solution to Equation (1.8.19), and hence solve Equation (1.8.18).

59. A differential equation of the form

$$y' + p(x)y + q(x)y^2 = r(x) \qquad (1.8.20)$$

is called a **Riccati** equation.

(a) If $y = Y(x)$ is a known solution to Equation (1.8.20), show that the substitution

$$y = Y(x) + v^{-1}(x)$$

reduces it to the linear equation

$$v' - [p(x) + 2Y(x)q(x)]v = q(x).$$

(b) Find the general solution to the Riccati equation

$$x^2 y' - xy - x^2 y^2 = 1, \quad x > 0,$$

given that $y = -x^{-1}$ is a solution.

60. Consider the Riccati equation

$$y' + 2x^{-1}y - y^2 = -2x^{-2}, \quad x > 0. \qquad (1.8.21)$$

(a) Determine the values of the constants a and r such that $y(x) = ax^r$ is a solution to Equation (1.8.21).

(b) Use the result from part (a) of the previous problem to determine the general solution to Equation (1.8.21).

61. (a) Show that the change of variables $y = x^{-1} + w$ transforms the Riccati differential equation

$$y' + 7x^{-1}y - 3y^2 = 3x^{-2} \qquad (1.8.22)$$

into the Bernoulli equation

$$w' + x^{-1}w = 3w^2. \qquad (1.8.23)$$

(b) Solve Equation (1.8.23), and hence determine the general solution to (1.8.22).

62. Consider the differential equation

$$y^{-1}y' + p(x)\ln y = q(x), \qquad (1.8.24)$$

where $p(x)$ and $q(x)$ are continuous functions on some interval (a, b). Show that the change of variables $u = \ln y$ reduces Equation (1.8.24) to the linear differential equation

$$u' + p(x)u = q(x),$$

and hence show that the general solution to Equation (1.8.24) is

$$y(x) = \exp\left\{ I^{-1}\left[\int I(x)q(x)\, dx + c \right] \right\},$$

where

$$I = e^{\int p(x)dx} \qquad (1.8.25)$$

and c is an arbitrary constant.

63. Use the technique derived in the previous problem to solve the initial-value problem

$$y^{-1}y' - 2x^{-1}\ln y = x^{-1}(1 - 2\ln x),$$

$$y(1) = e.$$

64. Consider the differential equation

$$f'(y)\frac{dy}{dx} + p(x)f(y) = q(x), \qquad (1.8.26)$$

where p and q are continuous functions on some interval (a, b), and f is an invertible function. Show that Equation (1.8.26) can be written as

$$\frac{du}{dx} + p(x)u = q(x),$$

where $u = f(y)$, and hence show that the general solution to Equation (1.8.26) is

$$y(x) = f^{-1}\left\{ I^{-1}\left[\int I(x)q(x)\,dx + c\right]\right\},$$

where I is given in (1.8.25), f^{-1} is the inverse of f,

and c is an arbitrary constant.

65. Solve

$$\sec^2 y\, \frac{dy}{dx} + \frac{1}{2\sqrt{1+x}}\tan y = \frac{1}{2\sqrt{1+x}}.$$

1.9 Exact Differential Equations

For the next technique it is best to consider first-order differential equations written in differential form

$$M(x, y)\, dx + N(x, y)\, dy = 0, \tag{1.9.1}$$

where M and N are given functions, assumed to be sufficiently smooth.[8] The method that we will consider is based on the idea of a differential. Recall from a previous calculus course that if $\phi = \phi(x, y)$ is a function of two variables, x and y, then the differential of ϕ, denoted $d\phi$, is defined by

$$d\phi = \frac{\partial \phi}{\partial x}\, dx + \frac{\partial \phi}{\partial y}\, dy. \tag{1.9.2}$$

Example 1.9.1 Solve

$$2x \sin y\, dx + x^2 \cos y\, dy = 0. \tag{1.9.3}$$

Solution: This equation is separable, but we will use a different technique to solve it. By inspection, we notice that

$$2x \sin y\, dx + x^2 \cos y\, dy = d(x^2 \sin y).$$

Consequently, Equation (1.9.3) can be written as $d(x^2 \sin y) = 0$, which implies that $x^2 \sin y$ is constant, hence the general solution to Equation (1.9.3) is

$$\sin y = \frac{c}{x^2},$$

where c is an arbitrary constant. $\square$

In the foregoing example we were able to write the given differential equation in the form $d\phi(x, y) = 0$, and hence obtain its solution. However, we cannot always do this. Indeed we see by comparing Equation (1.9.1) with (1.9.2) that the differential equation

$$M(x, y)\, dx + N(x, y)\, dy = 0$$

can be written as $d\phi = 0$ if and only if

$$M = \frac{\partial \phi}{\partial x} \quad \text{and} \quad N = \frac{\partial \phi}{\partial y}$$

for some function ϕ. This motivates the following definition:

---

[8]This means we assume that the functions M and N have continuous derivatives of sufficiently high order.

DEFINITION   1.9.2

The differential equation

$$M(x, y)\,dx + N(x, y)\,dy = 0$$

is said to be **exact** in a region R of the xy-plane if there exists a function $\phi(x, y)$ such that

$$\frac{\partial \phi}{\partial x} = M, \qquad \frac{\partial \phi}{\partial y} = N, \tag{1.9.4}$$

for all (x, y) in R.

Any function ϕ satisfying (1.9.4) is called a **potential function** for the differential equation

$$M(x, y)\,dx + N(x, y)\,dy = 0.$$

We emphasize that if such a function exists, then the preceding differential equation can be written as

$$d\phi = 0.$$

This is why such a differential equation is called an exact differential equation. From the previous example, a potential function for the differential equation

$$2x \sin y\,dx + x^2 \cos y\,dy = 0$$

is

$$\phi(x, y) = x^2 \sin y.$$

We now show that if a differential equation is exact and we can find a potential function ϕ, its solution can be written down immediately.

Theorem 1.9.3   The general solution to an exact equation

$$M(x, y)\,dx + N(x, y)\,dy = 0$$

is defined implicitly by

$$\phi(x, y) = c,$$

where ϕ satisfies (1.9.4) and c is an arbitrary constant.

Proof   We rewrite the differential equation in the form

$$M(x, y) + N(x, y)\frac{dy}{dx} = 0.$$

Since the differential equation is exact, there exists a potential function ϕ (see (1.9.4)) such that

$$\frac{\partial \phi}{\partial x} + \frac{\partial \phi}{\partial y}\frac{dy}{dx} = 0.$$

But this implies that $\partial \phi / \partial x = 0$. Consequently, $\phi(x, y)$ is a function of y only. By a similar argument, which we leave to the reader, we can deduce that $\phi(x, y)$ is a function of x only. We conclude therefore that $\phi(x, y) = c$, where c is a constant. ∎

Remarks

1. The potential function ϕ is a function of two variables x and y, and we interpret the relationship $\phi(x, y) = c$ as defining y implicitly as a function of x. The preceding theorem states that this relationship defines the general solution to the differential equation for which ϕ is a potential function.

2. Geometrically, Theorem 1.9.3 says that the solution curves of an exact differential equation are the family of curves $\phi(x, y) = k$, where k is a constant. These are called the **level curves** of the function $\phi(x, y)$.

The following two questions now arise:

1. How can we tell whether a given differential equation is exact?

2. If we have an exact equation, how do we find a potential function?

The answers are given in the next theorem and its proof.

Theorem 1.9.4   **(Test for Exactness)** Let M, N, and their first partial derivatives M_y and N_x, be continuous in a (simply connected[9]) region R of the xy-plane. Then the differential equation

$$M(x, y)\, dx + N(x, y)\, dy = 0$$

is exact for all x, y in R if and only if

$$\frac{\partial M}{\partial y} = \frac{\partial N}{\partial x}. \tag{1.9.5}$$

Proof We first prove that exactness implies the validity of Equation (1.9.5). If the differential equation is exact, then by definition there exists a potential function $\phi(x, y)$ such that $\phi_x = M$ and $\phi_y = N$. Thus, taking partial derivatives, $\phi_{xy} = M_y$ and $\phi_{yx} = N_x$. Since M_y and N_x are continuous in R, it follows that ϕ_{xy} and ϕ_{yx} are continuous in R. But, from multivariable calculus, this implies that $\phi_{xy} = \phi_{yx}$ and hence that $M_y = N_x$.

We now prove the converse. Thus we assume that Equation (1.9.5) holds and must prove that there exists a potential function ϕ such that

$$\frac{\partial \phi}{\partial x} = M \tag{1.9.6}$$

and

$$\frac{\partial \phi}{\partial y} = N. \tag{1.9.7}$$

The proof is constructional. That is, we will actually find a potential function ϕ. We begin by integrating Equation (1.9.6) with respect to x, holding y fixed (this is a partial integration) to obtain

$$\phi(x, y) = \int^x M(s, y)\, ds + h(y), \tag{1.9.8}$$

---

[9]Roughly speaking, simply connected means that the interior of any closed curve drawn in the region also lies in the region. For example, the interior of a circle is a simply connected region, although the region between two concentric circles is not.

where $h(y)$ is an arbitrary function of y (this is the integration "constant" that we must allow to depend on y, since we held y fixed in performing the integration[10]). We now show how to determine $h(y)$ so that the function f defined in (1.9.8) also satisfies Equation (1.9.7). Differentiating (1.9.8) partially with respect to y yields

$$\frac{\partial \phi}{\partial y} = \frac{\partial}{\partial y} \int^x M(s, y)\, ds + \frac{dh}{dy}.$$

In order that ϕ satisfy Equation (1.9.7) we must choose $h(y)$ to satisfy

$$\frac{\partial}{\partial y} \int^x M(s, y)\, ds + \frac{dh}{dy} = N(x, y).$$

That is,

$$\frac{dh}{dy} = N(x, y) - \frac{\partial}{\partial y} \int^x M(s, y)\, ds. \tag{1.9.9}$$

Since the left-hand side of this expression is a function of y only, we must show, for consistency, that the right-hand side also depends only on y. Taking the derivative of the right-hand side with respect to x yields

$$\begin{aligned}
\frac{\partial}{\partial x}\left(N - \frac{\partial}{\partial y} \int^x M(s, y)\, ds \right) &= \frac{\partial N}{\partial x} - \frac{\partial^2}{\partial x \partial y} \int^x M(s, y)\, ds \\
&= \frac{\partial N}{\partial x} - \frac{\partial}{\partial y}\left(\frac{\partial}{\partial x} \int^x M(s, y)\, ds \right) \\
&= \frac{\partial N}{\partial x} - \frac{\partial M}{\partial y}.
\end{aligned}$$

Thus, using (1.9.5), we have

$$\frac{\partial}{\partial x}\left(N - \frac{\partial}{\partial y} \int^x M(s, y)\, ds \right) = 0,$$

so that the right-hand side of Equation (1.9.9) does depend only on y. It follows that (1.9.9) is a consistent equation, and hence we can integrate both sides with respect to y to obtain

$$h(y) = \int^y N(x, t)\, dt - \int^y \frac{\partial}{\partial t}\left(\int^x M(s, t)\, ds \right) dt.$$

Finally, substituting into (1.9.8) yields the potential function

$$\phi(x, y) = \int^x M(s, y)\, dx + \int^y N(x, t)\, dt - \int^y \frac{\partial}{\partial t}\left(\int^x M(s, t)\, ds \right) dt. \quad \blacksquare$$

Remark   There is no need to memorize the final result for ϕ. For each particular problem, one can construct an appropriate potential function from first principles. This is illustrated in Examples 1.9.6 and 1.9.7.

---

[10]Throughout the text, $\int^x f(t)\, dt$ means "evaluate the indefinite integral $\int f(t)\, dt$ and replace t with x in the result."

Example 1.9.5 Determine whether the given differential equation is exact.

1. $[1 + \ln{(xy)}]\,dx + (x/y)\,dy = 0$.

2. $x^2 y\,dx - (xy^2 + y^3)\,dy = 0$.

Solution:

1. In this case, $M = 1 + \ln{(xy)}$ and $N = x/y$, so that $M_y = 1/y = N_x$. It follows from the previous theorem that the differential equation is exact.

2. In this case, we have $M = x^2 y$, $N = -(xy^2 + y^3)$, so that $M_y = x^2$, whereas $N_x = -y^2$. Since $M_y \neq N_x$, the differential equation is not exact. ☐

Example 1.9.6 Find the general solution to $2xe^y\,dx + (x^2 e^y + \cos y)\,dy = 0$.

Solution: We have

$$M(x, y) = 2xe^y, \qquad N(x, y) = x^2 e^y + \cos y,$$

so that

$$M_y = 2xe^y = N_x.$$

Hence the given differential equation is exact, and so there exists a potential function ϕ such that (see Definition 1.9.2)

$$\frac{\partial \phi}{\partial x} = 2xe^y, \tag{1.9.10}$$

$$\frac{\partial \phi}{\partial y} = x^2 e^y + \cos y. \tag{1.9.11}$$

Integrating Equation (1.9.10) with respect to x, holding y fixed, yields

$$\phi(x, y) = x^2 e^y + h(y), \tag{1.9.12}$$

where h is an arbitrary function of y. We now determine $h(y)$ such that (1.9.12) also satisfies Equation (1.9.11). Taking the derivative of (1.9.12) with respect to y yields

$$\frac{\partial \phi}{\partial y} = x^2 e^y + \frac{dh}{dy}. \tag{1.9.13}$$

Equations (1.9.11) and (1.9.13) give two expressions for $\partial \phi / \partial y$. This allows us to determine h. Subtracting Equation (1.9.11) from Equation (1.9.13) gives the consistency requirement

$$\frac{dh}{dy} = \cos y,$$

which implies, upon integration, that

$$h(y) = \sin y,$$

where we have set the integration constant equal to zero without loss of generality, since we require only one potential function. Substitution into (1.9.12) yields the potential function

$$\phi(x, y) = x^2 e^y + \sin y.$$

Consequently, the given differential equation can be written as

$$d(x^2 e^y + \sin y) = 0,$$

and so, from Theorem 1.9.3, the general solution is

$$x^2 e^y + \sin y = c. \qquad \square$$

Notice that the solution obtained in the preceding example is an implicit solution. Owing to the nature of the way in which the potential function for an exact equation is obtained, this is usually the case.

Example 1.9.7   Find the general solution to

$$\left[\sin(xy) + xy \cos(xy) + 2x\right] dx + \left[x^2 \cos(xy) + 2y\right] dy = 0.$$

Solution: We have

$$M(x, y) = \sin(xy) + xy \cos(xy) + 2x \qquad \text{and} \qquad N(x, y) = x^2 \cos(xy) + 2y.$$

Thus,

$$M_y = 2x \cos(xy) - x^2 y \sin(xy) = N_x,$$

and so the differential equation is exact. Hence there exists a potential function $\phi(x, y)$ such that

$$\frac{\partial \phi}{\partial x} = \sin(xy) + xy \cos(xy) + 2x, \tag{1.9.14}$$

$$\frac{\partial \phi}{\partial y} = x^2 \cos(xy) + 2y. \tag{1.9.15}$$

In this case, Equation (1.9.15) is the simpler equation, and so we integrate it with respect to y, holding x fixed, to obtain

$$\phi(x, y) = x \sin(xy) + y^2 + g(x), \tag{1.9.16}$$

where $g(x)$ is an arbitrary function of x. We now determine $g(x)$, and hence ϕ, from (1.9.14) and (1.9.16). Differentiating (1.9.16) partially with respect to x yields

$$\frac{\partial \phi}{\partial x} = \sin(xy) + xy \cos(xy) + \frac{dg}{dx}. \tag{1.9.17}$$

Equations (1.9.14) and (1.9.17) are consistent if and only if

$$\frac{dg}{dx} = 2x.$$

Hence, upon integrating,

$$g(x) = x^2,$$

where we have once more set the integration constant to zero without loss of generality, since we require only one potential function. Substituting into (1.9.16) gives the potential function

$$\phi(x, y) = x \sin xy + x^2 + y^2.$$

The original differential equation can therefore be written as

$$d(x \sin xy + x^2 + y^2) = 0,$$

and hence the general solution is

$$x \sin xy + x^2 + y^2 = c. \qquad \qquad \square$$

Remark   At first sight the above procedure appears to be quite complicated. However, with a little bit of practice, the steps are seen to be, in fact, fairly straightforward. As we have shown in Theorem 1.9.4, the method works in general, provided one starts with an exact differential equation.

Integrating Factors

Usually a given differential equation will not be exact. However, sometimes it is possible to multiply the differential equation by a nonzero function to obtain an exact equation that can then be solved using the technique we have described in this section. Notice that the solution to the resulting exact equation will be the same as that of the original equation, since we multiply by a nonzero function.

> ### DEFINITION   1.9.8
>
> A nonzero function $I(x, y)$ is called an **integrating factor** for the differential equation $M(x, y)dx + N(x, y)dy = 0$ if the differential equation
>
> $$I(x, y)M(x, y) \, dx + I(x, y)N(x, y) \, dy = 0$$
>
> is exact.

Example 1.9.9   Show that $I = x^2 y$ is an integrating factor for the differential equation

$$(3y^2 + 5x^2 y) \, dx + (3xy + 2x^3) \, dy = 0. \qquad (1.9.18)$$

Solution: Multiplying the given differential equation (which is not exact) by $x^2 y$ yields

$$(3x^2 y^3 + 5x^4 y^2) \, dx + (3x^3 y^2 + 2x^5 y) \, dy = 0. \qquad (1.9.19)$$

Thus,

$$M_y = 9x^2 y^2 + 10x^4 y = N_x,$$

so that the differential equation (1.9.19) is exact, and hence $I = x^2 y$ is an integrating factor for Equation (1.9.18). Indeed we leave it as an exercise to verify that (1.9.19) can be written as

$$d(x^3 y^3 + x^5 y^2) = 0,$$

so that the general solution to Equation (1.9.19) (and hence the general solution to Equation (1.9.18)) is defined implicitly by

$$x^3 y^3 + x^5 y^2 = c.$$

That is,

$$x^3 y^2 (y + x^2) = c. \qquad \square$$

As shown in the next theorem, using the test for exactness, it is straightforward to determine the conditions that a function $I(x, y)$ must satisfy in order to be an integrating factor for the differential equation $M(x, y)\, dx + N(x, y)\, dy = 0$.

Theorem 1.9.10   The function $I(x, y)$ is an integrating factor for

$$M(x, y)\, dx + N(x, y)\, dy = 0 \qquad (1.9.20)$$

if and only if it is a solution to the partial differential equation

$$N\frac{\partial I}{\partial x} - M\frac{\partial I}{\partial y} = \left(\frac{\partial M}{\partial y} - \frac{\partial N}{\partial x}\right) I. \qquad (1.9.21)$$

Proof   Multiplying Equation (1.9.20) by I yields

$$IM\, dx + IN\, dy = 0.$$

This equation is exact if and only if

$$\frac{\partial}{\partial y}(IM) = \frac{\partial}{\partial x}(IN),$$

that is, if and only if

$$\frac{\partial I}{\partial y}M + I\frac{\partial M}{\partial y} = \frac{\partial I}{\partial x}N + I\frac{\partial N}{\partial x}.$$

Rearranging the terms in this equation yields Equation (1.9.21). ∎

The preceding theorem is not too useful in general, since it is usually no easier to solve the partial differential equation (1.9.21) to find I than it is to solve the original Equation (1.9.20). However, it sometimes happens that an integrating factor exists that depends only on one variable. We now show that Theorem 1.9.10 can be used to determine when such an integrating factor exists and also to actually find a corresponding integrating factor.

Theorem 1.9.11   Consider the differential equation $M(x, y)\, dx + N(x, y)\, dy = 0$.

1. There exists an integrating factor that is dependent only on x if and only if $(M_y - N_x)/N = f(x)$, a function of x only. In such a case, an integrating factor is

$$I(x) = e^{\int f(x)\, dx}.$$

2. There exists an integrating factor that is dependent only on y if and only if $(M_y - N_x)/M = g(y)$, a function of y only. In such a case, an integrating factor is

$$I(y) = e^{-\int g(y)\, dy}.$$

Proof For part 1 of the theorem, we begin by assuming that $I = I(x)$ is an integrating factor for $M(x, y)\, dx + N(x, y)\, dy = 0$. Then $\partial I/\partial y = 0$, and so, from (1.9.21), I is a solution to

$$\frac{dI}{dx} N = (M_y - N_x)I.$$

That is,

$$\frac{1}{I} \frac{dI}{dx} = \frac{M_y - N_x}{N}.$$

Since, by assumption, I is a function of x only, it follows that the left-hand side of this expression depends only on x and hence also the right-hand side.

Conversely, suppose that $(M_y - N_x)/N = f(x)$, a function of x only. Then, dividing (1.9.21) by N, it follows that I is an integrating factor for $M(x, y)\, dx + N(x, y)\, dy = 0$ if and only if it is a solution to

$$\frac{\partial I}{\partial x} - \frac{M}{N} \frac{\partial I}{\partial y} = If(x). \tag{1.9.22}$$

We must show that this differential equation has a solution I that depends on x only. We do this by explicitly integrating the differential equation under the assumption that $I = I(x)$. Indeed, if $I = I(x)$, then Equation (1.9.22) reduces to

$$\frac{dI}{dx} = If(x),$$

which is a separable equation with solution

$$I(x) = e^{\int f(x)\, dx}$$

The proof of part 2 is similar, and so we leave it as an exercise (see Problem 30).   ∎

Example 1.9.12   Solve

$$(2x - y^2)\, dx + xy\, dy = 0, \qquad x > 0. \tag{1.9.23}$$

Solution: The equation is not exact ($M_y \neq N_x$). However,

$$\frac{M_y - N_x}{N} = \frac{-2y - y}{xy} = -\frac{3}{x},$$

which is a function of x only. It follows from part 1 of the preceding theorem that an integrating factor for Equation (1.9.23) is

$$I(x) = e^{-\int (3/x)dx} = e^{-3\ln x} = x^{-3}.$$

Multiplying Equation (1.9.23) by I yields the exact equation

$$(2x^{-2} - x^{-3}y^2)\, dx + x^{-2}y\, dy = 0. \tag{1.9.24}$$

(The reader should check that this is exact, although it must be, by the previous theorem.) We leave it as an exercise to verify that a potential function for Equation (1.9.24) is

$$\phi(x, y) = \frac{1}{2}x^{-2}y^2 - 2x^{-1},$$

and hence the general solution to (1.9.23) is given implicitly by

$$\frac{1}{2}x^{-2}y^2 - 2x^{-1} = c,$$

or equivalently,

$$y^2 - 4x = c_1 x^2. \qquad \square$$

Exercises for 1.9

Key Terms

Exact differential equation, Potential function, Integrating factor.

Skills

- Be able to determine whether or not a given differential equation is exact.

- Given the partial derivatives $\partial\phi/\partial x$ and $\partial\phi/\partial y$ of a potential function $\phi(x, y)$, be able to determine $\phi(x, y)$.

- Be able to find the general solution to an exact differential equation.

- When circumstances allow, be able to use an integrating factor to convert a given differential equation into an exact differential equation with the same solution set.

True-False Review

For Questions 1–9, decide if the given statement is **true** or **false**, and give a brief justification for your answer. If true,

you can quote a relevant definition or theorem from the text. If false, provide an example, illustration, or brief explanation of why the statement is false.

1. The differential equation $M(x, y)\, dx + N(x, y)\, dy = 0$ is exact in a simply connected region R if M_x and N_y are continuous partial derivatives with $M_x = N_y$.

2. The solution to an exact differential equation is called a potential function.

3. If $M(x)$ and $N(y)$ are continuous functions, then the differential equation $M(x)\, dx + N(y)\, dy = 0$ is exact.

4. If $(M_y - N_x)/N(x, y)$ is a function of x only, then the differential equation $M(x, y)\, dx + N(x, y)\, dy = 0$ becomes exact when it is multiplied through by

$$I(x) = \exp\left(\int (M_y - N_x)/N(x, y)\, dx \right).$$

5. There is a unique potential function for an exact differential equation $M(x, y)\, dx + N(x, y)\, dy = 0$.

6. The differential equation

$$(2ye^{2x} - \sin y)\, dx + (e^{2x} - x \cos y)\, dy = 0$$

is exact.

7. The differential equation

$$\frac{-2xy}{(x^2 + y)^2}\, dx + \frac{x^2}{(x^2 + y)^2}\, dy = 0$$

is exact.

8. The differential equation

$$(y^2 + \cos x)\, dx + 2xy^2\, dy = 0$$

is exact.

9. The differential equation

$$(e^{x \sin y} \sin y)\, dx + (e^{x \sin y} \cos y)\, dy = 0$$

is exact.

Problems

For Problems 1–3, determine whether the given differential equation is exact.

1. $(y + 3x^2)\, dx + xdy = 0.$

2. $[\cos(xy) - xy \sin(xy)]\, dx - x^2 \sin(xy)\, dy = 0.$

3. $ye^{xy}\, dx + (2y - xe^{xy})\, dy = 0.$

For Problems 4–12, solve the given differential equation.

4. $2xy\, dx + (x^2 + 1)\, dy = 0.$

5. $(y^2 + \cos x)\, dx + (2xy + \sin y)\, dy = 0.$

6. $x^{-1}(xy - 1)\, dx + y^{-1}(xy + 1)\, dy = 0.$

7. $(4e^{2x} + 2xy - y^2)\, dx + (x - y)^2\, dy = 0.$

8. $(y^2 - 2x)\, dx + 2xy\, dy = 0.$

9. $\left(\dfrac{1}{x} - \dfrac{y}{x^2 + y^2}\right) dx + \dfrac{x}{x^2 + y^2}\, dy = 0.$

10. $[1 + \ln(xy)]\, dx + xy^{-1}\, dy = 0.$

11. $[y \cos(xy) - \sin x]\, dx + x \cos(xy)\, dy = 0.$

12. $(2xy + \cos y)\, dx + (x^2 - x \sin y - 2y)\, dy = 0.$

For Problems 13–15, solve the given initial-value problem.

13. $(3x^2 \ln x + x^2 - y)\, dx - xdy = 0,\quad y(1) = 5.$

14. $2x^2 y' + 4xy = 3 \sin x,\quad y(2\pi) = 0.$

15. $(ye^{xy} + \cos x)\, dx + xe^{xy}\, dy = 0,\quad y(\pi/2) = 0.$

16. Show that if $\phi(x, y)$ is a potential function for $M(x, y)\, dx + N(x, y)\, dy = 0$, then so is $\phi(x, y) + c$, where c is an arbitrary constant. This shows that potential functions are uniquely defined only up to an additive constant.

For Problems 17–19, determine whether the given function is an integrating factor for the given differential equation.

17. $I(x, y) = \cos(xy),\ [\tan(xy) + xy]\, dx + x^2\, dy = 0.$

18. $I(x) = \sec x,\ [2x - (x^2 + y^2) \tan x]\, dx + 2y\, dy = 0.$

19. $I(x, y) = y^{-2}e^{-x/y},\ y(x^2 - 2xy)\, dx - x^3\, dy = 0.$

For Problems 20–26, determine an integrating factor for the given differential equation, and hence find the general solution.

20. $(xy - 1)\, dx + x^2\, dy = 0.$

21. $y\, dx - (2x + y^4)\, dy = 0.$

22. $x^2 y\, dx + y(x^3 + e^{-3y} \sin y)\, dy = 0.$

23. $(y - x^2)\, dx + 2xdy = 0,\quad x > 0.$

24. $xy[2 \ln(xy) + 1]\, dx + x^2\, dy = 0,\quad x > 0.$

25. $\dfrac{dy}{dx} + \dfrac{2x}{1 + x^2}y = \dfrac{1}{(1 + x^2)^2}.$

26. $(3xy - 2y^{-1})\, dx + x(x + y^{-2})\, dy = 0.$

For Problems 27–29, determine the values of the constants r and s such that $I(x, y) = x^r y^s$ is an integrating factor for the given differential equation.

27. $(y^{-1} - x^{-1})\, dx + (xy^{-2} - 2y^{-1})\, dy = 0.$

28. $y(5xy^2 + 4)\, dx + x(xy^2 - 1)\, dy = 0.$

29. $2y(y + 2x^2)\, dx + x(4y + 3x^2)\, dy = 0.$

30. Prove that if $(M_y - N_x)/M = g(y)$, a function of y only, then an integrating factor for

$$M(x, y)\, dx + N(x, y)\, dy = 0$$

is $I(y) = e^{-\int g(y)\, dy}.$

31. Consider the general first-order *linear* differential equation

$$\frac{dy}{dx} + p(x)y = q(x), \qquad (1.9.25)$$

where $p(x)$ and $q(x)$ are continuous functions on some interval (a, b).

(a) Rewrite Equation (1.9.25) in differential form, and show that an integrating factor for the resulting equation is

$$I(x) = e^{\int p(x)dx}. \qquad (1.9.26)$$

(b) Show that the general solution to Equation (1.9.25) can be written in the form

$$y(x) = I^{-1}\left\{ \int^{x} I(t)q(t)\, dt + c \right\},$$

where I is given in Equation (1.9.26), and c is an arbitrary constant.

1.10   Numerical Solution to First-Order Differential Equations

So far in this chapter we have investigated first-order differential equations geometrically via slope fields, and analytically by trying to construct exact solutions to certain types of differential equations. Certainly, for most first-order differential equations, it simply is not possible to find analytic solutions, since they will not fall into the few classes for which solution techniques are available. Our final approach to analyzing first-order differential equations is to look at the possibility of constructing a numerical approximation to the unique solution to the initial-value problem

$$\frac{dy}{dx} = f(x, y), \qquad y(x_0) = y_0. \qquad (1.10.1)$$

We consider three techniques that give varying levels of accuracy. In each case, we generate a sequence of approximations $y_1, y_2, \ldots$ to the value of the exact solution at the points $x_1, x_2, \ldots$, where $x_{n+1} = x_n + h, n = 0, 1, \ldots$, and h is a real number. We emphasize that numerical methods do not generate a formula for the solution to the differential equation. Rather they generate a sequence of approximations to the value of the solution at specified points. Furthermore, if we use a sufficient number of points, then by plotting the points (x_i, y_i) and joining them with straight-line segments, we are able to obtain an overall approximation to the solution curve corresponding to the solution of the given initial-value problem. This is how the approximate solution curves were generated in the preceding sections via the computer algebra system Maple. There are many subtle ideas associated with constructing numerical solutions to initial-value problems that are beyond the scope of this text. Indeed, a full discussion of the application of numerical methods to differential equations is best left for a future course in numerical analysis.

Euler's Method

Suppose we wish to approximate the solution to the initial-value problem (1.10.1) at $x = x_1 = x_0 + h$, where h is small. The idea behind Euler's method is to use the tangent line to the solution curve through (x_0, y_0) to obtain such an approximation. (See Figure 1.10.1.)

The equation of the tangent line through (x_0, y_0) is

$$y(x) = y_0 + m(x - x_0),$$

where m is the slope of the curve at (x_0, y_0). From Equation (1.10.1), $m = f(x_0, y_0)$, so

$$y(x) = y_0 + f(x_0, y_0)(x - x_0).$$

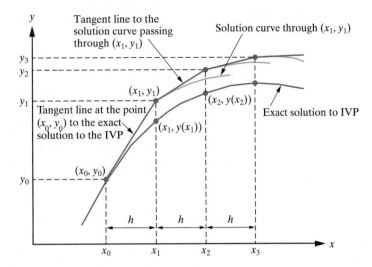

Figure 1.10.1: Euler's method for approximating the solution to the initial-value problem $dy/dx = f(x, y)$, $y(x_0) = y_0$.

Setting $x = x_1$ in this equation yields the Euler approximation to the exact solution at x_1, namely,

$$y_1 = y_0 + f(x_0, y_0)(x_1 - x_0),$$

which we write as

$$y_1 = y_0 + hf(x_0, y_0).$$

Now suppose we wish to obtain an approximation to the exact solution to the initial-value problem (1.10.1) at $x_2 = x_1 + h$. We can use the same idea, except we now use the tangent line to the solution curve through (x_1, y_1). From (1.10.1), the slope of this tangent line is $f(x_1, y_1)$, so that the equation of the required tangent line is

$$y(x) = y_1 + f(x_1, y_1)(x - x_1).$$

Setting $x = x_2$ yields the approximation

$$y_2 = y_1 + hf(x_1, y_1),$$

where we have substituted for $x_2 - x_1 = h$, to the solution to the initial-value problem at $x = x_2$. Continuing in this manner, we determine the sequence of approximations

$$y_{n+1} = y_n + hf(x_n, y_n), \qquad n = 0, 1, \ldots$$

to the solution to the initial-value problem (1.10.1) at the points $x_{n+1} = x_n + h$.

In summary, **Euler's method** for approximating the solution to the initial-value problem

$$y' = f(x, y), \qquad y(x_0) = y_0$$

at the points $x_{n+1} = x_0 + nh$ ($n = 0, 1, \ldots$) is

$$y_{n+1} = y_n + hf(x_n, y_n), \qquad n = 0, 1, \ldots . \tag{1.10.2}$$

Example 1.10.1   Consider the initial-value problem

$$y' = y - x, \quad y(0) = \tfrac{1}{2}.$$

Use Euler's method with (a) $h = 0.1$ and (b) $h = 0.05$ to obtain an approximation to $y(1)$. Given that the exact solution to the initial-value problem is

$$y(x) = x + 1 - \tfrac{1}{2}e^x,$$

compare the errors in the two approximations to $y(1)$.

Solution: In this problem we have

$$f(x, y) = y - x, \qquad x_0 = 0, \qquad y_0 = \tfrac{1}{2}.$$

(a) Setting $h = 0.1$ in (1.10.2) yields

$$y_{n+1} = y_n + 0.1(y_n - x_n).$$

Hence,

$$y_1 = y_0 + 0.1(y_0 - x_0) = 0.5 + 0.1(0.5 - 0) = 0.55,$$
$$y_2 = y_1 + 0.1(y_1 - x_1) = 0.55 + 0.1(0.55 - 0.1) = 0.595.$$

Continuing in this manner, we generate the approximations listed in Table 1.10.1, where we have rounded the calculations to six decimal places.

n	x_n	y_n	Exact Solution	Absolute Error
1	0.1	0.55	0.547414	0.002585
2	0.2	0.595	0.589299	0.005701
3	0.3	0.6345	0.625070	0.009430
4	0.4	0.66795	0.654088	0.013862
5	0.5	0.694745	0.675639	0.019106
6	0.6	0.714219	0.688941	0.025278
7	0.7	0.725641	0.693124	0.032518
8	0.8	0.728205	0.687229	0.040976
9	0.9	0.721026	0.670198	0.050828
10	1.0	0.703129	0.640859	0.062270

Table 1.10.1: The results of applying Euler's method with $h = 0.1$ to the initial-value problem in Example 1.10.1.

We have also listed the values of the exact solution and the absolute value of the error. In this case, the approximation to $y(1)$ is $y_{10} = 0.703129$, with an absolute error of

$$|y(1) - y_{10}| = 0.062270. \tag{1.10.3}$$

(b) When $h = 0.05$, Euler's method gives

$$y_{n+1} = y_n + 0.05(y_n - x_n), \qquad n = 0, 1, \ldots, 19,$$

which generates the approximations given in Table 1.10.2, where we have listed only every other intermediate approximation. We see that the approximation to $y(1)$ is

$$y_{20} = 0.673351$$

and that the absolute error in this approximation is

$$|y(1) - y_{20}| = 0.032492.$$

n	x_n	y_n	Exact Solution	Absolute Error
2	0.1	0.54875	0.547414	0.001335
4	0.2	0.592247	0.589299	0.002948
6	0.3	0.629952	0.625070	0.004881
8	0.4	0.661272	0.654088	0.007185
10	0.5	0.685553	0.675639	0.009913
12	0.6	0.702072	0.688941	0.013131
14	0.7	0.710034	0.693124	0.016910
16	0.8	0.708563	0.687229	0.021333
18	0.9	0.696690	0.670198	0.026492
20	1.0	0.686525	0.640859	0.032492

Table 1.10.2: The results of applying Euler's method with $h = 0.05$ to the initial-value problem in Example 1.10.1.

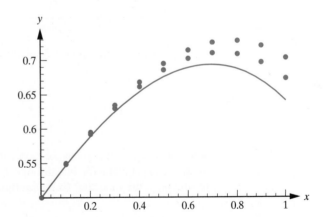

Figure 1.10.2: The exact solution to the initial-value problem considered in Example 1.10.1 and the two approximations obtained using Euler's method.

Comparing this with (1.10.3), we see that the smaller step size has led to a better approximation. In fact, it has almost halved the error at $y(1)$. In Figure 1.10.2 we have plotted the exact solution and the Euler approximations just obtained.   □

In the preceding example we saw that halving the step size had the effect of essentially halving the error. However, even then the accuracy was not as good as we probably would have liked. Of course we could just keep decreasing the step size (provided we did not take h to be so small that round-off errors started to play a role) to increase the accuracy, but then the number of steps we would have to take would make the calculations very cumbersome. A better approach is to derive methods that have a higher order of accuracy. We will consider two such methods.

Modified Euler Method (Heun's Method)

The method that we consider here is an example of what is called a **predictor-corrector method**. The idea is to use the formula from Euler's method to obtain a first approximation to the solution $y(x_{n+1})$. We denote this approximation by y_{n+1}^*, so that

$$y_{n+1}^* = y_n + hf(x_n, y_n).$$

We now improve (or "correct") this approximation by once more applying Euler's method. But this time, we use the average of the slopes of the solution curves through (x_n, y_n) and (x_{n+1}, y_{n+1}^*). This gives

$$y_{n+1} = y_n + \tfrac{1}{2}h[f(x_n, y_n) + f(x_{n+1}, y_{n+1}^*)].$$

As illustrated in Figure 1.10.3 for the case $n = 1$, we can interpret the modified Euler approximations as arising from first stepping to the point

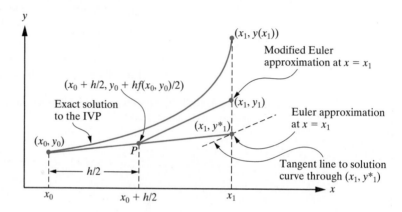

Figure 1.10.3: Derivation of the first step in the modified Euler method.

$$P\left(x_n + \frac{h}{2},\ y_n + \frac{hf(x_n, y_n)}{2}\right)$$

along the tangent line to the solution curve through (x_n, y_n) and then stepping from P to (x_{n+1}, y_{n+1}) along the line through P whose slope is $f(x_n, y_n^*)$.

In summary, the **modified Euler method** for approximating the solution to the initial-value problem

$$y' = f(x, y), \quad y(x_0) = y_0$$

at the points $x_{n+1} = x_0 + nh$ $(n = 0, 1, \dots)$ is

$$y_{n+1} = y_n + \tfrac{1}{2}h\left[f(x_n, y_n) + f(x_{n+1}, y_{n+1}^*)\right],$$

where

$$y_{n+1}^* = y_n + hf(x_n, y_n), \qquad n = 0, 1, \dots.$$

Example 1.10.2   Apply the modified Euler method with $h = 0.1$ to determine an approximation to the solution to the initial-value problem

$$y' = y - x, \qquad y(0) = \tfrac{1}{2}$$

at $x = 1$.

Solution: Taking $h = 0.1$ and $f(x, y) = y - x$ in the modified Euler method yields

$$y_{n+1}^* = y_n + 0.1(y_n - x_n),$$
$$y_{n+1} = y_n + 0.05(y_n - x_n + y_{n+1}^* - x_{n+1}).$$

Hence,

$$y_{n+1} = y_n + 0.05 \{y_n - x_n + [y_n + 0.1(y_n - x_n)] - x_{n+1}\}.$$

That is,

$$y_{n+1} = y_n + 0.05(2.1y_n - 1.1x_n - x_{n+1}), \qquad n = 0, 1, \ldots, 9.$$

When $n = 0$,

$$y_1 = y_0 + 0.05(2.1y_0 - 1.1x_0 - x_1) = 0.5475,$$

and when $n = 1$,

$$y_2 = y_1 + 0.05(2.1y_1 - 1.1x_1 - x_2) = 0.5894875.$$

n	x_n	y_n	Exact Solution	Absolute Error
1	0.1	0.5475	0.547414	0.000085
2	0.2	0.589487	0.589299	0.000189
3	0.3	0.625384	0.625070	0.000313
4	0.4	0.654549	0.654088	0.000461
5	0.5	0.676277	0.675639	0.000637
6	0.6	0.689786	0.688941	0.000845
7	0.7	0.694213	0.693124	0.001089
8	0.8	0.688605	0.687229	0.001376
9	0.9	0.671909	0.670198	0.001711
10	1.0	0.642959	0.640859	0.002100

Table 1.10.3: The results of applying the modified Euler method with $h = 0.1$ to the initial-value problem in Example 1.10.2.

Continuing in this manner, we generate the results displayed in Table 1.10.3. From this table, we see that the approximation to $y(1)$ according to the modified Euler method is

$$y_{10} = 0.642960.$$

As seen in the previous example, the value of the exact solution at $x = 1$ is

$$y(1) = 0.640859.$$

Consequently, the absolute error in the approximation at $x = 1$ using the modified Euler approximation with $h = 0.1$ is

$$|y(1) - y_{10}| = 0.002100.$$

Comparing this with the results of the previous example, we see that the modified Euler method has picked up approximately one decimal place of accuracy when using a step size $h = 0.1$. This is indicative of the general result that the error in the modified Euler method behaves as order h^2 as compared to the order h behavior of the Euler method. In Figure 1.10.4 we have sketched the exact solution to the differential equation and the modified Euler approximation with $h = 0.1$.

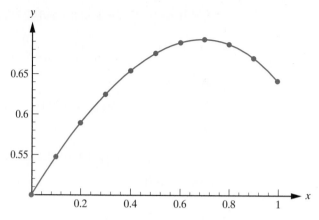

Figure 1.10.4: The exact solution to the initial-value problem in Example 1.10.2 and the approximations obtained using the modified Euler method with $h = 0.1$.

☐

Runge-Kutta Method of Order Four

The final method that we consider is somewhat more tedious to use in hand calculations, but is very easily programmed into a calculator or computer. It is a fourth-order method, which, in the case of a differential equation of the form $y' = f(x)$, reduces to Simpson's rule (which the reader has probably studied in a calculus course) for numerically evaluating definite integrals. Without justification, we state the algorithm.

The **fourth-order Runge-Kutta method** for approximating the solution to the initial-value problem

$$y' = f(x, y), \quad y(x_0) = y_0$$

at the points $x_{n+1} = x_0 + nh \ (n = 0, 1, \ldots)$ is

$$y_{n+1} = y_n + \tfrac{1}{6}(k_1 + 2k_2 + 2k_3 + k_4),$$

where

$$k_1 = hf(x_n, y_n), \quad k_2 = hf(x_n + \tfrac{1}{2}h, y_n + \tfrac{1}{2}k_1), \quad k_3 = hf(x_n + \tfrac{1}{2}h, y_n + \tfrac{1}{2}k_2),$$

$$k_4 = hf(x_{n+1}, y_n + k_3),$$

$n = 0, 1, 2, \ldots$.

Remark   In the previous sections, we used Maple to generate slope fields and approximate solution curves for first-order differential equations. The solution curves were in fact generated using a Runge-Kutta approximation.

Example 1.10.3   Apply the fourth-order Runge-Kutta method with $h = 0.1$ to determine an approximation to the solution to the initial-value problem below at $x = 1$:

$$y' = y - x, \quad y(0) = \tfrac{1}{2}$$

Solution: We take $h = 0.1$, and $f(x, y) = y - x$ in the fourth-order Runge-Kutta method, and we need to determine y_{10}. First we determine k_1, k_2, k_3, k_4.

$$k_1 = 0.1 f(x_n, y_n) = 0.1(y_n - x_n),$$
$$k_2 = 0.1 f(x_n + 0.05, y_n + 0.5k_1) = 0.1(y_n + 0.5k_1 - x_n - 0.05),$$
$$k_3 = 0.1 f(x_n + 0.05, y_n + 0.5k_2) = 0.1(y_n + 0.5k_2 - x_n - 0.05),$$
$$k_4 = 0.1 f(x_{n+1}, y_n + k_3) = 0.1(y_n + k_3 - x_{n+1}).$$

When $n = 0$,

$$k_1 = 0.1(0.5) = 0.05,$$
$$k_2 = 0.1[0.5 + (0.5)(0.05) - 0.05] = 0.0475,$$
$$k_3 = 0.1[0.5 + (0.5)(0.0475) - 0.05] = 0.047375,$$
$$k_4 = 0.1(0.5 + 0.047375 - 0.1) = 0.0447375,$$

so that

$$y_1 = y_0 + \tfrac{1}{6}(k_1 + 2k_2 + 2k_3 + k_4) = 0.5 + \tfrac{1}{6}(0.2844875) = 0.54741458,$$

rounded to eight decimal places. Continuing in this manner, we obtain the results displayed in Table 1.10.4.

n	x_n	y_n	Exact Solution	Absolute Error
1	0.1	0.54741458	0.54741454	0.00000004
2	0.2	0.58929871	0.58929862	0.00000009
3	0.3	0.62507075	0.62507060	0.00000015
4	0.4	0.65408788	0.65408765	0.00000022
5	0.5	0.67563968	0.67563936	0.00000032
6	0.6	0.68894102	0.68894060	0.00000042
7	0.7	0.69312419	0.69312365	0.00000054
8	0.8	0.68723022	0.68722954	0.00000068
9	0.9	0.67019929	0.67019844	0.00000085
10	1.0	0.64086013	0.64085909	0.00000104

Table 1.10.4: The results of applying the fourth-order Runge-Kutta method with $h = 0.1$ to the initial-value problem in Example 1.10.3.

In particular, we see that the fourth-order Runge-Kutta method approximation to $y(1)$ is

$$y_{10} = 0.64086013,$$

so that

$$|y(1) - y_{10}| = 0.00000104.$$

Clearly this is an excellent approximation. If we increase the step size to $h = 0.2$, the corresponding approximation to $y(1)$ becomes

$$y_5 = 0.640874,$$

with absolute error

$$|y(1) - y_5| = 0.000015,$$

which is still very impressive.   □

Exercises for 1.10

Key Terms

Euler's method, Predictor-corrector method, Modified Euler method (Heun's method), Fourth-order Runge-Kutta method.

Skills

- Be able to apply Euler's method to approximate the solution to an initial-value problem at a point near the initial value x_0.

- Be able to use the modified Euler method (Heun's method) to approximate the solution to an initial-value problem at a point near the initial value x_0.

- Be able to use the fourth-order Runge-Kutta method to approximate the solution to an initial-value problem at a point near the initial value x_0.

True-False Review

For Questions 1–4, decide if the given statement is **true** or **false**, and give a brief justification for your answer. If true, you can quote a relevant definition or theorem from the text. If false, provide an example, illustration, or brief explanation of why the statement is false.

1. Generally speaking, the smaller the step size in Euler's method, the more accurate the approximation to the solution of an initial-value problem at a point near the initial value x_0.

2. Euler's method is based on the equation of a tangent line to a curve at a given point (x_0, y_0).

3. With each additional step that is taken in Euler's method, the error in the approximation obtained from the method can only grow in size.

4. At each step of length h, Heun's method requires two applications of Euler's method with step size $h/2$.

Problems

For Problems 1–5, use Euler's method with the specified step size to determine the solution to the given initial-value problem at the specified point.

1. $y' = 4y - 1$, $y(0) = 1$, $h = 0.05$, $y(0.5)$.

2. $y' = -\dfrac{2xy}{1 + x^2}$, $y(0) = 1$, $h = 0.1$, $y(1)$.

3. $y' = x - y^2$, $y(0) = 2$, $h = 0.05$, $y(0.5)$.

4. $y' = -x^2 y$, $y(0) = 1$, $h = 0.2$, $y(1)$.

5. $y' = 2xy^2$, $y(0) = 0.5$, $h = 0.1$, $y(1)$.

For Problems 6–10, use the modified Euler method with the specified step size to determine the solution to the given initial-value problem at the specified point. In each case, compare your answer to that obtained using Euler's method.

6. The initial-value problem in Problem 1.

7. The initial-value problem in Problem 2.

8. The initial-value problem in Problem 3.

9. The initial-value problem in Problem 4.

10. The initial-value problem in Problem 5.

For Problems 11–15, use the fourth-order Runge-Kutta method with the specified step size to determine the solution to the given initial-value problem at the specified point. In each case, compare your answer to that obtained using Euler's method.

11. The initial-value problem in Problem 1.

12. The initial-value problem in Problem 2.

13. The initial-value problem in Problem 3.

14. The initial-value problem in Problem 4.

15. The initial-value problem in Problem 5.

16. ◇ Use the fourth-order Runge-Kutta method with $h = 0.5$ to approximate the solution to the initial-value problem

$$y' + \tfrac{1}{10}y = e^{-x/10}\cos x, \quad y(0) = 0$$

at the points $x = 0.5, 1.0, \ldots, 25$. Plot these points and describe the behavior of the corresponding solution.

1.11   Some Higher-Order Differential Equations

So far we have developed analytical techniques only for solving special types of first-order differential equations. The methods that we have discussed do not apply directly to higher-order differential equations, and so the solution to such equations usually requires the derivation of new techniques. One approach is to replace a higher-order differential equation by an equivalent system of first-order equations. (This will be developed further in Chapter 7.) For example, any second-order differential equation that can be written in the form

$$\frac{d^2y}{dx^2} = F\left(x, y, \frac{dy}{dx}\right) \tag{1.11.1}$$

where F is a known function, can be replaced by an equivalent pair of first-order differential equations as follows. We let $v = dy/dx$. Then $d^2y/dx^2 = dv/dx$, and so solving Equation (1.11.1) is equivalent to solving the following two first-order differential equations

$$\frac{dy}{dx} = v, \tag{1.11.2}$$

$$\frac{dv}{dx} = F(x, y, v). \tag{1.11.3}$$

In general the differential equation (1.11.3) cannot be solved directly, since it involves three variables, x, y, and v. However, for certain forms of the function F, Equation (1.11.3) will involve only two variables, and then we can sometimes solve it for v using one of our previous techniques. Having obtained v, we can then substitute into Equation (1.11.2) to obtain a first-order differential equation for y. We now discuss two forms of F for which this is certainly the case.

Case 1: Second-Order Equations with the Dependent Variable Missing

If y does not occur explicitly in the function F, then Equation (1.11.1) assumes the form

$$\frac{d^2y}{dx^2} = F\left(x, \frac{dy}{dx}\right). \tag{1.11.4}$$

Substituting $v = dy/dx$ and $dv/dx = d^2y/dx^2$ into this equation allows us to replace it with the two first-order equations

$$\frac{dy}{dx} = v, \tag{1.11.5}$$

$$\frac{dv}{dx} = F(x, v). \tag{1.11.6}$$

Thus, to solve Equation (1.11.4), we first solve Equation (1.11.6) for v in terms of x and then solve Equation (1.11.5) for y as a function of x.

Example 1.11.1   Find the general solution to

$$\frac{d^2y}{dx^2} = \frac{1}{x}\left(\frac{dy}{dx} + x^2\cos x\right), \qquad x > 0. \tag{1.11.7}$$

Solution: In Equation (1.11.7) the dependent variable is missing, and so we let $v = dy/dx$, which implies that $d^2y/dx^2 = dv/dx$. Substituting into Equation (1.11.7) yields the following equivalent first-order system:

$$\frac{dy}{dx} = v, \tag{1.11.8}$$

$$\frac{dv}{dx} = \frac{1}{x}(v + x^2 \cos x). \tag{1.11.9}$$

Equation (1.11.9) is a first-order linear differential equation with standard form

$$\frac{dv}{dx} - x^{-1}v = x \cos x. \tag{1.11.10}$$

An appropriate integrating factor is

$$I(x) = e^{-\int x^{-1}dx} = e^{-\ln x} = x^{-1}.$$

Multiplying Equation (1.11.10) by x^{-1} reduces it to

$$\frac{d}{dx}(x^{-1}v) = \cos x,$$

which can be integrated directly to obtain

$$x^{-1}v = \sin x + c.$$

Thus,

$$v = x \sin x + cx. \tag{1.11.11}$$

Substituting the expression for v from (1.11.11) into Equation (1.11.8) gives

$$\frac{dy}{dx} = x \sin x + cx$$

which we can integrate to obtain

$$y(x) = -x \cos x + \sin x + c_1 x^2 + c_2. \qquad \square$$

Case 2: Second-Order Equations with the Independent Variable Missing

If x does not occur explicitly in the function F in Equation (1.11.1), then we must solve a differential equation of the form

$$\frac{d^2y}{dx^2} = F\left(y, \frac{dy}{dx}\right). \tag{1.11.12}$$

In this case, we still let

$$v = \frac{dy}{dx},$$

as previously, but now we use the chain rule to express d^2y/dx^2 in terms of dv/dy. Specifically, we have

$$\frac{d^2y}{dx^2} = \frac{dv}{dx} = \frac{dv}{dy}\frac{dy}{dx} = v\frac{dv}{dy}.$$

Substituting for dy/dx and d^2y/dx^2 into Equation (1.11.12) reduces the second-order equation to the equivalent first-order system

$$\frac{dy}{dx} = v, \qquad\qquad (1.11.13)$$

$$\frac{dv}{dy} = F(y, v). \qquad\qquad (1.11.14)$$

In this case, we first solve Equation (1.11.14) for v as a function of y and then solve Equation (1.11.13) for y as a function of x.

Example 1.11.2   Find the general solution to

$$\frac{d^2y}{dx^2} = -\frac{2}{1-y}\left(\frac{dy}{dx}\right)^2. \qquad\qquad (1.11.15)$$

Solution: In this differential equation, the independent variable does not occur explicitly. Therefore, we let $v = dy/dx$ and use the chain rule to obtain

$$\frac{d^2y}{dx^2} = \frac{dv}{dx} = \frac{dv}{dy}\frac{dy}{dx} = v\frac{dv}{dy}.$$

Substituting into Equation (1.11.15) results in the equivalent system

$$\frac{dy}{dx} = v, \qquad\qquad (1.11.16)$$

$$v\frac{dv}{dy} = -\frac{2}{1-y}v^2. \qquad\qquad (1.11.17)$$

Separating the variables in the differential equation (1.11.17) gives

$$\frac{1}{v}dv = -\frac{2}{1-y}dy, \qquad\qquad (1.11.18)$$

which can be integrated to obtain

$$\ln|v| = 2\ln|1-y| + c.$$

Combining the logarithm terms and exponentiating yields

$$v(y) = c_1(1-y)^2, \qquad\qquad (1.11.19)$$

where we have set $c_1 = \pm e^c$. Notice that in solving Equation (1.11.17), we implicitly assumed that $v \neq 0$, since we divided by it to obtain Equation (1.11.18). However, the general form (1.11.19) does include the solution $v = 0$, provided we allow c_1 to equal zero. Substituting for v into Equation (1.11.16) yields

$$\frac{dy}{dx} = c_1(1-y)^2.$$

Separating the variables and integrating, we obtain

$$(1-y)^{-1} = c_1 x + d_1.$$

That is,

$$1 - y = \frac{1}{c_1 x + d_1}.$$

Solving for y gives

$$y(x) = \frac{c_1 x + (d_1 - 1)}{c_1 x + d_1}, \tag{1.11.20}$$

which can be written in the simpler form

$$y(x) = \frac{x + a}{x + b}, \tag{1.11.21}$$

where the constants a and b are defined by $a = (d_1 - 1)/c_1$ and $b = d_1/c_1$. Notice that the form (1.11.21) does not include the solution $y = $ constant, which is contained in (1.11.20) (set $c_1 = 0$). This is because in dividing by c_1, we implicitly assumed that $c_1 \neq 0$. Thus in specifying the solution in the form (1.11.21), we should also include the statement that any constant function $y = k$ (k a constant) is a solution.   □

Example 1.11.3   Determine the displacement at time t of a simple harmonic oscillator that is extended a distance A units from its equilibrium position and released from rest at $t = 0$.

Solution: According to the derivation in Section 1.1, the motion of the simple harmonic oscillator is governed by the initial-value problem

$$\frac{d^2 y}{dt^2} = -\omega^2 y, \tag{1.11.22}$$

$$y(0) = A, \quad \frac{dy}{dt}(0) = 0, \tag{1.11.23}$$

where ω is a positive constant. The differential equation (1.11.22) has the independent variable t missing. We therefore let $v = dy/dt$ and use the chain rule to write

$$\frac{d^2 y}{dt^2} = v \frac{dv}{dy}.$$

It then follows that Equation (1.11.22) can be replaced by the equivalent first-order system

$$\frac{dy}{dt} = v, \tag{1.11.24}$$

$$v \frac{dv}{dy} = -\omega^2 y. \tag{1.11.25}$$

Separating the variables and integrating Equation (1.11.25) yields

$$\frac{1}{2} v^2 = -\frac{1}{2} \omega^2 y^2 + c,$$

which implies that

$$v = \pm \sqrt{c_1 - \omega^2 y^2}$$

where $c_1 = 2c$. Substituting for v into Equation (1.11.24) yields

$$\frac{dy}{dt} = \pm \sqrt{c_1 - \omega^2 y^2}. \tag{1.11.26}$$

Setting $t = 0$ in this equation and using the initial conditions (1.11.23), we find that $c_1 = \omega^2 A^2$. Equation (1.11.26) therefore gives

$$\frac{dy}{dt} = \pm\omega\sqrt{A^2 - y^2}.$$

By separating the variables and integrating, we obtain

$$\arcsin(y/A) = \pm\omega t + b,$$

where b is an integration constant. Thus,

$$y(t) = A\sin(b \pm \omega t).$$

The initial condition $y(0) = A$ implies that $\sin b = 1$, and so we can choose $b = \pi/2$. We therefore have

$$y(t) = A\sin(\pi/2 \pm \omega t)$$

That is,

$$y(t) = A\cos\omega t.$$

Consequently the predicted motion is that the mass oscillates between $\pm A$ for all t. This solution makes sense physically, since the simple harmonic oscillator does not include dissipative forces that would slow the motion. $\square$

Remark   In Chapter 6 we will see how to solve the initial-value problem (1.11.22), (1.11.23) in just a few lines of work without requiring any integration!

Exercises for 1.11

Skills

- Be familiar with the strategy of solving a higher-order differential equation by replacing it with an equivalent system of first-order differential equations, and be able to carry out this strategy in particular instances.

Problems

For Problems 1–13, solve the given differential equation.

1. $y'' = 2x^{-1}y' + 4x^2$.

2. $(x-1)(x-2)y'' = y' - 1$.

3. $y'' + 2y^{-1})(y')^2 = y'$.

4. $y'' = (y')^2\tan y$.

5. $y'' + y'\tan x = (y')^2$.

6. $\dfrac{d^2x}{dt^2} = \left(\dfrac{dx}{dt}\right)^2 + 2\dfrac{dx}{dt}$.

7. $y'' - 2x^{-1}y' = 6x^4$.

8. $t\dfrac{d^2x}{dt^2} = 2(t + \dfrac{dx}{dt})$.

9. $y'' - \alpha(y')^2 - \beta y' = 0$, where α and β are nonzero constants.

10. $y'' - 2x^{-1}y' = 18x^4$.

11. $(1+x^2)y'' = -2xy'$.

12. $y'' + y^{-1}(y')^2 = ye^{-y}(y')^3$.

13. $y'' - y'\tan x = 1, \quad 0 \le x < \pi/2$.

In Problems 14–15, solve the given initial-value problem.

14. $yy'' = 2(y')^2 + y^2, \quad y(0) = 1, \quad y'(0) = 0$.

15. $y'' = \omega^2 y, \quad y(0) = a, \quad y'(0) = 0$, where ω, a are positive constants.

16. The following initial-value problem arises in the analysis of a cable suspended between two fixed points

$$y'' = \frac{1}{a}\sqrt{1 + (y')^2}, \quad y(0) = a, \quad y'(0) = 0,$$

where a is a nonzero constant. Solve this initial-value problem for $y(x)$. The corresponding solution curve is called a **catenary**.

17. Consider the general second-order linear differential equation with dependent variable missing:

$$y'' + p(x)y' = q(x).$$

Replace this differential equation with an equivalent pair of first-order equations and express the solution in terms of integrals.

18. Consider the general third-order differential equation of the form

$$y''' = F(x, y''). \tag{1.11.27}$$

(a) Show that Equation (1.11.27) can be replaced by the equivalent first-order system

$$\frac{du_1}{dx} = u_2, \quad \frac{du_2}{dx} = u_3, \quad \frac{du_3}{dx} = F(x, u_3),$$

where the variables u_1, u_2, u_3 are defined by

$$u_1 = y, \quad u_2 = y', \quad u_3 = y''.$$

(b) Solve $y''' = x^{-1}(y'' - 1)$.

19. A simple pendulum consists of a particle of mass m supported by a piece of string of length L. Assuming that the pendulum is displaced through an angle θ_0 radians from the vertical and then released from rest, the resulting motion is described by the initial-value problem

$$\frac{d^2\theta}{dt^2} + \frac{g}{L}\sin\theta = 0, \quad \theta(0) = \theta_0, \quad \frac{d\theta}{dt}(0) = 0. \tag{1.11.28}$$

(a) For small oscillations, $\theta \ll 1$, we can use the approximation $\sin\theta \approx \theta$ in Equation (1.11.28) to obtain the linear equation

$$\frac{d^2\theta}{dt^2} + \frac{g}{L}\theta = 0, \quad \theta(0) = \theta_0, \quad \frac{d\theta}{dt}(0) = 0.$$

Solve this initial-value problem for θ as a function of t. Is the predicted motion reasonable?

(b) Obtain the following first integral of (1.11.28):

$$\frac{d\theta}{dt} = \pm\sqrt{\frac{2g}{L}(\cos\theta - \cos\theta_0)}. \tag{1.11.29}$$

(c) Show from Equation (1.11.29) that the time T (equal to one-fourth of the period of motion) required for θ to go from 0 to θ_0 is given by the *elliptic integral of the first kind*

$$T = \sqrt{\frac{L}{2g}}\int_0^{\theta_0}\frac{1}{\sqrt{\cos\theta - \cos\theta_0}}\,d\theta. \tag{1.11.30}$$

(d) Show that (1.11.30) can be written as

$$T = \sqrt{\frac{L}{g}}\int_0^{\pi/2}\frac{1}{\sqrt{1 - k^2\sin^2 u}}\,du,$$

where $k = \sin(\theta_0/2)$. [**Hint:** First express $\cos\theta$ and $\cos\theta_0$ in terms of $\sin^2(\theta/2)$ and $\sin^2(\theta_0/2)$.]

1.12   Chapter Review

Basic Theory of Differential Equations

This chapter has provided an introduction to the theory of differential equations. A **differential equation** involves one or more derivatives of an unknown function, and the highest-order derivative is the **order** of the differential equation.

For an nth-order differential equation, the **general solution** contains n arbitrary constants, and all solutions can be obtained by assigning appropriate values to the constants. This chapter is concerned mainly with **first-order differential equations**, which may be written in the form

$$\frac{dy}{dx} = f(x, y), \tag{1.12.1}$$

for some given function f. If we impose an **initial condition** specifying the value of a solution $y(x)$ to the differential equation (1.12.1) at a particular point x_0, say $y_0 = y(x_0)$, then we have an **initial-value problem**:

$$\frac{dy}{dx} = f(x, y), \quad y(x_0) = y_0. \tag{1.12.2}$$

To solve an initial-value problem of the form (1.12.2), the first step is to determine the general solution to the differential equation (1.12.1), and then use the initial condition to determine the specific value of the arbitrary constant appearing in the general solution.

Solution Techniques for First-Order Differential Equations

One of our main goals in this chapter is to find solutions to first-order differential equations of the form (1.12.1). There are various ways in which we can seek these solutions:

1. **Geometrically:** The function $f(x, y)$ gives the slope of the tangent line to the solution curves of the differential equation (1.12.1) at the point (x, y). Thus, by computing $f(x, y)$ for various points (x, y), we can draw small line segments through the point (x, y) with slope $f(x, y)$ to depict how a solution curve would pass through (x, y). The resulting picture of line segments is called the **slope field** of the differential equation, and any solution curves to the differential equation in the xy-plane must be tangent to the slope field at all points.

   For example, the differential equation $dy/dx = -x/y$ determines a slope field consisting of small line segments that encircle the origin. Indeed, the solutions to this differential equation consist of concentric circles centered at the origin.

   One piece of theory is that different solution curves for the same differential equation can never cross (this essentially tells us that an initial-value problem cannot have multiple solutions). Thus, for example, if we find a solution to the differential equation (1.12.1) of the form $y(x) = y_0$, for some constant y_0 (recall that such a solution is called an **equilibrium solution**), then all other solution curves to the differential equation must lie entirely above the line $y = y_0$ or entirely below it.

2. **Numerically:** Suppose we wish to approximate the solution to the initial-value problem (1.12.2) at the point $x = x_1 = x_0 + h$, where h is small. **Euler's method** uses the slope of the solution at (x_0, y_0), which is $f(x_0, y_0)$, to use a tangent line approximation to the solution:

$$y(x) = y_0 + f(x_0, y_0)(x - x_0).$$

Therefore, we approximate

$$y(x_1) = y_0 + f(x_0, y_0)(x_1 - x_0) = y_0 + hf(x_0, y_0).$$

Now, starting from the point $(x_1, y(x_1))$, we can repeat the process to find approximations to the solutions at other points $x_2, x_3, \ldots$. The conclusion is that the approximation to the solution to the initial-value problem (1.12.2) at the points $x_{n+1} = x_0 + nh$ $(n = 0, 1, \ldots)$ is

$$y_{n+1} = y_n + hf(x_n, y_n), \quad n = 0, 1, \ldots$$

In Section 1.10, other modifications to Euler's method are also discussed.

3. **Analytically:** In some situations, we can explicitly obtain an equation for the general solution to the differential equation (1.12.1). These include situations in which the differential equation is separable, first-order linear, first-order homogeneous, Bernoulli, and/or exact. Table 1.12.1 shows the types of differential equations we can solve analytically and summarizes the solution techniques. If a given differential equation cannot be written in one of these forms, then the next step is to try to determine an integrating factor. If that fails, then we might try to find a change of variables that would reduce the differential equation to one of the above types.

Type	Standard Form	Technique
Separable	$p(y)y' = q(x)$	Separate the variables and integrate.
First-order linear	$y' + p(x)y = q(x)$	Rewrite as $\frac{d}{dx}(I \cdot y) = I \cdot q(x)$, where $I = e^{\int p(x)dx}$, and integrate with respect to x.
First-order homogeneous	$y' = f(x, y)$ where $f(tx, ty) = f(x, y)$	Change variables: $y = xV(x)$, and reduce to a separable equation.
Bernoulli	$y' + p(x)y = q(x)y^n$	Divide by y^n and make the change of variables $u = y^{1-n}$. This reduces the differential equation to a linear equation.
Exact	$M\,dx + N\,dy = 0$, with $M_y = N_x$	The solution is $\phi(x, y) = c$, where ϕ is determined by integrating $\phi_x = M$, $\phi_y = N$.

Table 1.12.1: A summary of the basic solution techniques for $y' = f(x, y)$.

Example 1.12.1   Determine which of the above types, if any, the following differential equation falls into:

$$\frac{dy}{dx} = -\frac{(8x^5 + 3y^4)}{4xy^3}.$$

Solution: Since the given differential equation is written in the form $dy/dx = f(x, y)$, we first check whether it is separable or homogeneous. By inspection, we see that it is neither of these. We next check to see whether it is a linear or a Bernoulli equation. We therefore rewrite the equation in the equivalent form

$$\frac{dy}{dx} + \frac{3}{4x}y = -2x^4y^{-3}, \tag{1.12.3}$$

which we recognize as a Bernoulli equation with $n = -3$. We could therefore solve the equation using the appropriate technique. Owing to the y^{-3} term in Equation (1.12.3), it follows that the equation is not a linear equation. Finally, we check for exactness. The natural differential form to try for the given differential equation is

$$(8x^5 + 3y^4)\,dx + 4xy^3\,dy = 0. \tag{1.12.4}$$

In this form, we have

$$M_y = 12y^3, \qquad N_x = 4y^3,$$

so that the equation is not exact. However, we see that

$$(M_y - N_x)/N = 2x^{-1},$$

so that according to Theorem 1.9.11, $I(x) = x^2$ is an integrating factor. Therefore, we could multiply Equation (1.12.4) by x^2 and then solve it as an exact equation.   □

Examples of First-Order Differential Equations

There are numerous real-world examples of first-order differential equations. Among the applications discussed in this chapter are Newton's law of cooling, families of orthogonal trajectories, Malthusian and logistic population models, mixing problems, electric circuits, and others.

Additional Problems

1. A racquetball player standing at the back wall of the court hits the ball from a height of 2 feet horizontally toward the front wall at 80 miles per hour. The length of a regulation racquetball court is 40 feet. Does the ball reach the front wall before hitting the ground? Neglect air resistance, and assume the acceleration of gravity is 32 feet/sec^2.

2. A boy 2 meters tall shoots a toy rocket straight up from head level at 10 meters per second. Assume the acceleration of gravity is 9.8 meters/sec^2.

   (a) What is the highest point above the ground reached by the rocket?

   (b) When does the rocket hit the ground?

In Problems 3–6, find the equation of the orthogonal trajectories to the given family of curves.

3. $y = cx^3$.

4. $y^2 = cx^3$.

5. $y = \ln(cx)$.

6. $x^4 + y^4 = c$.

7. Consider the family of curves

$$x^2 + 3y^2 = 2cy, \qquad (1.12.5)$$

   (a) Show that the differential equation of this family is
   $$\frac{dy}{dx} = \frac{2xy}{x^2 - 3y^2}.$$

   (b) Determine the orthogonal trajectories to the family (1.12.5).

In Problems 8–9, sketch the slope field and some representative solution curves for the given differential equation.

8. $y' = \sin x$.

9. $y' = y/x^2$.

10. At time t the velocity, $v(t)$, of an object is governed by the differential equation

$$\frac{dv}{dt} = \frac{1}{2}(25 - v), \quad t > 0.$$

   (a) Verify that $v(t) = 25$ is a solution to this differential equation.

   (b) Sketch the slope field for $0 \le v \le 25$. What happens to $v(t)$ as $t \to \infty$?

11. An object of mass m is released from rest in a medium in which the frictional forces are proportional to the square of the velocity. The initial-value problem that governs the subsequent motion is

$$mv\frac{dv}{dy} = mg - kv^2, \quad v(0) = 0, \qquad (1.12.6)$$

where $v(t)$ denotes the velocity of the object at time t, $y(t)$ denotes the distance traveled by the object at time t as measured from the point at which the object was released, and k is a positive constant.

(a) Solve (1.12.6) and show that

$$v^2 = \frac{mg}{k}(1 - e^{-2ky/m}).$$

(b) Make a sketch of v^2 as a function of y.

In Problems 12–37, determine which of the five types of differential equations we have studied the given equation falls into (see Table 1.12.1), and use an appropriate technique to find the general solution.

12. $\dfrac{dy}{dx} = \dfrac{2\ln x}{xy}$.

13. $xy' - 2y = 2x^2 \ln x$.

14. $\dfrac{dy}{dx} = -\dfrac{2xy}{x^2 + 2y}$.

15. $(y^2 + 3xy + x^2)\,dx - x^2\,dy = 0$.

16. $y' + y(\tan x + y\sin x) = 0$.

17. $\dfrac{dy}{dx} + \dfrac{2e^{2x}}{1 + e^{2x}}y = \dfrac{1}{e^{2x} - 1}$.

18. $y' - x^{-1}y = x^{-1}\sqrt{x^2 - y^2}$.

19. $\dfrac{dy}{dx} = \dfrac{\sin y + y\cos x + 1}{1 - x\cos y - \sin x}$.

20. $\dfrac{dy}{dx} + \dfrac{1}{x}y = \dfrac{25x^2 \ln x}{2y}$.

21. $e^{2x+y}\,dy - e^{x-y}\,dx = 0$.

22. $y' + y\cot x = \sec x$.

23. $\dfrac{dy}{dx} + \dfrac{2e^x}{1 + e^x}y = 2\sqrt{y}e^{-x}$.

24. $y[\ln(y/x) + 1]dx - x\,dy = 0$.

25. $(1 + 2xe^y)\,dx - (e^y + x)\,dy = 0$.

26. $y' + y\sin x = \sin x$.

27. $(3y^2 + x^2)\,dx - 2xy\,dy = 0$.

28. $2x(\ln x)y' - y = -9x^3 y^3 \ln x$.

29. $(1 + x)y' = y(2 + x)$.

30. $(x^2 - 1)(y' - 1) + 2y = 0$.

31. $x\sec^2(xy)\,dy = -\left[y\sec^2(xy) + 2x\right]dx$.

32. $\dfrac{dy}{dx} - x^2 y = \sqrt{y}$.

33. $\dfrac{dy}{dx} = \dfrac{x^2}{x^2 - y^2} + \dfrac{y}{x}$.

34. $[\ln(xy) + 1]\,dx + \left(\dfrac{x}{y} + 2y\right)dy = 0$.

35. $y' + \dfrac{y}{x} = \dfrac{25\ln x}{2x^3 y}$.

36. $(x + xy^2)y' = x^3 y e^{x-y}$.

37. $y' = \cos x(y\csc x - 1), \quad 0 < x < \dfrac{\pi}{2}$.

For Problems 38–41, determine which of the five types of differential equations we have studied the given differential equation falls into, and use an appropriate technique to find the solution to the initial-value problem.

38. $y' - x^2 y = x^2, \quad y(0) = 5$.

39. $e^{-3x+2y}\,dx + e^{x-4y}\,dy = 0, \quad y(0) = 0$.

40. $(3x^2 + 2xy^2)\,dx + (2x^2 y)\,dy = 0, \quad y(1) = 3$.

41. $\dfrac{dy}{dx} - (\sin x)y = e^{-\cos x}, \quad y(0) = \dfrac{1}{e}$.

42. Determine all values of the constants m and n, if there are any, for which the differential equation

$$(x^5 + y^m)\,dx - x^n y^3\,dy = 0$$

is each of the following:

(a) Exact.

(b) Separable.

(c) Homogeneous.

(d) Linear.

(e) Bernoulli.

43. A man's sandals are moved from poolside (80°F) to a sauna (180°F) to warm and dry them. If they are 100°F after 3 minutes in the sauna, how much time is required in the sauna to increase their temperature to 140°F, according to Newton's law of cooling?

44. A hot plate (150°F) is placed on a countertop in a room kept at 70°F. If the plate cools 25°F in the first 10 minutes, when does the plate reach 100°F, according to Newton's law of cooling?

45. A simple nonlinear law of cooling states that the rate of change of temperature of an object is proportional to the *square* of the temperature difference between the object and its surrounding medium (you may assume that the temperature of the surrounding medium is constant). Set up and solve the initial-value problem that governs this cooling process if the initial temperature is T_0. What happens to the temperature of the object as $t \to \infty$?

46. The temperature of an object at time t is governed by the *linear* differential equation

$$\frac{dT}{dt} = -k(T - 5\cos 2t).$$

At $t = 0$, the temperature of the object is $0°F$ and is, at that time, increasing at a rate of $5°F/min$.

(a) Determine the value of the constant k.

(b) Determine the temperature of the object at time t.

(c) Describe the behavior of the temperature of the object for large values of t.

47. Each spring, sandhill cranes migrate through the Platte River valley in central Nebraska. An estimated maximum of a half-million of these birds reach the region by April 1 each year. If there are only 100,000 sandhill cranes 15 days later and the sandhill cranes leave the Platte River valley at a rate proportional to the number of them still in the valley at the time,

(a) How many sandhill cranes remain in the valley 30 days after April 1?

(b) How many sandhill cranes remain in the valley 35 days after April 1?

(c) How many days after April 1 will there be fewer than 1000 sandhill cranes in the valley?

48. A city's population in the year 2000 was 200,000, in 2003 it was 230,000, and in 2006 it was 250,000. Using the logistic model of population, predict the population in 2010 and 2020.

49. Consider an RC circuit with $R = 4\ \Omega$, $C = \frac{1}{5}$ F, and $E(t) = 6\cos 2t$ V. If $q(0) = 3$ C, determine the current in the circuit for $t \geq 0$.

50. Consider an RL circuit with $R = 3\ \Omega$, $L = 0.3$ H, and $E(t) = 10$ V. If $i(0) = 3$ A, determine the current in the circuit for $t \geq 0$.

51. A solution containing 3 g/L of a salt solution pours into a tank, initially half full of water, at a rate of 6 L/min. The well-stirred mixture flows out at a rate of 4 L/min. If the tank holds 60 L, find the amount of salt (in grams) in the tank when the solution overflows.

In Problems 52–53, use Euler's method with the specified step size to determine the solution to the given initial-value problem at the specified point.

52. $y' = x^2 + 2y^2$, $y(0) = -3$, $h = 0.1$, $y(1)$.

53. $y' = \frac{3x}{y} + 2$, $y(1) = 2$, $h = 0.05$, $y(1.5)$.

In Problems 54–55, use the modified Euler method with the specified step size to determine the solution to the given initial-value problem at the specified point. In each case, compare your answer to that determined by using Euler's method.

54. The initial-value problem in Problem 52.

55. The initial-value problem in Problem 53.

In Problems 56–57, use the fourth-order Runge-Kutta method with the specified step size to determine the solution to the given initial-value problem at the specified point. In each case, compare your answer to that determined by using Euler's method.

56. The initial-value problem in Problem 52.

57. The initial-value problem in Problem 53.

Project: A Cylindrical Tank Problem

Consider an open cylindrical tank of height h_0 meters and radius r meters that is filled with water. A circular hole of radius l meters in the bottom of the tank allows the water to flow out under the influence of gravity. According to Torricelli's law, the water flows out with the same speed that it would acquire in falling freely from the water level in the tank to the hole.

1. Use Torricelli's law to derive the following equation for the rate of change of volume of water in the tank,

$$\frac{dV}{dt} = -a\sqrt{2gh}$$

where $h(t)$ denotes the height of water in the tank at time t, a denotes the area of the hole, and g denotes the acceleration due to gravity. [**Hint:** First show that an object that is released from rest at a height h hits the ground with a speed $\sqrt{2gh}$. Then consider the change in the volume of water in the tank in a time interval Δt.]

2. Show that the rate of change of volume of water in the tank is also given by

$$\frac{dV}{dt} = \pi r^2 \frac{dh}{dt}.$$

3. Using the results from problems (1) and (2), determine the height of the water in the tank at time t, and show that the tank will empty when $t = t_e$ where

$$t_e = \frac{\pi r^2}{a} \sqrt{\frac{2h_0}{g}}.$$

4. Suppose now that starting at $t = 0$ chemical is added to the water in the tank at a rate of w grams/second. Derive the following differential equation governing the amount of chemical, $A(t)$, in the tank at time t:

$$\frac{dA}{dt} - \frac{2}{t - t_e} A = w, \quad 0 < t < t_e. \tag{1.12.7}$$

5. Solve the differential equation (1.12.7). Determine the time when $A(t)$ is a maximum.

6. By making an appropriate change of variables in the differential equation (1.12.7), derive a differential equation for the concentration $c(t)$ of chemical in the tank at time t. Solve your differential equation and verify that you get the same expression for $c(t)$ as you do by dividing the expression for $A(t)$ obtained in the previous problem by $V(t)$.

7. In the particular case when $h_0 = 16$ m, $r = 5$ m, $l = 0.1$ m, and $w = 15$ g/s, determine t_e, and the time when the concentration of chemical in the tank reaches 1 g/L.

CHAPTER 2

Matrices and Systems of Linear Equations

Algebra is the intellectual instrument which has been created for rendering clear the quantitative aspects of the world. — Alfred North Whitehead

We will see in the later chapters that most problems in linear algebra can be reduced to questions regarding the solutions of systems of linear equations. In preparation for this, the next two chapters provide a detailed introduction to the theory and solution techniques for such systems. An example of a linear system of equations in the unknowns x_1, x_2, x_3 is

$$3x_1 + 4x_2 - 7x_3 = 5,$$
$$2x_1 - 3x_2 + 9x_3 = 7,$$
$$7x_1 + 2x_2 - 3x_3 = 4.$$

We see that this system is completely determined by the array of numbers

$$\begin{bmatrix} 3 & 4 & -7 & 5 \\ 2 & -3 & 9 & 7 \\ 7 & 2 & -3 & 4 \end{bmatrix},$$

which contains the coefficients of the unknowns on the left-hand side of the system and the numbers appearing on the right-hand side of the system. Such an array is an example of a matrix. In this chapter we see that, in general, linear systems of equations are best represented in terms of matrices and that, once such a representation has been made, the set of all solutions to the system can be easily determined. In the first few sections of this chapter we therefore introduce the basics of matrix algebra. We then apply matrices to solve systems of linear equations. In Chapter 7, we will see how matrices also give a natural framework for formulating and solving systems of linear differential equations.

111

2.1   Matrices: Definitions and Notation

We begin our discussion of matrices with a definition.

DEFINITION 2.1.1

An $m \times n$ (read "m by n") **matrix** is a rectangular array of numbers arranged in m horizontal rows and n vertical columns. Matrices are usually denoted by uppercase letters, such as A and B. The entries in the matrix are called the **elements** of the matrix.

Example 2.1.2   The following are examples of a 2×3 and a 3×3 matrix, respectively:

$$A = \begin{bmatrix} \frac{3}{2} & \frac{5}{4} & \frac{1}{5} \\ 0 & -\frac{3}{7} & \frac{5}{9} \end{bmatrix}, \qquad B = \begin{bmatrix} 2 & -1 & 3 \\ 1 & 1 & -1 \\ 0 & 0 & 1 \end{bmatrix}.$$

$\square$

We will use the index notation to denote the elements of a matrix. According to this notation, the element in the ith row and jth column of the matrix A will be denoted a_{ij}. Thus, for the matrices in the previous example we have

$$a_{13} = \tfrac{1}{5}, \qquad a_{22} = -\tfrac{3}{7}, \qquad b_{23} = -1, \qquad \text{and so on.}$$

Using the index notation, a general $m \times n$ matrix A is written

$$A = \begin{bmatrix} a_{11} & a_{12} & \dots & a_{1n} \\ a_{21} & a_{22} & \dots & a_{2n} \\ \vdots & \vdots & & \vdots \\ a_{m1} & a_{m2} & \dots & a_{mn} \end{bmatrix},$$

or, in a more abbreviated form, $A = [a_{ij}]$.

Remark   The expression $m \times n$ representing the number of rows and columns of a general matrix A is sometimes informally called the **size** of the matrix A. The numbers m and n themselves are sometimes called the **dimensions**[1] of the matrix A.

Next we define what is meant by equality of matrices.

DEFINITION 2.1.3

Two matrices A and B are **equal**, written $A = B$, if

1. They both have the same size, $m \times n$.
2. All corresponding elements in the matrices are equal: $a_{ij} = b_{ij}$ for all i and j with $1 \le i \le m$ and $1 \le j \le n$.

---

[1] Be careful not to confuse this usage of the term with the *dimension* of a vector space, which will be introduced in Chapter 4.

According to Definition 2.1.3, even though the matrices

$$A = \begin{bmatrix} 1 & 2 & 3 \\ 4 & 5 & 6 \end{bmatrix} \quad \text{and} \quad B = \begin{bmatrix} 4 & 2 \\ 3 & 6 \\ 1 & 5 \end{bmatrix}$$

contain the same six numbers, and therefore store the same basic information, they are not equal as matrices.

Row Vectors and Column Vectors

Of particular interest to us in the future will be $1 \times n$ and $n \times 1$ matrices. For this reason we give them special names.

DEFINITION   2.1.4

A $1 \times n$ matrix is called a **row n-vector**. An $n \times 1$ matrix is called a **column n-vector**. The elements of a row or column n-vector are called the **components** of the vector.

Remarks

1. We can refer to the objects just defined simply as row vectors and column vectors if the value of n is clear from the context.

2. We will see later in this chapter that when a system of linear equations is written using matrices, the basic unknown in the reformulated system is a column vector. A similar formulation will also be given in Chapter 7 for systems of differential equations.

Example 2.1.5   The matrix $\mathbf{a} = \begin{bmatrix} \frac{2}{3} & -\frac{1}{5} & \frac{4}{7} \end{bmatrix}$ is a row 3-vector and

$$\mathbf{b} = \begin{bmatrix} 1 \\ -1 \\ 3 \\ 4 \end{bmatrix}$$

is a column 4-vector.   □

As indicated here, we usually denote a row or column vector by a lowercase letter in **bold** print.

Associated with any $m \times n$ matrix are m row n-vectors and n column m-vectors. These are referred to as the **row vectors** of the matrix and the **column vectors** of the matrix, respectively.

Example 2.1.6   Associated with the matrix

$$A = \begin{bmatrix} -2 & 1 & 3 & 4 \\ 1 & 2 & 1 & 1 \\ 3 & -1 & 2 & 5 \end{bmatrix}$$

are the row 4-vectors

$$\begin{bmatrix} -2 & 1 & 3 & 4 \end{bmatrix}, \quad \begin{bmatrix} 1 & 2 & 1 & 1 \end{bmatrix}, \quad \text{and} \quad \begin{bmatrix} 3 & -1 & 2 & 5 \end{bmatrix},$$

and the column 3-vectors

$$\begin{bmatrix} -2 \\ 1 \\ 3 \end{bmatrix}, \quad \begin{bmatrix} 1 \\ 2 \\ -1 \end{bmatrix}, \quad \begin{bmatrix} 3 \\ 1 \\ 2 \end{bmatrix}, \quad \text{and} \quad \begin{bmatrix} 4 \\ 1 \\ 5 \end{bmatrix}.$$ □

Conversely, if $\mathbf{a}_1, \mathbf{a}_2, \ldots, \mathbf{a}_n$ are each column m-vectors, then we let $[\mathbf{a}_1, \mathbf{a}_2, \ldots, \mathbf{a}_n]$ denote the $m \times n$ matrix whose column vectors are $\mathbf{a}_1, \mathbf{a}_2, \ldots, \mathbf{a}_n$. Similarly, if $\mathbf{b}_1, \mathbf{b}_2, \ldots, \mathbf{b}_m$ are each row n-vectors, then we write

$$\begin{bmatrix} \mathbf{b}_1 \\ \mathbf{b}_2 \\ \vdots \\ \mathbf{b}_m \end{bmatrix}$$

for the $m \times n$ matrix with row vectors $\mathbf{b}_1, \mathbf{b}_2, \ldots, \mathbf{b}_m$. The reader should observe that a list of vectors arranged in a row will always consist of column vectors, while a list of vectors arranged in a column will always consist of row vectors.

Example 2.1.7 If $\mathbf{a}_1 = \begin{bmatrix} \frac{1}{5} \\ \frac{2}{3} \end{bmatrix}$, $\mathbf{a}_2 = \begin{bmatrix} \frac{4}{7} \\ \frac{5}{9} \end{bmatrix}$, and $\mathbf{a}_3 = \begin{bmatrix} -\frac{1}{3} \\ \frac{3}{11} \end{bmatrix}$, then

$$[\mathbf{a}_1, \mathbf{a}_2, \mathbf{a}_3] = \begin{bmatrix} \frac{1}{5} & \frac{4}{7} & -\frac{1}{3} \\ \frac{2}{3} & \frac{5}{9} & \frac{3}{11} \end{bmatrix}.$$ □

DEFINITION   2.1.8

If we interchange the row vectors and column vectors in an $m \times n$ matrix A, we obtain an $n \times m$ matrix called the **transpose** of A. We denote this matrix by A^T. In index notation, the (i, j)th element of A^T, denoted a_{ij}^T, is given by

$$a_{ij}^T = a_{ji}.$$

Example 2.1.9 If

$$A = \begin{bmatrix} 1 & 2 & 6 & 2 \\ 0 & 3 & 4 & 7 \end{bmatrix},$$

then

$$A^T = \begin{bmatrix} 1 & 0 \\ 2 & 3 \\ 6 & 4 \\ 2 & 7 \end{bmatrix}.$$

If

$$A = \begin{bmatrix} 1 & 3 & 5 \\ 2 & 0 & 7 \\ 3 & 4 & 9 \end{bmatrix},$$

then

$$A^T = \begin{bmatrix} 1 & 2 & 3 \\ 3 & 0 & 4 \\ 5 & 7 & 9 \end{bmatrix}.$$

□

Square Matrices

An $n \times n$ matrix is called a **square matrix**, since it has the same number of rows as columns. If A is a square matrix, then the elements a_{ii}, $1 \le i \le n$, make up the **main diagonal**, or **leading diagonal**, of the matrix. (See Figure 2.1.1 for the 3×3 case.)

$$\begin{bmatrix} a_{11} & a_{12} & a_{13} \\ a_{21} & a_{22} & a_{23} \\ a_{31} & a_{32} & a_{33} \end{bmatrix}$$

Figure 2.1.1: The main diagonal of a 3×3 matrix.

The sum of the main diagonal elements of an $n \times n$ matrix A is called the **trace** of A and is denoted $\text{tr}(A)$. Thus,

$$\text{tr}(A) = a_{11} + a_{22} + \cdots + a_{nn}.$$

An $n \times n$ matrix A is said to be **lower triangular** if $a_{ij} = 0$ whenever $i < j$ (zeros everywhere above (i.e.. "northeast of") the main diagonal), and it is said to be **upper triangular** if $a_{ij} = 0$ whenever $i > j$ (zeros everywhere below (i.e., "southwest of") the main diagonal). The following are examples of an upper triangular and lower triangular matrix, respectively:

$$\begin{bmatrix} 1 & -8 & 5 \\ 0 & -3 & 9 \\ 0 & 0 & 4 \end{bmatrix}, \qquad \begin{bmatrix} 2 & 0 & 0 \\ 0 & 1 & 0 \\ -6 & 7 & -3 \end{bmatrix}.$$

Observe that the transpose of a lower (upper) triangular matrix is an upper (lower) triangular matrix.

If every element on the main diagonal of a lower (upper) triangular matrix is a 1, the matrix is called a **unit lower (upper) triangular matrix**.

An $n \times n$ matrix $D = [d_{ij}]$ that has all *off-diagonal* elements equal to zero is called a **diagonal matrix**. Note that a matrix D is a diagonal matrix if and only if D is simultaneously upper and lower triangular. Such a matrix is completely determined by giving its main diagonal elements, since $d_{ij} = 0$ whenever $i \neq j$. Consequently, we can specify a diagonal matrix in the compact form

$$D = \text{diag}(d_1, d_2, \ldots, d_n),$$

where d_i denotes the diagonal element d_{ii}.

Example 2.1.10

The 4×4 diagonal matrix $D = \text{diag}(1, 2, 0, 3)$ is

$$D = \begin{bmatrix} 1 & 0 & 0 & 0 \\ 0 & 2 & 0 & 0 \\ 0 & 0 & 0 & 0 \\ 0 & 0 & 0 & 3 \end{bmatrix}.$$

□

The transpose naturally picks out two important types of square matrices as follows.

DEFINITION 2.1.11

1. A square matrix A satisfying $A^T = A$ is called a **symmetric matrix**.
2. If $A = [a_{ij}]$, then we let $-A$ denote the matrix with elements $-a_{ij}$. A square matrix A satisfying $A^T = -A$ is called a **skew-symmetric** (or **anti-symmetric**) **matrix**.

Example 2.1.12   The matrix

$$A = \begin{bmatrix} 1 & -1 & 1 & 5 \\ -1 & 2 & 2 & 6 \\ 1 & 2 & 3 & 4 \\ 5 & 6 & 4 & 9 \end{bmatrix}$$

is symmetric, whereas

$$B = \begin{bmatrix} 0 & -1 & -5 & 3 \\ 1 & 0 & 1 & -2 \\ 5 & -1 & 0 & 7 \\ -3 & 2 & -7 & 0 \end{bmatrix}$$

is skew-symmetric.   □

Notice that the main diagonal elements of the skew-symmetric matrix in the preceding example are all zero. This is true in general, since if A is a skew-symmetric matrix, then $a_{ij} = -a_{ji}$, which implies that when $i = j$, $a_{ii} = -a_{ii}$, so that $a_{ii} = 0$.

Matrix and Vector Functions

Later in the text we will be concerned with systems of two or more differential equations. The most effective way to study such systems, as it turns out, is to represent the system using matrices and vectors. However, we will need to allow the elements of the matrices and vectors that arise to contain *functions* of a single variable, not just real or complex numbers. This leads to the following definition, reminiscent of Definition 2.1.1.

DEFINITION 2.1.13

An $m \times n$ **matrix function** A is a rectangular array with m rows and n columns whose elements are functions of a single real variable t.

Example 2.1.14   Here are two examples of matrix functions:

$$A(t) = \begin{bmatrix} t^3 & t - \cos t & 5 \\ e^{t^2} & \ln(t+1) & te^t \end{bmatrix} \quad \text{and} \quad B(t) = \begin{bmatrix} 5 - t + t^2 & \sin(e^{2t}) \\ -1 & \tan t \\ 6 & 6 - t \end{bmatrix}.$$

A matrix function $A(t)$ is defined only for real values of t such that *all* elements in $A(t)$ assume a well-defined value. The function A is defined only for real values of t with

$t > -1$, since $\ln(t + 1)$ is defined only for $t > -1$. The reader should determine the values of t for which the matrix function B is defined. $\square$

Remark It is possible, of course, to consider matrix functions of more than one variable. However, this will not be particularly relevant for our purposes in this text.

Finally in this section, we have the following special type of matrix function.

DEFINITION 2.1.15

An $n \times 1$ matrix function is called a **column n-vector function**.

For instance, $\begin{bmatrix} t^2 \\ -6te^t \end{bmatrix}$ is a column 2-vector function.[2]

Exercises for 2.1

Key Terms

Matrices, Elements, Size (dimensions) of a matrix, Row vector, Column vector, Square matrix, Main diagonal, Trace, Lower (Upper) triangular matrix, Unit lower (upper) triangular matrix, Diagonal matrix, Symmetric matrix, Skew-symmetric matrix, Matrix function, Column n-vector function.

Skills

- Be able to determine the elements of a matrix.

- Be able to identify the size (i.e., dimensions) of a matrix.

- Be able to identify the row and column vectors of a matrix.

- Be able to determine the components of a row or column vector.

- Be able to say whether or not two given matrices are equal.

- Be able to find the transpose of a matrix.

- Be able to compute the trace of a square matrix.

- Be able to recognize square matrices that are upper triangular, lower triangular, or diagonal.

- Be able to recognize square matrices that are symmetric or skew-symmetric.

- Be able to determine the values of the variable t such that a matrix function A is defined.

True-False Review

For Questions 1–10, decide if the given statement is **true** or **false**, and give a brief justification for your answer. If true, you can quote a relevant definition or theorem from the text. If false, provide an example, illustration, or brief explanation of why the statement is false.

1. A diagonal matrix must be both upper triangular and lower triangular.

2. An $m \times n$ matrix has m column vectors and n row vectors.

3. If A is a symmetric matrix, then so is A^T.

4. The trace of a matrix is the product of the elements along the main diagonal.

5. A skew-symmetric matrix must have zeros along the main diagonal.

6. A matrix that is both symmetric and skew-symmetric cannot contain any nonzero elements.

7. The matrix functions

$$\begin{bmatrix} \sqrt{t} & 3t^2 \\ \dfrac{1}{|t|} & \sin 2t \end{bmatrix} \quad \text{and} \quad \begin{bmatrix} -2 + t & \ln t \\ e^{\sin t} & -3 \end{bmatrix}$$

are defined for exactly the same values of t.

---

[2]We could, of course, also speak of **row n-vector functions** as the $1 \times n$ matrix functions, but we will not need them in this text.

8. The matrix function

$$\begin{bmatrix} \cos t & t^2 \\ -2 & -t \\ e^t & \dfrac{1}{\sqrt{t-3}} \end{bmatrix}$$

is defined for all positive real numbers t.

9. Any matrix of numbers is a matrix function defined for all real values of the variable t.

10. If A and B are matrix functions such that the matrices $A(0)$ and $B(0)$ are the same, then we should consider A and B to be the same matrix function.

Problems

1. If

$$A = \begin{bmatrix} 1 & -2 & 3 & 2 \\ 7 & -6 & 5 & -1 \\ 0 & 2 & -3 & 4 \end{bmatrix},$$

determine $a_{31}, a_{24}, a_{14}, a_{32}, a_{21}$, and a_{34}.

For Problems 2–6, write the matrix with the given elements. In each case, specify the dimensions of the matrix.

2. $a_{11} = 1, a_{21} = -1, a_{12} = 5, a_{22} = 3$.

3. $a_{11} = 2, a_{12} = 1, a_{13} = -1, a_{21} = 0, a_{22} = 4, a_{23} = -2$.

4. $a_{11} = -1, a_{41} = -5, a_{31} = 1, a_{21} = 1$.

5. $a_{11} = 1, a_{31} = 2, a_{42} = -1, a_{32} = 7, a_{13} = -2, a_{23} = 0, a_{33} = 4, a_{21} = 3, a_{41} = -4, a_{12} = -3, a_{22} = 6, a_{43} = 5$.

6. $a_{12} = -1, a_{13} = 2, a_{23} = 3, a_{ji} = -a_{ij}, 1 \le i \le 3, 1 \le j \le 3$.

For Problems 7–9, determine $\text{tr}(A)$ for the given matrix.

7. $A = \begin{bmatrix} 1 & 0 \\ 2 & 3 \end{bmatrix}$.

8. $A = \begin{bmatrix} 1 & 2 & -1 \\ 3 & 2 & -2 \\ 7 & 5 & -3 \end{bmatrix}$.

9. $A = \begin{bmatrix} 2 & 0 & 1 \\ 3 & 2 & 5 \\ 0 & 1 & -5 \end{bmatrix}$.

For Problems 10–12, write the column vectors and row vectors of the given matrix.

10. $A = \begin{bmatrix} 1 & -1 \\ 3 & 5 \end{bmatrix}$.

11. $A = \begin{bmatrix} 1 & 3 & -4 \\ -1 & -2 & 5 \\ 2 & 6 & 7 \end{bmatrix}$.

12. $A = \begin{bmatrix} 2 & 10 & 6 \\ 5 & -1 & 3 \end{bmatrix}$.

13. If $\mathbf{a}_1 = [1 \quad 2], \mathbf{a}_2 = [3 \quad 4]$, and $\mathbf{a}_3 = [5 \quad 1]$, write the matrix

$$A = \begin{bmatrix} \mathbf{a}_1 \\ \mathbf{a}_2 \\ \mathbf{a}_3 \end{bmatrix},$$

and determine the column vectors of A.

14. If

$$\mathbf{b}_1 = \begin{bmatrix} 2 \\ -1 \\ 4 \end{bmatrix}, \quad \mathbf{b}_2 = \begin{bmatrix} 5 \\ 7 \\ -6 \end{bmatrix},$$

$$\mathbf{b}_3 = \begin{bmatrix} 0 \\ 0 \\ 0 \end{bmatrix}, \quad \mathbf{b}_4 = \begin{bmatrix} 1 \\ 2 \\ 3 \end{bmatrix},$$

write the matrix $B = [\mathbf{b}_1, \mathbf{b}_2, \mathbf{b}_3, \mathbf{b}_4]$ and determine the row vectors of B.

15. If $\mathbf{a}_1, \mathbf{a}_2, \ldots, \mathbf{a}_p$ are each column q-vectors, what are the dimensions of the matrix that has $\mathbf{a}_1, \mathbf{a}_2, \ldots, \mathbf{a}_p$ as its column vectors?

For Problems 16–20, give an example of a matrix of the specified form.

16. 3×3 diagonal matrix.

17. 4×4 upper triangular matrix.

18. 4×4 skew-symmetric matrix.

19. 3×3 upper triangular symmetric matrix.

20. 3×3 lower triangular skew-symmetric matrix.

For Problems 21–24, give an example of a matrix function of the specified form.

21. 2×3 matrix function defined only for values of t with $-2 \le t < 3$.

22. 4×2 matrix function A such that

$$A(0) = A(1) \neq A(2).$$

23. 1×5 matrix function A that is nonconstant such that all elements of $A(t)$ are positive for all t in $\mathbb{R}$.

24. 2×1 matrix function A that is nonconstant such that all elements of $A(t)$ are in $[0, 1]$ for every t in $\mathbb{R}$.

25. Construct *distinct* matrix functions A and B defined on all of $\mathbb{R}$ such that $A(0) = B(0)$ and $A(1) = B(1)$.

26. Prove that a symmetric upper triangular matrix is diagonal.

27. Determine all elements of the 3×3 skew-symmetric matrix A with $a_{21} = 1, a_{31} = 3, a_{23} = -1$.

2.2   Matrix Algebra

In the previous section we introduced the general idea of a matrix. The next step is to develop the algebra of matrices. Unless otherwise stated, we assume that all elements of the matrices that appear are real or complex numbers.

Addition and Subtraction of Matrices and Multiplication of a Matrix by a Scalar

Addition and subtraction of matrices is defined only for matrices with the same dimensions. We begin with addition.

DEFINITION   2.2.1

If A and B are both $m \times n$ matrices, then we define **addition** (or the **sum**) of A and B, denoted by $A + B$, to be the $m \times n$ matrix whose elements are obtained by adding corresponding elements of A and B. In index notation, if $A = [a_{ij}]$ and $B = [b_{ij}]$, then $A + B = [a_{ij} + b_{ij}]$.

Example 2.2.2   We have

$$\begin{bmatrix} 2 & -1 & 3 \\ 4 & -5 & 0 \end{bmatrix} + \begin{bmatrix} -1 & 0 & 5 \\ -5 & 2 & 7 \end{bmatrix} = \begin{bmatrix} 1 & -1 & 8 \\ -1 & -3 & 7 \end{bmatrix}. \qquad \square$$

Properties of Matrix Addition: If A and B are both $m \times n$ matrices, then

$$A + B = B + A \qquad \text{(matrix addition is commutative)},$$
$$A + (B + C) = (A + B) + C \qquad \text{(matrix addition is associative)}.$$

Both of these properties follow directly from Definition 2.2.1.

In order that we can model oscillatory physical phenomena, in much of the later work we will need to use complex as well as real numbers. Throughout the text we will use the term **scalar** to mean a real or complex number.

DEFINITION   2.2.3

If A is an $m \times n$ matrix and s is a scalar, then we let sA denote the matrix obtained by multiplying every element of A by s. This procedure is called **scalar multiplication**. In index notation, if $A = [a_{ij}]$, then $sA = [sa_{ij}]$.

Example 2.2.4   If $A = \begin{bmatrix} 2 & -1 \\ 4 & 6 \end{bmatrix}$, then $5A = \begin{bmatrix} 10 & -5 \\ 20 & 30 \end{bmatrix}.$ $\qquad \square$

Example 2.2.5   If $A = \begin{bmatrix} 1+i & i \\ 2+3i & 4 \end{bmatrix}$ and $s = 1 - 2i$, where $i = \sqrt{-1}$, find sA.

Solution: We have

$$sA = \begin{bmatrix} (1-2i)(1+i) & (1-2i)i \\ (1-2i)(2+3i) & (1-2i)4 \end{bmatrix} = \begin{bmatrix} 3-i & 2+i \\ 8-i & 4-8i \end{bmatrix}.$$ □

DEFINITION   2.2.6

We define **subtraction** of two matrices with the *same dimensions* by

$$A - B = A + (-1)B.$$

In index notation, $A - B = [a_{ij} - b_{ij}]$. That is, we subtract corresponding elements.

Further properties satisfied by the operations of matrix addition and multiplication of a matrix by a scalar are as follows:

Properties of Scalar Multiplication: For any scalars s and t, and for any matrices A and B of the same size,

$$\begin{aligned} 1A &= A && \text{(unit property)}, \\ s(A+B) &= sA + sB && \text{(distributivity of scalars over matrix addition)}, \\ (s+t)A &= sA + tA && \text{(distributivity of scalar addition over matrices)}, \\ s(tA) = (st)A &= (ts)A = t(sA) && \text{(associativity of scalar multiplication)}. \end{aligned}$$

The $m \times n$ **zero matrix**, denoted $0_{m \times n}$ (or simply 0, if the dimensions are clear), is the $m \times n$ matrix whose elements are all zeros. In the case of the $n \times n$ zero matrix, we may write 0_n. We now collect a few properties of the zero matrix. The first of these below indicates that the zero matrix plays a similar role in matrix addition to that played by the number zero in the addition of real numbers.

Properties of the Zero Matrix: For all matrices A and the zero matrix of the same size, we have

$$A + 0 = A, \qquad A - A = 0, \qquad \text{and} \qquad 0A = 0.$$

Note that in the last property here, the zero on the left side of the equation is a scalar, while the zero on the right side of the equation is a matrix.

Multiplication of Matrices

The definition we introduced above for how to multiply a matrix by a scalar is essentially the only possibility if, in the case when s is a positive integer, we want sA to be the same matrix as the one obtained when A is added to itself s times. We now define how to multiply two matrices together. In this case the multiplication operation is by no means obvious. However, in Chapter 5 when we study linear transformations, the motivation for the matrix multiplication procedure we are defining here will become quite transparent (see Theorem 5.5.7).

We will build up to the general definition of matrix multiplication in three stages.

Case 1: Product of a row n-vector and a column n-vector.   We begin by generalizing a concept from elementary calculus. If $\mathbf{a}$ and $\mathbf{b}$ are either row or column n-vectors, with components $a_1, a_2, \ldots, a_n$, and $b_1, b_2, \ldots, b_n$, respectively, then their **dot product**, denoted $\mathbf{a} \cdot \mathbf{b}$, is the *number*

$$\mathbf{a} \cdot \mathbf{b} = a_1 b_1 + a_2 b_2 + \cdots + a_n b_n.$$

As we will see, this is the key formula in defining the product of two matrices. Now let $\mathbf{a}$ be a *row n-vector*, and let $\mathbf{x}$ be a *column n-vector*. Then their **matrix product ax** is defined to be the 1×1 matrix whose single element is obtained by taking the dot product of the row vectors $\mathbf{a}$ and $\mathbf{x}^T$. Thus,

$$\mathbf{ax} = \begin{bmatrix} a_1 & a_1 & \cdots & a_n \end{bmatrix} \begin{bmatrix} x_1 \\ x_2 \\ \vdots \\ x_n \end{bmatrix} = [a_1 x_1 + a_2 x_2 + \cdots + a_n x_n].$$

Example 2.2.7   If $\mathbf{a} = \begin{bmatrix} 2 & -1 & 3 & 5 \end{bmatrix}$ and $\mathbf{x} = \begin{bmatrix} 3 \\ 2 \\ -3 \\ 4 \end{bmatrix}$, then

$$\mathbf{ax} = \begin{bmatrix} 2 & -1 & 3 & 5 \end{bmatrix} \begin{bmatrix} 3 \\ 2 \\ -3 \\ 4 \end{bmatrix} = [(2)(3) + (-1)(2) + (3)(-3) + (5)(4)] = [15]. \quad \square$$

Case 2: Product of an $m \times n$ matrix and a column n-vector.   If A is an $m \times n$ matrix and $\mathbf{x}$ is a column n-vector, then the product $A\mathbf{x}$ is defined to be the $m \times 1$ matrix whose *i*th element is obtained by taking the dot product of the *i*th row vector of A with $\mathbf{x}$. (See Figure 2.2.1.)

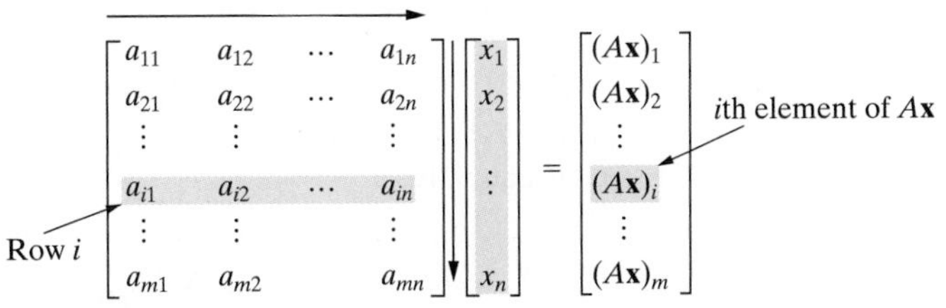

Figure 2.2.1: Multiplication of an $m \times n$ matrix with a column n-vector.

The *i*th row vector of A, $\mathbf{a}_i$, is

$$\mathbf{a}_i = \begin{bmatrix} a_{i1} & a_{i2} & \cdots & a_{in} \end{bmatrix},$$

so that $A\mathbf{x}$ has *i*th element

$$(A\mathbf{x})_i = a_{i1} x_1 + a_{i2} x_2 + \cdots + a_{in} x_n.$$

Consequently the column vector $A\mathbf{x}$ has elements

$$(A\mathbf{x})_i = \sum_{k=1}^{n} a_{ik} x_k, \qquad 1 \le i \le m. \tag{2.2.1}$$

As illustrated in the next example, in practice, we do not use the formula (2.2.1); rather, we explicitly take the matrix products of the row vectors of A with the column vector $\mathbf{x}$.

Example 2.2.8   Find $A\mathbf{x}$ if $A = \begin{bmatrix} 2 & 3 & -1 \\ 1 & 4 & -6 \\ 5 & -2 & 0 \end{bmatrix}$ and $\mathbf{x} = \begin{bmatrix} 7 \\ -3 \\ 1 \end{bmatrix}$.

Solution: We have

$$A\mathbf{x} = \begin{bmatrix} 2 & 3 & -1 \\ 1 & 4 & -6 \\ 5 & -2 & 0 \end{bmatrix} \begin{bmatrix} 7 \\ -3 \\ 1 \end{bmatrix} = \begin{bmatrix} 4 \\ -11 \\ 41 \end{bmatrix}.$$ □

The following result regarding multiplication of a column vector by a matrix will be used repeatedly in later chapters.

Theorem 2.2.9   If $A = \begin{bmatrix} \mathbf{a}_1, \mathbf{a}_2, \ldots, \mathbf{a}_n \end{bmatrix}$ is an $m \times n$ matrix and $\mathbf{c} = \begin{bmatrix} c_1 \\ c_2 \\ \vdots \\ c_n \end{bmatrix}$ is a column n-vector, then

$$A\mathbf{c} = c_1\mathbf{a}_1 + c_2\mathbf{a}_2 + \cdots + c_n\mathbf{a}_n. \tag{2.2.2}$$

Proof   The element a_{ik} of A is the ith component of the column m-vector $\mathbf{a}_k$, so

$$a_{ik} = (\mathbf{a}_k)_i.$$

Applying formula (2.2.1) for multiplication of a column vector by a matrix yields

$$(A\mathbf{c})_i = \sum_{k=1}^{n} a_{ik}c_k = \sum_{k=1}^{n} (\mathbf{a}_k)_i c_k = \sum_{k=1}^{n} (c_k\mathbf{a}_k)_i.$$

Consequently,

$$A\mathbf{c} = \sum_{k=1}^{n} c_k\mathbf{a}_k = c_1\mathbf{a}_1 + c_2\mathbf{a}_2 + \cdots + c_n\mathbf{a}_n$$

as required.   ■

If $\mathbf{x}_1, \mathbf{x}_2, \ldots, \mathbf{x}_n$ are column m-vectors and $c_1, c_2, \ldots, c_n$ are scalars, then an expression of the form

$$c_1\mathbf{x}_1 + c_2\mathbf{x}_2 + \cdots + c_n\mathbf{x}_n$$

is called a **linear combination** of the column vectors. Therefore, from Equation (2.2.2), we see that the vector $A\mathbf{c}$ is obtained by taking a linear combination of the column vectors of A. For example, if

$$A = \begin{bmatrix} 2 & -1 \\ 4 & 3 \end{bmatrix} \qquad \text{and} \qquad \mathbf{c} = \begin{bmatrix} 5 \\ -1 \end{bmatrix},$$

then

$$Ac = c_1 a_1 + c_2 a_2 = 5 \begin{bmatrix} 2 \\ 4 \end{bmatrix} + (-1) \begin{bmatrix} -1 \\ 3 \end{bmatrix} = \begin{bmatrix} 11 \\ 17 \end{bmatrix}.$$

Case 3: Product of an $m \times n$ matrix and an $n \times p$ matrix.   If A is an $m \times n$ matrix and B is an $n \times p$ matrix, then the product AB has columns defined by multiplying the matrix A by the respective column vectors of B, as described in Case 2. That is, if $B = [b_1, b_2, \ldots, b_p]$, then AB is the $m \times p$ matrix defined by

$$AB = [Ab_1, Ab_2, \ldots, Ab_p].$$

Example 2.2.10   If $A = \begin{bmatrix} 1 & 4 & 2 \\ 3 & 5 & 7 \end{bmatrix}$ and $B = \begin{bmatrix} 2 & 3 \\ 5 & -2 \\ 8 & 4 \end{bmatrix}$, determine AB.

Solution: We have

$$AB = \begin{bmatrix} 1 & 4 & 2 \\ 3 & 5 & 7 \end{bmatrix} \begin{bmatrix} 2 & 3 \\ 5 & -2 \\ 8 & 4 \end{bmatrix}$$

$$= \begin{bmatrix} [(1)(2) + (4)(5) + (2)(8)] & [(1)(3) + (4)(-2) + (2)(4)] \\ [(3)(2) + (5)(5) + (7)(8)] & [(3)(3) + (5)(-2) + (7)(4)] \end{bmatrix} = \begin{bmatrix} 38 & 3 \\ 87 & 27 \end{bmatrix}. \quad \square$$

Example 2.2.11   If $A = \begin{bmatrix} 2 \\ -1 \\ 3 \end{bmatrix}$ and $B = \begin{bmatrix} 2 & 4 \end{bmatrix}$, determine AB.

Solution: We have

$$AB = \begin{bmatrix} 2 \\ -1 \\ 3 \end{bmatrix} \begin{bmatrix} 2 & 4 \end{bmatrix} = \begin{bmatrix} (2)(2) & (2)(4) \\ (-1)(2) & (-1)(4) \\ (3)(2) & (3)(4) \end{bmatrix} = \begin{bmatrix} 4 & 8 \\ -2 & -4 \\ 6 & 12 \end{bmatrix}. \quad \square$$

Another way to describe AB is to note that the element $(AB)_{ij}$ is obtained by computing the matrix product of the ith row vector of A and the jth column vector of B. That is,

$$(AB)_{ij} = a_{i1} b_{1j} + a_{i2} b_{2j} + \cdots + a_{in} b_{nj}.$$

Expressing this using the summation notation yields the following result:

DEFINITION   2.2.12

If $A = [a_{ij}]$ is an $m \times n$ matrix, $B = [b_{ij}]$ is an $n \times p$ matrix, and $C = AB$, then

$$c_{ij} = \sum_{k=1}^{n} a_{ik} b_{kj}, \qquad 1 \leq i \leq m, \quad 1 \leq j \leq p. \qquad (2.2.3)$$

This is called the **index form** of the matrix product.

The formula (2.2.3) for the ijth element of AB is very important and will often be required in the future. The reader should memorize it.

In order for the product AB to be defined, we see that A and B must satisfy number of columns of A = number of rows of B.

In such a case, if C represents the product matrix AB, then the relationship between the dimensions of the matrices is

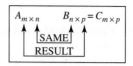

Now we give some further examples of matrix multiplication.

Example 2.2.13   If $A = \begin{bmatrix} 1 & 3 \\ 2 & 4 \end{bmatrix}$ and $B = \begin{bmatrix} 2 & -2 & 0 \\ 1 & 5 & 3 \end{bmatrix}$, then

$$AB = \begin{bmatrix} 1 & 3 \\ 2 & 4 \end{bmatrix} \begin{bmatrix} 2 & -2 & 0 \\ 1 & 5 & 3 \end{bmatrix} = \begin{bmatrix} 5 & 13 & 9 \\ 8 & 16 & 12 \end{bmatrix}.$$ □

Example 2.2.14   If $A = \begin{bmatrix} 1 & 2 & -1 \end{bmatrix}$ and $B = \begin{bmatrix} -1 & 1 \\ 0 & 1 \\ 1 & 2 \end{bmatrix}$, then

$$AB = \begin{bmatrix} 1 & 2 & -1 \end{bmatrix} \begin{bmatrix} -1 & 1 \\ 0 & 1 \\ 1 & 2 \end{bmatrix} = \begin{bmatrix} -2 & 1 \end{bmatrix}.$$ □

Example 2.2.15   If $A = \begin{bmatrix} 2 \\ -1 \\ 3 \end{bmatrix}$ and $B = \begin{bmatrix} 1 & 4 & -6 \end{bmatrix}$, then

$$AB = \begin{bmatrix} 2 \\ -1 \\ 3 \end{bmatrix} \begin{bmatrix} 1 & 4 & -6 \end{bmatrix} = \begin{bmatrix} 2 & 8 & -12 \\ -1 & -4 & 6 \\ 3 & 12 & -18 \end{bmatrix}.$$ □

Example 2.2.16   If $A = \begin{bmatrix} 1-i & i \\ 2+i & 1+i \end{bmatrix}$ and $B = \begin{bmatrix} 3+2i & 1+4i \\ i & -1+2i \end{bmatrix}$, then

$$AB = \begin{bmatrix} 1-i & i \\ 2+i & 1+i \end{bmatrix} \begin{bmatrix} 3+2i & 1+4i \\ i & -1+2i \end{bmatrix} = \begin{bmatrix} 4-i & 3+2i \\ 3+8i & -5+10i \end{bmatrix}.$$ □

Notice that in Examples 2.2.13 and 2.2.14 above, the product BA is not defined, since the number of columns of the matrix B does not agree with the number of rows of the matrix A.

We can now establish some basic properties of matrix multiplication.

Theorem 2.2.17   If A, B and C have appropriate dimensions for the operations to be performed, then

$$A(BC) = (AB)C \qquad \text{(associativity of matrix multiplication)}, \qquad (2.2.4)$$
$$A(B + C) = AB + AC \qquad \text{(left distributivity of matrix multiplication)}, \qquad (2.2.5)$$
$$(A + B)C = AC + BC \qquad \text{(right distributivity of matrix multiplication)}. \qquad (2.2.6)$$

Proof The idea behind the proof of each of these results is to use the definition of matrix multiplication to show that the ijth element of the matrix on the left-hand side of each equation is equal to the ijth element of the matrix on the right-hand side. We illustrate by proving (2.2.6), but we leave the proofs of (2.2.4) and (2.2.5) as exercises. Suppose that A and B are $m \times n$ matrices and that C is an $n \times p$ matrix. Then, from Equation (2.2.3),

$$[(A + B)C]_{ij} = \sum_{k=1}^{n}(a_{ik} + b_{ik})c_{kj} = \sum_{k=1}^{n} a_{ik}c_{kj} + \sum_{k=1}^{n} b_{ik}c_{kj}$$
$$= (AC)_{ij} + (BC)_{ij}$$
$$= (AC + BC)_{ij}, \quad 1 \le i \le m, \quad 1 \le j \le p.$$

Consequently,

$$(A + B)C = AC + BC. \qquad \blacksquare$$

Theorem 2.2.17 states that matrix multiplication is associative and distributive (over addition). We now consider the question of commutativity of matrix multiplication. If A is an $m \times n$ matrix and B is an $n \times m$ matrix, we can form both of the products AB and BA, which are $m \times m$ and $n \times n$, respectively. In the first of these, we say that B has been **premultiplied** by A, whereas in the second, we say that B has been **postmultiplied** by A. If $m \ne n$, then the matrices AB and BA will have different dimensions, so they cannot be equal. It is important to realize, however, that even if $m = n$, in general (that is, except for special cases)

$$AB \ne BA.$$

This is the statement that

matrix multiplication is not commutative.

With a little bit of thought this should not be too surprising, in view of the fact that the ijth element of AB is obtained by taking the matrix product of the ith row vector of A with the jth column vector of B, whereas the ijth element of BA is obtained by taking the matrix product of the ith row vector of B with the jth column vector of A. We illustrate with an example.

Example 2.2.18   If $A = \begin{bmatrix} 1 & 2 \\ -1 & 3 \end{bmatrix}$ and $B = \begin{bmatrix} 3 & 1 \\ 2 & -1 \end{bmatrix}$, find AB and BA.

Solution: We have

$$AB = \begin{bmatrix} 1 & 2 \\ -1 & 3 \end{bmatrix}\begin{bmatrix} 3 & 1 \\ 2 & -1 \end{bmatrix} = \begin{bmatrix} 7 & -1 \\ 3 & -4 \end{bmatrix} \quad \text{and} \quad BA = \begin{bmatrix} 3 & 1 \\ 2 & -1 \end{bmatrix}\begin{bmatrix} 1 & 2 \\ -1 & 3 \end{bmatrix} = \begin{bmatrix} 2 & 9 \\ 3 & 1 \end{bmatrix}.$$

Thus we see that in this example, $AB \ne BA$.   □

As an exercise, the reader can calculate the matrix BA in Examples 2.2.15 and 2.2.16 and again see that $AB \ne BA$.

For an $n \times n$ matrix we use the usual power notation to denote the operation of multiplying A by itself. Thus,

$$A^2 = AA, \qquad A^3 = AAA, \qquad \text{and so on.}$$

The **identity matrix**, I_n (or just I if the dimensions are obvious), is the $n \times n$ matrix with ones on the main diagonal and zeros elsewhere. For example,

$$I_2 = \begin{bmatrix} 1 & 0 \\ 0 & 1 \end{bmatrix} \quad \text{and} \quad I_3 = \begin{bmatrix} 1 & 0 & 0 \\ 0 & 1 & 0 \\ 0 & 0 & 1 \end{bmatrix}.$$

DEFINITION 2.2.19

The elements of I_n can be represented by the **Kronecker delta symbol**, δ_{ij}, defined by

$$\delta_{ij} = \begin{cases} 1, & \text{if } i = j, \\ 0, & \text{if } i \neq j. \end{cases}$$

Then,

$$I_n = [\delta_{ij}].$$

The following properties of the identity matrix indicate that it plays the same role in matrix multiplication as the number 1 does in the multiplication of real numbers.

Properties of the Identity Matrix:

1. $A_{m \times n} I_n = A_{m \times n}$.

2. $I_m A_{m \times p} = A_{m \times p}$.

Proof We establish property 1 and leave the proof of property 2 as an exercise (Problem 25). Using the index form of the matrix product, we have

$$(AI)_{ij} = \sum_{k=1}^{n} a_{ik} \delta_{kj} = a_{i1} \delta_{1j} + a_{i2} \delta_{2j} + \cdots + a_{ij} \delta_{jj} + \cdots + a_{in} \delta_{nj}.$$

But, from the definition of the Kronecker delta symbol, we see that all terms in the summation with $k \neq j$ vanish, so that we are left with

$$(AI)_{ij} = a_{ij} \delta_{jj} = a_{ij}, \qquad 1 \leq i \leq m, \quad 1 \leq j \leq n. \qquad \blacksquare$$

The next example illustrates property 2 of the identity matrix.

Example 2.2.20 If $A = \begin{bmatrix} 2 & -1 \\ 3 & 5 \\ 0 & -2 \end{bmatrix}$, verify that $I_3 A = A$.

Solution: We have

$$I_3 A = \begin{bmatrix} 1 & 0 & 0 \\ 0 & 1 & 0 \\ 0 & 0 & 1 \end{bmatrix} \begin{bmatrix} 2 & -1 \\ 3 & 5 \\ 0 & -2 \end{bmatrix} = \begin{bmatrix} 2 & -1 \\ 3 & 5 \\ 0 & -2 \end{bmatrix} = A. \qquad \square$$

Properties of the Transpose

The operation of taking the transpose of a matrix was introduced in the previous section. The next theorem gives three important properties satisfied by the transpose. These should be memorized.

Theorem 2.2.21   Let A and C be $m \times n$ matrices, and let B be an $n \times p$ matrix. Then

1. $(A^T)^T = A$.
2. $(A + C)^T = A^T + C^T$.
3. $(AB)^T = B^T A^T$.

Proof For all three statements, our strategy is again to show that the (i, j)-elements of each side of the equation are the same. We prove statement 3 and leave the proofs of 1 and 2 for the exercises (Problem 24). From the definition of the transpose and the index form of the matrix product, we have

$$[(AB)^T]_{ij} = (AB)_{ji} \qquad \text{(definition of the transpose)}$$

$$= \sum_{k=1}^{n} a_{jk} b_{ki} \qquad \text{(index form of the matrix product)}$$

$$= \sum_{k=1}^{n} b_{ki} a_{jk} = \sum_{k=1}^{n} b_{ik}^T a_{kj}^T$$

$$= (B^T A^T)_{ij}.$$

Consequently,

$$(AB)^T = B^T A^T. \qquad \blacksquare$$

Results for Triangular Matrices

Upper and lower triangular matrices play a significant role in the analysis of linear systems of equations. The following theorem and its corollary will be needed in Section 2.7.

Theorem 2.2.22   The product of two lower (upper) triangular matrices is a lower (upper) triangular matrix.

Proof Suppose that A and B are $n \times n$ lower triangular matrices. Then, $a_{ik} = 0$ whenever $i < k$, and $b_{kj} = 0$ whenever $k < j$. If we let $C = AB$, then we must prove that

$$c_{ij} = 0 \quad \text{whenever} \quad i < j.$$

Using the index form of the matrix product, we have

$$c_{ij} = \sum_{k=1}^{n} a_{ik} b_{kj} = \sum_{k=j}^{n} a_{ik} b_{kj} \qquad \text{(since } b_{kj} = 0 \text{ if } k < j\text{).} \qquad (2.2.7)$$

We now impose the condition that $i < j$. Then, since $k \geq j$ in (2.2.7), it follows that $k > i$. However, this implies that $a_{ik} = 0$ (since A is lower triangular), and hence, from (2.2.7), that

$$c_{ij} = 0 \quad \text{whenever} \quad i < j.$$

as required.

To establish the result for upper triangular matrices, either we can give an argument similar to that presented above for lower triangular matrices, or we can use the fact that the transpose of a lower triangular matrix is an upper triangular matrix, and vice versa. Hence, if A and B are $n \times n$ upper triangular matrices, then A^T and B^T are lower triangular, and therefore by what we proved above, $(AB)^T = B^T A^T$ remains lower triangular. Thus, AB is upper triangular.   ∎

Corollary 2.2.23   The product of two unit lower (upper) triangular matrices is a unit lower (upper) triangular matrix.

Proof Let A and B be unit lower triangular $n \times n$ matrices. We know from Theorem 2.2.22 that $C = AB$ is a lower triangular matrix. We must establish that $c_{ii} = 1$ for each i. The elements on the main diagonal of C can be obtained by setting $j = i$ in (2.2.7):

$$c_{ii} = \sum_{k=i}^{n} a_{ik} b_{ki}. \tag{2.2.8}$$

Since $a_{ik} = 0$ whenever $k > i$, the only nonzero term in the summation in (2.2.8) occurs when $k = i$. Consequently,

$$c_{ii} = a_{ii} b_{ii} = 1 \cdot 1 = 1, \qquad i = 1, 2, \ldots, n.$$

The proof for unit upper triangular matrices is similar and left as an exercise.   ∎

The Algebra and Calculus of Matrix Functions

By and large, the algebra of matrix and vector functions is the same as that for matrices and vectors of real or complex numbers. Since vector functions are a special case of matrix functions, we focus here on matrix functions. The main comment here pertains to scalar multiplication. In the description of scalar multiplication of matrices of numbers, the scalars were required to be real or complex numbers. However, for matrix functions, we can scalar multiply by any *scalar function* $s(t)$.

Example 2.2.24   If $s(t) = e^t$ and $A(t) = \begin{bmatrix} -2+t & e^{2t} \\ 4 & \cos t \end{bmatrix}$, then

$$s(t)A(t) = \begin{bmatrix} e^t(-2+t) & e^{3t} \\ 4e^t & e^t \cos t \end{bmatrix}.$$ □

Example 2.2.25   Referring to A and B from Example 2.1.14, find $2A - tB^T$.

Solution: We have

$$2A - tB^T = \begin{bmatrix} 2t^3 & 2t - 2\cos t & 10 \\ 2e^{t^2} & 2\ln(t+1) & 2te^t \end{bmatrix} - \begin{bmatrix} 5t - t^2 + t^3 & -t & 6t \\ t\sin(e^{2t}) & t\tan t & 6t - t^2 \end{bmatrix}$$

$$= \begin{bmatrix} t^3 + t^2 - 5t & 3t - 2\cos t & 10 - 6t \\ 2e^{t^2} - t\sin(e^{2t}) & 2\ln(t+1) - t\tan t & 2te^t + t^2 - 6t \end{bmatrix}. \qquad \square$$

We can also perform calculus operations on matrix functions. In particular we can differentiate and integrate them. The rules for doing so are as follows:

1. The derivative of a matrix function is obtained by differentiating *every* element of the matrix. Thus, if $A(t) = [a_{ij}(t)]$, then

$$\frac{dA}{dt} = \left[\frac{da_{ij}(t)}{dt}\right],$$

provided that each of the a_{ij} is differentiable.

2. It follows from (1) and the index form of the matrix product that if A and B are both differentiable and the product AB is defined, then

$$\frac{d}{dt}(AB) = A\frac{dB}{dt} + \frac{dA}{dt}B.$$

The key point to notice is that the order of the multiplication must be preserved.

3. If $A(t) = [a_{ij}(t)]$, where each $a_{ij}(t)$ is integrable on an interval $[a, b]$, then

$$\int_a^b A(t)\, dt = \left[\int_a^b a_{ij}(t)\, dt\right].$$

Example 2.2.26   If $A(t) = \begin{bmatrix} 2t & 1 \\ 6t^2 & 4e^{2t} \end{bmatrix}$, determine dA/dt and $\int_0^1 A(t)\, dt$.

Solution: We have

$$\frac{dA}{dt} = \begin{bmatrix} 2 & 0 \\ 12t & 8e^{2t} \end{bmatrix},$$

whereas

$$\int_0^1 A(t)\, dt = \begin{bmatrix} \int_0^1 2t\, dt & \int_0^1 1\, dt \\ \int_0^1 6t^2\, dt & \int_0^1 4e^{2t}\, dt \end{bmatrix} = \begin{bmatrix} 1 & 1 \\ 2 & 2(e^2 - 1) \end{bmatrix}. \qquad \square$$

Exercises for 2.2

Key Terms

Matrix addition and subtraction, Scalar multiplication, Matrix multiplication, Dot product, Linear combination of column vectors, Index form, Premultiplication, Postmultiplication, Zero matrix, Identity matrix, Kronecker delta symbol.

Skills

• Be able to perform matrix addition, subtraction, and multiplication.

• Know the basic relationships between the dimensions of two matrices A and B in order for $A + B$ to be defined, and in order for AB to be defined.

• Be able to multiply a matrix by a scalar.

• Be able to express the product $A\mathbf{x}$ of a matrix and a column vector as a linear combination of the columns of A.

- Be familiar with all of the basic properties of matrix addition, matrix multiplication, scalar multiplication, the zero matrix, the identity matrix, the transpose of a matrix, and lower (upper) triangular matrices.

- Know the basic technique for showing formally that two matrices are equal.

- Be able to perform algebra and calculus operations on matrix functions.

True-False Review

For Questions 1–12, decide if the given statement is **true** or **false**, and give a brief justification for your answer. If true, you can quote a relevant definition or theorem from the text. If false, provide an example, illustration, or brief explanation of why the statement is false.

1. For all matrices A, B, and C of the appropriate dimensions, we have

$$(AB)C = (CA)B.$$

2. If A is an $m \times n$ matrix, B is an $n \times p$ matrix, and C is a $p \times q$ matrix, then ABC is an $m \times q$ matrix.

3. If A and B are symmetric $n \times n$ matrices, then so is $A + B$.

4. If A and B are skew-symmetric $n \times n$ matrices, then AB is a symmetric matrix.

5. For $n \times n$ matrices A and B, we have

$$(A + B)^2 = A^2 + 2AB + B^2.$$

6. If $AB = 0$, then either $A = 0$ or $B = 0$.

7. If A and B are square matrices such that AB is upper triangular, then A and B must both be upper triangular.

8. If A is a square matrix such that $A^2 = A$, then A must be the zero matrix or the identity matrix.

9. If A is a matrix of numbers, then if we consider A as a matrix function, its derivative is the zero matrix.

10. If A and B are matrix functions whose product AB is defined, then

$$\frac{d}{dt}(AB) = A\frac{dB}{dt} + B\frac{dA}{dt}.$$

11. If A is an $n \times n$ matrix function such that A and dA/dt are the same function, then $A = ce^t I_n$ for some constant c.

12. If A and B are matrix functions whose product AB is defined, then the matrix functions $(AB)^T$ and $B^T A^T$ are the same.

Problems

1. If

$$A = \begin{bmatrix} 1 & 2 & -1 \\ 3 & 5 & 2 \end{bmatrix}, \qquad B = \begin{bmatrix} 2 & -1 & 3 \\ 1 & 4 & 5 \end{bmatrix},$$

find $2A$, $-3B$, $A - 2B$, and $3A + 4B$.

2. If

$$A = \begin{bmatrix} 2 & -1 & 0 \\ 3 & 1 & 2 \\ -1 & 1 & 1 \end{bmatrix}, \qquad B = \begin{bmatrix} 1 & -1 & 2 \\ 3 & 0 & 1 \\ -1 & 1 & 0 \end{bmatrix},$$

$$C = \begin{bmatrix} -1 & -1 & 1 \\ 1 & 2 & 3 \\ -1 & 1 & 0 \end{bmatrix},$$

find the matrix D such that $2A + B - 3C + 2D = A + 4C$.

3. Let

$$A = \begin{bmatrix} 1 & -1 & 2 \\ 3 & 1 & 4 \end{bmatrix}, \qquad B = \begin{bmatrix} 2 & -1 & 3 \\ 5 & 1 & 2 \\ 4 & 6 & -2 \end{bmatrix},$$

$$C = \begin{bmatrix} 1 \\ -1 \\ 2 \end{bmatrix}, \qquad D = \begin{bmatrix} 2 & -2 & 3 \end{bmatrix}.$$

Find, if possible, AB, BC, CA, DC, DB, AD, and CD.

For Problems 4–6, determine AB for the given matrices. In these problems i denotes $\sqrt{-1}$.

4. $A = \begin{bmatrix} 2-i & 1+i \\ -i & 2+4i \end{bmatrix}$, $B = \begin{bmatrix} i & 1-3i \\ 0 & 4+i \end{bmatrix}$.

5. $A = \begin{bmatrix} 3+2i & 2-4i \\ 5+i & -1+3i \end{bmatrix}$, $B = \begin{bmatrix} -1+i & 3+2i \\ 4-3i & 1+i \end{bmatrix}$.

6. $A = \begin{bmatrix} 3-2i & i \\ -i & 1 \end{bmatrix}$, $B = \begin{bmatrix} -1+i & 2-i & 0 \\ 1+5i & 0 & 3-2i \end{bmatrix}$.

7. Let

$$A = \begin{bmatrix} 1 & -1 & 2 & 3 \\ -2 & 3 & 4 & 6 \end{bmatrix}, \qquad B = \begin{bmatrix} 3 & 2 \\ 1 & 5 \\ 4 & -3 \\ -1 & 6 \end{bmatrix},$$

$$C = \begin{bmatrix} -3 & 2 \\ 1 & -4 \end{bmatrix}.$$

Find ABC and CAB.

8. If

$$A = \begin{bmatrix} 1 & -2 \\ 3 & 1 \end{bmatrix}, \qquad B = \begin{bmatrix} -1 & 2 \\ 5 & 3 \end{bmatrix}, \qquad C = \begin{bmatrix} 3 \\ -1 \end{bmatrix},$$

find $(2A - 3B)C$.

For Problems 9–11, determine $A\mathbf{c}$ by computing an appropriate linear combination of the column vectors of A.

9. $A = \begin{bmatrix} 1 & 3 \\ -5 & 4 \end{bmatrix}, \quad \mathbf{c} = \begin{bmatrix} 6 \\ -2 \end{bmatrix}.$

10. $A = \begin{bmatrix} 3 & -1 & 4 \\ 2 & 1 & 5 \\ 7 & -6 & 3 \end{bmatrix}, \quad \mathbf{c} = \begin{bmatrix} 2 \\ 3 \\ -4 \end{bmatrix}.$

11. $A = \begin{bmatrix} -1 & 2 \\ 4 & 7 \\ 5 & -4 \end{bmatrix}, \quad \mathbf{c} = \begin{bmatrix} 5 \\ -1 \end{bmatrix}.$

12. If A is an $m \times n$ matrix and C is an $r \times s$ matrix, what must be the dimensions of B in order for the product ABC to be defined? Write an expression for the (i, j)th element of ABC in terms of the elements of A, B and C.

13. Find A^2, A^3, and A^4 if

(a) $A = \begin{bmatrix} 1 & -1 \\ 2 & 3 \end{bmatrix}.$

(b) $A = \begin{bmatrix} 0 & 1 & 0 \\ -2 & 0 & 1 \\ 4 & -1 & 0 \end{bmatrix}.$

14. If A and B are $n \times n$ matrices, prove that

(a) $(A + B)^2 = A^2 + AB + BA + B^2.$

(b) $(A - B)^2 = A^2 - AB - BA + B^2.$

15. If

$$A = \begin{bmatrix} 3 & -1 \\ -5 & -1 \end{bmatrix},$$

calculate A^2 and verify that A satisfies $A^2 - 2A - 8I_2 = 0_2$.

16. Find a matrix

$$A = \begin{bmatrix} 1 & x & z \\ 0 & 1 & y \\ 0 & 0 & 1 \end{bmatrix}$$

such that

$$A^2 + \begin{bmatrix} 0 & -1 & 0 \\ 0 & 0 & -1 \\ 0 & 0 & 0 \end{bmatrix} = I_3.$$

17. If

$$A = \begin{bmatrix} x & 1 \\ -2 & y \end{bmatrix},$$

determine all values of x and y for which $A^2 = A$.

18. The Pauli spin matrices σ_1, σ_2, and σ_3 are defined by

$$\sigma_1 = \begin{bmatrix} 0 & 1 \\ 1 & 0 \end{bmatrix}, \qquad \sigma_2 = \begin{bmatrix} 0 & -i \\ i & 0 \end{bmatrix},$$

and

$$\sigma_3 = \begin{bmatrix} 1 & 0 \\ 0 & -1 \end{bmatrix}.$$

Verify that they satisfy

$$\sigma_1\sigma_2 = i\sigma_3, \qquad \sigma_2\sigma_3 = i\sigma_1, \qquad \sigma_3\sigma_1 = i\sigma_2.$$

If A and B are $n \times n$ matrices, we define their **commutator**, denoted $[A, B]$, by

$$[A, B] = AB - BA.$$

Thus, $[A, B] = 0$ if and only if A and B commute. That is, $AB = BA$. Problems 19–22 require the commutator.

19. If

$$A = \begin{bmatrix} 1 & -1 \\ 2 & 1 \end{bmatrix}, \qquad B = \begin{bmatrix} 3 & 1 \\ 4 & 2 \end{bmatrix},$$

find $[A, B]$.

20. If

$$A_1 = \begin{bmatrix} 1 & 0 \\ 0 & 1 \end{bmatrix}, \quad A_2 = \begin{bmatrix} 0 & 1 \\ 0 & 0 \end{bmatrix}, \quad A_3 = \begin{bmatrix} 0 & 0 \\ 1 & 0 \end{bmatrix},$$

compute all of the commutators $[A_i, A_j]$, and determine which of the matrices commute.

21. If

$$A_1 = \frac{1}{2}\begin{bmatrix} 0 & i \\ i & 0 \end{bmatrix}, \quad A_2 = \frac{1}{2}\begin{bmatrix} 0 & -1 \\ 1 & 0 \end{bmatrix},$$

$$A_3 = \frac{1}{2}\begin{bmatrix} i & 0 \\ 0 & -i \end{bmatrix},$$

verify that $[A_1, A_2] = A_3$, $[A_2, A_3] = A_1$, $[A_3, A_1] = A_2$.

22. If A, B and C are $n \times n$ matrices, find $[A, [B, C]]$ and prove the *Jacobi identity*

$$[A, [B, C]] + [B, [C, A]] + [C, [A, B]] = 0.$$

23. Use the index form of the matrix product to prove properties (2.2.4) and (2.2.5).

24. Prove parts 1 and 2 of Theorem 2.2.21.

25. Prove property 2 of the identity matrix.

26. If A and B are $n \times n$ matrices, prove that $\text{tr}(AB) = \text{tr}(BA)$.

27. If

$$A = \begin{bmatrix} 1 & -1 & 1 & 4 \\ 2 & 0 & 2 & -3 \\ 3 & 4 & -1 & 0 \end{bmatrix}, \qquad B = \begin{bmatrix} 0 & 1 \\ -1 & 2 \\ 1 & 1 \\ 2 & 1 \end{bmatrix},$$

find A^T, B^T, AA^T, AB and $B^T A^T$.

28. Let $A = \begin{bmatrix} 2 & 2 & 1 \\ 2 & 5 & 2 \\ 1 & 2 & 2 \end{bmatrix}$, and let S be the matrix with column vectors

$$\mathbf{s}_1 = \begin{bmatrix} -x \\ 0 \\ x \end{bmatrix}, \qquad \mathbf{s}_2 = \begin{bmatrix} -y \\ y \\ -y \end{bmatrix}, \qquad \mathbf{s}_3 = \begin{bmatrix} z \\ 2z \\ z \end{bmatrix},$$

where x, y, z are constants.

(a) Show that $AS = [\mathbf{s}_1, \mathbf{s}_2, 7\mathbf{s}_3]$.

(b) Find all values of x, y, z such that $S^T AS = \text{diag}(1, 1, 7)$.

29. A matrix that is a multiple of I_n is called an $n \times n$ **scalar matrix**.

(a) Determine the 4×4 scalar matrix whose trace is 8.

(b) Determine the 3×3 scalar matrix such that the product of the elements on the main diagonal is 343.

30. Prove that for each positive integer n, there is a unique scalar matrix whose trace is a given constant k.

If A is an $n \times n$ matrix, then the matrices S and T defined by

$$S = \tfrac{1}{2}(A + A^T), \qquad T = \tfrac{1}{2}(A - A^T)$$

are referred to as the symmetric and skew-symmetric parts of A, respectively. Problems 31–34 investigate properties of S and T.

31. Use the properties of the transpose to show that S and T are symmetric and skew-symmetric, respectively.

32. Find S and T for the matrix

$$A = \begin{bmatrix} 1 & -5 & 3 \\ 3 & 2 & 4 \\ 7 & -2 & 6 \end{bmatrix}.$$

33. If A is an $n \times n$ symmetric matrix, show that $T = 0$. What is the corresponding result for skew-symmetric matrices?

34. Show that every $n \times n$ matrix can be written as the sum of a symmetric and a skew-symmetric matrix.

35. Prove that if A is an $n \times p$ matrix and $D = \text{diag}(d_1, d_2, \ldots, d_n)$, then DA is the matrix obtained by multiplying the ith row vector of A by d_i ($1 \leq i \leq n$).

36. Use the properties of the transpose to prove that

(a) AA^T is a symmetric matrix.

(b) $(ABC)^T = C^T B^T A^T$.

For Problems 37–40, determine the derivative of the given matrix function.

37. $A(t) = \begin{bmatrix} e^{-2t} \\ \sin t \end{bmatrix}$.

38. $A(t) = \begin{bmatrix} t & \sin t \\ \cos t & 4t \end{bmatrix}$.

39. $A(t) = \begin{bmatrix} e^t & e^{2t} & t^2 \\ 2e^t & 4e^{2t} & 5t^2 \end{bmatrix}$.

40. $A(t) = \begin{bmatrix} \sin t & \cos t & 0 \\ -\cos t & \sin t & t \\ 0 & 3t & 1 \end{bmatrix}$.

41. Let $A = [a_{ij}(t)]$ be an $m \times n$ matrix function and let $B = [b_{ij}(t)]$ be an $n \times p$ matrix function. Use the definition of matrix multiplication to prove that

$$\frac{d}{dt}(AB) = A\frac{dB}{dt} + \frac{dA}{dt}B.$$

For Problems 42–45, determine $\int_a^b A(t)\, dt$ for the given matrix function.

42. $A(t) = \begin{bmatrix} \cos t \\ \sin t \end{bmatrix}$, $a = 0$, $b = \pi/2$.

43. $A(t) = \begin{bmatrix} e^t & e^{-t} \\ 2e^t & 5e^{-t} \end{bmatrix}$, $a = 0$, $b = 1$.

44. $A(t) = \begin{bmatrix} e^{2t} & \sin 2t \\ t^2 - 5 & te^t \\ \sec^2 t & 3t - \sin t \end{bmatrix}$, $a = 0, b = 1$.

45. The matrix function $A(t)$ in Problem 39, with $a = 0$ and $b = 1$.

Integration of matrix functions given in the text was done with definite integrals, but one can naturally compute indefinite integrals of matrix functions as well, by performing indefinite integrals for each element of the matrix function.

In Problems 46–49, evaluate the indefinite integral $\int A(t)\, dt$ for the given matrix function. You may assume that the constants of all indefinite integrations are zero.

46. $A(t) = \begin{bmatrix} 2t \\ 3t^2 \end{bmatrix}$.

47. The matrix function $A(t)$ in Problem 40.

48. The matrix function $A(t)$ in Problem 43.

49. The matrix function $A(t)$ in Problem 44.

2.3   Terminology for Systems of Linear Equations

As we mentioned in Section 2.1, a main aim of this chapter is to apply matrices to determine the solution properties of any system of linear equations. We are now in a position to pursue that aim. We begin by introducing some notation and terminology.

DEFINITION   2.3.1

The general $m \times n$ system of linear equations is of the form

$$
\begin{aligned}
a_{11}x_1 + a_{12}x_2 + \cdots + a_{1n}x_n &= b_1, \\
a_{21}x_1 + a_{22}x_2 + \cdots + a_{2n}x_n &= b_2, \\
&\ \ \vdots \\
a_{m1}x_1 + a_{m2}x_2 + \cdots + a_{mn}x_n &= b_m,
\end{aligned}
\tag{2.3.1}
$$

where the **system coefficients** a_{ij} and the **system constants** b_j are given scalars and $x_1, x_2, \ldots, x_n$ denote the unknowns in the system. If $b_i = 0$ for all i, then the system is called **homogeneous**; otherwise it is called **nonhomogeneous**.

DEFINITION   2.3.2

By a **solution** to the system (2.3.1) we mean an ordered n-tuple of scalars, $(c_1, c_2, \ldots, c_n)$, which, when substituted for $x_1, x_2, \ldots, x_n$ into the left-hand side of system (2.3.1), yield the values on the right-hand side. The set of all solutions to system (2.3.1) is called the **solution set** to the system.

Remarks

1. Usually the a_{ij} and b_j will be real numbers, and we will then be interested in determining only the real solutions to system (2.3.1). However, many of the problems that arise in the later chapters will require the solution to systems with complex coefficients, in which case the corresponding solutions will also be complex.

2. If $(c_1, c_2, \ldots, c_n)$ is a solution to the system (2.3.1), we will sometimes specify this solution by writing $x_1 = c_1, x_2 = c_2, \ldots, x_n = c_n$. For example, the ordered pair of numbers $(1, 2)$ is a solution to the system

$$
\begin{aligned}
x_1 + x_2 &= 3, \\
3x_1 - 2x_2 &= -1,
\end{aligned}
$$

and we could express this solution in the equivalent form $x_1 = 1, x_2 = 2$.

At this point, we pause to introduce some important notation that will be used frequently throughout the remainder of the text.

Notation 2.3.3   The set of all ordered n-tuples of real numbers $(c_1, c_2, \ldots, c_n)$ will be denoted by $\mathbb{R}^n$. Therefore, the set of all real solutions to the linear system (2.3.1) forms a subset of $\mathbb{R}^n$. In like manner, the set of all ordered n-tuples of complex numbers will be denoted by $\mathbb{C}^n$, and the solution set for a linear system (2.3.1) containing complex coefficients can be viewed as a subset of $\mathbb{C}^n$.

Notice that when we restrict all scalar values to be real, we have a natural correspondence between elements of $\mathbb{R}^n$, row n-vectors, and column n-vectors:

$$(x_1, x_2, \ldots, x_n) \longleftrightarrow [x_1 \quad x_2 \quad \ldots \quad x_n] \longleftrightarrow \begin{bmatrix} x_1 \\ x_2 \\ \vdots \\ x_n \end{bmatrix}.$$

Therefore, we may use the operations of addition, subtraction, and scalar multiplication of row n-vectors and column n-vectors to naturally equip $\mathbb{R}^n$ with these same operations. Therefore, just as we can perform addition and scalar multiplication of row or column vectors, so too can we perform these operations on n-tuples of scalars. In fact, we will often treat ordered n-tuples of scalars, row n-vectors, and column n-vectors as if they were just different representations of the same basic object.

Of course, if we allow all scalars in question to assume complex values, then the correspondence is between elements of $\mathbb{C}^n$, row n-vectors, and column n-vectors. We will have much more to say about the sets $\mathbb{R}^n$ and $\mathbb{C}^n$ in Chapter 4.

Returning to the general discussion of system (2.3.1), we will consider some fundamental questions:

1. Does the system (2.3.1) have a solution?

2. If the answer to question 1 is yes, then how many solutions are there?

3. How do we determine all of the solutions?

To obtain an idea of the answer to questions 1 and 2, consider the special case of a system of three equations in three unknowns. The linear system (2.3.1) then reduces to

$$\begin{aligned} a_{11}x_1 + a_{12}x_2 + a_{13}x_3 &= b_1, \\ a_{21}x_1 + a_{22}x_2 + a_{23}x_3 &= b_2, \\ a_{31}x_1 + a_{32}x_2 + a_{33}x_3 &= b_3, \end{aligned}$$

which can be interpreted as defining three planes in space. An ordered triple (c_1, c_2, c_3) is a solution to this system if and only if it corresponds to the coordinates of a point of intersection of the three planes. There are precisely four possibilities:

1. The planes have no intersection point.

2. The planes intersect in just one point.

3. The planes intersect in a line.

4. The planes are all identical.

In case 1, the corresponding system has no solution, whereas in case 2, the system has just one solution. Finally, in cases 3 and 4, every point on the line or plane (respectively) is a solution to the linear system and hence the system has an infinite number of solutions. Cases 1,2 and 3 are illustrated in Figure 2.3.1.

Three parallel planes (no intersection): no solution

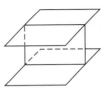

No common intersection: no solution

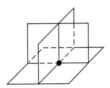

Planes intersect at a point: a unique solution

Planes intersect in a line: an infinite number of solutions

Figure 2.3.1: Possible intersection points for three planes in space.

We have therefore proved, geometrically, that there are precisely three possibilities for the solutions of a system of three equations in three unknowns. The system either has no solution, it has just one solution, or it has an infinite number of solutions. In Section 2.5, we will establish that these are the only possibilities for the general $m \times n$ system (2.3.1).

DEFINITION   2.3.4

A system of equations that has at least one solution is said to be **consistent**, whereas a system that has no solution is called **inconsistent**.

Our problem will be to determine whether a given system is consistent and then, if it is, to find its solution set.

DEFINITION   2.3.5

Naturally associated with the system (2.3.1) are the following two matrices:

1. The **matrix of coefficients** $A = \begin{bmatrix} a_{11} & a_{12} & \cdots & a_{1n} \\ a_{21} & a_{22} & \cdots & a_{2n} \\ & & \vdots & \\ a_{m1} & a_{m2} & \cdots & a_{mn} \end{bmatrix}.$

2. The **augmented matrix** $A^{\#} = \begin{bmatrix} a_{11} & a_{12} & \cdots & a_{1n} & b_1 \\ a_{21} & a_{22} & \cdots & a_{2n} & b_2 \\ & & \vdots & & \vdots \\ a_{m1} & a_{m2} & \cdots & a_{mn} & b_m \end{bmatrix}.$

The augmented matrix completely characterizes a system of equations, since it contains all of the system coefficients and system constants. We will see in the subsequent sections that the relationship between A and $A^{\#}$ determines the solution properties of a linear system. Notice that the matrix of coefficients is the matrix consisting of the first n columns of $A^{\#}$.

Example 2.3.6   Write the system of equations with the following augmented matrix:

$$\begin{bmatrix} 1 & 2 & 9 & -1 & 1 \\ 2 & -3 & 7 & 4 & 2 \\ 1 & 3 & 5 & 0 & -1 \end{bmatrix}.$$

Solution: The appropriate system is

$$\begin{aligned} x_1 + 2x_2 + 9x_3 - x_4 &= 1, \\ 2x_1 - 3x_2 + 7x_3 + 4x_4 &= 2, \\ x_1 + 3x_2 + 5x_3 &= -1. \end{aligned}$$

$\square$

Vector Formulation

We next show that the matrix product described in the preceding section can be used to write a linear system as a single equation involving the matrix of coefficients and column vectors. For example, the system

$$\begin{aligned} x_1 + 3x_2 - 4x_3 &= 1, \\ 2x_1 + 5x_2 - x_3 &= 5, \\ x_1 + 6x_3 &= 3 \end{aligned}$$

can be written as the vector equation

$$\begin{bmatrix} 1 & 3 & -4 \\ 2 & 5 & -1 \\ 1 & 0 & 6 \end{bmatrix} \begin{bmatrix} x_1 \\ x_2 \\ x_3 \end{bmatrix} = \begin{bmatrix} 1 \\ 5 \\ 3 \end{bmatrix},$$

since this vector equation is satisfied if and only if

$$\begin{bmatrix} x_1 + 3x_2 - 4x_3 \\ 2x_1 + 5x_2 - x_3 \\ x_1 + 6x_3 \end{bmatrix} = \begin{bmatrix} 1 \\ 5 \\ 3 \end{bmatrix};$$

that is, if and only if each equation of the given system is satisfied.

Similarly, the general $m \times n$ system of linear equations

$$\begin{aligned} a_{11}x_1 + a_{12}x_2 + \cdots + a_{1n}x_n &= b_1, \\ a_{21}x_1 + a_{22}x_2 + \cdots + a_{2n}x_n &= b_2, \\ &\vdots \\ a_{m1}x_1 + a_{m2}x_2 + \cdots + a_{mn}x_n &= b_m, \end{aligned}$$

can be written as the **vector equation**

$$A\mathbf{x} = \mathbf{b},$$

where A is the $m \times n$ matrix of coefficients and

$$\mathbf{x} = \begin{bmatrix} x_1 \\ x_2 \\ \vdots \\ x_n \end{bmatrix} \quad \text{and} \quad \mathbf{b} = \begin{bmatrix} b_1 \\ b_2 \\ \vdots \\ b_m \end{bmatrix}.$$

We will refer to the column n-vector $\mathbf{x}$ as the **vector of unknowns**, and to the column m-vector $\mathbf{b}$ as the **right-hand-side vector**. Assuming that all elements in the system are real, we can view $\mathbf{b}$ as an element of $\mathbb{R}^m$ and $\mathbf{x}$ as an element of $\mathbb{R}^n$. We can denote these statements by $\mathbf{b} \in \mathbb{R}^m$ and $\mathbf{x} \in \mathbb{R}^n$, respectively.[3] Therefore, the set of all real solutions to the system $A\mathbf{x} = \mathbf{b}$ is

$$S = \{\mathbf{x} \in \mathbb{R}^n : A\mathbf{x} = \mathbf{b}\},$$

which is a subset of $\mathbb{R}^n$.

Example 2.3.7   It can be shown, using the techniques of the next two sections, that the solution set of the linear system

$$\begin{aligned} x_1 + x_2 + 2x_3 - x_4 &= 0, \\ 3x_1 - 2x_2 + x_3 + 2x_4 &= 0, \\ 5x_1 + 3x_2 + 3x_3 - 2x_4 &= 0, \end{aligned}$$

is the subset of $\mathbb{R}^4$ defined by

$$S = \{(-t, 4t, t, 5t) : t \in \mathbb{R}\}. \qquad \square$$

A similar vector formulation for systems of differential equations can be used not only in developing the theory for such systems, but also in deriving solution techniques. As an example of this formulation, consider the system of differential equations

$$\begin{aligned} \frac{dx_1}{dt} &= 3tx_1 + 9x_2 + 6e^t, \\ \frac{dx_2}{dt} &= 2x_1 - 7x_2 + 3e^t. \end{aligned}$$

Using matrix and vector functions, this system can be written as the vector equation

$$\frac{d\mathbf{x}}{dt} = A(t)\mathbf{x}(t) + \mathbf{b}(t),$$

where

$$\mathbf{x}(t) = \begin{bmatrix} x_1(t) \\ x_2(t) \end{bmatrix}, \quad \frac{d\mathbf{x}}{dt} = \begin{bmatrix} dx_1/dt \\ dx_2/dt \end{bmatrix}, \quad A = \begin{bmatrix} 3t & 9 \\ 2 & -7 \end{bmatrix}, \quad \text{and} \quad \mathbf{b}(t) = \begin{bmatrix} 6e^t \\ 3e^t \end{bmatrix}.$$

In this formulation, the basic unknown is the column 2-vector function $\mathbf{x}(t)$.

Example 2.3.8   Give the vector formulation for the system of equations

$$\begin{aligned} x_1' &= 3x_1 + (\sin t)x_2 + e^t, \\ x_2' &= 7tx_1 + t^2x_2 - 4e^{-t}. \end{aligned}$$

---

[3]The symbol $\in$ is the set-theoretic notation declaring membership in a set; it will be often encountered in the text.

Solution: We have

$$\begin{bmatrix} x_1' \\ x_2' \end{bmatrix} = \begin{bmatrix} 3 & \sin t \\ 7t & t^2 \end{bmatrix} \begin{bmatrix} x_1 \\ x_2 \end{bmatrix} + \begin{bmatrix} e^t \\ -4e^{-t} \end{bmatrix}.$$

That is,

$$\mathbf{x}'(t) = A(t)\mathbf{x}(t) + \mathbf{b}(t),$$

where

$$\mathbf{x}(t) = \begin{bmatrix} x_1(t) \\ x_2(t) \end{bmatrix}, \quad A(t) = \begin{bmatrix} 3 & \sin t \\ 7t & t^2 \end{bmatrix}, \quad \mathbf{b}(t) = \begin{bmatrix} e^t \\ -4e^{-t} \end{bmatrix}.$$

□

Exercises for 2.3

Key Terms

System coefficients, System constants, Homogeneous system, Nonhomogeneous system, Solution, Solution set, Consistent system, Inconsistent system, Matrix of coefficients, Augmented matrix, Vector of unknowns, Right-hand-side vector.

Skills

- Be able to write a linear system of equations as a vector equation, and identify the matrix of coefficients, the right-hand-side vector, and the augmented matrix.

- Given a matrix of coefficients and a right-hand-side vector, or an augmented matrix, be able to write the corresponding linear system.

- Understand the geometric difference between a consistent linear system and an inconsistent one.

- Be able to verify that the components of a given vector provide a solution to a linear system.

- Be able to give the vector formulation for a system of differential equations.

True-False Review

For Questions 1–6, decide if the given statement is **true** or **false**, and give a brief justification for your answer. If true, you can quote a relevant definition or theorem from the text. If false, provide an example, illustration, or brief explanation of why the statement is false.

1. If a linear system of equations has an $m \times n$ augmented matrix, then the system has m equations and n unknowns.

2. A linear system that contains three distinct planes can have at most one solution.

3. If the matrix of coefficients of a linear system is an $m \times n$ matrix, then the right-hand-side vector must have n components.

4. It is impossible for a linear system of equations to have exactly two solutions.

5. If a linear system has an $m \times n$ coefficient matrix, then the augmented matrix for the linear system is $m \times (n + 1)$.

6. If A is an $n \times n$ matrix, then the linear systems $A\mathbf{x} = \mathbf{0}$ and $A^T\mathbf{x} = \mathbf{0}$ have the same solution set.

Problems

For Problems 1–2, verify that the given triple of real numbers is a solution to the given system.

1. $(1, -1, 2)$;

$$\begin{aligned} 2x_1 - 3x_2 + 4x_3 &= 13, \\ x_1 + x_2 - x_3 &= -2, \\ 5x_1 + 4x_2 + x_3 &= 3. \end{aligned}$$

2. $(2, -3, 1)$;

$$\begin{aligned} x_1 + x_2 - 2x_3 &= -3, \\ 3x_1 - x_2 - 7x_3 &= 2, \\ x_1 + x_2 + x_3 &= 0, \\ 2x_1 + 2x_2 - 4x_3 &= -6. \end{aligned}$$

3. Verify that for all values of t,

$$(1 - t, 2 + 3t, 3 - 2t)$$

is a solution to the linear system

$$\begin{aligned} x_1 + x_2 + x_3 &= 6, \\ x_1 - x_2 - 2x_3 &= -7, \\ 5x_1 + x_2 - x_3 &= 4. \end{aligned}$$

4. Verify that for all values of s and t,

$$(s, s - 2t, 2s + 3t, t)$$

is a solution to the linear system

$$\begin{aligned} x_1 + x_2 - x_3 + 5x_4 &= 0, \\ 2x_2 - x_3 + 7x_4 &= 0, \\ 4x_1 + 2x_2 - 3x_3 + 13x_4 &= 0. \end{aligned}$$

5. By making a sketch in the xy-plane, prove that the following linear system has no solution:

$$\begin{aligned} 2x + 3y &= 1, \\ 2x + 3y &= 2. \end{aligned}$$

For Problems 6–8, determine the coefficient matrix, A, the right-hand-side vector, $\mathbf{b}$, and the augmented matrix $A^{\#}$ of the given system.

6.
$$\begin{aligned} x_1 + 2x_2 - 3x_3 &= 1, \\ 2x_1 + 4x_2 - 5x_3 &= 2, \\ 7x_1 + 2x_2 - x_3 &= 3. \end{aligned}$$

7.
$$\begin{aligned} x + y + z - w &= 3, \\ 2x + 4y - 3z + 7w &= 2. \end{aligned}$$

8.
$$\begin{aligned} x_1 + 2x_2 - x_3 &= 0, \\ 2x_1 + 3x_2 - 2x_3 &= 0, \\ 5x_1 + 6x_2 - 5x_3 &= 0. \end{aligned}$$

For Problems 9–10, write the system of equations with the given coefficient matrix and right-hand-side vector.

9. $A = \begin{bmatrix} 1 & -1 & 2 & 3 \\ 1 & 1 & -2 & 6 \\ 3 & 1 & 4 & 2 \end{bmatrix}, \mathbf{b} = \begin{bmatrix} 1 \\ -1 \\ 2 \end{bmatrix}.$

10. $A = \begin{bmatrix} 2 & 1 & 3 \\ 4 & -1 & 2 \\ 7 & 6 & 3 \end{bmatrix}, \mathbf{b} = \begin{bmatrix} 3 \\ 1 \\ -5 \end{bmatrix}.$

11. Consider the $m \times n$ homogeneous system of linear equations

$$A\mathbf{x} = \mathbf{0}. \tag{2.3.2}$$

(a) If $\mathbf{x} = [x_1\ x_2\ \dots\ x_n]^T$ and $\mathbf{y} = [y_1\ y_2\ \dots\ y_n]^T$ are solutions to (2.3.2), show that

$$\mathbf{z} = \mathbf{x} + \mathbf{y} \quad \text{and} \quad \mathbf{w} = c\mathbf{x}$$

are also solutions, where c is an arbitrary scalar.

(b) Is the result of (a) true when $\mathbf{x}$ and $\mathbf{y}$ are solutions to the nonhomogeneous system $A\mathbf{x} = \mathbf{b}$? Explain.

For Problems 12–15, write the vector formulation for the given system of differential equations.

12. $x_1' = -4x_1 + 3x_2 + 4t, \quad x_2' = 6x_1 - 4x_2 + t^2.$

13. $x_1' = t^2 x_1 - tx_2, \quad x_2' = (-\sin t)x_1 + x_2.$

14. $x_1' = e^{2t}x_2, \quad x_2' + (\sin t)x_1 = 1.$

15. $x_1' = (-\sin t)x_2 + x_3 + t, \ x_2' = -e^t x_1 + t^2 x_3 + t^3,$
$x_3' = -tx_1 + t^2 x_2 + 1.$

For Problems 16–17 verify that the given vector function $\mathbf{x}$ defines a solution to $\mathbf{x}' = A\mathbf{x} + \mathbf{b}$ for the given A and $\mathbf{b}$.

16. $\mathbf{x}(t) = \begin{bmatrix} e^{4t} \\ -2e^{4t} \end{bmatrix}, \quad A = \begin{bmatrix} 2 & -1 \\ -2 & 3 \end{bmatrix}, \quad \mathbf{b}(t) = \begin{bmatrix} 0 \\ 0 \end{bmatrix}.$

17. $\mathbf{x}(t) = \begin{bmatrix} 4e^{-2t} + 2\sin t \\ 3e^{-2t} - \cos t \end{bmatrix}, \quad A = \begin{bmatrix} 1 & -4 \\ -3 & 2 \end{bmatrix},$
$\mathbf{b}(t) = \begin{bmatrix} -2(\cos t + \sin t) \\ 7\sin t + 2\cos t \end{bmatrix}.$

2.4   Elementary Row Operations and Row-Echelon Matrices

In the next section we will develop methods for solving a system of linear equations. These methods will consist of reducing a given system of equations to a new system that has the same solution set as the given system but is easier to solve. In this section we introduce the requisite mathematical results.

Elementary Row Operations

The first step in deriving systematic procedures for solving a linear system is to determine what operations can be performed on such a system without altering its solution set.

Example 2.4.1   Consider the system of equations

$$x_1 + 2x_2 + 4x_3 = 2, \tag{2.4.1}$$
$$2x_1 - 5x_2 + 3x_3 = 6, \tag{2.4.2}$$
$$4x_1 + 6x_2 - 7x_3 = 8. \tag{2.4.3}$$

Solution: If we permute (i.e., interchange), say, Equations (2.4.1) and (2.4.2), the resulting system is

$$2x_1 - 5x_2 + 3x_3 = 6,$$
$$x_1 + 2x_2 + 4x_3 = 2,$$
$$4x_1 + 6x_2 - 7x_3 = 8,$$

which certainly has the same solution set as the original system. Returning to the original system, if we multiply, say, Equation (2.4.2) by 5, we obtain the system

$$x_1 + 2x_2 + 4x_3 = 2,$$
$$10x_1 - 25x_2 + 15x_3 = 30,$$
$$4x_1 + 6x_2 - 7x_3 = 8,$$

which again has the same solution set as the original system. Finally, if we add, say, twice Equation (2.4.1) to Equation (2.4.3), we obtain the system

$$x_1 + 2x_2 + 4x_3 = 2, \tag{2.4.4}$$
$$2x_1 - 5x_2 + 3x_3 = 6, \tag{2.4.5}$$
$$(4x_1 + 6x_2 - 7x_3) + 2(x_1 + 2x_2 + 4x_3) = 8 + 2(2). \tag{2.4.6}$$

We can verify that, if (2.4.4)–(2.4.6) are satisfied, then so are (2.4.1)–(2.4.3), and vice versa. It follows that the system of equations (2.4.4)–(2.4.6) has the same solution set as the original system of equations (2.4.1)–(2.4.3).   □

More generally, similar reasoning can be used to show that the following three operations can be performed on any $m \times n$ system of linear equations without altering the solution set:

1. Permute equations.

2. Multiply an equation by a nonzero constant.

3. Add a multiple of one equation to another equation.

Since these operations involve changes only in the system coefficients and constants (and not changes in the variables), they can be represented by the following operations on the rows of the augmented matrix of the system:

1. Permute rows.

2. Multiply a row by a nonzero constant.

3. Add a multiple of one row to another row.

These three operations, called **elementary row operations**, will be a basic computational tool throughout the text, even in cases when the matrix under consideration is not derived from a system of linear equations. The following notation will be used to describe elementary row operations performed on a matrix A.

1. P_{ij}:   **P**ermute the ith and jth rows in A.

2. $M_i(k)$: **M**ultiply every element of the ith row of A by a nonzero scalar k.

3. $A_{ij}(k)$: **A**dd to the elements of the jth row of A the scalar k times the corresponding elements of the ith row of A.

Furthermore, the notation $A \sim B$ will mean that matrix B has been obtained from matrix A by a sequence of elementary row operations. To reference a particular elementary row operation used in, say, the nth step of the sequence of elementary row operations, we will write $\overset{n}{\sim} B$.

Example 2.4.2   The one-step operations performed on the system in Example 2.4.1 can be described as follows using elementary row operations on the augmented matrix of the system:

$$\begin{bmatrix} 1 & 2 & 4 & 2 \\ 2 & -5 & 3 & 6 \\ 4 & 6 & -7 & 8 \end{bmatrix} \overset{1}{\sim} \begin{bmatrix} 2 & -5 & 3 & 6 \\ 1 & 2 & 4 & 2 \\ 4 & 6 & -7 & 8 \end{bmatrix}$$ 1. P_{12}.   Permute (2.4.1) and (2.4.2).

$$\begin{bmatrix} 1 & 2 & 4 & 2 \\ 2 & -5 & 3 & 6 \\ 4 & 6 & -7 & 8 \end{bmatrix} \overset{1}{\sim} \begin{bmatrix} 1 & 2 & 4 & 2 \\ 10 & -25 & 15 & 30 \\ 4 & 6 & -7 & 8 \end{bmatrix}$$ 1. $M_2(5)$.   Multiply (2.4.2) by 5.

$$\begin{bmatrix} 1 & 2 & 4 & 2 \\ 2 & -5 & 3 & 6 \\ 4 & 6 & -7 & 8 \end{bmatrix} \overset{1}{\sim} \begin{bmatrix} 1 & 2 & 4 & 2 \\ 2 & -5 & 3 & 6 \\ 6 & 10 & 1 & 12 \end{bmatrix}$$ 1. $A_{13}(2)$.   Add 2 times (2.4.1) to (2.4.3).   □

It is important to realize that each elementary row operation is reversible; we can "undo" a given elementary row operation by another elementary row operation to bring the modified linear system back into its original form. Specifically, in terms of the notation introduced above, the reverse operations are determined as follows (ERO refers here to "elementary row operation"):

ERO Applied to A $A \sim B$	Reverse ERO Applied to B $B \sim A$
P_{ij}	P_{ji}: Permute row j and i in B.
$M_i(k)$	$M_i(1/k)$: Multiply the ith row of B by $1/k$.
$A_{ij}(k)$	$A_{ij}(-k)$: Add to the elements of the jth row of B the scalar $-k$ times the corresponding elements of the ith row of B

We introduce a special term for matrices that are related via elementary row operations.

DEFINITION   2.4.3

Let A be an $m \times n$ matrix. Any matrix obtained from A by a finite sequence of elementary row operations is said to be **row-equivalent** to A.

Thus, all of the matrices in the previous example are row-equivalent. Since elementary row operations do not alter the solution set of a linear system, we have the next theorem.

Theorem 2.4.4   Systems of linear equations with row-equivalent augmented matrices have the same solution sets.

Row-Echelon Matrices

Our methods for solving a system of linear equations will consist of using elementary row operations to reduce the augmented matrix of the given system to a simple form. But how simple a form should we aim for? In order to answer this question, consider the system

$$x_1 + x_2 - x_3 = 4, \tag{2.4.7}$$
$$x_2 - 3x_3 = 5, \tag{2.4.8}$$
$$x_3 = 2. \tag{2.4.9}$$

This system can be solved most easily as follows. From Equation (2.4.9), $x_3 = 2$. Substituting this value into Equation (2.4.8) and solving for x_2 yields $x_2 = 5 + 6 = 11$. Finally, substituting for x_3 and x_2 into Equation (2.4.7) and solving for x_1, we obtain $x_1 = -5$. Thus, the solution to the given system of equations is $(-5, 11, 2)$, a single vector in $\mathbb{R}^3$. This technique is called **back substitution** and could be used because the given system has a simple form. The augmented matrix of the system is

$$\begin{bmatrix} 1 & 1 & -1 & 4 \\ 0 & 1 & -3 & 5 \\ 0 & 0 & 1 & 2 \end{bmatrix}$$

We see that the submatrix consisting of the first three columns (which corresponds to the matrix of coefficients) is an upper triangular matrix with the leftmost nonzero entry in each row equal to 1. The back-substitution method will work on any system of linear equations with an augmented matrix of this form. Unfortunately, not all systems of equations have augmented matrices that can be reduced to such a form. However, there is a simple type of matrix to which any matrix can be reduced by elementary row operations, and which also represents a system of equations that can be solved (if it has a solution) by back substitution. This is called a *row-echelon matrix* and is defined as follows:

DEFINITION 2.4.5

An $m \times n$ matrix is called a **row-echelon matrix** if it satisfies the following three conditions:

1. If there are any rows consisting entirely of zeros, they are grouped together at the bottom of the matrix.

2. The first nonzero element in any nonzero row[4] is a 1 (called a **leading 1**).

3. The leading 1 of any row below the first row is to the right of the leading 1 of the row above it.

---

[4] A *nonzero row* (*nonzero column*) is any row (column) that does not consist entirely of zeros.

Example 2.4.6 Examples of row-echelon matrices are

$$
\begin{bmatrix} 1 & -2 & 3 & 7 \\ 0 & 1 & 5 & 0 \\ 0 & 0 & 0 & 1 \end{bmatrix}, \quad
\begin{bmatrix} 0 & 0 & 1 \\ 0 & 0 & 0 \\ 0 & 0 & 0 \end{bmatrix}, \quad \text{and} \quad
\begin{bmatrix} 1 & -1 & 6 & 5 & 9 \\ 0 & 0 & 1 & 2 & 5 \\ 0 & 0 & 0 & 1 & 0 \\ 0 & 0 & 0 & 0 & 0 \end{bmatrix},
$$

whereas

$$
\begin{bmatrix} 1 & 0 & -1 \\ 0 & 1 & 2 \\ 0 & 1 & -1 \end{bmatrix} \quad \text{and} \quad
\begin{bmatrix} 1 & 0 & 0 \\ 0 & 0 & 0 \\ 0 & 1 & -1 \\ 0 & 0 & 1 \end{bmatrix}
$$

are not row-echelon matrices. □

The basic result that will allow us to determine the solution set to any system of linear equations is stated in the next theorem.

Theorem 2.4.7 Any matrix is row-equivalent to a row-echelon matrix.

According to this theorem, by applying an appropriate sequence of elementary row operations to any $m \times n$ matrix, we can always reduce it to a row-echelon matrix. When a matrix A has been reduced to a row-echelon matrix in this way, we say that it has been reduced to **row-echelon form** and refer to the resulting matrix as a row-echelon form of A. The proof of Theorem 2.4.7 consists of giving an algorithm that will reduce an arbitrary $m \times n$ matrix to a row-echelon matrix after a finite sequence of elementary row operations. Before presenting such an algorithm, we first illustrate the result with an example.

Example 2.4.8 Use elementary row operations to reduce $\begin{bmatrix} 2 & 1 & -1 & 3 \\ 1 & -1 & 2 & 1 \\ -4 & 6 & -7 & 1 \\ 2 & 0 & 1 & 3 \end{bmatrix}$ to row-echelon form.

Solution: We show each step in detail.
Step 1: Put a leading 1 in the (1, 1) position.
This is most easily accomplished by permuting rows 1 and 2.

$$
\begin{bmatrix} 2 & 1 & -1 & 3 \\ 1 & -1 & 2 & 1 \\ -4 & 6 & -7 & 1 \\ 2 & 0 & 1 & 3 \end{bmatrix}
\overset{1}{\sim}
\begin{bmatrix} 1 & -1 & 2 & 1 \\ 2 & 1 & -1 & 3 \\ -4 & 6 & -7 & 1 \\ 2 & 0 & 1 & 3 \end{bmatrix}
$$

Step 2: Use the leading 1 to put zeros beneath it in column 1.
This is accomplished by adding appropriate multiples of row 1 to the remaining rows.

$$
\overset{2}{\sim}
\begin{bmatrix} 1 & -1 & 2 & 1 \\ 0 & 3 & -5 & 1 \\ 0 & 2 & 1 & 5 \\ 0 & 2 & -3 & 1 \end{bmatrix}
$$

Step 2 row operations: $\begin{cases} \text{Add } -2 \text{ times row 1 to row 2.} \\ \text{Add } 4 \text{ times row 1 to row 3.} \\ \text{Add } -2 \text{ times row 1 to row 4.} \end{cases}$

Step 3: Put a leading 1 in the (2, 2) position.
We could accomplish this by multiplying row 2 by 1/3. However, this would introduce fractions into the matrix and thereby complicate the remaining computations. In

hand calculations, fewer algebraic errors result if we avoid the use of fractions. In this case, we can obtain a leading 1 without the use of fractions by adding -1 times row 3 to row 2.

$$\underset{\sim}{3} \begin{bmatrix} 1 & -1 & 2 & 1 \\ 0 & 1 & -6 & -4 \\ 0 & 2 & 1 & 5 \\ 0 & 2 & -3 & 1 \end{bmatrix} \qquad \text{Step 3 row operation: \quad Add } -1 \text{ times row 3 to row 2.}$$

Step 4: Use the leading 1 in the (2, 2) position to put zeros beneath it in column 2.

We now add appropriate multiples of row 2 to the rows *beneath* it. For row-echelon form, we need not be concerned about the row above it, however.

$$\underset{\sim}{4} \begin{bmatrix} 1 & -1 & 2 & 1 \\ 0 & 1 & -6 & -4 \\ 0 & 0 & 13 & 13 \\ 0 & 0 & 9 & 9 \end{bmatrix} \qquad \text{Step 4 row operations:} \quad \begin{cases} \text{Add } -2 \text{ times row 2 to row 3.} \\ \text{Add } -2 \text{ times row 2 to row 4.} \end{cases}$$

Step 5: Put a leading 1 in the (3, 3) position.

This can be accomplished by multiplying row 3 by 1/13.

$$\underset{\sim}{5} \begin{bmatrix} 1 & -1 & 2 & 1 \\ 0 & 1 & -6 & -4 \\ 0 & 0 & 1 & 1 \\ 0 & 0 & 9 & 9 \end{bmatrix}$$

Step 6: Use the leading 1 in the (3, 3) position to put zeros beneath it in column 3.

The appropriate row operation is to add -9 times row 3 to row 4.

$$\underset{\sim}{6} \begin{bmatrix} 1 & -1 & 2 & 1 \\ 0 & 1 & -6 & -4 \\ 0 & 0 & 1 & 1 \\ 0 & 0 & 0 & 0 \end{bmatrix}$$

This is a row-echelon matrix, hence the given matrix has been reduced to row-echelon form. The specific operations used at each step are given next, using the notation introduced previously in this section. In future examples, we will simply indicate briefly the elementary row operation used at each step. The following shows this description for the present example.

---

1. P_{12} **2.** $A_{12}(-2)$, $A_{13}(4)$, $A_{14}(-2)$ **3.** $A_{32}(-1)$

4. $A_{23}(-2)$, $A_{24}(-2)$ **5.** $M_3(1/13)$ **6.** $A_{34}(-9)$

Remarks

1. Notice that in steps 2 and 4 of the preceding example we have performed multiple elementary row operations of the type $A_{ij}(k)$ in a single step. With this one exception, the reader is strongly advised not to combine multiple elementary row operations into a single step, particularly when they are of *different* types. This is a common source of calculation errors.

2. The reader may have noticed that the particular steps taken in the preceding example are not uniquely determined. For instance, we could have achieved a leading 1 in the (1, 1) position in step 1 by multiplying the first row by $1/2$, rather than permuting the first two rows. Therefore, we may have multiple strategies for reducing a matrix to row-echelon form, and indeed, many possible row-echelon forms for a given matrix A. In this particular case, we chose not to multiply the first row by $1/2$ in order to avoid introducing fractions into the calculations.

The reader is urged to study the foregoing example very carefully, since it illustrates the general procedure for reducing an $m \times n$ matrix to row-echelon form using elementary row operations. This procedure will be used repeatedly throughout the text. The idea behind reduction to row-echelon form is to start at the upper left-hand corner of the matrix and proceed downward and to the right in the matrix. The following algorithm formalizes the steps that reduce any $m \times n$ matrix to row-echelon form using a finite number of elementary row operations and thereby provides a proof of Theorem 2.4.7. An illustration of this algorithm is given in Figure 2.4.1.

Algorithm for Reducing an $m \times n$ Matrix A to Row-Echelon Form

1. Start with an $m \times n$ matrix A. If $A = 0$, go to step 7.
2. Determine the leftmost nonzero column (this is called a **pivot column**, and the topmost position in this column is called a **pivot position**).
3. Use elementary row operations to put a 1 in the pivot position.
4. Use elementary row operations to put zeros below the pivot position.
5. If there are no more nonzero rows below the pivot position go to step 7, otherwise go to step 6.
6. Apply steps 2 through 5 to the submatrix consisting of the rows that lie below the pivot position.
7. The matrix is a row-echelon matrix.

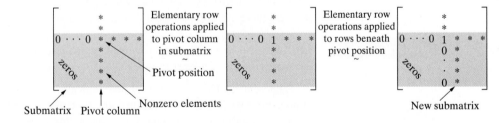

Figure 2.4.1: Illustration of an algorithm for reducing an $m \times n$ matrix to row-echelon form.

Remark   In order to obtain a row-echelon matrix, we put a 1 in each pivot position. However, many algorithms for solving systems of linear equations numerically are based around the preceding algorithm, except that in step 3 we place a nonzero number (not necessarily a 1) in the pivot position. Of course, the matrix resulting from an application of this algorithm differs from a row-echelon matrix, since it will have arbitrary nonzero elements in the pivot positions.

Example 2.4.9   Reduce $\begin{bmatrix} 3 & 2 & -5 & 2 \\ 1 & 1 & -2 & 1 \\ 1 & 0 & -3 & 4 \end{bmatrix}$ to row-echelon form.

Solution: Applying the row-reduction algorithm leads to the following sequence of elementary row operations. The specific row operations used at each step are given at the end of the process.

$$
\begin{array}{c}
\text{Pivot position} \\
\longrightarrow
\end{array}
\begin{bmatrix} ③ & 2 & -5 & 2 \\ 1 & 1 & -1 & 1 \\ 1 & 0 & -3 & 4 \end{bmatrix}
\overset{1}{\sim}
\begin{bmatrix} 1 & 1 & -1 & 1 \\ 3 & 2 & -5 & 2 \\ 1 & 0 & -3 & 4 \end{bmatrix}
\overset{2}{\sim}
\begin{bmatrix} 1 & 1 & -1 & 1 \\ 0 & ⊝1 & -2 & -1 \\ 0 & -1 & -2 & 3 \end{bmatrix}
$$

$$
\overset{3}{\sim}
\begin{bmatrix} 1 & 1 & -1 & 1 \\ 0 & 1 & 2 & 1 \\ 0 & -1 & -2 & 3 \end{bmatrix}
\overset{4}{\sim}
\begin{bmatrix} 1 & 1 & -1 & 1 \\ 0 & 1 & 2 & 1 \\ 0 & 0 & 0 & ④ \end{bmatrix}
\overset{5}{\sim}
\begin{bmatrix} 1 & 1 & -1 & 1 \\ 0 & 1 & 2 & 1 \\ 0 & 0 & 0 & 1 \end{bmatrix}
$$

Pivot position .

This is a row-echelon matrix and hence we are done. The row operations used are summarized here:

> **1.** P_{12} **2.** $A_{12}(-3), \ A_{13}(-1)$ **3.** $M_2(-1)$ **4.** $A_{23}(1)$ **5.** $M_3(1/4)$ □

The Rank of a Matrix

We now derive some further results on row-echelon matrices that will be required in the next section to develop the theory for solving systems of linear equations.

We first observe that a row-echelon form for a matrix A is not unique. Given one row-echelon form for A, we can always obtain a different one by taking the first row-echelon form for A and adding some multiple of a given row to any rows above it. The result is still in row-echelon form.

However, even though the row-echelon form of A is not unique, we do have the following theorem (in Chapter 4 we will see how the proof of this theorem arises naturally from the more sophisticated ideas from linear algebra yet to be introduced).

Theorem 2.4.10   Let A be an $m \times n$ matrix. All row-echelon matrices that are row-equivalent to A have the same number of nonzero rows.

Theorem 2.4.10 associates a number with any $m \times n$ matrix A—namely, the number of nonzero rows in any row-echelon form of A. As we will see in the next section, this number is fundamental in determining the solution properties of linear systems, and indeed it plays a central role in linear algebra in general. For this reason, we give it a special name.

DEFINITION   2.4.11

The number of nonzero rows in any row-echelon form of a matrix A is called the **rank** of A and is denoted rank(A).

Example 2.4.12   Determine rank(A) if $A = \begin{bmatrix} 3 & 1 & 4 \\ 4 & 3 & 5 \\ 2 & -1 & 3 \end{bmatrix}$.

Solution: In order to determine rank(A), we must first reduce A to row-echelon form.

$$\begin{bmatrix} 3 & 1 & 4 \\ 4 & 3 & 5 \\ 2 & -1 & 3 \end{bmatrix} \overset{1}{\sim} \begin{bmatrix} 1 & 2 & 1 \\ 4 & 3 & 5 \\ 2 & -1 & 3 \end{bmatrix} \overset{2}{\sim} \begin{bmatrix} 1 & 2 & 1 \\ 0 & -5 & 1 \\ 0 & -5 & 1 \end{bmatrix} \overset{3}{\sim} \begin{bmatrix} 1 & 2 & 1 \\ 0 & 1 & -\frac{1}{5} \\ 0 & -5 & 1 \end{bmatrix} \overset{4}{\sim} \begin{bmatrix} 1 & 2 & 1 \\ 0 & 1 & -\frac{1}{5} \\ 0 & 0 & 0 \end{bmatrix}.$$

Since there are two nonzero rows in this row-echelon form of A, it follows from Definition 2.4.11 that rank(A) = 2.

> **1.** $A_{31}(-1)$ **2.** $A_{12}(-4)$, $A_{13}(-2)$ **3.** $M_2(-1/5)$ **4.** $A_{23}(5)$ $\square$

In the preceding example, the original matrix A had three nonzero rows, whereas any row-echelon form of A has only two nonzero rows. We can interpret this geometrically as follows. The three row vectors of A can be considered as vectors in $\mathbb{R}^3$ with components

$$\mathbf{a}_1 = (3, 1, 4), \quad \mathbf{a}_2 = (4, 3, 5), \quad \mathbf{a}_3 = (2, -1, 3).$$

In performing elementary row operations on A, we are taking combinations of these vectors in the following way:

$$c_1 \mathbf{a}_1 + c_2 \mathbf{a}_2 + c_3 \mathbf{a}_3,$$

and thus the rows of a row-echelon form of A are all of this form. We have been combining the vectors *linearly*. The fact that we obtained a row of zeros in the row-echelon form means that there are values of the constants c_1, c_2, c_3 such that

$$c_1 \mathbf{a}_1 + c_2 \mathbf{a}_2 + c_3 \mathbf{a}_3 = \mathbf{0},$$

where $\mathbf{0}$ denotes the zero vector $(0, 0, 0)$. Equivalently, one of the vectors can be written in terms of the other two vectors, and therefore the three vectors lie in a plane. Reducing the matrix to row-echelon form has uncovered this relationship among the three vectors. We shall have much more to say about this in Chapter 4.

Remark   If A is an $m \times n$ matrix, then rank(A) $\leq m$ and rank(A) $\leq n$. This is because the number of nonzero rows in a row-echelon form of A is equal to the number of pivots in a row-echelon form of A, which cannot exceed the number of rows or columns of A, since there can be at most one pivot per row and per column.

Reduced Row-Echelon Matrices

In the future we will need to consider the special row-echelon matrices that arise when zeros are placed above, as well as beneath, each leading 1. Any such matrix is called a **reduced row-echelon matrix** and is defined precisely as follows.

DEFINITION   2.4.13

An $m \times n$ matrix is called a **reduced row-echelon matrix** if it satisfies the following conditions:

1. It is a row-echelon matrix.

2. Any *column* that contains a leading 1 has zeros everywhere else.

Example 2.4.14   The following are examples of reduced row-echelon matrices:

$$\begin{bmatrix} 1 & 3 & 0 & 0 \\ 0 & 0 & 1 & 0 \\ 0 & 0 & 0 & 1 \end{bmatrix}, \quad \begin{bmatrix} 1 & -1 & 7 & 0 \\ 0 & 0 & 0 & 1 \end{bmatrix}, \quad \begin{bmatrix} 1 & 0 & 5 & 3 \\ 0 & 1 & 2 & 1 \\ 0 & 0 & 0 & 0 \end{bmatrix}, \quad \text{and} \quad \begin{bmatrix} 1 & 0 & 0 \\ 0 & 1 & 0 \\ 0 & 0 & 1 \end{bmatrix}. \qquad \square$$

Although an $m \times n$ matrix A does not have a unique row-echelon form, in reducing A to a reduced row-echelon matrix we are making a particular choice of row-echelon matrix, since we arrange that all elements above each leading 1 are zeros. In view of this, the following theorem should not be too surprising:

Theorem 2.4.15   An $m \times n$ matrix is row-equivalent to a *unique* reduced row-echelon matrix.

The unique reduced row-echelon matrix to which a matrix A is row-equivalent will be called *the* reduced row-echelon form of A. As illustrated in the next example, the row-reduction algorithm is easily extended to determine the reduced row-echelon form of A—we just put zeros above and beneath each leading 1.

Example 2.4.16   Determine the reduced row-echelon form of $A = \begin{bmatrix} 3 & -1 & 22 \\ -1 & 5 & 2 \\ 2 & 4 & 24 \end{bmatrix}$.

Solution: We apply the row-reduction algorithm, but put 0s above and below the leading 1s. In so doing, it is immaterial whether we first reduce A to row-echelon form and then arrange 0s above the leading 1s, or arrange 0s both above and below the leading 1s as we proceed from left to right.

$$A = \begin{bmatrix} 3 & -1 & 22 \\ -1 & 5 & 2 \\ 2 & 4 & 24 \end{bmatrix} \overset{1}{\sim} \begin{bmatrix} 1 & 9 & 26 \\ -1 & 5 & 2 \\ 2 & 4 & 24 \end{bmatrix} \overset{2}{\sim} \begin{bmatrix} 1 & 9 & 26 \\ 0 & 14 & 28 \\ 0 & -14 & -28 \end{bmatrix} \overset{3}{\sim} \begin{bmatrix} 1 & 9 & 26 \\ 0 & 1 & 2 \\ 0 & -14 & -28 \end{bmatrix} \overset{4}{\sim} \begin{bmatrix} 1 & 0 & 8 \\ 0 & 1 & 2 \\ 0 & 0 & 0 \end{bmatrix}$$

which is the reduced row-echelon form of A.

> **1.** $A_{21}(2)$ **2.** $A_{12}(1),\ A_{13}(-2)$ **3.** $M_2(1/14)$ **4.** $A_{21}(-9),\ A_{23}(14)$ $\square$

Exercises for 2.4

Key Terms

Elementary row operations, Row-equivalent matrices, Back substitution, Row-echelon matrix, Row-echelon form, Leading 1, Pivot, Rank of a matrix, Reduced row-echelon matrix.

Skills

- Be able to perform elementary row operations on a matrix.

- Be able to determine a row-echelon form or reduced row-echelon form for a matrix.

- Be able to find the rank of a matrix.

True-False Review

For Questions 1–9, decide if the given statement is **true** or **false**, and give a brief justification for your answer. If true, you can quote a relevant definition or theorem from the text. If false, provide an example, illustration, or brief explanation of why the statement is false.

1. A matrix A can have many row-echelon forms but only one reduced row-echelon form.

2. Any upper triangular $n \times n$ matrix is in row-echelon form.

3. Any $n \times n$ matrix in row-echelon form is upper triangular.

4. If a matrix A has more rows than a matrix B, then $\text{rank}(A) \geq \text{rank}(B)$.

5. For any matrices A and B of the same dimensions,

$$\text{rank}(A + B) = \text{rank}(A) + \text{rank}(B).$$

6. For any matrices A and B of the appropriate dimensions,

$$\text{rank}(AB) = \text{rank}(A) \cdot \text{rank}(B).$$

7. If a matrix has rank zero, then it must be the zero matrix.

8. The matrices A and $2A$ must have the same rank.

9. The matrices A and $2A$ must have the same reduced row-echelon form.

Problems

For Problems 1–8, determine whether the given matrices are in reduced row-echelon form, row-echelon form but not reduced row-echelon form, or neither.

1. $\begin{bmatrix} 1 & 0 & -1 & 0 \\ 0 & 0 & 1 & 2 \\ 0 & 0 & 0 & 0 \end{bmatrix}$.

2. $\begin{bmatrix} 1 & 0 & 2 & 5 \\ 1 & 0 & 0 & 2 \\ 0 & 1 & 1 & 0 \end{bmatrix}$.

3. $\begin{bmatrix} 1 & 0 & 0 & 0 \\ 0 & 0 & 0 & 1 \end{bmatrix}$.

4. $\begin{bmatrix} 0 & 1 \\ 1 & 0 \end{bmatrix}$.

5. $\begin{bmatrix} 1 & 1 \\ 0 & 0 \end{bmatrix}$.

6. $\begin{bmatrix} 1 & 0 & 1 & 2 \\ 0 & 0 & 1 & 1 \\ 0 & 0 & 0 & 1 \\ 0 & 0 & 0 & 0 \end{bmatrix}$.

7. $\begin{bmatrix} 0 & 0 & 0 & 0 \\ 0 & 0 & 0 & 0 \\ 0 & 0 & 0 & 0 \end{bmatrix}$.

8. $\begin{bmatrix} 0 & 1 & 0 & 0 \\ 0 & 0 & 1 & 0 \\ 0 & 0 & 0 & 0 \end{bmatrix}$.

For Problems 9–18, use elementary row operations to reduce the given matrix to row-echelon form, and hence determine the rank of each matrix.

9. $\begin{bmatrix} 2 & 1 \\ 1 & -3 \end{bmatrix}$.

10. $\begin{bmatrix} 2 & -4 \\ -4 & 8 \end{bmatrix}$.

11. $\begin{bmatrix} 2 & 1 & 4 \\ 2 & -3 & 4 \\ 3 & -2 & 6 \end{bmatrix}$.

12. $\begin{bmatrix} 0 & 1 & 3 \\ 0 & 1 & 4 \\ 0 & 3 & 5 \end{bmatrix}$.

13. $\begin{bmatrix} 2 & -1 \\ 3 & 2 \\ 2 & 5 \end{bmatrix}$.

14. $\begin{bmatrix} 2 & -1 & 3 \\ 3 & 1 & -2 \\ 2 & -2 & 1 \end{bmatrix}$.

15. $\begin{bmatrix} 2 & -1 & 3 & 4 \\ 1 & -2 & 1 & 3 \\ 1 & -5 & 0 & 5 \end{bmatrix}$.

16. $\begin{bmatrix} 2 & -2 & -1 & 3 \\ 3 & -2 & 3 & 1 \\ 1 & -1 & 1 & 0 \\ 2 & -1 & 2 & 2 \end{bmatrix}$.

17. $\begin{bmatrix} 4 & 7 & 4 & 7 \\ 3 & 5 & 3 & 5 \\ 2 & -2 & 2 & -2 \\ 5 & -2 & 5 & -2 \end{bmatrix}$.

18. $\begin{bmatrix} 2 & 1 & 3 & 4 & 2 \\ 1 & 0 & 2 & 1 & 3 \\ 2 & 3 & 1 & 5 & 7 \end{bmatrix}$.

For Problems 19–25, reduce the given matrix to reduced row-echelon form and hence determine the rank of each matrix.

19. $\begin{bmatrix} 3 & 2 \\ 1 & -1 \end{bmatrix}$.

20. $\begin{bmatrix} 3 & 7 & 10 \\ 2 & 3 & -1 \\ 1 & 2 & 1 \end{bmatrix}$.

21. $\begin{bmatrix} 3 & -3 & 6 \\ 2 & -2 & 4 \\ 6 & -6 & 12 \end{bmatrix}$.

22. $\begin{bmatrix} 3 & 5 & -12 \\ 2 & 3 & -7 \\ -2 & -1 & 1 \end{bmatrix}$.

23. $\begin{bmatrix} 1 & -1 & -1 & 2 \\ 3 & -2 & 0 & 7 \\ 2 & -1 & 2 & 4 \\ 4 & -2 & 3 & 8 \end{bmatrix}$.

24. $\begin{bmatrix} 1 & -2 & 1 & 3 \\ 3 & -6 & 2 & 7 \\ 4 & -8 & 3 & 10 \end{bmatrix}$.

25. $\begin{bmatrix} 0 & 1 & 2 & 1 \\ 0 & 3 & 1 & 2 \\ 0 & 2 & 0 & 1 \end{bmatrix}$.

Many forms of technology have commands for performing elementary row operations on a matrix A. For example, in the linear algebra package of Maple, the three elementary row operations are

- swaprow(A, i, j): permute rows i and j
- mulrow(A, i, k): multiply row i by k
- addrow(A, i, j): add k times row i to row j

◇ For Problems 26–28, use some form of technology to determine a row-echelon form of the given matrix.

26. The matrix in Problem 14.

27. The matrix in Problem 15.

28. The matrix in Problem 18.

◇ Many forms of technology also have built-in functions for directly determining the reduced row-echelon form of a given matrix A. For example, in the linear algebra package of Maple, the appropriate command is rref(A). In Problems 29–31, use technology to determine directly the reduced row-echelon form of the given matrix.

29. The matrix in Problem 21.

30. The matrix in Problem 24.

31. The matrix in Problem 25.

2.5   Gaussian Elimination

We now illustrate how elementary row-operations applied to the augmented matrix of a system of linear equations can be used first to determine whether the system is consistent, and second, if the system is consistent, to find all of its solutions. In doing so, we will develop the general theory for linear systems of equations.

Example 2.5.1   Determine the solution set to

$$\begin{aligned} 3x_1 - 2x_2 + 2x_3 &= 9, \\ x_1 - 2x_2 + x_3 &= 5, \\ 2x_1 - x_2 - 2x_3 &= -1. \end{aligned} \qquad (2.5.1)$$

Solution: We first use elementary row operations to reduce the augmented matrix of the system to row-echelon form.

$$\begin{bmatrix} 3 & -2 & 2 & 9 \\ 1 & -2 & 1 & 5 \\ 2 & -1 & -2 & -1 \end{bmatrix} \overset{1}{\sim} \begin{bmatrix} 1 & -2 & 1 & 5 \\ 3 & -2 & 2 & 9 \\ 2 & -1 & -2 & -1 \end{bmatrix} \overset{2}{\sim} \begin{bmatrix} 1 & -2 & 1 & 5 \\ 0 & 4 & -1 & -6 \\ 0 & 3 & -4 & -11 \end{bmatrix}$$

$$\overset{3}{\sim} \begin{bmatrix} 1 & -2 & 1 & 5 \\ 0 & 1 & 3 & 5 \\ 0 & 3 & -4 & -11 \end{bmatrix} \overset{4}{\sim} \begin{bmatrix} 1 & -2 & 1 & 5 \\ 0 & 1 & 3 & 5 \\ 0 & 0 & -13 & -26 \end{bmatrix} \overset{5}{\sim} \begin{bmatrix} 1 & -2 & 1 & 5 \\ 0 & 1 & 3 & 5 \\ 0 & 0 & 1 & 2 \end{bmatrix}.$$

> **1.** P_{12} **2.** $A_{12}(-3)$, $A_{13}(-2)$ **3.** $A_{32}(-1)$ **4.** $A_{23}(-3)$ **5.** $M_3(-1/13)$

The system corresponding to this row-echelon form of the augmented matrix is

$$x_1 - 2x_2 + x_3 = 5, \tag{2.5.2}$$
$$x_2 + 3x_3 = 5, \tag{2.5.3}$$
$$x_3 = 2, \tag{2.5.4}$$

which can be solved by *back substitution*. From Equation (2.5.4), $x_3 = 2$. Substituting into Equation (2.5.3) and solving for x_2, we find that $x_2 = -1$. Finally, substituting into Equation (2.5.2) for x_3 and x_2 and solving for x_1 yields $x_1 = 1$. Thus, our original system of equations has the unique solution $(1, -1, 2)$, and the solution set to the system is

$$S = \{(1, -1, 2)\},$$

which is a subset of $\mathbb{R}^3$. ☐

The process of reducing the augmented matrix to row-echelon form and then using back substitution to solve the equivalent system is called **Gaussian elimination**. The particular case of Gaussian elimination that arises when the augmented matrix is reduced to reduced row-echelon form is called **Gauss-Jordan elimination**.

Example 2.5.2 Use Gauss-Jordan elimination to determine the solution set to

$$\begin{aligned} x_1 + 2x_2 - x_3 &= 1, \\ 2x_1 + 5x_2 - x_3 &= 3, \\ x_1 + 3x_2 + 2x_3 &= 6. \end{aligned}$$

Solution: In this case, we first reduce the augmented matrix of the system to reduced row-echelon form.

$$\begin{bmatrix} 1 & 2 & -1 & 1 \\ 2 & 5 & -1 & 3 \\ 1 & 3 & 2 & 6 \end{bmatrix} \overset{1}{\sim} \begin{bmatrix} 1 & 2 & -1 & 1 \\ 0 & 1 & 1 & 1 \\ 0 & 1 & 3 & 5 \end{bmatrix} \overset{2}{\sim} \begin{bmatrix} 1 & 0 & -3 & -1 \\ 0 & 1 & 1 & 1 \\ 0 & 0 & 2 & 4 \end{bmatrix} \overset{3}{\sim} \begin{bmatrix} 1 & 0 & -3 & -1 \\ 0 & 1 & 1 & 1 \\ 0 & 0 & 1 & 2 \end{bmatrix} \overset{4}{\sim} \begin{bmatrix} 1 & 0 & 0 & 5 \\ 0 & 1 & 0 & -1 \\ 0 & 0 & 1 & 2 \end{bmatrix}$$

> **1.** $A_{12}(-2)$, $A_{13}(-1)$ **2.** $A_{21}(-2)$, $A_{23}(-1)$ **3.** $M_3(1/2)$ **4.** $A_{31}(3)$, $A_{32}(-1)$

The augmented matrix is now in reduced row-echelon form. The equivalent system is

$$\begin{aligned} x_1 &= 5, \\ x_2 &= -1, \\ x_3 &= 2. \end{aligned}$$

and the solution can be read off directly as $(5, -1, 2)$. Consequently, the given system has solution set

$$S = \{(5, -1, 2)\}$$

in $\mathbb{R}^3$. ☐

We see from the preceding two examples that the advantage of Gauss-Jordan elimination over Gaussian elimination is that it does not require back substitution. However, the disadvantage is that reducing the augmented matrix to reduced row-echelon form requires more elementary row operations than reduction to row-echelon form. It can be shown, in fact, that in general, Gaussian elimination is the more computationally efficient technique. As we will see in the next section, the main reason for introducing the Gauss-Jordan method is its application to the computation of the inverse of an $n \times n$ matrix.

Remark   The Gaussian elimination method is so systematic that it can be programmed easily on a computer. Indeed, many large-scale programs for solving linear systems are based on the row-reduction method.

In both of the preceding examples,

$$\text{rank}(A) = \text{rank}(A^\#) = \text{number of unknowns in the system}$$

and the system had a unique solution. More generally, we have the following lemma:

Lemma 2.5.3   Consider the $m \times n$ linear system $A\mathbf{x} = \mathbf{b}$. Let $A^\#$ denote the augmented matrix of the system. If $\text{rank}(A) = \text{rank}(A^\#) = n$, then the system has a unique solution.

Proof If $\text{rank}(A) = \text{rank}(A^\#) = n$, then there are n leading ones in any row-echelon form of A, hence back substitution gives a unique solution. The form of the row-echelon form of $A^\#$ is shown below, with $m - n$ rows of zeros at the bottom of the matrix omitted and where the $*$'s denote unknown elements of the row-echelon form.

$$\begin{bmatrix} 1 & * & * & * & \ldots & * & * \\ 0 & 1 & * & * & \ldots & * & * \\ 0 & 0 & 1 & * & \ldots & * & * \\ \vdots & \vdots & \vdots & \vdots & \ldots & \vdots & \vdots \\ 0 & 0 & 0 & 0 & \ldots & 1 & * \end{bmatrix}$$ ■

Note that $\text{rank}(A)$ cannot exceed $\text{rank}(A^\#)$. Thus, there are only two possibilities for the relationship between $\text{rank}(A)$ and $\text{rank}(A^\#)$: $\text{rank}(A) < \text{rank}(A^\#)$ or $\text{rank}(A) = \text{rank}(A^\#)$. We now consider what happens in these cases.

Example 2.5.4   Determine the solution set to

$$\begin{aligned} x_1 + x_2 - x_3 + x_4 &= 1, \\ 2x_1 + 3x_2 + x_3 &= 4, \\ 3x_1 + 5x_2 + 3x_3 - x_4 &= 5. \end{aligned}$$

Solution: We use elementary row operations to reduce the augmented matrix:

$$\begin{bmatrix} 1 & 1 & -1 & 1 & 1 \\ 2 & 3 & 1 & 0 & 4 \\ 3 & 5 & 3 & -1 & 5 \end{bmatrix} \overset{1}{\sim} \begin{bmatrix} 1 & 1 & -1 & 1 & 1 \\ 0 & 1 & 3 & -2 & 2 \\ 0 & 2 & 6 & -4 & 2 \end{bmatrix} \overset{2}{\sim} \begin{bmatrix} 1 & 1 & -1 & 1 & 1 \\ 0 & 1 & 3 & -2 & 2 \\ 0 & 0 & 0 & 0 & -2 \end{bmatrix}$$

---

1. $A_{12}(-2)$, $A_{13}(-3)$ **2.** $A_{23}(-2)$

The last row tells us that the system of equations has no solution (that is, it is inconsistent), since it requires

$$0x_1 + 0x_2 + 0x_3 + 0x_4 = -2,$$

which is clearly impossible. The solution set to the system is thus the empty set $\emptyset$.   ☐

In the previous example, $\text{rank}(A) = 2$, whereas $\text{rank}(A^\#) = 3$. Thus, $\text{rank}(A) < \text{rank}(A^\#)$, and the corresponding system has no solution. Next we establish that this result is true in general.

Lemma 2.5.5 Consider the $m \times n$ linear system $A\mathbf{x} = \mathbf{b}$. Let $A^\#$ denote the augmented matrix of the system. If $\text{rank}(A) < \text{rank}(A^\#)$, then the system is inconsistent.

Proof If $\text{rank}(A) < \text{rank}(A^\#)$, then there will be one row in the reduced row-echelon form of the augmented matrix whose first nonzero element arises in the last column. Such a row corresponds to an equation of the form

$$0x_1 + 0x_2 + \cdots + 0x_n = 1,$$

which has no solution. Consequently, the system is inconsistent. ∎

Finally, we consider the case when $\text{rank}(A) = \text{rank}(A^\#)$. If $\text{rank}(A) = n$, we have already seen in Lemma 2.5.3 that the system has a unique solution. We now consider an example in which $\text{rank}(A) < n$.

Example 2.5.6 Determine the solution set to

$$\begin{aligned} 5x_1 - 6x_2 + x_3 &= 4, \\ 2x_1 - 3x_2 + x_3 &= 1, \\ 4x_1 - 3x_2 - x_3 &= 5. \end{aligned} \tag{2.5.5}$$

Solution: We begin by reducing the augmented matrix of the system.

$$\begin{bmatrix} 5 & -6 & 1 & 4 \\ 2 & -3 & 1 & 1 \\ 4 & -3 & -1 & 5 \end{bmatrix} \overset{1}{\sim} \begin{bmatrix} 1 & -3 & 2 & -1 \\ 2 & -3 & 1 & 1 \\ 4 & -3 & -1 & 5 \end{bmatrix} \overset{2}{\sim} \begin{bmatrix} 1 & -3 & 2 & -1 \\ 0 & 3 & -3 & 3 \\ 0 & 9 & -9 & 9 \end{bmatrix}$$
$$\overset{3}{\sim} \begin{bmatrix} 1 & -3 & 2 & -1 \\ 0 & 1 & -1 & 1 \\ 0 & 9 & -9 & 9 \end{bmatrix} \overset{4}{\sim} \begin{bmatrix} 1 & -3 & 2 & -1 \\ 0 & 1 & -1 & 1 \\ 0 & 0 & 0 & 0 \end{bmatrix}$$

> **1.** $A_{31}(-1)$ **2.** $A_{12}(-2)$, $A_{13}(-4)$ **3.** $M_2(1/3)$ **4.** $A_{23}(-9)$

The augmented matrix is now in row-echelon form, and the equivalent system is

$$x_1 - 3x_2 + 2x_3 = -1, \tag{2.5.6}$$
$$x_2 - x_3 = 1. \tag{2.5.7}$$

Since we have three variables, but only two equations relating them, we are free to specify one of the variables arbitrarily. The variable that we choose to specify is called a **free variable** or **free parameter**. The remaining variables are then determined by the system of equations and are called **bound variables** or **bound parameters**. In the foregoing system, we take x_3 as the free variable and set

$$x_3 = t,$$

where t can assume any real value[5]. It follows from (2.5.7) that

$$x_2 = 1 + t.$$

---

[5]When considering systems of equations with complex coefficients, we allow free variables to assume *complex* values as well.

Further, from Equation (2.5.6),

$$x_1 = -1 + 3(1 + t) - 2t = 2 + t.$$

Thus the solution set to the given system of equations is the following subset of $\mathbb{R}^3$:

$$S = \{(2 + t, 1 + t, t) : t \in \mathbb{R}\}.$$

The system has an infinite number of solutions, obtained by allowing the parameter t to assume all real values. For example, two particular solutions of the system are

$$(2, 1, 0) \qquad \text{and} \qquad (0, -1, -2),$$

corresponding to $t = 0$ and $t = -2$, respectively. Note that we can also write the solution set S above in the form

$$S = \{(2, 1, 0) + t(1, 1, 1) : t \in \mathbb{R}\}. \qquad \square$$

Remark   The geometry of the foregoing solution is as follows. The given system (2.5.5) can be interpreted as consisting of three planes in 3-space. Any solution to the system gives the coordinates of a point of intersection of the three planes. In the preceding example the planes intersect in a line whose parametric equations are

$$x_1 = 2 + t, \qquad x_2 = 1 + t, \qquad x_3 = t.$$

(See Figure 2.3.1.)

In general, the solution to a consistent $m \times n$ system of linear equations may involve more than one free variable. Indeed, the number of free variables will depend on how many nonzero rows arise in any row-echelon form of the augmented matrix, $A^\#$, of the system; that is, it will depend on the rank of $A^\#$. More precisely, if $\text{rank}(A^\#) = r^\#$, then the equivalent system will have only $r^\#$ relationships between the n variables. Consequently, provided the system is consistent,

$$\text{number of free variables} = n - r^\#.$$

We therefore have the following lemma.

Lemma 2.5.7   Consider the $m \times n$ linear system $A\mathbf{x} = \mathbf{b}$. Let $A^\#$ denote the augmented matrix of the system and let $r^\# = \text{rank}(A^\#)$. If $r^\# = \text{rank}(A) < n$, then the system has an infinite number of solutions, indexed by $n - r^\#$ free variables.

Proof As discussed before, any row-echelon equivalent system will have only $r^\#$ equations involving the n variables, and so there will be $n - r^\# > 0$ free variables. If we assign arbitrary values to these free variables, then the remaining $r^\#$ variables will be uniquely determined, by back substitution, from the system. Since each free variable can assume infinitely many values, in this case there are an infinite number of solutions to the system.   ∎

Example 2.5.8   Use Gaussian elimination to solve

$$\begin{aligned}
x_1 - 2x_2 + 2x_3 - x_4 &= 3, \\
3x_1 + x_2 + 6x_3 + 11x_4 &= 16, \\
2x_1 - x_2 + 4x_3 + 4x_4 &= 9.
\end{aligned}$$

Solution: A row-echelon form of the augmented matrix of the system is

$$\begin{bmatrix} 1 & -2 & 2 & -1 & 3 \\ 0 & 1 & 0 & 2 & 1 \\ 0 & 0 & 0 & 0 & 0 \end{bmatrix},$$

so that we have two free variables. The equivalent system is

$$x_1 - 2x_2 + 2x_3 - x_4 = 3, \tag{2.5.8}$$
$$x_2 \phantom{{}+{}} + 2x_4 = 1. \tag{2.5.9}$$

Notice that we cannot choose any two variables freely. For example, from Equation (2.5.9), we cannot specify both x_2 and x_4 independently. The bound variables should be taken as those that correspond to leading 1s in the row-echelon form of $A^{\#}$, since these are the variables that can always be determined by back substitution (they appear as the leftmost variable in some equation of the system corresponding to the row echelon form of the augmented matrix).

> Choose as free variables those variables that
> **do not** correspond to a leading 1 in a row-echelon form of $A^{\#}$.

Applying this rule to Equations (2.5.8) and (2.5.9), we choose x_3 and x_4 as free variables and therefore set

$$x_3 = s, \quad x_4 = t.$$

It then follows from Equation (2.5.9) that

$$x_2 = 1 - 2t.$$

Substitution into (2.5.8) yields

$$x_1 = 5 - 2s - 3t,$$

so that the solution set to the given system is the following subset of $\mathbb{R}^4$:

$$S = \{(5 - 2s - 3t, 1 - 2t, s, t) : s, t \in \mathbb{R}\}.$$
$$= \{(5, 1, 0, 0) + s(-2, 0, 1, 0) + t(-3, -2, 0, 1) : s, t \in \mathbb{R}\}. \qquad \square$$

Lemmas 2.5.3, 2.5.5, and 2.5.7 completely characterize the solution properties of an $m \times n$ linear system. Combining the results of these three lemmas gives the next theorem.

Theorem 2.5.9   Consider the $m \times n$ linear system $A\mathbf{x} = \mathbf{b}$. Let r denote the rank of A, and let $r^{\#}$ denote the rank of the augmented matrix of the system. Then

1. If $r < r^{\#}$, the system is inconsistent.
2. If $r = r^{\#}$, the system is consistent and

   (a) There exists a unique solution if and only if $r^{\#} = n$.
   (b) There exists an infinite number of solutions if and only if $r^{\#} < n$.

Homogeneous Linear Systems

Many problems that we will meet in the future will require the solution to a homogeneous system of linear equations. The general form for such a system is

$$
\begin{aligned}
a_{11}x_1 + a_{12}x_2 + \cdots + a_{1n}x_n &= 0, \\
a_{21}x_1 + a_{22}x_2 + \cdots + a_{2n}x_n &= 0, \\
&\ \ \vdots \\
a_{m1}x_1 + a_{m2}x_2 + \cdots + a_{mn}x_n &= 0,
\end{aligned}
\tag{2.5.10}
$$

or, in matrix form, $A\mathbf{x} = \mathbf{0}$, where A is the coefficient matrix of the system and $\mathbf{0}$ denotes the m-vector whose elements are all zeros.

Corollary 2.5.10   The homogeneous linear system $A\mathbf{x} = \mathbf{0}$ is consistent for any coefficient matrix A, with a solution given by $\mathbf{x} = \mathbf{0}$.

Proof We can see immediately from (2.5.10) that if $\mathbf{x} = \mathbf{0}$, then $A\mathbf{x} = \mathbf{0}$, so $\mathbf{x} = \mathbf{0}$ is a solution to the homogeneous linear system.

Alternatively, we can deduce the consistency of this system from Theorem 2.5.9 as follows. The augmented matrix $A^{\#}$ of a homogeneous linear system differs from that of the coefficient matrix A only by the addition of a column of zeros, a feature that does not affect the rank of the matrix. Consequently, for a homogeneous system, we have $\text{rank}(A^{\#}) = \text{rank}(A)$, and therefore, from Theorem 2.5.9, such a system is necessarily consistent.   ∎

Remarks

1. The solution $\mathbf{x} = \mathbf{0}$ is referred to as the **trivial solution**. Consequently, from Theorem 2.5.9, a homogeneous system either has *only* the trivial solution or has an infinite number of solutions (one of which must be the trivial solution).

2. Once more it is worth mentioning the geometric interpretation of Corollary 2.5.10 in the case of a homogeneous system with three unknowns. We can regard each equation of such a system as defining a plane. Owing to the homogeneity, each plane passes through the origin, hence the planes intersect at least at the origin.

Often we will be interested in determining whether a given homogeneous system has an infinite number of solutions, and not in actually obtaining the solutions. The following corollary to Theorem 2.5.9 can sometimes be used to determine by inspection whether a given homogeneous system has nontrivial solutions:

Corollary 2.5.11   A homogeneous system of m linear equations in n unknowns, with $m < n$, has an infinite number of solutions.

Proof Let r and $r^{\#}$ be as in Theorem 2.5.9. Using the fact that $r = r^{\#}$ for a homogeneous system, we see that since $r^{\#} \leq m < n$, Theorem 2.5.9 implies that the system has an infinite number of solutions.   ∎

Remark   If $m \geq n$, then we may or may not have nontrivial solutions, depending on whether the rank of the augmented matrix, $r^{\#}$, satisfies $r^{\#} < n$ or $r^{\#} = n$, respectively. We encourage the reader to construct linear systems that illustrate each of these two possibilities.

Example 2.5.12   Determine the solution set to $A\mathbf{x} = \mathbf{0}$, if $A = \begin{bmatrix} 0 & 2 & 3 \\ 0 & 1 & -1 \\ 0 & 3 & 7 \end{bmatrix}$.

Solution: The augmented matrix of the system is

$$\begin{bmatrix} 0 & 2 & 3 & 0 \\ 0 & 1 & -1 & 0 \\ 0 & 3 & 7 & 0 \end{bmatrix},$$

with reduced row-echelon form

$$\begin{bmatrix} 0 & 1 & 0 & 0 \\ 0 & 0 & 1 & 0 \\ 0 & 0 & 0 & 0 \end{bmatrix}.$$

The equivalent system is

$$x_2 = 0,$$
$$x_3 = 0.$$

It is tempting, but incorrect, to conclude from this that the solution to the system is $x_1 = x_2 = x_3 = 0$. Since x_1 does not occur in the system, it is a free variable and therefore *not necessarily* zero. Consequently, the correct solution to the foregoing system is $(r, 0, 0)$, where r is a free variable, and the solution set is $\{(r, 0, 0) : r \in \mathbb{R}\}$. $\square$

The linear systems that we have so far encountered have all had real coefficients, and we have considered corresponding real solutions. The techniques that we have developed for solving linear systems are also applicable to the case when our system has complex coefficients. The corresponding solutions will also be complex.

Remark   In general, the simplest method of putting a leading 1 in a position that contains the complex number $a + ib$ is to multiply the corresponding row by the scalar $\left(\frac{1}{a^2+b^2}\right)(a - ib)$. This is illustrated in steps 1 and 4 in the next example. If difficulties are encountered, consultation of Appendix A is in order.

Example 2.5.13   Determine the solution set to

$$\begin{aligned} (1 + 2i)x_1 + \quad\quad 4x_2 + (3 + i)x_3 &= 0, \\ (2 - i)x_1 + (1 + i)x_2 + \quad\quad 3x_3 &= 0, \\ 5ix_1 + (7 - i)x_2 + (3 + 2i)x_3 &= 0. \end{aligned}$$

Solution: We reduce the augmented matrix of the system.

$$\begin{bmatrix} 1+2i & 4 & 3+i & 0 \\ 2-i & 1+i & 3 & 0 \\ 5i & 7-i & 3+2i & 0 \end{bmatrix} \overset{1}{\sim} \begin{bmatrix} 1 & \frac{4}{5}(1-2i) & 1-i & 0 \\ 2-i & 1+i & 3 & 0 \\ 5i & 7-i & 3+2i & 0 \end{bmatrix}$$

$$\overset{2}{\sim} \begin{bmatrix} 1 & \frac{4}{5}(1-2i) & 1-i & 0 \\ 0 & (1+i) - \frac{4}{5}(1-2i)(2-i) & 3 - (1-i)(2-i) & 0 \\ 0 & (7-i) - 4i(1-2i) & (3+2i) - 5i(1-i) & 0 \end{bmatrix}$$

$$= \begin{bmatrix} 1 & \frac{4}{5}(1-2i) & 1-i & 0 \\ 0 & 1+5i & 2+3i & 0 \\ 0 & -1-5i & -2-3i & 0 \end{bmatrix} \overset{3}{\sim} \begin{bmatrix} 1 & \frac{4}{5}(1-2i) & 1-i & 0 \\ 0 & 1+5i & 2+3i & 0 \\ 0 & 0 & 0 & 0 \end{bmatrix}$$

$$\overset{4}{\sim} \begin{bmatrix} 1 & \frac{4}{5}(1-2i) & 1-i & 0 \\ 0 & 1 & \frac{1}{26}(17-7i) & 0 \\ 0 & 0 & 0 & 0 \end{bmatrix}.$$

---

1. $M_1((1-2i)/5)$ **2.** $A_{12}(-(2-i))$, $A_{13}(-5i)$ **3.** $A_{23}(1)$ **4.** $M_2((1-5i)/26)$

This matrix is now in row-echelon form. The equivalent system is

$$x_1 + \tfrac{4}{5}(1-2i)x_2 + (1-i)x_3 = 0,$$
$$x_2 + \tfrac{1}{26}(17-7i)x_3 = 0.$$

There is one free variable, which we take to be $x_3 = t$, where t can assume any *complex* value. Applying back substitution yields

$$x_2 = \tfrac{1}{26}t(-17+7i)$$
$$x_1 = -\tfrac{2}{65}t(1-2i)(-17+7i) - t(1-i)$$
$$= -\tfrac{1}{65}t(59+17i)$$

so that the solution set to the system is the subset of $\mathbb{C}^3$

$$\left\{ \left(-\tfrac{1}{65}t(59+17i), \tfrac{1}{26}t(-17+7i), t \right) : t \in \mathbb{C} \right\}. \qquad \square$$

Exercises for 2.5

Key Terms

Gaussian elimination, Gauss-Jordan elimination, Free variables, Bound (or leading) variables, Trivial solution.

Skills

- Be able to solve a linear system of equations by Gaussian elimination and by Gauss-Jordan elimination.

- Be able to identify free variables and bound variables and know how they are used to construct the solution set to a linear system.

- Understand the relationship between the ranks of A and $A^{\#}$, and how this affects the number of solutions to a linear system.

True-False Review

For Questions 1–6, decide if the given statement is **true** or **false**, and give a brief justification for your answer. If true, you can quote a relevant definition or theorem from the text. If false, provide an example, illustration, or brief explanation of why the statement is false.

1. The process by which a matrix is brought via elementary row operations to row-echelon form is known as Gauss-Jordan elimination.

2. A homogeneous linear system of equations is always consistent.

3. For a linear system $A\mathbf{x} = \mathbf{b}$, every column of the row-echelon form of A corresponds to either a bound variable or a free variable, but not both, of the linear system.

4. A linear system $A\mathbf{x} = \mathbf{b}$ is consistent if and only if the last column of the row-echelon form of the augmented matrix $[A \ \mathbf{b}]$ is not a pivot column.

5. A linear system is consistent if and only if there are free variables in the row-echelon form of the corresponding augmented matrix.

6. The columns of the row-echelon form of $A^{\#}$ that contain the leading 1s correspond to the free variables.

Problems

For Problems 1–9, use Gaussian elimination to determine the solution set to the given system.

1.
$$\begin{aligned} x_1 + 2x_2 + x_3 &= 1, \\ 3x_1 + 5x_2 + x_3 &= 3, \\ 2x_1 + 6x_2 + 7x_3 &= 1. \end{aligned}$$

2.
$$\begin{aligned} 3x_1 - x_2 &= 1, \\ 2x_1 + x_2 + 5x_3 &= 4, \\ 7x_1 - 5x_2 - 8x_3 &= -3. \end{aligned}$$

3.
$$\begin{aligned} 3x_1 + 5x_2 - x_3 &= 14, \\ x_1 + 2x_2 + x_3 &= 3, \\ 2x_1 + 5x_2 + 6x_3 &= 2. \end{aligned}$$

4.
$$\begin{aligned} 6x_1 - 3x_2 + 3x_3 &= 12, \\ 2x_1 - x_2 + x_3 &= 4, \\ -4x_1 + 2x_2 - 2x_3 &= -8. \end{aligned}$$

5.
$$\begin{aligned} 2x_1 - x_2 + 3x_3 &= 14, \\ 3x_1 + x_2 - 2x_3 &= -1, \\ 7x_1 + 2x_2 - 3x_3 &= 3, \\ 5x_1 - x_2 - 2x_3 &= 5. \end{aligned}$$

6.
$$\begin{aligned} 2x_1 - x_2 - 4x_3 &= 5, \\ 3x_1 + 2x_2 - 5x_3 &= 8, \\ 5x_1 + 6x_2 - 6x_3 &= 20, \\ x_1 + x_2 - 3x_3 &= -3. \end{aligned}$$

7.
$$\begin{aligned} x_1 + 2x_2 - x_3 + x_4 &= 1, \\ 2x_1 + 4x_2 - 2x_3 + 2x_4 &= 2, \\ 5x_1 + 10x_2 - 5x_3 + 5x_4 &= 5. \end{aligned}$$

8.
$$\begin{aligned} x_1 + 2x_2 - x_3 + x_4 &= 1, \\ 2x_1 - 3x_2 + x_3 - x_4 &= 2, \\ x_1 - 5x_2 + 2x_3 - 2x_4 &= 1, \\ 4x_1 + x_2 - x_3 + x_4 &= 3. \end{aligned}$$

9.
$$\begin{aligned} x_1 + 2x_2 + x_3 + x_4 - 2x_5 &= 3, \\ x_3 + 4x_4 - 3x_5 &= 2, \\ 2x_1 + 4x_2 - x_3 - 10x_4 + 5x_5 &= 0. \end{aligned}$$

For Problems 10–15, use Gauss-Jordan elimination to determine the solution set to the given system.

10.
$$\begin{aligned} 2x_1 - x_2 - x_3 &= 2, \\ 4x_1 + 3x_2 - 2x_3 &= -1, \\ x_1 + 4x_2 + x_3 &= 4. \end{aligned}$$

11.
$$\begin{aligned} 3x_1 + x_2 + 5x_3 &= 2, \\ x_1 + x_2 - x_3 &= 1, \\ 2x_1 + x_2 + 2x_3 &= 3. \end{aligned}$$

12.
$$\begin{aligned} x_1 - 2x_3 &= -3, \\ 3x_1 - 2x_2 - 4x_3 &= -9, \\ x_1 - 4x_2 + 2x_3 &= -3. \end{aligned}$$

13.
$$\begin{aligned} 2x_1 - x_2 + 3x_3 - x_4 &= 3, \\ 3x_1 + 2x_2 + x_3 - 5x_4 &= -6, \\ x_1 - 2x_2 + 3x_3 + x_4 &= 6. \end{aligned}$$

14.
$$\begin{aligned} x_1 + x_2 + x_3 - x_4 &= 4, \\ x_1 - x_2 - x_3 - x_4 &= 2, \\ x_1 + x_2 - x_3 + x_4 &= -2, \\ x_1 - x_2 + x_3 + x_4 &= -8. \end{aligned}$$

15.
$$\begin{aligned} 2x_1 - x_2 + 3x_3 + x_4 - x_5 &= 11, \\ x_1 - 3x_2 - 2x_3 - x_4 - 2x_5 &= 2, \\ 3x_1 + x_2 - 2x_3 - x_4 + x_5 &= -2, \\ x_1 + 2x_2 + x_3 + 2x_4 + 3x_5 &= -3, \\ 5x_1 - 3x_2 - 3x_3 + x_4 + 2x_5 &= 2. \end{aligned}$$

For Problems 16–20, determine the solution set to the system $A\mathbf{x} = \mathbf{b}$ for the given coefficient matrix A and right-hand side vector $\mathbf{b}$.

16. $A = \begin{bmatrix} 1 & -3 & 1 \\ 5 & -4 & 1 \\ 2 & 4 & -3 \end{bmatrix}$, $\mathbf{b} = \begin{bmatrix} 8 \\ 15 \\ -4 \end{bmatrix}$.

17. $A = \begin{bmatrix} 1 & 0 & 5 \\ 3 & -2 & 11 \\ 2 & -2 & 6 \end{bmatrix}$, $\mathbf{b} = \begin{bmatrix} 0 \\ 2 \\ 2 \end{bmatrix}$.

18. $A = \begin{bmatrix} 0 & 1 & -1 \\ 0 & 5 & 1 \\ 0 & 2 & 1 \end{bmatrix}$, $\mathbf{b} = \begin{bmatrix} -2 \\ 8 \\ 5 \end{bmatrix}$.

19. $A = \begin{bmatrix} 1 & -1 & 0 & -1 \\ 2 & 1 & 3 & 7 \\ 3 & -2 & 1 & 0 \end{bmatrix}$, $\mathbf{b} = \begin{bmatrix} 2 \\ 2 \\ 4 \end{bmatrix}$.

20. $A = \begin{bmatrix} 1 & 1 & 0 & 1 \\ 3 & 1 & -2 & 3 \\ 2 & 3 & 1 & 2 \\ -2 & 3 & 5 & -2 \end{bmatrix}$, $\mathbf{b} = \begin{bmatrix} 2 \\ 8 \\ 3 \\ -9 \end{bmatrix}$.

21. Determine all values of the constant k for which the following system has **(a)** no solution, **(b)** an infinite number of solutions, and **(c)** a unique solution.

$$\begin{aligned} x_1 + 2x_2 - x_3 &= 3, \\ 2x_1 + 5x_2 + x_3 &= 7, \\ x_1 + x_2 - k^2 x_3 &= -k. \end{aligned}$$

22. Determine all values of the constant k for which the following system has **(a)** no solution, **(b)** an infinite number of solutions, and **(c)** a unique solution.

$$\begin{aligned} 2x_1 + x_2 - x_3 + x_4 &= 0, \\ x_1 + x_2 + x_3 - x_4 &= 0, \\ 4x_1 + 2x_2 - x_3 + x_4 &= 0, \\ 3x_1 - x_2 + x_3 + kx_4 &= 0. \end{aligned}$$

23. Determine all values of the constants a and b for which the following system has **(a)** no solution, **(b)** an infinite number of solutions, and **(c)** a unique solution.

$$\begin{aligned} x_1 + x_2 - 2x_3 &= 4, \\ 3x_1 + 5x_2 - 4x_3 &= 16, \\ 2x_1 + 3x_2 - ax_3 &= b. \end{aligned}$$

24. Determine all values of the constants a and b for which the following system has **(a)** no solution, **(b)** an infinite number of solutions, and **(c)** a unique solution.

$$\begin{aligned} x_1 - ax_2 &= 3, \\ 2x_1 + x_2 &= 6, \\ -3x_1 + (a+b)x_2 &= 1. \end{aligned}$$

25. Show that the system

$$\begin{aligned} x_1 + x_2 + x_3 &= y_1, \\ 2x_1 + 3x_2 + x_3 &= y_2, \\ 3x_1 + 5x_2 + x_3 &= y_3, \end{aligned}$$

has an infinite number of solutions, provided that (y_1, y_2, y_3) lies on the plane whose equation is $y_1 - 2y_2 + y_3 = 0$.

26. Consider the system of linear equations

$$\begin{aligned} a_{11}x_1 + a_{12}x_2 &= b_1, \\ a_{21}x_1 + a_{22}x_2 &= b_2. \end{aligned}$$

Define Δ, Δ_1, and Δ_2 by

$$\Delta = a_{11}a_{22} - a_{12}a_{21},$$
$$\Delta_1 = a_{22}b_1 - a_{12}b_2, \qquad \Delta_2 = a_{11}b_2 - a_{12}b_1.$$

(a) Show that the given system has a unique solution if and only if $\Delta \neq 0$, and that the unique solution in this case is $x_1 = \Delta_1/\Delta$, $x_2 = \Delta_2/\Delta$.

(b) If $\Delta = 0$ and $a_{11} \neq 0$, determine the conditions on Δ_2 that would guarantee that the system has **(i)** no solution, **(ii)** an infinite number of solutions.

(c) Interpret your results in terms of intersections of straight lines.

Gaussian elimination with *partial pivoting* uses the following algorithm to reduce the augmented matrix:

1. Start with augmented matrix $A^\#$.
2. Determine the leftmost nonzero column.
3. Permute rows to put the element of largest absolute value in the pivot position.
4. Use elementary row operations to put zeros beneath the pivot position.
5. If there are no more nonzero rows below the pivot position, go to 7, otherwise go to 6.
6. Apply (2)–(5) to the submatrix consisting of the rows that lie below the pivot position.
7. The matrix is in reduced form.[6]

In Problems 27–30, use the preceding algorithm to reduce $A^\#$ and then apply back substitution to solve the equivalent system. Technology might be useful in performing the required row operations.

27. The system in Problem 1.

28. The system in Problem 5.

29. The system in Problem 6.

30. The system in Problem 10.

---

[6] Notice that this reduced form is *not* a row-echelon matrix.

31. (a) An $n \times n$ system of linear equations whose matrix of coefficients is a lower triangular matrix is called a **lower triangular system**. Assuming that $a_{ii} \neq 0$ for each i, devise a method for solving such a system that is analogous to the back-substitution method.

(b) Use your method from (a) to solve

$$
\begin{aligned}
x_1 &&&= 2, \\
2x_1 - 3x_2 &&&= 1, \\
3x_1 + x_2 - x_3 &&&= 8.
\end{aligned}
$$

32. Find all solutions to the following nonlinear system of equations:

$$
\begin{aligned}
4x_1^3 + 2x_2^2 + 3x_3 &= 12, \\
x_1^3 - x_2^2 + x_3 &= 2, \\
3x_1^3 + x_2^2 - x_3 &= 2.
\end{aligned}
$$

Does your answer contradict Theorem 2.5.9? Explain.

For Problems 33–43, determine the solution set to the given system.

33.
$$
\begin{aligned}
3x_1 + 2x_2 - x_3 &= 0, \\
2x_1 + x_2 + x_3 &= 0, \\
5x_1 - 4x_2 + x_3 &= 0.
\end{aligned}
$$

34.
$$
\begin{aligned}
2x_1 + x_2 - x_3 &= 0, \\
3x_1 - x_2 + 2x_3 &= 0, \\
x_1 - x_2 - x_3 &= 0, \\
5x_1 + 2x_2 - 2x_3 &= 0.
\end{aligned}
$$

35.
$$
\begin{aligned}
2x_1 - x_2 - x_3 &= 0, \\
5x_1 - x_2 + 2x_3 &= 0, \\
x_1 + x_2 + 4x_3 &= 0.
\end{aligned}
$$

36.
$$
\begin{aligned}
(1 + 2i)x_1 + (1 - i)x_2 + x_3 &= 0, \\
ix_1 + (1 + i)x_2 - ix_3 &= 0, \\
2ix_1 + x_2 + (1 + 3i)x_3 &= 0.
\end{aligned}
$$

37.
$$
\begin{aligned}
3x_1 + 2x_2 + x_3 &= 0, \\
6x_1 - x_2 + 2x_3 &= 0, \\
12x_1 + 6x_2 + 4x_3 &= 0.
\end{aligned}
$$

38.
$$
\begin{aligned}
2x_1 + x_2 - 8x_3 &= 0, \\
3x_1 - 2x_2 - 5x_3 &= 0, \\
5x_1 - 6x_2 - 3x_3 &= 0, \\
3x_1 - 5x_2 + x_3 &= 0.
\end{aligned}
$$

39.
$$
\begin{aligned}
x_1 + (1 + i)x_2 + (1 - i)x_3 &= 0, \\
ix_1 + x_2 + ix_3 &= 0, \\
(1 - 2i)x_1 - (1 - i)x_2 + (1 - 3i)x_3 &= 0.
\end{aligned}
$$

40.
$$
\begin{aligned}
x_1 - x_2 + x_3 &= 0, \\
3x_2 + 2x_3 &= 0, \\
3x_1 - x_3 &= 0, \\
5x_1 + x_2 - x_3 &= 0.
\end{aligned}
$$

41.
$$
\begin{aligned}
2x_1 - 4x_2 + 6x_3 &= 0, \\
3x_1 - 6x_2 + 9x_3 &= 0, \\
x_1 - 2x_2 + 3x_3 &= 0, \\
5x_1 - 10x_2 + 15x_3 &= 0.
\end{aligned}
$$

42.
$$
\begin{aligned}
4x_1 - 2x_2 - x_3 - x_4 &= 0, \\
3x_1 + x_2 - 2x_3 + 3x_4 &= 0, \\
5x_1 - x_2 - 2x_3 + x_4 &= 0.
\end{aligned}
$$

43.
$$
\begin{aligned}
2x_1 + x_2 - x_3 + x_4 &= 0, \\
x_1 + x_2 + x_3 - x_4 &= 0, \\
3x_1 - x_2 + x_3 - 2x_4 &= 0, \\
4x_1 + 2x_2 - x_3 + x_4 &= 0.
\end{aligned}
$$

For Problems 44–54, determine the solution set to the system $A\mathbf{x} = \mathbf{0}$ for the given matrix A.

44. $A = \begin{bmatrix} 2 & -1 \\ 3 & 4 \end{bmatrix}$.

45. $A = \begin{bmatrix} 1 - i & 2i \\ 1 + i & -2 \end{bmatrix}$.

46. $A = \begin{bmatrix} 1 + i & 1 - 2i \\ -1 + i & 2 + i \end{bmatrix}$.

47. $A = \begin{bmatrix} 1 & 2 & 3 \\ 2 & -1 & 0 \\ 1 & 1 & 1 \end{bmatrix}$.

48. $A = \begin{bmatrix} 1 & 1 & 1 & -1 \\ -1 & 0 & -1 & 2 \\ 1 & 3 & 2 & 2 \end{bmatrix}$.

49. $A = \begin{bmatrix} 2 - 3i & 1 + i & i - 1 \\ 3 + 2i & -1 + i & -1 - i \\ 5 - i & 2i & -2 \end{bmatrix}$.

50. $A = \begin{bmatrix} 1 & 3 & 0 \\ -2 & -3 & 0 \\ 1 & 4 & 0 \end{bmatrix}$.

51. $A = \begin{bmatrix} 1 & 0 & 3 \\ 3 & -1 & 7 \\ 2 & 1 & 8 \\ 1 & 1 & 5 \\ -1 & 1 & -1 \end{bmatrix}$.

52. $A = \begin{bmatrix} 1 & -1 & 0 & 1 \\ 3 & -2 & 0 & 5 \\ -1 & 2 & 0 & 1 \end{bmatrix}$.

53. $A = \begin{bmatrix} 1 & 0 & -3 & 0 \\ 3 & 0 & -9 & 0 \\ -2 & 0 & 6 & 0 \end{bmatrix}$.

54. $A = \begin{bmatrix} 2+i & i & 3-2i \\ i & 1-i & 4+3i \\ 3-i & 1+i & 1+5i \end{bmatrix}$.

2.6   The Inverse of a Square Matrix

In this section we investigate the situation when, for a given $n \times n$ matrix A, there exists a matrix B satisfying

$$AB = I_n \qquad \text{and} \qquad BA = I_n \qquad\qquad (2.6.1)$$

and derive an efficient method for determining B (when it does exist). As a possible application of the existence of such a matrix B, consider the $n \times n$ linear system

$$A\mathbf{x} = \mathbf{b}. \qquad\qquad (2.6.2)$$

Premultiplying both sides of (2.6.2) by an $n \times n$ matrix B yields

$$(BA)\mathbf{x} = B\mathbf{b}.$$

Assuming that $BA = I_n$, this reduces to

$$\mathbf{x} = B\mathbf{b}. \qquad\qquad (2.6.3)$$

Thus, we have determined a solution to the system (2.6.2) by a matrix multiplication. Of course, this depends on the existence of a matrix B satisfying (2.6.1), and even if such a matrix B does exist, it will turn out that using (2.6.3) to solve $n \times n$ systems is not very efficient computationally. Therefore it is generally not used in practice to solve $n \times n$ systems. However, from a theoretical point of view, a formula such as (2.6.3) is very useful. We begin the investigation by establishing that there can be at most one matrix B satisfying (2.6.1) for a given $n \times n$ matrix A.

Theorem 2.6.1   Let A be an $n \times n$ matrix. Suppose B and C are both $n \times n$ matrices satisfying

$$AB = BA = I_n, \qquad\qquad (2.6.4)$$
$$AC = CA = I_n, \qquad\qquad (2.6.5)$$

respectively. Then $B = C$.

Proof   From (2.6.4), it follows that

$$C = CI_n = C(AB).$$

That is,

$$C = (CA)B = I_n B = B,$$

where we have used (2.6.5) to replace CA by I_n in the second step.   ∎

Since the identity matrix I_n plays the role of the number 1 in the multiplication of matrices, the properties given in (2.6.1) are the analogs for matrices of the properties

$$xx^{-1} = 1, \quad x^{-1}x = 1,$$

which holds for all (nonzero) numbers x. It is therefore natural to denote the matrix B in (2.6.1) by A^{-1} and to call it the inverse of A. The following definition introduces the appropriate terminology.

> ### DEFINITION   2.6.2
>
> Let A be an $n \times n$ matrix. If there exists an $n \times n$ matrix A^{-1} satisfying
>
> $$AA^{-1} = A^{-1}A = I_n,$$
>
> then we call A^{-1} *the* matrix **inverse** to A, or just *the* inverse of A. We say that A is **invertible** if A^{-1} exists.

Invertible matrices are sometimes called **nonsingular**, while matrices that are not invertible are sometimes called **singular**.

Remark   It is important to realize that A^{-1} denotes the matrix that satisfies

$$AA^{-1} = A^{-1}A = I_n.$$

It does *not* mean $1/A$, which has no meaning whatsoever.

Example 2.6.3   If $A = \begin{bmatrix} 1 & -1 & 2 \\ 2 & -3 & 3 \\ 1 & -1 & 1 \end{bmatrix}$, verify that $B = \begin{bmatrix} 0 & -1 & 3 \\ 1 & -1 & 1 \\ 1 & 0 & -1 \end{bmatrix}$ is the inverse of A.

Solution: By direct multiplication, we find that

$$AB = \begin{bmatrix} 1 & -1 & 2 \\ 2 & -3 & 3 \\ 1 & -1 & 1 \end{bmatrix} \begin{bmatrix} 0 & -1 & 3 \\ 1 & -1 & 1 \\ 1 & 0 & -1 \end{bmatrix} = \begin{bmatrix} 1 & 0 & 0 \\ 0 & 1 & 0 \\ 0 & 0 & 1 \end{bmatrix} = I_3$$

and

$$BA = \begin{bmatrix} 0 & -1 & 3 \\ 1 & -1 & 1 \\ 1 & 0 & -1 \end{bmatrix} \begin{bmatrix} 1 & -1 & 2 \\ 2 & -3 & 3 \\ 1 & -1 & 1 \end{bmatrix} = \begin{bmatrix} 1 & 0 & 0 \\ 0 & 1 & 0 \\ 0 & 0 & 1 \end{bmatrix} = I_3.$$

Consequently, (2.6.1) is satisfied, hence B is indeed the inverse of A. We therefore write

$$A^{-1} = \begin{bmatrix} 0 & -1 & 3 \\ 1 & -1 & 1 \\ 1 & 0 & -1 \end{bmatrix}.$$ $\square$

We now return to the $n \times n$ system of Equations (2.6.2).

Theorem 2.6.4   If A^{-1} exists, then the $n \times n$ system of linear equations

$$Ax = b$$

has the *unique* solution

$$x = A^{-1}b$$

for every b in $\mathbb{R}^n$.

Proof We can verify by direct substitution that $\mathbf{x} = A^{-1}\mathbf{b}$ is indeed a solution to the linear system. The uniqueness of this solution is contained in the calculation leading from (2.6.2) to (2.6.3).   ∎

Our next theorem establishes when A^{-1} exists, and it also uncovers an efficient method for computing A^{-1}.

Theorem 2.6.5

An $n \times n$ matrix A is invertible if and only if $\text{rank}(A) = n$.

Proof If A^{-1} exists, then by Theorem 2.6.4, any $n \times n$ linear system $A\mathbf{x} = \mathbf{b}$ has a unique solution. Hence, Theorem 2.5.9 implies that $\text{rank}(A) = n$.

Conversely, suppose $\text{rank}(A) = n$. We must establish that there exists an $n \times n$ matrix X satisfying

$$AX = I_n = XA.$$

Let $\mathbf{e}_1, \mathbf{e}_2, \ldots, \mathbf{e}_n$ denote the column vectors of the identity matrix I_n. Since $\text{rank}(A) = n$, Theorem 2.5.9 implies that each of the linear systems

$$A\mathbf{x}_i = \mathbf{e}_i, \qquad i = 1, 2, \ldots, n \tag{2.6.6}$$

has a unique solution[7] $\mathbf{x}_i$. Consequently, if we let $X = [\mathbf{x}_1, \mathbf{x}_2, \ldots, \mathbf{x}_n]$, where $\mathbf{x}_1, \mathbf{x}_2, \ldots, \mathbf{x}_n$ are the unique solutions of the systems in (2.6.6), then

$$A[\mathbf{x}_1, \mathbf{x}_2, \ldots, \mathbf{x}_n] = [A\mathbf{x}_1, A\mathbf{x}_2, \ldots, A\mathbf{x}_n] = [\mathbf{e}_1, \mathbf{e}_2, \ldots, \mathbf{e}_n];$$

that is,

$$AX = I_n. \tag{2.6.7}$$

We must also show that, for the same matrix X,

$$XA = I_n.$$

Postmultiplying both sides of (2.6.7) by A yields

$$(AX)A = A.$$

That is,

$$A(XA - I_n) = 0_n. \tag{2.6.8}$$

Now let $\mathbf{y}_1, \mathbf{y}_2, \ldots, \mathbf{y}_n$ denote the column vectors of the $n \times n$ matrix $XA - I_n$. Equating corresponding column vectors on either side of (2.6.8) implies that

$$A\mathbf{y}_i = \mathbf{0}, \qquad i = 1, 2, \ldots, n. \tag{2.6.9}$$

But, by assumption, $\text{rank}(A) = n$, and so each system in (2.6.9) has a unique solution that, since the systems are homogeneous, must be the trivial solution. Consequently, each $\mathbf{y}_i$ is the zero vector, and thus

$$XA - I_n = 0_n.$$

Therefore,

$$XA = I_n. \tag{2.6.10}$$

---

[7]Notice that for an $n \times n$ system $A\mathbf{x} = \mathbf{b}$, if $\text{rank}(A) = n$, then $\text{rank}(A^{\#}) = n$.

Equations (2.6.7) and (2.6.10) imply that $X = A^{-1}$. ∎

We now have the following converse to Theorem 2.6.4.

Corollary 2.6.6 Let A be an $n \times n$ matrix. If $A\mathbf{x} = \mathbf{b}$ has a unique solution for some column n-vector $\mathbf{b}$, then A^{-1} exists.

Proof If $A\mathbf{x} = \mathbf{b}$ has a unique solution, then from Theorem 2.5.9, rank$(A) = n$, and so from the previous theorem, A^{-1} exists. ∎

Remark In particular, the above corollary tells us that if the homogeneous linear system $A\mathbf{x} = \mathbf{0}$ has only the trivial solution $\mathbf{x} = \mathbf{0}$, then A^{-1} exists.

Other criteria for deciding whether or not an $n \times n$ matrix A has an inverse will be developed in the next three chapters, but our goal at present is to develop a method for finding A^{-1}, should it exist.

Assuming that rank$(A) = n$, let $\mathbf{x}_1, \mathbf{x}_2, \ldots, \mathbf{x}_n$ denote the column vectors of A^{-1}. Then, from (2.6.6), these column vectors can be obtained by solving each of the $n \times n$ systems

$$A\mathbf{x}_i = \mathbf{e}_i, \qquad i = 1, 2, \ldots, n.$$

As we now show, some computation can be saved if we employ the Gauss-Jordan method in solving these systems. We first illustrate the method when $n = 3$. In this case, from (2.6.6), the column vectors of A^{-1} are determined by solving the three linear systems

$$A\mathbf{x}_1 = \mathbf{e}_1, \qquad A\mathbf{x}_2 = \mathbf{e}_2, \qquad A\mathbf{x}_3 = \mathbf{e}_3.$$

The augmented matrices of these systems can be written as

$$\left[A \begin{array}{c} 1 \\ 0 \\ 0 \end{array} \right], \left[A \begin{array}{c} 0 \\ 1 \\ 0 \end{array} \right], \left[A \begin{array}{c} 0 \\ 0 \\ 1 \end{array} \right],$$

respectively. Furthermore, since rank$(A) = 3$ by assumption, the reduced row-echelon form of A is I_3. Consequently, using elementary row operations to reduce the augmented matrix of the first system to reduced row-echelon form will yield, schematically,

$$\left[A \begin{array}{c} 1 \\ 0 \\ 0 \end{array} \right] \sim \underset{\cdots}{\text{ERO}} \sim \left[\begin{array}{ccc|c} 1 & 0 & 0 & a_1 \\ 0 & 1 & 0 & a_2 \\ 0 & 0 & 1 & a_3 \end{array} \right],$$

which implies that the first column vector of A^{-1} is

$$\mathbf{x}_1 = \left[\begin{array}{c} a_1 \\ a_2 \\ a_3 \end{array} \right].$$

Similarly, for the second system, the reduction

$$\left[A \begin{array}{c} 0 \\ 1 \\ 0 \end{array} \right] \sim \underset{\cdots}{\text{ERO}} \sim \left[\begin{array}{ccc|c} 1 & 0 & 0 & b_1 \\ 0 & 1 & 0 & b_2 \\ 0 & 0 & 1 & b_3 \end{array} \right]$$

implies that the second column vector of A^{-1} is

$$\mathbf{x}_2 = \begin{bmatrix} b_1 \\ b_2 \\ b_3 \end{bmatrix}.$$

Finally, for the third system, the reduction

$$\begin{bmatrix} \mathbf{A} & \begin{matrix} 0 \\ 0 \\ 1 \end{matrix} \end{bmatrix} \sim \overset{\text{ERO}}{\cdots} \sim \begin{bmatrix} 1 & 0 & 0 & c_1 \\ 0 & 1 & 0 & c_2 \\ 0 & 0 & 1 & c_3 \end{bmatrix}$$

implies that the third column vector of A^{-1} is

$$\mathbf{x}_3 = \begin{bmatrix} c_1 \\ c_2 \\ c_3 \end{bmatrix}.$$

Consequently,

$$A^{-1} = [\mathbf{x}_1, \mathbf{x}_2, \mathbf{x}_3] = \begin{bmatrix} a_1 & b_1 & c_1 \\ a_2 & b_2 & c_2 \\ a_3 & b_3 & c_3 \end{bmatrix}.$$

The key point to notice is that in solving for $\mathbf{x}_1, \mathbf{x}_2, \mathbf{x}_3$ we use the *same* elementary row operations to reduce A to I_3. We can therefore save a significant amount of work by combining the foregoing operations as follows:

$$\begin{bmatrix} \mathbf{A} & \begin{matrix} 1 & 0 & 0 \\ 0 & 1 & 0 \\ 0 & 0 & 1 \end{matrix} \end{bmatrix} \sim \overset{\text{ERO}}{\cdots} \sim \begin{bmatrix} 1 & 0 & 0 & a_1 & b_1 & c_1 \\ 0 & 1 & 0 & a_2 & b_2 & c_2 \\ 0 & 0 & 1 & a_3 & b_3 & c_3 \end{bmatrix}.$$

The generalization to the $n \times n$ case is immediate. We form the $n \times 2n$ matrix $[A \quad I_n]$ and reduce A to I_n using elementary row operations. Schematically,

$$[A \quad I_n] \sim \overset{\text{ERO}}{\cdots} \sim [I_n \quad A^{-1}].$$

This method of finding A^{-1} is called the **Gauss-Jordan technique.**

Remark   Notice that if we are given an $n \times n$ matrix A, we likely will not know from the outset whether $\text{rank}(A) = n$, hence we will not know whether A^{-1} exists. However, if at any stage in the row reduction of $[A \quad I_n]$ we find that $\text{rank}(A) < n$, then it will follow from Theorem 2.6.5 that A is not invertible.

Example 2.6.7   Find A^{-1} if $A = \begin{bmatrix} 1 & 1 & 3 \\ 0 & 1 & 2 \\ 3 & 5 & -1 \end{bmatrix}.$

Solution: Using the Gauss-Jordan technique, we proceed as follows.

$$\begin{bmatrix} 1 & 1 & 3 & 1 & 0 & 0 \\ 0 & 1 & 2 & 0 & 1 & 0 \\ 3 & 5 & -1 & 0 & 0 & 1 \end{bmatrix} \overset{1}{\sim} \begin{bmatrix} 1 & 1 & 3 & 1 & 0 & 0 \\ 0 & 1 & 2 & 0 & 1 & 0 \\ 0 & 2 & -10 & -3 & 0 & 1 \end{bmatrix} \overset{2}{\sim} \begin{bmatrix} 1 & 0 & 1 & 1 & -1 & 0 \\ 0 & 1 & 2 & 0 & 1 & 0 \\ 0 & 0 & -14 & -3 & -2 & 1 \end{bmatrix}$$

$$\overset{3}{\sim} \begin{bmatrix} 1 & 0 & 1 & 1 & -1 & 0 \\ 0 & 1 & 2 & 0 & 1 & 0 \\ 0 & 0 & 1 & \frac{3}{14} & \frac{1}{7} & -\frac{1}{14} \end{bmatrix} \overset{4}{\sim} \begin{bmatrix} 1 & 0 & 0 & \frac{11}{14} & -\frac{8}{7} & \frac{1}{14} \\ 0 & 1 & 0 & -\frac{3}{7} & \frac{5}{7} & \frac{1}{7} \\ 0 & 0 & 1 & \frac{3}{14} & \frac{1}{7} & -\frac{1}{14} \end{bmatrix}.$$

Thus,

$$A^{-1} = \begin{bmatrix} \frac{11}{14} & -\frac{8}{7} & \frac{1}{14} \\[2mm] -\frac{3}{7} & \frac{5}{7} & \frac{1}{7} \\[2mm] \frac{3}{14} & \frac{1}{7} & -\frac{1}{14} \end{bmatrix}.$$

We leave it as an exercise to confirm that $AA^{-1} = A^{-1}A = I_3$.

| **1.** $A_{13}(-3)$ | **2.** $A_{21}(-1)$, $A_{23}(-2)$ | **3.** $M_3(-1/14)$ | **4.** $A_{31}(-1)$, $A_{32}(-2)$ |

□

Example 2.6.8   Continuing the previous example, use A^{-1} to solve the system

$$\begin{aligned} x_1 + \ x_2 + 3x_3 &= 2, \\ x_2 + 2x_3 &= 1, \\ 3x_1 + 5x_2 - \ x_3 &= 4. \end{aligned}$$

Solution: The system can be written as

$$A\mathbf{x} = \mathbf{b},$$

where A is the matrix in the previous example, and

$$\mathbf{b} = \begin{bmatrix} 2 \\ 1 \\ 4 \end{bmatrix}.$$

Since A is invertible, the system has a unique solution that can be written as $\mathbf{x} = A^{-1}\mathbf{b}$. Thus, from the previous example we have

$$\mathbf{x} = \begin{bmatrix} \frac{11}{14} & -\frac{8}{7} & \frac{1}{14} \\[2mm] -\frac{3}{7} & \frac{5}{7} & \frac{1}{7} \\[2mm] \frac{3}{14} & \frac{1}{7} & -\frac{1}{14} \end{bmatrix} \begin{bmatrix} 2 \\ 1 \\ 4 \end{bmatrix} = \begin{bmatrix} \frac{5}{7} \\[2mm] \frac{3}{7} \\[2mm] \frac{2}{7} \end{bmatrix}.$$

Consequently, $x_1 = \frac{5}{7}$, $x_2 = \frac{3}{7}$, and $x_3 = \frac{2}{7}$, so that the solution to the system is $\left(\frac{5}{7}, \frac{3}{7}, \frac{2}{7}\right)$.

□

We now return to more theoretical information pertaining to the inverse of a matrix.

Properties of the Inverse

The inverse of an $n \times n$ matrix satisfies the properties stated in the following theorem, which should be committed to memory:

Theorem 2.6.9   Let A and B be invertible $n \times n$ matrices. Then

1. A^{-1} is invertible and $(A^{-1})^{-1} = A$.
2. AB is invertible and $(AB)^{-1} = B^{-1}A^{-1}$.
3. A^T is invertible and $(A^T)^{-1} = (A^{-1})^T$.

Proof The proof of each result consists of verifying that the appropriate matrix products yield the identity matrix.

1. We must verify that

$$A^{-1}A = I_n \qquad \text{and} \qquad AA^{-1} = I_n.$$

Both of these follow directly from Definition 2.6.2.

2. We must verify that

$$(AB)(B^{-1}A^{-1}) = I_n \qquad \text{and} \qquad (B^{-1}A^{-1})(AB) = I_n.$$

We establish the first equality, leaving the second equation as an exercise. We have

$$(AB)(B^{-1})(A^{-1}) = A(BB^{-1})A^{-1} = AI_nA^{-1} = AA^{-1} = I_n.$$

3. We must verify that

$$A^T(A^{-1})^T = I_n \qquad \text{and} \qquad (A^{-1})^T A^T = I_n.$$

Again, we prove the first part, leaving the second part as an exercise.
First recall from Theorem 2.2.21 that $A^T B^T = (BA)^T$. Using this property with $B = A^{-1}$ yields

$$A^T(A^{-1})^T = (A^{-1}A)^T = I_n^T = I_n. \qquad \blacksquare$$

The proof of property 2 of Theorem 2.6.9 can easily be extended to a statement about invertibility of a product of an arbitrary finite number of matrices. More precisely, we have the following.

Corollary 2.6.10   Let $A_1, A_2, \ldots, A_k$ be invertible $n \times n$ matrices. Then $A_1 A_2 \cdots A_k$ is invertible, and

$$(A_1 A_2 \cdots A_k)^{-1} = A_k^{-1} A_{k-1}^{-1} \cdots A_1^{-1}.$$

Proof The proof is left as an exercise (Problem 28). $\blacksquare$

Some Further Theoretical Results

Finally, in this section, we establish two results that will be required in Section 2.7 and also in a proof that arises in Section 3.2.

Theorem 2.6.11   Let A and B be $n \times n$ matrices. If $AB = I_n$, then both A and B are invertible and $B = A^{-1}$.

Proof Let $\mathbf{b}$ be an arbitrary column n-vector. Then, since $AB = I_n$, we have

$$A(B\mathbf{b}) = I_n\mathbf{b} = \mathbf{b}.$$

Consequently, for *every* $\mathbf{b}$, the system $A\mathbf{x} = \mathbf{b}$ has the solution $\mathbf{x} = B\mathbf{b}$. But this implies that $\text{rank}(A) = n$. To see why, suppose that $\text{rank}(A) < n$, and let A^* denote a row-echelon form of A. Note that the last row of A^* is zero. Choose $\mathbf{b}^*$ to be any column

n-vector whose last component is nonzero. Then, since $\text{rank}(A) < n$, it follows that the system

$$A^*\mathbf{x} = \mathbf{b}^*$$

is inconsistent. But, applying to the augmented matrix $[A^* \quad \mathbf{b}^*]$ the inverse row operations that reduced A to row-echelon form yields $[A \quad \mathbf{b}]$ for some $\mathbf{b}$. Since $A\mathbf{x} = \mathbf{b}$ has the same solution set as $A^*\mathbf{x} = \mathbf{b}^*$, it follows that $A\mathbf{x} = \mathbf{b}$ is inconsistent. We therefore have a contradiction, and so it must be the case that $\text{rank}(A) = n$, and therefore that A is invertible by Theorem 2.6.5.

We now establish that[8] $A^{-1} = B$. Since $AB = I_n$ by assumption, we have

$$A^{-1} = A^{-1}I_n = A^{-1}(AB) = (A^{-1}A)B = I_nB = B,$$

as required. It now follows directly from property 1 of Theorem 2.6.9 that B is invertible with inverse A. ∎

Corollary 2.6.12   Let A and B be $n \times n$ matrices. If AB is invertible, then both A and B are invertible.

Proof If we let $C = B(AB)^{-1}$ and $D = AB$, then

$$AC = AB(AB)^{-1} = DD^{-1} = I_n.$$

It follows from Theorem 2.6.11 that A is invertible. Similarly, if we let $C = (AB)^{-1}A$, then

$$CB = (AB)^{-1}AB = I_n.$$

Once more we can apply Theorem 2.6.11 to conclude that B is invertible. ∎

Exercises for 2.6

Key Terms

Inverse, Invertible, Singular, Nonsingular, Gauss-Jordan technique.

Skills

- Be able to check directly whether or not two matrices A and B are inverses of each other.

- Be able to find the inverse of an invertible matrix via the Gauss-Jordan technique.

- Be able to use the inverse of a coefficient matrix of a linear system in order to solve the system.

- Know the basic properties related to how the inverse operation behaves with respect to itself, multiplication, and transpose (Theorem 2.6.9).

True-False Review

For Questions 1–10, decide if the given statement is **true** or **false**, and give a brief justification for your answer. If true, you can quote a relevant definition or theorem from the text. If false, provide an example, illustration, or brief explanation of why the statement is false.

1. An invertible matrix is also known as a singular matrix.

---

[8]Note that it now makes sense to speak of A^{-1}, whereas prior to proving in the preceding paragraph that A is invertible, it would not have been legal to use the notation A^{-1}.

2. Every square matrix that does not contain a row of zeros is invertible.

3. A linear system $Ax = b$ with an $n \times n$ invertible coefficient matrix A has a unique solution.

4. If A is a matrix such that there exists a matrix B with $AB = I_n$, then A is invertible.

5. If A and B are invertible $n \times n$ matrices, then so is $A + B$.

6. If A and B are invertible $n \times n$ matrices, then so is AB.

7. If A is an invertible matrix such that $A^2 = A$, then A is the identity matrix.

8. If A is an $n \times n$ invertible matrix and B and C are $n \times n$ matrices such that $AB = AC$, then $B = C$.

9. If A is a 5×5 matrix of rank 4, then A is not invertible.

10. If A is a 6×6 matrix of rank 6, then A is invertible.

Problems

For Problems 1–3 verify by direct multiplication that the given matrices are inverses of one another.

1. $A = \begin{bmatrix} 2 & -1 \\ 3 & -1 \end{bmatrix}$, $A^{-1} = \begin{bmatrix} -1 & 1 \\ -3 & 2 \end{bmatrix}$.

2. $A = \begin{bmatrix} 4 & 9 \\ 3 & 7 \end{bmatrix}$, $A^{-1} = \begin{bmatrix} 7 & -9 \\ -3 & 4 \end{bmatrix}$.

3. $A = \begin{bmatrix} 3 & 5 & 1 \\ 1 & 2 & 1 \\ 2 & 6 & 7 \end{bmatrix}$, $A^{-1} = \begin{bmatrix} 8 & -29 & 3 \\ -5 & 19 & -2 \\ 2 & -8 & 1 \end{bmatrix}$.

For Problems 4–16, determine A^{-1}, if possible, using the Gauss-Jordan method. If A^{-1} exists, check your answer by verifying that $AA^{-1} = I_n$.

4. $A = \begin{bmatrix} 1 & 2 \\ 1 & 3 \end{bmatrix}$.

5. $A = \begin{bmatrix} 1 & 1+i \\ 1-i & 1 \end{bmatrix}$.

6. $A = \begin{bmatrix} 1 & -i \\ -1+i & 2 \end{bmatrix}$.

7. $A = \begin{bmatrix} 0 & 0 \\ 0 & 0 \end{bmatrix}$.

8. $A = \begin{bmatrix} 1 & -1 & 2 \\ 2 & 1 & 11 \\ 4 & -3 & 10 \end{bmatrix}$.

9. $A = \begin{bmatrix} 3 & 5 & 1 \\ 1 & 2 & 1 \\ 2 & 6 & 7 \end{bmatrix}$.

10. $A = \begin{bmatrix} 0 & 1 & 0 \\ 0 & 0 & 1 \\ 0 & 1 & 2 \end{bmatrix}$.

11. $A = \begin{bmatrix} 4 & 2 & -13 \\ 2 & 1 & -7 \\ 3 & 2 & 4 \end{bmatrix}$.

12. $A = \begin{bmatrix} 1 & 2 & -3 \\ 2 & 6 & -2 \\ -1 & 1 & 4 \end{bmatrix}$.

13. $A = \begin{bmatrix} 1 & i & 2 \\ 1+i & -1 & 2i \\ 2 & 2i & 5 \end{bmatrix}$.

14. $A = \begin{bmatrix} 2 & 1 & 3 \\ 1 & -1 & 2 \\ 3 & 3 & 4 \end{bmatrix}$.

15. $A = \begin{bmatrix} 1 & -1 & 2 & 3 \\ 2 & 0 & 3 & -4 \\ 3 & -1 & 7 & 8 \\ 1 & 0 & 3 & 5 \end{bmatrix}$.

16. $A = \begin{bmatrix} 0 & -2 & -1 & -3 \\ 2 & 0 & 2 & 1 \\ 1 & -2 & 0 & 2 \\ 3 & -1 & -2 & 0 \end{bmatrix}$.

17. Let

$$A = \begin{bmatrix} 2 & -1 & 4 \\ 5 & 1 & 2 \\ 1 & -1 & 3 \end{bmatrix}.$$

Find the second column vector of A^{-1} without determining the whole inverse.

For Problems 18–22, use A^{-1} to find the solution to the given system.

18.
$$\begin{aligned} x_1 + 3x_2 &= 1, \\ 2x_1 + 5x_2 &= 3. \end{aligned}$$

19.
$$\begin{aligned} x_1 + x_2 - 2x_3 &= -2, \\ x_2 + x_3 &= 3, \\ 2x_1 + 4x_2 - 3x_3 &= 1. \end{aligned}$$

20.
$$\begin{aligned} x_1 - 2ix_2 &= 2, \\ (2-i)x_1 + 4ix_2 &= -i. \end{aligned}$$

21.
$$3x_1 + 4x_2 + 5x_3 = 1,$$
$$2x_1 + 10x_2 + x_3 = 1,$$
$$4x_1 + x_2 + 8x_3 = 1.$$

22.
$$x_1 + x_2 + 2x_3 = 12,$$
$$x_1 + 2x_2 - x_3 = 24,$$
$$2x_1 - x_2 + x_3 = -36.$$

An $n \times n$ matrix A is called **orthogonal** if $A^T = A^{-1}$. For Problems 23–26, show that the given matrices are orthogonal.

23. $A = \begin{bmatrix} 0 & 1 \\ -1 & 0 \end{bmatrix}$.

24. $A = \begin{bmatrix} \sqrt{3}/2 & 1/2 \\ -1/2 & \sqrt{3}/2 \end{bmatrix}$.

25. $A = \begin{bmatrix} \cos\alpha & \sin\alpha \\ -\sin\alpha & \cos\alpha \end{bmatrix}$.

26. $A = \dfrac{1}{1+2x^2} \begin{bmatrix} 1 & -2x & 2x^2 \\ 2x & 1-2x^2 & -2x \\ 2x^2 & 2x & 1 \end{bmatrix}$.

27. Complete the proof of Theorem 2.6.9 by verifying the remaining properties in parts 2 and 3.

28. Prove Corollary 2.6.10.

For Problems 29–30, use properties of the inverse to prove the given statement.

29. If A is an $n \times n$ invertible *symmetric* matrix, then A^{-1} is symmetric.

30. If A is an $n \times n$ invertible *skew-symmetric* matrix, then A^{-1} is skew-symmetric.

31. Let A be an $n \times n$ matrix with $A^4 = 0$. Prove that $I_n - A$ is invertible with
$$(I_n - A)^{-1} = I_n + A + A^2 + A^3.$$

32. Prove that if A, B, C are $n \times n$ matrices satisfying $BA = I_n$ and $AC = I_n$, then $B = C$.

33. If A, B, C are $n \times n$ matrices satisfying $BA = I_n$ and $CA = I_n$, does it follow that $B = C$? Justify your answer.

34. Consider the general 2×2 matrix
$$A = \begin{bmatrix} a_{11} & a_{12} \\ a_{21} & a_{22} \end{bmatrix}$$
and let $\Delta = a_{11}a_{22} - a_{12}a_{21}$ with $a_{11} \neq 0$. Show that if $\Delta \neq 0$,
$$A^{-1} = \frac{1}{\Delta} \begin{bmatrix} a_{22} & -a_{12} \\ -a_{21} & a_{11} \end{bmatrix}.$$

The quantity Δ defined above is referred to as the determinant of A. We will investigate determinants in more detail in the next chapter.

35. Let A be an $n \times n$ matrix, and suppose that we have to solve the p linear systems
$$A\mathbf{x}_i = \mathbf{b}_i, \qquad i = 1, 2, \ldots, p$$
where the $\mathbf{b}_i$ are given. Devise an efficient method for solving these systems.

36. Use your method from the previous problem to solve the three linear systems
$$A\mathbf{x}_i = \mathbf{b}_i, \qquad i = 1, 2, 3$$
if
$$A = \begin{bmatrix} 1 & -1 & 1 \\ 2 & -1 & 4 \\ 1 & 1 & 6 \end{bmatrix}, \qquad \mathbf{b}_1 = \begin{bmatrix} 1 \\ 1 \\ -1 \end{bmatrix},$$
$$\mathbf{b}_2 = \begin{bmatrix} -1 \\ 2 \\ 5 \end{bmatrix}, \qquad \mathbf{b}_3 = \begin{bmatrix} 2 \\ 3 \\ 2 \end{bmatrix}.$$

37. Let A be an $m \times n$ matrix with $m \leq n$.

(a) If $\text{rank}(A) = m$, prove that there exists a matrix B satisfying $AB = I_m$. Such a matrix is called a **right inverse** of A.

(b) If
$$A = \begin{bmatrix} 1 & 3 & 1 \\ 2 & 7 & 4 \end{bmatrix},$$
determine all right inverses of A.

⋄ For Problems 38–39, reduce the matrix $[A \quad I_n]$ to reduced row-echelon form and thereby determine, if possible, the inverse of A.

38. $A = \begin{bmatrix} 5 & 9 & 17 \\ 7 & 21 & 13 \\ 27 & 16 & 8 \end{bmatrix}$.

39. A is a randomly generated 4×4 matrix.

⋄ For Problems 40–42, use built-in functions of some form of technology to determine $\text{rank}(A)$ and, if possible, A^{-1}.

40. $A = \begin{bmatrix} 3 & 5 & -7 \\ 2 & 5 & 9 \\ 13 & -11 & 22 \end{bmatrix}$.

41. $A = \begin{bmatrix} 7 & 13 & 15 & 21 \\ 9 & -2 & 14 & 23 \\ 17 & -27 & 22 & 31 \\ 19 & -42 & 21 & 33 \end{bmatrix}$.

42. A is a randomly generated 5×5 matrix.

43. ◇ For the system in Problem 21, determine A^{-1} and use it to solve the system.

44. ◇ Consider the $n \times n$ **Hilbert** matrix

$$H_n = \left[\frac{1}{i + j - 1} \right], \qquad 1 \le i, \quad j \le n.$$

(a) Determine H_4 and show that it is invertible.

(b) Find H_4^{-1} and use it to solve $H_4 \mathbf{x} = \mathbf{b}$ if $\mathbf{b} = [2, -1, 3, 5]^T$.

2.7   Elementary Matrices and the LU Factorization

We now introduce some matrices that can be used to perform elementary row operations on a matrix. Although they are of limited computational use, they do play a significant role in linear algebra and its applications.

DEFINITION   2.7.1

Any matrix obtained by performing a single elementary row operation on the identity matrix is called an **elementary matrix**.

In particular, an elementary matrix is always a square matrix. In general we will denote elementary matrices by E. If we are describing a specific elementary matrix, then in keeping with the notation introduced previously for elementary row operations, we will use the following notation for the three types of elementary matrices:

Type 1: P_{ij}—permute rows i and j in I_n.
Type 2: $M_i(k)$—multiply row i of I_n by the nonzero scalar k.
Type 3: $A_{ij}(k)$—add k times row i of I_n to row j of I_n.

Example 2.7.2   Write all 2×2 elementary matrices.

Solution: From Definition 2.7.1 and using the notation introduced above, we have

1. Permutation matrix: $P_{12} = \begin{bmatrix} 0 & 1 \\ 1 & 0 \end{bmatrix}$.

2. Scaling matrices: $M_1(k) = \begin{bmatrix} k & 0 \\ 0 & 1 \end{bmatrix}$, $M_2(k) = \begin{bmatrix} 1 & 0 \\ 0 & k \end{bmatrix}$.

3. Row combinations: $A_{12}(k) = \begin{bmatrix} 1 & 0 \\ k & 1 \end{bmatrix}$, $A_{21}(k) = \begin{bmatrix} 1 & k \\ 0 & 1 \end{bmatrix}$.

□

We leave it as an exercise to verify that the $n \times n$ elementary matrices have the following structure:

P_{ij}: ones along main diagonal except (i, i) and (j, j), ones in the (i, j) and (j, i) positions, and zeros elsewhere.

$M_i(k)$: the diagonal matrix $\text{diag}(1, 1, \ldots, k, \ldots, 1)$, where k appears in the (i, i) position.

$A_{ij}(k)$: ones along the main diagonal, k in the (j, i) position, and zeros elsewhere.

A key point to note about elementary matrices is the following:

> Premultiplying an $n \times p$ matrix A by an $n \times n$ elementary matrix E has the effect of performing the corresponding elementary row operation on A.

Rather than proving this statement, which we leave as an exercise, we illustrate with an example.

Example 2.7.3   If $A = \begin{bmatrix} 3 & -1 & 4 \\ 2 & 7 & 5 \end{bmatrix}$, then, for example,

$$M_1(k)A = \begin{bmatrix} k & 0 \\ 0 & 1 \end{bmatrix}\begin{bmatrix} 3 & -1 & 4 \\ 2 & 7 & 5 \end{bmatrix} = \begin{bmatrix} 3k & -k & 4k \\ 2 & 7 & 5 \end{bmatrix}.$$

Similarly,

$$A_{21}(k)A = \begin{bmatrix} 1 & k \\ 0 & 1 \end{bmatrix}\begin{bmatrix} 3 & -1 & 4 \\ 2 & 7 & 5 \end{bmatrix} = \begin{bmatrix} 3+2k & -1+7k & 4+5k \\ 2 & 7 & 5 \end{bmatrix}. \qquad \square$$

Since elementary row operations can be performed on a matrix by premultiplication by an appropriate elementary matrix, it follows that any matrix A can be reduced to row-echelon form by multiplication by a sequence of elementary matrices. Schematically we can therefore write

$$E_k E_{k-1} \cdots E_2 E_1 A = U,$$

where U denotes a row-echelon form of A and the E_i are elementary matrices.

Example 2.7.4   Determine elementary matrices that reduce $A = \begin{bmatrix} 2 & 3 \\ 1 & 4 \end{bmatrix}$ to row-echelon form.

Solution: We can reduce A to row-echelon form using the following sequence of elementary row operations:

$$\begin{bmatrix} 2 & 3 \\ 1 & 4 \end{bmatrix} \overset{1}{\sim} \begin{bmatrix} 1 & 4 \\ 2 & 3 \end{bmatrix} \overset{2}{\sim} \begin{bmatrix} 1 & 4 \\ 0 & -5 \end{bmatrix} \overset{3}{\sim} \begin{bmatrix} 1 & 4 \\ 0 & 1 \end{bmatrix}.$$

> **1.** P_{12} **2.** $A_{12}(-2)$ **3.** $M_2(-\tfrac{1}{5})$

Consequently,

$$M_2(-\tfrac{1}{5})A_{12}(-2)P_{12}A = \begin{bmatrix} 1 & 4 \\ 0 & 1 \end{bmatrix},$$

which we can verify by direct multiplication:

$$M_2(-\tfrac{1}{5})A_{12}(-2)P_{12}A = \begin{bmatrix} 1 & 0 \\ 0 & -\tfrac{1}{5} \end{bmatrix}\begin{bmatrix} 1 & 0 \\ -2 & 1 \end{bmatrix}\begin{bmatrix} 0 & 1 \\ 1 & 0 \end{bmatrix}\begin{bmatrix} 2 & 3 \\ 1 & 4 \end{bmatrix}$$

$$= \begin{bmatrix} 1 & 0 \\ 0 & -\tfrac{1}{5} \end{bmatrix}\begin{bmatrix} 1 & 0 \\ -2 & 1 \end{bmatrix}\begin{bmatrix} 1 & 4 \\ 2 & 3 \end{bmatrix}$$

$$= \begin{bmatrix} 1 & 0 \\ 0 & -\tfrac{1}{5} \end{bmatrix}\begin{bmatrix} 1 & 4 \\ 0 & -5 \end{bmatrix} = \begin{bmatrix} 1 & 4 \\ 0 & 1 \end{bmatrix}. \qquad \square$$

Since any elementary row operation is reversible, it follows that each elementary matrix is invertible. Indeed, in the 2×2 case it is easy to see that

$$P_{12}^{-1} = \begin{bmatrix} 0 & 1 \\ 1 & 0 \end{bmatrix}, \qquad M_1(k)^{-1} = \begin{bmatrix} 1/k & 0 \\ 0 & 1 \end{bmatrix}, \qquad M_2(k)^{-1} = \begin{bmatrix} 1 & 0 \\ 0 & 1/k \end{bmatrix},$$

$$A_{12}(k)^{-1} = \begin{bmatrix} 1 & 0 \\ -k & 1 \end{bmatrix}, \qquad A_{21}(k)^{-1} = \begin{bmatrix} 1 & -k \\ 0 & 1 \end{bmatrix}.$$

We leave it as an exercise to verify that in the $n \times n$ case, we have:

$$\boxed{M_i(k)^{-1} = M_i(1/k), \qquad P_{ij}^{-1} = P_{ij}, \qquad A_{ij}(k)^{-1} = A_{ij}(-k)}$$

Now consider an *invertible* $n \times n$ matrix A. Since the unique reduced row-echelon form of such a matrix is the identity matrix I_n, it follows from the preceding discussion that there exist elementary matrices $E_1, E_2, \ldots, E_k$ such that

$$E_k E_{k-1} \cdots E_2 E_1 A = I_n.$$

But this implies that

$$A^{-1} = E_k E_{k-1} \cdots E_2 E_1,$$

and hence,

$$A = (A^{-1})^{-1} = (E_k \cdots E_2 E_1)^{-1} = E_1^{-1} E_2^{-1} \cdots E_k^{-1},$$

which is a product of elementary matrices. So any invertible matrix is a product of elementary matrices. Conversely, since elementary matrices are invertible, a product of elementary matrices is a product of invertible matrices, hence is invertible by Corollary 2.6.10. Therefore, we have established the following.

Theorem 2.7.5   Let A be an $n \times n$ matrix. Then A is invertible if and only if A is a product of elementary matrices.

The LU Decomposition of an Invertible Matrix [9]

For the remainder of this section, we restrict our attention to invertible $n \times n$ matrices. In reducing such a matrix to row-echelon form, we have always placed leading ones on the main diagonal in order that we obtain a row-echelon matrix. We now lift the requirement that the main diagonal of the row-echelon form contain ones. As a consequence, the matrix that results from row reduction will be an upper triangular matrix but will not necessarily be in row-echelon form. Furthermore, reduction to such an upper triangular form can be accomplished without the use of Type 2 row operations.

[9] The material in the remainder of this section is not used elsewhere in the text.

Example 2.7.6   Use elementary row operations to reduce the matrix

$$A = \begin{bmatrix} 2 & 5 & 3 \\ 3 & 1 & -2 \\ -1 & 2 & 1 \end{bmatrix}$$

to upper triangular form.

Solution: The given matrix can be reduced to upper triangular form using the following sequence of elementary row operations:

$$\begin{bmatrix} 2 & 5 & 3 \\ 3 & 1 & -2 \\ -1 & 2 & 1 \end{bmatrix} \overset{1}{\sim} \begin{bmatrix} 2 & 5 & 2 \\ 0 & -\frac{13}{2} & -\frac{13}{2} \\ 0 & \frac{9}{2} & \frac{5}{2} \end{bmatrix} \overset{2}{\sim} \begin{bmatrix} 2 & 5 & 3 \\ 0 & -\frac{13}{2} & -\frac{13}{2} \\ 0 & 0 & -2 \end{bmatrix}.$$

1. $A_{12}(-\frac{3}{2})$, $A_{13}(\frac{1}{2})$ **2.** $A_{23}(\frac{9}{13})$ □

When using elementary row operations of Type 3, the multiple of a specific row that is *subtracted* from row i to put a zero in the (i, j) position is called a **multiplier** and denoted m_{ij}. Thus, in the preceding example, there are three multipliers—namely,

$$m_{21} = \tfrac{3}{2}, \qquad m_{31} = -\tfrac{1}{2}, \qquad m_{32} = -\tfrac{9}{13}.$$

The multipliers will be used in the forthcoming discussion.

In Example 2.7.6 we were able to reduce A to upper triangular form using only row operations of Type 3. This is not always the case. For example, the matrix

$$\begin{bmatrix} 0 & 5 \\ 3 & 2 \end{bmatrix}$$

requires that the two rows be permuted to obtain an upper triangular form. *For the moment, however, we will restrict our attention to invertible matrices A for which the reduction to upper triangular form can be accomplished without permuting rows.* In this case, we can therefore reduce A to upper triangular form using row operations of Type 3 only. Furthermore, throughout the reduction process, we can restrict ourselves to Type 3 operations that add multiples of a row to rows *beneath* that row, by simply performing the row operations column by column, from left to right. According to our description of the elementary matrices $A_{ij}(k)$, our reduction process therefore uses only elementary matrices that are *unit lower triangular*. More specifically, in terms of elementary matrices we have

$$E_k E_{k-1} \cdots E_2 E_1 A = U,$$

where E_k, E_{k-1}, ..., E_2, E_1 are unit lower triangular Type 3 elementary matrices and U is an upper triangular matrix. Since each elementary matrix is invertible, we can write the preceding equation as

$$A = E_1^{-1} E_2^{-1} \cdots E_k^{-1} U. \tag{2.7.1}$$

But, as we have already argued, each of the elementary matrices in (2.7.1) is a unit lower triangular matrix, and we know from Corollary 2.2.23 that the product of two unit lower triangular matrices is also a unit lower triangular matrix. Consequently, (2.7.1) can be written as

$$A = LU, \tag{2.7.2}$$

where

$$L = E_1^{-1} E_2^{-1} \cdots E_k^{-1} \tag{2.7.3}$$

is a unit lower triangular matrix and U is an upper triangular matrix. Equation (2.7.2) is referred to as the **LU factorization of** A. It can be shown (Problem 29) that this LU factorization is unique.

Example 2.7.7　Determine the LU factorization of the matrix

$$A = \begin{bmatrix} 2 & 5 & 3 \\ 3 & 1 & -2 \\ -1 & 2 & 1 \end{bmatrix}.$$

Solution:　Using the results of Example 2.7.6, we can write

$$E_3 E_2 E_1 A = \begin{bmatrix} 2 & 5 & 3 \\ 0 & -\frac{13}{2} & -\frac{13}{2} \\ 0 & 0 & -2 \end{bmatrix},$$

where

$$E_1 = A_{12}(-\tfrac{3}{2}), \qquad E_2 = A_{13}(\tfrac{1}{2}), \qquad \text{and} \qquad E_3 = A_{23}(\tfrac{9}{13}).$$

Therefore,

$$U = \begin{bmatrix} 2 & 5 & 3 \\ 0 & -\frac{13}{2} & -\frac{13}{2} \\ 0 & 0 & -2 \end{bmatrix}$$

and from (2.7.3),

$$L = E_1^{-1} E_2^{-1} \cdots E_k^{-1}. \tag{2.7.4}$$

Computing the inverses of the elementary matrices, we have

$$E_1^{-1} = A_{12}(\tfrac{3}{2}), \qquad E_2^{-1} = A_{13}(-\tfrac{1}{2}), \qquad \text{and} \qquad E_3^{-1} = A_{23}(-\tfrac{9}{13}).$$

Substituting these results into (2.7.4) yields

$$L = \begin{bmatrix} 1 & 0 & 0 \\ \frac{3}{2} & 1 & 0 \\ 0 & 0 & 1 \end{bmatrix} \begin{bmatrix} 1 & 0 & 0 \\ 0 & 1 & 0 \\ -\frac{1}{2} & 0 & 1 \end{bmatrix} \begin{bmatrix} 1 & 0 & 0 \\ 0 & 1 & 0 \\ 0 & -\frac{9}{13} & 1 \end{bmatrix} = \begin{bmatrix} 1 & 0 & 0 \\ \frac{3}{2} & 1 & 0 \\ -\frac{1}{2} & -\frac{9}{13} & 1 \end{bmatrix}.$$

Consequently,

$$A = \begin{bmatrix} 1 & 0 & 0 \\ \frac{3}{2} & 1 & 0 \\ -\frac{1}{2} & -\frac{9}{13} & 1 \end{bmatrix} \begin{bmatrix} 2 & 5 & 3 \\ 0 & -\frac{13}{2} & -\frac{13}{2} \\ 0 & 0 & -2 \end{bmatrix}$$

which is easily verified by a matrix multiplication.　　　　　□

Computing the lower triangular matrix L in the LU factorization of A using (2.7.3) can require a significant amount of work. However, if we look carefully at the matrix L in Example 2.7.7, we see that the elements beneath the leading diagonal are just the corresponding multipliers. That is, if l_{ij} denotes the (i, j) element of the matrix L, then

$$l_{ij} = m_{ij}, \qquad i > j. \tag{2.7.5}$$

Furthermore, it can be shown that this relationship holds in general. Consequently, we do not need to use (2.7.3) to obtain L. Instead we use row operations of Type 3 to reduce A to upper triangular form, and then we can use (2.7.5) to obtain L directly.

Example 2.7.8　Determine the LU decomposition for the matrix

$$A = \begin{bmatrix} 2 & -3 & 1 & 2 \\ 5 & -1 & 2 & 1 \\ 3 & 2 & 6 & -5 \\ -1 & 1 & 3 & 2 \end{bmatrix}.$$

Solution:　To determine U, we reduce A to upper triangular form using only row operations of Type 3 in which we add multiples of a given row only to rows *below* the given row.

$$A \overset{1}{\sim} \begin{bmatrix} 2 & -3 & 1 & 2 \\ 0 & \frac{13}{2} & -\frac{1}{2} & -4 \\ 0 & \frac{13}{2} & \frac{9}{2} & -8 \\ 0 & -\frac{1}{2} & \frac{7}{2} & 3 \end{bmatrix} \overset{2}{\sim} \begin{bmatrix} 2 & -3 & 1 & 2 \\ 0 & \frac{13}{2} & -\frac{1}{2} & -4 \\ 0 & 0 & 5 & -4 \\ 0 & 0 & \frac{45}{13} & \frac{35}{13} \end{bmatrix} \overset{3}{\sim} \begin{bmatrix} 2 & -3 & 1 & 2 \\ 0 & \frac{13}{2} & -\frac{1}{2} & -4 \\ 0 & 0 & 5 & -4 \\ 0 & 0 & 0 & \frac{71}{13} \end{bmatrix} = U.$$

Row Operations	**Corresponding Multipliers**
(1) $A_{12}(-\frac{5}{2})$, $A_{13}(-\frac{3}{2})$, $A_{14}(\frac{1}{2})$	$m_{21} = \frac{5}{2}$,　$m_{31} = \frac{3}{2}$,　$m_{41} = -\frac{1}{2}$
(2) $A_{23}(-1)$, $A_{24}(\frac{1}{13})$	$m_{32} = 1$,　$m_{42} = -\frac{1}{13}$
(3) $A_{34}(-\frac{9}{13})$	$m_{43} = \frac{9}{13}$

Consequently, from (2.7.4),

$$L = \begin{bmatrix} 1 & 0 & 0 & 0 \\ \frac{5}{2} & 1 & 0 & 0 \\ \frac{3}{2} & 1 & 1 & 0 \\ -\frac{1}{2} & -\frac{1}{13} & \frac{9}{13} & 1 \end{bmatrix}.$$

We leave it as an exercise to verify that $LU = A$.　□

The question undoubtedly in the reader's mind is: What is the use of the LU decomposition? In order to answer this question, consider the $n \times n$ system of linear equation $A\mathbf{x} = \mathbf{b}$, where $A = LU$. If we write the system as

$$LU\mathbf{x} = \mathbf{b}$$

and let $U\mathbf{x} = \mathbf{y}$, then solving $A\mathbf{x} = \mathbf{b}$ is equivalent to solving the pair of equations

$$L\mathbf{y} = \mathbf{b},$$
$$U\mathbf{x} = \mathbf{y}.$$

Owing to the triangular form of each of the coefficient matrices L and U, these systems can be solved easily—the first by "forward" substitution and the second by back substitution. In the case when we have a single right-hand-side vector $\mathbf{b}$, the LU factorization for solving the system has no advantage over Gaussian elimination. However, if we require the solution of several systems of equations with the same coefficient matrix A, say

$$A\mathbf{x}_i = \mathbf{b}_i, \qquad i = 1, 2, \ldots, p$$

then it is more efficient to compute the LU factorization of A once, and then successively solve the triangular systems

$$\left.\begin{array}{c} L\mathbf{y}_i = \mathbf{b}_i, \\ U\mathbf{x}_i = \mathbf{y}_i. \end{array}\right\} \qquad i = 1, 2, \ldots, p.$$

Example 2.7.9   Use the LU decomposition of

$$A = \begin{bmatrix} 2 & -3 & 1 & 2 \\ 5 & -1 & 2 & 1 \\ 3 & 2 & 6 & -5 \\ -1 & 1 & 3 & 2 \end{bmatrix}$$

to solve the system $A\mathbf{x} = \mathbf{b}$ if $\mathbf{b} = \begin{bmatrix} 2 \\ -3 \\ 5 \\ 7 \end{bmatrix}$.

Solution: We have shown in the previous example that $A = LU$ where

$$L = \begin{bmatrix} 1 & 0 & 0 & 0 \\ \frac{5}{2} & 1 & 0 & 0 \\ \frac{3}{2} & 1 & 1 & 0 \\ -\frac{1}{2} & -\frac{1}{13} & \frac{9}{13} & 1 \end{bmatrix} \quad \text{and} \quad U = \begin{bmatrix} 2 & -3 & 1 & 2 \\ 0 & \frac{13}{2} & -\frac{1}{2} & -4 \\ 0 & 0 & 5 & -4 \\ 0 & 0 & 0 & \frac{71}{13} \end{bmatrix}.$$

We now solve the two triangular systems $L\mathbf{y} = \mathbf{b}$ and $U\mathbf{x} = \mathbf{y}$. Using forward substitution on the first of these systems, we have

$$y_1 = 2, \qquad y_2 = -3 - \tfrac{5}{2}y_1 = -8,$$
$$y_3 = 5 - \tfrac{3}{2}y_1 - y_2 = 5 - 3 + 8 = 10,$$
$$y_4 = 7 + \tfrac{1}{2}y_1 + \tfrac{1}{13}y_2 - \tfrac{9}{13}y_3 = 8 - \tfrac{8}{13} - \tfrac{90}{13} = \tfrac{6}{13}.$$

Solving $U\mathbf{x} = \mathbf{y}$ via back substitution yields

$$x_4 = \tfrac{13}{71}y_4 = \tfrac{6}{71}, \qquad x_3 = \tfrac{1}{5}(y_3 + 4x_4) = \tfrac{1}{5}\left(10 + \tfrac{24}{71}\right) = \tfrac{734}{355},$$

$$x_2 = \tfrac{2}{13}\left(y_2 + \tfrac{1}{2}x_3 + 4x_4\right) = \tfrac{2}{13}\left(-8 + \tfrac{367}{355} + \tfrac{24}{71}\right) = -\tfrac{362}{355},$$

$$x_1 = \tfrac{1}{2}\left(y_1 + 3x_2 - x_3 - 2x_4\right) = \tfrac{1}{2}\left(2 - \tfrac{1086}{355} - \tfrac{734}{355} - \tfrac{12}{71}\right) = -\tfrac{117}{71}.$$

Consequently,

$$\mathbf{x} = \left(-\tfrac{117}{71}, -\tfrac{362}{355}, \tfrac{734}{355}, \tfrac{6}{71}\right).$$

$\square$

In the more general case when row interchanges are required to reduce an invertible matrix A to upper triangular form, it can be shown that A has a factorization of the form

$$A = PLU, \tag{2.7.6}$$

where P is an appropriate product of elementary permutation matrices, L is a unit lower triangular matrix, and U is an upper triangular matrix. From the properties of the elementary permutation matrices, it follows (see Problem 27), that $P^{-1} = P^T$. Using (2.7.6) the linear system $A\mathbf{x} = \mathbf{b}$ can be written as

$$PLU\mathbf{x} = \mathbf{b},$$

or equivalently,

$$LU\mathbf{x} = P^T\mathbf{b}.$$

Consequently, to solve $A\mathbf{x} = \mathbf{b}$ in this case we can solve the two triangular systems

$$\begin{cases} L\mathbf{y} = P^T\mathbf{b}, \\ U\mathbf{x} = \mathbf{y}. \end{cases}$$

For a full discussion of this and other factorizations of $n \times n$ matrices, and their applications, the reader is referred to more advanced texts on linear algebra or numerical analysis [for example, B. Noble and J. W.Daniel, *Applied Linear Algebra* (Englewood Cliffs, N.J., Prentice Hall, 1988); J. Ll. Morris, *Computational Methods in Elementary Numerical Analysis* (New York: Wiley, 1983)].

Exercises for 2.7

Key Terms

Elementary matrix, Multiplier, LU Factorization of a matrix.

Skills

- Be able to determine whether or not a given matrix is an elementary matrix.

- Know the form for the permutation matrices, scaling matrices, and row combination matrices.

- Be able to write down the inverse of an elementary matrix without any computation.

- Be able to determine elementary matrices that reduce a given matrix to row-echelon form.

- Be able to express an invertible matrix as a product of elementary matrices.

- Be able to determine the multipliers of a matrix.

- Be able to determine the LU factorization of a matrix.

- Be able to use the LU factorization of a matrix A to solve a linear system $A\mathbf{x} = \mathbf{b}$.

True-False Review

For Questions 1–10, decide if the given statement is **true** or **false**, and give a brief justification for your answer. If true, you can quote a relevant definition or theorem from the text. If false, provide an example, illustration, or brief explanation of why the statement is false.

1. Every elementary matrix is invertible.

2. A product of elementary matrices is an elementary matrix.

3. Every matrix can be expressed as a product of elementary matrices.

4. If A is an $m \times n$ matrix and E is an $m \times m$ elementary matrix, then the matrices A and EA have the same rank.

5. If P_{ij} is a permutation matrix, then $P_{ij}^2 = P_{ij}$.

6. If E_1 and E_2 are $n \times n$ elementary matrices, then $E_1 E_2 = E_2 E_1$.

7. If E_1 and E_2 are $n \times n$ elementary matrices *of the same type*, then $E_1 E_2 = E_2 E_1$.

8. Every matrix has an LU factorization.

9. In the LU factorization of a matrix A, the matrix L is a unit lower triangular matrix and the matrix U is a unit upper triangular matrix.

10. A 4×4 matrix A that has an LU factorization has 10 multipliers.

Problems

1. Write all 3×3 elementary matrices and their inverses.

For Problems 2–5, determine elementary matrices that reduce the given matrix to row-echelon form.

2. $\begin{bmatrix} 3 & 5 \\ 1 & -2 \end{bmatrix}$.

3. $\begin{bmatrix} 5 & 8 & 2 \\ 1 & 3 & -1 \end{bmatrix}$.

4. $\begin{bmatrix} 3 & -1 & 4 \\ 2 & 1 & 3 \\ 1 & 3 & 2 \end{bmatrix}$.

5. $\begin{bmatrix} 1 & 2 & 3 & 4 \\ 2 & 3 & 4 & 5 \\ 3 & 4 & 5 & 6 \end{bmatrix}$.

For Problems 6–12, express the matrix A as a product of elementary matrices.

6. $A = \begin{bmatrix} 1 & 2 \\ 1 & 3 \end{bmatrix}$.

7. $A = \begin{bmatrix} -2 & -3 \\ 5 & 7 \end{bmatrix}$.

8. $A = \begin{bmatrix} 3 & -4 \\ -1 & 2 \end{bmatrix}$.

9. $A = \begin{bmatrix} 4 & -5 \\ 1 & 4 \end{bmatrix}$.

10. $A = \begin{bmatrix} 1 & -1 & 0 \\ 2 & 2 & 2 \\ 3 & 1 & 3 \end{bmatrix}$.

11. $A = \begin{bmatrix} 0 & -4 & -2 \\ 1 & -1 & 3 \\ -2 & 2 & 2 \end{bmatrix}$.

12. $A = \begin{bmatrix} 1 & 2 & 3 \\ 0 & 8 & 0 \\ 3 & 4 & 5 \end{bmatrix}$.

13. Determine elementary matrices $E_1, E_2, \ldots, E_k$ that reduce
$$A = \begin{bmatrix} 2 & -1 \\ 1 & 3 \end{bmatrix}$$
to reduced row-echelon form. Verify by direct multiplication that $E_1 E_2 \cdots E_k A = I_2$.

14. Determine a Type 3 lower triangular elementary matrix E_1 that reduces
$$A = \begin{bmatrix} 3 & -2 \\ -1 & 5 \end{bmatrix}$$
to upper triangular form. Use Equation (2.7.3) to determine L and verify Equation (2.7.2).

For Problems 15–20, determine the LU factorization of the given matrix. Verify your answer by computing the product LU.

15. $A = \begin{bmatrix} 2 & 3 \\ 5 & 1 \end{bmatrix}$.

16. $A = \begin{bmatrix} 3 & 1 \\ 5 & 2 \end{bmatrix}$.

17. $A = \begin{bmatrix} 3 & -1 & 2 \\ 6 & -1 & 1 \\ -3 & 5 & 2 \end{bmatrix}$.

18. $A = \begin{bmatrix} 5 & 2 & 1 \\ -10 & -2 & 3 \\ 15 & 2 & -3 \end{bmatrix}$.

19. $A = \begin{bmatrix} 1 & -1 & 2 & 3 \\ 2 & 0 & 3 & -4 \\ 3 & -1 & 7 & 8 \\ 1 & 3 & 4 & 5 \end{bmatrix}$.

20. $A = \begin{bmatrix} 2 & -3 & 1 & 2 \\ 4 & -1 & 1 & 1 \\ -8 & 2 & 2 & -5 \\ 6 & 1 & 5 & 2 \end{bmatrix}$.

For Problems 21–24, use the LU factorization of A to solve the system $A\mathbf{x} = \mathbf{b}$.

21. $A = \begin{bmatrix} 1 & 2 \\ 2 & 3 \end{bmatrix}$, $\mathbf{b} = \begin{bmatrix} 3 \\ -1 \end{bmatrix}$.

22. $A = \begin{bmatrix} 1 & -3 & 5 \\ 3 & 2 & 2 \\ 2 & 5 & 2 \end{bmatrix}$, $\mathbf{b} = \begin{bmatrix} 1 \\ 5 \\ -1 \end{bmatrix}$.

23. $A = \begin{bmatrix} 2 & 2 & 1 \\ 6 & 3 & -1 \\ -4 & 2 & 2 \end{bmatrix}$, $\mathbf{b} = \begin{bmatrix} 1 \\ 0 \\ 2 \end{bmatrix}$.

24. $A = \begin{bmatrix} 4 & 3 & 0 & 0 \\ 8 & 1 & 2 & 0 \\ 0 & 5 & 3 & 6 \\ 0 & 0 & -5 & 7 \end{bmatrix}$, $\mathbf{b} = \begin{bmatrix} 2 \\ 3 \\ 0 \\ 5 \end{bmatrix}$.

25. Use the LU factorization of

$$A = \begin{bmatrix} 2 & -1 \\ -8 & 3 \end{bmatrix}$$

to solve each of the systems $A\mathbf{x}_i = \mathbf{b}_i$ if

$$\mathbf{b}_1 = \begin{bmatrix} 3 \\ -1 \end{bmatrix}, \qquad \mathbf{b}_2 = \begin{bmatrix} 2 \\ 7 \end{bmatrix}, \qquad \mathbf{b}_3 = \begin{bmatrix} 5 \\ -9 \end{bmatrix}.$$

26. Use the LU factorization of

$$A = \begin{bmatrix} -1 & 4 & 2 \\ 3 & 1 & 4 \\ 5 & -7 & 1 \end{bmatrix}$$

to solve each of the systems $A\mathbf{x}_i = \mathbf{e}_i$ and thereby determine A^{-1}.

27. If $P = P_1 P_2 \cdots P_k$, where each P_i is an elementary permutation matrix, show that $P^{-1} = P^T$.

28. Prove that

(a) The inverse of an invertible upper triangular matrix is upper triangular. Repeat for an invertible lower triangular matrix.

(b) The inverse of a unit upper triangular matrix is unit upper triangular. Repeat for a unit lower triangular matrix.

29. In this problem, we prove that the LU decomposition of an invertible $n \times n$ matrix is unique in the sense that, if $A = L_1 U_1$ and $A = L_2 U_2$, where L_1, L_2 are unit lower triangular matrices and U_1, U_2 are upper triangular matrices, then $L_1 = L_2$ and $U_1 = U_2$.

(a) Apply Corollary 2.6.12 to conclude that L_2 and U_1 are invertible, and then use the fact that $L_1 U_1 = L_2 U_2$ to establish that $L_2^{-1} L_1 = U_2 U_1^{-1}$.

(b) Use the result from (a) together with Theorem 2.2.22 and Corollary 2.2.23 to prove that $L_2^{-1} L_1 = I_n$ and $U_2 U_1^{-1} = I_n$, from which the required result follows.

30. QR Factorization: It can be shown that any invertible $n \times n$ matrix has a factorization of the form

$$A = QR,$$

where Q and R are invertible, R is upper triangular, and Q satisfies $Q^T Q = I_n$ (i.e., Q is **orthogonal**). Determine an algorithm for solving the linear system $A\mathbf{x} = \mathbf{b}$ using this QR factorization.

⋄ For Problems 31–33, use some form of technology to determine the LU factorization of the given matrix. Verify the factorization by computing the product LU.

31. $A = \begin{bmatrix} 3 & 5 & -2 \\ 2 & 7 & 9 \\ -5 & 5 & 11 \end{bmatrix}$.

32. $A = \begin{bmatrix} 27 & -19 & 32 \\ 15 & -16 & 9 \\ 23 & -13 & 51 \end{bmatrix}$.

33. $A = \begin{bmatrix} 34 & 13 & 19 & 22 \\ 53 & 17 & -71 & 20 \\ 21 & 37 & 63 & 59 \\ 81 & 93 & -47 & 39 \end{bmatrix}$.

2.8 The Invertible Matrix Theorem I

In Section 2.6, we defined an $n \times n$ invertible matrix A to be a matrix such that there exists an $n \times n$ matrix B satisfying $AB = BA = I_n$. There are, however, many other important and useful viewpoints on invertibility of matrices. Some of these we have already encountered in the preceding two sections, while others await us in later chapters. It is worthwhile to begin collecting a list of conditions on an $n \times n$ matrix A that are mathematically equivalent to its invertibility. We refer to this theorem as the Invertible Matrix Theorem. As we have indicated, this result is somewhat a "work in progress," and we shall return to it later in Sections 3.2 and 4.10.

Theorem 2.8.1   **(Invertible Matrix Theorem)**

Let A be an $n \times n$ matrix with real elements. The following conditions on A are equivalent:

(a) A is invertible.

(b) The equation $A\mathbf{x} = \mathbf{b}$ has a unique solution for every $\mathbf{b}$ in $\mathbb{R}^n$.

(c) The equation $A\mathbf{x} = \mathbf{0}$ has only the trivial solution $\mathbf{x} = \mathbf{0}$.

(d) rank$(A) = n$.

(e) A can be expressed as a product of elementary matrices.

(f) A is row-equivalent to I_n.

Proof The equivalence of (a), (b), and (d) has already been established in Section 2.6 in Theorems 2.6.4 and 2.6.5, as well as in Corollary 2.6.6. Moreover, the equivalence of (a) and (e) was already established in Theorem 2.7.5.

Next we establish that (c) is an equivalent statement by proving that (b) $\Longrightarrow$ (c) $\Longrightarrow$ (d). Assuming that (b) holds, we can conclude that the linear system $A\mathbf{x} = \mathbf{0}$ has a unique solution. However, one solution is evidently $\mathbf{x} = \mathbf{0}$, hence this is the unique solution to $A\mathbf{x} = \mathbf{0}$, which establishes (c). Next, assume that (c) holds. The fact that $A\mathbf{x} = \mathbf{0}$ has only the trivial solution means that, in reducing A to row-echelon form, we find no free parameters. Thus, every column (and hence every row) of A contains a pivot, which means that the row-echelon form of A has n nonzero rows; that is, rank$(A) = n$, which is (d).

Finally, we prove that (e) $\Longrightarrow$ (f) $\Longrightarrow$ (a). If (e) holds, we can left multiply I_n by a product of elementary matrices (corresponding to a sequence of elementary row operations applied to I_n) to obtain A. This means that A is row-equivalent to I_n, which is (f). Last, if A is row-equivalent to I_n, we can write A as a product of elementary matrices, each of which is invertible. Since a product of invertible matrices is invertible (by Corollary 2.6.10), we conclude that A is invertible, as needed. ∎

Exercises for 2.8

Skills

- Know the list of characterizations of invertible matrices given in the Invertible Matrix Theorem.

- Be able to use the Invertible Matrix Theorem to draw conclusions related to the invertibility of a matrix.

True-False Review

For Questions 1–4, decide if the given statement is **true** or **false**, and give a brief justification for your answer. If true, you can quote a relevant definition or theorem from the text. If false, provide an example, illustration, or brief explanation of why the statement is false.

1. If the linear system $A\mathbf{x} = \mathbf{0}$ has a nontrivial solution, then A can be expressed as a product of elementary matrices.

2. A 4×4 matrix A with rank$(A) = 4$ is row-equivalent to I_4.

3. If A is a 3×3 matrix with rank$(A) = 2$, then the linear system $A\mathbf{x} = \mathbf{b}$ must have infinitely many solutions.

4. Any $n \times n$ upper triangular matrix is row-equivalent to I_n.

Problems

1. Use part (c) of the Invertible Matrix Theorem to prove that if A is an invertible matrix and B and C are matrices of the same size as A such that $AB = AC$, then $B = C$. [**Hint:** Consider $AB - AC = 0$.]

2. Give a direct proof of the fact that (d) $\implies$ (c) in the Invertible Matrix Theorem.

3. Give a direct proof of the fact that (c) $\implies$ (b) in the Invertible Matrix Theorem.

4. Use the equivalence of (a) and (e) in the Invertible Matrix Theorem to prove that if A and B are invertible $n \times n$ matrices, then so is AB.

5. Use the equivalence of (a) and (c) in the Invertible Matrix Theorem to prove that if A and B are invertible $n \times n$ matrices, then so is AB.

2.9 Chapter Review

In this chapter we have investigated linear systems of equations. Matrices provide a convenient mathematical representation for linear systems, and whether or not a linear system has a solution (and if so, how many) can be determined entirely from the matrix for the linear system.

An $m \times n$ **matrix** $A = [a_{ij}]$ is a rectangular array of numbers arranged in m rows and n columns. The entry in the ith row and jth column is written a_{ij}. More generally, such an array, whose entries are allowed to depend on an indeterminate t, is known as a **matrix function**. Matrix functions can be used to formulate systems of differential equations.

If $m = n$, the matrix (or matrix function) is called a **square matrix**.

Concepts Related to Square Matrices

- **Main diagonal:** the entries $a_{11}, a_{22}, \ldots, a_{nn}$ in the matrix.

- **Trace:** the sum of the entries on the main diagonal.

- **Upper triangular matrix:** $a_{ij} = 0$ for $i > j$.

- **Lower triangular matrix:** $a_{ij} = 0$ for $i < j$.

- **Diagonal matrix:** $a_{ij} = 0$ for $i \neq j$.

- **Transpose:** applying to any $m \times n$ matrix A, this is the $n \times m$ matrix A^T obtained from A by interchanging its rows and columns

- **Symmetric matrix:** $A^T = A$; that is, $a_{ij} = a_{ji}$.

- **Skew-symmetric matrix:** $A^T = -A$; that is, $a_{ij} = -a_{ji}$. In particular, $a_{ii} = 0$ for each i.

Matrix Algebra

Given two matrices A and B of the same size $m \times n$, we can perform the following operations:

- **Addition/subtraction $A \pm B$:** add/subtract the corresponding elements of A and B.

- **Scalar multiplication rA:** multiply each entry of A by the real (or complex) scalar r.

If A is $m \times n$ and B is $n \times p$, we can form their product AB, which is an $m \times p$ matrix whose (i, j)-entry is computed by taking the dot product of the ith row vector of A with the jth column vector of B. Note that, in general, $AB \neq BA$.

Linear Systems

The general $m \times n$ system of linear equations is of the form

$$
\begin{aligned}
a_{11}x_1 + a_{12}x_2 + \cdots + a_{1n}x_n &= b_1, \\
a_{21}x_1 + a_{22}x_2 + \cdots + a_{2n}x_n &= b_2, \\
&\vdots \\
a_{m1}x_1 + a_{m2}x_2 + \cdots + a_{mn}x_n &= b_m.
\end{aligned}
$$

If each $b_i = 0$, the system is called **homogeneous**. There are two useful ways to formulate the above linear system:

1. **Augmented matrix:**

$$
A^{\#} = \begin{bmatrix}
a_{11} & a_{12} & \dots & a_{1n} & b_1 \\
a_{21} & a_{22} & \dots & a_{2n} & b_2 \\
& & \vdots & & \vdots \\
a_{m1} & a_{m2} & \dots & a_{mn} & b_m
\end{bmatrix}.
$$

2. **Vector form:**

$$
A\mathbf{x} = \mathbf{b},
$$

where

$$
A = \begin{bmatrix}
a_{11} & a_{12} & \dots & a_{1n} \\
a_{21} & a_{22} & \dots & a_{2n} \\
& & \vdots & \\
a_{m1} & a_{m2} & \dots & a_{mn}
\end{bmatrix}, \quad
\mathbf{x} = \begin{bmatrix}
x_1 \\
x_2 \\
\vdots \\
x_n
\end{bmatrix}, \quad
\mathbf{b} = \begin{bmatrix}
b_1 \\
b_2 \\
\vdots \\
b_m
\end{bmatrix}.
$$

Elementary Row Operations and Row Echelon Form

There are three types of elementary row operations on a matrix A:

1. P_{ij}: Permute the ith and jth rows in A.

2. $M_i(k)$: Multiply the entries in the ith row of A by the nonzero scalar k.

3. $A_{ij}(k)$: Add to the elements of the jth row of A the scalar k times the corresponding elements of the ith row of A.

By performing elementary row operations on the augmented matrix above, we can determine solutions, if any, to the linear system. The strategy is to apply elementary row operations in such a way that A is transformed into row-echelon form—a process known as **Gaussian elimination**. By applying back substitution to the linear system corresponding to the row-echelon form obtained, we find the solution. This solution agrees with the solution to the original linear system. If necessary, free parameters may be used to express this solution. A leading one in the far right-hand column of the row-echelon form indicates that the system has no solution.

A row-echelon form matrix is one in which

- All rows consisting entirely of zeros are placed at the bottom of the matrix.

- All other rows begin with a (leading) "1", called a **pivot**.

- The leading ones occur in columns strictly to the right of the leading ones in the rows above.

Invertible Matrices

An $n \times n$ matrix A is invertible if there exists an $n \times n$ matrix B such that $AB = I_n = BA$, where I_n is the $n \times n$ identity matrix (ones on the main diagonal, zeros elsewhere). We write A^{-1} for the (unique) inverse B of A. One procedure for determining A^{-1}, if it exists, is the **Gauss-Jordan technique**:

$$[A|I_n] \sim \underset{\cdots}{\overset{\text{ERO}}{\sim}} \sim [I_n|A^{-1}].$$

Invertible matrices A share all of the following equivalent properties:

- A can be reduced to I_n via a sequence of elementary row operations.

- The linear system $A\mathbf{x} = \mathbf{b}$ has a unique solution $\mathbf{x}$.

- The linear system $A\mathbf{x} = \mathbf{0}$ has only the trivial solution $\mathbf{x} = \mathbf{0}$.

- A can be expressed as a product of **elementary matrices** that are obtained from the identity matrix by applying exactly one elementary row operation.

Additional Problems

Let

$$A = \begin{bmatrix} -2 & 4 & 2 & 6 \\ -1 & -1 & 5 & 0 \end{bmatrix}, \quad B = \begin{bmatrix} -3 & 0 \\ 2 & 2 \\ 1 & -3 \\ 0 & 1 \end{bmatrix}, \quad C = \begin{bmatrix} -5 \\ -6 \\ 3 \\ 1 \end{bmatrix},$$

and $r = -4$. For Problems 1–6, compute the given expression, if possible.

1. $rA - B^T$.

2. AB and $\mathrm{tr}(AB)$.

3. $(AC)(AC)^T$.

4. $(rB)A$.

5. $(AB)^{-1}$.

6. $C^T C$ and $\mathrm{tr}(C^T C)$.

7. Let

$$A = \begin{bmatrix} 1 & 2 & 3 \\ 2 & 5 & 7 \end{bmatrix} \quad \text{and} \quad B = \begin{bmatrix} 3 & b \\ -4 & a \\ a & b \end{bmatrix}.$$

(a) Compute AB and determine the values of a and b such that $AB = I_2$.

(b) Using the values of a and b obtained in (a), compute BA.

8. Let A be an $m \times n$ matrix and let B be an $p \times n$ matrix. Use the index form of the matrix product to prove that $(AB^T)^T = BA^T$.

9. Let A be an $n \times n$ matrix.

(a) Use the index form of the matrix product to write the ijth element of A^2.

(b) In the case when A is a symmetric matrix, show that A^2 is also symmetric.

10. Let A and B be $n \times n$ matrices. If A is *skew-symmetric*, use properties of the transpose to establish that $B^T AB$ is also skew-symmetric.

An $n \times n$ matrix A is called **nilpotent** if $A^p = 0$ for some positive integer p. For Problems 11–12, show that the given matrix is nilpotent.

11. $A = \begin{bmatrix} 3 & 9 \\ -1 & -3 \end{bmatrix}$.

12. $A = \begin{bmatrix} 0 & 1 & 1 \\ 0 & 0 & 1 \\ 0 & 0 & 0 \end{bmatrix}$.

For Problems 13–16, let

$$A(t) = \begin{bmatrix} e^{-3t} & -\sec^2 t \\ 2t^3 & \cos t \\ 6\ln t & 36 - 5t \end{bmatrix}$$

and

$$B(t) = \begin{bmatrix} -7 & t^2 \\ 6-t & 3t^3 + 6t^2 \\ 1+t & \cos(\pi t/2) \\ e^t & 1 - t^3 \end{bmatrix}.$$

Compute the given expression, if possible.

13. $A'(t)$.

14. $\int_0^1 B(t)\, dt$.

15. $t^3 \cdot A(t) - \sin t \cdot B(t)$.

16. $B'(t) - e^t A(t)$.

For Problems 17–23, determine the solution set to the given linear system of equations.

17.
$$\begin{aligned} x_1 + 5x_2 + 2x_3 &= -6, \\ 4x_2 - 7x_3 &= 2, \\ 5x_3 &= 0. \end{aligned}$$

18.
$$\begin{aligned} 5x_1 - x_2 + 2x_3 &= 7, \\ -2x_1 + 6x_2 + 9x_3 &= 0, \\ -7x_1 + 5x_2 - 3x_3 &= -7. \end{aligned}$$

19.
$$\begin{aligned} x + 2y - z &= 1, \\ x \quad\quad + z &= 5, \\ 4x + 4y &= 12. \end{aligned}$$

20.
$$\begin{aligned} x_1 - 2x_2 - x_3 + 3x_4 &= 0, \\ -2x_1 + 4x_2 + 5x_3 - 5x_4 &= 3, \\ 3x_1 - 6x_2 - 6x_3 + 8x_4 &= 2. \end{aligned}$$

21.
$$\begin{aligned} 3x_1 \quad\quad - x_3 + 2x_4 - x_5 &= 1, \\ x_1 + 3x_2 + x_3 - 3x_4 + 2x_5 &= -1, \\ 4x_1 - 2x_2 - 3x_3 + 6x_4 - x_5 &= 5. \\ x_4 + 4x_5 &= -2. \end{aligned}$$

22.
$$\begin{aligned} x_1 + x_2 + x_3 + x_4 - 3x_5 &= 6, \\ x_1 + x_2 + x_3 + 2x_4 - 5x_5 &= 8, \\ 2x_1 + 3x_2 + x_3 + 4x_4 - 9x_5 &= 17, \\ 2x_1 + 2x_2 + 2x_3 + 3x_4 - 8x_5 &= 14. \end{aligned}$$

23.
$$\begin{aligned} x_1 - 3x_2 + 2ix_3 &= 1, \\ -2ix_1 + 6x_2 + 2x_3 &= -2. \end{aligned}$$

For Problems 24–27, determine all values of k for which the given linear system has (a) no solution, (b) a unique solution, and (c) infinitely many solutions.

24.
$$\begin{aligned} x_1 - kx_2 &= 6, \\ 2x_1 + 3x_2 &= k. \end{aligned}$$

25.
$$\begin{aligned} kx_1 + 2x_2 - x_3 &= 2, \\ kx_2 + x_3 &= 2. \end{aligned}$$

26.
$$\begin{aligned} 10x_1 + kx_2 - x_3 &= 0, \\ kx_1 + x_2 - x_3 &= 0, \\ 2x_1 + x_2 - x_3 &= 0. \end{aligned}$$

27.
$$\begin{aligned} x_1 - kx_2 + k^2x_3 &= 0, \\ x_1 \quad\quad + kx_3 &= 0, \\ x_2 - x_3 &= 1. \end{aligned}$$

28. Do the three planes $x_1 + 2x_2 + x_3 = 4$, $x_2 - x_3 = 1$, and $x_1 + 3x_2 = 0$ have at least one common point of intersection? Explain.

For Problems 29–34, (a) find a row-echelon form of the given matrix A, (b) determine rank(A), and (c) use the Gauss-Jordan technique to determine the inverse of A, if it exists.

29. $A = \begin{bmatrix} 4 & 7 \\ -2 & 5 \end{bmatrix}$.

30. $A = \begin{bmatrix} 2 & -7 \\ -4 & 14 \end{bmatrix}$.

31. $A = \begin{bmatrix} 3 & -1 & 6 \\ 0 & 2 & 3 \\ 3 & -5 & 0 \end{bmatrix}$.

32. $A = \begin{bmatrix} 2 & 1 & 0 & 0 \\ 1 & 2 & 0 & 0 \\ 0 & 0 & 3 & 4 \\ 0 & 0 & 4 & 3 \end{bmatrix}$.

33. $A = \begin{bmatrix} 3 & 0 & 0 \\ 0 & 2 & -1 \\ 1 & -1 & 2 \end{bmatrix}$.

34. $A = \begin{bmatrix} -2 & -3 & 1 \\ 1 & 4 & 2 \\ 0 & 5 & 3 \end{bmatrix}$.

35. Let

$$A = \begin{bmatrix} 1 & -1 & 3 \\ 4 & -3 & 13 \\ 1 & 1 & 4 \end{bmatrix}.$$

Solve each of the systems

$$A\mathbf{x}_i = \mathbf{e}_i, \quad i = 1, 2, 3$$

where $\mathbf{e}_i$ denote the column vectors of the identity matrix I_3.

36. Solve each of the systems $A\mathbf{x}_i = \mathbf{b}_i$ if

$$A = \begin{bmatrix} 2 & 5 \\ 7 & -2 \end{bmatrix}, \qquad \mathbf{b}_1 = \begin{bmatrix} 1 \\ 2 \end{bmatrix},$$

$$\mathbf{b}_2 = \begin{bmatrix} 4 \\ 3 \end{bmatrix}, \qquad \mathbf{b}_3 = \begin{bmatrix} -2 \\ 5 \end{bmatrix}.$$

37. Let A and B be invertible matrices.

(a) By computing an appropriate matrix product, verify that $(A^{-1}B)^{-1} = B^{-1}A$.

(b) Use properties of the inverse to *derive* $(A^{-1}B)^{-1} = B^{-1}A$.

38. Let S be an invertible $n \times n$ matrix and let k be a nonnegative integer. If $A = SDS^{-1}$, prove that $A^k = SD^kS^{-1}$.

For Problems 39–42, (a) express the given matrix as a product of elementary matrices, and (b) determine the LU decomposition of the matrix.

39. The matrix in Problem 29.

40. The matrix in Problem 32.

41. The matrix in Problem 33.

42. The matrix in Problem 34.

43. (a) Prove that if A and B are $n \times n$ matrices, then

$$(A + B)^3 = A^3 + A^2B + ABA + BA^2$$
$$+ AB^2 + BAB + B^2A + B^3.$$

(b) How does the formula change for $(A - B)^3$?

(c) Can you make a conjecture about the number of terms in the expansion of $(A + B)^k$, in terms of k?

44. Suppose that A and B are invertible matrices. Prove that the block matrix

$$\begin{bmatrix} A & 0 \\ 0 & B^{-1} \end{bmatrix}$$

is invertible.

45. In many different positions can two leading ones of a row-echelon form of a 2×4 matrix occur? How about three leading ones for a 3×4 matrix? How about four leading ones for a 4×6 matrix? How about m leading ones for an $m \times n$ matrix with $m \leq n$?

46. If the inverse of A^2 is the matrix B, what is the inverse of the matrix A^{10}? Prove your answer.

Project: Circles and Spheres via Gaussian Elimination

Part 1: Circles　In this part, we shall see that any three noncollinear points in the plane can be found on a unique circle, and we will use Gaussian elimination to find the center and radius of this circle.

(a) Show geometrically that three noncollinear points in the plane must lie on a unique circle. [**Hint:** The radius must lie on the line that passes through the midpoint of two of the three points and that is perpendicular to the segment connecting the two points.]

(b) A circle in the plane has an equation that can be given in the form

$$(x - a)^2 + (y - b)^2 = r^2,$$

where (a, b) is the center and r is the radius. By expanding the formula, we may write the equation of the circle in the form

$$x^2 + y^2 + cx + dy = k,$$

for constants c, d, and k. Using this latter formula together with Gaussian elimination, determine c, d, and k for each set of points below. Then solve for (a, b) and r to write the equation of the circle.

(i) $(2, -1), (3, 3), (4, -1)$.

(ii) $(-1, 0), (1, 2), (2, 2)$.

Part 2: Spheres   In this part, we shall extend the ideas of Part 1 and consider four noncoplanar points in 3-space. Any three of these four points lie in a plane but are noncollinear (why?). A sphere in 3-space has an equation that can be given in the form

$$(x - a)^2 + (y - b)^2 + (z - c)^2 = r^2,$$

where (a, b, c) is the center and r is the radius. By expanding the formula, we may write the equation of the sphere in the form

$$x^2 + y^2 + z^2 + ux + vy + wz = k,$$

for constants u, v, w, and k.

(a) Using the latter formula above together with Gaussian elimination, determine u, v, w, and k for each set of points below. Then solve for (a, b, c) and r to write the equation of the sphere.

    (i) $(1, -1, 2)$, $(2, -1, 4)$, $(-1, -1, -1)$, $(1, 4, 1)$.

    (ii) $(2, 0, 0)$, $(0, 3, 0)$, $(0, 0, 4)$, $(0, 0, 6)$.

(b) What goes wrong with the procedure in (a) if the points lie on a single plane? Choose four points of your own and carry out the procedure in part (a) to see what happens. Can you describe circumstances under which the four coplanar points *will* lie on a sphere?

CHAPTER
3

Determinants

Mathematics is the gate and key to the sciences. — Roger Bacon

In this chapter, we introduce a basic tool in applied mathematics, namely the determinant of a square matrix. The determinant is a number, associated with an $n \times n$ matrix A, whose value characterizes when the linear system $A\mathbf{x} = \mathbf{b}$ has a unique solution (or, equivalently, when A^{-1} exists). Determinants enjoy a wide range of applications, including coordinate geometry and function theory.

Sections 3.1–3.3 give a detailed introduction to determinants, their properties, and their applications. Alternatively, Section 3.4, "Summary of Determinants," can provide a nonrigorous and much more abbreviated introduction to the fundamental results required in the remainder of the text. We will see in later chapters that determinants are invaluable in the theory of eigenvalues and eigenvectors of a matrix, as well as in solution techniques for linear systems of differential equations.

3.1 The Definition of the Determinant

We will give a criterion shortly (Theorem 3.2.4) for the invertibility of a square matrix A in terms of the determinant of A, written $\det(A)$, which is a number determined directly from the elements of A. This criterion will provide a first extension of the Invertible Matrix Theorem introduced in Section 2.8.

To motivate the definition of the determinant of an $n \times n$ matrix A, we begin with the special cases $n = 1$, $n = 2$, and $n = 3$.

Case 1: $n = 1$. According to Theorem 2.6.5, the 1×1 matrix $A = [a_{11}]$ is invertible if and only if rank$(A) = 1$, if and only if the 1×1 determinant, $\det(A)$, defined by

$$\boxed{\det(A) = a_{11}}$$

is nonzero.

Case 2: $n = 2$.   According to Theorem 2.6.5, the 2×2 matrix

$$A = \begin{bmatrix} a_{11} & a_{12} \\ a_{21} & a_{22} \end{bmatrix}$$

is invertible if and only if rank$(A) = 2$, if and only if the row-echelon form of A has two nonzero rows. Provided that $a_{11} \neq 0$, we can reduce A to row-echelon form as follows:

$$\begin{bmatrix} a_{11} & a_{12} \\ a_{21} & a_{22} \end{bmatrix} \overset{1}{\sim} \begin{bmatrix} a_{11} & a_{12} \\ 0 & a_{22} - \dfrac{a_{12}a_{21}}{a_{11}} \end{bmatrix}.$$

$$\boxed{1. \ A_{12}\left(-\dfrac{a_{21}}{a_{11}}\right)}$$

For A to be invertible, it is necessary that $a_{22} - \dfrac{a_{12}a_{21}}{a_{11}} \neq 0$, or that $a_{11}a_{22} - a_{12}a_{21} \neq 0$. Thus, for A to be invertible, it is necessary that the 2×2 determinant, $\det(A)$, defined by

$$\boxed{\det(A) = a_{11}a_{22} - a_{12}a_{21}} \tag{3.1.1}$$

be nonzero. We will see in the next section that this condition is also sufficient for the 2×2 matrix A to be invertible.

Case 3: $n = 3$.   According to Theorem 2.6.5, the 3×3 matrix

$$A = \begin{bmatrix} a_{11} & a_{12} & a_{13} \\ a_{21} & a_{22} & a_{23} \\ a_{31} & a_{32} & a_{33} \end{bmatrix}$$

is invertible if and only if rank$(A) = 3$, if and only if the row-echelon form of A has three nonzero rows. Reducing A to row-echelon form as in Case 2, we find that it is necessary that the 3×3 determinant defined by

$$\boxed{\det(A) = a_{11}a_{22}a_{33} + a_{12}a_{23}a_{31} + a_{13}a_{21}a_{32} - a_{11}a_{23}a_{32} - a_{12}a_{21}a_{33} - a_{13}a_{22}a_{31}}$$

$$\tag{3.1.2}$$

be nonzero. Again, in the next section we will prove that this condition on $\det(A)$ is also sufficient for the 3×3 matrix A to be invertible.

To generalize the foregoing formulas for the determinant of an $n \times n$ matrix A, we take a closer look at their structure. Each determinant above consists of a sum of $n!$ products, where each product term contains precisely one element from each row and each column of A. Furthermore, each possible choice of one element from each row and each column of A does in fact occur as a term of the summation. Finally, each term is assigned a plus or a minus sign. Based on these observations, the appropriate way in which to define $\det(A)$ for an $n \times n$ matrix would seem to be to add up all possible products consisting of one element from each row and each column of A, with some condition on which products are taken with a plus sign and which with a minus sign. To describe this condition, we digress to discuss permutations.

Permutations

Consider the first n positive integers $1, 2, 3, \ldots, n$. Any arrangement of these integers in a specific order, say, $(p_1, p_2, \ldots, p_n)$, is called a **permutation**.

Example 3.1.1   There are precisely six distinct permutations of the integers 1, 2 and 3:

$$(1, 2, 3), \quad (1, 3, 2), \quad (2, 1, 3), \quad (2, 3, 1), \quad (3, 1, 2), \quad (3, 2, 1). \qquad \square$$

More generally, we have the following result:

Theorem 3.1.2   There are precisely $n!$ distinct permutations of the integers $1, 2, \ldots, n$.

The proof of this result is left as an exercise.

The elements in the permutation $(1, 2, \ldots, n)$ are said to be in their natural increasing order. We now introduce a number that describes how far a given permutation is from its natural order. For $i \neq j$, the pair of elements p_i and p_j in the permutation $(p_1, p_2, \ldots, p_n)$ are said to be **inverted** if they are out of their natural order; that is, if $p_i > p_j$ with $i < j$. If this is the case, we say that (p_i, p_j) is an **inversion**. For example, in the permutation $(4, 2, 3, 1)$, the pairs $(4, 2), (4, 3), (4, 1), (2, 1)$, and $(3, 1)$ are all out of their natural order. Consequently, there are a total of five inversions in this permutation. In general we let $N(p_1, p_2, \ldots, p_n)$ denote the total number of inversions in the permutation $(p_1, p_2, \ldots, p_n)$.

Example 3.1.3   Find the number of inversions in the permutations $(1, 3, 2, 4, 5)$ and $(2, 4, 5, 3, 1)$.

Solution: The only pair of elements in the permutation $(1, 3, 2, 4, 5)$ that is out of natural order is $(3, 2)$, so $N(1, 3, 2, 4, 5) = 1$.

The permutation $(2, 4, 5, 3, 1)$ has the following pairs of elements out of natural order: $(2, 1), (4, 3), (4, 1), (5, 3), (5, 1)$, and $(3, 1)$. Thus, $N(2, 4, 5, 3, 1) = 6$. $\square$

It can be shown that the number of inversions gives the minimum number of adjacent interchanges of elements in the permutation that are required to restore the permutation to its natural increasing order. This justifies the claim that the number of inversions describes how far from natural order a given permutation is. For example, $N(3, 2, 1) = 3$, and the permutation $(3, 2, 1)$ can be restored to its natural order by the following sequence of adjacent interchanges:

$$(3, 2, 1) \rightarrow (3, 1, 2) \rightarrow (1, 3, 2) \rightarrow (1, 2, 3).$$

The number of inversions enables us to distinguish two different types of permutations as follows.

DEFINITION   3.1.4

1. If $N(p_1, p_2, \ldots, p_n)$ is an even integer (or zero), we say $(p_1, p_2, \ldots, p_n)$ is an **even permutation**. We also say that $(p_1, p_2, \ldots, p_n)$ has **even parity**.

2. If $N(p_1, p_2, \ldots, p_n)$ is an odd integer, we say $(p_1, p_2, \ldots, p_n)$ is an **odd permutation**. We also say that $(p_1, p_2, \ldots, p_n)$ has **odd parity**.

Example 3.1.5   The permutation $(4, 1, 3, 2)$ has even parity, since we have $N(4, 1, 3, 2) = 4$, whereas $(3, 2, 1, 4)$ is an odd permutation since $N(3, 2, 1, 4) = 3$. □

We associate a plus or a minus sign with a permutation, depending on whether it has even or odd parity, respectively. The sign associated with the permutation $(p_1, p_2, \ldots, p_n)$ can be specified by the indicator $\sigma(p_1, p_2, \ldots, p_n)$, defined in terms of the number of inversions as follows:

$$\sigma(p_1, p_2, \ldots, p_n) = \begin{cases} +1 & \text{if } (p_1, p_2, \ldots, p_n) \text{ has even parity,} \\ -1 & \text{if } (p_1, p_2, \ldots, p_n) \text{ has odd parity.} \end{cases}$$

Hence,

$$\boxed{\sigma(p_1, p_2, \ldots, p_n) = (-1)^{N(p_1, p_2, \ldots, p_n)}.}$$

Example 3.1.6   It follows from Example 3.1.3 that

$$\sigma(1, 3, 2, 4, 5) = (-1)^1 = -1,$$

whereas

$$\sigma(2, 4, 5, 3, 1) = (-1)^6 = 1.$$ □

The proofs of some of our later results will depend upon the next theorem.

Theorem 3.1.7   If any two elements in a permutation are interchanged, then the parity of the resulting permutation is opposite to that of the original permutation.

Proof   We first show that interchanging two adjacent terms in a permutation changes its parity. Consider an arbitrary permutation $(p_1, \ldots, p_k, p_{k+1}, \ldots, p_n)$, and suppose we interchange the adjacent elements p_k and p_{k+1}. Then

- If $p_k > p_{k+1}$, then

$$N(p_1, p_2, \ldots, p_{k+1}, p_k, \ldots, p_n) = N(p_1, p_2, \ldots, p_k, p_{k+1}, \ldots, p_n) - 1,$$

- If $p_k < p_{k+1}$, then

$$N(p_1, p_2, \ldots, p_{k+1}, p_k, \ldots, p_n) = N(p_1, p_2, \ldots, p_k, p_{k+1}, \ldots, p_n) + 1,$$

so that the parity is changed in both cases.

Now suppose we interchange the elements p_i and p_k in the permutation $(p_1, p_2, \ldots, p_i, \ldots, p_k, \ldots, p_n)$. Note that $k - i > 0$. We can accomplish this by successively interchanging adjacent elements. In moving p_k to the ith position, we perform $k - i$ interchanges involving adjacent terms, and the resulting permutation is

$$(p_1, p_2, \ldots, p_k, p_i, \ldots, p_{k-1}, p_{k+1}, \ldots, p_n).$$

Next we move p_i to the kth position. A moment's thought shows that this requires $(k - i) - 1$ interchanges of adjacent terms. Thus, the total number of adjacent interchanges involved in interchanging the elements p_i and p_k is $2(k - i) - 1$, which is always an odd integer. Since each adjacent interchange changes the parity, the permutation resulting from an odd number of adjacent interchanges has opposite parity to the original permutation.   ■

At this point, we are ready to see how permutations can facilitate the definition of the determinant. From the expression (3.1.2) for the 3×3 determinant, we see that the row indices of each term have been arranged in their natural increasing order and that the column indices are each a permutation (p_1, p_2, p_3) of 1, 2, 3. Further, the sign attached to each term coincides with the sign of the permutation of the corresponding column indices; that is, $\sigma(p_1, p_2, p_3)$. These observations motivate the following general definition of the determinant of an $n \times n$ matrix:

DEFINITION   3.1.8

Let $A = [a_{ij}]$ be an $n \times n$ matrix. The **determinant of** A, denoted $\det(A)$, is defined as follows:

$$\det(A) = \sum \sigma(p_1, p_2, \ldots, p_n) a_{1p_1} a_{2p_2} a_{3p_3} \cdots a_{np_n}, \qquad (3.1.3)$$

where the summation is over the $n!$ distinct permutations $(p_1, p_2, \ldots, p_n)$ of the integers $1, 2, 3, \ldots, n$. The determinant of an $n \times n$ matrix is said to have **order** n.

We sometimes denote $\det(A)$ by

$$\begin{vmatrix} a_{11} & a_{12} & \cdots & a_{1n} \\ a_{21} & a_{22} & \cdots & a_{2n} \\ \vdots & \vdots & \vdots & \vdots \\ a_{n1} & a_{n2} & \cdots & a_{nn} \end{vmatrix}.$$

Thus, for example, from (3.1.1), we have

$$\begin{vmatrix} a_{11} & a_{12} \\ a_{21} & a_{22} \end{vmatrix} = a_{11}a_{22} - a_{12}a_{21}.$$

Example 3.1.9   Use Definition 3.1.8 to derive the expression for the determinant of order 3.

Solution: When $n = 3$, (3.1.3) reduces to

$$\det(A) = \sum \sigma(p_1, p_2, p_3) a_{1p_1} a_{2p_2} a_{3p_3},$$

where the summation is over the $3! = 6$ permutations of 1, 2, 3. It follows that the six terms in this summation are

$$a_{11}a_{22}a_{33}, \quad a_{11}a_{23}a_{32}, \quad a_{12}a_{21}a_{33}, \quad a_{12}a_{23}a_{31}, \quad a_{13}a_{21}a_{32}, \quad a_{13}a_{22}a_{31},$$

so that

$$\det(A) = \sigma(1, 2, 3)a_{11}a_{22}a_{33} + \sigma(1, 3, 2)a_{11}a_{23}a_{32} + \sigma(2, 1, 3)a_{12}a_{21}a_{33}$$
$$+ \sigma(2, 3, 1)a_{12}a_{23}a_{31} + \sigma(3, 1, 2)a_{13}a_{21}a_{32} + \sigma(3, 2, 1)a_{13}a_{22}a_{31}.$$

To obtain the values of each $\sigma(p_1, p_2, p_3)$, we determine the parity for each permutation (p_1, p_2, p_3). We find that

$$\sigma(1, 2, 3) = +1, \qquad \sigma(1, 3, 2) = -1, \qquad \sigma(2, 1, 3) = -1,$$
$$\sigma(2, 3, 1) = +1, \qquad \sigma(3, 1, 2) = +1, \qquad \sigma(3, 2, 1) = -1.$$

Hence,

$$\det(A) = \begin{vmatrix} a_{11} & a_{12} & a_{13} \\ a_{21} & a_{22} & a_{23} \\ a_{31} & a_{32} & a_{33} \end{vmatrix}$$

$$= a_{11}a_{22}a_{33} + a_{12}a_{23}a_{31} + a_{13}a_{21}a_{32} - a_{11}a_{23}a_{32} - a_{12}a_{21}a_{33} - a_{13}a_{22}a_{31}.$$

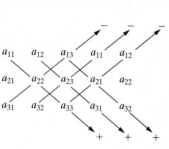

Figure 3.1.1: A schematic for obtaining the determinant of a 3×3 matrix $A = [a_{ij}]$.

A simple schematic for obtaining the terms in the determinant of order 3 is given in Figure 3.1.1. By taking the product of the elements joined by each arrow and attaching the indicated sign to the result, we obtain the six terms in the determinant of the 3×3 matrix $A = [a_{ij}]$. Note that this technique for obtaining the terms in a determinant *does not* generalize to determinants of $n \times n$ matrices with $n > 3$.

Example 3.1.10   Evaluate

$$\text{(a)} \quad |-3|. \qquad \text{(b)} \quad \begin{vmatrix} 3 & -2 \\ 1 & 4 \end{vmatrix}. \qquad \text{(c)} \quad \begin{vmatrix} 1 & 2 & -3 \\ 4 & -1 & 2 \\ 0 & 3 & 1 \end{vmatrix}.$$

Solution:

(a) $|-3| = -3$. In the case of a 1×1 matrix, the reader is cautioned not to confuse the vertical bars notation for the determinant with absolute value bars.

(b) $\begin{vmatrix} 3 & -2 \\ 1 & 4 \end{vmatrix} = (3)(4) - (-2)(1) = 14.$

(c) In this case, the schematic in Figure 3.1.1 is

$$\begin{array}{ccccc} 1 & 2 & -3 & 1 & 2 \\ 4 & -1 & 2 & 4 & -1 \\ 0 & 3 & 1 & 0 & 3 \end{array}$$

so that

$$\begin{vmatrix} 1 & 2 & -3 \\ 4 & -1 & 2 \\ 0 & 3 & 1 \end{vmatrix} = (1)(-1)(1) + (2)(2)(0)$$

$$+ (-3)(4)(3) - (0)(-1)(-3) - (3)(2)(1) - (1)(4)(2) = -51.$$

We now to turn to some geometric applications of the determinant.

Geometric Interpretation of the Determinants of Orders Two and Three

If $\mathbf{a}$ and $\mathbf{b}$ are two vectors in space, we recall that their dot product is the scalar

$$\mathbf{a} \cdot \mathbf{b} = ||\mathbf{a}|| \, ||\mathbf{b}|| \cos \theta, \qquad (3.1.4)$$

where θ is the angle between $\mathbf{a}$ and $\mathbf{b}$, and $||\mathbf{a}||$ and $||\mathbf{b}||$ denote the lengths of $\mathbf{a}$ and $\mathbf{b}$, respectively. On the other hand, the cross product of $\mathbf{a}$ and $\mathbf{b}$ is the vector

$$\mathbf{a} \times \mathbf{b} = ||\mathbf{a}|| \, ||\mathbf{b}|| \sin \theta \, \mathbf{n}, \qquad (3.1.5)$$

where **n** denotes a unit vector[1] that is perpendicular to the plane of **a** and **b** and chosen in such a way that {**a**, **b**, **n**} is a right-handed set of vectors. If **i**, **j**, **k** denote the unit vectors pointing along the positive x-, y- and z-axes, respectively, of a rectangular Cartesian coordinate system and $\mathbf{a} = a_1\mathbf{i} + a_2\mathbf{j} + a_3\mathbf{k}$, $\mathbf{b} = b_1\mathbf{i} + b_2\mathbf{j} + b_3\mathbf{k}$, then Equation (3.1.5) can be expressed in component form as

$$\mathbf{a} \times \mathbf{b} = (a_2b_3 - a_3b_2)\mathbf{i} + (a_3b_1 - a_1b_3)\mathbf{j} + (a_1b_2 - a_2b_1)\mathbf{k}. \tag{3.1.6}$$

This can be remembered most easily in the compact form

$$\mathbf{a} \times \mathbf{b} = \begin{vmatrix} \mathbf{i} & \mathbf{j} & \mathbf{k} \\ a_1 & a_2 & a_3 \\ b_1 & b_2 & b_3 \end{vmatrix},$$

whose validity is readily checked by using the schematic in Figure 3.1.1. We will use the equations above to establish the following theorem.

Theorem 3.1.11

1. The area of a parallelogram with sides determined by the vectors $\mathbf{a} = a_1\mathbf{i} + a_2\mathbf{j}$ and $\mathbf{b} = b_1\mathbf{i} + b_2\mathbf{j}$ is

$$\text{Area} = |\det(A)|,$$

where $A = \begin{bmatrix} a_1 & a_2 \\ b_1 & b_2 \end{bmatrix}$.

2. The volume of a parallelepiped determined by the vectors $\mathbf{a} = a_1\mathbf{i} + a_2\mathbf{j} + a_3\mathbf{k}$, $\mathbf{b} = b_1\mathbf{i} + b_2\mathbf{j} + b_3\mathbf{k}$, $\mathbf{c} = c_1\mathbf{i} + c_2\mathbf{j} + c_3\mathbf{k}$ is

$$\text{Volume} = |\det(A)|,$$

where $A = \begin{bmatrix} a_1 & a_2 & a_3 \\ b_1 & b_2 & b_3 \\ c_1 & c_2 & c_3 \end{bmatrix}$.

Before presenting the proof of this theorem, we make some remarks and give two examples.

Remarks

1. The vertical bars appearing in the formulas in Theorem 3.1.11 denote the absolute value of the number $\det(A)$.

2. We see from the expression for the volume of a parallelepiped that the condition for three vectors to lie in the same plane (i.e., the parallelepiped has zero volume) is that $\det(A) = 0$. This will be a useful result in the next chapter.

Example 3.1.12   Find the area of the parallelogram containing the points $(0, 0)$, $(1, 2)$, $(3, 4)$ and $(4, 6)$.

Solution: The sides of the parallelogram are determined by the vectors $\mathbf{a} = \mathbf{i} + 2\mathbf{j}$ and $\mathbf{b} = 3\mathbf{i} + 4\mathbf{j}$. According to part 1 of Theorem 3.1.11, the area of the parallelogram is

$$\left| \det \begin{bmatrix} 1 & 2 \\ 3 & 4 \end{bmatrix} \right| = |(1)(4) - (2)(3)| = |-2| = 2. \qquad \square$$

---

[1] A unit vector is a vector of length 1.

Example 3.1.13   Determine whether or not the vectors $\mathbf{a} = \mathbf{i} + 2\mathbf{j} + 3\mathbf{k}$, $\mathbf{b} = 4\mathbf{i} + 5\mathbf{j} + 6\mathbf{k}$, and $\mathbf{c} = -5\mathbf{i} + (-7)\mathbf{j} + (-9)\mathbf{k}$ lie in a single plane in 3-space.

Solution: By Remark 2 above, it suffices to determine whether or not the volume of the parallelepiped determined by the three vectors is zero or not. To do this, we use part 2 of Theorem 3.1.11:

$$
\text{Volume} = \left| \det \begin{bmatrix} 1 & 2 & 3 \\ 4 & 5 & 6 \\ -5 & -7 & -9 \end{bmatrix} \right|
$$

$$
= \left| \begin{array}{l} (1)(5)(-9) + (2)(6)(-5) + (3)(4)(-7) \\ -(-5)(5)(3) - (-7)(6)(1) - (-9)(4)(2) \end{array} \right| = 0,
$$

which shows that the three vectors do lie in a single plane.   □

Now we turn to the

Proof of Theorem 3.1.11:

1. The area of the parallelogram is

$$
\text{area} = (\text{length of base}) \times (\text{perpendicular height}).
$$

From Figure 3.1.2, this can be written as

$$
\text{Area} = ||\mathbf{a}||h = ||\mathbf{a}||\,||\mathbf{b}||\,|\sin\theta| = ||\mathbf{a} \times \mathbf{b}||. \tag{3.1.7}
$$

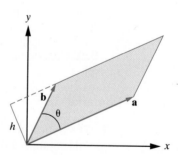

Figure 3.1.2: Determining the area of a parallelogram.

Since the $\mathbf{k}$ components of $\mathbf{a}$ and $\mathbf{b}$, a_3 and b_3, are both zero (since the vectors lie in the xy-plane), substitution from Equation (3.1.6) yields

$$
\text{Area} = ||(a_1 b_2 - a_2 b_1)\mathbf{k}|| = |a_1 b_2 - a_2 b_1| = |\det(A)|.
$$

2. The volume of the parallelepiped is

$$
\text{Volume} = (\text{area of base}) \times (\text{perpendicular height}).
$$

The base is determined by the vectors $\mathbf{b}$ and $\mathbf{c}$ (see Figure 3.1.3), and its area can be written as $||\mathbf{b} \times \mathbf{c}||$, in similar fashion to what was done in (3.1.7). From Figure 3.1.3 and Equation (3.1.4), we therefore have

$$\text{Volume} = ||\mathbf{b} \times \mathbf{c}||h = ||\mathbf{b} \times \mathbf{c}|| \, ||\mathbf{a}|| |\cos \psi| = ||\mathbf{b} \times \mathbf{c}|| \, |\mathbf{a} \cdot \mathbf{n}|,$$

where $\mathbf{n}$ is a unit vector that is perpendicular to the plane containing $\mathbf{b}$ and $\mathbf{c}$. We can now use Equations (3.1.5) and (3.1.6) to obtain

$$\text{Volume} = ||\mathbf{b} \times \mathbf{c}|| \, ||\mathbf{a}|| \, |\cos \psi| = |\mathbf{a} \cdot (\mathbf{b} \times \mathbf{c})|$$

$$= \left| (a_1\mathbf{i} + a_2\mathbf{j} + a_3\mathbf{k}) \cdot \left[(b_2c_3 - b_3c_2)\mathbf{i} + (b_3c_1 - b_1c_3)\mathbf{j} + (b_1c_2 - b_2c_1)\mathbf{k} \right] \right|$$

$$= |a_1(b_2c_3 - b_3c_2) + a_2(b_3c_1 - b_1c_3) + a_3(b_1c_2 - b_2c_1)|$$

$$= |\det(A)|,$$

as required.

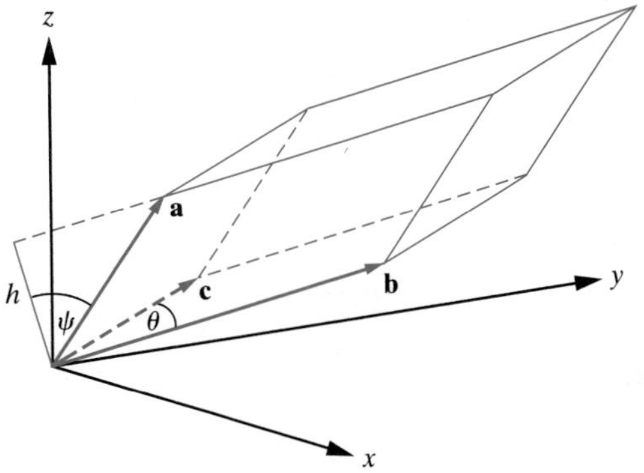

Figure 3.1.3: Determining the volume of a parallelepiped.

Exercises for 3.1

Key Terms

Permutation, Inversion, Parity, Determinant, Order, Dot product, Cross product.

Skills

- Be able to compute determinants by using Definition 3.1.8.

- Be able to list permutations of $1, 2, \ldots, n$.

- Be able to find the number of inversions of a given permutation and thus determine its parity.

- Be able to compute the area of a parallelogram with sides determined by vectors in $\mathbb{R}^2$.

- Be able to compute the volume of a parallelogram with sides determined by vectors in $\mathbb{R}^3$.

True-False Review

For Questions 1–8, decide if the given statement is **true** or **false**, and give a brief justification for your answer. If true, you can quote a relevant definition or theorem from the text. If false, provide an example, illustration, or brief explanation of why the statement is false.

1. If A is a 2×2 lower triangular matrix, then $\det(A)$ is the product of the elements on the main diagonal of A.

2. If A is a 3×3 upper triangular matrix, then $\det(A)$ is the product of the elements on the main diagonal of A.

3. The volume of the parallelepiped whose sides are determined by the vectors $\mathbf{a}$, $\mathbf{b}$, and $\mathbf{c}$ is given by $\det(A)$, where $A = [\mathbf{a}, \mathbf{b}, \mathbf{c}]$.

4. There are the same number of permutations of $\{1, 2, 3, 4\}$ of even parity as there are of odd parity.

5. If A and B are 2×2 matrices, then $\det(A + B) = \det(A) + \det(B)$.

6. The determinant of a matrix whose elements are all positive must be positive.

7. A matrix containing a row of zeros must have zero determinant.

8. Three vectors $\mathbf{v}_1$, $\mathbf{v}_2$, and $\mathbf{v}_3$ in $\mathbb{R}^3$ are coplanar if and only if the determinant of the 3×3 matrix $[\mathbf{v}_1, \mathbf{v}_2, \mathbf{v}_3]$ is zero.

Problems

For Problems 1–6, determine the parity of the given permutation.

1. $(2, 1, 3, 4)$.

2. $(1, 3, 2, 4)$.

3. $(1, 4, 3, 5, 2)$.

4. $(5, 4, 3, 2, 1)$.

5. $(1, 5, 2, 4, 3)$.

6. $(2, 4, 6, 1, 3, 5)$.

7. Use the definition of a determinant to derive the general expression for the determinant of A if

$$A = \begin{bmatrix} a_{11} & a_{12} \\ a_{21} & a_{22} \end{bmatrix}.$$

For Problems 8–15, evaluate the determinant of the given matrix.

8. $A = \begin{bmatrix} 1 & -1 \\ 2 & 3 \end{bmatrix}$.

9. $A = \begin{bmatrix} 2 & -1 \\ 6 & -3 \end{bmatrix}$.

10. $A = \begin{bmatrix} -4 & 10 \\ -1 & 8 \end{bmatrix}$.

11. $A = \begin{bmatrix} 1 & -1 & 0 \\ 2 & 3 & 6 \\ 0 & 2 & -1 \end{bmatrix}$.

12. $A = \begin{bmatrix} 2 & 1 & 5 \\ 4 & 2 & 3 \\ 9 & 5 & 1 \end{bmatrix}$.

13. $A = \begin{bmatrix} 0 & 0 & 2 \\ 0 & -4 & 1 \\ -1 & 5 & -7 \end{bmatrix}$.

14. $A = \begin{bmatrix} 1 & 2 & 3 & 4 \\ 0 & 5 & 6 & 7 \\ 0 & 0 & 8 & 9 \\ 0 & 0 & 0 & 10 \end{bmatrix}$.

15. $A = \begin{bmatrix} 0 & 0 & 2 & 0 \\ 5 & 0 & 0 & 0 \\ 0 & 0 & 0 & 3 \\ 0 & 2 & 0 & 0 \end{bmatrix}$.

For Problems 16–21, evaluate the given determinant.

16. $\begin{vmatrix} \pi & \pi^2 \\ \sqrt{2} & 2\pi \end{vmatrix}$.

17. $\begin{vmatrix} 2 & 3 & -1 \\ 1 & 4 & 1 \\ 3 & 1 & 6 \end{vmatrix}$.

18. $\begin{vmatrix} 3 & 2 & 6 \\ 2 & 1 & -1 \\ -1 & 1 & 4 \end{vmatrix}$.

19. $\begin{vmatrix} 2 & 3 & 6 \\ 0 & 1 & 2 \\ 1 & 5 & 0 \end{vmatrix}$.

20. $\begin{vmatrix} \sqrt{\pi} & e^2 & e^{-1} \\ \sqrt{67} & 1/30 & 2001 \\ \pi & \pi^2 & \pi^3 \end{vmatrix}$.

21. $\begin{vmatrix} e^{2t} & e^{3t} & e^{-4t} \\ 2e^{2t} & 3e^{3t} & -4e^{-4t} \\ 4e^{2t} & 9e^{3t} & 16e^{-4t} \end{vmatrix}$.

In Problems 22–23, we explore a relationship between determinants and solutions to a differential equation. The 3×3 matrix consisting of solutions to a differential equation and their derivatives is called the **Wronskian** and, as we will see in later chapters, plays a pivotal role in the theory of differential equations.

22. Verify that $y_1(x) = \cos 2x$, $y_2(x) = \sin 2x$, and $y_3(x) = e^x$ are solutions to the differential equation

$$y''' - y'' + 4y' - 4y = 0,$$

and show that $\begin{vmatrix} y_1 & y_2 & y_3 \\ y_1' & y_2' & y_3' \\ y_1'' & y_2'' & y_3'' \end{vmatrix}$ is nonzero on any interval.

23. (a) Verify that $y_1(x) = e^x$, $y_2(x) = \cosh x$, and $y_3(x) = \sinh x$ are solutions to the differential equation

$$y''' - y'' - y' + y = 0,$$

and show that $\begin{vmatrix} y_1 & y_2 & y_3 \\ y_1' & y_2' & y_3' \\ y_1'' & y_2'' & y_3'' \end{vmatrix}$ is identically zero.

(b) Determine nonzero constants d_1, d_2, and d_3 such that

$$d_1 y_1 + d_2 y_2 + d_3 y_3 = 0.$$

24. (a) Write all 24 distinct permutations of the integers $1, 2, 3, 4$.

(b) Determine the parity of each permutation in part (a).

(c) Use parts (a) and (b) to derive the expression for a determinant of order 4.

For Problems 25–27, use the previous problem to compute the determinant of A.

25. $A = \begin{bmatrix} 1 & -1 & 0 & 1 \\ 3 & 0 & 2 & 5 \\ 2 & 1 & 0 & 3 \\ 9 & -1 & 2 & 1 \end{bmatrix}$.

26. $A = \begin{bmatrix} 1 & 1 & 0 & 1 \\ 3 & 1 & -2 & 3 \\ 2 & 3 & 1 & 2 \\ -2 & 3 & 5 & -2 \end{bmatrix}$.

27. $A = \begin{bmatrix} 0 & 1 & 2 & 3 \\ 2 & 0 & 3 & 4 \\ 3 & 4 & 0 & 5 \\ 4 & 5 & 6 & 0 \end{bmatrix}$.

28. Use Problem 27 to find the determinant of A, where

$$A = \begin{bmatrix} 0 & 1 & 2 & 3 & 0 \\ 2 & 0 & 3 & 4 & 0 \\ 3 & 4 & 0 & 5 & 0 \\ 4 & 5 & 6 & 0 & 0 \\ 0 & 0 & 0 & 0 & 7 \end{bmatrix}.$$

29. (a) If $A = \begin{bmatrix} a_{11} & a_{12} \\ a_{21} & a_{22} \end{bmatrix}$ and c is a constant, verify that

$$\det(cA) = c^2 \det(A).$$

(b) Use the definition of a determinant to prove that if A is an $n \times n$ matrix and c is a constant, then

$$\det(cA) = c^n \det(A).$$

For Problems 30–33, determine whether the given expression is a term in the determinant of order 5. If it is, determine whether the permutation of the column indices has even or odd parity and hence find whether the term has a plus or a minus sign attached to it.

30. $a_{11}a_{25}a_{33}a_{42}a_{54}$.

31. $a_{11}a_{23}a_{34}a_{43}a_{52}$.

32. $a_{13}a_{25}a_{31}a_{44}a_{42}$.

33. $a_{11}a_{32}a_{24}a_{43}a_{55}$.

For Problems 34–37, determine the values of the indices p and q such that the following are terms in a determinant of order 4. In each case, determine the number of inversions in the permutation of the column indices and hence find the appropriate sign that should be attached to each term.

34. $a_{13}a_{p4}a_{32}a_{2q}$.

35. $a_{21}a_{3q}a_{p2}a_{43}$.

36. $a_{3q}a_{p4}a_{13}a_{42}$.

37. $a_{pq}a_{34}a_{13}a_{42}$.

38. The *alternating symbol* ϵ_{ijk} is defined by

$$\epsilon_{ijk} = \begin{cases} 1, & \text{if } (ijk) \text{ is an even permutation of } 1, 2, 3, \\ -1, & \text{if } (ijk) \text{ is an odd permutation of } 1, 2, 3, \\ 0, & \text{otherwise.} \end{cases}$$

(a) Write all nonzero ϵ_{ijk}, for $1 \le i \le 3$, $1 \le j \le 3$, $1 \le k \le 3$.

(b) If $A = [a_{ij}]$ is a 3×3 matrix, verify that

$$\det(A) = \sum_{i=1}^{3} \sum_{j=1}^{3} \sum_{k=1}^{3} \epsilon_{ijk} a_{1i} a_{2j} a_{3k}.$$

39. If A is the general $n \times n$ matrix, determine the sign attached to the term

$$a_{1n} a_{2\ n-1} a_{3\ n-2} \cdots a_{n1},$$

which arises in $\det(A)$.

40. ◇ Use some form of technology to evaluate the determinants in Problems 16–21.

41. ◇ Let A be an arbitrary 4×4 matrix. By experimenting with various elementary row operations, conjecture how elementary row operations applied to A affect the value of $\det(A)$.

42. ◇ Verify that $y_1(x) = e^{-2x} \cos 3x$, $y_2(x) = e^{-2x} \sin 3x$, and $y_3(x) = e^{-4x}$ are solutions to the differential equation

$$y''' + 8y'' + 29y' + 52y = 0,$$

and show that $\begin{vmatrix} y_1 & y_2 & y_3 \\ y_1' & y_2' & y_3' \\ y_1'' & y_2'' & y_3'' \end{vmatrix}$ is nonzero on any interval.

3.2   Properties of Determinants

For large values of n, evaluating a determinant of order n using the definition given in the previous section is not very practical, since the number of terms is $n!$ (for example, a determinant of order 10 contains 3,628,800 terms). In the next two sections, we develop better techniques for evaluating determinants. The following theorem suggests one way to proceed.

Theorem 3.2.1   If A is an $n \times n$ upper or lower triangular matrix, then

$$\det(A) = a_{11}a_{22}a_{33} \cdots a_{nn} = \prod_{i=1}^{n} a_{ii}.$$

Proof We use the definition of the determinant to prove the result in the upper triangular case. From Equation (3.1.3),

$$\det(A) = \sum \sigma(p_1, p_2, \ldots, p_n) a_{1p_1} a_{2p_2} a_{3p_3} \cdots a_{np_n}. \tag{3.2.1}$$

If A is upper triangular, then $a_{ij} = 0$ whenever $i > j$, and therefore the only nonzero terms in the preceding summation are those with $p_i \geq i$ for all i. Since all the p_i must be distinct, the only possibility is (by applying $p_i \geq i$ to $i = n, n-1, \ldots, 2, 1$ in turn)

$$p_i = i, \qquad i = 1, 2, \ldots, n,$$

and so Equation (3.2.1) reduces to the single term

$$\det(A) = \sigma(1, 2, \ldots, n) a_{11} a_{22} \cdots a_{nn}.$$

Since $\sigma(1, 2, \ldots, n) = 1$, it follows that

$$\det(A) = a_{11} a_{22} \cdots a_{nn}.$$

The proof in the lower triangular case is left as an exercise (Problem 47).   ■

Example 3.2.2   According to the previous theorem,

$$\begin{vmatrix} 2 & 5 & -1 & 3 \\ 0 & -1 & 0 & 4 \\ 0 & 0 & 7 & 8 \\ 0 & 0 & 0 & 5 \end{vmatrix} = (2)(-1)(7)(5) = -70. \qquad \square$$

Theorem 3.2.1 shows that it is easy to compute the determinant of an upper or lower triangular matrix. Recall from Chapter 2 that *any* matrix can be reduced to row-echelon form by a sequence of elementary row operations. In the case of an $n \times n$ matrix, any row-echelon form will be upper triangular. Theorem 3.2.1 suggests, therefore, that we should consider how elementary row operations performed on a matrix A alter the value of $\det(A)$.

Elementary Row Operations and Determinants

Let A be an $n \times n$ matrix.

P1. If B is the matrix obtained by permuting two rows of A, then

$$\det(B) = - \det(A).$$

P2. If B is the matrix obtained by multiplying one row of A by any[2] scalar k, then

$$\det(B) = k \, \det(A).$$

P3. If B is the matrix obtained by adding a multiple of any row of A to a different row of A, then

$$\det(B) = \det(A).$$

The proofs of these properties are given at the end of this section.

Remark   The main use of P2 is that it enables us to factor a common multiple of the entries of a particular row out of the determinant. For example, if

$$A = \begin{bmatrix} -1 & 4 \\ 3 & -2 \end{bmatrix} \quad \text{and} \quad B = \begin{bmatrix} -5 & 20 \\ 3 & -2 \end{bmatrix},$$

where B is obtained from A by multiplying the first row of A by 5, then we have

$$\det(B) = 5 \det(A) = 5[(-1)(-2) - (3)(4)] = 5(-10) = -50.$$

We now illustrate how the foregoing properties P1–P3, together with Theorem 3.2.1, can be used to evaluate a determinant. The basic idea is the same as that for Gaussian elimination. We use elementary row operations to reduce the determinant to upper triangular form and then use Theorem 3.2.1 to evaluate the resulting determinant.

Warning: When using the properties P1–P3 to simplify a determinant, one must remember to take account of any change that arises in the value of the determinant from the operations that have been performed on it.

Example 3.2.3   Evaluate $\begin{vmatrix} 2 & -1 & 3 & 7 \\ 1 & -2 & 4 & 3 \\ 3 & 4 & 2 & -1 \\ 2 & -2 & 8 & -4 \end{vmatrix}$.

---

[2]This statement is even true if $k = 0$.

Solution: We have

$$
\begin{vmatrix} 2 & -1 & 3 & 7 \\ 1 & -2 & 4 & 3 \\ 3 & 4 & 2 & -1 \\ 2 & -2 & 8 & -4 \end{vmatrix} \overset{1}{=} 2 \begin{vmatrix} 2 & -1 & 3 & 7 \\ 1 & -2 & 4 & 3 \\ 3 & 4 & 2 & -1 \\ 1 & -1 & 4 & -2 \end{vmatrix} \overset{2}{=} -2 \begin{vmatrix} 1 & -2 & 4 & 3 \\ 2 & -1 & 3 & 7 \\ 3 & 4 & 2 & -1 \\ 1 & -1 & 4 & -2 \end{vmatrix} \overset{3}{=} -2 \begin{vmatrix} 1 & -2 & 4 & 3 \\ 0 & 3 & -5 & 1 \\ 0 & 10 & -10 & -10 \\ 0 & 1 & 0 & -5 \end{vmatrix}
$$

$$
\overset{4}{=} 2 \begin{vmatrix} 1 & -2 & 4 & 3 \\ 0 & 1 & 0 & -5 \\ 0 & 10 & -10 & -10 \\ 0 & 3 & -5 & 1 \end{vmatrix} \overset{5}{=} 20 \begin{vmatrix} 1 & -2 & 4 & 3 \\ 0 & 1 & 0 & -5 \\ 0 & 1 & -1 & -1 \\ 0 & 3 & -5 & 1 \end{vmatrix} \overset{6}{=} 20 \begin{vmatrix} 1 & -2 & 4 & 3 \\ 0 & 1 & 0 & -5 \\ 0 & 0 & -1 & 4 \\ 0 & 0 & -5 & 16 \end{vmatrix}
$$

$$
\overset{7}{=} 20 \begin{vmatrix} 1 & -2 & 4 & 3 \\ 0 & 1 & 0 & -5 \\ 0 & 0 & -1 & 4 \\ 0 & 0 & 0 & -4 \end{vmatrix} = 80.
$$

---

1. $M_4(\frac{1}{2})$ **2.** P_{12} **3.** $A_{12}(-2),\ A_{13}(-3),\ A_{14}(-1)$ **4.** P_{24}

5. $M_3(\frac{1}{10})$ **6.** $A_{23}(-1),\ A_{24}(-3)$ **7.** $A_{34}(-5)$

Theoretical Results for $n \times n$ Matrices and $n \times n$ Linear Systems

In Section 2.8, we established several conditions on an $n \times n$ matrix A that are equivalent to saying that A is invertible. At this point, we are ready to give one additional characterization of invertible matrices in terms of determinants.

Theorem 3.2.4   Let A be an $n \times n$ matrix with real elements. The following conditions on A are equivalent.

    (a) A is invertible.

    (g) $\det(A) \neq 0$.

Proof Let A^* denote the reduced row-echelon form of A. Recall from Chapter 2 that A is invertible if and only if $A^* = I_n$. Since A^* is obtained from A by performing a sequence of elementary row operations, properties P1–P3 of determinants imply that $\det(A)$ is just a *nonzero* multiple of $\det(A^*)$. If A is invertible, then $\det(A^*) = \det(I_n) = 1$, so that $\det(A)$ is nonzero.

Conversely, if $\det(A) \neq 0$, then $\det(A^*) \neq 0$. This implies that $A^* = I_n$, hence A is invertible. ∎

According to Theorem 2.5.9 in the previous chapter, any linear system $A\mathbf{x} = \mathbf{b}$ has either no solution, exactly one solution, or infinitely many solutions. Recall from the Invertible Matrix Theorem that the linear system $A\mathbf{x} = \mathbf{b}$ has a unique solution for every $\mathbf{b}$ in $\mathbb{R}^n$ if and only if A is invertible. Thus, for an $n \times n$ linear system, Theorem 3.2.4 tells us that, for each $\mathbf{b}$ in $\mathbb{R}^n$, the system $A\mathbf{x} = \mathbf{b}$ has a unique solution $\mathbf{x}$ if and only if $\det(A) \neq 0$.

Next, we consider the homogeneous $n \times n$ linear system $A\mathbf{x} = \mathbf{0}$.

Corollary 3.2.5   The homogeneous $n \times n$ linear system $A\mathbf{x} = \mathbf{0}$ has an infinite number of solutions if and only if $\det(A) = 0$, and has only the trivial solution if and only if $\det(A) \neq 0$.

Proof The system $A\mathbf{x} = \mathbf{0}$ clearly has the trivial solution $\mathbf{x} = \mathbf{0}$ under any circumstances. By our remarks above, this must be the unique solution if and only if $\det(A) \neq 0$. The only other possibility, which occurs if and only if $\det(A) = 0$, is that the system has infinitely many solutions.   ∎

Remark   The preceding corollary is *very* important, since we are often interested only in determining the solution properties of a homogeneous linear system and not actually in finding the solutions themselves. We will refer back to this corollary on many occasions throughout the remainder of the text.

Example 3.2.6   Verify that the matrix

$$A = \begin{bmatrix} 1 & -1 & 3 \\ 2 & 4 & -2 \\ 3 & 5 & 7 \end{bmatrix}$$

is invertible. What can be concluded about the solution to $A\mathbf{x} = \mathbf{0}$?

Solution: It is easily shown that $\det(A) = 52 \neq 0$. Consequently, A is invertible. It follows from Corollary 3.2.5 that the homogeneous system $A\mathbf{x} = \mathbf{0}$ has only the trivial solution $(0, 0, 0)$.   □

Example 3.2.7   Verify that the matrix

$$A = \begin{bmatrix} 1 & 0 & 1 \\ 0 & 1 & 0 \\ -3 & 0 & -3 \end{bmatrix}$$

is not invertible and determine a set of real solutions to the system $A\mathbf{x} = \mathbf{0}$.

Solution: By the row operation $A_{13}(3)$, we see that A is row equivalent to the upper triangular matrix

$$B = \begin{bmatrix} 1 & 0 & 1 \\ 0 & 1 & 0 \\ 0 & 0 & 0 \end{bmatrix}.$$

By Theorem 3.2.1, $\det(B) = 0$, and hence B and A are not invertible. We illustrate Corollary 3.2.5 by finding an infinite number of solutions (x_1, x_2, x_3) to $A\mathbf{x} = \mathbf{0}$. Working with the upper triangular matrix B, we may set $x_3 = t$, a free parameter. The second row of the matrix system requires that $x_2 = 0$ and the first row requires that $x_1 + x_3 = 0$, so $x_1 = -x_3 = -t$. Hence, the set of solutions is $\{(-t, 0, t) : t \in \mathbb{R}\}$.   □

Further Properties of Determinants

In addition to elementary row operations, the following properties can also be useful in evaluating determinants.

Let A and B be $n \times n$ matrices.

P4. $\det(A^T) = \det(A)$.

P5. Let $\mathbf{a}_1, \mathbf{a}_2, \ldots, \mathbf{a}_n$ denote the row vectors of A. If the ith row vector of A is the sum of two row vectors, say $\mathbf{a}_i = \mathbf{b}_i + \mathbf{c}_i$, then $\det(A) = \det(B) + \det(C)$, where

$$B = \begin{bmatrix} \mathbf{a}_1 \\ \vdots \\ \mathbf{a}_{i-1} \\ \mathbf{b}_i \\ \mathbf{a}_{i+1} \\ \vdots \\ \mathbf{a}_n \end{bmatrix} \quad \text{and} \quad C = \begin{bmatrix} \mathbf{a}_1 \\ \vdots \\ \mathbf{a}_{i-1} \\ \mathbf{c}_i \\ \mathbf{a}_{i+1} \\ \vdots \\ \mathbf{a}_n \end{bmatrix}.$$

The corresponding property is also true for columns.

P6. If A has a row (or column) of zeros, then $\det(A) = 0$.

P7. If two rows (or columns) of A are the same, then $\det(A) = 0$.

P8. $\det(AB) = \det(A)\det(B)$.

The proofs of these properties are given at the end of the section. The main importance of P4 is the implication that any results regarding determinants that hold for the rows of a matrix also hold for the columns of a matrix. In particular, the properties P1–P3 regarding the effects that elementary row operations have on the determinant can be translated to corresponding statements on the effects that "elementary column operations" have on the determinant. We will use the notations

$$\text{CP}_{ij}, \qquad \text{CM}_i(k), \qquad \text{and} \qquad \text{CA}_{ij}(k)$$

to denote the three types of elementary column operations.

Example 3.2.8   Use only column operations to evaluate

$$\begin{vmatrix} 3 & 6 & -1 & 2 \\ 6 & 10 & 3 & 4 \\ 9 & 20 & 5 & 4 \\ 15 & 34 & 3 & 8 \end{vmatrix}.$$

Solution: We have

$$\begin{vmatrix} 3 & 6 & -1 & 2 \\ 6 & 10 & 3 & 4 \\ 9 & 20 & 5 & 4 \\ 15 & 34 & 3 & 8 \end{vmatrix} \overset{1}{=} 3 \cdot 2^2 \begin{vmatrix} 1 & 3 & -1 & 1 \\ 2 & 5 & 3 & 2 \\ 3 & 10 & 5 & 2 \\ 5 & 17 & 3 & 4 \end{vmatrix} \overset{2}{=} 12 \begin{vmatrix} 1 & 0 & 0 & 0 \\ 2 & -1 & 5 & 0 \\ 3 & 1 & 8 & -1 \\ 5 & 2 & 8 & -1 \end{vmatrix} \overset{3}{=} 12 \begin{vmatrix} 1 & 0 & 0 & 0 \\ 2 & -1 & 0 & 0 \\ 3 & 1 & 13 & -1 \\ 5 & 2 & 18 & -1 \end{vmatrix}$$

$$\overset{4}{=} 12 \begin{vmatrix} 1 & 0 & 0 & 0 \\ 2 & -1 & 0 & 0 \\ 3 & 1 & 13 & 0 \\ 5 & 2 & 18 & \frac{5}{13} \end{vmatrix} = 12(-5) = -60,$$

where we have once more used Theorem 3.2.1.

1. $\text{CM}_1(\frac{1}{3})$, $\text{CM}_2(\frac{1}{2})$, $\text{CM}_4(\frac{1}{2})$ 2. $\text{CA}_{12}(-3)$, $\text{CA}_{13}(1)$, $\text{CA}_{14}(-1)$

3. $\text{CA}_{23}(5)$ 4. $\text{CA}_{34}(\frac{1}{13})$

The property that often gives the most difficulty is P5. We explicitly illustrate its use with an example.

Example 3.2.9   Use property P5 to express

$$\begin{vmatrix} a_1 + b_1 & c_1 + d_1 \\ a_2 + b_2 & c_2 + d_2 \end{vmatrix}$$

as a sum of four determinants.

Solution: Applying P5 to row 1 yields:

$$\begin{vmatrix} a_1 + b_1 & c_1 + d_1 \\ a_2 + b_2 & c_2 + d_2 \end{vmatrix} = \begin{vmatrix} a_1 & c_1 \\ a_2 + b_2 & c_2 + d_2 \end{vmatrix} + \begin{vmatrix} b_1 & d_1 \\ a_2 + b_2 & c_2 + d_2 \end{vmatrix}.$$

Now we apply P5 to row 2 of both of the determinants on the right-hand side to obtain

$$\begin{vmatrix} a_1 + b_1 & c_1 + d_1 \\ a_2 + b_2 & c_2 + d_2 \end{vmatrix} = \begin{vmatrix} a_1 & c_1 \\ a_2 & c_2 \end{vmatrix} + \begin{vmatrix} a_1 & c_1 \\ b_2 & d_2 \end{vmatrix} + \begin{vmatrix} b_1 & d_1 \\ a_2 & c_2 \end{vmatrix} + \begin{vmatrix} b_1 & d_1 \\ b_2 & d_2 \end{vmatrix}.$$

Notice that we could also have applied P5 to the columns of the given determinant.   □

Warning   In view of P5, it may be tempting to believe that if A, B, and C are $n \times n$ matrices such that $A = B + C$, then $\det(A) = \det(B) + \det(C)$. This is not true! Examples abound to show the failure of this equation. For instance, if we take $B = I_2$ and $C = -I_2$, then $\det(A) = \det(0_2) = 0$, while $\det(B) = \det(C) = 1$. Thus, $\det(B) + \det(C) = 1 + 1 = 2 \neq 0$.

Next, we supply some examples of the last two properties, P7 and P8.

Example 3.2.10   Evaluate

(a) $\begin{vmatrix} 1 & 2 & -3 & 1 \\ -2 & 4 & 6 & 2 \\ -3 & -6 & 9 & 3 \\ 2 & 11 & -6 & 4 \end{vmatrix}.$

(b) $\begin{vmatrix} 2 - 4x & -4 & 2 \\ 5 + 3x & 3 & -3 \\ 1 - 2x & -2 & 1 \end{vmatrix}.$

Solution:

(a)   We have

$$\begin{vmatrix} 1 & 2 & -3 & 1 \\ -2 & 4 & 6 & 2 \\ -3 & -6 & 9 & 3 \\ 2 & 11 & -6 & 4 \end{vmatrix} \overset{1}{=} -3 \begin{vmatrix} 1 & 2 & 1 & 1 \\ -2 & 4 & -2 & 2 \\ -3 & -6 & -3 & 3 \\ 2 & 11 & 2 & 4 \end{vmatrix} = 0,$$

since the first and third columns of the latter matrix are identical (see P7).

$$\boxed{\text{1. } CM_3\left(-\frac{1}{3}\right)}$$

(b) Applying P5 to the first column, we have

$$\begin{vmatrix} 2-4x & -4 & 2 \\ 5+3x & 3 & -3 \\ 1-2x & -2 & 1 \end{vmatrix} = \begin{vmatrix} 2 & -4 & 2 \\ 5 & 3 & -3 \\ 1 & -2 & 1 \end{vmatrix} + \begin{vmatrix} -4x & -4 & 2 \\ 3x & 3 & -3 \\ -2x & -2 & 1 \end{vmatrix}$$

$$= 2\begin{vmatrix} 1 & -2 & 1 \\ 5 & 3 & -3 \\ 1 & -2 & 1 \end{vmatrix} + x\begin{vmatrix} -4 & -4 & 2 \\ 3 & 3 & -3 \\ -2 & -2 & 1 \end{vmatrix} = 0 + 0 = 0,$$

since the first and third rows of the first matrix agree, and the first and second columns of the second matrix agree.   □

Example 3.2.11   If

$$A = \begin{bmatrix} \sin\phi & \cos\phi \\ -\cos\phi & \sin\phi \end{bmatrix} \quad \text{and} \quad B = \begin{bmatrix} \cos\theta & -\sin\theta \\ \sin\theta & \cos\theta \end{bmatrix},$$

show that $\det(AB) = 1$.

Solution: Using P8, we have

$$\det(AB) = \det(A)\det(B) = (\sin^2\phi + \cos^2\phi)(\cos^2\theta + \sin^2\theta) = 1 \cdot 1 = 1. \quad □$$

Example 3.2.12   Find all x satisfying

$$\begin{vmatrix} x^2 & x & 1 \\ 1 & 1 & 1 \\ 4 & 2 & 1 \end{vmatrix} = 0.$$

Solution: If we expanded this determinant according to Definition 3.1.8 (or using the schematic in Figure 3.1.1), then we would have a quadratic equation in x. Thus, there are at most two distinct values of x that satisfy the equation. By inspection, the determinant vanishes when $x = 1$ (since the first two rows of the matrix coincide in this case), and it vanishes when $x = 2$ (since the first and third rows of the matrix coincide in this case). Consequently, the two values of x satisfying the given equation are $x = 1$ and $x = 2$.   □

Proofs of the Properties of Determinants

We now prove the properties P1–P8.

Proof of P1:   Let B be the matrix obtained by interchanging row r with row s in A. Then the elements of B are related to those of A as follows:

$$b_{ij} = \begin{cases} a_{ij} & \text{if } i \neq r, s, \\ a_{sj} & \text{if } i = r, \\ a_{rj} & \text{if } i = s. \end{cases}$$

Thus, from Definition 3.1.8,

$$\det(B) = \sum \sigma(p_1, p_2, \cdots, p_r, \cdots, p_s, \cdots, p_n) b_{1p_1} b_{2p_2} \cdots b_{rp_r} \cdots b_{sp_s} \cdots b_{np_n}$$

$$= \sum \sigma(p_1, p_2, \cdots, p_r, \cdots, p_s, \cdots, p_n) a_{1p_1} a_{2p_2} \cdots a_{sp_r} \cdots a_{rp_s} \cdots a_{np_n}.$$

Interchanging p_r and p_s in $\sigma(p_1, p_2, \ldots, p_r, \ldots, p_s, \ldots, p_n)$ and recalling from Theorem 3.1.7 that such an interchange has the effect of changing the parity of the permutation, we obtain

$$\det(B) = -\sum \sigma(p_1, p_2, \cdots, p_s, \cdots, p_r, \cdots, p_n)a_{1p_1}a_{2p_2}\cdots a_{rp_s}\cdots a_{sp_r}\cdots a_{np_n},$$

where we have also rearranged the terms so that the row indices are in their natural increasing order. The sum on the right-hand side of this equation is just $\det(A)$, so that

$$\det(B) = -\det(A). \qquad \blacksquare$$

Proof of P2: Let B be the matrix obtained by multiplying the ith row of A through by any scalar k. Then $b_{ij} = ka_{ij}$ for each j. Then

$$\det(B) = \sum \sigma(p_1, p_2, \cdots, p_n)b_{1p_1}b_{2p_2}\cdots b_{np_n}$$

$$= \sum \sigma(p_1, p_2, \cdots, p_n)a_{1p_1}a_{2p_2}\cdots (ka_{ip_i})\cdots a_{np_n} = k\det(A). \qquad \blacksquare$$

We prove properties P5 and P7 next, since they simplify the proof of P3.

Proof of P5: The elements of A are

$$a_{kj} = \begin{cases} a_{kj}, & \text{if } k \neq i, \\ b_{ij} + c_{ij}, & \text{if } k = i. \end{cases}$$

Thus, from Definition 3.1.8,

$$\det(A) = \sum \sigma(p_1, p_2, \cdots, p_n)a_{1p_1}a_{2p_2}\cdots a_{np_n}$$

$$= \sum \sigma(p_1, p_2, \cdots, p_n)a_{1p_1}a_{2p_2}\cdots a_{i-1p_{i-1}}(b_{ip_i} + c_{ip_i})a_{i+1p_{i+1}}\cdots a_{np_n}$$

$$= \sum \sigma(p_1, p_2, \cdots, p_n)a_{1p_1}a_{2p_2}\cdots a_{i-1p_{i-1}}b_{ip_i}a_{i+1p_{i+1}}\cdots a_{np_n}$$

$$+ \sum \sigma(p_1, p_2, \cdots, p_n)a_{1p_1}a_{2p_2}\cdots a_{i-1p_{i-1}}c_{ip_i}a_{i+1p_{i+1}}\cdots a_{np_n}$$

$$= \det(B) + \det(C). \qquad \blacksquare$$

Proof of P7: Suppose rows i and j in A are the same. Then if we interchange these rows, the matrix, and hence its determinant, are unaltered. However, according to P1, the determinant of the resulting matrix is $-\det(A)$. Therefore,

$$\det(A) = -\det(A),$$

which implies that

$$\det(A) = 0. \qquad \blacksquare$$

Proof of P3: Let $A = [\mathbf{a}_1, \mathbf{a}_2, \ldots, \mathbf{a}_n]^T$, and let B be the matrix obtained from A when k times row j of A is added to row i of A. Then

$$B = [\mathbf{a}_1, \mathbf{a}_2, \ldots, \mathbf{a}_i + k\mathbf{a}_j, \ldots, \mathbf{a}_n]^T$$

so that, using P5,

$$\det(B) = \det([\mathbf{a}_1, \mathbf{a}_2, \ldots, \mathbf{a}_i + k\mathbf{a}_j, \ldots, \mathbf{a}_n]^T)$$
$$= \det([\mathbf{a}_1, \mathbf{a}_2, \ldots, \mathbf{a}_n]^T) + \det([\mathbf{a}_1, \mathbf{a}_2, \ldots, k\mathbf{a}_j, \ldots, \mathbf{a}_n]^T).$$

By P2, we can factor out k from row i of the second determinant on the right-hand side. If we do this, it follows that row i and row j of the resulting determinant are the same, and so, from P7, the value of the second determinant is zero. Thus,

$$\det(B) = \det([\mathbf{a}_1, \mathbf{a}_2, \ldots, \mathbf{a}_n]^T) = \det(A),$$

as required. ∎

Proof of P4: Using Definition 3.1.8, we have

$$\det(A^T) = \sum \sigma(p_1, p_2, \ldots, p_n) a_{p_1 1} a_{p_2 2} a_{p_3 3} \cdots a_{p_n n}. \qquad (3.2.2)$$

Since $(p_1, p_2, \ldots, p_n)$ is a permutation of $1, 2, \ldots, n$, it follows that, by rearranging terms,

$$a_{p_1 1} a_{p_2 2} a_{p_3 3} \cdots a_{p_n n} = a_{1 q_1} a_{2 q_2} a_{3 q_3} \cdots a_{n q_n}, \qquad (3.2.3)$$

for appropriate values of $q_1, q_2, \ldots, q_n$. Furthermore,

$$N(p_1, \ldots, p_n) = \text{\# of interchanges in changing } (1, 2, \ldots, n) \text{ to } (p_1, p_2, \ldots, p_n)$$
$$= \text{\# of interchanges in changing } (p_1, p_2, \ldots, p_n) \text{ to } (1, 2, \ldots, n)$$

and by (3.2.3), this number is

$$= \text{\# of interchanges in changing } (1, 2, \ldots, n) \text{ to } (q_1, q_2, \ldots, q_n)$$
$$= N(q_1, \ldots, q_n).$$

Thus,

$$\sigma(p_1, p_2, \ldots, p_n) = \sigma(q_1, q_2, \ldots, q_n). \qquad (3.2.4)$$

Substituting Equations (3.2.3) and (3.2.4) into Equation (3.2.2), we have

$$\det(A^T) = \sum \sigma(q_1, q_2, \ldots, q_n) a_{1 q_1} a_{2 q_2} a_{3 q_3} \cdots a_{n q_n}$$
$$= \det(A). \qquad ∎$$

Proof of P6: Since each term $\sigma(p_1, p_2, \ldots, p_n) a_{1 p_1} a_{2 p_2} \cdots a_{n p_n}$ in the formula for $\det(A)$ contains a factor from the row (or column) of zeros, each such term is zero. Thus, $\det(A) = 0$. ∎

Proof of P8: Let E denote an elementary matrix. We leave it as an exercise (Problem 51) to verify that

$$\det(E) = \begin{cases} -1, & \text{if } E \text{ permutes rows,} \\ +1, & \text{if } E \text{ adds a multiple of one row to another row,} \\ k, & \text{if } E \text{ scales a row by } k. \end{cases}$$

It then follows from properties P1–P3 that in each case

$$\det(EA) = \det(E)\det(A). \qquad (3.2.5)$$

Now consider a general product AB. We need to distinguish two cases.

Case 1: If A is not invertible, then from Corollary 2.6.12, so is AB. Consequently, applying Theorem 3.2.4,

$$\det(AB) = 0 = \det(A)\det(B).$$

Case 2: If A is invertible, then from Section 2.7, we know that it can be expressed as the product of elementary matrices, say, $A = E_1 E_2 \cdots E_r$. Hence, repeatedly applying (3.2.5) gives

$$\det(AB) = \det(E_1 E_2 \cdots E_r B) = \det(E_1)\det(E_2 \cdots E_r B)$$
$$= \det(E_1)\det(E_2)\cdots\det(E_r)\det(B)$$
$$= \det(E_1 E_2 \cdots E_r)\det(B) = \det(A)\det(B). \qquad\blacksquare$$

Exercises for 3.2

Skills

- Be able to compute the determinant of an upper or lower triangular matrix "at a glance" (Theorem 3.2.1).

- Know the effects that elementary row operations have on the determinant of a matrix.

- Likewise, be comfortable with the effects that column operations have on the determinant of a matrix.

- Be able to use the determinant to decide if a matrix is invertible (Theorem 3.2.4).

- Know how the determinant is affected by matrix multiplication and by matrix transpose.

True-False Review

For Questions 1–6, decide if the given statement is **true** or **false**, and give a brief justification for your answer. If true, you can quote a relevant definition or theorem from the text. If false, provide an example, illustration, or brief explanation of why the statement is false.

1. If each element of an $n \times n$ matrix is doubled, then the determinant of the matrix also doubles.

2. Multiplying a row of an $n \times n$ matrix through by a scalar c has the same effect on the determinant as multiplying a column of the matrix through by c.

3. If A is an $n \times n$ matrix, then $\det(A^5) = (\det A)^5$.

4. If A is a real $n \times n$ matrix, then $\det(A^2)$ cannot be negative.

5. The matrix $\begin{bmatrix} x^2 & x \\ y^2 & y \end{bmatrix}$ is not invertible if and only if $x = 0$ or $y = 0$.

6. If A and B are $n \times n$ matrices, then $\det(AB) = \det(BA)$.

Problems

For Problems 1–12, reduce the given determinant to upper triangular form and then evaluate.

1. $\begin{vmatrix} 1 & 2 & 3 \\ 2 & 6 & 4 \\ 3 & -5 & 2 \end{vmatrix}$.

2. $\begin{vmatrix} 2 & -1 & 4 \\ 3 & 2 & 1 \\ -2 & 1 & 4 \end{vmatrix}$.

3. $\begin{vmatrix} 2 & 1 & 3 \\ -1 & 2 & 6 \\ 4 & 1 & 12 \end{vmatrix}$.

4. $\begin{vmatrix} 0 & 1 & -2 \\ -1 & 0 & 3 \\ 2 & -3 & 0 \end{vmatrix}$.

5. $\begin{vmatrix} 3 & 7 & 1 \\ 5 & 9 & -6 \\ 2 & 1 & 3 \end{vmatrix}$.

6. $\begin{vmatrix} 1 & -1 & 2 & 4 \\ 3 & 1 & 2 & 4 \\ -1 & 1 & 3 & 2 \\ 2 & 1 & 4 & 2 \end{vmatrix}$.

7. $\begin{vmatrix} 2 & 32 & 1 & 4 \\ 26 & 104 & 26 & -13 \\ 2 & 56 & 2 & 7 \\ 1 & 40 & 1 & 5 \end{vmatrix}$.

8. $\begin{vmatrix} 0 & 1 & -1 & 1 \\ -1 & 0 & 1 & 1 \\ 1 & -1 & 0 & 1 \\ -1 & -1 & -1 & 0 \end{vmatrix}.$

9. $\begin{vmatrix} 2 & 1 & 3 & 5 \\ 3 & 0 & 1 & 2 \\ 4 & 1 & 4 & 3 \\ 5 & 2 & 5 & 3 \end{vmatrix}.$

10. $\begin{vmatrix} 2 & -1 & 3 & 4 \\ 7 & 1 & 2 & 3 \\ -2 & 4 & 8 & 6 \\ 6 & -6 & 18 & -24 \end{vmatrix}.$

11. $\begin{vmatrix} 7 & -1 & 3 & 4 \\ 14 & 2 & 4 & 6 \\ 21 & 1 & 3 & 4 \\ -7 & 4 & 5 & 8 \end{vmatrix}.$

12. $\begin{vmatrix} 3 & 7 & 1 & 2 & 3 \\ 1 & 1 & -1 & 0 & 1 \\ 4 & 8 & -1 & 6 & 6 \\ 3 & 7 & 0 & 9 & 4 \\ 8 & 16 & -1 & 8 & 12 \end{vmatrix}.$

For Problems 13–19, use Theorem 3.2.4 to determine whether the given matrix is invertible or not.

13. $\begin{bmatrix} 2 & 1 \\ 3 & 2 \end{bmatrix}.$

14. $\begin{bmatrix} -1 & 1 \\ 1 & -1 \end{bmatrix}.$

15. $\begin{bmatrix} 2 & 6 & -1 \\ 3 & 5 & 1 \\ 2 & 0 & 1 \end{bmatrix}.$

16. $\begin{bmatrix} -1 & 2 & 3 \\ 5 & -2 & 1 \\ 8 & -2 & 5 \end{bmatrix}.$

17. $\begin{bmatrix} 1 & 0 & 2 & -1 \\ 3 & -2 & 1 & 4 \\ 2 & 1 & 6 & 2 \\ 1 & -3 & 4 & 0 \end{bmatrix}.$

18. $\begin{bmatrix} 1 & 1 & 1 & 1 \\ -1 & 1 & -1 & 1 \\ 1 & 1 & -1 & -1 \\ -1 & 1 & 1 & -1 \end{bmatrix}.$

19. $\begin{bmatrix} 1 & 2 & -3 & 5 \\ -1 & 2 & -3 & 6 \\ 2 & 3 & -1 & 4 \\ 1 & -2 & 3 & -6 \end{bmatrix}.$

20. Determine all values of the constant k for which the given system has a unique solution

$$x_1 + kx_2 = b_1,$$
$$kx_1 + 4x_2 = b_2.$$

21. Determine all values of the constant k for which the given system has an infinite number of solutions.

$$x_1 + 2x_2 + kx_3 = 0,$$
$$2x_1 - kx_2 + x_3 = 0,$$
$$3x_1 + 6x_2 + x_3 = 0.$$

22. Determine all values of k for which the given system has an infinite number of solutions.

$$x_1 + 2x_2 + x_3 = kx_1,$$
$$2x_1 + x_2 + x_3 = kx_2,$$
$$x_1 + x_2 + 2x_3 = kx_3.$$

23. Determine all values of k for which the given system has a unique solution.

$$x_1 + kx_2 \qquad\quad = 2,$$
$$kx_1 + x_2 + x_3 = 1,$$
$$x_1 + x_2 + x_3 = 1.$$

24. If

$$A = \begin{bmatrix} 1 & -1 & 2 \\ 3 & 1 & 4 \\ 0 & 1 & 3 \end{bmatrix},$$

find $\det(A)$, and use properties of determinants to find $\det(A^T)$ and $\det(-2A)$.

25. If

$$A = \begin{bmatrix} 1 & -1 \\ 2 & 3 \end{bmatrix} \quad \text{and} \quad B = \begin{bmatrix} 1 & 2 \\ -2 & 4 \end{bmatrix},$$

evaluate $\det(AB)$ and verify P8.

26. If

$$A = \begin{bmatrix} \cosh x & \sinh x \\ \sinh x & \cosh x \end{bmatrix} \quad \text{and} \quad B = \begin{bmatrix} \cosh y & \sinh y \\ \sinh y & \cosh y \end{bmatrix},$$

evaluate $\det(AB)$.

For Problems 27–29, use properties of determinants to show that $\det(A) = 0$ for the given matrix A.

27. $A = \begin{bmatrix} 3 & 2 & 1 \\ 6 & 4 & -1 \\ 9 & 6 & 2 \end{bmatrix}.$

28. $A = \begin{bmatrix} 1 & -3 & 1 \\ 2 & -1 & 7 \\ 3 & 1 & 13 \end{bmatrix}$.

29. $A = \begin{bmatrix} 1+3a & 1 & 3 \\ 1+2a & 1 & 2 \\ 2 & 2 & 0 \end{bmatrix}$.

For Problems 30–32, let $A = \begin{bmatrix} a & b \\ c & d \end{bmatrix}$ and assume $\det(A) = 1$. Find $\det(B)$.

30. $B = \begin{bmatrix} 3c & 3d \\ 4a & 4b \end{bmatrix}$.

31. $B = \begin{bmatrix} -2a & -2c \\ 3a+b & 3c+d \end{bmatrix}$.

32. $B = \begin{bmatrix} -b & -a \\ d-4b & c-4a \end{bmatrix}$.

For Problems 33–35, let

$$A = \begin{bmatrix} a & b & c \\ d & e & f \\ g & h & i \end{bmatrix}$$

and assume $\det(A) = -6$. Find $\det(B)$.

33. $B = \begin{bmatrix} -4d & -4e & -4f \\ g+5a & h+5b & i+5c \\ a & b & c \end{bmatrix}$.

34. $B = \begin{bmatrix} d & e & f \\ -3a & -3b & -3c \\ g-4d & h-4e & i-4f \end{bmatrix}$.

35. $B = \begin{bmatrix} 2a & 2d & 2g \\ b-c & e-f & h-i \\ c-a & f-d & i-g \end{bmatrix}$.

For Problems 36–40, let A and B be 4×4 matrices such that $\det(A) = 5$ and $\det(B) = 3$. Compute the determinant of the given matrix.

36. AB^T.

37. $A^2 B^5$.

38. $(A^{-1}B^2)^3$.

39. $((2B)^{-1}(AB)^T)$.

40. $(5A)(2B)$.

41. Let

$$A = \begin{bmatrix} 1 & 2 & 4 \\ 3 & 1 & 6 \\ k & 3 & 2 \end{bmatrix}$$.

(a) In terms of k, find the volume of the parallelepiped determined by the row vectors of the matrix A.

(b) Does your answer to (a) change if we instead consider the volume of the parallelepiped determined by the *column* vectors of the matrix A? Why or why not?

(c) For what value(s) of k, if any, is A invertible?

42. Without expanding the determinant, determine all values of x for which $\det(A) = 0$ if

$$A = \begin{bmatrix} 1 & -1 & x \\ 2 & 1 & x^2 \\ 4 & -1 & x^3 \end{bmatrix}$$.

43. Use *only* properties P5, P1, and P2 to show that

$$\begin{vmatrix} \alpha x - \beta y & \beta x - \alpha y \\ \beta x + \alpha y & \alpha x + \beta y \end{vmatrix} = (x^2 + y^2)\begin{vmatrix} \alpha & \beta \\ \beta & \alpha \end{vmatrix}.$$

44. Use *only* properties P5, P1, and P2 to find the value of $\alpha\beta\gamma$ such that

$$\begin{vmatrix} a_1 + \beta b_1 & b_1 + \gamma c_1 & c_1 + \alpha a_1 \\ a_2 + \beta b_2 & b_2 + \gamma c_2 & c_2 + \alpha a_2 \\ a_3 + \beta b_3 & b_3 + \gamma c_3 & c_3 + \alpha a_3 \end{vmatrix} = 0$$

for all values of a_i, b_i, c_i.

45. Use *only* properties P3 and P7 to prove property P6.

46. An $n \times n$ matrix A that satisfies $A^T = A^{-1}$ is called an **orthogonal matrix**. Show that if A is an orthogonal matrix, then $\det(A) = \pm 1$.

47. **(a)** Use the definition of a determinant to prove that if A is an $n \times n$ lower triangular matrix, then

$$\det(A) = a_{11}a_{22}a_{33} \cdots a_{nn} = \prod_{i=1}^{n} a_{ii}.$$

(b) Evaluate the following determinant by first reducing it to lower triangular form and then using the result from (a):

$$\begin{vmatrix} 2 & -1 & 3 & 5 \\ 1 & 2 & 2 & 1 \\ 3 & 0 & 1 & 4 \\ 1 & 2 & 0 & 1 \end{vmatrix}.$$

48. Use determinants to prove that if A is invertible and B and C are matrices with $AB = AC$, then $B = C$.

49. If A and S are $n \times n$ matrices with S invertible, show that $\det(S^{-1}AS) = \det(A)$. [**Hint:** Since $S^{-1}S = I_n$, how are $\det(S^{-1})$ and $\det(S)$ related?]

50. If $\det(A^3) = 0$, is it possible for A to be invertible? Justify your answer.

51. Let E be an elementary matrix. Verify the formula for $\det(E)$ given in the text at the beginning of the proof of P8.

52. Show that
$$\begin{vmatrix} x & y & 1 \\ x_1 & y_1 & 1 \\ x_2 & y_2 & 1 \end{vmatrix} = 0$$
represents the equation of the straight line through the distinct points (x_1, y_1) and (x_2, y_2).

53. Without expanding the determinant, show that
$$\begin{vmatrix} 1 & x & x^2 \\ 1 & y & y^2 \\ 1 & z & z^2 \end{vmatrix} = (y - z)(z - x)(x - y).$$

54. If A is an $n \times n$ *skew-symmetric* matrix and n is odd, prove that $\det(A) = 0$.

55. Let $A = [\mathbf{a}_1, \mathbf{a}_2, \dots, \mathbf{a}_n]$ be an $n \times n$ matrix, and let $\mathbf{b} = c_1\mathbf{a}_1 + c_2\mathbf{a}_2 + \cdots + c_n\mathbf{a}_n$, where $c_1, c_2, \dots, c_n$ are constants. If B_k denotes the matrix obtained from A by replacing the kth column vector by $\mathbf{b}$, prove that
$$\det(B_k) = c_k \det(A), \qquad k = 1, 2, \dots, n.$$

56. ◇ Let A be the general 4×4 matrix.

(a) Verify property P1 of determinants in the case when the first two rows of A are permuted.

(b) Verify property P2 of determinants in the case when row 1 of A is divided by k.

(c) Verify property P3 of determinants in the case when k times row 2 is added to row 1.

57. ◇ For a randomly generated 5×5 matrix, verify that $\det(A^T) = \det(A)$.

58. ◇ Determine all values of a for which
$$\begin{bmatrix} 1 & 2 & 3 & 4 & a \\ 2 & 1 & 2 & 3 & 4 \\ 3 & 2 & 1 & 2 & 3 \\ 4 & 3 & 2 & 1 & 2 \\ a & 4 & 3 & 2 & 1 \end{bmatrix}$$
is invertible.

59. ◇ If
$$A = \begin{bmatrix} 1 & 4 & 1 \\ 3 & 2 & 1 \\ 3 & 4 & -1 \end{bmatrix},$$
determine all values of the constant k for which the linear system $(A - kI_3)\mathbf{x} = \mathbf{0}$ has an infinite number of solutions, and find the corresponding solutions.

60. ◇ Use the determinant to show that
$$A = \begin{bmatrix} 1 & 2 & 3 & 4 \\ 2 & 1 & 2 & 3 \\ 3 & 2 & 1 & 2 \\ 4 & 3 & 2 & 1 \end{bmatrix}$$
is invertible, and use A^{-1} to solve $A\mathbf{x} = \mathbf{b}$ if $\mathbf{b} = [3, 7, 1, -4]^T$.

3.3   Cofactor Expansions

We now obtain an alternative method for evaluating determinants. The basic idea is that we can reduce a determinant of order n to a sum of determinants of order $n-1$. Continuing in this manner, it is possible to express any determinant as a sum of determinants of order 2. This method is the one most frequently used to evaluate a determinant by hand, although the procedure introduced in the previous section whereby we use elementary row operations to reduce the matrix to upper triangular form involves less work in general. When A is invertible, the technique we derive leads to formulas for both A^{-1} and the unique solution to $A\mathbf{x} = \mathbf{b}$. We first require two preliminary definitions.

DEFINITION   3.3.1

Let A be an $n \times n$ matrix. The **minor**, M_{ij}, of the element a_{ij}, is the determinant of the matrix obtained by deleting the ith row vector and jth column vector of A.

Remark   Notice that if A is an $n \times n$ matrix, then M_{ij} is a determinant of order $n - 1$. By convention, if $n = 1$, we define the "empty" determinant M_{11} to be 1.

Example 3.3.2   If

$$A = \begin{bmatrix} a_{11} & a_{12} & a_{13} \\ a_{21} & a_{22} & a_{23} \\ a_{31} & a_{32} & a_{33} \end{bmatrix},$$

then, for example,

$$M_{23} = \begin{vmatrix} a_{11} & a_{12} \\ a_{31} & a_{32} \end{vmatrix} \quad \text{and} \quad M_{31} = \begin{vmatrix} a_{12} & a_{13} \\ a_{22} & a_{23} \end{vmatrix}. \qquad \square$$

Example 3.3.3   Determine the minors M_{11}, M_{23}, and M_{31} for

$$A = \begin{bmatrix} 2 & 1 & 3 \\ -1 & 4 & -2 \\ 3 & 1 & 5 \end{bmatrix}.$$

Solution: Using Definition 3.3.1, we have

$$M_{11} = \begin{vmatrix} 4 & -2 \\ 1 & 5 \end{vmatrix} = 22, \qquad M_{23} = \begin{vmatrix} 2 & 1 \\ 3 & 1 \end{vmatrix} = -1, \qquad M_{31} = \begin{vmatrix} 1 & 3 \\ 4 & -2 \end{vmatrix} = -14. \qquad \square$$

DEFINITION   3.3.4

Let A be an $n \times n$ matrix. The **cofactor**, C_{ij}, of the element a_{ij}, is defined by

$$C_{ij} = (-1)^{i+j} M_{ij},$$

where M_{ij} is the minor of a_{ij}.

From Definition 3.3.4, we see that the cofactor of a_{ij} and the minor of a_{ij} are the same if $i + j$ is even, and they differ by a minus sign if $i + j$ is odd. The appropriate sign in the cofactor C_{ij} is easy to remember, since it alternates in the following manner:

$$\begin{vmatrix} + & - & + & - & + & \cdots \\ - & + & - & + & - & \cdots \\ + & - & + & - & + & \cdots \\ \vdots & \vdots & \vdots & \vdots & \vdots & \end{vmatrix}.$$

Example 3.3.5   Determine the cofactors C_{11}, C_{23}, and C_{31} for the matrix in Example 3.3.3.

Solution: We have already obtained the minors M_{11}, M_{23}, and M_{31} in Example 3.3.3, so it follows that

$$C_{11} = +M_{11} = 22, \qquad C_{23} = -M_{23} = 1, \qquad C_{31} = +M_{31} = -14. \qquad \square$$

Example 3.3.6 If $A = \begin{bmatrix} a_{11} & a_{12} \\ a_{21} & a_{22} \end{bmatrix}$, verify that $\det(A) = a_{11}C_{11} + a_{12}C_{12}$.

Solution: In this case,

$$C_{11} = +\det[a_{22}] = a_{22}, \qquad C_{12} = -\det[a_{12}] = -a_{12},$$

so that

$$a_{11}C_{11} + a_{12}C_{12} = a_{11}a_{22} + a_{12}(-a_{21}) = \det(A). \qquad \square$$

The preceding example is a special case of the following important theorem.

Theorem 3.3.7 **(Cofactor Expansion Theorem)**

Let A be an $n \times n$ matrix. If we multiply the elements in any row (or column) of A by their cofactors, then the sum of the resulting products is $\det(A)$. Thus,

1. If we expand along row i,

$$\det(A) = a_{i1}C_{i1} + a_{i2}C_{i2} + \cdots + a_{in}C_{in} = \sum_{k=1}^{n} a_{ik}C_{ik}.$$

2. If we expand along column j,

$$\det(A) = a_{1j}C_{1j} + a_{2j}C_{2j} + \cdots + a_{nj}C_{nj} = \sum_{k=1}^{n} a_{kj}C_{kj}.$$

The expressions for $\det(A)$ appearing in this theorem are known as **cofactor expansions**. Notice that a cofactor expansion can be formed along *any* row or column of A. Regardless of the chosen row or column, the cofactor expansion will always yield the determinant of A. However, sometimes the calculation is simpler if the row or column of expansion is wisely chosen. We will illustrate this in the examples below. The proof of the Cofactor Expansion Theorem will be presented after some examples.

Example 3.3.8 Use the Cofactor Expansion Theorem along (a) row 1, (b) column 3 to find

$$\begin{vmatrix} 2 & 3 & 4 \\ 1 & -1 & 1 \\ 6 & 3 & 0 \end{vmatrix}.$$

Solution:

(a) We have

$$\begin{vmatrix} 2 & 3 & 4 \\ 1 & -1 & 1 \\ 6 & 3 & 0 \end{vmatrix} = 2 \begin{vmatrix} -1 & 1 \\ 3 & 0 \end{vmatrix} - 3 \begin{vmatrix} 1 & 1 \\ 6 & 0 \end{vmatrix} + 4 \begin{vmatrix} 1 & -1 \\ 6 & 3 \end{vmatrix} = -6 + 18 + 36 = 48.$$

(b) We have

$$\begin{vmatrix} 2 & 3 & 4 \\ 1 & -1 & 1 \\ 6 & 3 & 0 \end{vmatrix} = 4 \begin{vmatrix} 1 & -1 \\ 6 & 3 \end{vmatrix} - 1 \begin{vmatrix} 2 & 3 \\ 6 & 3 \end{vmatrix} + 0 = 36 + 12 + 0 = 48. \qquad \square$$

Notice that (b) was easier than (a) in the previous example, because of the zero in column 3. Whenever one uses the cofactor expansion method to evaluate a determinant, it is usually best to select a row or column containing as many zeros as possible in order to minimize the amount of computation required.

Example 3.3.9 Evaluate

$$\begin{vmatrix} 0 & 3 & -1 & 0 \\ 5 & 0 & 8 & 2 \\ 7 & 2 & 5 & 4 \\ 6 & 1 & 7 & 0 \end{vmatrix}.$$

Solution: In this case, it is easiest to use either row 1 or column 4. Choosing row 1, we have

$$\begin{vmatrix} 0 & 3 & -1 & 0 \\ 5 & 0 & 8 & 2 \\ 7 & 2 & 5 & 4 \\ 6 & 1 & 7 & 0 \end{vmatrix} = -3 \begin{vmatrix} 5 & 8 & 2 \\ 7 & 5 & 4 \\ 6 & 7 & 0 \end{vmatrix} + (-1) \begin{vmatrix} 5 & 0 & 2 \\ 7 & 2 & 4 \\ 6 & 1 & 0 \end{vmatrix}$$

$$= -3 [2(49 - 30) - 4(35 - 48) + 0] - [5(0 - 4) - 0 + 2(7 - 12)]$$
$$= -240.$$

In evaluating the determinants of order 3 on the right side of the first equality, we have used cofactor expansion along column 3 and row 1, respectively. For additional practice, the reader may wish to verify our result here by cofactor expansion along a different row or column. $\qquad \square$

Now we turn to the

Proof of the Cofactor Expansion Theorem: It follows from the definition of the determinant that $\det(A)$ can be written in the form

$$\det(A) = a_{i1}\hat{C}_{i1} + a_{12}\hat{C}_{i2} + \cdots + a_{in}\hat{C}_{in} \tag{3.3.1}$$

where the coefficients $\hat{C}_{ij}$ contain no elements from row i or column j. We must show that

$$\hat{C}_{ij} = C_{ij}$$

where C_{ij} is the cofactor of a_{ij}.

Consider first a_{11}. From Definition 3.1.8, the terms of $\det(A)$ that contain a_{11} are given by

$$a_{11} \sum \sigma(1, p_2, p_3, \ldots, p_n) a_{2p_2} a_{3p_3} \cdots a_{np_n},$$

where the summation is over the $(n-1)!$ distinct permutations of $2, 3, \ldots, n$. Thus,

$$\hat{C}_{11} = \sum \sigma(1, p_2, p_3, \ldots, p_n) a_{2p_2} a_{3p_3} \cdots a_{np_n}.$$

However, this summation is just the minor M_{11}, and since $C_{11} = M_{11}$, we have shown the coefficient of a_{11} in $\det(A)$ is indeed the cofactor C_{11}.

Now consider the element a_{ij}. By successively interchanging adjacent rows and columns of A, we can move a_{ij} into the $(1, 1)$ position *without altering the relative positions of the other rows and columns of A.* We let A' denote the resulting matrix. Obtaining A' from A requires $i - 1$ row interchanges and $j - 1$ column interchanges. Therefore, the total number of interchanges required to obtain A' from A is $i + j - 2$. Consequently,

$$\det(A) = (-1)^{i+j-2} \det(A') = (-1)^{i+j} \det(A').$$

Now for the key point. The coefficient of a_{ij} in $\det(A)$ must be $(-1)^{i+j}$ times the coefficient of a_{ij} in $\det(A')$. But, a_{ij} occurs in the $(1, 1)$ position of A', and so, as we have previously shown, its coefficient in $\det(A')$ is M_{11}'. Since the relative positions of the remaining rows in A have not altered, it follows that $M_{11}' = M_{ij}$, and therefore the coefficient of a_{ij} in $\det(A')$ is M_{ij}. Consequently, the coefficient of a_{ij} in $\det(A)$ is $(-1)^{i+j} M_{ij} = C_{ij}$. Applying this result to the elements $a_{i1}, a_{i2}, \ldots, a_{in}$ and comparing with (3.3.1) yields

$$\hat{C}_{ij} = C_{ij}, \qquad j = 1, 2, \ldots, n,$$

which establishes the theorem for expansion along a row. The result for expansion along a column follows directly, since $\det(A^T) = \det(A)$. ∎

We now have two computational methods for evaluating determinants: the use of elementary row operations given in the previous section to reduce the matrix in question to upper triangular form, and the Cofactor Expansion Theorem. In evaluating a given determinant by hand, it is usually most efficient (and least error prone) to use a combination of the two techniques. More specifically, we use elementary row operations to set all except one element in a row or column equal to zero and then use the Cofactor Expansion Theorem on that row or column. We illustrate with an example.

Example 3.3.10   Evaluate

$$\begin{vmatrix} 2 & 1 & 8 & 6 \\ 1 & 4 & 1 & 3 \\ -1 & 2 & 1 & 4 \\ 1 & 3 & -1 & 2 \end{vmatrix}.$$

Solution: We have

$$\begin{vmatrix} 2 & 1 & 8 & 6 \\ 1 & 4 & 1 & 3 \\ -1 & 2 & 1 & 4 \\ 1 & 3 & -1 & 2 \end{vmatrix} \overset{1}{=} \begin{vmatrix} 0 & -7 & 6 & 0 \\ 1 & 4 & 1 & 3 \\ 0 & 6 & 2 & 7 \\ 0 & -1 & -2 & -1 \end{vmatrix} \overset{2}{=} -\begin{vmatrix} -7 & 6 & 0 \\ 6 & 2 & 7 \\ -1 & -2 & -1 \end{vmatrix} \overset{3}{=} -\begin{vmatrix} -7 & 6 & 0 \\ -1 & -12 & 0 \\ -1 & -2 & -1 \end{vmatrix} \overset{4}{=} 90.$$

---

1. $A_{21}(-2)$, $A_{23}(1)$, $A_{24}(-1)$ **2.** Cofactor expansion along column 1
3. $A_{32}(7)$ **4.** Cofactor expansion along column 3

Example 3.3.11 Determine all values of k for which the system

$$
\begin{aligned}
10x_1 + kx_2 - x_3 &= 0, \\
kx_1 + x_2 - x_3 &= 0, \\
2x_1 + x_2 - 3x_3 &= 0,
\end{aligned}
$$

has nontrivial solutions.

Solution: We will apply Corollary 3.2.5. The determinant of the matrix of coefficients of the system is

$$
\det(A) = \begin{vmatrix} 10 & k & -1 \\ k & 1 & -1 \\ 2 & 1 & -3 \end{vmatrix} \overset{1}{=} \begin{vmatrix} 10 & k & -1 \\ k-10 & 1-k & 0 \\ -28 & 1-3k & 0 \end{vmatrix} \overset{2}{=} -\begin{vmatrix} k-10 & 1-k \\ -28 & 1-3k \end{vmatrix}
$$

$$
= -[(k-10)(1-3k) - (-28)(1-k)] = 3k^2 - 3k - 18 = 3(k^2 - k - 6)
$$

$$
= 3(k-3)(k+2).
$$

1. $A_{12}(-1)$, $A_{13}(-3)$	**2.** Cofactor expansion along column 3.

From Corollary 3.2.5, the system has nontrivial solutions if and only if $\det(A) = 0$; that is, if and only if $k = 3$ or $k = -2$. $\square$

The Adjoint Method for A^{-1}

We next establish two corollaries to the Cofactor Expansion Theorem that, in the case of an invertible matrix A, lead to a method for expressing the elements of A^{-1} in terms of determinants.

Corollary 3.3.12 If the elements in the ith row (or column) of an $n \times n$ matrix A are multiplied by the cofactors of a different row (or column), then the sum of the resulting products is zero. That is,

1. If we use the elements of row i and the cofactors of row j,

$$
\sum_{k=1}^{n} a_{ik} C_{jk} = 0, \qquad i \neq j. \tag{3.3.2}
$$

2. If we use the elements of column i and the cofactors of column j,

$$
\sum_{k=1}^{n} a_{ki} C_{kj} = 0, \qquad i \neq j. \tag{3.3.3}
$$

Proof We prove (3.3.2). Let B be the matrix obtained from A by adding row i to row j ($i \neq j$) in the matrix A. By P3, $\det(B) = \det(A)$. Cofactor expansion of B along row j gives

$$
\det(A) = \det(B) = \sum_{k=1}^{n} (a_{jk} + a_{ik}) C_{jk} = \sum_{k=1}^{n} a_{jk} C_{jk} + \sum_{k=1}^{n} a_{ik} C_{jk}.
$$

That is,

$$\det(A) = \det(A) + \sum_{k=1}^{n} a_{ik} C_{jk},$$

since by the Cofactor Expansion Theorem the first summation on the right-hand side is simply $\det(A)$. It follows immediately that

$$\sum_{k=1}^{n} a_{ik} C_{jk} = 0, \qquad i \neq j.$$

Equation (3.3.3) can be proved similarly (Problem 47). ∎

The Cofactor Expansion Theorem and the above corollary can be combined into the following corollary.

Corollary 3.3.13　Let A be an $n \times n$ matrix. If δ_{ij} is the Kronecker delta symbol (see Definition 2.2.19), then

$$\sum_{k=1}^{n} a_{ik} C_{jk} = \delta_{ij} \det(A), \qquad \sum_{k=1}^{n} a_{ki} C_{kj} = \delta_{ij} \det(A). \tag{3.3.4}$$

The formulas in (3.3.4) should be reminiscent of the index form of the matrix product. Combining this with the fact that the Kronecker delta gives the elements of the identity matrix, we might suspect that (3.3.4) is telling us something about the inverse of A. Before establishing that this suspicion is indeed correct, we need a definition.

DEFINITION　3.3.14

If every element in an $n \times n$ matrix A is replaced by its cofactor, the resulting matrix is called the **matrix of cofactors** and is denoted M_C. The transpose of the matrix of cofactors, M_C^T, is called the **adjoint** of A and is denoted $\text{adj}(A)$. Thus, the elements of $\text{adj}(A)$ are

$$\text{adj}(A)_{ij} = C_{ji}.$$

Example 3.3.15　Determine $\text{adj}(A)$ if

$$A = \begin{bmatrix} 2 & 0 & -3 \\ -1 & 5 & 4 \\ 3 & -2 & 0 \end{bmatrix}.$$

Solution:　We first determine the cofactors of A:

$$C_{11} = 8, \quad C_{12} = 12, \quad C_{13} = -13, \quad C_{21} = 6, \quad C_{22} = 9, \quad C_{23} = 4,$$
$$C_{31} = 15, \quad C_{32} = -5, \quad C_{33} = 10.$$

Thus,

$$M_C = \begin{bmatrix} 8 & 12 & -13 \\ 6 & 9 & 4 \\ 15 & -5 & 10 \end{bmatrix},$$

so that

$$\text{adj}(A) = M_C^T = \begin{bmatrix} 8 & 6 & 15 \\ 12 & 9 & -5 \\ -13 & 4 & 10 \end{bmatrix}.$$ □

We can now prove the next theorem.

Theorem 3.3.16 **(The Adjoint Method for Computing A^{-1})**

If $\det(A) \neq 0$, then

$$A^{-1} = \frac{1}{\det(A)} \text{adj}(A).$$

Proof Let $B = \dfrac{1}{\det(A)} \text{adj}(A)$. Then we must establish that $AB = I_n = BA$. But, using the index form of the matrix product,

$$(AB)_{ij} = \sum_{k=1}^{n} a_{ik} b_{kj} = \sum_{k=1}^{n} a_{ik} \cdot \frac{1}{\det(A)} \cdot \text{adj}(A)_{kj} = \frac{1}{\det(A)} \sum_{k=1}^{n} a_{ik} C_{jk} = \delta_{ij},$$

where we have used Equation (3.3.4) in the last step. Consequently, $AB = I_n$. We leave it as an exercise (Problem 53) to verify that $BA = I_n$ also.   ■

Example 3.3.17   For the matrix in Example 3.3.15,

$$\det(A) = 55,$$

so that

$$A^{-1} = \frac{1}{55} \begin{bmatrix} 8 & 6 & 15 \\ 12 & 9 & -5 \\ -13 & 4 & 10 \end{bmatrix}.$$ □

For square matrices of relatively small size, the adjoint method for computing A^{-1} is often easier than using elementary row operations to reduce A to upper triangular form.

In Chapter 7, we will find that the solution of a system of *differential* equations can be expressed naturally in terms of matrix functions. Certain problems will require us to find the inverse of such matrix functions. For 2×2 systems, the adjoint method is very quick.

Example 3.3.18   Find A^{-1} if $A = \begin{bmatrix} e^{2t} & e^{-t} \\ 3e^{2t} & 6e^{-t} \end{bmatrix}$.

Solution: In this case,

$$\det(A) = (e^{2t})(6e^{-t}) - (3e^{2t})(e^{-t}) = 3e^t,$$

and

$$\text{adj}(A) = \begin{bmatrix} 6e^{-t} & -e^{-t} \\ -3e^{2t} & e^{2t} \end{bmatrix},$$

so that

$$A^{-1} = \begin{bmatrix} 2e^{-2t} & -\frac{1}{3}e^{-2t} \\ -e^t & \frac{1}{3}e^t \end{bmatrix}.$$ □

Cramer's Rule

We now derive a technique that enables us, in the case when $\det(A) \neq 0$, to express the unique solution of an $n \times n$ linear system

$$A\mathbf{x} = \mathbf{b}$$

directly in terms of determinants. Let B_k denote the matrix obtained by replacing the kth column vector of A with $\mathbf{b}$. Thus,

$$B_k = \begin{bmatrix} a_{11} & a_{12} & \ldots & b_1 & \ldots & a_{1n} \\ a_{21} & a_{22} & \ldots & b_2 & \ldots & a_{2n} \\ \vdots & \vdots & & \vdots & & \vdots \\ a_{n1} & a_{n2} & \ldots & b_n & \ldots & a_{nn} \end{bmatrix}.$$

The key point to notice is that the cofactors of the elements in the kth column of B_k coincide with the corresponding cofactors of A. Thus, expanding $\det(B_k)$ along the kth column using the Cofactor Expansion Theorem yields

$$\det(B_k) = b_1 C_{1k} + b_2 C_{2k} + \cdots + b_n C_{nk} = \sum_{i=1}^{n} b_i C_{ik}, \quad k = 1, 2, \ldots, n, \quad (3.3.5)$$

where the C_{ij} are the cofactors of A. We can now prove Cramer's rule.

Theorem 3.3.19 **(Cramer's Rule)**

If $\det(A) \neq 0$, the unique solution to the $n \times n$ system $A\mathbf{x} = \mathbf{b}$ is $(x_1, x_2, \ldots, x_n)$, where

$$x_k = \frac{\det(B_k)}{\det(A)}, \quad k = 1, 2, \ldots, n. \quad (3.3.6)$$

Proof If $\det(A) \neq 0$, then the system $A\mathbf{x} = \mathbf{b}$ has the unique solution

$$\mathbf{x} = A^{-1}\mathbf{b}, \quad (3.3.7)$$

where, from Theorem 3.3.16, we can write

$$A^{-1} = \frac{1}{\det(A)} \operatorname{adj}(A). \quad (3.3.8)$$

If we let

$$\mathbf{x} = \begin{bmatrix} x_1 \\ x_2 \\ \vdots \\ x_n \end{bmatrix} \quad \text{and} \quad \mathbf{b} = \begin{bmatrix} b_1 \\ b_2 \\ \vdots \\ b_n \end{bmatrix}$$

and recall that $\operatorname{adj}(A)_{ij} = C_{ji}$, then substitution from (3.3.8) into (3.3.7) and use of the index form of the matrix product yields

$$x_k = \sum_{i=1}^{n} (A^{-1})_{ki} b_i = \sum_{i=1}^{n} \frac{1}{\det(A)} \operatorname{adj}(A)_{ki} b_i$$

$$= \frac{1}{\det(A)} \sum_{i=1}^{n} C_{ik} b_i, \qquad k = 1, 2, \ldots, n.$$

Using (3.3.5), we can write this as

$$x_k = \frac{\det(B_k)}{\det(A)}, \qquad k = 1, 2, \ldots, n$$

as required.   ■

Remark   In general, Cramer's rule requires more work than the Gaussian elimination method, and it is restricted to $n \times n$ systems whose coefficient matrix is invertible. However, it is a powerful theoretical tool, since it gives us a formula for the solution of an $n \times n$ system, provided $\det(A) \neq 0$.

Example 3.3.20   Solve

$$\begin{aligned} 3x_1 + 2x_2 - x_3 &= 4, \\ x_1 + x_2 - 5x_3 &= -3, \\ -2x_1 - x_2 + 4x_3 &= 0. \end{aligned}$$

Solution: The following determinants are easily evaluated:

$$\det(A) = \begin{vmatrix} 3 & 2 & -1 \\ 1 & 1 & -5 \\ -2 & -1 & 4 \end{vmatrix} = 8, \qquad \det(B_1) = \begin{vmatrix} 4 & 2 & -1 \\ -3 & 1 & -5 \\ 0 & -1 & 4 \end{vmatrix} = 17,$$

$$\det(B_2) = \begin{vmatrix} 3 & 4 & -1 \\ 1 & -3 & -5 \\ -2 & 0 & 4 \end{vmatrix} = -6, \qquad \det(B_3) = \begin{vmatrix} 3 & 2 & 4 \\ 1 & 1 & -3 \\ -2 & -1 & 0 \end{vmatrix} = 7.$$

Inserting these results into (3.3.6) yields $x_1 = \frac{17}{8}$, $x_2 = -\frac{6}{8} = -\frac{3}{4}$, and $x_3 = \frac{7}{8}$, so that the solution to the system is $(\frac{17}{8}, -\frac{3}{4}, \frac{7}{8})$.   □

Exercises for 3.3

Key Terms

Minor, Cofactor, Cofactor expansion, Matrix of cofactors, Adjoint, Cramer's rule.

Skills

• Be able to compute the minors and cofactors of a matrix.

• Understand the difference between M_{ij} and C_{ij}.

• Be able to compute the determinant of a matrix via cofactor expansion.

• Be able to compute the matrix of cofactors and the adjoint of a matrix.

• Be able to use the adjoint of an invertible matrix A to compute A^{-1}.

• Be able to use Cramer's rule to solve a linear system of equations.

True-False Review

For Questions 1–7, decide if the given statement is **true** or **false**, and give a brief justification for your answer. If true, you can quote a relevant definition or theorem from the text. If false, provide an example, illustration, or brief explanation of why the statement is false.

1. The $(2, 3)$-minor of a matrix is the same as the $(2, 3)$-cofactor of the matrix.

2. We have $A \cdot \text{adj}(A) = \det(A) \cdot I_n$ for all $n \times n$ matrices A.

3. Cofactor expansion of a matrix along any row or column will yield the same result, although the individual terms in the expansion along different rows or columns can vary.

4. If A is an $n \times n$ matrix and c is a scalar, then

$$\text{adj}(cA) = c \cdot \text{adj}(A).$$

5. If A and B are 2×2 matrices, then

$$\text{adj}(A + B) = \text{adj}(A) + \text{adj}(B).$$

6. If A and B are 2×2 matrices, then

$$\text{adj}(AB) = \text{adj}(A) \cdot \text{adj}(B).$$

7. For every n, $\text{adj}(I_n) = I_n$.

Problems

For Problems 1–3, determine all minors and cofactors of the given matrix.

1. $A = \begin{bmatrix} 1 & -3 \\ 2 & 4 \end{bmatrix}$.

2. $A = \begin{bmatrix} 1 & -1 & 2 \\ 3 & -1 & 4 \\ 2 & 1 & 5 \end{bmatrix}$.

3. $A = \begin{bmatrix} 2 & 10 & 3 \\ 0 & -1 & 0 \\ 4 & 1 & 5 \end{bmatrix}$.

4. If

$$A = \begin{bmatrix} 1 & 3 & -1 & 2 \\ 3 & 4 & 1 & 2 \\ 7 & 1 & 4 & 6 \\ 5 & 0 & 1 & 2 \end{bmatrix},$$

determine the minors $M_{12}, M_{31}, M_{23}, M_{42}$, and the corresponding cofactors.

For Problems 5–10, use the Cofactor Expansion Theorem to evaluate the given determinant along the specified row or column.

5. $\begin{vmatrix} 1 & -2 \\ 1 & 3 \end{vmatrix}$, row 1.

6. $\begin{vmatrix} -1 & 2 & 3 \\ 1 & 4 & -2 \\ 3 & 1 & 4 \end{vmatrix}$, column 3.

7. $\begin{vmatrix} 2 & 1 & -4 \\ 7 & 1 & 3 \\ 1 & 5 & -2 \end{vmatrix}$, row 2.

8. $\begin{vmatrix} 3 & 1 & 4 \\ 7 & 1 & 2 \\ 2 & 3 & -5 \end{vmatrix}$, column 1.

9. $\begin{vmatrix} 0 & 2 & -3 \\ -2 & 0 & 5 \\ 3 & -5 & 0 \end{vmatrix}$, row 3.

10. $\begin{vmatrix} 1 & -2 & 3 & 0 \\ 4 & 0 & 7 & -2 \\ 0 & 1 & 3 & 4 \\ 1 & 5 & -2 & 0 \end{vmatrix}$, column 4.

For Problems 11–19, evaluate the given determinant using the techniques of this section.

11. $\begin{vmatrix} 1 & 0 & -2 \\ 3 & 1 & -1 \\ 7 & 2 & 5 \end{vmatrix}$.

12. $\begin{vmatrix} -1 & 2 & 3 \\ 0 & 1 & 4 \\ 2 & -1 & 3 \end{vmatrix}$.

13. $\begin{vmatrix} 2 & -1 & 3 \\ 5 & 2 & 1 \\ 3 & -3 & 7 \end{vmatrix}$.

14. $\begin{vmatrix} 0 & -2 & 1 \\ 2 & 0 & -3 \\ -1 & 3 & 0 \end{vmatrix}$.

15. $\begin{vmatrix} 1 & 0 & -1 & 0 \\ 0 & 1 & 0 & -1 \\ -1 & 0 & -1 & 0 \\ 0 & 1 & 0 & 1 \end{vmatrix}$.

16. $\begin{vmatrix} 2 & -1 & 3 & 1 \\ 1 & 4 & -2 & 3 \\ 0 & 2 & -1 & 0 \\ 1 & 3 & -2 & 4 \end{vmatrix}$.

17. $\begin{vmatrix} 3 & 5 & 2 & 6 \\ 2 & 3 & 5 & -5 \\ 7 & 5 & -3 & -16 \\ 9 & -6 & 27 & -12 \end{vmatrix}$.

18. $\begin{vmatrix} 2 & -7 & 4 & 3 \\ 5 & 5 & -3 & 7 \\ 6 & 2 & 6 & 3 \\ 4 & 2 & -4 & 5 \end{vmatrix}$.

19. $\begin{vmatrix} 2 & 0 & -1 & 3 & 0 \\ 0 & 3 & 0 & 1 & 2 \\ 0 & 1 & 3 & 0 & 4 \\ 1 & 0 & 1 & -1 & 0 \\ 3 & 0 & 2 & 0 & 5 \end{vmatrix}$.

20. If

$$A = \begin{bmatrix} 0 & x & y & z \\ -x & 0 & 1 & -1 \\ -y & -1 & 0 & 1 \\ -z & 1 & -1 & 0 \end{bmatrix},$$

show that $\det(A) = (x + y + z)^2$.

21. **(a)** Consider the 3×3 *Vandermonde* determinant $V(r_1, r_2, r_3)$ defined by

$$V(r_1, r_2, r_3) = \begin{vmatrix} 1 & 1 & 1 \\ r_1 & r_2 & r_3 \\ r_1^2 & r_2^2 & r_3^2 \end{vmatrix}.$$

Show that

$$V(r_1, r_2, r_3) = (r_2 - r_1)(r_3 - r_1)(r_3 - r_2).$$

(b) More generally, show that the $n \times n$ Vandermonde determinant

$$V(r_1, r_2, \ldots, r_n) = \begin{vmatrix} 1 & 1 & \cdots & 1 \\ r_1 & r_2 & \cdots & r_n \\ r_1^2 & r_2^2 & \cdots & r_n^2 \\ \vdots & \vdots & & \vdots \\ r_1^{n-1} & r_2^{n-1} & \cdots & r_n^{n-1} \end{vmatrix}$$

has value

$$V(r_1, r_2, \ldots, r_n) = \prod_{1 \le i < m \le n} (r_m - r_i).$$

For Problems 22–31, find **(a)** $\det(A)$, **(b)** the matrix of cofactors M_C, **(c)** $\text{adj}(A)$, and, if possible, **(d)** A^{-1}.

22. $A = \begin{bmatrix} 3 & 1 \\ 4 & 5 \end{bmatrix}$.

23. $A = \begin{bmatrix} -1 & -2 \\ 4 & 1 \end{bmatrix}$.

24. $A = \begin{bmatrix} 5 & 2 \\ -15 & -6 \end{bmatrix}$.

25. $A = \begin{bmatrix} 2 & -3 & 0 \\ 2 & 1 & 5 \\ 0 & -1 & 2 \end{bmatrix}$.

26. $A = \begin{bmatrix} -2 & 3 & -1 \\ 2 & 1 & 5 \\ 0 & 2 & 3 \end{bmatrix}$.

27. $A = \begin{bmatrix} 1 & -1 & 2 \\ 3 & -1 & 4 \\ 5 & 1 & 7 \end{bmatrix}$.

28. $A = \begin{bmatrix} 0 & 1 & 2 \\ -1 & -1 & 3 \\ 1 & -2 & 1 \end{bmatrix}$.

29. $A = \begin{bmatrix} 2 & -3 & 5 \\ 1 & 2 & 1 \\ 0 & 7 & -1 \end{bmatrix}$.

30. $A = \begin{bmatrix} 1 & 1 & 1 & 1 \\ -1 & 1 & -1 & 1 \\ 1 & 1 & -1 & -1 \\ -1 & 1 & 1 & -1 \end{bmatrix}$.

31. $A = \begin{bmatrix} 1 & 0 & 3 & 5 \\ -2 & 1 & 1 & 3 \\ 3 & 9 & 0 & 2 \\ 2 & 0 & 3 & -1 \end{bmatrix}$.

32. Let $A = \begin{bmatrix} 1 & -2x & 2x^2 \\ 2x & 1 - 2x^2 & -2x \\ 2x^2 & 2x & 1 \end{bmatrix}$.

(a) Show that $\det(A) = (1 + 2x^2)^3$.

(b) Use the adjoint method to find A^{-1}.

In Problems 33–35, find the specified element in the *inverse* of the given matrix. Do not use elementary row operations.

33. $A = \begin{bmatrix} 1 & 1 & 1 \\ 1 & 2 & 2 \\ 1 & 2 & 3 \end{bmatrix}$; $(3, 2)$-element.

34. $A = \begin{bmatrix} 2 & 0 & -1 \\ 2 & 1 & 1 \\ 3 & -1 & 0 \end{bmatrix}$; $(3, 1)$-element.

35. $A = \begin{bmatrix} 1 & 0 & 1 & 0 \\ 2 & -1 & 1 & 3 \\ 0 & 1 & -1 & 2 \\ -1 & 1 & 2 & 0 \end{bmatrix}$; $(2, 3)$-element.

In Problems 36–38, find A^{-1}.

36. $A = \begin{bmatrix} 3e^t & e^{2t} \\ 2e^t & 2e^{2t} \end{bmatrix}$.

37. $A = \begin{bmatrix} e^t \sin 2t & -e^{-t} \cos 2t \\ e^t \cos 2t & e^{-t} \sin 2t \end{bmatrix}$.

38. $A = \begin{bmatrix} e^t & te^t & e^{-2t} \\ e^t & 2te^t & e^{-2t} \\ e^t & te^t & 2e^{-2t} \end{bmatrix}$.

39. If

$$A = \begin{bmatrix} 1 & 2 & 3 \\ 3 & 4 & 5 \\ 4 & 5 & 6 \end{bmatrix},$$

compute the matrix product $A \cdot \text{adj}(A)$. What can you conclude about $\det(A)$?

For Problems 40–43, use Cramer's rule to solve the given linear system.

40.
$$\begin{aligned} 2x_1 - 3x_2 &= 2, \\ x_1 + 2x_2 &= 4. \end{aligned}$$

41.
$$\begin{aligned} 3x_1 - 2x_2 + x_3 &= 4, \\ x_1 + x_2 - x_3 &= 2, \\ x_1 \quad\quad + x_3 &= 1. \end{aligned}$$

42.
$$\begin{aligned} x_1 - 3x_2 + x_3 &= 0, \\ x_1 + 4x_2 - x_3 &= 0, \\ 2x_1 + x_2 - 3x_3 &= 0. \end{aligned}$$

43.
$$\begin{aligned} x_1 - 2x_2 + 3x_3 - x_4 &= 1, \\ 2x_1 \quad\quad + x_3 &= 2, \\ x_1 + x_2 \quad\quad - x_4 &= 0, \\ x_2 - 2x_3 + x_4 &= 3. \end{aligned}$$

44. Use Cramer's rule to determine x_1 and x_2 if

$$\begin{aligned} e^t x_1 + e^{-2t} x_2 &= 3\sin t, \\ e^t x_1 - 2e^{-2t} x_2 &= 4\cos t. \end{aligned}$$

45. Determine the value of x_2 such that

$$\begin{aligned} x_1 + 4x_2 - 2x_3 + x_4 &= 2, \\ 2x_1 + 9x_2 - 3x_3 - 2x_4 &= 5, \\ x_1 + 5x_2 + x_3 - x_4 &= 3, \\ 3x_1 + 14x_2 + 7x_3 - 2x_4 &= 6. \end{aligned}$$

46. Find all solutions to the system

$$\begin{aligned} (b+c)x_1 + a(x_2 + x_3) &= a, \\ (c+a)x_1 + b(x_3 + x_1) &= b, \\ (a+b)x_1 + c(x_1 + x_2) &= c, \end{aligned}$$

where a, b, c are constants. Make sure you consider all cases (that is, those when there is a unique solution, an infinite number of solutions, and no solutions).

47. Prove Equation (3.3.3).

48. ◇ Let A be a randomly generated invertible 4×4 matrix. Verify the Cofactor Expansion Theorem for expansion along row 1.

49. ◇ Let A be a randomly generated 4×4 matrix. Verify Equation (3.3.3) when $i = 2$ and $j = 4$.

50. ◇ Let A be a randomly generated 5×5 matrix. Determine $\text{adj}(A)$ and compute $A \cdot \text{adj}(A)$. Use your result to determine $\det(A)$.

51. ◇ Solve the system of equations

$$\begin{aligned} 1.21x_1 + 3.42x_2 + 2.15x_3 &= 3.25, \\ 5.41x_1 + 2.32x_2 + 7.15x_3 &= 4.61, \\ 21.63x_1 + 3.51x_2 + 9.22x_3 &= 9.93. \end{aligned}$$

Round answers to two decimal places.

52. ◇ Use Cramer's rule to solve the system $A\mathbf{x} = \mathbf{b}$ if

$$A = \begin{bmatrix} 1 & 2 & 3 & 4 & 4 \\ 2 & 1 & 2 & 3 & 4 \\ 3 & 2 & 1 & 2 & 3 \\ 4 & 3 & 2 & 1 & 2 \\ 4 & 4 & 3 & 2 & 1 \end{bmatrix}, \quad \text{and} \quad \mathbf{b} = \begin{bmatrix} 68 \\ -72 \\ -87 \\ 79 \\ 43 \end{bmatrix}.$$

53. Verify that $BA = I_n$ in the proof of Theorem 3.3.16.

3.4   Summary of Determinants

The primary aim of this section is to serve as a stand-alone introduction to determinants for readers who desire only a cursory review of the major facts pertaining to determinants. It may also be used as a review of the results derived in Sections 3.1–3.3.

Formulas for the Determinant

The determinant of an $n \times n$ matrix A, denoted $\det(A)$, is a scalar whose value can be obtained in the following manner.

1. If $A = [a_{11}]$, then $\det(A) = a_{11}$.

2. If $A = \begin{bmatrix} a_{11} & a_{12} \\ a_{21} & a_{22} \end{bmatrix}$, then $\det(A) = a_{11}a_{22} - a_{12}a_{21}$.

3. For $n > 2$, the determinant of A can be computed using either of the following formulas:

$$\det(A) = a_{i1}C_{i1} + a_{i2}C_{i2} + \cdots + a_{in}C_{in}, \tag{3.4.1}$$

$$\det(A) = a_{1j}C_{1j} + a_{2j}C_{2j} + \cdots + a_{nj}C_{nj}, \tag{3.4.2}$$

where $C_{ij} = (-1)^{i+j}M_{ij}$, and M_{ij} is the determinant of the matrix obtained by deleting the ith row and jth column of A. The formulas (3.4.1) and (3.4.2) are referred to as cofactor expansion along the ith row and cofactor expansion along the jth column, respectively. The determinants M_{ij} and C_{ij} are called the **minors** and **cofactors** of A, respectively. We also denote $\det(A)$ by

$$\begin{vmatrix} a_{11} & a_{12} & \cdots & a_{1n} \\ a_{21} & a_{22} & \cdots & a_{2n} \\ \vdots & \vdots & & \vdots \\ a_{n1} & a_{n2} & \cdots & a_{nn} \end{vmatrix}.$$

As an example, consider the general 3×3 matrix

$$A = \begin{bmatrix} a_{11} & a_{12} & a_{13} \\ a_{21} & a_{22} & a_{23} \\ a_{31} & a_{32} & a_{33} \end{bmatrix}.$$

Using cofactor expansion along row 1, we have

$$\det(A) = a_{11}C_{11} + a_{12}C_{12} + a_{13}C_{13}. \tag{3.4.3}$$

We next compute the required cofactors:

$$C_{11} = +M_{11} = \begin{vmatrix} a_{22} & a_{23} \\ a_{32} & a_{33} \end{vmatrix} = a_{22}a_{33} - a_{23}a_{32},$$

$$C_{12} = -M_{12} = -\begin{vmatrix} a_{21} & a_{23} \\ a_{31} & a_{33} \end{vmatrix} = -(a_{21}a_{33} - a_{23}a_{31}),$$

$$C_{13} = +M_{13} = \begin{vmatrix} a_{21} & a_{22} \\ a_{31} & a_{32} \end{vmatrix} = a_{21}a_{32} - a_{22}a_{31}.$$

Inserting these expressions for the cofactors into Equation (3.4.3) yields

$$\det(A) = a_{11}(a_{22}a_{33} - a_{23}a_{32}) - a_{12}(a_{21}a_{33} - a_{23}a_{31}) + a_{13}(a_{21}a_{32} - a_{22}a_{31}),$$

which can be written as

$$\det(A) = a_{11}a_{22}a_{33} + a_{12}a_{23}a_{31} + a_{13}a_{21}a_{32} - a_{11}a_{23}a_{32} - a_{12}a_{21}a_{33} - a_{13}a_{22}a_{31}.$$

Although we chose to use cofactor expansion along the first row to obtain the preceding formula, according to (3.4.1) and (3.4.2), the same result would have been obtained if we had chosen to expand along any row or column of A. A simple schematic for obtaining the terms in the determinant of a 3×3 matrix is given in Figure 3.4.1. By taking the product of the elements joined by each arrow and attaching the indicated sign to the result, we obtain the six terms in the determinant of the 3×3 matrix $A = [a_{ij}]$. Note that this technique for obtaining the terms in a 3×3 determinant *does not* generalize to determinants of larger matrices.

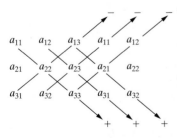

Figure 3.4.1: A schematic for obtaining the determinant of a 3×3 matrix $A = [a_{ij}]$.

Example 3.4.1   Evaluate

$$\begin{vmatrix} 2 & -1 & 1 \\ 3 & 4 & 2 \\ 7 & 5 & 8 \end{vmatrix}.$$

Solution: In this case, the schematic given in Figure 3.4.1 is

$$\begin{matrix} 2 & -1 & 1 & 2 & -1 \\ 3 & 4 & 2 & 3 & 4 \\ 7 & 5 & 8 & 7 & 5 \end{matrix}$$

so that

$$\begin{vmatrix} 2 & -1 & 1 \\ 3 & 4 & 2 \\ 7 & 5 & 8 \end{vmatrix} = (2)(4)(8) + (-1)(2)(7) + (1)(3)(5) - (7)(4)(1) - (5)(2)(2) - (8)(3)(-1)$$

$$= 41.$$ □

Properties of Determinants

Let A and B be $n \times n$ matrices. The determinant has the following properties:

P1. If B is obtained by permuting two rows (or columns) of A, then

$$\det(B) = -\det(A).$$

P2. If B is obtained by multiplying any row (or column) of A by a scalar k, then

$$\det(B) = k \det(A).$$

P3. If B is obtained by adding a multiple of any row (or column) of A to another row (or column) of A, then

$$\det(B) = \det(A).$$

P4. $\det(A^T) = \det(A).$

P5. Let $\mathbf{a}_1, \mathbf{a}_2, \ldots, \mathbf{a}_n$ denote the row vectors of A. If the ith row vector of A is the sum of two row vectors, say $\mathbf{a}_i = \mathbf{b}_i + \mathbf{c}_i$, then

$$\det(A) = \det(B) + \det(C),$$

where

$$B = [\mathbf{a}_1, \mathbf{a}_2, \ldots, \mathbf{a}_{i-1}, \mathbf{b}_i, \mathbf{a}_{i+1}, \ldots, \mathbf{a}_n]^T$$

and

$$C = [\mathbf{a}_1, \mathbf{a}_2, \ldots, \mathbf{a}_{i-1}, \mathbf{c}_i, \mathbf{a}_{i+1}, \ldots, \mathbf{a}_n]^T.$$

The corresponding property for columns is also true.

P6. If A has a row (or column) of zeros, then $\det(A) = 0.$

P7. If two rows (or columns) of A are the same, then $\det(A) = 0.$

P8. $\det(AB) = \det(A)\det(B).$

The first three properties tell us how elementary row operations and elementary column operations performed on a matrix A alter the value of $\det(A)$. They can be very helpful in reducing the amount of work required to evaluate a determinant, since we can use elementary row operations to put several zeros in a row or column of A and then use cofactor expansion along that row or column. We illustrate with an example.

Example 3.4.2 Evaluate

$$\begin{vmatrix} 2 & 1 & 3 & 2 \\ -1 & 1 & -2 & 2 \\ 5 & 1 & -2 & 1 \\ -2 & 3 & 1 & 1 \end{vmatrix}.$$

Solution: Before performing a cofactor expansion, we first use elementary row operations to simplify the determinant:

$$\begin{bmatrix} 2 & 1 & 3 & 2 \\ -1 & 1 & -2 & 2 \\ 5 & 1 & -2 & 1 \\ -2 & 3 & 1 & 1 \end{bmatrix} \overset{1}{\sim} \begin{bmatrix} 0 & 3 & -1 & 6 \\ -1 & 1 & -2 & 2 \\ 0 & 6 & -12 & 11 \\ 0 & 1 & 5 & -3 \end{bmatrix}$$

According to P3, the determinants of the two matrices above are the same. To evaluate the determinant of the matrix on the right, we use cofactor expansion along the first column.

$$\begin{vmatrix} 0 & 3 & -1 & 6 \\ -1 & 1 & -2 & 2 \\ 0 & 6 & -12 & 11 \\ 0 & 1 & 5 & -3 \end{vmatrix} = -(-1)\begin{vmatrix} 3 & -1 & 6 \\ 6 & -12 & 11 \\ 1 & 5 & -3 \end{vmatrix}$$

To evaluate the determinant of the 3×3 matrix on the right, we can use the schematic given in Figure 3.4.1, or, we can continue to use elementary row operations to introduce zeros into the matrix:

$$\begin{vmatrix} 3 & -1 & 6 \\ 6 & -12 & 11 \\ 1 & 5 & -3 \end{vmatrix} \overset{2}{=} \begin{vmatrix} 0 & -16 & 15 \\ 0 & -42 & 29 \\ 1 & 5 & -3 \end{vmatrix} = \begin{vmatrix} -16 & 15 \\ -42 & 29 \end{vmatrix} = 166.$$

Here, we have reduced the 3×3 determinant to a 2×2 determinant by using cofactor expansion along the first column of the 3×3 matrix.

> **1.** $A_{21}(2),\ A_{23}(5),\ A_{24}(-2)$ **2.** $A_{31}(-3),\ A_{32}(-6)$

Basic Theoretical Results

The determinant is a useful theoretical tool in linear algebra. We list next the major results that will be needed in the remainder of the text.

1. The volume of the parallelepiped determined by the vectors

$$\mathbf{a} = a_1\mathbf{i} + a_2\mathbf{j} + a_3\mathbf{k}, \quad \mathbf{b} = b_1\mathbf{i} + b_2\mathbf{j} + b_3\mathbf{k}, \quad \mathbf{c} = c_1\mathbf{i} + c_2\mathbf{j} + c_3\mathbf{k}$$

is

$$\boxed{\text{Volume} = |\det(A)|,}$$

where $A = \begin{bmatrix} a_1 & a_2 & a_3 \\ b_1 & b_2 & b_3 \\ c_1 & c_2 & c_3 \end{bmatrix}.$

2. An $n \times n$ matrix is invertible if and only if $\det(A) \neq 0$.

3. An $n \times n$ linear system $A\mathbf{x} = \mathbf{b}$ has a unique solution if and only if $\det(A) \neq 0$.

4. An $n \times n$ *homogeneous* linear system $A\mathbf{x} = \mathbf{0}$ has an infinite number of solutions if and only if $\det(A) = 0$.

We see, for example, that according to (2), the matrices in Examples 3.4.1 and 3.4.2 are both invertible.

If A is an $n \times n$ matrix with $\det(A) \neq 0$, then the following two methods can be derived for obtaining the inverse of A and for finding the unique solution to the linear system $A\mathbf{x} = \mathbf{b}$, respectively.

1. Adjoint Method for A^{-1}: If A is invertible, then

$$A^{-1} = \frac{1}{\det(A)} \text{adj}(A),$$

where adj(A) denotes the transpose of the matrix obtained by replacing each element in A by its cofactor.

2. Cramer's Rule: If $\det(A) \neq 0$, then the unique solution to $A\mathbf{x} = \mathbf{b}$ is $\mathbf{x} = (x_1, x_2, \ldots, x_n)$, where

$$x_k = \frac{\det(B_k)}{\det(A)}, \qquad k = 1, 2, \ldots, n,$$

and B_k denotes the matrix obtained when the kth column vector of A is replaced by $\mathbf{b}$.

Example 3.4.3   Use the adjoint method to find A^{-1} if $A = \begin{bmatrix} 2 & -1 & 1 \\ 3 & 4 & 2 \\ 7 & 5 & 8 \end{bmatrix}$.

Solution: We have already shown in Example 3.4.1 that $\det(A) = 41$, so that A is invertible. Replacing each element in A with its cofactor yields the **matrix of cofactors**

$$M_C = \begin{bmatrix} 22 & -10 & -13 \\ 13 & 9 & -17 \\ -6 & -1 & 11 \end{bmatrix},$$

so that

$$\text{adj}(A) = M_C^T = \begin{bmatrix} 22 & 13 & -6 \\ -10 & 9 & -1 \\ -13 & -17 & 11 \end{bmatrix}.$$

Consequently,

$$A^{-1} = \frac{1}{\det(A)} \text{adj}(A) = \begin{bmatrix} \frac{22}{41} & \frac{13}{41} & -\frac{6}{41} \\ -\frac{10}{41} & \frac{9}{41} & -\frac{1}{41} \\ -\frac{13}{41} & -\frac{17}{41} & \frac{11}{41} \end{bmatrix}. \qquad \square$$

Example 3.4.4   Use Cramer's rule to solve the linear system

$$\begin{aligned} 2x_1 - x_2 + x_3 &= 2, \\ 3x_1 + 4x_2 + 2x_3 &= 5, \\ 7x_1 + 5x_2 + 8x_3 &= 3. \end{aligned}$$

Solution: The matrix of coefficients is

$$A = \begin{bmatrix} 2 & -1 & 1 \\ 3 & 4 & 2 \\ 7 & 5 & 8 \end{bmatrix}.$$

We have already shown in Example 3.4.1 that $\det(A) = 41$. Consequently, Cramer's rule can indeed be applied. In this problem, we have

$$\det(B_1) = \begin{vmatrix} 2 & -1 & 1 \\ 5 & 4 & 2 \\ 3 & 5 & 8 \end{vmatrix} = 91,$$

$$\det(B_2) = \begin{vmatrix} 2 & 2 & 1 \\ 3 & 5 & 2 \\ 7 & 3 & 8 \end{vmatrix} = 22,$$

$$\det(B_3) = \begin{vmatrix} 2 & -1 & 2 \\ 3 & 4 & 5 \\ 7 & 5 & 3 \end{vmatrix} = -78.$$

It therefore follows from Cramer's rule that

$$x_1 = \frac{\det(B_1)}{\det(A)} = \frac{91}{41}, \qquad x_2 = \frac{\det(B_2)}{\det(A)} = \frac{22}{41}, \qquad x_3 = \frac{\det(B_3)}{\det(A)} = -\frac{78}{41}. \qquad \square$$

Exercises for 3.4

Skills

- Be able to compute the determinant of an $n \times n$ matrix.

- Know the effects that elementary row operations and elementary column operations have on the determinant of a matrix.

- Be able to use the determinant to decide if a matrix is invertible.

- Know how the determinant is affected by matrix multiplication and by matrix transpose.

- Be able to compute the adjoint of a matrix and use it to find A^{-1} for an invertible matrix A.

Problems

For Problems 1–7, evaluate the given determinant.

1. $\begin{vmatrix} 5 & -1 \\ 3 & 7 \end{vmatrix}.$

2. $\begin{vmatrix} 3 & 5 & 7 \\ -1 & 2 & 4 \\ 6 & 3 & -2 \end{vmatrix}.$

3. $\begin{vmatrix} 5 & 1 & 4 \\ 6 & 1 & 3 \\ 14 & 2 & 7 \end{vmatrix}.$

4. $\begin{vmatrix} 2.3 & 1.5 & 7.9 \\ 4.2 & 3.3 & 5.1 \\ 6.8 & 3.6 & 5.7 \end{vmatrix}.$

5. $\begin{vmatrix} a & b & c \\ b & c & a \\ c & a & b \end{vmatrix}.$

6. $\begin{vmatrix} 3 & 5 & -1 & 2 \\ 2 & 1 & 5 & 2 \\ 3 & 2 & 5 & 7 \\ 1 & -1 & 2 & 1 \end{vmatrix}.$

7. $\begin{vmatrix} 7 & 1 & 2 & 3 \\ 2 & -2 & 4 & 6 \\ 3 & -1 & 5 & 4 \\ 18 & 9 & 27 & 54 \end{vmatrix}.$

For Problems 8–12, find $\det(A)$. If A is invertible, use the adjoint method to find A^{-1}.

8. $A = \begin{bmatrix} 3 & 5 \\ 2 & 7 \end{bmatrix}$.

9. $A = \begin{bmatrix} 1 & 2 & 3 \\ 2 & 3 & 1 \\ 3 & 1 & 2 \end{bmatrix}$.

10. $A = \begin{bmatrix} 3 & 4 & 7 \\ 2 & 6 & 1 \\ 3 & 14 & -1 \end{bmatrix}$.

11. $A = \begin{bmatrix} 2 & 5 & 7 \\ 4 & -3 & 2 \\ 6 & 9 & 11 \end{bmatrix}$.

12. $A = \begin{bmatrix} 5 & -1 & 2 & 1 \\ 3 & -1 & 4 & 5 \\ 1 & -1 & 2 & 1 \\ 5 & 9 & -3 & 2 \end{bmatrix}$.

For Problems 13–17, use Cramer's rule to determine the unique solution to the system $A\mathbf{x} = \mathbf{b}$ for the given matrix and vector.

13. $A = \begin{bmatrix} 3 & 5 \\ 6 & 2 \end{bmatrix}$, $\mathbf{b} = \begin{bmatrix} 4 \\ 9 \end{bmatrix}$.

14. $A = \begin{bmatrix} \cos t & \sin t \\ \sin t & -\cos t \end{bmatrix}$, $\mathbf{b} = \begin{bmatrix} e^{-t} \\ 3e^{-t} \end{bmatrix}$.

15. $A = \begin{bmatrix} 4 & 1 & 3 \\ 2 & -1 & 5 \\ 2 & 3 & 1 \end{bmatrix}$, $\mathbf{b} = \begin{bmatrix} 5 \\ 7 \\ 2 \end{bmatrix}$.

16. $A = \begin{bmatrix} 5 & 3 & 6 \\ 2 & 4 & -7 \\ 2 & 5 & 9 \end{bmatrix}$, $\mathbf{b} = \begin{bmatrix} 3 \\ -1 \\ 4 \end{bmatrix}$.

17. $A = \begin{bmatrix} 3.1 & 3.5 & 7.1 \\ 2.2 & 5.2 & 6.3 \\ 1.4 & 8.1 & 0.9 \end{bmatrix}$, $\mathbf{b} = \begin{bmatrix} 3.6 \\ 2.5 \\ 9.3 \end{bmatrix}$.

18. If A is an invertible $n \times n$ matrix, prove that

$$\det(A^{-1}) = \frac{1}{\det(A)}.$$

19. Let A and B be 3×3 matrices with $\det(A) = 3$ and $\det(B) = -4$. Determine

$$\det(2A), \quad \det(A^{-1}), \quad \det(A^T B),$$
$$\det(B^5), \quad \det(B^{-1}AB).$$

3.5   Chapter Review

This chapter has laid out a basic introduction to the theory of determinants.

Determinants and Elementary Row Operations

For a square matrix A, one approach for computing the determinant of A, $\det(A)$, is to use elementary row operations to reduce A to row-echelon form. The effects of the various types of elementary row operations on $\det(A)$ are as follows:

- P_{ij}: permuting two rows of A alters the determinant by a factor of -1.

- $M_i(k)$: multiplying the ith row of A by k multiplies the determinant of the matrix by a factor of k.

- $A_{ij}(k)$: adding a multiple of one row of A to another has no effect whatsoever on $\det(A)$.

A crucial fact in this approach is the following:

Theorem 3.5.1   If A is an $n \times n$ upper (or lower) triangular matrix, its determinant is

$$\det(A) = a_{11}a_{22} \cdots a_{nn}.$$

Therefore, since the row-echelon form of A is upper triangular, we can compute $\det(A)$ by using Theorem 3.5.1 and by keeping track of the elementary row operations involved in the row-reduction process.

Cofactor Expansion

Another way to compute $\det(A)$ is via the Cofactor Expansion Theorem: For $n \geq 2$, the determinant of A can be computed using either of the following formulas:

$$\det(A) = a_{i1}C_{i1} + a_{i2}C_{i2} + \cdots + a_{in}C_{in}, \tag{3.5.1}$$

$$\det(A) = a_{1j}C_{1j} + a_{2j}C_{2j} + \cdots + a_{nj}C_{nj}, \tag{3.5.2}$$

where $C_{ij} = (-1)^{i+j}M_{ij}$, and M_{ij} is the determinant of the matrix obtained by deleting the ith row and jth column of A. The formulas (3.5.1) and (3.5.2) are referred to as cofactor expansion along the ith row and cofactor expansion along the jth column, respectively. The determinants M_{ij} and C_{ij} are called the **minors** and **cofactors** of A, respectively.

Adjoint Method and Cramer's Rule

If A is an $n \times n$ matrix with $\det(A) \neq 0$, then the following two methods can be derived for obtaining the inverse of A and for finding the unique solution to the linear system $A\mathbf{x} = \mathbf{b}$, respectively.

1. **Adjoint Method for A^{-1}:** If A is invertible, then

$$A^{-1} = \frac{1}{\det(A)} \, \text{adj}(A),$$

   where $\text{adj}(A)$ denotes the transpose of the matrix obtained by replacing each element in A by its cofactor.

2. **Cramer's Rule:** If $\det(A) \neq 0$, then the unique solution to $A\mathbf{x} = \mathbf{b}$ is $\mathbf{x} = (x_1, x_2, \ldots, x_n)$, where

$$x_k = \frac{\det(B_k)}{\det(A)}, \qquad k = 1, 2, \ldots, n,$$

   and B_k denotes the matrix obtained when the kth column vector of A is replaced by $\mathbf{b}$.

Additional Problems

For Problems 1–6, evaluate the determinant of the given matrix A by using **(a)** the definition, **(b)** elementary row operations to reduce A to an upper triangular matrix, and **(c)** the Cofactor Expansion Theorem.

1. $A = \begin{bmatrix} -7 & -2 \\ 1 & -5 \end{bmatrix}$.

2. $A = \begin{bmatrix} 6 & 6 \\ -2 & 1 \end{bmatrix}$.

3. $A = \begin{bmatrix} -1 & 4 & 1 \\ 0 & 2 & 2 \\ 2 & 2 & -3 \end{bmatrix}$.

4. $A = \begin{bmatrix} 2 & 3 & -5 \\ -4 & 0 & 2 \\ 6 & -3 & 3 \end{bmatrix}$.

5. $A = \begin{bmatrix} 3 & -1 & -2 & 1 \\ 0 & 0 & 1 & 4 \\ 0 & 2 & 1 & -1 \\ 0 & 0 & 0 & -4 \end{bmatrix}$.

6. $A = \begin{bmatrix} 0 & 0 & 0 & -2 \\ 0 & 0 & -5 & 1 \\ 0 & 1 & -4 & 1 \\ -3 & -3 & -3 & -3 \end{bmatrix}$.

For Problems 7–10, suppose that

$$A = \begin{bmatrix} a & b & c \\ d & e & f \\ g & h & i \end{bmatrix}, \text{ and } \det(A) = 4.$$

Compute the determinant of each matrix below.

7. $\begin{bmatrix} g & h & i \\ -4a & -4b & -4c \\ 2d & 2e & 2f \end{bmatrix}.$

8. $\begin{bmatrix} a-5d & b-5e & c-5f \\ 3g & 3h & 3i \\ -d+3g & -e+3h & -f+3i \end{bmatrix}.$

9. $\begin{bmatrix} 3b & 3e & 3h \\ c-2a & f-2d & i-2g \\ -a & -d & -g \end{bmatrix}.$

10. $3 \begin{bmatrix} a-d & b-e & c-f \\ 2g & 2h & 2i \\ -d & -e & -f \end{bmatrix}.$

For Problems 11–14, suppose that A and B are 4×4 invertible matrices. If $\det(A) = -2$ and $\det(B) = 3$, compute each determinant below.

11. $\det(AB)$.

12. $\det(B^2 A^{-1})$.

13. $\det(((A^{-1}B)^T)(2B^{-1}))$.

14. $\det((-A)^3(2B^2))$.

15. Let

$$A = \begin{bmatrix} 1 & 2 & -1 \\ 2 & 1 & 4 \end{bmatrix}, \quad B = \begin{bmatrix} 2 & 1 \\ 5 & -2 \\ 4 & 7 \end{bmatrix}, \quad C = \begin{bmatrix} 1 & 0 & 5 \\ 3 & -1 & 4 \\ 2 & -2 & 6 \end{bmatrix}.$$

Determine, if possible,

$\det(A)$, $\quad$ $\det(B)$, $\quad$ $\det(C)$,
$\det(C^T)$, $\quad$ $\det(AB)$, $\quad$ $\det(BA)$,
$\det(B^T A^T)$, $\quad$ $\det(BAC)$, $\quad$ $\det(ACB)$.

16. Let

$$A = \begin{bmatrix} 1 & 2 \\ 3 & 4 \end{bmatrix}, \quad \text{and} \quad B = \begin{bmatrix} 5 & 4 \\ 1 & 1 \end{bmatrix}.$$

Use the adjoint method to find B^{-1} and then determine $(A^{-1}B^T)^{-1}$.

For Problems 17–21, use the adjoint method to determine A^{-1} for the given matrix A.

17. $A = \begin{bmatrix} 2 & -1 & 1 \\ 0 & 5 & -1 \\ 1 & 1 & 3 \end{bmatrix}.$

18. $A = \begin{bmatrix} 0 & -3 & 2 & 2 \\ 0 & 1 & 1 & 1 \\ 1 & 2 & 3 & -4 \\ 1 & 0 & 0 & 5 \end{bmatrix}.$

19. $A = \begin{bmatrix} 0 & 0 & 0 & 1 \\ 0 & 1 & 3 & -3 \\ -2 & -3 & -5 & 2 \\ 4 & -4 & 4 & 6 \end{bmatrix}.$

20. $A = \begin{bmatrix} 5 & 8 & 16 \\ 4 & 1 & 8 \\ -4 & -4 & -11 \end{bmatrix}.$

21. $A = \begin{bmatrix} 2 & 6 & 6 \\ 2 & 7 & 6 \\ 2 & 7 & 7 \end{bmatrix}.$

22. Add one row to the matrix

$$A = \begin{bmatrix} 4 & -1 & 0 \\ 5 & 1 & 4 \end{bmatrix}$$

so as to create a 3×3 matrix B with $\det(B) = 10$.

23. **True or False:** Given any real number r and any 3×3 matrix A whose entries are all nonzero, it is always possible to change at most one entry of A to get a matrix B with $\det(B) = r$.

24. Let $A = \begin{bmatrix} 1 & 2 & 4 \\ 3 & 1 & 6 \\ k & 3 & 2 \end{bmatrix}.$

(a) Find all value(s) of k for which the matrix A fails to be invertible.

(b) In terms of k, determine the volume of the parallelepiped determined by the row vectors of the matrix A. Is that the same as the volume of the parallelepiped determined by the column vectors of the matrix A? Explain how you know this without any calculation.

25. Repeat the preceding problem for the matrix

$$A = \begin{bmatrix} k+1 & 2 & 1 \\ 0 & 3 & k \\ 1 & 1 & 1 \end{bmatrix}.$$

26. Repeat the preceding problem for the matrix

$$A = \begin{bmatrix} 2 & k-3 & k^2 \\ 2 & 1 & 4 \\ 1 & k & 0 \end{bmatrix}.$$

27. Let A and B be $n \times n$ matrices such that $AB = -BA$. Use determinants to prove that if n is odd, then A and B cannot both be invertible.

28. A real $n \times n$ matrix A is called *orthogonal* if $AA^T = A^T A = I_n$. If A is an orthogonal matrix, prove that $\det(A) = \pm 1$.

For Problems 29–31, use Cramer's rule to solve the given linear system.

29.
$$-3x_1 + x_2 = 3,$$
$$x_1 + 2x_2 = 1.$$

30.
$$2x_1 - x_2 + x_3 = 2,$$
$$4x_1 + 5x_2 + 3x_3 = 0,$$
$$4x_1 - 3x_2 + 3x_3 = 2.$$

31.
$$3x_1 + x_2 + 2x_3 = -1,$$
$$2x_1 - x_2 + x_3 = -1,$$
$$5x_2 + 5x_3 = -5.$$

Project: Volume of a Tetrahedron

In this project, we use determinants and vectors to derive the formula for the volume of a tetrahedron with vertices $A = (x_1, y_1, z_1)$, $B = (x_2, y_2, z_2)$, $C = (x_3, y_3, z_3)$, and $D = (x_4, y_4, z_4)$.

Let h denote the distance from A to the plane determined by B, C, and D. From geometry, the volume of the tetrahedron is given by

$$\text{Volume} = \tfrac{1}{3}h(\text{area of triangle } BCD). \tag{3.5.3}$$

(a) Express the area of triangle BCD in terms of a cross product of vectors.

(b) Use trigonometry to express h in terms of the distance from A to B and the angle between the vector $\overrightarrow{AB}$ and the segment connecting A to the base BCD at a right angle.

(c) Combining (a) and (b) with the volume of the tetrahedron given above, express the volume of the tetrahedron in terms of dot products and cross products of vectors.

(d) Following the proof of part 2 of Theorem 3.1.11, express the volume of the tetrahedron in terms of a determinant with entries in terms of the x_i, y_i, and z_i for $1 \leq i \leq 4$.

(e) Show that the expression in part (d) is the same as

$$\text{Volume} = \frac{1}{6} \begin{vmatrix} x_1 & y_1 & z_1 & 1 \\ x_2 & y_2 & z_2 & 1 \\ x_3 & y_3 & z_3 & 1 \\ x_4 & y_4 & z_4 & 1 \end{vmatrix}. \tag{3.5.4}$$

(f) For each set of four points below, determine the volume of the tetrahedron with those points as vertices by using (3.5.3) and by using (3.5.4). Both formulas should yield the same answer.

    (i) $(0, 0, 0), (1, 0, 0), (0, 1, 0), (0, 0, 1)$.

    (ii) $(-1, 1, 2), (0, 3, 3), (1, -1, 2), (0, 0, 1)$.

CHAPTER

4

Vector Spaces

To criticize mathematics for its abstraction is to miss the point entirely. Abstraction is what makes mathematics work. — Ian Stewart

The main aim of this text is to study linear mathematics. In Chapter 2 we studied systems of linear equations, and the theory underlying the solution of a system of linear equations can be considered as a special case of a general mathematical framework for linear problems. To illustrate this framework, we discuss an example.

Consider the homogeneous linear system $A\mathbf{x} = \mathbf{0}$, where

$$A = \begin{bmatrix} 1 & -1 & 2 \\ 2 & -2 & 4 \\ 3 & -3 & 6 \end{bmatrix}.$$

It is straightforward to show that this system has solution set

$$S = \{(r - 2s, r, s) : r, s \in \mathbb{R}\}.$$

Geometrically we can interpret each solution as defining the coordinates of a point in space or, equivalently, as the geometric vector with components

$$\mathbf{v} = (r - 2s, r, s).$$

Using the standard operations of vector addition and multiplication of a vector by a real number, it follows that $\mathbf{v}$ can be written in the form

$$\mathbf{v} = r(1, 1, 0) + s(-2, 0, 1).$$

We see that every solution to the given linear problem can be expressed as a linear combination of the two basic solutions (see Figure 4.0.1):

$$\mathbf{v}_1 = (1, 1, 0) \qquad \text{and} \qquad \mathbf{v}_2 = (-2, 0, 1).$$

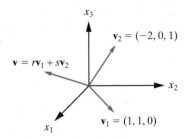

Figure 4.0.1: Two basic solutions to $A\mathbf{x} = \mathbf{0}$ and an example of an arbitrary solution to the system.

We will observe a similar phenomenon in Chapter 6, when we establish that every solution to the homogeneous second-order linear differential equation

$$y'' + a_1 y' + a_2 y = 0$$

can be written in the form

$$y(x) = c_1 y_1(x) + c_2 y_2(x),$$

where $y_1(x)$ and $y_2(x)$ are two nonproportional solutions to the differential equation on the interval of interest.

In each of these problems, we have a set of "vectors" V (in the first problem the vectors are ordered triples of numbers, whereas in the second, they are functions that are at least twice differentiable on an interval I) and a linear vector equation. Further, in both cases, all solutions to the given equation can be expressed as a linear combination of two particular solutions.

In the next two chapters we develop this way of formulating linear problems in terms of an abstract set of vectors, V, and a linear vector equation with solutions in V. We will find that many problems fit into this framework and that the solutions to these problems can be expressed as linear combinations of a certain number (not necessarily two) of basic solutions. The importance of this result cannot be overemphasized. It reduces the search for all solutions to a given problem to that of finding a finite number of solutions. As specific applications, we will derive the theory underlying linear differential equations and linear systems of differential equations as special cases of the general framework.

Before proceeding further, we give a word of encouragement to the more application-oriented reader. It will probably seem at times that the ideas we are introducing are rather esoteric and that the formalism is pure mathematical abstraction. However, in addition to its inherent mathematical beauty, the formalism incorporates ideas that pervade many areas of applied mathematics, particularly engineering mathematics and mathematical physics, where the problems under investigation are very often linear in nature. Indeed, the linear algebra introduced in the next two chapters should be considered an extremely important addition to one's mathematical repertoire, certainly on a par with the ideas of elementary calculus.

4.1 Vectors in $\mathbb{R}^n$

In this section, we use some familiar ideas about geometric vectors to motivate the more general and abstract idea of a vector space, which will be introduced in the next section. We begin by recalling that a geometric vector can be considered mathematically as a directed line segment (or arrow) that has both a magnitude (length) and a direction attached to it. In calculus courses, we define **vector addition** according to the parallelogram law (see Figure 4.1.1); namely, the sum of the vectors $\mathbf{x}$ and $\mathbf{y}$ is the diagonal of the parallelogram formed by $\mathbf{x}$ and $\mathbf{y}$. We denote the sum by $\mathbf{x} + \mathbf{y}$. It can then be shown geometrically that for all vectors $\mathbf{x}$, $\mathbf{y}$, $\mathbf{z}$,

$$\mathbf{x} + \mathbf{y} = \mathbf{y} + \mathbf{x} \tag{4.1.1}$$

and

$$\mathbf{x} + (\mathbf{y} + \mathbf{z}) = (\mathbf{x} + \mathbf{y}) + \mathbf{z}. \tag{4.1.2}$$

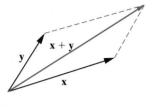

Figure 4.1.1: Parallelogram law of vector addition.

These are the statements that the vector addition operation is commutative and associative. The *zero vector*, denoted $\mathbf{0}$, is defined as the vector satisfying

$$\mathbf{x} + \mathbf{0} = \mathbf{x}, \tag{4.1.3}$$

for all vectors $\mathbf{x}$. We consider the zero vector as having zero magnitude and arbitrary direction. Geometrically, we picture the zero vector as corresponding to a point in space. Let $-\mathbf{x}$ denote the vector that has the same magnitude as $\mathbf{x}$, but the opposite direction. Then according to the parallelogram law of addition,

$$\mathbf{x} + (-\mathbf{x}) = \mathbf{0}. \tag{4.1.4}$$

The vector $-\mathbf{x}$ is called the *additive inverse* of $\mathbf{x}$. Properties (4.1.1)–(4.1.4) are the fundamental properties of vector addition.

The basic algebra of vectors is completed when we also define the operation of multiplication of a vector by a real number. Geometrically, if $\mathbf{x}$ is a vector and k is a real number, then $k\mathbf{x}$ is defined to be the vector whose magnitude is $|k|$ times the magnitude of $\mathbf{x}$ and whose direction is the same as $\mathbf{x}$ if $k > 0$, and opposite to $\mathbf{x}$ if $k < 0$. (See Figure 4.1.2.) If $k = 0$, then $k\mathbf{x} = \mathbf{0}$. This **scalar multiplication** operation has several important properties that we now list. Once more, each of these can be established geometrically using only the foregoing definitions of vector addition and scalar multiplication.

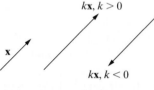

Figure 4.1.2: Scalar multiplication of $\mathbf{x}$ by k.

For all vectors $\mathbf{x}$ and $\mathbf{y}$, and all real numbers r, s and t,

$$1\mathbf{x} = \mathbf{x}, \tag{4.1.5}$$
$$(st)\mathbf{x} = s(t\mathbf{x}), \tag{4.1.6}$$
$$r(\mathbf{x} + \mathbf{y}) = r\mathbf{x} + r\mathbf{y}, \tag{4.1.7}$$
$$(s + t)\mathbf{x} = s\mathbf{x} + t\mathbf{x}. \tag{4.1.8}$$

It is important to realize that, in the foregoing development, we *have not* defined a "multiplication of vectors." In Chapter 3 we discussed the idea of a dot product and cross product of two vectors in space (see Equations (3.1.4) and (3.1.5)), but for the purposes of discussing abstract vector spaces we will essentially ignore the dot product and cross product. We will revisit the dot product in Section 4.11, when we develop inner product spaces.

We will see in the next section how the concept of a vector space arises as a direct generalization of the ideas associated with geometric vectors. Before performing this abstraction, we want to recall some further features of geometric vectors and give one specific and important extension.

We begin by considering vectors in the plane. Recall that $\mathbb{R}^2$ denotes the set of all ordered pairs of real numbers; thus,

$$\mathbb{R}^2 = \{(x, y) : x \in \mathbb{R}, y \in \mathbb{R}\}.$$

The elements of this set are called *vectors in* $\mathbb{R}^2$, and we use the usual vector notation to denote these elements. Geometrically we identify the vector $\mathbf{v} = (x, y)$ in $\mathbb{R}^2$ with

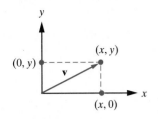

Figure 4.1.3: Identifying vectors in $\mathbb{R}^2$ with geometric vectors in the plane.

the geometric vector **v** directed from the origin of a Cartesian coordinate system to the point with coordinates (x, y). This identification is illustrated in Figure 4.1.3. The numbers x and y are called the **components** of the geometric vector **v**. The geometric vector addition and scalar multiplication operations are consistent with the addition and scalar multiplication operations defined in Chapter 2 via the correspondence with row (or column) vectors for $\mathbb{R}^2$:

If $\mathbf{v} = (x_1, y_1)$ and $\mathbf{w} = (x_2, y_2)$, and k is an arbitrary real number, then

$$\mathbf{v} + \mathbf{w} = (x_1, y_1) + (x_2, y_2) = (x_1 + x_2, y_1 + y_2), \qquad (4.1.9)$$
$$k\mathbf{v} = k(x_1, y_1) = (kx_1, ky_1). \qquad (4.1.10)$$

These are the algebraic statements of the parallelogram law of vector addition and the scalar multiplication law, respectively. (See Figure 4.1.4.) Using the parallelogram law of vector addition and Equations (4.1.9) and (4.1.10), it follows that any vector $\mathbf{v} = (x, y)$ can be written as

$$\mathbf{v} = x\mathbf{i} + y\mathbf{j} = x(1, 0) + y(0, 1),$$

where $\mathbf{i} = (1, 0)$ and $\mathbf{j} = (0, 1)$ are the unit vectors pointing along the positive x- and y-coordinate axes, respectively.

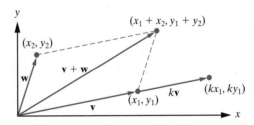

Figure 4.1.4: Vector addition and scalar multiplication in $\mathbb{R}^2$.

The properties (4.1.1)–(4.1.8) are now easily verified for vectors in $\mathbb{R}^2$. In particular, the zero vector in $\mathbb{R}^2$ is the vector

$$\mathbf{0} = (0, 0).$$

Furthermore, Equation (4.1.9) implies that

$$(x, y) + (-x, -y) = (0, 0) = \mathbf{0},$$

so that the additive inverse of the general vector $\mathbf{v} = (x, y)$ is $-\mathbf{v} = (-x, -y)$.

It is straightforward to extend these ideas to vectors in 3-space. We recall that

$$\mathbb{R}^3 = \{(x, y, z) : x \in \mathbb{R}, y \in \mathbb{R}, z \in \mathbb{R}\}.$$

As illustrated in Figure 4.1.5, each vector $\mathbf{v} = (x, y, z)$ in $\mathbb{R}^3$ can be identified with the geometric vector **v** that joins the origin of a Cartesian coordinate system to the point with coordinates (x, y, z). We call x, y, and z the components of **v**.

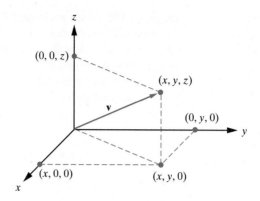

Figure 4.1.5: Identifying vectors in $\mathbb{R}^3$ with geometric vectors in space.

Recall that if $\mathbf{v} = (x_1, y_1, z_1)$, $\mathbf{w} = (x_2, y_2, z_2)$, and k is an arbitrary real number, then addition and scalar multiplication were given in Chapter 2 by

$$\mathbf{v} + \mathbf{w} = (x_1, y_1, z_1) + (x_2, y_2, z_2) = (x_1 + x_2, y_1 + y_2, z_1 + z_2), \qquad (4.1.11)$$
$$k\mathbf{v} = k(x_1, y_1, z_1) = (kx_1, ky_1, kz_1). \qquad (4.1.12)$$

Once more, these are, respectively, the component forms of the laws of vector addition and scalar multiplication for geometric vectors. It follows that an arbitrary vector $\mathbf{v} = (x, y, z)$ can be written as

$$\mathbf{v} = x\mathbf{i} + y\mathbf{j} + z\mathbf{k} = x(1, 0, 0) + y(0, 1, 0) + z(0, 0, 1),$$

where $\mathbf{i} = (1, 0, 0)$, $\mathbf{j} = (0, 1, 0)$, and $\mathbf{k} = (0, 0, 1)$ denote the unit vectors which point along the positive x-, y-, and z-coordinate axes, respectively.

We leave it as an exercise to check that the properties (4.1.1)–(4.1.8) are satisfied by vectors in $\mathbb{R}^3$, where

$$\mathbf{0} = (0, 0, 0),$$

and the additive inverse of $\mathbf{v} = (x, y, z)$ is $-\mathbf{v} = (-x, -y, -z)$.

We now come to our first major abstraction. Whereas the sets $\mathbb{R}^2$ and $\mathbb{R}^3$ and their associated algebraic operations arise naturally from our experience with Cartesian geometry, the motivation behind the algebraic operations in $\mathbb{R}^n$ for larger values of n does not come from geometry. Rather, we can view the addition and scalar multiplication operations in $\mathbb{R}^n$ for $n > 3$ as the natural extension of the component forms of addition and scalar multiplication in $\mathbb{R}^2$ and $\mathbb{R}^3$ in (4.1.9)–(4.1.12). Therefore, in $\mathbb{R}^n$ we have that if $\mathbf{v} = (x_1, x_2, \ldots, x_n)$, $\mathbf{w} = (y_1, y_2, \ldots, y_n)$, and k is an arbitrary real number, then

$$\mathbf{v} + \mathbf{w} = (x_1 + y_1, x_2 + y_2, \ldots, x_n + y_n), \qquad (4.1.13)$$
$$k\mathbf{v} = (kx_1, kx_2, \ldots, kx_n). \qquad (4.1.14)$$

Again, these definitions are direct generalizations of the algebraic operations defined in $\mathbb{R}^2$ and $\mathbb{R}^3$, but there is no geometric analogy when $n > 3$. It is easily established that these operations satisfy properties (4.1.1)–(4.1.8), where the **zero vector** in $\mathbb{R}^n$ is

$$\mathbf{0} = (0, 0, \ldots, 0),$$

and the **additive inverse** of the vector $\mathbf{v} = (x_1, x_2, \ldots, x_n)$ is

$$-\mathbf{v} = (-x_1, -x_2, \ldots, -x_n).$$

The verification of this is left as an exercise.

| **Example 4.1.1** | If $\mathbf{v} = (1.2, 3.5, 2, 0)$ and $\mathbf{w} = (12.23, 19.65, 23.22, 9.76)$, then |

$$\mathbf{v} + \mathbf{w} = (1.2, 3.5, 2, 0) + (12.23, 19.65, 23.22, 9.76) = (13.43, 23.15, 25.22, 9.76)$$

and

$$2.35\mathbf{v} = (2.82, 8.225, 4.7, 0). \qquad \square$$

Exercises for 4.1

Key Terms

Vectors in $\mathbb{R}^n$, Vector addition, Scalar multiplication, Zero vector, Additive inverse, Components of a vector.

Skills

- Be able to perform vector addition and scalar multiplication for vectors in $\mathbb{R}^n$ given in component form.

- Understand the geometric perspective on vector addition and scalar multiplication in the cases of $\mathbb{R}^2$ and $\mathbb{R}^3$.

- Be able to formally verify the axioms (4.1.1)–(4.1.8) for vectors in $\mathbb{R}^n$.

True-False Review

For Questions 1–12, decide if the given statement is **true** or **false**, and give a brief justification for your answer. If true, you can quote a relevant definition or theorem from the text. If false, provide an example, illustration, or brief explanation of why the statement is false.

1. The vector (x, y) in $\mathbb{R}^2$ is the same as the vector $(x, y, 0)$ in $\mathbb{R}^3$.

2. Each vector (x, y, z) in $\mathbb{R}^3$ has exactly one additive inverse.

3. The solution set to a linear system of 4 equations and 6 unknowns consists of a collection of vectors in $\mathbb{R}^6$.

4. For every vector $(x_1, x_2, \ldots, x_n)$ in $\mathbb{R}^n$, the vector $(-1) \cdot (x_1, x_2, \ldots, x_n)$ is an additive inverse.

5. A vector whose components are all positive is called a "positive vector."

6. If s and t are scalars and $\mathbf{x}$ and $\mathbf{y}$ are vectors in $\mathbb{R}^n$, then $(s + t)(\mathbf{x} + \mathbf{y}) = s\mathbf{x} + t\mathbf{y}$.

7. For every vector $\mathbf{x}$ in $\mathbb{R}^n$, the vector $0\mathbf{x}$ is the zero vector of $\mathbb{R}^n$.

8. The parallelogram whose sides are determined by vectors $\mathbf{x}$ and $\mathbf{y}$ in $\mathbb{R}^2$ have diagonals determined by the vectors $\mathbf{x} + \mathbf{y}$ and $\mathbf{x} - \mathbf{y}$.

9. If $\mathbf{x}$ is a vector in the first quadrant of $\mathbb{R}^2$, then any scalar multiple $k\mathbf{x}$ of $\mathbf{x}$ is still a vector in the first quadrant of $\mathbb{R}^2$.

10. The vector $5\mathbf{i} - 6\mathbf{j} + \sqrt{2}\mathbf{k}$ in $\mathbb{R}^3$ is the same as $(5, -6, \sqrt{2})$.

11. Three vectors $\mathbf{x}$, $\mathbf{y}$, and $\mathbf{z}$ in $\mathbb{R}^3$ always determine a 3-dimensional solid region in $\mathbb{R}^3$.

12. If $\mathbf{x}$ and $\mathbf{y}$ are vectors in $\mathbb{R}^2$ whose components are even integers and k is a scalar, then $\mathbf{x} + \mathbf{y}$ and $k\mathbf{x}$ are also vectors in $\mathbb{R}^2$ whose components are even integers.

Problems

1. If $\mathbf{x} = (3, 1), \mathbf{y} = (-1, 2)$, determine the vectors $\mathbf{v}_1 = 2\mathbf{x}, \mathbf{v}_2 = 3\mathbf{y}, \mathbf{v}_3 = 2\mathbf{x} + 3\mathbf{y}$. Sketch the corresponding points in the xy-plane and the equivalent geometric vectors.

2. If $\mathbf{x} = (-1, -4)$ and $\mathbf{y} = (-5, 1)$, determine the vectors $\mathbf{v}_1 = 3\mathbf{x}, \mathbf{v}_2 = -4\mathbf{y}, \mathbf{v}_3 = 3\mathbf{x} + (-4)\mathbf{y}$. Sketch the corresponding points in the xy-plane and the equivalent geometric vectors.

3. If $\mathbf{x} = (3, -1, 2, 5), \mathbf{y} = (-1, 2, 9, -2)$, determine $\mathbf{v} = 5\mathbf{x} + (-7)\mathbf{y}$ and its additive inverse.

4. If $\mathbf{x} = (1, 2, 3, 4, 5)$ and $\mathbf{z} = (-1, 0, -4, 1, 2)$, find $\mathbf{y}$ in $\mathbb{R}^5$ such that $2\mathbf{x} + (-3)\mathbf{y} = -\mathbf{z}$.

5. Verify the commutative law of addition for vectors in $\mathbb{R}^4$.

6. Verify the associative law of addition for vectors in $\mathbb{R}^4$.

7. Verify properties (4.1.5)–(4.1.8) for vectors in $\mathbb{R}^3$.

8. Show with examples that if $\mathbf{x}$ is a vector in the first quadrant of $\mathbb{R}^2$ (i.e., both coordinates of $\mathbf{x}$ are positive) and $\mathbf{y}$ is a vector in the third quadrant of $\mathbb{R}^2$ (i.e., both coordinates of $\mathbf{y}$ are negative), then the sum $\mathbf{x} + \mathbf{y}$ could occur in any of the four quadrants.

4.2   Definition of a Vector Space

In the previous section, we showed how the set $\mathbb{R}^n$ of all ordered n-tuples of real numbers, together with the addition and scalar multiplication operations defined on it, has the same algebraic properties as the familiar algebra of geometric vectors. We now push this abstraction one step further and introduce the idea of a vector space. Such an abstraction will enable us to develop a mathematical framework for studying a broad class of linear problems, such as systems of linear equations, linear differential equations, and systems of linear differential equations, which have far-reaching applications in all areas of applied mathematics, science, and engineering.

Let V be a nonempty set. For our purposes, it is useful to call the elements of V vectors and use the usual vector notation $\mathbf{u}, \mathbf{v}, \ldots$, to denote these elements. For example, if V is the set of all 2×2 matrices, then the vectors in V are 2×2 matrices, whereas if V is the set of all positive integers, then the vectors in V are positive integers. We will be interested only in the case when the set V has an addition operation and a scalar multiplication operation defined on its elements in the following senses:

Vector Addition: A rule for combining any two vectors in V. We will use the usual $+$ sign to denote an addition operation, and the result of adding the vectors $\mathbf{u}$ and $\mathbf{v}$ will be denoted $\mathbf{u} + \mathbf{v}$.

Real (or Complex) Scalar Multiplication: A rule for combining each vector in V with any real (or complex) number. We will use the usual notation $k\mathbf{v}$ to denote the result of scalar multiplying the vector $\mathbf{v}$ by the real (or complex) number k.

To combine the two types of scalar multiplication, we let F denote the set of scalars for which the operation is defined. Thus, for us, F is either the set of all real numbers or the set of all complex numbers. For example, if V is the set of all 2×2 matrices with complex elements and F denotes the set of all complex numbers, then the usual operation of matrix addition is an addition operation on V, and the usual method of multiplying a matrix by a scalar is a scalar multiplication operation on V. Notice that the result of applying either of these operations is always another vector (2×2 matrix) in V.

As a further example, let V be the set of positive integers, and let F be the set of all real numbers. Then the usual operations of addition and multiplication within the real numbers define addition and scalar multiplication operations on V. Note in this case, however, that the scalar multiplication operation, in general, will not yield another vector in V, since when we multiply a positive integer by a real number, the result is not, in general, a positive integer.

We are now in a position to give a precise definition of a vector space.

DEFINITION   4.2.1

Let V be a nonempty set (whose elements are called vectors) on which are defined an addition operation and a scalar multiplication operation with scalars in F. We call V a **vector space over** F, provided the following ten conditions are satisfied:

A1. *Closure under addition:* For each pair of vectors $\mathbf{u}$ and $\mathbf{v}$ in V, the sum $\mathbf{u} + \mathbf{v}$ is also in V. We say that V is **closed under addition**.

A2. *Closure under scalar multiplication:* For each vector $\mathbf{v}$ in V and each scalar k in F, the scalar multiple $k\mathbf{v}$ is also in V. We say that V is **closed under scalar multiplication**.

A3. *Commutativity of addition:* For all $\mathbf{u}, \mathbf{v} \in V$, we have

$$\mathbf{u} + \mathbf{v} = \mathbf{v} + \mathbf{u}.$$

A4. *Associativity of addition:* For all $\mathbf{u}, \mathbf{v}, \mathbf{w} \in V$, we have

$$(\mathbf{u} + \mathbf{v}) + \mathbf{w} = \mathbf{u} + (\mathbf{v} + \mathbf{w}).$$

A5. *Existence of a zero vector in V:* In V there is a vector, denoted $\mathbf{0}$, satisfying

$$\mathbf{v} + \mathbf{0} = \mathbf{v}, \qquad \text{for all } \mathbf{v} \in V.$$

A6. *Existence of additive inverses in V:* For each vector $\mathbf{v}$ In V, there is a vector, denoted $-\mathbf{v}$, in V such that

$$\mathbf{v} + (-\mathbf{v}) = \mathbf{0}.$$

A7. *Unit property:* For all $\mathbf{v} \in V$,

$$1\mathbf{v} = \mathbf{v}.$$

A8. *Associativity of scalar multiplication:* For all $\mathbf{v} \in V$ and all scalars $r, s \in F$,

$$(rs)\mathbf{v} = r(s\mathbf{v}).$$

A9. *Distributive property of scalar multiplication over vector addition:* For all $\mathbf{u}, \mathbf{v} \in V$ and all scalars $r \in F$,

$$r(\mathbf{u} + \mathbf{v}) = r\mathbf{u} + r\mathbf{v}.$$

A10. *Distributive property of scalar multiplication over scalar addition:* For all $\mathbf{v} \in V$ and all scalars $r, s \in F$,

$$(r + s)\mathbf{v} = r\mathbf{v} + s\mathbf{v}.$$

Remarks

1. A key point to note is that in order to define a vector space, we must start with all of the following:

   (a) A nonempty set of vectors V.

   (b) A set of scalars F (either $\mathbb{R}$ or $\mathbb{C}$).

(c) An addition operation defined on V.

(d) A scalar multiplication operation defined on V.

Then we must check that the axioms A1–A10 are satisfied.

2. Terminology: A vector space over the real numbers will be referred to as a **real vector space**, whereas a vector space over the complex numbers will be called a **complex vector space**.

3. As indicated in Definition 4.2.1, we will use boldface to denote vectors in a general vector space. In handwriting, it is strongly advised that vectors be denoted either as $\vec{v}$ or as $\underset{\sim}{v}$. This will avoid any confusion between vectors in V and scalars in F.

4. When we deal with a familiar vector space, we will use the usual notation for vectors in the space. For example, as seen below, the set $\mathbb{R}^n$ of ordered n-tuples is a vector space, and we will denote vectors here in the form $(x_1, x_2, \ldots, x_n)$, as in the previous section. As another illustration, it is shown below that the set of all real-valued functions defined on an interval is a vector space, and we will denote the vectors in this vector space by $f, g, \ldots$.

Examples of Vector Spaces

1. The set of all real numbers, together with the usual operations of addition and multiplication, is a real vector space.

2. The set of all complex numbers is a complex vector space when we use the usual operations of addition and multiplication by a complex number. It is also possible to restrict the set of scalars to $\mathbb{R}$, in which case the set of complex numbers becomes a real vector space.

3. The set $\mathbb{R}^n$, together with the operations of addition and scalar multiplication defined in (4.1.13) and (4.1.14), is a real vector space. As we saw in the previous section, the zero vector in $\mathbb{R}^n$ is the n-tuple of zeros $(0, 0, \ldots, 0)$, and the additive inverse of the vector $\mathbf{v} = (x_1, x_2, \ldots, x_n)$ is $-\mathbf{v} = (-x_1, -x_2, \ldots, -x_n)$.

Strictly speaking, for each of the examples above it is necessary to verify all of the axioms A1–A10 of a vector space. However, in these examples, the axioms hold immediately as well-known properties of real and complex numbers and n-tuples.

Example 4.2.2   Let V be the set of all 2×2 matrices with real elements. Show that V, together with the usual operations of matrix addition and multiplication of a matrix by a real number, is a real vector space.

Solution: We must verify the axioms A1–A10. If A and B are in V (that is, A and B are 2×2 matrices with real entries), then $A + B$ and kA are in V for all real numbers k. Consequently, V is closed under addition and scalar multiplication, and therefore Axioms A1 and A2 of the vector space definition hold.

A3. Given two 2×2 matrices

$$A = \begin{bmatrix} a_1 & a_2 \\ a_3 & a_4 \end{bmatrix} \quad \text{and} \quad B = \begin{bmatrix} b_1 & b_2 \\ b_3 & b_4 \end{bmatrix},$$

we have

$$A + B = \begin{bmatrix} a_1 & a_2 \\ a_3 & a_4 \end{bmatrix} + \begin{bmatrix} b_1 & b_2 \\ b_3 & b_4 \end{bmatrix} = \begin{bmatrix} a_1 + b_1 & a_2 + b_2 \\ a_3 + b_3 & a_4 + b_4 \end{bmatrix}$$

$$= \begin{bmatrix} b_1 + a_1 & b_2 + a_2 \\ b_3 + a_3 & b_4 + a_4 \end{bmatrix} = \begin{bmatrix} b_1 & b_2 \\ b_3 & b_4 \end{bmatrix} + \begin{bmatrix} a_1 & a_2 \\ a_3 & a_4 \end{bmatrix} = B + A.$$

A4. Given three 2×2 matrices

$$A = \begin{bmatrix} a_1 & a_2 \\ a_3 & a_4 \end{bmatrix}, \qquad B = \begin{bmatrix} b_1 & b_2 \\ b_3 & b_4 \end{bmatrix}, \qquad C = \begin{bmatrix} c_1 & c_2 \\ c_3 & c_4 \end{bmatrix},$$

we have

$$(A + B) + C = \left(\begin{bmatrix} a_1 & a_2 \\ a_3 & a_4 \end{bmatrix} + \begin{bmatrix} b_1 & b_2 \\ b_3 & b_4 \end{bmatrix} \right) + \begin{bmatrix} c_1 & c_2 \\ c_3 & c_4 \end{bmatrix}$$

$$= \begin{bmatrix} a_1 + b_1 & a_2 + b_2 \\ a_3 + b_3 & a_4 + b_4 \end{bmatrix} + \begin{bmatrix} c_1 & c_2 \\ c_3 & c_4 \end{bmatrix}$$

$$= \begin{bmatrix} (a_1 + b_1) + c_1 & (a_2 + b_2) + c_2 \\ (a_3 + b_3) + c_3 & (a_4 + b_4) + c_4 \end{bmatrix}$$

$$= \begin{bmatrix} a_1 + (b_1 + c_1) & a_2 + (b_2 + c_2) \\ a_3 + (b_3 + c_3) & a_4 + (b_4 + c_4) \end{bmatrix}$$

$$= \begin{bmatrix} a_1 & a_2 \\ a_3 & a_4 \end{bmatrix} + \begin{bmatrix} b_1 + c_1 & b_2 + c_2 \\ b_3 + c_3 & b_4 + c_4 \end{bmatrix}$$

$$= \begin{bmatrix} a_1 & a_2 \\ a_3 & a_4 \end{bmatrix} + \left(\begin{bmatrix} b_1 & b_2 \\ b_3 & b_4 \end{bmatrix} + \begin{bmatrix} c_1 & c_2 \\ c_3 & c_4 \end{bmatrix} \right) = A + (B + C).$$

A5. If A is any matrix in V, then

$$A + \begin{bmatrix} 0 & 0 \\ 0 & 0 \end{bmatrix} = A.$$

Thus, 0_2 is the zero vector in V.

A6. The additive inverse of $A = \begin{bmatrix} a & b \\ c & d \end{bmatrix}$ is $-A = \begin{bmatrix} -a & -b \\ -c & -d \end{bmatrix}$, since

$$A + (-A) = \begin{bmatrix} a + (-a) & b + (-b) \\ c + (-c) & d + (-d) \end{bmatrix} = \begin{bmatrix} 0 & 0 \\ 0 & 0 \end{bmatrix} = 0_2.$$

A7. If A is any matrix in V, then

$$1A = A,$$

thus verifying the unit property.

A8. Given a matrix $A = \begin{bmatrix} a & b \\ c & d \end{bmatrix}$ and scalars r and s, we have

$$(rs)A = \begin{bmatrix} (rs)a & (rs)b \\ (rs)c & (rs)d \end{bmatrix} = \begin{bmatrix} r(sa) & r(sb) \\ r(sc) & r(sd) \end{bmatrix} = r \begin{bmatrix} sa & sb \\ sc & sd \end{bmatrix} = r(sA),$$

as required.

A9. Given matrices $A = \begin{bmatrix} a_1 & a_2 \\ a_3 & a_4 \end{bmatrix}$ and $B = \begin{bmatrix} b_1 & b_2 \\ b_3 & b_4 \end{bmatrix}$ and a scalar r, we have

$$r(A + B) = r \left(\begin{bmatrix} a_1 & a_2 \\ a_3 & a_4 \end{bmatrix} + \begin{bmatrix} b_1 & b_2 \\ b_3 & b_4 \end{bmatrix} \right)$$

$$= r \begin{bmatrix} a_1 + b_1 & a_2 + b_2 \\ a_3 + b_3 & a_4 + b_4 \end{bmatrix} = \begin{bmatrix} r(a_1 + b_1) & r(a_2 + b_2) \\ r(a_3 + b_3) & r(a_4 + b_4) \end{bmatrix}$$

$$= \begin{bmatrix} ra_1 + rb_1 & ra_2 + rb_2 \\ ra_3 + rb_3 & ra_4 + rb_4 \end{bmatrix} = \begin{bmatrix} ra_1 & ra_2 \\ ra_3 & ra_4 \end{bmatrix} + \begin{bmatrix} rb_1 & rb_2 \\ rb_3 & rb_4 \end{bmatrix} = rA + rB.$$

A10. Given A, r, and s as in A8 above, we have

$$(r+s)A = \begin{bmatrix} (r+s)a & (r+s)b \\ (r+s)c & (r+s)d \end{bmatrix} = \begin{bmatrix} ra+sa & rb+sb \\ rc+sc & rd+sd \end{bmatrix}$$

$$= \begin{bmatrix} ra & rb \\ rc & rd \end{bmatrix} + \begin{bmatrix} sa & sb \\ sc & sd \end{bmatrix} = rA + sA,$$

as required.

Thus V, together with the given operations, is a real vector space.   □

Remark   In a manner similar to the previous example, it is easily established that the set of all $m \times n$ matrices with real entries is a real vector space when we use the usual operations of addition of matrices and multiplication of matrices by a real number. We will denote the vector space of all $m \times n$ matrices with real elements by $M_{m \times n}(\mathbb{R})$, and we denote the vector space of all $n \times n$ matrices with real elements by $M_n(\mathbb{R})$.

Example 4.2.3   Let V be the set of all real-valued functions defined on an interval I. Define addition and scalar multiplication in V as follows. If f and g are in V and k is any real number, then $f + g$ and kf are defined by

$$(f+g)(x) = f(x) + g(x) \qquad \text{for all } x \in I,$$
$$(kf)(x) = kf(x) \qquad \text{for all } x \in I.$$

Show that V, together with the given operations of addition and scalar multiplication, is a real vector space.

Solution: It follows from the given definitions of addition and scalar multiplication that if f and g are in V, and k is any real number, then $f + g$ and kf are both real-valued functions on I and are therefore in V. Consequently, the closure axioms A1 and A2 hold. We now check the remaining axioms.

A3. Let f and g be arbitrary functions in V. From the definition of function addition, we have

$$(f+g)(x) = f(x) + g(x) = g(x) + f(x) = (g+f)(x),$$

for all $x \in I$. (The middle step here follows from the fact that $f(x)$ and $g(x)$ are real numbers associated with evaluating f and g at the input x, and real number addition commutes.) Consequently, $f + g = g + f$ (since the values of $f + g$ and $g + f$ agree for every $x \in I$), and so addition in V is commutative.

A4. Let $f, g, h \in V$. Then for all $x \in I$, we have

$$[(f+g)+h](x) = (f+g)(x) + h(x) = [f(x)+g(x)] + h(x)$$
$$= f(x) + [g(x)+h(x)] = f(x) + (g+h)(x)$$
$$= [f+(g+h)](x).$$

Consequently, $(f+g)+h = f+(g+h)$, so that addition in V is indeed associative.

A5. If we define the zero function, O, by $O(x) = 0$, for all $x \in I$, then

$$(f+O)(x) = f(x) + O(x) = f(x) + 0 = f(x),$$

for all $f \in V$ and all $x \in I$, which implies that $f + O = f$. Hence, O is the zero vector in V. (See Figure 4.2.1.)

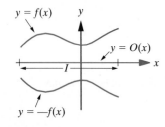

$y = f(x)$

$y = O(x)$

$y = -f(x)$

Figure 4.2.1: In the vector space of all functions defined on an interval I, the additive inverse of a function f is obtained by reflecting the graph of f about the x-axis. The zero vector is the zero function $O(x)$.

A6. If $f \in V$, then $-f$ is defined by $(-f)(x) = -f(x)$ for all $x \in I$, since

$$[f + (-f)](x) = f(x) + (-f)(x) = f(x) - f(x) = 0$$

for all $x \in I$. This implies that $f + (-f) = O$.

A7. Let $f \in V$. Then, by definition of the scalar multiplication operation, for all $x \in I$, we have

$$(1f)(x) = 1f(x) = f(x).$$

Consequently, $1f = f$.

A8. Let $f \in V$, and let $r, s \in \mathbb{R}$. Then, for all $x \in I$,

$$[(rs)f](x) = (rs)f(x) = r[sf(x)] = r[(sf)(x)].$$

Hence, the functions $(rs)f$ and $r(sf)$ agree on every $x \in I$, and hence $(rs)f = r(sf)$, as required.

A9. Let $f, g \in V$ and let $r \in \mathbb{R}$. Then, for all $x \in I$,

$$[r(f + g)](x) = r[(f + g)(x)] = r[f(x) + g(x)] = rf(x) + rg(x)$$
$$= (rf)(x) + (rg)(x) = (rf + rg)(x).$$

Hence, $r(f + g) = rf + rg$.

A10. Let $f \in V$, and let $r, s \in \mathbb{R}$. Then for all $x \in I$,

$$[(r+s)f](x) = (r+s)f(x) = rf(x)+sf(x) = (rf)(x)+(sf)(x) = (rf+sf)(x),$$

which proves that $(r + s)f = rf + sf$.

Since all parts of Definition 4.2.1 are satisfied, it follows that V, together with the given operations of addition and scalar multiplication, is a real vector space. □

Remark   As the previous two examples indicate, a full verification of the vector space definition can be somewhat tedious and lengthy, although it is usually straightforward. Be careful to not leave out any important steps in such a verification.

The Vector Space $\mathbb{C}^n$

We now introduce the most important complex vector space. Let $\mathbb{C}^n$ denote the set of all ordered n-tuples of *complex numbers*. Thus,

$$\mathbb{C}^n = \{(z_1, z_2, \ldots, z_n) : z_1, z_2, \ldots, z_n \in \mathbb{C}\}.$$

We refer to the elements of $\mathbb{C}^n$ as **vectors in** $\mathbb{C}^n$. A typical vector in $\mathbb{C}^n$ is $(z_1, z_2, \ldots, z_n)$, where each z_k is a complex number.

Example 4.2.4   The following are examples of vectors in $\mathbb{C}^2$ and $\mathbb{C}^4$, respectively:
$$\mathbf{u} = (2.1 - 3i, -1.5 + 3.9i), \qquad \mathbf{v} = (5 + 7i, 2 - i, 3 + 4i, -9 - 17i). \quad □$$

In order to obtain a vector space, we must define appropriate operations of "vector addition" and "multiplication by a scalar" on the set of vectors in question. In the case of $\mathbb{C}^n$, we are motivated by the corresponding operations in $\mathbb{R}^n$ and thus define the addition and scalar multiplication operations componentwise. Thus, if $\mathbf{u} = (u_1, u_2, \ldots, u_n)$ and $\mathbf{v} = (v_1, v_2, \ldots, v_n)$ are vectors in $\mathbb{C}^n$ and k is an arbitrary *complex number*, then

$$\mathbf{u} + \mathbf{v} = (u_1 + v_1, u_2 + v_2, \ldots, u_n + v_n),$$
$$k\mathbf{u} = (ku_1, ku_2, \ldots, ku_n).$$

Example 4.2.5   If $\mathbf{u} = (1 - 3i, 2 + 4i)$, $\mathbf{v} = (-2 + 4i, 5 - 6i)$, and $k = 2 + i$, find $\mathbf{u} + k\mathbf{v}$.

Solution: We have

$$
\begin{aligned}
\mathbf{u} + k\mathbf{v} &= (1 - 3i, 2 + 4i) + (2 + i)(-2 + 4i, 5 - 6i) \\
&= (1 - 3i, 2 + 4i) + (-8 + 6i, 16 - 7i) = (-7 + 3i, 18 - 3i). \quad \square
\end{aligned}
$$

It is straightforward to show that $\mathbb{C}^n$, together with the given operations of addition and scalar multiplication, is a *complex* vector space.

Further Properties of Vector Spaces

The main reason for formalizing the definition of an abstract vector space is that any results that we can prove based solely on the definition will then apply to *all* vector spaces we care to examine; that is, we do not have to prove separate results for geometric vectors, $m \times n$ matrices, vectors in $\mathbb{R}^n$ or $\mathbb{C}^n$, or real-valued functions, and so on. The next theorem lists some results that can be proved using the vector space axioms.

Theorem 4.2.6   Let V be a vector space over F.

1. The zero vector is unique.
2. $0\mathbf{u} = \mathbf{0}$ for all $\mathbf{u} \in V$.
3. $k\mathbf{0} = \mathbf{0}$ for all scalars $k \in F$.
4. The additive inverse of each element in V is unique.
5. For all $\mathbf{u} \in V$, $-\mathbf{u} = (-1)\mathbf{u}$.
6. If k is a scalar and $\mathbf{u} \in V$ such that $k\mathbf{u} = \mathbf{0}$, then either $k = 0$ or $\mathbf{u} = \mathbf{0}$.

Proof   **1.** Suppose there were two zero vectors in V, denoted $\mathbf{0}_1$ and $\mathbf{0}_2$. Then, for any $\mathbf{v} \in V$, we would have

$$\mathbf{v} + \mathbf{0}_1 = \mathbf{v} \tag{4.2.1}$$

and

$$\mathbf{v} + \mathbf{0}_2 = \mathbf{v}. \tag{4.2.2}$$

We must prove that $\mathbf{0}_1 = \mathbf{0}_2$. But, applying (4.2.1) with $\mathbf{v} = \mathbf{0}_2$, we have

$$
\begin{aligned}
\mathbf{0}_2 &= \mathbf{0}_2 + \mathbf{0}_1 \\
&= \mathbf{0}_1 + \mathbf{0}_2 \qquad \text{(Axiom A3)} \\
&= \mathbf{0}_1 \qquad\quad \text{(from (4.2.2) with } \mathbf{v} = \mathbf{0}_1 \text{)}.
\end{aligned}
$$

Consequently, $\mathbf{0}_1 = \mathbf{0}_2$, so the zero vector is unique in a vector space.

2. Let **u** be an arbitrary element in a vector space V. Since $0 = 0 + 0$, we have

$$0\mathbf{u} = (0 + 0)\mathbf{u} = 0\mathbf{u} + 0\mathbf{u},$$

by Axiom A10. Now Axiom A6 implies that the vector $-(0\mathbf{u})$ exists, and adding it to both sides of the previous equation yields

$$0\mathbf{u} + [-(0\mathbf{u})] = (0\mathbf{u} + 0\mathbf{u}) + [-(0\mathbf{u})].$$

Thus, since addition in a vector space is associative (Axiom A4),

$$0\mathbf{u} + [-(0\mathbf{u})] = 0\mathbf{u} + (0\mathbf{u} + [-(0\mathbf{u})]).$$

Applying Axiom A6 on both sides and then using Axiom A5, this becomes

$$\mathbf{0} = 0\mathbf{u} + \mathbf{0} = 0\mathbf{u},$$

and this completes the verification of (2).

3. Using the fact that $\mathbf{0} = \mathbf{0} + \mathbf{0}$ (by Axiom A5), the proof here proceeds along the same lines as the proof of result 2. We leave the verification to the reader as an exercise (Problem 21).

4. Let $\mathbf{u} \in V$ be an arbitrary vector, and suppose that there were two additive inverses, say **v** and **w**, for **u**. According to Axiom A6, this implies that

$$\mathbf{u} + \mathbf{v} = \mathbf{0} \tag{4.2.3}$$

and

$$\mathbf{u} + \mathbf{w} = \mathbf{0}. \tag{4.2.4}$$

We wish to show that $\mathbf{v} = \mathbf{w}$. Now, Axiom A6 implies that a vector $-\mathbf{v}$ exists, so adding it on the right to both sides of (4.2.3) yields

$$(\mathbf{u} + \mathbf{v}) + (-\mathbf{v}) = \mathbf{0} + (-\mathbf{v}) = -\mathbf{v}.$$

Applying Axioms A4 and A6 on the left side, we simplify this to

$$\mathbf{u} = -\mathbf{v}.$$

Substituting this into (4.2.4) yields

$$-\mathbf{v} + \mathbf{w} = \mathbf{0}.$$

Adding **v** to the left of both sides and applying Axioms A4 and A6 once more yields $\mathbf{v} = \mathbf{w}$, as desired.

5. To verify that $-\mathbf{u} = (-1)\mathbf{u}$ for all $\mathbf{u} \in V$, we note that

$$\mathbf{0} = 0\mathbf{u} = (1 + (-1))\mathbf{u} = 1\mathbf{u} + (-1)\mathbf{u} = \mathbf{u} + (-1)\mathbf{u},$$

where we have used property 2 and Axioms A10 and A7. The equation above proves that $(-1)\mathbf{u}$ is an additive inverse of **u**, and by the uniqueness of additive inverses that we just proved, we conclude that $(-1)\mathbf{u} = -\mathbf{u}$, as desired.

Finally, we leave the proof of result 6 in Theorem 4.2.6 as an exercise (Problem 22).

∎

Remark   The proof of Theorem 4.2.6 involved a number of tedious and seemingly obvious steps. It is important to remember, however, that in an abstract vector space we are not allowed to rely on past experience in deriving results for the first time. For instance, the statement "$\mathbf{0} + \mathbf{0} = \mathbf{0}$" may seem intuitively clear, but in our newly developed mathematical structure, we must appeal specifically to the rules A1–A10 given for a vector space. Hence, the statement "$\mathbf{0} + \mathbf{0} = \mathbf{0}$" should be viewed as a consequence of Axiom A5 and nothing else. Once we have proved these basic results, of course, then we are free to use them in *any* vector space context where they are needed. This is the whole advantage to working in the general vector space setting.

We end this section with a list of the most important vector spaces that will be required throughout the remainder of the text. In each case the addition and scalar multiplication operations are the usual ones associated with the set of vectors.

- $\mathbb{R}^n$, the (real) vector space of all ordered n-tuples of real numbers.

- $\mathbb{C}^n$, the (complex) vector space of all ordered n-tuples of complex numbers.

- $M_{m \times n}(\mathbb{R})$, the (real) vector space of all $m \times n$ matrices with real elements.

- $M_n(\mathbb{R})$, the (real) vector space of all $n \times n$ matrices with real elements.

- $C^k(I)$, the vector space of all real-valued functions that are continuous and have (at least) k continuous derivatives on I. We will show that this set of vectors is a (real) vector space in the next section.

- P_n, the (real) vector space of all real-valued polynomials of degree $\leq n$ with real coefficients. That is,

$$P_n = \{a_0 + a_1 x + a_2 x^2 + \cdots + a_n x^n : a_0, a_1, \ldots, a_n \in \mathbb{R}\}.$$

We leave the verification that P_n is a (real) vector space as an exercise (Problem 23).

Exercises for 4.2

Key Terms

Vector space (real or complex), Closure under addition, Closure under scalar multiplication, Commutativity of addition, Associativity of addition, Existence of zero vector, Existence of additive inverses, Unit property, Associativity of scalar multiplication, Distributive properties, Examples: $\mathbb{R}^n$, $\mathbb{C}^n$, $M_n(\mathbb{R})$, $C^k(I)$, P_n.

Skills

- Be able to define a vector space. Specifically, be able to identify and list the ten axioms A1–A10 governing the vector space operations.

- Know each of the standard examples of vector spaces given at the end of the section, and know how to perform the vector operations in these vector spaces.

- Be able to check whether or not each of the axioms A1–A10 holds for specific examples V. This includes, if possible, closure of V under vector addition and scalar multiplication, as well as identification of the zero vector and the additive inverse of each vector in the set V.

- Be able to prove basic properties that hold generally for vector spaces V (see Theorem 4.2.6).

True-False Review

For Questions 1–8, decide if the given statement is **true** or **false**, and give a brief justification for your answer. If true, you can quote a relevant definition or theorem from the text. If false, provide an example, illustration, or brief explanation of why the statement is false.

1. The zero vector in a vector space V is unique.

2. If $\mathbf{v}$ is a vector in a vector space V, and r and s are scalars such that $r\mathbf{v} = s\mathbf{v}$, then $r = s$.

3. The set $\mathbb{Z}$ of integers, together with the usual operations of addition and scalar multiplication, forms a vector space.

4. If $\mathbf{x}$ and $\mathbf{y}$ are vectors in a vector space V, then the additive inverse of $\mathbf{x} + \mathbf{y}$ is $(-\mathbf{x}) + (-\mathbf{y})$.

5. The additive inverse of a vector $\mathbf{v}$ in a vector space V is unique.

6. The set $\{0\}$, with the usual operations of addition and scalar multiplication, forms a vector space.

7. The set $\{0, 1\}$, with the usual operations of addition and scalar multiplication, forms a vector space.

8. The set of positive real numbers, with the usual operations of addition and scalar multiplication, forms a vector space.

Problems

For Problems 1–5, determine whether the given set of vectors is closed under addition and closed under scalar multiplication. In each case, take the set of scalars to be the set of all real numbers.

1. The set of all rational numbers.

2. The set of all upper triangular $n \times n$ matrices with real elements.

3. The set of all solutions to the differential equation $y' + 9y = 4x^2$. (Do not solve the differential equation.)

4. The set of all solutions to the differential equation $y' + 9y = 0$. (Do not solve the differential equation.)

5. The set of all solutions to the homogeneous linear system $A\mathbf{x} = \mathbf{0}$.

6. Let
$$S = \{A \in M_2(\mathbb{R}) : \det(A) = 0\}.$$

(a) Is the zero vector from $M_2(\mathbb{R})$ in S?

(b) Give an explicit example illustrating that S is not closed under matrix addition.

(c) Is S closed under scalar multiplication? Justify your answer.

7. Let $\mathbb{N} = \{1, 2, \dots\}$ denote the set of all positive integers. Give three reasons why $\mathbb{N}$, together with the usual operations of addition and scalar multiplication, is not a real vector space.

8. We have defined the set $\mathbb{R}^2 = \{(x, y) : x, y \in \mathbb{R}\}$, together with the addition and scalar multiplication operations as follows:
$$(x_1, y_1) + (x_2, y_2) = (x_1 + x_2, y_1 + y_2),$$
$$k(x_1, y_1) = (kx_1, ky_1).$$

Give a complete verification that each of the vector space axioms is satisfied.

9. Determine the zero vector in the vector space $M_{2\times3}(\mathbb{R})$, and the additive inverse of a general element. (Note that the vector space axioms A1–A4 and A7–A10 follow directly from matrix algebra.)

10. Generalize the previous exercise to find the zero vector and the additive inverse of a general element of $M_{m\times n}(\mathbb{R})$.

11. Let P denote the set of all polynomials whose degree is exactly 2. Is P a vector space? Justify your answer.

12. On $\mathbb{R}^+$, the set of *positive* real numbers, define the operations of addition and scalar multiplication as follows:
$$x + y = xy,$$
$$c \cdot x = x^c.$$

Note that the multiplication and exponentiation appearing on the right side of these formulas refer to the ordinary operations on real numbers. Determine whether $\mathbb{R}^+$, together with these algebraic operations, is a vector space.

13. On $\mathbb{R}^2$, define the operation of addition and multiplication by a real number as follows:
$$(x_1, y_1) + (x_2, y_2) = (x_1 - x_2, y_1 - y_2),$$
$$k(x_1, y_1) = (-kx_1, -ky_1).$$

Which of the axioms for a vector space are satisfied by $\mathbb{R}^2$ with these algebraic operations?

14. On $\mathbb{R}^2$, define the operation of addition by

$$(x_1, y_1) + (x_2, y_2) = (x_1 x_2, y_1 y_2).$$

Do axioms A5 and A6 in the definition of a vector space hold? Justify your answer.

15. On $M_2(\mathbb{R})$, define the operation of addition by

$$A + B = AB,$$

and use the usual scalar multiplication operation. Determine which axioms for a vector space are satisfied by $M_2(\mathbb{R})$ with the above operations.

16. On $M_2(\mathbb{R})$, define the operations of addition and multiplication by a real number ($\oplus$ and $\cdot$, respectively) as follows:

$$A \oplus B = -(A + B),$$
$$k \cdot A = -kA,$$

where the operations on the right-hand sides of these equations are the usual ones associated with $M_2(\mathbb{R})$.

Determine which of the axioms for a vector space are satisfied by $M_2(\mathbb{R})$ with the operations $\oplus$ and $\cdot$.

For Problems 17–18, verify that the given set of objects together with the usual operations of addition and scalar multiplication is a *complex* vector space.

17. $\mathbb{C}^2$.

18. $M_2(\mathbb{C})$, the set of all 2×2 matrices with complex entries.

19. Is $\mathbb{C}^3$ a *real* vector space? Explain.

20. Is $\mathbb{R}^3$ a *complex* vector space? Explain.

21. Prove part 3 of Theorem 4.2.6.

22. Prove part 6 of Theorem 4.2.6.

23. Prove that P_n is a vector space.

4.3   Subspaces

Let us try to make contact between the abstract vector space idea and the solution of an applied problem. Vector spaces generally arise as the sets containing the unknowns in a given problem. For example, if we are solving a differential equation, then the basic unknown is a function, and therefore any solution to the differential equation will be an element of the vector space V of all functions defined on an appropriate interval. Consequently, the solution set of a differential equation is a subset of V. Similarly, consider the system of linear equations $A\mathbf{x} = \mathbf{b}$, where A is an $m \times n$ matrix with real elements. The basic unknown in this system, $\mathbf{x}$, is a column n-vector, or equivalently a vector in $\mathbb{R}^n$. Consequently, the solution set to the system is a subset of the vector space $\mathbb{R}^n$. As these examples illustrate, the solution set of an applied problem is generally a subset of vectors from an appropriate vector space (schematically represented in Figure 4.3.1). The question we will need to answer in the future is whether this subset of vectors is a vector space in its own right. The following definition introduces the terminology we will use:

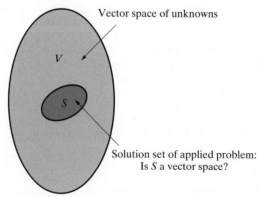

Figure 4.3.1: The solution set S of an applied problem is a subset of the vector space V of unknowns in the problem.

DEFINITION   4.3.1

Let S be a nonempty subset of a vector space V. If S is itself a vector space under the same operations of addition and scalar multiplication as used in V, then we say that S is a **subspace** of V.

In establishing that a given subset S of vectors from a vector space V is a subspace of V, it would appear as though we must check that each axiom in the vector space definition is satisfied when we restrict our attention to vectors lying only in S. The first and most important theorem of the section tells us that all we need do, in fact, is check the closure axioms A1 and A2. If these are satisfied, then the remaining axioms necessarily hold in S. This is a very useful theorem that will be applied on several occasions throughout the remainder of the text.

Theorem 4.3.2   Let S be a nonempty subset of a vector space V. Then S is a subspace of V if and only if S is closed under the operations of addition and scalar multiplication in V.

Proof   If S is a subspace of V, then it is a vector space, and hence, it is certainly closed under addition and scalar multiplication. Conversely, assume that S is closed under addition and scalar multiplication. We must prove that Axioms A3–A10 of Definition 4.2.1 hold when we restrict to vectors in S. Consider first the axioms A3, A4, and A7–A10. These are properties of the addition and scalar multiplication operations, hence since we use the same operations in S as in V, these axioms are all inherited from V by the subset S. Finally, we establish A5 and A6: Choose any vector[1] $\mathbf{u}$ in S. Since S is closed under scalar multiplication, both $0\mathbf{u}$ and $(-1)\mathbf{u}$ are in S. But by Theorem 4.2.6, $0\mathbf{u} = \mathbf{0}$ and $(-1)\mathbf{u} = -\mathbf{u}$, hence $\mathbf{0}$ and $-\mathbf{u}$ are both in S. Therefore, A5 and A6 are satisfied.   ■

The idea behind Theorem 4.3.2 is that once we have a vector space V in place, then any nonempty subset S, equipped with the same addition and scalar multiplication operations, will inherit all of the axioms that involve those operations. The only possible concern we have for S is whether or not it satisfies the closure axioms A1 and A2. Of course, we presumably had to carry out the full verification of A1–A10 for the vector space V in the first place, before gaining the shortcut of Theorem 4.3.2 for the subset S.

In determining whether a subset S of a vector space V is a subspace of V, we must keep clear in our minds what the given vector space is and what conditions on the vectors in V restrict them to lie in the subset S. This is most easily done by expressing S in set notation as follows:

$$S = \{\mathbf{v} \in V : \text{conditions on } \mathbf{v}\}.$$

We illustrate with an example.

Example 4.3.3   Verify that the set of all real solutions to the following linear system is a subspace of $\mathbb{R}^3$:

$$\begin{aligned} x_1 + 2x_2 - x_3 &= 0, \\ 2x_1 + 5x_2 - 4x_3 &= 0. \end{aligned}$$

Solution: The reduced row-echelon form of the augmented matrix of the system is

$$\begin{bmatrix} 1 & 0 & 3 & 0 \\ 0 & 1 & -2 & 0 \end{bmatrix},$$

---

[1]This is possible since S is assumed to be nonempty.

so that the solution set of the system is

$$S = \{\mathbf{x} \in \mathbb{R}^3 : \mathbf{x} = (-3r, 2r, r), \ r \in \mathbb{R}\},$$

which is a nonempty subset of $\mathbb{R}^3$. We now use Theorem 4.3.2 to verify that S is a subspace of $\mathbb{R}^3$: If $\mathbf{x} = (-3r, 2r, r)$ and $\mathbf{y} = (-3s, 2s, s)$ are any two vectors in S, then

$$\mathbf{x} + \mathbf{y} = (-3r, 2r, r) + (-3s, 2s, s) = (-3(r + s), 2(r + s), r + s) = (-3t, 2t, t),$$

where $t = r + s$. Thus, $\mathbf{x} + \mathbf{y}$ meets the required form for elements of S, and consequently, if we add two vectors in S, the result is another vector in S. Similarly, if we multiply an arbitrary vector $\mathbf{x} = (-3r, 2r, r)$ in S by a real number k, the resulting vector is

$$k\mathbf{x} = k(-3r, 2r, r) = (-3kr, 2kr, kr) = (-3w, 2w, w),$$

where $w = kr$. Hence, $k\mathbf{x}$ again has the proper form for membership in the subset S, and so S is closed under scalar multiplication. By Theorem 4.3.2, S is a subspace of $\mathbb{R}^3$. Note, of course, that our application of Theorem 4.3.2 hinges on our prior knowledge that $\mathbb{R}^3$ is a vector space.

Geometrically, the vectors in S lie along the line of intersection of the planes with the given equations. This is the line through the origin in the direction of the vector $\mathbf{v} = (-3, 2, 1)$. (See Figure 4.3.2.)

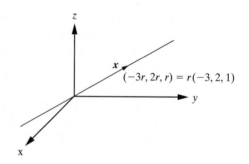

Figure 4.3.2: The solution set to the homogeneous system of linear equations in Example 4.3.3 is a subspace of $\mathbb{R}^3$.

□

Example 4.3.4   Verify that $S = \{\mathbf{x} \in \mathbb{R}^2 : \mathbf{x} = (r, -3r + 1), \ r \in \mathbb{R}\}$ is not a subspace of $\mathbb{R}^2$.

Solution: One approach here, according to Theorem 4.3.2, is to demonstrate the failure of closure under addition or scalar multiplication. For example, if we start with two vectors in S, say $\mathbf{x} = (r, -3r + 1)$ and $\mathbf{y} = (s, -3s + 1)$, then

$$\mathbf{x} + \mathbf{y} = (r, -3r + 1) + (s, -3s + 1) = (r + s, -3(r + s) + 2) = (w, -3w + 2),$$

where $w = r + s$. We see that $\mathbf{x} + \mathbf{y}$ does not have the required form for membership in S. Hence, S is not closed under addition and therefore fails to be a subspace of $\mathbb{R}^2$. Alternatively, we can show similarly that S is not closed under scalar multiplication.

Observant readers may have noticed another reason that S cannot form a subspace. Geometrically, the points in S correspond to those points that lie on the line with Cartesian equation $y = -3x + 1$. Since this line does not pass through the origin, S does not contain the zero vector $\mathbf{0} = (0, 0)$, and therefore we know S cannot be a subspace.   □

Remark   In general, we have the following important observation.

> If a subset S of a vector space V fails to contain the zero vector $\mathbf{0}$, then it cannot form a subspace.

This observation can often be made more quickly than deciding whether or not S is closed under addition and closed under scalar multiplication. However, we caution that if the zero vector *does* belong to S, then the observation is inconclusive and further investigation is required to determine whether or not S forms a subspace of V.

Example 4.3.5   Let S denote the set of all real symmetric $n \times n$ matrices. Verify that S is a subspace of $M_n(\mathbb{R})$.

Solution: The subset of interest is

$$S = \{A \in M_n(\mathbb{R}) : A^T = A\}.$$

Note that S is nonempty, since, for example, it contains the zero matrix 0_n. We now verify closure of S under addition and scalar multiplication. Let A and B be in S. Then

$$A^T = A \quad \text{and} \quad B^T = B.$$

Using these conditions and the properties of the transpose yields

$$(A + B)^T = A^T + B^T = A + B$$

and

$$(kA)^T = kA^T = kA$$

for all real values of k. Consequently $A + B$ and kA are both symmetric matrices, so they are elements of S. Hence S is closed under both addition and scalar multiplication and so is indeed a subspace of $M_n(\mathbb{R})$.   □

Remark   Notice in Example 4.3.5 that it was not necessary to actually write out the matrices A and B in terms of their elements $[a_{ij}]$ and $[b_{ij}]$, respectively. This shows the advantage of using simple abstract notation to describe the elements of the subset S in some situations.

Example 4.3.6   Let V be the vector space of all real-valued functions defined on an interval $[a, b]$, and let S denote the set of all functions in V that satisfy $f(a) = 0$. Verify that S is a subspace of V.

Solution: We have

$$S = \{f \in V : f(a) = 0\},$$

which is nonempty since it contains, for example, the zero function

$$O(x) = 0 \quad \text{for all } x \text{ in } [a, b].$$

Assume that f and g are in S, so that $f(a) = 0$ and $g(a) = 0$. We now check for closure of S under addition and scalar multiplication. We have

$$(f + g)(a) = f(a) + g(a) = 0 + 0 = 0,$$

which implies that $f + g \in S$. Hence, S is closed under addition. Further, if k is any real number,

$$(kf)(a) = kf(a) = k0 = 0,$$

so that S is also closed under scalar multiplication. Theorem 4.3.2 therefore implies that S is a subspace of V. Some representative functions from S are sketched in Figure 4.3.3. $\qquad \square$

In the next theorem, we establish that the subset $\{\mathbf{0}\}$ of a vector space V is in fact a subspace of V. We call this subspace the **trivial subspace** of V.

Theorem 4.3.7   Let V be a vector space with zero vector $\mathbf{0}$. Then $S = \{\mathbf{0}\}$ is a subspace of V.

Proof Note that S is nonempty. Further, the closure of S under addition and scalar multiplication follow, respectively, from

$$\mathbf{0} + \mathbf{0} = \mathbf{0} \qquad \text{and} \qquad k\mathbf{0} = \mathbf{0},$$

where the second statement follows from Theorem 4.2.6. $\qquad \blacksquare$

We now use Theorem 4.3.2 to establish an important result pertaining to homogeneous systems of linear equations that has already been illustrated in Example 4.3.3.

Theorem 4.3.8   Let A be an $m \times n$ matrix. The solution set of the homogeneous system of linear equations $A\mathbf{x} = \mathbf{0}$ is a subspace of $\mathbb{C}^n$.

Proof Let S denote the solution set of the homogeneous linear system. Then we can write

$$S = \{\mathbf{x} \in \mathbb{C}^n : A\mathbf{x} = \mathbf{0}\},$$

a subset of $\mathbb{C}^n$. Since a homogeneous system always admits the trivial solution $\mathbf{x} = \mathbf{0}$, we know that S is nonempty. If $\mathbf{x}_1$ and $\mathbf{x}_2$ are in S, then

$$A\mathbf{x}_1 = \mathbf{0} \qquad \text{and} \qquad A\mathbf{x}_2 = \mathbf{0}.$$

Using properties of the matrix product, we have

$$A(\mathbf{x}_1 + \mathbf{x}_2) = A\mathbf{x}_1 + A\mathbf{x}_2 = \mathbf{0} + \mathbf{0} = \mathbf{0},$$

so that $\mathbf{x}_1 + \mathbf{x}_2$ also solves the system and therefore is in S. Furthermore, if k is any complex scalar, then

$$A(k\mathbf{x}) = kA\mathbf{x} = k\mathbf{0} = \mathbf{0},$$

so that $k\mathbf{x}$ is also a solution of the system and therefore is in S. Since S is closed under both addition and scalar multiplication, it follows from Theorem 4.3.2 that S is a subspace of $\mathbb{C}^n$. $\qquad \blacksquare$

The preceding theorem has established that the solution set to any homogeneous linear system of equations is a vector space. Owing to the importance of this vector space, it is given a special name.

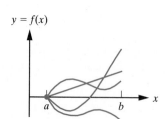

$y = f(x)$

Figure 4.3.3: Representative functions in the subspace S given in Example 4.3.6. Each function in S satisfies $f(a) = 0$.

DEFINITION   4.3.9

Let A be an $m \times n$ matrix. The solution set to the corresponding homogeneous linear system $A\mathbf{x} = \mathbf{0}$ is called the **null space of** A and is denoted nullspace(A). Thus,

$$\text{nullspace}(A) = \{\mathbf{x} \in \mathbb{C}^n : A\mathbf{x} = \mathbf{0}\}.$$

Remarks

1. If the matrix A has real elements, then we will consider only the corresponding real solutions to $A\mathbf{x} = \mathbf{0}$. Consequently, in this case,

$$\text{nullspace}(A) = \{\mathbf{x} \in \mathbb{R}^n : A\mathbf{x} = \mathbf{0}\},$$

a subspace of $\mathbb{R}^n$.

2. The previous theorem *does not* hold for the solution set of a *nonhomogeneous* linear system $A\mathbf{x} = \mathbf{b}$, for $\mathbf{b} \neq \mathbf{0}$, since $\mathbf{x} = \mathbf{0}$ is not in the solution set of the system.

Next we introduce the vector space of primary importance in the study of linear differential equations. This vector space arises as a subspace of the vector space of all functions that are defined on an interval I.

Example 4.3.10   Let V denote the vector space of all functions that are defined on an interval I, and let $C^k(I)$ denote the set of all functions that are continuous and have (at least) k continuous derivatives on the interval I, for a fixed non-negative integer k. Show that $C^k(I)$ is a subspace of V.

Solution: In this case

$$C^k(I) = \{f \in V : f, f', f'', \ldots, f^{(k)} \text{ exist and are continuous on } I\}.$$

This set is nonempty, as the zero function $O(x) = 0$ for all $x \in I$ is an element of $C^k(I)$. Moreover, it follows from the properties of derivatives that if we add two functions in $C^k(I)$, the result is a function in $C^k(I)$. Similarly, if we multiply a function in $C^k(I)$ by a scalar, then the result is a function in $C^k(I)$. Thus, Theorem 4.3.2 implies that $C^k(I)$ is a subspace of V. $\square$

Our final result in this section ties together the ideas introduced here with the theory of differential equations.

Theorem 4.3.11   The set of all solutions to the homogeneous linear differential equation

$$y'' + a_1(x)y' + a_2(x)y = 0 \tag{4.3.1}$$

on an interval I is a vector space.

Proof Let S denote the set of all solutions to the given differential equation. Then S is a nonempty subset of $C^2(I)$, since the identically zero function $y = 0$ is a solution to the differential equation. We establish that S is in fact a subspace of[2] $C^k(I)$. Let y_1 and y_2 be in S, and let k be a scalar. Then we have the following:

$$y_1'' + a_1(x)y_1' + a_2(x)y_1 = 0 \quad \text{and} \quad y_2'' + a_1(x)y_2' + a_2(x)y_2 = 0. \tag{4.3.2}$$

Now, if $y(x) = y_1(x) + y_2(x)$, then

$$
\begin{aligned}
y'' + a_1 y' + a_2 y &= (y_1 + y_2)'' + a_1(x)(y_1 + y_2)' + a_2(x)(y_1 + y_2) \\
&= [y_1'' + a_1(x)y_1' + a_2(x)y_1] + [y_2'' + a_1(x)y_2' + a_2(x)y_2] \\
&= 0 + 0 = 0,
\end{aligned}
$$

---

[2] It is important at this point that we have already established Example 4.3.10, so that S is a subset of a set that is indeed a vector space.

where we have used (4.3.2). Consequently, $y(x) = y_1(x) + y_2(x)$ is a solution to the differential equation (4.3.1). Moreover, if $y(x) = ky_1(x)$, then

$$y'' + a_1 y' + a_2 y = (ky_1)'' + a_1(x)(ky_1)' + a_2(x)(ky_1)$$
$$= k[y_1'' + a_1(x)y_1' + a_2(x)y_1] = 0,$$

where we have once more used (4.3.2). This establishes that $y(x) = ky_1(x)$ is a solution to Equation (4.3.1). Therefore, S is closed under both addition and scalar multiplication. Consequently, the set of all solutions to Equation (4.3.1) is a subspace of $C^2(I)$.   ∎

We will refer to the set of all solutions to a differential equation of the form (4.3.1) as the **solution space** of the differential equation. A key theoretical result that we will establish in Chapter 6 regarding the homogeneous linear differential equation (4.3.1) is that *every* solution to the differential equation has the form

$$y(x) = c_1 y_1(x) + c_2 y_2(x),$$

where y_1, y_2 are any two nonproportional solutions. The power of this result is impressive: It reduces the search for all solutions to Equation (4.3.1) to the search for just two nonproportional solutions. In vector space terms, the result can be restated as follows:

> Every vector in the solution space to the differential equation (4.3.1) can be written as a linear combination of any two nonproportional solutions y_1 and y_2.

We say that the solution space is **spanned** by y_1 and y_2. Moreover, two nonproportional solutions are referred to as **linearly independent**. For example, we saw in Example 1.2.16 that the set of all solutions to the differential equation

$$y'' + \omega^2 y = 0$$

is spanned by $y_1(x) = \cos \omega x$, and $y_2(x) = \sin \omega x$, and y_1 and y_2 are linearly independent. We now begin our investigation as to whether this type of idea will work more generally when the solution set to a problem is a vector space. For example, what about the solution set to a homogeneous linear system $A\mathbf{x} = \mathbf{0}$? We might suspect that if there are k free variables defining the vectors in nullspace(A), then every solution to $A\mathbf{x} = \mathbf{0}$ can be expressed as a linear combination of k basic solutions. We will establish that this is indeed the case in Section 4.9. The two key concepts we need to generalize are (1) spanning a general vector space with a set of vectors, and (2) linear independence in a general vector space. These will be addressed in turn in the next two sections.

Exercises for 4.3

Key Terms

Subspace, Trivial subspace, Null space of a matrix A.

Skills

- Be able to check whether or not a subset S of a vector space V is a subspace of V.

- Be able to compute the null space of an $m \times n$ matrix A.

True-False Review

For Questions 1–8, decide if the given statement is **true** or **false**, and give a brief justification for your answer. If true,

you can quote a relevant definition or theorem from the text. If false, provide an example, illustration, or brief explanation of why the statement is false.

1. The null space of an $m \times n$ matrix A with real elements is a subspace of $\mathbb{R}^m$.

2. The solution set of any linear system of m equations in n variables forms a subspace of $\mathbb{C}^n$.

3. The points in $\mathbb{R}^2$ that lie on the line $y = mx + b$ form a subspace of $\mathbb{R}^2$ if and only if $b = 0$.

4. If $m < n$, then $\mathbb{R}^m$ is a subspace of $\mathbb{R}^n$.

5. A nonempty set S of a vector space V that is closed under scalar multiplication contains the zero vector of V.

6. If $V = \mathbb{R}$ is a vector space under the usual operations of addition and scalar multiplication, then the subset $\mathbb{R}^+$ of *positive* real numbers, together with the operations defined in Problem 12 of Section 4.2, forms a subspace of V.

7. If $V = \mathbb{R}^3$ and S consists of all points on the xy-plane, the xz-plane, and the yz-plane, then S is a subspace of V.

8. If V is a vector space, then two different subspaces of V can contain no common vectors other than $\mathbf{0}$.

Problems

1. Let $S = \{\mathbf{x} \in \mathbb{R}^2 : \mathbf{x} = (2k, -3k), k \in \mathbb{R}\}$.

   (a) Establish that S is a subspace of $\mathbb{R}^2$.

   (b) Make a sketch depicting the subspace S in the Cartesian plane.

2. Let $S = \{\mathbf{x} \in \mathbb{R}^3 : \mathbf{x} = (r - 2s, 3r + s, s), r, s \in \mathbb{R}\}$.

   (a) Establish that S is a subspace of $\mathbb{R}^3$.

   (b) Show that the vectors in S lie on the plane with equation $3x - y + 7z = 0$.

For Problems 3–19, express S in set notation and determine whether it is a subspace of the given vector space V.

3. $V = \mathbb{R}^2$, and S is the set of all vectors (x, y) in V satisfying $3x + 2y = 0$.

4. $V = \mathbb{R}^4$, and S is the set of all vectors of the form $(x_1, 0, x_3, 2)$.

5. $V = \mathbb{R}^3$, and S is the set of all vectors (x, y, z) in V satisfying $x + y + z = 1$.

6. $V = \mathbb{R}^n$, and S is the set of all solutions to the nonhomogeneous linear system $A\mathbf{x} = \mathbf{b}$, where A is a fixed $m \times n$ matrix and $\mathbf{b} (\neq \mathbf{0})$ is a fixed vector.

7. $V = \mathbb{R}^2$, and S consists of all vectors (x, y) satisfying $x^2 - y^2 = 0$.

8. $V = M_2(\mathbb{R})$, and S is the subset of all 2×2 matrices with $\det(A) = 1$.

9. $V = M_n(\mathbb{R})$, and S is the subset of all $n \times n$ lower triangular matrices.

10. $V = M_n(\mathbb{R})$, and S is the subset of all $n \times n$ invertible matrices.

11. $V = M_2(\mathbb{R})$, and S is the subset of all 2×2 symmetric matrices.

12. $V = M_2(\mathbb{R})$, and S is the subset of all 2×2 skew-symmetric matrices.

13. V is the vector space of all real-valued functions defined on the interval $[a, b]$, and S is the subset of V consisting of all functions satisfying $f(a) = f(b)$.

14. V is the vector space of all real-valued functions defined on the interval $[a, b]$, and S is the subset of V consisting of all functions satisfying $f(a) = 1$.

15. V is the vector space of all real-valued functions defined on the interval $(-\infty, \infty)$, and S is the subset of V consisting of all functions satisfying $f(-x) = f(x)$ for all $x \in (-\infty, \infty)$.

16. $V = P_2$, and S is the subset of P_2 consisting of all polynomials of the form $p(x) = ax^2 + b$.

17. $V = P_2$, and S is the subset of P_2 consisting of all polynomials of the form $p(x) = ax^2 + 1$.

18. $V = C^2(I)$, and S is the subset of V consisting of those functions satisfying the differential equation

$$y'' + 2y' - y = 0$$

on I.

19. $V = C^2(I)$, and S is the subset of V consisting of those functions satisfying the differential equation

$$y'' + 2y' - y = 1$$

on I.

For Problems 20–22, determine the null space of the given matrix A.

20. $A = \begin{bmatrix} 1 & -2 & 1 \\ 4 & -7 & -2 \\ -1 & 3 & 4 \end{bmatrix}$.

21. $A = \begin{bmatrix} 1 & 3 & -2 & 1 \\ 3 & 10 & -4 & 6 \\ 2 & 5 & -6 & -1 \end{bmatrix}$.

22. $A = \begin{bmatrix} 1 & i & -2 \\ 3 & 4i & -5 \\ -1 & -3i & i \end{bmatrix}$.

23. Show that the set of all solutions to the nonhomogeneous differential equation

$$y'' + a_1 y' + a_2 y = F(x),$$

where $F(x)$ is nonzero on an interval I, is not a subspace of $C^2(I)$.

24. Let S_1 and S_2 be subspaces of a vector space V. Let

$$S_1 \cup S_2 = \{\mathbf{v} \in V : \mathbf{v} \in S_1 \text{ or } \mathbf{v} \in S_2\},$$
$$S_1 \cap S_2 = \{\mathbf{v} \in V : \mathbf{v} \in S_1 \text{ and } \mathbf{v} \in S_2\},$$

and let

$$S_1 + S_2 = \{\mathbf{v} \in V :$$
$$\mathbf{v} = \mathbf{x} + \mathbf{y} \text{ for some } \mathbf{x} \in S_1 \text{ and } \mathbf{y} \in S_2\}.$$

(a) Show that, in general, $S_1 \cup S_2$ is not a subspace of V.

(b) Show that $S_1 \cap S_2$ is a subspace of V.

(c) Show that $S_1 + S_2$ is a subspace of V.

4.4   Spanning Sets

The only algebraic operations that are defined in a vector space V are those of addition and scalar multiplication. Consequently, the most general way in which we can combine the vectors $\mathbf{v}_1, \mathbf{v}_2, \ldots, \mathbf{v}_k$ in V is

$$c_1 \mathbf{v}_1 + c_2 \mathbf{v}_2 + \cdots + c_k \mathbf{v}_k, \qquad (4.4.1)$$

where $c_1, c_2, \ldots, c_k$ are scalars. An expression of the form (4.4.1) is called a **linear combination** of $\mathbf{v}_1, \mathbf{v}_2, \ldots, \mathbf{v}_k$. Since V is closed under addition and scalar multiplication, it follows that the foregoing linear combination is itself a vector in V. One of the questions we wish to answer is whether every vector in a vector space can be obtained by taking linear combinations of a finite set of vectors. The following terminology is used in the case when the answer to this question is affirmative:

DEFINITION 4.4.1

If *every* vector in a vector space V can be written as a linear combination of $\mathbf{v}_1, \mathbf{v}_2, \ldots, \mathbf{v}_k$, we say that V is **spanned** or **generated** by $\mathbf{v}_1, \mathbf{v}_2, \ldots, \mathbf{v}_k$ and call the set of vectors $\{\mathbf{v}_1, \mathbf{v}_2, \ldots, \mathbf{v}_k\}$ a **spanning set** for V. In this case, we also say that $\{\mathbf{v}_1, \mathbf{v}_2, \ldots, \mathbf{v}_k\}$ **spans** V.

This spanning idea was introduced in the preceding section within the framework of differential equations. In addition, we are all used to representing geometric vectors in $\mathbb{R}^3$ in terms of their components as (see Section 4.1)

$$\mathbf{v} = a\mathbf{i} + b\mathbf{j} + c\mathbf{k},$$

where $\mathbf{i}$, $\mathbf{j}$, and $\mathbf{k}$ denote the unit vectors pointing along the positive x-, y-, and z-axes, respectively, of a rectangular Cartesian coordinate system. Using the above terminology, we say that $\mathbf{v}$ has been expressed as a linear combination of the vectors $\mathbf{i}, \mathbf{j}$, and $\mathbf{k}$, and that the vector space of all geometric vectors is spanned by $\mathbf{i}, \mathbf{j}$, and $\mathbf{k}$.

We now consider several examples to illustrate the spanning concept in different vector spaces.

Example 4.4.2   Show that $\mathbb{R}^2$ is spanned by the vectors

$$\mathbf{v}_1 = (1, 1) \quad \text{and} \quad \mathbf{v}_2 = (2, -1).$$

Solution: We must establish that for every $\mathbf{v} = (x_1, x_2)$ in $\mathbb{R}^2$, there exist constants c_1 and c_2 such that

$$\mathbf{v} = c_1\mathbf{v}_1 + c_2\mathbf{v}_2. \tag{4.4.2}$$

That is, in component form,

$$(x_1, x_2) = c_1(1, 1) + c_2(2, -1).$$

Equating corresponding components in this equation yields the following linear system:

$$c_1 + 2c_2 = x_1,$$
$$c_1 - c_2 = x_2.$$

In this system, we view x_1 and x_2 as fixed, while the variables we must solve for are c_1 and c_2. The determinant of the matrix of coefficients of this system is

$$\begin{vmatrix} 1 & 2 \\ 1 & -1 \end{vmatrix} = -3.$$

Since this is nonzero regardless of the values of x_1 and x_2, the matrix of coefficients is invertible, and hence for all $(x_1, x_2) \in \mathbb{R}^2$, the system has a (unique) solution according to Theorem 2.6.4. Thus, Equation (4.4.2) can be satisfied for every vector $\mathbf{v} \in \mathbb{R}^2$, so the given vectors do span $\mathbb{R}^2$. Indeed, solving the linear system yields

$$c_1 = \tfrac{1}{3}(x_1 + 2x_2), \qquad c_2 = \tfrac{1}{3}(x_1 - x_2).$$

Hence,

$$(x_1, x_2) = \tfrac{1}{3}(x_1 + 2x_2)\mathbf{v}_1 + \tfrac{1}{3}(x_1 - x_2)\mathbf{v}_2.$$

For example, if $\mathbf{v} = (2, 1)$, then $c_1 = \tfrac{4}{3}$ and $c_2 = \tfrac{1}{3}$, so that $\mathbf{v} = \tfrac{4}{3}\mathbf{v}_1 + \tfrac{1}{3}\mathbf{v}_2$. This is illustrated in Figure 4.4.1.   □

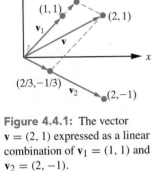

Figure 4.4.1: The vector $\mathbf{v} = (2, 1)$ expressed as a linear combination of $\mathbf{v}_1 = (1, 1)$ and $\mathbf{v}_2 = (2, -1)$.

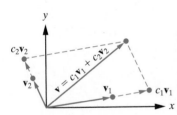

Figure 4.4.2: Any two noncollinear vectors in $\mathbb{R}^2$ span $\mathbb{R}^2$.

More generally, any two nonzero and noncolinear vectors $\mathbf{v}_1$ and $\mathbf{v}_2$ in $\mathbb{R}^2$ span $\mathbb{R}^2$, since, as illustrated geometrically in Figure 4.4.2, every vector in $\mathbb{R}^2$ can be written as a linear combination of $\mathbf{v}_1$ and $\mathbf{v}_2$.

Example 4.4.3

Determine whether the vectors $\mathbf{v}_1 = (1, -1, 4)$, $\mathbf{v}_2 = (-2, 1, 3)$, and $\mathbf{v}_3 = (4, -3, 5)$ span $\mathbb{R}^3$.

Solution: Let $\mathbf{v} = (x_1, x_2, x_3)$ be an arbitrary vector in $\mathbb{R}^3$. We must determine whether there are real numbers c_1, c_2, c_3 such that

$$\mathbf{v} = c_1\mathbf{v}_1 + c_2\mathbf{v}_2 + c_3\mathbf{v}_3 \tag{4.4.3}$$

or, in component form,

$$(x_1, x_2, x_3) = c_1(1, -1, 4) + c_2(-2, 1, 3) + c_3(4, -3, 5).$$

Equating corresponding components on either side of this vector equation yields

$$c_1 - 2c_2 + 4c_3 = x_1,$$
$$-c_1 + c_2 - 3c_3 = x_2,$$
$$4c_1 + 3c_2 + 5c_3 = x_3.$$

Reducing the augmented matrix of this system to row-echelon form, we obtain

$$\begin{bmatrix} 1 & -2 & 4 & x_1 \\ 0 & 1 & -1 & -x_1 - x_2 \\ 0 & 0 & 0 & 7x_1 + 11x_2 + x_3 \end{bmatrix}.$$

It follows that the system is consistent if and only if x_1, x_2, x_3 satisfy

$$7x_1 + 11x_2 + x_3 = 0. \tag{4.4.4}$$

Consequently, Equation (4.4.3) holds only for those vectors $\mathbf{v} = (x_1, x_2, x_3)$ in $\mathbb{R}^3$ whose components satisfy Equation (4.4.4). Hence, $\mathbf{v}_1$, $\mathbf{v}_2$, and $\mathbf{v}_3$ *do not* span $\mathbb{R}^3$. Geometrically, Equation (4.4.4) is the equation of a plane through the origin in space, and so by taking linear combinations of the given vectors, we can obtain only those vectors which lie on this plane. We leave it as an exercise to verify that indeed the three given vectors lie in the plane with Equation (4.4.4). It is worth noting that this plane forms a subspace S of $\mathbb{R}^3$, and that while V is not spanned by the vectors $\mathbf{v}_1$, $\mathbf{v}_2$, and $\mathbf{v}_3$, S is.   □

The reason that the vectors in the previous example did not span $\mathbb{R}^3$ was because they were coplanar. In general, any three noncoplanar vectors $\mathbf{v}_1$, $\mathbf{v}_2$, and $\mathbf{v}_3$ in $\mathbb{R}^3$ span $\mathbb{R}^3$, since, as illustrated in Figure 4.4.3, every vector in $\mathbb{R}^3$ can be written as a linear combination of $\mathbf{v}_1$, $\mathbf{v}_2$, and $\mathbf{v}_3$. In subsequent sections we will make this same observation from a more algebraic point of view.

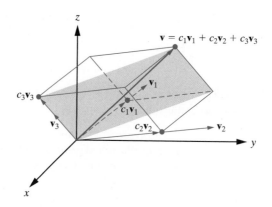

Figure 4.4.3: Any three noncoplanar vectors in $\mathbb{R}^3$ span $\mathbb{R}^3$.

Notice in the previous example that the linear combination (4.4.3) can be written as the matrix equation

$$A\mathbf{c} = \mathbf{v},$$

where the columns of A are the given vectors $\mathbf{v}_1$, $\mathbf{v}_2$, and $\mathbf{v}_3$: $A = [\mathbf{v}_1, \mathbf{v}_2, \mathbf{v}_3]$. Thus, the question of whether or not the vectors $\mathbf{v}_1$, $\mathbf{v}_2$, and $\mathbf{v}_3$ span $\mathbb{R}^3$ can be formulated as follows: Does the system $A\mathbf{c} = \mathbf{v}$ have a solution $\mathbf{c}$ for every $\mathbf{v}$ in $\mathbb{R}^3$? If so, then the column vectors of A span $\mathbb{R}^3$, and if not, then the column vectors of A do not span $\mathbb{R}^3$. This reformulation applies more generally to vectors in $\mathbb{R}^n$, and we state it here for the record.

Theorem 4.4.4   Let $\mathbf{v}_1, \mathbf{v}_2, \ldots, \mathbf{v}_k$ be vectors in $\mathbb{R}^n$. Then $\{\mathbf{v}_1, \mathbf{v}_2, \ldots, \mathbf{v}_k\}$ spans $\mathbb{R}^n$ if and only if, for the matrix $A = [\mathbf{v}_1, \mathbf{v}_2, \ldots, \mathbf{v}_k]$, the linear system $A\mathbf{c} = \mathbf{v}$ is consistent for every $\mathbf{v}$ in $\mathbb{R}^n$.

Proof Rewriting the system $A\mathbf{c} = \mathbf{v}$ as the linear combination

$$c_1\mathbf{v}_1 + c_2\mathbf{v}_2 + \cdots + c_k\mathbf{v}_k = \mathbf{v},$$

we see that the existence of a solution $(c_1, c_2, \ldots, c_k)$ to this vector equation for each $\mathbf{v}$ in $\mathbb{R}^n$ is equivalent to the statement that $\{\mathbf{v}_1, \mathbf{v}_2, \ldots, \mathbf{v}_k\}$ spans $\mathbb{R}^n$.   ∎

Next, we consider a couple of examples involving vector spaces other than $\mathbb{R}^n$.

Example 4.4.5   Verify that

$$A_1 = \begin{bmatrix} 1 & 0 \\ 0 & 0 \end{bmatrix}, \qquad A_2 = \begin{bmatrix} 1 & 1 \\ 0 & 0 \end{bmatrix}, \qquad A_3 = \begin{bmatrix} 1 & 1 \\ 1 & 0 \end{bmatrix}, \qquad A_4 = \begin{bmatrix} 1 & 1 \\ 1 & 1 \end{bmatrix}$$

span $M_2(\mathbb{R})$.

Solution: An arbitrary vector in $M_2(\mathbb{R})$ is of the form

$$A = \begin{bmatrix} a & b \\ c & d \end{bmatrix}.$$

If we write

$$c_1 A_1 + c_2 A_2 + c_3 A_3 + c_4 A_4 = A,$$

then equating the elements of the matrices on each side of the equation yields the system

$$\begin{aligned} c_1 + c_2 + c_3 + c_4 &= a, \\ c_2 + c_3 + c_4 &= b, \\ c_3 + c_4 &= c, \\ c_4 &= d. \end{aligned}$$

Solving this by back substitution gives

$$c_1 = a - b, \qquad c_2 = b - c, \qquad c_3 = c - d, \qquad c_4 = d.$$

Hence, we have

$$A = (a - b)A_1 + (b - c)A_2 + (c - d)A_3 + dA_4.$$

Consequently every vector in $M_2(\mathbb{R})$ can be written as a linear combination of A_1, A_2, A_3, and A_4, and therefore these matrices do indeed span $M_2(\mathbb{R})$.   □

Remark   The most natural spanning set for $M_2(\mathbb{R})$ is

$$\left\{ \begin{bmatrix} 1 & 0 \\ 0 & 0 \end{bmatrix}, \begin{bmatrix} 0 & 1 \\ 0 & 0 \end{bmatrix}, \begin{bmatrix} 0 & 0 \\ 1 & 0 \end{bmatrix}, \begin{bmatrix} 0 & 0 \\ 0 & 1 \end{bmatrix} \right\},$$

a fact that we leave to the reader as an exercise.

Example 4.4.6   Determine a spanning set for P_2, the vector space of all polynomials of degree 2 or less.

Solution: The general polynomial in P_2 is

$$p(x) = a_0 + a_1 x + a_2 x^2.$$

If we let

$$p_0(x) = 1, \qquad p_1(x) = x, \qquad p_2(x) = x^2,$$

then

$$p(x) = a_0 p_0(x) + a_1 p_1(x) + a_2 p_2(x).$$

Thus, every vector in P_2 is a linear combination of 1, x, and x^2, and so a spanning set for P_2 is $\{1, x, x^2\}$. For practice, the reader might show that $\{x^2, x + x^2, 1 + x + x^2\}$ is another spanning set for P_2, by making the appropriate modifications to the calculations in this example.   □

The Linear Span of a Set of Vectors

Now let $v_1, v_2, \ldots, v_k$ be vectors in a vector space V. Forming all possible linear combinations of $v_1, v_2, \ldots, v_k$ generates a subset of V called the **linear span** of $\{v_1, v_2, \ldots, v_k\}$, denoted span$\{v_1, v_2, \ldots, v_k\}$. We have

$$\text{span}\{v_1, v_2, \ldots, v_k\} = \{v \in V : v = c_1 v_1 + c_2 v_2 + \cdots + c_k v_k, c_1, c_2, \ldots, c_k \in F\}.$$
(4.4.5)

For example, suppose $V = C^2(I)$, and let $y_1(x) = \sin x$ and $y_2(x) = \cos x$. Then

$$\text{span}\{y_1, y_2\} = \{y \in C^2(I) : y(x) = c_1 \cos x + c_2 \sin x, c_1, c_2 \in \mathbb{R}\}.$$

From Example 1.2.16, we recognize y_1 and y_2 as being nonproportional solutions to the differential equation $y'' + y = 0$. Consequently, in this example, the linear span of the given functions coincides with the set of all solutions to the differential equation $y'' + y = 0$ and therefore is a *subspace* of V. Our next theorem generalizes this to show that any linear span of vectors in any vector space forms a subspace.

Theorem 4.4.7   Let $v_1, v_2, \ldots, v_k$ be vectors in a vector space V. Then span$\{v_1, v_2, \ldots, v_k\}$ is a subspace of V.

Proof Let $S = \text{span}\{v_1, v_2, \ldots, v_k\}$. Then $\mathbf{0} \in S$ (corresponding to $c_1 = c_2 = \cdots = c_k = 0$ in (4.4.5)), so S is nonempty. We now verify closure of S under addition and scalar multiplication. If u and v are in S, then, from Equation (4.4.5),

$$u = a_1 v_1 + a_2 v_2 + \cdots + a_k v_k \qquad \text{and} \qquad v = b_1 v_1 + b_2 v_2 + \cdots + b_k v_k,$$

for some scalars a_i, b_i. Thus,

$$\begin{aligned} u + v &= (a_1 v_1 + a_2 v_2 + \cdots + a_k v_k) + (b_1 v_1 + b_2 v_2 + \cdots + b_k v_k) \\ &= (a_1 + b_1)v_1 + (a_2 + b_2)v_2 + \cdots + (a_k + b_k)v_k \\ &= c_1 v_1 + c_2 v_2 + \cdots + c_k v_k, \end{aligned}$$

where $c_i = a_i + b_i$ for each $i = 1, 2, \ldots, k$. Consequently, $u + v$ has the proper form for membership in S according to (4.4.5), so S is closed under addition. Further, if r is any scalar, then

$$\begin{aligned} ru &= r(a_1 v_1 + a_2 v_2 + \cdots + a_k v_k) \\ &= (ra_1)v_1 + (ra_2)v_2 + \cdots + (ra_k)v_k \\ &= d_1 v_1 + d_2 v_2 + \cdots + d_k v_k, \end{aligned}$$

where $d_i = ra_i$ for each $i = 1, 2, \ldots, k$. Consequently, $ru \in S$, and so S is also closed under scalar multiplication. Hence, $S = \text{span}\{v_1, v_2, \ldots, v_k\}$ is a subspace of V.   ■

Remarks

1. We will also refer to span$\{v_1, v_2, \ldots, v_k\}$ as the **subspace of** V **spanned by** $v_1, v_2, \ldots, v_k$.

2. As a special case, we will declare that span$(\emptyset) = \{\mathbf{0}\}$.

Example 4.4.8   If $V = \mathbb{R}^2$ and $\mathbf{v}_1 = (-1, 1)$, determine span$\{\mathbf{v}_1\}$.

Solution: We have

$$\begin{aligned}
\text{span}\{\mathbf{v}_1\} &= \{\mathbf{v} \in \mathbb{R}^2 : \mathbf{v} = c_1\mathbf{v}_1, \ c_1 \in \mathbb{R}\} \\
&= \{\mathbf{v} \in \mathbb{R}^2 : \mathbf{v} = c_1(-1, 1), \ c_1 \in \mathbb{R}\} \\
&= \{\mathbf{v} \in \mathbb{R}^2 : \mathbf{v} = (-c_1, c_1), \ c_1 \in \mathbb{R}\}.
\end{aligned}$$

Geometrically, this is the line through the origin with parametric equations $x = -c_1$, $y = c_1$, so that the Cartesian equation of the line is $y = -x$. (See Figure 4.4.4.)

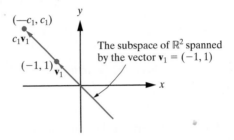

Figure 4.4.4: The subspace of $\mathbb{R}^2$ spanned by $\mathbf{v}_1 = (-1, 1)$.

Example 4.4.9   If $V = \mathbb{R}^3$, $\mathbf{v}_1 = (1, 0, 1)$, and $\mathbf{v}_2 = (0, 1, 1)$, determine the subspace of $\mathbb{R}^3$ spanned by $\mathbf{v}_1$ and $\mathbf{v}_2$. Does $\mathbf{w} = (1, 1, -1)$ lie in this subspace?

Solution: We have

$$\begin{aligned}
\text{span}\{\mathbf{v}_1, \mathbf{v}_2\} &= \{\mathbf{v} \in \mathbb{R}^3 : \mathbf{v} = c_1\mathbf{v}_1 + c_2\mathbf{v}_2, \ c_1, c_2 \in \mathbb{R}\} \\
&= \{\mathbf{v} \in \mathbb{R}^3 : \mathbf{v} = c_1(1, 0, 1) + c_2(0, 1, 1), \ c_1, c_2 \in \mathbb{R}\} \\
&= \{\mathbf{v} \in \mathbb{R}^3 : \mathbf{v} = (c_1, c_2, c_1 + c_2), \ c_1, c_2 \in \mathbb{R}\}.
\end{aligned}$$

Since the vector $\mathbf{w} = (1, 1, -1)$ is not of the form $(c_1, c_2, c_1 + c_2)$, it does not lie in span$\{\mathbf{v}_1, \mathbf{v}_2\}$. Geometrically, span$\{\mathbf{v}_1, \mathbf{v}_2\}$ is the plane through the origin determined by the two given vectors $\mathbf{v}_1$ and $\mathbf{v}_2$. It has parametric equations $x = c_1$, $y = c_2$, $z = c_1 + c_2$, which implies that its Cartesian equation is $z = x + y$. Thus, the fact that $\mathbf{w}$ is not in span$\{\mathbf{v}_1, \mathbf{v}_2\}$ means that $\mathbf{w}$ does not lie in this plane. The subspace is depicted in Figure 4.4.5.

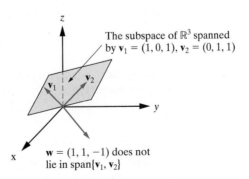

Figure 4.4.5: The subspace of $\mathbb{R}^3$ spanned by $\mathbf{v}_1 = (1, 0, 1)$ and $\mathbf{v}_2 = (0, 1, 1)$ is the plane with Cartesian equation $z = x + y$.

Example 4.4.10   Let

$$A_1 = \begin{bmatrix} 1 & 0 \\ 0 & 0 \end{bmatrix}, \qquad A_2 = \begin{bmatrix} 0 & 1 \\ 1 & 0 \end{bmatrix}, \qquad A_3 = \begin{bmatrix} 0 & 0 \\ 0 & 1 \end{bmatrix}$$

in $M_2(\mathbb{R})$. Determine $\text{span}\{A_1, A_2, A_3\}$.

Solution: By definition we have

$$\text{span}\{A_1, A_2, A_3\} = \{A \in M_2(\mathbb{R}) : A = c_1 A_1 + c_2 A_2 + c_3 A_3, \ c_1, c_2, c_3 \in \mathbb{R}\}$$
$$= \left\{ A \in M_2(\mathbb{R}) : A = c_1 \begin{bmatrix} 1 & 0 \\ 0 & 0 \end{bmatrix} + c_2 \begin{bmatrix} 0 & 1 \\ 1 & 0 \end{bmatrix} + c_3 \begin{bmatrix} 0 & 0 \\ 0 & 1 \end{bmatrix} \right\}$$
$$= \left\{ A \in M_2(\mathbb{R}) : A = \begin{bmatrix} c_1 & c_2 \\ c_2 & c_3 \end{bmatrix}, \ c_1, c_2, c_3 \in \mathbb{R} \right\}.$$

This is the set of all real 2×2 *symmetric* matrices.   □

Example 4.4.11   Determine the subspace of P_2 spanned by

$$p_1(x) = 1 + 3x, \qquad p_2(x) = x + x^2,$$

and decide whether $\{p_1, p_2\}$ is a spanning set for P_2.

Solution: We have

$$\text{span}\{p_1, p_2\} = \{p \in P_2 : p(x) = c_1 p_1(x) + c_2 p_2(x), \ c_1, c_2 \in \mathbb{R}\}$$
$$= \{p \in P_2 : p(x) = c_1(1 + 3x) + c_2(x + x^2), \ c_1, c_2 \in \mathbb{R}\}$$
$$= \{p \in P_2 : p(x) = c_1 + (3c_1 + c_2)x + c_2 x^2, \ c_1, c_2 \in \mathbb{R}\}.$$

Next, we will show that $\{p_1, p_2\}$ is not a spanning set for P_2. To establish this, we need give only one example of a polynomial in P_2 that is not in $\text{span}\{p_1, p_2\}$. There are many such choices here, but suppose we consider $p(x) = 1 + x$. If this polynomial were in $\text{span}\{p_1, p_2\}$, then we would have to be able to find values of c_1 and c_2 such that

$$1 + x = c_1 + (3c_1 + c_2)x + c_2 x^2. \tag{4.4.6}$$

Since there is no x^2 term on the left-hand side of this expression, we must set $c_2 = 0$. But then (4.4.6) would reduce to

$$1 + x = c_1(1 + 3x).$$

Equating the constant terms on each side of this forces $c_1 = 1$, but then the coefficients of x do not match. Hence, such an equality is impossible. Consequently, there are no values of c_1 and c_2 such that the Equation (4.4.6) holds, and therefore, $\text{span}\{p_1, p_2\} \neq P_2$.   □

Remark   In the previous example, the reader may well wonder why we knew from the beginning to select $p(x) = 1 + x$ as a vector that would be outside of $\text{span}\{p_1, p_2\}$. In truth, we only need to find a polynomial that does not have the form $p(x) = c_1 + (3c_1 + c_2)x + c_2 x^2$ and in fact, "most" of the polynomials in P_2 would have achieved the desired result here.

Exercises for 4.4

Key Terms

Linear combination, Linear span, Spanning set.

Skills

- Be able to determine whether a given set of vectors S spans a vector space V, and be able to prove your answer mathematically.

- Be able to determine the linear span of a set of vectors. For vectors in $\mathbb{R}^n$, be able to give a geometric description of the linear span.

- If S is a spanning set for a vector space V, be able to write any vector in V as a linear combination of the elements of S.

- Be able to construct a spanning set for a vector space V. As a special case, be able to determine a spanning set for the null space of an $m \times n$ matrix.

- Be able to determine whether a particular vector $\mathbf{v}$ in a vector space V lies in the linear span of a set S of vectors in V.

True-False Review

For Questions 1–12, decide if the given statement is **true** or **false**, and give a brief justification for your answer. If true, you can quote a relevant definition or theorem from the text. If false, provide an example, illustration, or brief explanation of why the statement is false.

1. The linear span of a set of vectors in a vector space V forms a subspace of V.

2. If some vector $\mathbf{v}$ in a vector space V is a linear combination of vectors in a set S, then S spans V.

3. If S is a spanning set for a vector space V and W is a subspace of V, then S is a spanning set for W.

4. If S is a spanning set for a vector space V, then every vector $\mathbf{v}$ in V must be uniquely expressible as a linear combination of the vectors in S.

5. A set S of vectors in a vector space V spans V if and only if the linear span of S is V.

6. The linear span of two vectors in $\mathbb{R}^3$ is a plane through the origin.

7. Every vector space V has a finite spanning set.

8. If S is a spanning set for a vector space V, then any proper subset S' of S is not a spanning set for V.

9. The vector space of 3×3 upper triangular matrices is spanned by the matrices E_{ij} where $1 \le i \le j \le 3$.

10. A spanning set for the vector space P_2 must contain a polynomial of each degree 0, 1, and 2.

11. If $m < n$, then any spanning set for $\mathbb{R}^n$ must contain more vectors than any spanning set for $\mathbb{R}^m$.

12. The vector space P of all polynomials with real coefficients cannot be spanned by a finite set S.

Problems

For Problems 1–3, determine whether the given set of vectors spans $\mathbb{R}^2$.

1. $\{(1, -1), (2, -2), (2, 3)\}$.

2. $\{(2, 5), (0, 0)\}$.

3. $\{(6, -2), (-2, 2/3), (3, -1)\}$.

Recall that three vectors $\mathbf{v}_1, \mathbf{v}_2, \mathbf{v}_3$ in $\mathbb{R}^3$ are coplanar if and only if

$$\det([\mathbf{v}_1, \mathbf{v}_2, \mathbf{v}_3]) = 0.$$

For Problems 4–6, use this result to determine whether the given set of vectors spans $\mathbb{R}^3$.

4. $\{(1, -1, 1), (2, 5, 3), (4, -2, 1)\}$.

5. $\{(1, -2, 1), (2, 3, 1), (0, 0, 0), (4, -1, 2)\}$.

6. $\{(2, -1, 4), (3, -3, 5), (1, 1, 3)\}$.

7. Show that the set of vectors

$$\{(1, 2, 3), (3, 4, 5), (4, 5, 6)\}$$

does not span $\mathbb{R}^3$, but that it does span the subspace of $\mathbb{R}^3$ consisting of all vectors lying in the plane with equation $x - 2y + z = 0$.

8. Show that $\mathbf{v}_1 = (2, -1)$, $\mathbf{v}_2 = (3, 2)$ span $\mathbb{R}^2$, and express the vector $\mathbf{v} = (5, -7)$ as a linear combination of $\mathbf{v}_1, \mathbf{v}_2$.

9. Show that $\mathbf{v}_1 = (-1, 3, 2)$, $\mathbf{v}_2 = (1, -2, 1)$, $\mathbf{v}_3 = (2, 1, 1)$ span $\mathbb{R}^3$, and express $\mathbf{v} = (x, y, z)$ as a linear combination of $\mathbf{v}_1, \mathbf{v}_2, \mathbf{v}_3$.

10. Show that $\mathbf{v}_1 = (1, 1)$, $\mathbf{v}_2 = (-1, 2)$, $\mathbf{v}_3 = (1, 4)$ span $\mathbb{R}^2$. Do $\mathbf{v}_1$, $\mathbf{v}_2$ alone span $\mathbb{R}^2$ also?

11. Let S be the subspace of $\mathbb{R}^3$ consisting of all vectors of the form $\mathbf{v} = (c_1, c_2, c_2 - 2c_1)$. Show that S is spanned by $\mathbf{v}_1 = (1, 0, -2)$, $\mathbf{v}_2 = (0, 1, 1)$.

12. Let S be the subspace of $\mathbb{R}^4$ consisting of all vectors of the form $\mathbf{v} = (c_1, c_2, c_2 - c_1, c_1 - 2c_2)$. Determine a set of vectors that spans S.

13. Let S be the subspace of $\mathbb{R}^3$ consisting of all solutions to the linear system

$$x - 2y - z = 0.$$

Determine a set of vectors that spans S.

For Problems 14–15, determine a spanning set for the null space of the given matrix A.

14. $A = \begin{bmatrix} 1 & 2 & 3 \\ 3 & 4 & 5 \\ 5 & 6 & 7 \end{bmatrix}$.

15. $A = \begin{bmatrix} 1 & 2 & 3 & 5 \\ 1 & 3 & 4 & 2 \\ 2 & 4 & 6 & -1 \end{bmatrix}$.

16. Let S be the subspace of $M_2(\mathbb{R})$ consisting of all symmetric 2×2 matrices with real elements. Show that S is spanned by the matrices

$$A_1 = \begin{bmatrix} 1 & 0 \\ 0 & 0 \end{bmatrix}, \quad A_2 = \begin{bmatrix} 0 & 0 \\ 0 & 1 \end{bmatrix}, \quad A_3 = \begin{bmatrix} 0 & 1 \\ 1 & 0 \end{bmatrix}.$$

17. Let S be the subspace of $M_2(\mathbb{R})$ consisting of all skew-symmetric 2×2 matrices with real elements. Determine a matrix that spans S.

18. Let S be the subset of $M_2(\mathbb{R})$ consisting of all upper triangular 2×2 matrices.

   (a) Verify that S is a subspace of $M_2(\mathbb{R})$.

   (b) Determine a set of 2×2 matrices that spans S.

For Problems 19–20, determine span$\{\mathbf{v}_1, \mathbf{v}_2\}$ for the given vectors in $\mathbb{R}^3$, and describe it geometrically.

19. $\mathbf{v}_1 = (1, -1, 2)$, $\mathbf{v}_2 = (2, -1, 3)$.

20. $\mathbf{v}_1 = (1, 2, -1)$, $\mathbf{v}_2 = (-2, -4, 2)$.

21. Let S be the subspace of $\mathbb{R}^3$ spanned by the vectors $\mathbf{v}_1 = (1, 1, -1)$, $\mathbf{v}_2 = (2, 1, 3)$, $\mathbf{v}_3 = (-2, -2, 2)$. Show that S also is spanned by $\mathbf{v}_1$ and $\mathbf{v}_2$ only.

For Problems 22–24, determine whether the given vector $\mathbf{v}$ lies in span$\{\mathbf{v}_1, \mathbf{v}_2\}$.

22. $\mathbf{v} = (3, 3, 4)$, $\mathbf{v}_1 = (1, -1, 2)$, $\mathbf{v}_2 = (2, 1, 3)$ in $\mathbb{R}^3$.

23. $\mathbf{v} = (5, 3, -6)$, $\mathbf{v}_1 = (-1, 1, 2)$, $\mathbf{v}_2 = (3, 1, -4)$ in $\mathbb{R}^3$.

24. $\mathbf{v} = (1, 1, -2)$, $\mathbf{v}_1 = (3, 1, 2)$, $\mathbf{v}_2 = (-2, -1, 1)$ in $\mathbb{R}^3$.

25. If $p_1(x) = x - 4$ and $p_2(x) = x^2 - x + 3$, determine whether $p(x) = 2x^2 - x + 2$ lies in span$\{p_1, p_2\}$.

26. Consider the vectors

$$A_1 = \begin{bmatrix} 1 & -1 \\ 2 & 0 \end{bmatrix}, \quad A_2 = \begin{bmatrix} 0 & 1 \\ -2 & 1 \end{bmatrix}, \quad A_3 = \begin{bmatrix} 3 & 0 \\ 1 & 2 \end{bmatrix}$$

in $M_2(\mathbb{R})$. Determine span$\{A_1, A_2, A_3\}$.

27. Consider the vectors

$$A_1 = \begin{bmatrix} 1 & 2 \\ -1 & 3 \end{bmatrix}, \quad A_2 = \begin{bmatrix} -2 & 1 \\ 1 & -1 \end{bmatrix}$$

in $M_2(\mathbb{R})$. Find span$\{A_1, A_2\}$, and determine whether or not

$$B = \begin{bmatrix} 3 & 1 \\ -2 & 4 \end{bmatrix}$$

lies in this subspace.

28. Let $V = C^\infty(I)$ and let S be the subspace of V spanned by the functions

$$f(x) = \cosh x, \qquad g(x) = \sinh x.$$

   (a) Give an expression for a general vector in S.

   (b) Show that S is also spanned by the functions

$$h(x) = e^x, \qquad j(x) = e^{-x}.$$

For Problems 29–32, give a geometric description of the subspace of $\mathbb{R}^3$ spanned by the given set of vectors.

29. $\{\mathbf{0}\}$.

30. $\{\mathbf{v}_1\}$, where $\mathbf{v}_1$ is any nonzero vector in $\mathbb{R}^3$.

31. $\{\mathbf{v}_1, \mathbf{v}_2\}$, where $\mathbf{v}_1$, $\mathbf{v}_2$ are nonzero and noncollinear vectors in $\mathbb{R}^3$.

32. $\{\mathbf{v}_1, \mathbf{v}_2\}$, where $\mathbf{v}_1, \mathbf{v}_2$ are collinear vectors in $\mathbb{R}^3$.

33. Prove that if S and S' are subsets of a vector space V such that S is a subset of S', then span(S) is a subset of span(S').

34. Prove that

$$\text{span}\{\mathbf{v}_1, \mathbf{v}_2, \mathbf{v}_3\} = \text{span}\{\mathbf{v}_1, \mathbf{v}_2\}$$

if and only if $\mathbf{v}_3$ can be written as a linear combination of $\mathbf{v}_1$ and $\mathbf{v}_2$.

4.5   Linear Dependence and Linear Independence

As indicated in the previous section, in analyzing a vector space we will be interested in determining a spanning set. The reader has perhaps already noticed that a vector space V can have many such spanning sets.

Example 4.5.1   Observe that $\{(1, 0), (0, 1)\}$, $\{(1, 0), (1, 1)\}$, and $\{(1, 0), (0, 1), (1, 2)\}$ are all spanning sets for $\mathbb{R}^2$.   □

As another illustration, two different spanning sets for $V = M_2(\mathbb{R})$ were given in Example 4.4.5 and the remark that followed. Given the abundance of spanning sets available for a given vector space V, we are faced with a natural question: Is there a "best class of" spanning sets to use? The answer, to a large degree, is "yes". For instance, in Example 4.5.1, the spanning set $\{(1, 0), (0, 1), (1, 2)\}$ contains an "extra" vector, $(1, 2)$, which seems to be unnecessary for spanning $\mathbb{R}^2$, since $\{(1, 0), (0, 1)\}$ is already a spanning set. In some sense, $\{(1, 0), (0, 1)\}$ is a more efficient spanning set. It is what we call a ***minimal* spanning set**, since it contains the minimum number of vectors needed to span the vector space.[3]

But how will we know if we have found a minimal spanning set (assuming one exists)? Returning to the example above, we have seen that

$$\text{span}\{(1, 0), (0, 1)\} = \text{span}\{(1, 0), (0, 1), (1, 2)\} = \mathbb{R}^2.$$

Observe that the vector $(1, 2)$ is already a linear combination of $(1, 0)$ and $(0, 1)$, and therefore it does not add any new vectors to the linear span of $\{(1, 0), (0, 1)\}$.

As a second example, consider the vectors $\mathbf{v}_1 = (1, 1, 1)$, $\mathbf{v}_2 = (3, -2, 1)$, and $\mathbf{v}_3 = 4\mathbf{v}_1 + \mathbf{v}_2 = (7, 2, 5)$. It is easily verified that $\det([\mathbf{v}_1, \mathbf{v}_2, \mathbf{v}_3]) = 0$. Consequently, the three vectors lie in a plane (see Figure 4.5.1) and therefore, since they are not collinear, the linear span of these three vectors is the whole of this plane. Furthermore, the same plane is generated if we consider the linear span of $\mathbf{v}_1$ and $\mathbf{v}_2$ alone. As in the previous example, the reason that $\mathbf{v}_3$ does not add any new vectors to the linear span of $\{\mathbf{v}_1, \mathbf{v}_2\}$ is that it is already a linear combination of $\mathbf{v}_1$ and $\mathbf{v}_2$. It is not possible, however, to generate all vectors in the plane by taking linear combinations of just one vector, as we could generate only a line lying in the plane in that case. Consequently, $\{\mathbf{v}_1, \mathbf{v}_2\}$ is a minimal spanning set for the subspace of $\mathbb{R}^3$ consisting of all points lying on the plane.

As a final example, recall from Example 1.2.16 that the solution space to the differential equation

$$y'' + y = 0$$

---

[3] Since a single (nonzero) vector in $\mathbb{R}^2$ spans only the line through the origin along which it points, it cannot span all of $\mathbb{R}^2$; hence, the minimum number of vectors required to span $\mathbb{R}^2$ is 2.

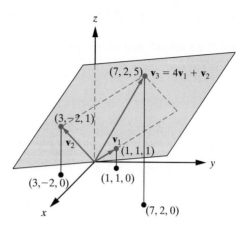

Figure 4.5.1: $\mathbf{v}_3 = 4\mathbf{v}_1 + \mathbf{v}_2$ lies in the plane through the origin containing $\mathbf{v}_1$ and $\mathbf{v}_2$, and so, span$\{\mathbf{v}_1, \mathbf{v}_2, \mathbf{v}_3\} = $ span$\{\mathbf{v}_1, \mathbf{v}_2\}$.

can be written as span$\{y_1, y_2\}$, where $y_1(x) = \cos x$ and $y_2(x) = \sin x$. However, if we let $y_3(x) = 3\cos x - 2\sin x$, for instance, then $\{y_1, y_2, y_3\}$ is also a spanning set for the solution space of the differential equation, since

$$\begin{aligned}
\text{span}\{y_1, y_2, y_3\} &= \{c_1 \cos x + c_2 \sin x + c_3(3\cos x - 2\sin x) : \ c_1, c_2, c_3 \in \mathbb{R}\} \\
&= \{(c_1 + 3c_3)\cos x + (c_2 - 2c_3)\sin x : \ c_1, c_2, c_3 \in \mathbb{R}\} \\
&= \{d_1 \cos x + d_2 \sin x : \ d_1, d_2 \in \mathbb{R}\} \\
&= \text{span}\{y_1, y_2\}.
\end{aligned}$$

The reason that $\{y_1, y_2, y_3\}$ is not a *minimal* spanning set for the solution space is that y_3 is a linear combination of y_1 and y_2, and therefore, as we have just shown, it does not add any new vectors to the linear span of $\{\cos x, \sin x\}$.

More generally, it is not too difficult to extend the argument used in the preceding examples to establish the following general result.

Theorem 4.5.2   Let $\{\mathbf{v}_1, \mathbf{v}_2, \ldots, \mathbf{v}_k\}$ be a set of at least two vectors in a vector space V. If one of the vectors in the set is a linear combination of the other vectors in the set, then that vector can be deleted from the given set of vectors and the linear span of the resulting set of vectors will be the same as the linear span of $\{\mathbf{v}_1, \mathbf{v}_2, \ldots, \mathbf{v}_k\}$.

Proof The proof of this result is left for the exercises (Problem 48).   ■

For instance, if $\mathbf{v}_1$ is a linear combination of $\mathbf{v}_2, \mathbf{v}_3, \ldots, \mathbf{v}_k$, then Theorem 4.5.2 says that

$$\text{span}\{\mathbf{v}_1, \mathbf{v}_2, \ldots, \mathbf{v}_k\} = \text{span}\{\mathbf{v}_2, \mathbf{v}_3, \ldots, \mathbf{v}_k\}.$$

In this case, the set $\{\mathbf{v}_1, \mathbf{v}_2, \ldots, \mathbf{v}_k\}$ is not a minimal spanning set.

To determine a minimal spanning set, the problem we face in view of Theorem 4.5.2 is that of determining when a vector in $\{\mathbf{v}_1, \mathbf{v}_2, \ldots, \mathbf{v}_k\}$ can be expressed as a linear combination of the remaining vectors in the set. The correct formulation for solving this problem requires the concepts of linear dependence and linear independence, which we are now ready to introduce. First we consider linear dependence.

DEFINITION 4.5.3

A finite nonempty set of vectors $\{v_1, v_2, \ldots, v_k\}$ in a vector space V is said to be **linearly dependent** if there exist scalars $c_1, c_2, \ldots, c_k$, *not all zero*, such that

$$c_1 v_1 + c_2 v_2 + \cdots + c_k v_k = 0.$$

Such a nontrivial linear combination of vectors is sometimes referred to as a **linear dependency** among the vectors $v_1, v_2, \ldots, v_k$.

A set of vectors that is not linearly dependent is called linearly independent. This can be stated mathematically as follows:

DEFINITION 4.5.4

A finite, nonempty set of vectors $\{v_1, v_2, \ldots, v_k\}$ in a vector space V is said to be **linearly independent** if the *only* values of the scalars $c_1, c_2, \ldots, c_k$ for which

$$c_1 v_1 + c_2 v_2 + \cdots + c_k v_k = 0$$

are $c_1 = c_2 = \cdots = c_k = 0$.

Remarks

1. It follows immediately from the preceding two definitions that a nonempty set of vectors in a vector space V is linearly independent if and only if it is not linearly dependent.

2. If $\{v_1, v_2, \ldots, v_k\}$ is a linearly independent set of vectors, we sometimes informally say that the vectors $v_1, v_2, \ldots, v_k$ are themselves linearly independent. The same remark applies to the linearly dependent condition as well.

Consider the simple case of a set containing a single vector v. If $v = 0$, then $\{v\}$ is linearly dependent, since for any nonzero scalar c_1,

$$c_1 0 = 0.$$

On the other hand, if $v \neq 0$, then the only value of the scalar c_1 for which

$$c_1 v = 0$$

is $c_1 = 0$. Consequently, $\{v\}$ is linearly independent. We can therefore state the next theorem.

Theorem 4.5.5   A set consisting of a single vector v in a vector space V is linearly dependent if and only if $v = 0$. Therefore, any set consisting of a single *nonzero* vector is linearly independent.

We next establish that linear dependence of a set containing at least two vectors is equivalent to the property that we are interested in—namely, that at least one vector in the set can be expressed as a linear combination of the remaining vectors in the set.

Theorem 4.5.6 Let $\{v_1, v_2, \ldots, v_k\}$ be a set of at least two vectors in a vector space V. Then $\{v_1, v_2, \ldots, v_k\}$ is linearly dependent if and only if at least one of the vectors in the set can be expressed as a linear combination of the others.

Proof If $\{v_1, v_2, \ldots, v_k\}$ is linearly dependent, then according to Definition 4.5.3, there exist scalars $c_1, c_2, \ldots, c_k$, not all zero, such that

$$c_1 v_1 + c_2 v_2 + \cdots + c_k v_k = 0.$$

Suppose that $c_i \neq 0$. Then we can express v_i as a linear combination of the other vectors as follows:

$$v_i = -\frac{1}{c_i}(c_1 v_1 + c_2 v_2 + \cdots + c_{i-1} v_{i-1} + c_{i+1} v_{i+1} + \cdots + c_k v_k).$$

Conversely, suppose that one of the vectors, say, v_j, can be expressed as a linear combination of the remaining vectors. That is,

$$v_j = c_1 v_1 + c_2 v_2 + \cdots + c_{j-1} v_{j-1} + c_{j+1} v_{j+1} + \cdots + c_k v_k.$$

Adding $(-1)v_j$ to both sides of this equation yields

$$c_1 v_1 + c_2 v_2 + \cdots + c_{j-1} v_{j-1} - v_j + c_{j+1} v_{j+1} + \cdots + c_k v_k = 0.$$

Since the coefficient of v_j is $-1 \neq 0$, the set of vectors $\{v_1, v_2, \ldots, v_k\}$ is linearly dependent. ∎

As far as the minimal-spanning-set idea is concerned, Theorems 4.5.6 and 4.5.2 tell us that a linearly dependent spanning set for a (nontrivial) vector space V cannot be a minimal spanning set. On the other hand, we will see in the next section that a linearly independent spanning set for V *must* be a minimal spanning set for V. For the remainder of this section, however, we focus more on the mechanics of determining whether a given set of vectors is linearly independent or linearly dependent. Sometimes this can be done by inspection. For example, Figure 4.5.2 illustrates that any set of three vectors in $\mathbb{R}^2$ is linearly dependent.

As another example, let V be the vector space of all functions defined on an interval I. If

$$f_1(x) = 1, \qquad f_2(x) = 2 \sin^2 x, \qquad f_3(x) = -5 \cos^2 x,$$

then $\{f_1, f_2, f_3\}$ is linearly dependent in V, since the identity $\sin^2 x + \cos^2 x = 1$ implies that for all $x \in I$,

$$f_1(x) = \tfrac{1}{2} f_2(x) - \tfrac{1}{5} f_3(x).$$

We can therefore conclude from Theorem 4.5.2 that

$$\text{span}\{1, 2 \sin^2 x, -5 \cos^2 x\} = \text{span}\{2 \sin^2 x, -5 \cos^2 x\}.$$

In relatively simple examples, the following general results can be applied. They are a direct consequence of the definition of linearly dependent vectors and are left for the exercises (Problem 49).

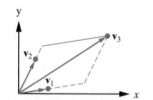

Figure 4.5.2: The set of vectors $\{v_1, v_2, v_3\}$ is linearly dependent in $\mathbb{R}^2$, since v_3 is a linear combination of v_1 and v_2.

Proposition 4.5.7 Let V be a vector space.

1. Any set of *two* vectors in V is linearly dependent if and only if the vectors are proportional.

2. Any set of vectors in V containing the zero vector is linearly dependent.

Remark   We emphasize that the first result in Proposition 4.5.7 holds only for the case of two vectors. It cannot be applied to sets containing more than two vectors.

Example 4.5.8   If $v_1 = (1, 2, -9)$ and $v_2 = (-2, -4, 18)$, then $\{v_1, v_2\}$ is linearly dependent in $\mathbb{R}^3$, since $v_2 = -2v_1$. Geometrically, v_1 and v_2 lie on the same line. $\square$

Example 4.5.9   If

$$A_1 = \begin{bmatrix} 2 & 1 \\ 3 & 4 \end{bmatrix}, \qquad A_2 = \begin{bmatrix} 0 & 0 \\ 0 & 0 \end{bmatrix}, \qquad A_3 = \begin{bmatrix} 2 & 5 \\ -3 & 2 \end{bmatrix},$$

then $\{A_1, A_2, A_3\}$ is linearly dependent in $M_2(\mathbb{R})$, since it contains the zero vector from $M_2(\mathbb{R})$. $\square$

For more complicated situations, we must resort to Definitions 4.5.3 and 4.5.4, although conceptually it is always helpful to keep in mind that the essence of the problem we are solving is to determine whether a vector in a given set can be expressed as a linear combination of the remaining vectors in the set. We now give some examples to illustrate the use of Definitions 4.5.3 and 4.5.4.

Example 4.5.10   If $v_1 = (1, 2, -1)$ $v_2 = (2, -1, 1)$, and $v_3 = (8, 1, 1)$, show that $\{v_1, v_2, v_3\}$ is linearly dependent in $\mathbb{R}^3$, and determine the linear dependency relationship.

Solution: We must first establish that there are values of the scalars c_1, c_2, c_3, not all zero, such that

$$c_1 v_1 + c_2 v_2 + c_3 v_3 = 0. \tag{4.5.1}$$

Substituting for the given vectors yields

$$c_1(1, 2, -1) + c_2(2, -1, 1) + c_3(8, 1, 1) = (0, 0, 0).$$

That is,

$$(c_1 + 2c_2 + 8c_3, 2c_1 - c_2 + c_3, -c_1 + c_2 + c_3) = (0, 0, 0).$$

Equating corresponding components on either side of this equation yields

$$\begin{aligned} c_1 + 2c_2 + 8c_3 &= 0, \\ 2c_1 - c_2 + c_3 &= 0, \\ -c_1 + c_2 + c_3 &= 0. \end{aligned}$$

The reduced row-echelon form of the augmented matrix of this system is

$$\begin{bmatrix} 1 & 0 & 2 & 0 \\ 0 & 1 & 3 & 0 \\ 0 & 0 & 0 & 0 \end{bmatrix}.$$

Consequently, the system has an infinite number of solutions for c_1, c_2, c_3, so the vectors are linearly dependent.

In order to determine a specific linear dependency relationship, we proceed to find c_1, c_2, and c_3. Setting $c_3 = t$, we have $c_2 = -3t$ and $c_1 = -2t$. Taking $t = 1$ and substituting these values for c_1, c_2, c_3 into (4.5.1), we obtain the linear dependency relationship

$$-2\mathbf{v}_1 - 3\mathbf{v}_2 + \mathbf{v}_3 = \mathbf{0},$$

or equivalently,

$$\mathbf{v}_1 = -\tfrac{3}{2}\mathbf{v}_2 + \tfrac{1}{2}\mathbf{v}_3,$$

which can be easily verified using the given expressions for $\mathbf{v}_1$, $\mathbf{v}_2$, and $\mathbf{v}_3$. It follows from Theorem 4.5.2 that

$$\text{span}\{\mathbf{v}_1, \mathbf{v}_2, \mathbf{v}_3\} = \text{span}\{\mathbf{v}_2, \mathbf{v}_3\}.$$

Geometrically, we can conclude that $\mathbf{v}_1$ lies in the plane determined by the vectors $\mathbf{v}_2$ and $\mathbf{v}_3$.   □

Example 4.5.11   Determine whether the following matrices are linearly dependent or linearly independent in $M_2(\mathbb{R})$:

$$A_1 = \begin{bmatrix} 1 & -1 \\ 2 & 0 \end{bmatrix}, \qquad A_2 = \begin{bmatrix} 2 & 1 \\ 0 & 3 \end{bmatrix}, \qquad A_3 = \begin{bmatrix} 1 & -1 \\ 2 & 1 \end{bmatrix}.$$

Solution: The condition for determining whether these vectors are linearly dependent or linearly independent,

$$c_1 A_1 + c_2 A_2 + c_3 A_3 = 0_2,$$

is equivalent in this case to

$$c_1 \begin{bmatrix} 1 & -1 \\ 2 & 0 \end{bmatrix} + c_2 \begin{bmatrix} 2 & 1 \\ 0 & 3 \end{bmatrix} + c_3 \begin{bmatrix} 1 & -1 \\ 2 & 1 \end{bmatrix} = \begin{bmatrix} 0 & 0 \\ 0 & 0 \end{bmatrix},$$

which is satisfied if and only if

$$\begin{aligned} c_1 + 2c_2 + c_3 &= 0, \\ -c_1 + c_2 - c_3 &= 0, \\ 2c_1 \phantom{{}+ 2c_2} + 2c_3 &= 0, \\ 3c_2 + c_3 &= 0. \end{aligned}$$

The reduced row-echelon form of the augmented matrix of this homogeneous system is

$$\begin{bmatrix} 1 & 0 & 0 & 0 \\ 0 & 1 & 0 & 0 \\ 0 & 0 & 1 & 0 \\ 0 & 0 & 0 & 0 \end{bmatrix},$$

which implies that the system has only the trivial solution $c_1 = c_2 = c_3 = 0$. It follows from Definition 4.5.4 that $\{A_1, A_2, A_3\}$ is linearly independent.   □

As a corollary to Theorem 4.5.2, we establish the following result.

Corollary 4.5.12   Any nontrivial, finite set of linearly dependent vectors in a vector space V contains a linearly independent subset that has the same linear span as the given set of vectors.

Proof Since the given set is linearly dependent, at least one of the vectors in the set is a linear combination of the remaining vectors, by Theorem 4.5.6. Thus, by Theorem 4.5.2, we can delete that vector from the set, and the resulting set of vectors will span the same subspace of V as the original set. If the resulting set is linearly independent, then we are done. If not, then we can repeat the procedure to eliminate another vector in the set. Continuing in this manner (with a finite number of iterations), we will obtain a linearly independent set that spans the same subspace of V as the subspace spanned by the original set of vectors.   ∎

Remark   Corollary 4.5.12 is actually true even if the set of vectors in question is infinite, but we shall not need to consider that case in this text. In the case of an infinite set of vectors, other techniques are required for the proof.

Note that the linearly independent set obtained using the procedure given in the previous theorem is not unique, and therefore the question arises whether the number of vectors in any resulting linearly independent set is independent of the manner in which the procedure is applied. We will give an affirmative answer to this question in Section 4.6.

Example 4.5.13   Let $v_1 = (1, 2, 3)$, $v_2 = (-1, 1, 4)$, $v_3 = (3, 3, 2)$, and $v_4 = (-2, -4, -6)$. Determine a linearly independent set of vectors that spans the same subspace of $\mathbb{R}^3$ as span$\{v_1, v_2, v_3, v_4\}$.

Solution: Setting

$$c_1 v_1 + c_2 v_2 + c_3 v_3 + c_4 v_4 = 0$$

requires that

$$c_1(1, 2, 3) + c_2(-1, 1, 4) + c_3(3, 3, 2) + c_4(-2, -4, -6) = (0, 0, 0),$$

leading to the linear system

$$\begin{aligned} c_1 - c_2 + 3c_3 - 2c_4 &= 0, \\ 2c_1 + c_2 + 3c_3 - 4c_4 &= 0, \\ 3c_1 + 4c_2 + 2c_3 - 6c_4 &= 0. \end{aligned}$$

The augmented matrix of this system is

$$\begin{bmatrix} 1 & -1 & 3 & -2 & 0 \\ 2 & 1 & 3 & -4 & 0 \\ 3 & 4 & 2 & -6 & 0 \end{bmatrix}$$

and the reduced row-echelon form of the augmented matrix of this system is

$$\begin{bmatrix} 1 & 0 & 2 & -2 & 0 \\ 0 & 1 & -1 & 0 & 0 \\ 0 & 0 & 0 & 0 & 0 \end{bmatrix}.$$

The system has two free variables, $c_3 = s$ and $c_4 = t$, and so $\{v_1, v_2, v_3, v_4\}$ is linearly dependent. Then $c_2 = s$ and $c_1 = 2t - 2s$. So the general form of the solution is

$$(2t - 2s, s, s, t) = s(-2, 1, 1, 0) + t(2, 0, 0, 1).$$

Setting $s = 1$ and $t = 0$ yields the linear combination

$$-2v_1 + v_2 + v_3 = 0, \tag{4.5.2}$$

and setting $s = 0$ and $t = 1$ yields the linear combination

$$2\mathbf{v}_1 + \mathbf{v}_4 = \mathbf{0}. \tag{4.5.3}$$

We can solve (4.5.2) for $\mathbf{v}_3$ in terms of $\mathbf{v}_1$ and $\mathbf{v}_2$, and we can solve (4.5.3) for $\mathbf{v}_4$ in terms of $\mathbf{v}_1$. Hence, according to Theorem 4.5.2, we have

$$\text{span}\{\mathbf{v}_1, \mathbf{v}_2, \mathbf{v}_3, \mathbf{v}_4\} = \text{span}\{\mathbf{v}_1, \mathbf{v}_2\}.$$

By Proposition 4.5.7, $\mathbf{v}_1$ and $\mathbf{v}_2$ are linearly independent, so $\{\mathbf{v}_1, \mathbf{v}_2\}$ is the linearly independent set we are seeking. Geometrically, the subspace of $\mathbb{R}^3$ spanned by $\mathbf{v}_1$ and $\mathbf{v}_2$ is a plane, and the vectors $\mathbf{v}_3$ and $\mathbf{v}_4$ lie in this plane.   ☐

Linear Dependence and Linear Independence in $\mathbb{R}^n$

Let $\{\mathbf{v}_1, \mathbf{v}_2, \ldots, \mathbf{v}_k\}$ be a set of vectors in $\mathbb{R}^n$, and let A denote the matrix that has $\mathbf{v}_1, \mathbf{v}_2, \ldots, \mathbf{v}_k$ as *column* vectors. Thus,

$$A = [\mathbf{v}_1, \mathbf{v}_2, \ldots, \mathbf{v}_k]. \tag{4.5.4}$$

Since each of the given vectors is in $\mathbb{R}^n$, it follows that A has n rows and is therefore an $n \times k$ matrix.

The linear combination $c_1\mathbf{v}_1 + c_2\mathbf{v}_2 + \cdots + c_k\mathbf{v}_k = \mathbf{0}$ can be written in matrix form as (see Theorem 2.2.9)

$$A\mathbf{c} = \mathbf{0}, \tag{4.5.5}$$

where A is given in Equation (4.5.4) and $\mathbf{c} = [c_1 \, c_2 \, \ldots \, c_k]^T$. Consequently, we can state the following theorem and corollary:

Theorem 4.5.14   Let $\mathbf{v}_1, \mathbf{v}_2, \ldots, \mathbf{v}_k$ be vectors in $\mathbb{R}^n$ and $A = [\mathbf{v}_1, \mathbf{v}_2, \ldots, \mathbf{v}_k]$. Then $\{\mathbf{v}_1, \mathbf{v}_2, \ldots, \mathbf{v}_k\}$ is linearly dependent if and only if the linear system $A\mathbf{c} = \mathbf{0}$ has a nontrivial solution.

Corollary 4.5.15   Let $\mathbf{v}_1, \mathbf{v}_2, \ldots, \mathbf{v}_k$ be vectors in $\mathbb{R}^n$ and $A = [\mathbf{v}_1, \mathbf{v}_2, \ldots, \mathbf{v}_k]$.

1. If $k > n$, then $\{\mathbf{v}_1, \mathbf{v}_2, \ldots, \mathbf{v}_k\}$ is linearly dependent.

2. If $k = n$, then $\{\mathbf{v}_1, \mathbf{v}_2, \ldots, \mathbf{v}_k\}$ is linearly dependent if and only if $\det(A) = 0$.

Proof If $k > n$, the system (4.5.5) has an infinite number of solutions (see Corollary 2.5.11), hence the vectors are linearly dependent by Theorem 4.5.14.

On the other hand, if $k = n$, the system (4.5.5) is $n \times n$, and hence, from Corollary 3.2.5, it has an infinite number of solutions if and only if $\det(A) = 0$.   ∎

Example 4.5.16   Determine whether the given vectors are linearly dependent or linearly independent in $\mathbb{R}^4$.

1. $\mathbf{v}_1 = (1, 3, -1, 0)$, $\mathbf{v}_2 = (2, 9, -1, 3)$, $\mathbf{v}_3 = (4, 5, 6, 11)$, $\mathbf{v}_4 = (1, -1, 2, 5)$, $\mathbf{v}_5 = (3, -2, 6, 7)$.

2. $\mathbf{v}_1 = (1, 4, 1, 7)$, $\mathbf{v}_2 = (3, -5, 2, 3)$, $\mathbf{v}_3 = (2, -1, 6, 9)$, $\mathbf{v}_4 = (-2, 3, 1, 6)$.

Solution:

1. Since we have five vectors in $\mathbb{R}^4$, Corollary 4.5.15 implies that $\{\mathbf{v}_1, \mathbf{v}_2, \mathbf{v}_3, \mathbf{v}_4, \mathbf{v}_5\}$ is necessarily linearly dependent.

2. In this case, we have four vectors in $\mathbb{R}^4$, and therefore, we can use the determinant:

$$\det(A) = \det[\mathbf{v}_1, \mathbf{v}_2, \mathbf{v}_3, \mathbf{v}_4] = \begin{vmatrix} 1 & 3 & 2 & -2 \\ 4 & -5 & -1 & 3 \\ 1 & 2 & 6 & 1 \\ 7 & 3 & 9 & 6 \end{vmatrix} = -462.$$

Since the determinant is nonzero, it follows from Corollary 4.5.15 that the given set of vectors is linearly independent. $\square$

Linear Independence of Functions

We now consider the general problem of determining whether or not a given set of functions is linearly independent or linearly dependent. We begin by specializing the general Definition 4.5.4 to the case of a set of functions defined on an interval I.

DEFINITION   4.5.17

The set of functions $\{f_1, f_2, \ldots, f_k\}$ is **linearly independent on an interval** I if and only if the only values of the scalars $c_1, c_2, \ldots, c_k$ such that

$$c_1 f_1(x) + c_2 f_2(x) + \cdots + c_k f_k(x) = 0, \qquad \text{for all } x \in I, \qquad (4.5.6)$$

are $c_1 = c_2 = \cdots = c_k = 0$.

The main point to notice is that the condition (4.5.6) must hold for all x in I.

A key tool in deciding whether or not a collection of functions is linearly independent on an interval I is the Wronskian. As we will see in Chapter 6, we can draw particularly sharp conclusions from the Wronskian about the linear dependence or independence of a family of *solutions to a linear homogeneous differential equation*.

DEFINITION   4.5.18

Let $f_1, f_2, \ldots, f_k$ be functions in $C^{k-1}(I)$. The **Wronskian** of these functions is the order k determinant defined by

$$W[f_1, f_2, \ldots, f_k](x) = \begin{vmatrix} f_1(x) & f_2(x) & \cdots & f_k(x) \\ f_1'(x) & f_2'(x) & \cdots & f_k'(x) \\ \vdots & \vdots & & \vdots \\ f_1^{(k-1)}(x) & f_2^{(k-1)}(x) & \cdots & f_k^{(k-1)}(x) \end{vmatrix}.$$

Remark   Notice that the Wronskian is a function defined on I. Also note that this function depends on the order of the functions in the Wronskian. For example, using properties of determinants,

$$W[f_2, f_1, \ldots, f_k](x) = -W[f_1, f_2, \ldots, f_k](x).$$

Example 4.5.19   If $f_1(x) = \sin x$ and $f_2(x) = \cos x$ on $(-\infty, \infty)$, then

$$W[f_1, f_2](x) = \begin{vmatrix} \sin x & \cos x \\ \cos x & -\sin x \end{vmatrix} = (\sin x)(-\sin x) - (\cos x)(\cos x)$$

$$= -(\sin^2 x + \cos^2 x) = -1. \qquad \square$$

Example 4.5.20   If $f_1(x) = x$, $f_2(x) = x^2$, and $f_3(x) = x^3$ on $(-\infty, \infty)$, then

$$W[f_1, f_2, f_3](x) = \begin{vmatrix} x & x^2 & x^3 \\ 1 & 2x & 3x^2 \\ 0 & 2 & 6x \end{vmatrix} = x(12x^2 - 6x^2) - (6x^3 - 2x^3) = 2x^3. \qquad \square$$

We can now state and prove the main result about the Wronskian.

Theorem 4.5.21   Let $f_1, f_2, \ldots, f_k$ be functions in $C^{k-1}(I)$. If $W[f_1, f_2, \ldots, f_k]$ is nonzero at some point x_0 in I, then $\{f_1, f_2, \ldots, f_k\}$ is linearly independent on I.

Proof   To apply Definition 4.5.17, assume that

$$c_1 f_1(x) + c_2 f_2(x) + \cdots + c_k f_k(x) = 0,$$

for all x in I. Then, differentiating $k - 1$ times yields the linear system

$$\begin{aligned} c_1 f_1(x) &+ c_2 f_2(x) &+ \cdots + c_k f_k(x) &= 0, \\ c_1 f_1'(x) &+ c_2 f_2'(x) &+ \cdots + c_k f_k'(x) &= 0, \\ &&&\vdots \\ c_1 f_1^{(k-1)}(x) &+ c_2 f_2^{(k-1)}(x) &+ \cdots + c_k f_k^{(k-1)}(x) &= 0, \end{aligned}$$

where the unknowns in the system are $c_1, c_2, \ldots, c_k$. We wish to show that $c_1 = c_2 = \cdots = c_k = 0$. The determinant of the matrix of coefficients of this system is just $W[f_1, f_2, \ldots, f_k](x)$. Consequently, if $W[f_1, f_2, \ldots, f_k](x_0) \neq 0$ for some x_0 in I, then the determinant of the matrix of coefficients of the system is nonzero at that point, and therefore the only solution to the system is the trivial solution $c_1 = c_2 = \cdots = c_k = 0$. That is, the given set of functions is linearly independent on I.   ■

Remarks

1. Notice that it is only necessary for $W[f_1, f_2, \ldots, f_k](x)$ to be nonzero at one point in I for $\{f_1, f_2, \ldots, f_k\}$ to be linearly independent on I.

2. Theorem 4.5.21 *does not say* that if $W[f_1, f_2, \ldots, f_k](x) = 0$ for every x in I, then $\{f_1, f_2, \ldots, f_k\}$ is linearly dependent on I. As we will see in the next example below, the Wronskian of a linearly independent set of functions on an interval I can be identically zero on I. Instead, the logical equivalent of the preceding theorem is: If $\{f_1, f_2, \ldots, f_k\}$ is linearly dependent on I, then $W[f_1, f_2, \ldots, f_k](x) = 0$ at every point of I.

> If $W[f_1, f_2, \ldots, f_k](x) = 0$ for all x in I, Theorem 4.5.21 gives no information as to the linear dependence or independence of $\{f_1, f_2, \ldots, f_k\}$ on I.

Example 4.5.22   Determine whether the following functions are linearly dependent or linearly independent on $I = (-\infty, \infty)$.

(a) $f_1(x) = e^x$, $f_2(x) = x^2 e^x$.

(b) $f_1(x) = x$, $f_2(x) = x + x^2$, $f_3(x) = 2x - x^2$.

(c) $f_1(x) = x^2$, $f_2(x) = \begin{cases} 2x^2, & \text{if } x \geq 0, \\ -x^2, & \text{if } x < 0. \end{cases}$

Solution:

(a)
$$W[f_1, f_2](x) = \begin{vmatrix} e^x & x^2 e^x \\ e^x & e^x(x^2 + 2x) \end{vmatrix} = e^{2x}(x^2 + 2x) - x^2 e^{2x} = 2x e^{2x}.$$

Since $W[f_1, f_2](x) \neq 0$ (except at $x = 0$), the functions are linearly independent on $(-\infty, \infty)$.

(b)
$$W[f_1, f_2, f_3](x) = \begin{vmatrix} x & x + x^2 & 2x - x^2 \\ 1 & 1 + 2x & 2 - 2x \\ 0 & 2 & -2 \end{vmatrix}$$
$$= x\left[(-2)(1 + 2x) - 2(2 - 2x)\right]$$
$$- \left[(-2)(x + x^2) - 2(2x - x^2)\right] = 0.$$

Thus, no conclusion can be drawn from Theorem 4.5.21. However, a closer inspection of the functions reveals, for example, that

$$f_2 = 3f_1 - f_3.$$

Consequently, the functions are linearly dependent on $(-\infty, \infty)$.

(c) If $x \geq 0$, then
$$W[f_1, f_2](x) = \begin{vmatrix} x^2 & 2x^2 \\ 2x & 4x \end{vmatrix} = 0,$$

whereas if $x < 0$, then

$$W[f_1, f_2](x) = \begin{vmatrix} x^2 & -x^2 \\ 2x & -2x \end{vmatrix} = 0.$$

Thus, $W[f_1, f_2](x) = 0$ for all x in $(-\infty, \infty)$, so no conclusion can be drawn from Theorem 4.5.21. Again we take a closer look at the given functions. They are sketched in Figure 4.5.3. In this case, we see that on the interval $(-\infty, 0)$, the functions are linearly dependent, since

$$f_1 + f_2 = 0.$$

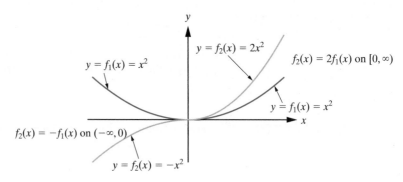

Figure 4.5.3: Two functions that are linearly independent on $(-\infty, \infty)$, but whose Wronskian is identically zero on that interval.

They are also linearly dependent on $[0, \infty)$, since on this interval we have

$$2f_1 - f_2 = 0.$$

The key point is to realize that there is no set of *nonzero* constants c_1, c_2 for which

$$c_1 f_1 + c_2 f_2 = 0$$

holds *for all* x in $(-\infty, \infty)$. Hence, the given functions are linearly independent on $(-\infty, \infty)$. This illustrates our second remark following Theorem 4.5.21, and it emphasizes the importance of the role played by the interval I when discussing linear dependence and linear independence of functions. A collection of functions may be linearly independent on an interval I_1, but linearly dependent on another interval I_2.   □

It might appear at this stage that the usefulness of the Wronskian is questionable, since if $W[f_1, f_2, \ldots, f_k]$ vanishes on an interval I, then no conclusion can be drawn as to the linear dependence or linear independence of the functions $f_1, f_2, \ldots, f_k$ on I. However, the real power of the Wronskian is in its application to solutions of linear differential equations of the form

$$y^{(n)} + a_1(x)y^{(n-1)} + \cdots + a_{n-1}(x)y' + a_n(x)y = 0. \tag{4.5.7}$$

In Chapter 6, we will establish that if we have n functions that are *solutions of an equation of the form* (4.5.7) on an interval I, then if the Wronskian of these functions is identically zero on I, the functions are indeed linearly dependent on I. Thus, the Wronskian does completely characterize the linear dependence or linear independence of solutions of such equations. This is a fundamental result in the theory of linear differential equations.

Exercises for 4.5

Key Terms

Linearly dependent set, Linear dependency, Linearly independent set, Minimal spanning set, Wronskian of a set of functions.

Skills

- Be able to determine whether a given finite set of vectors is linearly dependent or linearly independent. For sets of one or two vectors, you should be able to do this *at a glance*. If the set is linearly dependent, be able to determine a linear dependency relationship among the vectors.

- Be able to take a linearly dependent set of vectors and remove vectors until it becomes a linearly independent set of vectors with the same span as the original set.

- Be able to produce a linearly independent set of vectors that spans a given subspace of a vector space V.

- Be able to conclude immediately that a set of k vectors in $\mathbb{R}^n$ is linearly dependent if $k > n$, and know what can be said in the case where $k = n$ as well.

- Know what information the Wronskian does (and does not) give about the linear dependence or linear independence of a set of functions on an interval I.

True-False Review

For Questions 1–9, decide if the given statement is **true** or **false**, and give a brief justification for your answer. If true, you can quote a relevant definition or theorem from the text. If false, provide an example, illustration, or brief explanation of why the statement is false.

1. Every vector space V possesses a unique minimal spanning set.

2. The set of column vectors of a 5×7 matrix A must be linearly dependent.

3. The set of column vectors of a 7×5 matrix A must be linearly independent.

4. Any nonempty subset of a linearly independent set of vectors is linearly independent.

5. If the Wronskian of a set of functions is nonzero at some point x_0 in an interval I, then the set of functions is linearly independent.

6. If it is possible to express one of the vectors in a set S as a linear combination of the others, then S is a linearly dependent set.

7. If a set of vectors S in a vector space V contains a linearly dependent subset, then S is itself a linearly dependent set.

8. A set of three vectors in a vector space V is linearly dependent if and only if all three vectors are proportional to one another.

9. If the Wronskian of a set of functions is identically zero at every point of an interval I, then the set of functions is linearly dependent.

Problems

For Problems 1–9, determine whether the given set of vectors is linearly independent or linearly dependent in $\mathbb{R}^n$. In the case of linear dependence, find a dependency relationship.

1. $\{(1, -1), (1, 1)\}$.

2. $\{(2, -1), (3, 2), (0, 1)\}$.

3. $\{(1, -1, 0), (0, 1, -1), (1, 1, 1)\}$.

4. $\{(1, 2, 3), (1, -1, 2), (1, -4, 1)\}$.

5. $\{(-2, 4, -6), (3, -6, 9)\}$.

6. $\{(1, -1, 2), (2, 1, 0)\}$.

7. $\{(-1, 1, 2), (0, 2, -1), (3, 1, 2), (-1, -1, 1)\}$.

8. $\{(1, -1, 2, 3), (2, -1, 1, -1), (-1, 1, 1, 1)\}$.

9. $\{(2, -1, 0, 1), (1, 0, -1, 2), (0, 3, 1, 2),$
   $(-1, 1, 2, 1)\}$.

10. Let $\mathbf{v}_1 = (1, 2, 3)$, $\mathbf{v}_2 = (4, 5, 6)$, $\mathbf{v}_3 = (7, 8, 9)$. Determine whether $\{\mathbf{v}_1, \mathbf{v}_2, \mathbf{v}_3\}$ is linearly independent in $\mathbb{R}^3$. Describe
$$\text{span}\{\mathbf{v}_1, \mathbf{v}_2, \mathbf{v}_3\}$$
geometrically.

11. Consider the vectors $\mathbf{v}_1 = (2, -1, 5)$,
$\mathbf{v}_2 = (1, 3, -4)$, $\mathbf{v}_3 = (-3, -9, 12)$ in $\mathbb{R}^3$.

   (a) Show that $\{\mathbf{v}_1, \mathbf{v}_2, \mathbf{v}_3\}$ is linearly dependent.

   (b) Is $\mathbf{v}_1 \in \text{span}\{\mathbf{v}_2, \mathbf{v}_3\}$? Draw a picture illustrating your answer.

12. Determine all values of the constant k for which the vectors $(1, 1, k)$, $(0, 2, k)$, and $(1, k, 6)$ are linearly dependent in $\mathbb{R}^3$.

For Problems 13–14, determine all values of the constant k for which the given set of vectors is linearly independent in $\mathbb{R}^4$.

13. $\{(1, 0, 1, k), (-1, 0, k, 1), (2, 0, 1, 3)\}$.

14. $\{(1, 1, 0, -1), (1, k, 1, 1), (2, 1, k, 1), (-1, 1, 1, k)\}$.

For Problems 15–17, determine whether the given set of vectors is linearly independent in $M_2(\mathbb{R})$.

15. $A_1 = \begin{bmatrix} 1 & 1 \\ 0 & 1 \end{bmatrix}$, $A_2 = \begin{bmatrix} 2 & -1 \\ 0 & 1 \end{bmatrix}$, $A_3 = \begin{bmatrix} 3 & 6 \\ 0 & 4 \end{bmatrix}$.

16. $A_1 = \begin{bmatrix} 2 & -1 \\ 3 & 4 \end{bmatrix}$, $A_2 = \begin{bmatrix} -1 & 2 \\ 1 & 3 \end{bmatrix}$.

17. $A_1 = \begin{bmatrix} 1 & 0 \\ 1 & 2 \end{bmatrix}$, $A_2 = \begin{bmatrix} -1 & 1 \\ 2 & 1 \end{bmatrix}$, $A_3 = \begin{bmatrix} 2 & 1 \\ 5 & 7 \end{bmatrix}$.

For Problems 18–19, determine whether the given set of vectors is linearly independent in P_1.

18. $p_1(x) = 1 - x$, $\quad p_2(x) = 1 + x$.

19. $p_1(x) = 2 + 3x$, $\quad p_2(x) = 4 + 6x$.

20. Show that the vectors

$$p_1(x) = a + bx \quad \text{and} \quad p_2(x) = c + dx$$

are linearly independent in P_1 if and only if the constants a, b, c, d satisfy $ad - bc \neq 0$.

21. If $f_1(x) = \cos 2x$, $f_2(x) = \sin^2 x$, $f_3(x) = \cos^2 x$, determine whether $\{f_1, f_2, f_3\}$ is linearly dependent or linearly independent in $C^\infty(-\infty, \infty)$.

For Problems 22–28, determine a linearly independent set of vectors that spans the same subspace of V as that spanned by the original set of vectors.

22. $V = \mathbb{R}^3$, $\{(1, 2, 3), (-3, 4, 5), (1, -\frac{4}{3}, -\frac{5}{3})\}$.

23. $V = \mathbb{R}^3$, $\{(3, 1, 5), (0, 0, 0), (1, 2, -1), (-1, 2, 3)\}$.

24. $V = \mathbb{R}^3$, $\{(1, 1, 1), (1, -1, 1), (1, -3, 1), (3, 1, 2)\}$.

25. $V = \mathbb{R}^4$,
$\{(1, 1, -1, 1), (2, -1, 3, 1), (1, 1, 2, 1), (2, -1, 2, 1)\}$.

26. $V = M_2(\mathbb{R})$,

$$\left\{ \begin{bmatrix} 1 & 2 \\ 3 & 4 \end{bmatrix}, \begin{bmatrix} -1 & 2 \\ 5 & 7 \end{bmatrix}, \begin{bmatrix} 3 & 2 \\ 1 & 1 \end{bmatrix} \right\}.$$

27. $V = P_1$, $\{2 - 5x, 3 + 7x, 4 - x\}$.

28. $V = P_2$, $\{2 + x^2, 4 - 2x + 3x^2, 1 + x\}$.

For Problems 29–33, use the Wronskian to show that the given functions are linearly independent on the given interval I.

29. $f_1(x) = 1$, $f_2(x) = x$, $f_3(x) = x^2$, $I = (-\infty, \infty)$.

30. $f_1(x) = \sin x$, $f_2(x) = \cos x$, $f_3(x) = \tan x$, $I = (-\pi/2, \pi/2)$.

31. $f_1(x) = 1$, $f_2(x) = 3x$, $f_3(x) = x^2 - 1$, $I = (-\infty, \infty)$.

32. $f_1(x) = e^{2x}$, $f_2(x) = e^{3x}$, $f_3(x) = e^{-x}$, $I = (-\infty, \infty)$.

33.

$$f_1(x) = \begin{cases} x^2, & \text{if } x \geq 0, \\ 3x^3, & \text{if } x < 0, \end{cases}$$

$$f_2(x) = 7x^2, \ I = (-\infty, \infty).$$

For Problems 34–36, show that the Wronskian of the given functions is identically zero on $(-\infty, \infty)$. Determine whether the functions are linearly independent or linearly dependent on that interval.

34. $f_1(x) = 1$, $f_2(x) = x$, $f_3(x) = 2x - 1$.

35. $f_1(x) = e^x$, $f_2(x) = e^{-x}$, $f_3(x) = \cosh x$.

36. $f_1(x) = 2x^3$,

$$f_2(x) = \begin{cases} 5x^3, & \text{if } x \geq 0, \\ -3x^3, & \text{if } x < 0. \end{cases}$$

37. Consider the functions $f_1(x) = x$,

$$f_2(x) = \begin{cases} x, & \text{if } x \geq 0, \\ -x, & \text{if } x < 0. \end{cases}$$

(a) Show that f_2 is not in $C^1(-\infty, \infty)$.

(b) Show that $\{f_1, f_2\}$ is linearly dependent on the intervals $(-\infty, 0)$ and $[0, \infty)$, while it is linearly independent on the interval $(-\infty, \infty)$. Justify your results by making a sketch showing both of the functions.

38. Determine whether the functions $f_1(x) = x$,

$$f_2(x) = \begin{cases} x, & \text{if } x \neq 0, \\ 1, & \text{if } x = 0. \end{cases}$$

are linearly dependent or linearly independent on $I = (-\infty, \infty)$.

39. Show that the functions

$$f_1(x) = \begin{cases} x - 1, & \text{if } x \geq 1, \\ 2(x - 1), & \text{if } x < 1, \end{cases}$$

$f_2(x) = 2x$, $f_3(x) = 3$ form a linearly independent set on $(-\infty, \infty)$. Determine all intervals on which $\{f_1, f_2, f_3\}$ is linearly dependent.

40. (a) Show that $\{1, x, x^2, x^3\}$ is linearly independent on every interval.

(b) If $f_k(x) = x^k$ for $k = 0, 1, \ldots, n$, show that $\{f_0, f_1, \ldots, f_n\}$ is linearly independent on every interval for all fixed n.

41. (a) Show that the functions

$$f_1(x) = e^{r_1 x}, \quad f_2(x) = e^{r_2 x}, \quad f_3(x) = e^{r_3 x}$$

have Wronskian

$$W[f_1, f_2, f_3](x) = e^{(r_1+r_2+r_3)x}\begin{vmatrix} 1 & 1 & 1 \\ r_1 & r_2 & r_3 \\ r_1^2 & r_2^2 & r_3^2 \end{vmatrix}$$

$$= e^{(r_1+r_2+r_3)x}(r_3 - r_1)(r_3 - r_2)(r_2 - r_1),$$

and hence determine the conditions on r_1, r_2, r_3 such that $\{f_1, f_2, f_3\}$ is linearly independent on every interval.

(b) More generally, show that the set of functions

$$\{e^{r_1 x}, e^{r_2 x}, \ldots, e^{r_n x}\}$$

is linearly independent on every interval if and only if all of the r_i are distinct. [**Hint:** Show that the Wronskian of the given functions is a multiple of the $n \times n$ Vandermonde determinant, and then use Problem 21 in Section 3.3.]

42. Let $\{v_1, v_2\}$ be a linearly independent set in a vector space V, and let $v = \alpha v_1 + v_2$, $w = v_1 + \alpha v_2$, where α is a constant. Use Definition 4.5.4 to determine all values of α for which $\{v, w\}$ is linearly independent.

43. If v_1 and v_2 are vectors in a vector space V, and u_1, u_2, u_3 are each linear combinations of them, prove that $\{u_1, u_2, u_3\}$ is linearly dependent.

44. Let $v_1, v_2, \ldots, v_m$ be a set of linearly independent vectors in a vector space V and suppose that the vectors $u_1, u_2, \ldots, u_n$ are each linear combinations of them. It follows that we can write

$$u_k = \sum_{i=1}^{m} a_{ik} v_i, \quad k = 1, 2, \ldots, n,$$

for appropriate constants a_{ik}.

(a) If $n > m$, prove that $\{u_1, u_2, \ldots, u_n\}$ is linearly dependent on V.

(b) If $n = m$, prove that $\{u_1, u_2, \ldots, u_n\}$ is linearly independent in V if and only if $\det[a_{ij}] \neq 0$.

(c) If $n < m$, prove that $\{u_1, u_2, \ldots, u_n\}$ is linearly independent in V if and only if $\text{rank}(A) = n$, where $A = [a_{ij}]$.

(d) Which result from this section do these results generalize?

45. Prove from the definition of "linearly independent" that if $\{v_1, v_2, \ldots, v_n\}$ is linearly independent and if A is an invertible $n \times n$ matrix, then the set $\{Av_1, Av_2, \ldots, Av_n\}$ is linearly independent.

46. Prove that if $\{v_1, v_2\}$ is linearly independent and v_3 is not in $\text{span}\{v_1, v_2\}$, then $\{v_1, v_2, v_3\}$ is linearly independent.

47. Generalizing the previous exercise, prove that if $\{v_1, v_2, \ldots, v_k\}$ is linearly independent and v_{k+1} is not in $\text{span}\{v_1, v_2, \ldots, v_k\}$, then $\{v_1, v_2, \ldots, v_{k+1}\}$ is linearly independent.

48. Prove Theorem 4.5.2.

49. Prove Proposition 4.5.7.

50. Prove that if $\{v_1, v_2, \ldots, v_k\}$ spans a vector space V, then for every vector v in V, $\{v, v_1, v_2, \ldots, v_k\}$ is linearly dependent.

51. Prove that if $V = P_n$ and $S = \{p_1, p_2, \ldots, p_k\}$ is a set of vectors in V each of a different degree, then S is linearly independent. [**Hint:** Assume without loss of generality that the polynomials are ordered in descending degree: $\deg(p_1) > \deg(p_2) > \cdots > \deg(p_k)$. Assuming that $c_1 p_1 + c_2 p_2 + \cdots + c_k p_k = 0$, first show that c_1 is zero by examining the highest degree. Then repeat for lower degrees to show successively that $c_2 = 0$, $c_3 = 0$, and so on.]

4.6 Bases and Dimension

The results of the previous section show that if a minimal spanning set exists in a (nontrivial) vector space V, it cannot be linearly dependent. Therefore if we are looking for minimal spanning sets for V, we should focus our attention on spanning sets that are linearly independent. One of the results of this section establishes that *every* spanning set for V that is linearly independent is indeed a minimal spanning set. Such a set will be called a *basis*. This is one of the most important concepts in this text and a cornerstone of linear algebra.

DEFINITION   4.6.1

A set of vectors $\{\mathbf{v}_1, \mathbf{v}_2, \ldots, \mathbf{v}_k\}$ in a vector space V is called a **basis**[4] for V if

(a) The vectors are linearly independent.

(b) The vectors span V.

Notice that if we have a finite spanning set for a vector space, then we can always, in principle, determine a basis for V by using the technique of Corollary 4.5.12. Furthermore, the computational aspects of determining a basis have been covered in the previous two sections, since all we are really doing is combining the two concepts of linear independence and linear span. Consequently, this section is somewhat more theoretically oriented than the preceding ones. The reader is encouraged not to gloss over the theoretical aspects, as these really are fundamental results in linear algebra.

There do exist vector spaces V for which it is impossible to find a finite set of linearly independent vectors that span V. The vector space $C^n(I)$, $n \geq 1$, is such an example (Example 4.6.19). Such vector spaces are called **infinite-dimensional vector spaces**. Our primary interest in this text, however, will be vector spaces that contain a *finite* spanning set of linearly independent vectors. These are known as **finite-dimensional vector spaces**, and we will encounter numerous examples of them throughout the remainder of this section.

We begin with the vector space $\mathbb{R}^n$. In $\mathbb{R}^2$, the most natural basis, denoted $\{\mathbf{e}_1, \mathbf{e}_2\}$, consists of the two vectors

$$\mathbf{e}_1 = (1, 0), \qquad \mathbf{e}_2 = (0, 1), \tag{4.6.1}$$

and in $\mathbb{R}^3$, the most natural basis, denoted $\{\mathbf{e}_1, \mathbf{e}_2, \mathbf{e}_3\}$, consists of the three vectors

$$\mathbf{e}_1 = (1, 0, 0), \qquad \mathbf{e}_2 = (0, 1, 0), \qquad \mathbf{e}_3 = (0, 0, 1). \tag{4.6.2}$$

The verification that the sets (4.6.1) and (4.6.2) are indeed bases of $\mathbb{R}^2$ and $\mathbb{R}^3$, respectively, is straightforward and left as an exercise.[5] These bases are referred to as the **standard basis** on $\mathbb{R}^2$ and $\mathbb{R}^3$, respectively. In the case of the standard basis for $\mathbb{R}^3$ given in (4.6.2), we recognize the vectors $\mathbf{e}_1, \mathbf{e}_2, \mathbf{e}_3$ as the familiar unit vectors $\mathbf{i}, \mathbf{j}, \mathbf{k}$ pointing along the positive x-, y-, and z-axes of the rectangular Cartesian coordinate system.

More generally, consider the set of vectors $\{\mathbf{e}_1, \mathbf{e}_2, \ldots, \mathbf{e}_n\}$ in $\mathbb{R}^n$ defined by

$$\mathbf{e}_1 = (1, 0, \ldots, 0), \qquad \mathbf{e}_2 = (0, 1, \ldots, 0), \qquad \ldots, \qquad \mathbf{e}_n = (0, 0, \ldots, 1).$$

These vectors are linearly independent by Corollary 4.5.15, since

$$\det([\mathbf{e}_1, \mathbf{e}_2, \ldots, \mathbf{e}_n]) = \det(I_n) = 1 \neq 0.$$

Furthermore, the vectors span $\mathbb{R}^n$, since an arbitrary vector $\mathbf{v} = (x_1, x_2, \ldots, x_n)$ in $\mathbb{R}^n$ can be written as

$$\begin{aligned} \mathbf{v} &= x_1(1, 0, \ldots, 0) + x_2(0, 1, \ldots, 0) + \cdots + x_n(0, 0, \ldots, 1) \\ &= x_1\mathbf{e}_1 + x_2\mathbf{e}_2 + \cdots + x_n\mathbf{e}_n. \end{aligned}$$

---

[4] The plural of basis is bases.

[5] Alternatively, the verification is a special case of that given shortly for the general case of $\mathbb{R}^n$.

Consequently, $\{e_1, e_2, \ldots, e_n\}$ is a basis for $\mathbb{R}^n$. We refer to this basis as the **standard basis** for $\mathbb{R}^n$.

The general vector in $\mathbb{R}^n$ has n components, and the standard basis vectors arise as the n vectors that are obtained by sequentially setting one component to the value 1 and the other components to 0. In general, this is how we obtain standard bases in vector spaces whose vectors are determined by the specification of n independent constants. We illustrate with some examples.

Example 4.6.2   Determine the standard basis for $M_2(\mathbb{R})$.

Solution: The general matrix in $M_2(\mathbb{R})$ is

$$\begin{bmatrix} a & b \\ c & d \end{bmatrix}.$$

Consequently, there are four independent parameters that give rise to four special vectors in $M_2(\mathbb{R})$. Sequentially setting one of these parameters to the value 1 and the others to 0 generates the following four matrices:

$$A_1 = \begin{bmatrix} 1 & 0 \\ 0 & 0 \end{bmatrix}, \quad A_2 = \begin{bmatrix} 0 & 1 \\ 0 & 0 \end{bmatrix}, \quad A_3 = \begin{bmatrix} 0 & 0 \\ 1 & 0 \end{bmatrix}, \quad A_4 = \begin{bmatrix} 0 & 0 \\ 0 & 1 \end{bmatrix}.$$

We see that $\{A_1, A_2, A_3, A_4\}$ is a spanning set for $M_2(\mathbb{R})$. Furthermore,

$$c_1 A_1 + c_2 A_2 + c_3 A_3 + c_4 A_4 = 0_2$$

holds if and only if

$$c_1 \begin{bmatrix} 1 & 0 \\ 0 & 0 \end{bmatrix} + c_2 \begin{bmatrix} 0 & 1 \\ 0 & 0 \end{bmatrix} + c_3 \begin{bmatrix} 0 & 0 \\ 1 & 0 \end{bmatrix} + c_4 \begin{bmatrix} 0 & 0 \\ 0 & 1 \end{bmatrix} = \begin{bmatrix} 0 & 0 \\ 0 & 0 \end{bmatrix}$$

—that is, if and only if $c_1 = c_2 = c_3 = c_4 = 0$. Consequently, $\{A_1, A_2, A_3, A_4\}$ is a linearly independent spanning set for $M_2(\mathbb{R})$, hence it is a basis. This is the standard basis for $M_2(\mathbb{R})$.   □

Remark   More generally, consider the vector space of all $m \times n$ matrices with real entries, $M_{m \times n}(\mathbb{R})$. If we let E_{ij} denote the $m \times n$ matrix with value 1 in the (i, j)-position and zeros elsewhere, then we can show routinely that

$$\{E_{ij} : 1 \le i \le m, \ 1 \le j \le n\}$$

is a basis for $M_{m \times n}(\mathbb{R})$, and it is the standard basis.

Example 4.6.3   Determine the standard basis for P_2.

Solution: We have

$$P_2 = \{a_0 + a_1 x + a_2 x^2 : a_0, a_1, a_2 \in \mathbb{R}\},$$

so that the vectors in P_2 are determined by specifying values for the three parameters a_0, a_1, and a_2. Sequentially setting one of these parameters to the value 1 and the other two to the value 0 yields the following vectors in P_2:

$$p_0(x) = 1, \qquad p_1(x) = x, \qquad p_2(x) = x^2.$$

We have shown in Example 4.4.6 that $\{p_0, p_1, p_2\}$ is a spanning set for P_2. Furthermore,

$$W[p_0, p_1, p_2](x) = \begin{vmatrix} 1 & x & x^2 \\ 0 & 1 & 2x \\ 0 & 0 & 2 \end{vmatrix} = 2 \neq 0,$$

which implies that $\{p_0, p_1, p_2\}$ is linearly independent on any interval.[6] Consequently, $\{p_0, p_1, p_2\}$ is a basis for P_2. This is the standard basis for P_2.   ☐

Remark   More generally, the reader can check that the standard basis for the vector space of all polynomials of degree n or less, P_n, is

$$\{1, x, x^2, \ldots, x^n\}.$$

Dimension of a Finite-Dimensional Vector Space

The reader has probably realized that there can be many different bases for a given vector space V. In addition to the standard basis $\{e_1, e_2, e_3\}$ on $\mathbb{R}^3$, for example, it can be checked[7] that $\{(1, 2, 3), (4, 5, 6), (7, 8, 8)\}$ and $\{(1, 0, 0), (1, 1, 0), (1, 1, 1)\}$ are also bases for $\mathbb{R}^3$. And there are countless others as well.

Despite the multitude of different bases available for a vector space V, they all share one common feature: the *number* of vectors in each basis for V is the same. This fact will be deduced as a corollary of our next theorem, a fundamental result in the theory of vector spaces.

Theorem 4.6.4   If a finite-dimensional vector space has a basis consisting of m vectors, then any set of more than m vectors is linearly dependent.

Proof Let $\{v_1, v_2, \ldots, v_m\}$ be a basis for V, and consider an arbitrary set of vectors in V, say, $\{u_1, u_2, \ldots, u_n\}$, with $n > m$. We wish to prove that $\{u_1, u_2, \ldots, u_n\}$ is necessarily linearly dependent. Since $\{v_1, v_2, \ldots, v_m\}$ is a basis for V, it follows that each u_j can be written as a linear combination of $v_1, v_2, \ldots, v_m$. Thus, there exist constants a_{ij} such that

$$\begin{aligned} u_1 &= a_{11}v_1 + a_{21}v_2 + \cdots + a_{m1}v_m, \\ u_2 &= a_{12}v_1 + a_{22}v_2 + \cdots + a_{m2}v_m, \\ &\vdots \\ u_n &= a_{1n}v_1 + a_{2n}v_2 + \cdots + a_{mn}v_m. \end{aligned}$$

To prove that $\{u_1, u_2, \ldots, u_n\}$ is linearly dependent, we must show that there exist scalars $c_1, c_2, \ldots, c_n$, not all zero, such that

$$c_1 u_1 + c_2 u_2 + \cdots + c_n u_n = 0. \tag{4.6.3}$$

Inserting the expressions for $u_1, u_2, \ldots, u_n$ into Equation (4.6.3) yields

$$c_1(a_{11}v_1 + a_{21}v_2 + \cdots + a_{m1}v_m) + c_2(a_{12}v_1 + a_{22}v_2 + \cdots + a_{m2}v_m)$$
$$+ \cdots + c_n(a_{1n}v_1 + a_{2n}v_2 + \cdots + a_{mn}v_m) = 0.$$

---

[6] Alternatively, we can start with the equation $c_0 p_0(x) + c_1 p_1(x) + c_2 p_2(x) = 0$ for all x in $\mathbb{R}$ and show readily that $c_0 = c_1 = c_2 = 0$.

[7] The reader desiring extra practice at the computational aspects of verifying a basis is encouraged to pause here to check these examples.

Rearranging terms, we have

$$(a_{11}c_1 + a_{12}c_2 + \cdots + a_{1n}c_n)\mathbf{v}_1 + (a_{21}c_1 + a_{22}c_2 + \cdots + a_{2n}c_n)\mathbf{v}_2$$
$$+ \cdots + (a_{m1}c_1 + a_{m2}c_2 + \cdots + a_{mn}c_n)\mathbf{v}_m = \mathbf{0}.$$

Since $\{\mathbf{v}_1, \mathbf{v}_2, \ldots, \mathbf{v}_m\}$ is linearly independent, we can conclude that

$$\begin{aligned}
a_{11}c_1 + a_{12}c_2 + \cdots + a_{1n}c_n &= 0, \\
a_{21}c_1 + a_{22}c_2 + \cdots + a_{2n}c_n &= 0, \\
&\;\;\vdots \\
a_{m1}c_1 + a_{m2}c_2 + \cdots + a_{mn}c_n &= 0.
\end{aligned}$$

This is an $m \times n$ homogeneous system of linear equations with $m < n$, and hence, from Corollary 2.5.11, it has nontrivial solutions for $c_1, c_2, \ldots, c_n$. It therefore follows from Equation (4.6.3) that $\{\mathbf{u}_1, \mathbf{u}_2, \ldots, \mathbf{u}_n\}$ is linearly dependent. ∎

Corollary 4.6.5   All bases in a finite-dimensional vector space V contain the same number of vectors.

Proof Suppose $\{\mathbf{v}_1, \mathbf{v}_2, \ldots, \mathbf{v}_n\}$ and $\{\mathbf{u}_1, \mathbf{u}_2, \ldots, \mathbf{u}_m\}$ are two bases for V. From Theorem 4.6.4 we know that we cannot have $m > n$ (otherwise $\{\mathbf{u}_1, \mathbf{u}_2, \ldots, \mathbf{u}_m\}$ would be a linearly dependent set and hence could not be a basis for V). Nor can we have $n > m$ (otherwise $\{\mathbf{v}_1, \mathbf{v}_2, \ldots, \mathbf{v}_n\}$ would be a linearly dependent set and hence could not be a basis for V). Thus, it follows that we must have $m = n$. ∎

We can now prove that any basis provides a minimal spanning set for V.

Corollary 4.6.6   If a finite-dimensional vector space V has a basis consisting of n vectors, then any spanning set must contain at least n vectors.

Proof If the spanning set contained fewer than n vectors, then there would be a subset of less than n linearly independent vectors that spanned V; that is, there would be a basis consisting of less than n vectors. But this would contradict the previous corollary. ∎

The number of vectors in a basis for a finite-dimensional vector space is clearly a fundamental property of the vector space, and by Corollary 4.6.5 it is independent of the particular chosen basis. We call this number the *dimension* of the vector space.

DEFINITION 4.6.7

The **dimension** of a finite-dimensional vector space V, written dim$[V]$, is the number of vectors in any basis for V. If V is the trivial vector space, $V = \{\mathbf{0}\}$, then we define its dimension to be zero.

Remark   We say that the dimension of the world we live in is three for the very reason that the maximum number of independent directions that we can perceive is three. If a vector space has a basis containing n vectors, then from Theorem 4.6.4, the maximum number of vectors in any linearly independent set is n. Thus, we see that the terminology *dimension* used in an arbitrary vector space is a generalization of a familiar idea.

Example 4.6.8   It follows from our examples earlier in this section that $\dim[\mathbb{R}^3] = 3$, $\dim[M_2(\mathbb{R})] = 4$, and $\dim[P_2] = 3$. ☐

More generally, the following dimensions should be remembered:

$$\dim[\mathbb{R}^n] = n, \quad \dim[M_{m \times n}(\mathbb{R})] = mn, \quad \dim[M_n(\mathbb{R})] = n^2, \quad \dim[P_n] = n + 1.$$

These values have essentially been established previously in our discussion of standard bases. The standard basis for $\mathbb{R}^n$ is $\{\mathbf{e}_1, \mathbf{e}_2, \ldots, \mathbf{e}_n\}$, where $\mathbf{e}_i$ is the n-tuple with value 1 in the ith position and value 0 elsewhere. Thus, this basis contains n vectors. The standard basis for $M_{m \times n}(\mathbb{R})$ is the set of matrices E_{ij} ($1 \leq i \leq m$, $1 \leq j \leq n$) with value 1 in the (i, j) position and value 0 elsewhere. There are mn such matrices in this standard basis. The case of $M_n(\mathbb{R})$ is just a special case of $M_{m \times n}(\mathbb{R})$ in which $m = n$. Finally, the standard basis for P_n is $\{1, x, x^2, \ldots, x^n\}$, a set of $n + 1$ vectors.

Next, let us return once more to Example 1.2.16 to cast its results in terms of the basis concept.

Example 4.6.9   Determine a basis for the solution space to the differential equation

$$y'' + y = 0$$

on any interval I.

Solution: Our results from Example 1.2.16 tell us that all solutions to the given differential equation are of the form

$$y(x) = c_1 \cos x + c_2 \sin x.$$

Consequently, $\{\cos x, \sin x\}$ is a linearly independent spanning set for the solution space of the differential equation and therefore is a basis.   □

More generally, we will show in Chapter 6 that all solutions to the differential equation

$$y'' + a_1(x)y' + a_2(x)y = 0$$

on the interval I have the form

$$y(x) = c_1 y_1(x) + c_2 y_2(x),$$

where $\{y_1, y_2\}$ is any linearly independent set of solutions to the differential equation. Using the terminology introduced in this section, it will therefore follow that:

> The set of all solutions to $y'' + a_1(x)y' + a_2(x)y = 0$ on an interval I is a vector space of dimension two.

If a vector space has dimension n, then from Theorem 4.6.4, the maximum number of vectors in any linearly independent set is n. On the other hand, from Corollary 4.6.6, the minimum number of vectors that can span V is also n. Thus, a basis for V must be a linearly independent set of n vectors. Our next theorem establishes that *any* set of n linearly independent vectors is a basis for V.

Theorem 4.6.10   If $\dim[V] = n$, then any set of n linearly independent vectors in V is a basis for V.

Proof Let $v_1, v_2, \ldots, v_n$ be n linearly independent vectors in V. We need to show that they span V. To do this, let v be an arbitrary vector in V. From Theorem 4.6.4, the set of vectors $\{v, v_1, v_2, \ldots, v_n\}$ is linearly dependent, and so there exist scalars $c_0, c_1, \ldots, c_n$, not all zero, such that

$$c_0 v + c_1 v_1 + \cdots + c_n v_n = 0. \tag{4.6.4}$$

If $c_0 = 0$, then the linear independence of $\{v_1, v_2, \ldots, v_n\}$ and (4.6.4) would imply that $c_0 = c_1 = \cdots = c_n = 0$, a contradiction. Hence, $c_0 \neq 0$, and so, from Equation (4.6.4),

$$v = -\frac{1}{c_0}(c_1 v_1 + c_2 v_2 + \cdots + c_n v_n).$$

Thus v, and hence any vector in V, can be written as a linear combination of $v_1, v_2, \ldots, v_n$, and hence, $\{v_1, v_2, \ldots, v_n\}$ spans V, in addition to being linearly independent. Hence it is a basis for V, as required. ∎

Theorem 4.6.10 is one of the most important results of the section. In Chapter 6, we will explicitly construct a basis for the solution space to the differential equation

$$y^{(n)} + a_1(x)y^{(n-1)} + \cdots + a_{n-1}(x)y' + a_n(x)y = 0$$

consisting of n vectors. That is, we will show that the solution space to this differential equation is n-dimensional. It will then follow immediately from Theorem 4.6.10 that every solution to this differential equation is of the form

$$y(x) = c_1 y_1(x) + c_2 y_2(x) + \cdots + c_n y_n(x),$$

where $\{y_1, y_2, \ldots, y_n\}$ is *any* linearly independent set of n solutions to the differential equation. Therefore, determining all solutions to the differential equation will be reduced to determining any linearly independent set of n solutions. A similar application of the theorem will be used to develop the theory for systems of differential equations in Chapter 7.

More generally, Theorem 4.6.10 says that if we know in advance that the dimension of the vector space V is n, then n linearly independent vectors in V are already guaranteed to form a basis for V without the need to explicitly verify that these n vectors also span V. This represents a useful reduction in the work required to verify a basis. Here is an example:

Example 4.6.11   Verify that $\{1 + x, 2 - 2x + x^2, 1 + x^2\}$ is a basis for P_2.

Solution: Since $\dim[P_2] = 3$, Theorem 4.6.10 will guarantee that the three given vectors are a basis, once we confirm only that they are linearly independent. The polynomials

$$p_1(x) = 1 + x, \qquad p_2(x) = 2 - 2x + x^2, \qquad p_3(x) = 1 + x^2$$

have Wronskian

$$W[p_1, p_2, p_3](x) = \begin{vmatrix} 1 + x & 2 - 2x + x^2 & 1 + x^2 \\ 1 & -2 + 2x & 2x \\ 0 & 2 & 2 \end{vmatrix} = -6 \neq 0.$$

Since the Wronskian is nonzero, the given set of vectors is linearly independent on any interval. Consequently, $\{1 + x, 2 - 2x + x^2, 1 + x^2\}$ is indeed a basis for P_2. □

There is a notable parallel result to Theorem 4.6.10 which can also cut down the work required to verify that a set of vectors in V is a basis for V, provided that we know the dimension of V in advance.

Theorem 4.6.12   If $\dim[V] = n$, then any set of n vectors in V that spans V is a basis for V.

Proof Let $\mathbf{v}_1, \mathbf{v}_2, \ldots, \mathbf{v}_n$ be n vectors in V that span V. To confirm that $\{\mathbf{v}_1, \mathbf{v}_2, \ldots, \mathbf{v}_n\}$ is a basis for V, we need only show that this is a linearly independent set of vectors. Suppose, to the contrary, that $\{\mathbf{v}_1, \mathbf{v}_2, \ldots, \mathbf{v}_n\}$ is a linearly dependent set. By Corollary 4.5.12, there is a linearly independent subset of $\{\mathbf{v}_1, \mathbf{v}_2, \ldots, \mathbf{v}_n\}$, with fewer than n vectors, which also spans V. But this implies that V contains a basis with fewer than n vectors, a contradiction. ∎

Putting the results of Theorems 4.6.10 and 4.6.12 together, the following result is immediate.

Corollary 4.6.13   If $\dim[V] = n$ and $S = \{\mathbf{v}_1, \mathbf{v}_2, \ldots, \mathbf{v}_n\}$ is a set of n vectors in V, the following statements are equivalent:

1. S is a basis for V.
2. S is linearly independent.
3. S spans V.

We emphasize once more the importance of this result. It means that if we have a set S of $\dim[V]$ vectors in V, then to determine whether or not S is a basis for V, we need only check if S is linearly independent *or* if S spans V, *not both*.

We next establish another corollary to Theorem 4.6.10.

Corollary 4.6.14   Let S be a subspace of a finite-dimensional vector space V. If $\dim[V] = n$, then

$$\dim[S] \leq n.$$

Furthermore, if $\dim[S] = n$, then $S = V$.

Proof Suppose that $\dim[S] > n$. Then any basis for S would contain more than n linearly independent vectors, and therefore we would have a linearly independent set of more than n vectors in V. This would contradict Theorem 4.6.4. Thus, $\dim[S] \leq n$.

Now consider the case when $\dim[S] = n = \dim[V]$. In this case, any basis for S consists of n linearly independent vectors in S and hence n linearly independent vectors in V. Thus, by Theorem 4.6.10, these vectors also form a basis for V. Hence, every vector in V is spanned by the basis vectors for S, and hence, every vector in V lies in S. Thus, $V = S$. ∎

Example 4.6.15   Give a geometric description of the subspaces of $\mathbb{R}^3$ of dimensions 0, 1, 2, 3.

Solution: *Zero-dimensional subspace:* This corresponds to the subspace $\{(0, 0, 0)\}$, and therefore it is represented geometrically by the origin of a Cartesian coordinate system.

One-dimensional subspace: These are subspaces generated by a single (nonzero) basis vector. Consequently, they correspond geometrically to lines through the origin.

Two-dimensional subspace: These are the subspaces generated by any two noncollinear vectors and correspond geometrically to planes through the origin.

Three-dimensional subspace: Since $\dim[\mathbb{R}^3] = 3$, it follows from Corollary 4.6.14 that the only three-dimensional subspace of $\mathbb{R}^3$ is $\mathbb{R}^3$ itself.   □

Example 4.6.16   Determine a basis for the subspace of $\mathbb{R}^3$ consisting of all solutions to the equation $x_1 + 2x_2 - x_3 = 0$.

Solution: We can solve this problem geometrically. The given equation is that of a plane through the origin and therefore is a two-dimensional subspace of $\mathbb{R}^3$. In order to determine a basis for this subspace, we need only choose two linearly independent (i.e., noncollinear) vectors that lie in the plane. A simple choice of vectors is[8] $\mathbf{v}_1 = (1, 0, 1)$ and $\mathbf{v}_2 = (2, -1, 0)$. Thus, a basis for the subspace is $\{(1, 0, 1), (2, -1, 0)\}$.   □

Corollary 4.6.14 has shown that if S is a subspace of a finite-dimensional vector space V with $\dim[S] = \dim[V]$, then $S = V$. Our next result establishes that, in general, a basis for a subspace of a finite-dimensional vector space V can be extended to a basis for V. This result will be required in the next section and also in Chapter 5.

Theorem 4.6.17   Let S be a subspace of a finite-dimensional vector space V. Any basis for S is part of a basis for V.

Proof   Suppose $\dim[V] = n$ and $\dim[S] = k$. By Corollary 4.6.14, $k \leq n$. If $k = n$, then $S = V$, so that any basis for S is a basis for V. Suppose now that $k < n$, and let $\{\mathbf{v}_1, \mathbf{v}_2, \ldots, \mathbf{v}_k\}$ be a basis for S. These basis vectors are linearly independent, but they fail to span V (otherwise they would form a basis for V, contradicting $k < n$). Thus, there is at least one vector, say $\mathbf{v}_{k+1}$, in V that is not in $\mathrm{span}\{\mathbf{v}_1, \mathbf{v}_2, \ldots, \mathbf{v}_k\}$. Hence, $\{\mathbf{v}_1, \mathbf{v}_2, \ldots, \mathbf{v}_k, \mathbf{v}_{k+1}\}$ is linearly independent. If $k+1 = n$, then we have a basis for V by Theorem 4.6.10, and we are done. Otherwise, we can repeat the procedure to obtain the linearly independent set $\{\mathbf{v}_1, \mathbf{v}_2, \ldots, \mathbf{v}_k, \mathbf{v}_{k+1}, \mathbf{v}_{k+2}\}$. The process will terminate when we have a linearly independent set containing n vectors, including the original vectors $\mathbf{v}_1, \mathbf{v}_2, \ldots, \mathbf{v}_k$ in the basis for S. This proves the theorem.   ■

Remark   The process used in proving the previous theorem is referred to as **extending a basis**.

Example 4.6.18   Let S denote the subspace of $M_2(\mathbb{R})$ consisting of all symmetric 2×2 matrices. Determine a basis for S, and find $\dim[S]$. Extend this basis for S to obtain a basis for $M_2(\mathbb{R})$.

Solution: We first express S in set notation as

$$S = \{A \in M_2(\mathbb{R}) : A^T = A\}.$$

In order to determine a basis for S, we need to obtain the element form of the matrices in S. We can write

$$S = \left\{ \begin{bmatrix} a & b \\ b & c \end{bmatrix} : a, b, c \in \mathbb{R} \right\}.$$

Since

$$\begin{bmatrix} a & b \\ b & c \end{bmatrix} = a \begin{bmatrix} 1 & 0 \\ 0 & 0 \end{bmatrix} + b \begin{bmatrix} 0 & 1 \\ 1 & 0 \end{bmatrix} + c \begin{bmatrix} 0 & 0 \\ 0 & 1 \end{bmatrix},$$

it follows that

$$S = \mathrm{span}\left\{ \begin{bmatrix} 1 & 0 \\ 0 & 0 \end{bmatrix}, \begin{bmatrix} 0 & 1 \\ 1 & 0 \end{bmatrix}, \begin{bmatrix} 0 & 0 \\ 0 & 1 \end{bmatrix} \right\}.$$

---

[8]There are many others, of course.

Furthermore, it is easily shown that the matrices in this spanning set are linearly independent. Consequently, a basis for S is

$$\left\{ \begin{bmatrix} 1 & 0 \\ 0 & 0 \end{bmatrix}, \begin{bmatrix} 0 & 1 \\ 1 & 0 \end{bmatrix}, \begin{bmatrix} 0 & 0 \\ 0 & 1 \end{bmatrix} \right\},$$

so that $\dim[S] = 3$. Since $\dim[M_2(\mathbb{R})] = 4$, in order to extend the basis for S to a basis for $M_2(\mathbb{R})$, we need to add one additional matrix from $M_2(\mathbb{R})$ such that the resulting set is linearly independent. We must choose a nonsymmetric matrix, for any symmetric matrix can be expressed as a linear combination of the three basis vectors for S, and this would create a linear dependency among the matrices. A simple choice of nonsymmetric matrix (although this is certainly not the *only* choice) is

$$\begin{bmatrix} 0 & 1 \\ 0 & 0 \end{bmatrix}.$$

Adding this vector to the basis for S yields the linearly independent set

$$\left\{ \begin{bmatrix} 1 & 0 \\ 0 & 0 \end{bmatrix}, \begin{bmatrix} 0 & 1 \\ 1 & 0 \end{bmatrix}, \begin{bmatrix} 0 & 0 \\ 0 & 1 \end{bmatrix}, \begin{bmatrix} 0 & 1 \\ 0 & 0 \end{bmatrix} \right\}. \tag{4.6.5}$$

Since $\dim[M_2(\mathbb{R})] = 4$, Theorem 4.6.10 implies that (4.6.5) is a basis for $M_2(\mathbb{R})$. $\square$

It is important to realize that not all vector spaces are finite dimensional. Some are infinite-dimensional. In an infinite-dimensional vector space, we can find an arbitrarily large number of linearly independent vectors. We now give an example of an infinite-dimensional vector space that is of primary importance in the theory of differential equations, $C^n(I)$.

Example 4.6.19   Show that the vector space $C^n(I)$ is an infinite-dimensional vector space.

Solution: Consider the functions $1, x, x^2, \ldots, x^k$ in $C^n(I)$. Of course, each of these functions is in $C^k(I)$ as well, and for each fixed k, the Wronskian of these functions is nonzero (the reader can check that the matrix involved in this calculation is upper triangular, with nonzero entries on the main diagonal). Hence, the functions are linearly independent on I by Theorem 4.5.21. Since we can choose k arbitrarily, it follows that there are an arbitrarily large number of linearly independent vectors in $C^n(I)$, hence $C^n(I)$ is infinite-dimensional. $\square$

In this example we showed that $C^n(I)$ is an infinite-dimensional vector space. Consequently, the use of our finite-dimensional vector space theory in the analysis of differential equations appears questionable. However, the key theoretical result that we will establish in Chapter 6 is that the *solution set* of certain linear differential equations is a *finite-dimensional subspace* of $C^n(I)$, and therefore our basis results will be applicable to this solution set.

Exercises for 4.6

Key Terms

Basis, Standard basis, Infinite-dimensional, Finite-dimensional, Dimension, Extension of a subspace basis.

Skills

- Be able to determine whether a given set of vectors forms a basis for a vector space V.

- Be able to construct a basis for a given vector space V.

- Be able to extend a basis for a subspace of V to V itself.

- Be familiar with the standard bases on $\mathbb{R}^n$, $M_{m \times n}(\mathbb{R})$, and P_n.

- Be able to give the dimension of a vector space V.

- Be able to draw conclusions about the properties of a set of vectors in a vector space (i.e., spanning or linear independence) based solely on the size of the set.

- Understand the usefulness of Theorems 4.6.10 and 4.6.12.

True-False Review

For Questions 1–11, decide if the given statement is **true** or **false**, and give a brief justification for your answer. If true, you can quote a relevant definition or theorem from the text. If false, provide an example, illustration, or brief explanation of why the statement is false.

1. A basis for a vector space V is a set S of vectors that spans V.

2. If V and W are vector spaces of dimensions n and m, respectively, and if $n > m$, then W is a subspace of V.

3. A vector space V can have many different bases.

4. $\dim[P_n] = \dim[\mathbb{R}^n]$.

5. If V is an n-dimensional vector space, then any set S of m vectors with $m > n$ must span V.

6. Five vectors in P_3 must be linearly dependent.

7. Two vectors in P_3 must be linearly independent.

8. Ten vectors in $M_3(\mathbb{R})$ must be linearly dependent.

9. If V is an n-dimensional vector space, then every set S with fewer than n vectors can be extended to a basis for V.

10. Every set of vectors that spans a finite-dimensional vector space V contains a subset which forms a basis for V.

11. The set of all 3×3 *upper triangular* matrices forms a three-dimensional subspace of $M_3(\mathbb{R})$.

Problems

For Problems 1–5, determine whether the given set of vectors is a basis for $\mathbb{R}^n$.

1. $\{(1, 1), (-1, 1)\}$.

2. $\{(1, 2, 1), (3, -1, 2), (1, 1, -1)\}$.

3. $\{(1, -1, 1), (2, 5, -2), (3, 11, -5)\}$.

4. $\{(1, 1, -1, 2), (1, 0, 1, -1), (2, -1, 1, -1)\}$.

5. $\{(1, 1, 0, 2), (2, 1, 3, -1), (-1, 1, 1, -2), (2, -1, 1, 2)\}$.

6. Determine all values of the constant k for which the set of vectors $\{(0, -1, 0, k), (1, 0, 1, 0), (0, 1, 1, 0), (k, 0, 2, 1)\}$ is a basis for $\mathbb{R}^4$.

7. Determine a basis S for P_3, and hence, prove that $\dim[P_3] = 4$. Be sure to prove that S is a basis.

8. Determine a basis S for P_3 whose elements all have the same degree. Be sure to prove that S is a basis.

For Problems 9–12, find the dimension of the null space of the given matrix A.

9. $A = \begin{bmatrix} 1 & 3 \\ -2 & -6 \end{bmatrix}$.

10. $A = \begin{bmatrix} 0 & 0 & 0 \\ 0 & 0 & 0 \\ 0 & 1 & 0 \end{bmatrix}$.

11. $A = \begin{bmatrix} 1 & -1 & 4 \\ 2 & 3 & -2 \\ 1 & 2 & -2 \end{bmatrix}$.

12. $A = \begin{bmatrix} 1 & -1 & 2 & 3 \\ 2 & -1 & 3 & 4 \\ 1 & 0 & 1 & 1 \\ 3 & -1 & 4 & 5 \end{bmatrix}$.

13. Let S be the subspace of $\mathbb{R}^3$ that consists of all solutions to the equation $x - 3y + z = 0$. Determine a basis for S, and hence, find $\dim[S]$.

14. Let S be the subspace of $\mathbb{R}^3$ consisting of all vectors of the form $(r, r - 2s, 3s - 5r)$, where r and s are real numbers. Determine a basis for S, and hence, find $\dim[S]$.

15. Let S be the subspace of $M_2(\mathbb{R})$ consisting of all 2×2 upper triangular matrices. Determine a basis for S, and hence, find $\dim[S]$.

16. Let S be the subspace of $M_2(\mathbb{R})$ consisting of all 2×2 matrices with trace zero. Determine a basis for S, and hence, find $\dim[S]$.

17. Let S be the subspace of $\mathbb{R}^3$ spanned by the vectors $\mathbf{v}_1 = (1, 0, 1)$, $\mathbf{v}_2 = (0, 1, 1)$, $\mathbf{v}_3 = (2, 0, 2)$. Determine a basis for S, and hence, find $\dim[S]$.

18. Let S be the vector space consisting of the set of all linear combinations of the functions $f_1(x) = e^x$, $f_2(x) = e^{-x}$, $f_3(x) = \sinh(x)$. Determine a basis for S, and hence, find $\dim[S]$.

19. Determine a basis for the subspace of $M_2(\mathbb{R})$ spanned by

$$\begin{bmatrix} 1 & 3 \\ -1 & 2 \end{bmatrix}, \quad \begin{bmatrix} 0 & 0 \\ 0 & 0 \end{bmatrix}, \quad \begin{bmatrix} -1 & 4 \\ 1 & 1 \end{bmatrix}, \quad \begin{bmatrix} 5 & -6 \\ -5 & 1 \end{bmatrix}.$$

20. Let $\mathbf{v}_1 = (1, 1)$ and $\mathbf{v}_2 = (-1, 1)$.

(a) Show that $\{\mathbf{v}_1, \mathbf{v}_2\}$ spans $\mathbb{R}^2$.

(b) Show that $\{\mathbf{v}_1, \mathbf{v}_2\}$ is linearly independent.

(c) Conclude from (a) *or* (b) that $\{\mathbf{v}_1, \mathbf{v}_2\}$ is a basis for $\mathbb{R}^2$. What theorem in this section allows you to draw this conclusion from either (a) or (b), without proving both?

21. Let $\mathbf{v}_1 = (2, 1)$ and $\mathbf{v}_2 = (3, 1)$.

(a) Show that $\{\mathbf{v}_1, \mathbf{v}_2\}$ spans $\mathbb{R}^2$.

(b) Show that $\{\mathbf{v}_1, \mathbf{v}_2\}$ is linearly independent.

(c) Conclude from (a) *or* (b) that $\{\mathbf{v}_1, \mathbf{v}_2\}$ is a basis for $\mathbb{R}^2$. What theorem in this section allows you to draw this conclusion from either (a) or (b), without proving both?

22. Let $\mathbf{v}_1 = (0, 6, 3)$, $\mathbf{v}_2 = (3, 0, 3)$, and $\mathbf{v}_3 = (6, -3, 0)$. Show that $\{\mathbf{v}_1, \mathbf{v}_2, \mathbf{v}_3\}$ is a basis for $\mathbb{R}^3$. [**Hint:** You need not show that the set is both linearly independent and a spanning set for P_2. Use a theorem from this section to shorten your work.]

23. Determine all values of the constant α for which $\{1 + \alpha x^2, 1 + x + x^2, 2 + x\}$ is a basis for P_2.

24. Let $p_1(x) = 1 + x$, $p_2(x) = x(x - 1)$, $p_3(x) = 1 + 2x^2$. Show that $\{p_1, p_2, p_3\}$ is a basis for P_2. [**Hint:** You need not show that the set is both linearly independent and a spanning set for P_2. Use a theorem from this section to shorten your work.]

25. The **Legendre polynomial of degree** n, $p_n(x)$, is defined to be the polynomial solution of the differential equation

$$(1 - x^2)y'' - 2xy' + n(n + 1)y = 0,$$

which has been normalized so that $p_n(1) = 1$. The first three Legendre polynomials are $p_0(x) = 1$, $p_1(x) = x$, and $p_2(x) = \frac{1}{2}(3x^2 - 1)$. Show that $\{p_0, p_1, p_2\}$ is a basis for P_2. [The hint for the previous problem applies again.]

26. Let

$$A_1 = \begin{bmatrix} -1 & 1 \\ 0 & 1 \end{bmatrix}, \quad A_2 = \begin{bmatrix} 1 & 3 \\ -1 & 0 \end{bmatrix},$$

$$A_3 = \begin{bmatrix} 1 & 0 \\ 1 & 2 \end{bmatrix}, \quad A_4 = \begin{bmatrix} 0 & -1 \\ 2 & 3 \end{bmatrix}.$$

(a) Show that $\{A_1, A_2, A_3, A_4\}$ is a basis for $M_2(\mathbb{R})$. [The hint on the previous problems applies again.]

(b) Express the vector

$$\begin{bmatrix} 5 & 6 \\ 7 & 8 \end{bmatrix}$$

as a linear combination of the basis in (a).

27. Let

$$A = \begin{bmatrix} 1 & 1 & -1 & 1 \\ 2 & -3 & 5 & -6 \\ 5 & 0 & 2 & -3 \end{bmatrix},$$

and let $\mathbf{v}_1 = (-2, 7, 5, 0)$ and $\mathbf{v}_2 = (3, -8, 0, 5)$.

(a) Show that $\{\mathbf{v}_1, \mathbf{v}_2\}$ is a basis for the null space of A.

(b) Using the basis in part (a), write an expression for an arbitrary vector (x, y, z, w) in the null space of A.

28. Let $V = M_3(\mathbb{R})$ and let S be the subset of all vectors in V such that the sum of the entries in each row and in each column is zero.

(a) Find a basis and the dimension of S.

(b) Extend the basis in (a) to a basis for V.

29. Let $V = M_3(\mathbb{R})$ and let S be the subset of all vectors in V such that the sum of the entries in each row and in each column is the same.

(a) Find a basis and the dimension of S.

(b) Extend the basis in (a) to a basis for V.

For Problems 30–31, $Sym_n(\mathbb{R})$ and $Skew_n(\mathbb{R})$ denote the vector spaces consisting of all real $n \times n$ matrices that are symmetric and skew-symmetric, respectively.

30. Find a basis for $Sym_2(\mathbb{R})$ and $Skew_2(\mathbb{R})$, and show that

$$\dim[Sym_2(\mathbb{R})] + \dim[Skew_2(\mathbb{R})] = \dim[M_2(\mathbb{R})].$$

31. Determine the dimensions of $Sym_n(\mathbb{R})$ and $Skew_n(\mathbb{R})$, and show that

$$\dim[Sym_n(\mathbb{R})] + \dim[Skew_n(\mathbb{R})] = \dim[M_n(\mathbb{R})].$$

For Problems 32–34, a subspace S of a vector space V is given. Determine a basis for S and extend your basis for S to obtain a basis for V.

32. $V = \mathbb{R}^3$, S is the subspace consisting of all points lying on the plane with Cartesian equation

$$x + 4y - 3z = 0.$$

33. $V = M_2(\mathbb{R})$, S is the subspace consisting of all matrices of the form

$$\begin{bmatrix} a & b \\ b & a \end{bmatrix}.$$

34. $V = P_2$, S is the subspace consisting of all polynomials of the form $(2a_1 + a_2)x^2 + (a_1 + a_2)x + (3a_1 - a_2)$.

35. Let S be a basis for P_{n-1}. Prove that $S \cup \{x^n\}$ is a basis for P_n.

36. Generalize the previous problem as follows. Let S be a basis for P_{n-1}, and let p be any polynomial of degree n. Prove that $S \cup \{p\}$ is a basis for P_n.

37. **(a)** What is the dimension of $\mathbb{C}^n$ as a real vector space? Determine a basis.

(b) What is the dimension of $\mathbb{C}^n$ as a complex vector space? Determine a basis.

4.7 Change of Basis

Throughout this section, we restrict our attention to vector spaces that are *finite-dimensional*. If we have a (finite) basis for such a vector space V, then, since the vectors in a basis span V, any vector in V can be expressed as a linear combination of the basis vectors. The next theorem establishes that there is only one way in which we can do this.

Theorem 4.7.1 If V is a vector space with basis $\{v_1, v_2, \ldots, v_n\}$, then every vector $v \in V$ can be written **uniquely** as a linear combination of $v_1, v_2, \ldots, v_n$.

Proof Since $v_1, v_2, \ldots, v_n$ span V, every vector $v \in V$ can be expressed as

$$v = a_1v_1 + a_2v_2 + \cdots + a_nv_n, \tag{4.7.1}$$

for some scalars $a_1, a_2, \ldots, a_n$. Suppose also that

$$v = b_1v_1 + b_2v_2 + \cdots + b_nv_n, \tag{4.7.2}$$

for some scalars $b_1, b_2, \ldots, b_n$. We will show that $a_i = b_i$ for each i, which will prove the uniqueness assertion of this theorem. Subtracting Equation (4.7.2) from Equation (4.7.1) yields

$$(a_1 - b_1)v_1 + (a_2 - b_2)v_2 + \cdots + (a_n - b_n)v_n = 0. \tag{4.7.3}$$

But $\{v_1, v_2, \ldots, v_n\}$ is linearly independent, and so Equation (4.7.3) implies that

$$a_1 - b_1 = 0, \qquad a_2 - b_2 = 0, \qquad \ldots, \qquad a_n - b_n = 0.$$

That is, $a_i = b_i$ for each $i = 1, 2, \ldots, n$. ∎

Remark The converse of Theorem 4.7.1 is also true. That is, if every vector v in a vector space V can be written uniquely as a linear combination of the vectors in $\{v_1, v_2, \ldots, v_n\}$, then $\{v_1, v_2, \ldots, v_n\}$ is a basis for V. The proof of this fact is left as an exercise (Problem 38).

Up to this point, we have not paid particular attention to the order in which the vectors of a basis are listed. However, in the remainder of this section, this will become an important consideration. By an **ordered basis** for a vector space, we mean a basis in which we are keeping track of the order in which the basis vectors are listed.

DEFINITION 4.7.2

If $B = \{v_1, v_2, \ldots, v_n\}$ is an ordered basis for V and v is a vector in V, then the scalars $c_1, c_2, \ldots, c_n$ in the unique n-tuple $(c_1, c_2, \ldots, c_n)$ such that

$$v = c_1 v_1 + c_2 v_2 + \cdots + c_n v_n$$

are called the **components of v relative to the ordered basis** $B = \{v_1, v_2, \ldots, v_n\}$. We denote the *column* vector consisting of the components of v relative to the ordered basis B by $[v]_B$, and we call $[v]_B$ the **component vector of v relative to B**.

Example 4.7.3   Determine the components of the vector $v = (1, 7)$ relative to the ordered basis $B = \{(1, 2), (3, 1)\}$.

Solution: If we let $v_1 = (1, 2)$ and $v_2 = (3, 1)$, then since these vectors are not collinear, $B = \{v_1, v_2\}$ is a basis for $\mathbb{R}^2$. We must determine constants c_1, c_2 such that

$$c_1 v_1 + c_2 v_2 = v.$$

We write

$$c_1(1, 2) + c_2(3, 1) = (1, 7).$$

This requires that

$$c_1 + 3c_2 = 1 \qquad \text{and} \qquad 2c_1 + c_2 = 7.$$

The solution to this system is $(4, -1)$, which gives the components of v relative to the ordered basis $B = \{v_1, v_2\}$. (See Figure 4.7.1.) Thus,

$$v = 4v_1 - v_2.$$

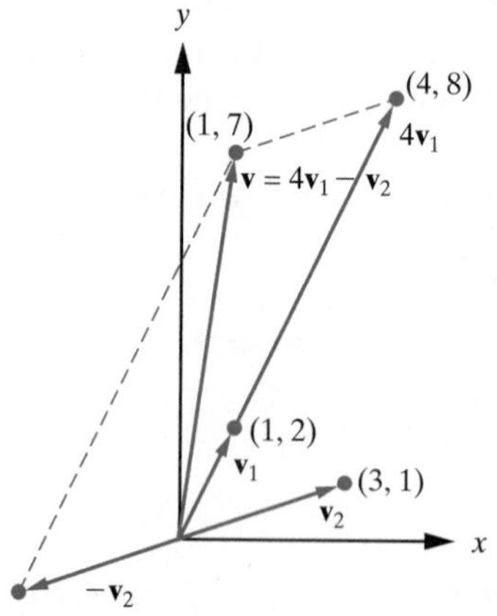

Therefore, we have

$$[v]_B = \begin{bmatrix} 4 \\ -1 \end{bmatrix}. \qquad \square$$

Figure 4.7.1: The components of the vector $v = (1, 7)$ relative to the basis $\{(1, 2), (3, 1)\}$.

Remark   In the preceding example, the component vector of $v = (1, 7)$ relative to the ordered basis $B' = \{(3, 1), (1, 2)\}$ is

$$[v]_{B'} = \begin{bmatrix} -1 \\ 4 \end{bmatrix}.$$

Thus, even though the bases B and B' contain the same vectors, the fact that the vectors are listed in different order affects the components of the vectors in the vector space.

Example 4.7.4   In P_2, determine the component vector of $p(x) = 5 + 7x - 3x^2$ relative to the following:

(a) The standard (ordered) basis $B = \{1, x, x^2\}$.

(b) The ordered basis $C = \{1 + x, 2 + 3x, 5 + x + x^2\}$.

Solution:

(a) The given polynomial is already written as a linear combination of the standard basis vectors. Consequently, the components of $p(x) = 5 + 7x - 3x^2$ relative to the standard basis B are 5, 7, and -3. We write

$$[p(x)]_B = \begin{bmatrix} 5 \\ 7 \\ -3 \end{bmatrix}.$$

(b) The components of $p(x) = 5 + 7x - 3x^2$ relative to the ordered basis

$$C = \{1 + x, 2 + 3x, 5 + x + x^2\}$$

are c_1, c_2, and c_3, where

$$c_1(1 + x) + c_2(2 + 3x) + c_3(5 + x + x^2) = 5 + 7x - 3x^2.$$

That is,

$$(c_1 + 2c_2 + 5c_3) + (c_1 + 3c_2 + c_3)x + c_3 x^2 = 5 + 7x - 3x^2.$$

Hence, c_1, c_2, and c_3 satisfy

$$\begin{array}{rcl} c_1 + 2c_2 + 5c_3 & = & 5, \\ c_1 + 3c_2 + c_3 & = & 7, \\ c_3 & = & -3. \end{array}$$

The augmented matrix of this system has reduced row-echelon form

$$\begin{bmatrix} 1 & 0 & 0 & 40 \\ 0 & 1 & 0 & -10 \\ 0 & 0 & 1 & -3 \end{bmatrix},$$

so that the system has solution $(40, -10, -3)$, which gives the required components. Hence, we can write

$$5 + 7x - 3x^2 = 40(1 + x) - 10(2 + 3x) - 3(5 + x + x^2).$$

Therefore,

$$[p(x)]_C = \begin{bmatrix} 40 \\ -10 \\ -3 \end{bmatrix}. \qquad \square$$

Change-of-Basis Matrix

The preceding example naturally motivates the following question: If we are given two different ordered bases for an n-dimensional vector space V, say

$$B = \{\mathbf{v}_1, \mathbf{v}_2, \ldots, \mathbf{v}_n\} \quad \text{and} \quad C = \{\mathbf{w}_1, \mathbf{w}_2, \ldots, \mathbf{w}_n\}, \qquad (4.7.4)$$

and a vector $\mathbf{v}$ in V, how are $[\mathbf{v}]_B$ and $[\mathbf{v}]_C$ related? In practical terms, we may know the components of $\mathbf{v}$ relative to B and wish to know the components of $\mathbf{v}$ relative to a different ordered basis C. This question actually arises quite often, since different bases are advantageous in different circumstances, so it is useful to be able to convert components of a vector relative to one basis to components relative to another basis. The tool we need in order to do this efficiently is the change-of-basis matrix. Before we describe this matrix, we pause to record the linearity properties satisfied by the components of a vector. These properties will facilitate the discussion that follows.

Lemma 4.7.5   Let V be a vector space with ordered basis $B = \{\mathbf{v}_1, \mathbf{v}_2, \ldots, \mathbf{v}_n\}$, let $\mathbf{x}$ and $\mathbf{y}$ be vectors in V, and let c be a scalar. Then we have

(a) $[\mathbf{x} + \mathbf{y}]_B = [\mathbf{x}]_B + [\mathbf{y}]_B$.

(b) $[c\mathbf{x}]_B = c[\mathbf{x}]_B$.

Proof Write

$$\mathbf{x} = a_1\mathbf{v}_1 + a_2\mathbf{v}_2 + \cdots + a_n\mathbf{v}_n \quad \text{and} \quad \mathbf{y} = b_1\mathbf{v}_1 + b_2\mathbf{v}_2 + \cdots + b_n\mathbf{v}_n,$$

so that

$$\mathbf{x} + \mathbf{y} = (a_1 + b_1)\mathbf{v}_1 + (a_2 + b_2)\mathbf{v}_2 + \cdots + (a_n + b_n)\mathbf{v}_n.$$

Hence,

$$[\mathbf{x} + \mathbf{y}]_B = \begin{bmatrix} a_1 + b_1 \\ a_2 + b_2 \\ \vdots \\ a_n + b_n \end{bmatrix} = \begin{bmatrix} a_1 \\ a_2 \\ \vdots \\ a_n \end{bmatrix} + \begin{bmatrix} b_1 \\ b_2 \\ \vdots \\ b_n \end{bmatrix} = [\mathbf{x}]_B + [\mathbf{y}]_B,$$

which establishes (a). The proof of (b) is left as an exercise (Problem 37).   ■

DEFINITION   4.7.6

Let V be an n-dimensional vector space with ordered bases B and C given in (4.7.4). We define the **change-of-basis matrix from B to C** by

$$P_{C \leftarrow B} = \left[[\mathbf{v}_1]_C, [\mathbf{v}_2]_C, \ldots, [\mathbf{v}_n]_C \right]. \tag{4.7.5}$$

In words, we determine the components of each vector in the "old basis" B with respect the "new basis" C and write the component vectors in the columns of the change-of-basis matrix.

Remark   Of course, there is also a change-of-basis matrix from C to B, given by

$$P_{B \leftarrow C} = \left[[\mathbf{w}_1]_B, [\mathbf{w}_2]_B, \ldots, [\mathbf{w}_n]_B \right].$$

We will see shortly that the matrices $P_{B \leftarrow C}$ and $P_{C \leftarrow B}$ are intimately related.

Our first order of business at this point is to see why the matrix in (4.7.5) converts the components of a vector relative to B into components relative to C. Let $\mathbf{v}$ be a vector in V and write

$$\mathbf{v} = a_1\mathbf{v}_1 + a_2\mathbf{v}_2 + \cdots + a_n\mathbf{v}_n.$$

Then

$$[\mathbf{v}]_B = \begin{bmatrix} a_1 \\ a_2 \\ \vdots \\ a_n \end{bmatrix}.$$

Hence, using Theorem 2.2.9 and Lemma 4.7.5, we have

$$P_{C \leftarrow B}[\mathbf{v}]_B = a_1[\mathbf{v}_1]_C + a_2[\mathbf{v}_2]_C + \cdots + a_n[\mathbf{v}_n]_C = [a_1\mathbf{v}_1 + a_2\mathbf{v}_2 + \cdots + a_n\mathbf{v}_n]_C = [\mathbf{v}]_C.$$

This calculation shows that premultiplying the component vector of $\mathbf{v}$ relative to B by the change of basis matrix $P_{C \leftarrow B}$ yields the component vector of $\mathbf{v}$ relative to C:

$$\boxed{[\mathbf{v}]_C = P_{C \leftarrow B}[\mathbf{v}]_B.} \tag{4.7.6}$$

Example 4.7.7   Let $V = \mathbb{R}^2$, $B = \{(1, 2), (3, 4)\}$, $C = \{(7, 3), (4, 2)\}$, and $\mathbf{v} = (1, 0)$. It is routine to verify that B and C are bases for V.

(a) Determine $[\mathbf{v}]_B$ and $[\mathbf{v}]_C$.

(b) Find $P_{C \leftarrow B}$ and $P_{B \leftarrow C}$.

(c) Use (4.7.6) to compute $[\mathbf{v}]_C$, and compare your answer with (a).

Solution:

(a) Solving $(1, 0) = a_1(1, 2) + a_2(3, 4)$, we find $a_1 = -2$ and $a_2 = 1$. Hence,

$$[\mathbf{v}]_B = \begin{bmatrix} -2 \\ 1 \end{bmatrix}.$$

Likewise, setting $(1, 0) = b_1(7, 3) + b_2(4, 2)$, we find $b_1 = 1$ and $b_2 = -1.5$. Hence,

$$[\mathbf{v}]_C = \begin{bmatrix} 1 \\ -1.5 \end{bmatrix}.$$

(b) A short calculation shows that

$$[(1, 2)]_C = \begin{bmatrix} -3 \\ 5.5 \end{bmatrix} \quad \text{and} \quad [(3, 4)]_C = \begin{bmatrix} -5 \\ 9.5 \end{bmatrix}.$$

Thus, we have

$$P_{C \leftarrow B} = \begin{bmatrix} -3 & -5 \\ 5.5 & 9.5 \end{bmatrix}.$$

Likewise, another short calculation shows that

$$[(7, 3)]_B = \begin{bmatrix} -9.5 \\ 5.5 \end{bmatrix} \quad \text{and} \quad [(4, 2)]_B = \begin{bmatrix} -5 \\ 3 \end{bmatrix}.$$

Hence,

$$P_{B \leftarrow C} = \begin{bmatrix} -9.5 & -5 \\ 5.5 & 3 \end{bmatrix}.$$

(c) We compute as follows:

$$P_{C \leftarrow B}[\mathbf{v}]_B = \begin{bmatrix} -3 & -5 \\ 5.5 & 9.5 \end{bmatrix} \begin{bmatrix} -2 \\ 1 \end{bmatrix} = \begin{bmatrix} 1 \\ -1.5 \end{bmatrix} = [\mathbf{v}]_C,$$

as we found in part (a).   □

The reader may have noticed a close resemblance between the two matrices $P_{C \leftarrow B}$ and $P_{B \leftarrow C}$ computed in part (b) of the preceding example. In fact, a brief calculation shows that

$$P_{C \leftarrow B} P_{B \leftarrow C} = I_2 = P_{B \leftarrow C} P_{C \leftarrow B}.$$

The two change-of-basis matrices are inverses of each other. This turns out to be always true. To see why, consider again Equation (4.7.6). If we premultiply both sides of (4.7.6) by the matrix $P_{B \leftarrow C}$, we get

$$P_{B \leftarrow C}[\mathbf{v}]_C = P_{B \leftarrow C} P_{C \leftarrow B}[\mathbf{v}]_B. \tag{4.7.7}$$

Rearranging the roles of B and C in (4.7.6), the left side of (4.7.7) is simply $[\mathbf{v}]_B$. Thus,

$$P_{B \leftarrow C} P_{C \leftarrow B}[\mathbf{v}]_B = [\mathbf{v}]_B.$$

Since this is true for any vector $[\mathbf{v}]_B$ in $\mathbb{R}^n$, this implies that

$$P_{B \leftarrow C} P_{C \leftarrow B} = I_n,$$

the $n \times n$ identity matrix. Likewise, a similar calculation shows that

$$P_{C \leftarrow B} P_{B \leftarrow C} = I_n.$$

Thus, we have proved that

The matrices $P_{C \leftarrow B}$ and $P_{B \leftarrow C}$ are inverses of one another.

Example 4.7.8   Let $V = P_2$, and let $B = \{1, 1+x, 1+x+x^2\}$, and $C = \{2+x+x^2, x+x^2, x\}$. It is routine to verify that B and C are bases for V. Find the change-of-basis matrix from B to C, and use it to calculate the change-of-basis matrix from C to B.

Solution: We set $1 = a_1(2+x+x^2) + a_2(x+x^2) + a_3 x$. With a quick calculation, we find that $a_1 = 0.5$, $a_2 = -0.5$, and $a_3 = 0$. Next, we set $1 + x = b_1(2+x+x^2) + b_2(x+x^2) + b_3 x$, and we find that $b_1 = 0.5$, $b_2 = -0.5$, and $b_3 = 1$. Finally, we set $1 + x + x^2 = c_1(2+x+x^2) + c_2(x+x^2) + c_3 x$, from which it follows that $c_1 = 0.5$, $c_2 = 0.5$, and $c_3 = 0$. Hence, we have

$$P_{C \leftarrow B} = \begin{bmatrix} a_1 & b_1 & c_1 \\ a_2 & b_2 & c_2 \\ a_3 & b_3 & c_3 \end{bmatrix} = \begin{bmatrix} 0.5 & 0.5 & 0.5 \\ -0.5 & -0.5 & 0.5 \\ 0 & 1 & 0 \end{bmatrix}.$$

Thus, we have

$$P_{B \leftarrow C} = (P_{C \leftarrow B})^{-1} = \begin{bmatrix} 1 & -1 & -1 \\ 0 & 0 & 1 \\ 1 & 1 & 0 \end{bmatrix}.$$ □

In much the same way that we showed above that the matrices $P_{C \leftarrow B}$ and $P_{B \leftarrow C}$ are inverses of one another, we can make the following observation.

Theorem 4.7.9   Let V be a vector space with ordered bases A, B, and C. Then

$$P_{C \leftarrow A} = P_{C \leftarrow B} P_{B \leftarrow A}. \qquad (4.7.8)$$

Proof Using (4.7.6), for every $\mathbf{v} \in V$, we have

$$P_{C \leftarrow B} P_{B \leftarrow A}[\mathbf{v}]_A = P_{C \leftarrow B}[\mathbf{v}]_B = [\mathbf{v}]_C = P_{C \leftarrow A}[\mathbf{v}]_A,$$

so that premultiplication of $[\mathbf{v}]_A$ by either matrix in (4.7.8) yields the same result. Hence, the matrices on either side of (4.7.8) are the same. ∎

We conclude this section by using Theorem 4.7.9 to show how an arbitrary change-of-basis matrix $P_{C \leftarrow B}$ in $\mathbb{R}^n$ can be expressed as a product of change-of-basis matrices involving the standard basis $E = \{\mathbf{e}_1, \mathbf{e}_2, \ldots, \mathbf{e}_n\}$ of $\mathbb{R}^n$. Let $B = \{\mathbf{v}_1, \mathbf{v}_2, \ldots, \mathbf{v}_n\}$ and $C = \{\mathbf{w}_1, \mathbf{w}_2, \ldots, \mathbf{w}_n\}$ be arbitrary ordered bases for $\mathbb{R}^n$. Since $[\mathbf{v}]_E = \mathbf{v}$ for all column vectors $\mathbf{v}$ in $\mathbb{R}^n$, the matrices

$$P_{E \leftarrow B} = [[\mathbf{v}_1]_E, [\mathbf{v}_2]_E, \ldots, [\mathbf{v}_n]_E] = [\mathbf{v}_1, \mathbf{v}_2, \ldots, \mathbf{v}_n]$$

and

$$P_{E \leftarrow C} = [[\mathbf{w}_1]_E, [\mathbf{w}_2]_E, \ldots, [\mathbf{w}_n]_E] = [\mathbf{w}_1, \mathbf{w}_2, \ldots, \mathbf{w}_n]$$

can be written down immediately. Using these matrices, together with Theorem 4.7.9, we can compute the arbitrary change-of-basis matrix $P_{C \leftarrow B}$ with ease:

$$P_{C \leftarrow B} = P_{C \leftarrow E} P_{E \leftarrow B} = (P_{E \leftarrow C})^{-1} P_{E \leftarrow B}.$$

Exercises for 4.7

Key Terms

Ordered basis, Components of a vector relative to an ordered basis, Change-of-basis matrix.

Skills

- Be able to find the components of a vector relative to a given ordered basis for a vector space V.

- Be able to compute the change-of-basis matrix for a vector space V from one ordered basis B to another ordered basis C.

- Be able to use the change-of-basis matrix from B to C to determine the components of a vector relative to C from the components of the vector relative to B.

- Be familiar with the relationship between the two change-of-basis matrices $P_{C \leftarrow B}$ and $P_{B \leftarrow C}$.

True-False Review

For Questions 1–8, decide if the given statement is **true** or **false**, and give a brief justification for your answer. If true, you can quote a relevant definition or theorem from the text. If false, provide an example, illustration, or brief explanation of why the statement is false.

1. Every vector in a finite-dimensional vector space V can be expressed uniquely as a linear combination of vectors comprising a basis for V.

2. The change-of-basis matrix $P_{B \leftarrow C}$ acts on the component vector of a vector $\mathbf{v}$ relative to the basis C and produces the component vector of $\mathbf{v}$ relative to the basis B.

3. A change-of-basis matrix is always a square matrix.

4. A change-of-basis matrix is always invertible.

5. For any vectors $\mathbf{v}$ and $\mathbf{w}$ in a finite-dimensional vector space V with basis B, we have $[\mathbf{v}-\mathbf{w}]_B = [\mathbf{v}]_B - [\mathbf{w}]_B$.

6. If the bases B and C for a vector space V contain the same set of vectors, then $[\mathbf{v}]_B = [\mathbf{v}]_C$ for every vector $\mathbf{v}$ in V.

7. If B and C are bases for a finite-dimensional vector space V, and $\mathbf{v}$ and $\mathbf{w}$ are in V such that $[\mathbf{v}]_B = [\mathbf{w}]_C$, then $\mathbf{v} = \mathbf{w}$.

8. The matrix $P_{B \leftarrow B}$ is the identity matrix for any basis B for V.

Problems

For Problems 1–13, determine the component vector of the given vector in the vector space V relative to the given ordered basis B.

1. $V = \mathbb{R}^2$; $B = \{(2, -2), (1, 4)\}$; $\mathbf{v} = (5, -10)$.

2. $V = \mathbb{R}^2$; $B = \{(-1, 3), (3, 2)\}$; $\mathbf{v} = (8, -2)$.

3. $V = \mathbb{R}^3$; $B = \{(1, 0, 1), (1, 1, -1), (2, 0, 1)\}$; $\mathbf{v} = (-9, 1, -8)$.

4. $V = \mathbb{R}^3$; $B = \{(1, -6, 3), (0, 5, -1), (3, -1, -1)\}$; $\mathbf{v} = (1, 7, 7)$.

5. $V = \mathbb{R}^3$; $B = \{(3, -1, -1), (1, -6, 3), (0, 5, -1)\}$; $\mathbf{v} = (1, 7, 7)$.

6. $V = \mathbb{R}^3$; $B = \{(-1, 0, 0), (0, 0, -3), (0, -2, 0)\}$; $\mathbf{v} = (5, 5, 5)$.

7. $V = P_2$; $B = \{x^2 + x, 2 + 2x, 1\}$; $p(x) = -4x^2 + 2x + 6$.

8. $V = P_2$; $B = \{5 - 3x, 1, 1 + 2x^2\}$; $p(x) = 15 - 18x - 30x^2$.

9. $V = P_3$; $B = \{1, 1+x, 1+x+x^2, 1+x+x^2+x^3\}$; $p(x) = 4 - x + x^2 - 2x^3$.

10. $V = P_3$; $B = \{x^3 + x^2, x^3 - 1, x^3 + 1, x^3 + x\}$; $p(x) = 8 + x + 6x^2 + 9x^3$.

11. $V = M_2(\mathbb{R})$;
$$B = \left\{ \begin{bmatrix} 1 & 1 \\ 1 & 1 \end{bmatrix}, \begin{bmatrix} 1 & 1 \\ 1 & 0 \end{bmatrix}, \begin{bmatrix} 1 & 1 \\ 0 & 0 \end{bmatrix}, \begin{bmatrix} 1 & 0 \\ 0 & 0 \end{bmatrix} \right\};$$
$$A = \begin{bmatrix} -3 & -2 \\ -1 & 2 \end{bmatrix}.$$

12. $V = M_2(\mathbb{R})$;
$$B = \left\{ \begin{bmatrix} 2 & -1 \\ 3 & 5 \end{bmatrix}, \begin{bmatrix} 0 & 4 \\ -1 & 1 \end{bmatrix}, \begin{bmatrix} 1 & 1 \\ 1 & 1 \end{bmatrix}, \begin{bmatrix} 3 & -1 \\ 2 & 5 \end{bmatrix} \right\};$$
$$A = \begin{bmatrix} -10 & 16 \\ -15 & -14 \end{bmatrix}.$$

13. $V = M_2(\mathbb{R})$;
$$B = \left\{ \begin{bmatrix} -1 & 1 \\ 0 & 1 \end{bmatrix}, \begin{bmatrix} 1 & 3 \\ -1 & 0 \end{bmatrix}, \begin{bmatrix} 1 & 0 \\ 1 & 2 \end{bmatrix}, \begin{bmatrix} 0 & -1 \\ 2 & 3 \end{bmatrix} \right\};$$
$$A = \begin{bmatrix} 5 & 6 \\ 7 & 8 \end{bmatrix}.$$

14. Let $\mathbf{v}_1 = (0, 6, 3)$, $\mathbf{v}_2 = (3, 0, 3)$, and $\mathbf{v}_3 = (6, -3, 0)$. Determine the component vector of an arbitrary vector $\mathbf{v} = (x, y, z)$ relative to the ordered basis $\{\mathbf{v}_1, \mathbf{v}_2, \mathbf{v}_3\}$.

15. Let $p_1(x) = 1 + x$, $p_2(x) = x(x - 1)$, and $p_3(x) = 1 + 2x^2$. Determine the component vector of an arbitrary polynomial $p(x) = a_0 + a_1 x + a_2 x^2$ relative to the ordered basis $\{p_1, p_2, p_3\}$.

For Problems 16–25, find the change-of-basis matrix $P_{C \leftarrow B}$ from the given ordered basis B to the given ordered basis C of the vector space V.

16. $V = \mathbb{R}^2$; $B = \{(9, 2), (4, -3)\}$; $C = \{(2, 1), (-3, 1)\}$.

17. $V = \mathbb{R}^2$; $B = \{(-5, -3), (4, 28)\}$; $C = \{(6, 2), (1, -1)\}$.

18. $V = \mathbb{R}^3$; $B = \{(2, -5, 0), (3, 0, 5), (8, -2, -9)\}$; $C = \{(1, -1, 1), (2, 0, 1), (0, 1, 3)\}$.

19. $V = \mathbb{R}^3$; $B = \{(-7, 4, 4), (4, 2, -1), (-7, 5, 0)\}$; $C = \{(1, 1, 0), (0, 1, 1), (3, -1, -1)\}$.

20. $V = P_1$; $B = \{7 - 4x, 5x\}$; $C = \{1 - 2x, 2 + x\}$.

21. $V = P_2$; $B = \{-4 + x - 6x^2, 6 + 2x^2, -6 - 2x + 4x^2\}$; $C = \{1 - x + 3x^2, 2, 3 + x^2\}$.

22. $V = P_3$;
$B = \{-2 + 3x + 4x^2 - x^3, 3x + 5x^2 + 2x^3, -5x^2 - 5x^3, 4 + 4x + 4x^2\}$; $C = \{1 - x^3, 1 + x, x + x^2, x^2 + x^3\}$.

23. $V = P_2$; $B = \{2 + x^2, -1 - 6x + 8x^2, -7 - 3x - 9x^2\}$; $C = \{1 + x, -x + x^2, 1 + 2x^2\}$.

24. $V = M_2(\mathbb{R})$;
$$B = \left\{ \begin{bmatrix} 1 & 0 \\ -1 & -2 \end{bmatrix}, \begin{bmatrix} 0 & -1 \\ 3 & 0 \end{bmatrix}, \begin{bmatrix} 3 & 5 \\ 0 & 0 \end{bmatrix}, \begin{bmatrix} -2 & -4 \\ 0 & 0 \end{bmatrix} \right\};$$
$$C = \left\{ \begin{bmatrix} 1 & 1 \\ 1 & 1 \end{bmatrix}, \begin{bmatrix} 1 & 1 \\ 1 & 0 \end{bmatrix}, \begin{bmatrix} 1 & 1 \\ 0 & 0 \end{bmatrix}, \begin{bmatrix} 1 & 0 \\ 0 & 0 \end{bmatrix} \right\}.$$

25. $V = M_2(\mathbb{R})$; $B = \{E_{12}, E_{22}, E_{21}, E_{11}\}$;
$C = \{E_{22}, E_{11}, E_{21}, E_{12}\}$.

For Problems 26–31, find the change-of-basis matrix $P_{B \leftarrow C}$ from the given basis C to the given basis B of the vector space V.

26. V, B, and C from Problem 16.

27. V, B, and C from Problem 17.

28. V, B, and C from Problem 18.

29. V, B, and C from Problem 20.

30. V, B, and C from Problem 22.

31. V, B, and C from Problem 25.

For Problems 32–36, verify Equation (4.7.6) for the given vector.

32. $\mathbf{v} = (-5, 3)$; V, B, and C from Problem 16.

33. $\mathbf{v} = (-1, 2, 0)$; V, B, and C from Problem 19.

34. $p(x) = 6 - 4x$; V, B, and C from Problem 20.

35. $p(x) = 5 - x + 3x^2$; V, B, and C from Problem 21.

36. $A = \begin{bmatrix} -1 & -1 \\ -4 & 5 \end{bmatrix}$; V, B, and C from Problem 24.

37. Prove part (b) of Lemma 4.7.5.

38. Prove that if every vector $\mathbf{v}$ in a vector space V can be written uniquely as a linear combination of the vectors in $\{\mathbf{v}_1, \mathbf{v}_2, \ldots, \mathbf{v}_n\}$, then $\{\mathbf{v}_1, \mathbf{v}_2, \ldots, \mathbf{v}_n\}$ is a basis for V.

39. Show that if B is a basis for a finite-dimensional vector space V, and C is a basis obtained by reordering the vectors in B, then the matrices $P_{C \leftarrow B}$ and $P_{B \leftarrow C}$ each contain exactly one 1 in each row and column, and zeros elsewhere.

4.8   Row Space and Column Space

In this section, we consider two vector spaces that can be associated with any $m \times n$ matrix. For simplicity, we will assume that the matrices have real entries, although the results that we establish can easily be extended to matrices with complex entries.

Row Space

Let $A = [a_{ij}]$ be an $m \times n$ real matrix. The row vectors of this matrix are row n-vectors, and therefore they can be associated with vectors in $\mathbb{R}^n$. The subspace of $\mathbb{R}^n$ spanned by these vectors is called the **row space** of A and denoted rowspace(A). For example, if

$$A = \begin{bmatrix} 2 & -1 & 3 \\ 5 & 9 & -7 \end{bmatrix},$$

then

$$\text{rowspace}(A) = \text{span}\{(2, -1, 3), (5, 9, -7)\}.$$

For a general $m \times n$ matrix A, how can we obtain a basis for rowspace(A)? By its very definition, the row space of A is spanned by the row vectors of A, but these may not be linearly independent, hence the row vectors of A do not necessarily form a basis for rowspace(A). We wish to determine a systematic and efficient method for obtaining a basis for the row space. Perhaps not surprisingly, it involves the use of elementary row operations.

If we perform elementary row operations on A, then we are merely taking linear combinations of vectors in rowspace(A), and we therefore might suspect that the row space of the resulting matrix coincides with the row space of A. This is the content of the following theorem.

Theorem 4.8.1   If A and B are row-equivalent matrices, then

$$\text{rowspace}(A) = \text{rowspace}(B).$$

Proof We establish that the matrix that results from performing any of the three elementary row operations on a matrix A has the same row space as the row space of A. If we interchange two rows of A, then clearly we have not altered the row space, since we still have the same set of row vectors (listed in a different order).

Now let $\mathbf{a}_1, \mathbf{a}_2, \ldots, \mathbf{a}_m$ denote the row vectors of A. We combine the remaining two types of elementary row operations by considering the result of replacing $\mathbf{a}_i$ by the vector $r\mathbf{a}_i + s\mathbf{a}_j$, where $r \, (\neq 0)$ and s are real numbers. If $s = 0$, then this corresponds to scaling $\mathbf{a}_i$ by a factor of r, whereas if $r = 1$ and $s \neq 0$, this corresponds to adding a multiple of row j to row i. If B denotes the resulting matrix, then

$$
\begin{aligned}
\text{rowspace}(B) &= \{c_1\mathbf{a}_1 + c_2\mathbf{a}_2 + \cdots + c_i(r\mathbf{a}_i + s\mathbf{a}_j) + \cdots + c_m\mathbf{a}_m\} \\
&= \{c_1\mathbf{a}_1 + c_2\mathbf{a}_2 + \cdots + (rc_i)\mathbf{a}_i + \cdots + (c_j + sc_i)\mathbf{a}_j + \cdots + c_m\mathbf{a}_m\} \\
&= \{c_1\mathbf{a}_1 + c_2\mathbf{a}_2 + \cdots + d_i\mathbf{a}_i + \cdots + d_j\mathbf{a}_j + \cdots + c_m\mathbf{a}_m\},
\end{aligned}
$$

where $d_i = rc_i$ and $d_j = c_j + sc_i$. Note that d_i and d_j can take on arbitrary values, hence the vectors in $\text{rowspace}(B)$ consist precisely of arbitrary linear combinations of $\mathbf{a}_1, \mathbf{a}_2, \ldots, \mathbf{a}_m$. That is,

$$
\text{rowspace}(B) = \text{span}\{\mathbf{a}_1, \mathbf{a}_2, \ldots, \mathbf{a}_m\} = \text{rowspace}(A).
$$

■

The previous theorem is the key to determining a basis for $\text{rowspace}(A)$. The idea we use is to reduce A to row-echelon form. If $\mathbf{d}_1, \mathbf{d}_2, \ldots, \mathbf{d}_k$ denote the nonzero row vectors in this row-echelon form, then from the previous theorem,

$$
\text{rowspace}(A) = \text{span}\{\mathbf{d}_1, \mathbf{d}_2, \ldots, \mathbf{d}_k\}.
$$

We now establish that $\{\mathbf{d}_1, \mathbf{d}_2, \ldots, \mathbf{d}_k\}$ is linearly independent. Consider

$$
c_1\mathbf{d}_1 + c_2\mathbf{d}_2 + \cdots + c_k\mathbf{d}_k = \mathbf{0}. \tag{4.8.1}
$$

Owing to the positioning of the leading ones in a row-echelon matrix, each of the row vectors $\mathbf{d}_1, \mathbf{d}_2, \ldots, \mathbf{d}_{k-1}$ will have a leading one in a position where each succeeding row vector in the row-echelon form has a zero. Hence, Equation (4.8.1) is satisfied only if

$$
c_1 = c_2 = \cdots = c_{k-1} = 0,
$$

and therefore, it reduces to

$$
c_k\mathbf{d}_k = \mathbf{0}.
$$

However, $\mathbf{d}_k$ is a nonzero vector, and so we must have $c_k = 0$. Consequently, all of the constants in Equation (4.8.1) must be zero, and therefore $\{\mathbf{d}_1, \mathbf{d}_2, \ldots, \mathbf{d}_k\}$ not only spans $\text{rowspace}(A)$, but also is linearly independent. Hence, $\{\mathbf{d}_1, \mathbf{d}_2, \ldots, \mathbf{d}_k\}$ is a basis for $\text{rowspace}(A)$. We have therefore established the next theorem.

Theorem 4.8.2   The set of nonzero row vectors in any row-echelon form of an $m \times n$ matrix A is a basis for $\text{rowspace}(A)$.

As a consequence of the preceding theorem, we can conclude that *all* row-echelon forms of A have the same number of nonzero rows. For if this were not the case, then we could find two bases for $\text{rowspace}(A)$ containing a different number of vectors, which would contradict Corollary 4.6.5. We can therefore consider Theorem 2.4.10 as a direct consequence of Theorem 4.8.2.

Example 4.8.3   Determine a basis for the row space of

$$A = \begin{bmatrix} 1 & -1 & 1 & 3 & 2 \\ 2 & -1 & 1 & 5 & 1 \\ 3 & -1 & 1 & 7 & 0 \\ 0 & 1 & -1 & -1 & -3 \end{bmatrix}.$$

Solution: We first reduce A to row-echelon form:

$$A \overset{1}{\sim} \begin{bmatrix} 1 & -1 & 1 & 3 & 2 \\ 0 & 1 & -1 & -1 & -3 \\ 0 & 2 & -2 & -2 & -6 \\ 0 & 1 & -1 & -1 & -3 \end{bmatrix} \overset{2}{\sim} \begin{bmatrix} 1 & -1 & 1 & 3 & 2 \\ 0 & 1 & -1 & -1 & -3 \\ 0 & 0 & 0 & 0 & 0 \\ 0 & 0 & 0 & 0 & 0 \end{bmatrix}.$$

> **1.** $A_{12}(-2)$, $A_{13}(-3)$ **2.** $A_{23}(-2)$, $A_{24}(-1)$

Consequently, a basis for rowspace(A) is $\{(1, -1, 1, 3, 2), (0, 1, -1, -1, -3)\}$, and therefore rowspace(A) is a two-dimensional subspace of $\mathbb{R}^5$.   □

Theorem 4.8.2 also gives an efficient method for determining a basis for the subspace of $\mathbb{R}^n$ spanned by a given set of vectors. If we let A be the matrix whose row vectors are the given vectors from $\mathbb{R}^n$, then rowspace(A) coincides with the subspace of $\mathbb{R}^n$ spanned by those vectors. Consequently, the nonzero row vectors in any row-echelon form of A will be a basis for the subspace spanned by the given set of vectors.

Example 4.8.4   Determine a basis for the subspace of $\mathbb{R}^4$ spanned by $\{(1,2,3,4), (4,5,6,7), (7,8,9,10)\}$.

Solution: We first let A denote the matrix that has the given vectors as row vectors. Thus,

$$A = \begin{bmatrix} 1 & 2 & 3 & 4 \\ 4 & 5 & 6 & 7 \\ 7 & 8 & 9 & 10 \end{bmatrix}.$$

We now reduce A to row-echelon form:

$$A \overset{1}{\sim} \begin{bmatrix} 1 & 2 & 3 & 4 \\ 0 & -3 & -6 & -9 \\ 0 & -6 & -12 & -18 \end{bmatrix} \overset{2}{\sim} \begin{bmatrix} 1 & 2 & 3 & 4 \\ 0 & 1 & 2 & 3 \\ 0 & -6 & -12 & -18 \end{bmatrix} \overset{3}{\sim} \begin{bmatrix} 1 & 2 & 3 & 4 \\ 0 & 1 & 2 & 3 \\ 0 & 0 & 0 & 0 \end{bmatrix}.$$

> **1.** $A_{12}(-4)$, $A_{13}(-7)$ **2.** $M_2(-\frac{1}{3})$ **3.** $A_{23}(6)$

Consequently, a basis for the subspace of $\mathbb{R}^4$ spanned by the given vectors is $\{(1, 2, 3, 4), (0, 1, 2, 3)\}$. We see that the given vectors span a two-dimensional subspace of $\mathbb{R}^4$.   □

Column Space

If A is an $m \times n$ matrix, the column vectors of A are column m-vectors and therefore can be associated with vectors in $\mathbb{R}^m$. The subspace of $\mathbb{R}^m$ spanned by these vectors is called the **column space** of A and denoted colspace(A).

Example 4.8.5   For the matrix

$$A = \begin{bmatrix} 2 & -1 & 3 \\ 5 & 9 & -7 \end{bmatrix},$$

we have colspace(A) = span$\{(2, 5), (-1, 9), (3, -7)\}$.   ☐

We now consider the problem of determining a basis for the column space of an $m \times n$ matrix A. Since the column vectors of A coincide with the row vectors of A^T, it follows that

$$\text{colspace}(A) = \text{rowspace}(A^T).$$

Hence one way to obtain a basis for colspace(A) would be to reduce A^T to row-echelon form, and then the nonzero row vectors in the resulting matrix would form a basis for colspace(A).

There is, however, a better method for determining a basis for colspace(A) directly from any row-echelon form of A. The derivation of this technique is somewhat involved and will require full attention.

We begin by determining the column space of an $m \times n$ *reduced row-echelon matrix*. In order to introduce the basic ideas, consider the particular reduced row-echelon matrix

$$E = \begin{bmatrix} 1 & 2 & 0 & 3 & 0 \\ 0 & 0 & 1 & 5 & 0 \\ 0 & 0 & 0 & 0 & 1 \\ 0 & 0 & 0 & 0 & 0 \end{bmatrix}.$$

In this case, we see that the first, third, and fifth column vectors, which are the column vectors containing the leading ones, coincide with the first three standard basis vectors in $\mathbb{R}^4$ (written as column vectors):

$$\mathbf{e}_1 = \begin{bmatrix} 1 \\ 0 \\ 0 \\ 0 \end{bmatrix}, \qquad \mathbf{e}_2 = \begin{bmatrix} 0 \\ 1 \\ 0 \\ 0 \end{bmatrix}, \qquad \mathbf{e}_3 = \begin{bmatrix} 0 \\ 0 \\ 1 \\ 0 \end{bmatrix}.$$

Consequently, these column vectors are linearly independent. Furthermore, the remaining column vectors in E (those that do not contain leading ones) are both linear combinations of $\mathbf{e}_1$ and $\mathbf{e}_2$, columns that *do* contain leading ones. Therefore $\{\mathbf{e}_1, \mathbf{e}_2, \mathbf{e}_3\}$ is a linearly independent set of vectors that spans colspace(E), and so a basis for colspace(E) is

$$\{(1, 0, 0, 0), (0, 1, 0, 0), (0, 0, 1, 0)\}.$$

Clearly, the same arguments apply to any reduced row-echelon matrix E. Thus, if E contains k (necessarily $\leq n$) leading ones, a basis for colspace(E) is $\{\mathbf{e}_1, \mathbf{e}_2, \ldots, \mathbf{e}_k\}$.

Now consider an arbitrary $m \times n$ matrix A, and let E denote the reduced row-echelon form of A. Recall from Chapter 2 that performing elementary row operations on a linear system does not alter its solution set. Hence, the two homogeneous systems of equations

$$A\mathbf{c} = \mathbf{0} \qquad \text{and} \qquad E\mathbf{c} = \mathbf{0} \tag{4.8.2}$$

have the same solution sets. If we write A and E in column-vector form as $A = [\mathbf{a}_1, \mathbf{a}_2, \ldots, \mathbf{a}_n]$ and $E = [\mathbf{d}_1, \mathbf{d}_2, \ldots, \mathbf{d}_n]$, respectively, then the two systems in (4.8.2) can be written as

$$c_1\mathbf{a}_1 + c_2\mathbf{a}_2 + \cdots + c_n\mathbf{a}_n = \mathbf{0},$$
$$c_1\mathbf{d}_1 + c_2\mathbf{d}_2 + \cdots + c_n\mathbf{d}_n = \mathbf{0},$$

respectively. Thus, the fact that these two systems have the same solution set means that a linear dependence relationship will hold between the column vectors of E if and only if precisely the same linear dependence relation holds between the corresponding column vectors of A. In particular, since our previous work shows that the column vectors in E that contain leading ones give a basis for colspace(E), they give a maximal linearly independent set in colspace(E). Therefore, the corresponding column vectors in A will also be a maximal linearly independent set in colspace(A). Consequently, this set of vectors from A will be a basis for colspace(A).

We have therefore shown that the set of column vectors of A corresponding to those column vectors containing leading ones in the reduced row-echelon form of A is a basis for colspace(A). But do we have to reduce A to reduced row-echelon form? The answer is no. We need only reduce A to row-echelon form. The reason is that going further to reduce a matrix from row-echelon form to reduced row-echelon form does not alter the position or number of leading ones in a matrix, and therefore the column vectors containing leading ones in any row-echelon form of A will correspond to the column vectors containing leading ones in the reduced row-echelon form of A. Consequently, we have established the following result.

Theorem 4.8.6   Let A be an $m \times n$ matrix. The set of column vectors of A corresponding to those column vectors containing leading ones in any row-echelon form of A is a basis for colspace(A).

Example 4.8.7   Determine a basis for colspace(A) if

$$A = \begin{bmatrix} 1 & 2 & -1 & -2 & -1 \\ 2 & 4 & -2 & -3 & -1 \\ 5 & 10 & -5 & -3 & -1 \\ -3 & -6 & 3 & 2 & 1 \end{bmatrix}.$$

Solution: We first reduce A to row-echelon form:

$$A \overset{1}{\sim} \begin{bmatrix} 1 & 2 & -1 & -2 & -1 \\ 0 & 0 & 0 & 1 & 1 \\ 0 & 0 & 0 & 7 & 4 \\ 0 & 0 & 0 & -4 & -2 \end{bmatrix} \overset{2}{\sim} \begin{bmatrix} 1 & 2 & -1 & -2 & -1 \\ 0 & 0 & 0 & 1 & 1 \\ 0 & 0 & 0 & 0 & -3 \\ 0 & 0 & 0 & 0 & 2 \end{bmatrix}$$

$$\overset{3}{\sim} \begin{bmatrix} 1 & 2 & -1 & -2 & -1 \\ 0 & 0 & 0 & 1 & 1 \\ 0 & 0 & 0 & 0 & 1 \\ 0 & 0 & 0 & 0 & 2 \end{bmatrix} \overset{4}{\sim} \begin{bmatrix} 1 & 2 & -1 & -2 & -1 \\ 0 & 0 & 0 & 1 & 1 \\ 0 & 0 & 0 & 0 & 1 \\ 0 & 0 & 0 & 0 & 0 \end{bmatrix}.$$

---

1. $A_{12}(-2)$, $A_{13}(-5)$, $A_{14}(3)$ **2.** $A_{23}(-7)$, $A_{24}(4)$ **3.** $M_3(-\frac{1}{3})$ **4.** $A_{34}(-2)$

Since the first, fourth, and fifth column vectors in this row-echelon form of A contain the leading ones, it follows from Theorem 4.8.6 that the set of *corresponding* column vectors in A is a basis for colspace(A). Consequently, a basis for colspace(A) is

$$\{(1, 2, 5, -3), (-2, -3, -3, 2), (-1, -1, -1, 1)\}.$$

Hence, colspace(A) is a three-dimensional subspace of $\mathbb{R}^4$. Notice from the row-echelon form of A that a basis for rowspace(A) is $\{(1, 2, -1, -2, -1), (0, 0, 0, 1, 1), (0, 0, 0, 0, 1)\}$ so that rowspace(A) is a three-dimensional subspace of $\mathbb{R}^5$.   □

We now summarize the discussion of row space and column space.

Summary: Let A be an $m \times n$ matrix. In order to determine a basis for rowspace(A) and a basis for colspace(A), we reduce A to row-echelon form.

---

1. The row vectors containing the leading ones in the row-echelon form give a basis for rowspace(A) (a subspace of $\mathbb{R}^n$).
2. The column vectors of A corresponding to the column vectors containing the leading ones in the row-echelon form give a basis for colspace(A) (a subspace of $\mathbb{R}^m$).

Since the number of vectors in a basis for rowspace(A) or in a basis for colspace(A) is equal to the number of leading ones in any row-echelon form of A, it follows that

$$\dim[\text{rowspace}(A)] = \dim[\text{colspace}(A)] = \text{rank}(A).$$

However, we emphasize that rowspace(A) and colspace(A) are, in general, subspaces of *different* vector spaces. In Example 4.8.7, for instance, rowspace(A) is a subspace of $\mathbb{R}^5$, while colspace(A) is a subspace of $\mathbb{R}^4$. For an $m \times n$ matrix, rowspace(A) is a subspace of $\mathbb{R}^n$, whereas colspace(A) is a subspace of $\mathbb{R}^m$.

Exercises for 4.8

Key Terms

Row space, Column space.

Skills

- Be able to compute a basis for the row space of a matrix.

- Be able to compute a basis for the column space of a matrix.

True-False Review

For Questions 1–6, decide if the given statement is **true** or **false**, and give a brief justification for your answer. If true, you can quote a relevant definition or theorem from the text. If false, provide an example, illustration, or brief explanation of why the statement is false.

1. If A is an $m \times n$ matrix such that rowspace(A) = colspace(A), then $m = n$.

2. A basis for the row space of a matrix A consists of the row vectors of any row-echelon form of A.

3. The nonzero column vectors of a row-echelon form of a matrix A form a basis for colspace(A).

4. The sets rowspace(A) and colspace(A) have the same dimension.

5. If A is an $n \times n$ *invertible* matrix, then rowspace(A) = $\mathbb{R}^n$.

6. If A is an $n \times n$ *invertible* matrix, then colspace(A) = $\mathbb{R}^n$.

Problems

For Problems 1–6, determine a basis for rowspace(A) and a basis for colspace(A).

1. $A = \begin{bmatrix} 1 & -2 \\ -3 & 6 \end{bmatrix}$.

2. $A = \begin{bmatrix} 1 & 1 & -3 & 2 \\ 3 & 4 & -11 & 7 \end{bmatrix}$.

3. $A = \begin{bmatrix} 1 & 2 & 3 \\ 5 & 6 & 7 \\ 9 & 10 & 11 \end{bmatrix}$.

4. $A = \begin{bmatrix} 0 & 3 & 1 \\ 0 & -6 & -2 \\ 0 & 12 & 4 \end{bmatrix}$.

5. $A = \begin{bmatrix} 1 & 2 & -1 & 3 \\ 3 & 6 & -3 & 5 \\ 1 & 2 & -1 & -1 \\ 5 & 10 & -5 & 7 \end{bmatrix}$.

6. $A = \begin{bmatrix} 1 & -1 & 2 & 3 \\ 1 & 1 & -2 & 6 \\ 3 & 1 & 4 & 2 \end{bmatrix}$.

For Problems 7–10, use the ideas in this section to determine a basis for the subspace of $\mathbb{R}^n$ spanned by the given set of vectors.

7. $\{(1, -1, 2), (5, -4, 1), (7, -5, -4)\}$.

8. $\{(1, 3, 3), (1, 5, -1), (2, 7, 4), (1, 4, 1)\}$.

9. $\{(1, 1, -1, 2), (2, 1, 3, -4), (1, 2, -6, 10)\}$.

10. $\{(1, 4, 1, 3), (2, 8, 3, 5), (1, 4, 0, 4), (2, 8, 2, 6)\}$.

11. Let
$$A = \begin{bmatrix} -3 & 9 \\ 1 & -3 \end{bmatrix}.$$

Find a basis for rowspace(A) and colspace(A). Make a sketch to show each subspace in the xy-plane.

12. Let $A = \begin{bmatrix} 1 & 2 & 4 \\ 5 & 11 & 21 \\ 3 & 7 & 13 \end{bmatrix}$.

(a) Find a basis for rowspace(A) and colspace(A).

(b) Show that rowspace(A) corresponds to the plane with Cartesian equation $2x + y - z = 0$, whereas colspace(A) corresponds to the plane with Cartesian equation $2x - y + z = 0$.

13. Give examples to show how each type of elementary row operation applied to a matrix can change the column space of the matrix.

14. Give an example of a square matrix A whose row space and column space have no nonzero vectors in common.

4.9   The Rank-Nullity Theorem

In Section 4.3, we defined the null space of a real $m \times n$ matrix A to be the set of all real solutions to the associated homogeneous linear system $A\mathbf{x} = \mathbf{0}$. Thus,

$$\text{nullspace}(A) = \{\mathbf{x} \in \mathbb{R}^n : A\mathbf{x} = \mathbf{0}\}.$$

The dimension of nullspace(A) is referred to as the **nullity** of A and is denoted nullity(A). In order to find nullity(A), we need to determine a basis for nullspace(A). Recall that if rank(A) $= r$, then any row-echelon form of A contains r leading ones, which correspond to the bound variables in the linear system. Thus, there are $n - r$ columns without leading ones, which correspond to free variables in the solution of the system $A\mathbf{x} = \mathbf{0}$. Hence, there are $n - r$ free variables in the solution of the system $A\mathbf{x} = \mathbf{0}$. We might therefore suspect that nullity(A) $= n - r$. Our next theorem, often referred to as the Rank-Nullity Theorem, establishes that this is indeed the case.

Theorem 4.9.1   **(Rank-Nullity Theorem)**

For any $m \times n$ matrix A,

$$\text{rank}(A) + \text{nullity}(A) = n. \tag{4.9.1}$$

Proof   If rank(A) $= n$, then by the Invertible Matrix Theorem, the only solution to $A\mathbf{x} = \mathbf{0}$ is the trivial solution $\mathbf{x} = \mathbf{0}$. Hence, in this case, nullspace(A) $= \{\mathbf{0}\}$, so nullity(A) $= 0$ and Equation (4.9.1) holds.

Now suppose rank(A) $= r < n$. In this case, there are $n - r > 0$ free variables in the solution to $A\mathbf{x} = \mathbf{0}$. Let $t_1, t_2, \ldots, t_{n-r}$ denote these free variables (chosen as those variables not attached to a leading one in any row-echelon form of A), and let $\mathbf{x}_1, \mathbf{x}_2, \ldots, \mathbf{x}_{n-r}$ denote the solutions obtained by sequentially setting each free variable to 1 and the remaining free variables to zero. Note that $\{\mathbf{x}_1, \mathbf{x}_2, \ldots, \mathbf{x}_{n-r}\}$ is linearly independent. Moreover, every solution to $A\mathbf{x} = \mathbf{0}$ is a linear combination of $\mathbf{x}_1, \mathbf{x}_2, \ldots, \mathbf{x}_{n-r}$:

$$\mathbf{x} = t_1\mathbf{x}_1 + t_2\mathbf{x}_2 + \cdots + t_{n-r}\mathbf{x}_{n-r},$$

which shows that $\{\mathbf{x}_1, \mathbf{x}_2, \ldots, \mathbf{x}_{n-r}\}$ spans nullspace(A). Thus, $\{\mathbf{x}_1, \mathbf{x}_2, \ldots, \mathbf{x}_{n-r}\}$ is a basis for nullspace(A), and nullity(A) $= n - r$. ∎

Example 4.9.2 If

$$
A = \begin{bmatrix} 1 & 1 & 2 & 3 \\ 3 & 4 & -1 & 2 \\ -1 & -2 & 5 & 4 \end{bmatrix},
$$

find a basis for nullspace(A) and verify Theorem 4.9.1.

Solution: We must find all solutions to $A\mathbf{x} = \mathbf{0}$. Reducing the augmented matrix of this system yields

$$
A^{\#} \stackrel{1}{\sim} \begin{bmatrix} 1 & 1 & 2 & 3 & 0 \\ 0 & 1 & -7 & -7 & 0 \\ 0 & -1 & 7 & 7 & 0 \end{bmatrix} \stackrel{2}{\sim} \begin{bmatrix} 1 & 1 & 2 & 3 & 0 \\ 0 & 1 & -7 & -7 & 0 \\ 0 & 0 & 0 & 0 & 0 \end{bmatrix}.
$$

> **1.** $A_{12}(-3),\ A_{13}(1)$ **2.** $A_{23}(1)$

Consequently, there are two free variables, $x_3 = t_1$ and $x_4 = t_2$, so that

$$
x_2 = 7t_1 + 7t_2, \qquad x_1 = -9t_1 - 10t_2.
$$

Hence,

$$
\begin{aligned}
\text{nullspace}(A) &= \{(-9t_1 - 10t_2, 7t_1 + 7t_2, t_1, t_2) : t_1, t_2 \in \mathbb{R}\} \\
&= \{t_1(-9, 7, 1, 0) + t_2(-10, 7, 0, 1) : t_1, t_2 \in \mathbb{R}\} \\
&= \text{span}\{(-9, 7, 1, 0), (-10, 7, 0, 1)\}.
\end{aligned}
$$

Since the two vectors in this spanning set are not proportional, they are linearly independent. Consequently, a basis for nullspace(A) is $\{(-9, 7, 1, 0), (-10, 7, 0, 1)\}$, so that nullity($A$) = 2. In this problem, A is a 3×4 matrix, and so, in the Rank-Nullity Theorem, $n = 4$. Further, from the foregoing row-echelon form of the augmented matrix of the system $A\mathbf{x} = \mathbf{0}$, we see that rank(A) = 2. Hence,

$$
\text{rank}(A) + \text{nullity}(A) = 2 + 2 = 4 = n,
$$

and the Rank-Nullity Theorem is verified.   □

Systems of Linear Equations

We now examine the linear structure of the solution set to the linear system $A\mathbf{x} = \mathbf{b}$ in terms of the concepts introduced in the last few sections. First we consider the homogeneous case $\mathbf{b} = \mathbf{0}$.

Corollary 4.9.3 Let A be an $m \times n$ matrix, and consider the corresponding homogeneous linear system $A\mathbf{x} = \mathbf{0}$.

1. If rank(A) = n, then $A\mathbf{x} = \mathbf{0}$ has only the trivial solution, so nullspace(A) = $\{\mathbf{0}\}$.

2. If rank(A) = $r < n$, then $A\mathbf{x} = \mathbf{0}$ has an infinite number of solutions, all of which can be obtained from

$$
\mathbf{x} = c_1\mathbf{x}_1 + c_2\mathbf{x}_2 + \cdots + c_{n-r}\mathbf{x}_{n-r}, \tag{4.9.2}
$$

where $\{\mathbf{x}_1, \mathbf{x}_2, \ldots, \mathbf{x}_{n-r}\}$ is any linearly independent set of $n - r$ solutions to $A\mathbf{x} = \mathbf{0}$.

Proof Note that part 1 is a restatement of previous results, or can be quickly deduced from the Rank-Nullity Theorem. Now for part 2, assume that $\text{rank}(A) = r < n$. By the Rank-Nullity Theorem, $\text{nullity}(A) = n - r$. Thus, from Theorem 4.6.10, if $\{\mathbf{x}_1, \mathbf{x}_2, \ldots, \mathbf{x}_{n-r}\}$ is any set of $n - r$ linearly independent solutions to $A\mathbf{x} = \mathbf{0}$, it is a basis for $\text{nullspace}(A)$, and so all vectors in $\text{nullspace}(A)$ can be written as

$$\mathbf{x} = c_1\mathbf{x}_1 + c_2\mathbf{x}_2 + \cdots + c_{n-r}\mathbf{x}_{n-r},$$

for appropriate values of the constants $c_1, c_2, \ldots, c_{n-r}$.   ∎

Remark   The expression (4.9.2) is referred to as the **general solution** to the system $A\mathbf{x} = \mathbf{0}$.

We now turn our attention to nonhomogeneous linear systems. We begin by formulating Theorem 2.5.9 in terms of $\text{colspace}(A)$.

Theorem 4.9.4   Let A be an $m \times n$ matrix and consider the linear system $A\mathbf{x} = \mathbf{b}$.

1. If $\mathbf{b}$ is not in $\text{colspace}(A)$, then the system is inconsistent.
2. If $\mathbf{b} \in \text{colspace}(A)$, then the system is consistent and has the following:

   (a) a unique solution if and only if $\dim[\text{colspace}(A)] = n$.
   (b) an infinite number of solutions if and only if $\dim[\text{colspace}(A)] < n$.

Proof If we write A in terms of its column vectors as $A = [\mathbf{a}_1, \mathbf{a}_2, \ldots, \mathbf{a}_n]$, then the linear system $A\mathbf{x} = \mathbf{b}$ can be written as

$$x_1\mathbf{a}_1 + x_2\mathbf{a}_2 + \cdots + x_n\mathbf{a}_n = \mathbf{b}.$$

Consequently, the linear system is consistent if and only if the vector $\mathbf{b}$ is a linear combination of the column vectors of A. Thus, the system is consistent if and only if $\mathbf{b} \in \text{colspace}(A)$. This proves part 1. Parts 2(a) and 2(b) follow directly from Theorem 2.5.9, since $\text{rank}(A) = \dim[\text{colspace}(A)]$.   ∎

The set of all solutions to a nonhomogeneous linear system is not a vector space, since, for example, it does not contain the zero vector, but the linear structure of $\text{nullspace}(A)$ can be used to determine the general form of the solution of a nonhomogeneous system.

Theorem 4.9.5   Let A be an $m \times n$ matrix. If $\text{rank}(A) = r < n$ and $\mathbf{b} \in \text{colspace}(A)$, then all solutions to $A\mathbf{x} = \mathbf{b}$ are of the form

$$\mathbf{x} = c_1\mathbf{x}_1 + c_2\mathbf{x}_2 + \cdots + c_{n-r}\mathbf{x}_{n-r} + \mathbf{x}_p, \tag{4.9.3}$$

where $\mathbf{x}_p$ is any particular solution to $A\mathbf{x} = \mathbf{b}$, and $\{\mathbf{x}_1, \mathbf{x}_2, \ldots, \mathbf{x}_{n-r}\}$ is a basis for $\text{nullspace}(A)$.

Proof Since $\mathbf{x}_p$ is a solution to $A\mathbf{x} = \mathbf{b}$, we have

$$A\mathbf{x}_p = \mathbf{b}. \tag{4.9.4}$$

Let $\mathbf{x} = \mathbf{u}$ be an arbitrary solution to $A\mathbf{x} = \mathbf{b}$. Then we also have

$$A\mathbf{u} = \mathbf{b}. \tag{4.9.5}$$

Subtracting (4.9.4) from (4.9.5) yields

$$A\mathbf{u} - A\mathbf{x}_p = \mathbf{0},$$

or equivalently,

$$A(\mathbf{u} - \mathbf{x}_p) = \mathbf{0}.$$

Consequently, the vector $\mathbf{u} - \mathbf{x}_p$ is in nullspace(A), and so there exist scalars $c_1, c_2,$ $\ldots, c_{n-r}$ such that

$$\mathbf{u} - \mathbf{x}_p = c_1\mathbf{x}_1 + c_2\mathbf{x}_2 + \cdots + c_{n-r}\mathbf{x}_{n-r},$$

since $\{\mathbf{x}_1, \mathbf{x}_2, \ldots, \mathbf{x}_{n-r}\}$ is a basis for nullspace(A). Hence,

$$\mathbf{u} = c_1\mathbf{x}_1 + c_2\mathbf{x}_2 + \cdots + c_{n-r}\mathbf{x}_{n-r} + \mathbf{x}_p,$$

as required.   ∎

Remark   The expression given in Equation (4.9.3) is called the **general solution** to $A\mathbf{x} = \mathbf{b}$. It has the structure

$$\mathbf{x} = \mathbf{x}_c + \mathbf{x}_p,$$

where

$$\mathbf{x}_c = c_1\mathbf{x}_1 + c_2\mathbf{x}_2 + \cdots + c_{n-r}\mathbf{x}_{n-r}$$

is the general solution of the associated homogeneous system and $\mathbf{x}_p$ is one particular solution of the nonhomogeneous system. In later chapters, we will see that this structure is also apparent in the solution of all linear differential equations and in all linear systems of differential equations. It is a result of the linearity inherent in the problem, rather than the specific problem that we are studying. The unifying concept, in addition to the vector space, is the idea of a linear transformation, which we will study in the next chapter.

Example 4.9.6   Let

$$A = \begin{bmatrix} 1 & 1 & 2 & 3 \\ 3 & 4 & -1 & 2 \\ -1 & -2 & 5 & 4 \end{bmatrix} \quad \text{and} \quad \mathbf{b} = \begin{bmatrix} 3 \\ 10 \\ -4 \end{bmatrix}.$$

Verify that $\mathbf{x}_p = (1, 1, -1, 1)$ is a particular solution to $A\mathbf{x} = \mathbf{b}$, and use Theorem 4.9.5 to determine the general solution to the system.

Solution: For the given $\mathbf{x}_p$, we have

$$A\mathbf{x}_p = \begin{bmatrix} 1 & 1 & 2 & 3 \\ 3 & 4 & -1 & 2 \\ -1 & -2 & 5 & 4 \end{bmatrix} \begin{bmatrix} 1 \\ 1 \\ -1 \\ 1 \end{bmatrix} = \begin{bmatrix} 3 \\ 10 \\ -4 \end{bmatrix} = \mathbf{b}.$$

Consequently, $\mathbf{x}_p = (1, 1, -1, 1)$ is a particular solution to $A\mathbf{x} = \mathbf{b}$. Further, from Example 4.9.2, a basis for nullspace(A) is $\{\mathbf{x}_1, \mathbf{x}_2\}$, where $\mathbf{x}_1 = (-9, 7, 1, 0)$ and $\mathbf{x}_2 = (-10, 7, 0, 1)$. Thus, the general solution to $A\mathbf{x} = \mathbf{0}$ is

$$\mathbf{x}_c = c_1\mathbf{x}_1 + c_2\mathbf{x}_2,$$

and therefore, from Theorem 4.9.5, the general solution to $A\mathbf{x} = \mathbf{b}$ is

$$\mathbf{x} = c_1\mathbf{x}_1 + c_2\mathbf{x}_2 + \mathbf{x}_p = c_1(-9, 7, 1, 0) + c_2(-10, 7, 0, 1) + (1, 1, -1, 1),$$

which can be written as

$$\mathbf{x} = (-9c_1 - 10c_2 + 1, 7c_1 + 7c_2 + 1, c_1 - 1, c_2 + 1). \qquad \square$$

Exercises for 4.9

Skills

- For a given matrix A, be able to determine the rank from the nullity, or the nullity from the rank.

- Know the relationship between the rank of a matrix A and the consistency of a linear system $A\mathbf{x} = \mathbf{b}$.

- Know the relationship between the column space of a matrix A and the consistency of a linear system $A\mathbf{x} = \mathbf{b}$.

- Be able to formulate the solution set to a linear system $A\mathbf{x} = \mathbf{b}$ in terms of the solution set to the corresponding homogeneous linear equation.

True-False Review

For Questions 1–9, decide if the given statement is **true** or **false**, and give a brief justification for your answer. If true, you can quote a relevant definition or theorem from the text. If false, provide an example, illustration, or brief explanation of why the statement is false.

1. For an $m \times n$ matrix A, the nullity of A must be at least $|m - n|$.

2. If A is a 7×9 matrix with nullity$(A) = 2$, then rowspace$(A) = \mathbb{R}^7$.

3. If A is a 9×7 matrix with nullity$(A) = 0$, then rowspace$(A) = \mathbb{R}^7$.

4. The nullity of an $n \times n$ *upper triangular* matrix A is simply the number of zeros appearing on the main diagonal of A.

5. An $n \times n$ matrix A for which nullspace$(A) = $ colspace(A) cannot be invertible.

6. For all $m \times n$ matrices A and B, nullity$(A + B) = $ nullity$(A) + $ nullity(B).

7. For all $n \times n$ matrices A and B, nullity$(AB) = $ nullity$(A) \cdot$ nullity(B).

8. For all $n \times n$ matrices A and B, nullity$(AB) \geq$ nullity(B).

9. If $\mathbf{x}_p$ is a solution to the linear system $A\mathbf{x} = \mathbf{b}$, then $\mathbf{y} + \mathbf{x}_p$ is also a solution for any $\mathbf{y}$ in nullspace(A).

Problems

For Problems 1–4, determine the null space of A and verify the Rank-Nullity Theorem.

1. $A = \begin{bmatrix} 1 & 0 & -6 & -1 \end{bmatrix}$.

2. $A = \begin{bmatrix} 2 & -1 \\ -4 & 2 \end{bmatrix}$.

3. $A = \begin{bmatrix} 1 & 1 & -1 \\ 3 & 4 & 4 \\ 1 & 1 & 0 \end{bmatrix}$.

4. $A = \begin{bmatrix} 1 & 4 & -1 & 3 \\ 2 & 9 & -1 & 7 \\ 2 & 8 & -2 & 6 \end{bmatrix}$.

For Problems 5–8, determine the nullity of A "by inspection" by appealing to the Rank-Nullity Theorem. Avoid computations.

5. $A = \begin{bmatrix} 2 & -3 \\ 0 & 0 \\ -4 & 6 \\ 22 & -33 \end{bmatrix}$.

6. $A = \begin{bmatrix} 1 & 3 & -3 & 2 & 5 \\ -4 & -12 & 12 & -8 & -20 \\ 0 & 0 & 0 & 0 & 0 \\ 1 & 3 & -3 & 2 & 6 \end{bmatrix}$.

7. $A = \begin{bmatrix} 0 & 1 & 0 \\ 0 & 1 & 0 \\ 0 & 0 & 1 \\ 0 & 0 & 1 \end{bmatrix}$.

8. $A = \begin{bmatrix} 0 & 0 & 0 & -2 \end{bmatrix}$.

For Problems 9–12, determine the solution set to $A\mathbf{x} = \mathbf{b}$, and show that all solutions are of the form (4.9.3).

9. $A = \begin{bmatrix} 1 & 3 & -1 \\ 2 & 7 & 9 \\ 1 & 5 & 21 \end{bmatrix}$, $\mathbf{b} = \begin{bmatrix} 4 \\ 11 \\ 10 \end{bmatrix}$.

10. $A = \begin{bmatrix} 2 & -1 & 1 & 4 \\ 1 & -1 & 2 & 3 \\ 1 & -2 & 5 & 5 \end{bmatrix}$, $\mathbf{b} = \begin{bmatrix} 5 \\ 6 \\ 13 \end{bmatrix}$.

11. $A = \begin{bmatrix} 1 & 1 & -2 \\ 3 & -1 & -7 \\ 1 & 1 & 1 \\ 2 & 2 & -4 \end{bmatrix}$, $\mathbf{b} = \begin{bmatrix} -3 \\ 2 \\ 0 \\ -6 \end{bmatrix}$.

12. $A = \begin{bmatrix} 1 & 1 & -1 & 5 \\ 0 & 2 & -1 & 7 \\ 4 & 2 & -3 & 13 \end{bmatrix}$, $\mathbf{b} = \begin{bmatrix} 0 \\ 0 \\ 0 \end{bmatrix}$.

13. Show that a 3×7 matrix A with nullity$(A) = 4$ must have colspace$(A) = \mathbb{R}^3$. Is rowspace$(A) = \mathbb{R}^3$?

14. Show that a 6×4 matrix A with nullity$(A) = 0$ must have rowspace$(A) = \mathbb{R}^4$. Is colspace$(A) = \mathbb{R}^4$?

15. Prove that if rowspace$(A) = $ nullspace(A), then A contains an even number of columns.

16. Show that a 5×7 matrix A must have $2 \le$ nullity$(A) \le 7$. Give an example of a 5×7 matrix A with nullity$(A) = 2$ and an example of a 5×7 matrix A with nullity$(A) = 7$.

17. Show that 3×8 matrix A must have $5 \le$ nullity$(A) \le 8$. Give an example of a 3×8 matrix A with nullity$(A) = 5$ and an example of a 3×8 matrix A with nullity$(A) = 8$.

18. Prove that if A and B are $n \times n$ matrices and A is invertible, then

$$\text{nullity}(AB) = \text{nullity}(B).$$

[**Hint:** $B\mathbf{x} = \mathbf{0}$ if and only if $AB\mathbf{x} = \mathbf{0}$.]

4.10   The Invertible Matrix Theorem II

In Section 2.8, we gave a list of characterizations of invertible matrices (Theorem 2.8.1). In view of the concepts introduced in this chapter, we are now in a position to add to the list that was begun there.

Theorem 4.10.1   **(Invertible Matrix Theorem)**

Let A be an $n \times n$ matrix with real elements. The following conditions on A are equivalent:

(a) A is invertible.

(h) nullity$(A) = 0$.

(i) nullspace$(A) = \{\mathbf{0}\}$.

(j) The columns of A form a linearly independent set of vectors in $\mathbb{R}^n$.

(k) colspace$(A) = \mathbb{R}^n$ (that is, the columns of A span $\mathbb{R}^n$).

(l) The columns of A form a basis for $\mathbb{R}^n$.

(m) The rows of A form a linearly independent set of vectors in $\mathbb{R}^n$.

(n) rowspace$(A) = \mathbb{R}^n$ (that is, the rows of A span $\mathbb{R}^n$).

(o) The rows of A form a basis for $\mathbb{R}^n$.

(p) A^T is invertible.

Proof The equivalence of (a) and (h) follows at once from Theorem 2.8.1(d) and the Rank-Nullity Theorem (Theorem 4.9.1). The equivalence of (h) and (i) is immediately clear. The equivalence of (a) and (j) is immediate from Theorem 2.8.1(c) and Theorem 4.5.14. Since the dimension of colspace(A) is simply rank(A), the equivalence of (a) and (k) is immediate from Theorem 2.8.1(d). Next, from the definition of a basis, we see that (j) and (k) are logically equivalent to (l). Moreover, since the row space and column space of A have the same dimension, (k) and (n) are equivalent. Since $\text{rowspace}(A) = \text{colspace}(A^T)$, the equivalence of (k) and (n) proves that (a) and (p) are equivalent. Finally, the equivalence of (a) and (p) proves that (j) is equivalent to (m) and that (l) is equivalent to (o). ∎

Example 4.10.2 Do the rows of the matrix below span $\mathbb{R}^4$?

$$A = \begin{bmatrix} -2 & -2 & 1 & 3 \\ 3 & 3 & 0 & -1 \\ -1 & -1 & -2 & -5 \\ 2 & 2 & 1 & 1 \end{bmatrix}$$

Solution: We see by inspection that the columns of A are linearly dependent, since the first two columns are identical. Therefore, by the equivalence of (j) and (n) in the Invertible Matrix Theorem, the rows of A do not span $\mathbb{R}^4$. □

Example 4.10.3 If A is an $n \times n$ matrix such that the linear system $A^T\mathbf{x} = \mathbf{0}$ has no nontrivial solution $\mathbf{x}$, then $\text{nullspace}(A^T) = \{\mathbf{0}\}$, and thus A^T is invertible by the equivalence of (a) and (i) in the Invertible Matrix Theorem. Thus, by the same theorem, we can conclude that the columns of A form a linearly independent set. □

Despite the lengthy list of characterizations of invertible matrices that we have been able to develop so far, this list is still by no means complete. In the next chapter, we will use linear transformations and eigenvalues to provide further characterizations of invertible matrices.

Exercises for 4.10

Skills

- Be well familiar with all of the conditions (a)–(p) in the Invertible Matrix Theorem that characterize invertible matrices.

True-False Review

For Questions 1–10, decide if the given statement is **true** or **false**, and give a brief justification for your answer. If true, you can quote a relevant definition or theorem from the text. If false, provide an example, illustration, or brief explanation of why the statement is false.

1. The set of all row vectors of an invertible matrix is linearly independent.

2. An $n \times n$ matrix can have n linearly independent rows and n linearly dependent columns.

3. The set of all row vectors of an $n \times n$ matrix can be linearly dependent while the set of all columns is linearly independent.

4. If A is an $n \times n$ matrix with $\det(A) = 0$, then the columns of A must form a basis for $\mathbb{R}^n$.

5. If A and B are row-equivalent $n \times n$ matrices such that $\text{rowspace}(A) \neq \mathbb{R}^n$, then $\text{colspace}(B) \neq \mathbb{R}^n$.

6. If E is an $n \times n$ elementary matrix and A is an $n \times n$ matrix with $\text{nullspace}(A) = \{\mathbf{0}\}$, then $\det(EA) = 0$.

7. If A and B are $n \times n$ invertible matrices, then $\text{nullity}([A|B]) = 0$, where $[A|B]$ is the $n \times 2n$ matrix with the blocks A and B as shown.

8. A matrix of the form

$$\begin{bmatrix} 0 & a & 0 \\ b & 0 & c \\ 0 & d & 0 \end{bmatrix}$$

cannot be invertible.

9. A matrix of the form

$$\begin{bmatrix} 0 & a & 0 & b \\ c & 0 & d & 0 \\ 0 & e & 0 & f \\ g & 0 & h & 0 \end{bmatrix}$$

cannot be invertible.

10. A matrix of the form

$$\begin{bmatrix} a & b & c \\ d & e & f \\ g & h & i \end{bmatrix}$$

such that $ae - bd = 0$ cannot be invertible.

4.11   Inner Product Spaces

We now extend the familiar idea of a dot product for geometric vectors to an arbitrary vector space V. This enables us to associate a magnitude with each vector in V and also to define the angle between two vectors in V. The major reason that we want to do this is that, as we will see in the next section, it enables us to construct orthogonal bases in a vector space, and the use of such a basis often simplifies the representation of vectors. We begin with a brief review of the dot product.

Let $\mathbf{x} = (x_1, x_2, x_3)$ and $\mathbf{y} = (y_1, y_2, y_3)$ be two arbitrary vectors in $\mathbb{R}^3$, and consider the corresponding geometric vectors

$$\mathbf{x} = x_1\mathbf{i} + x_2\mathbf{j} + x_3\mathbf{k}, \qquad \mathbf{y} = y_1\mathbf{i} + y_2\mathbf{j} + y_3\mathbf{k}.$$

The dot product of $\mathbf{x}$ and $\mathbf{y}$ can be defined in terms of the components of these vectors as

$$\mathbf{x} \cdot \mathbf{y} = x_1 y_1 + x_2 y_2 + x_3 y_3. \tag{4.11.1}$$

An equivalent geometric definition of the dot product is

$$\mathbf{x} \cdot \mathbf{y} = ||\mathbf{x}||\, ||\mathbf{y}|| \cos\theta, \tag{4.11.2}$$

where $||\mathbf{x}||, ||\mathbf{y}||$ denote the lengths of $\mathbf{x}$ and $\mathbf{y}$ respectively, and $0 \le \theta \le \pi$ is the angle between them. (See Figure 4.11.1.)

Taking $\mathbf{y} = \mathbf{x}$ in Equations (4.11.1) and (4.11.2) yields

$$||\mathbf{x}||^2 = \mathbf{x} \cdot \mathbf{x} = x_1^2 + x_2^2 + x_3^2,$$

so that the length of a geometric vector is given in terms of the dot product by

$$||\mathbf{x}|| = \sqrt{\mathbf{x} \cdot \mathbf{x}} = \sqrt{x_1^2 + x_2^2 + x_3^2}.$$

Furthermore, from Equation (4.11.2), the angle between any two nonzero vectors $\mathbf{x}$ and $\mathbf{y}$ is

$$\cos\theta = \frac{\mathbf{x} \cdot \mathbf{y}}{||\mathbf{x}||\, ||\mathbf{y}||}, \tag{4.11.3}$$

which implies that $\mathbf{x}$ and $\mathbf{y}$ are orthogonal (perpendicular) if and only if

$$\mathbf{x} \cdot \mathbf{y} = 0.$$

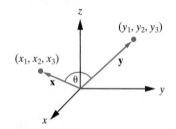

Figure 4.11.1: Defining the dot product in $\mathbb{R}^3$.

In a general vector space, we do not have a geometrical picture to guide us in defining the dot product, hence our definitions must be purely algebraic. We begin by considering the vector space $\mathbb{R}^n$, since there is a natural way to extend Equation (4.11.1) in this case. Before proceeding, we note that from now on we will use the standard terms *inner product* and *norm* in place of dot product and length, respectively.

DEFINITION   4.11.1

Let $\mathbf{x} = (x_1, x_2, \ldots, x_n)$ and $\mathbf{y} = (y_1, y_2, \ldots, y_n)$ be vectors in $\mathbb{R}^n$. We define the **standard inner product in $\mathbb{R}^n$**, denoted $\langle \mathbf{x}, \mathbf{y} \rangle$, by

$$\langle \mathbf{x}, \mathbf{y} \rangle = x_1 y_1 + x_2 y_2 + \cdots + x_n y_n.$$

The **norm** of $\mathbf{x}$ is

$$\|\mathbf{x}\| = \sqrt{\langle \mathbf{x}, \mathbf{x} \rangle} = \sqrt{x_1^2 + x_2^2 + \cdots + x_n^2}.$$

Example 4.11.2   If $\mathbf{x} = (1, -1, 0, 2, 4)$ and $\mathbf{y} = (2, 1, 1, 3, 0)$ in $\mathbb{R}^5$, then

$$\langle \mathbf{x}, \mathbf{y} \rangle = (1)(2) + (-1)(1) + (0)(1) + (2)(3) + (4)(0) = 7,$$
$$\|\mathbf{x}\| = \sqrt{1^2 + (-1)^2 + 0^2 + 2^2 + 4^2} = \sqrt{22},$$
$$\|\mathbf{y}\| = \sqrt{2^2 + 1^2 + 1^2 + 3^2 + 0^2} = \sqrt{15}. \qquad \square$$

Basic Properties of the Standard Inner Product in $\mathbb{R}^n$

In the case of $\mathbb{R}^n$, the definition of the standard inner product was a natural extension of the familiar dot product in $\mathbb{R}^3$. To generalize this definition further to an arbitrary vector space, we isolate the most important properties of the standard inner product in $\mathbb{R}^n$ and use them as the defining criteria for a general notion of an inner product. Let us examine the inner product in $\mathbb{R}^n$ more closely. We view it as a mapping that associates with any two vectors $\mathbf{x} = (x_1, x_2, \ldots, x_n)$ and $\mathbf{y} = (y_1, y_2, \ldots, y_n)$ in $\mathbb{R}^n$ the real number

$$\langle \mathbf{x}, \mathbf{y} \rangle = x_1 y_1 + x_2 y_2 + \cdots + x_n y_n.$$

This mapping has the following properties:

For all $\mathbf{x}$, $\mathbf{y}$, and $\mathbf{z}$ in $\mathbb{R}^n$ and all real numbers k,

1. $\langle \mathbf{x}, \mathbf{x} \rangle \geq 0$. Furthermore, $\langle \mathbf{x}, \mathbf{x} \rangle = 0$ if and only if $\mathbf{x} = \mathbf{0}$.
2. $\langle \mathbf{y}, \mathbf{x} \rangle = \langle \mathbf{x}, \mathbf{y} \rangle$.
3. $\langle k\mathbf{x}, \mathbf{y} \rangle = k \langle \mathbf{x}, \mathbf{y} \rangle$.
4. $\langle \mathbf{x} + \mathbf{y}, \mathbf{z} \rangle = \langle \mathbf{x}, \mathbf{z} \rangle + \langle \mathbf{y}, \mathbf{z} \rangle$.

These properties are easily established using Definition 4.11.1. For example, to prove property 1, we proceed as follows. From Definition 4.11.1,

$$\langle \mathbf{x}, \mathbf{x} \rangle = x_1^2 + x_2^2 + \cdots + x_n^2.$$

Since this is a sum of squares of real numbers, it is necessarily nonnegative. Further, $\langle \mathbf{x}, \mathbf{x} \rangle = 0$ if and only if $x_1 = x_2 = \cdots = x_n = 0$—that is, if and only if $\mathbf{x} = \mathbf{0}$. Similarly, for property 2, we have

$$\langle \mathbf{y}, \mathbf{x} \rangle = y_1 x_1 + y_2 x_2 + \cdots + y_n x_n = x_1 y_1 + x_2 y_2 + \cdots + x_n y_n = \langle \mathbf{x}, \mathbf{y} \rangle.$$

We leave the verification of properties 3 and 4 for the reader.

Definition of a Real Inner Product Space

We now use properties 1–4 as the basic defining properties of an inner product in a real vector space.

DEFINITION   4.11.3

Let V be a real vector space. A mapping that associates with each pair of vectors $\mathbf{u}$ and $\mathbf{v}$ in V a real number, denoted $\langle \mathbf{u}, \mathbf{v} \rangle$, is called an **inner product** in V, provided it satisfies the following properties. For all $\mathbf{u}$, $\mathbf{v}$, and $\mathbf{w}$ in V, and all real numbers k,

1. $\langle \mathbf{u}, \mathbf{u} \rangle \geq 0$. Furthermore, $\langle \mathbf{u}, \mathbf{u} \rangle = 0$ if and only if $\mathbf{u} = \mathbf{0}$.
2. $\langle \mathbf{v}, \mathbf{u} \rangle = \langle \mathbf{u}, \mathbf{v} \rangle$.
3. $\langle k\mathbf{u}, \mathbf{v} \rangle = k \langle \mathbf{u}, \mathbf{v} \rangle$.
4. $\langle \mathbf{u} + \mathbf{v}, \mathbf{w} \rangle = \langle \mathbf{u}, \mathbf{w} \rangle + \langle \mathbf{v}, \mathbf{w} \rangle$.

The **norm** of $\mathbf{u}$ is defined in terms of an inner product by

$$\|\mathbf{u}\| = \sqrt{\langle \mathbf{u}, \mathbf{u} \rangle}.$$

A real vector space together with an inner product defined in it is called a **real inner product space**.

Remarks

1. Observe that $\|\mathbf{u}\| = \sqrt{\langle \mathbf{u}, \mathbf{u} \rangle}$ takes a well-defined nonnegative real value, since property 1 of an inner product guarantees that the norm evaluates the square root of a nonnegative real number.

2. It follows from the discussion above that $\mathbb{R}^n$ together with the inner product defined in Definition 4.11.1 is an example of a real inner product space.

One of the fundamental inner products arises in the vector space $C^0[a, b]$ of all real-valued functions that are *continuous* on the interval $[a, b]$. In this vector space, we define the mapping $\langle f, g \rangle$ by

$$\langle f, g \rangle = \int_a^b f(x)g(x)\, dx, \tag{4.11.4}$$

for all f and g in $C^0[a, b]$. We establish that this mapping defines an inner product in $C^0[a, b]$ by verifying properties 1–4 of Definition 4.11.3. If f is in $C^0[a, b]$, then

$$\langle f, f \rangle = \int_a^b [f(x)]^2\, dx.$$

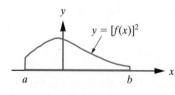

Figure 4.11.2: $\langle f, f \rangle$ gives the area between the graph of $y = [f(x)]^2$ and the x-axis, lying over the interval $[a, b]$.

Since the integrand, $[f(x)]^2$, is a nonnegative continuous function, it follows that $\langle f, f \rangle$ measures the area between the graph $y = [f(x)]^2$ and the x-axis on the interval $[a, b]$. (See Figure 4.11.2.)

Consequently, $\langle f, f \rangle \geq 0$. Furthermore, $\langle f, f \rangle = 0$ if and only if there is zero area between the graph $y = [f(x)]^2$ and the x-axis—that is, if and only if

$$[f(x)]^2 = 0 \qquad \text{for all } x \text{ in } [a, b].$$

Hence, $\langle f, f \rangle = 0$ if and only if $f(x) = 0$, for all x in $[a, b]$, so f must be the zero function. (See Figure 4.11.3.) Consequently, property 1 of Definition 4.11.3 is satisfied. Now let f, g, and h be in $C^0[a, b]$, and let k be an arbitrary real number. Then

$$\langle g, f \rangle = \int_a^b g(x)f(x)\,dx = \int_a^b f(x)g(x)\,dx = \langle f, g \rangle.$$

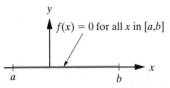

$f(x) = 0$ for all x in $[a,b]$

Figure 4.11.3: $\langle f, f \rangle = 0$ if and only if f is the zero function.

Hence, property 2 of Definition 4.11.3 is satisfied. For property 3, we have

$$\langle kf, g \rangle = \int_a^b (kf)(x)g(x)\,dx = \int_a^b kf(x)g(x)\,dx = k\int_a^b f(x)g(x)\,dx = k\langle f, g \rangle,$$

as needed. Finally,

$$\langle f + g, h \rangle = \int_a^b (f+g)(x)h(x)\,dx = \int_a^b [f(x) + g(x)]h(x)\,dx$$
$$= \int_a^b f(x)h(x)\,dx + \int_a^b g(x)h(x)\,dx = \langle f, h \rangle + \langle g, h \rangle,$$

so that property (4) of Definition 4.11.3 is satisfied. We can now conclude that Equation (4.11.4) does define an inner product in the vector space $C^0[a, b]$.

Example 4.11.4   Use Equation (4.11.4) to determine the inner product of the following functions in $C^0[0, 1]$:

$$f(x) = 8x, \qquad g(x) = x^2 - 1.$$

Also find $\|f\|$ and $\|g\|$.

Solution: From Equation (4.11.4),

$$\langle f, g \rangle = \int_0^1 8x(x^2 - 1)\,dx = \left[2x^4 - 4x^2 \right]_0^1 = -2.$$

Moreover, we have

$$\|f\| = \sqrt{\int_0^1 64x^2\,dx} = \frac{8}{\sqrt{3}}$$

and

$$\|g\| = \sqrt{\int_0^1 (x^2 - 1)^2\,dx} = \sqrt{\int_0^1 (x^4 - 2x^2 + 1)\,dx} = \sqrt{\frac{8}{15}}.\qquad\square$$

We have already seen that the norm concept generalizes the length of a geometric vector. Our next goal is to show how an inner product enables us to define the angle between two vectors in an abstract vector space. The key result is the *Cauchy-Schwarz inequality* established in the next theorem.

Theorem 4.11.5   **(Cauchy-Schwarz Inequality)**

Let $\mathbf{u}$ and $\mathbf{v}$ be arbitrary vectors in a real inner product space V. Then

$$|\langle \mathbf{u}, \mathbf{v} \rangle| \leq \|\mathbf{u}\|\,\|\mathbf{v}\|. \qquad (4.11.5)$$

Proof Let k be an arbitrary real number. For the vector $\mathbf{u} + k\mathbf{v}$, we have

$$0 \le ||\mathbf{u} + k\mathbf{v}||^2 = \langle \mathbf{u} + k\mathbf{v}, \mathbf{u} + k\mathbf{v} \rangle. \tag{4.11.6}$$

But, using the properties of a real inner product,

$$
\begin{aligned}
\langle \mathbf{u} + k\mathbf{v}, \mathbf{u} + k\mathbf{v} \rangle &= \langle \mathbf{u}, \mathbf{u} + k\mathbf{v} \rangle + \langle k\mathbf{v}, \mathbf{u} + k\mathbf{v} \rangle \\
&= \langle \mathbf{u} + k\mathbf{v}, \mathbf{u} \rangle + \langle \mathbf{u} + k\mathbf{v}, k\mathbf{v} \rangle \\
&= \langle \mathbf{u}, \mathbf{u} \rangle + \langle k\mathbf{v}, \mathbf{u} \rangle + \langle \mathbf{u}, k\mathbf{v} \rangle + \langle k\mathbf{v}, k\mathbf{v} \rangle \\
&= \langle \mathbf{u}, \mathbf{u} \rangle + 2\langle k\mathbf{v}, \mathbf{u} \rangle + k\langle \mathbf{v}, k\mathbf{v} \rangle \\
&= \langle \mathbf{u}, \mathbf{u} \rangle + 2\langle k\mathbf{v}, \mathbf{u} \rangle + k\langle k\mathbf{v}, \mathbf{v} \rangle \\
&= \langle \mathbf{u}, \mathbf{u} \rangle + 2\langle k\mathbf{v}, \mathbf{u} \rangle + k^2\langle \mathbf{v}, \mathbf{v} \rangle \\
&= ||\mathbf{u}||^2 + 2k\langle \mathbf{v}, \mathbf{u} \rangle + k^2||\mathbf{v}||^2.
\end{aligned}
$$

Consequently, (4.11.6) implies that

$$||\mathbf{v}||^2 k^2 + 2\langle \mathbf{u}, \mathbf{v} \rangle k + ||\mathbf{u}||^2 \ge 0. \tag{4.11.7}$$

The left-hand side of this inequality defines the quadratic expression

$$P(k) = ||\mathbf{v}||^2 k^2 + 2\langle \mathbf{u}, \mathbf{v} \rangle k + ||\mathbf{u}||^2.$$

The discriminant of this quadratic is

$$\Delta = 4(\langle \mathbf{u}, \mathbf{v} \rangle)^2 - 4||\mathbf{u}||^2 ||\mathbf{v}||^2.$$

If $\Delta > 0$, then $P(k)$ has two real and distinct roots. This would imply that the graph of P crosses the k-axis and, therefore, P would assume negative values, contrary to (4.11.7). Consequently, we must have $\Delta \le 0$. That is,

$$4(\langle \mathbf{u}, \mathbf{v} \rangle)^2 - 4||\mathbf{u}||^2 ||\mathbf{v}||^2 \le 0,$$

or equivalently,

$$(\langle \mathbf{u}, \mathbf{v} \rangle)^2 \le ||\mathbf{u}||^2 ||\mathbf{v}||^2.$$

Hence,

$$|\langle \mathbf{u}, \mathbf{v} \rangle| \le ||\mathbf{u}|| \, ||\mathbf{v}||. \qquad \blacksquare$$

If $\mathbf{u}$ and $\mathbf{v}$ are arbitrary vectors in a real inner product space V, then $\langle \mathbf{u}, \mathbf{v} \rangle$ is a real number, and so (4.11.5) can be written in the equivalent form

$$-||\mathbf{u}|| \, ||\mathbf{v}|| \le \langle \mathbf{u}, \mathbf{v} \rangle \le ||\mathbf{u}|| \, ||\mathbf{v}||.$$

Consequently, provided that $\mathbf{u}$ and $\mathbf{v}$ are nonzero vectors, we have

$$-1 \le \frac{\langle \mathbf{u}, \mathbf{v} \rangle}{||\mathbf{u}|| \, ||\mathbf{v}||} \le 1.$$

Thus, each pair of nonzero vectors in a *real* inner product space V determines a unique angle θ by

$$\cos \theta = \frac{\langle \mathbf{u}, \mathbf{v} \rangle}{||\mathbf{u}|| \, ||\mathbf{v}||}, \qquad 0 \le \theta \le \pi. \tag{4.11.8}$$

We call θ the angle between $\mathbf{u}$ and $\mathbf{v}$. In the case when $\mathbf{u}$ and $\mathbf{v}$ are geometric vectors, the formula (4.11.8) coincides with Equation (4.11.3).

Example 4.11.6   Determine the angle between the vectors $\mathbf{u} = (1, -1, 2, 3)$ and $\mathbf{v} = (-2, 1, 2, -2)$ in $\mathbb{R}^4$.

Solution: Using the standard inner product in $\mathbb{R}^4$ yields

$$\langle \mathbf{u}, \mathbf{v} \rangle = -5, \qquad ||\mathbf{u}|| = \sqrt{15}, \qquad ||\mathbf{v}|| = \sqrt{13},$$

so that the angle between $\mathbf{u}$ and $\mathbf{v}$ is given by

$$\cos\theta = -\frac{5}{\sqrt{15}\sqrt{13}} = -\frac{\sqrt{195}}{39}, \qquad 0 \le \theta \le \pi.$$

Hence,

$$\theta = \arccos\left(-\frac{\sqrt{195}}{39}\right) \approx 1.937 \text{ radians} \approx 110° \, 58'. \qquad \square$$

Example 4.11.7   Use the inner product (4.11.4) to determine the angle between the functions $f_1(x) = \sin 2x$ and $f_2(x) = \cos 2x$ on the interval $[-\pi, \pi]$.

Solution: Using the inner product (4.11.4), we have

$$\langle f_1, f_2 \rangle = \int_{-\pi}^{\pi} \sin 2x \, \cos 2x \, dx = \frac{1}{2} \int_{-\pi}^{\pi} \sin 4x \, dx = \frac{1}{8}\left(-\cos 4x\right)\Big|_{-\pi}^{\pi} = 0.$$

Consequently, the angle between the two functions satisfies

$$\cos\theta = 0, \qquad 0 \le \theta \le \pi,$$

which implies that $\theta = \pi/2$. We say that the functions are *orthogonal* on the interval $[-\pi, \pi]$, relative to the inner product (4.11.4). In the next section we will have much more to say about orthogonality of vectors. $\qquad \square$

Complex Inner Products[9]

The preceding discussion has been concerned with real vector spaces. In order to generalize the definition of an inner product to a complex vector space, we first consider the case of $\mathbb{C}^n$. By analogy with Definition 4.11.1, one might think that the natural inner product in $\mathbb{C}^n$ would be obtained by summing the products of corresponding components of vectors in $\mathbb{C}^n$ in exactly the same manner as in the standard inner product for $\mathbb{R}^n$. However, one reason for introducing an inner product is to obtain a concept of "length" of a vector. In order for a quantity to be considered a reasonable measure of length, we would want it to be a nonnegative real number that vanishes if and only if the vector itself is the zero vector (property 1 of a real inner product). But, if we apply the inner product in $\mathbb{R}^n$ given in Definition 4.11.1 to vectors in $\mathbb{C}^n$, then, since the components of vectors in $\mathbb{C}^n$ are complex numbers, it follows that the resulting norm of a vector in

---

[9]In the remainder of the text, the only complex inner product that we will require is the standard inner product in $\mathbb{C}^n$, and this is needed only in Section 5.10.

$\mathbb{C}^n$ would be a complex number also. Furthermore, applying the $\mathbb{R}^2$ inner product to, for example, the vector $\mathbf{u} = (1 - i, 1 + i)$, we obtain

$$||\mathbf{u}||^2 = (1 - i)^2 + (1 + i)^2 = 0,$$

which means that a nonzero vector would have zero "length." To rectify this situation, we must define an inner product in $\mathbb{C}^n$ more carefully. We take advantage of complex conjugation to do this, as the definition shows.

DEFINITION   4.11.8

If $\mathbf{u} = (u_1, u_2, \ldots, u_n)$ and $\mathbf{v} = (v_1, v_2, \ldots, v_n)$ are vectors in $\mathbb{C}^n$, we define the **standard inner product in $\mathbb{C}^n$ by**[10]

$$\langle \mathbf{u}, \mathbf{v} \rangle = u_1\overline{v}_1 + u_2\overline{v}_2 + \cdots + u_n\overline{v}_n.$$

The **norm** of $\mathbf{u}$ is defined to be the *real number*

$$||\mathbf{u}|| = \sqrt{\langle \mathbf{u}, \mathbf{u} \rangle} = \sqrt{|u_1|^2 + |u_2|^2 + \cdots + |u_n|^2}.$$

The preceding inner product is a mapping that associates with the two vectors $\mathbf{u} = (u_1, u_2, \ldots, u_n)$ and $\mathbf{v} = (v_1, v_2, \ldots, v_n)$ in $\mathbb{C}^n$ the *scalar*

$$\langle \mathbf{u}, \mathbf{v} \rangle = u_1\overline{v}_1 + u_2\overline{v}_2 + \cdots + u_n\overline{v}_n.$$

In general, $\langle \mathbf{u}, \mathbf{v} \rangle$ will be nonreal (i.e., it will have a nonzero imaginary part). The key point to notice is that the norm of $\mathbf{u}$ is always a *real* number, even though the separate components of $\mathbf{u}$ are complex numbers.

Example 4.11.9   If $\mathbf{u} = (1 + 2i, 2 - 3i)$ and $\mathbf{v} = (2 - i, 3 + 4i)$, find $\langle \mathbf{u}, \mathbf{v} \rangle$ and $||\mathbf{u}||$.

Solution: Using Definition 4.11.8,

$$\langle \mathbf{u}, \mathbf{v} \rangle = (1 + 2i)(2 + i) + (2 - 3i)(3 - 4i) = 5i - 6 - 17i = -6 - 12i,$$
$$||\mathbf{u}|| = \sqrt{\langle \mathbf{u}, \mathbf{u} \rangle} = \sqrt{(1 + 2i)(1 - 2i) + (2 - 3i)(2 + 3i)} = \sqrt{5 + 13} = 3\sqrt{2}. \quad \square$$

The standard inner product in $\mathbb{C}^n$ satisfies properties (1), (3), and (4), but not property (2). We now derive the appropriate generalization of property (2) when using the standard inner product in $\mathbb{C}^n$. Let $\mathbf{u} = (u_1, u_2, \ldots, u_n)$ and $\mathbf{v} = (v_1, v_2, \ldots, v_n)$ be vectors in $\mathbb{C}^n$. Then, from Definition 4.11.8,

$$\langle \mathbf{v}, \mathbf{u} \rangle = v_1\overline{u}_1 + v_2\overline{u}_2 + \cdots + v_n\overline{u}_n = \overline{u_1\overline{v}_1 + u_2\overline{v}_2 + \cdots + u_n\overline{v}_n} = \overline{\langle \mathbf{u}, \mathbf{v} \rangle}.$$

Thus,

$$\langle \mathbf{v}, \mathbf{u} \rangle = \overline{\langle \mathbf{u}, \mathbf{v} \rangle}.$$

We now use the properties satisfied by the standard inner product in $\mathbb{C}^n$ to define an inner product in an arbitrary (that is, real or complex) vector space.

---

[10]Recall that if $z = a + ib$, then $\overline{z} = a - ib$ and $|z|^2 = z\overline{z} = (a + ib)(a - ib) = a^2 + b^2$.

DEFINITION 4.11.10

Let V be a (real or complex) vector space. A mapping that associates with each pair of vectors $\mathbf{u}$, $\mathbf{v}$ in V a scalar, denoted $\langle \mathbf{u}, \mathbf{v} \rangle$, is called an **inner product** in V, provided it satisfies the following properties. For all $\mathbf{u}$, $\mathbf{v}$ and $\mathbf{w}$ in V and all (real or complex) scalars k,

1. $\langle \mathbf{u}, \mathbf{u} \rangle \geq 0$. Furthermore, $\langle \mathbf{u}, \mathbf{u} \rangle = 0$ if and only if $\mathbf{u} = \mathbf{0}$.
2. $\langle \mathbf{v}, \mathbf{u} \rangle = \overline{\langle \mathbf{u}, \mathbf{v} \rangle}$.
3. $\langle k\mathbf{u}, \mathbf{v} \rangle = k \langle \mathbf{u}, \mathbf{v} \rangle$.
4. $\langle \mathbf{u} + \mathbf{v}, \mathbf{w} \rangle = \langle \mathbf{u}, \mathbf{w} \rangle + \langle \mathbf{v}, \mathbf{w} \rangle$.

The **norm** of $\mathbf{u}$ is defined in terms of the inner product by

$$\|\mathbf{u}\| = \sqrt{\langle \mathbf{u}, \mathbf{u} \rangle}.$$

Remark   Notice that the properties in the preceding definition reduce to those in Definition 4.11.3 in the case that V is a *real* vector space, since in such a case the complex conjugates are unnecessary. Thus, this definition is a consistent extension of Definition 4.11.3.

Example 4.11.11   Use properties 2 and 3 of Definition 4.11.10 to prove that in an inner product space

$$\langle \mathbf{u}, k\mathbf{v} \rangle = \overline{k} \langle \mathbf{u}, \mathbf{v} \rangle$$

for all vectors $\mathbf{u}$, $\mathbf{v}$ and all scalars k.

Solution: From properties 2 and 3, we have

$$\langle \mathbf{u}, k\mathbf{v} \rangle = \overline{\langle k\mathbf{v}, \mathbf{u} \rangle} = \overline{k \langle \mathbf{v}, \mathbf{u} \rangle} = \overline{k} \; \overline{\langle \mathbf{v}, \mathbf{u} \rangle} = \overline{k} \langle \mathbf{u}, \mathbf{v} \rangle.$$

Notice that in the particular case of a real vector space, the foregoing result reduces to

$$\langle \mathbf{u}, k\mathbf{v} \rangle = k \langle \mathbf{u}, \mathbf{v} \rangle,$$

since in such a case the scalars are real numbers. $\square$

Exercises for 4.11

Key Terms

Inner product, Axioms of an inner product, Real (complex) inner product space, Norm, Angle, Cauchy-Schwarz inequality.

Skills

- Know the four inner product space axioms.

- Be able to check whether or not a proposed inner product on a vector space V satisfies the inner product space axioms.

- Be able to compute the inner product of two vectors in an inner product space.

- Be able to find the norm of a vector in an inner product space.

- Be able to find the angle between two vectors in an inner product space.

True-False Review

For Questions 1–7, decide if the given statement is **true** or **false**, and give a brief justification for your answer. If true, you can quote a relevant definition or theorem from the text. If false, provide an example, illustration, or brief explanation of why the statement is false.

1. If $\mathbf{v}$ and $\mathbf{w}$ are linearly independent vectors in an inner product space V, then $\langle \mathbf{v}, \mathbf{w} \rangle = 0$.

2. In any inner product space V, we have

$$\langle k\mathbf{v}, k\mathbf{w} \rangle = k \langle \mathbf{v}, \mathbf{w} \rangle.$$

3. If $\langle \mathbf{v}_1, \mathbf{w} \rangle = \langle \mathbf{v}_2, \mathbf{w} \rangle = 0$ in an inner product space V, then

$$\langle c_1 \mathbf{v}_1 + c_2 \mathbf{v}_2, \mathbf{w} \rangle = 0.$$

4. In any inner product space V, $\langle \mathbf{x} + \mathbf{y}, \mathbf{x} - \mathbf{y} \rangle < 0$ if and only if $||\mathbf{x}|| < ||\mathbf{y}||$.

5. In any vector space V, there is at most one valid inner product $\langle \, , \, \rangle$ that can be defined on V.

6. The angle between the vectors $\mathbf{v}$ and $\mathbf{w}$ in an inner product space V is the same as the angle between the vectors $-2\mathbf{v}$ and $-2\mathbf{w}$.

7. If $p(x) = a_0 + a_1 x + a_2 x^2$ and $q(x) = b_0 + b_1 x + b_2 x^2$, then we can define an inner product on P_2 via $\langle p, q \rangle = a_0 b_0$.

Problems

1. Use the standard inner product in $\mathbb{R}^4$ to determine the angle between the vectors $\mathbf{v} = (1, 3, -1, 4)$ and $\mathbf{w} = (-1, 1, -2, 1)$.

2. If $f(x) = \sin x$ and $g(x) = x$ on $[0, \pi]$, use the function inner product defined in the text to determine the angle between f and g.

3. If $\mathbf{v} = (2+i, 3-2i, 4+i)$ and $\mathbf{w} = (-1+i, 1-3i, 3-i)$, use the standard inner product in $\mathbb{C}^3$ to determine, $\langle \mathbf{v}, \mathbf{w} \rangle$, $||\mathbf{v}||$, and $||\mathbf{w}||$.

4. Let

$$A = \begin{bmatrix} a_{11} & a_{12} \\ a_{21} & a_{22} \end{bmatrix}, \qquad B = \begin{bmatrix} b_{11} & b_{12} \\ b_{21} & b_{22} \end{bmatrix}$$

be vectors in $M_2(\mathbb{R})$. Show that the mapping

$$\langle A, B \rangle = a_{11}b_{11} + a_{12}b_{12} + a_{21}b_{21} + a_{22}b_{22} \tag{4.11.9}$$

defines an inner product in $M_2(\mathbb{R})$.

5. Referring to A and B in the previous problem, show that the mapping

$$\langle A, B \rangle = a_{11}b_{22} + a_{12}b_{21} + a_{21}b_{12} + a_{22}b_{11}$$

does *not* define a valid inner product on $M_2(\mathbb{R})$.

For Problems 6–7, use the inner product (4.11.9) to determine $\langle A, B \rangle$, $||A||$, and $||B||$.

6. $A = \begin{bmatrix} 2 & -1 \\ 3 & 5 \end{bmatrix}$, $B = \begin{bmatrix} 3 & 1 \\ -1 & 2 \end{bmatrix}$.

7. $A = \begin{bmatrix} 3 & 2 \\ -2 & 4 \end{bmatrix}$, $B = \begin{bmatrix} 1 & 1 \\ -2 & 1 \end{bmatrix}$.

8. Let $p_1(x) = a + bx$ and $p_2(x) = c + dx$ be vectors in P_1. Determine a mapping $\langle p_1, p_2 \rangle$ that defines an inner product on P_1.

Consider the vector space $\mathbb{R}^2$. Define the mapping $\langle \, , \, \rangle$ by

$$\langle \mathbf{v}, \mathbf{w} \rangle = 2v_1 w_1 + v_1 w_2 + v_2 w_1 + 2v_2 w_2 \tag{4.11.10}$$

for all vectors $\mathbf{v} = (v_1, v_2)$ and $\mathbf{w} = (w_1, w_2)$ in $\mathbb{R}^2$. This mapping is required for Problems 9–12.

9. Verify that Equation (4.11.10) defines an inner product on $\mathbb{R}^2$.

For Problems 10–12, determine the inner product of the given vectors using (a) the inner product (4.11.10), (b) the standard inner product in $\mathbb{R}^2$.

10. $\mathbf{v} = (1, 0)$, $\mathbf{w} = (-1, 2)$.

11. $\mathbf{v} = (2, -1)$, $\mathbf{w} = (3, 6)$.

12. $\mathbf{v} = (1, -2)$, $\mathbf{w} = (2, 1)$.

13. Consider the vector space $\mathbb{R}^2$. Define the mapping $\langle \, , \, \rangle$ by

$$\langle \mathbf{v}, \mathbf{w} \rangle = v_1 w_1 - v_2 w_2, \tag{4.11.11}$$

for all vectors $\mathbf{v} = (v_1, v_2)$ and $\mathbf{w} = (w_1, w_2)$. Verify that all of the properties in Definition 4.11.3 except (1) are satisfied by (4.11.11).

The mapping (4.11.11) is called a **pseudo-inner product** in $\mathbb{R}^2$ and, when generalized to $\mathbb{R}^4$, is of fundamental importance in Einstein's special relativity theory.

14. Using Equation (4.11.11), determine all nonzero vectors satisfying $\langle \mathbf{v}, \mathbf{v} \rangle = 0$. Such vectors are called **null** vectors.

15. Using Equation (4.11.11), determine all vectors satisfying $\langle \mathbf{v}, \mathbf{v} \rangle < 0$. Such vectors are called **timelike** vectors.

16. Using Equation (4.11.11), determine all vectors satisfying $\langle \mathbf{v}, \mathbf{v} \rangle > 0$. Such vectors are called **spacelike** vectors.

17. Make a sketch of $\mathbb{R}^2$ and indicate the position of the null, timelike, and spacelike vectors.

18. Consider the vector space $\mathbb{R}^n$, and let $\mathbf{v} = (v_1, v_2, \ldots, v_n)$ and $\mathbf{w} = (w_1, w_2, \ldots, w_n)$ be vectors in $\mathbb{R}^n$. Show that the mapping $\langle , \rangle$ defined by

$$\langle \mathbf{v}, \mathbf{w} \rangle = k_1 v_1 w_1 + k_2 v_2 w_2 + \cdots + k_n v_n w_n$$

is a valid inner product on $\mathbb{R}^n$ if and only if the constants $k_1, k_2, \ldots, k_n$ are all positive.

19. Prove from the inner product axioms that, in any inner product space V, $\langle \mathbf{v}, \mathbf{0} \rangle = 0$ for all $\mathbf{v}$ in V.

20. Let V be a real inner product space.

(a) Prove that for all $\mathbf{v}, \mathbf{w} \in V$,

$$\|\mathbf{v} + \mathbf{w}\|^2 = \|\mathbf{v}\|^2 + 2\langle \mathbf{v}, \mathbf{w} \rangle + \|\mathbf{w}\|^2.$$

[**Hint:** $\|\mathbf{v} + \mathbf{w}\|^2 = \langle \mathbf{v} + \mathbf{w}, \mathbf{v} + \mathbf{w} \rangle$.]

(b) Two vectors $\mathbf{v}$ and $\mathbf{w}$ in an inner product space V are called *orthogonal* if $\langle \mathbf{v}, \mathbf{w} \rangle = 0$. Use (a) to prove the general **Pythagorean theorem:** If $\mathbf{v}$ and $\mathbf{w}$ are orthogonal in an inner product space V, then

$$\|\mathbf{v} + \mathbf{w}\|^2 = \|\mathbf{v}\|^2 + \|\mathbf{w}\|^2.$$

(c) Prove that for all $\mathbf{v}, \mathbf{w}$ in V,

(i) $\|\mathbf{v} + \mathbf{w}\|^2 - \|\mathbf{v} - \mathbf{w}\|^2 = 4\langle \mathbf{v}, \mathbf{w} \rangle$.

(ii) $\|\mathbf{v} + \mathbf{w}\|^2 + \|\mathbf{v} - \mathbf{w}\|^2 = 2(\|\mathbf{v}\|^2 + \|\mathbf{w}\|^2)$.

21. Let V be a complex inner product space. Prove that for all $\mathbf{v}, \mathbf{w}$ in V,

$$\|\mathbf{v} + \mathbf{w}\|^2 = \|\mathbf{v}\|^2 + 2\mathrm{Re}(\langle \mathbf{v}, \mathbf{w} \rangle) + \|\mathbf{v}\|^2,$$

where Re denotes the real part of a complex number.

4.12 Orthogonal Sets of Vectors and the Gram-Schmidt Process

The discussion in the previous section has shown how an inner product can be used to define the angle between two nonzero vectors. In particular, if the inner product of two nonzero vectors is zero, then the angle between those two vectors is $\pi/2$ radians, and therefore it is natural to call such vectors orthogonal (perpendicular). The following definition extends the idea of orthogonality into an arbitrary inner product space.

DEFINITION 4.12.1

Let V be an inner product space.

1. Two vectors $\mathbf{u}$ and $\mathbf{v}$ in V are said to be **orthogonal** if $\langle \mathbf{u}, \mathbf{v} \rangle = 0$.

2. A set of nonzero vectors $\{\mathbf{v}_1, \mathbf{v}_2, \ldots, \mathbf{v}_k\}$ in V is called an **orthogonal set** of vectors if
$$\langle \mathbf{v}_i, \mathbf{v}_j \rangle = 0, \qquad \text{whenever } i \neq j.$$

(That is, every vector is orthogonal to every other vector in the set.)

3. A vector $\mathbf{v}$ in V is called a **unit vector** if $\|\mathbf{v}\| = 1$.

4. An *orthogonal* set of *unit* vectors is called an **orthonormal set** of vectors. Thus, $\{\mathbf{v}_1, \mathbf{v}_2, \ldots, \mathbf{v}_k\}$ in V is an orthonormal set if and only if

(a) $\langle \mathbf{v}_i, \mathbf{v}_j \rangle = 0$ whenever $i \neq j$.

(b) $\langle \mathbf{v}_i, \mathbf{v}_i \rangle = 1$ for all $i = 1, 2, \ldots, k$.

Remarks

1. The conditions in (4a) and (4b) can be written compactly in terms of the Kronecker delta symbol as

$$\langle \mathbf{v}_i, \mathbf{v}_j \rangle = \delta_{ij}, \qquad i, j = 1, 2, \ldots, k.$$

2. Note that the inner products occurring in Definition 4.12.1 will depend upon which inner product space we are working in.

3. If $\mathbf{v}$ is any nonzero vector, then $\dfrac{1}{||\mathbf{v}||} \mathbf{v}$ is a unit vector, since the properties of an inner product imply that

$$\left\langle \frac{1}{||\mathbf{v}||} \mathbf{v}, \frac{1}{||\mathbf{v}||} \mathbf{v} \right\rangle = \frac{1}{||\mathbf{v}||^2} \langle \mathbf{v}, \mathbf{v} \rangle = \frac{1}{||\mathbf{v}||^2} ||\mathbf{v}||^2 = 1.$$

Using Remark 3 above, we can take an orthogonal set of vectors $\{\mathbf{v}_1, \mathbf{v}_2, \ldots, \mathbf{v}_k\}$ and create a new set $\{\mathbf{u}_1, \mathbf{u}_2, \ldots, \mathbf{u}_k\}$, where $\mathbf{u}_i = \dfrac{1}{||\mathbf{v}_i||} \mathbf{v}_i$ is a unit vector for each i. Using the properties of an inner product, it is easy to see that the new set $\{\mathbf{u}_1, \mathbf{u}_2, \ldots, \mathbf{u}_k\}$ is an orthonormal set (see Problem 31). The process of replacing the $\mathbf{v}_i$ by the $\mathbf{u}_i$ is called **normalization**.

Example 4.12.2   Verify that $\{(-2, 1, 3, 0), (0, -3, 1, -6), (-2, -4, 0, 2)\}$ is an orthogonal set of vectors in $\mathbb{R}^4$, and use it to construct an orthonormal set of vectors in $\mathbb{R}^4$.

Solution: Let $\mathbf{v}_1 = (-2, 1, 3, 0)$, $\mathbf{v}_2 = (0, -3, 1, -6)$, and $\mathbf{v}_3 = (-2, -4, 0, 2)$. Then

$$\langle \mathbf{v}_1, \mathbf{v}_2 \rangle = 0, \qquad \langle \mathbf{v}_1, \mathbf{v}_3 \rangle = 0, \qquad \langle \mathbf{v}_2, \mathbf{v}_3 \rangle = 0,$$

so that the given set of vectors is an orthogonal set. Dividing each vector in the set by its norm yields the following orthonormal set:

$$\left\{ \frac{1}{\sqrt{14}} \mathbf{v}_1, \frac{1}{\sqrt{46}} \mathbf{v}_2, \frac{1}{2\sqrt{6}} \mathbf{v}_3 \right\}.$$ □

Example 4.12.3   Verify that the functions $f_1(x) = 1$, $f_2(x) = \sin x$, and $f_3(x) = \cos x$ are orthogonal in $C^0[-\pi, \pi]$, and use them to construct an orthonormal set of functions in $C^0[-\pi, \pi]$.

Solution: In this case, we have

$$\langle f_1, f_2 \rangle = \int_{-\pi}^{\pi} \sin x \, dx = 0, \qquad \langle f_1, f_3 \rangle = \int_{-\pi}^{\pi} \cos x \, dx = 0,$$

$$\langle f_2, f_3 \rangle = \int_{-\pi}^{\pi} \sin x \cos x \, dx = \left[\frac{1}{2} \sin^2 x \right]_{-\pi}^{\pi} = 0,$$

so that the functions are indeed orthogonal on $[-\pi, \pi]$. Taking the norm of each function, we obtain

$$||f_1|| = \sqrt{\int_{-\pi}^{\pi} 1 \, dx} = \sqrt{2\pi},$$

$$||f_2|| = \sqrt{\int_{-\pi}^{\pi} \sin^2 x \, dx} = \sqrt{\int_{-\pi}^{\pi} \frac{1}{2}(1 - \cos 2x) \, dx} = \sqrt{\pi},$$

$$||f_3|| = \sqrt{\int_{-\pi}^{\pi} \cos^2 x \, dx} = \sqrt{\int_{-\pi}^{\pi} \frac{1}{2}(1 + \cos 2x) \, dx} = \sqrt{\pi}.$$

Thus an orthonormal set of functions on $[-\pi, \pi]$ is

$$\left\{ \frac{1}{\sqrt{2\pi}}, \frac{1}{\sqrt{\pi}} \sin x, \frac{1}{\sqrt{\pi}} \cos x \right\}.$$

□

Orthogonal and Orthonormal Bases

In the analysis of geometric vectors in elementary calculus courses, it is usual to use the standard basis $\{\mathbf{i}, \mathbf{j}, \mathbf{k}\}$. Notice that this set of vectors is in fact an orthonormal set. The introduction of an inner product in a vector space opens up the possibility of using similar bases in a general finite-dimensional vector space. The next definition introduces the appropriate terminology.

DEFINITION   4.12.4

A basis $\{\mathbf{v}_1, \mathbf{v}_2, \ldots, \mathbf{v}_n\}$ for a (finite-dimensional) inner product space is called an **orthogonal** basis if

$$\langle \mathbf{v}_i, \mathbf{v}_j \rangle = 0 \qquad \text{whenever } i \neq j,$$

and it is called an **orthonormal** basis if

$$\langle \mathbf{v}_i, \mathbf{v}_j \rangle = \delta_{ij}, \qquad i, j = 1, 2, \ldots, n.$$

There are two natural questions at this point: (1) How can we obtain an orthogonal or orthonormal basis for an inner product space V? (2) Why is it beneficial to work with an orthogonal or orthonormal basis of vectors? We address the second question first.

In light of our work in previous sections of this chapter, the importance of our next theorem should be self-evident.

Theorem 4.12.5   If $\{\mathbf{v}_1, \mathbf{v}_2, \ldots, \mathbf{v}_k\}$ is an *orthogonal* set of *nonzero* vectors in an inner product space V, then $\{\mathbf{v}_1, \mathbf{v}_2, \ldots, \mathbf{v}_k\}$ is linearly independent.

Proof Assume that

$$c_1 \mathbf{v}_1 + c_2 \mathbf{v}_2 + \cdots + c_k \mathbf{v}_k = \mathbf{0}. \tag{4.12.1}$$

We will show that $c_1 = c_2 = \cdots = c_k = 0$. Taking the inner product of each side of (4.12.1) with $\mathbf{v}_i$, we find that

$$\langle c_1 \mathbf{v}_1 + c_2 \mathbf{v}_2 + \cdots + c_k \mathbf{v}_k, \mathbf{v}_i \rangle = \langle \mathbf{0}, \mathbf{v}_i \rangle = 0.$$

Using the inner product properties on the left side, we have

$$c_1 \langle \mathbf{v}_1, \mathbf{v}_i \rangle + c_2 \langle \mathbf{v}_2, \mathbf{v}_i \rangle + \cdots + c_k \langle \mathbf{v}_k, \mathbf{v}_i \rangle = 0.$$

Finally, using the fact that for all $j \neq i$, we have $\langle \mathbf{v}_j, \mathbf{v}_i \rangle = 0$, we conclude that

$$c_i \langle \mathbf{v}_i, \mathbf{v}_i \rangle = 0.$$

Since $\mathbf{v}_i \neq \mathbf{0}$, it follows that $c_i = 0$, and this holds for each i with $1 \leq i \leq k$.   ∎

Example 4.12.6   Let $V = M_2(\mathbb{R})$, let W be the subspace of all 2×2 *symmetric* matrices, and let

$$S = \left\{ \begin{bmatrix} 2 & -1 \\ -1 & 0 \end{bmatrix}, \begin{bmatrix} 1 & 1 \\ 1 & 2 \end{bmatrix}, \begin{bmatrix} 2 & 2 \\ 2 & -3 \end{bmatrix} \right\}.$$

Define an inner product on V via[11]

$$\left\langle \begin{bmatrix} a_{11} & a_{12} \\ a_{21} & a_{22} \end{bmatrix}, \begin{bmatrix} b_{11} & b_{12} \\ b_{21} & b_{22} \end{bmatrix} \right\rangle = a_{11}b_{11} + a_{12}b_{12} + a_{21}b_{21} + a_{22}b_{22}.$$

Show that S is an orthogonal basis for W.

Solution: According to Example 4.6.18, we already know that $\dim[W] = 3$. Using the given inner product, it can be directly shown that S is an orthogonal set, and hence, Theorem 4.12.5 implies that S is linearly independent. Therefore, by Theorem 4.6.10, S is a basis for W. □

Let V be a (finite-dimensional) inner product space, and suppose that we have an *orthogonal* basis $\{\mathbf{v}_1, \mathbf{v}_2, \ldots, \mathbf{v}_n\}$ for V. As we saw in Section 4.7, any vector $\mathbf{v}$ in V can be written *uniquely* in the form

$$\mathbf{v} = c_1\mathbf{v}_1 + c_2\mathbf{v}_2 + \cdots + c_n\mathbf{v}_n, \tag{4.12.2}$$

where the unique n-tuple $(c_1, c_2, \ldots, c_n)$ consists of the components of $\mathbf{v}$ relative to the given basis. It is easier to determine the components c_i in the case of an orthogonal basis than it is for other bases, because we can simply form the inner product of both sides of (4.12.2) with $\mathbf{v}_i$ as follows:

$$\begin{aligned} \langle \mathbf{v}, \mathbf{v}_i \rangle &= \langle c_1\mathbf{v}_1 + c_2\mathbf{v}_2 + \cdots + c_n\mathbf{v}_n, \mathbf{v}_i \rangle \\ &= c_1\langle \mathbf{v}_1, \mathbf{v}_i \rangle + c_2\langle \mathbf{v}_2, \mathbf{v}_i \rangle + \cdots + c_n\langle \mathbf{v}_n, \mathbf{v}_i \rangle \\ &= c_i\|\mathbf{v}_i\|^2, \end{aligned}$$

where the last step follows from the orthogonality properties of the basis $\{\mathbf{v}_1, \mathbf{v}_2, \ldots, \mathbf{v}_n\}$. Therefore, we have proved the following theorem.

Theorem 4.12.7   Let V be a (finite-dimensional) inner product space with orthogonal basis $\{\mathbf{v}_1, \mathbf{v}_2, \ldots, \mathbf{v}_n\}$. Then any vector $\mathbf{v} \in V$ may be expressed in terms of the basis as

$$\mathbf{v} = \left(\frac{\langle \mathbf{v}, \mathbf{v}_1 \rangle}{\|\mathbf{v}_1\|^2} \right) \mathbf{v}_1 + \left(\frac{\langle \mathbf{v}, \mathbf{v}_2 \rangle}{\|\mathbf{v}_2\|^2} \right) \mathbf{v}_2 + \cdots + \left(\frac{\langle \mathbf{v}, \mathbf{v}_n \rangle}{\|\mathbf{v}_n\|^2} \right) \mathbf{v}_n.$$

Theorem 4.12.7 gives a simple formula for writing an arbitrary vector in an inner product space V as a linear combination of vectors in an orthogonal basis for V. Let us illustrate with an example.

Example 4.12.8   Let V, W, and S be as in Example 4.12.6. Find the components of the vector

$$\mathbf{v} = \begin{bmatrix} 0 & -1 \\ -1 & 2 \end{bmatrix}$$

relative to S.

Solution: From the formula given in Theorem 4.12.7, we have

$$\mathbf{v} = \frac{2}{6}\begin{bmatrix} 2 & -1 \\ -1 & 0 \end{bmatrix} + \frac{2}{7}\begin{bmatrix} 1 & 1 \\ 1 & 2 \end{bmatrix} - \frac{10}{21}\begin{bmatrix} 2 & 2 \\ 2 & -3 \end{bmatrix},$$

---

[11]This defines a valid inner product on V by Problem 4 in Section 4.11.

so the components of $\mathbf{v}$ relative to S are

$$\left(\frac{1}{3}, \frac{2}{7}, -\frac{10}{21}\right).$$

□

If the orthogonal basis $\{\mathbf{v}_1, \mathbf{v}_2, \ldots, \mathbf{v}_n\}$ for V is in fact orthonormal, then since $||\mathbf{v}_i|| = 1$ for each i, we immediately deduce the following corollary of Theorem 4.12.7.

Corollary 4.12.9   Let V be a (finite-dimensional) inner product space with an orthonormal basis $\{\mathbf{v}_1, \mathbf{v}_2, \ldots, \mathbf{v}_n\}$. Then any vector $\mathbf{v} \in V$ may be expressed in terms of the basis as

$$\mathbf{v} = \langle \mathbf{v}, \mathbf{v}_1 \rangle \mathbf{v}_1 + \langle \mathbf{v}, \mathbf{v}_2 \rangle \mathbf{v}_2 + \cdots + \langle \mathbf{v}, \mathbf{v}_n \rangle \mathbf{v}_n.$$

Remark   Corollary 4.12.9 tells us that the components of a given vector $\mathbf{v}$ relative to the orthonormal basis $\{\mathbf{v}_1, \mathbf{v}_2, \ldots, \mathbf{v}_n\}$ are precisely the numbers $\langle \mathbf{v}, \mathbf{v}_i \rangle$, for $1 \leq i \leq n$. Thus, by working with an orthonormal basis for a vector space, we have a simple method for getting the components of any vector in the vector space.

Example 4.12.10   We can write an arbitrary vector in $\mathbb{R}^n$, $\mathbf{v} = (a_1, a_2, \ldots, a_n)$, in terms of the standard basis $\{\mathbf{e}_1, \mathbf{e}_2, \ldots, \mathbf{e}_n\}$ by noting that $\langle \mathbf{v}, \mathbf{e}_i \rangle = a_i$. Thus, $\mathbf{v} = a_1 \mathbf{e}_1 + a_2 \mathbf{e}_2 + \cdots + a_n \mathbf{e}_n$.

□

Example 4.12.11   We can equip the vector space P_1 of all polynomials of degree ≤ 1 with inner product

$$\langle p, q \rangle = \int_{-1}^{1} p(x) q(x) \, dx,$$

thus making P_1 into an inner product space. Verify that the vectors $p_0 = 1/\sqrt{2}$ and $p_1 = \sqrt{1.5}x$ form an orthonormal basis for P_1 and use Corollary 4.12.9 to write the vector $q = 1 + x$ as a linear combination of p_0 and p_1.

Solution: We have

$$\langle p_0, p_1 \rangle = \int_{-1}^{1} \frac{1}{\sqrt{2}} \cdot \sqrt{1.5}x \, dx = 0,$$

$$||p_0|| = \sqrt{\langle p_0, p_0 \rangle} = \sqrt{\int_{-1}^{1} p_0^2 \, dx} = \sqrt{\int_{-1}^{1} \frac{1}{2} \, dx} = \sqrt{1} = 1,$$

$$||p_1|| = \sqrt{\langle p_1, p_1 \rangle} = \sqrt{\int_{-1}^{1} p_1^2 \, dx} = \sqrt{\int_{-1}^{1} \frac{3}{2} x^2 \, dx} = \sqrt{\frac{1}{2} x^3 \Big|_{-1}^{1}} = \sqrt{1} = 1.$$

Thus, $\{p_0, p_1\}$ is an orthonormal (and hence linearly independent) set of vectors in P_1. Since $\dim[P_1] = 2$, Theorem 4.6.10 shows that $\{p_0, p_1\}$ is an (orthonormal) basis for P_1.

Finally, we wish to write $q = 1 + x$ as a linear combination of p_0 and p_1, by using Corollary 4.12.9. We leave it to the reader to verify that $\langle q, p_0 \rangle = \sqrt{2}$ and $\langle q, p_1 \rangle = \sqrt{\frac{2}{3}}$. Thus, we have

$$1 + x = \sqrt{2} \, p_0 + \sqrt{\frac{2}{3}} \, p_1 = \sqrt{2} \cdot \frac{1}{\sqrt{2}} + \sqrt{\frac{2}{3}} \cdot \left(\sqrt{\frac{3}{2}} x\right).$$

So the component vector of $1 + x$ relative to $\{p_0, p_1\}$ is $(\sqrt{2}, \sqrt{\frac{2}{3}})^T$.

□

The Gram-Schmidt Process

Next, we return to address the first question we raised earlier: How can we obtain an orthogonal or orthonormal basis for an inner product space V? The idea behind the process is to begin with *any* basis for V, say $\{x_1, x_2, \ldots, x_n\}$, and to successively replace these vectors with vectors $v_1, v_2, \ldots, v_n$ that are orthogonal to one another, and to ensure that, throughout the process, the span of the vectors remains unchanged. This is known as the **Gram-Schmidt process**. To describe it, we shall once more appeal to a look at geometric vectors.

If v and w are any two linearly independent (noncollinear) geometric vectors, then the **orthogonal projection** of w on v is the vector $P(w, v)$ shown in Figure 4.12.1. We see from the figure that an orthogonal basis for the subspace (plane) of 3-space spanned by v and w is $\{v_1, v_2\}$, where

$$v_1 = v \quad \text{and} \quad v_2 = w - P(w, v).$$

In order to generalize this result to an arbitrary inner product space, we need to derive an expression for $P(w, v)$ in terms of the dot product. We see from Figure 4.12.1 that the norm of $P(w, v)$ is

$$\|P(w, v)\| = \|w\| \cos\theta,$$

where θ is the angle between v and w. Thus

$$P(w, v) = \|w\| \cos\theta \frac{v}{\|v\|},$$

which we can write as

$$P(w, v) = \left(\frac{\|w\| \, \|v\|}{\|v\|^2} \cos\theta \right) v. \tag{4.12.3}$$

Recalling that the dot product of the vectors w and v is defined by

$$w \cdot v = \|w\| \, \|v\| \cos\theta,$$

it follows from Equation (4.12.3) that

$$P(w, v) = \frac{(w \cdot v)}{\|v\|^2} v,$$

or equivalently, using the notation for the inner product introduced in the previous section,

$$P(w, v) = \frac{\langle w, v \rangle}{\|v\|^2} v.$$

Now let x_1 and x_2 be linearly independent vectors in an arbitrary inner product space V. We show next that the foregoing formula can also be applied in V to obtain an orthogonal basis $\{v_1, v_2\}$ for the subspace of V spanned by $\{x_1, x_2\}$. Let

$$v_1 = x_1$$

and

$$v_2 = x_2 - P(x_2, v_1) = x_2 - \frac{\langle x_2, v_1 \rangle}{\|v_1\|^2} v_1. \tag{4.12.4}$$

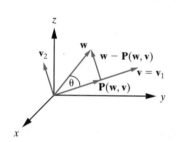

Figure 4.12.1: Obtaining an orthogonal basis for a two-dimensional subspace of $\mathbb{R}^3$.

Note from (4.12.4) that $\mathbf{v}_2$ can be written as a linear combination of $\{\mathbf{x}_1, \mathbf{x}_2\}$, and hence, $\mathbf{v}_2 \in \text{span}\{\mathbf{x}_1, \mathbf{x}_2\}$. Since we also have that $\mathbf{x}_2 \in \text{span}\{\mathbf{v}_1, \mathbf{v}_2\}$, it follows that $\text{span}\{\mathbf{v}_1, \mathbf{v}_2\} = \text{span}\{\mathbf{x}_1, \mathbf{x}_2\}$. Next we claim that $\mathbf{v}_2$ is orthogonal to $\mathbf{v}_1$. We have

$$\langle \mathbf{v}_2, \mathbf{v}_1 \rangle = \Big\langle \mathbf{x}_2 - \frac{\langle \mathbf{x}_2, \mathbf{v}_1 \rangle}{||\mathbf{v}_1||^2}\mathbf{v}_1, \mathbf{v}_1 \Big\rangle = \langle \mathbf{x}_2, \mathbf{v}_1 \rangle - \Big\langle \frac{\langle \mathbf{x}_2, \mathbf{v}_1 \rangle}{||\mathbf{v}_1||^2}\mathbf{v}_1, \mathbf{v}_1 \Big\rangle$$

$$= \langle \mathbf{x}_2, \mathbf{v}_1 \rangle - \frac{\langle \mathbf{x}_2, \mathbf{v}_1 \rangle}{||\mathbf{v}_1||^2}\langle \mathbf{v}_1, \mathbf{v}_1 \rangle = 0,$$

which verifies our claim. We have shown that $\{\mathbf{v}_1, \mathbf{v}_2\}$ is an orthogonal set of vectors which spans the same subspace of V as $\mathbf{x}_1$ and $\mathbf{x}_2$.

The calculations just presented can be generalized to prove the following useful result (see Problem 32).

Lemma 4.12.12   Let $\{\mathbf{v}_1, \mathbf{v}_2, \ldots, \mathbf{v}_k\}$ be an orthogonal set of vectors in an inner product space V. If $\mathbf{x} \in V$, then the vector

$$\mathbf{x} - P(\mathbf{x}, \mathbf{v}_1) - P(\mathbf{x}, \mathbf{v}_2) - \cdots - P(\mathbf{x}, \mathbf{v}_k)$$

is orthogonal to $\mathbf{v}_i$ for each i.

Now suppose we are given a linearly independent set of vectors $\{\mathbf{x}_1, \mathbf{x}_2, \ldots, \mathbf{x}_m\}$ in an inner product space V. Using Lemma 4.12.12, we can construct an orthogonal basis for the subspace of V spanned by these vectors. We begin with the vector $\mathbf{v}_1 = \mathbf{x}_1$ as above, and we define $\mathbf{v}_i$ by subtracting off appropriate projections of $\mathbf{x}_i$ on $\mathbf{v}_1, \mathbf{v}_2, \ldots, \mathbf{v}_{i-1}$. The resulting procedure is called the **Gram-Schmidt orthogonalization procedure**. The formal statement of the result is as follows.

Theorem 4.12.13   **(Gram-Schmidt Process)**

Let $\{\mathbf{x}_1, \mathbf{x}_2, \ldots, \mathbf{x}_m\}$ be a linearly independent set of vectors in an inner product space V. Then an *orthogonal basis* for the subspace of V spanned by these vectors is $\{\mathbf{v}_1, \mathbf{v}_2, \ldots, \mathbf{v}_m\}$, where

$$\mathbf{v}_1 = \mathbf{x}_1$$

$$\mathbf{v}_2 = \mathbf{x}_2 - \frac{\langle \mathbf{x}_2, \mathbf{v}_1 \rangle}{||\mathbf{v}_1||^2}\mathbf{v}_1$$

$$\mathbf{v}_3 = \mathbf{x}_3 - \frac{\langle \mathbf{x}_3, \mathbf{v}_1 \rangle}{||\mathbf{v}_1||^2}\mathbf{v}_1 - \frac{\langle \mathbf{x}_3, \mathbf{v}_2 \rangle}{||\mathbf{v}_2||^2}\mathbf{v}_2$$

$$\vdots$$

$$\mathbf{v}_i = \mathbf{x}_i - \sum_{k=1}^{i-1} \frac{\langle \mathbf{x}_i, \mathbf{v}_k \rangle}{||\mathbf{v}_k||^2}\mathbf{v}_k$$

$$\vdots$$

$$\mathbf{v}_m = \mathbf{x}_m - \sum_{k=1}^{m-1} \frac{\langle \mathbf{x}_m, \mathbf{v}_k \rangle}{||\mathbf{v}_k||^2}\mathbf{v}_k.$$

Proof Lemma 4.12.12 shows that $\{\mathbf{v}_1, \mathbf{v}_2, \ldots, \mathbf{v}_m\}$ is an orthogonal set of vectors. Thus, both $\{\mathbf{v}_1, \mathbf{v}_2, \ldots, \mathbf{v}_m\}$ and $\{\mathbf{x}_1, \mathbf{x}_2, \ldots, \mathbf{x}_m\}$ are linearly independent sets, and hence

$$\text{span}\{\mathbf{v}_1, \mathbf{v}_2, \ldots, \mathbf{v}_m\} \qquad \text{and} \qquad \text{span}\{\mathbf{x}_1, \mathbf{x}_2, \ldots, \mathbf{x}_m\}$$

are m-dimensional subspaces of V. (Why?) Moreover, from the formulas given in Theorem 4.12.13, we see that each $x_i \in \text{span}\{v_1, v_2, \ldots, v_m\}$, and so $\text{span}\{x_1, x_2, \ldots, x_m\}$ is a subset of $\text{span}\{v_1, v_2, \ldots, v_m\}$. Thus, by Corollary 4.6.14,

$$\text{span}\{v_1, v_2, \ldots, v_m\} = \text{span}\{x_1, x_2, \ldots, x_m\}.$$

We conclude that $\{v_1, v_2, \ldots, v_m\}$ is a basis for the subspace of V spanned by $x_1, x_2, \ldots, x_m$.   ■

Example 4.12.14   Obtain an orthogonal basis for the subspace of $\mathbb{R}^4$ spanned by

$$x_1 = (1, 0, 1, 0), \qquad x_2 = (1, 1, 1, 1), \qquad x_3 = (-1, 2, 0, 1).$$

Solution: Following the Gram-Schmidt process, we set $v_1 = x_1 = (1, 0, 1, 0)$. Next, we have

$$v_2 = x_2 - \frac{\langle x_2, v_1 \rangle}{||v_1||^2} v_1 = (1, 1, 1, 1) - \frac{2}{2}(1, 0, 1, 0) = (0, 1, 0, 1)$$

and

$$\begin{aligned} v_3 &= x_3 - \frac{\langle x_3, v_1 \rangle}{||v_1||^2} v_1 - \frac{\langle x_3, v_2 \rangle}{||v_2||^2} v_2 \\ &= (-1, 2, 0, 1) + \frac{1}{2}(1, 0, 1, 0) - \frac{3}{2}(0, 1, 0, 1) \\ &= \left(-\frac{1}{2}, \frac{1}{2}, \frac{1}{2}, -\frac{1}{2}\right). \end{aligned}$$

The orthogonal basis so obtained is

$$\left\{ (1, 0, 1, 0), (0, 1, 0, 1), \left(-\frac{1}{2}, \frac{1}{2}, \frac{1}{2}, -\frac{1}{2}\right) \right\}.$$ □

Of course, once an orthogonal basis $\{v_1, v_2, \ldots, v_m\}$ is obtained for a subspace of V, we can normalize this basis by setting $u_i = \dfrac{v_i}{||v_i||}$ to obtain an orthonormal basis $\{u_1, u_2, \ldots, u_m\}$. For instance, an orthonormal basis for the subspace of $\mathbb{R}^4$ in the preceding example is

$$\left\{ \left(\frac{1}{\sqrt{2}}, 0, \frac{1}{\sqrt{2}}, 0\right), \left(0, \frac{1}{\sqrt{2}}, 0, \frac{1}{\sqrt{2}}\right), \left(-\frac{1}{2}, \frac{1}{2}, \frac{1}{2}, -\frac{1}{2}\right) \right\}.$$

Example 4.12.15   Determine an orthogonal basis for the subspace of $C^0[-1, 1]$ spanned by the functions $f_1(x) = x$, $f_2(x) = x^3$, $f_3(x) = x^5$, using the same inner product introduced in the previous section.

Solution: In this case, we let $\{g_1, g_2, g_3\}$ denote the orthogonal basis, and we apply the Gram-Schmidt process. Thus, $g_1(x) = x$, and

$$g_2(x) = f_2(x) - \frac{\langle f_2, g_1 \rangle}{||g_1||^2} g_1(x). \tag{4.12.5}$$

We have

$$\langle f_2, g_1 \rangle = \int_{-1}^{1} f_2(x)g_1(x)\, dx = \int_{-1}^{1} x^4\, dx = \tfrac{2}{5} \quad \text{and}$$

$$\|g_1\|^2 = \langle g_1, g_1 \rangle = \int_{-1}^{1} x^2\, dx = \tfrac{2}{3}.$$

Substituting into Equation (4.12.5) yields

$$g_2(x) = x^3 - \tfrac{3}{5}x = \tfrac{1}{5}x(5x^2 - 3).$$

We now compute $g_3(x)$. According to the Gram-Schmidt process,

$$g_3(x) = f_3(x) - \frac{\langle f_3, g_1 \rangle}{\|g_1\|^2} g_1(x) - \frac{\langle f_3, g_2 \rangle}{\|g_2\|^2} g_2(x). \tag{4.12.6}$$

We first evaluate the required inner products:

$$\langle f_3, g_1 \rangle = \int_{-1}^{1} f_3(x)g_1(x)\, dx = \int_{-1}^{1} x^6\, dx = \tfrac{2}{7},$$

$$\langle f_3, g_2 \rangle = \int_{-1}^{1} f_3(x)g_2(x)\, dx = \frac{1}{5} \int_{-1}^{1} x^6(5x^2 - 3)\, dx = \tfrac{1}{5}\left(\tfrac{10}{9} - \tfrac{6}{7}\right) = \tfrac{16}{315},$$

$$\|g_2\|^2 = \int_{-1}^{1} [g_2(x)]^2\, dx = \tfrac{1}{25} \int_{-1}^{1} x^2(5x^2 - 3)^2\, dx$$

$$= \tfrac{1}{25} \int_{-1}^{1} (25x^6 - 30x^4 + 9x^2)\, dx = \tfrac{8}{175}.$$

Substituting into Equation (4.12.6) yields

$$g_3(x) = x^5 - \tfrac{3}{7}x - \tfrac{2}{9}x(5x^2 - 3) = \tfrac{1}{63}(63x^5 - 70x^3 + 15x).$$

Thus, an orthogonal basis for the subspace of $C^0[-1, 1]$ spanned by f_1, f_2, and f_3 is

$$\left\{x, \tfrac{1}{5}x(5x^2 - 3), \tfrac{1}{63}x(63x^4 - 70x^2 + 15)\right\}. \qquad \square$$

Exercises for 4.12

Key Terms

Orthogonal vectors, Orthogonal set, Unit vector, Orthonormal vectors, Orthonormal set, Normalization, Orthogonal basis, Orthonormal basis, Gram-Schmidt process, Orthogonal projection.

Skills

- Be able to determine whether a given set of vectors are orthogonal and/or orthonormal.

- Be able to determine whether a given set of vectors forms an orthogonal and/or orthonormal basis for an inner product space.

- Be able to replace an orthogonal set with an orthonormal set via normalization.

- Be able to readily compute the components of a vector $\mathbf{v}$ in an inner product space V relative to an orthogonal (or orthonormal) basis for V.

- Be able to compute the orthogonal projection of one vector $\mathbf{w}$ along another vector $\mathbf{v}$: $\mathbf{P}(\mathbf{w}, \mathbf{v})$.

- Be able to carry out the Gram-Schmidt process to replace a basis for V with an orthogonal (or orthonormal) basis for V.

True-False Review

For Questions 1–7, decide if the given statement is **true** or **false**, and give a brief justification for your answer. If true, you can quote a relevant definition or theorem from the text. If false, provide an example, illustration, or brief explanation of why the statement is false.

1. Every orthonormal basis for an inner product space V is also an orthogonal basis for V.

2. Every linearly independent set of vectors in an inner product space V is orthogonal.

3. With the inner product $\langle f, g \rangle = \int_0^\pi f(t)g(t)\, dt$, the functions $f(x) = \cos x$ and $g(x) = \sin x$ are an orthogonal basis for span$\{\cos x, \sin x\}$.

4. The Gram-Schmidt process applied to the vectors $\{\mathbf{x}_1, \mathbf{x}_2, \mathbf{x}_3\}$ yields the same basis as the Gram-Schmidt process applied to the vectors $\{\mathbf{x}_3, \mathbf{x}_2, \mathbf{x}_1\}$.

5. In expressing the vector $\mathbf{v}$ as a linear combination of the orthogonal basis $\{\mathbf{v}_1, \mathbf{v}_2, \ldots, \mathbf{v}_n\}$ for an inner product space V, the coefficient of $\mathbf{v}_i$ is

$$c_i = \frac{\langle \mathbf{v}, \mathbf{v}_i \rangle}{||\mathbf{v}_i||^2}.$$

6. If $\mathbf{u}$ and $\mathbf{v}$ are orthogonal vectors and $\mathbf{w}$ is any vector, then

$$P(P(\mathbf{w}, \mathbf{v}), \mathbf{u}) = \mathbf{0}.$$

7. If $\mathbf{w}_1$, $\mathbf{w}_2$, and $\mathbf{v}$ are vectors in an inner product space V, then

$$P(\mathbf{w}_1 + \mathbf{w}_2, \mathbf{v}) = P(\mathbf{w}_1, \mathbf{v}) + P(\mathbf{w}_2, \mathbf{v}).$$

Problems

For Problems 1–4, determine whether the given set of vectors is an orthogonal set in $\mathbb{R}^n$. For those that are, determine a corresponding orthonormal set of vectors.

1. $\{(2, -1, 1), (1, 1, -1), (0, 1, 1)\}$.

2. $\{(1, 3, -1, 1), (-1, 1, 1, -1), (1, 0, 2, 1)\}$

3. $\{(1, 2, -1, 0), (1, 0, 1, 2), (-1, 1, 1, 0), (1, -1, -1, 0)\}$.

4. $\{(1, 2, -1, 0, 3), (1, 1, 0, 2, -1), (4, 2, -4, -5, -4)\}$

5. Let $\mathbf{v}_1 = (1, 2, 3)$, $\mathbf{v}_2 = (1, 1, -1)$. Determine all nonzero vectors $\mathbf{w}$ such that $\{\mathbf{v}_1, \mathbf{v}_2, \mathbf{w}\}$ is an orthogonal set. Hence obtain an orthonormal set of vectors in $\mathbb{R}^3$.

For Problems 6–7, show that the given set of vectors is an orthogonal set in $\mathbb{C}^n$, and hence obtain an orthonormal set of vectors in $\mathbb{C}^n$ in each case.

6. $\{(1 - i, 3 + 2i), (2 + 3i, 1 - i)\}$.

7. $\{(1 - i, 1 + i, i), (0, i, 1 - i), (-3 + 3i, 2 + 2i, 2i)\}$.

8. Consider the vectors $\mathbf{v} = (1 - i, 1 + 2i)$, $\mathbf{w} = (2 + i, z)$ in $\mathbb{C}^2$. Determine the complex number z such that $\{\mathbf{v}, \mathbf{w}\}$ is an orthogonal set of vectors, and hence obtain an orthonormal set of vectors in $\mathbb{C}^2$.

For Problems 9–10, show that the given functions in $C^0[-1, 1]$ are orthogonal, and use them to construct an orthonormal set of functions in $C^0[-1, 1]$.

9. $f_1(x) = 1$, $f_2(x) = \sin \pi x$, $f_3(x) = \cos \pi x$.

10. $f_1(x) = 1$, $f_2(x) = x$, $f_3(x) = \frac{1}{2}(3x^2 - 1)$. These are the **Legendre polynomials** that arise as solutions of the Legendre differential equation

$$(1 - x^2)y'' - 2xy' + n(n + 1)y = 0,$$

when $n = 0, 1, 2$, respectively.

For Problems 11–12, show that the given functions are orthonormal on $[-1, 1]$.

11. $f_1(x) = \sin \pi x$, $f_2(x) = \sin 2\pi x$, $f_3(x) = \sin 3\pi x$. [**Hint:** The trigonometric identity

$$\sin a \sin b = \frac{1}{2}[\cos(a + b) - \cos(a - b)]$$

will be useful.]

12. $f_1(x) = \cos \pi x$, $f_2(x) = \cos 2\pi x$, $f_3(x) = \cos 3\pi x$.

13. Let

$$A_1 = \begin{bmatrix} 1 & 1 \\ -1 & 2 \end{bmatrix}, \quad A_2 = \begin{bmatrix} -1 & 1 \\ 2 & 1 \end{bmatrix}, \quad \text{and}$$

$$A_3 = \begin{bmatrix} -1 & -3 \\ 0 & 2 \end{bmatrix}.$$

Use the inner product

$$\langle A, B \rangle = a_{11}b_{11} + a_{12}b_{12} + a_{21}b_{21} + a_{22}b_{22}$$

to find all matrices

$$A_4 = \begin{bmatrix} a & b \\ c & d \end{bmatrix}$$

such that $\{A_1, A_2, A_3, A_4\}$ is an orthogonal set of matrices in $M_2(\mathbb{R})$.

For Problems 14–19, use the Gram-Schmidt process to determine an *orthonormal* basis for the subspace of $\mathbb{R}^n$ spanned by the given set of vectors.

14. $\{(1, -1, -1), (2, 1, -1)\}$.

15. $\{(2, 1, -2), (1, 3, -1)\}$.

16. $\{(-1, 1, 1, 1), (1, 2, 1, 2)\}$.

17. $\{(1, 0, -1, 0), (1, 1, -1, 0), (-1, 1, 0, 1)\}$

18. $\{(1, 2, 0, 1), (2, 1, 1, 0), (1, 0, 2, 1)\}$.

19. $\{(1, 1, -1, 0), (-1, 0, 1, 1), (2, -1, 2, 1)\}$.

20. If

$$A = \begin{bmatrix} 3 & 1 & 4 \\ 1 & -2 & 1 \\ 1 & 5 & 2 \end{bmatrix},$$

determine an orthogonal basis for rowspace(A).

For Problems 21–22, determine an *orthonormal* basis for the subspace of $\mathbb{C}^3$ spanned by the given set of vectors. Make sure that you use the appropriate inner product in $\mathbb{C}^3$.

21. $\{(1 - i, 0, i), (1, 1 + i, 0)\}$.

22. $\{(1 + i, i, 2 - i), (1 + 2i, 1 - i, i)\}$.

For Problems 23–25, determine an orthogonal basis for the subspace of $C^0[a, b]$ spanned by the given vectors, for the given interval $[a, b]$.

23. $f_1(x) = 1$, $f_2(x) = x$, $f_3(x) = x^2$, $a = 0, b = 1$.

24. $f_1(x) = 1$, $f_2(x) = x^2$, $f_3(x) = x^4$, $a = -1, b = 1$.

25. $f_1(x) = 1$, $f_2(x) = \sin x$, $f_3(x) = \cos x$, $a = -\pi/2$, $b = \pi/2$.

On $M_2(\mathbb{R})$ define the inner product $\langle A, B \rangle$ by

$$\langle A, B \rangle = 5a_{11}b_{11} + 2a_{12}b_{12} + 3a_{21}b_{21} + 5a_{22}b_{22}$$

for all matrices $A = [a_{ij}]$ and $B = [b_{ij}]$. For Problems 26–27, use this inner product in the Gram-Schmidt procedure to determine an orthogonal basis for the subspace of $M_2(\mathbb{R})$ spanned by the given matrices.

26. $A_1 = \begin{bmatrix} 1 & -1 \\ 2 & 1 \end{bmatrix}$, $A_2 = \begin{bmatrix} 2 & -3 \\ 4 & 1 \end{bmatrix}$.

27. $A_1 = \begin{bmatrix} 0 & 1 \\ 1 & 0 \end{bmatrix}$, $A_2 = \begin{bmatrix} 0 & 1 \\ 1 & 1 \end{bmatrix}$, $A_3 = \begin{bmatrix} 1 & 1 \\ 1 & 0 \end{bmatrix}$.

Also identify the subspace of $M_2(\mathbb{R})$ spanned by $\{A_1, A_2, A_3\}$.

On P_n, define the inner product $\langle p_1, p_2 \rangle$ by

$$\langle p_1, p_2 \rangle = a_0b_0 + a_1b_1 + \cdots + a_nb_n$$

for all polynomials

$$p_1(x) = a_0 + a_1x + \cdots + a_nx^n,$$
$$p_2(x) = b_0 + b_1x + \cdots + b_nx^n.$$

For Problems 28–29, use this inner product to determine an orthogonal basis for the subspace of P_n spanned by the given polynomials.

28. $p_1(x) = 1 - 2x + 2x^2$, $p_2(x) = 2 - x - x^2$.

29. $p_1(x) = 1 + x^2$, $p_2(x) = 2 - x + x^3$, $p_3(x) = 2x^2 - x$.

30. Let $\{\mathbf{u}_1, \mathbf{u}_2, \mathbf{v}\}$ be linearly independent vectors in an inner product space V, and suppose that $\mathbf{u}_1$ and $\mathbf{u}_2$ are orthogonal. Define the vector $\mathbf{u}_3$ in V by

$$\mathbf{u}_3 = \mathbf{v} + \lambda\mathbf{u}_1 + \mu\mathbf{u}_2,$$

where λ, μ are scalars. Derive the values of λ and μ such that $\{\mathbf{u}_1, \mathbf{u}_2, \mathbf{u}_3\}$ is an orthogonal basis for the subspace of V spanned by $\{\mathbf{u}_1, \mathbf{u}_2, \mathbf{v}\}$.

31. Prove that if $\{\mathbf{v}_1, \mathbf{v}_2, \ldots, \mathbf{v}_k\}$ is an orthogonal set of vectors in an inner product space V and if $\mathbf{u}_i = \dfrac{1}{||\mathbf{v}_i||}\mathbf{v}_i$ for each i, then $\{\mathbf{u}_1, \mathbf{u}_2, \ldots, \mathbf{u}_k\}$ form an orthonormal set of vectors.

32. Prove Lemma 4.12.12.

Let V be an inner product space, and let W be a subspace of V. Set

$$W^\perp = \{\mathbf{v} \in V : \langle \mathbf{v}, \mathbf{w} \rangle = 0 \text{ for all } \mathbf{w} \in W\}.$$

The set $W^\perp$ is called the **orthogonal complement** of W in V. Problems 33–38 explore this concept in some detail. Deeper applications can be found in Project 1 at the end of this chapter.

33. Prove that $W^\perp$ is a subspace of V.

34. Let $V = \mathbb{R}^3$ and let

$$W = \text{span}\{(1, 1, -1)\}.$$

Find $W^\perp$.

35. Let $V = \mathbb{R}^4$ and let

$$W = \text{span}\{(0, 1, -1, 3), (1, 0, 0, 3)\}.$$

Find $W^\perp$.

36. Let $V = M_2(\mathbb{R})$ and let W be the subspace of 2×2 symmetric matrices. Compute $W^\perp$.

37. Prove that $W \cap W^\perp = 0$. (That is, W and $W^\perp$ have no nonzero elements in common.)

38. Prove that if W_1 is a subset of W_2, then $(W_2)^\perp$ is a subset of $(W_1)^\perp$.

39. The subject of Fourier series is concerned with the representation of a 2π-periodic function f as the following *infinite* linear combination of the set of functions $\{1, \sin nx, \cos nx\}_{n=1}^\infty$:

$$f(x) = \tfrac{1}{2}a_0 + \sum_{n=1}^\infty (a_n \cos nx + b_n \sin nx).$$
$$(4.12.7)$$

In this problem, we investigate the possibility of performing such a representation.

(a) Use appropriate trigonometric identities, or some form of technology, to verify that the set of functions
$$\{1, \sin nx, \cos nx\}_{n=1}^\infty$$
is orthogonal on the interval $[-\pi, \pi]$.

(b) By multiplying (4.12.7) by $\cos mx$ and integrating over the interval $[-\pi, \pi]$, show that
$$a_0 = \frac{1}{\pi} \int_{-\pi}^{\pi} f(x)\, dx$$

and
$$a_m = \frac{1}{\pi} \int_{-\pi}^{\pi} f(x) \cos mx\, dx.$$

[**Hint**: You may assume that interchange of the infinite summation with the integral is permissible.]

(c) Use a similar procedure to show that

$$b_m = \frac{1}{\pi} \int_{-\pi}^{\pi} f(x) \sin mx\, dx.$$

It can be shown that if f is in $C^1(-\pi, \pi)$, then Equation (4.12.7) holds for each $x \in (-\pi, \pi)$. The series appearing on the right-hand side of (4.12.7) is called the **Fourier series of** f, and the constants in the summation are called the **Fourier coefficients for** f.

(d) Show that the Fourier coefficients for the function $f(x) = x$, $-\pi < x \le \pi$, $f(x + 2\pi) = f(x)$, are

$$a_n = 0, \qquad\qquad n = 0, 1, 2, \ldots,$$
$$b_n = -\frac{2}{n} \cos n\pi, \qquad n = 1, 2, \ldots,$$

and thereby determine the Fourier series of f.

(e) $\diamond$ Using some form of technology, sketch the approximations to $f(x) = x$ on the interval $(-\pi, \pi)$ obtained by considering the first three terms, first five terms, and first ten terms in the Fourier series for f. What do you conclude?

4.13 Chapter Review

In this chapter we have derived some basic results in linear algebra regarding vector spaces. These results form the framework for much of linear mathematics. Following are listed some of the chapter highlights.

The Definition of a Vector Space

A vector space consists of four different components:

1. A set of vectors V.

2. A set of scalars F (either the set of real numbers $\mathbb{R}$, or the set of complex numbers $\mathbb{C}$).

3. A rule, $+$, for adding vectors in V.

4. A rule, $\cdot$, for multiplying vectors in V by scalars in F.

Then $(V, +, \cdot)$ is a vector space over F if and only if axioms A1–A10 of Definition 4.2.1 are satisfied. If F is the set of all real numbers, then $(V, +, \cdot)$ is called a *real* vector space, whereas if F is the set of all complex numbers, then $(V, +, \cdot)$ is called a *complex* vector space. Since it is usually quite clear what the addition and scalar multiplication operations are, we usually specify a vector space by giving only the set of vectors V. The major vector spaces we have dealt with are the following:

$\mathbb{R}^n$ the (real) vector space of all ordered n-tuples of real numbers.

$\mathbb{C}^n$ the (complex) vector space of all ordered n-tuples of complex numbers.

$M_n(\mathbb{R})$ the (real) vector space of all $n \times n$ matrices with real elements.

$C^k(I)$ the vector space of all real-valued functions that are continuous and have (at least) k continuous derivatives on I.

P_n the vector space of all polynomials of degree $\leq n$ with real coefficients.

Subspaces

Usually the vector space V that underlies a given problem is known. It is often one that appears in the list above. However, the solution of a given problem in general involves only a subset of vectors from this vector space. The question that then arises is whether this subset of vectors is itself a vector space under the same operations of addition and scalar multiplication as in V. In order to answer this question, Theorem 4.3.2 tells us that *a nonempty subset of a vector space V is a subspace of V if and only if the subset is closed under addition and closed under scalar multiplication.*

Spanning Sets

A set of vectors $\{\mathbf{v}_1, \mathbf{v}_2, \ldots, \mathbf{v}_k\}$ in a vector space V is said to *span* V if *every* vector in V can be written as a linear combination of $\mathbf{v}_1, \mathbf{v}_2, \ldots, \mathbf{v}_k$—that is, if for every $\mathbf{v} \in V$, there exist scalars $c_1, c_2, \ldots, c_k$ such that

$$\mathbf{v} = c_1\mathbf{v}_1 + c_2\mathbf{v}_2 + \cdots + c_k\mathbf{v}_k.$$

Given a set of vectors $\{\mathbf{v}_1, \mathbf{v}_2, \ldots, \mathbf{v}_k\}$ in a vector space V, we can form the set of *all* vectors that can be written as a linear combination of $\mathbf{v}_1, \mathbf{v}_2, \ldots, \mathbf{v}_k$. This collection of vectors is a subspace of V called the *subspace spanned by* $\{\mathbf{v}_1, \mathbf{v}_2, \ldots, \mathbf{v}_k\}$, and denoted span$\{\mathbf{v}_1, \mathbf{v}_2, \ldots, \mathbf{v}_k\}$. Thus,

$$\text{span}\{\mathbf{v}_1, \mathbf{v}_2, \ldots, \mathbf{v}_k\} = \{\mathbf{v} \in V : \mathbf{v} = c_1\mathbf{v}_1 + c_2\mathbf{v}_2 + \cdots + c_k\mathbf{v}_k\}.$$

Linear Dependence and Linear Independence

Let $\{\mathbf{v}_1, \mathbf{v}_2, \ldots, \mathbf{v}_k\}$ be a set of vectors in a vector space V, and consider the vector equation

$$c_1\mathbf{v}_1 + c_2\mathbf{v}_2 + \cdots + c_k\mathbf{v}_k = \mathbf{0}. \tag{4.13.1}$$

Clearly this equation will hold if $c_1 = c_2 = \cdots = c_k = 0$. The question of interest is whether there are nonzero values of some or all of the scalars $c_1, c_2, \ldots, c_k$ such that (4.13.1) holds. This leads to the following two ideas:

Linear dependence: There exist scalars $c_1, c_2, \ldots, c_k$, not all zero, such that (4.13.1) holds.

Linear independence: The *only* values of the scalars $c_1, c_2, \ldots, c_k$ such that (4.13.1) holds are $c_1 = c_2 = \cdots = c_k = 0$.

To determine whether a set of vectors is linearly dependent or linearly independent we usually have to use (4.13.1). However, if the vectors are from $\mathbb{R}^n$, then we can use Corollary 4.5.15, whereas for vectors in $C^{k-1}(I)$ the Wronskian can be useful.

Bases and Dimension

A linearly independent set of vectors that spans a vector space V is called a *basis* for V. If $\{\mathbf{v}_1, \mathbf{v}_2, \ldots, \mathbf{v}_k\}$ is a basis for V, then any vector in V can be written uniquely as

$$\mathbf{v} = c_1 \mathbf{v}_1 + c_2 \mathbf{v}_2 + \cdots + c_k \mathbf{v}_k,$$

for appropriate values of the scalars $c_1, c_2, \ldots, c_k$.

1. All bases in a finite-dimensional vector space V contain the same number of vectors, and this number is called the *dimension* of V, denoted $\dim[V]$.

2. We can view the dimension of a finite-dimensional vector space V in two different ways. First, it gives the minimum number of vectors that span V. Alternatively, we can regard $\dim[V]$ as determining the maximum number of vectors that a linearly independent set in V can contain.

3. If $\dim[V] = n$, then *any* linearly independent set of n vectors in V is a basis for V. Alternatively, *any* set of n vectors that spans V is a basis for V.

Inner Product Spaces

An inner product is a mapping that associates, with any two vectors $\mathbf{u}$ and $\mathbf{v}$ in a vector space V, a scalar that we denote by $\langle \mathbf{u}, \mathbf{v} \rangle$. This mapping must satisfy the properties given in Definition 4.11.10. The main reason for introducing the idea of an inner product is that it enables us to extend the familiar idea of orthogonality and length of vectors in $\mathbb{R}^3$ to a general vector space. Thus $\mathbf{u}$ and $\mathbf{v}$ are said to be orthogonal in an inner product space if and only if

$$\langle \mathbf{u}, \mathbf{v} \rangle = 0.$$

The Gram-Schmidt Orthonormalization Process

The Gram-Schmidt procedure is a process that takes a linearly independent set of vectors $\{\mathbf{x}_1, \mathbf{x}_2, \ldots, \mathbf{x}_m\}$ in an inner product space V and returns an *orthogonal* basis $\{\mathbf{v}_1, \mathbf{v}_2, \ldots, \mathbf{v}_m\}$ for $\text{span}\{\mathbf{x}_1, \mathbf{x}_2, \ldots, \mathbf{x}_m\}$.

Additional Problems

For Problems 1–2, let r and s denote scalars and let $\mathbf{v}$ and $\mathbf{w}$ denote vectors in $\mathbb{R}^5$.

1. Prove that $(r + s)\mathbf{v} = r\mathbf{v} + s\mathbf{v}$.

2. Prove that $r(\mathbf{v} + \mathbf{w}) = r\mathbf{v} + r\mathbf{w}$.

For Problems 3–13, determine whether the given set (together with the usual operations on that set) forms a vector space over $\mathbb{R}$. In all cases, justify your answer carefully.

3. The set of polynomials of degree 5 or less whose coefficients are even integers.

4. The set of all polynomials of degree 5 or less whose coefficients of x^2 and x^3 are zero.

5. The set of solutions to the linear system

$$\begin{aligned} -2x_2 + 5x_3 &= 7, \\ 4x_1 - 6x_2 + 3x_3 &= 0. \end{aligned}$$

6. The set of solutions to the linear system

$$4x_1 - 7x_2 + 2x_3 = 0,$$
$$5x_1 - 2x_2 + 9x_3 = 0.$$

7. The set of 2×2 real matrices whose entries are either all zero or all nonzero.

8. The set of 2×2 real matrices that commute with the matrix

$$\begin{bmatrix} 1 & 2 \\ 0 & 2 \end{bmatrix}.$$

9. The set of all functions $f : [0, 1] \to [0, 1]$ such that $f(0) = f(\frac{1}{4}) = f(\frac{1}{2}) = f(\frac{3}{4}) = f(1) = 0$.

10. The set of all functions $f : [0, 1] \to [0, 1]$ such that $f(x) \leq x$ for all x in $[0, 1]$.

11. The set of $n \times n$ matrices A such that A^2 is symmetric.

12. The set of all points (x, y) in $\mathbb{R}^2$ that are equidistant from $(-1, 2)$ and $(1, -2)$.

13. The set of all points (x, y, z) in $\mathbb{R}^3$ that are a distance 5 from the point $(0, -3, 4)$.

14. Let

$$V = \{(a_1, a_2) : a_1, a_2 \in \mathbb{R}, a_2 > 0\}.$$

Define addition and scalar multiplication on V as follows:

$$(a_1, a_2) + (b_1, b_2) = (a_1 + b_1, a_2 b_2),$$
$$k(a_1, a_2) = (ka_1, a_2^k), \quad k \in \mathbb{R}.$$

Explicitly verify that V is a vector space over $\mathbb{R}$.

15. Show that

$$W = \{(a, 2^a) : a \in \mathbb{R}\}$$

is a subspace of the vector space V given in the preceding problem.

16. Show that $\{(1, 2), (3, 8)\}$ is a linearly dependent set in the vector space V in Problem 14.

17. Show that $\{(1, 4), (2, 1)\}$ is a basis for the vector space V in Problem 14.

18. What is the dimension of the subspace of P_2 given by

$$W = \text{span}\{2 + x^2, 4 - 2x + 3x^2, 1 + x\}?$$

For Problems 19–24, decide (with justification) whether W is a subspace of V.

19. $V = \mathbb{R}^2$, $W = \{(x, y) : x^2 - y = 0\}$.

20. $V = \mathbb{R}^2$, $W = \{(x, x^3) : x \in \mathbb{R}\}$.

21. $V = M_2(\mathbb{R})$, $W = \{2 \times 2 \text{ orthogonal matrices}\}$. [An $n \times n$ matrix A is *orthogonal* if it is invertible and $A^{-1} = A^T$.]

22. $V = C[a, b]$, $W = \{f \in V : f(a) = 2f(b)\}$.

23. $V = C[a, b]$, $W = \{f \in V : \int_a^b f(x) \, dx = 0\}$.

24. $V = M_{3 \times 2}(\mathbb{R})$,

$$W = \left\{ \begin{bmatrix} a & b \\ c & d \\ e & f \end{bmatrix} : a + b = c + f \text{ and } a - c = e - f - d \right\}.$$

For Problems 25–32, decide (with justification) whether or not the given set S of vectors (a) spans V, and (b) is linearly independent.

25. $V = \mathbb{R}^3$, $S = \{(5, -1, 2), (7, 1, 1)\}$.

26. $V = \mathbb{R}^3$, $S = \{(6, -3, 2), (1, 1, 1), (1, -8, -1)\}$.

27. $V = \mathbb{R}^4$, $S = \{(6, -3, 2, 0), (1, 1, 1, 0), (1, -8, -1, 0)\}$.

28. $V = \mathbb{R}^3$, $S = \{(10, -6, 5), (3, -3, 2), (0, 0, 0), (6, 4, -1), (7, 7, -2)\}$.

29. $V = P_3$, $S = \{2x - x^3, 1 + x + x^2, 3, x\}$.

30. $V = P_4$, $S = \{x^4 + x^2 + 1, x^2 + x + 1, x + 1, x^4 + 2x + 3\}$.

31. $V = M_{2 \times 3}(\mathbb{R})$,

$$S = \left\{ \begin{bmatrix} -1 & 0 & 0 \\ 0 & 1 & 1 \end{bmatrix}, \begin{bmatrix} 3 & 2 & 1 \\ 1 & 2 & 3 \end{bmatrix}, \begin{bmatrix} -1 & -2 & -3 \\ 3 & 2 & 1 \end{bmatrix}, \begin{bmatrix} -11 & -6 & -5 \\ 1 & -2 & -5 \end{bmatrix} \right\}.$$

32. $V = M_2(\mathbb{R})$,

$$S = \left\{ \begin{bmatrix} 1 & 2 \\ 2 & 1 \end{bmatrix}, \begin{bmatrix} 3 & 4 \\ 4 & 3 \end{bmatrix}, \begin{bmatrix} -2 & -1 \\ -1 & -2 \end{bmatrix}, \begin{bmatrix} -3 & 0 \\ 0 & 3 \end{bmatrix}, \begin{bmatrix} 2 & 0 \\ 0 & 0 \end{bmatrix} \right\}.$$

33. Prove that if $\{\mathbf{v}_1, \mathbf{v}_2, \mathbf{v}_3\}$ is linearly independent and $\mathbf{v}_4$ is not in $\text{span}\{\mathbf{v}_1, \mathbf{v}_2, \mathbf{v}_3\}$, then $\{\mathbf{v}_1, \mathbf{v}_2, \mathbf{v}_3, \mathbf{v}_4\}$ is linearly independent.

34. Let A be an $m \times n$ matrix, let $\mathbf{v} \in \text{colspace}(A)$ and let $\mathbf{w} \in \text{nullspace}(A^T)$. Prove that $\mathbf{v}$ and $\mathbf{w}$ are orthogonal.

35. Let W denote the set of all 3×3 skew-symmetric matrices.

    (a) Show that W is a subspace of $M_3(\mathbb{R})$.

    (b) Find a basis and the dimension of W.

    (c) Extend the basis you constructed in part (b) to a basis for $M_3(\mathbb{R})$.

36. Let W denote the set of all 3×3 matrices whose rows and columns add up to zero.

    (a) Show that W is a subspace of $M_3(\mathbb{R})$.

    (b) Find a basis and the dimension of W.

    (c) Extend the basis you constructed in part (b) to a basis for $M_3(\mathbb{R})$.

37. Let $(V, +_V, \cdot_V)$ and $(W, +_W, \cdot_W)$ be vector spaces and define

$$V \oplus W = \{(\mathbf{v}, \mathbf{w}) : \mathbf{v} \in V \text{ and } \mathbf{w} \in W\}.$$

Prove that

    (a) $V \oplus W$ is a vector space, under componentwise operations.

    (b) Via the identification $\mathbf{v} \mapsto (\mathbf{v}, 0)$, V is a subspace of $V \oplus W$, and likewise for W.

    (c) If $\dim[V] = n$ and $\dim[W] = m$, then $\dim[V \oplus W] = m + n$. [**Hint:** Write a basis for $V \oplus W$ in terms of bases for V and W.]

38. Show that a basis for P_3 need not contain a polynomial of each degree 0, 1, 2, 3.

39. Prove that if A is a matrix whose nullspace and column space are the same, then A must have an even number of columns.

40. Let

$$B = \begin{bmatrix} b_1 \\ b_2 \\ \vdots \\ b_n \end{bmatrix} \quad \text{and} \quad C = \begin{bmatrix} c_1 & c_2 & \cdots & c_n \end{bmatrix}.$$

    Prove that if all entries $b_1, b_2, \ldots, b_n$ and $c_1, c_2, \ldots, c_n$ are nonzero, then the $n \times n$ matrix $A = BC$ has nullity $n - 1$.

For Problems 41–44, find a basis and the dimension for the row space, column space, and null space of the given matrix A.

41. $A = \begin{bmatrix} -3 & -6 \\ -6 & -12 \end{bmatrix}$.

42. $A = \begin{bmatrix} -1 & 6 & 2 & 0 \\ 3 & 3 & 1 & 5 \\ 7 & 21 & 7 & 15 \end{bmatrix}$.

43. $A = \begin{bmatrix} -4 & 0 & 3 \\ 0 & 10 & 13 \\ 6 & 5 & 2 \\ -2 & 5 & 10 \end{bmatrix}$.

44. $A = \begin{bmatrix} 3 & 5 & 5 & 2 & 0 \\ 1 & 0 & 2 & 2 & 1 \\ 1 & 1 & 1 & -2 & -2 \\ -2 & 0 & -4 & -2 & -2 \end{bmatrix}$.

For Problems 45–46, find an orthonormal basis for the row space, column space, and null space of the given matrix A.

45. $A = \begin{bmatrix} 1 & 2 & 6 \\ 2 & 1 & 6 \\ 0 & 1 & 2 \\ 1 & 0 & 2 \end{bmatrix}$.

46. $A = \begin{bmatrix} 1 & 3 & 5 \\ -1 & -3 & 1 \\ 0 & 2 & 3 \\ 1 & 5 & 2 \\ 1 & 5 & 8 \end{bmatrix}$.

For Problems 47–50, find an orthogonal basis for the span of the set S, where S is given in

47. Problem 25.

48. Problem 26.

49. Problem 29, using $p \cdot q = \int_0^1 p(t)q(t)\, dt$.

50. Problem 32, using the inner product defined in Problem 4 of Section 4.11.

For Problems 51–54, determine the angle between the given vectors $\mathbf{u}$ and $\mathbf{v}$ using the standard inner product on $\mathbb{R}^n$.

51. $\mathbf{u} = (2, 3)$ and $\mathbf{v} = (4, -1)$.

52. $\mathbf{u} = (-2, -1, 2, 4)$ and $\mathbf{v} = (-3, 5, 1, 1)$.

53. Repeat Problems 51–52 for the inner product on $\mathbb{R}^n$ given by

$$\langle \mathbf{u}, \mathbf{v} \rangle = 2u_1 v_1 + u_2 v_2 + u_3 v_3 + \cdots + u_n v_n.$$

54. Let $t_0, t_1, \ldots, t_n$ be real numbers. For p and q in P_n, define

$$p \cdot q = p(t_0)q(t_0) + p(t_1)q(t_1) + \cdots + p(t_n)q(t_n).$$

   (a) Prove that $p \cdot q$ defines a valid inner product on P_n.

   (b) Let $t_0 = -3$, $t_1 = -1$, $t_2 = 1$, and $t_3 = 3$. Let $p_0(t) = 1$, $p_1(t) = t$, and $p_2(t) = t^2$. Find a polynomial q that is orthogonal to p_0 and p_1, such that $\{p_0, p_1, q\}$ is an orthogonal basis for span$\{p_0, p_1, p_2\}$.

55. Find the distance from the point $(2, 3, 4)$ to the line in $\mathbb{R}^3$ passing through $(0, 0, 0)$ and $(6, -1, -4)$.

56. Let V be an inner product space with basis $\{\mathbf{v}_1, \mathbf{v}_2, \ldots, \mathbf{v}_n\}$. If $\mathbf{x}$ and $\mathbf{y}$ are vectors in V such that $\mathbf{x} \cdot \mathbf{v}_i = \mathbf{y} \cdot \mathbf{v}_i$ for each $i = 1, 2, \ldots, n$, prove that $\mathbf{x} = \mathbf{y}$.

57. State as many conditions as you can on an $n \times n$ matrix A that are equivalent to its invertibility.

Project I: Orthogonal Complement

Let V be an inner product space and let W be a subspace of V.

Part 1 Definition   Let

$$W^{\perp} = \{\mathbf{v} \in V : \langle \mathbf{v}, \mathbf{w} \rangle = 0 \text{ for all } \mathbf{w} \in W\}.$$

Show that $W^{\perp}$ is a subspace of V and that $W^{\perp}$ and W share only the zero vector: $W^{\perp} \cap W = \{\mathbf{0}\}$.

Part 2 Examples

   (a) Let $V = M_2(\mathbb{R})$ with inner product

$$\left\langle \begin{bmatrix} a_{11} & a_{12} \\ a_{21} & a_{22} \end{bmatrix}, \begin{bmatrix} b_{11} & b_{12} \\ b_{21} & b_{22} \end{bmatrix} \right\rangle = a_{11}b_{11} + a_{12}b_{12} + a_{21}b_{21} + a_{22}b_{22}.$$

Find the orthogonal complement of the set W of 2×2 *symmetric matrices*.

   (b) Let A be an $m \times n$ matrix. Show that

$$(\text{rowspace}(A))^{\perp} = \text{nullspace}(A)$$

and

$$(\text{colspace}(A))^{\perp} = \text{nullspace}(A^T).$$

Use this to find the orthogonal complement of the row space and column space of the matrices below:

   (i) $A = \begin{bmatrix} 3 & 1 & -1 \\ 6 & 0 & -4 \end{bmatrix}$.

   (ii) $A = \begin{bmatrix} -1 & 0 & 6 & 2 \\ 3 & -1 & 0 & 4 \\ 1 & 1 & 1 & -1 \end{bmatrix}$.

   (c) Find the orthogonal complement of

     (i) the line in $\mathbb{R}^3$ containing the points $(0, 0, 0)$ and $(2, -1, 3)$.

     (ii) the plane $2x + 3y - 4z = 0$ in $\mathbb{R}^3$.

Part 3 Some Theoretical Results   Let W be a subspace of a finite-dimensional inner product space V.

(a) Show that every vector in V can be written *uniquely* in the form $\mathbf{w} + \mathbf{w}^\perp$, where $\mathbf{w} \in W$ and $\mathbf{w}^\perp \in W^\perp$. [**Hint:** By Gram-Schmidt, $\mathbf{v}$ can be projected onto the subspace W as, say, $\text{proj}_W(\mathbf{v})$, and so $\mathbf{v} = \text{proj}_W(\mathbf{v}) + \mathbf{w}^\perp$, where $w^\perp \in W^\perp$. For the uniqueness, use the fact that $W \cap W^\perp = \{\mathbf{0}\}$.]

(b) Use part (a) to show that

$$\dim[V] = \dim[W] + \dim[W^\perp].$$

(c) Show that

$$(W^\perp)^\perp = W.$$

Project II: Line-Fitting Data Points

Suppose data points (x_1, y_1), (x_2, y_2), ..., (x_n, y_n) in the xy-plane have been collected. Unless these data points are collinear, there will be no line that contains all of them. We wish to find a line, commonly known as a **least-squares line**, that approximates the data points as closely as possible.

How do we go about finding such a line? The approach we take[12] is to write the line as $y = mx + b$, where m and b are unknown constants.

Part 1 Derivation of the Least-Squares Line

(a) By substituting the data points (x_i, y_i) for x and y in the equation $y = mx + b$, show that the matrix equation $A\mathbf{x} = \mathbf{y}$ is obtained, where

$$A = \begin{bmatrix} x_1 & 1 \\ x_2 & 1 \\ \vdots & \vdots \\ x_n & 1 \end{bmatrix}, \qquad \mathbf{x} = \begin{bmatrix} m \\ b \end{bmatrix}, \qquad \text{and} \qquad \mathbf{y} = \begin{bmatrix} y_1 \\ y_2 \\ \vdots \\ y_n \end{bmatrix}.$$

Unless the data points are collinear, the system $A\mathbf{x} = \mathbf{y}$ obtained in part (a) has no solution for $\mathbf{x}$. In other words, the vector $\mathbf{y}$ does not lie in the column space of A. The goal then becomes to find $\mathbf{x}_0$ such that the distance $||\mathbf{y} - A\mathbf{x}_0||$ is as small as possible. This will happen precisely when $\mathbf{y} - A\mathbf{x}_0$ is perpendicular to the column space of A. In other words, for all $\mathbf{x} \in \mathbb{R}^2$, we must have

$$(A\mathbf{x}) \cdot (\mathbf{y} - A\mathbf{x}_0) = 0.$$

(b) Using the fact that the dot product of vectors $\mathbf{u}$ and $\mathbf{v}$ can be written as a matrix multiplication,

$$\mathbf{u} \cdot \mathbf{v} = \mathbf{u}^T \mathbf{v},$$

show that

$$(A\mathbf{x}) \cdot (\mathbf{y} - A\mathbf{x}_0) = \mathbf{x} \cdot (A^T \mathbf{y} - A^T A\mathbf{x}_0).$$

(c) Conclude that

$$A^T \mathbf{y} = A^T A\mathbf{x}_0.$$

Provided that A has linearly independent columns, the matrix $A^T A$ is invertible (see Problem 34, in Section 4.13).

---

[12]We can also obtain the least-squares line by using optimization techniques from multivariable calculus, but the goal here is to illustrate the use of linear systems and projections.

(d) Show that the least-squares solution is

$$\mathbf{x}_0 = (A^T A)^{-1} A^T \mathbf{y}$$

and therefore,

$$A\mathbf{x}_0 = A(A^T A)^{-1} A^T \mathbf{y}$$

is the point in the column space of A that is closest to $\mathbf{y}$. Therefore, it is the **projection** of $\mathbf{y}$ onto the column space of A, and we write

$$A\mathbf{x}_0 = A(A^T A)^{-1} A^T \mathbf{y} = P\mathbf{y},$$

where

$$P = A(A^T A)^{-1} A^T \tag{4.13.2}$$

is called a **projection matrix**. If A is $m \times n$, what are the dimensions of P?

(e) Referring to the projection matrix P in (4.13.2), show that $PA = A$ and $P^2 = P$. Geometrically, why are these facts to be expected? Also show that P is a symmetric matrix.

Part 2 Some Applications   In parts (a)–(d) below, find the equation of the least-squares line to the given data points.

(a) $(0, -2), (1, -1), (2, 1), (3, 2), (4, 2)$.

(b) $(-1, 5), (1, 1), (2, 1), (3, -3)$.

(c) $(-4, -1), (-3, 1), (-2, 3), (0, 7)$.

(d) $(-3, 1), (-2, 0), (-1, 1), (0, -1), (2, -1)$.

In parts (e)–(f), by using the ideas in this project, find the distance from the point P to the given plane.

(e) $P(0, 0, 0); 2x - y + 3z = 6$.

(f) $P(-1, 3, 5); -x + 3y + 3z = 8$.

Part 3 A Further Generalization   Instead of fitting data points to a least-squares line, one could also attempt to do a parabolic approximation of the form $ax^2 + bx + c$. By following the outline in Part 1 above, try to determine a procedure for finding the best parabolic approximation to a set of data points. Then try out your procedure on the data points given in Part 2, (a)–(d).

5

Linear Transformations

There is nothing mysterious, as some have tried to maintain, about the
applicability of mathematics. What we get by abstraction from something
can be returned. — R. L. Wilder

In the preceding chapter, we began building a general framework for studying linear problems. This framework was the mathematical concept of a vector space. Heretofore, our attention at any given moment has been focused on the properties of a single vector space V. However, a rich mastery of linear algebra also requires a working knowledge of the *relationships between different vector spaces*.

Many problems in linear algebra involve the simultaneous consideration of more than one vector space. For example, let A be an $m \times n$ matrix and consider the linear system

$$A\mathbf{x} = \mathbf{0}.$$

In this case, note that $\mathbf{x}$ is a vector in $\mathbb{R}^n$, while the right-hand-side vector $\mathbf{0}$ is a vector in $\mathbb{R}^m$. In fact, the matrix A can be viewed as a mapping T that accepts inputs $\mathbf{x}$ from the vector space $V = \mathbb{R}^n$ and yields outputs

$$T(\mathbf{x}) = A\mathbf{x}$$

in the vector space $W = \mathbb{R}^m$. In terms of this mapping, the solution set to the homogeneous linear system $A\mathbf{x} = \mathbf{0}$, for example, consists of all vectors $\mathbf{x}$ in $\mathbb{R}^n$ with the property that $T(\mathbf{x}) = \mathbf{0}$.

We can use the general mapping notation T in other problems as well. For example, consider the second-order differential equation

$$y'' + y = 0 \tag{5.0.1}$$

and the associated "mapping of functions" T defined by

$$T(y) = y'' + y.$$

Given a function y, T maps y to the function $y'' + y$. For example,

$$T(x^2) = (x^2)'' + (x^2) = 2 + x^2,$$

$$T(\ln x) = (\ln x)'' + (\ln x) = -\frac{1}{x^2} + \ln x.$$

In terms of the mapping T, the solution set S to the differential equation (5.0.1) consists of all those functions y that are mapped to the zero function:

$$S = \{y : T(y) = 0\}.$$

The point that we are making is that a variety of problems we have studied to this point in the text, both in linear algebra and in differential equations, can be viewed as special cases of the general problem of finding all vectors $\mathbf{v}$ in a vector space with the property that $T(\mathbf{v}) = \mathbf{0}$, where T is a mapping from a vector space V into a vector space W.

The mappings satisfy T discussed above satisfy the linearity properties

$$T(\mathbf{u} + \mathbf{v}) = T(\mathbf{u}) + T(\mathbf{v}) \qquad \text{for all } \mathbf{u}, \mathbf{v} \in V,$$
$$T(c\,\mathbf{v}) = c\,T(\mathbf{v}) \qquad \text{for all } \mathbf{v} \in V \text{ and all scalars } c.$$

Any mapping that satisfies these properties is called a linear function, or a linear transformation. We will see in this chapter that the general linear framework we have been aiming for is indeed completed once an appropriate linear transformation is defined with inputs from the vector space of unknowns in the problem we are studying. We will show that the set of all solutions to the corresponding homogeneous linear problem $T(\mathbf{v}) = \mathbf{0}$ is a subspace of the vector space from which we are mapping. Consequently, once we have determined the dimension of that solution space, we will know how many linearly independent solutions to $T(\mathbf{v}) = \mathbf{0}$ are required to determine all of its solutions.

5.1   Definition of a Linear Transformation

We begin with a precise definition of a mapping between two vector spaces.

DEFINITION   5.1.1

Let V and W be vector spaces. A **mapping** T from V into W is a rule that assigns to each vector $\mathbf{v}$ in V precisely one vector $\mathbf{w} = T(\mathbf{v})$ in W. We denote such a mapping by $T : V \to W$.

Example 5.1.2   The following are examples of mappings between vector spaces:

1. $T : M_n(\mathbb{R}) \to M_n(\mathbb{R})$ defined by $T(A) = A^T$.
2. $T : M_n(\mathbb{R}) \to \mathbb{R}$ defined by $T(A) = \det(A)$.
3. $T : P_1 \to P_2$ defined by $T(a_0 + a_1 x) = 2a_0 + a_1 + (a_0 + 3a_1)x + 4a_1 x^2$.
4. $T : C^0[a, b] \to \mathbb{R}$ defined by $T(f) = \int_a^b f(x)\, dx$. $\square$

The basic operations of addition and scalar multiplication in a vector space V enable us to form only linear combinations of vectors in V. In keeping with the aim of studying linear mathematics, it is natural to restrict attention to mappings $T : V \to W$ that preserve such linear combinations of vectors in the sense that

$$T(c_1 \mathbf{v}_1 + c_2 \mathbf{v}_2) = c_1 T(\mathbf{v}_1) + c_2 T(\mathbf{v}_2),$$

for all vectors $\mathbf{v}_1$, $\mathbf{v}_2$ in V and all scalars c_1, c_2. The most general type of mapping that does this is called a linear transformation.

DEFINITION 5.1.3

Let V and W be vector spaces.[1] A mapping $T : V \to W$ is called a **linear transformation** from V to W if it satisfies the following properties:

1. $T(\mathbf{u} + \mathbf{v}) = T(\mathbf{u}) + T(\mathbf{v})$ for *all* $\mathbf{u}, \mathbf{v} \in V$.
2. $T(c\,\mathbf{v}) = c\,T(\mathbf{v})$ for *all* $\mathbf{v} \in V$ and *all* scalars c.

We refer to these properties as the **linearity properties**. The vector space V is called the **domain** of T, while the vector space W is called the **codomain** of T.

Observe that the additive operation appearing on the left side of property 1 refers to addition in V, while the additive operation appearing on the right side refers to addition in W. Although we use the same symbol for addition in each vector space, it is important to realize that they need not be the same operation. The same remarks apply to the scalar multiplication operations appearing in property 2.

A mapping $T : V \to W$ that does not satisfy Definition 5.1.3 is called a **nonlinear transformation**. For instance, in Example 5.1.2, the mapping in part 2 is a nonlinear transformation if $n > 1$, since, for example

$$T(2A) = \det(2A) = 2^n \det(A) \neq 2 \, \det(A) = 2\,T(A),$$

unless A is not invertible. On the other hand, the mapping in part 1 is a linear transformation, since for all $n \times n$ matrices A, B and scalars c, we have

$$T(A + B) = (A + B)^T = A^T + B^T = T(A) + T(B),$$

and

$$T(cA) = (cA)^T = cA^T = c\,T(A).$$

Likewise, for the mapping in part 3, we can consider two polynomials $a_0 + a_1 x$ and $b_0 + b_1 x$, and a scalar c. Then we have

$$
\begin{aligned}
T((a_0 + a_1 x) + (b_0 + b_1 x)) &= T((a_0 + b_0) + (a_1 + b_1)x) \\
&= 2(a_0 + b_0) + (a_1 + b_1) + ((a_0 + b_0) \\
&\quad + 3(a_1 + b_1))x + 4(a_1 + b_1)x^2 \\
&= (2a_0 + a_1 + (a_0 + 3a_1)x + 4a_1 x^2) + (2b_0 + b_1) \\
&\quad + (b_0 + 3b_1)x + 4b_1 x^2 \\
&= T(a_0 + a_1 x) + T(b_0 + b_1 x),
\end{aligned}
$$

and

$$
\begin{aligned}
T(c(a_0 + a_1 x)) = T(ca_0 + ca_1 x) &= 2ca_0 + ca_1 + (ca_0 + 3ca_1)x + 4ca_1 x^2 \\
&= c(2a_0 + a_1 + (a_0 + 3a_1)x + 4a_1 x^2) \\
&= c\,T(a_0 + a_1 x).
\end{aligned}
$$

---

[1] The vector spaces V and W must either be both real vector spaces or both complex vector spaces in order that we use the same scalars in both spaces.

Similarly, it can be shown that the mapping in part 4 is a linear transformation. We now give some further examples.

Example 5.1.4 | Define $T: C^1(I) \rightarrow C^0(I)$ by $T(f) = f'$. Verify that T is a linear transformation.

Solution: If f and g are in $C^1(I)$ and c is a real number, then

$$T(f + g) = (f + g)' = f' + g' = T(f) + T(g)$$

and

$$T(cf) = (cf)' = cf' = c\,T(f).$$

Thus, T satisfies both properties of Definition 5.1.3 and is a linear transformation. ☐

Example 5.1.5 | Define $T: C^2(I) \rightarrow C^0(I)$ by $T(y) = y'' + y$. Verify that T is a linear transformation.

Solution: If y_1 and y_2 are in $C^2(I)$, then

$$\begin{aligned}
T(y_1 + y_2) &= (y_1 + y_2)'' + (y_1 + y_2) \\
&= y_1'' + y_2'' + y_1 + y_2 \\
&= y_1'' + y_1 + y_2'' + y_2 \\
&= T(y_1) + T(y_2).
\end{aligned}$$

Furthermore, if c is an arbitrary real number, then

$$T(cy_1) = (cy_1)'' + (cy_1) = cy_1'' + cy_1 = c(y_1'' + y_1) = c\,T(y_1).$$

Consequently, since both properties of Definition 5.1.3 are satisfied, T is a linear transformation. ☐

Example 5.1.6 | Define $S: M_2(\mathbb{R}) \rightarrow M_2(\mathbb{R})$ by

$$S\left(\begin{bmatrix} a & b \\ c & d \end{bmatrix}\right) = \begin{bmatrix} a - 2b & 0 \\ 3a + 4d & a + b - c \end{bmatrix}.$$

Verify that S is a linear transformation.

Solution: Let $A = \begin{bmatrix} a & b \\ c & d \end{bmatrix}$ and $B = \begin{bmatrix} a' & b' \\ c' & d' \end{bmatrix}$. Then

$$\begin{aligned}
S(A + B) &= S\left(\begin{bmatrix} a + a' & b + b' \\ c + c' & d + d' \end{bmatrix}\right) \\
&= \begin{bmatrix} (a + a') - 2(b + b') & 0 \\ 3(a + a') + 4(d + d') & (a + a') + (b + b') - (c + c') \end{bmatrix} \\
&= \begin{bmatrix} a - 2b & 0 \\ 3a + 4d & a + b - c \end{bmatrix} + \begin{bmatrix} a' - 2b' & 0 \\ 3a' + 4d' & a' + b' - c' \end{bmatrix} = S(A) + S(B).
\end{aligned}$$

Moreover, for any scalar k, we have

$$\begin{aligned}
S(kA) &= S\left(\begin{bmatrix} ka & kb \\ kc & kd \end{bmatrix}\right) = \begin{bmatrix} ka - 2kb & 0 \\ 3ka + 4kd & ka + kb - kc \end{bmatrix} \\
&= k \begin{bmatrix} a - 2b & 0 \\ 3a + 4d & a + b - c \end{bmatrix} = kS(A).
\end{aligned}$$

Once more, both conditions of Definition 5.1.3 are satisfied, hence S is a linear transformation.   □

Remark   The codomain of the linear transformation S given in Example 5.1.6 is $M_2(\mathbb{R})$. However, since all output values $S(A)$ are lower triangular 2×2 matrices, we can view the transformation as $S': M_2(\mathbb{R}) \to W$, where W is the vector space of all 2×2 lower triangular matrices.

Theorem 5.1.7   A mapping $T: V \to W$ is a linear transformation if and only if

$$T(c_1\mathbf{v}_1 + c_2\mathbf{v}_2) = c_1 T(\mathbf{v}_1) + c_2 T(\mathbf{v}_2), \tag{5.1.1}$$

for all $\mathbf{v}_1, \mathbf{v}_2$ in V and all scalars c_1, c_2.

Proof   Suppose T satisfies Equation (5.1.1). Then property (1) of Definition 5.1.3 arises as the special case $c_1 = c_2 = 1, \mathbf{v}_1 = \mathbf{u}, \mathbf{v}_2 = \mathbf{v}$. Further, property (2) of Definition 5.1.3 is the special case $c_1 = c, c_2 = 0, \mathbf{v}_1 = \mathbf{v}$. Since properties 1 and 2 of Definition 5.1.3 are both satisfied, T is a linear transformation.

Conversely, if T is a linear transformation, then using properties 1 and 2 from Definition 5.1.3 yields

$$T(c_1\mathbf{v}_1 + c_2\mathbf{v}_2) = T(c_1\mathbf{v}_1) + T(c_2\mathbf{v}_2) = c_1 T(\mathbf{v}_1) + c_2 T(\mathbf{v}_2),$$

so that Equation (5.1.1) is satisfied.   ■

Repeated application of the linearity properties can now be used to establish that if $T: V \to W$ is a linear transformation, then for all $\mathbf{v}_1, \mathbf{v}_2, \ldots, \mathbf{v}_k$ in V and all scalars $c_1, c_2, \ldots, c_k$, we have

$$T(c_1\mathbf{v}_1 + c_2\mathbf{v}_2 + \cdots + c_k\mathbf{v}_k) = c_1 T(\mathbf{v}_1) + c_2 T(\mathbf{v}_2) + \cdots + c_k T(\mathbf{v}_k). \tag{5.1.2}$$

In particular, if $\{\mathbf{v}_1, \mathbf{v}_2, \ldots, \mathbf{v}_k\}$ is a basis for V, then any vector in V can be written as

$$\mathbf{v} = c_1\mathbf{v}_1 + c_2\mathbf{v}_2 + \cdots + c_k\mathbf{v}_k,$$

for appropriate scalars $c_1, c_2, \ldots, c_k$, so that from Equation (5.1.2),

$$T(\mathbf{v}) = T(c_1\mathbf{v}_1 + c_2\mathbf{v}_2 + \cdots + c_k\mathbf{v}_k) = c_1 T(\mathbf{v}_1) + c_2 T(\mathbf{v}_2) + \cdots + c_k T(\mathbf{v}_k).$$

Consequently, if we know $T(\mathbf{v}_1), T(\mathbf{v}_2), \ldots, T(\mathbf{v}_k)$, then we know how every vector in V transforms. This once more emphasizes the importance of the basis in studying vector spaces.

Example 5.1.8   Let $T: P_2 \to P_2$ be a *linear transformation* satisfying

$$T(1) = 2 - 3x, \qquad T(x) = 2x + 5x^2, \qquad T(x^2) = 3 - x + x^2.$$

For an arbitrary vector $p = a_0 + a_1 x + a_2 x^2$ in P_2, determine $T(p)$.

Solution: Since we have been given the transformation of the (standard) basis $\{1, x, x^2\}$ for P_2, we can determine the transformation of all vectors in P_2. Using the linearity properties in Definition 5.1.3, it follows that

$$
\begin{aligned}
T(a_0 + a_1 x + a_2 x^2) &= T(a_0) + T(a_1 x) + T(a_2 x^2) \\
&= a_0 T(1) + a_1 T(x) + a_2 T(x^2) \\
&= a_0(2 - 3x) + a_1(2x + 5x^2) + a_2(3 - x + x^2) \\
&= 2a_0 + 3a_2 + (-3a_0 + 2a_1 - a_2)x + (5a_1 + a_2)x^2. \quad \square
\end{aligned}
$$

The next theorem lists two basic properties of linear transformations. In this theorem we distinguish the zero vector in V, denoted $\mathbf{0}_V$, from the zero vector in W, denoted $\mathbf{0}_W$.

Theorem 5.1.9   Let $T : V \to W$ be a linear transformation. Then

1. $T(\mathbf{0}_V) = \mathbf{0}_W$,
2. $T(-\mathbf{v}) = -T(\mathbf{v})$ for all $\mathbf{v} \in V$.

Proof   1. If $\mathbf{v}$ is a vector in V, then $0 \cdot \mathbf{v} = \mathbf{0}_V$, by Theorem 4.2.6 (2). Consequently,

$$
T(\mathbf{0}_V) = T(0 \cdot \mathbf{v}) = 0 \cdot T(\mathbf{v}) = \mathbf{0}_W.
$$

2. We know that $-\mathbf{v} = (-1)\mathbf{v}$ for all $\mathbf{v} \in V$, by Theorem 4.2.6 (5). Consequently,

$$
T(-\mathbf{v}) = T((-1)\mathbf{v}) = (-1)T(\mathbf{v}) = -T(\mathbf{v}). \quad \blacksquare
$$

Linear Transformations from $\mathbb{R}^n$ to $\mathbb{R}^m$

Linear transformations between the vector spaces $\mathbb{R}^n$ and $\mathbb{R}^m$ play a very fundamental role in linear algebra and its applications. We now investigate some of their properties.

Example 5.1.10   Define $T : \mathbb{R}^3 \to \mathbb{R}^2$ as follows: If $\mathbf{x} = (x_1, x_2, x_3)$, then

$$
T(\mathbf{x}) = (2x_1 + x_2, x_1 + x_2 - x_3).
$$

Verify that T is a linear transformation from $\mathbb{R}^3$ to $\mathbb{R}^2$.

Solution: Let $\mathbf{x} = (x_1, x_2, x_3)$ and $\mathbf{y} = (y_1, y_2, y_3)$ be arbitrary vectors in $\mathbb{R}^3$. Then, using vector addition in $\mathbb{R}^3$, we have $\mathbf{x} + \mathbf{y} = (x_1 + y_1, x_2 + y_2, x_3 + y_3)$. Consequently,

$$
\begin{aligned}
T(\mathbf{x} + \mathbf{y}) &= T((x_1 + y_1), (x_2 + y_2), (x_3 + y_3)) \\
&= (2(x_1 + y_1) + (x_2 + y_2), (x_1 + y_1) + (x_2 + y_2) - (x_3 + y_3)) \\
&= (2x_1 + x_2, x_1 + x_2 - x_3) + (2y_1 + y_2, y_1 + y_2 - y_3) \\
&= T(\mathbf{x}) + T(\mathbf{y}).
\end{aligned}
$$

Further, if c is any real number, then $c\mathbf{x} = (cx_1, cx_2, cx_3)$, so that

$$
\begin{aligned}
T(c\mathbf{x}) &= T(cx_1, cx_2, cx_3) = (2cx_1 + cx_2, cx_1 + cx_2 - cx_3) \\
&= c(2x_1 + x_2, x_1 + x_2 - x_3) \\
&= cT(\mathbf{x}).
\end{aligned}
$$

Hence, properties 1 and 2 of Definition 5.1.3 are satisfied, and so T is a linear transformation from $\mathbb{R}^3$ to $\mathbb{R}^2$. $\square$

Our next theorem introduces how linear transformations from $\mathbb{R}^n$ to $\mathbb{R}^m$ arise.

Theorem 5.1.11   Let A be an $m \times n$ real matrix, and define $T : \mathbb{R}^n \to \mathbb{R}^m$ by $T(\mathbf{x}) = A\mathbf{x}$. Then T is a linear transformation.

Proof We need only verify the two linearity properties in Definition 5.1.3. Let **x** and **y** be arbitrary vectors in $\mathbb{R}^n$, and let c be an arbitrary real number. Then

$$T(\mathbf{x} + \mathbf{y}) = A(\mathbf{x} + \mathbf{y}) = A\mathbf{x} + A\mathbf{y} = T(\mathbf{x}) + T(\mathbf{y}),$$
$$T(c\mathbf{x}) = A(c\mathbf{x}) = cA\mathbf{x} = c\,T(\mathbf{x}).$$

■

A linear transformation $T : \mathbb{R}^n \to \mathbb{R}^m$ defined by $T(\mathbf{x}) = A\mathbf{x}$, where A is an $m \times n$ matrix, is called a **matrix transformation**.

Example 5.1.12   Determine the matrix transformation $T : \mathbb{R}^3 \to \mathbb{R}^2$ if

$$A = \begin{bmatrix} 2 & 1 & 0 \\ 1 & 1 & -1 \end{bmatrix}.$$

Solution: We have

$$T(\mathbf{x}) = A\mathbf{x} = \begin{bmatrix} 2 & 1 & 0 \\ 1 & 1 & -1 \end{bmatrix} \begin{bmatrix} x_1 \\ x_2 \\ x_3 \end{bmatrix} = \begin{bmatrix} 2x_1 + x_2 \\ x_1 + x_2 - x_3 \end{bmatrix},$$

which we write as

$$T(x_1, x_2, x_3) = (2x_1 + x_2, x_1 + x_2 - x_3).$$

□

If we compare the previous two examples, we see that they contain the same linear transformation, but defined in two different ways. This leads to the question as to whether we can always describe a linear transformation from $\mathbb{R}^n$ to $\mathbb{R}^m$ as a matrix transformation. The following theorem answers this in the affirmative:

Theorem 5.1.13   Let $T : \mathbb{R}^n \to \mathbb{R}^m$ be a linear transformation. Then T is described by the matrix transformation

$$T(\mathbf{x}) = A\mathbf{x},$$

where A is the $m \times n$ matrix

$$A = [T(\mathbf{e}_1), T(\mathbf{e}_2), \ldots, T(\mathbf{e}_n)]$$

and $\mathbf{e}_1, \mathbf{e}_2, \ldots, \mathbf{e}_n$ denote the standard basis vectors in $\mathbb{R}^n$.

Proof Any vector $\mathbf{x} = (x_1, x_2, \ldots, x_n)$ in $\mathbb{R}^n$ can be expressed in terms of the standard basis as

$$\mathbf{x} = x_1\mathbf{e}_1 + x_2\mathbf{e}_2 + \cdots + x_n\mathbf{e}_n.$$

Thus, since $T : \mathbb{R}^n \to \mathbb{R}^m$ is a linear transformation, properties 1 and 2 of Definition 5.1.3 imply that

$$\begin{aligned} T(\mathbf{x}) &= T(x_1\mathbf{e}_1 + x_2\mathbf{e}_2 + \cdots + x_n\mathbf{e}_n) \\ &= x_1 T(\mathbf{e}_1) + x_2 T(\mathbf{e}_2) + \cdots + x_n T(\mathbf{e}_n). \end{aligned}$$

Consequently, $T(\mathbf{x})$ is a linear combination of the vectors $T(\mathbf{e}_1), T(\mathbf{e}_2), \ldots, T(\mathbf{e}_n)$, each of which is a vector in $\mathbb{R}^m$. Therefore, the preceding expression for $T(\mathbf{x})$ can be written as the matrix product

$$T(\mathbf{x}) = [T(\mathbf{e}_1), T(\mathbf{e}_2), \ldots, T(\mathbf{e}_n)] \begin{bmatrix} x_1 \\ x_2 \\ \vdots \\ x_n \end{bmatrix} = A\mathbf{x},$$

where $A = [T(\mathbf{e}_1), T(\mathbf{e}_2), \ldots, T(\mathbf{e}_n)]$. ∎

DEFINITION 5.1.14

If $T: \mathbb{R}^n \to \mathbb{R}^m$ is a linear transformation, then the $m \times n$ matrix

$$A = [T(\mathbf{e}_1), T(\mathbf{e}_2), \ldots, T(\mathbf{e}_n)]$$

is called the **matrix of T**.

Example 5.1.15 Determine the matrix of the linear transformation $T: \mathbb{R}^4 \to \mathbb{R}^3$ defined by

$$T(x_1, x_2, x_3, x_4) = (2x_1 + 3x_2 + x_4, 5x_1 + 9x_3 - x_4, 4x_1 + 2x_2 - x_3 + 7x_4).$$
$$(5.1.3)$$

Solution: The standard basis vectors in $\mathbb{R}^4$ are

$$\mathbf{e}_1 = (1, 0, 0, 0), \qquad \mathbf{e}_2 = (0, 1, 0, 0), \qquad \mathbf{e}_3 = (0, 0, 1, 0), \qquad \mathbf{e}_4 = (0, 0, 0, 1).$$

Consequently, from (5.1.3),

$$T(\mathbf{e}_1) = (2, 5, 4), \quad T(\mathbf{e}_2) = (3, 0, 2), \quad T(\mathbf{e}_3) = (0, 9, -1), \quad T(\mathbf{e}_4) = (1, -1, 7),$$

so that the matrix of the transformation is

$$A = [T(\mathbf{e}_1), T(\mathbf{e}_2), T(\mathbf{e}_3), T(\mathbf{e}_4)] = \begin{bmatrix} 2 & 3 & 0 & 1 \\ 5 & 0 & 9 & -1 \\ 4 & 2 & -1 & 7 \end{bmatrix}. \qquad \square$$

Example 5.1.16 Determine the linear transformation $T: \mathbb{R}^2 \to \mathbb{R}^3$ satisfying

$$T(1, 0) = (2, 3, -1), \qquad T(0, 1) = (5, -4, 7).$$

Solution: The matrix of the transformation is

$$A = [T(\mathbf{e}_1), T(\mathbf{e}_2)] = \begin{bmatrix} 2 & 5 \\ 3 & -4 \\ -1 & 7 \end{bmatrix},$$

so that

$$T(\mathbf{x}) = \begin{bmatrix} 2 & 5 \\ 3 & -4 \\ -1 & 7 \end{bmatrix} \begin{bmatrix} x_1 \\ x_2 \end{bmatrix} = \begin{bmatrix} 2x_1 + 5x_2 \\ 3x_1 - 4x_2 \\ -x_1 + 7x_2 \end{bmatrix},$$

which we write as

$$T(x_1, x_2) = (2x_1 + 5x_2, 3x_1 - 4x_2, -x_1 + 7x_2).$$ □

As the above examples illustrate, linear transformations $T : \mathbb{R}^n \to \mathbb{R}^m$ are completely determined from the matrix of T. Therefore, to answer questions regarding such linear transformations, it is almost always desirable to calculate the matrix of T. Here is one more example.

Example 5.1.17 Suppose $T : \mathbb{R}^3 \to \mathbb{R}^2$ is a linear transformation with $T(1, 0, 0) = (4, 5)$, $T(0, 1, 0) = (-1, 1)$, and $T(2, 1, -3) = (7, -1)$. Find $T(x_1, x_2, x_3)$.

Solution: We need to compute the matrix of T, which is $A = \left[\, T(\mathbf{e}_1), T(\mathbf{e}_2), T(\mathbf{e}_3)\,\right]$. We are already given $T(1, 0, 0)$ and $T(0, 1, 0)$, but to find $T(0, 0, 1)$, we must write $(0, 0, 1)$ as a linear combination of the vectors $(1, 0, 0)$, $(0, 1, 0)$, and $(2, 1, -3)$. A short calculation shows that

$$(0, 0, 1) = \tfrac{2}{3}(1, 0, 0) + \tfrac{1}{3}(0, 1, 0) - \tfrac{1}{3}(2, 1, -3),$$

which shows that

$$T(0, 0, 1) = \tfrac{2}{3}T(1, 0, 0) + \tfrac{1}{3}T(0, 1, 0) - \tfrac{1}{3}T(2, 1, -3)$$
$$= \tfrac{2}{3}(4, 5) + \tfrac{1}{3}(-1, 1) - \tfrac{1}{3}(7, -1) = (0, 4).$$

Thus, the matrix of T is

$$A = \begin{bmatrix} 4 & -1 & 0 \\ 5 & 1 & 4 \end{bmatrix}.$$

Thus, we have

$$T(x_1, x_2, x_3) = \begin{bmatrix} 4 & -1 & 0 \\ 5 & 1 & 4 \end{bmatrix} \begin{bmatrix} x_1 \\ x_2 \\ x_3 \end{bmatrix} = (4x_1 - x_2, 5x_1 + x_2 + 4x_3).$$ □

Finally in this section, consider the mapping $T : \mathbb{R}^2 \to \mathbb{R}^2$, which rotates each point in the plane through an angle θ in the counterclockwise direction, where $0 \le \theta \le 2\pi$. As Figure 5.1.1 illustrates, T is a *linear* transformation.

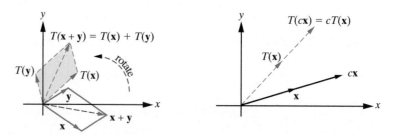

Figure 5.1.1: Rotation in the plane satisfies the basic linearity properties and therefore is a linear transformation.

In order to determine the matrix of this transformation, all we need to do is obtain $T(\mathbf{e}_1)$ and $T(\mathbf{e}_2)$. From Figure 5.1.2, we see that

$$T(\mathbf{e}_1) = (\cos\theta, \ \sin\theta), \qquad T(\mathbf{e}_2) = (-\sin\theta, \ \cos\theta).$$

Consequently, the matrix of the transformation is

$$T(\theta) = \begin{bmatrix} \cos\theta & -\sin\theta \\ \sin\theta & \cos\theta \end{bmatrix}.$$

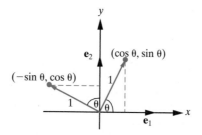

Figure 5.1.2: Determining the transformation matrix corresponding to a rotation in the xy-plane.

Exercises for 5.1

Key Terms

Mapping, Linear transformation, Linearity properties, Nonlinear transformation, Matrix transformation, Matrix of T.

Skills

- Be able to determine and verify whether a given mapping $T : V \to W$ is a linear or nonlinear transformation.

- Be able to determine the matrix of a linear transformation $T : \mathbb{R}^n \to \mathbb{R}^m$.

- Given a linear transformation $T : V \to W$ and values $T(\mathbf{v}_1), T(\mathbf{v}_2), \dots, T(\mathbf{v}_k)$ for a basis $\{\mathbf{v}_1, \mathbf{v}_2, \dots, \mathbf{v}_k\}$ of V, be able to find $T(\mathbf{v})$ for any vector $\mathbf{v}$ in V.

True-False Review

For Questions 1–6, decide if the given statement is **true** or **false**, and give a brief justification for your answer. If true, you can quote a relevant definition or theorem from the text. If false, provide an example, illustration, or brief explanation of why the statement is false.

1. A linear transformation $T : V \to W$ is a mapping that satisfies the conditions $T(\mathbf{u} + \mathbf{v}) = T(\mathbf{u}) + T(\mathbf{v})$ and $T(c \cdot \mathbf{v}) = c \cdot T(\mathbf{v})$ for some vectors $\mathbf{u}, \mathbf{v}$ in V and for some scalar c.

2. A linear transformation $T : \mathbb{R}^n \to \mathbb{R}^m$ can be represented by the formula $T(\mathbf{x}) = A\mathbf{x}$ for some $n \times m$ matrix A.

3. The formula $T(\mathbf{0}_V) = \mathbf{0}_W$ holds for any mapping $T : V \to W$.

4. The matrix of a linear transformation $T : \mathbb{R}^n \to \mathbb{R}^m$ is the matrix $A = \begin{bmatrix} T(\mathbf{e}_1), T(\mathbf{e}_2), \dots, T(\mathbf{e}_n) \end{bmatrix}$.

5. If $T : V \to W$ is a linear transformation, the formula $T(-\mathbf{v}) = -T(\mathbf{v})$ holds for every $\mathbf{v}$ in V.

6. A linear transformation $T : V \to W$ must satisfy

$$T((c + d)\mathbf{v}) = cT(\mathbf{v}) + dT(\mathbf{v})$$

for every vector $\mathbf{v}$ in V and for all scalars c and d.

Problems

For Problems 1–8, show that the given mapping is a linear transformation.

1. $T : \mathbb{R}^2 \to \mathbb{R}^2$ defined by

$$T(x_1, x_2) = (x_1 + 2x_2, 2x_1 - x_2).$$

2. $T : \mathbb{R}^3 \to \mathbb{R}^2$ defined by

$$T(x_1, x_2, x_3) = (x_1 + 3x_2 + x_3, x_1 - x_2).$$

3. $T : C^2(I) \to C^0(I)$ defined by

$$T(y) = y'' - 16y.$$

4. $T: C^2(I) \rightarrow C^0(I)$ defined by

$$T(y) = y'' + a_1 y' + a_2 y,$$

where a_1 and a_2 are functions defined on I.

5. $T: C^0[a, b] \rightarrow \mathbb{R}$ defined by

$$T(f) = \int_a^b f(x)\, dx.$$

6. $T: M_n(\mathbb{R}) \rightarrow M_n(\mathbb{R})$ defined by

$$T(A) = AB - BA,$$

where B is a fixed $n \times n$ matrix.

7. $S: M_n(\mathbb{R}) \rightarrow M_n(\mathbb{R})$ defined by

$$S(A) = A + A^T.$$

8. $T: M_n(\mathbb{R}) \rightarrow \mathbb{R}$ defined by $T(A) = \text{tr}(A)$, where $\text{tr}(A)$ denotes the trace of A.

For Problems 9–10, show that the given mapping is a non-linear transformation.

9. $T: \mathbb{R}^2 \rightarrow \mathbb{R}^2$ defined by

$$T(x_1, x_2) = (x_1 + x_2, 2).$$

10. $T: M_2(\mathbb{R}) \rightarrow \mathbb{R}$ defined by

$$T(A) = \det(A).$$

For Problems 11–15, determine the matrix of the given transformation

$$T: \mathbb{R}^n \rightarrow \mathbb{R}^m.$$

11. $T(x_1, x_2) = (3x_1 - 2x_2, x_1 + 5x_2).$

12. $T(x_1, x_2) = (x_1 + 3x_2, 2x_1 - 7x_2, x_1).$

13. $T(x_1, x_2, x_3) = (x_1 - x_2 + x_3, x_3 - x_1).$

14. $T(x_1, x_2, x_3) = x_1 + 5x_2 - 3x_3.$

15. $T(x_1, x_2, x_3) = (x_3 - x_1, -x_1, 3x_1 + 2x_3, 0).$

For Problems 16–20, determine the linear transformation $T: \mathbb{R}^n \rightarrow \mathbb{R}^m$ that has the given matrix.

16. $A = \begin{bmatrix} 1 & 3 \\ -4 & 7 \end{bmatrix}.$

17. $A = \begin{bmatrix} 2 & -1 & 5 \\ 3 & 1 & -2 \end{bmatrix}.$

18. $A = \begin{bmatrix} 2 & 2 & -3 \\ 4 & -1 & 2 \\ 5 & 7 & -8 \end{bmatrix}.$

19. $A = \begin{bmatrix} -3 \\ -2 \\ 0 \\ 1 \end{bmatrix}.$

20. $A = \begin{bmatrix} 1 & -4 & -6 & 0 & 2 \end{bmatrix}.$

21. Let V be a real inner product space, and let $\mathbf{u}$ be a fixed (nonzero) vector in V. Define $T: V \rightarrow \mathbb{R}$ by

$$T(\mathbf{v}) = \langle \mathbf{u}, \mathbf{v} \rangle.$$

Use properties of the inner product to show that T is a linear transformation.

22. Let V be a real inner product space, and let $\mathbf{u}_1$ and $\mathbf{u}_2$ be fixed (nonzero) vectors in V. Define $T: V \rightarrow \mathbb{R}^2$ by

$$T(\mathbf{v}) = (\langle \mathbf{u}_1, \mathbf{v} \rangle, \langle \mathbf{u}_2, \mathbf{v} \rangle).$$

Use properties of the inner product to show that T is a linear transformation.

23. **(a)** Let $\mathbf{v}_1 = (1, 1)$ and $\mathbf{v}_2 = (1, -1)$. Show that $\{\mathbf{v}_1, \mathbf{v}_2\}$ is a basis for $\mathbb{R}^2$.

(b) Let $T: \mathbb{R}^2 \rightarrow \mathbb{R}^2$ be the *linear* transformation satisfying

$$T(\mathbf{v}_1) = (2, 3), \qquad T(\mathbf{v}_2) = (-1, 1),$$

where $\mathbf{v}_1$ and $\mathbf{v}_2$ are the basis vectors given in (a). Find $T(x_1, x_2)$ for an arbitrary vector (x_1, x_2) in $\mathbb{R}^2$. What is $T(4, -2)$?

For Problems 24–27, assume that T defines a linear transformation and use the given information to find the matrix of T.

24. $T: \mathbb{R}^2 \rightarrow \mathbb{R}^4$ such that $T(-1, 1) = (1, 0, -2, 2)$ and $T(1, 2) = (-3, 1, 1, 1)$.

25. $T: \mathbb{R}^4 \rightarrow \mathbb{R}^2$ such that $T(1, 0, 0, 0) = (3, -2)$, $T(1, 1, 0, 0) = (5, 1)$, $T(1, 1, 1, 0) = (-1, 0)$, and $T(1, 1, 1, 1) = (2, 2)$.

26. $T: \mathbb{R}^3 \rightarrow \mathbb{R}^3$ such that $T(1, 2, 0) = (2, -1, 1)$, $T(0, 1, 1) = (3, -1, -1)$ and $T(0, 2, 3) = (6, -5, 4)$.

27. $T: \mathbb{R}^3 \to \mathbb{R}^4$ such that $T(0, -1, 4) = (2, 5, -2, 1)$, $T(0, 3, 3) = (-1, 0, 0, 5)$, and $T(4, 4, -1) = (-3, 1, 1, 3)$.

28. Let $T: P_2 \to P_2$ be the *linear* transformation satisfying

$$T(1) = x + 1, \quad T(x) = x^2 - 1, \quad T(x^2) = 3x + 2.$$

Determine $T(ax^2 + bx + c)$, where a, b, and c are arbitrary real numbers.

29. Let $T: V \to V$ be a linear transformation, and suppose that

$$T(2\mathbf{v}_1 + 3\mathbf{v}_2) = \mathbf{v}_1 + \mathbf{v}_2,$$
$$T(\mathbf{v}_1 + \mathbf{v}_2) = 3\mathbf{v}_1 - \mathbf{v}_2.$$

Find $T(\mathbf{v}_1)$ and $T(\mathbf{v}_2)$.

30. Let $T: P_2 \to P_2$ be the *linear* transformation satisfying:

$$T(x^2 - 1) = x^2 + x - 3, \qquad T(2x) = 4x,$$
$$T(3x^2 + 2) = 2(x + 3).$$

Find $T(1)$, $T(x)$, $T(x^2)$, and hence, show that

$$T(ax^2 + bx + c) = ax^2 - (a - 2b + 2c)x + 3c,$$

where a, b, and c are arbitrary real numbers.

31. Let $\{\mathbf{v}_1, \mathbf{v}_2\}$ be a basis for the vector space V. If $T: V \to V$ is the linear transformation satisfying

$$T(\mathbf{v}_1) = 3\mathbf{v}_1 - \mathbf{v}_2, \qquad T(\mathbf{v}_2) = \mathbf{v}_1 + 2\mathbf{v}_2,$$

find $T(\mathbf{v})$ for an arbitrary vector in V.

32. Let $T: V \to W$ and $S: V \to W$ be linear transformations, and assume that $\{\mathbf{v}_1, \mathbf{v}_2, \ldots, \mathbf{v}_k\}$ spans V. Prove that if $T(\mathbf{v}_i) = S(\mathbf{v}_i)$ for each $i = 1, 2, \ldots, k$, then $T = S$; that is, $T(\mathbf{v}) = S(\mathbf{v})$ for each $\mathbf{v} \in V$.

33. Let V be a vector space with basis $\{\mathbf{v}_1, \mathbf{v}_2, \ldots, \mathbf{v}_k\}$ and suppose $T: V \to W$ is a linear transformation such that $T(\mathbf{v}_i) = \mathbf{0}$ for each $i = 1, 2, \ldots, k$. Prove that T is the zero transformation; that is, $T(\mathbf{v}) = \mathbf{0}$ for each $\mathbf{v} \in V$.

Let $T_1: V \to W$ and $T_2: V \to W$ be linear transformations, and let c be a scalar. We define the **sum** $T_1 + T_2$ and the **scalar product** cT_1 by

$$(T_1 + T_2)(\mathbf{v}) = T_1(\mathbf{v}) + T_2(\mathbf{v})$$

and

$$(cT_1)(\mathbf{v}) = cT_1(\mathbf{v})$$

for all $\mathbf{v} \in V$. The remaining problems in this section consider the properties of these mappings.

34. Verify that $T_1 + T_2$ and cT_1 are *linear transformations*.

35. Let $T_1: \mathbb{R}^2 \to \mathbb{R}^2$ and $T_2: \mathbb{R}^2 \to \mathbb{R}^2$ be the linear transformations with matrices

$$A = \begin{bmatrix} 3 & 1 \\ -1 & 2 \end{bmatrix}, \qquad B = \begin{bmatrix} 2 & 5 \\ 3 & -4 \end{bmatrix}.$$

Find $T_1 + T_2$ and cT_1.

36. Let $T_1: \mathbb{R}^n \to \mathbb{R}^m$ and $T_2: \mathbb{R}^n \to \mathbb{R}^m$ be the linear transformations with matrices A and B, respectively. Show that $T_1 + T_2$ and cT_1 are the linear transformations with matrices $A + B$ and cA, respectively.

37. Let V and W be vector spaces, and let $L(V, W)$ denote the set of all linear transformations from V into W. Verify that $L(V, W)$ together with the operations of addition and scalar multiplication just defined for linear transformations, is a vector space.

---

*5.2   Transformations of $\mathbb{R}^2$

To gain some geometric insight into linear transformations, we consider the particular case of linear transformations from $\mathbb{R}^2 \to \mathbb{R}^2$. Any such transformation is called a **transformation of $\mathbb{R}^2$**. Geometrically, the action of a transformation of $\mathbb{R}^2$ can be represented by its effect on an arbitrary point in the Cartesian plane. We first establish that transformations of $\mathbb{R}^2$ map lines into lines. Recall that the parametric equations of a line passing through the point $\mathbf{x}_1 = (x_1, y_1)$ in the direction of the vector $\mathbf{v}$ with components (a, b) are

$$x = x_1 + at, \qquad y = y_1 + bt,$$

---

*This section can be omitted without loss of continuity.

which can be written as

$$\mathbf{x} = \mathbf{x}_1 + t\mathbf{v}. \tag{5.2.1}$$

The transformation $T(\mathbf{x}) = A\mathbf{x}$ therefore transforms the points along this line into

$$T(\mathbf{x}) = A(\mathbf{x}_1 + t\mathbf{v}) = A\mathbf{x}_1 + tA\mathbf{v}. \tag{5.2.2}$$

Consequently, the transformed points lie along the line with parametric equations

$$\mathbf{x} = \mathbf{y}_1 + t\mathbf{w},$$

where $\mathbf{y}_1 = A\mathbf{x}_1$ and $\mathbf{w} = A\mathbf{v}$. This is illustrated in Figure 5.2.1. It follows that if we know how two points in $\mathbb{R}^2$ transform, then we can determine the transformation of all points along the line joining those two points. Further, the linearity properties (1) and (2) of Definition 5.1.3 are the statement that the parallelogram law for vector addition is preserved under the transformation. This is illustrated in Figure 5.2.2.

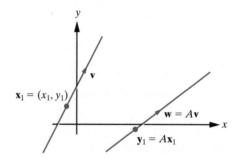

Figure 5.2.1: A transformation of $\mathbb{R}^2$ maps the line with vector parametric equation $\mathbf{x} = \mathbf{x}_1 + t\mathbf{v}$ into the line $\mathbf{x} = A\mathbf{x}_1 + tA\mathbf{v} = \mathbf{y}_1 + t\mathbf{w}$.

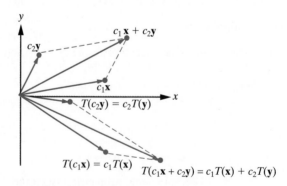

Figure 5.2.2: A transformation of $\mathbb{R}^2$ preserves the parallelogram law of vector addition in the sense that $T(c_1\mathbf{x} + c_2\mathbf{y}) = c_1 T(\mathbf{x}) + c_2 T(\mathbf{y})$.

From Equations (5.2.1) and (5.2.2), we see that *any* line with direction vector $\mathbf{v}$ is mapped into a line with direction vector $A\mathbf{v}$, so that parallel lines are mapped to parallel lines by a transformation of $\mathbb{R}^2$. In describing specific transformations of $\mathbb{R}^2$, it is often useful to determine the effect of such a transformation on the points lying inside a rectangle. Because a transformation of $\mathbb{R}^2$ maps parallel lines into parallel lines, it follows that the transform of a rectangle will be the parallelogram whose vertices are the transforms of the vertices of the rectangle. This is illustrated in Figure 5.2.3.

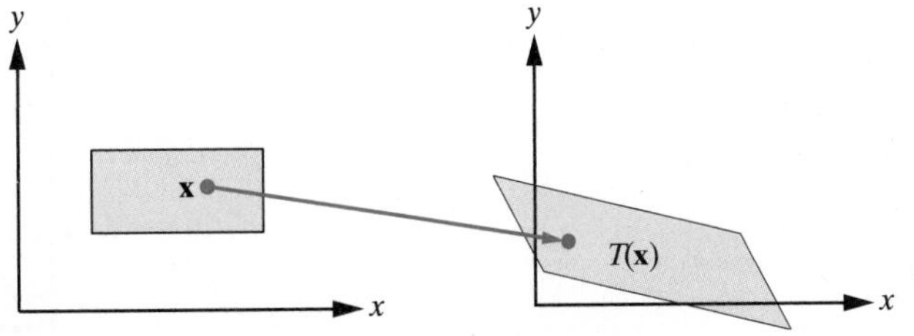

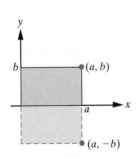

Figure 5.2.3: Transformations of $\mathbb{R}^2$ map rectangles into parallelograms.

Simple Transformations

We next introduce some simple transformations of $\mathbb{R}^2$ and their geometrical interpretation. We then show how more complicated transformations can be considered as a combination of these simple ones.

I: Reflections. Consider the transformation of $\mathbb{R}^2$ with matrix

$$R_x = \begin{bmatrix} 1 & 0 \\ 0 & -1 \end{bmatrix}.$$

If $\mathbf{v} = (x, y)$ is an arbitrary point in $\mathbb{R}^2$, then

$$T(\mathbf{v}) = R_x(\mathbf{v}) = \begin{bmatrix} 1 & 0 \\ 0 & -1 \end{bmatrix} \begin{bmatrix} x \\ y \end{bmatrix} = \begin{bmatrix} x \\ -y \end{bmatrix}.$$

Thus,

$$T(x, y) = (x, -y).$$

We see that each point in the Cartesian plane is mapped to its mirror image in the x-axis. Hence, R_x describes the reflection in the x-axis. This can be illustrated by considering the rectangle in the xy-plane through the points $(0, 0)$, $(a, 0)$, $(0, b)$, and (a, b). In order to determine the transform of this rectangle, all we need to do is determine the transform of the vertices. (See Figure 5.2.4.) We leave it as an exercise to show that the transformation of $\mathbb{R}^2$ with matrix

$$R_y = \begin{bmatrix} -1 & 0 \\ 0 & 1 \end{bmatrix}$$

describes a reflection in the y-axis. Now consider the transformation of $\mathbb{R}^2$ with matrix

$$R_{xy} = \begin{bmatrix} 0 & 1 \\ 1 & 0 \end{bmatrix}.$$

If $\mathbf{v} = (x, y)$, then

$$T(\mathbf{v}) = R_{xy}\mathbf{v} = \begin{bmatrix} 0 & 1 \\ 1 & 0 \end{bmatrix} \begin{bmatrix} x \\ y \end{bmatrix} = \begin{bmatrix} y \\ x \end{bmatrix}.$$

Thus,

$$T(x, y) = (y, x).$$

We see that the x- and y- coordinates have been interchanged. Geometrically, all points along the line $y = x$ remain fixed, and all other points are transformed into their mirror image in the line $y = x$. Consequently R_{xy} describes a reflection in the line $y = x$. This is illustrated in Figure 5.2.5.

Figure 5.2.4: Reflection in the x-axis.

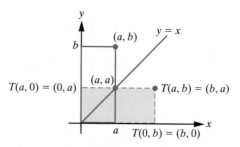

Figure 5.2.5: Reflection in the line $y = x$. The rectangle with vertices $(0, 0)$, $(a, 0)$, $(0, b)$, (a, b) is transformed into the rectangle with vertices $(0, 0)$, $(0, a)$, $(b, 0)$, (b, a).

II: Stretches. The next transformations we analyze are those with matrices

$$\textbf{(a)} \quad LS_x = \begin{bmatrix} k & 0 \\ 0 & 1 \end{bmatrix}, \quad k > 0 \qquad \textbf{(b)} \quad LS_y = \begin{bmatrix} 1 & 0 \\ 0 & k \end{bmatrix}, \quad k > 0.$$

Consider first (a). If $\mathbf{v} = (x, y)$, then

$$T(\mathbf{x}) = LS_x \mathbf{v} = \begin{bmatrix} k & 0 \\ 0 & 1 \end{bmatrix} \begin{bmatrix} x \\ y \end{bmatrix} = \begin{bmatrix} kx \\ y \end{bmatrix}.$$

Thus,

$$T(x, y) = (kx, y).$$

We see that the x-coordinate of each point in the plane is scaled by a factor k, whereas the y-coordinate is unaltered. Thus, points along the y-axis ($x = 0$) remain fixed, whereas points with positive (negative) x-coordinate are moved horizontally to the right (left). This transformation is called a **linear stretch in the x-direction** (hence the notation LS_x). The effect of this transformation is illustrated in Figure 5.2.6. If $k > 1$, then the transformation is an **expansion**, whereas if $0 < k < 1$, we have a **compression**. If $k = 1$, then $LS_x = I_2$, and all points remain fixed. This is the **identity transformation**.

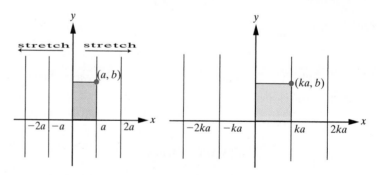

Figure 5.2.6: The effect of a linear stretch in the x-direction. Points along the y-axis remain fixed. All other points are moved parallel to the x-axis.

In a similar manner, it is easily verified that the transformation with matrix LS_y corresponds to a **linear stretch in the y-direction**.

III: Shears. Now consider the transformations of $\mathbb{R}^2$ with matrices

$$\textbf{(a)} \quad S_x = \begin{bmatrix} 1 & k \\ 0 & 1 \end{bmatrix}, \qquad \textbf{(b)} \quad S_y = \begin{bmatrix} 1 & 0 \\ k & 1 \end{bmatrix}.$$

For the matrix in (a), if $\mathbf{v} = (x, y)$, then

$$T(\mathbf{v}) = S_x \mathbf{v} = \begin{bmatrix} 1 & k \\ 0 & 1 \end{bmatrix} \begin{bmatrix} x \\ y \end{bmatrix} = \begin{bmatrix} x + ky \\ y \end{bmatrix},$$

so that

$$T(x, y) = (x + ky, y).$$

In this case, each point in the plane is moved parallel to the x-axis a distance proportional to its y-coordinate (points along the x-axis remain fixed). This is referred to as a **shear parallel to the x-axis** and is illustrated in Figure 5.2.7 for the case $k > 0$. We leave it as an exercise to verify that the transformation of $\mathbb{R}^2$ with matrix S_y corresponds to a **shear parallel to the y-axis**.

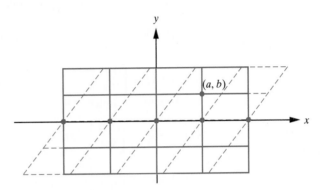

Figure 5.2.7: A shear parallel to the x-axis. Points on the x-axis remain fixed. Rectangles are transformed into parallelograms.

Invertible Transformations of $\mathbb{R}^2$

We now show that any transformation of $\mathbb{R}^2$ with an *invertible* matrix can be obtained by combining the basic transformations I–III just described. To do so, we recall from Section 2.7 that any matrix A can be reduced to reduced row-echelon form through multiplication by an appropriate sequence of elementary matrices. In the case of an invertible 2×2 matrix, the reduced row-echelon form is I_2, so that, denoting the elementary matrices by $E_1, E_2, \ldots, E_n$, we can write

$$E_n E_{n-1} \cdots E_2 E_1 A = I_2.$$

Equivalently, since each of the elementary matrices is invertible,

$$A = E_1^{-1} E_2^{-1} \cdots E_n^{-1}. \tag{5.2.3}$$

Now let T be any transformation of $\mathbb{R}^2$ with *invertible* matrix A. It follows from Equation (5.2.3) that

$$T(\mathbf{v}) = A\mathbf{v} = E_1^{-1} E_2^{-1} \cdots E_n^{-1} \mathbf{v}. \tag{5.2.4}$$

Since, as we have shown in Section 2.7, the inverse of an elementary matrix is also an elementary matrix, (5.2.4) implies that a general transformation of $\mathbb{R}^2$ with invertible matrix can be obtained by applying a sequence of transformations corresponding to appropriate elementary matrices. Now for the key point. A closer look at the elementary matrices shows that the relationships listed next hold between them and the matrices of the simple transformations introduced previously in this section.

(1) $P_{12} = \begin{bmatrix} 0 & 1 \\ 1 & 0 \end{bmatrix}$ corresponds to a reflection in the line $y = x$.

(2a) $M_1(k) = \begin{bmatrix} k & 0 \\ 0 & 1 \end{bmatrix}$, $k > 0$ corresponds to a linear stretch in the x-direction.

(2b) $M_1(k) = \begin{bmatrix} k & 0 \\ 0 & 1 \end{bmatrix}$, $k < 0$. In this case, we can write

$$\begin{bmatrix} k & 0 \\ 0 & 1 \end{bmatrix} = \begin{bmatrix} -k & 0 \\ 0 & 1 \end{bmatrix}\begin{bmatrix} -1 & 0 \\ 0 & 1 \end{bmatrix},$$

which corresponds to a reflection in the y-axis followed by a linear stretch (stretch factor $-k > 0$) in the x-direction.

(3a) $M_2(k) = \begin{bmatrix} 1 & 0 \\ 0 & k \end{bmatrix}$, $k > 0$ corresponds to a linear stretch in the y-direction.

(3b) $M_2(k) = \begin{bmatrix} 1 & 0 \\ 0 & k \end{bmatrix}$, $k < 0$. This corresponds to a reflection in the x-axis followed by a linear stretch in the y-direction.

(4) $A_{12}(k) = \begin{bmatrix} 1 & 0 \\ k & 1 \end{bmatrix}$ corresponds to a shear parallel to the y-axis.

(5) $A_{21}(k) = \begin{bmatrix} 1 & k \\ 0 & 1 \end{bmatrix}$ corresponds to a shear parallel to the x-axis.

We can therefore conclude that any transformation of $\mathbb{R}^2$ with invertible matrix can be obtained by applying an appropriate sequence of reflections, shears, and stretches.

Example 5.2.1   Let $T: \mathbb{R}^2 \to \mathbb{R}^2$ be the transformation of $\mathbb{R}^2$ with invertible matrix

$$A = \begin{bmatrix} 3 & 9 \\ 1 & 2 \end{bmatrix}.$$

Describe T as a combination of reflections, shears, and stretches.

Solution: Reducing A to reduced row-echelon form in the usual manner yields

$$\begin{bmatrix} 3 & 9 \\ 1 & 2 \end{bmatrix} \overset{1}{\sim} \begin{bmatrix} 1 & 2 \\ 3 & 9 \end{bmatrix} \overset{2}{\sim} \begin{bmatrix} 1 & 2 \\ 0 & 3 \end{bmatrix} \overset{3}{\sim} \begin{bmatrix} 1 & 2 \\ 0 & 1 \end{bmatrix} \overset{4}{\sim} \begin{bmatrix} 1 & 0 \\ 0 & 1 \end{bmatrix}.$$

> **1.** P_{12} **2.** $A_{13}(-3)$ **3.** $M_2(\frac{1}{3})$ **4.** $A_{21}(-2)$

The corresponding elementary matrices that accomplish this reduction are

$$P_{12} = \begin{bmatrix} 0 & 1 \\ 1 & 0 \end{bmatrix}, \quad A_{12}(-3) = \begin{bmatrix} 1 & 0 \\ -3 & 1 \end{bmatrix}, \quad M_2(1/3) = \begin{bmatrix} 1 & 0 \\ 0 & \frac{1}{3} \end{bmatrix}, \quad A_{21}(-2) = \begin{bmatrix} 1 & -2 \\ 0 & 1 \end{bmatrix},$$

with inverses

$$P_{12}, \quad A_{12}(3), \quad M_2(3), \quad A_{21}(2),$$

respectively. Consequently,

$$A = \begin{bmatrix} 0 & 1 \\ 1 & 0 \end{bmatrix}\begin{bmatrix} 1 & 0 \\ 3 & 1 \end{bmatrix}\begin{bmatrix} 1 & 0 \\ 0 & 3 \end{bmatrix}\begin{bmatrix} 1 & 2 \\ 0 & 1 \end{bmatrix}.$$

We can therefore write

$$T(\mathbf{v}) = A\mathbf{v} = \begin{bmatrix} 0 & 1 \\ 1 & 0 \end{bmatrix}\begin{bmatrix} 1 & 0 \\ 3 & 1 \end{bmatrix}\begin{bmatrix} 1 & 0 \\ 0 & 3 \end{bmatrix}\begin{bmatrix} 1 & 2 \\ 0 & 1 \end{bmatrix}\mathbf{v}.$$

We see that T consists of a shear parallel to the x-axis, followed by a stretch in the y-direction, followed by a shear parallel to the y-axis, followed by a reflection in $y = x$. $\square$

In the previous section, we derived the matrix of the transformation of $\mathbb{R}^2$ that corresponds to a rotation through an angle θ in the counterclockwise direction, namely,

$$T(\theta) = \begin{bmatrix} \cos\theta & -\sin\theta \\ \sin\theta & \cos\theta \end{bmatrix}.$$

Since this is an invertible matrix for any value of θ, it follows from the analysis in this section that a rotation is an appropriate combination of reflections, shears, and stretches. Indeed, we leave it as an exercise to verify that, for $\theta \neq \pi/2, 3\pi/2$,

$$T(\theta) = \begin{bmatrix} \cos\theta & 0 \\ 0 & 1 \end{bmatrix}\begin{bmatrix} 1 & 0 \\ \sin\theta & 1 \end{bmatrix}\begin{bmatrix} 1 & 0 \\ 0 & \sec\theta \end{bmatrix}\begin{bmatrix} 1 & -\tan\theta \\ 0 & 1 \end{bmatrix}. \tag{5.2.5}$$

Exercises for 5.2

Key Terms

Transformation of $\mathbb{R}^2$, Reflection, Stretch (expansion and compression), Shear, Invertible transformation of $\mathbb{R}^2$.

Skills

- Be able to describe the relationships between elementary matrices and the matrices representing reflections, stretches, and shears.

- Be able to express any transformation of $\mathbb{R}^2$ with invertible matrix as a composition of reflections, stretches, and shears.

- Be able to recognize reflections, stretches, and shears of $\mathbb{R}^2$ by looking at the matrix of the transformation.

True-False Review

For Questions 1–6, decide if the given statement is **true** or **false**, and give a brief justification for your answer. If true, you can quote a relevant definition or theorem from the text. If false, provide an example, illustration, or brief explanation of why the statement is false.

1. Any transformation of $\mathbb{R}^2$ maps a line in the plane onto another line.

2. The matrix of a reflection, stretch, or shear of $\mathbb{R}^2$ is an elementary matrix.

3. A composition of two shears is a shear.

4. Every invertible transformation of $\mathbb{R}^2$ is a composition of reflections, stretches, and shears.

5. A composition of two reflections is a stretch.

6. A composition of two stretches is a stretch.

Problems

For Problems 1–4, for the transformation of $\mathbb{R}^2$ with the given matrix, sketch the transform of the square with vertices $(1, 1)$, $(2, 1)$, $(2, 2)$, and $(1, 2)$.

1. $A = \begin{bmatrix} 0 & 1 \\ -1 & 0 \end{bmatrix}$.

2. $A = \begin{bmatrix} 1 & -1 \\ 1 & 2 \end{bmatrix}$.

3. $A = \begin{bmatrix} 1 & 1 \\ -1 & 1 \end{bmatrix}$.

4. $A = \begin{bmatrix} -2 & -2 \\ -2 & 0 \end{bmatrix}$.

For Problems 5–12, describe the transformation of $\mathbb{R}^2$ with the given matrix as a product of reflections, stretches, and shears.

5. $A = \begin{bmatrix} 1 & 2 \\ 0 & 1 \end{bmatrix}$.

6. $A = \begin{bmatrix} 0 & 2 \\ 2 & 0 \end{bmatrix}$.

7. $A = \begin{bmatrix} 1 & 0 \\ 3 & 1 \end{bmatrix}$.

8. $A = \begin{bmatrix} -1 & 0 \\ 0 & -1 \end{bmatrix}$.

9. $A = \begin{bmatrix} 1 & -3 \\ -2 & 8 \end{bmatrix}$.

10. $A = \begin{bmatrix} 1 & 2 \\ 3 & 4 \end{bmatrix}$.

11. $A = \begin{bmatrix} 1 & 0 \\ 0 & -2 \end{bmatrix}$.

12. $A = \begin{bmatrix} -1 & -1 \\ -1 & 0 \end{bmatrix}$.

13. Consider the transformation of $\mathbb{R}^2$ corresponding to a counterclockwise rotation through angle θ ($0 \leq \theta < 2\pi$). For $\theta \neq \pi/2, 3\pi/2$, verify that the matrix of the transformation is given by (5.2.5), and describe it in terms of reflections, stretches, and shears.

14. Express the transformation of $\mathbb{R}^2$ corresponding to a counterclockwise rotation through an angle $\theta = \pi/2$ as a product of reflections, stretches, and shears. Repeat for the case $\theta = 3\pi/2$.

5.3   The Kernel and Range of a Linear Transformation

If $T: V \to W$ is any linear transformation, there is an associated homogeneous linear vector equation, namely,

$$T(\mathbf{v}) = \mathbf{0}.$$

The solution set to this vector equation is a subset of V called the kernel of T.

DEFINITION 5.3.1

Let $T: V \to W$ be a linear transformation. The set of *all* vectors $\mathbf{v} \in V$ such that $T(\mathbf{v}) = \mathbf{0}$ is called the **kernel** of T and is denoted Ker(T). Thus,

$$\text{Ker}(T) = \{\mathbf{v} \in V : T(\mathbf{v}) = \mathbf{0}\}.$$

Example 5.3.2   Determine Ker(T) for the linear transformation $T: C^2(I) \to C^0(I)$ defined by $T(y) = y'' + y$.

Solution: We have

$$\text{Ker}(T) = \{y \in C^2(I) : T(y) = 0\} = \{y \in C^2(I) : y'' + y = 0 \text{ for all } x \in I\}.$$

Hence, in this case, Ker(T) is the solution set to the differential equation

$$y'' + y = 0.$$

Since this differential equation has general solution $y(x) = c_1 \cos x + c_2 \sin x$, we have

$$\text{Ker}(T) = \{y \in C^2(I) : y(x) = c_1 \cos x + c_2 \sin x\}.$$

This is the subspace of $C^2(I)$ spanned by $\{\cos x, \sin x\}$.   □

The set of all vectors in W that we map onto when T is applied to *all* vectors in V is called the range of T. We can think of the range of T as being the set of function output values. A formal definition follows.

DEFINITION   5.3.3

The **range** of the linear transformation $T : V \to W$ is the subset of W consisting of all transformed vectors from V. We denote the range of T by $\text{Rng}(T)$. Thus,

$$\text{Rng}(T) = \{T(\mathbf{v}) : \mathbf{v} \in V\}.$$

A schematic representation of $\text{Ker}(T)$ and $\text{Rng}(T)$ is given in Figure 5.3.1.

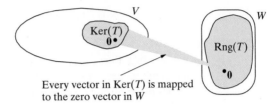

Every vector in $\text{Ker}(T)$ is mapped
to the zero vector in W

Figure 5.3.1: Schematic representation of the kernel and range of a linear transformation.

Let us now focus on matrix transformations, say $T : \mathbb{R}^n \to \mathbb{R}^m$. In this particular case,

$$\text{Ker}(T) = \{\mathbf{x} \in \mathbb{R}^n : T(\mathbf{x}) = \mathbf{0}\}.$$

If we let A denote the matrix of T, then $T(\mathbf{x}) = A\mathbf{x}$, so that

$$\text{Ker}(T) = \{\mathbf{x} \in \mathbb{R}^n : A\mathbf{x} = \mathbf{0}\}.$$

Consequently,

> If $T : \mathbb{R}^n \to \mathbb{R}^m$ is the linear transformation with matrix A, then $\text{Ker}(T)$ is the solution set to the homogeneous linear system $A\mathbf{x} = \mathbf{0}$.

In Section 4.3, we defined the solution set to $A\mathbf{x} = \mathbf{0}$ to be $\text{nullspace}(A)$. Therefore, we have

$$\text{Ker}(T) = \text{nullspace}(A), \tag{5.3.1}$$

from which it follows directly that[2] $\text{Ker}(T)$ is a *subspace* of $\mathbb{R}^n$. Furthermore, for a linear transformation $T : \mathbb{R}^n \to \mathbb{R}^m$,

$$\text{Rng}(T) = \{T(\mathbf{x}) : \mathbf{x} \in \mathbb{R}^n\}.$$

If $A = [\mathbf{a}_1, \mathbf{a}_2, \ldots, \mathbf{a}_n]$ denotes the matrix of T, then

$$\begin{aligned}
\text{Rng}(T) &= \{A\mathbf{x} : \mathbf{x} \in \mathbb{R}^n\} \\
&= \{x_1\mathbf{a}_1 + x_2\mathbf{a}_2 + \cdots + x_n\mathbf{a}_n : x_1, x_2, \ldots, x_n \in \mathbb{R}\} \\
&= \text{colspace}(A).
\end{aligned}$$

Consequently, $\text{Rng}(T)$ is a *subspace* of $\mathbb{R}^m$. We illustrate these results with an example.

Example 5.3.4   Let $T : \mathbb{R}^3 \to \mathbb{R}^2$ be the linear transformation with matrix

$$A = \begin{bmatrix} 1 & -2 & 5 \\ -2 & 4 & -10 \end{bmatrix}.$$

---

[2]It is also easy to verify this fact directly by using Definition 5.3.1 and Theorem 4.3.2.

Determine Ker(T) and Rng(T).

Solution: To determine Ker(T), (5.3.1) implies that we need to find the solution set to the system $A\mathbf{x} = \mathbf{0}$. The reduced row-echelon form of the augmented matrix of this system is

$$\begin{bmatrix} 1 & -2 & 5 & 0 \\ 0 & 0 & 0 & 0 \end{bmatrix},$$

so that there are two free variables. Setting $x_2 = r$ and $x_3 = s$, it follows that $x_1 = 2r - 5s$, so that $\mathbf{x} = (2r - 5s, r, s)$. Hence,

$$\begin{aligned} \text{Ker}(T) &= \{\mathbf{x} \in \mathbb{R}^3 : \mathbf{x} = (2r - 5s, r, s) : r, s \in \mathbb{R}\} \\ &= \{\mathbf{x} \in \mathbb{R}^3 : \mathbf{x} = r(2, 1, 0) + s(-5, 0, 1), \ r, s \in \mathbb{R}\}. \end{aligned}$$

We see that Ker(T) is the two-dimensional subspace of $\mathbb{R}^3$ spanned by the linearly independent vectors $(2, 1, 0)$ and $(-5, 0, 1)$, and therefore it consists of all points lying on the plane through the origin that contains these vectors. We leave it as an exercise to verify that the equation of this plane is $x_1 - 2x_2 + 5x_3 = 0$. The linear transformation T maps all points lying on this plane to the zero vector in $\mathbb{R}^2$. (See Figure 5.3.2.)

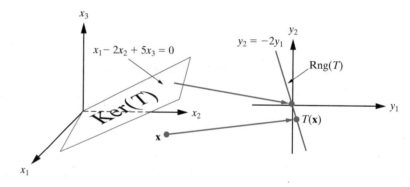

Figure 5.3.2: The kernel and range of the linear transformation in Example 5.3.6.

Turning our attention to Rng(T), recall that, since T is a matrix transformation,

$$\text{Rng}(T) = \text{colspace}(A).$$

From the foregoing reduced row-echelon form of A we see that colspace(A) is generated by the first column vector in A. Consequently,

$$\text{Rng}(T) = \{\mathbf{y} \in \mathbb{R}^2 : \ \mathbf{y} = r(1, -2), \ r \in \mathbb{R}\}.$$

Hence, the points in Rng(T) lie along the line through the origin in $\mathbb{R}^2$ whose direction is determined by $\mathbf{v} = (1, -2)$. The Cartesian equation of this line is $y_2 = -2y_1$. Consequently, T maps all points in $\mathbb{R}^3$ onto this line, and therefore Rng(T) is a one-dimensional subspace of $\mathbb{R}^2$. This is illustrated in Figure 5.3.2. $\square$

To summarize, any matrix transformation $T : \mathbb{R}^n \rightarrow \mathbb{R}^m$ with $m \times n$ matrix A has natural subspaces

$$\begin{aligned} \text{Ker}(T) &= \text{nullspace}(A) \quad \text{(subspace of } \mathbb{R}^n) \\ \text{Rng}(T) &= \text{colspace}(A) \quad \text{(subspace of } \mathbb{R}^m) \end{aligned}$$

Now let us return to arbitrary linear transformations. The preceding discussion has shown that both the kernel and range of any linear transformation from $\mathbb{R}^n$ to $\mathbb{R}^m$ are subspaces of $\mathbb{R}^n$ and $\mathbb{R}^m$, respectively. Our next result, which is fundamental, establishes that this is true in general.

Theorem 5.3.5   If $T : V \rightarrow W$ is a linear transformation, then

1. $\text{Ker}(T)$ is a subspace of V.

2. $\text{Rng}(T)$ is a subspace of W.

Proof In this proof, we once more denote the zero vector in W by $\mathbf{0}_W$. Both $\text{Ker}(T)$ and $\text{Rng}(T)$ are necessarily nonempty, since, as we verified in Section 5.1, any linear transformation maps the zero vector in V to the zero vector in W. We must now establish that $\text{Ker}(T)$ and $\text{Rng}(T)$ are both closed under addition and closed under scalar multiplication in the appropriate vector space.

1. If $\mathbf{v}_1$ and $\mathbf{v}_2$ are in $\text{Ker}(T)$, then $T(\mathbf{v}_1) = \mathbf{0}_W$ and $T(\mathbf{v}_2) = \mathbf{0}_W$. We must show that $\mathbf{v}_1 + \mathbf{v}_2$ is in $\text{Ker}(T)$; that is, $T(\mathbf{v}_1 + \mathbf{v}_2) = \mathbf{0}_W$. But we have

$$T(\mathbf{v}_1 + \mathbf{v}_2) = T(\mathbf{v}_1) + T(\mathbf{v}_2) = \mathbf{0}_W + \mathbf{0}_W = \mathbf{0}_W,$$

so that $\text{Ker}(T)$ is closed under addition. Further, if c is any scalar,

$$T(c\mathbf{v}_1) = cT(\mathbf{v}_1) = c\mathbf{0}_W = \mathbf{0}_W,$$

which shows that $c\mathbf{v}_1$ is in $\text{Ker}(T)$, and so $\text{Ker}(T)$ is also closed under scalar multiplication. Thus, $\text{Ker}(T)$ is a subspace of V.

2. If $\mathbf{w}_1$ and $\mathbf{w}_2$ are in $\text{Rng}(T)$, then $\mathbf{w}_1 = T(\mathbf{v}_1)$ and $\mathbf{w}_2 = T(\mathbf{v}_2)$ for some $\mathbf{v}_1$ and $\mathbf{v}_2$ in V. Thus,

$$\mathbf{w}_1 + \mathbf{w}_2 = T(\mathbf{v}_1) + T(\mathbf{v}_2) = T(\mathbf{v}_1 + \mathbf{v}_2).$$

This says that $\mathbf{w}_1 + \mathbf{w}_2$ arises as an output of the transformation T; that is, $\mathbf{w}_1 + \mathbf{w}_2$ is in $\text{Rng}(T)$. Thus, $\text{Rng}(T)$ is closed under addition. Further, if c is any scalar, then

$$c\mathbf{w}_1 = cT(\mathbf{v}_1) = T(c\mathbf{v}_1),$$

so that $c\mathbf{w}_1$ is the transform of $c\mathbf{v}_1$, and therefore $c\mathbf{w}_1$ is in $\text{Rng}(T)$. Consequently, $\text{Rng}(T)$ is a subspace of W.   ■

Remark   We can interpret the first part of the preceding theorem as telling us that if T is a linear transformation, then the solution set to the corresponding linear homogeneous problem

$$T(\mathbf{v}) = \mathbf{0}$$

is a vector space. Consequently, if we can determine the dimension of this vector space, then we know how many linearly independent solutions are required to build every solution to the problem. This is the formulation for linear problems that we have been looking for.

Example 5.3.6   Find $\text{Ker}(S)$, $\text{Rng}(S)$, and their dimensions for the linear transformation $S : M_2(\mathbb{R}) \rightarrow M_2(\mathbb{R})$ defined by

$$S(A) = A - A^T.$$

Solution: In this case,

$$\text{Ker}(S) = \{A \in M_2(\mathbb{R}) : S(A) = 0\} = \{A \in M_2(\mathbb{R}) : A - A^T = 0_2\}.$$

Thus, $\text{Ker}(S)$ is the solution set of the matrix equation

$$A - A^T = 0_2,$$

so that the matrices in $\text{Ker}(S)$ satisfy

$$A^T = A.$$

Hence, $\text{Ker}(S)$ is the subspace of $M_2(\mathbb{R})$ consisting of all symmetric 2×2 matrices. We have shown previously that a basis for this subspace is

$$\left\{ \begin{bmatrix} 1 & 0 \\ 0 & 0 \end{bmatrix}, \begin{bmatrix} 0 & 1 \\ 1 & 0 \end{bmatrix}, \begin{bmatrix} 0 & 0 \\ 0 & 1 \end{bmatrix} \right\},$$

so that $\dim[\text{Ker}(S)] = 3$. We now determine the range of S:

$$\text{Rng}(S) = \{S(A) : A \in M_2(\mathbb{R})\} = \{A - A^T : A \in M_2(\mathbb{R})\}$$
$$= \left\{ \begin{bmatrix} a & b \\ c & d \end{bmatrix} - \begin{bmatrix} a & c \\ b & d \end{bmatrix} : a, b, c, d \in \mathbb{R} \right\}$$
$$= \left\{ \begin{bmatrix} 0 & b-c \\ -(b-c) & 0 \end{bmatrix} : b, c \in \mathbb{R} \right\}.$$

Thus,

$$\text{Rng}(S) = \left\{ \begin{bmatrix} 0 & e \\ -e & 0 \end{bmatrix} : e \in \mathbb{R} \right\} = \text{span}\left\{ \begin{bmatrix} 0 & 1 \\ -1 & 0 \end{bmatrix} \right\}.$$

Consequently, $\text{Rng}(S)$ consists of all skew-symmetric 2×2 matrices with real elements. Since $\text{Rng}(S)$ is generated by the single nonzero matrix

$$\begin{bmatrix} 0 & 1 \\ -1 & 0 \end{bmatrix},$$

it follows that a basis for $\text{Rng}(S)$ is

$$\left\{ \begin{bmatrix} 0 & 1 \\ -1 & 0 \end{bmatrix} \right\},$$

so that $\dim[\text{Rng}(S)] = 1$.   □

Example 5.3.7   Let $T : P_1 \rightarrow P_2$ be the linear transformation defined by

$$T(a + bx) = (2a - 3b) + (b - 5a)x + (a + b)x^2.$$

Find $\text{Ker}(T)$, $\text{Rng}(T)$, and their dimensions.

Solution: From Definition 5.3.1,

$$\text{Ker}(T) = \{p \in P_1 : T(p) = 0\}$$
$$= \{a + bx : (2a - 3b) + (b - 5a)x + (a + b)x^2 = 0 \text{ for all } x\}$$
$$= \{a + bx : a + b = 0, \quad b - 5a = 0, \quad 2a - 3b = 0\}.$$

But the only values of a and b that satisfy the conditions

$$a + b = 0, \qquad b - 5a = 0, \qquad 2a - 3b = 0$$

are

$$a = b = 0.$$

Consequently, $\text{Ker}(T)$ contains only the zero polynomial. Hence, we write

$$\text{Ker}(T) = \{\mathbf{0}\}.$$

It follows from Definition 4.6.7 that $\dim[\text{Ker}(T)] = 0$. Furthermore,

$$\text{Rng}(T) = \{T(a + bx) : a, b \in \mathbb{R}\} = \{(2a - 3b) + (b - 5a)x + (a + b)x^2 : a, b \in \mathbb{R}\}.$$

To determine a basis for $\text{Rng}(T)$, we write this as

$$\text{Rng}(T) = \{a(2 - 5x + x^2) + b(-3 + x + x^2) : a, b \in \mathbb{R}\}$$
$$= \text{span}\{2 - 5x + x^2, -3 + x + x^2\}.$$

Thus, $\text{Rng}(T)$ is spanned by $p_1(x) = 2 - 5x + x^2$ and $p_2(x) = -3 + x + x^2$. Since p_1 and p_2 are not proportional to one another, they are linearly independent on any interval. Consequently, a basis for $\text{Rng}(T)$ is $\{2 - 5x + x^2, -3 + x + x^2\}$, so that $\dim[\text{Rng}(T)] = 2$. □

The General Rank-Nullity Theorem

In concluding this section, we consider a fundamental theorem for linear transformations $T : V \rightarrow W$ that gives a relationship between the dimensions of $\text{Ker}(T)$, $\text{Rng}(T)$, and V. This is a generalization of the Rank-Nullity Theorem considered in Section 4.9. The theorem here, Theorem 5.3.8, reduces to the previous result, Theorem 4.9.1, in the case when T is a linear transformation from $\mathbb{R}^n$ to $\mathbb{R}^m$ with $m \times n$ matrix A. Suppose that $\dim[V] = n$ and that $\dim[\text{Ker}(T)] = k$. Then k-dimensions worth of the vectors in V are all mapped onto the zero vector in W. Consequently, we only have $n - k$ dimensions worth of vectors left to map onto the remaining vectors in W. This *suggests* that

$$\dim[\text{Rng}(T)] = \dim[V] - \dim[\text{Ker}(T)].$$

This is indeed correct, although a rigorous proof is somewhat involved. We state the result as a theorem here.

Theorem 5.3.8   **(General Rank-Nullity Theorem)**

If $T : V \rightarrow W$ is a linear transformation and V is finite-dimensional, then

$$\dim[\text{Ker}(T)] + \dim[\text{Rng}(T)] = \dim[V].$$

Before presenting the proof of this theorem, we give a few applications and examples. The general Rank-Nullity Theorem is useful for checking that we have the correct dimensions when determining the kernel and range of a linear transformation. Furthermore, it can also be used to determine the dimension of $\text{Rng}(T)$, once we know the dimension of $\text{Ker}(T)$, or vice versa. For example, consider the linear transformation discussed in Example 5.3.6. Theorem 5.3.8 tells us that

$$\dim[\text{Ker}(S)] + \dim[\text{Rng}(S)] = \dim[M_2(\mathbb{R})],$$

so that once we have determined that $\dim[\text{Ker}(S)] = 3$, it immediately follows that

$$3 + \dim[\text{Rng}(S)] = 4$$

so that

$$\dim[\text{Rng}(S)] = 1.$$

As another illustration, consider the matrix transformation in Example 5.3.4 with

$$A = \begin{bmatrix} 1 & -2 & 5 \\ -2 & 4 & -10 \end{bmatrix}.$$

By inspection, we can see that $\dim[\text{Rng}(T)] = \dim[\text{colspace}(A)] = 1$, so the Rank-Nullity Theorem implies that

$$\dim[\text{Ker}(T)] = 3 - 1 = 2,$$

with no additional calculation. Of course, to obtain a basis for $\text{Ker}(T)$, it becomes necessary to carry out the calculations presented in Example 5.3.4.

We close this section with a proof of Theorem 5.3.8.

Proof of Theorem 5.3.8: Suppose that $\dim[V] = n$. We consider three cases:

Case 1: If $\dim[\text{Ker}(T)] = n$, then by Corollary 4.6.14 we conclude that $\text{Ker}(T) = V$. This means that $T(\mathbf{v}) = \mathbf{0}$ for every vector $\mathbf{v} \in V$. In this case

$$\text{Rng}(T) = \{T(\mathbf{v}) : \mathbf{v} \in V\} = \{\mathbf{0}\},$$

hence $\dim[\text{Rng}(T)] = 0$. Thus, we have

$$\dim[\text{Ker}(T)] + \dim[\text{Rng}(T)] = n + 0 = n = \dim[V],$$

as required.

Case 2: Assume $\dim[\text{Ker}(T)] = k$, where $0 < k < n$. Let $\{\mathbf{v}_1, \mathbf{v}_2, \ldots, \mathbf{v}_k\}$ be a basis for $\text{Ker}(T)$. Then, using Theorem 4.6.17, we can extend this basis to a basis for V, which we denote by $\{\mathbf{v}_1, \mathbf{v}_2, \ldots, \mathbf{v}_k, \mathbf{v}_{k+1}, \ldots, \mathbf{v}_n\}$.

We prove that $\{T(\mathbf{v}_{k+1}), T(\mathbf{v}_{k+2}), \ldots, T(\mathbf{v}_n)\}$ is a basis for $\text{Rng}(T)$. To do this, we first prove that $\{T(\mathbf{v}_{k+1}), T(\mathbf{v}_{k+2}), \ldots, T(\mathbf{v}_n)\}$ spans $\text{Rng}(T)$. Let $\mathbf{w}$ be any vector in $\text{Rng}(T)$. Then $\mathbf{w} = T(\mathbf{v})$, for some $\mathbf{v} \in V$. Using the basis for V, we have $\mathbf{v} = c_1\mathbf{v}_1 + c_2\mathbf{v}_2 + \cdots + c_n\mathbf{v}_n$ for some scalars $c_1, c_2, \ldots, c_n$. Hence,

$$\mathbf{w} = T(\mathbf{v}) = T(c_1\mathbf{v}_1 + c_2\mathbf{v}_2 + \cdots + c_n\mathbf{v}_n) = c_1T(\mathbf{v}_1) + c_2T(\mathbf{v}_2) + \cdots + c_nT(\mathbf{v}_n).$$

Since $\mathbf{v}_1, \mathbf{v}_2, \ldots, \mathbf{v}_k$ are in $\text{Ker}(T)$, this reduces to

$$\mathbf{w} = c_{k+1}T(\mathbf{v}_{k+1}) + c_{k+2}T(\mathbf{v}_{k+2}) + \cdots + c_nT(\mathbf{v}_n).$$

Thus,

$$\text{Rng}(T) = \text{span}\{T(\mathbf{v}_{k+1}), T(\mathbf{v}_{k+2}), \ldots, T(\mathbf{v}_n)\}.$$

Next we show that $\{T(\mathbf{v}_{k+1}), T(\mathbf{v}_{k+2}), \ldots, T(\mathbf{v}_n)\}$ is linearly independent. Suppose that

$$d_{k+1}T(\mathbf{v}_{k+1}) + d_{k+2}T(\mathbf{v}_{k+2}) + \cdots + d_nT(\mathbf{v}_n) = \mathbf{0}, \tag{5.3.2}$$

where $d_{k+1}, d_{k+2}, \ldots, d_n$ are scalars. Then, using the linearity of T,

$$T(d_{k+1}\mathbf{v}_{k+1} + d_{k+2}\mathbf{v}_{k+2} + \cdots + d_n\mathbf{v}_n) = \mathbf{0},$$

which implies that the vector $d_{k+1}\mathbf{v}_{k+1} + d_{k+2}\mathbf{v}_{k+2} + \cdots + d_n\mathbf{v}_n$ is in $\text{Ker}(T)$. Consequently, there exist scalars $d_1, d_2, \ldots, d_k$ such that

$$d_{k+1}\mathbf{v}_{k+1} + d_{k+2}\mathbf{v}_{k+2} + \cdots + d_n\mathbf{v}_n = d_1\mathbf{v}_1 + d_2\mathbf{v}_2 + \cdots + d_k\mathbf{v}_k,$$

which means that

$$d_1\mathbf{v}_1 + d_2\mathbf{v}_2 + \cdots + d_k\mathbf{v}_k - (d_{k+1}\mathbf{v}_{k+1} + d_{k+2}\mathbf{v}_{k+2} + \cdots + d_n\mathbf{v}_n) = \mathbf{0}.$$

Since the set of vectors $\{\mathbf{v}_1, \mathbf{v}_2, \ldots, \mathbf{v}_k, \mathbf{v}_{k+1}, \ldots, \mathbf{v}_n\}$ is linearly independent, we must have

$$d_1 = d_2 = \cdots = d_k = d_{k+1} = \cdots = d_n = 0.$$

Thus, from Equation (5.3.2), $\{T(\mathbf{v}_{k+1}), T(\mathbf{v}_{k+2}), \ldots, T(\mathbf{v}_n)\}$ is linearly independent.

By the work in the last two paragraphs, $\{T(\mathbf{v}_{k+1}), T(\mathbf{v}_{k+2}), \ldots, T(\mathbf{v}_n)\}$ is a basis for $\text{Rng}(T)$. Since there are $n - k$ vectors in this basis, it follows that $\dim[\text{Rng}(T)] = n - k$. Consequently,

$$\dim[\text{Ker}(T)] + \dim[\text{Rng}(T)] = k + (n - k) = n = \dim[V],$$

as required.

Case 3:   If $\dim[\text{Ker}(T)] = 0$, then $\text{Ker}(T) = \{\mathbf{0}\}$. Let $\{\mathbf{v}_1, \mathbf{v}_2, \ldots, \mathbf{v}_n\}$ be any basis for V. By a similar argument to that used in Case 2 above, it can be shown that (see Problem 18) $\{T(\mathbf{v}_1), T(\mathbf{v}_2), \ldots, T(\mathbf{v}_n)\}$ is a basis for $\text{Rng}(T)$, and so again we have

$$\dim[\text{Ker}(T)] + \dim[\text{Rng}(T)] = n. \qquad \blacksquare$$

Exercises for 5.3

Key Terms

Kernel and range of a linear transformation, Rank-Nullity Theorem.

Skills

- Be able to find the kernel of a linear transformation $T: V \to W$ and give a basis and the dimension of $\text{Ker}(T)$.

- Be able to find the range of a linear transformation $T: V \to W$ and give a basis and the dimension of $\text{Rng}(T)$.

- Be able to show that the kernel (resp. range) of a linear transformation $T: V \to W$ is a subspace of V (resp. W).

- Be able to verify the Rank-Nullity Theorem for a given linear transformation $T: V \to W$.

- Be able to utilize the Rank-Nullity Theorem to help find the dimensions of the kernel and range of a linear transformation $T: V \to W$.

True-False Review

For Questions 1–6, decide if the given statement is **true** or **false**, and give a brief justification for your answer. If true, you can quote a relevant definition or theorem from the text. If false, provide an example, illustration, or brief explanation of why the statement is false.

1. If $T: V \to W$ is a linear transformation and W is finite-dimensional, then

$$\dim[\text{Ker}(T)] + \dim[\text{Rng}(T)] = \dim[W].$$

2. If $T: P_4 \to \mathbb{R}^7$ is a linear transformation, then $\text{Ker}(T)$ must be at least two-dimensional.

3. If $T: \mathbb{R}^n \to \mathbb{R}^m$ is a linear transformation with matrix A, then Rng(T) is the solution set to the homogeneous linear system $A\mathbf{x} = \mathbf{0}$.

4. The range of a linear transformation $T: V \to W$ is a subspace of V.

5. If $T: M_{23} \to P_7$ is a linear transformation with

$$T\begin{bmatrix} 1 & 1 & 1 \\ 0 & 0 & 0 \end{bmatrix} = 0, \qquad T\begin{bmatrix} 1 & 2 & 3 \\ 4 & 5 & 6 \end{bmatrix} = 0,$$

then Rng(T) is at most four-dimensional.

6. If $T: \mathbb{R}^n \to \mathbb{R}^m$ is a linear transformation with matrix A, then Rng(T) is the column space of A.

Problems

1. Consider $T: \mathbb{R}^3 \to \mathbb{R}^2$ defined by $T(\mathbf{x}) = A\mathbf{x}$, where

$$A\begin{bmatrix} 1 & -1 & 2 \\ 1 & -2 & -3 \end{bmatrix}.$$

Find $T(\mathbf{x})$ and thereby determine whether $\mathbf{x}$ is in Ker(T).

(a) $\mathbf{x} = (7, 5, -1)$.

(b) $\mathbf{x} = (-21, -15, 2)$.

(c) $\mathbf{x} = (35, 25, -5)$.

For Problems 2–6, find Ker(T) and Rng(T) and give a geometrical description of each. Also, find dim[Ker(T)] and dim[Rng(T)] and verify Theorem 5.3.8.

2. $T: \mathbb{R}^2 \to \mathbb{R}^2$ defined by $T(\mathbf{x}) = A\mathbf{x}$, where

$$A = \begin{bmatrix} 3 & 6 \\ 1 & 2 \end{bmatrix}.$$

3. $T: \mathbb{R}^3 \to \mathbb{R}^3$ defined by $T(\mathbf{x}) = A\mathbf{x}$, where

$$A = \begin{bmatrix} 1 & -1 & 0 \\ 0 & 1 & 2 \\ 2 & -1 & 1 \end{bmatrix}.$$

4. $T: \mathbb{R}^3 \to \mathbb{R}^3$ defined by $T(\mathbf{x}) = A\mathbf{x}$, where

$$A = \begin{bmatrix} 1 & -2 & 1 \\ 2 & -3 & -1 \\ 5 & -8 & -1 \end{bmatrix}.$$

5. $T: \mathbb{R}^3 \to \mathbb{R}^2$ defined by $T(\mathbf{x}) = A\mathbf{x}$, where

$$A = \begin{bmatrix} 1 & -1 & 2 \\ -3 & 3 & -6 \end{bmatrix}.$$

6. $T: \mathbb{R}^3 \to \mathbb{R}^2$ defined by $T(\mathbf{x}) = A\mathbf{x}$, where

$$A = \begin{bmatrix} 1 & 3 & 2 \\ 2 & 6 & 5 \end{bmatrix}.$$

For Problems 7–10, compute Ker(T) and Rng(T).

7. The linear transformation T defined in Problem 24 in Section 5.1.

8. The linear transformation T defined in Problem 25 in Section 5.1.

9. The linear transformation T defined in Problem 26 in Section 5.1.

10. The linear transformation T defined in Problem 27 in Section 5.1.

11. Consider the linear transformation $T: \mathbb{R}^3 \to \mathbb{R}$ defined by

$$T(\mathbf{v}) = \langle \mathbf{u}, \mathbf{v} \rangle,$$

where $\mathbf{u}$ is a fixed nonzero vector in $\mathbb{R}^3$.

(a) Find Ker(T) and dim[Ker(T)], and interpret this geometrically.

(b) Find Rng(T) and dim[Rng(T)].

12. Consider the linear transformation $S: M_n(\mathbb{R}) \to M_n(\mathbb{R})$ defined by $S(A) = A + A^T$, where A is a fixed $n \times n$ matrix.

(a) Find Ker(S) and describe it. What is dim[Ker(S)]?

(b) In the particular case when A is a 2×2 matrix, determine a basis for Ker(S), and hence, find its dimension.

13. Consider the linear transformation $T: M_n(\mathbb{R}) \to M_n(\mathbb{R})$ defined by

$$T(A) = AB - BA,$$

where B is a fixed $n \times n$ matrix. Find Ker(T) and describe it.

14. Consider the *linear* transformation $T: P_2 \to P_2$ defined by

$$T(ax^2+bx+c) = ax^2+(a+2b+c)x+(3a-2b-c),$$

where $a, b,$ and c are arbitrary constants.

(a) Show that Ker(T) consists of all polynomials of the form $b(x - 2)$, and hence, find its dimension.

(b) Find Rng(T) and its dimension.

15. Consider the *linear* transformation $T: P_2 \to P_1$ defined by

$$T(ax^2 + bx + c) = (a + b) + (b - c)x,$$

where $a, b,$ and c are arbitrary real numbers. Determine $\text{Ker}(T)$, $\text{Rng}(T)$, and their dimensions.

16. Consider the *linear* transformation $T: P_1 \to P_2$ defined by

$$T(ax + b) = (b - a) + (2b - 3a)x + bx^2.$$

Determine $\text{Ker}(T)$, $\text{Rng}(T)$, and their dimensions.

17. Let $\{\mathbf{v}_1, \mathbf{v}_2, \mathbf{v}_3\}$ and $\{\mathbf{w}_1, \mathbf{w}_2\}$ be bases for real vector spaces V and W, respectively, and let $T: V \to W$ be the linear transformation satisfying

$$T(\mathbf{v}_1) = 2\mathbf{w}_1 - \mathbf{w}_2, \qquad T(\mathbf{v}_2) = \mathbf{w}_1 - \mathbf{w}_2,$$

$$T(\mathbf{v}_3) = \mathbf{w}_1 + 2\mathbf{w}_2.$$

Find $\text{Ker}(T)$, $\text{Rng}(T)$, and their dimensions.

18. **(a)** Let $T: V \to W$ be a linear transformation, and suppose that $\dim[V] = n$. If $\text{Ker}(T) = \{\mathbf{0}\}$ and $\{\mathbf{v}_1, \mathbf{v}_2, \ldots, \mathbf{v}_n\}$ is any basis for V, prove that

$$\{T(\mathbf{v}_1), T(\mathbf{v}_2), \ldots, T(\mathbf{v}_n)\}$$

is a basis for $\text{Rng}(T)$. (This fills in the missing details in the proof of Theorem 5.3.8.)

(b) Show that the conclusion from part (a) fails to hold if $\text{Ker}(T) \neq \{\mathbf{0}\}$.

5.4 Additional Properties of Linear Transformations

One aim of this section is to establish that all real vector spaces of (finite) dimension n are essentially the same as $\mathbb{R}^n$. In order to do so, we need to consider the composition of linear transformations.

DEFINITION 5.4.1

Let $T_1: U \to V$ and $T_2: V \to W$ be two linear transformations.[3] We define the **composition**, or **product**, $T_2 T_1: U \to W$ by

$$(T_2 T_1)(\mathbf{u}) = T_2(T_1(\mathbf{u})) \qquad \text{for all } \mathbf{u} \in U.$$

The composition is illustrated in Figure 5.4.1.

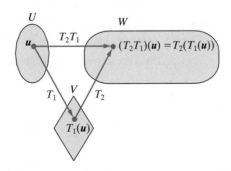

Figure 5.4.1: The composition of two linear transformations.

Our first result establishes that $T_2 T_1$ is a *linear* transformation.

Theorem 5.4.2 Let $T_1: U \to V$ and $T_2: V \to W$ be linear transformations. Then $T_2 T_1: U \to W$ is a linear transformation.

---

[3]We assume that $U, V,$ and W are either all real vector spaces or all complex vector spaces.

Proof Let $\mathbf{u}_1, \mathbf{u}_2$ be arbitrary vectors in U, and let c be a scalar. We must prove that

$$(T_2 T_1)(\mathbf{u}_1 + \mathbf{u}_2) = (T_2 T_1)(\mathbf{u}_1) + (T_2 T_1)(\mathbf{u}_2) \qquad (5.4.1)$$

and that

$$(T_2 T_1)(c\mathbf{u}_1) = c(T_2 T_1)(\mathbf{u}_1). \qquad (5.4.2)$$

Consider first Equation (5.4.1). We have

$$
\begin{aligned}
(T_2 T_1)(\mathbf{u}_1 + \mathbf{u}_2) &= T_2(T_1(\mathbf{u}_1 + \mathbf{u}_2)) &&\text{(definition of } T_2 T_1) \\
&= T_2(T_1(\mathbf{u}_1) + T_1(\mathbf{u}_2)) &&\text{(using the linearity of } T_1) \\
&= T_2(T_1(\mathbf{u}_1)) + T_2(T_1(\mathbf{u}_2)) &&\text{(using the linearity of } T_2) \\
&= (T_2 T_1)(\mathbf{u}_1) + (T_2 T_1)(\mathbf{u}_2), &&\text{(definition of } T_2 T_1)
\end{aligned}
$$

so that (5.4.1) is satisfied. We leave the proof of (5.4.2) as an exercise (Problem 30). ∎

Example 5.4.3   Let $T_1 \colon \mathbb{R}^n \to \mathbb{R}^m$ and $T_2 \colon \mathbb{R}^m \to \mathbb{R}^p$ be linear transformations with matrices A and B, respectively. Determine the linear transformation $T_2 T_1 \colon \mathbb{R}^n \to \mathbb{R}^p$.

Solution: From Definition 5.4.1, for any vector $\mathbf{x}$ in $\mathbb{R}^n$, we have

$$(T_2 T_1)(\mathbf{x}) = T_2(T_1(\mathbf{x})) = T_2(A\mathbf{x}) = B(A\mathbf{x}) = (BA)\mathbf{x}.$$

Consequently, $T_2 T_1$ is the linear transformation with matrix BA. Note that A is an $m \times n$ matrix and B is a $p \times m$ matrix, so that the matrix product BA is defined, with size $p \times n$. $\square$

Example 5.4.4   Let $T_1 \colon M_n(\mathbb{R}) \to M_n(\mathbb{R})$ and $T_2 \colon M_n(\mathbb{R}) \to \mathbb{R}$ be the linear transformations defined by

$$T_1(A) = A + A^T, \qquad T_2(A) = \operatorname{tr}(A).$$

Determine $T_2 T_1$.

Solution: In this case, $T_2 T_1 \colon M_n(\mathbb{R}) \to \mathbb{R}$ is defined by

$$(T_2 T_1)(A) = T_2(T_1(A)) = T_2(A + A^T) = \operatorname{tr}(A + A^T).$$

This can be written in the equivalent form

$$(T_2 T_1)(A) = 2\operatorname{tr}(A).$$

As a specific example, consider the matrix $A = \begin{bmatrix} 2 & -1 \\ -3 & 6 \end{bmatrix}$. We have

$$T_1(A) = A + A^T = \begin{bmatrix} 4 & -4 \\ -4 & 12 \end{bmatrix}$$

so that

$$T_2(T_1(A)) = \operatorname{tr}\left(\begin{bmatrix} 4 & -4 \\ -4 & 12 \end{bmatrix} \right) = 16. \qquad \square$$

We now extend the definitions of *one-to-one* and *onto*, which should be familiar in the case of a function of a single variable $f \colon \mathbb{R} \to \mathbb{R}$, to the case of arbitrary linear transformations.

DEFINITION   5.4.5

A linear transformation $T: V \to W$ is said to be

1. **one-to-one** if distinct elements in V are mapped via T to distinct elements in W; that is, whenever $\mathbf{v}_1 \neq \mathbf{v}_2$ in V, we have $T(\mathbf{v}_1) \neq T(\mathbf{v}_2)$.

2. **onto** if the range of T is the whole of W; that is, if *every* $\mathbf{w} \in W$ is the image under T of at least one vector $\mathbf{v} \in V$.

Example 5.4.6   The linear transformation $T: M_2(\mathbb{R}) \to \mathbb{R}$ defined by $T(A) = \text{tr}(A)$ is not one-to-one, since it is possible to find two distinct matrices A and B for which $T(A) = T(B)$. For instance, we can take

$$A = \begin{bmatrix} 0 & 1 \\ 0 & 0 \end{bmatrix} \quad \text{and} \quad B = \begin{bmatrix} 0 & 2 \\ 0 & 0 \end{bmatrix},$$

but $\text{tr}(A) = \text{tr}(B) = 0$. Many other choices of A and B can also be used to illustrate this.

On the other hand, T is onto, since every real number $w \in \mathbb{R}$ is the image of some matrix in $M_2(\mathbb{R})$. For example, take

$$A = \begin{bmatrix} w & 0 \\ 0 & 0 \end{bmatrix}.$$

Then $T(A) = w$, as desired. Again, many other choices for A are possible here.   □

The following theorem can be helpful in determining whether a given linear transformation is one-to-one.

Theorem 5.4.7   Let $T: V \to W$ be a linear transformation. Then T is one-to-one if and only if $\text{Ker}(T) = \{\mathbf{0}\}$.

Proof   Since T is a linear transformation, we have $T(\mathbf{0}) = \mathbf{0}$. Thus, if T is one-to-one, there can be no other vector $\mathbf{v}$ in V satisfying $T(\mathbf{v}) = \mathbf{0}$, so $\text{Ker}(T) = \{\mathbf{0}\}$.

Conversely, suppose that $\text{Ker}(T) = \{\mathbf{0}\}$. If $\mathbf{v}_1 \neq \mathbf{v}_2$, then $\mathbf{v}_1 - \mathbf{v}_2 \neq \mathbf{0}$, and therefore, since $\text{Ker}(T) = \{\mathbf{0}\}$, $T(\mathbf{v}_1 - \mathbf{v}_2) \neq \mathbf{0}$. Hence, by the linearity of T, $T(\mathbf{v}_1) - T(\mathbf{v}_2) \neq \mathbf{0}$, or equivalently, $T(\mathbf{v}_1) \neq T(\mathbf{v}_2)$. Thus, if $\text{Ker}(T) = \{\mathbf{0}\}$, then T is one-to-one.   ■

For instance, our calculations in Example 5.3.7 showed that $\text{Ker}(T) = \{\mathbf{0}\}$, so the linear transformation T in that example is one-to-one.

We now have the following characterization of one-to-one and onto in terms of the kernel and range of T:

> The linear transformation $T: V \to W$ is
> 1. one-to-one if and only if $\text{Ker}(T) = \{\mathbf{0}\}$.
> 2. onto if and only if $\text{Rng}(T) = W$.

Example 5.4.8   If $T: \mathbb{R}^7 \to \mathbb{R}^5$ is a linear transformation and $\text{Ker}(T)$ is two-dimensional, then by Theorem 5.3.8, we have that $\text{Rng}(T)$ is five-dimensional, and hence $\text{Rng}(T) = W$. That is, T is onto.   □

Example 5.4.9   If $T: \mathbb{R}^3 \to \mathbb{R}^6$ is a linear transformation and $\text{Rng}(T)$ is three-dimensional, then by Theorem 5.3.8, $\text{Ker}(T) = \{\mathbf{0}\}$, so T must be one-to-one.   □

Example 5.4.10   If $T: P_6 \to \mathbb{R}^5$ is a linear transformation with $\dim[\text{Ker}(T)] = 2$, then T is onto, since $\text{Rng}(T)$ is five-dimensional, by Theorem 5.3.8.   □

Example 5.4.11   Consider the linear transformation $T: P_2 \to P_2$ defined by

$$T(a + bx + cx^2) = (2a - b + c) + (b - 2a)x + cx^2.$$

Determine whether T is one-to-one, onto, both, or neither.

Solution: To determine whether T is one-to-one, we find $\text{Ker}(T)$. For the given transformation, we have

$$
\begin{aligned}
\text{Ker}(T) &= \{p \in P_2: T(p) = 0\} \\
&= \{a + bx + cx^2: T(a + bx + cx^2) = 0 \text{ for all } x\} \\
&= \{a + bx + cx^2: (2a - b + c) + (b - 2a)x + cx^2 = 0 \text{ for all } x\}.
\end{aligned}
$$

But,

$$(2a - b + c) + (b - 2a)x + cx^2 = 0,$$

for all real x, if and only if

$$c = 0, \qquad b - 2a = 0, \qquad 2a - b + c = 0.$$

These equations are satisfied if and only if

$$c = 0, \qquad b = 2a,$$

so that

$$
\begin{aligned}
\text{Ker}(T) &= \{a + bx + cx^2: c = 0, b = 2a\} \\
&= \{a(1 + 2x): a \in \mathbb{R}\}.
\end{aligned}
$$

Since the kernel of T contains nonzero vectors, Theorem 5.4.7 implies that T is *not* one-to-one. To determine whether T is onto, we can check whether or not $\text{Rng}(T) = P_2$. However, there is a shorter method, using Theorem 5.3.8. Since the vectors in $\text{Ker}(T)$ consist of all scalar multiples of the nonzero polynomial $p_1(x) = 1 + 2x$, $\text{Ker}(T) = \text{span}\{1 + 2x\}$, and so $\dim[\text{Ker}(T)] = 1$. Further, we have seen in Chapter 4 that $\dim[P_2] = 3$. Therefore, from Theorem 5.3.8, we have

$$1 + \dim[\text{Rng}(T)] = 3,$$

which implies that

$$\dim[\text{Rng}(T)] = 2.$$

Thus, $\text{Rng}(T)$ is a two-dimensional subspace of the three-dimensional vector space P_2, and so $\text{Rng}(T) \neq P_2$. Hence, T is *not* onto.   □

Let us explore the relationship between the one-to-one and onto properties of a linear transformation $T: V \to W$ and the dimensions of V and W a bit further. We have the following useful result.

Corollary 5.4.12   Let $T: V \to W$ be a linear transformation, and assume that V and W are both finite-dimensional. Then

1. If T is one-to-one, then $\dim[V] \leq \dim[W]$.

2. If T is onto, then $\dim[V] \geq \dim[W]$.

3. If T is one-to-one and onto, then $\dim[V] = \dim[W]$.

Proof We appeal once more to Theorem 5.3.8. To prove part 1, assume that $T : V \to W$ is one-to-one. Then by Theorem 5.4.7, we see that $\dim[\mathrm{Ker}(T)] = 0$. Thus, Theorem 5.3.8 implies that $\dim[V] = \dim[\mathrm{Rng}(T)]$. But $\mathrm{Rng}(T)$ is a subspace of W, and so by Corollary 4.6.14, $\dim[\mathrm{Rng}(T)] \leq \dim[W]$. Hence, $\dim[V] \leq \dim[W]$.

To prove part 2, assume that T is onto. Then $\mathrm{Rng}(T) = W$, so Theorem 5.3.8 can be rewritten as

$$\dim[V] = \dim[\mathrm{Ker}(T)] + \dim[W],$$

which immediately shows that $\dim[W] \leq \dim[V]$.

Finally, part 3 is an immediate consequence of parts 1 and 2. ∎

In many situations, Corollary 5.4.12 can be used in contrapositive form to eliminate the possibility of a one-to-one or onto linear transformation from V to W. For instance, if $\dim[V] < \dim[W]$, then part 2 of Corollary 5.4.12 implies at once that there can be no onto linear transformation from V to W. For a specific example, no linear transformation $T : M_2(\mathbb{R}) \to \mathbb{R}^5$ can be onto, since this would imply that $\dim[M_2(\mathbb{R})] \geq \dim[\mathbb{R}^5]$, a contradiction.

In a similar way, if $\dim[V] > \dim[W]$, then part 1 of Corollary 5.4.12 implies at once that there can be no one-to-one linear transformation from V to W. For a specific example, no linear transformation $T : M_3(\mathbb{R}) \to \mathbb{R}^5$ can be one-to-one, since this would imply that $\dim[M_3(\mathbb{R})] \leq \dim[\mathbb{R}^5]$, a contradiction. Likewise, we could have used this type of reasoning to arrive at the conclusion that the linear transformation in Example 5.4.6 cannot be one-to-one.

From part 3 of Corollary 5.4.12, it follows that a necessary condition on $T : V \to W$ for T to be both one-to-one and onto is that $\dim[V] = \dim[W]$. The reader must be careful not to take this conclusion too far. In particular, if two vector spaces V and W *do* have the same dimension, this does not guarantee that any given linear transformation $T : V \to W$ is both one-to-one and onto. This point is well illustrated by Example 5.4.11. However, if we know in advance that two finite-dimensional vector spaces V and W have the same dimension, then we can draw the following conclusion regarding a linear transformation between them.

Proposition 5.4.13   Assume V and W are finite-dimensional vector spaces with $\dim[V] = \dim[W]$. If $T : V \to W$ is a linear transformation, then T is one-to-one if and only if T is onto.

The proof of Proposition 5.4.13 uses Theorem 5.3.8 as was done in the proof of Corollary 5.4.12 and is left as an exercise (Problem 31). The utility of Proposition 5.4.13 is that, for a linear transformation T between vector spaces V and W of the same dimension, if we show that T is one-to-one, it automatically follows that T is also onto. Alternatively, if we show that T is onto, then it automatically follows that T is one-to-one. In this way, only one of the two properties, one-to-one *or* onto, needs to be explicitly verified.

If $T : V \to W$ is both one-to-one and onto, then for each $\mathbf{w} \in W$, there is a *unique* $\mathbf{v} \in V$ such that $T(\mathbf{v}) = \mathbf{w}$. We can therefore define a mapping $T^{-1} : W \to V$ by

$$T^{-1}(\mathbf{w}) = \mathbf{v} \quad \text{if and only if} \quad \mathbf{w} = T(\mathbf{v}).$$

This mapping satisfies the basic properties of an inverse—namely,

$$T^{-1}(T(\mathbf{v})) = \mathbf{v} \quad \text{for all} \quad \mathbf{v} \in V$$

and

$$T(T^{-1}(\mathbf{w})) = \mathbf{w} \qquad \text{for all } \mathbf{w} \in W.$$

We call T^{-1} the **inverse** transformation to T. Again we stress that T^{-1} exists *if and only if* T is both one-to-one *and* onto, in which case we call T an **invertible** linear transformation. We leave it as an exercise to verify that T^{-1} is a *linear* transformation (Problem 32).

DEFINITION   5.4.14

Let $T : V \to W$ be a linear transformation. If T is both one-to-one and onto, then the linear transformation $T^{-1} : W \to V$ defined by

$$T^{-1}(\mathbf{w}) = \mathbf{v} \text{ if and only if } \mathbf{w} = T(\mathbf{v})$$

is called the **inverse transformation** to T.

For linear transformations $T : \mathbb{R}^n \to \mathbb{R}^n$, the following theorem characterizes the existence of an inverse transformation.

Theorem 5.4.15   Let $T : \mathbb{R}^n \to \mathbb{R}^n$ be a linear transformation with matrix A. Then T^{-1} exists if and only if $\det(A) \neq 0$. Furthermore, $T^{-1} : \mathbb{R}^n \to \mathbb{R}^n$ is a linear transformation with matrix A^{-1}.

Proof   We have

$$\text{Ker}(T) = \{\mathbf{x} \in \mathbb{R}^n : A\mathbf{x} = \mathbf{0}\}.$$

Hence, T is one-to-one if and only if the homogeneous linear system $A\mathbf{x} = \mathbf{0}$ has only the trivial solution. But this is true if and only if $\det(A) \neq 0$. Furthermore,

$$\text{Rng}(T) = \{A\mathbf{x} : \mathbf{x} \in \mathbb{R}^n\} = \text{colspace}(A).$$

Consequently,

$$\begin{aligned}
T \text{ is onto} &\iff \text{colspace}(A) = \mathbb{R}^n \\
&\iff \text{the column vectors of } A \text{ span } \mathbb{R}^n \\
&\iff \text{the columns of } A \text{ are linearly independent} \\
&\iff \det(A) \neq 0,
\end{aligned}$$

where we have used the Invertible Matrix Theorem (Theorem 4.10.1) for the last two equivalences. Hence, T is both one-to-one and onto if and only if $\det(A) \neq 0$. That is, T^{-1} exists if and only if $\det(A) \neq 0$.

Finally, if $\det(A) \neq 0$, then A^{-1} exists, so that

$$T(\mathbf{x}) = \mathbf{y} \iff A\mathbf{x} = \mathbf{y} \iff \mathbf{x} = A^{-1}\mathbf{y}.$$

Consequently, the inverse transformation is

$$T^{-1}(\mathbf{y}) = A^{-1}\mathbf{y},$$

from which it follows that T^{-1} is itself a *linear* transformation with matrix A^{-1}.   ∎

Example 5.4.16   If $T : \mathbb{R}^3 \to \mathbb{R}^3$ has matrix

$$A = \begin{bmatrix} 2 & 3 & 1 \\ -1 & 2 & 3 \\ 4 & 1 & 6 \end{bmatrix},$$

show that T^{-1} exists and find it.

Solution: It is easily shown that $\det(A) = 63 \neq 0$, so that A is invertible. Consequently, T^{-1} exists and is given by

$$T^{-1}(\mathbf{x}) = A^{-1}\mathbf{x}.$$

Using the Gauss-Jordan technique or the adjoint method, we find that

$$A^{-1} = \begin{bmatrix} \frac{1}{7} & -\frac{17}{63} & \frac{1}{9} \\ \frac{2}{7} & \frac{8}{63} & -\frac{1}{9} \\ -\frac{1}{7} & \frac{10}{63} & \frac{1}{9} \end{bmatrix}.$$

Hence,

$$T^{-1}(x_1, x_2, x_3) = (\tfrac{1}{7}x_1 - \tfrac{17}{63}x_2 + \tfrac{1}{9}x_3, \tfrac{2}{7}x_1 + \tfrac{8}{63}x_2 - \tfrac{1}{9}x_3, -\tfrac{1}{7}x_1 + \tfrac{10}{63}x_2 + \tfrac{1}{9}x_3). \quad \square$$

Isomorphism

Now let V be a (real) vector space of finite dimension n, and let $\{\mathbf{v}_1, \mathbf{v}_2, \ldots, \mathbf{v}_n\}$ be a basis for V. We define a mapping $T : \mathbb{R}^n \to V$ via

$$T(c_1, c_2, \ldots, c_n) = c_1\mathbf{v}_1 + c_2\mathbf{v}_2 + \cdots + c_n\mathbf{v}_n,$$

where $c_1, c_2, \ldots, c_n \in \mathbb{R}$. It is easy to check that T is a linear transformation. Moreover, since $\{\mathbf{v}_1, \mathbf{v}_2, \ldots, \mathbf{v}_n\}$ is linearly independent, it follows that $\text{Ker}(T) = \{\mathbf{0}\}$, so that T is one-to-one. Furthermore, $\text{Rng}(T) = \text{span}\{\mathbf{v}_1, \mathbf{v}_2, \ldots, \mathbf{v}_n\} = V$, so that T is also onto. Since T is both one-to-one and onto, *every* vector in V occurs as the image of exactly one vector in $\mathbb{R}^n$. This transformation has therefore matched up vectors in $\mathbb{R}^n$ with vectors in V in such a manner that linear combinations are preserved under the mapping. Such a transformation is called an *isomorphism* between $\mathbb{R}^n$ and V. We say that $\mathbb{R}^n$ and V are *isomorphic*.

DEFINITION   5.4.17

Let V and W be vector spaces.[4] If there exists a linear transformation $T : V \to W$ that is both one-to-one and onto, we call T an **isomorphism**, and we say that V and W are **isomorphic** vector spaces, written $V \cong W$.

Remarks

1. If V and W are vector spaces with $\dim[V] \neq \dim[W]$, then V and W cannot be isomorphic, by Corollary 5.4.12.

2. Our discussion above proves that *all* n-dimensional (real) vector spaces are isomorphic to $\mathbb{R}^n$. Hence, if V and W are vector spaces with $\dim[V] = \dim[W] = n$, then

$$V \cong \mathbb{R}^n \cong W.$$

---

[4] We assume that V and W are either both real vector spaces or both complex vector spaces.

Thus, all (real) n-dimensional vector spaces are isomorphic to one another. Hence, in studying properties of the vector space $\mathbb{R}^n$, we are really studying *all* (real) vector spaces of dimension n. This illustrates the importance of the vector space $\mathbb{R}^n$.

Example 5.4.18   Determine an isomorphism between $\mathbb{R}^3$ and the vector space P_2.

Solution: An arbitrary vector in P_2 can be expressed relative to the standard basis as

$$a_0 + a_1x + a_2x^2.$$

Consequently, an isomorphism between $\mathbb{R}^3$ and P_2 is $T: \mathbb{R}^3 \to P_2$, defined by

$$T(a_0, a_1, a_2) = a_0 + a_1x + a_2x^2.$$

It is straightforward to verify that T is one-to-one, onto, and a linear transformation. $\square$

Example 5.4.19   Determine an isomorphism between $\mathbb{R}^4$ and $M_2(\mathbb{R})$.

Solution: An arbitrary vector

$$A = \begin{bmatrix} a & b \\ c & d \end{bmatrix}$$

in $M_2(\mathbb{R})$ can be written relative to the standard basis as

$$A = a \begin{bmatrix} 1 & 0 \\ 0 & 0 \end{bmatrix} + b \begin{bmatrix} 0 & 1 \\ 0 & 0 \end{bmatrix} + c \begin{bmatrix} 0 & 0 \\ 1 & 0 \end{bmatrix} + d \begin{bmatrix} 0 & 0 \\ 0 & 1 \end{bmatrix}.$$

Hence, we can define an isomorphism $T: \mathbb{R}^4 \to M_2(\mathbb{R})$ by

$$T(a, b, c, d) = \begin{bmatrix} a & b \\ c & d \end{bmatrix}.$$ $\square$

Remark   In the preceding two examples, note that the given isomorphism T is not unique. For instance, in Example 5.4.19, we could also define an isomorphism $S: \mathbb{R}^4 \to M_2(\mathbb{R})$ via

$$S(a, b, c, d) = \begin{bmatrix} d & c \\ b & a \end{bmatrix}.$$

In general, isomorphisms are not unique.

Finally in this section, we can extend our list of criteria for an $n \times n$ matrix A to be invertible as follows.

Theorem 5.4.20   Let A be an $n \times n$ matrix with real elements, and let $T: \mathbb{R}^n \to \mathbb{R}^n$ be the matrix transformation defined by $T(\mathbf{x}) = A\mathbf{x}$. The following conditions are equivalent:

    (a) A is invertible.

    (q) T is one-to-one.

    (r) T is onto.

    (s) T is an isomorphism.

Proof By the Invertible Matrix Theorem, A is invertible if and only if nullspace$(A) = \{\mathbf{0}\}$. According to Equation (5.3.1), this is equivalent to the statement that Ker$(T) = \{\mathbf{0}\}$, and Theorem 5.4.7 shows that this is equivalent to the statement that T is one-to-one. Hence, (a) and (q) are equivalent. Now (q) and (r) are equivalent by Proposition 5.4.13, and (q) and (r) together are equivalent to (s) by the definition of an isomorphism. ∎

Exercises for 5.4

Key Terms

Composition of linear transformations, One-to-one and onto properties, Inverse transformation, Invertible linear transformation, Isomorphism, Isomorphic vector spaces.

Skills

- Be able to determine the composition of two (or more) given linear transformations.

- Be able to determine whether a given linear transformation $T : V \to W$ is one-to-one, onto, both, or neither.

- Be able to use the one-to-one and onto properties of a linear transformation $T : V \to W$ to draw conclusions about the relationship between dim$[V]$ and dim$[W]$.

- Conversely, given information about the dimensions of vector spaces V and W, decide quickly whether it is possible for a linear transformation $T : V \to W$ to be one-to-one and/or onto.

- Be able to determine whether a given linear transformation $T : V \to W$ has an inverse transformation, and if so, be able to find it.

- Be able to determine whether two vector spaces V and W are isomorphic, and if so, be able to construct an isomorphism $T : V \to W$.

True-False Review

For Questions 1–12, decide if the given statement is **true** or **false**, and give a brief justification for your answer. If true, you can quote a relevant definition or theorem from the text. If false, provide an example, illustration, or brief explanation of why the statement is false.

1. There is no one-to-one linear transformation $T : P_3 \to M_{32}$.

2. If V denotes the set of all 3×3 upper triangular matrices, then $V \cong M_{32}$.

3. If $T_1 : V_1 \to V_2$ and $T_2 : V_2 \to V_3$, then Ker(T_1) is a subspace of Ker$(T_2 T_1)$.

4. If $T : M_{22} \to P_2$ is a linear transformation and Ker(T) is one-dimensional, then T is onto but not one-to-one.

5. There is no onto linear transformation $T : M_2(\mathbb{R}) \to P_4$.

6. If $T_1 : V_1 \to V_2$ and $T_2 : V_2 \to V_3$ are both one-to-one, then so is the composition $T_2 T_1 : V_1 \to V_3$.

7. The linear transformation $T : C^1[a, b] \to C^0[a, b]$ given by $T(f) = f'$ is both one-to-one and onto.

8. If $T : P_3 \to M_{23}$ is a linear transformation and Rng(T) is four-dimensional, then T is one-to-one but not onto.

9. There is no isomorphism $T : \mathbb{R}^n \to \mathbb{R}^m$ unless $m = n$.

10. Every real vector space is isomorphic to $\mathbb{R}^n$ for some positive integer n.

11. If $T_1 : V_1 \to V_2$ and $T_2 : V_2 \to V_3$ are linear transformations such that T_2 is onto, then $T_2 T_1$ is onto.

12. If $T : \mathbb{R}^8 \to \mathbb{R}^3$ is an onto linear transformation, then Ker(T) is five-dimensional.

Problems

1. Let $T_1 : \mathbb{R}^2 \to \mathbb{R}^2$ and $T_2 : \mathbb{R}^2 \to \mathbb{R}^2$ be the linear transformations with matrices

$$A = \begin{bmatrix} -1 & 2 \\ 3 & 1 \end{bmatrix}, \qquad B = \begin{bmatrix} 1 & 5 \\ -2 & 0 \end{bmatrix},$$

respectively. Find $T_1 T_2$ and $T_2 T_1$. Does $T_1 T_2 = T_2 T_1$?

2. Let $T_1 \colon \mathbb{R}^2 \to \mathbb{R}^2$ and $T_2 \colon \mathbb{R}^2 \to \mathbb{R}$ be the linear transformations with matrices

$$A = \begin{bmatrix} 1 & -1 \\ 3 & 2 \end{bmatrix}, \qquad B = [-1 \ \ 1],$$

respectively. Find $T_2 T_1$. Does $T_1 T_2$ exist? Explain.

3. Let $T_1 \colon \mathbb{R}^2 \to \mathbb{R}^2$ and $T_2 \colon \mathbb{R}^2 \to \mathbb{R}^2$ be the linear transformations with matrices

$$A = \begin{bmatrix} 1 & -1 \\ 2 & -2 \end{bmatrix}, B = \begin{bmatrix} 2 & 1 \\ 3 & -1 \end{bmatrix},$$

respectively. Find $\text{Ker}(T_1)$, $\text{Ker}(T_2)$, $\text{Ker}(T_1 T_2)$, and $\text{Ker}(T_2 T_1)$.

4. Let $T_1 \colon M_n(\mathbb{R}) \to M_n(\mathbb{R})$ and $T_2 \colon M_n(\mathbb{R}) \to M_n(\mathbb{R})$ be the linear transformations defined by $T_1(A) = A - A^T$ and $T_2(A) = A + A^T$. Show that $T_2 T_1$ is the zero transformation.

5. Define $T_1 \colon C^1[a, b] \to C^0[a, b]$ and $T_2 \colon C^0[a, b] \to C^1[a, b]$ by

$$T_1(f) = f',$$

$$[T_2(f)](x) = \int_a^x f(t)\, dt, \qquad a \le x \le b.$$

(a) If $f(x) = \sin(x - a)$, find

$$[T_1(f)](x) \qquad \text{and} \qquad [T_2(f)](x),$$

and show that, for the given function,

$$[T_1 T_2](f) = [T_2 T_1](f) = f.$$

(b) Show that for general functions f and g,

$$[T_1 T_2](f) = f, \quad \{[T_2 T_1](g)\}(x) = g(x) - g(a).$$

6. Let $\{v_1, v_2\}$ be a basis for the vector space V, and suppose that $T_1 \colon V \to V$ and $T_2 \colon V \to V$ are the linear transformations satisfying

$$T_1(v_1) = v_1 - v_2, \qquad T_1(v_2) = 2v_1 + v_2$$
$$T_2(v_1) = v_1 + 2v_2, \qquad T_2(v_2) = 3v_1 - v_2.$$

Determine $(T_2 T_1)(v)$ for an arbitrary vector in V.

7. Repeat Problem 6 under the assumption

$$T_1(v_1) = 3v_1 + v_2, \qquad T_1(v_2) = 0$$
$$T_2(v_1) = -5v_2, \qquad T_2(v_2) = -v_1 + 6v_2.$$

For Problems 8–11, find $\text{Ker}(T)$ and $\text{Rng}(T)$, and hence, determine whether the given transformation is one-to-one, onto, both, or neither. If T^{-1} exists, find it.

8. $T(\mathbf{x}) = A\mathbf{x}$, where $A = \begin{bmatrix} 4 & 2 \\ 1 & 3 \end{bmatrix}$.

9. $T(\mathbf{x}) = A\mathbf{x}$, where $A = \begin{bmatrix} 1 & 2 \\ -2 & -4 \end{bmatrix}$.

10. $T(\mathbf{x}) = A\mathbf{x}$, where $A = \begin{bmatrix} 1 & 2 & -1 \\ 2 & 5 & 1 \end{bmatrix}$.

11. $T(\mathbf{x}) = A\mathbf{x}$, where $A = \begin{bmatrix} 0 & 1 & 2 \\ 3 & 4 & 5 \\ 5 & 4 & 3 \\ 2 & 1 & 0 \end{bmatrix}$.

12. Suppose $T \colon \mathbb{R}^3 \to \mathbb{R}^2$ is a linear transformation such that $T(1, 0, 0) = (4, 5)$, $T(0, 1, 0) = (-1, 1)$, and $T(2, 1, -3) = (7, -1)$.

(a) Find the matrix of T.

(b) Is T one-to-one, onto, both, or neither? Explain briefly.

13. Let V be a vector space and define $T \colon V \to V$ by $T(\mathbf{x}) = \lambda \mathbf{x}$, where λ is a nonzero scalar. Show that T is a linear transformation that is one-to-one and onto, and find T^{-1}.

14. Define $T \colon P_1 \to P_1$ by

$$T(ax + b) = (2b - a)x + (b + a).$$

Show that T is both one-to-one and onto, and find T^{-1}.

15. Define $T \colon P_2 \to P_1$ by

$$T(ax^2 + bx + c) = (a - b)x + c.$$

Determine whether T is one-to-one, onto, both, or neither. Find T^{-1} or explain why it does not exist.

16. Let $\{v_1, v_2\}$ be a basis for the vector space V, and suppose that $T \colon V \to V$ is a linear transformation. If $T(v_1) = v_1 + 2v_2$ and $T(v_2) = 2v_1 - 3v_2$, determine whether T is one-to-one, onto, both, or neither. Find T^{-1} or explain why it does not exist.

17. Let v_1 and v_2 be a basis for the vector space V, and suppose that $T_1 \colon V \to V$ and $T_2 \colon V \to V$ are the linear transformations satisfying

$$T_1(v_1) = v_1 + v_2, \qquad T_1(v_2) = v_1 - v_2,$$
$$T_2(v_1) = \tfrac{1}{2}(v_1 + v_2), \qquad T_2(v_2) = \tfrac{1}{2}(v_1 - v_2).$$

Find $(T_1 T_2)(v)$ and $(T_2 T_1)(v)$ for an arbitrary vector in V and show that $T_2 = T_1^{-1}$.

18. Determine an isomorphism between $\mathbb{R}^2$ and the vector space P_1.

19. Determine an isomorphism between $\mathbb{R}^3$ and the subspace of $M_2(\mathbb{R})$ consisting of all upper triangular matrices.

20. Determine an isomorphism between $\mathbb{R}$ and the subspace of $M_2(\mathbb{R})$ consisting of all skew-symmetric matrices.

21. Determine an isomorphism between $\mathbb{R}^3$ and the subspace of $M_2(\mathbb{R})$ consisting of all symmetric matrices.

22. Let V denote the vector space of all 4×4 upper triangular matrices. Find n such that $V \cong \mathbb{R}^n$, and construct an isomorphism.

23. Let V denote the subspace of P_8 consisting of all polynomials whose odd coefficients are zero. Find n such that $V \cong \mathbb{R}^n$, and construct an isomorphism.

For Problems 24–27, an invertible linear transformation $\mathbb{R}^n \to \mathbb{R}^n$ is given. Find a formula for the inverse linear transformation.

24. $T_1: \mathbb{R}^2 \to \mathbb{R}^2$ defined by $T_1(\mathbf{x}) = A\mathbf{x}$, where

$$A = \begin{bmatrix} 1 & 1 \\ 2 & 3 \end{bmatrix}.$$

25. $T_2: \mathbb{R}^2 \to \mathbb{R}^2$ defined by $T_2(\mathbf{x}) = A\mathbf{x}$, where

$$A = \begin{bmatrix} -4 & -1 \\ 2 & 2 \end{bmatrix}.$$

26. $T_3: \mathbb{R}^3 \to \mathbb{R}^3$ defined by $T_3(\mathbf{x}) = A\mathbf{x}$, where

$$A = \begin{bmatrix} 1 & 1 & 3 \\ 0 & 1 & 2 \\ 3 & 5 & -1 \end{bmatrix}.$$

27. $T_4: \mathbb{R}^3 \to \mathbb{R}^3$ defined by $T_4(\mathbf{x}) = A\mathbf{x}$, where

$$A = \begin{bmatrix} 3 & 5 & 1 \\ 1 & 2 & 1 \\ 2 & 6 & 7 \end{bmatrix}.$$

28. Referring to Problems 24–25, compute the matrix of $T_2 T_1$ and the matrix of $T_1 T_2$.

29. Referring to Problems 26–27, compute the matrix of $T_4 T_3$ and the matrix of $T_3 T_4$.

30. Complete the proof of Theorem 5.4.2 by verifying Equation (5.4.2).

31. Prove Proposition 5.4.13.

32. If $T: V \to W$ is an invertible linear transformation (that is, T^{-1} exists), show that $T^{-1}: W \to V$ is also a linear transformation.

33. Prove that if $T: V \to V$ is a one-to-one linear transformation, and V is finite-dimensional, then T^{-1} exists.

34. Prove that if $T: V \to W$ is a one-to-one linear transformation and $\{\mathbf{v}_1, \mathbf{v}_2, \ldots, \mathbf{v}_k\}$ is a linearly independent set of vectors in V, then

$$\{T(\mathbf{v}_1), T(\mathbf{v}_2), \ldots, T(\mathbf{v}_k)\}$$

is a linearly independent set of vectors in W.

35. Suppose $T: V \to W$ is a linear transformation and $\{\mathbf{w}_1, \mathbf{w}_2, \ldots, \mathbf{w}_m\}$ spans W. If there exist vectors $\mathbf{v}_1, \mathbf{v}_2, \ldots, \mathbf{v}_m$ in V such that $T(\mathbf{v}_i) = \mathbf{w}_i$ for each $i = 1, 2, \ldots, m$, prove that T is onto.

36. Prove that if $T: V \to W$ is a linear transformation with $\dim[W] = n = \dim[\mathrm{Rng}(T)]$, then T is onto.

37. Let $T_1: V \to V$ and $T_2: V \to V$ be linear transformations and suppose that T_2 is one-to-one. If

$$(T_1 T_2)(\mathbf{v}) = \mathbf{v} \text{ for all } \mathbf{v} \text{ in } V,$$

prove that

$$(T_2 T_1)(\mathbf{v}) = \mathbf{v} \text{ for all } \mathbf{v} \text{ in } V.$$

38. Prove that if $T: V \to V$ is a linear transformation such that $T^2 = 0$ (that is, $T(T(\mathbf{v})) = \mathbf{0}$ for all $\mathbf{v} \in V$), then $\mathrm{Rng}(T)$ is a subspace of $\mathrm{Ker}(T)$.

39. Let $T_1: V_1 \to V_2$ and $T_2: V_2 \to V_3$ be linear transformations.

    (a) Prove that if T_1 and T_2 are both one-to-one, then so is $T_2 T_1: V_1 \to V_3$.

    (b) Prove that if T_1 and T_2 are both onto, then so is $T_2 T_1: V_1 \to V_3$.

    (c) Prove that if T_1 and T_2 are both isomorphisms, then so is $T_2 T_1: V_1 \to V_3$.

*5.5　The Matrix of a Linear Transformation

In Section 5.1 we associated an $m \times n$ matrix with any linear transformation $T : \mathbb{R}^n \to \mathbb{R}^m$ (see Definition 5.1.14). In an effort to generalize this, let V and W be vector spaces with $\dim[V] = n$ and $\dim[W] = m$. If we fix ordered bases B and C on V and W, respectively, we will see that we can uniquely associate an $m \times n$ matrix to any linear transformation $T : V \to W$. All of the essential information about the linear transformation $T : V \to W$ can be found in the associated matrix, and therefore all of the ideas we have been developing for linear transformations between finite-dimensional vector spaces can be expressed entirely in the language of matrices. Before we begin, recall from Section 4.7 that the component vector of a vector $\mathbf{v}$ relative to an ordered basis B is denoted $[\mathbf{v}]_B$. We proceed as follows.

DEFINITION　5.5.1

Let V and W be vector spaces with ordered bases $B = \{\mathbf{v}_1, \mathbf{v}_2, \ldots, \mathbf{v}_n\}$ and $C = \{\mathbf{w}_1, \mathbf{w}_2, \ldots, \mathbf{w}_m\}$, respectively, and let $T : V \to W$ be a linear transformation. The $m \times n$ matrix

$$[T]_B^C = \left[[T(\mathbf{v}_1)]_C, [T(\mathbf{v}_2)]_C, \ldots, [T(\mathbf{v}_n)]_C \right]$$

is called the **matrix representation of T relative to the bases B and C.** In case $V = W$ and $B = C$, we refer to $[T]_B^B$ simply as the **matrix representation of T relative to the basis B.**

Remarks

1. If $V = \mathbb{R}^n$ and $W = \mathbb{R}^m$ are each equipped with the standard bases, then $[T]_B^C$ is the same as the matrix of T introduced in Definition 5.1.14.

2. If $V = W$ and $T(\mathbf{v}) = \mathbf{v}$ for all $\mathbf{v}$ in V (i.e., T is the identity transformation), then $[T]_B^C$ is just the change-of-basis matrix from B to C described in Section 4.7.

Example 5.5.2　Recall the linear transformation $T : P_1 \to P_2$ defined by

$$T(a + bx) = (2a - 3b) + (b - 5a)x + (a + b)x^2$$

in Example 5.3.7. Determine the matrix representation of T relative to the given ordered bases B and C.

(a) $B = \{1, x\}$ and $C = \{1, x, x^2\}$.

(b) $B = \{1, x + 5\}$ and $C = \{1, 1 + x, 1 + x^2\}$.

Solution:

(a) We have $T(1) = 2 - 5x + x^2$ and $T(x) = -3 + x + x^2$, so

$$[T(1)]_C = \begin{bmatrix} 2 \\ -5 \\ 1 \end{bmatrix} \qquad \text{and} \qquad [T(x)]_C = \begin{bmatrix} -3 \\ 1 \\ 1 \end{bmatrix}.$$

---

*This section can be omitted without loss of continunity

Thus,

$$[T]_B^C = \begin{bmatrix} 2 & -3 \\ -5 & 1 \\ 1 & 1 \end{bmatrix}.$$

(b) We have $T(1) = 2 - 5x + x^2$ and $T(x + 5) = 7 - 24x + 6x^2$. Writing

$$2 - 5x + x^2 = a_1(1) + a_2(1 + x) + a_3(1 + x^2) = (a_1 + a_2 + a_3) + a_2 x + a_3 x^2,$$

we find that $a_3 = 1$, $a_2 = -5$, and $a_1 = 6$. Thus, we have

$$[T(1)]_C = \begin{bmatrix} 6 \\ -5 \\ 1 \end{bmatrix}.$$

Next, we write

$$7 - 24x + 6x^2 = (b_1 + b_2 + b_3) + b_2 x + b_3 x^2,$$

from which it follows that $b_3 = 6$, $b_2 = -24$, and $b_1 = 25$. Thus, we have

$$[T(x + 5)]_C = \begin{bmatrix} 25 \\ -24 \\ 6 \end{bmatrix}.$$

Thus,

$$[T]_B^C = \begin{bmatrix} 6 & 25 \\ -5 & -24 \\ 1 & 6 \end{bmatrix}. \qquad \square$$

Example 5.5.3   Let $T : M_2(\mathbb{R}) \rightarrow \mathbb{R}^2$ be defined by

$$T \begin{bmatrix} a & b \\ c & d \end{bmatrix} = (a + d, -b + 3c + d).$$

Find $[T]_B^C$, where

$$B = \left\{ \begin{bmatrix} 1 & 0 \\ -1 & 1 \end{bmatrix}, \begin{bmatrix} 0 & 0 \\ 0 & 2 \end{bmatrix}, \begin{bmatrix} 1 & 1 \\ 0 & 0 \end{bmatrix}, \begin{bmatrix} 0 & 1 \\ 0 & 0 \end{bmatrix} \right\} \qquad \text{and} \qquad C = \{(1, -1), (1, 1)\}.$$

Solution: We have

$$\left[T \begin{bmatrix} 1 & 0 \\ -1 & 1 \end{bmatrix} \right]_C = [(2, -2)]_C = \begin{bmatrix} 2 \\ 0 \end{bmatrix},$$

$$\left[T \begin{bmatrix} 0 & 0 \\ 0 & 2 \end{bmatrix} \right]_C = [(2, 2)]_C = \begin{bmatrix} 0 \\ 2 \end{bmatrix},$$

$$\left[T \begin{bmatrix} 1 & 1 \\ 0 & 0 \end{bmatrix} \right]_C = [(1, -1)]_C = \begin{bmatrix} 1 \\ 0 \end{bmatrix},$$

$$\left[T \begin{bmatrix} 0 & 1 \\ 0 & 0 \end{bmatrix} \right]_C = [(0, -1)]_C = \begin{bmatrix} 0.5 \\ -0.5 \end{bmatrix}.$$

Thus,

$$[T]_B^C = \begin{bmatrix} 2 & 0 & 1 & 0.5 \\ 0 & 2 & 0 & -0.5 \end{bmatrix}. \qquad \square$$

Given the matrix $[T]_B^C$ representing $T: V \to W$ relative to the bases B and C, we can completely recover the formula for $T(\mathbf{v})$, for any vector $\mathbf{v}$ in V. The next theorem gives us a way of doing this.

Theorem 5.5.4 Let V and W be vector spaces with ordered bases B and C, respectively. If $T: V \to W$ is a linear transformation and $\mathbf{v}$ is any vector in V, then we have

$$[T(\mathbf{v})]_C = [T]_B^C[\mathbf{v}]_B. \tag{5.5.1}$$

Proof Let $B = \{\mathbf{v}_1, \mathbf{v}_2, \dots, \mathbf{v}_n\}$, and consider a vector $\mathbf{v}$ in V. Writing $\mathbf{v} = a_1\mathbf{v}_1 + a_2\mathbf{v}_2 + \cdots + a_n\mathbf{v}_n$, we have

$$[T]_B^C[\mathbf{v}]_B = a_1[T(\mathbf{v}_1)]_C + a_2[T(\mathbf{v}_2)]_C + \cdots + a_n[T(\mathbf{v}_n)]_C$$
$$= [T(a_1\mathbf{v}_1 + a_2\mathbf{v}_2 + \cdots + a_n\mathbf{v}_n)]_C = [T(\mathbf{v})]_C,$$

where we have used the linearity of the components of a vector (see Lemma 4.7.5). ∎

Example 5.5.5 Let us verify that the matrix $[T]_B^C$ found in Example 5.5.3 above does indeed contain all of the information needed to compute $T(\mathbf{v})$ for any vector $\mathbf{v}$ in V. Our first step is to find the components of a general vector

$$\mathbf{v} = \begin{bmatrix} a & b \\ c & d \end{bmatrix}$$

relative to the basis B above. Writing

$$\begin{bmatrix} a & b \\ c & d \end{bmatrix} = k_1\begin{bmatrix} 1 & 0 \\ -1 & 1 \end{bmatrix} + k_2\begin{bmatrix} 0 & 0 \\ 0 & 2 \end{bmatrix} + k_3\begin{bmatrix} 1 & 1 \\ 0 & 0 \end{bmatrix} + k_4\begin{bmatrix} 0 & 1 \\ 0 & 0 \end{bmatrix},$$

we can solve for k_1, k_2, k_3, and k_4:

$$k_1 = -c, \qquad k_2 = \frac{c+d}{2}, \qquad k_3 = a+c, \qquad k_4 = b-a-c.$$

Thus,

$$[\mathbf{v}]_B = \begin{bmatrix} -c \\ (c+d)/2 \\ a+c \\ b-a-c \end{bmatrix}.$$

Applying Theorem 5.5.4, we obtain the components of $T(\mathbf{v})$ relative to C:

$$[T(\mathbf{v})]_C = \begin{bmatrix} 2 & 0 & 1 & 0.5 \\ 0 & 2 & 0 & -0.5 \end{bmatrix}\begin{bmatrix} -c \\ (c+d)/2 \\ a+c \\ b-a-c \end{bmatrix} = \begin{bmatrix} \frac{1}{2}a + \frac{1}{2}b - \frac{3}{2}c \\ \frac{1}{2}a - \frac{1}{2}b + \frac{3}{2}c + d \end{bmatrix}.$$

Hence, we have

$$T(\mathbf{v}) = \left(\tfrac{1}{2}a + \tfrac{1}{2}b - \tfrac{3}{2}c\right)(1, -1) + \left(\tfrac{1}{2}a - \tfrac{1}{2}b + \tfrac{3}{2}c + d\right)(1, 1) = (a+d, -b+3c+d),$$

so that we have recovered the formula for the linear transformation T.   □

We next consider the special case of a linear transformation from the n-dimensional vector space V to itself, $T : V \rightarrow V$. If we let B and C denote two ordered bases on V, we pose the following natural question: Does there exist any relationship between the matrices $[T]_B^B$ and $[T]_C^C$? This is really a question of change-of-basis for linear transformations, and our next theorem provides the answer. Recall from Section 4.7 that $P_{C \leftarrow B}$ denotes the change-of-basis matrix from B to C.

Theorem 5.5.6   Let V be a vector space with ordered bases B and C. If $T : V \rightarrow V$ is a linear transformation, then we have

$$[T]_C^C = P_{C \leftarrow B}[T]_B^B P_{B \leftarrow C}.$$

Proof Let $\mathbf{v}$ be an arbitrary vector in V. On the one hand, we have

$$[T]_C^C[\mathbf{v}]_C = [T(\mathbf{v})]_C$$

by Theorem 5.5.4. On the other hand,

$$P_{C \leftarrow B}[T]_B^B P_{B \leftarrow C}[\mathbf{v}]_C = P_{C \leftarrow B}[T]_B^B[\mathbf{v}]_B = P_{C \leftarrow B}[T(\mathbf{v})]_B = [T(\mathbf{v})]_C.$$

Thus, the two matrices $[T]_C^C$ and $P_{C \leftarrow B}[T]_B^B P_{B \leftarrow C}$ have the same effect on the component vector $[\mathbf{v}]_C$. Since this holds for every $\mathbf{v}$ in V, the matrices must be identical. ■

We have seen that the matrix representation of a linear transformation between vector spaces with fixed ordered bases fosters a natural relationship between linear transformations and matrices. The added perspective we have gained in this section now enables us to better understand both of these types of objects. This is one of many instances in mathematics in which one's knowledge and understanding of two or more topics can be considerably enhanced through an understanding of how those topics are connected to one another. This remarkably important realization about the nature of mathematics cannot be too strongly emphasized. Therefore, for the remainder of this section we look for additional insight about linear transformations that is available through the use of matrix representations.

Linear Transformation Concepts in Terms of Matrix Representations

In the preceding sections of this chapter, we have explored several key concepts pertaining to linear transformations, including composition of linear transformations, kernels and ranges of linear transformations, and one-to-one, onto, and invertible linear transformations. Let us see what additional light can be brought to each of these concepts through the machinery of matrix representations. We begin with composition of linear transformations.

Composition of Linear Transformations: Let U, V, and W be vector spaces with ordered bases A, B, and C. Let $T_1 : U \rightarrow V$ and $T_2 : V \rightarrow W$ be linear transformations.

The next theorem determines the matrix representation of $T_2T_1 : U \to W$ relative to the bases A and C in terms of the matrix representations of T_1 and T_2 (with respect to the appropriate bases) separately. The theorem is a generalization of Theorem 4.7.9.

Theorem 5.5.7   If U, V, and W are vector spaces with ordered bases A, B, and C, respectively, and $T_1 : U \to V$ and $T_2 : V \to W$ are linear transformations, then we have

$$[T_2T_1]_A^C = [T_2]_B^C[T_1]_A^B. \tag{5.5.2}$$

Proof It suffices to show that when we premultiply any column vector of the form $[\mathbf{u}]_A$, where $\mathbf{u}$ is a vector in U, by the matrices appearing on either side of (5.5.2), we obtain the same result. Using Equation (5.5.1), we have

$$[T_2]_B^C[T_1]_A^B[\mathbf{u}]_A = [T_2]_B^C[T_1(\mathbf{u})]_B = [T_2(T_1(\mathbf{u}))]_C = [(T_2T_1)\mathbf{u}]_C = [T_2T_1]_A^C[\mathbf{u}]_A,$$

as required. ∎

Theorem 5.5.7 is the extremely powerful statement that multiplication of matrix representations corresponds to composition of the associated linear transformations. At first glance, the procedure for multiplying matrices that was presented in Section 2.2 may have seemed somewhat unmotivated, but now we see that it is precisely that matrix multiplication scheme that allows Theorem 5.5.7 to work. Therefore, we now have clear evidence and motivation for why we multiply matrices the way we do.

Example 5.5.8   Define

$$T_1 : P_2 \to M_2(\mathbb{R}) \qquad \text{via} \qquad T_1(a + bx + cx^2) = \begin{bmatrix} a - 2c & a + 3b \\ a - b - c & -4b \end{bmatrix}$$

and

$$T_2 : M_2(\mathbb{R}) \to \mathbb{R}^2 \qquad \text{via} \qquad T_2\left(\begin{bmatrix} a & b \\ c & d \end{bmatrix}\right) = (-2a, a + b + c + d).$$

Use the matrix representation of T_2T_1 relative to the standard bases on P_2 and $\mathbb{R}^2$ to find the formula for the composition $T_2T_1 : P_2 \to \mathbb{R}^2$.

Solution: Let A, B, and C be the standard bases on the vector spaces P_2, $M_2(\mathbb{R})$, and $\mathbb{R}^2$, respectively. We have

$$[T_2T_1]_A^C = [T_2]_B^C[T_1]_A^B = \begin{bmatrix} -2 & 0 & 0 & 0 \\ 1 & 1 & 1 & 1 \end{bmatrix} \begin{bmatrix} 1 & 0 & -2 \\ 1 & 3 & 0 \\ 1 & -1 & -1 \\ 0 & -4 & 0 \end{bmatrix} = \begin{bmatrix} -2 & 0 & 4 \\ 3 & -2 & -3 \end{bmatrix}.$$

From this matrix, we conclude that

$$(T_2T_1)(a + bx + cx^2) = (-2a + 4c, 3a - 2b - 3c).$$

This can also be verified directly by composing the two linear transformations. □

Kernels and Ranges of Linear Transformations: Because of the close relationship between the matrix representation of a linear transformation and the linear transformation itself, the following theorem is to be expected.

Theorem 5.5.9   Let $T : V \to W$ be a linear transformation, and let B and C be ordered bases for V and W, respectively. Then

(a) For all $\mathbf{v}$ in V, $\mathbf{v}$ belongs to $\text{Ker}(T)$ if and only if $[\mathbf{v}]_B$ belongs to nullspace $([T]_B^C)$.

(b) For all $\mathbf{w}$ in W, $\mathbf{w}$ belongs to $\text{Rng}(T)$ if and only if $[\mathbf{w}]_C$ belongs to colspace$([T]_B^C)$.

Proof We prove part (a) and leave (b) as an exercise (Problem 21). Given $\mathbf{v}$ in V, we have $T(\mathbf{v}) = \mathbf{0}$ if and only if $[T(\mathbf{v})]_C$ is the zero vector in $\mathbb{R}^n$, where $n = \dim[W]$. But since $[T(\mathbf{v})]_C = [T]_B^C[\mathbf{v}]_B$ by Equation (5.5.1), this is equivalent to saying that $[\mathbf{v}]_B$ is in nullspace$([T]_B^C)$, as required. ∎

One-to-one, Onto, and Invertible Linear Transformations: We saw in the previous section that a linear transformation $T : V \to W$ is one-to-one if and only if $\text{Ker}(T) = \{\mathbf{0}\}$, and T is onto if and only if $\text{Rng}(T) = W$. Combining this with Theorem 5.5.9, we have the following.

Corollary 5.5.10

Let $T : V \to W$ be a linear transformation, and let B and C be ordered bases for V and W, respectively. Then

(a) T is one-to-one if and only if nullspace$([T]_B^C) = \{\mathbf{0}\}$.

(b) T is onto if and only if colspace$([T]_B^C) = \mathbb{R}^n$, where $n = \dim[W]$.

Finally, a linear transformation $T : V \to V$ is invertible if and only if T is both one-to-one and onto. If $\dim[V] = n$, then the matrix $[T]_B^C$ is an $n \times n$ matrix, and from Corollary 5.5.10 we see that

> T is invertible if and only if $[T]_B^C$ is an invertible matrix for all ordered bases B and C of V if and only if $[T]_B^C$ is an invertible matrix for some ordered bases B and C of V.

This is the natural generalization of Theorem 5.4.15. It says that, in order to check that T is invertible, we need only show that $[T]_B^C$ is an invertible matrix for *one choice* of ordered bases B and C. However, if we know already that T is invertible, then the matrix $[T]_B^C$ will be invertible for *all* choices of ordered bases B and C.

Given an invertible linear transformation T with matrix representation $[T]_B^C$, we can now use matrices to determine the inverse linear transformation T^{-1}. According to Theorem 5.5.7, we have

$$[T^{-1}]_C^B[T]_B^C = [I]_B^B, \tag{5.5.3}$$

and the latter matrix is simply the identity matrix. Likewise,

$$[T]_B^C[T^{-1}]_C^B = [I]_C^C \tag{5.5.4}$$

is the identity matrix. Therefore, (5.5.3) and (5.5.4) imply that

$$\boxed{\left([T]_B^C\right)^{-1} = [T^{-1}]_C^B.}$$

Therefore, we can use the matrix representation $[T]_B^C$ of T to determine the matrix representation $[T^{-1}]_C^B$ of T^{-1}, thus enabling us to determine T^{-1} directly from matrix algebra. We illustrate with an example.

Example 5.5.11   Let $T : P_2 \to P_2$ be defined via

$$T(a + bx + cx^2) = (3a - b + c) + (a - c)x + (4b + c)x^2.$$

(a) Find the matrix representation of T relative to the standard basis $B = \{1, x, x^2\}$ on P_2.

(b) Use the matrix in part (a) to prove that T is invertible.

(c) Determine the linear transformation $T^{-1} : P_2 \to P_2$ by using the matrix representation of T^{-1} relative to $B = \{1, x, x^2\}$.

Solution:

(a) We have $T(1) = 3 + x$, $T(x) = -1 + 4x^2$, and $T(x^2) = 1 - x + x^2$. Finding the components of these polynomials relative to the standard basis on P_2 enables us to quickly write down

$$[T]_B^B = \begin{bmatrix} 3 & -1 & 1 \\ 1 & 0 & -1 \\ 0 & 4 & 1 \end{bmatrix}.$$

(b) We can verify easily that $[T]_B^B$ is invertible (for example, its determinant is nonzero), and hence T is invertible.

(c) We have that

$$[T^{-1}]_B^B = \left([T]_B^B\right)^{-1} = \begin{bmatrix} 3 & -1 & 1 \\ 1 & 0 & -1 \\ 0 & 4 & 1 \end{bmatrix}^{-1} = \tfrac{1}{17} \begin{bmatrix} 4 & 5 & 1 \\ -1 & 3 & 4 \\ 4 & -12 & 1 \end{bmatrix}.$$

Thus,

$$T^{-1}(a + bx + cx^2) = \tfrac{1}{17}[(4a + 5b + c) + (-a + 3b + 4c)x \\ + (4a - 12b + c)x^2].$$ □

Exercises for 5.5

Key Terms

Matrix representation of T relative to bases B and C.

Skills

- Be able to determine the matrix representation of a linear transformation $T : V \to W$ relative to bases B and C for vector spaces V and W, respectively.

- Be able to use the matrix representation of a linear transformation $T : V \to W$ and components of vec-

tors v in V to determine the action of T on v—that is, to determine $T(v)$.

- Be aware of the rich relationship between composition of linear transformations and multiplication of corresponding matrix representations.

- Be able to rephrase the concepts of kernel and range of a linear transformation in terms of matrix representations, and likewise for one-to-one, onto, and invertible linear transformations.

True-False Review

For Questions 1–6, decide if the given statement is **true** or **false**, and give a brief justification for your answer. If true, you can quote a relevant definition or theorem from the text. If false, provide an example, illustration, or brief explanation of why the statement is false.

1. If $\dim[V] = n$ and $\dim[W] = m$, then any matrix representation $[T]_B^C$ of $T: V \to W$ relative to ordered bases B and C (for V and W, respectively) is an $n \times m$ matrix.

2. If $T: V \to W$ is any linear transformation, and $[T]_B^C$ is a matrix representation of T, then we have another matrix representation for T given by $[T]_C^B$.

3. If $U, V,$ and W are vector spaces with ordered bases A, B, and C, respectively, and $T_1: U \to V$ and $T_2: V \to W$ are linear transformations, then $[T_2 T_1]_A^C = [T_2]_C^B [T_1]_B^A$.

4. If $T: V \to V$ is an invertible linear transformation, and B and C are ordered bases for V, then $([T]_B^C)^{-1} = [T^{-1}]_B^C$.

5. Two different linear transformations $T_1: V_1 \to W_1$ and $T_2: V_2 \to W_2$ (with bases B_1, C_1, B_2, C_2 for V_1, W_1, V_2, W_2, respectively) can have the same matrix representations: $[T]_{B_1}^{C_1} = [T]_{B_2}^{C_2}$.

6. If $T: V \to W$ is an onto linear transformation (with $\dim[V] = n$ and $\dim[W] = m$), then $\text{colspace}([T]_B^C) = \mathbb{R}^m$ for any choice of ordered bases B and C for V and W, respectively.

Problems

For Problems 1–8, determine the matrix representation $[T]_B^C$ for the given linear transformation T and ordered bases B and C.

1. $T: P_2 \to \mathbb{R}^2$ given by
$$T(a + bx + cx^2) = (a - 3c, 2a + b - 2c),$$
   (a) $B = \{1, x, x^2\}; C = \{(1, 0), (0, 1)\}$.
   (b) $B = \{1, 1+x, 1+x+x^2\}; C = \{(1, -1), (2, 1)\}$.

2. $T: M_2(\mathbb{R}) \to P_3$ given by
$$T\left(\begin{bmatrix} a & b \\ c & d \end{bmatrix}\right) = (a - d) + 3bx^2 + (c - a)x^3,$$
   (a) $B = \{E_{11}, E_{12}, E_{21}, E_{22}\}; C = \{1, x, x^2, x^3\}$.
   (b) $B = \{E_{21}, E_{11}, E_{22}, E_{12}\}; C = \{x, 1, x^3, x^2\}$.

3. $T: \mathbb{R}^3 \to \text{span}\{\cos x, \sin x\}$ given by
$$T(a, b, c) = (a - 2c) \cos x + (3b + c) \sin x,$$
   (a) $B = \{(1, 0, 0), (0, 1, 0), (0, 0, 1)\}$;
   $C = \{\cos x, \sin x\}$.
   (b) $B = \{(2, -1, -1), (1, 3, 5), (0, 4, -1)\}$;
   $C = \{\cos x - \sin x, \cos x + \sin x\}$.

4. $T: P_2 \to P_3$ given by $T(p(x)) = (x + 1)p(x)$,
   (a) $B = \{1, x, x^2\}; C = \{1, x, x^2, x^3\}$.
   (b) $B = \{1, x - 1, (x - 1)^2\}$;
   $C = \{1, x - 1, (x - 1)^2, (x - 1)^3\}$.

5. $T: P_3 \to P_2$ given by $T(p(x)) = p'(x)$,
   (a) $B = \{1, x, x^2, x^3\}; C = \{1, x, x^2\}$.
   (b) $B = \{x^3, x^3 + 1, x^3 + x, x^3 + x^2\}$;
   $C = \{1, 1 + x, 1 + x + x^2\}$.

6. $T: M_2(\mathbb{R}) \to \mathbb{R}^2$ given by
$$T(A) = (\text{tr}(A), \text{tr}(A)),$$
   (a) $B = \{E_{11}, E_{12}, E_{21}, E_{22}\}; C = \{(1, 0), (0, 1)\}$.
   (b) $B = \left\{\begin{bmatrix} -1 & -2 \\ -2 & -3 \end{bmatrix}, \begin{bmatrix} 1 & 1 \\ 2 & 2 \end{bmatrix}, \begin{bmatrix} 0 & -3 \\ 2 & -2 \end{bmatrix}, \begin{bmatrix} 0 & 4 \\ 1 & 0 \end{bmatrix}\right\}$;
   $C = \{(1, 0), (0, 1)\}$.

7. $T: M_2(\mathbb{R}) \to M_2(\mathbb{R})$ given by
$$T(A) = 2A - A^T,$$
   (a) $B = C = \{E_{11}, E_{12}, E_{21}, E_{22}\}$.
   (b) $B = \left\{\begin{bmatrix} -1 & -2 \\ -2 & -3 \end{bmatrix}, \begin{bmatrix} 1 & 1 \\ 2 & 2 \end{bmatrix}, \begin{bmatrix} 0 & -3 \\ 2 & -2 \end{bmatrix}, \begin{bmatrix} 0 & 4 \\ 1 & 0 \end{bmatrix}\right\}$;
   $C = \{E_{11}, E_{12}, E_{21}, E_{22}\}$.

8. $T: V \to V$ (where $V = \text{span}\{e^{2x}, e^{-3x}\}$) given by
$$T(f) = f',$$
   (a) $B = C = \{e^{2x}, e^{-3x}\}$.
   (b) $B = \{e^{2x} - 3e^{-3x}, 2e^{-3x}\}$;
   $C = \{e^{2x} + e^{-3x}, -e^{2x}\}$.

For Problems 9–15, determine $T(\mathbf{v})$ for the given linear transformation T and vector $\mathbf{v}$ in V by

   (a) computing $[T]_B^C$ and $[\mathbf{v}]_B$ and using Theorem 5.5.4.
   (b) direct calculation.

9. $T: P_1 \rightarrow M_2(\mathbb{R})$ via

$$T(a + bx) = \begin{bmatrix} a - b & 0 \\ -2b & -a + b \end{bmatrix},$$

relative to the standard bases B and C; $p(x) = -2 + 3x$.

10. $T: \mathbb{R}^3 \rightarrow P_3$ via

$$T(a, b, c) = 2a - (a + b - c)x + (2c - a)x^3,$$

relative to the standard bases B and C; $\mathbf{v} = (2, -1, 5)$.

11. $T: M_2(\mathbb{R}) \rightarrow M_2(\mathbb{R})$ via

$$T\left(\begin{bmatrix} a & b \\ c & d \end{bmatrix}\right) = \begin{bmatrix} 2a - b + d & -a + 3d \\ 0 & -a - b + 3c \end{bmatrix},$$

relative to the standard basis $B = C$;

$$A = \begin{bmatrix} -7 & 2 \\ 1 & -3 \end{bmatrix}.$$

12. $T: P_2 \rightarrow P_4$ via $T(p(x)) = x^2 p(x)$, relative to the standard bases B and C; $p(x) = -1 + 5x - 6x^2$.

13. $T: M_3(\mathbb{R}) \rightarrow \mathbb{R}$ via $T(A) = \text{tr}(A)$, relative to the standard bases B and C;

$$A = \begin{bmatrix} 2 & -6 & 0 \\ 1 & 4 & -4 \\ 0 & 0 & -3 \end{bmatrix}.$$

14. $T: P_4 \rightarrow P_3$ via $T(p(x)) = p'(x)$, relative to the standard bases B and C; $p(x) = 3 - 4x + 6x^2 + 6x^3 - 2x^4$.

15. $T: P_3 \rightarrow \mathbb{R}$ via $T(p(x)) = p(2)$, relative to the standard bases B and C; $p(x) = 2x - 3x^2$.

16. Let T_1 be the linear transformation from Problem 14 and T_2 be the linear transformation from Problem 15.

(a) Find the matrix representation of $T_2 T_1$ relative to the standard bases.

(b) Verify Theorem 5.5.7 by comparing part (a) with the product of the matrices in Problems 14 and 15.

(c) Use the matrix representation found in (a) to determine $(T_2 T_1)(2 + 5x - x^2 + 3x^4)$. Verify your answer by computing this directly.

17. Let T_1 be the linear transformation from Problem 9 and let T_2 be the linear transformation from Problem 6.

(a) Find the matrix representation of $T_2 T_1$ relative to the standard bases.

(b) Verify Theorem 5.5.7 by comparing part (a) with the product of the matrices in Problems 6 and 9.

(c) Use the matrix representation found in (a) to determine $(T_2 T_1)(-3 + 8x)$. Verify your answer by computing this directly.

18. Let T_1 be the linear transformation from Problem 4 and let T_2 be the linear transformation from Problem 5.

(a) Find the matrix representation of $T_2 T_1$ relative to the standard bases.

(b) Verify Theorem 5.5.7 by comparing part (a) with the product of the matrices in Problems 4 and 5.

(c) Use the matrix representation found in (a) to determine $(T_2 T_1)(7 - x + 2x^2)$. Verify your answer by computing it directly.

(d) Is $T_2 T_1$ invertible? Use matrix representations to explain your answer.

19. Is the linear transformation in Problem 2 invertible? Use matrix representations to explain your answer.

20. Let $T: P_2 \rightarrow \mathbb{R}^3$ be given by

$$T(p(x)) = (p(0), p(1), p(2)).$$

Is T invertible? Find the matrix representation of T with respect to the standard bases and use it to support your answer.

21. Supply a proof of part (b) of Theorem 5.5.9.

5.6   The Eigenvalue/Eigenvector Problem

In order to motivate the problem to be studied in the next several sections, we recall from Chapter 1 that the differential equation

$$\frac{dx}{dt} = ax, \tag{5.6.1}$$

where a is a constant, has general solution

$$x(t) = ce^{at}. \tag{5.6.2}$$

Then in Section 2.3 we considered a system of two differential equations

$$\frac{dx_1}{dt} = a_{11}x_1 + a_{12}x_2,$$

$$\frac{dx_2}{dt} = a_{21}x_1 + a_{22}x_2,$$

where x_1 and x_2 are functions of the independent variable t, and the a_{ij} are constants. We wrote this more succinctly as the vector equation

$$\frac{d\mathbf{x}}{dt} = A\mathbf{x}(t), \tag{5.6.3}$$

where

$$\mathbf{x}(t) = \begin{bmatrix} x_1(t) \\ x_2(t) \end{bmatrix}, \qquad A = [a_{ij}], \qquad \frac{d\mathbf{x}}{dt} = \begin{bmatrix} dx_1/dt \\ dx_2/dt \end{bmatrix}.$$

Based on the solution (5.6.2) to the differential equation (5.6.1), we might suspect that the system (5.6.3) may have solutions of the form

$$\mathbf{x}(t) = \begin{bmatrix} e^{\lambda t} v_1 \\ e^{\lambda t} v_2 \end{bmatrix} = e^{\lambda t} \mathbf{v},$$

where λ, v_1, and v_2 are constants and

$$\mathbf{v} = \begin{bmatrix} v_1 \\ v_2 \end{bmatrix}.$$

Indeed, substituting this expression for $\mathbf{x}(t)$ into (5.6.3) yields

$$\lambda e^{\lambda t} \mathbf{v} = A(e^{\lambda t} \mathbf{v}),$$

or equivalently,

$$A\mathbf{v} = \lambda \mathbf{v}. \tag{5.6.4}$$

We have therefore shown that

$$\mathbf{x}(t) = e^{\lambda t} \mathbf{v}$$

is a solution to the system of differential equations (5.6.3), provided that λ and $\mathbf{v}$ satisfy Equation (5.6.4).

In Chapter 7 we will pursue this technique for determining solutions to general linear systems of differential equations. For the remainder of the present section, however, we focus our attention on the mathematical problem of finding all scalars λ and all nonzero vectors $\mathbf{v}$ satisfying Equation (5.6.4) for a given $n \times n$ matrix A. Although we used systems of differential equations to motivate the study of this problem, it is important to realize that it arises also in many other areas of applications.

We begin by introducing the appropriate terminology.

DEFINITION   5.6.1

Let A be an $n \times n$ matrix. Any values of λ for which

$$A\mathbf{v} = \lambda \mathbf{v} \tag{5.6.5}$$

has *nontrivial* solutions $\mathbf{v}$ are called **eigenvalues** of A. The corresponding *nonzero* vectors $\mathbf{v}$ are called **eigenvectors** of A.

Remark   Eigenvalues and eigenvectors are also sometimes referred to as **characteristic values** and **characteristic vectors** of A.

In order to formulate the eigenvalue/eigenvector problem within the vector-space framework, we will interpret A as the matrix of a linear transformation $T : \mathbb{C}^n \to \mathbb{C}^n$ in the usual manner; that is, $T(\mathbf{v}) = A\mathbf{v}$. In many of our problems, A and λ will both be real, which will enable us to restrict attention to $\mathbb{R}^n$, although we will often require the complex vector space $\mathbb{C}^n$. Indeed, we will see in the later chapters that complex eigenvalues and eigenvectors are required to describe linear physical systems that exhibit oscillatory behavior.

It is always helpful to have a geometric interpretation of the problem under consideration. According to Equation (5.6.5), the eigenvectors of A are those nonzero vectors that are mapped into a constant scalar multiple of themselves by the linear transformation $T(\mathbf{v}) = A\mathbf{v}$. Geometrically, this means that the linear transformation leaves the direction of $\mathbf{v}$ unchanged[5] and stretches the vector $\mathbf{v}$ by a factor of λ. This is illustrated for the case $\mathbb{R}^2$ in Figure 5.6.1.

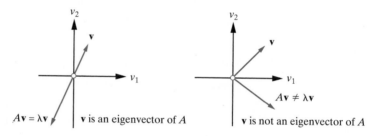

Figure 5.6.1: A geometrical description of the eigenvalue/eigenvector problem in $\mathbb{R}^2$.

Note that if $A\mathbf{v} = \lambda\mathbf{v}$ and c is an arbitrary scalar, then

$$A(c\mathbf{v}) = cA\mathbf{v} = c(\lambda\mathbf{v}) = \lambda(c\mathbf{v}).$$

Consequently, if $\mathbf{v}$ is an eigenvector of A, then so is $c\mathbf{v}$ for any nonzero scalar c.

Example 5.6.2   Verify that $\mathbf{v} = (1, 3)$ is an eigenvector of the matrix

$$A = \begin{bmatrix} 1 & 1 \\ -3 & 5 \end{bmatrix}$$

corresponding to the eigenvalue $\lambda = 4$.

Solution: For the given vector, we have the following:[6]

$$A\mathbf{v} = \begin{bmatrix} 1 & 1 \\ -3 & 5 \end{bmatrix} \begin{bmatrix} 1 \\ 3 \end{bmatrix} = \begin{bmatrix} 4 \\ 12 \end{bmatrix} = 4 \begin{bmatrix} 1 \\ 3 \end{bmatrix} = 4\mathbf{v}.$$

Consequently, $\mathbf{v}$ is an eigenvector of A corresponding to the eigenvalue $\lambda = 4$.   □

---

[5]If $\lambda < 0$, then the transformed vector points in the direction opposite to $\mathbf{v}$, due to the minus sign.
[6]Notice that once more we will switch between vectors in $\mathbb{R}^n$ and column n-vectors.

Solution of the Problem

The solution of the eigenvalue/eigenvector problem hinges on the observation that (5.6.5) can be written in the equivalent form

$$(A - \lambda I)\mathbf{v} = \mathbf{0}, \qquad (5.6.6)$$

where I denotes the identity matrix. Consequently, the eigenvalues of A are those values of λ for which the $n \times n$ linear system (5.6.6) has nontrivial solutions, and the eigenvectors are the corresponding solutions. But, according to Corollary 3.2.5, the system (5.6.6) has nontrivial solutions if and only if

$$\det(A - \lambda I) = 0.$$

To solve the eigenvalue/eigenvector problem, we therefore proceed as follows:

Solution to the Eigenvalue/Eigenvector Problem

> **1.** Find all scalars λ with $\det(A - \lambda I) = 0$. These are the eigenvalues of A.
> **2.** If $\lambda_1, \lambda_2, \ldots, \lambda_k$ are the *distinct* eigenvalues obtained in step 1, then solve the k systems of linear equations
> $$(A - \lambda_i I)\mathbf{v}_i = 0, \qquad i = 1, 2, \ldots, k$$
> to find all eigenvectors $\mathbf{v}_i$ corresponding to each eigenvalue.

DEFINITION 5.6.3

For a given $n \times n$ matrix A, the polynomial $p(\lambda)$ defined by

$$p(\lambda) = \det(A - \lambda I)$$

is called the **characteristic polynomial** of A, and the equation

$$p(\lambda) = 0$$

is called the **characteristic equation** of A.

From the definition of the determinant, Definition 3.1.8, we can verify that[7] $p(\lambda)$ is a polynomial in λ of degree n. Also, it follows from step 1 above that the eigenvalues of A are the roots of the characteristic equation. Using this fact, we can deduce the following fundamental result (Problem 35).

Proposition 5.6.4

An $n \times n$ matrix A is invertible if and only if 0 is not an eigenvalue of A.

Some Examples

We now consider three examples to illustrate some of the possibilities that can arise in the eigenvalue/eigenvector problem and also to motivate some of the theoretical results that will be established in the next section.

---

[7] Alternatively, this can be verified by induction on n, using the Cofactor Expansion Theorem.

Example 5.6.5   Find all eigenvalues and eigenvectors of

$$A = \begin{bmatrix} 5 & -4 \\ 8 & -7 \end{bmatrix}.$$

Solution: The linear system for determining the eigenvalues and eigenvectors is $(A - \lambda I)\mathbf{v} = \mathbf{0}$; that is,

$$\begin{bmatrix} 5 - \lambda & -4 \\ 8 & -7 - \lambda \end{bmatrix} \begin{bmatrix} v_1 \\ v_2 \end{bmatrix} = \begin{bmatrix} 0 \\ 0 \end{bmatrix}. \tag{5.6.7}$$

This system has nontrivial solutions if and only if λ satisfies the characteristic equation

$$\begin{vmatrix} 5 - \lambda & -4 \\ 8 & -7 - \lambda \end{vmatrix} = 0.$$

Expanding the determinant yields

$$(\lambda - 5)(\lambda + 7) + 32 = 0.$$

That is,

$$\lambda^2 + 2\lambda - 3 = 0,$$

which has factorization

$$(\lambda + 3)(\lambda - 1) = 0.$$

Consequently, the eigenvalues of A are

$$\lambda_1 = -3, \qquad \lambda_2 = 1.$$

The corresponding eigenvectors are obtained by successively substituting the foregoing eigenvalues into (5.6.7) and solving the resulting system.
Eigenvalue $\lambda_1 = -3$: We have

$$A - \lambda_1 I = \begin{bmatrix} 8 & -4 \\ 8 & -4 \end{bmatrix},$$

so the augmented matrix of the system $(A - \lambda_1 I)\mathbf{v} = \mathbf{0}$ is

$$\begin{bmatrix} 8 & -4 & 0 \\ 8 & -4 & 0 \end{bmatrix},$$

with reduced row-echelon form

$$\begin{bmatrix} 1 & -\frac{1}{2} & 0 \\ 0 & 0 & 0 \end{bmatrix}.$$

The solution to the system can therefore be written in the form $\mathbf{v} = (r, 2r)$, where r is a free variable. It follows that the eigenvectors corresponding to $\lambda_1 = -3$ are those vectors in $\mathbb{R}^2$ of the form

$$\mathbf{v} = r(1, 2),$$

where r is any *nonzero* real number. Notice that there is only one linearly independent eigenvector corresponding to the eigenvalue $\lambda = -3$, which we may choose as $\mathbf{v}_1 = (1, 2)$. All other eigenvectors (corresponding to $\lambda_1 = -3$) are scalar multiples of $\mathbf{v}_1$.
Eigenvalue $\lambda_2 = 1$: The augmented matrix of the system $(A - \lambda_2 I)\mathbf{v} = \mathbf{0}$ is

$$\begin{bmatrix} 4 & -4 & 0 \\ 8 & -8 & 0 \end{bmatrix},$$

with reduced row-echelon form

$$\begin{bmatrix} 1 & -1 & 0 \\ 0 & 0 & 0 \end{bmatrix}.$$

Consequently, the system has solutions $\mathbf{v} = (s, s)$, where s is a free variable. It follows that the eigenvectors corresponding to $\lambda_2 = 1$ are those vectors in $\mathbb{R}^2$ of the form

$$\mathbf{v} = s(1, 1),$$

where s is any *nonzero* real number. Once more there is only one linearly independent eigenvector corresponding to the eigenvalue $\lambda_2 = 1$, which we may choose as $\mathbf{v}_2 = (1, 1)$. All other eigenvectors (corresponding to $\lambda_2 = 1$) are scalar multiples of $\mathbf{v}_2$.

Notice that the eigenvectors $\mathbf{v}_1 = (1, 2)$ and $\mathbf{v}_2 = (1, 1)$ (which correspond to the two different eigenvalues here) are nonproportional and therefore are linearly independent in $\mathbb{R}^2$. The matrix A has therefore picked out a basis for $\mathbb{R}^2$. This is illustrated in Figure 5.6.2. □

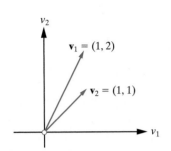

Figure 5.6.2: Two linearly independent eigenvectors for the matrix in Example 5.6.5.

Example 5.6.6   Find all eigenvalues and eigenvectors of

$$A = \begin{bmatrix} 5 & 12 & -6 \\ -3 & -10 & 6 \\ -3 & -12 & 8 \end{bmatrix}.$$

Solution: The system $(A - \lambda I)\mathbf{v} = \mathbf{0}$ has nontrivial solutions if and only if $\det(A - \lambda I) = 0$. For the given matrix,

$$\det(A - \lambda I) = \begin{vmatrix} 5 - \lambda & 12 & -6 \\ -3 & -10 - \lambda & 6 \\ -3 & -12 & 8 - \lambda \end{vmatrix}.$$

Using the Cofactor Expansion Theorem along row 1 yields

$$\begin{aligned} \det(A - \lambda I) &= (5 - \lambda)[(\lambda - 8)(\lambda + 10) + 72] - 12(3\lambda - 6) - 6(6 - 3\lambda) \\ &= (5 - \lambda)(\lambda^2 + 2\lambda - 8) + 18(2 - \lambda) \\ &= (5 - \lambda)(\lambda - 2)(\lambda + 4) + 18(2 - \lambda) \\ &= (2 - \lambda)[(\lambda - 5)(\lambda + 4) + 18] = (2 - \lambda)(\lambda^2 - \lambda - 2) \\ &= (2 - \lambda)(\lambda - 2)(\lambda + 1) = -(\lambda - 2)^2(\lambda + 1). \end{aligned}$$

Consequently, A has eigenvalues

$$\lambda_1 = 2 \quad \text{(repeated twice)}, \qquad \lambda_2 = -1.$$

To determine the corresponding eigenvectors of A, we must solve each of the homogeneous linear systems

$$(A - \lambda_1 I)\mathbf{v} = \mathbf{0}, \qquad (A - \lambda_2 I)\mathbf{v} = \mathbf{0}.$$

Eigenvalue $\lambda_1 = 2$: The augmented matrix of the linear system $(A - \lambda_1 I)\mathbf{v} = \mathbf{0}$ is

$$\begin{bmatrix} 3 & 12 & -6 & 0 \\ -3 & -12 & 6 & 0 \\ -3 & -12 & 6 & 0 \end{bmatrix},$$

with reduced row-echelon form

$$\begin{bmatrix} 1 & 4 & -2 & 0 \\ 0 & 0 & 0 & 0 \\ 0 & 0 & 0 & 0 \end{bmatrix}.$$

Hence, the eigenvectors of A corresponding to the eigenvalue $\lambda_1 = 2$ are

$$\mathbf{v} = (-4r + 2s, r, s),$$

where r and s are free variables that cannot be simultaneously zero. (Why not?) Writing $\mathbf{v}$ in the equivalent form

$$\mathbf{v} = r(-4, 1, 0) + s(2, 0, 1),$$

we see that there are two linearly independent eigenvectors corresponding to $\lambda_1 = 2$, which we may choose as $\mathbf{v}_1 = (-4, 1, 0)$ and $\mathbf{v}_2 = (2, 0, 1)$. All other eigenvectors corresponding to $\lambda_1 = 2$ are obtained by taking nontrivial linear combinations of $\mathbf{v}_1, \mathbf{v}_2$. Therefore, geometrically, these eigenvectors all lie in the plane through the origin of a Cartesian coordinate system that contains $\mathbf{v}_1$ and $\mathbf{v}_2$.

Eigenvalue $\lambda_2 = -1$: The augmented matrix of the linear system $(A - \lambda_2 I)\mathbf{v} = \mathbf{0}$ is

$$\begin{bmatrix} 6 & 12 & -6 & 0 \\ -3 & -9 & 6 & 0 \\ -3 & -12 & 9 & 0 \end{bmatrix},$$

with reduced row-echelon form

$$\begin{bmatrix} 1 & 0 & 1 & 0 \\ 0 & 1 & -1 & 0 \\ 0 & 0 & 0 & 0 \end{bmatrix}.$$

Consequently, the eigenvectors of A corresponding to the eigenvalue $\lambda_2 = -1$ are those nonzero vectors in $\mathbb{R}^3$ of the form

$$\mathbf{v} = t(-1, 1, 1),$$

where t is a *nonzero* free variable. We see that there is only one linearly independent eigenvector corresponding to $\lambda_2 = -1$, which we may take to be $\mathbf{v}_3 = (-1, 1, 1)$. All other eigenvectors of A corresponding to $\lambda_2 = -1$ are scalar multiples of $\mathbf{v}_3$. Geometrically, these eigenvectors lie on the line with direction vector $\mathbf{v}_3$ which passes through the origin of a Cartesian coordinate system.

We note that the eigenvectors of A have given rise to the set of vectors $\{(-4, 1, 0), (2, 0, 1), (-1, 1, 1)\}$. Furthermore, since

$$\det([\mathbf{v}_1, \mathbf{v}_2, \mathbf{v}_3]) = \begin{vmatrix} -4 & 2 & -1 \\ 1 & 0 & 1 \\ 0 & 1 & 1 \end{vmatrix} = 1 \neq 0,$$

the set of eigenvectors $\{(-4, 1, 0), (2, 0, 1), (-1, 1, 1)\}$ is linearly independent and therefore is a basis for $\mathbb{R}^3$.   □

The reader may have noticed a couple of common features of the two examples just given. First, the number of linearly independent eigenvectors associated with a given eigenvalue λ is equal to the number of times λ occurs as a root of the characteristic equation. Second, the total number of linearly independent eigenvectors obtained from all of the eigenvalues agrees with the number of rows and columns of the matrix A.

Are these observations always to be expected? The answer is no. The next example illustrates this, and in the next section we will have much more to say about this issue from a theoretical perspective.

Example 5.6.7   Find all eigenvalues and eigenvectors of

$$A = \begin{bmatrix} 1 & 1 \\ 0 & 1 \end{bmatrix}.$$

Solution: The system $(A - \lambda I)\mathbf{v} = \mathbf{0}$ has nontrivial solutions if and only if $\det(A - \lambda I) = 0$. For the given matrix,

$$\det(A - \lambda I) = \begin{vmatrix} 1 - \lambda & 1 \\ 0 & 1 - \lambda \end{vmatrix} = (\lambda - 1)^2.$$

Hence, A has eigenvalue $\lambda_1 = 1$ (repeated twice).

To determine the corresponding eigenvectors of A, we must solve the homogeneous linear system

$$(A - \lambda_1 I)\mathbf{v} = \mathbf{0}$$

which, for $\lambda_1 = 1$, has augmented matrix

$$\begin{bmatrix} 0 & 1 & 0 \\ 0 & 0 & 0 \end{bmatrix}.$$

Hence, the eigenvectors of A corresponding to $\lambda_1 = 1$ are

$$\mathbf{v} = (r, 0),$$

where r is a *nonzero* free variable. It follows that the eigenvectors corresponding to $\lambda_1 = 1$ are those vectors in $\mathbb{R}^2$ of the form

$$\mathbf{v} = r(1, 0).$$

Notice that there is only one linearly independent eigenvector here,

$$\mathbf{v}_1 = (1, 0).$$

In this case, therefore, the number of linearly independent eigenvectors associated with λ falls short of the number of times λ occurs as a root of the characteristic equation, as well as the number of rows and columns of A. In the next section we will refer to such a matrix as **defective**, since it fails to have "enough" linearly independent eigenvectors to match its size.   □

The matrices and eigenvalues in the previous examples have all been *real*, and this enabled us to regard the eigenvectors as vectors in $\mathbb{R}^n$. We now consider the case when some or all of the eigenvalues and eigenvectors are complex. The steps in determining the eigenvalues and eigenvectors do not change, although they can be more complicated algebraically, since the equations determining the eigenvectors will have complex coefficients. For the majority of matrices that we consider, the matrix A will have only *real* elements. In these cases, the following theorem can save some work in determining any complex eigenvectors.

Theorem 5.6.8   Let A be an $n \times n$ matrix with *real* elements. If λ is a complex eigenvalue of A with corresponding eigenvector $\mathbf{v}$, then $\bar{\lambda}$ is an eigenvalue of A with corresponding eigenvector $\bar{\mathbf{v}}$.

Proof If $A\mathbf{v} = \lambda\mathbf{v}$, then $\overline{A\mathbf{v}} = \overline{\lambda\mathbf{v}}$, which implies that $A\overline{\mathbf{v}} = \overline{\lambda}\overline{\mathbf{v}}$, since A has real entries. ∎

Remark According to the previous theorem, if we find the eigenvectors of a real matrix A corresponding to a complex eigenvalue λ, then we can obtain the eigenvectors corresponding to the eigenvalue $\overline{\lambda}$ *without having to solve a linear system.*

Example 5.6.9 Find all eigenvalues and eigenvectors of

$$A = \begin{bmatrix} -2 & -6 \\ 3 & 4 \end{bmatrix}.$$

Solution: The characteristic polynomial of A is

$$p(\lambda) = \begin{vmatrix} -2 - \lambda & -6 \\ 3 & 4 - \lambda \end{vmatrix} = \lambda^2 - 2\lambda + 10.$$

Consequently, the eigenvalues of A are

$$\lambda_1 = 1 + 3i, \qquad \lambda_2 = \overline{\lambda_1} = 1 - 3i.$$

Since these are complex eigenvalues, we take the underlying vector space as $\mathbb{C}^2$, hence any scalars that arise in the solution of the problem will be complex.

Eigenvalue $\lambda_1 = 1 + 3i$: The augmented matrix of the system $(A - \lambda_1 I)\mathbf{v} = \mathbf{0}$ reduces to

$$\begin{bmatrix} -3 - 3i & -6 & 0 \\ 3 & 3 - 3i & 0 \end{bmatrix} \overset{1}{\sim} \begin{bmatrix} 0 & 0 & 0 \\ 3 & 3 - 3i & 0 \end{bmatrix} \overset{2}{\sim} \begin{bmatrix} 3 & 3 - 3i & 0 \\ 0 & 0 & 0 \end{bmatrix} \overset{3}{\sim} \begin{bmatrix} 1 & 1 - i & 0 \\ 0 & 0 & 0 \end{bmatrix},$$

> **1.** $A_{21}(1 + i)$ **2.** P_{12} **3.** $M_1(\frac{1}{3})$

so that the eigenvectors of A corresponding to $\lambda_1 = 1 + 3i$ are those vectors in $\mathbb{C}^2$ of the form

$$\mathbf{v} = r(-(1 - i), 1),$$

where r is an arbitrary *nonzero complex* number.

Eigenvalue $\lambda_2 = 1 - 3i$: From Theorem 5.6.8, the eigenvectors in this case are those vectors in $\mathbb{C}^2$ of the form $\mathbf{v} = s(-(1 + i), 1)$, where s is an arbitrary *nonzero complex* number.

Notice that the eigenvectors corresponding to different eigenvalues are linearly independent vectors in $\mathbb{C}^2$. For example, a linearly independent set of eigenvectors is $\{(-(1 - i), 1), (-(1 + i), 1)\}$. Once more the eigenvectors have determined a basis in the underlying vector space (in this case $\mathbb{C}^2$). ☐

When first encountering the eigenvalue/eigenvector problem, students often focus so much attention on the computational aspects of finding the eigenvalues and eigenvectors that they lose sight of the original equation that defines the eigenvalues and eigenvectors of A. The following examples illustrate the importance of the defining equation when establishing theoretical results.

Example 5.6.10 Let λ be an eigenvalue of the matrix A with corresponding eigenvector $\mathbf{v}$. Prove that λ^2 is an eigenvalue of A^2 with corresponding eigenvector $\mathbf{v}$.

Solution: We are given that

$$A\mathbf{v} = \lambda\mathbf{v}, \tag{5.6.8}$$

and we must establish that

$$A^2\mathbf{v} = \lambda^2\mathbf{v}.$$

From Equation (5.6.8), we have

$$A^2\mathbf{v} = A(A\mathbf{v}) = A(\lambda\mathbf{v}) = \lambda(A\mathbf{v}) = \lambda(\lambda\mathbf{v}) = \lambda^2\mathbf{v},$$

and the result is established.   □

Example 5.6.11   Let λ and $\mathbf{v}$ be an eigenvalue/eigenvector pair for the $n \times n$ matrix A. If k is an arbitrary real number, prove that $\mathbf{v}$ is also an eigenvector of the matrix $A - kI$ corresponding to the eigenvalue $\lambda - k$.

Solution: Once more, the only information that we are given is that

$$A\mathbf{v} = \lambda\mathbf{v}.$$

If we let $B = A - kI$, then we must establish that

$$B\mathbf{v} = (\lambda - k)\mathbf{v}.$$

But,

$$B\mathbf{v} = (A - kI)\mathbf{v} = A\mathbf{v} - k\mathbf{v} = \lambda\mathbf{v} - k\mathbf{v} = (\lambda - k)\mathbf{v},$$

as required.   □

Exercises for 5.6

Key Terms

Eigenvalue, Eigenvector, Characteristic polynomial, Characteristic equation.

Skills

- Be able to determine whether a given scalar λ and vector $\mathbf{v}$ form an eigenvalue/eigenvector pair for a given matrix A.

- Be able to determine eigenvectors that correspond to a given eigenvalue of A.

- Be able to determine the eigenvalue that corresponds to a given eigenvector of A.

- For 2×2 and 3×3 matrices, be able to provide a geometric interpretation of the eigenvalue/eigenvector problem. For special cases, you should be able to determine eigenvalue-eigenvector pairs by arguing geometrically.

- Be able to compute the characteristic polynomial for a given matrix A and use it to find the eigenvalues of A.

- Be able to prove basic facts about eigenvalue/eigenvector pairs from the definitions in this section. (See Examples 5.6.10 and 5.6.11.)

- Be able to find all eigenvalues and corresponding eigenvectors for a given matrix A.

True-False Review

For Questions 1–9, decide if the given statement is **true** or **false**, and give a brief justification for your answer. If true, you can quote a relevant definition or theorem from the text. If false, provide an example, illustration, or brief explanation of why the statement is false.

1. An eigenvector corresponding to the eigenvalue λ of a matrix A is any vector $\mathbf{v}$ such that $A\mathbf{v} = \lambda\mathbf{v}$.

2. The eigenvalues of an upper or lower triangular matrix A are the entries appearing on the main diagonal of A.

3. If two matrices A and B have the same characteristic polynomial, then A and B must have exactly the same set of eigenvalues.

4. If two matrices A and B have exactly the same characteristic polynomial, then A and B must have exactly the same set of eigenvectors.

5. If A is the 2×2 matrix of the linear transformation $T: \mathbb{R}^2 \to \mathbb{R}^2$ that rotates points of the xy-plane counterclockwise by 90 degrees, then A has no real eigenvalues.

6. If A is an $n \times n$ matrix, then A has n eigenvalues, including possible repeated eigenvalues and complex eigenvalues.

7. A linear combination of a set of eigenvectors of a matrix A is again an eigenvector of A.

8. If $\lambda = a + ib$ $(b \neq 0)$ is a complex eigenvalue of a matrix A, then so is $\bar{\lambda} = a - ib$.

9. If λ is an eigenvalue of the matrix A, then λ^2 is an eigenvalue of A^2.

Problems

For Problems 1–3, use Equation (5.6.5) to verify that λ and $\mathbf{v}$ are an eigenvalue/eigenvector pair for the given matrix A.

1. $\lambda = 4$, $\mathbf{v} = (1, 1)$,
$$A = \begin{bmatrix} 1 & 3 \\ 2 & 2 \end{bmatrix}.$$

2. $\lambda = 3$, $\mathbf{v} = (2, 1, -1)$,
$$A = \begin{bmatrix} 1 & -2 & -6 \\ -2 & 2 & -5 \\ 2 & 1 & 8 \end{bmatrix}.$$

3. $\lambda = -2$, $\mathbf{v} = c_1(1, 0, -3) + c_2(4, -3, 0)$,
$$A = \begin{bmatrix} 1 & 4 & 1 \\ 3 & 2 & 1 \\ 3 & 4 & -1 \end{bmatrix},$$
where c_1 and c_2 are constants.

4. Given that $\mathbf{v}_1 = (1, -2)$ and $\mathbf{v}_2 = (1, 1)$ are eigenvectors of
$$A = \begin{bmatrix} 4 & 1 \\ 2 & 3 \end{bmatrix},$$
determine the eigenvalues of A.

5. The effect of the linear transformation $T: \mathbb{R}^2 \to \mathbb{R}^2$ with matrix
$$A = \begin{bmatrix} 1 & 0 \\ 0 & -1 \end{bmatrix}$$
is to reflect each vector in the x-axis. By arguing geometrically, determine all eigenvalues and eigenvectors of A.

6. The effect of the linear transformation $T: \mathbb{R}^2 \to \mathbb{R}^2$ with matrix
$$A = \begin{bmatrix} 0 & 1 \\ 1 & 0 \end{bmatrix}$$
is to reflect each vector across the line $y = x$. By arguing geometrically, determine all eigenvalues and eigenvectors of A.

7. The linear transformation $T: \mathbb{R}^2 \to \mathbb{R}^2$ with matrix
$$A = \begin{bmatrix} \cos\theta & -\sin\theta \\ \sin\theta & \cos\theta \end{bmatrix}$$
rotates vectors in the xy-plane counterclockwise through an angle θ, where $0 \leq \theta < 2\pi$. By arguing geometrically, determine all values of θ for which A has *real* eigenvalues. Find the real eigenvalues and the corresponding eigenvectors.

8. The linear transformation $T: \mathbb{R}^3 \to \mathbb{R}^3$ with matrix
$$A = \begin{bmatrix} 0 & 0 & 0 \\ 0 & 1 & 0 \\ 0 & 0 & 0 \end{bmatrix}$$
takes vectors (x, y, z) in $\mathbb{R}^3$ and moves them to the corresponding point $(0, y, 0)$ on the y-axis. By arguing geometrically, determine all eigenvalues and eigenvector of A.

For Problems 9–27, determine all eigenvalues and corresponding eigenvectors of the given matrix.

9. $\begin{bmatrix} 3 & -1 \\ -5 & -1 \end{bmatrix}$.

10. $\begin{bmatrix} 1 & 6 \\ 2 & -3 \end{bmatrix}$.

11. $\begin{bmatrix} 7 & 4 \\ -1 & 3 \end{bmatrix}$.

12. $\begin{bmatrix} 2 & 0 \\ 0 & 2 \end{bmatrix}$.

13. $\begin{bmatrix} 3 & -2 \\ 4 & -1 \end{bmatrix}$.

14. $\begin{bmatrix} 2 & 3 \\ -3 & 2 \end{bmatrix}$.

15. $\begin{bmatrix} 10 & -12 & 8 \\ 0 & 2 & 0 \\ -8 & 12 & -6 \end{bmatrix}$.

16. $\begin{bmatrix} 3 & 0 & 0 \\ 0 & 2 & -1 \\ 1 & -1 & 2 \end{bmatrix}$.

17. $\begin{bmatrix} 1 & 0 & 0 \\ 0 & 3 & 2 \\ 2 & -2 & -1 \end{bmatrix}$.

18. $\begin{bmatrix} 6 & 3 & -4 \\ -5 & -2 & 2 \\ 0 & 0 & -1 \end{bmatrix}$.

19. $\begin{bmatrix} 7 & -8 & 6 \\ 8 & -9 & 6 \\ 0 & 0 & -1 \end{bmatrix}$.

20. $\begin{bmatrix} 0 & 1 & -1 \\ 0 & 2 & 0 \\ 2 & -1 & 3 \end{bmatrix}$.

21. $\begin{bmatrix} 1 & 0 & 0 \\ 0 & 0 & 1 \\ 0 & -1 & 0 \end{bmatrix}$.

22. $\begin{bmatrix} -2 & 1 & 0 \\ 1 & -1 & -1 \\ 1 & 3 & -3 \end{bmatrix}$.

23. $\begin{bmatrix} 2 & -1 & 3 \\ 3 & 1 & 0 \\ 2 & -1 & 3 \end{bmatrix}$.

24. $\begin{bmatrix} 5 & 0 & 0 \\ 0 & 5 & 0 \\ 0 & 0 & 5 \end{bmatrix}$.

25. $\begin{bmatrix} 0 & 2 & 2 \\ 2 & 0 & 2 \\ 2 & 2 & 0 \end{bmatrix}$.

26. $\begin{bmatrix} 1 & 2 & 3 & 4 \\ 4 & 3 & 2 & 1 \\ 4 & 5 & 6 & 7 \\ 7 & 6 & 5 & 4 \end{bmatrix}$.

27. $\begin{bmatrix} 0 & 1 & 0 & 0 \\ -1 & 0 & 0 & 0 \\ 0 & 0 & 0 & -1 \\ 0 & 0 & 1 & 0 \end{bmatrix}$.

28. Find all eigenvalues and corresponding eigenvectors of

$$A = \begin{bmatrix} 1+i & 0 & 0 \\ 2-2i & 1-3i & 0 \\ 2i & 0 & 1 \end{bmatrix}.$$

Note that the eigenvectors do not occur in complex conjugate pairs. Does this contradict Theorem 5.6.8? Explain.

29. Consider the matrix

$$A = \begin{bmatrix} 1 & -1 \\ 2 & 4 \end{bmatrix}.$$

    (a) Show that the characteristic polynomial of A is $p(\lambda) = \lambda^2 - 5\lambda + 6$.

    (b) Show that A satisfies its characteristic equation. That is, $A^2 - 5A + 6I_2 = 0_2$. (This result is known as the Cayley-Hamilton theorem and is true for general $n \times n$ matrices.

    (c) Use the result from (b) to find A^{-1}. [**Hint:** Multiply the equation in (b) by A^{-1}.]

30. Let $A = \begin{bmatrix} 1 & 2 \\ 2 & -2 \end{bmatrix}$.

    (a) Determine all eigenvalues of A.

    (b) Reduce A to row-echelon form and determine the eigenvalues of the resulting matrix. Are these the same as the eigenvalues of A?

31. If $\mathbf{v}_1 = (1, -1)$ and $\mathbf{v}_2 = (2, 1)$ are eigenvectors of the matrix A corresponding to the eigenvalues $\lambda_1 = 2, \lambda_2 = -3$, respectively, find $A(3\mathbf{v}_1 - \mathbf{v}_2)$.

32. Let $\mathbf{v}_1 = (1, -1, 1)$, $\mathbf{v}_2 = (2, 1, 3)$, and $\mathbf{v}_3 = (-1, -1, 2)$ be eigenvectors of the matrix A corresponding to the eigenvalues $\lambda_1 = 2, \lambda_2 = -2$, and $\lambda_3 = 3$, respectively, and let $\mathbf{v} = (5, 0, 3)$.

    (a) Express $\mathbf{v}$ as a linear combination of $\mathbf{v}_1, \mathbf{v}_2$, and $\mathbf{v}_3$.

    (b) Find $A\mathbf{v}$.

33. If $\mathbf{v}_1, \mathbf{v}_2$, and $\mathbf{v}_3$ are eigenvectors of A corresponding to the eigenvalue λ, and c_1, c_2, and c_3 are scalars (not all zero), show that $c_1\mathbf{v}_1 + c_2\mathbf{v}_2 + c_3\mathbf{v}_3$ is also an eigenvector of A corresponding to the eigenvalue λ.

34. Prove that the eigenvalues of an upper (or lower) triangular matrix are just the diagonal elements of the matrix.

35. Prove Proposition 5.6.4.

36. Let A be an $n \times n$ invertible matrix. Prove that if λ is an eigenvalue of A, then $1/\lambda$ is an eigenvalue of A^{-1}. [**Note:** By Proposition 5.6.4, $\lambda \neq 0$ here.]

37. Let A and B be $n \times n$ matrices, and assume that $\mathbf{v}$ in $\mathbb{R}^n$ is an eigenvector of A corresponding to the eigenvalue λ and also an eigenvector of B corresponding to the eigenvalue μ.

   (a) Prove that $\mathbf{v}$ is an eigenvector of the matrix AB. What is the corresponding eigenvalue?

   (b) Prove that $\mathbf{v}$ is an eigenvector of the matrix $A+B$. What is the corresponding eigenvalue?

38. Let A be an $n \times n$ matrix. Prove that A and A^T have the same eigenvalues. [**Hint:** Show that $\det(A^T - \lambda I) = \det(A - \lambda I)$.]

39. Let A be an $n \times n$ real matrix with complex eigenvalue $\lambda = a + ib$, where $b \neq 0$, and let $\mathbf{v} = \mathbf{r} + i\mathbf{s}$ be a corresponding eigenvector of A.

   (a) Prove that $\mathbf{r}$ and $\mathbf{s}$ are *nonzero* vectors in $\mathbb{R}^n$.

   (b) Prove that $\{\mathbf{r}, \mathbf{s}\}$ is linearly independent in $\mathbb{R}^n$.

For Problems 40–45, use some form of technology to determine the eigenvalues and eigenvectors of A in the following manner:

   (a) Form the matrix $A - \lambda I$.

   (b) Solve the characteristic equation $\det(A - \lambda I) = 0$ to determine the eigenvalues of A.

   (c) For each eigenvalue λ_i found in (2), solve the system $(A - \lambda_i I)\mathbf{v} = \mathbf{0}$ to determine the eigenvectors of A.

40. $\diamond A = \begin{bmatrix} 3 & 1 \\ 2 & 4 \end{bmatrix}$.

41. $\diamond A = \begin{bmatrix} 5 & 34 & -41 \\ 4 & 17 & -23 \\ 5 & 24 & -31 \end{bmatrix}$.

42. $\diamond A = \begin{bmatrix} 4 & 1 & 1 \\ 1 & 4 & 1 \\ 1 & 1 & 4 \end{bmatrix}$.

43. $\diamond A = \begin{bmatrix} 1 & 1 & 1 \\ 3 & -1 & 2 \\ 3 & 1 & 4 \end{bmatrix}$.

44. $\diamond A = \begin{bmatrix} 0 & 1 & -2 \\ -1 & 0 & 2 \\ 2 & -2 & 0 \end{bmatrix}$.

45. $\diamond A = \begin{bmatrix} 0 & 1 & 1 & 1 & 1 \\ 1 & 0 & 1 & 1 & 1 \\ 1 & 1 & 0 & 1 & 1 \\ 1 & 1 & 1 & 0 & 1 \\ 1 & 1 & 1 & 1 & 0 \end{bmatrix}$.

For Problems 46–51, use some form of technology to directly determine the eigenvalues and eigenvectors of the given matrix.

46. $\diamond$ The matrix in Problem 40.

47. $\diamond$ The matrix in Problem 41.

48. $\diamond$ The matrix in Problem 42.

49. $\diamond$ The matrix in Problem 43.

50. $\diamond$ The matrix in Problem 44.

51. $\diamond$ The matrix in Problem 45.

5.7 General Results for Eigenvalues and Eigenvectors

In this section, we look more closely at the relationship between the eigenvalues and eigenvectors of an $n \times n$ matrix. Our aim is to formalize several of the ideas introduced via the examples of the previous section.

For a given $n \times n$ matrix $A = [a_{ij}]$, the characteristic polynomial $p(\lambda)$ assumes the form

$$p(\lambda) = \det(A - \lambda I) = \begin{vmatrix} a_{11} - \lambda & a_{12} & \cdots & a_{1n} \\ a_{21} & a_{22} - \lambda & \cdots & a_{2n} \\ \vdots & \vdots & & \vdots \\ a_{n1} & a_{n2} & \cdots & a_{nn} - \lambda \end{vmatrix}.$$

Expanding this determinant yields a polynomial of degree n in λ with leading coefficient $(-1)^n$. It follows that $p(\lambda)$ can be written in the form

$$p(\lambda) = (-1)^n \lambda^n + b_1 \lambda^{n-1} + b_2 \lambda^{n-2} + \cdots + b_n,$$

where $b_1, b_2, \ldots, b_n$ are scalars. Since we consider the underlying vector space to be $\mathbb{C}^n$, the Fundamental Theorem of Algebra guarantees that $p(\lambda)$ will have precisely n zeros (not necessarily distinct), hence A will have n eigenvalues. If we let $\lambda_1, \lambda_2, \ldots, \lambda_k$ denote the *distinct* eigenvalues of A, then $p(\lambda)$ can be factored as

$$p(\lambda) = (-1)^n (\lambda - \lambda_1)^{m_1} (\lambda - \lambda_2)^{m_2} (\lambda - \lambda_3)^{m_3} \cdots (\lambda - \lambda_k)^{m_k},$$

where, since $p(\lambda)$ has degree n,

$$m_1 + m_2 + \cdots + m_k = n.$$

Thus, associated with each eigenvalue λ_i is a number m_i, called the **multiplicity** of λ_i.
    We now focus our attention on the eigenvectors of A.

DEFINITION   5.7.1

Let A be an $n \times n$ matrix. For a given eigenvalue λ_i, let E_i denote the set of *all* vectors $\mathbf{v}$ satisfying $A\mathbf{v} = \lambda_i \mathbf{v}$. Then E_i is called the **eigenspace** of A corresponding to the eigenvalue λ_i. Thus, E_i is the solution set to the linear system $(A - \lambda_i I)\mathbf{v} = \mathbf{0}$.

Remarks

1. Equivalently, we can say that the eigenspace E_i is the kernel of the linear transformation $T_i \colon \mathbb{C}^n \to \mathbb{C}^n$ defined by $T_i(\mathbf{v}) = (A - \lambda_i I)\mathbf{v}$.

2. It is important to notice that there is one eigenspace associated with each eigenvalue of A.

3. Also note that the only difference between the eigenspace corresponding to a specific eigenvalue, and the set of all eigenvectors corresponding to that eigenvalue is that the eigenspace includes the zero vector.

Example 5.7.2   Determine all eigenspaces for the matrix

$$A = \begin{bmatrix} 5 & -4 \\ 8 & -7 \end{bmatrix}.$$

Solution: We have already computed the eigenvalues and eigenvectors of A in Example 5.6.5. The eigenvalues of A are $\lambda_1 = -3$ and $\lambda_2 = 1$. The eigenvectors corresponding to $\lambda_1 = -3$ are all nonzero vectors of the form $\mathbf{v} = r(1, 2)$, where r is a constant. Thus, the eigenspace corresponding to $\lambda_1 = -3$ is

$$E_1 = \{\mathbf{v} \in \mathbb{R}^2 : \mathbf{v} = r(1, 2), \ r \in \mathbb{R}\}.$$

The eigenvectors corresponding to the eigenvalue $\lambda_2 = 1$ are of the form $\mathbf{v} = s(1, 1)$, where $s \neq 0$, so that the eigenspace corresponding to $\lambda_2 = 1$ is

$$E_2 = \{\mathbf{v} \in \mathbb{R}^2 : \mathbf{v} = s(1, 1), \ s \in \mathbb{R}\}.$$

$\square$

We have one main result for eigenspaces.

Theorem 5.7.3   Let λ_i be an eigenvalue of A of multiplicity m_i and let E_i denote the corresponding eigenspace. Then

1. For each i, E_i is a subspace of $\mathbb{C}^n$.

2. If n_i denotes the dimension of E_i, then $1 \leq n_i \leq m_i$ for each i. In words, the dimension of the eigenspace corresponding to λ_i is at most the multiplicity of λ_i.

Proof

1. From Definition 5.7.1, E_i is the null space of the matrix $A - \lambda_i I$ and hence is a subspace of $\mathbb{C}^n$. Alternatively, Remark 1 above, coupled with Theorem 5.3.5, provides an immediate proof.

2. The proof of this result requires some more advanced ideas about linear transformations than we have developed and is therefore omitted. [See, for example, G. E. Shilov, *Linear Algebra* (New York: Dover Publications, 1977). ■

Remark   The numbers m_i and n_i are called the **algebraic** multiplicity and the **geometric** multiplicity of the eigenvalue λ_i, respectively.

Example 5.7.4   Determine all eigenspaces and their dimensions for the matrix

$$A = \begin{bmatrix} 3 & -1 & 0 \\ 0 & 2 & 0 \\ -1 & 1 & 2 \end{bmatrix}.$$

Solution: A straightforward calculation yields the characteristic polynomial

$$p(\lambda) = -(\lambda - 2)^2(\lambda - 3),$$

so that the eigenvalues of A are $\lambda_1 = 2$ (with algebraic multiplicity 2) and $\lambda_2 = 3$ (with algebraic multiplicity 1). The eigenvectors corresponding to $\lambda_1 = 2$ are determined by solving the linear system $(A - \lambda_1 I)\mathbf{v} = \mathbf{0}$ for $\mathbf{v}$. The augmented matrix for the system is

$$\begin{bmatrix} 1 & -1 & 0 & 0 \\ 0 & 0 & 0 & 0 \\ -1 & 1 & 0 & 0 \end{bmatrix},$$

which has reduced row-echelon form

$$\begin{bmatrix} 1 & -1 & 0 & 0 \\ 0 & 0 & 0 & 0 \\ 0 & 0 & 0 & 0 \end{bmatrix}.$$

The general solution to this system is therefore

$$\mathbf{v} = (r, r, s) = r(1, 1, 0) + s(0, 0, 1).$$

Thus, the eigenspace corresponding to $\lambda_1 = 2$ is

$$E_1 = \{\mathbf{v} \in \mathbb{R}^3 : \mathbf{v} = r(1, 1, 0) + s(0, 0, 1), \ r, s \in \mathbb{R}\} = \text{span}\{(1, 1, 0), (0, 0, 1)\}.$$

We see that the linearly independent set $\{(1, 1, 0), (0, 0, 1)\}$ is a basis for the eigenspace E_1, and hence, $\dim[E_1] = 2$; that is, $n_1 = 2$.

It is easily shown that the eigenvectors corresponding to the eigenvalue $\lambda_2 = 3$ are of the form

$$\mathbf{v} = (t, 0, -t) = t(1, 0, -1),$$

where t is a nonzero real number. Thus, the eigenspace corresponding to $\lambda_2 = 3$ is

$$E_2 = \{\mathbf{v} \in \mathbb{R}^3 : \mathbf{v} = t(1, 0, -1), \ t \in \mathbb{R}\} = \text{span}\{(1, 0, -1)\}.$$

Consequently, $\{(1, 0, -1)\}$ is a basis for this eigenspace E_2, and hence, $\dim[E_2] = 1$; that is, $n_2 = 1$. The eigenspaces E_1 and E_2 are sketched in Figure 5.7.1.

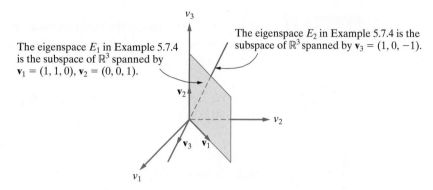

The eigenspace E_1 in Example 5.7.4 is the subspace of $\mathbb{R}^3$ spanned by $\mathbf{v}_1 = (1, 1, 0)$, $\mathbf{v}_2 = (0, 0, 1)$.

The eigenspace E_2 in Example 5.7.4 is the subspace of $\mathbb{R}^3$ spanned by $\mathbf{v}_3 = (1, 0, -1)$.

Figure 5.7.1: Geometrical description of the eigenspaces determined in Example 5.7.4.

We now consider the relationship between eigenvectors corresponding to *distinct* eigenvalues. There is one key theorem which has already been illustrated by the examples in the previous section.

Theorem 5.7.5   Eigenvectors corresponding to *distinct* eigenvalues are linearly independent.

Proof   We use induction to prove the result. Let $\lambda_1, \lambda_2, \ldots, \lambda_m$ be distinct eigenvalues of A with corresponding eigenvectors $\mathbf{v}_1, \mathbf{v}_2, \ldots, \mathbf{v}_m$. It is certainly true that $\{\mathbf{v}_1\}$ is linearly independent. Now suppose that $\{\mathbf{v}_1, \mathbf{v}_2, \ldots, \mathbf{v}_k\}$ is linearly independent for some $k < m$, and consider the set $\{\mathbf{v}_1, \mathbf{v}_2, \ldots, \mathbf{v}_k, \mathbf{v}_{k+1}\}$. We wish to show that this set of vectors is linearly independent. Consider

$$c_1\mathbf{v}_1 + c_2\mathbf{v}_2 + \cdots + c_k\mathbf{v}_k + c_{k+1}\mathbf{v}_{k+1} = \mathbf{0}. \tag{5.7.1}$$

Premultiplying both sides of this equation by A and using $A\mathbf{v}_i = \lambda_i\mathbf{v}_i$ yields

$$c_1\lambda_1\mathbf{v}_1 + c_2\lambda_2\mathbf{v}_2 + \cdots + c_k\lambda_k\mathbf{v}_k + c_{k+1}\lambda_{k+1}\mathbf{v}_{k+1} = \mathbf{0}. \tag{5.7.2}$$

But, from Equation (5.7.1),

$$c_{k+1}\mathbf{v}_{k+1} = -(c_1\mathbf{v}_1 + c_2\mathbf{v}_2 + \cdots + c_k\mathbf{v}_k),$$

so that Equation (5.7.2) can be written as

$$c_1\lambda_1\mathbf{v}_1 + c_2\lambda_2\mathbf{v}_2 + \cdots + c_k\lambda_k\mathbf{v}_k - \lambda_{k+1}(c_1\mathbf{v}_1 + c_2\mathbf{v}_2 + \cdots + c_k\mathbf{v}_k) = \mathbf{0}.$$

That is,

$$c_1(\lambda_1 - \lambda_{k+1})\mathbf{v}_1 + c_2(\lambda_2 - \lambda_{k+1})\mathbf{v}_2 + \cdots + c_k(\lambda_k - \lambda_{k+1})\mathbf{v}_k = \mathbf{0}.$$

Since $\mathbf{v}_1, \mathbf{v}_2, \ldots, \mathbf{v}_k$ are linearly independent, this implies that

$$c_1(\lambda_1 - \lambda_{k+1}) = 0, \qquad c_2(\lambda_2 - \lambda_{k+1}) = 0, \qquad \ldots, \qquad c_k(\lambda_k - \lambda_{k+1}) = 0,$$

and hence, since the λ_i are distinct, $c_1 = c_2 = \cdots = c_k = 0$. But now, since $\mathbf{v}_{k+1} \neq \mathbf{0}$, it follows from Equation (5.7.1) that $c_{k+1} = 0$ also, and so $\{\mathbf{v}_1, \mathbf{v}_2, \ldots, \mathbf{v}_k, \mathbf{v}_{k+1}\}$ is linearly independent. We have therefore shown that the desired result is true for $\{\mathbf{v}_1, \mathbf{v}_2, \ldots, \mathbf{v}_k, \mathbf{v}_{k+1}\}$ whenever it is true for $\{\mathbf{v}_1, \mathbf{v}_2, \ldots, \mathbf{v}_k\}$, and, since the result is true for a single eigenvector, it is true for $\{\mathbf{v}_1, \mathbf{v}_2, \ldots, \mathbf{v}_k\}$, $1 \leq k \leq m$. ∎

Corollary 5.7.6   Let $E_1, E_2, \ldots, E_k$ denote the eigenspaces of an $n \times n$ matrix A. In each eigenspace, choose a set of linearly independent eigenvectors, and let $\{\mathbf{v}_1, \mathbf{v}_2, \ldots, \mathbf{v}_r\}$ denote the union of the linearly independent sets. Then $\{\mathbf{v}_1, \mathbf{v}_2, \ldots, \mathbf{v}_r\}$ is linearly independent.

Proof We argue by contradiction. Suppose that $\{\mathbf{v}_1, \mathbf{v}_2, \ldots, \mathbf{v}_r\}$ is linearly dependent. Then there exist scalars $c_1, c_2, \ldots, c_r$, not all zero, such that

$$c_1\mathbf{v}_1 + c_2\mathbf{v}_2 + \cdots + c_r\mathbf{v}_r = \mathbf{0}, \tag{5.7.3}$$

which can be written as

$$\mathbf{w}_1 + \mathbf{w}_2 + \cdots + \mathbf{w}_k = \mathbf{0},$$

where $\mathbf{w}_i$ is the sum of those terms in (5.7.3) that involve vectors in E_i. Note that some $\mathbf{w}_i \neq \mathbf{0}$; otherwise, since some $c_j \neq 0$, we would have a nontrivial linear combination of the vectors in E_j resulting in $\mathbf{w}_j = \mathbf{0}$, which contradicts our choice of the set of linearly independent eigenvectors. But if $\mathbf{w}_i \neq \mathbf{0}$, then this implies that $\{\mathbf{w}_1, \mathbf{w}_2, \ldots, \mathbf{w}_k\}$ is linearly dependent, which contradicts Theorem 5.7.5. Consequently, all of the scalars in Equation (5.7.3) must be zero, and so $\{\mathbf{v}_1, \mathbf{v}_2, \ldots, \mathbf{v}_r\}$ is indeed linearly independent. ∎

Since the dimension of $\mathbb{R}^n$ (or $\mathbb{C}^n$) is n, the maximum number of linearly independent eigenvectors that A can have is n. In such a case, we say that A is nondefective. The following definition introduces the appropriate terminology.

DEFINITION   5.7.7

An $n \times n$ matrix A that has n linearly independent eigenvectors is called **nondefective**. In such a case, we say that A has a **complete set of eigenvectors**. If A has less than n linearly independent eigenvectors, it is called **defective**.

If A is nondefective, then any set of n linearly independent eigenvectors of A is a basis for $\mathbb{R}^n$ (or $\mathbb{C}^n$). Such a basis is referred to as an **eigenbasis** of A.

Example 5.7.8   For the matrix in Example 5.7.4, $\{(1, 1, 0), (0, 0, 1), (1, 0, -1)\}$ is a complete set of eigenvectors. Consequently, the matrix is nondefective. Likewise, the matrices in Examples 5.6.5, 5.6.6, and 5.6.9 are all nondefective, while the matrix in Example 5.6.7 is defective. □

Example 5.7.9   Determine whether $A = \begin{bmatrix} 4 & -1 \\ 1 & 2 \end{bmatrix}$ is defective or not.

Solution: The characteristic polynomial of A is

$$p(\lambda) = (4 - \lambda)(2 - \lambda) + 1 = \lambda^2 - 6\lambda + 9 = (\lambda - 3)^2.$$

Thus, $\lambda_1 = 3$ is an eigenvalue of multiplicity 2. The eigenvectors of A are easily found from the augmented matrix of the system $(A - \lambda_1 I)\mathbf{v} = \mathbf{0}$, which is

$$\begin{bmatrix} 1 & -1 & 0 \\ 1 & -1 & 0 \end{bmatrix}.$$

We find that the set of eigenvectors take the form

$$\mathbf{v} = (r, r) = r(1, 1),$$

where r is a nonzero real number. So the eigenspace corresponding to $\lambda_1 = 3$ is

$$E_1 = \{\mathbf{v} \in \mathbb{R}^2 : \mathbf{v} = r(1, 1), \ r \in \mathbb{R}\} = \text{span}\{(1, 1)\}.$$

Thus, $\dim[E_1] = 1 < 2$, and hence A is defective. □

The next result is a direct consequence of Theorem 5.7.5.

Corollary 5.7.10 If an $n \times n$ matrix A has n *distinct* eigenvalues, then it is nondefective.

Proof Denote the n distinct eigenvalues of A by $\lambda_1, \lambda_2, \ldots, \lambda_n$, and denote the corresponding eigenvectors by $\mathbf{v}_1, \mathbf{v}_2, \ldots, \mathbf{v}_n$, respectively. By Theorem 5.7.5, $\{\mathbf{v}_1, \mathbf{v}_2, \ldots, \mathbf{v}_n\}$ is linearly independent. Thus, A is nondefective. ∎

Note that if A does *not* have n distinct eigenvalues, it *may still* be nondefective. For instance, in Example 5.6.6, there are only two distinct eigenvalues, but there are three linearly independent eigenvectors, which gives a complete set. The general result is as follows.

Theorem 5.7.11 An $n \times n$ matrix A is nondefective if and only if the dimension of each eigenspace is the same as the algebraic multiplicity m_i of the corresponding eigenvalue—that is, if and only if $\dim[E_i] = m_i$ for each i.

Proof Suppose that A is nondefective, with eigenspaces $E_1, E_2, \ldots, E_k$ of dimensions $n_1, n_2, \ldots, n_k$, respectively. Since A is nondefective, $n_1 + n_2 + \cdots + n_k = n$. If $n_i < m_i$ for some i, then since $n_i \leq m_i$ for each i, we have

$$n = n_1 + n_2 + \cdots + n_k < m_1 + m_2 + \cdots + m_k = n,$$

a contradiction. Thus, $n_i = m_i$ for each i; that is, the dimension of each eigenspace is the same as the algebraic multiplicity of the corresponding eigenvalue.

Conversely, if $n_i = m_i$ for each i, then

$$n = m_1 + m_2 + \cdots + m_k = n_1 + n_2 + \cdots + n_k,$$

which means that the union of the linearly independent eigenvectors that span each eigenspace consists of n eigenvectors of A, and this union is linearly independent by Corollary 5.7.6. Thus, A has n linearly independent eigenvectors. ∎

Theorem 5.7.11 really says that, in order for a matrix A to be nondefective, each eigenvalue of A must "pull its weight" in the sense that it must have a corresponding eigenspace "large enough" in terms of dimension to match its multiplicity. In Example 5.6.7, for instance, the eigenvalue $\lambda_1 = 1$ has multiplicity 2, but the corresponding eigenspace is only one dimensional. Therefore, the matrix in that example is defective.

Exercises for 5.7

Key Terms

Algebraic multiplicity, Geometric multiplicity, Eigenspace of A corresponding to λ, Defective matrix, Nondefective matrix, Complete set of eigenvectors, eigenbasis.

Skills

- Be able to compute the algebraic and geometric multiplicities of an eigenvalue λ of a square matrix A.

- For a square matrix A, be able to compute its eigenspaces and find bases and the dimension of each.

- Be able to determine if a given square matrix is defective or nondefective.

True-False Review

For Questions 1–7, decide if the given statement is **true** or **false**, and give a brief justification for your answer. If true, you can quote a relevant definition or theorem from the text. If false, provide an example, illustration, or brief explanation of why the statement is false.

1. An $n \times n$ matrix A is nondefective if it has n linearly independent eigenvectors.

2. Each eigenspace of an $n \times n$ matrix is a subspace of $\mathbb{R}^n$.

3. If A has an eigenvalue λ of algebraic multiplicity 3, then eigenspace E_λ cannot be more than three dimensional.

4. If S is a set consisting of exactly one nonzero vector from each eigenspace of a matrix A, then S is linearly independent.

5. If the eigenvalues of a 3×3 matrix A are $\lambda = -1, 2, 6$, then A is nondefective.

6. If a matrix A is has a repeated eigenvalue, then it is defective.

7. If $(\lambda_1, \mathbf{v}_1)$ and $(\lambda_2, \mathbf{v}_2)$ are two eigenvalue/eigenvector pairs of a matrix A with $\lambda_1 \neq \lambda_2$, then $\{\mathbf{v}_1, \mathbf{v}_2\}$ is linearly independent.

Problems

For Problems 1–15, determine the multiplicity of each eigenvalue and a basis for each eigenspace of the given matrix. Hence, determine the dimension of each eigenspace and state whether the matrix is defective or nondefective.

1. $\begin{bmatrix} 1 & 4 \\ 2 & 3 \end{bmatrix}.$

2. $\begin{bmatrix} 3 & 0 \\ 0 & 3 \end{bmatrix}.$

3. $\begin{bmatrix} 1 & 2 \\ -2 & 5 \end{bmatrix}.$

4. $\begin{bmatrix} 5 & 5 \\ -2 & -1 \end{bmatrix}.$

5. $\begin{bmatrix} 3 & -4 & -1 \\ 0 & -1 & -1 \\ 0 & -4 & 2 \end{bmatrix}.$

6. $\begin{bmatrix} 4 & 0 & 0 \\ 0 & 2 & -3 \\ 0 & -2 & 1 \end{bmatrix}.$

7. $\begin{bmatrix} 3 & 1 & 0 \\ -1 & 5 & 0 \\ 0 & 0 & 4 \end{bmatrix}.$

8. $\begin{bmatrix} 3 & 0 & 0 \\ 2 & 0 & -4 \\ 1 & 4 & 0 \end{bmatrix}.$

9. $\begin{bmatrix} 4 & 1 & 6 \\ -4 & 0 & -7 \\ 0 & 0 & -3 \end{bmatrix}.$

10. $\begin{bmatrix} 2 & 0 & 0 \\ 0 & 2 & 0 \\ 0 & 0 & 2 \end{bmatrix}.$

11. $\begin{bmatrix} 7 & -8 & 6 \\ 8 & -9 & 6 \\ 0 & 0 & -1 \end{bmatrix}.$

12. $\begin{bmatrix} 2 & 2 & -1 \\ 2 & 1 & -1 \\ 2 & 3 & -1 \end{bmatrix}.$

13. $\begin{bmatrix} 1 & -1 & 2 \\ 1 & -1 & 2 \\ 1 & -1 & 2 \end{bmatrix}.$

14. $\begin{bmatrix} 2 & 3 & 0 \\ -1 & 0 & 1 \\ -2 & -1 & 4 \end{bmatrix}.$

15. $\begin{bmatrix} 0 & -1 & -1 \\ -1 & 0 & -1 \\ -1 & -1 & 0 \end{bmatrix}.$

For Problems 16–20, determine whether the given matrix is defective or nondefective.

16. $A = \begin{bmatrix} 2 & 3 \\ 2 & 1 \end{bmatrix}$;

characteristic polynomial $p(\lambda) = (\lambda + 1)(\lambda - 4)$.

17. $A = \begin{bmatrix} 6 & 5 \\ -5 & -4 \end{bmatrix}$;

characteristic polynomial $p(\lambda) = (\lambda - 1)^2$.

18. $A = \begin{bmatrix} 1 & -2 \\ 5 & 3 \end{bmatrix}$;

characteristic polynomial $p(\lambda) = \lambda^2 - 4\lambda + 13$.

19. $A = \begin{bmatrix} 1 & -3 & 1 \\ -1 & -1 & 1 \\ -1 & -3 & 3 \end{bmatrix}$;

characteristic polynomial $p(\lambda) = -(\lambda - 2)^2(\lambda + 1)$.

20. $A = \begin{bmatrix} -1 & 2 & 2 \\ -4 & 5 & 2 \\ -4 & 2 & 5 \end{bmatrix}$;

characteristic polynomial $p(\lambda) = (3 - \lambda)^3$.

For Problems 21–25, determine a basis for each eigenspace of A and sketch the eigenspaces.

21. $A = \begin{bmatrix} 2 & 1 \\ 3 & 4 \end{bmatrix}$.

22. $A = \begin{bmatrix} 2 & 3 \\ 0 & 2 \end{bmatrix}$.

23. $A = \begin{bmatrix} 5 & 0 \\ 0 & 5 \end{bmatrix}$.

24. $A = \begin{bmatrix} 3 & 1 & -1 \\ 1 & 3 & -1 \\ -1 & -1 & 3 \end{bmatrix}$;

characteristic polynomial $p(\lambda) = (5 - \lambda)(\lambda - 2)^2$.

25. $A = \begin{bmatrix} -3 & 1 & 0 \\ -1 & -1 & 2 \\ 0 & 0 & -2 \end{bmatrix}$.

26. The matrix

$$A = \begin{bmatrix} 2 & -2 & 3 \\ 1 & -1 & 3 \\ 1 & -2 & 4 \end{bmatrix}$$

has eigenvalues $\lambda_1 = 1$ and $\lambda_2 = 3$.

(a) Determine a basis for the eigenspace E_1 and then use the Gram-Schmidt procedure to obtain an orthogonal basis for E_1.

(b) Are the vectors in E_1 orthogonal to the vectors in E_2?

27. Repeat the previous question for

$$A = \begin{bmatrix} 1 & -1 & 1 \\ -1 & 1 & 1 \\ 1 & 1 & 1 \end{bmatrix},$$

assuming that A has eigenvalues $\lambda_1 = 2$, $\lambda_2 = -1$.

28. The matrix

$$A = \begin{bmatrix} a & b & c \\ a & b & c \\ a & b & c \end{bmatrix}$$

has eigenvalues $0, 0$, and $a + b + c$. Determine all values of the constants a, b, and c for which A is non-defective.

29. Consider the characteristic polynomial of an $n \times n$ matrix A; namely,

$$p(\lambda) = \det(A - \lambda I)$$

$$= \begin{vmatrix} a_{11} - \lambda & a_{12} & \cdots & a_{1n} \\ a_{21} & a_{22} - \lambda & \cdots & a_{2n} \\ \vdots & \vdots & & \vdots \\ a_{n1} & a_{n2} & \cdots & a_{nn} - \lambda \end{vmatrix} \quad (5.7.4)$$

which can be written in either of the following equivalent forms:

$$p(\lambda) = (-1)^n \lambda^n + b_1 \lambda^{n-1} + \cdots + b_n, \quad (5.7.5)$$
$$p(\lambda) = (\lambda_1 - \lambda)(\lambda_2 - \lambda) \ldots (\lambda_n - \lambda), \quad (5.7.6)$$

where $\lambda_1, \lambda_2, \ldots, \lambda_n$ (not necessarily distinct) are the eigenvalues of A.

(a) Use Equations (5.7.4) and (5.7.5) to show that

$$b_1 = (-1)^{n-1}(a_{11} + a_{22} + \cdots + a_{nn}),$$
$$b_n = \det(A).$$

Recall that the quantity $a_{11} + a_{22} + \cdots + a_{nn}$ is called the *trace* of the matrix A, denoted $\text{tr}(A)$.

(b) Use Equations (5.7.5) and (5.7.6) to show that

$$b_1 = (-1)^{n-1}(\lambda_1 + \lambda_2 + \cdots + \lambda_n),$$
$$b_n = \lambda_1 \lambda_2 \ldots \lambda_n.$$

(c) Use your results from (a) and (b) to show that

$$\det(A) = \text{product of the eigenvalues of } A$$
$$\text{tr}(A) = \text{sum of the eigenvalues of } A$$

30. Use the result of Problem 29 to determine the sum and the product of the eigenvalues of each matrix A below:

(a) $A = \begin{bmatrix} 2 & 0 & 5 \\ 0 & -1 & 1 \\ 3 & -4 & 2 \end{bmatrix}$.

(b) $A = \begin{bmatrix} 0 & -3 & 1 & 1 \\ 0 & 2 & -1 & 3 \\ -1 & 1 & 1 & 1 \\ 1 & 0 & 5 & -2 \end{bmatrix}$.

(c) $A = \begin{bmatrix} 12 & 11 & 9 & -7 \\ 2 & 3 & -5 & 6 \\ 10 & 8 & 5 & 4 \\ 1 & 0 & 3 & 4 \end{bmatrix}$.

31. Let E_i denote the eigenspace of A corresponding to the eigenvalue λ_i. Use Theorem 4.3.2 to prove that E_i is a subspace of $\mathbb{C}^n$.

32. Let $\mathbf{v}_1$ and $\mathbf{v}_2$ be eigenvectors of A corresponding to the distinct eigenvalues λ_1 and λ_2, respectively. Prove that $\mathbf{v}_1$ and $\mathbf{v}_2$ are linearly independent. [**Hint:** Model your proof on the general case considered in Theorem 5.7.5.]

33. Let E_i denote the eigenspace of A corresponding to the eigenvalue λ_i. If $\{\mathbf{v}_i\}$ is a basis for E_1 and $\{\mathbf{v}_2, \mathbf{v}_3\}$ is a basis for E_2, prove that $\{\mathbf{v}_1, \mathbf{v}_2, \mathbf{v}_3\}$ is linearly independent. [**Hint:** Model your proof on the general case considered in Corollary 5.7.6.]

For Problems 34–38, use some form of technology to determine the eigenvalues and a basis for each eigenspace of the given matrix. Hence, determine the dimension of each eigenspace and state whether the matrix is defective or nondefective.

34. $\diamond$ $A = \begin{bmatrix} 1 & -3 & 3 \\ -1 & -2 & 3 \\ -1 & -3 & 4 \end{bmatrix}$.

35. $\diamond$ $A = \begin{bmatrix} 1 & 1 & 1 \\ 1 & 1 & 1 \\ 1 & 1 & 1 \end{bmatrix}$.

36. $\diamond$ $A = \begin{bmatrix} 3 & \sqrt{2} & 3 \\ \sqrt{2} & 3 & \sqrt{2} \\ 3 & \sqrt{2} & 3 \end{bmatrix}$.

37. $\diamond$ $A = \begin{bmatrix} 25 & -6 & 12 \\ 11 & 0 & 6 \\ -44 & 12 & -21 \end{bmatrix}$.

38. $\diamond$ $A = \begin{bmatrix} 1 & 2 & 1 & 2 \\ 2 & 1 & 2 & 1 \\ 1 & 2 & 1 & 2 \\ 2 & 1 & 2 & 1 \end{bmatrix}$.

For Problems 39–40, show that the given matrix is nondefective.

39. $\diamond$ $A = \begin{bmatrix} a & b & a \\ b & a & b \\ a & b & a \end{bmatrix}$.

40. $\diamond$ $A = \begin{bmatrix} a & a & b \\ a & 2a+b & a \\ b & a & a \end{bmatrix}$.

5.8 Diagonalization

A powerful application of the theory of eigenvalues and eigenvectors is diagonalization. As motivation for this application, we consider the linear system of differential equations

$$\frac{dx_1}{dt} = a_{11}x_1 + a_{12}x_2, \tag{5.8.1}$$

$$\frac{dx_2}{dt} = a_{21}x_1 + a_{22}x_2, \tag{5.8.2}$$

which we write as a vector equation

$$\mathbf{x}' = A\mathbf{x}, \tag{5.8.3}$$

where

$$\mathbf{x} = \begin{bmatrix} x_1 \\ x_2 \end{bmatrix}, \qquad \mathbf{x}' = \begin{bmatrix} x_1' \\ x_2' \end{bmatrix}, \qquad A = [a_{ij}].$$

The prime symbol denotes differentiation with respect to the independent variable t. In general, we cannot integrate the given system directly, because each equation involves both unknown functions x_1 and x_2. We say that the equations are *coupled*. Suppose, however, we make a *linear* change of variables defined by

$$\mathbf{x} = S\mathbf{y}, \tag{5.8.4}$$

where S is an invertible matrix of constants. Then

$$\mathbf{x}' = S\mathbf{y}',$$

so that (5.8.3) is transformed to the equivalent system

$$S\mathbf{y}' = AS\mathbf{y}.$$

Premultiplying by S^{-1} yields

$$\mathbf{y}' = B\mathbf{y}, \tag{5.8.5}$$

where $B = S^{-1}AS$. The question that now arises is whether it is possible to choose S such that the system (5.8.5) can be integrated. For if this is the case, then, upon performing the integration to find $\mathbf{y}$, the solution $\mathbf{x}$ to (5.8.3) can be determined from (5.8.4). The results of this section will establish that, provided A is nondefective, all of this is possible.

The aim of the section therefore is to investigate matrices that are related via $B = S^{-1}AS$. Of particular interest to us is the possibility of choosing S so that $S^{-1}AS$ has a simple structure. Of course, the question that needs answering is: How simple a form should we aim for? First we introduce some terminology and a helpful result.

DEFINITION 5.8.1

Let A and B be $n \times n$ matrices. We say A is **similar** to B if there exists an invertible matrix S such that $B = S^{-1}AS$.

Example 5.8.2 If

$$A = \begin{bmatrix} -4 & -10 \\ 3 & 7 \end{bmatrix} \quad \text{and} \quad B = \begin{bmatrix} 1 & -1 \\ 0 & 2 \end{bmatrix},$$

verify that $B = S^{-1}AS$, where

$$S = \begin{bmatrix} 2 & -3 \\ -1 & 2 \end{bmatrix}.$$

Solution: It is easily shown that

$$S^{-1} = \begin{bmatrix} 2 & 3 \\ 1 & 2 \end{bmatrix},$$

so that

$$S^{-1}AS = \begin{bmatrix} 2 & 3 \\ 1 & 2 \end{bmatrix} \begin{bmatrix} -4 & -10 \\ 3 & 7 \end{bmatrix} \begin{bmatrix} 2 & -3 \\ -1 & 2 \end{bmatrix}$$

$$= \begin{bmatrix} 2 & 3 \\ 1 & 2 \end{bmatrix} \begin{bmatrix} 2 & -8 \\ -1 & 5 \end{bmatrix} = \begin{bmatrix} 1 & -1 \\ 0 & 2 \end{bmatrix};$$

that is, $S^{-1}AS = B$. $\square$

Theorem 5.8.3 Similar matrices have the same eigenvalues (including multiplicities).

Proof If A is similar to B, then $B = S^{-1}AS$ for some invertible matrix S. Thus,

$$\det(B - \lambda I) = \det(S^{-1}AS - \lambda I) = \det(S^{-1}AS - \lambda S^{-1}S)$$
$$= \det(S^{-1}(A - \lambda I)S) = \det(S^{-1})\det(A - \lambda I)\det(S)$$
$$= \det(A - \lambda I),$$

where we have used the fact that $\det(S^{-1}) = 1/\det(S)$ in the final step. Consequently, A and B have the same characteristic polynomial and hence the same eigenvalues (and multiplicities). ∎

We now know from Theorem 5.8.3 that A and $S^{-1}AS$ have the same eigenvalues $\lambda_1, \lambda_2, \ldots, \lambda_n$. Furthermore, the simplest matrix that has these eigenvalues is the diagonal matrix $D = \text{diag}(\lambda_1, \lambda_2, \ldots, \lambda_n)$. Consequently, the simplest possible structure for $S^{-1}AS$ is

$$S^{-1}AS = \text{diag}(\lambda_1, \lambda_2, \ldots, \lambda_n).$$

We have therefore been led to the question:

> For an $n \times n$ matrix A, when does an invertible matrix S exist such that
> $$S^{-1}AS = \text{diag}(\lambda_1, \lambda_2, \ldots, \lambda_n)?$$

The answer is provided in the next crucial theorem.

Theorem 5.8.4

An $n \times n$ matrix A is similar to a diagonal matrix if and only if A is nondefective. In such a case, if $\mathbf{v}_1, \mathbf{v}_2, \ldots, \mathbf{v}_n$ denote n linearly independent eigenvectors of A and

$$S = [\mathbf{v}_1, \mathbf{v}_2, \ldots, \mathbf{v}_n],$$

then

$$S^{-1}AS = \text{diag}(\lambda_1, \lambda_2, \ldots, \lambda_n),$$

where $\lambda_1, \lambda_2, \ldots, \lambda_n$ are the eigenvalues of A (not necessarily distinct) corresponding to the eigenvectors $\mathbf{v}_1, \mathbf{v}_2, \ldots, \mathbf{v}_n$.

Proof If A is similar to a diagonal matrix, then there exists an invertible matrix $S = [\mathbf{v}_1, \mathbf{v}_2, \ldots, \mathbf{v}_n]$ such that

$$S^{-1}AS = D, \tag{5.8.6}$$

where $D = \text{diag}(\lambda_1, \lambda_2, \ldots, \lambda_n)$ and, from Theorem 5.8.3, $\lambda_1, \lambda_2, \ldots, \lambda_n$ are the eigenvalues of A. Premultiplying both sides of (5.8.6) by S yields

$$AS = SD,$$

or equivalently,

$$[A\mathbf{v}_1, A\mathbf{v}_2, \ldots, A\mathbf{v}_n] = [\lambda_1\mathbf{v}_1, \lambda_2\mathbf{v}_2, \ldots, \lambda_n\mathbf{v}_n].$$

Equating corresponding column vectors, we must have

$$A\mathbf{v}_1 = \lambda_1\mathbf{v}_1, \qquad A\mathbf{v}_2 = \lambda_2\mathbf{v}_2, \qquad \ldots, \qquad A\mathbf{v}_n = \lambda_n\mathbf{v}_n.$$

Consequently, $\mathbf{v}_1, \mathbf{v}_2, \ldots, \mathbf{v}_n$ are eigenvectors of A corresponding to the eigenvalues $\lambda_1, \lambda_2, \ldots, \lambda_n$. Further, since $\det(S) \neq 0$, the eigenvectors are linearly independent.

Conversely, suppose A is nondefective, and let $S = [\mathbf{v}_1, \mathbf{v}_2, \ldots, \mathbf{v}_n]$, where $\{\mathbf{v}_1, \mathbf{v}_2, \ldots, \mathbf{v}_n\}$ is any complete set of eigenvectors of A. Then

$$AS = A[\mathbf{v}_1, \mathbf{v}_2, \ldots, \mathbf{v}_n] = [A\mathbf{v}_1, A\mathbf{v}_2, \ldots, A\mathbf{v}_n] = [\lambda_1\mathbf{v}_1, \lambda_2\mathbf{v}_2, \ldots, \lambda_n\mathbf{v}_n].$$

This can be written in the equivalent form

$$AS = SD, \tag{5.8.7}$$

where $D = \text{diag}(\lambda_1, \lambda_2, \ldots, \lambda_n)$. Since the columns of S form a linearly independent set, $\det(S) \neq 0$, and hence S is invertible. Premultiplying both sides of (5.8.7) by S^{-1} yields

$$S^{-1}AS = D$$

so that A is indeed similar to a diagonal matrix. ■

DEFINITION 5.8.5

An $n \times n$ matrix that is similar to a diagonal matrix is said to be **diagonalizable**.

Remark   By Theorem 5.8.4, the term "diagonalizable" is synonymous with "nondefective." The matrices in Examples 5.6.5, 5.6.6, 5.6.9, and 5.7.4 are diagonalizable, while the matrix in Example 5.6.7 is not. As an exercise, the reader should write down an appropriate matrix S in each of Examples 5.6.5, 5.6.6, 5.6.9, and 5.7.4, together with diagonal matrices to which the given matrices are similar. To illustrate this, we offer the following example.

Example 5.8.6   Verify that

$$A = \begin{bmatrix} 3 & -2 & -2 \\ -3 & -2 & -6 \\ 3 & 6 & 10 \end{bmatrix}$$

is diagonalizable and find a matrix S such that $S^{-1}AS = \text{diag}(\lambda_1, \lambda_2, \lambda_3)$.

Solution: The characteristic polynomial for A is $p(\lambda) = -(\lambda - 4)^2(\lambda - 3)$ (this can be determined, for example, by cofactor expansion of $A - \lambda I$), so that the eigenvalues of A are $\lambda = 4, 4, 3$. Corresponding linearly independent eigenvectors are

$$\lambda = 4: \qquad \mathbf{v}_1 = (-2, 0, 1), \quad \mathbf{v}_2 = (-2, 1, 0),$$
$$\lambda = 3: \qquad \mathbf{v}_3 = (1, 3, -3).$$

Consequently, A is nondefective and therefore diagonalizable. If we set

$$S = \begin{bmatrix} -2 & -2 & 1 \\ 0 & 1 & 3 \\ 1 & 0 & -3 \end{bmatrix},$$

then, according to Theorem 5.8.4,

$$S^{-1}AS = \text{diag}(4, 4, 3).$$

It is important to note that the ordering of the eigenvalues in the diagonal matrix must be in correspondence with the ordering of the eigenvectors in the matrix S. For example, permuting columns 2 and 3 in S yields the matrix

$$S_1 = \begin{bmatrix} -2 & 1 & -2 \\ 0 & 3 & 1 \\ 1 & -3 & 0 \end{bmatrix}.$$

Since the column vectors of S_1 are eigenvectors of A, Theorem 5.8.4 implies that

$$S_1^{-1} A S_1 = \text{diag}(4, 3, 4). \qquad \square$$

Now we return to the system of differential equations (5.8.1) and (5.8.2) that motivated our discussion. We assume that A is nondefective and choose $S = [\mathbf{v}_1, \mathbf{v}_2]$ such that $S^{-1} A S = \text{diag}(\lambda_1, \lambda_2)$. Then the system of differential equations (5.8.5) reduces to

$$\mathbf{y}' = \text{diag}(\lambda_1, \lambda_2)\mathbf{y}.$$

That is,

$$\begin{bmatrix} y_1' \\ y_2' \end{bmatrix} = \begin{bmatrix} \lambda_1 & 0 \\ 0 & \lambda_2 \end{bmatrix} \begin{bmatrix} y_1 \\ y_2 \end{bmatrix},$$

so that

$$y_1' = \lambda_1 y_1, \qquad y_2' = \lambda_2 y_2.$$

We see that the system of differential equations has decoupled, and both of the resulting differential equations are easily integrated to obtain

$$y_1(t) = c_1 e^{\lambda_1 t}, \qquad y_2(t) = c_2 e^{\lambda_2 t}.$$

From (5.8.4) we see that the solution in the original variables is

$$\mathbf{x}(t) = S\mathbf{y}(t),$$

which can be written in the equivalent form

$$\mathbf{x}(t) = S\mathbf{y}(t) = [\mathbf{v}_1, \mathbf{v}_2] \begin{bmatrix} c_1 e^{\lambda_1 t} \\ c_2 e^{\lambda_2 t} \end{bmatrix}.$$

That is,

$$\mathbf{x}(t) = c_1 e^{\lambda_1 t} \mathbf{v}_1 + c_2 e^{\lambda_2 t} \mathbf{v}_2, \qquad (5.8.8)$$

where λ_1, λ_2 are the eigenvalues of A corresponding to the eigenvectors $\mathbf{v}_1, \mathbf{v}_2$. The formula (5.8.8) looks suspiciously like a statement about the set of all solutions to the system of differential equations being generated by taking all linear combinations of a certain set of basic solutions (in this case, two such solutions). As mentioned previously, a full vector space formulation for systems of linear differential equations will be given in Chapter 7.

Example 5.8.7   Use the ideas introduced in this section to determine all solutions to

$$x_1' = 6x_1 - x_2, \qquad x_2' = -5x_1 + 2x_2.$$

Solution: The given system can be written as

$$\mathbf{x}' = A\mathbf{x},$$

where $A = \begin{bmatrix} 6 & -1 \\ -5 & 2 \end{bmatrix}$. The transformed system is

$$\mathbf{y}' = (S^{-1}AS)\mathbf{y}, \tag{5.8.9}$$

where

$$\mathbf{x} = S\mathbf{y}. \tag{5.8.10}$$

To determine S, we need the eigenvalues and eigenvectors of A. The characteristic polynomial of A is

$$p(\lambda) = \begin{vmatrix} 6 - \lambda & -1 \\ -5 & 2 - \lambda \end{vmatrix} = (\lambda - 1)(\lambda - 7).$$

Hence, A is nondefective by Corollary 5.7.10. The eigenvectors are easily computed:

$$\lambda_1 = 1: \qquad \mathbf{v} = r(1, 5),$$
$$\lambda_2 = 7: \qquad \mathbf{v} = s(-1, 1).$$

We could now substitute into (5.8.8) to obtain the solution to the system, but it is more instructive to go through the steps that led to that equation. If we set

$$S = \begin{bmatrix} 1 & -1 \\ 5 & 1 \end{bmatrix},$$

then from Theorem 5.8.4,

$$S^{-1}AS = \text{diag}(1, 7),$$

so that the system (5.8.9) is

$$\begin{bmatrix} y_1' \\ y_2' \end{bmatrix} = \begin{bmatrix} 1 & 0 \\ 0 & 7 \end{bmatrix} \begin{bmatrix} y_1 \\ y_2 \end{bmatrix}.$$

Hence,

$$y_1' = y_1, \qquad y_2' = 7y_2.$$

Both of these equations can be integrated to obtain

$$y_1(t) = c_1 e^t, \qquad y_2(t) = c_2 e^{7t}.$$

Using (5.8.10) to return to the original variables, we have

$$\mathbf{x} = S\mathbf{y} = \begin{bmatrix} 1 & -1 \\ 5 & 1 \end{bmatrix} \begin{bmatrix} c_1 e^t \\ c_2 e^{7t} \end{bmatrix} = \begin{bmatrix} c_1 e^t - c_2 e^{7t} \\ 5c_1 e^t + c_2 e^{7t} \end{bmatrix}.$$

Consequently,

$$x_1(t) = c_1 e^t - c_2 e^{7t}, \qquad x_2(t) = 5c_1 e^t + c_2 e^{7t}. \qquad \square$$

In concluding this section, we note that if a matrix A is defective, then from Theorem 5.8.4, there does *not* exist a matrix S such that $S^{-1}AS = \text{diag}(\lambda_1, \lambda_2, \ldots, \lambda_n)$. To handle defective matrices, we try to find an invertible matrix S such that $S^{-1}AS$ is "close" to a diagonal matrix. This will be pursued in Section 5.11.

Exercises for 5.8

Key Terms

Similar matrices, Diagonalizable matrix.

Skills

- Be able to find matrices that are similar to a given matrix A.

- Be able to list properties shared by similar matrices and use these to help decide whether two given matrices are similar or not.

- Be able to determine if a given matrix is diagonalizable or not.

- Be able to find n linearly independent eigenvectors for an $n \times n$ diagonalizable matrix A and thus construct an $n \times n$ matrix S such that $S^{-1}AS$ is a diagonal matrix.

- Be able to solve linear systems of differential equations in which the coefficient matrix A is diagonalizable.

True-False Review

For Questions 1–8, decide if the given statement is **true** or **false**, and give a brief justification for your answer. If true, you can quote a relevant definition or theorem from the text. If false, provide an example, illustration, or brief explanation of why the statement is false.

1. A square matrix A is diagonalizable if and only if it is nondefective.

2. If A is an invertible, diagonalizable matrix, then so is A^{-1}.

3. If two matrices A and B have the same set of eigenvalues (including multiplicities), then they are similar.

4. An $n \times n$ matrix is diagonalizable if and only if it has n eigenvectors.

5. If the characteristic polynomial $p(\lambda)$ of a matrix A has no repeated roots, then A is diagonalizable.

6. If A is a diagonalizable matrix, then so is A^2.

7. A square matrix A is always similar to itself.

8. If A is an $n \times n$ matrix with n odd whose eigenspaces are all even dimensional, then A is not diagonalizable.

Problems

For Problems 1–12, determine whether the given matrix is diagonalizable. Where possible, find a matrix S such that

$$S^{-1}AS = \text{diag}(\lambda_1, \lambda_2, \ldots, \lambda_n).$$

1. $\begin{bmatrix} -1 & -2 \\ -2 & 2 \end{bmatrix}$.

2. $\begin{bmatrix} -7 & 4 \\ -4 & 1 \end{bmatrix}$.

3. $\begin{bmatrix} 1 & -8 \\ 2 & -7 \end{bmatrix}$.

4. $\begin{bmatrix} 0 & 4 \\ -4 & 0 \end{bmatrix}$.

5. $\begin{bmatrix} 1 & 0 & 0 \\ 0 & 3 & 7 \\ 1 & 1 & -3 \end{bmatrix}$.

6. $\begin{bmatrix} 1 & -2 & 0 \\ 2 & -3 & 0 \\ 2 & -2 & -1 \end{bmatrix}$.

7. $\begin{bmatrix} 0 & -2 & -2 \\ -2 & 0 & -2 \\ -2 & -2 & 0 \end{bmatrix}$.

8. $\begin{bmatrix} -2 & 1 & 4 \\ -2 & 1 & 4 \\ -2 & 1 & 4 \end{bmatrix}$.

9. $\begin{bmatrix} 2 & 0 & 0 \\ 0 & 1 & 0 \\ 2 & -1 & 1 \end{bmatrix}$.

10. $\begin{bmatrix} 4 & 0 & 0 \\ 3 & -1 & -1 \\ 0 & 2 & 1 \end{bmatrix}$.

11. $\begin{bmatrix} 0 & 2 & -1 \\ -2 & 0 & -2 \\ 1 & 2 & 0 \end{bmatrix}$.

12. $\begin{bmatrix} 1 & -2 & 0 \\ -2 & 1 & 0 \\ 0 & 0 & 3 \end{bmatrix}$.

◇ For Problems 13–14, use some form of technology to determine a complete set of eigenvectors for the given matrix A. Construct a matrix S that diagonalizes A and explicitly verify that $S^{-1}AS = \text{diag}(\lambda_1, \lambda_2, \ldots, \lambda_n)$.

13. $A = \begin{bmatrix} 1 & -3 & 3 \\ -2 & -4 & 6 \\ -2 & -6 & 8 \end{bmatrix}$.

14. $A = \begin{bmatrix} 3 & -2 & 3 & -2 \\ -2 & 3 & -2 & 3 \\ 3 & -2 & 3 & -2 \\ -2 & 3 & -2 & 3 \end{bmatrix}$.

For Problems 15–19, use the ideas introduced in this section to solve the given system of differential equations.

15. $x_1' = x_1 + 4x_2, \quad x_2' = 2x_1 + 3x_2$.

16. $x_1' = 6x_1 - 2x_2, \quad x_2' = -2x_1 + 6x_2$.

17. $x_1' = 9x_1 + 6x_2, \quad x_2' = -10x_1 - 7x_2$.

18. $x_1' = -12x_1 - 7x_2, \quad x_2' = 16x_1 + 10x_2$.

19. $x_1' = x_2, \quad x_2' = -x_1$.

For Problems 20–21, first write the given system of differential equations in matrix form, and then use the ideas from this section to determine all solutions.

20. $x_1' = 3x_1 - 4x_2 - x_3, x_2' = -x_2 - x_3,$
$x_3' = -4x_2 + 2x_3$.

21. $x_1' = x_1 + x_2 - x_3, x_2' = x_1 + x_2 + x_3,$
$x_3' = -x_1 + x_2 + x_3$.

22. Let A be a nondefective matrix. Then

$$S^{-1}AS = D,$$

where D is a diagonal matrix. This can be written as

$$A = SDS^{-1}.$$

Use this result to show that

$$A^2 = SD^2S^{-1},$$

and that for every positive integer k,

$$A^k = SD^kS^{-1}.$$

23. If $D = \text{diag}(\lambda_1, \lambda_2, \ldots, \lambda_n)$, show that for every positive integer k,

$$D^k = \text{diag}(\lambda_1^k, \lambda_2^k, \ldots, \lambda_n^k).$$

24. Use the results of the preceding two problems to determine A^3 and A^5, given that

$$A = \begin{bmatrix} -7 & -4 \\ 18 & 11 \end{bmatrix}.$$

25. We call a matrix B a **square root** of A if $B^2 = A$.

(a) Show that if $D = \text{diag}(\lambda_1, \lambda_2, \ldots, \lambda_n)$, then the matrix

$$\sqrt{D} = \text{diag}(\sqrt{\lambda_1}, \sqrt{\lambda_2}, \ldots, \sqrt{\lambda_n})$$

is a square root of D.

(b) Show that if A is a nondefective matrix with $S^{-1}AS = D$ for some invertible matrix S and diagonal matrix D, then $S\sqrt{D}S^{-1}$ is a square root of A.

(c) Find a square root for the matrix

$$A = \begin{bmatrix} 6 & -2 \\ -3 & 7 \end{bmatrix}.$$

26. Prove the following properties for similar matrices:

(a) A matrix A is always similar to itself.

(b) If A is similar to B, then B is similar to A.

(c) If A is similar to B and B is similar to C, then A is similar to C.

27. If A is similar to B, prove that A^T is similar to B^T.

28. In Theorem 5.8.3, we proved that similar matrices have the same eigenvalues. This problem investigates the relationship between their eigenvectors. Let $\mathbf{v}$ be an eigenvector of A corresponding to the eigenvalue λ. Prove that if $B = S^{-1}AS$, then $S^{-1}\mathbf{v}$ is an eigenvector of B corresponding to the eigenvalue λ.

29. Let A be a nondefective matrix and let S be a matrix such that $S^{-1}AS = \text{diag}(\lambda_1, \lambda_2, \ldots, \lambda_n)$, where all λ_i are nonzero.

(a) Prove that A is invertible.

(b) Prove that

$$S^{-1}A^{-1}S = \text{diag}\left(\frac{1}{\lambda_1}, \frac{1}{\lambda_2}, \ldots, \frac{1}{\lambda_n}\right).$$

30. Let A be a nondefective matrix and let S be a matrix such that $S^{-1}AS = \text{diag}(\lambda_1, \lambda_2, \ldots, \lambda_n)$.

(a) Prove that if $Q = (S^T)^{-1}$, then

$$Q^{-1}A^TQ = \text{diag}(\lambda_1, \lambda_2, \ldots, \lambda_n).$$

This establishes that A^T is also nondefective.

(b) If M_C denotes the matrix of cofactors of S, prove that the column vectors of M_C are linearly independent eigenvectors of A^T. [**Hint:** Use the adjoint method to determine S^{-1}.]

31. If

$$A = \begin{bmatrix} -2 & 4 \\ 1 & 1 \end{bmatrix},$$

determine S such that $S^{-1}AS = \text{diag}(-3, 2)$, and use the result from the previous problem to determine all eigenvectors of A^T.

Problems 32–34 deal with the generalization of the diagonalization problem to defective matrices. A complete discussion of this topic can be found in Section 5.11.

32. Let A be a 2×2 *defective* matrix. It follows from Theorem 5.8.4 that A is not diagonalizable. However, it can be shown that A is similar to the Jordan canonical form matrix

$$J_\lambda = \begin{bmatrix} \lambda & 1 \\ 0 & \lambda \end{bmatrix}.$$

Thus, there exists a matrix $S = [\mathbf{v}_1, \mathbf{v}_2]$, such that

$$S^{-1}AS = J_\lambda.$$

Prove that $\mathbf{v}_1$ and $\mathbf{v}_2$ must satisfy

$$(A - \lambda I)\mathbf{v}_1 = \mathbf{0}, \qquad (5.8.11)$$
$$(A - \lambda I)\mathbf{v}_2 = \mathbf{v}_1. \qquad (5.8.12)$$

Equation (5.8.11) is the statement that $\mathbf{v}_1$ must be an eigenvector of A corresponding to the eigenvalue λ. Any vectors that satisfy (5.8.12) are called *generalized eigenvectors* of A. The subject of generalized eigenvectors and Jordan canonical form matrices will be taken up in detail in Section 5.11.

33. Show that

$$A = \begin{bmatrix} 2 & 1 \\ -1 & 4 \end{bmatrix}$$

is defective and use the previous problem to determine a matrix S such that

$$S^{-1}AS = \begin{bmatrix} 3 & 1 \\ 0 & 3 \end{bmatrix}.$$

34. Let λ be an eigenvalue of the 3×3 matrix A of multiplicity 3, and suppose the corresponding eigenspace has dimension 1. It can be shown that, in this case, there exists a matrix $S = [\mathbf{v}_1, \mathbf{v}_2, \mathbf{v}_3]$ such that

$$S^{-1}AS = \begin{bmatrix} \lambda & 1 & 0 \\ 0 & \lambda & 1 \\ 0 & 0 & \lambda \end{bmatrix}.$$

Prove that $\mathbf{v}_1, \mathbf{v}_2, \mathbf{v}_3$ must satisfy

$$(A - \lambda I)\mathbf{v}_1 = \mathbf{0}$$
$$(A - \lambda I)\mathbf{v}_2 = \mathbf{v}_1$$
$$(A - \lambda I)\mathbf{v}_3 = \mathbf{v}_2.$$

35. In this problem, we establish that similar matrices describe the same linear transformation relative to different bases. Assume that $\{\mathbf{e}_1, \mathbf{e}_2, \ldots, \mathbf{e}_n\}$ and $\{\mathbf{f}_1, \mathbf{f}_2, \ldots, \mathbf{f}_n\}$ are bases for a vector space V and let $T: V \rightarrow V$ be a linear transformation. Define the $n \times n$ matrices $A = [a_{ik}]$ and $B = [b_{ik}]$ by

$$T(\mathbf{e}_k) = \sum_{i=1}^{n} a_{ik}\mathbf{e}_i, \qquad k = 1, 2, \ldots, n, \quad (5.8.13)$$

$$T(\mathbf{f}_k) = \sum_{i=1}^{n} b_{ik}\mathbf{f}_i, \qquad k = 1, 2, \ldots, n. \quad (5.8.14)$$

If we express each of the basis vectors $\mathbf{f}_1, \mathbf{f}_2, \ldots, \mathbf{f}_n$ in terms of the basis vectors $\mathbf{e}_1, \mathbf{e}_2, \ldots, \mathbf{e}_n$, we have that

$$\mathbf{f}_i = \sum_{j=1}^{n} s_{ji}\mathbf{e}_j, \qquad i = 1, 2, \ldots, n, \quad (5.8.15)$$

for appropriate scalars s_{ji}. Thus, the matrix $S = [s_{ji}]$ describes the relationship between the two bases.

(a) Prove that S is invertible. [**Hint:** Use the fact that $\mathbf{f}_1, \mathbf{f}_2, \ldots, \mathbf{f}_n$ are linearly independent.]

(b) Use (5.8.14) and (5.8.15) to show that, for $k = 1, 2, \ldots, n$, we have

$$T(\mathbf{f}_k) = \sum_{j=1}^{n}\left(\sum_{i=1}^{n} s_{ji}b_{ik}\right)\mathbf{e}_j,$$

or equivalently,

$$T(\mathbf{f}_k) = \sum_{i=1}^{n}\left(\sum_{j=1}^{n} s_{ij}b_{jk}\right)\mathbf{e}_i. \quad (5.8.16)$$

(c) Use (5.8.13) and (5.8.15) to show that, for $k = 1, 2, \ldots n$, we have

$$T(\mathbf{f}_k) = \sum_{i=1}^{n}\left(\sum_{j=1}^{n} a_{ij}s_{jk}\right)\mathbf{e}_i. \quad (5.8.17)$$

(d) Use (5.8.16) and (5.8.17) together with the fact that $\{e_1, e_2, \ldots, e_n\}$ is linearly independent to show that

$$\sum_{j=1}^{n} s_{ij}b_{jk} = \sum_{j=1}^{n} a_{ij}s_{jk}, \quad 1 \le i, k \le n,$$

and hence that

$$SB = AS.$$

Finally, conclude that A and B are related by

$$B = S^{-1}AS.$$

5.9   An Introduction to the Matrix Exponential Function

This section provides a brief introduction to the matrix exponential function. In Chapter 7, we will see that this function plays a valuable role in the analysis and solution of systems of differential equations.

DEFINITION   5.9.1

Let A be an $n \times n$ matrix of constants. We define the **matrix exponential function**, denoted e^{At}, by

$$e^{At} = I_n + At + \frac{1}{2!}(At)^2 + \frac{1}{3!}(At)^3 + \cdots + \frac{1}{k!}(At)^k + \cdots . \tag{5.9.1}$$

It can be shown that for all $n \times n$ matrices A and all values of $t \in (-\infty, \infty)$, the infinite series appearing on the right-hand-side of (5.9.1) converges to an $n \times n$ matrix. Consequently, e^{At} is a well-defined $n \times n$ matrix.

Properties of the Matrix Exponential Function

1. If A and B are $n \times n$ matrices satisfying $AB = BA$, then

$$\boxed{e^{(A+B)t} = e^{At}e^{Bt}.}$$

2. For all $n \times n$ matrices A, e^{At} is invertible and

$$\boxed{(e^{At})^{-1} = e^{(-A)t} = e^{-At}.}$$

That is,

$$e^{At}e^{-At} = I_n.$$

The proofs of these results require a precise definition of convergence of an infinite series of matrices. This would take us too far astray from the main focus of this text, hence the proofs are omitted. [See, for example, M. W. Hirsch and S. Smale, *Differential Equations, Dynamical Systems, and Linear Algebra* (New York: Academic Press, 1974).]

We now turn to the issue of computing e^{At}.

Example 5.9.2   Compute e^{At} if

$$A = \begin{bmatrix} 2 & 0 \\ 0 & -1 \end{bmatrix}.$$

Solution: In this case, we see that

$$At = \begin{bmatrix} 2t & 0 \\ 0 & -t \end{bmatrix},$$

so that for any positive integer k, we have

$$(At)^k = \begin{bmatrix} (2t)^k & 0 \\ 0 & (-t)^k \end{bmatrix}.$$

Thus,

$$e^{At} = \begin{bmatrix} 1 & 0 \\ 0 & 1 \end{bmatrix} + \begin{bmatrix} 2t & 0 \\ 0 & -t \end{bmatrix} + \frac{1}{2!} \begin{bmatrix} (2t)^2 & 0 \\ 0 & (-t)^2 \end{bmatrix} + \cdots + \frac{1}{k!} \begin{bmatrix} (2t)^k & 0 \\ 0 & (-t)^k \end{bmatrix} + \cdots$$

$$= \begin{bmatrix} \sum_{k=0}^{\infty} \frac{1}{k!}(2t)^k & 0 \\ 0 & \sum_{k=0}^{\infty} \frac{1}{k!}(-t)^k \end{bmatrix}.$$

Hence,

$$e^{At} = \begin{bmatrix} e^{2t} & 0 \\ 0 & e^{-t} \end{bmatrix}. \qquad \square$$

More generally, it can be shown (Problem 1) that

> If $A = \mathrm{diag}(d_1, d_2, \ldots, d_n)$, then $e^{At} = \mathrm{diag}(e^{d_1 t}, e^{d_2 t}, \ldots, e^{d_n t})$.

If A is not a diagonal matrix, then the computation of e^{At} is more involved. The next simplest case that can arise is when A is nondefective. In this case, as we have shown in the previous section, A is *similar* to a diagonal matrix, and we might suspect that this would lead to a simplification in the evaluation of e^{At}. We now show that this is indeed the case. Suppose that A has n linearly independent eigenvectors $\mathbf{v}_1, \mathbf{v}_2, \ldots, \mathbf{v}_n$, and define the $n \times n$ matrix S by

$$S = [\mathbf{v}_1, \mathbf{v}_2, \ldots, \mathbf{v}_n].$$

Then from Theorem 5.8.4,

$$S^{-1}AS = \mathrm{diag}(\lambda_1, \lambda_2, \ldots, \lambda_n), \tag{5.9.2}$$

where $\lambda_1, \lambda_2, \ldots, \lambda_n$ are the eigenvalues of A corresponding to the eigenvectors $\mathbf{v}_1, \mathbf{v}_2, \ldots, \mathbf{v}_n$. Premultiplying (5.9.2) by S and postmultiplying by S^{-1} yields

$$A = SDS^{-1},$$

where

$$D = \mathrm{diag}(\lambda_1, \lambda_2, \ldots, \lambda_n).$$

We now compute e^{At}. From Definition 5.9.1,

$$e^{At} = I_n + At + \frac{1}{2!}(At)^2 + \frac{1}{3!}(At)^3 + \cdots + \frac{1}{k!}(At)^k + \cdots$$

$$= SS^{-1} + (SDS^{-1})t + \frac{1}{2!}(SDS^{-1})^2 t^2 + \cdots + \frac{1}{k!}(SDS^{-1})^k t^k + \cdots .$$

A short exercise (see Problem 22 in Section 5.8) shows that for every positive integer k, we have $(SDS^{-1})^k = SD^kS^{-1}$. Substituting this into the preceding expression for e^{At} above, we get

$$e^{At} = S\left[I + Dt + \frac{1}{2!}(Dt)^2 + \cdots + \frac{1}{k!}(Dt)^k + \cdots\right]S^{-1}.$$

That is,

$$e^{At} = Se^{Dt}S^{-1}.$$

Consequently, we have established the next theorem.

Theorem 5.9.3   Let A be a nondefective $n \times n$ matrix with linearly independent eigenvectors $\mathbf{v}_1, \mathbf{v}_2, \ldots, \mathbf{v}_n$ and corresponding eigenvalues $\lambda_1, \lambda_2, \ldots, \lambda_n$. Then

$$e^{At} = Se^{Dt}S^{-1},$$

where $S = [\mathbf{v}_1, \mathbf{v}_2, \ldots, \mathbf{v}_n]$ and $D = \mathrm{diag}(\lambda_1, \lambda_2, \ldots, \lambda_n)$.

Example 5.9.4   Determine e^{At} if

$$A = \begin{bmatrix} 3 & 3 \\ 5 & 1 \end{bmatrix}.$$

Solution: The eigenvalues of A are $\lambda_1 = 6$ and $\lambda_2 = -2$, and therefore A is nondefective. A straightforward computation yields the following eigenvectors, which correspond respectively to λ_1 and λ_2:

$$\mathbf{v}_1 = (1, 1) \qquad \text{and} \qquad \mathbf{v}_2 = (-3, 5).$$

It follows from Theorem 5.9.3 that if we set

$$S = \begin{bmatrix} 1 & -3 \\ 1 & 5 \end{bmatrix} \qquad \text{and} \qquad D = \mathrm{diag}(6, -2),$$

then

$$e^{At} = Se^{Dt}S^{-1}.$$

That is,

$$e^{At} = S\begin{bmatrix} e^{6t} & 0 \\ 0 & e^{-2t} \end{bmatrix}S^{-1}. \tag{5.9.3}$$

It is easily shown that

$$S^{-1} = \frac{1}{8}\begin{bmatrix} 5 & 3 \\ -1 & 1 \end{bmatrix},$$

so that, substituting into Equation (5.9.3),

$$e^{At} = \frac{1}{8}\begin{bmatrix} 1 & -3 \\ 1 & 5 \end{bmatrix}\begin{bmatrix} e^{6t} & 0 \\ 0 & e^{-2t} \end{bmatrix}\begin{bmatrix} 5 & 3 \\ -1 & 1 \end{bmatrix} = \frac{1}{8}\begin{bmatrix} 1 & -3 \\ 1 & 5 \end{bmatrix}\begin{bmatrix} 5e^{6t} & 3e^{6t} \\ -e^{-2t} & e^{-2t} \end{bmatrix}.$$

Consequently,

$$e^{At} = \frac{1}{8}\begin{bmatrix} 5e^{6t} + 3e^{-2t} & 3(e^{6t} - e^{-2t}) \\ 5(e^{6t} - e^{-2t}) & 3e^{6t} + 5e^{-2t} \end{bmatrix}. \qquad \square$$

The computation of e^{At} when A is a defective matrix is best accomplished by relating e^{At} to the solution of a linear, homogeneous system of differential equations (see Section 7.8). Alternatively, one can use a generalization of the diagonalization procedure to defective matrices via the machinery of Jordan canonical forms (see Section 5.11).

Exercises for 5.9

Key Terms

Matrix exponential function.

Skills

- Be able to compute the matrix exponential function e^{At} for any nondefective matrix A.

True-False Review

For Questions 1–6, decide if the given statement is **true** or **false**, and give a brief justification for your answer. If true, you can quote a relevant definition or theorem from the text. If false, provide an example, illustration, or brief explanation of why the statement is false.

1. The matrix exponential function e^{At} is defined only for a square matrix A.

2. If $A^3 = 0$, then $e^A = I + A + A^2$.

3. The matrix exponential function e^{At} is invertible if and only if A is invertible.

4. The matrix exponential function e^{At} is an infinite series that converges for all values of t.

5. For any diagonal matrix D and invertible matrix S, we have $(SDS^{-1})^k = S^k D^k (S^{-1})^k$ for all positive integers k.

6. For all $n \times n$ matrices A, $e^{A^2 t} = (e^{At})^2$.

Problems

1. If $A = \text{diag}(d_1, d_2, \ldots, d_n)$, prove that

$$e^{At} = \text{diag}(e^{d_1 t}, e^{d_2 t}, e^{d_3 t}, \ldots, e^{d_n t}).$$

2. If

$$A = \begin{bmatrix} -3 & 0 \\ 0 & 5 \end{bmatrix},$$

determine e^{At} and e^{-At}.

3. Prove that for all values of the constant λ,

$$e^{\lambda I_n t} = e^{\lambda t} I_n.$$

4. Consider the matrix

$$A = \begin{bmatrix} a & b \\ 0 & a \end{bmatrix}.$$

We can write $A = B + C$, where

$$B = \begin{bmatrix} a & 0 \\ 0 & a \end{bmatrix}, \qquad C = \begin{bmatrix} 0 & b \\ 0 & 0 \end{bmatrix}.$$

(a) Verify that $BC = CB$.

(b) Verify that $C^2 = 0_2$, and determine e^{Ct}.

(c) Use property (1) of the matrix exponential function to find e^{At}.

5. If

$$A = \begin{bmatrix} a & b \\ -b & a \end{bmatrix},$$

use property 1 of the matrix exponential function and Definition 5.9.1 to show that

$$e^{At} = e^{at} \begin{bmatrix} \cos bt & \sin bt \\ -\sin bt & \cos bt \end{bmatrix}.$$

For Problems 6–12, show that A is nondefective and use Theorem 5.9.3 to find e^{At}.

6. $A = \begin{bmatrix} 1 & 2 \\ 0 & 3 \end{bmatrix}$.

7. $A = \begin{bmatrix} 3 & 1 \\ 1 & 3 \end{bmatrix}$.

8. $A = \begin{bmatrix} 0 & 2 \\ -2 & 0 \end{bmatrix}$.

9. $A = \begin{bmatrix} -1 & 3 \\ -3 & -1 \end{bmatrix}$.

10. $A = \begin{bmatrix} a & b \\ -b & a \end{bmatrix}$.

11. $A = \begin{bmatrix} 3 & -2 & -2 \\ 1 & 0 & -2 \\ 0 & 0 & 3 \end{bmatrix}$.

12. $A = \begin{bmatrix} 6 & -2 & -1 \\ 8 & -2 & -2 \\ 4 & -2 & 1 \end{bmatrix}$, and you may assume that $p(\lambda) = -(\lambda - 2)^2(\lambda - 1)$.

An $n \times n$ matrix A that satisfies $A^k = 0$ for some k is called **nilpotent**. For Problems 13–17, show that the given matrix is nilpotent, and use Definition 5.9.1 to determine e^{At}.

13. $A = \begin{bmatrix} -3 & 9 \\ -1 & 3 \end{bmatrix}$.

14. $A = \begin{bmatrix} 1 & 1 \\ -1 & -1 \end{bmatrix}$.

15. $A = \begin{bmatrix} 0 & 0 & 0 \\ 1 & 0 & 0 \\ 0 & 1 & 0 \end{bmatrix}$.

16. $A = \begin{bmatrix} -1 & -6 & -5 \\ 0 & -2 & -1 \\ 1 & 2 & 3 \end{bmatrix}$.

17. $A = \begin{bmatrix} 0 & 1 & 0 & 0 \\ 0 & 0 & 1 & 0 \\ 0 & 0 & 0 & 1 \\ 0 & 0 & 0 & 0 \end{bmatrix}$.

18. Let A be the $n \times n$ matrix whose only nonzero elements are
$$a_{i+1\, i} = 1, \quad i = 1, 2, \ldots, n - 1.$$
Determine e^{At}. (See Problem 15 for the case $n = 3$.)

19. If
$$A = \begin{pmatrix} A_0 & 0 \\ 0 & B_0 \end{pmatrix}$$
is a block diagonal matrix with matrices A_0 and B_0, prove that
$$e^{At} = \begin{pmatrix} e^{A_0 t} & 0 \\ 0 & e^{B_0 t} \end{pmatrix}.$$

---

*5.10   Orthogonal Diagonalization and Quadratic Forms

Symmetric matrices with real elements play an important role in many applications of linear algebra. In particular, they arise in the study of quadratic forms (defined below), and these appear in applications including geometry, statistics, mechanics, and electrical engineering.

In this section, we study the special properties satisfied by the eigenvectors of a real symmetric matrix and show how this simplifies the diagonalization problem introduced in the previous section. We also give a brief application of the theoretical results that are obtained to quadratic forms. We begin with a definition.

DEFINITION 5.10.1

A real $n \times n$ invertible matrix A is called **orthogonal** if
$$A^{-1} = A^T.$$

Example 5.10.2   Verify that the following matrix is an orthogonal matrix:
$$A = \begin{bmatrix} \frac{1}{3} & -\frac{2}{3} & \frac{2}{3} \\ \frac{2}{3} & -\frac{1}{3} & -\frac{2}{3} \\ \frac{2}{3} & \frac{2}{3} & \frac{1}{3} \end{bmatrix}.$$

Solution: For the given matrix, we have
$$AA^T = \begin{bmatrix} \frac{1}{3} & -\frac{2}{3} & \frac{2}{3} \\ \frac{2}{3} & -\frac{1}{3} & -\frac{2}{3} \\ \frac{2}{3} & \frac{2}{3} & \frac{1}{3} \end{bmatrix} \begin{bmatrix} \frac{1}{3} & \frac{2}{3} & \frac{2}{3} \\ -\frac{2}{3} & -\frac{1}{3} & \frac{2}{3} \\ \frac{2}{3} & -\frac{2}{3} & \frac{1}{3} \end{bmatrix} = I_3.$$

Similarly, $A^T A = I_3$, so that $A^T = A^{-1}$. Consequently, A is an orthogonal matrix.   □

---

*This section may be omitted without loss of continuity

If we look more closely at the preceding example, we see that the column vectors of A form an *orthonormal* set of vectors. The same can be said of the set of row vectors of A. The next theorem establishes that this is a basic characterizing property of all orthogonal matrices.[8]

Theorem 5.10.3   A real $n \times n$ matrix A is an orthogonal matrix if and only if the row (or column) vectors of A form an orthonormal set of vectors.

Proof We leave this proof as an exercise (Problem 26). ∎

We can now state the main results of this section.

Basic Results for Real Symmetric Matrices

Theorem 5.10.4   Let A be a real symmetric matrix. Then

1. All eigenvalues of A are real.
2. Real eigenvectors of A that correspond to *distinct* eigenvalues are *orthogonal*.
3. A is nondefective.
4. A has a complete set of *orthonormal* eigenvectors.
5. A can be diagonalized with an *orthogonal* matrix S such that $S^T A S$ is diagonal.

The proof of Theorem 5.10.4 is deferred to the end of the section. We emphasize that the orthogonal diagonalization of a real symmetric matrix A alluded to in part 5 requires that we use a complete set of *orthonormal* eigenvectors in constructing S.

Example 5.10.5   Find a complete set of orthonormal eigenvectors of

$$A = \begin{bmatrix} 2 & 2 & 1 \\ 2 & 5 & 2 \\ 1 & 2 & 2 \end{bmatrix},$$

and determine an orthogonal matrix that diagonalizes A.

Solution: Since A is real and symmetric, a complete orthonormal set of eigenvectors exists. To find it, we first seek the eigenvalues of A. The characteristic polynomial of A is

$$\det(A - \lambda I) = \begin{vmatrix} 2 - \lambda & 2 & 1 \\ 2 & 5 - \lambda & 2 \\ 1 & 2 & 2 - \lambda \end{vmatrix} = -(\lambda - 1)^2(\lambda - 7).$$

Thus, A has eigenvalues

$$\lambda_1 = 1 \text{ (with multiplicity 2)} , \qquad \lambda_2 = 7 \text{ (with multiplicity 1).}$$

Eigenvalue $\lambda_1 = 1$: In this case, the system $(A - \lambda_1 I)\mathbf{v} = \mathbf{0}$ reduces to the single equation

$$v_1 + 2v_2 + v_3 = 0,$$

---

[8] Because of this characterization of orthogonal matrices, it might be more appropriate to call them *orthonormal* matrices. However, this would run contrary to the vast literature that has established the fundamental term *orthogonal* matrix as we have done in Definition 5.10.1.

which has solution $(-2r - s, r, s)$, so that the corresponding eigenvectors are

$$\mathbf{v} = (-2r - s, r, s) = r(-2, 1, 0) + s(-1, 0, 1).$$

Two linearly independent eigenvectors corresponding to the eigenvalue $\lambda_1 = 1$ are

$$\mathbf{x}_1 = (-1, 0, 1), \qquad \mathbf{x}_2 = (-2, 1, 0).$$

Although $\{\mathbf{x}_1, \mathbf{x}_2\}$ is not an orthonormal set, we can apply the Gram-Schmidt process in Section 4.12 to replace this set with an orthonormal set $\{\mathbf{u}_1, \mathbf{u}_2\}$ as follows.

First, let $\mathbf{v}_1 = \mathbf{x}_1$ and $\mathbf{v}_2 = \mathbf{x}_2 - \mathbf{P}(\mathbf{x}_2, \mathbf{v}_1) = \mathbf{x}_2 - \mathbf{v}_1 = (-1, 1, -1)$. Now $\{\mathbf{v}_1, \mathbf{v}_2\}$ is an orthogonal set, and we can normalize each vector to get

$$\mathbf{u}_1 = \left(-\frac{1}{\sqrt{2}}, 0, \frac{1}{\sqrt{2}}\right) \quad \text{and} \quad \mathbf{u}_2 = \left(-\frac{1}{\sqrt{3}}, \frac{1}{\sqrt{3}}, -\frac{1}{\sqrt{3}}\right).$$

Thus, $\{\mathbf{u}_1, \mathbf{u}_2\}$ is an orthonormal set of eigenvectors of A corresponding to $\lambda_1 = 1$.

Eigenvalue $\lambda_2 = 7$: The reader can check that Gaussian elimination applied to the system $(A - \lambda_2 I)\mathbf{v} = \mathbf{0}$ yields the eigenvector

$$\mathbf{v}_3 = (1, 2, 1).$$

Replacing this with a unit vector, we have

$$\mathbf{u}_3 = \left(\frac{1}{\sqrt{6}}, \frac{2}{\sqrt{6}}, \frac{1}{\sqrt{6}}\right).$$

Notice that $\mathbf{u}_3$ is orthogonal to both $\mathbf{u}_1$ and $\mathbf{u}_2$, as guaranteed by Theorem 5.10.4 (2). It follows that a complete set of orthonormal eigenvectors for the matrix A is $\{\mathbf{u}_1, \mathbf{u}_2, \mathbf{u}_3\}$; that is,

$$\left\{\left(-\frac{1}{\sqrt{2}}, 0, \frac{1}{\sqrt{2}}\right), \left(-\frac{1}{\sqrt{3}}, \frac{1}{\sqrt{3}}, -\frac{1}{\sqrt{3}}\right), \left(\frac{1}{\sqrt{6}}, \frac{2}{\sqrt{6}}, \frac{1}{\sqrt{6}}\right)\right\}.$$

If we set

$$S = \begin{bmatrix} -\frac{1}{\sqrt{2}} & -\frac{1}{\sqrt{3}} & \frac{1}{\sqrt{6}} \\ 0 & \frac{1}{\sqrt{3}} & \frac{2}{\sqrt{6}} \\ \frac{1}{\sqrt{2}} & -\frac{1}{\sqrt{3}} & \frac{1}{\sqrt{6}} \end{bmatrix},$$

then S is an orthogonal matrix satisfying

$$S^T A S = S^{-1} A S = \text{diag}(1, 1, 7). \qquad \square$$

Notice that in this example, we took the linearly independent eigenvectors associated with each eigenvalue separately and applied Gram-Schmidt to them to get an orthonormal basis for each eigenspace separately. It is important to realize that this is a legal process to apply to *any* $n \times n$ matrix. However, the advantage of a real *symmetric* matrix is that, for such a matrix, the eigenvectors from one eigenspace and the eigenvectors from another eigenspace are already orthogonal. This is not true for an arbitrary matrix.

When we apply Gram-Schmidt to a set of eigenvectors *within a single eigenspace*, the vectors resulting from the process are still eigenvectors belonging to that eigenspace. However, applying Gram-Schmidt to vectors arising in different eigenspaces will result in vectors $\mathbf{w}_i$ and $\mathbf{u}_i$ that are not even eigenvectors at all! Hence, for an arbitrary matrix, we have no possibility of "orthogonalizing" eigenvectors occurring in distinct eigenspaces.

Quadratic Forms

If A is a *symmetric* $n \times n$ real matrix and $\mathbf{x}$ is a (column) vector in $\mathbb{R}^n$, then an expression of the form

$$\mathbf{x}^T A \mathbf{x}$$

is called a **quadratic form**. We can consider a quadratic form as defining a mapping from $\mathbb{R}^n$ to $\mathbb{R}$. For example, if $n = 2$, then

$$
\begin{aligned}
\mathbf{x}^T A \mathbf{x} &= [x_1 \ x_2] \begin{bmatrix} a_{11} & a_{12} \\ a_{12} & a_{22} \end{bmatrix} \begin{bmatrix} x_1 \\ x_2 \end{bmatrix} \\
&= [x_1 \ x_2] \begin{bmatrix} a_{11}x_1 + a_{12}x_2 \\ a_{12}x_1 + a_{22}x_2 \end{bmatrix} \\
&= a_{11}x_1^2 + 2a_{12}x_1x_2 + a_{22}x_2^2.
\end{aligned}
$$

We see that this is indeed quadratic in x_1 and x_2. Quadratic forms arise in many applications. For example, in geometry, the conic sections have Cartesian equations that can be expressed as

$$\mathbf{x}^T A \mathbf{x} = c,$$

whereas quadric surfaces have equations of this same form, where now A is a 3×3 matrix and $\mathbf{x}$ is a vector in $\mathbb{R}^3$. In mechanics, the kinetic energy K of a physical system with n degrees of freedom (and time-independent constraints) can be written as

$$K = \mathbf{x}^T A \mathbf{x},$$

where A is an $n \times n$ matrix and $\mathbf{x}$ is a vector of (generalized) velocities. The question we are going to address is whether it is possible to make a *linear* change of variables

$$\mathbf{x} = S\mathbf{y} \tag{5.10.1}$$

that enables us to simplify a quadratic form by eliminating the cross terms. Equivalently, we want to reduce a quadratic form to a sum of squares. Making the change of variables (5.10.1) in the general quadratic form yields

$$\mathbf{x}^T A \mathbf{x} = (S\mathbf{y})^T A (S\mathbf{y}) = \mathbf{y}^T (S^T A S)\mathbf{y}. \tag{5.10.2}$$

If we choose $S = [\mathbf{w}_1, \mathbf{w}_2, \ldots, \mathbf{w}_n]$, where $\{\mathbf{w}_1, \mathbf{w}_2, \ldots, \mathbf{w}_n\}$ is any complete set of orthonormal eigenvectors for A, then Theorem 5.10.4 implies that

$$S^T A S = \text{diag}(\lambda_1, \lambda_2, \ldots, \lambda_n),$$

where $\lambda_1, \lambda_2, \ldots, \lambda_n$ are the eigenvalues of A corresponding to $\mathbf{w}_1, \mathbf{w}_2, \ldots, \mathbf{w}_n$. With this choice of S, the right-hand side of Equation (5.10.2) reduces to

$$
[y_1 \ y_2 \ \cdots \ y_n] \begin{bmatrix} \lambda_1 & & & \\ & \lambda_2 & & \\ & & \ddots & \\ & & & \lambda_n \end{bmatrix} \begin{bmatrix} y_1 \\ y_2 \\ \vdots \\ y_n \end{bmatrix} = \lambda_1 y_1^2 + \lambda_2 y_2^2 + \cdots + \lambda_n y_n^2,
$$

and we have accomplished our goal of reducing the quadratic form to a sum of squares. This result is summarized in the following theorem.

Theorem 5.10.6   **(Principal Axes Theorem)**

Let A be an $n \times n$ real symmetric matrix. Then the quadratic form $\mathbf{x}^T A \mathbf{x}$ can be reduced to a sum of squares by the change of variables $\mathbf{x} = S\mathbf{y}$, where S is an orthogonal matrix whose column vectors $\{\mathbf{w}_1, \mathbf{w}_2, \ldots, \mathbf{w}_n\}$ are any complete orthonormal set of eigenvectors for A. The transformed quadratic form is

$$\lambda_1 y_1^2 + \lambda_2 y_2^2 + \cdots + \lambda_n y_n^2,$$

where $\lambda_1, \lambda_2, \ldots, \lambda_n$ are the eigenvalues of A corresponding to $\mathbf{w}_1, \mathbf{w}_2, \ldots, \mathbf{w}_n$.

The column vectors of the matrix S in the Principal Axes Theorem are called **principal axes** for the quadratic form $\mathbf{x}^T A \mathbf{x}$.

Example 5.10.7   By transforming to principal axes, reduce the quadratic form $\mathbf{x}^T A \mathbf{x}$ for

$$A = \begin{bmatrix} 2 & 2 & 1 \\ 2 & 5 & 2 \\ 1 & 2 & 2 \end{bmatrix}.$$

to a sum of squares.

Solution: In the previous example, we found an orthogonal matrix that diagonalizes A; namely,

$$S = \begin{bmatrix} -\frac{1}{\sqrt{2}} & -\frac{1}{\sqrt{3}} & \frac{1}{\sqrt{6}} \\ 0 & \frac{1}{\sqrt{3}} & \frac{2}{\sqrt{6}} \\ \frac{1}{\sqrt{2}} & -\frac{1}{\sqrt{3}} & \frac{1}{\sqrt{6}} \end{bmatrix}.$$

Furthermore, the corresponding eigenvalues of A are $(1, 1, 7)$. Consequently, the change of variables $\mathbf{x} = S\mathbf{y}$ reduces the quadratic form to

$$y_1^2 + y_2^2 + 7y_3^2.$$

This is quite a simplification from the quadratic form expressed in the original variables; namely,

$$\mathbf{x}^T A \mathbf{x} = [x_1 \ x_2 \ x_3] \begin{bmatrix} 2 & 2 & 1 \\ 2 & 5 & 2 \\ 1 & 2 & 2 \end{bmatrix} \begin{bmatrix} x_1 \\ x_2 \\ x_3 \end{bmatrix}$$
$$= 2x_1^2 + 5x_2^2 + 2x_3^2 + 4x_1x_2 + 2x_1x_3 + 4x_2x_3. \qquad \square$$

Example 5.10.8   By transforming to principal axes, identify the conic section with Cartesian equation

$$5x_1^2 + 2x_1x_2 + 5x_2^2 = 1.$$

Solution: We first write the quadratic form appearing on the left-hand side of the given equation in matrix form. If we set

$$A = \begin{bmatrix} 5 & 1 \\ 1 & 5 \end{bmatrix},$$

then the given equation is

$$\mathbf{x}^T A \mathbf{x} = 1. \qquad (5.10.3)$$

Observe that A has characteristic polynomial $p(\lambda) = (\lambda - 6)(\lambda - 4)$, so that the eigenvalues of A are $\lambda_1 = 6$ and $\lambda_2 = 4$, with corresponding orthonormal eigenvectors

$$\mathbf{w}_1 = \left(\frac{1}{\sqrt{2}}, \frac{1}{\sqrt{2}}\right) \quad \text{and} \quad \mathbf{w}_2 = \left(-\frac{1}{\sqrt{2}}, \frac{1}{\sqrt{2}}\right).$$

Therefore, if we set

$$S = \begin{bmatrix} \frac{1}{\sqrt{2}} & -\frac{1}{\sqrt{2}} \\ \frac{1}{\sqrt{2}} & \frac{1}{\sqrt{2}} \end{bmatrix},$$

then under the change of variables $\mathbf{x} = S\mathbf{y}$, Equation (5.10.3) reduces to

$$6y_1^2 + 4y_2^2 = 1, \tag{5.10.4}$$

which is the equation of an ellipse. Geometrically, vectors $\mathbf{w}_1$ and $\mathbf{w}_2$ are obtained by rotating the standard basis vectors $\mathbf{e}_1$ and $\mathbf{e}_2$, respectively, counterclockwise through an angle $\pi/4$ radians. Relative to the Cartesian coordinate system (y_1, y_2), corresponding to the principal axes $\mathbf{w}_1, \mathbf{w}_2$, the ellipse has the simple equation (5.10.4). Figure 5.10.1 shows how the conic looks relative to the standard Cartesian axes and the rotated principal axes. □

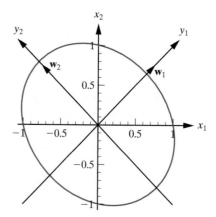

Figure 5.10.1: Relative to Cartesian axes (x_1, x_2), cross terms come into the equation for the ellipse. Rotating to the principal axes (y_1, y_2) eliminates these cross terms.

We conclude this section with a proof of Theorem 5.10.4. The following preliminary lemma will be useful for the proof.

Lemma 5.10.9   Let $\mathbf{v}_1$ and $\mathbf{v}_2$ be vectors in $\mathbb{R}^n$ (or $\mathbb{C}^n$), and let A be an $n \times n$ real symmetric matrix. Then

$$\langle A\mathbf{v}_1, \mathbf{v}_2 \rangle = \langle \mathbf{v}_1, A\mathbf{v}_2 \rangle.$$

Proof The key to the proof is to note that the inner product of two column vectors in $\mathbb{R}^n$ or $\mathbb{C}^n$ can be written in matrix form as

$$[\langle \mathbf{x}, \mathbf{y} \rangle] = \mathbf{x}^T \overline{\mathbf{y}}.$$

Applying this to the vectors $A\mathbf{v}_1$ and $\mathbf{v}_2$ yields

$$[\langle A\mathbf{v}_1, \mathbf{v}_2 \rangle] = (A\mathbf{v}_1)^T \overline{\mathbf{v}}_2 = \mathbf{v}_1^T A^T \overline{\mathbf{v}}_2 = \mathbf{v}_1^T \overline{(A\mathbf{v}_2)} = [\langle \mathbf{v}_1, A\mathbf{v}_2 \rangle],$$

from which the result follows directly. ■

Using this lemma, we can now give the

Proof of Theorem 5.10.4:

1. We must prove that every eigenvalue of a real symmetric matrix A is real. Let $(\lambda_1, \mathbf{v}_1)$ be an eigenvalue/eigenvector pair for A; that is

$$A\mathbf{v}_1 = \lambda_1\mathbf{v}_1. \tag{5.10.5}$$

We will show that $\lambda_1 = \overline{\lambda}_1$, which will prove part 1. Taking the inner product of both sides of (5.10.5) with the vector $\mathbf{v}_1$ yields

$$\langle A\mathbf{v}_1, \mathbf{v}_1\rangle = \langle \lambda_1\mathbf{v}_1, \mathbf{v}_1\rangle.$$

Using the properties of the inner product, we obtain

$$\langle A\mathbf{v}_1, \mathbf{v}_1\rangle = \lambda_1\|\mathbf{v}_1\|^2. \tag{5.10.6}$$

Taking the complex conjugate of (5.10.6) yields (remember that $\|\mathbf{v}_1\|$ is a *real* number)

$$\overline{\langle A\mathbf{v}_1, \mathbf{v}_1\rangle} = \overline{\lambda}_1\|\mathbf{v}_1\|^2.$$

Using the fact that $\overline{\langle \mathbf{u}, \mathbf{v}\rangle} = \langle \mathbf{v}, \mathbf{u}\rangle$,

$$\langle \mathbf{v}_1, A\mathbf{v}_1\rangle = \overline{\lambda}_1\|\mathbf{v}_1\|^2. \tag{5.10.7}$$

Subtracting (5.10.7) from (5.10.6) and using Lemma 5.10.9 yields

$$0 = (\lambda_1 - \overline{\lambda}_1)\|\mathbf{v}_1\|^2. \tag{5.10.8}$$

However, $\mathbf{v}_1 \neq \mathbf{0}$, since it is an eigenvector of A. Consequently, from Equation (5.10.8), we must have

$$\lambda_1 = \overline{\lambda}_1.$$

2. It follows from part 1 that all eigenvalues of A are necessarily real. We can therefore take the underlying vector space as $\mathbb{R}^n$, so that the corresponding eigenvectors of A will also be real. Let $(\lambda_1, \mathbf{v}_1)$ and $(\lambda_2, \mathbf{v}_2)$ be two such eigenvalue/eigenvector pairs with $\lambda_1 \neq \lambda_2$. Then, in addition to (5.10.5) we also have

$$A\mathbf{v}_2 = \lambda_2\mathbf{v}_2. \tag{5.10.9}$$

We must prove that $\langle \mathbf{v}_1, \mathbf{v}_2\rangle = 0$. Taking the inner product of (5.10.5) with $\mathbf{v}_2$ and the inner product of (5.10.9) with $\mathbf{v}_1$ and pulling the scalars out of the inner products, we get, respectively,

$$\langle A\mathbf{v}_1, \mathbf{v}_2\rangle = \lambda_1\langle \mathbf{v}_1, \mathbf{v}_2\rangle, \qquad \langle \mathbf{v}_1, A\mathbf{v}_2\rangle = \lambda_2\langle \mathbf{v}_1, \mathbf{v}_2\rangle.$$

Subtracting the second equation from the first and using Lemma 5.10.9, we obtain

$$0 = (\lambda_1 - \lambda_2)\langle \mathbf{v}_1, \mathbf{v}_2\rangle.$$

Since $\lambda_1 - \lambda_2 \neq 0$ by assumption, it follows that $\langle \mathbf{v}_1, \mathbf{v}_2\rangle = 0$.

3. The proof of this part utilizes some ideas from the next section, so we postpone the proof until the end of the next section.

4. Eigenvectors corresponding to distinct eigenvalues are orthogonal from part 2. Now suppose that the eigenvalue λ_i has multiplicity m_i. Then, since A is nondefective (from part 3), it follows from Theorem 5.7.11 that we can find m_i linearly independent eigenvectors corresponding to λ_i. These vectors span the eigenspace corresponding to λ_i, and hence, we can use the Gram-Schmidt process to find m_i *orthonormal* eigenvectors (corresponding to λ_i) that span this eigenspace. Proceeding in this manner for each eigenvalue, we obtain a complete set of orthonormal eigenvectors.

5. If we denote the orthonormal set of eigenvectors in (4) by $\{\mathbf{w}_1, \mathbf{w}_2, \ldots, \mathbf{w}_n\}$ and let $S = [\mathbf{w}_1, \mathbf{w}_2, \ldots, \mathbf{w}_n]$, then Theorem 5.10.3 implies that S is an orthogonal matrix. Consequently,

$$S^T A S = S^{-1} A S = \text{diag}(\lambda_1, \lambda_2, \ldots, \lambda_n),$$

where we have used Theorem 5.8.4.   ∎

Exercises for 5.10

Key Terms

Orthogonal matrix, Complete set of orthonormal eigenvectors, Quadratic forms, Principal axes.

Skills

- Be able to determine whether a given matrix A is orthogonal or not.

- Be able to call upon the properties of real symmetric matrices given in Theorem 5.10.4.

- Be able to construct a complete set of orthonormal eigenvectors for a given matrix A and construct an orthogonal matrix S such that $S^T A S$ is a diagonal matrix.

- Determine a set of principal axes for a given quadratic form and reduce it to a sum of squares by an appropriate change of variables.

True-False Review

For Questions 1–8, decide if the given statement is **true** or **false**, and give a brief justification for your answer. If true, you can quote a relevant definition or theorem from the text. If false, provide an example, illustration, or brief explanation of why the statement is false.

1. If A is an $n \times n$ orthogonal matrix, then A is invertible and $AA^T = A^T A = I_n$.

2. A real matrix A whose characteristic polynomial is $p(\lambda) = \lambda^3 + \lambda$ cannot be symmetric.

3. A real matrix A with eigenvectors

$$\mathbf{v}_1 = \begin{bmatrix} 1 \\ 2 \end{bmatrix} \quad \text{and} \quad \mathbf{v}_2 = \begin{bmatrix} 1 \\ 3 \end{bmatrix}$$

cannot be symmetric.

4. If A is an $n \times n$ real, symmetric matrix with eigenvalues $\lambda_1, \lambda_2, \ldots, \lambda_n$ corresponding to a complete orthonormal set of eigenvectors for A, then the quadratic form $\mathbf{x}^T A \mathbf{x}$ is transformed into $\lambda_1 y_1^2 + \lambda_2 y_2^2 + \cdots + \lambda_n y_n^2$.

5. For any $n \times n$ real, symmetric matrix A and vectors $\mathbf{v}$ and $\mathbf{w}$ in $\mathbb{R}^n$, we have $\langle A\mathbf{v}, \mathbf{w} \rangle = \langle \mathbf{v}, A\mathbf{w} \rangle$.

6. If A and B are $n \times n$ orthogonal matrices, then AB is also an orthogonal matrix.

7. A real $n \times n$ matrix A is orthogonal if and only if its row (or column) vectors are orthogonal unit vectors.

8. Any real matrix with a complete set of orthonormal eigenvectors is symmetric.

Problems

For Problems 1–13, determine an orthogonal matrix S such that $S^T A S = \text{diag}(\lambda_1, \lambda_2, \ldots, \lambda_n)$, where A denotes the given matrix.

1. $\begin{bmatrix} 2 & 2 \\ 2 & -1 \end{bmatrix}$.

2. $\begin{bmatrix} 4 & 6 \\ 6 & 9 \end{bmatrix}$.

3. $\begin{bmatrix} 1 & 2 \\ 2 & 1 \end{bmatrix}$.

4. $\begin{bmatrix} 0 & 0 & 3 \\ 0 & -2 & 0 \\ 3 & 0 & 0 \end{bmatrix}$.

5. $\begin{bmatrix} 1 & 2 & 1 \\ 2 & 4 & 2 \\ 1 & 2 & 1 \end{bmatrix}$.

6. $\begin{bmatrix} 2 & 0 & 0 \\ 0 & 3 & 1 \\ 0 & 1 & 3 \end{bmatrix}$.

7. $\begin{bmatrix} 0 & 1 & 0 \\ 1 & 0 & 0 \\ 0 & 0 & 1 \end{bmatrix}$.

8. $\begin{bmatrix} 1 & 1 & -1 \\ 1 & 1 & 1 \\ -1 & 1 & 1 \end{bmatrix}$.

9. $\begin{bmatrix} 1 & 0 & -1 \\ 0 & 1 & 1 \\ -1 & 1 & 0 \end{bmatrix}$.

10. $\begin{bmatrix} 3 & 3 & 4 \\ 3 & 3 & 0 \\ 4 & 0 & 3 \end{bmatrix}$.

You may assume that $p(\lambda) = (\lambda + 2)(\lambda - 3)(8 - \lambda)$.

11. $\begin{bmatrix} -3 & 2 & 2 \\ 2 & -3 & 2 \\ 2 & 2 & -3 \end{bmatrix}$.

You may assume that $p(\lambda) = (1 - \lambda)(\lambda + 5)^2$.

12. $\begin{bmatrix} 0 & 1 & 1 \\ 1 & 0 & 1 \\ 1 & 1 & 0 \end{bmatrix}$.

You may assume that $p(\lambda) = (\lambda + 1)^2(2 - \lambda)$.

For Problems 13–16, determine a set of principal axes for the given quadratic form, and reduce the quadratic form to a sum of squares.

13. $\mathbf{x}^T A \mathbf{x}$, $A = \begin{bmatrix} 1 & 3 \\ 3 & 1 \end{bmatrix}$.

14. $\mathbf{x}^T A \mathbf{x}$, $A = \begin{bmatrix} 5 & 2 \\ 2 & 5 \end{bmatrix}$.

15. $\mathbf{x}^T A \mathbf{x}$, $A = \begin{bmatrix} 1 & 1 & -1 \\ 1 & 1 & 1 \\ -1 & 1 & 1 \end{bmatrix}$.

16. $\diamond$ $\mathbf{x}^T A \mathbf{x}$, $A = \begin{bmatrix} 3 & 1 & 3 & 1 \\ 1 & 3 & 1 & 3 \\ 3 & 1 & 3 & 1 \\ 1 & 3 & 1 & 3 \end{bmatrix}$.

17. Consider the general 2×2 real symmetric matrix

$$A = \begin{bmatrix} a & b \\ b & c \end{bmatrix}.$$

Prove that A has an eigenvalue of multiplicity 2 if and only if it is a scalar matrix (that is, a matrix of the form $r I_2$, where r is a constant).

18. (a) Let A be an $n \times n$ real symmetric matrix. Prove that if λ is an eigenvalue of A of multiplicity n, then A is a scalar matrix. [**Hint:** Prove that there exists an orthogonal matrix S such that $S^T A S = \lambda I_n$, and then solve for A.]

(b) State and prove the corresponding result for general $n \times n$ matrices.

19. The 2×2 real symmetric matrix A has two distinct eigenvalues, λ_1 and λ_2. If $\mathbf{v}_1 = (1, 2)$ is an eigenvector of A corresponding to the eigenvalue λ_1, determine an eigenvector corresponding to λ_2.

20. The 2×2 real symmetric matrix A has two distinct eigenvalues λ_1 and λ_2.

(a) If $\mathbf{v}_1 = (a, b)$ is an eigenvector of A corresponding to the eigenvalue λ_1, determine an eigenvector corresponding to λ_2, and hence find an orthogonal matrix S such that $S^T A S = \text{diag}(\lambda_1, \lambda_2)$.

(b) Use your result from part (a) to find A. [Your answer will involve λ_1, λ_2, a and b.]

21. The 3×3 real symmetric matrix A has eigenvalues λ_1 and λ_2 (multiplicity 2).

(a) If $\mathbf{v}_1 = (1, -1, 1)$ spans the eigenspace E_1, determine a basis for E_2 and hence find an orthogonal matrix S, such that $S^T A S = \text{diag}(\lambda_1, \lambda_2, \lambda_2)$.

(b) Use your result from part (a) to find A.

Problems 22–24 deal with the eigenvalue/eigenvector problem for $n \times n$ real *skew-symmetric* matrices.

22. Let A be an $n \times n$ real skew-symmetric matrix.

(a) Prove that for all $\mathbf{v}_1$ and $\mathbf{v}_2$ in $\mathbb{C}^n$,

$$\langle A\mathbf{v}_1, \mathbf{v}_2 \rangle = -\langle \mathbf{v}_1, A\mathbf{v}_2 \rangle,$$

where $\langle \, , \rangle$ denotes the standard inner product in $\mathbb{C}^n$. [**Hint:** See Lemma 5.10.9.]

(b) Prove that all *nonzero* eigenvalues of A are pure imaginary ($\lambda = -\bar{\lambda}$). [**Hint:** Model your proof after that of part 1 in Theorem 5.10.4.]

23. It follows from the previous problem that the only real eigenvalue a real skew-symmetric matrix can possess is $\lambda = 0$. Use this to prove that if A is an $n \times n$ real skew-symmetric matrix, *with n odd*, then A necessarily has zero as one of its eigenvalues.

24. Determine all eigenvalues and corresponding eigenvectors of the matrix

$$A = \begin{bmatrix} 0 & 4 & -4 \\ -4 & 0 & -2 \\ 4 & 2 & 0 \end{bmatrix}.$$

25. Repeat Problem 24 for the matrix

$$A = \begin{bmatrix} 0 & -1 & -6 \\ 1 & 0 & 5 \\ 6 & -5 & 0 \end{bmatrix}.$$

26. Prove Theorem 5.10.3.

---

[*]**5.11**   Jordan Canonical Forms

The diagonalization of an $n \times n$ matrix A described in Section 5.8 is not possible when the matrix A is defective. With this in mind, the question at hand naturally becomes: For a defective matrix A, is an "approximation" to the diagonalization procedure available? The answer is affirmative, provided we allow the matrices arising in the theory to have complex entries. The approximation we are alluding to here gives rise to the *Jordan canonical form* of A, and the present section aims to give a few examples and introduce the theory of Jordan canonical forms.

For nondefective matrices A, one can construct an invertible matrix S and a diagonal matrix D such that $S^{-1}AS = D$. The matrix S is constructed by placing n linearly independent eigenvectors as its columns. For a defective matrix, we simply do not have "enough" linearly independent eigenvectors to form such a matrix S. The strategy then becomes to search for additional linearly independent vectors that are "close" to being eigenvectors of A. The definition below makes this more precise.

DEFINITION   5.11.1

Let A be an $n \times n$ matrix. A nonzero vector $\mathbf{v}$ is called a **generalized eigenvector** of A corresponding to the eigenvalue λ if

$$(A - \lambda I)^p \, \mathbf{v} = \mathbf{0}$$

for some positive integer p. That is, $\mathbf{v}$ belongs to the nullspace of the matrix $(A - \lambda I)^p$.

Remarks

1. By setting $p = 1$, we see that every eigenvector of A is a generalized eigenvector of A.

2. If p is the *smallest* positive integer such that $(A - \lambda I)^p \, \mathbf{v} = \mathbf{0}$, then the vector $(A - \lambda I)^{p-1} \, \mathbf{v}$ is a (regular) eigenvector of A corresponding to λ, since it belongs to the nullspace of $A - \lambda I$.

Example 5.11.2   Returning to the defective matrix

$$A = \begin{bmatrix} 4 & -1 \\ 1 & 2 \end{bmatrix}$$

in Example 5.7.9, note that $(A - 3I)^2 = 0_2$, which implies that *every* nonzero vector $\mathbf{v}$ is a generalized eigenvector of A corresponding to $\lambda_1 = 3$. For instance, along with the

[*]This section may omited without loss of continuity.

eigenvector $v_1 = (1, 1)$, we may choose the vector $v_2 = (1, 0)$ as a generalized eigenvector such that $\{v_1, v_2\}$ is linearly independent. Since $\{v_1, v_2\}$ is a linearly independent set of two vectors in $\mathbb{R}^2$, it is a basis of $\mathbb{R}^2$.   □

Example 5.11.3   Determine generalized eigenvectors for the matrix

$$B = \begin{bmatrix} 1 & 1 & 0 \\ 0 & 1 & 2 \\ 0 & 0 & 3 \end{bmatrix}.$$

Solution: Direct calculation shows that the characteristic polynomial for B is

$$p(\lambda) = (3 - \lambda)(1 - \lambda)^2,$$

so the eigenvalues of B are $\lambda_1 = 3$ and $\lambda_2 = 1$ (with multiplicity 2). We now look for eigenvectors corresponding to each of these eigenvalues.
Eigenvalue $\lambda_1 = 3$: The augmented matrix of the linear system $(B - \lambda_1 I)v = 0$ is

$$\begin{bmatrix} -2 & 1 & 0 & 0 \\ 0 & -2 & 2 & 0 \\ 0 & 0 & 0 & 0 \end{bmatrix},$$

and we quickly find a solution $v_1 = (1, 2, 2)$, which is an eigenvector of B corresponding to $\lambda_1 = 3$.

Eigenvalue $\lambda_2 = 1$: The augmented matrix of the linear system $(B - \lambda_2 I)v = 0$ is

$$\begin{bmatrix} 0 & 1 & 0 & 0 \\ 0 & 0 & 2 & 0 \\ 0 & 0 & 2 & 0 \end{bmatrix},$$

so only one linearly independent eigenvector, $v_2 = (1, 0, 0)$, is obtained by solving this system. Hence, B is defective. We therefore seek to use a generalized eigenvector that is not parallel to v_2 as a substitute for a second linearly independent eigenvector corresponding to λ_2. To do this, we compute

$$(B - \lambda_2 I)^2 = \begin{bmatrix} 0 & 0 & 2 \\ 0 & 0 & 4 \\ 0 & 0 & 4 \end{bmatrix},$$

which, in addition to multiples of v_2, contains the vector $v_3 = (0, 1, 0)$ in its nullspace. Therefore, v_3 is a generalized eigenvector of B corresponding to $\lambda_2 = 1$. It is worth noting, however, that in this case, any nonzero vector of the form $(a, b, 0)$ is a legitimate generalized eigenvector of B corresponding to $\lambda_2 = 1$. Computing the powers $(B - \lambda_2 I)^3$, $(B - \lambda_2 I)^4, \ldots$, we see that no further generalized eigenvectors of B can be found. Observe that $\{v_2, v_3\}$ is linearly independent and that any generalized eigenvector of B corresponding to $\lambda_2 = 1$ is a linear combination of v_2 and v_3. Note therefore that $\{v_1, v_2, v_3\}$ is a basis of $\mathbb{R}^3$ consisting of generalized eigenvectors.   □

In Examples 5.11.2 and 5.11.3, we see that by using generalized eigenvectors, we are able to obtain n linearly independent vectors. This is not merely a coincidence, and it turns out that for *any* $n \times n$ matrix A, one can always find a linearly independent set of n vectors in $\mathbb{C}^n$ (and hence a basis for $\mathbb{C}^n$ over $\mathbb{C}$) consisting of generalized eigenvectors.[9]

In view of the diagonalization procedure described in Section 5.8, we are tempted in Examples 5.11.2 and 5.11.3 above to construct a matrix S whose columns consist of a linearly independent set of generalized eigenvectors of A and compute[10] $S^{-1}AS$ and $S^{-1}BS$, respectively. Unlike the case of a diagonalizable matrix, where we can use any linearly independent set of eigenvectors, in this case the set of generalized eigenvectors must be carefully chosen. We will see how to make this choice below. In Example 5.11.2, for instance, we can construct the matrix

$$S = \begin{bmatrix} 1 & 1 \\ 1 & 0 \end{bmatrix}.$$

Computing $S^{-1}AS$ gives

$$J_1 = S^{-1}AS = \begin{bmatrix} 3 & 1 \\ 0 & 3 \end{bmatrix}. \tag{5.11.1}$$

Likewise, in Example 5.11.3, we can construct the matrix

$$S = \begin{bmatrix} 1 & 1 & 0 \\ 2 & 0 & 1 \\ 2 & 0 & 0 \end{bmatrix}.$$

Computing $S^{-1}BS$ gives

$$J_2 = S^{-1}BS = \begin{bmatrix} 3 & 0 & 0 \\ 0 & 1 & 1 \\ 0 & 0 & 1 \end{bmatrix}. \tag{5.11.2}$$

The matrices J_1 and J_2 appearing in (5.11.1) and (5.11.2) are almost diagonal matrices, with the eigenvalues along the main diagonal. The only discrepancy is the appearance of some 1's along the *superdiagonals* of these matrices, where the superdiagonal consists of the elements $a_{i,i+1}$ appearing on the diagonal line directly above and parallel to the main diagonal. This is a characteristic feature of matrices in Jordan canonical form. A formal description will be given momentarily, but one feature we observe here is that a matrix in Jordan canonical form consists of block matrices inside it, each containing an eigenvalue on the main diagonal and 1's on the superdiagonal. These block matrices are so important in this theory that they enjoy a special name.

DEFINITION   5.11.4

If λ is a real number, then a square matrix of the form

$$J_\lambda = \begin{bmatrix} \lambda & 1 & 0 & 0 & \ldots & 0 \\ 0 & \lambda & 1 & 0 & \ldots & 0 \\ 0 & 0 & \lambda & 1 & \ldots & 0 \\ \vdots & \vdots & \vdots & \ddots & \ddots & \vdots \\ 0 & 0 & \ldots & 0 & \lambda & 1 \\ 0 & 0 & 0 & \ldots & 0 & \lambda \end{bmatrix}$$

is called a **Jordan block corresponding to** λ.

---

[9]The proof of this fact is beyond the scope of this text, but can be found in more advanced texts on linear algebra, such as S. Friedberg, A. Insel, L. Spence, *Linear Algebra*, (Upper Saddle River, NJ: Prentice Hall, 4th ed. 2002).

[10]Here again note that such S is invertible, since its columns are linearly independent.

Example 5.11.5   The matrices

$$
\begin{bmatrix} 2 & 1 \\ 0 & 2 \end{bmatrix}, \quad [3], \quad \text{and} \quad
\begin{bmatrix}
5 & 1 & 0 & 0 & 0 \\
0 & 5 & 1 & 0 & 0 \\
0 & 0 & 5 & 1 & 0 \\
0 & 0 & 0 & 5 & 1 \\
0 & 0 & 0 & 0 & 5
\end{bmatrix}
$$

are Jordan blocks.   □

DEFINITION   5.11.6

Any square matrix consisting of Jordan blocks centered along the main diagonal and zeros elsewhere is said to be in **Jordan canonical form**.

Example 5.11.7   The matrix

$$
\begin{bmatrix}
2 & 1 & 0 & & & & & & \\
0 & 2 & 1 & & & & & & \\
0 & 0 & 2 & & & & & & \\
& & & 5 & 1 & & & & \\
& & & 0 & 5 & & & & \\
& & & & & 7 & 1 & & \\
& & & & & 0 & 7 & & \\
& & & & & & & 7 & \\
& & & & & & & & 9
\end{bmatrix},
$$

where the empty elements are all zero, is in Jordan canonical form. Moreover, the matrices J_1 and J_2 in (5.11.1) and (5.11.2) are in Jordan canonical form.   □

In general, since each Jordan block is an upper triangular matrix, every matrix in Jordan canonical form is upper triangular. Notice, too, that the size of the Jordan blocks in a Jordan canonical form can vary. For instance, the matrix in Example 5.11.7 contains one 3×3 block, two 2×2 blocks, and two 1×1 blocks.

It turns out that, by a careful selection and arrangement of n linearly independent generalized eigenvectors of an $n \times n$ matrix A in the columns of a matrix S, we will have $S^{-1}AS = J$, a matrix in Jordan canonical form. Therefore, we have the following.

Theorem 5.11.8   Every square matrix A is similar to a matrix J that is in Jordan canonical form.

Proof We refer the reader to S. Friedberg, A. Insel, L. Spence, *Linear Algebra*, Upper Saddle River, NJ: Prentice Hall, 4th ed. 2002.   ∎

Remark   It can be shown that the matrix J to which the square matrix A in Theorem 5.11.8 is similar is uniquely determined, up to a rearrangement of the Jordan blocks. Thus, we can write JCF(A) for the unique Jordan canonical form of A.

As we indicated above, considerable care must be exercised in selecting and arranging the generalized eigenvectors in the columns of S in order to ensure that $S^{-1}AS$ is in Jordan canonical form. To describe this, we need the following terminology.

DEFINITION 5.11.9

Let A be an $n \times n$ matrix, and let $\mathbf{v}$ be a generalized eigenvector of A corresponding to λ. If p is the smallest positive integer such that $(A - \lambda I)^p \mathbf{v} = \mathbf{0}$, then the ordered set

$$\{(A - \lambda I)^{p-1}\mathbf{v}, (A - \lambda I)^{p-2}\mathbf{v}, \dots, (A - \lambda I)\mathbf{v}, \mathbf{v}\} \tag{5.11.3}$$

is called a **cycle of generalized eigenvectors of A corresponding to λ**. The integer p is called the **length** of the cycle.

We often refer to the vector $\mathbf{v}$ in (5.11.3) as the **initial vector** of the cycle, and the vector $(A - \lambda I)^{p-1}\mathbf{v}$ as the **terminal vector** of the cycle.

Theorem 5.11.10   The cycle of generalized eigenvectors in (5.11.3) is linearly independent.

Proof Assume that

$$a_0\mathbf{v} + a_1(A - \lambda I)\mathbf{v} + a_2(A - \lambda I)^2\mathbf{v} + \cdots + a_{p-1}(A - \lambda I)^{p-1}\mathbf{v} = \mathbf{0}. \tag{5.11.4}$$

Multiplying (5.11.4) through by $(A - \lambda I)^{p-1}$ and using the fact that $(A - \lambda I)^p \mathbf{v} = \mathbf{0}$, we have

$$a_0(A - \lambda I)^{p-1}\mathbf{v} = \mathbf{0}.$$

Thus, $a_0 = 0$. Substituting this into (5.11.4) yields

$$a_1(A - \lambda I)\mathbf{v} + a_2(A - \lambda I)^2\mathbf{v} + \cdots + a_{p-1}(A - \lambda I)^{p-1}\mathbf{v} = \mathbf{0}. \tag{5.11.5}$$

Now multiply (5.11.5) through by $(A - \lambda I)^{p-2}$ to get

$$a_1(A - \lambda I)^{p-1}\mathbf{v} = \mathbf{0},$$

which implies $a_1 = 0$.

Continuing in this way, we deduce that $a_0 = a_1 = a_2 = \cdots = a_{p-1} = 0$, as needed. ∎

Notice that the terminal vector of (5.11.3) is the one (and only) vector in the cycle that is an eigenvector. Hence, every cycle of generalized eigenvectors of A can be associated with exactly one eigenvector of A. In fact, if λ is an eigenvalue of A of multiplicity m, then it can be shown[11] that there are m linearly independent generalized eigenvectors of A corresponding to λ.

The basic idea in constructing the matrix S so that $S^{-1}AS$ is in Jordan canonical form is to put vectors occurring in various cycles of the type (5.11.3) in adjacent columns and in the *same order* as listed in (5.11.3). For each cycle of generalized eigenvectors of length p corresponding to λ, the matrix $S^{-1}AS$ will contain a Jordan block of size $p \times p$ corresponding to λ.

We summarize the above observations as follows:

1. The number of Jordan blocks in JCF(A) is the number of linearly independent eigenvectors of A.
2. The size of a Jordan block is equal to the number of vectors in the corresponding cycle of generalized eigenvectors of A.

---

[11]Friedberg et al., op. cit.

Rather than prove these observations, which is best left for a higher level course on linear algebra, we content ourselves with some examples. The observations listed above can often simplify considerably the process of determining JCF(A).

Example 5.11.11   Suppose A is a 4×4 matrix with eigenvalues $\lambda, \lambda, \lambda, \lambda$. List the possible Jordan canonical forms of A.

Solution: Since JCF(A) is a 4×4 matrix, the possible Jordan block decompositions are as follows:

(a) one 4×4 block,

(b) one 3×3 block and one 1×1 block,

(c) two 2×2 blocks,

(d) one 2×2 block and two 1×1 blocks,

(e) four 1×1 blocks.

The Jordan canonical forms associated with these five cases are, respectively:

$$
\begin{bmatrix} \lambda & 1 & 0 & 0 \\ 0 & \lambda & 1 & 0 \\ 0 & 0 & \lambda & 1 \\ 0 & 0 & 0 & \lambda \end{bmatrix},
\begin{bmatrix} \lambda & 1 & 0 & 0 \\ 0 & \lambda & 1 & 0 \\ 0 & 0 & \lambda & 0 \\ 0 & 0 & 0 & \lambda \end{bmatrix},
\begin{bmatrix} \lambda & 1 & 0 & 0 \\ 0 & \lambda & 0 & 0 \\ 0 & 0 & \lambda & 1 \\ 0 & 0 & 0 & \lambda \end{bmatrix},
\begin{bmatrix} \lambda & 1 & 0 & 0 \\ 0 & \lambda & 0 & 0 \\ 0 & 0 & \lambda & 0 \\ 0 & 0 & 0 & \lambda \end{bmatrix},
\begin{bmatrix} \lambda & 0 & 0 & 0 \\ 0 & \lambda & 0 & 0 \\ 0 & 0 & \lambda & 0 \\ 0 & 0 & 0 & \lambda \end{bmatrix}
$$

By calculating a set of linearly independent eigenvectors of A, we can narrow down which of the five matrices listed above can be the JCF(A). For example, if A has only one linearly independent eigenvector, then JCF(A) is the first matrix, which has only one Jordan block. Likewise, if A has three linearly independent eigenvectors, then JCF(A) is the fourth matrix, and if A has four linearly independent eigenvectors (i.e., if A is nondefective), then JCF(A) is the last matrix, which is the same matrix we would obtain via the diagonalization procedure described in Section 5.8.

On the other hand, if A has two linearly independent eigenvectors, then either the second or third matrix could be JCF(A), since both of these matrices contain two Jordan blocks. To distinguish these two matrices, we examine the sizes of the Jordan blocks. This information comes from the length of the cycles of generalized eigenvectors. If we can find a cycle of generalized eigenvectors of length 3, $\{(A - \lambda I)^2 \mathbf{v}, (A - \lambda I)\mathbf{v}, \mathbf{v}\}$, then JCF($A$) would have to contain a 3×3 block. Otherwise it would not contain a 3×3 block. In this way, we could distinguish cases (b) and (c) above. $\square$

More generally, suppose the $n \times n$ matrix A has a single eigenvalue λ of multiplicity n. To determine whether or not A contains a cycle of generalized eigenvectors corresponding to λ of length p or more, it is necessary and sufficient to determine whether or not $(A - \lambda I)^{p-1} = 0_n$. If so, then no cycle of length p is possible. Otherwise, $(A - \lambda I)^{p-1} \neq 0_n$, and so we can find a vector $\mathbf{v}$ with $(A - \lambda I)^{p-1}\mathbf{v} \neq \mathbf{0}$, and hence we may use $\mathbf{v}$ as an initial vector for a cycle of generalized eigenvectors of length p or more.

Example 5.11.12   Let

$$
A = \begin{bmatrix} 0 & 0 & 1 \\ 0 & 0 & 0 \\ 0 & 0 & 0 \end{bmatrix}.
$$

Compute JCF(A).

Solution: The eigenvalues of A are $\lambda = 0, 0, 0$, so JCF(A) contains either (a) one 3×3 block, (b) one 2×2 block and one 1×1 block, or (c) three 1×1 blocks. To determine which, we simply need to know how many linearly independent eigenvectors A has. It is easily seen that the null space of the matrix $A - \lambda I$ is two-dimensional. Therefore, JCF(A) contains two Jordan blocks, which is case (b). Thus, we have

$$\text{JCF}(A) = \begin{bmatrix} 0 & 1 & 0 \\ 0 & 0 & 0 \\ 0 & 0 & 0 \end{bmatrix}. \qquad \square$$

Example 5.11.13   Let

$$A = \begin{bmatrix} 7 & 0 & 0 & 4 & 0 & 0 \\ 0 & 7 & 0 & 0 & 5 & 0 \\ 0 & 0 & 7 & 0 & 0 & 6 \\ 0 & 0 & 0 & 7 & 0 & 0 \\ 0 & 0 & 0 & 0 & 7 & 0 \\ 0 & 0 & 0 & 0 & 0 & 7 \end{bmatrix}.$$

Compute JCF(A).

Solution: The eigenvalues of A are $\lambda = 7, 7, 7, 7, 7, 7$. There are a number of possibilities for JCF(A). However, easy computation (or inspection) shows that the null space of $A - 7I$ is three-dimensional, so JCF(A) must contain three Jordan blocks. The possibilities are thus

(a) one 4×4 block and two 1×1 blocks,

(b) one 3×3 block, one 2×2 block, and one 1×1 block,

(c) three 2×2 blocks.

Observe, however, that $(A - 7I)^2 = 0$, so that it is impossible to build a cycle of generalized eigenvectors of length greater than 2. Thus, JCF(A) cannot have Jordan blocks of size larger than 2×2, and hence, case (c) is the answer:

$$\text{JCF}(A) = \begin{bmatrix} 7 & 1 & & & & \\ 0 & 7 & & & & \\ & & 7 & 1 & & \\ & & 0 & 7 & & \\ & & & & 7 & 1 \\ & & & & 0 & 7 \end{bmatrix}. \qquad \square$$

Example 5.11.14   Let

$$A = \begin{bmatrix} 0 & 1 & 0 \\ 0 & 0 & 1 \\ 1 & -3 & 3 \end{bmatrix}.$$

Find an invertible matrix S such that $S^{-1}AS$ is in Jordan canonical form, and determine JCF(A).

Solution: It is routine to check that the eigenvalues of A are $\lambda = 1, 1, 1$. Now

$$\text{nullspace}(A - I) = \begin{bmatrix} -1 & 1 & 0 \\ 0 & -1 & 1 \\ 1 & -3 & 2 \end{bmatrix},$$

which row reduces to

$$\begin{bmatrix} -1 & 1 & 0 \\ 0 & -1 & 1 \\ 0 & 0 & 0 \end{bmatrix},$$

and has only a one-dimensional solution. Thus, JCF(A) consists of just one 3×3 Jordan block:

$$\text{JCF}(A) = \begin{bmatrix} 1 & 1 & 0 \\ 0 & 1 & 1 \\ 0 & 0 & 1 \end{bmatrix}.$$

To find an invertible matrix S, we must produce a cycle of generalized eigenvectors of length 3 and place them into the columns of S. The cycle will take the form

$$\{(A - I)^2\mathbf{v}, (A - I)\mathbf{v}, \mathbf{v}\},$$

so we simply need to find a vector $\mathbf{v}$ such that $(A - I)^2\mathbf{v} \neq \mathbf{0}$. Direct computation shows that

$$(A - I)^2 = \begin{bmatrix} 1 & -2 & 1 \\ 1 & -2 & 1 \\ 1 & -2 & 1 \end{bmatrix},$$

and we may take $\mathbf{v}$ to be any vector that is not killed by this matrix—say

$$\mathbf{v} = \begin{bmatrix} 1 \\ 0 \\ 0 \end{bmatrix}.$$

Then

$$(A - I)\mathbf{v} = \begin{bmatrix} -1 \\ 0 \\ 1 \end{bmatrix} \quad \text{and} \quad (A - I)^2\mathbf{v} = \begin{bmatrix} 1 \\ 1 \\ 1 \end{bmatrix}.$$

Hence, we have

$$S = \begin{bmatrix} 1 & -1 & 1 \\ 1 & 0 & 0 \\ 1 & 1 & 0 \end{bmatrix}. \qquad \square$$

Example 5.11.15   Let

$$A = \begin{bmatrix} -2 & -1 & 1 \\ 2 & -5 & 2 \\ 1 & -1 & -2 \end{bmatrix}.$$

Find an invertible matrix S such that $S^{-1}AS$ is in Jordan canonical form, and determine JCF(A).

Solution: The reader can check that the eigenvalues of A are $\lambda = -3, -3, -3$. Since

$$A + 3I = \begin{bmatrix} 1 & -1 & 1 \\ 2 & -2 & 2 \\ 1 & -1 & 1 \end{bmatrix},$$

nullspace($A + 3I$) contains two linearly independent solutions. Thus, JCF(A) consists of two Jordan blocks (one of them of size 2×2 and the other of size 1×1):

$$\text{JCF}(A) = \begin{bmatrix} -3 & 1 & 0 \\ 0 & -3 & 0 \\ 0 & 0 & -3 \end{bmatrix}.$$

To find an invertible matrix S, let us first determine a cycle of generalized eigenvectors of length 2 (no such cycle of length 3 is possible in this case, since there is no 3×3 Jordan block in JCF(A)). A cycle of generalized eigenvectors of length 2 will take the form

$$\{(A + 3I)\mathbf{v}, \mathbf{v}\},$$

so we seek a vector $\mathbf{v}$ such that $(A + 3I)\mathbf{v} \neq \mathbf{0}$. We quickly see that the vector

$$\mathbf{v} = \begin{bmatrix} 1 \\ 0 \\ 0 \end{bmatrix}$$

will serve nicely, and thus

$$(A + 3I)\mathbf{v} = \begin{bmatrix} 1 \\ 2 \\ 1 \end{bmatrix}.$$

So we have a cycle of generalized eigenvectors

$$\left\{ \begin{bmatrix} 1 \\ 2 \\ 1 \end{bmatrix}, \begin{bmatrix} 1 \\ 0 \\ 0 \end{bmatrix} \right\}.$$

Finally, we need to find an additional eigenvector that is nonproportional to the eigenvector $(A + 3I)\mathbf{v}$. There are many possibilities, and the reader can check that

$$\mathbf{w} = \begin{bmatrix} 1 \\ 1 \\ 0 \end{bmatrix}$$

is workable. Hence, we have

$$S = \begin{bmatrix} 1 & 1 & 1 \\ 2 & 0 & 1 \\ 1 & 0 & 0 \end{bmatrix}. \qquad \square$$

The Jordan canonical form concept is a powerful tool in the study of matrices (and corresponding linear transformations). To reiterate, *every* $n \times n$ matrix has a unique Jordan canonical form (up to the order of the Jordan blocks) that it is similar to. Since similar matrices share many properties, such as characteristic polynomial, eigenvalues, determinant, trace, dimension of eigenspaces, and so on, many questions that one can ask about matrices in general can be reduced to questions about matrices that are in Jordan canonical form, a much more specialized class. An excellent illustration of this can be seen in Project II at the end of this chapter.

As a final application of the concepts in this section, we return once more to linear systems of differential equations. We saw in Section 5.8 that a linear system of differential equations with a diagonalizable coefficient matrix can be solved by transforming the given system to a diagonal system. Now, through the machinery of Jordan canonical forms, we can attempt a similar strategy on linear systems of differential equations in which the coefficient matrix is not diagonalizable. Consider once again the linear system $\mathbf{x}' = A\mathbf{x}$. Given an invertible matrix S of constants, recall the linear change of variables $\mathbf{x} = S\mathbf{y}$ from (5.8.4). This yields the transformed linear system

$$\mathbf{y}' = B\mathbf{y}, \tag{5.11.6}$$

where $B = S^{-1}AS$. In this case, we wish to choose S so that B is the Jordan canonical form of A. As such, the system (5.11.6) will consist of first-order linear differential equations, which we can solve by the technique of Section 1.6. We illustrate with an example.

Example 5.11.16   Solve the linear system

$$x_1' = -\ 9x_1 +\ 9x_2,$$
$$x_2' = -16x_1 + 15x_2.$$

Solution: The coefficient matrix for the system is

$$A = \begin{bmatrix} -9 & 9 \\ -16 & 15 \end{bmatrix}.$$

The eigenvalues of A are $\lambda = 3, 3$, and we find that the eigenspace $E_{\lambda=3}$ is only one-dimensional, with basis vector $\begin{bmatrix} 3 \\ 4 \end{bmatrix}$. We can find a generalized eigenvector $\mathbf{v}$ of A corresponding to $\lambda = 3$ by finding a vector $\mathbf{v}$ such that $(A - 3I)\mathbf{v} \neq \mathbf{0}$. Since

$$A - 3I = \begin{bmatrix} -12 & 9 \\ -16 & 12 \end{bmatrix},$$

we see by inspection that the vector $\begin{bmatrix} 0 \\ 1 \end{bmatrix}$ is a generalized eigenvector of A. Since

$$\begin{bmatrix} -12 & 9 \\ -16 & 12 \end{bmatrix} \begin{bmatrix} 0 \\ 1 \end{bmatrix} = \begin{bmatrix} 9 \\ 12 \end{bmatrix},$$

we have the cycle of generalized eigenvectors

$$\left\{ \begin{bmatrix} 9 \\ 12 \end{bmatrix}, \begin{bmatrix} 0 \\ 1 \end{bmatrix} \right\},$$

and thus we construct the matrices

$$S = \begin{bmatrix} 9 & 0 \\ 12 & 1 \end{bmatrix} \quad \text{and} \quad B = \begin{bmatrix} 3 & 1 \\ 0 & 3 \end{bmatrix}.$$

Via the substitution $\mathbf{x} = S\mathbf{y}$, the original system is transformed into

$$\mathbf{y}' = \begin{bmatrix} 3 & 1 \\ 0 & 3 \end{bmatrix} \mathbf{y},$$

which corresponds to the equations

$$y_1' = 3y_1 +\ y_2,$$
$$y_2' = \qquad 3y_2.$$

The solution to the second equation is

$$y_2(t) = c_1 e^{3t}. \tag{5.11.7}$$

Substituting (5.11.7) and rearranging, the first equation becomes

$$y_1' - 3y_1 = c_1 e^{3t}. \tag{5.11.8}$$

This is a first-order linear equation, with integrating factor $I(t) = e^{-3t}$. Multiplying (5.11.8) through by $I(t)$, it becomes

$$(y_1 \cdot e^{-3t})' = c_1.$$

Integrating both sides yields

$$y_1 \cdot e^{-3t} = c_1 t + c_2,$$

and thus,

$$y_1(t) = c_1 t e^{3t} + c_2 e^{3t}.$$

Thus, we have

$$\mathbf{y} = \begin{bmatrix} y_1(t) \\ y_2(t) \end{bmatrix} = \begin{bmatrix} c_1 t e^{3t} + c_2 e^{3t} \\ c_1 e^{3t} \end{bmatrix}. \tag{5.11.9}$$

Finally, we must substitute (5.11.9) into the equation $\mathbf{x} = S\mathbf{y}$ to solve for $\mathbf{x}$. We obtain

$$\mathbf{x} = \begin{bmatrix} 9 & 0 \\ 12 & 1 \end{bmatrix} \mathbf{y} = \begin{bmatrix} 9(c_1 t e^{3t} + c_2 e^{3t}) \\ 12(c_1 t e^{3t} + c_2 e^{3t}) + c_1 e^{3t} \end{bmatrix} = c_1 e^{3t} \begin{bmatrix} 9t \\ 12t + 1 \end{bmatrix} + c_2 e^{3t} \begin{bmatrix} 9 \\ 12 \end{bmatrix}.$$

Thus, the general solution to the linear system of differential equations is

$$x_1(t) = 9c_1 t e^{3t} + 9c_2 e^{3t}, \qquad x_2(t) = c_1 e^{3t}(12t + 1) + 12c_2 e^{3t}. \qquad \square$$

Concluding this section, we return to prove part 3 of Theorem 5.10.4 in Section 5.10, which states that every real symmetric matrix is diagonalizable.

Proof that every real symmetric matrix is nondefective (Theorem 5.10.4, part 3):

Suppose to the contrary that A is a real symmetric matrix that is defective. There exists an eigenvalue λ of A and a corresponding generalized eigenvector $\mathbf{v}$ with $(A - \lambda I)^2 \mathbf{v} = \mathbf{0}$, but $(A - \lambda I)\mathbf{v} \neq \mathbf{0}$. Then

$$\begin{aligned} 0 \neq \langle (A - \lambda I)\mathbf{v}, (A - \lambda I)\mathbf{v} \rangle \qquad & \text{since } (A - \lambda I)\mathbf{v} \neq \mathbf{0} \\ = \langle \mathbf{v}, (A - \lambda I)^2 \mathbf{v} \rangle \qquad & \text{by Lemma 5.10.9} \\ = \langle \mathbf{v}, \mathbf{0} \rangle = 0, \end{aligned}$$

a contradiction. ∎

Exercises for 5.11

Key Terms

Generalized eigenvector, Superdiagonal, Jordan block corresponding to λ, Jordan canonical form, Cycle of generalized eigenvectors corresponding to λ, Length of a cycle, Initial and terminal vectors.

Skills

- For a given matrix A and eigenvalue λ, be able to determine whether a vector $\mathbf{v}$ is an eigenvector, a generalized eigenvector, or neither.

- Be able to construct a cycle of generalized eigenvectors corresponding to an eigenvalue λ of a matrix A.

- Be able to compute the Jordan canonical form of a given matrix A, along with an invertible matrix S whose columns are comprised of the cycles of generalized eigenvectors.

- Be able to list the possible Jordan canonical forms for a matrix A, given only the multiplicities of its eigenvalues.

- Be able to solve linear systems of differential equations in which the coefficient matrix A is not necessarily diagonalizable.

True-False Review

For Questions 1–12, decide if the given statement is **true** or **false**, and give a brief justification for your answer. If true, you can quote a relevant definition or theorem from the text. If false, provide an example, illustration, or brief explanation of why the statement is false.

1. Every eigenvector is a generalized eigenvector.

2. The number of Jordan blocks in the Jordan canonical form of a matrix A is the number of linearly independent eigenvectors of A.

3. For every square matrix A, there is a unique invertible matrix S such that $S^{-1}AS$ is in Jordan canonical form.

4. If J_1 and J_2 are $n \times n$ matrices in Jordan canonical form, then the matrix $J_1 + J_2$ is in Jordan canonical form.

5. A generalized eigenvector of A corresponding to an eigenvalue λ is a member of the null space of $(A-\lambda I)^p$ for some positive integer p.

6. The dimension of K_λ, the vector space of generalized eigenvectors corresponding to an eigenvalue λ, is equal to the number of Jordan blocks corresponding to λ in the Jordan canonical form of A.

7. Every square matrix A is similar to a matrix J in Jordan canonical form.

8. If J_1 and J_2 are $n \times n$ matrices in Jordan canonical form, then the matrix $J_1 J_2$ is in Jordan canonical form.

9. The size of a Jordan block is equal to the number of vectors in the corresponding cycle of generalized eigenvectors of A.

10. If A is an $n \times n$ matrix with no cycles of generalized eigenvectors of length $p \geq 2$, then A is diagonalizable.

11. Similar matrices must have the same Jordan canonical form, up to rearrangement of the Jordan blocks.

12. If J is in Jordan canonical form and r is a scalar, then the matrix rJ is in Jordan canonical form.

Problems

For Problems 1–4, determine how many possible Jordan canonical forms are possible with the given eigenvalues (not counting rearrangements of the Jordan blocks) and list each of them.

1. A 3×3 matrix with eigenvalues $\lambda = 1, 1, 1$.

2. A 4×4 matrix with eigenvalues $\lambda = 1, 1, 3, 3$.

3. A 5×5 matrix with eigenvalues $\lambda = 2, 2, 2, 2, 2$.

4. A 6×6 matrix with eigenvalues $\lambda = 3, 3, 3, 3, 9, 9$.

For Problems 5–6, determine how many Jordan canonical forms are possible with the given eigenvalues (not counting rearrangements of the Jordan blocks). You do not need to list them.

5. An 11×11 matrix with eigenvalues $\lambda = 2, 2, 2, 2, 6, 6, 6, 6, 8, 8, 8$.

6. A 10×10 matrix with eigenvalues $\lambda = 2, 2, 2, 2, 5, 5, 5, 5, 5, 5$.

7. If it is known that $(A-5I)^2 = 0$ for the matrix in Problem 6, how many Jordan canonical form structures are possible for the matrix A?

8. Let A be a 5×5 matrix with eigenvalues $\lambda_1, \lambda_1, \lambda_1, \lambda_2, \lambda_2$, where $\lambda_1 \neq \lambda_2$.

    (a) Determine the complete list of possible Jordan canonical forms of A.

    (b) Assume further that $(A - \lambda_1 I)^2 \neq 0_5$. Among the matrices listed in part (a), which of them are the possible Jordan canonical form of A in light of this new information?

9. Suppose A is a 6×6 matrix with eigenvalue λ (of multiplicity 6). If it is known that $(A - \lambda I)^3 = 0$ but $(A - \lambda I)^2 \neq 0$, write down all possible Jordan canonical forms of A.

For Problems 10–13, the characteristic polynomial $p(\lambda)$ for a square matrix A is given. Write down a set S of matrices such that every square matrix with characteristic polynomial $p(\lambda)$ is guaranteed to be similar to exactly one of the matrices in the set S.

10. $p(\lambda) = (4 - \lambda)^2(-6 - \lambda)$.

11. $p(\lambda) = (4 - \lambda)^3(-1 - \lambda)^2$.

12. $p(\lambda) = (3 - \lambda)^2(-2 - \lambda)^3\lambda^2$.

13. $p(\lambda) = (-2 - \lambda)^2(6 - \lambda)^5$.

14. Which of the matrices in the set S in Problem 13 have a set of exactly five (and not more than five) linearly independent eigenvectors? Explain.

15. Give an example of a 2×2 matrix A that has a generalized eigenvector that is not an eigenvector, and exhibit such a generalized eigenvector.

16. Give an example of a 3×3 matrix A that has a generalized eigenvector that is not an eigenvector, and exhibit such a generalized eigenvector.

For Problems 17–26, find the Jordan canonical form J for the matrix A, and determine an invertible matrix S such that $S^{-1}AS = J$.

17. $A = \begin{bmatrix} 1 & 1 \\ -1 & 3 \end{bmatrix}$.

18. $A = \begin{bmatrix} 1 & 1 & 1 \\ 0 & 1 & 1 \\ 0 & 1 & 1 \end{bmatrix}$.

19. $A = \begin{bmatrix} 5 & 0 & -1 \\ 1 & 4 & -1 \\ 1 & 0 & 3 \end{bmatrix}$.

20. $A = \begin{bmatrix} 4 & -4 & 5 \\ -1 & 4 & 2 \\ -1 & 2 & 4 \end{bmatrix}$.

21. $A = \begin{bmatrix} -6 & 1 & 0 \\ -\frac{1}{2} & -\frac{9}{2} & \frac{1}{2} \\ -\frac{1}{2} & \frac{1}{2} & -\frac{11}{2} \end{bmatrix}$. [**Hint:** The eigenvalues of A are $\lambda = -5$ (with multiplicity 2) and $\lambda = -6$.]

22. $A = \begin{bmatrix} 2 & -2 & 14 \\ 0 & 3 & -7 \\ 0 & 0 & 2 \end{bmatrix}$.

23. $A = \begin{bmatrix} 7 & -2 & 2 \\ 0 & 4 & -1 \\ -1 & 1 & 4 \end{bmatrix}$.

24. $A = \begin{bmatrix} -1 & -1 & 0 \\ 0 & -1 & -2 \\ 0 & 0 & -1 \end{bmatrix}$.

25. $A = \begin{bmatrix} 2 & -1 & 0 & 1 \\ 0 & 3 & -1 & 0 \\ 0 & 1 & 1 & 0 \\ 0 & -1 & 0 & 3 \end{bmatrix}$.

26. $A = \begin{bmatrix} 2 & -4 & 2 & 2 \\ -2 & 0 & 1 & 3 \\ -2 & -2 & 3 & 3 \\ -2 & -6 & 3 & 7 \end{bmatrix}$. [The characteristic polynomial is $p(\lambda) = (2 - \lambda)^2(4 - \lambda)^2$.]

For Problems 27–29, find the Jordan canonical form J for the matrix A. You need not determine an invertible matrix S such that $S^{-1}AS = J$.

27. $A = \begin{bmatrix} 2 & 1 & 1 & 1 & 1 \\ 0 & 2 & 0 & 0 & 1 \\ 0 & 0 & 2 & 0 & 1 \\ 0 & 0 & 0 & 2 & 1 \\ 0 & 0 & 0 & 0 & 2 \end{bmatrix}$.

28. $A = \begin{bmatrix} 0 & 0 & 0 & 1 & 4 \\ 0 & 0 & 1 & 1 & 1 \\ 0 & 0 & 0 & 6 & 0 \\ 0 & 0 & 0 & 0 & 0 \\ 0 & 0 & 0 & 0 & 0 \end{bmatrix}$.

29. $A = \begin{bmatrix} 1 & 1 & 0 & 1 & 0 & 1 & 0 & 1 \\ 0 & 1 & 0 & 0 & 0 & 0 & 0 & 0 \\ 0 & 0 & 1 & 1 & 0 & 1 & 0 & 1 \\ 0 & 0 & 0 & 1 & 0 & 0 & 0 & 0 \\ 0 & 0 & 0 & 0 & 1 & 1 & 0 & 1 \\ 0 & 0 & 0 & 0 & 0 & 1 & 0 & 0 \\ 0 & 0 & 0 & 0 & 0 & 0 & 1 & 1 \\ 0 & 0 & 0 & 0 & 0 & 0 & 0 & 1 \end{bmatrix}$.

For Problems 30–31, use Jordan canonical forms to determine whether the given pair of matrices are similar.

30. $A = \begin{bmatrix} 7 & 1 & 0 \\ -1 & 5 & 0 \\ 1 & 0 & 6 \end{bmatrix}$; $B = \begin{bmatrix} 6 & -1 & 1 \\ 0 & 6 & 0 \\ 0 & 0 & 6 \end{bmatrix}$.

31. $A = \begin{bmatrix} 7 & -2 & 2 \\ 0 & 4 & -1 \\ -1 & 1 & 4 \end{bmatrix}$; $B = \begin{bmatrix} 3 & -1 & -2 \\ 1 & 6 & 1 \\ 1 & 0 & 6 \end{bmatrix}$.

For Problems 32–35, determine the general solution to the system $\mathbf{x}' = A\mathbf{x}$ for the given matrix A.

32. $A = \begin{bmatrix} -3 & -2 \\ 2 & 1 \end{bmatrix}$.

33. $A = \begin{bmatrix} 0 & 1 & 0 \\ 0 & 0 & 1 \\ 1 & 1 & -1 \end{bmatrix}$.

34. $A = \begin{bmatrix} -2 & 0 & 0 \\ 1 & -3 & -1 \\ -1 & 1 & -1 \end{bmatrix}$.

35. $A = \begin{bmatrix} 4 & 0 & 0 \\ 1 & 4 & 0 \\ 0 & 1 & 4 \end{bmatrix}$.

36. Solve the initial-value problem $\mathbf{x}' = A\mathbf{x}$, where
$$A = \begin{bmatrix} -2 & -1 \\ 1 & -4 \end{bmatrix}, \qquad \mathbf{x}(0) = \begin{bmatrix} 0 \\ -1 \end{bmatrix}.$$

37. Prove that if A and B are $n \times n$ matrices with the same Jordan canonical form, then A is similar to B.

38. Let A be a square matrix with characteristic polynomial $p(\lambda) = -\lambda^3$. Use Jordan canonical forms to prove that A is a nilpotent[12] matrix.

39. (a) Let J be a Jordan block. Prove that the Jordan canonical form of the matrix J^T is J.

(b) Let A be an $n \times n$ matrix. Prove that A and A^T have the same Jordan canonical form.

5.12   Chapter Review

Linear Transformations

In this chapter we have considered mappings $T : V \to W$ between vector spaces V and W that satisfy the basic linearity properties

$$T(\mathbf{x} + \mathbf{y}) = T(\mathbf{x}) + T(\mathbf{y}), \quad \text{for all } \mathbf{x} \text{ and } \mathbf{y} \text{ in } V,$$
$$T(c\mathbf{x}) = cT(\mathbf{x}), \quad \text{for all } \mathbf{x} \text{ in } V \text{ and all scalars } c.$$

We now list some of the key definitions and theorems for linear transformations.

We have identified the following two important subsets of vectors associated with a linear transformation:

1. The *kernel* of T, denoted Ker(T). This is the set of all vectors in V that are mapped to $\mathbf{0}_W$, the zero vector in W.

2. The *range* of T, denoted Rng(T). This is the set of vectors in W that we obtain when we allow T to act on every vector in V. Equivalently, Rng(T) is the set of all transformed vectors. The key results about Ker(T) and Rng(T) are as follows.

Let $T : V \to W$ be a linear transformation. Then,

1. Ker(T) is a subspace of V.

2. Rng(T) is a subspace of W.

3. If V is finite-dimensional, $\dim[\text{Ker}(T)] + \dim[\text{Rng}(T)] = \dim[V]$.

4. T is one-to-one if and only if Ker(T) = $\{\mathbf{0}\}$.

5. T is onto if and only if Rng(T) = W.

Finally, if $T : V \to W$ is a linear transformation, then the inverse transformation $T^{-1} : W \to V$ exists if and only if T is both one-to-one and onto, in which case V and W are *isomorphic* and T is called an *isomorphism*.

The Algebraic Eigenvalue/Eigenvector Problem

We next summarize the main results obtained in this chapter regarding the algebraic eigenvalue/eigenvector problem.

1. For a given $n \times n$ matrix A, the eigenvalue/eigenvector problem consists of determining all scalars λ and all *nonzero* vectors $\mathbf{v}$ such that

$$A\mathbf{v} = \lambda\mathbf{v}.$$

2. The eigenvalues of A are the roots of the characteristic equation

$$p(\lambda) = \det(A - \lambda I) = 0, \tag{5.12.1}$$

---

[12] Recall that an $n \times n$ matrix A is called *nilpotent* if $A^p = 0_n$ for some positive integer p.

and the eigenvectors of A are obtained by solving the linear systems

$$(A - \lambda I)\mathbf{v} = \mathbf{0}, \qquad (5.12.2)$$

when λ assumes the values obtained in (5.12.1).

3. If A is a matrix with real elements, then complex eigenvalues and eigenvectors occur in conjugate pairs.

4. Associated with each eigenvalue λ there is a vector space, called the eigenspace of λ. This is the set of all eigenvectors corresponding to λ together with the zero vector. Equivalently, it can be considered as the set of *all* solutions to the linear system (5.12.2).

5. If m denotes the multiplicity of the eigenvalue λ and n denotes the dimension of the corresponding eigenspace, then

$$1 \leq n \leq m.$$

6. Eigenvectors corresponding to distinct eigenvalues are linearly independent.

7. An $n \times n$ matrix that has n linearly independent eigenvectors is said to have a *complete set of eigenvectors*, and we call such a matrix *nondefective*.

8. Two $n \times n$ matrices A and B are said to be *similar* if there exists a matrix S such that

$$B = S^{-1}AS.$$

A matrix that is similar to a diagonal matrix is said to be diagonalizable. We have shown that A is diagonalizable if and only if it is nondefective.

9. If A is nondefective and $S = [\mathbf{v}_1, \mathbf{v}_2, \ldots, \mathbf{v}_n]$, where $\mathbf{v}_1, \mathbf{v}_2, \ldots, \mathbf{v}_n$ are linearly independent eigenvectors of A, then

$$S^{-1}AS = \text{diag}(\lambda_1, \lambda_2, \ldots, \lambda_n),$$

where $\lambda_1, \lambda_2, \ldots, \lambda_n$ are the eigenvalues of A corresponding to $\mathbf{v}_1, \mathbf{v}_2, \ldots, \mathbf{v}_n$.

10. If A is a *real symmetric* matrix, then

    (a) All eigenvalues of A are real.

    (b) Eigenvectors corresponding to different eigenvalues are *orthogonal*.

    (c) A is nondefective.

    (d) A has a complete set of orthonormal eigenvectors, say $\mathbf{w}_1, \mathbf{w}_2, \ldots, \mathbf{w}_n$.

    (e) If we let $S = [\mathbf{w}_1, \mathbf{w}_2, \ldots, \mathbf{w}_n]$, where $\mathbf{w}_1, \mathbf{w}_2, \ldots, \mathbf{w}_n$ are the orthonormal eigenvectors determined in (d), then S is an orthogonal matrix ($S^{-1} = S^T$), and hence, from result 9, for this matrix,

$$S^T AS = \text{diag}(\lambda_1, \lambda_2, \ldots, \lambda_n).$$

11. Every matrix A is similar to a matrix J in Jordan canonical form. Here are some facts about JCF(A):

    (a) The JCF(A) consists of Jordan blocks arranged along the main diagonal, and each Jordan block consists of an eigenvalue placed along its main diagonal with 1's appearing on the superdiagonal and zeros elsewhere.

(b) The number of Jordan blocks in JCF(A) is equal to the number of linearly independent eigenvectors of A.

(c) Each Jordan block coincides with a cycle of generalized eigenvectors. A generalized eigenvector $\mathbf{v}$ satisfies the equation $(A - \lambda I)^p \mathbf{v} = \mathbf{0}$ for some $p \geq 1$.

(d) The length of a cycle of generalized eigenvectors is the same as the size of the corresponding Jordan block.

Additional Problems

In Problems 1–10, decide whether or not the given mapping T is a linear transformation. Justify your answers. For each mapping that is a linear transformation, decide whether or not T is one-to-one, onto, both, or neither, and find a basis and dimension for Ker(T) and Rng(T).

1. $T: \mathbb{R}^2 \to \mathbb{R}^4$ defined by
$T(x, y) = (x + y, 0, x - y, xy)$.

2. $T: \mathbb{R}^3 \to \mathbb{R}^2$ defined by $T(x, y, z) = (2x - 3y, -x)$.

3. $T: \mathbb{R}^3 \to \mathbb{R}^2$ defined by
$T(x, y, z) = (-3z, 2x - y + 5z)$.

4. $T: P_4 \to \mathbb{R}^2$ defined by $T(p) = (p(0), p(1))$.

5. $T: \mathbb{R}^2 \to \mathbb{R}$ defined by $T(x, y) = \dfrac{x + y}{5}$.

6. $T: M_2(\mathbb{R}) \to \mathbb{R}^2$ defined by
$$T \begin{bmatrix} a & b \\ c & d \end{bmatrix} = (ac, bd).$$

7. $T: P_2 \to M_2(\mathbb{R})$ defined by
$$T(a + bx + cx^2) = \begin{bmatrix} -a - b & 0 \\ 3c - a & -2b \end{bmatrix}.$$

8. $T: M_2(\mathbb{R}) \to M_2(\mathbb{R})$ defined by
$$T(A) = A + A^T.$$

9. $T: \mathbb{R}^3 \to P_2$ defined by
$$T((a, b, c)) = ax^2 + (2b - c)x + (a - 2b + c).$$

10. $T: \mathbb{R}^3 \to M_2(\mathbb{R})$ defined by
$$T((x_1, x_2, x_3)) = \begin{bmatrix} 0 & x_1 - x_2 + x_3 \\ -x_1 + x_2 - x_3 & 0 \end{bmatrix}.$$

In Problems 11–15, determine a formula for the linear transformation meeting the given conditions.

11. $T: \mathbb{R}^3 \to \mathbb{R}^2$ given by $T(\mathbf{x}) = A\mathbf{x}$, where
$$A = \begin{bmatrix} -1 & 8 & 0 \\ 2 & -2 & -5 \end{bmatrix}.$$

12. $T: \mathbb{R}^2 \to \mathbb{R}^5$ given by $T(\mathbf{x}) = A\mathbf{x}$, where
$$A = \begin{bmatrix} -1 & 4 \\ 0 & 2 \\ 3 & -3 \\ 3 & -3 \\ 2 & -6 \end{bmatrix}.$$

13. $T: \mathbb{R} \to \mathbb{R}^4$ such that $T(2) = (-1, 5, 0, -2)$.

14. $T: M_2(\mathbb{R}) \to \mathbb{R}^2$ such that
$$T \begin{bmatrix} 1 & 0 \\ 0 & 1 \end{bmatrix} = (2, -5),$$
$$T \begin{bmatrix} 0 & 1 \\ 1 & 0 \end{bmatrix} = (0, -3),$$
$$T \begin{bmatrix} 1 & 0 \\ 0 & 0 \end{bmatrix} = (1, 1),$$
$$T \begin{bmatrix} 1 & 1 \\ 0 & 0 \end{bmatrix} = (-6, 2).$$

15. $T: P_2 \to M_2(\mathbb{R})$ such that
$$T(x^2 - x - 3) = \begin{bmatrix} -2 & 1 \\ -4 & -1 \end{bmatrix},$$
$$T(2x + 5) = \begin{bmatrix} 0 & 1 \\ 2 & -2 \end{bmatrix},$$
$$T(6) = \begin{bmatrix} 12 & 6 \\ 6 & 18 \end{bmatrix}.$$

16. If $T: P_5 \rightarrow M_2(\mathbb{R})$ is an onto linear transformation, what is $\dim[\mathrm{Ker}(T)]$?

17. If $T: M_{2\times 3}(\mathbb{R}) \rightarrow P_6$ is one-to-one, what is $\dim[\mathrm{Rng}(T)]$?

In Problems 18–23, decide whether or not the given matrix A is diagonalizable. If so, find an invertible matrix S and a diagonal matrix D such that $S^{-1}AS = D$.

18. $A = \begin{bmatrix} 3 & 0 \\ 16 & -1 \end{bmatrix}$.

19. $A = \begin{bmatrix} 13 & -9 \\ 25 & -17 \end{bmatrix}$.

20. $A = \begin{bmatrix} -4 & 3 & 0 \\ -6 & 5 & 0 \\ 3 & -3 & -1 \end{bmatrix}$.

21. $A = \begin{bmatrix} 1 & 1 & 0 \\ -4 & 5 & 0 \\ 17 & -11 & -2 \end{bmatrix}$.

22. $A = \begin{bmatrix} -1 & -1 & 3 \\ 4 & 4 & -4 \\ -1 & 0 & 3 \end{bmatrix}$. [**Hint:** The only eigenvalue of A is $\lambda = 2$.]

23. $A = \begin{bmatrix} 9 & 5 & -5 \\ 0 & -1 & 0 \\ 10 & 5 & -6 \end{bmatrix}$. [**Hint:** The eigenvalues of A are $\lambda = 4$ and $\lambda = -1$.]

◇ In Problems 24–27, use some form of technology to find a complete set of *orthonormal* eigenvectors for A and an *orthogonal* matrix S and a diagonal matrix D such that $S^{-1}AS = D$.

24. $A = \begin{bmatrix} 2 & 2 & 1 \\ 2 & 5 & 2 \\ 1 & 2 & 2 \end{bmatrix}$.

25. $A = \begin{bmatrix} 0 & -1 & 4 \\ -1 & 5 & 2 \\ 4 & 2 & 2 \end{bmatrix}$.

26. $A = \begin{bmatrix} -2 & 1 & 1 \\ 1 & 3 & 6 \\ 1 & 6 & -1 \end{bmatrix}$.

27. $A = \begin{bmatrix} 1 & 0 & 1 \\ 0 & 1 & 1 \\ 1 & 1 & 1 \end{bmatrix}$.

Find the Jordan canonical form of each matrix in Problems 28–29.

28. $A = \begin{bmatrix} 5 & 8 & 16 \\ 4 & 1 & 8 \\ -4 & -4 & -11 \end{bmatrix}$. [**Hint:** The eigenvalues of A are $\lambda = 1$ and $\lambda = -3$.]

29. $A = \begin{bmatrix} 2 & 1 & 1 \\ 2 & 1 & -2 \\ -1 & 0 & -2 \end{bmatrix}$. [**Hint:** The eigenvalues of A are $\lambda = -1$ and $\lambda = 3$.]

In Problems 30–33, write down all of the possible Jordan canonical form structures, up to a rearrangement of the blocks, for matrices of the specified type. For each Jordan canonical form structure, list the number of linearly independent eigenvectors of a matrix with that Jordan canonical form, and list the maximum length of a cycle of generalized eigenvectors of the matrix.

30. 4×4 matrices with eigenvalues $\lambda = -1, -1, -1, 2$.

31. 5×5 matrices with eigenvalues $\lambda = 4, 4, 4, 4, 4$.

32. 6×6 matrices with eigenvalues $\lambda = 6, 6, 6, 6, -3, -3$.

33. 7×7 matrices with eigenvalues $\lambda = 2, 2, 2, 2, -4, -4, -4$.

34. True or False: If A and B are $n \times n$ matrices with eigenvalues λ_A and λ_B, respectively, then $\lambda_A - \lambda_B$ is an eigenvalue of $A - B$. Explain.

35. True or False: If A and B are square matrices such that A^2 is similar to B^2, then A is similar to B. Explain.

If T_1 and T_2 are both linear transformations from V to W, then we can define a mapping $T_1 + T_2: V \rightarrow W$, given by $(T_1 + T_2)(\mathbf{v}) = T_1(\mathbf{v}) + T_2(\mathbf{v})$ for all $\mathbf{v}$ in V. The next three problems concern the mapping $T_1 + T_2$.

36. Let T_1 and T_2 be linear transformations from V to W. Prove that $T_1 + T_2$ is a linear transformation. Must there be any relationship between $\mathrm{Ker}(T_1)$, $\mathrm{Ker}(T_2)$, and $\mathrm{Ker}(T_1 + T_2)$?

37. True or False: Let T_1 and T_2 be linear transformations from V to W. If T_1 and T_2 are both onto, then $T_1 + T_2$ is onto.

38. True or False: Let T_1 and T_2 be linear transformations from V to W. If T_1 and T_2 are both one-to-one, then $T_1 + T_2$ is one-to-one.

39. Let V and W be vector spaces, and let $T: V \rightarrow W$ be a linear transformation with $\mathrm{Ker}(T) = \{\mathbf{0}\}$. If $\{\mathbf{v}_1, \mathbf{v}_2, \ldots, \mathbf{v}_n\}$ is a linearly independent subset of V, show that $\{T(\mathbf{v}_1), T(\mathbf{v}_2), \ldots, T(\mathbf{v}_n)\}$ is a linearly independent subset of W.

40. Prove that if V_1 is isomorphic to V_2 and V_2 is isomorphic to V_3, then V_1 is isomorphic to V_3.

41. Assume that $A_1, A_2, \ldots, A_k$ are $n \times n$ matrices and, for each i, a vector $\mathbf{v}$ is an eigenvector of A_i with corresponding eigenvalue λ_i. Show that $\mathbf{v}$ is also an eigenvector of the matrix $A_1 A_2 \cdots A_k$. What is the corresponding eigenvalue?

42. Fix an invertible $n \times n$ matrix S. Show that the function $T : M_n(\mathbb{R}) \to M_n(\mathbb{R})$ defined by $T(A) = S^{-1} A S$ is an isomorphism.

Project I: The Hungry Knights

King Arthur and his knights are sitting at the round table for a hearty breakfast of porridge. Each knight is served a portion of porridge (the servings are not necessarily equal portions). But the knights are greedy, and in the first minute of the meal, each knight (including King Arthur) steals half of the porridge from the knight on his left, and half of the porridge from the knight on his right. In the second minute, each knight again steals half the porridge from each neighbor. This process continues indefinitely. The knights are so busy stealing each other's porridge that nothing ever actually gets eaten.

Assume there are n knights (including King Arthur), and they receive initial distribution of porridge $(a_1, a_2, \ldots, a_n)$. The problem is to determine the long-term distribution of porridge. [Note that knight 1 and knight n are seated next to each other at the round table, and so they steal from each other.]

Part 1   Solve the above problem for $n = 3$ as follows:

(a) Regarding the above porridge distribution as a column vector $\mathbf{v}$, find a matrix A such that $A\mathbf{v}$ represents the porridge distribution after one minute, $A^2\mathbf{v}$ represents the porridge distribution after two minutes, and so on.

(b) Diagonalize A.

(c) Determine the distribution of porridge as time $\to \infty$.

(d) Do the amounts of porridge stabilize, oscillate, or does this depend on the initial distribution of porridge ? If so, how?

(e) Discuss the impact that the eigenvalues of A have on the limiting behavior. Explain how and why certain eigenvalues play a larger role in shaping the limit behavior of the distribution than others.

(f) Discuss any special cases you think are interesting (such as what happens in the case where all knights start with equal porridge, or in the case where one knight starts with all the porridge).

Part 2   Redo Part 1 for $n = 4$. [**Hint**: You can save yourself some trouble in computing eigenvalues if you find a couple of eigenvectors by inspection (trial and error) and use facts about $\mathrm{tr}(A)$ and $\det(A)$. In fact, with a little clever experimenting, you may be able to get all of the eigenvalues and eigenvectors by "inspection."]

Part 3   Based on your work above, try to guess what happens for larger values of n. Explore with larger values of n on a calculator or computer to check out your guess.

Project II: Square Roots of Matrices

In this project we will study which $n \times n$ matrices possess square roots. We begin with the formal definition.

DEFINITION 5.12.1

A **square root** for an $n \times n$ matrix A is an $n \times n$ matrix B such that $B^2 = A$.

Throughout this project we allow the matrices under consideration to have complex entries. We begin with diagonalizable matrices.

Part 1: Diagonalizable matrices

(a) In Problem 22 of Section 5.8, it was shown that $(SDS^{-1})^k = SD^k S^{-1}$ for all positive integers k. Extend this result to all positive fractions $k = p/q$, where p and q are positive integers.

(b) Applying part (a) with $k = \frac{1}{2}$, prove that all diagonalizable matrices have square roots.

(c) Find four different square roots for the matrix

$$A = \begin{bmatrix} -7 & -32 \\ 16 & 41 \end{bmatrix}.$$

(d) Generalizing part (b) above, prove that if A and B are similar matrices, then A has a square root if and only if B has a square root.

(e) Conclude that the classification of matrices that possess square roots is reduced to the classification of Jordan canonical form structures that have square roots.

Part 2: Invertible matrices

(a) Show that the Jordan canonical form of an invertible matrix A is an upper triangular matrix J with all entries along the main diagonal nonzero.

(b) Use the fact that the Jordan canonical form for a 2×2 matrix A is either a diagonal matrix or a single 2×2 Jordan block to prove that any invertible 2×2 matrix has a square root.

(c) Perform a similar analysis for the case of a 3×3 matrix A to prove that any invertible 3×3 matrix has a square root.[13]

Part 3: Matrices with all zero eigenvalues
By Part 1, the determination of which matrices with all zero eigenvalues have square roots is reduced to the determination of which Jordan canonical form structures whose blocks all correspond to the eigenvalue $\lambda = 0$ have square roots.

---

[13]In fact, it can be shown that *any* $n \times n$ invertible matrix has a square root.

(a) Prove that the Jordan block

$$J = \begin{bmatrix} 0 & 1 \\ 0 & 0 \end{bmatrix}$$

does not possess a square root. [**Hint**: Assume to the contrary that such a square root B *does* exist, and find the relationship between the Jordan canonical form of B and J to obtain a contradiction.]

(b) On the other hand, construct a square root for the 4×4 matrix J' consisting of two 2×2 Jordan blocks J as in part (a).

CHAPTER

6

Linear Differential Equations of Order n

There is no branch of mathematics, however abstract, which may not someday be applied to the phenomena of the real world. — Nicolai Lobachevsky

In Chapter 1 we developed a technique that enabled us to solve any first-order linear differential equation. We now turn our attention to linear differential equations of arbitrary order n. The general form for such a differential equation is

$$a_0(x)y^{(n)} + a_1(x)y^{(n-1)} + \cdots + a_{n-1}(x)y' + a_n(x)y = F(x), \qquad (6.0.1)$$

where $a_0, a_1, \ldots a_n, F$ are functions defined on an interval I. In this chapter we will apply the results obtained in the preceding two chapters to develop the underlying theory for the solution of (6.0.1). This will be accomplished in three steps.

1. Reformulate the problem of solving (6.0.1) in the equivalent form

$$Ly = F,$$

where L is an appropriate *linear transformation*.

2. Establish that the set of all solutions to the associated homogeneous differential equation

$$Ly = 0$$

is a vector space of *dimension n*, so that every solution to the homogeneous differential equation can be expressed as

$$y(x) = c_1 y_1(x) + c_2 y_2(x) + \cdots + c_n y_n(x),$$

where $\{y_1, y_2, \ldots, y_n\}$ is any linearly independent set of n solutions to $Ly = 0$.

3. Establish that every solution to the nonhomogeneous problem $Ly = F$ is of the form

$$y(x) = c_1 y_1(x) + c_2 y_2(x) + \cdots + c_n y_n(x) + y_p(x), \qquad (6.0.2)$$

where $y_p(x)$ is any particular solution to the nonhomogeneous equation.

The hard work put into understanding the material in Chapters 4 and 5 is really seen to pay off in Section 6.1, where the power of the vector space methods enables us to build this general theory very quickly.

Having obtained the general theory for linear differential equation of order n, we are concerned in the remainder of the chapter with obtaining the requisite number of solutions needed to build (6.0.2).

6.1   General Theory for Linear Differential Equations

Recall from Chapter 5 that the mapping $D : C^1(I) \rightarrow C^0(I)$ defined by $D(f) = f'$ is a *linear* transformation. We call D the **derivative operator**. Higher-order derivative operators can be defined by composition. Thus, $D^k : C^k(I) \rightarrow C^0(I)$ is defined by

$$D^k = D(D^{k-1}), \qquad k = 2, 3, \ldots,$$

so that

$$D^k(f) = \frac{d^k f}{dx^k}.$$

By taking a linear combination of the basic derivative operators, we obtain the general **linear differential operator of order** n,

$$L = D^n + a_1 D^{n-1} + \cdots + a_{n-1} D + a_n, \tag{6.1.1}$$

defined by

$$Ly = y^{(n)} + a_1 y^{(n-1)} + \cdots + a_{n-1} y' + a_n y,$$

where the a_i are, in general, functions of x. We leave it as an exercise (Problem 43) to verify that for all $y_1, y_2 \in C^n(I)$, and all scalars c,

$$L(y_1 + y_2) = Ly_1 + Ly_2,$$
$$L(cy_1) = cL(y_1).$$

Consequently, L is a linear transformation from $C^n(I)$ into $C^0(I)$.

Example 6.1.1   If $L = D^2 + 4xD - 3x$, then

$$Ly = y'' + 4xy' - 3xy,$$

so that, for example,

$$L(\sin x) = -\sin x + 4x \cos x - 3x \sin x,$$

whereas

$$L(x^2) = 2 + 8x^2 - 3x^3. \qquad \square$$

Now consider the general nth-order linear differential equation

$$a_0(x)y^{(n)} + a_1(x)y^{(n-1)} + \cdots + a_{n-1}(x)y' + a_n(x)y = F(x), \tag{6.1.2}$$

where $a_0(\neq 0), a_1, \ldots, a_n$ and F are functions specified on an interval I. If $F(x)$ is identically zero on I, then the differential equation (6.1.2) is called **homogeneous**. Otherwise, it is called **nonhomogeneous**. We will assume that $a_0(x)$ is nonzero on I, in which case we can divide Equation (6.1.2) by a_0 and redefine the remaining functions to obtain the following standard form:

$$y^{(n)} + a_1(x)y^{(n-1)} + \cdots + a_{n-1}(x)y' + a_n(x)y = F(x). \qquad (6.1.3)$$

This can be written in the equivalent form

$$Ly = F(x),$$

where L is given in Equation (6.1.1).

The key result that we require in developing the theory for linear differential equations is the following existence and uniqueness theorem.

Theorem 6.1.2   Let $a_1, a_2, \ldots, a_n$, and F be functions that are continuous on an interval I. Then, for any x_0 in I, the initial-value problem

$$Ly = F(x),$$
$$y(x_0) = y_0, \quad y'(x_0) = y_1, \quad \ldots, \quad y^{(n-1)}(x_0) = y_{n-1}$$

has a unique solution on I.

Proof   The proof of this theorem requires concepts from advanced calculus and is best left for a second course in differential equations. See, for example, E. A. Coddington, and N. Levinson, *Theory of Differential Equations* (New York: McGraw-Hill, 1955). ■

The differential equation (6.1.3) is said to be **regular** on I if the functions $a_1, a_2, \ldots, a_n$, and F are continuous on I. In developing the theory for linear differential equations, we will always assume that our differential equations are regular on the interval of interest, so that the existence and uniqueness theorem can be applied on that interval.

Homogeneous Linear Differential Equations

We first consider the nth-order linear homogeneous differential equation

$$y^{(n)} + a_1(x)y^{(n-1)} + \cdots + a_{n-1}(x)y' + a_n(x)y = 0 \qquad (6.1.4)$$

on an interval I. This differential equation can be written as the operator equation

$$Ly = 0,$$

where $L: C^n(I) \to C^0(I)$ is the nth-order linear differential operator

$$L = D^n + a_1 D^{n-1} + \cdots + a_{n-1}D + a_n.$$

If we let S denote the set of all solutions to the differential equation (6.1.4), then

$$S = \{y \in C^n(I) : Ly = 0\}.$$

That is,

$$S = \text{Ker}(L).$$

In Chapter 5, we proved that the kernel of any linear transformation $T: V \to W$ is a subspace of V. It follows directly from this result that S, the set of all solutions to (6.1.4), is a subspace of $C^n(I)$. We will refer to this subspace as the **solution space** of the differential equation. If we can determine the dimension of S, then we will know how many linearly independent solutions are required to span the solution space. This is dealt with in the following theorem.

Theorem 6.1.3   The set of all solutions to the regular nth-order homogeneous linear differential equation

$$y^{(n)} + a_1(x)y^{(n-1)} + \cdots + a_{n-1}(x)y' + a_n(x)y = 0$$

on an interval I is a vector space of dimension n.

Proof The given differential equation can be written in operator form as

$$Ly = 0.$$

We have already shown that the set of all solutions to this differential equation is a vector space. To prove that the dimension of the solution space is n, we must establish the existence of a basis consisting of n solutions. For simplicity, we provide the details only for the case $n = 2$.

Let y_1, y_2 be the unique solutions to the initial-value problems

$$Ly_1 = 0, \quad y_1(x_0) = 1, \quad y_1'(x_0) = 0, \tag{6.1.5}$$
$$Ly_2 = 0, \quad y_2(x_0) = 0, \quad y_2'(x_0) = 1, \tag{6.1.6}$$

where

$$L = D^2 + a_1(x)D + a_2(x).$$

The Wronskian of these solutions at $x_0 \in I$ is $W[y_1, y_2](x_0) = \det(I_2) = 1 \neq 0$, so that the solutions are linearly independent on I. To be a basis for the solution space, y_1 and y_2 must also span the solution space. Let $y = u(x)$ be any solution to the differential equation $Ly = 0$ on I, and suppose that

$$u(x_0) = c_1, \quad u'(x_0) = c_2,$$

where c_1, c_2 are constants. Then $y = u(x)$ is the unique solution to the initial-value problem

$$Ly = 0, \quad y(x_0) = c_1, \quad y'(x_0) = c_2. \tag{6.1.7}$$

However, if we define

$$w(x) = c_1 y_1(x) + c_2 y_2(x),$$

then, using the linearity of L,

$$Lw = L(c_1 y_1 + c_2 y_2) = c_1 L(y_1) + c_2 L(y_2) = 0.$$

Further, using the initial values in (6.1.5) and (6.1.6),

$$w(x_0) = c_1 y_1(x_0) + c_2 y_2(x_0) = c_1 \quad \text{and} \quad w'(x_0) = c_1 y_1'(x_0) + c_2 y_2'(x_0) = c_2.$$

Consequently, $w(x)$ also satisfies the initial-value problem (6.1.7). Thus, by uniqueness, we must have

$$u(x) = w(x).$$

That is,

$$u(x) = c_1 y_1(x) + c_2 y_2(x).$$

Therefore, we have shown that any solution to $Ly = 0$ can be written as a linear combination of the linearly independent solutions y_1, y_2, hence these solutions do span the solution space. It follows that $\{y_1, y_2\}$ is a basis for the solution space, and since the basis consists of two vectors, the dimension of this solution space is two. The extension of the foregoing proof to arbitrary n is left as an exercise (Problem 40). ∎

It follows from the previous theorem and Theorem 4.6.10 that *any* set of n linearly independent solutions, say $\{y_1, y_2, \ldots, y_n\}$, to

$$y^{(n)} + a_1(x)y^{(n-1)} + \cdots + a_{n-1}(x)y' + a_n(x)y = 0 \qquad (6.1.8)$$

is a basis for the solution space of this differential equation. Consequently, every solution to the differential equation can be written as

$$y(x) = c_1 y_1(x) + c_2 y_2(x) + \cdots + c_n y_n(x), \qquad (6.1.9)$$

for appropriate constants $c_1, c_2, \ldots, c_n$. We refer to (6.1.9) as the **general solution** to the differential equation (6.1.8).

Example 6.1.4   Determine all solutions to the differential equation $y'' + y' - 6y = 0$ of the form $y(x) = e^{rx}$, where r is a constant. Use your solutions to determine the general solution to the differential equation.

Solution: Substituting $y(x) = e^{rx}$ into the given differential equation yields

$$e^{rx}(r^2 + r - 6) = 0,$$

or equivalently,

$$(r + 3)(r - 2) = 0.$$

Hence, two solutions to the differential equation are

$$y_1(x) = e^{2x} \qquad \text{and} \qquad y_2(x) = e^{-3x}.$$

Furthermore, the Wronskian of these solutions is $W[y_1, y_2](x) = -4e^{-x} \neq 0$, so that they are linearly independent on any interval. It follows directly from Theorem 6.1.3 that a basis for the set of all solutions to the differential equation is $\{e^{2x}, e^{-3x}\}$, so that the general solution to the differential equation is

$$y(x) = c_1 e^{2x} + c_2 e^{-3x}. \qquad \square$$

In the previous example we were able to determine that the solutions y_1 and y_2 are linearly independent on any interval, since their Wronskian was nonzero. What would have happened if their Wronskian had been identically zero? Based on Theorem 4.5.21, we would not have been able to draw any conclusion as to the linear dependence or linear independence of the solutions. We now show, however, that *when dealing with solutions of an nth-order homogeneous linear differential equation*, if the Wronskian of the solutions is zero for at least one point in I, then the solutions are linearly dependent on I.

Theorem 6.1.5   Let $y_1, y_2, \ldots, y_n$ be solutions to the regular nth-order differential equation $Ly = 0$ on an interval I, and let $W[y_1, y_2, \ldots, y_n](x)$ denote their Wronskian. If $W[y_1, y_2, \ldots, y_n](x_0) = 0$ at any point x_0 in I, then $\{y_1, y_2, \ldots, y_n\}$ is *linearly dependent* on I.

Proof We provide details for the case $n = 2$ and leave the extension to arbitrary n as an exercise (Problem 41). Once more, the proof depends on the existence-uniqueness theorem. Let x_0 be a point in I at which $W[y_1, y_2](x_0) = 0$, and consider the linear system

$$c_1 y_1(x_0) + c_2 y_2(x_0) = 0,$$
$$c_1 y_1'(x_0) + c_2 y_2'(x_0) = 0,$$

where the unknowns are c_1, c_2. The determinant of the matrix of coefficients of this system is $W[y_1, y_2](x_0) = 0$, so that the system has nontrivial solutions. Let (α_1, α_2) be one such *nontrivial* solution, and define the function $u(x)$ by

$$u(x) = \alpha_1 y_1(x) + \alpha_2 y_2(x).$$

Then $y = u(x)$ satisfies the initial-value problem

$$Ly = 0, \qquad y(x_0) = 0, \qquad y'(x_0) = 0.$$

However, $y(x) = 0$ also satisfies the initial-value problem, hence by uniqueness we must have $u(x) = 0$; that is,

$$\alpha_1 y_1(x) + \alpha_2 y_2(x) = 0,$$

where at least one of α_1, α_2 is nonzero. Consequently, $\{y_1, y_2\}$ is linearly dependent on I. ∎

To summarize:

> The vanishing or nonvanishing of the Wronskian on an interval I completely characterizes whether *solutions to $Ly = 0$* are linearly dependent or linearly independent on I.

Example 6.1.6 Verify that $y_1(x) = \cos 2x$ and $y_2(x) = 3(1 - 2\sin^2 x)$ are solutions to the differential equation $y'' + 4y = 0$ on $(-\infty, \infty)$. Determine whether they are linearly independent on $(-\infty, \infty)$.

Solution: It is easily verified by direct substitution that

$$y_1'' + 4y_1 = 0 \quad \text{and} \quad y_2'' + 4y_2 = 0,$$

so that y_1 and y_2 are solutions to the given differential equation on $(-\infty, \infty)$. To determine whether they are linearly independent on $(-\infty, \infty)$, we compute their Wronskian.

$$W[y_1, y_2](x) = \begin{vmatrix} \cos 2x & 3(1 - 2\sin^2 x) \\ -2\sin 2x & -12\sin x \cos x \end{vmatrix}$$
$$= -6\sin 2x \cos 2x + 6\sin 2x \cos 2x,$$
$$= 0,$$

so that, from Theorem 6.1.5, the solutions are linearly dependent on $(-\infty, \infty)$. Indeed, since $\cos 2x = 1 - 2\sin^2 x$, the dependency relation is $3y_1 - y_2 = 0$. We leave it as an exercise to verify that a second linearly independent solution to the given differential equation is $y_3(x) = \sin 2x$, so that the general solution to $y'' + 4y = 0$ is

$$y(x) = c_1 \cos 2x + c_2 \sin 2x. \qquad \square$$

Nonhomogeneous Linear Differential Equations

We now consider the nonhomogeneous linear differential equation

$$y^{(n)} + a_1(x)y^{(n-1)} + \cdots + a_{n-1}(x)y' + a_n(x)y = F(x), \qquad (6.1.10)$$

where $F(x)$ is not identically zero on the interval of interest. If we set $F(x) = 0$ in (6.1.10), we obtain the **associated homogeneous equation**

$$y^{(n)} + a_1(x)y^{(n-1)} + \cdots + a_{n-1}(x)y' + a_n(x)y = 0. \qquad (6.1.11)$$

Equations (6.1.10) and (6.1.11) can be written in operator form as

$$Ly = F \quad \text{and} \quad Ly = 0,$$

respectively, where

$$L = D^n + a_1(x)D^{n-1} + \cdots + a_{n-1}(x)D + a_n(x).$$

The main theoretical result for nonhomogeneous linear differential equations is given in the following theorem:

Theorem 6.1.7

Let $\{y_1, y_2, \ldots, y_n\}$ be a *linearly independent* set of solutions to $Ly = 0$ on an interval I, and let $y = y_p$ be any *particular* solution to $Ly = F$ on I. Then every solution to $Ly = F$ on I is of the form

$$y = c_1 y_1 + c_2 y_2 + \cdots + c_n y_n + y_p,$$

for appropriate constants $c_1, c_2, \ldots, c_n$.

Proof Since $y = y_p$ satisfies Equation (6.1.10), we have

$$Ly_p = F. \qquad (6.1.12)$$

Let $y = u$ be any solution to Equation (6.1.10). Then we also have

$$Lu = F. \qquad (6.1.13)$$

Subtracting (6.1.12) from (6.1.13) and using the linearity of L yields

$$L(u - y_p) = 0.$$

Thus, $y = u - y_p$ is a solution to the associated homogeneous equation $Ly = 0$ and therefore can be written as

$$u - y_p = c_1 y_1 + c_2 y_2 + \cdots + c_n y_n,$$

for appropriately chosen constants $c_1, c_2, \ldots, c_n$. Consequently,

$$u = c_1 y_1 + c_2 y_2 + \cdots + c_n y_n + y_p. \qquad \blacksquare$$

According to the previous theorem, the general solution to the nonhomogeneous differential equation $Ly = F$ is of the form

$$y(x) = y_c(x) + y_p(x),$$

where

$$y_c(x) = c_1 y_1(x) + c_2 y_2(x) + \cdots + c_n y_n(x)$$

is the general solution to the associated homogeneous equation $Ly = 0$, and y_p is a particular solution to $Ly = F$. We refer to y_c as the **complementary function** for $Ly = F$.

Example 6.1.8   Verify that $y_p(x) = \frac{1}{3}e^{5x}$ is a particular solution to the differential equation

$$y'' + y' - 6y = 8e^{5x}$$

and determine the general solution.

Solution: For the given function, we have

$$y_p'' + y_p' - 6y_p = \frac{25}{3}e^{5x} + \frac{5}{3}e^{5x} - 2e^{5x} = 8e^{5x}.$$

Hence, $y_p(x) = \frac{1}{3}e^{5x}$ is a solution to the given differential equation. We have seen in Example 6.1.4 that the general solution to the associated homogeneous differential equation is

$$y_c(x) = c_1 e^{2x} + c_2 e^{-3x}.$$

So, Theorem 6.1.7 tells us that the general solution to the given differential equation is

$$y(x) = c_1 e^{2x} + c_2 e^{-3x} + \frac{1}{3}e^{5x}. \qquad \square$$

We need one final result regarding solutions to nonhomogeneous differential equations.

Theorem 6.1.9   If $y = u_p$ and $y = v_p$ are particular solutions of $Ly = f(x)$ and $Ly = g(x)$, respectively, then $y = u_p + v_p$ is a solution to $Ly = f(x) + g(x)$.

Proof We have $L(u_p + v_p) = L(u_p) + L(v_p) = f(x) + g(x)$.   ∎

We have now derived the fundamental theory for linear differential equations. In the remainder of the chapter we focus on developing techniques for finding the solutions whose existence is guaranteed by our theory.

Exercises for 6.1

Key Terms

Derivative operator, Linear differential operator of order n, Regular differential equation, nth-order linear homogeneous differential equation, Solution space of a differential equation, General solution to a differential equation, nth-order linear nonhomogeneous differential equation, Complementary function.

Skills

- Be able to evaluate a given linear differential operator L on a function y.

- Be able to compute the kernel of a linear differential operator L.

- Be able to write an nth-order linear differential equation as an operator equation.

- Be able to find solutions to a given nth-order linear homogeneous differential equation of a specified form.

- Be able to use basic solutions to an nth-order linear homogeneous differential equation to find the general solution to the differential equation.

- Be able to use the Wronskian to determine whether a collection of solutions to $Ly = 0$ are linearly dependent or linearly independent.

- Given a particular solution to an nth-order linear nonhomogeneous differential equation, be able to find the general solution to the differential equation.

True-False Review

For Questions 1–7, decide if the given statement is **true** or **false**, and give a brief justification for your answer. If true, you can quote a relevant definition or theorem from the text. If false, provide an example, illustration, or brief explanation of why the statement is false.

1. A regular nth-order linear homogeneous differential equation defined on an interval I always has n solutions that are linearly independent on the interval I.

2. If $y_1, y_2, \ldots, y_n$ are solutions to a regular nth-order linear homogeneous differential equation such that $W[y_1, y_2, \ldots, y_n](x)$ is zero at some points of I and nonzero at other points of I, then $\{y_1, y_2, \ldots, y_n\}$ is a linearly independent set of functions.

3. If L_1 and L_2 are linear differential operators, then $L_1 L_2 = L_2 L_1$.

4. If L_1 and L_2 are linear differential operators, then so is $L_1 + L_2$.

5. If L is a linear differential operator, then so is cL for all constants c.

6. If y_p is a particular solution to the differential equation $Ly = F$, then $y_p + u$ is also a solution to $Ly = F$ for every solution u of the corresponding homogeneous differential equation $Ly = 0$.

7. If y_1 is a solution to $Ly = F_1$ and y_2 is a solution to $Ly = F_2$, then $y_1 + y_2$ is a solution to $Ly = F_1 + F_2$.

Problems

For Problems 1–4, find Ly for the given differential operator if **(a)** $y(x) = 2x - 3e^{2x}$ and **(b)** $y(x) = 3\sin^2 x$.

1. $L = D - x$.

2. $L = D^2 - x^2 D + x$.

3. $L = D^3 - 2x D^2$.

4. $L = D^3 - D + 4$.

For Problems 5–8, verify that the given function is in the kernel of L.

5. $y(x) = x^{-2}, \quad L = x^2 D^2 + 2x D - 2$.

6. $y(x) = \sin(x^2), \quad L = D^2 - x^{-1} D + 4x^2$.

7. $y(x) = \sin x + \cos x, \quad L = D^3 + D^2 + D + 1$.

8. $y(x) = xe^x, \quad L = -D^2 + 2D - 1$.

For Problems 9–12, compute $\text{Ker}(L)$.

9. $L = D - 2x$.

10. $L = D^2 + 1$.

11. $L = D^2 + 2D - 15$. [**Hint:** Try for two solutions to $Ly = 0$ of the form e^{rx}.]

12. $L = x^2 D + x$.

For Problems 13–14, find $L_1 L_2$ and $L_2 L_1$ for the given differential operators, and determine whether $L_1 L_2 = L_2 L_1$.

13. $L_1 = D + 1, \quad L_2 = D - 2x^2$.

14. $L_1 = D + x, \quad L_2 = D + (2x - 1)$.

15. If $L_1 = D + a_1(x)$, determine all differential operators of the form $L_2 = D + b_1(x)$ such that $L_1 L_2 = L_2 L_1$.

For Problems 16–17, write the given nonhomogeneous differential equation as an operator equation, and give the associated homogeneous differential equation.

16. $y''' + x^2 y'' - (\sin x)y' + e^x y = x^3$.

17. $y'' + 4xy' - 6x^2 y = x^2 \sin x$.

18. Use the existence and uniqueness theorem to prove that the only solution to the initial-value problem

$$y'' + x^2 y + e^x y = 0,$$
$$y(0) = 0, \quad y'(0) = 0$$

is the trivial solution $y(x) = 0$.

19. Use the existence and uniqueness theorem to formulate and prove a general theorem regarding the solution to the initial-value problem

$$Ly = 0,$$
$$y(x_0) = 0, \quad y'(x_0) = 0, \quad \ldots, \quad y^{(n-1)}(x_0) = 0.$$

For Problems 20–23, determine two linearly independent solutions to the given differential equation of the form $y(x) = e^{rx}$, and thereby determine the general solution to the differential equation.

20. $y'' - 2y' - 3y = 0$.

21. $y'' + 7y' + 10y = 0$.

22. $y'' - 36y = 0$.

23. $y'' + 4y' = 0$.

For Problems 24–28, determine three linearly independent solutions to the given differential equation of the form $y(x) = e^{rx}$, and thereby determine the general solution to the differential equation.

24. $y''' - 3y'' - y' + 3y = 0.$

25. $y''' + 3y'' - 4y' - 12y = 0.$

26. $y''' + 3y'' - 18y' - 40y = 0.$

27. $y''' - y'' - 2y' = 0.$

28. $y''' + y'' - 10y' + 8y = 0.$

For Problems 29–30, determine four linearly independent solutions to the given differential equation of the form $y(x) = e^{rx}$, and thereby determine the general solution to the differential equation.

29. $y^{(iv)} - 2y''' - y'' + 2y' = 0.$

30. $y^{(iv)} - 13y'' + 36y = 0.$

   [**Hint:** Factor the fourth-degree equation you get as a product of two quadratic polynomials first.]

For Problems 31–32, determine two linearly independent solutions to the given differential equation of the form $y(x) = x^r$, and thereby determine the general solution to the differential equation on $(0, \infty)$.

31. $x^2 y'' + 3xy' - 8y = 0, \quad x > 0.$

32. $2x^2 y'' + 5xy' + y = 0, \quad x > 0.$

For Problems 33–34, determine three linearly independent solutions to the given differential equation of the form $y(x) = x^r$, and thereby determine the general solution to the differential equation on $(0, \infty)$.

33. $x^3 y''' + x^2 y'' - 2xy' + 2y = 0, \quad x > 0.$

34. $x^3 y''' + 3x^2 y'' - 6xy' = 0, \quad x > 0.$

35. Determine a particular solution to the given differential equation of the form $y_p(x) = A_0 e^{3x}$. Also find the general solution to the differential equation:

$$y'' + y' - 6y = 18e^{3x}.$$

36. Determine a particular solution to the given differential equation of the form

$$y_p(x) = A_0 + A_1 x + A_2 x^2.$$

Also find the general solution to the differential equation:

$$y'' + y' - 2y = 4x^2.$$

37. Determine a particular solution to the given differential equation of the form $y_p(x) = A_0 e^{3x}$. Also find the general solution to the differential equation:

$$y''' + 2y'' - y' - 2y = 4e^{3x}.$$

38. Determine a particular solution to the given differential equation of the form $y_p(x) = A_0 e^{-2x}$. Also find the general solution to the differential equation:

$$y''' + y'' - 10y' + 8y = e^{-2x}.$$

39. Determine a particular solution to the given differential equation of the form $y_p(x) = A_0 e^{4x}$. Also find the general solution to the differential equation:

$$y''' + 5y'' + 6y' = -3e^{4x}.$$

40. Extend the proof of Theorem 6.1.3 to an arbitrary positive integer n.

41. Extend the proof of Theorem 6.1.5 to an arbitrary positive integer n.

42. Let $T: V \rightarrow W$ be a linear transformation, and suppose that $\{v_1, v_2, \ldots, v_n\}$ is a basis for Ker(T). Prove that every solution to the operator equation

$$T(v) = w \tag{6.1.14}$$

is of the form

$$v = c_1 v_1 + c_2 v_2 + \cdots + c_n v_n + v_p,$$

where v_p is any particular solution to Equation (6.1.14).

43. Prove that the linear differential operator of order n,

$$L = D^n + a_1 D^{n-1} + \cdots + a_{n-1} D + a_n,$$

is a linear transformation from $C^n(I)$ to $C^0(I)$.

6.2   Constant-Coefficient Homogeneous Linear Differential Equations

In the next few sections we develop techniques for solving linear differential equations of order n that have constant coefficients. These are differential equations that can be written in the form

$$y^{(n)} + a_1 y^{(n-1)} + \cdots + a_{n-1} y' + a_n y = F(x),$$

where $a_1, a_2, \ldots, a_n$ are *constants*. To determine the general solution to this differential equation we will require the complementary function. Consequently, we begin by analyzing the associated homogeneous equation

$$y^{(n)} + a_1 y^{(n-1)} + \cdots + a_{n-1} y' + a_n y = 0.$$

In operator form, we write this as

$$P(D)y = 0,$$

where

$$P(D) = D^n + a_1 D^{n-1} + \cdots + a_{n-1} D + a_n.$$

The operator $P(D)$ is called a **polynomial differential operator**. It is the special case of the general linear differential operator introduced in the previous section that arises when the coefficients are constant. Associated with any polynomial differential operator is the real polynomial

$$P(r) = r^n + a_1 r^{n-1} + \cdots + a_{n-1} r + a_n,$$

referred to as the **auxiliary polynomial**. The corresponding polynomial equation

$$P(r) = 0$$

is called the **auxiliary equation**.

In general, the composition of two linear transformations is *not* commutative, so that, in particular, if L_1 and L_2 are two linear differential operators, then, in general, $L_1 L_2 \neq L_2 L_1$. According to the next theorem, however, commutativity *does* hold for *polynomial differential operators*. This is a key result in determining all solutions to $P(D)y = 0$.

Theorem 6.2.1   If $P(D)$ and $Q(D)$ are polynomial differential operators, then

$$P(D)Q(D) = Q(D)P(D).$$

Proof   The proof consists of a straightforward verification and is left for the exercises (Problem 42). ∎

Example 6.2.2   If $P(D) = D - 5$ and $Q(D) = D + 7$, verify that

$$P(D)Q(D) = Q(D)P(D).$$

Solution: For any twice-differentiable function f, we have

$$P(D)Q(D)f = (D - 5)(D + 7)f$$
$$= (D - 5)(f' + 7f) = f'' + 2f' - 35f$$
$$= (D^2 + 2D - 35)f.$$

Consequently,

$$P(D)Q(D) = D^2 + 2D - 35.$$

Similarly,

$$Q(D)P(D)f = (D + 7)(D - 5)f = (D + 7)(f' - 5f)$$
$$= f'' + 2f' - 35f = (D^2 + 2D - 35)f,$$

so that

$$Q(D)P(D) = D^2 + 2D - 35 = P(D)Q(D). \qquad \square$$

The importance of the preceding theorem is that it enables us to factor polynomial differential operators in the same way that we can factor real polynomials. More specifically, if $P(D)$ is a polynomial differential operator of degree n, then the auxiliary polynomial $P(r)$ can be factored as

$$P(r) = (r - r_1)^{m_1}(r - r_2)^{m_2} \cdots (r - r_k)^{m_k},$$

where m_i denotes the multiplicity of the root r_i, and

$$m_1 + m_2 + \cdots + m_k = n.$$

Consequently, $P(D)$ has the corresponding factorization

$$P(D) = (D - r_1)^{m_1}(D - r_2)^{m_2} \cdots (D - r_k)^{m_k},$$

and Theorem 6.2.1 tells us that the ordering of the terms in this factored form of $P(D)$ does not matter. It follows that the differential equation $P(D)y = 0$ can be written as

$$(D - r_1)^{m_1}(D - r_2)^{m_2} \cdots (D - r_k)^{m_k} y = 0. \tag{6.2.1}$$

The next step is to establish the following theorem.

Theorem 6.2.3   If $P(D) = P_1(D)P_2(D) \cdots P_k(D)$, where each $P_i(D)$ is a polynomial differential operator, then, for $1 \leq i \leq k$, any solution to $P_i(D)y = 0$ is also a solution to $P(D)y = 0$.

Proof Suppose $P_i(D)u = 0$ for some i satisfying $1 \leq i \leq k$. Then, since we can change the order of the factors in a polynomial differential operator (with constant coefficients), it follows that the expression for $P(D)$ can be rearranged as follows:

$$P(D) = P_1(D) \cdots P_{i-1}(D)P_{i+1}(D) \cdots P_k(D)P_i(D).$$

Hence,

$$P(D)u = P_1(D) \cdots P_{i-1}(D)P_{i+1}(D) \cdots P_k(D)P_i(D)u = 0. \qquad \blacksquare$$

Applying the preceding theorem to Equation (6.2.1), we see that any solutions to

$$(D - r_i)^{m_i} y = 0 \tag{6.2.2}$$

will also be solutions to the full differential equation (6.2.1). Therefore, we first focus our attention on differential equations of the form (6.2.2). Before determining the solution to this differential equation, we recall two properties of the complex exponential function. (See Appendix A for a fuller discussion of complex exponential functions.)

1. Euler's formula:

$$e^{(a+ib)x} = e^{ax}(\cos bx + i \sin bx). \tag{6.2.3}$$

2. If $r = a + ib$, then

$$\frac{d}{dx}(e^{rx}) = re^{rx}.$$

We now determine a set of linearly independent solutions to the differential equation (6.2.2).

Theorem 6.2.4   The differential equation $(D - r)^m y = 0$, where m is a positive integer and r is a real or complex number, has the following m solutions that are linearly independent on any interval:

$$e^{rx}, xe^{rx}, x^2 e^{rx}, \ldots, x^{m-1} e^{rx}.$$

Proof   Consider the effect of applying the differential operator $(D - r)^m$ to the function $e^{rx} u(x)$, where u is an arbitrary (but sufficiently smooth) function. When $m = 1$, we obtain

$$(D - r)(e^{rx} u) = e^{rx} u' + re^{rx} u - re^{rx} u.$$

Thus,

$$(D - r)(e^{rx} u) = e^{rx} u'.$$

Repeating this procedure yields

$$(D - r)^2 (e^{rx} u) = (D - r)(e^{rx} u') = e^{rx} u'',$$

so that in general,

$$(D - r)^m (e^{rx} u) = e^{rx} D^m(u). \tag{6.2.4}$$

Choosing $u(x) = x^k$ and using the fact that $D^m(x^k) = 0$, for $k = 0, 1, \ldots, m - 1$, it follows from (6.2.4) that

$$(D - r)^m (e^{rx} x^k) = 0, \qquad k = 0, 1, \ldots, m - 1;$$

hence $e^{rx}, xe^{rx}, x^2 e^{rx}, \ldots, x^{m-1} e^{rx}$ are solutions to the differential equation $(D - r)^m y = 0$. We now prove that these solutions are linearly independent on any interval. We must show that

$$c_1 e^{rx} + c_2 x e^{rx} + c_3 x^2 e^{rx} + \cdots + c_m x^{m-1} e^{rx} = 0,$$

for x in any interval if and only if $c_1 = c_2 = \cdots = c_m = 0$. Dividing by e^{rx}, we obtain the equivalent expression

$$c_1 + c_2 x + c_3 x^2 + \cdots + c_m x^{m-1} = 0.$$

Since the set of functions $\{1, x, x^2, \ldots, x^{m-1}\}$ is linearly independent on any interval (see Problem 40 in Section 4.5), it follows that $c_1 = c_2 = \cdots = c_m = 0$. Hence, the given functions are indeed linearly independent on any interval. ∎

We now apply the results of the previous two theorems to the differential equation

$$(D - r_1)^{m_1}(D - r_2)^{m_2} \cdots (D - r_k)^{m_k} y = 0. \tag{6.2.5}$$

The solutions obtained due to a term of the form $(D - r)^m$ depend on whether r is a real or complex number.

We consider the two cases separately.

1. Each factor of the form $(D - r)^m$, where r is *real*, contributes the m linearly independent solutions

$$e^{rx}, xe^{rx}, \ldots, x^{m-1}e^{rx}.$$

2. For each factor of the form $(D - r)^m$, where $r = a + ib$ ($b \neq 0$), the factor $(D - \bar{r})^m$ must also occur in Equation (6.2.5). These complex conjugate terms contribute the complex-valued solutions

$$e^{(a\pm ib)x}, xe^{(a\pm ib)x}, x^2e^{(a\pm ib)x}, \ldots, x^{m-1}e^{(a\pm ib)x}.$$

Real-valued solutions can be obtained in the following manner. Consider the two complex conjugate solutions

$$w_1(x) = x^k e^{(a+ib)x} = x^k e^{ax}(\cos bx + i \sin bx),$$
$$w_2(x) = x^k e^{(a-ib)x} = x^k e^{ax}(\cos bx - i \sin bx),$$

where $0 \leq k \leq m - 1$ and we have used Euler's formula (6.2.3). Since these are both solutions to a linear homogeneous equation, any linear combination of them is also a solution to the same equation. In particular, defining $y_1(x)$ and $y_2(x)$ by

$$y_1(x) = \frac{1}{2}[w_1(x) + w_2(x)] = x^k e^{ax} \cos bx$$
$$y_2(x) = \frac{1}{2i}[w_1(x) - w_2(x)] = x^k e^{ax} \sin bx$$

respectively, yields two corresponding *real-valued* solutions. Repeating this procedure for each value of k, we obtain the following $2m$ real-valued solutions to $(D - r)^m(D - \bar{r})^m y = 0$,

$$e^{ax} \cos bx, e^{ax} \sin bx, xe^{ax} \cos bx, xe^{ax} \sin bx, \ldots, x^{m-1}e^{ax} \cos bx, x^{m-1}e^{ax} \sin bx.$$

We leave the verification that these solutions are linearly independent on any interval for the exercises (Problems 43 and 44).

By considering each factor in Equation (6.2.5) successively, we can therefore obtain n real-valued solutions to $P(D)y = 0$. The proof that the resulting set of solutions is linearly independent on any interval is tedious and not particularly instructive. Consequently, this proof is omitted [see, for example, W.L. Kaplan, *Differential Equations* (Reading, MA: Addison-Wesley, 1958)].

We now summarize our results.

Theorem 6.2.5   Consider the differential equation

$$P(D)y = 0. \tag{6.2.6}$$

Let $r_1, r_2, \ldots, r_k$ be the distinct roots of the auxiliary equation, so that

$$P(r) = (r - r_1)^{m_1}(r - r_2)^{m_2} \cdots (r - r_k)^{m_k},$$

where m_i denotes the multiplicity of the root $r = r_i$.

1. If r_i is *real*, then the functions $e^{r_1 x}, xe^{r_1 x}, \ldots, x^{m_1-1}e^{r_1 x}$ are linearly independent solutions to Equation (6.2.6) on any interval.
2. If r_j is *complex*, say $r_j = a + ib$ (a and b are real, with $b \neq 0$), then the functions

$$e^{ax}\cos bx, \, xe^{ax}\cos bx, \ldots, x^{m_j-1}e^{ax}\cos bx$$
$$e^{ax}\sin bx, \, xe^{ax}\sin bx, \ldots, x^{m_j-1}e^{ax}\sin bx$$

corresponding to the conjugate roots $r = a \pm ib$ are linearly independent solutions to Equation (6.2.6) on any interval.
3. The n real-valued solutions $y_1, y_2, \ldots, y_n$ to Equation (6.2.6) that are obtained by considering the distinct roots $r_1, r_2, \ldots, r_k$ are linearly independent on any interval. Consequently, the general solution to Equation (6.2.6) is

$$y(x) = c_1 y_1(x) + c_2 y_2(x) + \cdots + c_n y_n(x).$$

Example 6.2.6   Determine the general solution to $y'' - y' - 2y = 0$.

Solution: The auxiliary polynomial is

$$P(r) = r^2 - r - 2 = (r - 2)(r + 1).$$

Therefore, the auxiliary equation has roots $r_1 = 2$, $r_2 = -1$, so that two linearly independent solutions to the given differential equation are

$$y_1(x) = e^{2x} \quad \text{and} \quad y_2(x) = e^{-x}.$$

Hence, the general solution to the differential equation is

$$y(x) = c_1 e^{2x} + c_2 e^{-x}.$$

Some representative solution curves are sketched in Figure 6.2.1.

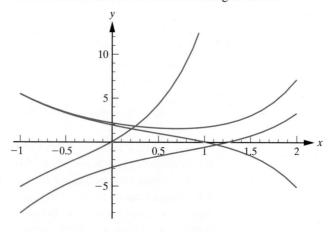

Figure 6.2.1: Representative solution curves for the differential equation in Example 6.2.6.

□

Example 6.2.7 Determine the general solution to $y'' + 6y' + 25y = 0$.

Solution: The auxiliary equation is

$$r^2 + 6r + 25 = 0,$$

with roots $r = -3 \pm 4i$. Consequently, two linearly independent real-valued solutions to the differential equation are

$$y_1(x) = e^{-3x} \cos 4x \quad \text{and} \quad y_2(x) = e^{-3x} \sin 4x$$

and the general solution to the differential equation is

$$y(x) = e^{-3x}(c_1 \cos 4x + c_2 \sin 4x).$$

We see that, owing to the presence of the trigonometric functions, the solutions are oscillatory. The negative exponential term implies that the amplitude of the oscillations decays as x increases. Some representative solution curves are given in Figure 6.2.2.

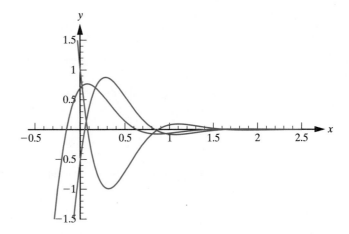

Figure 6.2.2: Representative solution curves for the differential equation in Example 6.2.7.

☐

Example 6.2.8 Solve the initial-value problem

$$y'' + 4y' + 4y = 0, \qquad y(0) = 1, \qquad y'(0) = 4.$$

Solution: The auxiliary polynomial is

$$P(r) = r^2 + 4r + 4 = (r + 2)^2.$$

Thus, $r = -2$ is a repeated root of the auxiliary equation, and therefore two linearly independent solutions to the given differential equation are

$$y_1(x) = e^{-2x}, \qquad y_2(x) = xe^{-2x}.$$

Consequently, the general solution is

$$y(x) = e^{-2x}(c_1 + c_2x).$$

Owing to the presence of the negative exponential term, it follows that all solutions approach zero as $x \to \infty$. The initial condition $y(0) = 1$ implies that $c_1 = 1$. Thus,

$$y(x) = e^{-2x}(1 + c_2 x).$$

Differentiating this expression yields

$$y'(x) = -2e^{-2x}(1 + c_2 x) + c_2 e^{-2x}$$

so that the second initial condition requires $c_2 = 6$. Hence, the unique solution to the given initial-value problem is

$$y(x) = e^{-2x}(1 + 6x).$$

Some solution curves are sketched in Figure 6.2.3. Which one corresponds to the given initial conditions?

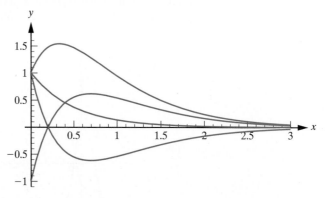

Figure 6.2.3: Representative solution curves for the differential equation in Example 6.2.8.

◻

Example 6.2.9 Determine the general solution to $y''' + 2y'' + 3y' + 2y = 0$.

Solution: The auxiliary polynomial is $P(r) = r^3 + 2r^2 + 3r + 2$, which can be factored as

$$P(r) = (r + 1)(r^2 + r + 2).$$

The roots of the auxiliary equation are therefore

$$r = -1 \quad \text{and} \quad r = \frac{-1 \pm i\sqrt{7}}{2}.$$

Hence, three linearly independent solutions to the given differential equation are

$$y_1(x) = e^{-x}, \qquad y_2(x) = e^{-x/2}\cos\left(\frac{\sqrt{7}x}{2}\right), \qquad y_3(x) = e^{-x/2}\sin\left(\frac{\sqrt{7}x}{2}\right),$$

so that the general solution is

$$y(x) = c_1 e^{-x} + c_2 e^{-x/2}\cos\left(\frac{\sqrt{7}x}{2}\right) + c_3 e^{-x/2}\sin\left(\frac{\sqrt{7}x}{2}\right).$$

◻

Example 6.2.10   Determine the general solution to

$$(D - 3)(D^2 + 2D + 2)^2 y = 0. \qquad (6.2.7)$$

Solution: The auxiliary polynomial is

$$P(r) = (r - 3)(r^2 + 2r + 2)^2,$$

so that the roots of the auxiliary equation are $r = 3$, and $r = -1 \pm i$ (multiplicity 2). The corresponding linearly independent solutions to Equation (6.2.7) are

$$y_1(x) = e^{3x},\, y_2(x) = e^{-x} \cos x,\, y_3(x) = xe^{-x} \cos x,$$
$$y_4(x) = e^{-x} \sin x,\, y_5(x) = xe^{-x} \sin x,$$

hence the general solution to (6.2.7) is

$$y(x) = c_1 e^{3x} + e^{-x}(c_2 \cos x + c_3 x \cos x + c_4 \sin x + c_5 x \sin x). \qquad \square$$

Example 6.2.11   Determine the general solution to

$$D^3(D - 2)^2(D^2 + 1)^2 y = 0. \qquad (6.2.8)$$

Solution: The auxiliary polynomial is $P(r) = r^3(r - 2)^2(r^2 + 1)^2$, with zeros $r = 0$ (multiplicity 3), $r = 2$ (multiplicity 2), and $r = \pm i$ (multiplicity 2). We therefore obtain the following linearly independent solutions to the given differential equation:

$$y_1(x) = 1, \quad y_2(x) = x, \quad y_3(x) = x^2, \quad y_4(x) = e^{2x}, \quad y_5(x) = xe^{2x},$$
$$y_6(x) = \cos x, \quad y_7(x) = x \cos x, \quad y_8(x) = \sin x, \quad y_9(x) = x \sin x.$$

Hence, the general solution to (6.2.8) is

$$y(x) = c_1 + c_2 x + c_3 x^2 + c_4 e^{2x} + c_5 xe^{2x} + (c_6 + c_7 x) \cos x + (c_8 + c_9 x) \sin x. \quad \square$$

Exercises for 6.2

Key Terms

Polynomial differential operator, Auxiliary polynomial, Auxiliary equation.

Skills

- Be able to express an nth-order constant-coefficient homogeneous linear differential equation in polynomial differential operator form.

- Be able to find the auxiliary polynomial and equation

associated with an nth-order constant-coefficient homogeneous linear differential equation.

- Be able to use the distinct roots (and their multiplicities) of the auxiliary equation to find n linearly independent solutions to an nth-order constant-coefficient homogeneous linear differential equation.

- Be able to find the general solution to an nth-order constant-coefficient homogeneous linear differential equation.

True-False Review

For Questions 1–8, decide if the given statement is **true** or **false**, and give a brief justification for your answer. If true, you can quote a relevant definition or theorem from the text. If false, provide an example, illustration, or brief explanation of why the statement is false.

1. An nth-order constant-coefficient homogeneous linear differential equation has n linearly independent solutions if and only if the corresponding auxiliary polynomial has n distinct roots.

2. Any two differential operators L_1 and L_2 commute; that is, $L_1L_2 = L_2L_1$.

3. The roots of the auxiliary polynomial of an nth-order constant-coefficient homogeneous linear differential equation have multiplicities that sum to n.

4. If 0 is a root of multiplicity 4 of the auxiliary polynomial of an nth-order constant-coefficient homogeneous linear differential equation, then any polynomial of degree 3 or less is a solution to the differential equation.

5. The general solution to the differential equation

$$D(D^2 - 4)^2 y = 0$$

is $y(x) = c_1 + c_2x + c_3e^{2x} + c_4e^{-2x} + c_5xe^{2x} + c_6xe^{-2x}.$

6. The general solution to the differential equation

$$(D + 3)^2(D^2 + 25)y = 0$$

is $y(x) = c_1e^{-3x} + c_2xe^{-3x} + c_3\cos 5x + c_4\sin 5x.$

7. The general solution to the differential equation

$$(D^2 - 4D + 5)^2 y = 0$$

is $y(x) = c_1e^{2x}\cos x + c_2e^{2x}\sin x + c_3xe^{2x}\cos x + c_4xe^{2x}\sin x.$

8. If $y(x)$ is the general solution to the differential equation $P(D)y = 0$, then $xy(x)$ is the general solution to the differential equation $(P(D))^2y = 0$.

Problems

For Problems 1–3, determine a basis for the solution space of the given differential equation.

1. $y'' + 2y' - 3y = 0.$

2. $y'' + 6y' + 9y = 0.$

3. $y'' - 6y' + 25y = 0.$

4. Let S denote the subspace of the solution space to the differential equation $y'' + 16y = 0$, with basis $\{\sin 4x + 5\cos 4x\}$. Write the general vector in S and extend the basis for S to a basis for the full solution space of the differential equation.

For Problems 5–31, find the general solution to the given differential equation.

5. $y'' - y' - 2y = 0.$

6. $y'' - 6y' + 9y = 0.$

7. $(D^2 + 6D + 25)y = 0.$

8. $(D + 1)(D - 5)y = 0.$

9. $(D + 2)^2 y = 0.$

10. $y'' - 6y' + 34y = 0.$

11. $y'' + 10y' + 25y = 0.$

12. $(D^2 - 2)y = 0.$

13. $y'' + 8y' + 20y = 0.$

14. $y'' + 2y' + 2y = 0.$

15. $(D - 4)(D + 2)y = 0.$

16. $y'' - 14y' + 58y = 0.$

17. $y''' - y'' + y' - y = 0.$

18. $y''' - 2y'' - 4y' + 8y = 0.$

19. $(D - 2)(D^2 - 16)y = 0.$

20. $(D^2 + 2D + 10)^2 y = 0.$

21. $(D^2 + 4)^2(D + 1)y = 0.$

22. $(D^2 + 3)(D + 1)^2 y = 0.$

23. $D^2(D - 1)y = 0.$

24. $y^{(iv)} - 8y'' + 16y = 0.$

25. $y^{(iv)} - 16y = 0.$

26. $y''' + 8y'' + 22y' + 20y = 0.$

27. $y^{(iv)} - 16y'' + 40y' - 25y = 0.$

28. $(D - 1)^3(D^2 + 9)y = 0.$

29. $(D^2 - 2D + 2)^2(D^2 - 1)y = 0.$

30. $(D + 3)(D - 1)(D + 5)^3 y = 0.$

31. $(D^2 + 9)^3 y = 0.$

For Problems 32–35, solve the given initial-value problem.

32. $y'' - 8y' + 16y = 0$, $y(0) = 2$, $y'(0) = 7$.

33. $y'' - 4y' + 5y = 0$, $y(0) = 3$, $y'(0) = 5$.

34. $y''' - y'' + y' - y = 0$, $y(0) = 0$, $y'(0) = 1$, $y''(0) = 2$.

35. $y''' + 2y'' - 4y' - 8y = 0$, $y(0) = 0$, $y'(0) = 6$, $y''(0) = 8$.

36. Solve the initial-value problem

$$y'' - 2my' + (m^2 + k^2)y = 0, \quad y(0) = 0, \quad y'(0) = k,$$

where m and k are positive constants.

37. Find the general solution to

$$y'' - 2my' + (m^2 - k^2)y = 0,$$

where m and k are positive constants. Show that the solution can be written in the form:

$$y(x) = e^{mx}(c_1 \cosh kx + c_2 \sinh kx).$$

38. An object of mass m is attached to one end of a spring, and the other end of the unstretched spring is attached to a fixed wall. (See Figure 6.2.4.)

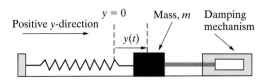

Figure 6.2.4: The spring–mass system considered in Problem 38.

The object is pulled to the right a distance y_0 and released from rest. Assuming that there is a damping force that is proportional to the velocity of the object, an application of Hooke's law and Newton's second law of motion yields an initial-value problem that can be written in the form (using appropriate units)

$$\frac{d^2 y}{dt^2} + 2c\frac{dy}{dt} + k^2 y = 0,$$

$$y(0) = y_0, \quad \frac{dy}{dt}(0) = 0,$$

where $y(t)$ denotes the displacement of the spring from its equilibrium position at time t, and c, k are positive constants.

(a) Assuming that $c^2 < k^2$, solve the preceding initial-value problem to obtain

$$y(t) = \left(\frac{y_0}{\omega}\right)e^{-ct}(\omega \cos \omega t + c \sin \omega t),$$

where $\omega = \sqrt{k^2 - c^2}$.

(b) Show that the solution in (a) can be written in the form

$$y(t) = \left(\frac{ky_0}{\omega}\right)e^{-ct} \sin(\omega t + \phi),$$

where $\phi = \tan^{-1}(\omega/c)$, and then sketch the graph of y against t. Is the predicted motion reasonable? Explain.

39. Consider the *partial differential equation* (**Laplace's equation**)

$$\frac{\partial^2 u}{\partial x^2} + \frac{\partial^2 u}{\partial y^2} = 0. \tag{6.2.9}$$

(a) Show that the substitution

$$u(x, y) = e^{x/\alpha} f(\xi),$$

where $\xi = \beta x - \alpha y$, (and α, β are positive constants) reduces Equation (6.2.9) to the differential equation

$$\frac{d^2 f}{d\xi^2} + 2p\frac{df}{d\xi} + \frac{q}{\alpha^2}f = 0, \tag{6.2.10}$$

where

$$p = \frac{\beta}{\alpha(\alpha^2 + \beta^2)}, \quad q = \frac{1}{\alpha^2 + \beta^2}. \tag{6.2.11}$$

[**Hint:** Use the chain rule, for example: $\dfrac{\partial f}{\partial x} = \dfrac{df}{d\xi} \cdot \dfrac{\partial \xi}{\partial x}$.]

(b) Solve Equation (6.2.10), and hence, find the corresponding solution to (6.2.9). [**Hint:** In solving Equation (6.2.10), you will need to use (6.2.11) in order to obtain a simple form of solution.]

40. Consider the differential equation

$$y'' + a_1 y' + a_2 y = 0, \tag{6.2.12}$$

where a_1, a_2 are constants.

(a) If the auxiliary equation has real roots r_1 and r_2, what conditions on these roots would guarantee that every solution to Equation (6.2.12) satisfies

$$\lim_{x \to +\infty} y(x) = 0?$$

(b) If the auxiliary equation has complex conjugate roots $r = a \pm ib$, what conditions on these roots would guarantee that every solution to Equation (6.2.12) satisfies

$$\lim_{x \to +\infty} y(x) = 0?$$

(c) If a_1, a_2 are positive, prove that $\lim_{x \to +\infty} y(x) = 0$, for every solution to Equation (6.2.12).

(d) If $a_1 > 0$ and $a_2 = 0$, prove that all solutions to Equation (6.2.12) approach a constant value as $x \to +\infty$.

(e) If $a_1 = 0$ and $a_2 > 0$, prove that all solutions to Equation (6.2.12) remain bounded as $x \to +\infty$.

41. Consider $P(D)y = 0$. What conditions on the roots of the auxiliary equation would guarantee that every solution to the differential equation satisfies

$$\lim_{x \to +\infty} y(x) = 0?$$

42. Prove Theorem 6.2.1.

43. For all constants a and b, prove that the set of functions $\{e^{ax} \cos bx, e^{ax} \sin bx, xe^{ax} \cos bx, xe^{ax} \sin bx\}$ is linearly independent on $(-\infty, \infty)$.

44. Generalizing the previous exercise, prove that the set of functions

$$\{e^{ax} \cos bx, xe^{ax} \cos bx, x^2 e^{ax} \cos bx, \ldots,$$
$$x^m e^{ax} \cos bx, e^{ax} \sin bx, xe^{ax} \sin bx,$$
$$x^2 e^{ax} \sin bx, \ldots, x^m e^{ax} \sin bx\}$$

is linearly independent on $(-\infty, \infty)$. [**Hint:** Show that the condition for determining linear dependence or linear independence can be written as

$$P(x) \cos bx + Q(x) \sin bx = 0,$$

where $P(x) = c_0 + c_1 x + \cdots + c_m x^m$ and $Q(x) = d_0 + d_1 x + \cdots + d_m x^m$. Then show that this implies $P(n\pi/b) = 0$ and $Q((2n+1)\pi/b) = 0$ for all integers n, which means that P and Q must both be the zero polynomial.]

For Problems 45–49, use some form of technology to factor the auxiliary polynomial of the given differential equation. Write the general solution to the differential equation.

45. $\diamond$ $y''' - 7y'' - 193y' - 665y = 0$.

46. $\diamond$ $y^{(iv)} + 4y''' - 3y'' - 64y' - 208y = 0$.

47. $\diamond$ $y^{(iv)} + 8y''' + 28y'' + 47y' + 36y = 0$.

48. $\diamond$ $y^{(v)} + 4y^{(iv)} + 50y''' + 200y'' + 625y' + 2500y = 0$.

49. $\diamond$ $y^{(vii)} + 3y^{(vi)} + 3y^{(v)} + 9y^{(iv)} + 3y''' + 9y'' + y' + 3y = 0$.

For Problems 50–51, use some form of technology to solve the given initial-value problem. Also, sketch the solution curve.

50. $\diamond$ Problem 34.

51. $\diamond$ Problem 35.

6.3   The Method of Undetermined Coefficients: Annihilators

According to Theorem 6.1.7, the general solution to the nonhomogeneous differential equation

$$P(D)y = F(x) \tag{6.3.1}$$

is of the form

$$y(x) = y_c(x) + y_p(x)$$

where y_c is the general solution to the associated homogeneous differential equation and y_p is one particular solution to (6.3.1). We have seen in the previous section how y_c can be obtained. We now turn our attention to determining a particular solution y_p. In this section we develop a method that can be applied whenever $F(x)$ has certain special forms. This technique can be introduced quite simply as follows. Consider the differential equation (6.3.1), and suppose that there is a polynomial differential operator $A(D)$ such that

$$A(D)F = 0.$$

Then, operating on (6.3.1) with $A(D)$ yields the homogeneous differential equation

$$A(D)P(D)y = 0 \qquad (6.3.2)$$

This is a constant-coefficient homogeneous linear differential equation and therefore can be solved using the technique of the previous section. The key point is the following. Any solution to (6.3.1) must also solve (6.3.2). Consequently, by choosing the arbitrary constants in the general solution to (6.3.2) appropriately, we must be able to obtain a particular solution to (6.3.1). We note that the general solution to (6.3.2) will contain the complementary function for (6.3.1), since $P(D)$ is part of the composed differential operator $A(D)P(D)$ in (6.3.2). Hence, we must be able to obtain a particular solution to (6.3.1) from that part of the general solution to (6.3.2) that does not include the complementary function.

Example 6.3.1   Determine the general solution to

$$(D + 1)(D - 1)y = 16e^{3x}. \qquad (6.3.3)$$

Solution: We first obtain the complementary function. The auxiliary polynomial is

$$P(r) = (r - 1)(r + 1),$$

so that

$$y_c(x) = c_1 e^x + c_2 e^{-x}.$$

The nonhomogeneous term in (6.3.3) is $F(x) = 16e^{3x}$, and so we need a polynomial differential operator $A(D)$ such that $A(D)F = 0$. Since $(D - a)(e^{ax}) = 0$, it follows that we can choose

$$A(D) = D - 3.$$

Operating on (6.3.3) with $A(D)$ yields the homogeneous differential equation

$$(D - 3)(D + 1)(D - 1)y = 0,$$

which has general solution

$$y(x) = c_1 e^x + c_2 e^{-x} + A_0 e^{3x}.$$

This solution must contain a particular solution to (6.3.3) for appropriate values of the constants c_1, c_2, and A_0. However, the first two terms coincide with the complementary function to (6.3.3) and therefore satisfy

$$(D + 1)(D - 1)(c_1 e^x + c_2 e^{-x}) = 0.$$

Consequently, (6.3.3) must have a solution of the form

$$y_p(x) = A_0 e^{3x}. \qquad (6.3.4)$$

We call $y_p(x)$ a **trial solution** for the differential equation (6.3.3). It contains one *undetermined coefficient*, A_0. In order to determine the appropriate value for A_0, we substitute the trial solution into (6.3.3). We have

$$(D + 1)(D - 1)(A_0 e^{3x}) = 16e^{3x}$$

—that is,

$$(D^2 - 1)(A_0 e^{3x}) = 16 e^{3x}$$

or equivalently,

$$A_0(9 e^{3x} - e^{3x}) = 16 e^{3x}.$$

We must therefore choose A_0 to satisfy

$$8 A_0 e^{3x} = 16 e^{3x},$$

so that $A_0 = 2$. Substituting this value for A_0 into (6.3.4) yields the following particular solution to (6.3.3):

$$y_p(x) = 2 e^{3x}.$$

Consequently, the general solution to (6.3.3) is

$$y(x) = y_c(x) + y_p(x) = c_1 e^x + c_2 e^{-x} + 2 e^{3x}. \qquad \square$$

This technique for obtaining a particular solution is called the **method of undetermined coefficients**. It is applicable only to differential equations that satisfy the following two conditions:

1. The differential equation has constant coefficients and therefore is of the form

$$P(D)y = F(x). \tag{6.3.5}$$

2. There exists a polynomial differential operator $A(D)$ such that

$$A(D)F = 0. \tag{6.3.6}$$

Any polynomial differential operator $A(D)$ that satisfies (6.3.6) is said to **annihilate** $F(x)$. The polynomial differential operator of lowest order that satisfies Equation (6.3.6) is called the **annihilator** of F.

Example 6.3.2   Show that $A(D) = D^2 + 4$ annihilates $F(x) = 5 \cos 2x$.

Solution: We have

$$A(D)(5 \cos 2x) = (D^2 + 4)(5 \cos 2x) = D^2(5 \cos 2x) + 20 \cos 2x$$
$$= -20 \cos 2x + 20 \cos 2x = 0. \qquad \square$$

More generally, a polynomial differential operator $A(D)$ annihilates F if and only if $y = F(x)$ is a solution to

$$A(D)y = 0.$$

Thus, the only types of functions that can be annihilated by a polynomial differential operator are those that arise as solutions to a homogeneous constant-coefficient linear differential equation. From our results of the previous section it follows that $F(x)$ must be one of the following forms:

1. $F(x) = cx^k e^{ax}$,

2. $F(x) = cx^k e^{ax} \sin bx$,

3. $F(x) = cx^k e^{ax} \cos bx$,

4. sums of equations 1–3,

where a, b, c are real numbers and k is a nonnegative integer. We next derive appropriate annihilators to cover any case that might arise. Consider first $F(x) = x^k e^{ax}$, where a is a real number. Since the differential equation

$$(D - a)^{k+1} y = 0,$$

where a is a real number and k is a nonnegative integer, has the real-valued solutions

$$e^{ax}, xe^{ax}, \ldots, x^k e^{ax},$$

it follows that

> 1. $A(D) = (D - a)^{k+1}$ annihilates each of the functions
> $$e^{ax}, xe^{ax}, \ldots, x^k e^{ax},$$
> and therefore it also annihilates
> $$F(x) = (a_0 + a_1 x + \cdots + a_k x^k) e^{ax},$$
> for all values of the constants $a_0, a_1, \ldots, a_k$.

Remark   Note the special case of statement 1 that arises when $a = 0$, namely

> $A(D) = D^{k+1}$ annihilates $F(x) = a_0 + a_1 x + \cdots + a_k x^k$.

Now consider the functions $e^{ax} \cos bx$ and $e^{ax} \sin bx$, where a and b are real numbers. These functions arise as linearly independent (real-valued) solutions to the differential equation

$$(D - \alpha)(D - \bar{\alpha}) y = 0,$$

where $\alpha = a + ib$. Expanding the polynomial differential operator, we have

$$(D^2 - 2aD + a^2 + b^2) y = 0.$$

Consequently,

> 2. $A(D) = D^2 - 2aD + a^2 + b^2$ annihilates both of the functions
> $$e^{ax} \cos bx \text{ and } e^{ax} \sin bx,$$
> and therefore it also annihilates
> $$F(x) = e^{ax}(a_0 \cos bx + b_0 \sin bx),$$
> for all values of the constants a_0, b_0. In particular,
> $$A(D) = D^2 + b^2$$
> annihilates the functions $\cos bx$ and $\sin bx$.

Further, the functions

$$e^{ax}\cos bx,\, xe^{ax}\cos bx,\, x^2 e^{ax}\cos bx,\, \ldots,\, x^k e^{ax}\cos bx,$$

$$e^{ax}\sin bx,\, xe^{ax}\sin bx,\, x^2 e^{ax}\sin bx,\, \ldots,\, x^k e^{ax}\sin bx,$$

arise as linearly independent (real-valued) solutions to the differential equation

$$(D^2 - 2aD + a^2 + b^2)^{k+1}y = 0.$$

Equivalently, we can state that

---

3. $A(D) = (D^2 - 2aD + a^2 + b^2)^{k+1}$ annihilates each of the functions

$$e^{ax}\cos bx, \quad xe^{ax}\cos bx, \quad x^2 e^{ax}\cos bx, \quad \ldots, \quad x^k e^{ax}\cos bx,$$
$$e^{ax}\sin bx, \quad xe^{ax}\sin bx, \quad x^2 e^{ax}\sin bx, \quad \ldots, \quad x^k e^{ax}\sin bx,$$

hence for all values of the constants $a_0, a_1, \ldots, a_k, b_0, b_1, \ldots, b_k$ it annihilates.

$$F(x) = (a_0 + a_1 x + \ldots + a_k x^k)e^{ax}\cos bx + (b_0 + b_1 x + \ldots + b_k x^k)e^{ax}\sin bx.$$

Finally, if $A_1(D)F_1 = 0$ and $A_2(D)F_2 = 0$, then

$$
\begin{aligned}
A_1(D)A_2(D)(F_1 + F_2) &= A_1(D)A_2(D)F_1 + A_1(D)A_2(D)F_2 \\
&= A_2(D)A_1(D)F_1 + A_1(D)(0) \\
&= A_2(D)(0) = 0.
\end{aligned}
$$

We can therefore state:

---

4. If $F(x)$ is a sum of functions of the forms given in statements 1–3, then $F(x)$ is annihilated by the corresponding *product* of the annihilators in statements 1–3.

The following examples give further illustrations of the annihilator technique.

Example 6.3.3   Determine the general solution to

$$(D - 4)(D + 1)y = 15e^{4x}. \tag{6.3.7}$$

Solution: The auxiliary polynomial for the given differential equation is $P(r) = (r - 4)(r + 1)$, so that the complementary function is

$$y_c(x) = c_1 e^{-x} + c_2 e^{4x}.$$

In this case $F(x) = 15e^{4x}$, which has annihilator $A(D) = D - 4$. Operating on the given differential equation with $A(D)$ yields the homogeneous differential equation

$$(D - 4)^2(D + 1)y = 0$$

with general solution

$$y(x) = c_1 e^{-x} + c_2 e^{4x} + A_0 x e^{4x}.$$

Since the first two terms coincide with the complementary function, an appropriate trial solution for (6.3.7) is

$$y_p(x) = A_0 x e^{4x}.$$

To determine A_0 we substitute this trial solution into (6.3.7). Differentiating y_p twice yields

$$y_p'(x) = A_0 e^{4x}(4x + 1), \qquad y_p'' = A_0 e^{4x}(16x + 8).$$

Thus, substituting y_p into $(D - 4)(D + 1)y = y'' - 3y' - 4y = 15e^{4x}$, it follows that A_0 must satisfy

$$A_0 e^{4x}[(16x + 8) - 3(4x + 1) - 4x] = 15e^{4x},$$

that is,

$$5A_0 = 15,$$

so that

$$A_0 = 3.$$

Consequently, a particular solution to (6.3.7) is

$$y_p(x) = 2xe^{4x}.$$

Hence, the general solution to (6.3.7) is

$$y(x) = c_1 e^{-x} + c_2 e^{4x} + 3xe^{4x}. \qquad \square$$

Example 6.3.4   Solve the initial-value problem

$$y'' - y' - 2y = 10 \sin x, \qquad y(0) = 0, \quad y'(0) = 1. \tag{6.3.8}$$

Solution: The auxiliary polynomial is

$$P(r) = r^2 - r - 2 = (r - 2)(r + 1),$$

so that the complementary function is

$$y_c(x) = c_1 e^{2x} + c_2 e^{-x}.$$

The annihilator for $F(x) = 10 \sin x$ is $A(D) = D^2 + 1$. The operator form of the differential equation in (6.3.8) is $(D - 2)(D + 1)y = 10 \sin x$, and operating on this with $A(D)$ therefore yields

$$(D^2 + 1)(D - 2)(D + 1)y = 0,$$

which has general solution

$$y(x) = c_1 e^{2x} + c_2 e^{-x} + A_0 \sin x + A_1 \cos x.$$

The first two terms coincide with the complementary function, so that an appropriate trial solution is

$$y_p(x) = A_0 \sin x + A_1 \cos x$$

Substituting this trial solution into Equation (6.3.8) yields

$$(-A_0 \sin x - A_1 \cos x) - (A_0 \cos x - A_1 \sin x) - 2(A_0 \sin x + A_1 \cos x) = 10 \sin x.$$

That is,

$$(-3A_0 + A_1) \sin x - (A_0 + 3A_1) \cos x = 10 \sin x.$$

This equation is satisfied for all x if and only if

$$-3A_0 + A_1 = 10 \quad \text{and} \quad A_0 + 3A_1 = 0.$$

The unique solution to this system of equations is

$$A_0 = -3 \quad \text{and} \quad A_1 = 1,$$

so that a particular solution to the differential equation in Equation (6.3.8) is

$$y_p(x) = -3 \sin x + \cos x.$$

Consequently the general solution is

$$y(x) = c_1 e^{2x} + c_2 e^{-x} - 3 \sin x + \cos x. \tag{6.3.9}$$

We now impose the initial conditions given in Equation (6.3.8). From Equation (6.3.9), $y(0) = 0$ if and only if

$$c_1 + c_2 = -1, \tag{6.3.10}$$

whereas $y'(0) = 1$ if and only if

$$2c_1 - c_2 = 4. \tag{6.3.11}$$

Solving Equations (6.3.10) and (6.3.11) yields

$$c_1 = 1 \quad \text{and} \quad c_2 = -2,$$

so that, from Equation (6.3.9), the unique solution to the given initial-value problem is

$$y(x) = e^{2x} - 2e^{-x} - 3 \sin x + \cos x. \qquad \square$$

Example 6.3.5   Determine the general solution to

$$(D^2 + 1)y = 3 \cos x + 4 \sin x. \tag{6.3.12}$$

Solution: The complementary function is

$$y_c(x) = c_1 \cos x + c_2 \sin x.$$

Furthermore, the annihilator for $F(x) = 3 \cos x + 4 \sin x$ is $A(D) = D^2 + 1$. Operating on the differential equation (6.3.12) with $A(D)$ yields

$$(D^2 + 1)^2 y = 0,$$

which has general solution

$$y(x) = c_1 \cos x + c_2 \sin x + x(A_0 \cos x + B_0 \sin x).$$

Hence, a trial solution for (6.3.12) is

$$y_p(x) = x(A_0 \cos x + B_0 \sin x).$$

Consequently,

$$y_p'(x) = x(-A_0 \sin x + B_0 \cos x) + A_0 \cos x + B_0 \sin x$$

and

$$y_p''(x) = -x(A_0 \cos x + B_0 \sin x) + 2(-A_0 \sin x + B_0 \cos x).$$

Substituting these results into the differential equation $y'' + y = 3 \cos x + 4 \sin x$ corresponding to (6.3.12) and simplifying yields

$$2(-A_0 \sin x + B_0 \cos x) = 3 \cos x + 4 \sin x,$$

so that $A_0 = -2$, and $B_0 = \frac{3}{2}$. Therefore, a particular solution to Equation (6.3.12) is

$$y_p(x) = x\left(-2 \cos x + \frac{3}{2} \sin x\right) = \frac{1}{2}x(3 \sin x - 4 \cos x),$$

and the general solution is

$$y(x) = c_1 \cos x + c_2 \sin x + \frac{1}{2}x(3 \sin x - 4 \cos x). \qquad \square$$

Example 6.3.6   Determine the general solution to

$$(D^2 - 4D + 5)y = 8xe^{2x} \cos x \qquad (6.3.13)$$

Solution: The auxiliary equation is $r^2 - 4r + 5 = 0$, with roots $r = 2 \pm i$. Therefore,

$$y_c(x) = e^{2x}(c_1 \cos x + c_2 \sin x).$$

An annihilator for $F(x) = 8xe^{2x} \cos x$ is $A(D) = (D^2 - 4D + 5)^2$. Operating on Equation (6.3.13) with $A(D)$ yields

$$(D^2 - 4D + 5)^3 y = 0,$$

which has general solution

$$y(x) = e^{2x}(c_1 \cos x + c_2 \sin x) + xe^{2x}(A_0 \cos x + B_0 \sin x) + x^2 e^{2x}(A_1 \cos x + B_1 \sin x).$$

Neglecting the contribution from the complementary function, we obtain the trial solution

$$y_p(x) = xe^{2x}(A_0 \cos x + B_0 \sin x) + x^2 e^{2x}(A_1 \cos x + B_1 \sin x).$$

Substituting into the differential equation $y'' - 4y' + 5y = 8xe^{2x} \cos x$ corresponding to (6.3.13) and simplifying yields

$$(-2A_0 + 2B_1 - 4xA_1) \sin x + (2B_0 + 2A_1 + 4xB_1) \cos x = 8x \cos x,$$

so that A_0, A_1, B_0, B_1 must satisfy

$$-2A_0 + 2B_1 = 0,$$
$$2A_1 + 2B_0 = 0,$$
$$-4A_1 = 0,$$
$$4B_1 = 8.$$

We see that $A_1 = 0 = B_0$ and $A_0 = 2 = B_1$, so that a particular solution to (6.3.13) is

$$y_p(x) = 2xe^{2x}(x \sin x + \cos x),$$

and the general solution is

$$y(x) = e^{2x}[c_1 \cos x + c_2 \sin x + 2x(x \sin x + \cos x)]. \qquad \square$$

Example 6.3.7   Use the annihilator technique to determine a trial solution for

$$(D + 1)(D^2 + 9)y = 4xe^{-x} + 5e^{2x} \cos 3x.$$

Solution: The complementary function is

$$y_c(x) = c_1 e^{-x} + c_2 \cos 3x + c_3 \sin 3x.$$

An annihilator for $F_1(x) = 4xe^{-x}$ is

$$A_1(D) = (D + 1)^2,$$

whereas an annihilator for $F_2(x) = 5e^{2x} \cos 3x$ is

$$A_2(D) = D^2 - 4D + 13.$$

Hence, operating on the given differential equation with $A(D) = (D^2 - 4D + 13)(D+1)^2$ yields the homogeneous differential equation

$$(D^2 - 4D + 13)(D + 1)^3(D^2 + 9)y = 0,$$

which has general solution

$$y(x) = c_1 e^{-x} + c_2 \cos 3x + c_3 \sin 3x + A_0 xe^{-x} + A_1 x^2 e^{-x} + e^{2x}(B_0 \cos 3x + B_1 \sin 3x).$$

Consequently, a trial solution for the given differential equation is

$$y_p(x) = A_0 xe^{-x} + A_1 x^2 e^{-x} + e^{2x}(B_0 \cos 3x + B_1 \sin 3x). \qquad \square$$

In the preceding examples, we have used annihilators to determine appropriate trial solutions on a case-by-case basis for the given nonhomogeneous linear constant-coefficient differential equation with $F(x)$ of one of the forms 1–4. As we now show, it is actually possible to derive generally the appropriate trial solutions without the need to make reference to annihilators. For example, consider

$$P(D)y = cx^k e^{ax}, \qquad (6.3.14)$$

and let y_c denote the complementary function. The appropriate annihilator for Equation (6.3.14) is $A(D) = (D - a)^{k+1}$, and so a trial solution for Equation (6.3.14) can be determined from the general solution to

$$A(D)P(D)y = 0. \qquad (6.3.15)$$

The following two cases arise:

Case 1: $r = a$ **is *not* a root of** $P(r) = 0$**:**   Then the general solution to Equation (6.3.15) will be of the form

$$y(x) = y_c(x) + e^{ax}(A_0 + A_1 x + \cdots + A_k x^k),$$

so that an appropriate trial solution is

$$y_p(x) = e^{ax}(A_0 + A_1 x + \cdots + A_k x^k).$$

This is the "usual" trial solution.

Case 2: $r = a$ **is a root of multiplicity** m **of** $P(r) = 0$**:**   Then the complementary function $y_c(x)$ will contain the terms

$$e^{ax}(c_0 + c_1 x + \cdots + c_{m-1} x^{m-1}).$$

The operator $A(D)P(D)$ will therefore contain the factor $(D - a)^{m+k+1}$, so that the terms in the general solution to Equation (6.3.15) that do not arise in the complementary function are

$$y_p(x) = e^{ax} x^m (A_0 + A_1 x + \cdots + A_k x^k),$$

which is the "modified" trial solution.

The derivation of appropriate trial solutions for $P(D)y = F(x)$ in the case when

$$P(x) = cx^k e^{ax} \cos bx \quad \text{and} \quad F(x) = cx^k e^{ax} \sin bx$$

is left as an exercise (Problem 40). We summarize the results in a table.

$F(x)$	**Usual Trial Solution**	**Modified Trial Solution**
$cx^k e^{ax}$	If $P(a) \neq 0$: $$y_p(x) = e^{ax}(A_0 + \cdots + A_k x^k)$$	If a is a root of $P(r) = 0$ of multiplicity m: $$y_p(x) = x^m e^{ax}(A_0 + \cdots + A_k x^k)$$
$cx^k e^{ax} \cos bx$ or $cx^k e^{ax} \sin bx$	If $P(a + ib) \neq 0$: $y_p(x) =$ $e^{ax}[A_0 \cos bx + B_0 \sin bx$ $+x(A_1 \cos bx + B_1 \sin bx)$ $+ \cdots +$ $x^k(A_k \cos bx + B_k \sin bx)]$	If $a + ib$ is a root of $P(r)$ of multiplicity m: $y_p(x) = x^m e^{ax}[A_0 \cos bx$ $+B_0 \sin bx+$ $x(A_1 \cos bx + B_1 \sin bx)$ $+ \cdots +$ $x^k(A_k \cos bx + B_k \sin bx)]$

If $F(x)$ is the sum of functions of the preceding form, then the appropriate trial solution is the corresponding sum.

In the following table, we have specialized the foregoing trial solutions to the cases that arise most often in applications:

$F(x)$	**Usual Trial Solution**	**Modified Trial Solution**
ce^{ax}	If $P(a) \neq 0$: $y_p(x) = A_0 e^{9x}$	If a is a root of $P(r) = 0$ of multiplicity m: $y_p(x) = A_0 x^m e^{ax}$
$c \cos bx$ or $c \sin bx$	If $P(ib) \neq 0$: $y_p(x) = A_0 \cos bx + B_0 \sin bx$	If ib is a root of $P(r) = 0$ of multiplicity m: $y_p(x) = x^m(A_0 \cos bx + B_0 \sin bx)$
cx^k	If $P(0) \neq 0$: $y_p(x) = A_0 + A_1 x + \cdots + A_k x^k$	If zero is a root of $P(r) = 0$ of multiplicity m: $y_p(x) = x^m(A_0 + A_1 x + \cdots + A_k x^k)$

Exercises for 6.3

Key Terms

Trial solution, Undetermined coefficients, Annihilate, Annihilator.

Skills

- Be able to determine the annihilator of a given function.

- Be able to use annihilators to derive an appropriate trial solution for the constant-coefficient differential equation $P(D)y = F(x)$.

- Be able to determine the general solution to an nth-order constant-coefficient nonhomogeneous differential equation $P(D)y = F(x)$ by using an appropriate trial solution.

True-False Review

For Questions 1–8, decide if the given statement is **true** or **false**, and give a brief justification for your answer. If true, you can quote a relevant definition or theorem from the text. If false, provide an example, illustration, or brief explanation of why the statement is false.

1. If $A_1(D)$ annihilates $F_1(x)$ and $A_2(D)$ annihilates $F_2(x)$, then $A_1(D) + A_2(D)$ annihilates $F_1(x) + F_2(x)$.

2. The annihilator of $F(x) = a_0 + a_1 x + \cdots + a_k x^k$ is $A(D) = D^k$.

3. The annihilator of $F(x) = xe^{ax}$ is $A(D) = (D-a)^2$.

4. Every function $F(x)$ has a unique annihilator $A(D)$.

5. If $A_1(D)A_2(D)$ annihilates $F(x)$, then either $A_1(D)$ annihilates $F(x)$ or $A_2(D)$ annihilates $F(x)$.

6. The appropriate trial solution for the fourth-order differential equation $D^2(D^2 + 4)y = 3 - 5x$ is $y_p(x) = A_0 + A_1 x + A_2 x^2 + A_3 x^3$.

7. The appropriate trial solution for the fifth-order differential equation $D^3(D^2 + 1)y = x^4$ is $y_p(x) = A_0 + A_1 x + A_2 x^2 + A_3 x^3 + A_4 x^4$.

8. The appropriate trial solution for the fifth-order differential equation $D^3(D^2 + 1)y = \cos x$ is $y_p(x) = x(A_0 \cos x + B_0 \sin x)$.

Problems

For Problems 1–16, determine the annihilator of the given function.

1. $F(x) = 2e^x - 3x$.

2. $F(x) = 5e^{-3x}$.

3. $F(x) = x^3 e^{7x} + 5 \sin 4x$.

4. $F(x) = \sin x + 3xe^{2x}$.

5. $F(x) = e^x \sin 2x + 3 \cos 2x$.

6. $F(x) = 4e^{-2x} \sin x$.

7. $F(x) = e^{5x}(2 - x^2) \cos x$.

8. $F(x) = (1 - 3x)e^{4x} + 2x^2$.

9. $F(x) = e^{4x}(x - 2 \sin 5x) + 3x - x^2 e^{-2x} \cos x$.

10. $F(x) = e^{-3x}(2 \sin x + 7 \cos x)$.

11. $F(x) = x \cos 3x$.

12. $F(x) = x^2 \sin x$.

13. $F(x) = \sin^4 x$. [**Hint:** Write $\sin^2 x = \dfrac{1 - \cos 2x}{2}$.]

14. $F(x) = \cos^2 x$. [**Hint:** Write $\cos^2 x = \dfrac{1 + \cos 2x}{2}$.]

15. $F(x) = \sin^2 x \cos^2 x \cos^2 2x$.

16. $F(x) = \sin x \cos^3 x$.

For Problems 17–27, determine the general solution to the given differential equation. Derive your trial solution using the annihilator technique.

17. $y'' + 4y' + 4y = 5xe^{-2x}$.

18. $y'' + y = 6e^x$.

19. $y'' - y' - 2y = 5e^{2x}$.

20. $y'' + 16y = 4\cos x$.

21. $(D - 2)(D - 3)y = 7e^{2x}$.

22. $y'' + 2y' + 5y = 3\sin 2x$.

23. $(D^2 + 6)y = \sin^2 x \cos^2 x$.

24. $(D + 3)(D - 1)y = \sin^2 x$.

25. $y''' + 2y'' - 5y' - 6y = 4x^2$.

26. $y''' - y'' + y' - y = 9e^{-x}$.

27. $y''' + 3y'' + 3y' + y = 2e^{-x} + 3e^{2x}$.

28. Solve the given initial-value problem:
$$(D - 1)(D - 2)(D - 3)y = 6e^{4x},$$
$$y(0) = 4, \qquad y'(0) = 10, \qquad y''(0) = 30.$$

For Problems 29–33, find the general solution to the given differential equation. Do not use annihilators in determining your trial solution.

29. $y''' - 2y'' - y' + 2y = 4e^{3x}$.

30. $y''' - 3y'' - 16y' + 48y = 6e^{3x}$.

31. $y''' + 3y'' - 4y' - 12y = 4\cos x$.

32. $D^2(D^2 + 1)y = 6 - 12x$.

33. $(D + 1)^3 y = 5e^{-x}$.

For Problems 34–39, determine an appropriate trial solution for the given differential equation. Do *not* solve for the constants that arise in your trial solution.

34. $(D + 1)(D^2 + 1)y = 4xe^x$.

35. $(D^2 + 4D + 13)^2 y = 5e^{-2x}\cos 3x$.

36. $(D^2 + 4)(D - 2)^3 y = 4x + 9xe^{2x}$.

37. $D^2(D - 1)(D^2 + 4)^2 y = 11e^x - \sin 2x$.

38. $(D^2 - 2D + 2)^3(D - 2)^2(D + 4)y = e^x \cos x - 3e^{2x}$.

39. $D(D^2 - 9)(D^2 - 4D + 5)y = 2e^{3x} + e^{2x}\sin x$.

40. Derive an appropriate trial solution for the differential equation
$$P(D)y = cx^k e^{ax} \cos bx.$$

6.4   Complex-Valued Trial Solutions

We now introduce an alternative method for solving constant-coefficient differential equations of the form

$$y'' + a_1 y' + a_2 y = F(x)$$

when $F(x)$ contains terms of the form $x^k e^{ax} \sin bx$ or $x^k e^{ax} \cos bx$. The technique is based on the observation that

$$x^k e^{ax} \cos bx = \text{Re}\{x^k e^{(a+ib)x}\} \qquad \text{and} \qquad x^k e^{ax} \sin bx = \text{Im}\{x^k e^{(a+ib)x}\}$$

where "Re" and "Im" denote the real part and the imaginary part of a complex-valued function, respectively. To see why this observation is useful, we need the next theorem.

Theorem 6.4.1   If $y(x) = u(x) + iv(x)$ is a complex-valued solution to

$$y'' + a_1 y' + a_2 y = F(x) + iG(x) \tag{6.4.1}$$

then

$$u'' + a_1 u' + a_2 u = F(x) \qquad \text{and} \qquad v'' + a_1 v' + a_2 v = G(x).$$

Proof If $y(x) = u(x) + iv(x)$, then

$$y'' + a_1 y' + a_2 y = [u(x) + iv(x)]'' + a_1[u(x) + iv(x)]' + a_2[u(x) + iv(x)]$$
$$= (u'' + a_1 u' + a_2 u) + i(v'' + a_1 v' + a_2 v).$$

Since y solves Equation (6.4.1), we must have

$$(u'' + a_1 u' + a_2 u) + i(v'' + a_1 v' + a_2 v) = F(x) + iG(x).$$

Equating real and imaginary parts on either side of this equation yields the desired result. ∎

Consequently, if we solve the complex equation

$$y'' + a_1 y' + a_2 y = cx^k e^{(a+ib)x}, \tag{6.4.2}$$

then, by taking the real and imaginary parts of the resulting complex-valued solution, we can directly determine solutions to

$$y'' + a_1 y' + a_2 y = cx^k e^{ax} \cos bx \quad \text{and} \quad y'' + a_1 y' + a_2 y = cx^k e^{ax} \sin bx. \tag{6.4.3}$$

The key point is that Equation (6.4.2) is a simpler equation to solve than its real counterparts given in (6.4.3). We illustrate the technique with some examples.

Example 6.4.2   Solve

$$y'' + y' - 6y = 4\cos 2x. \tag{6.4.4}$$

Solution: The complementary function for Equation (6.4.4) is

$$y_c(x) = c_1 e^{-3x} + c_2 e^{2x}.$$

In determining a particular solution, we consider the complex differential equation

$$z'' + z' - 6z = 4e^{2ix}. \tag{6.4.5}$$

An appropriate complex-valued trial solution for this differential equation is

$$z_p(x) = A_0 e^{2ix}, \tag{6.4.6}$$

where A_0 is a complex constant. The first two derivatives of z_p are

$$z_p'(x) = 2i A_0 e^{2ix} \qquad \text{and} \qquad z_p''(x) = -4 A_0 e^{2ix},$$

so that z_p is a solution to Equation (6.4.5) if and only if

$$(-4A_0 + 2i A_0 - 6A_0)e^{2ix} = 4e^{2ix}.$$

This is the case if and only if

$$A_0 = \frac{2}{-5+i} = -\frac{1}{13}(5+i).$$

Substituting this value of A_0 into (6.4.6) yields

$$z_p(x) = -\frac{1}{13}(5+i)e^{2ix} = -\frac{1}{13}(5+i)(\cos 2x + i \sin 2x)$$

$$= \frac{1}{13}(\sin 2x - 5\cos 2x) - \frac{1}{13}i(\cos 2x + 5\sin 2x).$$

Consequently, a particular solution to Equation (6.4.4) is

$$y_p(x) = \text{Re}\{z_p\} = \frac{1}{13}(\sin 2x - 5\cos 2x),$$

so that the general solution to Equation (6.4.4) is

$$y(x) = y_c(x) + y_p(x) = c_1 e^{-3x} + c_2 e^{2x} + \frac{1}{13}(\sin 2x - 5\cos 2x).$$

Notice that we can also write down the general solution to the differential equation

$$y'' + y' - 6y = 4\sin 2x,$$

since a particular solution will just be $\text{Im}\{z_p\}$.   □

Example 6.4.3   Solve

$$y'' - 2y' + 5y = 8e^x \sin 2x. \tag{6.4.7}$$

Solution: The complementary function is

$$y_c(x) = e^x(c_1 \cos 2x + c_2 \sin 2x).$$

In order to determine a particular solution to Equation (6.4.7), we consider the complex counterpart

$$z'' - 2z' + 5z = 8e^{(1+2i)x}. \tag{6.4.8}$$

Since $1 + 2i$ is a root of the auxiliary equation, an appropriate trial solution for Equation (6.4.8) is

$$z_p(x) = A_0 x e^{(1+2i)x}. \tag{6.4.9}$$

Differentiating with respect to x yields

$$\begin{cases} z_p'(x) = A_0 e^{(1+2i)x}[(1+2i)x + 1], \\ z_p''(x) = A_0 e^{(1+2i)x}[(1+2i)^2 x + 2(1+2i)] = A_0 e^{(1+2i)x}[(-3+4i)x + 2(1+2i)]. \end{cases}$$

Substituting into Equation (6.4.8) leads to the following condition on A_0:

$$A_0\,[(-3+4i)x + 2(1+2i) - 2(1+2i)x - 2 + 5x] = 8.$$

Hence,

$$A_0 = \frac{2}{i} = -2i.$$

It follows from Equation (6.4.9) that a complex-valued solution to Equation (6.4.8) is

$$z_p(x) = -2ixe^{(1+2i)x} = -2ixe^x(\cos 2x + i\sin 2x)$$

and so a particular solution to the original differential equation is

$$y_p(x) = \text{Im}\{z_p\} = -2xe^x\cos 2x.$$

Consequently, Equation (6.4.7) has general solution

$$y(x) = e^x(c_1\cos 2x + c_2\sin 2x) - 2xe^x\cos 2x.$$

◻

Exercises for 6.4

Skills

- Be able to use the method of Section 6.3 to solve differential equations of the form (6.4.3) by using a complex-valued trial solution.

True-False Review

For Questions 1–6, decide if the given statement is **true** or **false**, and give a brief justification for your answer. If true, you can quote a relevant definition or theorem from the text. If false, provide an example, illustration, or brief explanation of why the statement is false.

1. An appropriate complex-valued trial solution for the differential equation $y'' + y' = e^{-x}\cos 2x$ is $y_p(x) = A_0e^{(-1+2i)x}$.

2. An appropriate complex-valued trial solution for the differential equation $y'' + 2y' - 15y = xe^{2x}\sin 3x$ is $y_p(x) = A_0xe^{(2+3i)x}$.

3. An appropriate complex-valued trial solution for the differential equation $y'' + y = -\cos x$ is $y_p(x) = A_0e^{ix}$.

4. An appropriate complex-valued trial solution for the differential equation $y'' + y = 3\sin 2x$ is $y_p(x) = A_0e^{2ix}$.

5. An appropriate complex-valued trial solution for the differential equation $y'' - 4y' + 29y = xe^{2x}\sin 5x$ is $y_p(x) = A_0xe^{(2+5i)x}$.

6. An appropriate complex-valued trial solution for the differential equation $y'' + 9y = \cos 3x + \sin 4x$ is $y_p(x) = Axe^{3ix} + Be^{4ix}$.

Problems

For all problems below, use a complex-valued trial solution to determine a particular solution to the given differential equation.

1. $y'' + 2y' + y = 50\sin 3x$.

2. $y'' - y = 10e^{2x}\cos x$.

3. $y'' + 4y' + 4y = 169\sin 3x$.

4. $y'' - y' - 2y = 40\sin^2 x$.

5. $y'' + y = 3e^x\cos 2x$.

6. $y'' + 2y' + 2y = 2e^{-x}\sin x$.

7. $y'' - 4y = 100xe^x\sin x$.

8. $y'' + 2y' + 5y = 4e^{-x}\cos 2x$.

9. $y'' - 2y' + 10y = 24e^x\cos 3x$.

10. $y'' + 16y = 34e^x + 16\cos 4x - 8\sin 4x$.

11. $d^2y/dt^2 + \omega_0^2 y = F_0\cos\omega t$, where ω_0, ω are positive constants, and F_0 is an arbitrary constant. You will need to consider the cases $\omega \neq \omega_0$ and $\omega = \omega_0$ separately.

6.5 Oscillations of a Mechanical System

In this section we analyze in some detail the motion of a mechanical system consisting of a mass attached to a spring. We will see that even such a simple physical system has interesting and varied behavior. We begin by constructing an appropriate mathematical model.

Mathematical Formulation

Statement of the problem: A mass of m kilograms is attached to the end of a spring whose natural length is l_0 meters. At $t = 0$, the mass is displaced a distance y_0 meters from its equilibrium position and released with a velocity v_0 meters/second. We wish to determine the initial-value problem that governs the resulting motion.

Mathematical formulation of the problem: We assume that the motion takes place vertically and adopt the convention that distances are measured positive in the downward direction. In order to formulate the problem mathematically, we need to determine the forces acting on the mass. Consider first the static equilibrium position, in which the mass hangs freely from the spring with no motion. (See Figure 6.5.1.) The forces acting on the mass in this equilibrium position are

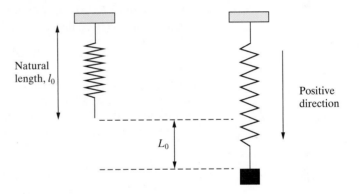

Figure 6.5.1: Spring–mass system in static equilibrium.

1. The force due to gravity

$$F_g = mg.$$

2. The spring force, F_s. According to Hooke's law (see Section 1.1),

$$F_s = -kL_0,$$

    where k is the (positive) spring constant and L_0 is the displacement of the spring from its equilibrium position.

    Since the system is in static equilibrium, these forces must exactly balance, so that $F_s + F_g = 0$. Hence,

$$mg = kL_0. \tag{6.5.1}$$

Now consider the situation when the mass has been set in motion. (See Figure 6.5.2.) We let $y(t)$ denote the position of the mass at time t and take $y = 0$ to coincide with

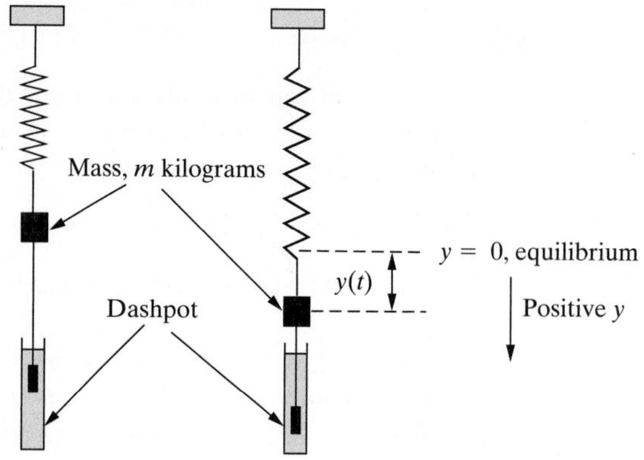

Figure 6.5.2: A simple model of a damped spring–mass system.

the equilibrium position of the system. The equation of motion of the mass can then be obtained from Newton's second law. The forces that now act on the mass are as follows:

1. The force due to gravity F_g. Once more this is

$$F_g = mg. \tag{6.5.2}$$

2. The spring force F_s. At time t the total displacement of the spring from its natural length is $L_0 + y(t)$, so that, according to Hooke's law,

$$F_s = -k[L_0 + y(t)]. \tag{6.5.3}$$

3. A damping force F_d. In general, the motion will be damped by, for example, air resistance or, as shown in Figure 6.5.2, an external damping system, such as a dashpot. We assume that any damping forces that are present are directly proportional to the velocity of the mass. Under this assumption, we have

$$F_d = -c\frac{dy}{dt} \tag{6.5.4}$$

where c is a *positive* constant called the **damping constant**. Note that the negative sign is inserted in Equation (6.5.4), since F_d always acts in the opposite direction to that of the motion.

4. Any external driving forces, $F(t)$, that are present. For example, the top of the spring or the mass itself may be subjected to an external force.

The total force acting on the system will be the sum of the preceding forces. Thus, using Newton's second law of motion, the differential equation governing the motion of the mass is

$$m\frac{d^2y}{dt^2} = F_g + F_s + F_d + F(t).$$

Substituting from Equations (6.5.2)–(6.5.4) yields

$$m\frac{d^2y}{dt^2} = mg - k(L_0 + y) - c\frac{dy}{dt} + F(t).$$

That is, using Equation (6.5.1), and rearranging terms,

$$\frac{d^2y}{dt^2} + \frac{c}{m}\frac{dy}{dt} + \frac{k}{m}y = \frac{1}{m}F(t).$$

In addition, we also have the initial conditions

$$y(0) = y_0 \quad \text{and} \quad \frac{dy}{dt}(0) = v_0,$$

where y_0 denotes the initial displacement of the mass from its equilibrium position, and v_0 denotes the initial velocity of the mass. The motion of the spring–mass system is therefore governed by the initial-value problem

$$\frac{d^2y}{dt^2} + \frac{c}{m}\frac{dy}{dt} + \frac{k}{m}y = \frac{1}{m}F(t), \qquad y(0) = y_0, \qquad \frac{dy}{dt}(0) = v_0. \tag{6.5.5}$$

Free Oscillations of a Mechanical System

We first consider the case when there are no external forces acting on the system. In the preceding formulation this corresponds to setting $F(t) = 0$, so that the initial-value problem (6.5.5) reduces to

$$\frac{d^2y}{dt^2} + \frac{c}{m}\frac{dy}{dt} + \frac{k}{m}y = 0, \qquad y(0) = y_0, \qquad \frac{dy}{dt}(0) = v_0.$$

For most of the discussion we will concentrate on the differential equation alone, since the initial conditions do not significantly affect the behavior of its solutions. We must therefore solve the constant-coefficient homogeneous differential equation

$$\frac{d^2y}{dt^2} + \frac{c}{m}\frac{dy}{dt} + \frac{k}{m}y = 0. \tag{6.5.6}$$

We divide the discussion of the solution to Equation (6.5.6) into several subcases.

Case 1: No Damping.   This is the simplest case that can arise and is of importance for understanding the more general situation. Setting $c = 0$ in Equation (6.5.6) yields

$$\frac{d^2y}{dt^2} + \omega_0^2 y = 0, \tag{6.5.7}$$

where

$$\omega_0 = \sqrt{k/m}.$$

Equation (6.5.7) has general solution

$$y(t) = c_1 \cos \omega_0 t + c_2 \sin \omega_0 t. \tag{6.5.8}$$

It is instructive to introduce two new constants A_0 and ϕ, defined in terms of c_1 and c_2 by (see Figure 6.5.3)

$$A_0 \cos \phi = c_1, \qquad A_0 \sin \phi = c_2. \tag{6.5.9}$$

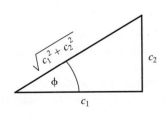

Figure 6.5.3: The definition of the phase angle ϕ.

That is,

$$A_0 = \sqrt{c_1^2 + c_2^2}, \qquad \phi = \arctan\left(\frac{c_2}{c_1}\right).$$

Substituting from Equation (6.5.9) into Equation (6.5.8) yields

$$y(t) = A_0(\cos \omega_0 t \cos \phi + \sin \omega_0 t \sin \phi).$$

Consequently,

$$y(t) = A_0 \cos(\omega_0 t - \phi). \qquad (6.5.10)$$

Clearly, the motion described by Equation (6.5.10) is periodic. We refer to such motion as **simple harmonic motion (SHM)**. Figure 6.5.4 depicts this motion for typical values of the constants A_0, ω_0, and ϕ. The standard names for the constants arising in the solution are as follows:

A_0: the **amplitude** of the motion.

ω_0: the **circular frequency** of the system.

ϕ: the **phase** of the motion.

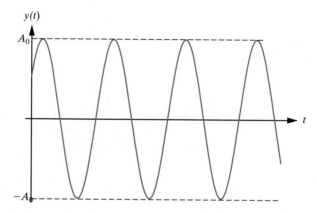

Figure 6.5.4: Simple harmonic motion. The mass continues to oscillate with a constant amplitude A_0.

The fundamental **period** of oscillation (that is, the time for the system to undergo one complete cycle), T, is

$$T = \frac{2\pi}{\omega_0} = 2\pi \sqrt{\frac{m}{k}}.$$

Consequently, the frequency of oscillation (number of oscillations per unit of time), f, is given by

$$f = \frac{1}{T} = \frac{\omega_0}{2\pi} = \frac{1}{2\pi} \sqrt{\frac{k}{m}}.$$

Notice that this is independent of the initial conditions. It is truly a property of the system.

Case 2: Damping. We now discuss the motion of the spring–mass system when the damping constant, c, is *nonzero*. In this case, the auxiliary polynomial for Equation (6.5.6) is

$$P(r) = r^2 + \frac{c}{m} r + \frac{k}{m}$$

with roots

$$r = \frac{-c \pm \sqrt{c^2 - 4km}}{2m}.$$

As we might expect, the behavior of the system is dependent on whether the auxiliary polynomial has distinct real roots, a repeated real root, or complex conjugate roots. These three situations will arise, depending on the magnitude of the (dimensionless) combination of the system variables $c^2/(4km)$. For a given spring and mass, only the damping can be altered, which leads to the following terminology. We say that the system is

(a) *underdamped* if $c^2/(4km) < 1$ (complex conjugate roots),

(b) *critically damped* if $c^2/(4km) = 1$ (repeated real root)

(c) *overdamped* if $c^2/(4km) > 1$ (two distinct real roots).

The corresponding solutions to Equation (6.5.6) are

$$y(t) = e^{-ct/(2m)}(c_1 \cos \mu t + c_2 \sin \mu t), \qquad \mu = \frac{\sqrt{4km - c^2}}{2m}, \qquad (6.5.11)$$

$$y(t) = e^{-ct/(2m)}(c_1 + c_2 t), \qquad (6.5.12)$$

$$y(t) = e^{-ct/(2m)}(c_1 e^{\mu t} + c_2 e^{-\mu t}), \qquad \mu = \frac{\sqrt{c^2 - 4km}}{2m}. \qquad (6.5.13)$$

As we shall discuss below, in all three cases (6.5.11)–(6.5.13) we have $\lim\limits_{t \to \infty} y(t) = 0$, which implies that the motion dies out for large t. This is certainly consistent with our everyday experience. We will discuss the different cases separately.

Case 2a: Underdamping.   In this case, the position of the mass at time t is given in (6.5.11), which reduces to SHM when $c = 0$. Once more it is convenient to introduce constants A_0 and ϕ, defined by

$$A_0 \cos \phi = c_1 \qquad \text{and} \qquad A_0 \sin \phi = c_2.$$

Now (6.5.11) can be written in the equivalent form

$$y(t) = A_0 e^{-ct/(2m)} \cos(\mu t - \phi). \qquad (6.5.14)$$

We see that the mass oscillates between $\pm A_0 e^{-ct/(2m)}$. The corresponding motion is depicted in Figure 6.5.5 for the case when $y(0) > 0$ and $dy/dt(0) > 0$.

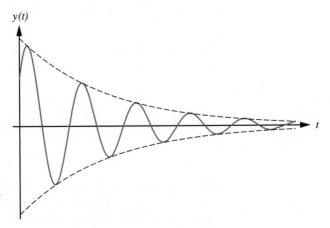

Figure 6.5.5: Underdamping: The mass oscillates between $\pm A_0 e^{-ct/(2m)}$.

In general the motion *is* oscillatory, but it is *not* periodic. The amplitude of the motion dies out exponentially with time, although the time interval, T, between successive maxima (or minima) of $y(t)$ has the constant value (see Problem 14)

$$T = \frac{2\pi}{\mu} = \frac{4\pi m}{\sqrt{4km - c^2}}.$$

This is called the **quasiperiod**.

Case 2b: Critical Damping. This case arises when $c^2/(4km) = 1$. From Equation (6.5.6), the motion is governed by the differential equation

$$\frac{d^2y}{dt^2} + \frac{c}{m}\frac{dy}{dt} + \frac{c^2}{4m^2}y = 0$$

with general solution

$$y(t) = e^{-ct/(2m)}(c_1 + c_2 t). \tag{6.5.15}$$

Now the damping is so severe that the system can pass through the equilibrium position at most once, and so we do not have oscillatory behavior. If we impose the initial conditions

$$y(0) = y_0 \quad \text{and} \quad \frac{dy}{dt}(0) = v_0,$$

then it is easily shown (see Problem 15) that (6.5.15) can be written in the form

$$y(t) = e^{-ct/(2m)}\left[y_0 + t\left(v_0 + \frac{c}{2m}y_0\right)\right].$$

Consequently, the system will pass through the equilibrium position, provided y_0 and $v_0 + c/2m\, y_0$ have opposite signs. A sketch of the motion described by (6.5.15) is given in Figure 6.5.6.

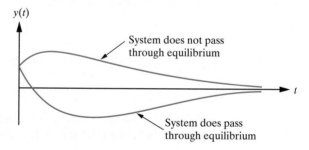

Figure 6.5.6: Critical damping: The system can pass through equilibrium at most once.

Case 2c: Overdamping. In this case we have $c^2/(4km) > 1$. The roots of the auxiliary equation corresponding to Equation (6.5.6) are

$$r_1 = \frac{-c + \sqrt{c^2 - 4km}}{2m} \quad \text{and} \quad r_2 = \frac{-c - \sqrt{c^2 - 4km}}{2m}$$

so that the general solution to Equation (6.5.6) is

$$y(t) = e^{-ct/(2m)}(c_1 e^{\mu t} + c_2 e^{-\mu t}), \quad \mu = \frac{\sqrt{c^2 - 4km}}{2m}.$$

Since c, k, and m are positive, it follows that both of the roots of the auxiliary equation are negative, which implies that both terms in $y(t)$ decay in time. Once more, we do not have oscillatory behavior. The motion is very similar to that of the critically damped case. The system can pass through the equilibrium position at most once. (The graphs given in Figure 6.5.6 are representative of this case also.)

Forced Oscillations

We now consider the case when an external force acts on the spring–mass system. As shown at the beginning of the section, the appropriate differential equation describing the motion of the system is

$$\frac{d^2 y}{dt^2} + \frac{c}{m}\frac{dy}{dt} + \frac{k}{m}y = \frac{F(t)}{m}.$$

The situation of most interest arises when the applied force is periodic in time, and we therefore restrict attention to a driving term of the form

$$F(t) = F_0 \cos \omega t,$$

where F_0 and ω are constants. Then the differential equation governing the motion is

$$\frac{d^2 y}{dt^2} + \frac{c}{m}\frac{dy}{dt} + \frac{k}{m}y = \frac{F_0}{m}\cos \omega t. \tag{6.5.16}$$

Once more we will divide our discussion into several cases.

Case 1: No Damping.   Setting $c = 0$ in Equation (6.5.16) yields

$$\frac{d^2 y}{dt^2} + \omega_0^2 y = \frac{F_0}{m}\cos \omega t, \tag{6.5.17}$$

where

$$\omega_0 = \sqrt{k/m}$$

denotes the circular frequency of the system. The complementary function for Equation (6.5.17) is

$$y_c(t) = c_1 \cos \omega_0 t + c_2 \sin \omega_0 t$$

which can be written in the form

$$y_c(t) = A_0 \cos(\omega_0 t - \phi),$$

for appropriate constants A_0 and ϕ. We therefore need to find a particular solution to Equation (6.5.17). The right-hand side of Equation (6.5.17) is of an appropriate form to use the method of undetermined coefficients, although the trial solution will depend on whether $\omega \neq \omega_0$ or $\omega = \omega_0$.

Case 1a: $\omega \neq \omega_0$.   In this case, the appropriate trial solution is

$$y_p(t) = A \cos \omega t + B \sin \omega t.$$

A straightforward calculation yields the particular solution for Equation (6.5.17) (see Problem 26)

$$y_p(t) = \frac{F_0}{m(\omega_0^2 - \omega^2)}\cos \omega t \tag{6.5.18}$$

so that the general solution to Equation (6.5.17) is

$$y(t) = A_0 \cos(\omega_0 t - \phi) + \frac{F_0}{m(\omega_0^2 - \omega^2)}\cos \omega t. \tag{6.5.19}$$

Comparing this with (6.5.10), we see that the resulting motion consists of a superposition of two simple harmonic oscillation modes. One of these modes has the circular frequency, ω_0, of the system, whereas the other has the frequency of the driving force. Consequently, the motion is oscillatory and bounded for all time, but in general it is not periodic. Indeed, it can be shown that the motion is periodic whenever the ratio ω/ω_0 is a rational number—say,

$$\frac{\omega}{\omega_0} = \frac{p}{q}, \tag{6.5.20}$$

where p and q are positive integers (see Problem 27). In such a case, the fundamental period of the motion is

$$T = \frac{2\pi q}{\omega_0} = \frac{2\pi p}{\omega}$$

where p and q are the smallest integers satisfying Equation (6.5.20). A typical (nonperiodic) motion of the form (6.5.19) is sketched in Figure 6.5.7.

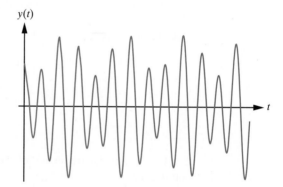

Figure 6.5.7: Forced harmonic oscillation.

An interesting occurrence arises when the driving frequency ω is close (but not equal to) the natural frequency of the system. To investigate this situation, we first impose the initial conditions $y(0) = 0$ and $dy/dt(0) = 0$ on the general solution (6.5.19). These conditions imply that A_0 and ϕ must satisfy

$$A_0 \cos \phi + \frac{F_0}{m(\omega_0^2 - \omega^2)} = 0, \qquad \omega_0 A_0 \sin \phi = 0.$$

Hence

$$A_0 = -\frac{F_0}{m(\omega_0^2 - \omega^2)}, \qquad \phi = 0.$$

Substituting these values into the general solution (6.5.19) gives

$$y(t) = \frac{F_0}{m(\omega_0^2 - \omega^2)} (\cos \omega t - \cos \omega_0 t).$$

We next use the trigonometric identity $2 \sin A \sin B = \cos(A - B) - \cos(A + B)$ with $A = (\omega_0 - \omega)t/2$ and $B = (\omega_0 + \omega)t/2$ to obtain

$$y(t) = \frac{2F_0}{m(\omega_0^2 - \omega^2)} \sin\left[\left(\frac{\omega_0 - \omega}{2}\right)t\right] \sin\left[\left(\frac{\omega_0 + \omega}{2}\right)t\right].$$

If ω and ω_0 are nearly equal, then $\sin[(\omega_0 - \omega)t/2)]$ is slowly varying compared to $\sin[(\omega_0 + \omega)t/2)]$. Thus, $y(t)$ behaves like a rapidly oscillating SHM mode whose amplitude is slowly varying in time. (See Figure 6.5.8.) One of the simplest occurrences of this phenomenon is when two tuning forks whose frequencies are nearly equal are struck simultaneously.

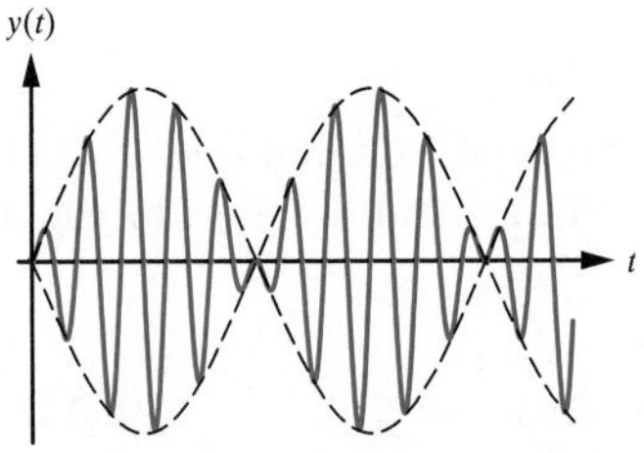

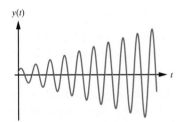

$y(t)$

Figure 6.5.8: When $\omega_0 \approx \omega$, the resulting motion can be interpreted as being simple harmonic with a slowly varying amplitude.

Case 1b: $\omega = \omega_0$ **(Resonance).**   When the frequency of the driving term coincides with the frequency of the system, we must solve

$$\frac{d^2 y}{dt^2} + \omega_0^2 y = \frac{F_0}{m} \cos \omega_0 t. \tag{6.5.21}$$

The complementary function can be written as

$$y_c(t) = A_0 \cos(\omega_0 t - \phi)$$

and an appropriate trial solution is

$$y_p(t) = t(A \cos \omega_0 t + B \sin \omega_0 t).$$

A straightforward application of the method of undetermined coefficients yields the particular solution (see Problem 26)

$$y_p(t) = \frac{F_0}{2m\omega_0} t \sin \omega_0 t \tag{6.5.22}$$

so that the general solution to Equation (6.5.21) is

$$y(t) = A_0 \cos(\omega_0 t - \phi) + \frac{F_0}{2m\omega_0} t \sin \omega_0 t.$$

We see that the motion is oscillatory, but we also see that the amplitude increases without bound as $t \to \infty$. This phenomenon, which occurs when the driving and natural frequencies coincide, is called **resonance**. Its physical consequences cannot be overemphasized. For example, the occurrence of resonance in the present situation would eventually lead to the spring's elastic limit being exceeded, and hence, the system would be destroyed. This situation is depicted in Figure 6.5.9.

$y(t)$

Figure 6.5.9: Resonance: The amplitude of the oscillation increases without bound as $t \to \infty$.

Case 2: Damping.   We now consider the general damped equation

$$\frac{d^2 y}{dt^2} + \frac{c}{m} \frac{dy}{dt} + \frac{k}{m} y = \frac{F_0}{m} \cos \omega t \tag{6.5.23}$$

where $c \neq 0$. An appropriate trial solution for this equation is

$$y_p(t) = A \cos \omega t + B \sin \omega t.$$

A fairly lengthy, but straightforward, computation yields the particular solution (see Problem 29)

$$y_p(t) = \frac{F_0}{(k - m\omega^2)^2 + c^2\omega^2} \left[(k - m\omega^2) \cos \omega t + c\omega \sin \omega t \right], \tag{6.5.24}$$

which can be written in the form

$$y_p(t) = \frac{F_0}{H} \cos(\omega t - \eta), \tag{6.5.25}$$

where

$$\cos \eta = \frac{m(\omega_0^2 - \omega^2)}{H}, \qquad \sin \eta = \frac{c\omega}{H}, \qquad H = \sqrt{m^2(\omega_0^2 - \omega^2)^2 + c^2\omega^2}$$

and

$$\omega_0 = \sqrt{k/m}.$$

Consider first the case of underdamping. Using the homogeneous solution from (6.5.14) and the foregoing particular solution, it follows that the general solution to Equation (6.5.23) in this case is

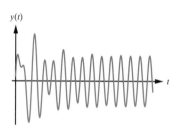

$$y(t) = A_0 e^{-ct/(2m)} \cos(\mu t - \phi) + \frac{F_0}{H} \cos(\omega t - \eta). \qquad (6.5.26)$$

For large t, we see that y_p is dominant. For this reason we refer to the complementary function as the **transient** part of the solution and call y_p the **steady-state** solution. We recognize Equation (6.5.26) as consisting of a superposition of two harmonic oscillations, one damped and the other undamped. The motion is eventually simple harmonic with a frequency coinciding with that of the driving term.

Figure 6.5.10: An example of forced motion with damping.

The cases for critical damping and overdamping are similar, since in both cases the complementary function (transient part of the solution) dies out exponentially and the steady-state solution (6.5.25) dominates. A typical motion of a forced mechanical system with damping is shown in Figure 6.5.10.

Exercises for 6.5

Key Terms

Spring–mass system, Static equilibrium, Spring force, Damping force, Damping constant, External driving force, Free oscillations, Simple harmonic motion, Amplitude, Circular frequency, Phase, Period, Underdamping, Quasi period, Critical damping, Overdamping, Forced oscillations, Resonance.

Skills

- Understand the statement and mathematical formulation of the problem of determining the motion of a spring in a mechanical system.

- Be familiar with the different forces that are present in the spring–mass system described in this section.

- Be able to write down and solve the initial-value problem governing the motion of the spring–mass system in the case when no external forces are present.

- Be able to determine the simple harmonic motion associated with the free oscillations of a system with no damping, including its amplitude, frequency, phase, and period.

- Be able to determine whether a damped spring–mass system with no external forces is underdamped, critically damped, or overdamped, and be able to determine the motion of the spring–mass system in each case.

- Be able to determine the motion of a spring–mass system subject to external forces, whether or not there is damping.

- Be able to determine the steady-state and transient solutions for the motion of a spring–mass system subject to a periodic external force and damping.

True-False Review

For Questions 1–9, decide if the given statement is **true** or **false**, and give a brief justification for your answer. If true, you can quote a relevant definition or theorem from the text. If false, provide an example, illustration, or brief explanation of why the statement is false.

1. The spring force on a mass acts in a direction opposite to the displacement of the mass from equilibrium.

2. The circular frequency of a spring–mass system is the spring constant k divided by the mass m.

3. For simple harmonic motion, the product of the frequency of oscillation and the period of oscillation is 1.

4. An underdamped spring–mass system tends to rest as $t \to \infty$.

5. Underdamped, critically damped, and overdamped spring–mass systems can all exhibit periodic motion.

6. Resonance occurs when the circular frequency of a spring–mass system agrees with the frequency of the driving force.

7. The air resistance experienced by a spring–mass system is an example of a damping force.

8. The larger the mass, the shorter the period of a spring–mass system that is undergoing simple harmonic motion.

9. In order for the amplitude of a spring–mass system to increase without bound, an external driving force must be present.

Problems

For Problems 1–2, consider the spring–mass system whose motion is governed by the given initial-value problem. Determine the circular frequency of the system and the amplitude, phase, and period of the motion.

1. $\dfrac{d^2y}{dt^2} + 4y = 0, \quad y(0) = 2, \quad \dfrac{dy}{dt}(0) = 4.$

2. $\dfrac{d^2y}{dt^2} + \omega_0^2 y = 0, \quad y(0) = y_0, \quad \dfrac{dy}{dt}(0) = v_0,$

   where ω_0, y_0, v_0 are constants.

3. A force of 3 N stretches a spring by 1 m.

   (a) Find the spring constant k.

   (b) A mass of 4 kg is attached to the spring. At $t = 0$, the mass is pulled down a distance 1 m from equilibrium and released with a downward velocity of 0.5 m/s. Assuming that damping is negligible, determine an expression for the position of the mass at time t. Find the circular frequency of the system and the amplitude, phase, and period of the motion.

For Problems 4–10, determine the motion of the spring–mass system governed by the given initial-value problem. In each case, state whether the motion is underdamped, critically damped, or overdamped, and make a sketch depicting the motion.

4. $\dfrac{d^2y}{dt^2} + 2\dfrac{dy}{dt} + 5y = 0, \quad y(0) = 1, \quad \dfrac{dy}{dt}(0) = 3.$

5. $\dfrac{d^2y}{dt^2} + 3\dfrac{dy}{dt} + 2y = 0, \quad y(0) = 1, \quad \dfrac{dy}{dt}(0) = 0.$

6. $4\dfrac{d^2y}{dt^2} + 12\dfrac{dy}{dt} + 5y = 0, \quad y(0) = 1, \quad \dfrac{dy}{dt}(0) = -3.$

7. $\dfrac{d^2y}{dt^2} + 2\dfrac{dy}{dt} + y = 0, \quad y(0) = -1, \quad \dfrac{dy}{dt}(0) = 2.$

8. $4\dfrac{d^2y}{dt^2} + 4\dfrac{dy}{dt} + y = 0, \quad y(0) = 4, \quad \dfrac{dy}{dt}(0) = -1.$

9. $\dfrac{d^2y}{dt^2} + 4\dfrac{dy}{dt} + 7y = 0, \quad y(0) = 2, \quad \dfrac{dy}{dt}(0) = 6.$

10. $\dfrac{d^2y}{dt^2} + 5\dfrac{dy}{dt} + 6y = 0, \quad y(0) = -1, \quad \dfrac{dy}{dt}(0) = 4.$

11. In the previous problem, find the time at which the mass passes through the equilibrium position, and determine the maximum positive displacement of the mass from equilibrium. Make a sketch depicting the motion.

12. Consider the spring–mass system whose motion is governed by the differential equation

$$\dfrac{d^2y}{dt^2} + 2\alpha\dfrac{dy}{dt} + y = 0.$$

Determine all values of the (positive) constant α for which the system is (a) underdamped, (b) critically damped, and (c) overdamped. In the case of overdamping, solve the system fully. If the initial velocity of the mass is zero, determine if the mass passes through equilibrium.

13. Consider the spring–mass system whose motion is governed by the initial-value problem

$$\dfrac{d^2y}{dt^2} + 3\dfrac{dy}{dt} + 2y = 0, \quad y(0) = 1, \quad \dfrac{dy}{dt}(0) = -3.$$

   (a) Determine the position of the mass at time t.

   (b) Determine the time when the mass passes through the equilibrium position.

   (c) Make a sketch depicting the general motion of the system.

14. Consider the general solution for an *underdamped* spring–mass system.

   (a) Show that the time between successive maxima (or minima) of $y(t)$ is

   $$T = \frac{2\pi}{\mu} = \frac{4\pi m}{\sqrt{4km - c^2}}.$$

   (b) Show that if $\dfrac{c^2}{4km} << 1$, then

   $$T \approx 2\pi \sqrt{\frac{m}{k}}.$$

   Is this result reasonable? Explain.

15. Show that the general solution for the motion of a *critically damped* spring–mass system, with initial displacement y_0 and initial velocity v_0, can be written in the form

   $$y(t) = e^{-ct/(2m)}\left[y_0 + t\left(v_0 + \frac{c}{2m}y_0\right)\right]$$

   and that the system can pass through the equilibrium position at most once.

16. A cylinder of side L meters lies one-quarter submerged and upright in a certain fluid. At $t = 0$, the cylinder is pushed down a distance of $L/2$ meters and released from rest. Show that the resulting motion is simple harmonic, and determine the circular frequency and period of the motion.[1]

A simple pendulum consists of a mass, m kilograms, attached to the end of a light rod of length L meters, whose other end is fixed. (See Figure 6.5.11.)

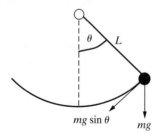

Figure 6.5.11: The simple pendulum

If we let θ radians denote the angle the rod is displaced from the vertical at time t, then the component of the velocity in the direction of motion is $v = L \cdot d\theta/dt$,

so that the component of the acceleration in the direction of motion is $L \cdot d^2\theta/dt^2$. Further, the tangential component of the force is $F_T = -mg \sin\theta$, so that, from Newton's second law, the equation of motion of the pendulum is

$$mL\frac{d^2\theta}{dt^2} = -mg\sin\theta.$$

That is,

$$\frac{d^2\theta}{dt^2} + \frac{g}{L}\sin\theta = 0. \qquad (6.5.27)$$

This is a nonlinear differential equation. However, if we recall the Maclaurin expansion for $\sin\theta$, namely,

$$\sin\theta = \theta - \frac{1}{3!}\theta^3 + \frac{1}{5!}\theta^5 - \cdots,$$

it follows that for small oscillations we can approximate $\sin\theta$ by θ. Then Equation (6.5.27) can be replaced to reasonable accuracy by the simple *linear* differential equation

$$\frac{d^2\theta}{dt^2} + \frac{g}{L}\theta = 0. \qquad (6.5.28)$$

Problems 17–20 deal with the simple pendulum whose motion is described by Equation (6.5.28).

17. A pendulum of length 0.5 m is displaced an angle 0.1 rad from the equilibrium position and released from rest. Determine the resulting motion.

18. A pendulum of length L meters is displaced an angle α radians from the vertical and released with an angular velocity of β radians/second. Determine the amplitude, phase, and period of the resulting motion.

19. Show that the period of the simple pendulum is $T = 2\pi\sqrt{L/g}$. Determine the length of a pendulum that takes one second to swing from its extreme position on the right to its extreme position on the left. Let $g = 9.8$ m/s^2.

20. A clock has a pendulum of length 90 centimeters. If the clock ticks each time the pendulum swings from its extreme position on the right to its extreme position on the left, determine the number of times the clock ticks in one minute. Let $g = 9.8$ m/s^2.

---

[1]According to Archimedes' principle, when an object is partially or wholly immersed in a fluid, it experiences an upward force equal to the weight of fluid displaced.

21. An object of mass m is attached to the midpoint of a light elastic string of natural length $6a$. When the ends of the string are fixed at the same level a distance $6a$ apart and the mass is allowed to hang in equilibrium, the length of the stretched string is $10a$. (See Figure 6.5.12.) The mass is pulled down a small vertical distance from equilibrium and released.

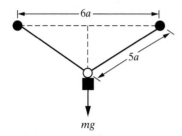

Figure 6.5.12: The static equilibrium position.

Show that, *for small oscillations*, the period of the resulting motion is

$$T = \frac{20\pi}{7} \cdot \sqrt{\frac{a}{g}}.$$

22. Repeat the previous problem if the string has natural length $2L_0$ and in equilibrium the stretched string has length $2L$.

23. Consider the damped spring–mass system whose motion is governed by

$$\frac{d^2y}{dt^2} + 2\frac{dy}{dt} + 5y = 17\sin 2t, \quad y(0) = -2, \quad \frac{dy}{dt}(0) = 0.$$

(a) Determine whether the motion is underdamped, overdamped, or critically damped.

(b) Find the solution to the given initial-value problem and identify the steady-state and transient parts.

24. Consider the spring–mass system whose motion is governed by

$$\frac{d^2y}{dt^2} + \omega_0^2 y = F_0 \sin \omega t, \quad y(0) = 0, \quad \frac{dy}{dt}(0) = 0.$$

Determine the solution if the system is resonating.

25. Consider the spring–mass system whose motion is governed by

$$\frac{d^2y}{dt^2} + 3\frac{dy}{dt} + 2y = 10\sin t.$$

Determine the steady-state solution, y_p, and express your answer in the form

$$y_p(t) = A_0 \sin(t - \phi),$$

for appropriate constants A_0 and ϕ.

26. Consider the forced undamped spring–mass system whose motion is governed by

$$\frac{d^2y}{dt^2} + \omega_0^2 y = \frac{F_0}{m}\cos \omega t.$$

Derive the particular solutions given in Equations (6.5.18) and (6.5.22). (You will need to consider $\omega \neq \omega_0$ and $\omega = \omega_0$ separately.)

27. The general solution to the forced undamped (nonresonating) spring–mass system is

$$y(t) = A_0 \cos(\omega_0 t - \phi) + \frac{F_0 \cos \omega t}{m(\omega_0^2 - \omega^2)}.$$

If $\omega/\omega_0 = p/q$, where p and q are integers, show that the motion is periodic with period $T = 2\pi q/\omega_0$.

28. Determine the period of the motion for the spring–mass system governed by the differential equation

$$\frac{d^2y}{dt^2} + \frac{9}{16}y = 55\cos 2t.$$

29. Consider the damped forced motion described by

$$\frac{d^2y}{dt^2} + \frac{c}{m}\frac{dy}{dt} + \frac{k}{m}y = \frac{F_0}{m}\cos \omega t.$$

Derive the steady-state solution (6.5.24) given in the text.

30. Consider the damped forced motion described by

$$\frac{d^2y}{dt^2} + \frac{c}{m}\frac{dy}{dt} + \frac{k}{m}y = \frac{F_0}{m}\cos \omega t.$$

We have shown that the steady-state solution can be written in the form

$$y_p(t) = \frac{F_0}{H}\cos(\omega t - \nu),$$

where

$$\cos \nu = \frac{m(\omega_0^2 - \omega^2)}{H},$$

$$\sin \nu = \frac{c\omega}{H},$$

$$\omega_0 = \sqrt{\frac{k}{m}},$$

$$H = \sqrt{m^2(\omega_0^2 - \omega^2)^2 + c^2\omega^2}.$$

Assuming that $c^2/(2m^2\omega_0^2) < 1$, show that the amplitude of the steady-state solution is a maximum when

$$\omega = \sqrt{\omega_0^2 - \frac{c^2}{2m^2}}.$$

[**Hint:** The maximum occurs at the value of ω that makes H a minimum. Assume that H is a function of ω, and determine the value of ω that minimizes H.]

31. Consider the damped spring–mass system with $m = 1$, $k = 5$, $c = 2$, and $F(t) = 8\cos\omega t$.

   (a) Determine the transient part of the solution and the steady-state solution.

   (b) Determine the value of ω that maximizes the amplitude of the steady-state solution and express the corresponding solution in the form

   $$y_p(t) = A_0 \cos(\omega t - v),$$

   for appropriate constants A_0, ω, and v.

32. Consider the spring–mass system whose motion is governed by the differential equation

$$\frac{d^2y}{dt^2} + 2\frac{dy}{dt} + 5y = 4e^{-t}\cos 2t.$$

   (a) Describe the variation with time of the applied external force.

   (b) Determine the motion of the mass. What happens as $t \to \infty$?

33. Consider the spring–mass system whose motion is governed by the differential equation

$$\frac{d^2y}{dt^2} + 16y = 130e^{-t}\cos t.$$

Determine the resulting motion, and identify any transient and steady-state parts of your solution.

6.6   RLC Circuits

In Section 1.7 we used Kirchoff's second law to derive the differential equation

$$\frac{di}{dt} + \frac{R}{L}i + \frac{1}{LC}q = \frac{1}{L}E(t), \tag{6.6.1}$$

which governs the behavior of the RLC circuit shown in Figure 6.6.1. Here, q is the charge on the capacitor at time t, the constants R, L, and C are the resistance, inductance, and capacitance of the circuit elements, respectively, and $E(t)$ denotes the driving electromotive force (EMF). The current in the circuit is related to the charge on the capacitor via

$$i(t) = \frac{dq}{dt}.$$

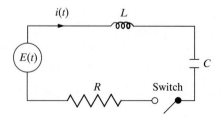

Figure 6.6.1: An RLC circuit.

Substituting this expression for $i(t)$ into Equation (6.6.1) yields the second-order constant-coefficient differential equation

$$\frac{d^2q}{dt^2} + \frac{R}{L}\frac{dq}{dt} + \frac{1}{LC}q = \frac{1}{L}E(t). \tag{6.6.2}$$

A comparison of Equation (6.6.2) with the basic differential equation governing the motion of a spring–mass system, namely

$$\frac{d^2y}{dt^2} + \frac{c}{m}\frac{dy}{dt} + \frac{k}{m}y = \frac{1}{m}F(t),$$

reveals that, although physically the two problems are distinct, from a purely mathematical standpoint they are identical. The correspondence between the variables and parameters in an RLC circuit and a spring–mass system is given in Table 6.6.1. It follows that the results derived in the previous section for a spring–mass system can be translated into corresponding results for RLC circuits. Rather than repeating these results, we will make some general observations and then consider one illustrative example. The full investigation of the behavior of an RLC circuit is left for the exercises.

RLC Circuit	Spring–Mass System
$q(t)$	$y(t)$
L	m
R	c
$1/C$	k
$E(t)$	$F(t)$

Table 6.6.1: Comparison of an RLC circuit and a spring–mass system.

Consider first the homogeneous differential equation

$$\frac{d^2q}{dt^2} + \frac{R}{L}\frac{dq}{dt} + \frac{1}{LC}q = 0. \tag{6.6.3}$$

This has auxiliary equation

$$r^2 + \frac{R}{L}r + \frac{1}{L}C = 0$$

with roots

$$r = \frac{-R \pm \sqrt{R^2 - 4L/C}}{2L}.$$

Three familiar cases arise. The circuit is said to be

1. *underdamped* if $R^2 < 4L/C$.
2. *critically damped* if $R^2 = 4L/C$.
3. *overdamped* if $R^2 > 4L/C$.

The corresponding solutions to Equation (6.6.3) are

$$\begin{aligned}
q(t) &= e^{-Rt/(2L)}(c_1 \cos \mu t + c_2 \sin \mu t), & \mu &= \frac{\sqrt{4L/C - R^2}}{2L}, \\
q(t) &= e^{-Rt/(2L)}(c_1 + c_2 t), & & \\
q(t) &= e^{-Rt/(2L)}(c_1 e^{\mu t} + c_2 e^{-\mu t}), & \mu &= \frac{\sqrt{R^2 - 4L/C}}{2L}.
\end{aligned} \tag{6.6.4}$$

In all cases of physical relevance, $R/L > 0$, so that

$$\lim_{t \to \infty} q(t) = 0.$$

Equivalently, we can state that the complementary function $q_c(t)$ for Equation (6.6.2) satisfies

$$\lim_{t \to \infty} q_c(t) = 0.$$

We refer to q_c as the **transient** part of the solution to Equation (6.6.2), since it decays exponentially with time. As a specific example, we consider the case of a periodic driving EMF in an underdamped circuit.

Example 6.6.1   Determine the current in the RLC circuit

$$\frac{d^2q}{dt^2} + \frac{R}{L}\frac{dq}{dt} + \frac{1}{LC}q = \frac{E_0}{L}\cos \omega t, \tag{6.6.5}$$

where E_0 and ω are positive constants and $R^2 < 4L/C$.

Solution: The complementary function given in Equation (6.6.4) can be written in phase-amplitude form as

$$q_c(t) = A_0 e^{-Rt/(2L)} \cos(\mu t - \phi),$$

where A_0 and ϕ are defined in the usual manner. A particular solution to Equation (6.6.5) can be obtained by using Table 6.6.1 to make the appropriate replacements in the solution (6.5.18) for the corresponding spring–mass system. The result is

$$q_p(t) = \frac{E_0}{H} \cos(\omega t - \eta),$$

where

$$H = \sqrt{L^2(\omega_0^2 - \omega^2)^2 + R^2 w^2}$$

and

$$\cos \eta = \frac{L(\omega_0^2 - \omega^2)}{H}, \quad \sin \eta = \frac{R\omega}{H}, \quad \omega_0 = \frac{1}{\sqrt{LC}}.$$

Consequently, the charge on the capacitor at time t is

$$q(t) = A_0 e^{-Rt/(2L)} \cos(\mu t - \phi) + \frac{E_0}{H} \cos(\omega t - \eta),$$

and the corresponding current in the circuit can be determined from

$$i(t) = \frac{dq}{dt}.$$

Rather than compute this derivative, we consider the behavior of $q(t)$ and the corresponding behavior of the current as $t \to \infty$. Since q_c tends to zero as $t \to \infty$, for large t the particular solution q_p will be the dominant part of $q(t)$. For this reason, we refer to

q_p as the **steady-state solution**. The corresponding **steady-state current** in the circuit, denoted i_S, is given by

$$i_S(t) = \frac{dq_p}{dt} = -\frac{\omega E_0}{H} \sin(\omega t - \eta).$$

We see that this is periodic and that the frequency of the oscillation coincides with the frequency of the driving EMF. The amplitude of the oscillation is

$$A = \frac{\omega E_0}{H}.$$

That is, upon substituting for H,

$$A = \frac{\omega E_0}{\sqrt{L^2(\omega_0^2 - \omega^2)^2 + R^2 w^2}}. \tag{6.6.6}$$

It is often required to determine the value of ω that maximizes this amplitude. In order to do so, we rewrite (6.6.6) in the equivalent form

$$A = \frac{E_0}{\sqrt{\omega^{-2} L^2(\omega_0^2 - \omega^2)^2 + R^2}}.$$

This will be a maximum when the term in parentheses vanishes, which occurs when

$$\omega^2 = \omega_0^2.$$

Substituting for $\omega_0^2 = 1/(LC)$, it follows that the amplitude of the steady-state current will be a maximum when $\omega = \omega_{max}$, where

$$\omega_{max} = \frac{1}{\sqrt{LC}}.$$

The corresponding value of A is

$$A_{max} = \frac{E_0}{R}.$$

The behavior of A as a function of ω for typical values of E_0, R, L, and C is shown in Figure 6.6.2. □

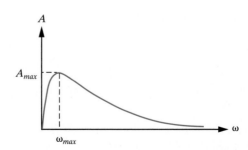

Figure 6.6.2: The behavior of the amplitude of the steady-state current as a function of the driving frequency.

Exercises for 6.6

Key Terms

RLC circuit: Underdamped, Critically damped, Overdamped, Transient solution, Steady-state solution.

Skills

- Be able to solve the differential equation arising from Kirchoff's second law in order to determine the charge on a capacitor or the current in an RLC circuit.

- Be able to determine the transient and steady-state parts of the solution $q(t)$ of Equation (6.6.2) and the solution $i(t)$ of Equation (6.6.1).

True-False Review

For Questions 1–6, decide if the given statement is **true** or **false**, and give a brief justification for your answer. If true, you can quote a relevant definition or theorem from the text. If false, provide an example, illustration, or brief explanation of why the statement is false.

1. If there is no driving electromotive force in an RLC circuit, then the charge on the capacitor and the current in the circuit tend to zero as $t \to \infty$.

2. If $R = 4\,\Omega$, $L = 4\,H$, and $C = \frac{1}{17}\,F$, then the RLC is underdamped.

3. An external driving force $E(t) = E_0 \cos \omega t$ produces a steady-state current of maximum amplitude when $\omega = 1/\sqrt{LC}$.

4. The amplitude of the steady-state current in an RLC circuit is proportional to the amplitude E_0 of the external driving force.

5. If the resistance R in an RLC circuit is doubled, then the current $i(t)$ in the circuit is reduced by one-half.

6. The charge on a capacitor in an RLC circuit with no driving force varies periodically with time.

Problems

1. Determine the steady-state current in the RLC circuit that has $R = \frac{3}{2}\,\Omega$, $L = \frac{1}{2}\,H$, $C = \frac{2}{3}\,F$, and $E(t) = 13\cos 3t$ V.

2. Determine the charge on the capacitor at time t in an RLC circuit that has $R = 4\,\Omega$, $L = 4\,H$, $C = \frac{1}{17}\,F$, and $E = E_0$ V, where E_0 is constant. What happens to the charge on the capacitor as $t \to +\infty$? Describe the behavior of the current in the circuit.

3. Consider the RLC circuit with $E(t) = E_0 \cos \omega t$ V, where E_0 and ω are constants. If there is no resistor in the circuit, show that the charge on the capacitor satisfies

$$\lim_{t \to \infty} q(t) = +\infty$$

if and only if $\omega = 1/\sqrt{LC}$. What happens to the current in the circuit as $t \to +\infty$?

4. Consider the RLC circuit with $R = 16\,\Omega$, $L = 8\,H$, $C = \frac{1}{40}\,F$, and $E(t) = 17\cos 2t$ V. Determine the current in the circuit for $t > 0$, given that at $t = 0$ the capacitor is uncharged and there is no current flowing.

5. Consider the RLC circuit with $R = 3\,\Omega$, $L = \frac{1}{2}\,H$, $C = \frac{1}{5}\,F$, and $E(t) = 2\cos \omega t$ V. Determine the current in the circuit at time t, and find the value of ω that maximizes the amplitude of the steady-state current.

6. Show that the differential equation governing the behavior of an RLC circuit can be written directly in terms of the current $i(t)$ as

$$\frac{d^2 i}{dt^2} + \frac{R}{L}\frac{di}{dt} + \frac{1}{LC}i = \frac{1}{L}\frac{dE}{dt}.$$

7. Determine the current in the general RLC circuit with $R^2 < 4L/C$, if $E(t) = E_0 e^{-at}$, where E_0, a are constants.

8. Consider the RLC circuit with $R = 2\,\Omega$, $L = \frac{1}{2}\,H$, $C = \frac{2}{5}\,F$. Initially the capacitor is uncharged, and no current is flowing in the circuit. Determine the current for $t > 0$, if the applied EMF is (see Figure 6.6.3)

$$E(t) = \begin{cases} 50t, & 0 \le t < \pi, \\ 50\pi, & t \ge \pi. \end{cases}$$

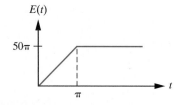

Figure 6.6.3: EMF for Problem 8.

6.7   The Variation-of-Parameters Method

The method of undetermined coefficients has two severe limitations. First, it is applicable only to differential equations with constant-coefficients, and second, it can be applied only to differential equations whose nonhomogeneous terms are of the form described in Section 6.3. For example, we could not use the method of undetermined coefficients to find a particular solution to the differential equation

$$y'' + 4y' - 6y = x^2 \ln x.$$

In this section we introduce a very powerful technique, the variation-of-parameters method, for obtaining particular solutions to n^{th}-order linear nonhomogeneous differential equations, assuming that we know the general solution to the associated homogeneous equation. Unlike the method of undetermined coefficients, the variation-of-parameters method is not restricted to differential equations with constant coefficients, and, at least in theory, the actual form of the nonhomogeneous term is immaterial. We will begin by considering the *second-order* case, since the generalization to *n*th-order will then be fairly straightforward.

Consider the second-order linear nonhomogeneous differential equation

$$y'' + a_1 y' + a_2 y = F, \tag{6.7.1}$$

where we assume that a_1, a_2, and F are continuous on an interval I. Suppose that $y = y_1(x)$ and $y = y_2(x)$ are two linearly independent solutions to the associated homogeneous equation

$$y'' + a_1 y' + a_2 y = 0 \tag{6.7.2}$$

on I, so that the general solution to Equation (6.7.2) on I is

$$y_c(x) = c_1 y_1(x) + c_2 y_2(x).$$

The variation-of-parameters method consists of replacing the constants c_1 and c_2 by functions $u_1(x)$ and $u_2(x)$ (that is, we allow the parameters c_1 and c_2 to vary), determined in such a way that the resulting function

$$y_p(x) = u_1(x)y_1(x) + u_2(x)y_2(x) \tag{6.7.3}$$

is a particular solution to Equation (6.7.1).

Differentiating Equation (6.7.3) with respect to x yields

$$y_p' = u_1' y_1 + u_1 y_1' + u_2' y_2 + u_2 y_2'.$$

It is tempting to differentiate this expression once more and then substitute into Equation (6.7.1) to determine u_1 and u_2. However, if we did this, the resulting expression for y_p'' would involve second derivatives of u_1 and u_2, hence we would have complicated our problem. Since y_p contains two unknown functions, whereas Equation (6.7.1) gives only one condition for determining them, we have the freedom to impose a further constraint on u_1 and u_2. In order to eliminate second derivatives of u_1 and u_2 arising in y_p'', we try for solutions of the form (6.7.3) satisfying the constraint

$$u_1' y_1 + u_2' y_2 = 0. \tag{6.7.4}$$

The expression for y_p' then reduces to

$$y_p' = u_1 y_1' + u_2 y_2',$$

so that

$$y_p'' = u_1'y_1' + u_1y_1'' + u_2'y_2' + u_2y_2''.$$

Substituting into Equation (6.7.1) and collecting terms yields

$$u_1(y_1'' + a_1y_1' + a_2y_1) + u_2(y_2'' + a_1y_2' + a_2y_2) + (u_1'y_1' + u_2'y_2') = F(x).$$

The terms multiplying u_1 and u_2 vanish, since y_1 and y_2 each solve $y'' + a_1y' + a_2y = 0$. We therefore require that

$$u_1'y_1' + u_2'y_2' = F. \tag{6.7.5}$$

Consequently, $y_p(x) = u_1(x)y_1(x) + u_2(x)y_2(x)$ is a solution to Equation (6.7.1), provided that u_1 and u_2 satisfy Equations (6.7.4) and (6.7.5). That is,

$$y_1u_1' + y_2u_2' = 0 \quad \text{and} \quad y_1'u_1' + y_2'u_2' = F. \tag{6.7.6}$$

This is a linear system of equations for the unknowns u_1' and u_2'. The matrix of coefficients of this system has determinant

$$\begin{vmatrix} y_1 & y_2 \\ y_1' & y_2' \end{vmatrix},$$

which is the Wronskian, $W[y_1, y_2](x)$, of y_1 and y_2. Since y_1 and y_2 are linearly independent on I, $W[y_1, y_2](x)$ is *nonzero* on I and hence the system (6.7.6) has a unique solution for u_1' and u_2'. Indeed, applying Cramer's rule to (6.7.6) yields

$$u_1'(x) = -\frac{y_2(x)F(x)}{W[y_1, y_2](x)}, \qquad u_2'(x) = \frac{y_1(x)F(x)}{W[y_1, y_2](x)},$$

which can be integrated directly to obtain

$$u_1(x) = -\int_{x_0}^{x} \frac{y_2(t)F(t)}{W[y_1, y_2](t)}\, dt, \qquad u_2(x) = \int_{x_0}^{x} \frac{y_1(t)F(t)}{W[y_1, y_2](t)}\, dt, \tag{6.7.7}$$

where $x_0 \in I$. We have therefore established the next theorem.

Theorem 6.7.1   **(Variation-of-Parameters Method)**

Consider

$$y'' + a_1y' + a_2y = F, \tag{6.7.8}$$

where a_1, a_2, and F are assumed to be (at least) continuous on the interval I. Let y_1 and y_2 be linearly independent solutions to the associated homogeneous equation

$$y'' + a_1y' + a_2y = 0$$

on I. Then a particular solution to Equation (6.7.8) is

$$y_p = u_1y_1 + u_2y_2$$

where u_1 and u_2 satisfy

$$y_1u_1' + y_2u_2' = 0 \quad \text{and} \quad y_1'u_1' + y_2'u_2' = F.$$

Example 6.7.2 Solve $y'' + y = \sec x$.

Solution: Two linearly independent solutions to the associated homogeneous equation are $y_1(x) = \cos x$ and $y_2(x) = \sin x$. Thus, a particular solution to the given differential equation is

$$y_p(x) = u_1 y_1 + u_2 y_2 = u_1 \cos x + u_2 \sin x, \tag{6.7.9}$$

where u_1 and u_2 satisfy

$$\cos x \, u_1' + \sin x \, u_2' = 0,$$
$$-\sin x \, u_1' + \cos x \, u_2' = \sec x.$$

Applying Cramer's rule, the solution to this system is

$$u_1' = \frac{\begin{vmatrix} 0 & \sin x \\ \sec x & \cos x \end{vmatrix}}{\begin{vmatrix} \cos x & \sin x \\ -\sin x & \cos x \end{vmatrix}} = -\sin x \sec x, \qquad u_2' = \frac{\begin{vmatrix} \cos x & 0 \\ -\sin x & \sec x \end{vmatrix}}{\begin{vmatrix} \cos x & \sin x \\ -\sin x & \cos x \end{vmatrix}} = \cos x \sec x = 1.$$

Consequently,

$$u_1(x) = -\int \sin x \sec x \, dx = -\int \frac{\sin x}{\cos x} \, dx = \ln|\cos x|$$

and

$$u_2(x) = \int 1 \cdot dx = x,$$

where we have set the integration constants to zero, since we require only one particular solution. Substitution into Equation (6.7.9) yields

$$y_p(x) = \cos x \cdot \ln|\cos x| + x \sin x$$

so that the general solution to the given differential equation is

$$y(x) = c_1 \cos x + c_2 \sin x + \cos x \cdot \ln|\cos x| + x \sin x. \qquad \square$$

Example 6.7.3 Solve $y'' + 4y' + 4y = e^{-2x} \ln x$, $x > 0$.

Solution: In this case, two linearly independent solutions to the associated homogeneous equation are $y_1(x) = e^{-2x}$ and $y_2(x) = xe^{-2x}$, hence we seek a particular solution to the given differential equation of the form

$$y_p(x) = u_1 e^{-2x} + u_2 x e^{-2x},$$

where u_1 and u_2 satisfy

$$e^{-2x} u_1' + xe^{-2x} u_2' = 0,$$
$$-2e^{-2x} u_1' + e^{-2x}(1 - 2x) u_2' = e^{-2x} \ln x.$$

The solution to this system is

$$u_1' = -x \ln x, \qquad u_2' = \ln x.$$

Integrating both of these expressions by parts (and setting the integration constants to zero), we obtain

$$u_1(x) = \tfrac{1}{4}x^2(1 - 2\ln x), \qquad u_2(x) = x(\ln x - 1).$$

Thus,

$$y_p(x) = \tfrac{1}{4}x^2 e^{-2x}(1 - 2\ln x) + x^2 e^{-2x}(\ln x - 1).$$

That is,

$$y_p(x) = \tfrac{1}{4}x^2 e^{-2x}(2\ln x - 3).$$

Consequently, the general solution to the given differential equation is

$$y(x) = e^{-2x}\left[c_1 + c_2 x + \tfrac{1}{4}x^2(2\ln x - 3)\right]. \qquad \square$$

Sometimes we can use a combination of the variation-of-parameters method and the method of undetermined coefficients to obtain a particular solution to a differential equation. We illustrate this with an example.

Example 6.7.4   Determine the general solution to the differential equation

$$y'' + 9y = 6\cot^2 3x + 5e^{2x}, \qquad 0 < x < \pi/6. \qquad (6.7.10)$$

Solution: The complementary function for the given differential equation is

$$y_c(x) = c_1 \cos 3x + c_2 \sin 3x.$$

Application of the variation-of-parameters technique directly to the differential equation (6.7.10) leads to some rather nasty integrals arising from the e^{2x} term in the right-hand side. However, if we determine a particular solution to each of the differential equations

$$y'' + 9y = 6\cot^2 3x \qquad (6.7.11)$$

and

$$y'' + 9y = 5e^{2x}, \qquad (6.7.12)$$

then Theorem 6.1.9 can be applied to conclude that the sum of these two solutions will itself be a particular solution to Equation (6.7.10). The key point is that Equation (6.7.12) can be solved easily using the method of undetermined coefficients, and therefore we have alleviated the problem of evaluating the integrals mentioned previously. Consider first Equation (6.7.11). According to the variation-of-parameters method, there is a particular solution to this differential equation of the form

$$y_{p_1}(x) = u_1 \cos 3x + u_2 \sin 3x,$$

where u_1 and u_2 satisfy

$$\cos 3x u_1' + \sin 3x u_2' = 0,$$
$$-\sin 3x u_1' + \cos 3x u_2' = 2\cot^2 3x.$$

Solving this system of equations yields

$$u_1' = -2\cot^2 3x \sin 3x, \qquad u_2' = 2\cot^2 3x \cos 3x.$$

Consequently, using the trigonometric identity $\cot^2 \theta = \csc^2 \theta - 1$ and integration formulas,

$$u_1 = -2 \int \cot^2 3x \sin 3x \, dx = -2 \int (\csc 3x - \sin 3x) \, dx$$

$$= -\tfrac{2}{3} [\ln \ (\csc 3x - \cot 3x) + \cos 3x]$$

and

$$u_2 = 2 \int \cot^2 3x \cos 3x \, dx = 2 \int \left(\frac{\cos 3x}{\sin^2 3x} - \cos 3x \right) dx$$

$$= -\tfrac{2}{3}(\csc 3x + \sin 3x).$$

Therefore,

$$y_{p_1}(x) = -\tfrac{2}{3} \cos 3x \ [\ln \ (\csc 3x - \cot 3x) + \cos 3x] - \tfrac{2}{3} \sin 3x (\csc 3x + \sin 3x),$$

which simplifies to

$$y_{p_1}(x) = -\tfrac{2}{3}[\cos 3x \ln \ (\csc 3x - \cot 3x) + 2].$$

Next consider Equation (6.7.12). An appropriate trial solution for this differential equation is

$$y_{p_2}(x) = A_0 e^{2x}$$

and substitution into Equation (6.7.12) yields $A_0 = \tfrac{5}{13}$. Consequently, a particular solution to Equation (6.7.12) is

$$y_{p_2}(x) = \tfrac{5}{13} e^{2x}.$$

It follows directly from Theorem 6.1.9 that a particular solution to Equation (6.7.10) is

$$y_p(x) = y_{p_1}(x) + y_{p_2}(x) = -\tfrac{2}{3} [\cos 3x \ln(\csc 3x - \cot 3x) + 2] + \tfrac{5}{13} e^{2x}.$$

The general solution to Equation (6.7.10) is therefore

$$y(x) = c_1 \cos 3x + c_2 \sin 3x - \tfrac{2}{3} [\cos 3x \ln(\csc 3x - \cot 3x) + 2] + \tfrac{5}{13} e^{2x}. \qquad \square$$

Green's Functions

According to Theorem 6.7.1, a particular solution to the differential equation

$$y'' + a_1 y' + a_2 y = F, \qquad x \in I, \tag{6.7.13}$$

where a_1, a_2, F are continuous on I, is

$$y_p(x) = u_1(x) y_1(x) + u_2(x) y_2(x). \tag{6.7.14}$$

Substituting the expressions for u_1 and u_2 obtained in (6.7.7) gives

$$y_p(x) = -y_1(x) \int_{x_0}^x \frac{y_2(t) F(t)}{W[y_1, y_2](t)} \, dt + y_2(x) \int_{x_0}^x \frac{y_1(t) F(t)}{W[y_1, y_2](t)} \, dt.$$

The two terms on the right-hand side of the preceding equation can be combined to obtain

$$y_p(x) = \int_{x_0}^x \left\{ \frac{y_1(t)y_2(x) - y_2(t)y_1(x)}{W[y_1, y_2](t)} \right\} F(t)\, dt,$$

which we write as

$$y_p(x) = \int_{x_0}^x K(x, t) F(t)\, dt, \qquad (6.7.15)$$

where

$$K(x, t) = \frac{y_1(t)y_2(x) - y_2(t)y_1(x)}{W[y_1, y_2](t)}. \qquad (6.7.16)$$

The function $K(x, t)$ is called a **Green's function** for the problem. We see that it depends only on the solutions to the associated homogeneous problem and not on the nonhomogeneous term $F(x)$.

Example 6.7.5   Use a Green's function to determine a particular solution to the differential equation

$$y'' + 16y = F(x).$$

Solution: Two linearly independent solutions to the associated homogeneous differential equation are

$$y_1(x) = \cos 4x \qquad \text{and} \qquad y_2(x) = \sin 4x$$

with Wronskian

$$W[y_1, y_2](x) = (\cos 4x)(4 \cos 4x) - (\sin 4x)(-4 \sin 4x) = 4.$$

Substitution into (6.7.16) yields

$$K(x, t) = \tfrac{1}{4}(\cos 4t \sin 4x - \sin 4t \cos 4x) = \tfrac{1}{4} \sin [4(x - t)].$$

Consequently, from (6.7.15),

$$y_p(x) = \tfrac{1}{4} \int_{x_0}^x \sin [4(x - t)]\, F(t)\, dt.$$

The general solution to the given differential equation can therefore be expressed as

$$y(x) = c_1 \cos 4x + c_2 \sin 4x + \tfrac{1}{4} \int_{x_0}^x \sin [4(x - t)]\, F(t)\, dt. \qquad \square$$

Generalization to Higher Order

We now consider the generalization of the variation-of-parameters method to linear nonhomogeneous differential equations of arbitrary order n. In this case, the basic equation is

$$y^{(n)} + a_1(x)y^{(n-1)} + \cdots + a_{n-1}(x)y' + a_n(x)y = F(x), \qquad (6.7.17)$$

where we assume that the functions $a_1, a_2, \ldots, a_n$ and F are at least continuous on the interval I. Let $\{y_1(x), y_2(x), \ldots, y_n(x)\}$ be a linearly independent set of solutions to the associated homogeneous equation

$$y^{(n)} + a_1(x)y^{(n-1)} + \cdots + a_{n-1}(x)y' + a_n(x)y = 0 \qquad (6.7.18)$$

on I, so that the general solution to Equation (6.7.18) on I is

$$y_c(x) = c_1 y_1(x) + c_2 y_2(x) + \cdots + c_n y_n(x).$$

We now look for a particular solution to Equation (6.7.17) of the form

$$y_p(x) = u_1(x)y_1(x) + u_2(x)y_2(x) + \cdots + u_n(x)y_n(x). \tag{6.7.19}$$

The idea is to substitute this expression for y_p into Equation (6.7.17) and choose the functions $u_1, u_2, \ldots, u_n$ so that the resulting y_p is indeed a solution. However, Equation (6.7.17) will give only one constraint on the functions $u_1, u_2, \ldots, u_n$ and their derivatives. Since we have n functions, we might expect that we can impose $n - 1$ further constraints on these functions. Following the steps taken in the second-order case, we differentiate y_p a total of n times, while imposing the constraint that the sum of the terms involving derivatives of the $u_1, u_2, \ldots, u_n$ that arise at each stage (except the last) should equal zero. For example, at the first stage, we obtain

$$y'_p = u_1 y'_1 + u'_1 y_1 + u_2 y'_2 + u'_2 y_2 + \cdots + u_n y'_n + u'_n y_n,$$

so we impose the constraint

$$u'_1 y_1 + u'_2 y_2 + \cdots + u'_n y_n = 0$$

in which case the foregoing expression for y'_p reduces to

$$y'_p = u_1 y'_1 + u_2 y'_2 + \cdots + u_n y'_n.$$

Continuing in this manner leads to the following expressions for y_p and its derivatives:

$$
\begin{aligned}
y_p &= u_1 y_1 + u_2 y_2 & + \cdots + u_n y_n, \\
y'_p &= u_1 y'_1 + u_2 y'_2 & + \cdots + u_n y_n, \\
&\ \vdots \\
y_p^{(n)} &= u_1 y_1^{(n)} + u_2 y_2^{(n)} + \cdots + u_n y_n^{(n)} + \left[u'_1 y_1^{(n-1)} + u'_2 y_2^{(n-1)} + \cdots + u'_n y_n^{(n-1)} \right],
\end{aligned}
$$
$$\tag{6.7.20}$$

together with the corresponding constraint conditions

$$
\begin{aligned}
u'_1 y_1 + u'_2 y_2 & + \cdots + u'_n y_n & = 0, \\
u'_1 y'_1 + u'_2 y'_2 & + \cdots + u'_n y'_n & = 0, \\
& \ \vdots \\
u'_1 y_1^{(n-2)} + u'_2 y_2^{(n-2)} & + \cdots + u'_n y_n^{(n-2)} & = 0.
\end{aligned}
\tag{6.7.21}
$$

Substitution from (6.7.20) into (6.7.17) yields the following condition in order for y_p to be a solution [simply multiply each equation in (6.7.20) by the appropriate a_i and add the elements in each column]:

$$
\begin{aligned}
u_1 &\left[y_1^{(n)} + a_1 y_1^{(n-1)} + \cdots + a_{n-1} y'_1 + a_n y_1 \right] \\
&+ u_2 \left[y_2^{(n)} + a_1 y_2^{(n-1)} + \cdots + a_{n-1} y'_2 + a_n y_2 \right] + \cdots \\
&+ u_n \left[y_n^{(n)} + a_1 y_n^{(n-1)} + \cdots + a_{n-1} y'_n + a_n y_n \right] \\
&+ \left[u'_1 y_1^{(n-1)} + u'_2 y_2^{(n-1)} + \cdots + u'_n y_n^{(n-1)} \right] = F(x).
\end{aligned}
$$

The terms in each of the brackets, except the last, vanish, since $y_1, y_2, \ldots, y_n$ are solutions of Equation (6.7.18). We are therefore left with the condition

$$u_1' y_1^{(n-1)} + u_2' y_2^{(n-1)} + \cdots + u_n' y_n^{(n-1)} = F(x).$$

Combining this with the constraints given in (6.7.21) leads to the following linear system of equations for determining $u_1', u_2', \ldots, u_n'$:

$$(6.7.22)$$
$$\begin{aligned}
y_1 u_1' + y_2 u_2' &+ \cdots + y_n u_n' = 0, \\
y_1' u_1' + y_2' u_2' &+ \cdots + y_n' u_n' = 0, \\
&\vdots \\
y_1^{(n-2)} u_1' + y_2^{(n-2)} u_2' &+ \cdots + y_n^{(n-2)} u_n' = 0, \\
y_1^{(n-1)} u_1' + y_2^{(n-1)} u_2' &+ \cdots + y_n^{(n-1)} u_n' = F(x).
\end{aligned}$$

The determinant of the matrix of coefficients of this system is the Wronskian of the functions $y_1, y_2, \ldots, y_n$, which is necessarily nonzero on I, since $y_1, y_2, \ldots, y_n$ are linearly independent on I. Consequently, the system (6.7.22) has a unique solution for the derivatives $u_1', u_2', \ldots, u_n'$, from which we can determine $u_1, u_2, \ldots, u_n$ by integration. Having found the functions $u_1, u_2, \ldots, u_n$, we can obtain y_p by substitution into Equation (6.7.19).

Theorem 6.7.6   **(Variation-of-Parameters Method)**

Consider

$$y^{(n)} + a_1(x)y^{(n-1)} + \cdots + a_{n-1}y' + a_n(x)y = F(x), \qquad (6.7.23)$$

where $a_1, a_2, \ldots, a_n, F$ are assumed to be (at least) continuous on the interval I. Let $\{y_1, y_2, \ldots, y_n\}$ be a *linearly independent* set of solutions to the associated homogeneous equation

$$y^{(n)} + a_1(x)y^{(n-1)} + \cdots + a_{n-1}(x)y' + a_n(x)y = 0$$

on I. Then a particular solution to Equation (6.7.23) is

$$y_p = u_1 y_1 + u_2 y_2 + \cdots + u_n y_n,$$

where the functions $u_1, u_2, \ldots, u_n$ satisfy (6.7.22).

Example 6.7.7   Determine the general solution to

$$y''' - 3y'' + 3y' - y = 36e^x \ln x. \qquad (6.7.24)$$

Solution: In this case, the auxiliary polynomial of the associated homogeneous equation is

$$P(r) = r^3 - 3r^2 + 3r - 1 = (r-1)^3,$$

so that three linearly independent solutions are

$$y_1(x) = e^x, \quad y_2(x) = xe^x, \quad y_3(x) = x^2 e^x.$$

According to the variation-of-parameters method, there is a particular solution to Equation (6.7.24) of the form

$$y_p(x) = e^x u_1(x) + xe^x u_2(x) + x^2 e^x u_3(x), \qquad (6.7.25)$$

where $u_1, u_2,$ and u_3 satisfy (6.7.22), which, in this case (after division by e^x), assumes the form

$$
\begin{aligned}
u_1' + && xu_2' + && x^2 u_3' &= 0, \\
u_1' + (x+1)u_2' + && (x^2 + 2x)u_3' &= 0, \\
u_1' + (x+2)u_2' + && (x^2 + 4x + 2)u_3' &= 36 \ln x.
\end{aligned}
$$

Since we have more than two variables in the system, it is more efficient to use Gaussian elimination, rather than Cramer's rule, to determine the solution. We therefore reduce the augmented matrix of the system to row-echelon form:

$$
\begin{bmatrix}
1 & x & x^2 & 0 \\
1 & x+1 & x^2+2x & 0 \\
1 & x+2 & x^2+4x+2 & 36 \ln x
\end{bmatrix}
\sim
\begin{bmatrix}
1 & x & x^2 & 0 \\
0 & 1 & 2x & 0 \\
0 & 2 & 4x+2 & 36 \ln x
\end{bmatrix}
$$

$$
\sim
\begin{bmatrix}
1 & x & x^2 & 0 \\
0 & 1 & 2x & 0 \\
0 & 0 & 2 & 36 \ln x
\end{bmatrix}
\sim
\begin{bmatrix}
1 & x & x^2 & 0 \\
0 & 1 & 2x & 0 \\
0 & 0 & 1 & 18 \ln x
\end{bmatrix}.
$$

Consequently,

$$
u_1' = 18x^2 \ln x, \quad u_2' = -36x \ln x, \quad u_3' = 18 \ln x.
$$

By integrating, we obtain

$$
u_1(x) = 18 \int x^2 \ln x \, dx = 2x^3 (3 \ln x - 1),
$$

$$
u_2(x) = -36 \int x \ln x \, dx = 9x^2 (1 - 2 \ln x),
$$

$$
u_3(x) = 18 \int \ln x \, dx = 18x(\ln x - 1),
$$

where we have set the integration constants to zero without loss of generality. Substituting these expressions for $u_1, u_2,$ and u_3 into Equation (6.7.25) yields the particular solution

$$
y_p(x) = x^3 e^x (6 \ln x - 11).
$$

The general solution to the given differential equation is therefore

$$
y(x) = e^x \left[c_1 + c_2 x + c_3 x^2 + x^3 (6 \ln x - 11) \right].
$$

☐

Exercises for 6.7

Key Terms

Variation-of-parameters, Green's function.

Skills

- Be able to use the variation-of-parameters method to find the general solution to an nth-order linear differential equation.

- Be able to use a Green's function to determine a particular solution to a given differential equation.

True-False Review

For Questions 1–3, decide if the given statement is **true** or **false**, and give a brief justification for your answer. If true, you can quote a relevant definition or theorem from the text. If false, provide an example, illustration, or brief explanation of why the statement is false.

1. The functions $u_1(x)$ and $u_2(x)$ arising in the particular solution (6.7.3) to (6.7.1) satisfy

$$u_1'(x)y_1'(x) + u_2'(x)y_2'(x) = F(x),$$

where y_1 and y_2 are linearly independent solutions to (6.7.2).

2. The variation-of-parameters method seeks a particular solution to (6.7.17) of the form

$$y_p(x) = u_1(x)y_1(x) + u_2(x)y_2(x) + \cdots + u_n(x)y_n(x),$$

where $y_1, y_2, \ldots, y_n$ are any n solutions to the associated homogeneous differential equation (6.7.18).

3. The functions $u_1(x), u_2(x), \ldots, u_n(x)$ arising in the particular solution (6.7.19) to the differential equation (6.7.17) are uniquely determined by the variation-of-parameters method.

Problems

For Problems 1–22, use the variation-of-parameters method to find the general solution to the given differential equation.

1. $y'' + 6y' + 9y = \dfrac{2e^{-3x}}{x^2 + 1}$.

2. $y'' - 4y = \dfrac{8}{e^{2x} + 1}$.

3. $y'' - 4y' + 5y = e^{2x} \tan x, \ 0 < x < \pi/2$.

4. $y'' - 6y' + 9y = 4e^{3x} \ln x, \quad x > 0$.

5. $y'' + 4y' + 4y = x^{-2}e^{-2x}, \quad x > 0$.

6. $y'' + 9y = 18\sec^3(3x), \quad |x| < \pi/6$.

7. $y'' - y = 2\tanh x$.

8. $y'' - 2my' + m^2 y = e^{mx}/(1 + x^2), \quad m$ constant.

9. $y'' - 2y' + y = 4e^x x^{-3} \ln x, \quad x > 0$.

10. $y'' + 2y' + y = \dfrac{e^{-x}}{\sqrt{4 - x^2}}, \quad |x| < 2$.

11. $y'' + 2y' + 17y = \dfrac{64e^{-x}}{3 + \sin^2(4x)}$.

12. $y'' + 9y = \dfrac{36}{4 - \cos^2(3x)}$.

13. $y'' - 10y' + 25y = \dfrac{2e^{5x}}{4 + x^2}$.

14. $y'' - 6y' + 13y = 4e^{3x}\sec^2(2x), \ |x| < \pi/4$.

15. $y'' + y = \sec x + 4e^x, \quad |x| < \pi/2$.

16. $y'' + y = \csc x + 2x^2 + 5x + 1, \ 0 < x < \pi$.

17. $y'' + 4y' + 4y = 15e^{-2x}\ln x + 25\cos x, \ x > 0$.

18. $y'' + 4y' + 4y = \dfrac{4e^{-2x}}{1 + x^2} + 2x^2 - 1$.

19. $y''' - 6y'' + 12y' - 8y = 36e^{2x}\ln x$.

20. $y''' - 3y'' + 3y' - y = 2x^{-2}e^x, \quad x > 0$.

21. $y''' + 3y'' + 3y' + y = \dfrac{2e^{-x}}{1 + x^2}$.

22. $y''' - 6y'' + 9y' = 12e^{3x}$. Suggest a better method for solving this problem.

For Problems 23–26, use a Green's function to determine a particular solution to the given differential equation.

23. $y'' - y = F(x)$.

24. $y'' + y' - 2y = F(x)$.

25. $y'' + 5y' + 4y = F(x)$.

26. $y'' + 4y' - 12y = F(x)$.

For Problems 27–28, use a Green's function to solve the given initial-value problem. [**Hint:** Choose $x_0 = 0$.]

27. $y'' + y = \sec x, \quad y(0) = 0, \quad y'(0) = 1$.

28. $y'' - 4y' + 4y = 5xe^{2x}, \quad y(0) = 1, \quad y'(0) = 0$.

29. Determine a Green's function for

$$y'' - 2ay' + a^2 y = F(x), \tag{6.7.26}$$

where a is a constant, and use it to find a particular solution to (6.7.26) when (in each case, α, β are constants):

(a) $F(x) = \dfrac{\alpha e^{ax}}{x^2 + \beta^2}$.

(b) $F(x) = \dfrac{\alpha e^{ax}}{\sqrt{\beta^2 - x^2}}, \quad |x| < \beta$.

(c) $F(x) = e^{ax} x^\alpha \ln x, \quad x > 0$.

30. Consider the differential equation $y'' + y = F(x)$, where F is continuous on the interval $[a, b]$. If $x_0 \in (a, b)$, show that the solution to the initial-value problem

$$\begin{cases} y'' + y = F(x), \\ y(x_0) = y_0, \qquad y'(x_0) = y_1 \end{cases}$$

is

$$y(x) = y_0 \cos(x - x_0) + y_1 \sin(x - x_0)$$
$$+ \int_{x_0}^{x} F(t) \sin(x - t)\, dt.$$

31. If F is continuous on the interval $[a, b]$, use the variation-of-parameters technique to show that a particular solution to

$$(D - r)^3 y = F(x), \qquad r \text{ constant}$$

is

$$y_p(x) = \tfrac{1}{2} \int_a^x F(t)(x - t)^2 e^{r(x-t)}\, dt.$$

32. (a) Use Cramer's rule to show that the solution to the system (6.7.22) can be written in the form

$$u'_k = \frac{F(x) W_k(x)}{W[y_1, y_2, \ldots, y_n](x)}, \qquad k = 1, 2, \ldots, n,$$

where $W_k(x)$ denotes the determinant that is obtained when the kth column of $W[y_1, y_2, \ldots, y_n](x)$ is replaced by

$$\begin{bmatrix} 0 \\ 0 \\ \vdots \\ 1 \end{bmatrix}.$$

(b) Use your result from (a) to show that a particular solution to Equation (6.7.23) is

$$y_p(x) = \int_{x_0}^{x} K(x, t) F(t)\, dt,$$

where K is given by

$$K(x, t) =$$
$$\frac{y_1(x) W_1(t) + y_2(x) W_2(t) + \cdots + y_n(x) W_n(t)}{W[y_1, y_2, \ldots, y_n](t)},$$

and x_0 is an arbitrary point in the interval of interest.

For Problems 33–37, use the result of the preceding problem to determine a particular solution to the given differential equation.

33. $(D + 3)(D - 3)(D + 5)y = F(x)$.

34. $(D + 1)(D^2 + 9)y = F(x)$.

35. $(D^2 + 8D + 16)(D - 2)y = F(x)$.

36. $(D^2 - 4D + 13)(D - 3)y = F(x)$.

37. $(D - 1)(D - 2)(D + 4)y = F(x)$.

6.8 A Differential Equation with Nonconstant Coefficients

In this section we consider a particular type of homogeneous differential equation that has nonconstant coefficients. Its solution will be useful in Chapter 9 and also will enable us to give a further illustration of the power of the variation-of-parameters technique introduced in the preceding section.

DEFINITION 6.8.1

A differential equation of the form

$$x^n \frac{d^n y}{dx^n} + a_1 x^{n-1} \frac{d^{n-1} y}{dx^{n-1}} + \cdots + a_{n-1} x \frac{dy}{dx} + a_n y = 0,$$

where $a_1, a_2, \ldots, a_n$ are constants, is called a **Cauchy-Euler** equation.

Notice that if we replace x by kx, where k is a constant, then the form of a Cauchy-Euler equation is unaltered. Such a rescaling of x can be interpreted as a dimensional change (for example, inches to centimeters), and so Cauchy-Euler equations are sometimes called **equidimensional** equations.

We begin our analysis by restricting attention to the second-order case and will assume that $x > 0$ (the extension to the interval $x < 0$ is left for the exercises). Thus, consider the differential equation

$$x^2 y'' + a_1 x y' + a_2 y = 0, \qquad x > 0, \tag{6.8.1}$$

where a_1 and a_2 are constants. The solution technique is based on the observation that if we substitute $y(x) = x^r$ into Equation (6.8.1), then each of the resulting terms on the left-hand side will be multiplied by the same power of x, which suggests that there may be solutions of the form

$$y(x) = x^r \tag{6.8.2}$$

for an appropriately chosen constant r. In order to investigate this possibility we differentiate (6.8.2) twice to obtain

$$y' = r x^{r-1}, \qquad y'' = r(r-1)x^{r-2}.$$

Substituting these expressions into Equation (6.8.1) yields the condition

$$x^r [r(r-1) + a_1 r + a_2] = 0$$

so that (6.8.2) is indeed a solution to Equation (6.8.1), provided that r satisfies

$$r(r-1) + a_1 r + a_2 = 0.$$

That is,

$$r^2 + (a_1 - 1)r + a_2 = 0. \tag{6.8.3}$$

This is referred to as the **indicial equation** associated with Equation (6.8.1). The roots of (6.8.3) are

$$r_1 = \frac{-(a_1 - 1) + \sqrt{(a_1 - 1)^2 - 4a_2}}{2}, \qquad r_2 = \frac{-(a_1 - 1) - \sqrt{(a_1 - 1)^2 - 4a_2}}{2}$$

so that there are three cases to consider.

1. r_1, r_2 real and distinct: In this case two solutions to Equation (6.8.1) are

$$y_1(x) = x^{r_1} \qquad \text{and} \qquad y_2(x) = x^{r_2}.$$

Further, since by assumption in this case $r_1 \neq r_2$,

$$\frac{y_2}{y_1} = x^{r_2 - r_1} \neq \text{constant}$$

so that y_1 and y_2 are linearly independent on $(0, \infty)$. Consequently, the general solution to Equation (6.8.1) in this case is

$$y(x) = c_1 x^{r_1} + c_2 x^{r_2}. \tag{6.8.4}$$

2. $r_1 = r_2 = -(a_1 - 1)/2$: In this case, we obtain only one solution to Equation (6.8.1), namely

$$y_1(x) = x^{r_1}.$$

We try for a second linearly independent solution to Equation (6.8.1) of the form

$$y_2(x) = x^{r_1} u(x).$$

where $u' \neq 0$. Differentiating with respect to x yields

$$y_2'(x) = x^{r_1} u' + r_1 x^{r_1-1} u, \quad y_2''(x) = x^{r_1} u'' + 2r_1 x^{r_1-1} u' + r_1(r_1 - 1) x^{r_1-2} u.$$

Substituting these expressions into Equation (6.8.1), we obtain the following equation for u:

$$x^2 \left[x^{r_1} u'' + 2r_1 x^{r_1-1} u' + r_1(r_1-1) x^{r_1-2} u \right] + a_1 x (x^{r_1} u' + r_1 x^{r_1-1} u) + a_2 x^{r_1} u = 0.$$

Equivalently,

$$x^{r_1+2} u'' + (2r_1 + a_1) x^{r_1+1} u' + x^{r_1} [r_1(r_1-1) + a_1 r_1 + a_2] u = 0.$$

The last term on the left-hand side vanishes, since r_1 is a root of the indicial Equation (6.8.3). Thus, substituting $r_1 = -(a_1 - 1)/2$ and dividing through by x^{r_1+2}, we see that u must satisfy

$$u'' + x^{-1} u' = 0$$

This separable equation can be written as

$$\frac{u''}{u'} = -\frac{1}{x}$$

which can be integrated directly to obtain

$$\ln |u'| = -\ln x + c.$$

We can therefore choose (by setting $c = 0$)

$$u' = x^{-1}.$$

Integrating once more yields

$$u(x) = \ln x$$

where we have set the integration constant to zero again. Consequently, a second solution to Equation (6.8.1) in this case is

$$y_2(x) = x^{r_1} \ln x.$$

Since

$$\frac{y_2}{y_1} = \ln x \neq \text{constant}$$

it follows that y_1 and y_2 are linearly independent on $(0, \infty)$. The general solution to Equation (6.8.1) is therefore given by

$$y(x) = c_1 x^{r_1} + c_2 x^{r_1} \ln x = x^{r_1} (c_1 + c_2 \ln x). \tag{6.8.5}$$

3. Complex conjugate roots, $r_1 = a + ib$, $r_2 = a - ib$, where $b \neq 0$: In this case, two complex-valued solutions to Equation (6.8.1) are

$$w_1(x) = x^{a+ib} = e^{(a+ib)\ln x} = x^a[\cos(b \ln x) + i \sin(b \ln x)],$$
$$w_2(x) = x^{a-ib} = e^{(a-ib)\ln x} = x^a[\cos(b \ln x) - i \sin(b \ln x)],$$

where we have used Euler's formula. Two corresponding real-valued solutions are

$$y_1(x) = \frac{1}{2}[w_1(x) + w_2(x)] = x^a \cos(b \ln x)$$

$$y_2(x) = \frac{1}{2i}[w_1(x) - w_2(x)] = x^a \sin(b \ln x).$$

Further,

$$\frac{y_2}{y_1} = \tan(b \ln x) \neq \text{constant}$$

since $b \neq 0$. Consequently, y_1 and y_2 are linearly independent on $(0, \infty)$, and so the general solution to Equation (6.8.1) in this case is

$$y(x) = x^a[c_1 \cos(b \ln x) + c_2 \sin(b \ln x)]. \tag{6.8.6}$$

The preceding discussion is summarized in Table 6.8.1.

Roots of Indicial Equation	Linearly Independent Solutions to Differential Equation
Real distinct: $r_1 \neq r_2$	$y_1(x) = x^{r_1}$, $y_2(x) = x^{r_2}$.
Real repeated: $r_1 = r_2$	$y_1(x) = x^{r_1}$, $y_2(x) = x^{r_1} \ln x$.
Complex conjugate: $r_1 = a + ib, r_2 = a - ib$	$y_1(x) = x^a \cos(b \ln x)$, $y_2(x) = x^a \sin(b \ln x)$.

Table 6.8.1: Linearly independent solutions to a Cauchy-Euler equation.

Example 6.8.2   Solve

$$x^2 y'' - xy' - 8y = 0, \qquad x > 0. \tag{6.8.7}$$

Solution: Inserting $y = x^r$ into Equation (6.8.7) yields the indicial equation

$$r(r-1) - r - 8 = 0.$$

That is,

$$r^2 - 2r - 8 = (r-4)(r+2) = 0.$$

Hence, two linearly independent solutions to Equation (6.8.7) are

$$y_1(x) = x^4 \qquad \text{and} \qquad y_2(x) = x^{-2}.$$

Consequently, Equation (6.8.7) has general solution

$$y(x) = c_1 x^4 + c_2 x^{-2}.$$ □

Example 6.8.3   Solve the initial-value problem

$$x^2 y'' - 3xy' + 13y = 0,$$ (6.8.8)

$$y(1) = 2, \quad y'(1) = -5.$$ (6.8.9)

Solution: Substituting $y = x^r$ into Equation (6.8.8) yields the indicial equation

$$r^2 - 4r + 13 = 0,$$

which has the complex conjugate roots

$$r = 2 \pm 3i.$$

It follows that two linearly independent solutions to Equation (6.8.8) are

$$y_1(x) = x^2 \cos(3 \ln x) \quad \text{and} \quad y_2(x) = x^2 \sin(3 \ln x),$$

so that the general solution is

$$y(x) = c_1 x^2 \cos(3 \ln x) + c_2 x^2 \sin(3 \ln x),$$

which we write as

$$y(x) = x^2 [c_1 \cos(3 \ln x) + c_2 \sin(3 \ln x)].$$

The first initial condition in (6.8.9) requires that

$$c_1 \cos 0 + c_2 \sin 0 = 2,$$

so that $c_1 = 2$. Inserting this value of c_1 into the general solution and differentiating with respect to x yields

$$y'(x) = 2x\,[2 \cos(3 \ln x) + c_2 \sin(3 \ln x)]$$
$$+ x^2 \left[-6x^{-1} \sin(3 \ln x) + 3x^{-1} c_2 \cos(3 \ln x) \right].$$

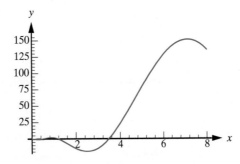

Figure 6.8.1: The solution to the initial-value problem in Example 6.8.3.

The second initial condition in (6.8.9) therefore requires

$$2(2 + 0) + (0 + 3c_2) = -5,$$

so that $c_2 = -3$. Consequently the solution to the initial-value problem is

$$y(x) = x^2[2\cos(3\ln x) - 3\sin(3\ln x)].$$

A sketch of the corresponding solution curve is given in Figure 6.8.1. Owing to the trigonometric terms, the solution is oscillatory. The amplitude of the oscillation is growing rapidly with x, owing to the multiplicative factor x^2. Furthermore, as $x \to 0^+$, the amplitude also approaches zero.   ☐

Now consider the nonhomogeneous equation

$$x^2 y'' + a_1 xy' + a_2 y = g(x), \tag{6.8.10}$$

where a_1 and a_2 are constants. Since the associated homogeneous equation is a Cauchy-Euler equation, we can determine the complementary function and then use the variation-of-parameters method to determine a particular solution. We must remember, however, that the formulas derived in the variation-of-parameters technique are based around a differential equation written in the standard form

$$y'' + a_1(x)y' + a_2(x)y = F(x).$$

Consequently, when applying the method to a differential equation of the form (6.8.10), the appropriate formulas for determining u_1 and u_2 are

$$y_1 u_1' + y_2 u_2' = 0,$$
$$y_1' u_1' + y_2' u_2' = x^{-2} g(x).$$

We illustrate with an example.

Example 6.8.4   Determine the general solution to

$$x^2 y'' - 3xy' + 4y = x^2 \ln x, \qquad x > 0. \tag{6.8.11}$$

Solution: The associated homogeneous equation is the Cauchy-Euler equation

$$x^2 y'' - 3xy' + 4y = 0. \tag{6.8.12}$$

Substituting $y = x^r$ into this equation yields the indicial equation

$$r^2 - 4r + 4 = 0.$$

That is,

$$(r - 2)^2 = 0.$$

Hence two linearly independent solutions to Equation (6.8.12) are

$$y_1(x) = x^2 \qquad \text{and} \qquad y_2(x) = x^2 \ln x.$$

According to the variation-of-parameters technique, a particular solution to Equation (6.8.11) is

$$y_p(x) = y_1(x)u_1(x) + y_2(x)u_2(x) = x^2 u_1 + x^2 \ln x \, u_2, \tag{6.8.13}$$

where u_1 and u_2 are determined from

$$x^2 u_1' + x^2 \ln x \, u_2' = 0, \qquad 2x u_1' + (2x \ln x + x) u_2' = \ln x.$$

Hence,

$$u_1' = -x^{-1} (\ln x)^2 \qquad \text{and} \qquad u_2' = x^{-1} \ln x,$$

which upon integration gives

$$u_1(x) = -\tfrac{1}{3} (\ln x)^3 \qquad \text{and} \qquad u_2(x) = \tfrac{1}{2} (\ln x)^2,$$

where we have set the integration constants to zero without loss of generality. Substitution into (6.8.13) yields

$$y_p(x) = -\tfrac{1}{3} x^2 (\ln x)^3 + \tfrac{1}{2} x^2 (\ln x)^3 = \tfrac{1}{6} x^2 (\ln x)^3.$$

Thus, Equation (6.8.11) has general solution

$$y(x) = c_1 x^2 + c_2 x^2 \ln x + \tfrac{1}{6} x^2 (\ln x)^3,$$

which can be written as

$$y(x) = \tfrac{1}{6} x^2 [c_3 + c_4 \ln x + (\ln x)^3],$$

where $c_3 = 6 c_1$ and $c_4 = 6 c_2$. $\square$

Generalization to Higher Order

Now consider the general Cauchy-Euler equation of order n,

$$x^n y^{(n)} + a_1 x^{n-1} y^{(n-1)} + \cdots + a_{n-1} x y' + a_n y = 0, \tag{6.8.14}$$

where $a_1, a_2, \ldots, a_n$ are constants, on the interval $x > 0$. We begin by substituting $y(x) = x^r$ into (6.8.14). The result is the indicial equation

$$\begin{aligned} r(r-1)(r-2) \cdots (r-n+1) + a_1 r(r-1) \cdots (r-n+2) + \cdots \\ + a_{n-1} r + a_n = 0. \end{aligned} \tag{6.8.15}$$

In the case that (6.8.15) has n distinct roots, say $r_1, r_2, \ldots, r_n$, we directly obtain the n solutions

$$x^{r_1}, x^{r_2}, \ldots, x^{r_n},$$

and it can be shown that these solutions are linearly independent on $(0, \infty)$. Of course, if some of the roots are complex, then we must take the real and imaginary parts of the corresponding complex-valued solutions in order to determine real-valued solutions. If, however, $r = r_1$ is a root of multiplicity k, then we cannot directly determine the appropriate number of solutions. Based on our experience in the second-order case, we might suspect that there are k solutions of the form

$$x^{r_1}, x^{r_1} \ln x, x^{r_1} (\ln x)^2, \ldots, x^{r_1} (\ln x)^{k-1}. \tag{6.8.16}$$

We now show that this is indeed correct. The key idea is that the change of variables $x = e^z$, or equivalently $z = \ln x$, transforms Equation (6.8.14) into a constant-coefficient equation, which can be solved via the technique of Section 6.2. To establish this we need the following lemma.

Lemma 6.8.5   If y is a sufficiently smooth function of x and $x = e^z$, then

$$x^k \frac{d^k y}{dx^k} = D(D-1)(D-2)\cdots(D-k+1)y, \quad k = 1, 2, \ldots, \tag{6.8.17}$$

where $D = d/dz$.

Proof   The proof of the result requires the chain rule and mathematical induction. Since $x = e^z$, we have $z = \ln x$, so that $dz/dx = 1/x$. Thus, by the chain rule,

$$\frac{dy}{dx} = \frac{dy}{dz}\frac{dz}{dx} = \frac{1}{x}\frac{dy}{dz},$$

which implies that

$$x\frac{dy}{dx} = Dy \tag{6.8.18}$$

where $D = d/dz$. Thus, (6.8.17) is true when $k = 1$. Now suppose that the result is true when $k = m$. That is,

$$x^m \frac{d^m y}{dx^m} = D(D-1)(D-2)\cdots(D-m+1)y. \tag{6.8.19}$$

We must show that this implies its validity when $k = m + 1$. We proceed as follows. Using the product rule, we have

$$x\frac{d}{dx}\left(x^m \frac{d^m y}{dx^m}\right) = x^{m+1}\frac{d^{m+1}y}{dx^{m+1}} + mx^m \frac{d^m y}{dx^m},$$

which can be rearranged to obtain

$$x^{m+1}\frac{d^{m+1}y}{dx^{m+1}} = x\frac{d}{dx}\left(x^m \frac{d^m y}{dx^m}\right) - mx^m \frac{d^m y}{dx^m}.$$

Substituting from (6.8.18) for $x(d/dx) = D$ and from (6.8.19) for $x^m(d^m y/dx^m)$ into this equation yields

$$x^{m+1}\frac{d^{m+1}y}{dx^{m+1}} = D[D(D-1)(D-2)\cdots(D-m+1)y]$$
$$- m[D(D-1)(D-2)\cdots(D-m+1)y];$$

that is, since we can interchange the order of the factors in a polynomial differential operator,

$$x^{m+1}\frac{d^{m+1}y}{dx^{m+1}} = D(D-1)(D-2)\cdots(D-m+1)(D-m)y.$$

We have therefore established that the validity of (6.8.17) when $k = m$ implies its validity when $k = m + 1$. Since the result is true when $k = 1$, it follows, by induction, that it is valid for all positive integers k. ■

Remark   Although the rule for transforming derivatives given in (6.8.17) looks quite formidable, it is very easy to remember. We write out the first three derivatives in order to elucidate this:

$$x\frac{dy}{dx} = Dy, \quad x^2\frac{d^2y}{dx^2} = D(D-1)y, \quad x^3\frac{d^3y}{dx^3} = D(D-1)(D-2)y,$$

where $D = d/dz$.

We can now establish the main result.

Theorem 6.8.6   The change of variables $x = e^z$ transforms the Cauchy-Euler equation

$$x^n\frac{d^ny}{dx^n} + a_1 x^{n-1}\frac{d^{n-1}y}{dx^{n-1}} + \cdots + a_{n-1}x\frac{dy}{dx} + a_n y = 0, \quad x > 0, \tag{6.8.20}$$

into the constant-coefficient equation

$$[D(D-1)(D-2) \cdots (D-n+1) + a_1 D(D-1) \\ \cdots (D-n+2) + \cdots + a_{n-1}D + a_n]y = 0. \tag{6.8.21}$$

Proof   Equation (6.8.21) follows directly by substituting for each of the terms $x^k(d^k y/ dx^k)$ in (6.8.20), using the previous lemma.   ∎

The auxiliary equation for the constant-coefficient equation (6.8.21) is

$$r(r-1)(r-2) \cdots (r-n+1) + a_1 r(r-1) \cdots (r-n+2) + \cdots + a_{n-1}r + a_n = 0,$$

which coincides with the indicial equation (6.8.15) of the original differential equation. If $r = r_1$ is a root of multiplicity k, then it follows from our results of Section 6.2 that the corresponding linearly independent solutions to (6.8.21), and hence (6.8.20), are

$$e^{r_1 z}, \quad ze^{r_1 z}, \quad \ldots \quad , z^{k-1}e^{r_1 z};$$

that is, since $z = \ln x$,

$$x^{r_1}, \quad x^{r_1}\ln x, \quad \ldots \quad , x^{r_1}(\ln x)^{k-1}.$$

If $r = a + ib$ is complex, then we can obtain the appropriate real-valued solutions by extracting the real and imaginary parts of the corresponding complex-valued solutions. Thus, corresponding to complex conjugate roots of the indicial equation of multiplicity k, we obtain the $2k$ real-valued solutions:

$$x^a \cos(b\ln x), \quad x^a \sin(b\ln x), \quad x^a \ln x \cos(b\ln x), \quad x^a \ln x \sin(b\ln x), \quad \ldots, \\ x^a(\ln x)^{k-1}\cos(b\ln x), \quad x^a(\ln x)^{k-1}\sin(b\ln x).$$

In summary, to solve a Cauchy-Euler equation we can substitute $y(x) = x^r$ into the differential equation to obtain the indicial equation. Corresponding to each root, r_i, we can then determine the appropriate number of linearly independent solutions, as indicated above.

Example 6.8.7   Determine the general solution to

$$x^3 y''' + 2x^2 y'' + 4xy' - 4y = 0, \quad x > 0. \tag{6.8.22}$$

Solution: Substituting $y(x) = x^r$ into (6.8.22) yields the indicial equation

$$r(r-1)(r-2) + 2r(r-1) + 4r - 4 = 0;$$

that is,

$$r^3 - r^2 + 4r - 4 = 0,$$

which can be factored as

$$(r-1)(r^2 + 4) = 0.$$

The roots of the indicial equation are, therefore,

$$r = 1, \quad r = \pm 2i,$$

so that three linearly independent solutions to (6.8.22) on $(0, \infty)$ are

$$y_1(x) = x, \quad y_2(x) = \cos(2 \ln x), \quad y_3(x) = \sin(2 \ln x).$$

Consequently, the general solution is

$$y(x) = c_1 x + c_2 \cos(2 \ln x) + c_3 \sin(2 \ln x). \qquad \square$$

Example 6.8.8   Determine the general solution to

$$x^3 y''' - 3x^2 y'' + 7xy' - 8y = 0, \quad x > 0.$$

Solution: In this case the substitution $y(x) = x^r$ yields the indicial equation

$$r(r-1)(r-2) - 3r(r-1) + 7r - 8 = 0;$$

that is,

$$r^3 - 6r^2 + 12r - 8 = 0.$$

By inspection we see that $r = 2$ is a root, and therefore the indicial equation can be written as

$$(r-2)(r^2 - 4r + 4) = 0;$$

that is,

$$(r-2)^3 = 0.$$

It follows that three linearly independent solutions to the given differential equation are

$$y_1(x) = x^2, \quad y_2(x) = x^2 \ln x, \quad y_3(x) = x^2 (\ln x)^2,$$

so that the general solution is

$$y(x) = x^2 [c_1 + c_2 \ln x + c_3 (\ln x)^2]. \qquad \square$$

Finally, we mention that on the interval $(-\infty, 0)$, the substitution $y(x) = (-x)^r$ can be used to obtain the indicial equation, and the solutions of the differential equation can be obtained by replacing x with $-x$ in the solutions obtained previously in this section. Consequently, if we use $|x|$ in place of x, we will obtain solutions to a Cauchy-Euler equation that are valid for all $x \neq 0$.

Exercises for 6.8

Key Terms

Cauchy-Euler equation, Equidimensional equation, Indicial equation.

Skills

- Be able to determine whether or not a given differential equation is a Cauchy-Euler equation.

- Be able to determine the indicial equation associated with a Cauchy-Euler equation.

- Be able to determine two linearly independent solutions to a Cauchy-Euler equation according to whether the associated indicial equation has real distinct roots, real repeated roots, or complex conjugate roots.

- Be able to find the general solution to a Cauchy-Euler equation.

True-False Review

For Questions 1–6, decide if the given statement is **true** or **false**, and give a brief justification for your answer. If true, you can quote a relevant definition or theorem from the text. If false, provide an example, illustration, or brief explanation of why the statement is false.

1. A Cauchy-Euler equation is a differential equation of the form
$$x^2 y'' + a_1 y' + a_2 y = 0,$$
where a_1 and a_2 are constants.

2. The Cauchy-Euler equation
$$x^2 y'' - 2xy' - 18y = 0$$
has two linearly independent solutions of the form $y(x) = x^r$.

3. The Cauchy-Euler equation
$$x^2 y'' + 9xy' + 16y = 0$$
has two linearly independent solutions of the form $y(x) = x^r$.

4. All solutions $y(x)$ to the Cauchy-Euler equation
$$x^2 y'' + 6xy' + 6y = 0$$
tend to 0 as $x \to +\infty$.

5. If $y(x) = \ln x / x$ is obtained by the method of this section as a solution to a Cauchy-Euler equation, then the differential equation is
$$x^2 y'' + 3xy' + y = 0.$$

6. All nontrivial solutions to the Cauchy-Euler equation
$$x^2 y'' - 5xy' - 7y = 0$$
are oscillatory as a function of x.

Problems

For Problems 1–8, determine the general solution to the given differential equation on $(0, \infty)$.

1. $x^2 y'' - xy' + 5y = 0.$

2. $x^2 y'' - 6y = 0.$

3. $x^2 y'' - 3xy' + 4y = 0.$

4. $x^2 y'' - 4xy' + 4y = 0.$

5. $x^2 y'' + 3xy' + y = 0.$

6. $x^2 y'' + 5xy' + 13y = 0.$

7. $x^2 y'' - xy' - 35y = 0.$

8. $x^2 y'' + xy' + 16y = 0.$

For Problems 9–11, find the solution to the Cauchy-Euler equation on the interval $(0, \infty)$. In each case, m and k are positive constants.

9. $x^2 y'' + xy' - m^2 y = 0.$

10. $x^2 y'' - x(2m - 1)y' + m^2 y = 0.$

11. $x^2 y'' - x(2m - 1)y' + (m^2 + k^2)y = 0.$

12. Consider the Cauchy-Euler equation
$$x^2 y'' + xa_1 y' + a_2 y = 0, \quad x > 0. \quad (6.8.23)$$

(a) Show that the change of independent variable defined by $x = e^z$ transforms Equation (6.8.23) into the constant-coefficient equation
$$\frac{d^2 y}{dz^2} + (a_1 - 1)\frac{dy}{dz} + a_2 y = 0. \quad (6.8.24)$$

(b) Show that if $y_1(z), y_2(z)$ are linearly independent solutions to Equation (6.8.24), then $y_1(\ln x), y_2(\ln x)$ are linearly independent solutions to Equation (6.8.23). [**Hint:** From (a), we already know that y_1, y_2 are solutions to Equation (6.8.23). To show that they are linearly independent, verify that

$$W[y_1, y_2](x) = \frac{dz}{dx} W[y_1, y_2](z).]$$

13. Consider the Cauchy-Euler equation

$$x^2 y'' + axy' + by = 0, \quad x < 0. \qquad (6.8.25)$$

Show that the substitution $y = (-x)^r$ yields the indicial equation

$$r^2 + (a - 1)r + b = 0.$$

Thus, linearly independent solutions to Equation (6.8.25) on $(-\infty, 0)$ can be determined by replacing x with $-x$ in (6.8.4), (6.8.5), and (6.8.6). As a consequence, if we replace x by $|x|$ in these solutions, we will obtain solutions to Equation (6.8.1) that are valid for all $x \neq 0$.

For Problems 14–21, solve the given differential equation on the interval $x > 0$. (Remember to put the equation in standard form.)

14. $x^2 y'' + 4xy' + 2y = 4\ln x$.

15. $x^2 y'' - 4xy' + 6y = x^4 \sin x$.

16. $x^2 y'' + 6xy' + 6y = 4e^{2x}$.

17. $x^2 y'' - 3xy' + 4y = \dfrac{x^2}{\ln x}$.

18. $x^2 y'' + 4xy' + 2y = \cos x$.

19. $x^2 y'' + xy' + 9y = 9\ln x$.

20. $x^2 y'' - xy' + 5y = 8x(\ln x)^2$.

21. $x^2 y'' - (2m - 1)xy' + m^2 y = x^m (\ln x)^k$, where m, k are constants.

22. (a) Solve the initial-value problem

$$x^2 y'' - xy' + 5y = 0,$$
$$y(1) = \sqrt{2}, \quad y'(1) = 3\sqrt{2},$$

and show that your solution can be written in the form

$$y(x) = 2x \cos(2\ln x - \frac{\pi}{4}).$$

(b) Determine all zeros of $y(x)$.

(c) ◇ Sketch the corresponding solution curve on the interval $[0.001, 16]$, and verify the zeros in the interval $[3, 16]$.

23. The motion of a physical system is governed by the initial-value problem

$$t^2 \frac{d^2 y}{dt^2} + t \frac{dy}{dt} + 25y = 0$$
$$y(1) = 3\frac{\sqrt{3}}{2}, \quad y'(1) = \frac{15}{2}.$$

(a) Solve the given initial-value problem, and show that your solution can be written in the form

$$y(t) = 3\cos(5\ln t - \pi/6).$$

(b) Determine all zeros of $y(t)$.

(c) ◇ Sketch the corresponding solution curve on the interval $[0.01, 2]$.

(d) Is the system performing simple harmonic motion? Justify your answer.

24. We have shown that in the case of complex conjugate roots, $r = a \pm ib, b \neq 0$, of the indicial equation, the general solution to the Cauchy-Euler equation

$$y'' + a_1 y' + a_2 y = 0, \quad x > 0$$

is

$$y(x) = x^a [c_1 \cos(b\ln x) + c_2 \sin(b\ln x)]. \quad (6.8.26)$$

(a) Show that (6.8.26) can be written in the form

$$y(x) = Ax^a \cos(b\ln x - \phi)$$

for appropriate constants A and ϕ.

(b) Determine all zeros of $y(x)$, and the distance between successive zeros. What happens to this distance as $x \to \infty$, and as $x \to 0^+$?

(c) Describe the behavior of the solution as $x \to \infty$ and as $x \to 0^+$ in each of the three cases $a > 0, a < 0$, and $a = 0$. In each case, give a general sketch of a generic solution curve.

6.9   Reduction of Order

Finally, in this chapter, we consider a powerful technique for determining the general solution to any second-order linear differential equation, assuming that we know just one solution to the associated homogeneous equation. This technique is usually referred to as **reduction of order**.

Consider

$$y'' + a_1(x)y' + a_2(x)y = F(x), \qquad (6.9.1)$$

where we assume that the functions a_1, a_2, $F(x)$ are continuous on an interval J. We know that the general solution to Equation (6.9.1) is of the form

$$y(x) = c_1 y_1(x) + c_2 y_2(x) + y_p(x),$$

where y_1 and y_2 are linearly independent (i.e., nonproportional) solutions to

$$y'' + a_1(x)y' + a_2(x)y = 0 \qquad (6.9.2)$$

on J . Suppose that we have found one solution, say, $y = y_1(x)$, to (6.9.2). We now replace the constant c with an arbitrary function $u(x)$ and try for a solution to (6.9.1) of the form

$$y(x) = u(x)y_1(x). \qquad (6.9.3)$$

The following derivation establishes that $u(x)$ can, in theory, always be determined from Equation (6.9.1). Differentiating (6.9.3) twice with respect to x yields

$$y' = u'y_1 + uy_1'$$
$$y'' = u''y_1 + 2u'y_1' + uy_1''.$$

Substituting into Equation (6.9.1) gives

$$(u''y_1 + 2u'y_1' + uy_1'') + a_1(x)(u'y_1 + uy_1') + a_2(x)(uy_1) = F(x)$$

so that (6.9.3) solves Equation (6.9.1), provided u satisfies

$$u[y_1'' + a_1(x)y_1' + a_2(x)y_1] + u''y_1 + u'[2y_1' + a_1(x)y_1] = F(x). \qquad (6.9.4)$$

Since $y = y_1(x)$ is a solution to Equation (6.9.2) the coefficient of u in this expression vanishes. Consequently, Equation (6.9.4) reduces to

$$u''y_1 + u'[2y_1' + a_1(x)y_1] = F(x)$$

which is a first-order linear differential equation for u'. We have therefore reduced the order of the differential equation, hence the name of the technique. If we let $w = u'$, then the preceding differential equation can be written as

$$w' + \left(\frac{2y_1'}{y_1} + a_1\right) w = \frac{F(x)}{y_1}. \qquad (6.9.5)$$

An integrating factor for (6.9.5) is

$$I(x) = e^{\int^x \left(\frac{2y_1'(s)}{y_1(s)} + a_1(s)\right) ds} = y_1^2(x) e^{\int^x a_1(s) ds}.$$

According to the technique developed in Section 1.6, multiplying Equation (6.9.5) by $I(x)$ reduces it to the integrable form

$$\frac{d}{dx}[I(x)w(x)] = \frac{I(x)F(x)}{y_1(x)},$$

Integrating the preceding equation and dividing by $I(x)$ yields

$$w(x) = \frac{1}{I(x)} \int^x \frac{I(s)F(s)}{y_1(s)} \, ds + \frac{c_1}{I(x)},$$

where c_1 is an integration constant. Since $w = u'$, we have

$$u'(x) = \frac{1}{I(x)} \int^x \frac{I(s)F(s)}{y_1(s)} \, ds + \frac{c_1}{I(x)}.$$

One more integration yields

$$u(x) = \int^x \frac{1}{I(t)} \int^t \frac{I(s)F(s)}{y_1(s)} \, ds \, dt + c_1 \int^x \frac{1}{I(s)} \, ds + c_2,$$

so that

$$\begin{aligned}
y(x) = u(x)y_1(x) &= c_1 y_1(x) \int^x \frac{1}{I(s)} \, ds \\
&+ c_2 y_1(x) + y_1(x) \int^x \frac{1}{I(t)} \int^t \frac{I(s)F(s)}{y_1(s)} \, ds \, dt
\end{aligned} \qquad (6.9.6)$$

Setting $F(s) = 0$ in this formula we can identify two nonproportional, and hence linearly independent, solutions to (6.9.2), namely

$$y(x) = y_1(x) \quad \text{and} \quad y(x) = y_1(x) \int^x \frac{1}{I(s)} \, ds.$$

Further, setting $c_1 = c_2 = 0$ in (6.9.6) yields the following particular solution to (6.9.1):

$$y_p(x) = y_1(x) \int^x \frac{1}{I(t)} \int^t \frac{I(s)F(s)}{y_1(s)} \, ds \, dt.$$

Consequently, Equation (6.9.6) gives the general solution to Equation (6.9.1). We have therefore established the following theorem.

Theorem 6.9.1   If $y = y_1(x)$ is a solution to

$$y'' + a_1(x)y' + a_2(x)y = 0$$

on an interval J, then substituting $y(x) = y_1(x)u(x)$ into

$$y'' + a_1(x)y' + a_2(x)y = F(x)$$

yields its general solution.

Example 6.9.2   Determine the general solution to

$$xy'' - 2y' + (2 - x)y = 0, \qquad x > 0, \qquad (6.9.7)$$

given that one solution is $y_1(x) = e^x$.

Solution: To determine the general solution we substitute

$$y(x) = y_1(x)u(x) = e^x u(x) \tag{6.9.8}$$

into (6.9.7). We first compute the appropriate derivatives.

$$y' = e^x (u' + u)$$
$$y'' = e^x (u'' + 2u' + u).$$

Substituting these expressions into Equation (6.9.7), we find that u must satisfy

$$x(u'' + 2u' + u) - 2(u' + u) + (2 - x)u = 0,$$

which simplifies to

$$xu'' + 2u'(x - 1) = 0,$$

or equivalently,

$$w' + \frac{2(x - 1)}{x} w = 0,$$

where $w = u'$. We can solve the preceding differential equation either as a linear equation or a separable equation. Separating the variables yields

$$\frac{w'}{w} = 2(x^{-1} - 1).$$

By integrating, we obtain

$$\ln |w| = 2(\ln x - x) + c,$$

which, upon exponentiation, can be written as

$$w = c_1 x^2 e^{-2x}.$$

Therefore,

$$u' = c_1 x^2 e^{-2x}.$$

Integration by parts gives

$$u(x) = -\tfrac{1}{4} c_1 e^{-2x} (1 + 2x + 2x^2) + c_2.$$

Substituting into (6.9.8) yields the general solution to (6.9.7), namely,

$$y(x) = c_1 e^{-x} (1 + 2x + 2x^2) + c_2 e^x.$$

where we have absorbed the factor of $-\tfrac{1}{4}$ into c_1.   □

Example 6.9.3   Determine the general solution to

$$x^2 y'' + 3xy' + y = 4 \ln x, \quad x > 0, \tag{6.9.9}$$

given that one solution to the associated homogeneous equation is $y(x) = x^{-1}$.

Solution: We let

$$y(x) = x^{-1}u(x) \tag{6.9.10}$$

where $u(x)$ is to be determined. Differentiating y twice yields

$$y' = x^{-1}u' - x^{-2}u, \qquad y'' = x^{-1}u'' - 2x^{-2}u' + 2x^{-3}u.$$

Substituting into Equation (6.9.9) and collecting terms, we obtain the following differential equation for u:

$$u'' + x^{-1}u' = 4x^{-1}\ln x,$$

or equivalently,

$$w' + x^{-1}w = 4x^{-1}\ln x. \tag{6.9.11}$$

where $w = u'$. An integrating factor for (6.9.11) is $I(x) = e^{\int x^{-1}dx} = x$, so that Equation (6.9.11) can be written in the equivalent form

$$\frac{d}{dx}(xw) = 4\ln x.$$

Integrating both sides with respect to x yields

$$xw = 4x(\ln x - 1) + c_1,$$

where c_1 is a constant. Thus,

$$w(x) = 4(\ln x - 1) + c_1 x^{-1}.$$

Consequently,

$$u'(x) = 4(\ln x - 1) + c_1 x^{-1},$$

which can be integrated directly to obtain

$$u(x) = 4x(\ln x - 2) + c_1 \ln x + c_2,$$

where c_2 is another integration constant. Inserting this expression for u into (6.9.10) yields

$$y(x) = 4(\ln x - 2) + c_1 x^{-1}\ln x + c_2 x^{-1}.$$

which is the general solution to Equation (6.9.9).   ☐

Exercises for 6.9

Skills

- Be able to carry out the reduction of order technique to find the general solution to a second-order linear differential equation.

Problems

For Problems 1–6, y_1 is a solution to the given differential equation. Use the method of reduction of order to determine a second linearly independent solution and write the general solution.

1. $x^2 y'' - 3xy' + 4y = 0$, $x > 0$, $y_1(x) = x^2$.

2. $x^2 y'' - 2xy' + (x^2 + 2)y = 0$, $x > 0$, $y_1(x) = x \sin x$.

3. $xy'' + (1-2x)y' + (x-1)y = 0$, $x > 0$, $y_1(x) = e^x$.

4. $y'' - x^{-1}y' + 4x^2 y = 0$, $x > 0$, $y_1(x) = \sin(x^2)$.

5. $(1 - x^2)y'' - 2xy' + 2y = 0$, $-1 < x < 1$, $y_1(x) = x$.

6. $4x^2y'' + 4xy' + (4x^2 - 1)y = 0, \quad x > 0, \quad y_1(x) = x^{-1/2}\sin x.$

7. Consider the **Cauchy-Euler equation**

$$x^2y'' - (2m-1)xy' + m^2y = 0, \quad x > 0,$$
$$(6.9.12)$$

where m is a constant.

   (a) Determine a particular solution to Equation (6.9.12) of the form $y_1(x) = x^r$.

   (b) Use your solution from (a) and the method of reduction of order to obtain a second linearly independent solution.

8. Determine the values of the constants $a_0, a_1,$ and a_2 such that

$$y(x) = a_0 + a_1 x + a_2 x^2$$

is a solution to

$$(4 + x^2)y'' - 2y = 0,$$

and use the reduction of order technique to find a second linearly independent solution.

9. Consider the differential equation

$$xy'' - (\alpha x + \beta)y' + \alpha\beta y = 0, \quad x > 0, \quad (6.9.13)$$

where α and β are constants.

   (a) Show that $y_1(x) = e^{\alpha x}$ is a solution to Equation (6.9.13).

   (b) Use reduction of order to derive the second linearly independent solution

$$y_2(x) = e^{\alpha x}\int x^\beta e^{-\alpha x}\,dx.$$

   (c) In the particular case when $\alpha = 1$ and β is a nonnegative integer, show that a second linearly independent solution to Equation (6.9.13) is

$$y_2(x) = 1 + x + \frac{1}{2!}x^2 + \cdots + \frac{1}{\beta!}x^\beta.$$

For Problems 10–15, y_1 is a solution to the associated homogeneous equation. Use the method of reduction of order to determine the general solution to the given differential equation.

10. $y'' - 6y' + 9y = 15e^{3x}\sqrt{x}, \quad x > 0, \quad y_1(x) = e^{3x}.$

11. $y'' - 4y' + 4y = 4e^{2x}\ln x, \quad x > 0, \quad y_1(x) = e^{2x}.$

12. $4x^2y'' + y = \sqrt{x}\ln x, \quad x > 0, \quad y_1(x) = x^{1/2}.$

13. $y'' + y = \csc x, \quad 0 < x < \pi, \quad y_1(x) = \sin x.$

14. $xy'' - (2x+1)y' + 2y = 8x^2e^{2x}, \quad x > 0, \quad y_1(x) = e^{2x}.$

15. $x^2y'' - 3xy' + 4y = 8x^4, \quad x > 0, \quad y_1(x) = x^2.$

16. Consider the differential equation

$$y'' + p(x)y' + q(x)y = r(x), \quad (6.9.14)$$

where p, q and r are continuous on an interval I. If $y = y_1(x)$ is a solution to the associated homogeneous equation, show that $y_2(x) = u(x)y_1(x)$ is a solution to Equation (6.9.14), provided $v = u'$ is a solution to the linear differential equation

$$v' + \left(2\frac{y_1'}{y_1} + p\right)v = \frac{r}{y_1}.$$

Express the solution to Equation (6.9.14) in terms of integrals. Identify two linearly independent solutions to the associated homogeneous equation and a particular solution to Equation (6.9.14).

6.10   Chapter Review

In this chapter we have studied the general theory for linear differential equations of arbitrary order n. To do so, we use the **linear differential operator of order** n,

$$L = D^n + a_1(x)D^{n-1} + \cdots + a_{n-1}(x)D + a_n(x),$$

where $D(y) = y'$ for a differentiable function y on an interval I. Thus,

$$Ly = y^{(n)} + a_1(x)y^{(n-1)} + \cdots + a_{n-1}(x)y' + a_n(x)y.$$

A homogeneous linear differential equation of order n can then be expressed as

$$Ly = 0,$$

and the solution set is an n-dimensional vector space. Therefore, the general solution to $Ly = 0$ can be constructed from any n linearly independent solutions to the differential equation.

If L is a polynomial differential operator $P(D)$ (with constant-coefficients), then a real root r of the equation $P(D) = 0$ of multiplicity m contributes linearly independent solutions

$$e^{rx}, \quad xe^{rx}, \quad \ldots, \quad x^{m-1}e^{rx}.$$

On the other hand, a complex root $r = a + ib$ ($b \neq 0$) of multiplicity m contributes $2m$ solutions

$$e^{ax}\cos bx, \quad xe^{ax}\cos bx, \quad \ldots, \quad x^{m-1}e^{ax}\cos bx,$$
$$e^{ax}\sin bx, \quad xe^{ax}\sin bx, \quad \ldots, \quad x^{m-1}e^{ax}\sin bx.$$

Note that $r = a - bi$ necessarily will occur with multiplicity m as well, but contributes the same $2m$ solutions as $r = a + ib$.

The general solution to a *nonhomogeneous* differential equation $Ly = F(x)$ has the form

$$y(x) = y_c(x) + y_p(x),$$

where $y_c(x)$ is the complementary function (which solves the corresponding homogeneous differential equation) and $y_p(x)$ is any particular solution to the original nonhomogeneous differential equation. We have discussed two basic methods for determining a particular solution $y_p(x)$, undetermined coefficients and variation-of-parameters.

Method of Undetermined Coefficients: Annihilators

For differential equations of the form $P(D)y = F(x)$ with

$$F(x) = \begin{cases} cx^k e^{ax}, \\ cx^k e^{ax}\sin bx, \\ cx^k e^{ax}\cos bx, \end{cases}$$

or sums of these types, we can find an **annihilator** $A(D)$ of F (i.e., such that $A(D)F = 0$). For instance, $A(D) = (D - a)^{k+1}$ annihilates $cx^k e^{ax}$, while $A(D) = (D^2 - 2aD + a^2 + b^2)^{k+1}$ annihilates $cx^k e^{ax}\cos bx$ and $cx^k e^{ax}\sin bx$. If $F(x)$ is given by a sum of the forms above, then it is annihilated by the corresponding *product* of annihilators.

By operating on the differential equation $P(D)y = F(x)$ by the annihilator $A(D)$ of F, we obtain the homogeneous differential equation $A(D)P(D)y = 0$, whose general solution can be readily obtained by determining the roots of the equation $A(D)P(D) = 0$. By removing terms of the general solution to $A(D)P(D)y = 0$ that coincide with the complementary function for the differential equation $P(D)y = F(x)$, the remaining terms provide the form of a particular solution y_p, also known as a **trial solution**, to the nonhomogeneous differential equation. This trial solution provides the correct form for a particular solution, but contains **undetermined coefficients** that can be computed by substituting the trial solution into the differential equation. The table at the end of Section 6.3 shows the appropriate trial solution to use if $F(x)$ is of any of the forms mentioned above.

Method of Variation-of-Parameters

Here we obtain a particular solution to obtain $Ly = F(x)$ of the form

$$y_p(x) = u_1(x)y_1(x) + u_2(x)y_2(x) + \cdots + u_n(x)y_n(x),$$

where n is the degree of L, $y_1, y_2, \ldots, y_n$ are n linearly independent solutions to the corresponding homogeneous differential equation $Ly = 0$, and $u_1, u_2, \ldots, u_n$ are chosen to satisfy Equations (6.7.22).

Applications

In Sections 6.5 and 6.6, we have considered some applications of second-order nonhomogeneous linear differential equations. First we explored the motion of a mechanical system consisting of a mass attached to a spring under the influence of a variety of possible forces. Next we saw that the mathematics of a spring–mass system coincides with that of an RLC circuit, given an appropriate renaming of the variables.

Cauchy-Euler Equations

A second-order Cauchy-Euler differential equation has the form

$$x^2 y'' + a_1 x y' + a_2 y = 0, \qquad x > 0,$$

and is an important example of a differential equation with nonconstant coefficients. Two linearly independent solutions to the Cauchy-Euler equation can be obtained by substituting $y(x) = x^r$ into the differential equation. The text also discusses a natural generalization to higher-order Cauchy-Euler equations.

Reduction of Order

In the final section of this chapter, a powerful method is used to determine the general solution to a second-order nonhomogeneous differential equation, once one solution to the associated homogeneous differential equation has been obtained. The procedure described actually reduces the problem to solving a first-order linear differential equation.

Additional Problems

In Problems 1–6, find Ly for the given differential operator L and the given function y.

1. $L = D^2 + 3, \quad y(x) = e^{x^3}$.

2. $L = 5, \quad y(x) = \dfrac{1}{1 + x^2}$.

3. $L = \dfrac{1}{x} D^2 + xD - 2, \quad y(x) = 4\sin x$.

4. $L = x^2 D^3 - \sin x D, \quad y(x) = e^{2x} + \cos x$.

5. $L = (x^2 + 1)D^3 - (\cos x)D + 5x^2, \quad y(x) = \ln x + 8x^5$.

6. $L = 4x^2 D, y(x) = \sin^2(x^2 + 1)$.

In Problems 7–13, find the general solution to the given differential equation.

7. $y''' + 3y'' - 4y = 0$.

8. $y''' + 11y'' + 36y' + 26y = 0$. [**Hint:** $r = -1$ is a root of the auxiliary polynomial.]

9. $y^{(iv)} + 13y'' + 36y = 0$.

10. $y''' + 10y'' + 25y' = 0$.

11. $(D + 3)^3(D^2 - 4D + 13)y = 0$.

12. $(D^2 - 2D + 2)^3 y = 0$.

13. $(D^2 + 4D + 4)(D - 3)y = 0$.

In Problems 14–17, determine the annihilator of the given function.

14. $F(x) = e^{-x} + x$.

15. $F(x) = e^{3x} \sin x$.

16. $F(x) = x^5 \cos 4x$.

17. $F(x) = x \sin x + e^{-2x}$.

In Problems 18–23, determine a trial solution for the given nonhomogeneous differential equation.

18. $y'' + 6y' + 9y = 4e^{-3x}$.

19. $y'' + 6y' + 9y = 4e^{-2x}$.

20. $y''' - 6y'' + 25y' = x^2$.

21. $y''' - 6y'' + 25y' = \sin 4x$.

22. $y''' + 9y'' + 24y' + 16y = 8e^{-x} + 1$.

23. $y^{(vi)} + 3y^{(iv)} + 3y'' + y = 2\sin x$.

For Problems 24–29, solve the given nonhomogeneous differential equation by using **(a)** the method of annihilators, and **(b)** the variation-of-parameters method.

24. The differential equation in Problem 19.

25. The differential equation in Problem 20.

26. The differential equation in Problem 21.

27. $y'' - 4y = 5e^x$.

28. $y'' + 2y' + y = 2xe^{-x}$.

29. $y'' - y = 4e^x$.

In Problems 30–39, state whether the annihilator method can be used to determine a particular solution to the given differential equation. If the technique cannot be used, state why not. If the technique can be used, then give an appropriate trial solution.

30. $y'' + xy = \sin x$.

31. $y'' + 4y = \ln x$.

32. $y'' + 2y' - 3y = 5e^x$.

33. $y'' + y = \tan x$.

34. $y'' + y = 4\cos 2x + 3e^x$.

35. $y'' - 8y' + 16y = 7e^{4x}$.

36. $x^2 y'' + 5xy' + 7y = 3e^x$.

37. $y'' - 2y' + 5y = 7e^x \cos x + \sin x$.

38. $y'' + 4y = 7\cos^2 x$.

39. $\dfrac{d^2 y}{dt^2} - 2a\dfrac{dy}{dt} + (a^2 + b^2)y = e^{at}(4t + \cos bt)$, where a and b are positive constants.

In Problems 40–45, use the annihilator method to solve the given differential equation.

40. $y''' + 4y = 7e^x$.

41. $y'' + 2y' - 3y = 2xe^{-3x}$.

42. $y'' + 4y' = 4x^2$.

43. $y'' + 4y = 8\cos 2x$.

44. $y'' - 8y' + 16y = 5e^{4x}$.

45. $y'' - y = 3e^{2x} + \sin x$.

46. Solve the initial-value problem.

$$y'' - y' - 2y = 15e^{2x},$$
$$y(0) = 0, \quad y'(0) = 8.$$

In Problems 47–51, use the variation-of-parameters method to solve the given differential equation.

47. $y'' + y = \dfrac{1}{\sin x}$.

48. $y'' + y = \tan x$.

49. $y'' - 2my' + m^2 y = e^{mx} \ln x, x > 0$, where m is a real constant.

50. $y'' + 2y' + y = x^{-1}e^x$.

51. $y'' - 2y' + y = e^x \ln x, x > 0$.

52. Solve Problem 49 by the *reduction of order* method, given that $y_1(x) = e^{mx}$ is a solution to the associated homogeneous differential equation.

For Problems 53–58, find the general solution to the given differential equation on the interval $(0, \infty)$.

53. $x^2 y'' + 9xy' + 16y = 0$.

54. $x^2 y'' + 9xy' + 15y = 0$.

55. $x^2 y'' - 11xy' + 37y = 0$.

56. $x^2 y'' + xy' + 25y = 0$.

57. $x^2 y'' - 2xy' - 18y = 0$.

58. $x^2 y'' - xy' = 0$.

For Problems 59–61, solve the given differential equation on the interval $x > 0$. Use the variation-of-parameters technique to obtain a particular solution.

59. $x^2 y'' + 9xy' + 16y = x^{-3}$.

60. $x^2 y'' - 3xy' - 12y = x^4 + 5x^2$.

61. $x^2 y'' - 5xy' + 10y = x^3$.

For Problems 62–66, find a particular solution to the given differential equation.

62. $y'' - 4y' - 5y = e^{3x} \sin 2x$.

63. $y'' + 4y' + 4y = \dfrac{\ln x}{xe^{2x}}$.

64. $y'' - 2y' + 26y = e^x \cos 5x$.

65. $y'' - 9y' + 20y = x^3 e^{5x}$.

66. $y'' - 8y' + 17y = e^{4x} \csc x$.

CHAPTER
7

Systems of Differential Equations

Go down deep enough into anything and you will find mathematics. —
Dean Schlicter

In practice, most applied problems involve more than one unknown function for their formulation and hence require the solution to a system of differential equations. Perhaps the simplest way to see how systems naturally arise is to consider the motion of an object in space. If this object has mass m and is moving under the influence of a force $\mathbf{F}(t) = (F_1(t), F_2(t), F_3(t))$, then, according to Newton's second law of motion, the position of the object at time t, $(x(t), y(t), z(t))$, is obtained by solving the system

$$m\frac{d^2x}{dt^2} = F_1, \qquad m\frac{d^2y}{dt^2} = F_2, \qquad m\frac{d^2z}{dt^2} = F_3.$$

In this chapter we consider the formulation and solution of systems of differential equations. The majority of the chapter is concerned with *linear* systems of differential equations. In this case, the following familiar questions need addressing:

Question 1: How can we formulate problems in a way suitable for solution?
Question 2: How many solutions, if any, does our linear system of differential equations possess?
Question 3: How do we find the solutions that arise in Question 2?

For Question 1, we have already seen some examples in Chapters 2 and 3 of how to formulate applied problems in a fruitful way by using vectors and matrices. Answering Question 2 will once more require the vector space techniques from Chapters 4 and 5, whereas, in the case when our linear systems have constant coefficients, we will find an elegant answer to Question 3 using eigenvalues and eigenvectors of appropriate matrices.

Before beginning the general development of the theory for systems of differential equations, we consider two physical problems that can be formulated mathematically in terms of such systems.

Consider the coupled spring–mass system that consists of two masses m_1, m_2 connected by two springs whose spring constants are k_1 and k_2, respectively. (See Figure 7.0.1.)

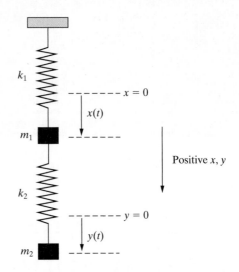

Figure 7.0.1: A coupled spring–mass system.

Let $x(t)$ and $y(t)$ denote the displacement of m_1 and m_2, respectively, from their positions when the system is in the static equilibrium position. Then, using Hooke's law and Newton's second law, it follows that the motion of the masses is governed by the system of differential equations

$$m_1 \frac{d^2x}{dt^2} = -k_1 x + k_2(y - x),$$

$$m_2 \frac{d^2y}{dt^2} = -k_2(y - x).$$

We would expect the problem to have a unique solution once we have specified the initial positions and velocities of the masses.

As a second example, consider the mixing problem depicted in Figure 7.0.2. Two tanks contain a solution consisting of chemical dissolved in water. A solution containing c grams/L of the chemical flows into tank 1 at a rate of r L/minute, and the solution in tank 2 flows out at the same rate. In addition, the solution flows into tank 1 from tank 2 at a rate of r_{12} L/minute and into tank 2 from tank 1 at a rate of r_{21} L/minute.

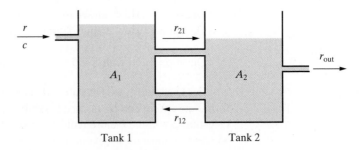

Figure 7.0.2: A mixing problem.

We wish to determine the amounts of chemical $A_1(t)$ and $A_2(t)$ in tanks 1 and 2 at any time t. Analysis similar to that used in Section 1.7 yields the following system of differential equations governing the behavior of A_1 and A_2:

$$\frac{dA_1}{dt} = -\frac{r_{21}}{V_1}A_1 + \frac{r_{12}}{V_2}A_2 + cr,$$

$$\frac{dA_2}{dt} = \frac{r_{21}}{V_1}A_1 - \frac{(r_{12} + r)}{V_2}A_2,$$

where V_1 and V_2 denote the volume of solution in each tank at time t.

We will give a full discussion of both of the foregoing problems in Section 7.7, once we have developed the theory and solution techniques for linear systems of differential equations.

7.1   First-Order Linear Systems

We first focus our attention on *linear* systems of differential equations, sometimes called **linear differential systems**. Once such a system has been appropriately formulated, vector space methods can be applied to derive the complete theory regarding their solution properties.

DEFINITION   7.1.1

A system of differential equations of the form

$$\begin{aligned}
\frac{dx_1}{dt} &= a_{11}(t)x_1(t) + a_{12}(t)x_2(t) + \cdots + a_{1n}(t)x_n(t) + b_1(t), \\
\frac{dx_2}{dt} &= a_{21}(t)x_1(t) + a_{22}(t)x_2(t) + \cdots + a_{2n}(t)x_n(t) + b_2(t), \\
&\ \vdots \\
\frac{dx_n}{dt} &= a_{n1}(t)x_1(t) + a_{n2}(t)x_2(t) + \cdots + a_{nn}(t)x_n(t) + b_n(t),
\end{aligned}$$

(7.1.1)

where the $a_{ij}(t)$ and $b_i(t)$ are specified functions on an interval I, is called a **first-order linear system**. If $b_1 = b_2 = \cdots = b_n = 0$, then the system is called **homogeneous**. Otherwise, it is called **nonhomogeneous**.

Remarks

1. It is important to notice the structure of a first-order linear system. The highest derivative occurring in such a system is a first derivative. Further, there is precisely one equation involving the derivative of each separate unknown function. Finally, the terms that appear on the right-hand side of the equations do *not* involve any derivatives and are linear in the unknown functions $x_1, x_2, \ldots, x_n$.

2. We will usually denote dx_i/dt by x_i'.

Example 7.1.2   An example of a nonhomogeneous first-order linear system is

$$\begin{aligned}
x_1' &= e^t x_1 + t^2 x_2 + \sin t, \\
x_2' &= t x_1 + 3x_2 - \cos t.
\end{aligned}$$

The associated homogeneous system is

$$\begin{aligned}
x_1' &= e^t x_1 + t^2 x_2, \\
x_2' &= t x_1 + 3x_2.
\end{aligned}$$

$\square$

> **DEFINITION 7.1.3**
>
> By a **solution** to the system (7.1.1) on an interval I we mean an ordered n-tuple of functions $x_1(t), x_2(t), \ldots, x_n(t)$, which, when substituted into both sides of the system, yield the same result for all t in I.

Example 7.1.4   Verify that

$$x_1(t) = -2e^{5t} + 4e^{-t}, \qquad x_2(t) = e^{5t} + e^{-t} \tag{7.1.2}$$

is a solution to the linear system of differential equations

$$x_1' = x_1 - 8x_2 \tag{7.1.3}$$
$$x_2' = -x_1 + 3x_2 \tag{7.1.4}$$

on $(-\infty, \infty)$.

Solution: From (7.1.2) it follows that the left-hand side of Equation (7.1.3) is

$$x_1'(t) = -10e^{5t} - 4e^{-t},$$

whereas the right-hand side is

$$x_1(t) - 8x_2(t) = (-2e^{5t} + 4e^{-t}) - 8(e^{5t} + e^{-t}) = -10e^{5t} - 4e^{-t}.$$

Consequently,

$$x_1' = x_1 - 8x_2,$$

so that Equation (7.1.3) is satisfied by the given functions for all $t \in (-\infty, \infty)$. Similarly, it is easily shown that Equation (7.1.4) is also satisfied. It follows that x_1 and x_2 do define a solution to the given system on $(-\infty, \infty)$.   □

We now derive a simple technique for solving the system (7.1.1) that can be used when the coefficients $a_{ij}(t)$ in the system are constants. Although we will develop a superior technique for such systems in the later sections, the method introduced here does have importance and will be useful in motivating some of the subsequent results. For simplicity, we will consider only $n = 2$. Under the assumption that all a_{ij} are constants, the system (7.1.1) reduces to

$$x_1' = a_{11}x_1 + a_{12}x_2 + b_1(t),$$
$$x_2' = a_{21}x_1 + a_{22}x_2 + b_2(t).$$

This system can be written in the equivalent form

$$(D - a_{11})x_1 - a_{12}x_2 = b_1(t), \tag{7.1.5}$$
$$-a_{21}x_1 + (D - a_{22})x_2 = b_2(t), \tag{7.1.6}$$

where D is the differential operation d/dt. The idea behind the solution technique is that we can now easily eliminate x_2 between these two equations by operating on Equation

(7.1.5) with $D - a_{22}$, multiplying Equation (7.1.6) by a_{12}, and adding the resulting equations. This yields a second-order constant-coefficient linear differential equation for x_1 only, which can be solved using the techniques of Chapter 6. Substituting the expression thereby obtained for x_1 into Equation (7.1.5) will then yield x_2.[1] We illustrate the technique with an example.

Example 7.1.5   Solve the system

$$x_1' = x_1 + 2x_2 \tag{7.1.7}$$
$$x_2' = 2x_1 - 2x_2. \tag{7.1.8}$$

Solution: We begin by rewriting the system in operator form as

$$(D - 1)x_1 - \quad 2x_2 = 0, \tag{7.1.9}$$
$$- 2x_1 + (D + 2)x_2 = 0. \tag{7.1.10}$$

To eliminate x_2 between these two equations, we first operate on Equation (7.1.9) with $D + 2$ to obtain

$$(D + 2)(D - 1)x_1 - 2(D + 2)x_2 = 0.$$

Adding twice Equation (7.1.10) to this equation eliminates x_2 and yields

$$(D + 2)(D - 1)x_1 - 4x_1 = 0.$$

That is,

$$(D^2 + D - 6)x_1 = 0.$$

This constant-coefficient differential equation has auxiliary polynomial

$$P(r) = r^2 + r - 6 = (r + 3)(r - 2).$$

Consequently,

$$x_1(t) = c_1 e^{-3t} + c_2 e^{2t}. \tag{7.1.11}$$

We now determine x_2. From Equation (7.1.9), we have

$$x_2(t) = \tfrac{1}{2}(D - 1)x_1.$$

Inserting the expression for x_1 from (7.1.11) into the previous equation yields

$$x_2(t) = \tfrac{1}{2}(Dx_1 - x_1) = \tfrac{1}{2}(-4c_1 e^{-3t} + c_2 e^{2t}).$$

Hence, the solution to the system of differential equations (7.1.7) and (7.1.8) is

$$x_1(t) = c_1 e^{-3t} + c_2 e^{2t}, \qquad x_2(t) = \tfrac{1}{2}(-4c_1 e^{-3t} + c_2 e^{2t}),$$

where c_1 and c_2 are arbitrary constants. $\square$

In solving an applied problem that is governed by a system of differential equations, we usually require the particular solution to the system that corresponds to the specific

---

[1] If $a_{12} = 0$, we can determine x_1 directly from Equation (7.1.5) and then determine x_2 from Equation (7.1.6).

problem of interest. Such a particular solution is obtained by specifying appropriate auxiliary conditions. This leads to the idea of an initial-value problem for linear systems.

DEFINITION 7.1.6

Solving the system (7.1.1) subject to n auxiliary conditions imposed at the *same* value of the independent variable is called an **initial-value problem**. Thus, the general form of the auxiliary conditions for an initial-value problem is:

$$x_1(t_0) = \alpha_1, \quad x_2(t_0) = \alpha_2, \quad \ldots, \quad x_n(t_0) = \alpha_n,$$

where $\alpha_1, \alpha_2, \ldots, \alpha_n$ are constants.

Example 7.1.7   Solve the initial-value problem

$$x_1' = x_1 + 2x_2, \qquad x_2' = 2x_1 - 2x_2,$$
$$x_1(0) = 1, \qquad x_2(0) = 0.$$

Solution: We have already seen in the previous example that the solution to the given system of differential equations is

$$x_1(t) = c_1 e^{-3t} + c_2 e^{2t}, \qquad x_2(t) = \tfrac{1}{2}(-4c_1 e^{-3t} + c_2 e^{2t}), \qquad (7.1.12)$$

where c_1 and c_2 are arbitrary constants. Imposing the two initial conditions yields the following equations for determining c_1 and c_2:

$$c_1 + c_2 = 1,$$
$$-4c_1 + c_2 = 0.$$

Consequently,

$$c_1 = \tfrac{1}{5} \quad \text{and} \quad c_2 = \tfrac{4}{5}.$$

Substituting for c_1 and c_2 into (7.1.12) yields the unique solution

$$x_1(t) = \tfrac{1}{5}(e^{-3t} + 4e^{2t}), \qquad x_2(t) = \tfrac{2}{5}(e^{2t} - e^{-3t}). \qquad \square$$

It might appear that restricting to *first-order* linear systems means that we are considering only very special types of linear differential equations. In fact this is incorrect, since most systems of k differential equations that are linear in k unknown functions and their derivatives can be rewritten as equivalent first-order systems by redefining the dependent variables. We illustrate with an example.

Example 7.1.8   Rewrite the linear system

$$\frac{d^2x}{dt^2} - 4y = e^t \qquad (7.1.13)$$

$$\frac{d^2y}{dt^2} + t^2 \frac{dx}{dt} = \sin t, \qquad (7.1.14)$$

as an equivalent first-order system.

Solution: We introduce new dependent variables relative to which Equations (7.1.13) and (7.1.14) reduce to first-order differential equations. Let

$$x_1 = x, \qquad x_2 = \frac{dx}{dt}, \qquad x_3 = y, \qquad x_4 = \frac{dy}{dt}. \qquad (7.1.15)$$

Then Equations (7.1.13) and (7.1.14) can be replaced by

$$\frac{dx_2}{dt} - 4x_3 = e^t, \qquad \frac{dx_4}{dt} + t^2 x_2 = \sin t.$$

These equations must also be supplemented with equations for x_1 and x_3. From (7.1.15), we see that

$$\frac{dx_1}{dt} = x_2, \qquad \frac{dx_3}{dt} = x_4.$$

Consequently, the given system of differential equations is equivalent to the first-order linear system

$$\frac{dx_1}{dt} = x_2, \qquad \frac{dx_2}{dt} = 4x_3 + e^t, \qquad \frac{dx_3}{dt} = x_4, \qquad \frac{dx_4}{dt} = -t^2 x_2 + \sin t. \quad \square$$

Finally, consider the general nth-order linear differential equation

$$y^{(n)} + a_1(t)y^{(n-1)} + \cdots + a_{n-1}(t)y' + a_n(t)y = F(t). \qquad (7.1.16)$$

If we introduce the new variables $x_1, x_2, \ldots, x_n$, defined by

$$x_1 = y, \qquad x_2 = y', \qquad \ldots, \qquad x_n = y^{(n-1)},$$

then Equation (7.1.16) can be replaced by the equivalent first-order linear system

$$x_1' = x_2, \qquad x_2' = x_3, \qquad \ldots, \qquad x_{n-1}' = x_n,$$

$$x_n' = -a_n(t)x_1 - a_{n-1}(t)x_2 - \cdots - a_1(t)x_n + F(t).$$

Consequently, any nth-order linear differential equation can be replaced by an equivalent system of *first-order* differential equations.

Example 7.1.9   Write the following differential equation as an equivalent first-order system:

$$\frac{d^2 y}{dt^2} + 4e^t \frac{dy}{dt} - 9t^2 y = 7t^2.$$

Solution: We introduce new variables x_1 and x_2 defined by

$$x_1 = y, \qquad x_2 = \frac{dy}{dt}.$$

Then the given differential equation can be replaced by the first-order system

$$\frac{dx_1}{dt} = x_2, \qquad \frac{dx_2}{dt} = 9t^2 x_1 - 4e^t x_2 + 7t^2. \qquad \square$$

Exercises for 7.1

Key Terms

First-order linear system, Homogeneous linear system, Non-homogeneous linear system, Solution to a first-order linear system, Initial-value problem.

Skills

- Be able to use differential operators to solve a first-order linear system of differential equations.

- Be able to use differential operators together with initial conditions to solve an initial-value problem consisting of a first-order linear system.

- Be able to convert higher-order linear systems of differential equations into a first-order linear system by introducing new variables.

- Be able to convert a first-order linear system of differential equations into a single higher-order differential equation that can be solved by the techniques of the previous chapter.

True-False Review

For Questions 1–10, decide if the given statement is **true** or **false**, and give a brief justification for your answer. If true, you can quote a relevant definition or theorem from the text. If false, provide an example, illustration, or brief explanation of why the statement is false.

1. The system of differential equations $x' = t^2x - ty$, $y' = (\sin t)x + 5y$, is a first-order linear system of differential equations.

2. The system of differential equations $x' = t^4x - e^t y + 4$, $y' = (t^2 + 3)x - t^2$, is a first-order linear system of differential equations.

3. The system of differential equations $x' = txy + 5y - 3$, $y' = x - y - 1$, is a first-order linear system of differential equations.

4. The system of differential equations $x' = -2x$, $y' = e^y x + e^t y$, is a first-order linear system of differential equations.

5. A third-order linear differential equation can be replaced by a first-order linear system consisting of three differential equations.

6. A first-order linear system of two differential equations can be solved by converting it into a second-order linear differential equation.

7. The first-order linear system $x' = x + y$, $y' = x - y$, with auxiliary conditions $x(0) = 0$, $y(1) = 1$, is an initial-value problem.

8. The first-order linear system $x' = -2x - 3y$, $y' = 5x - y$, with auxiliary conditions $x(0) = 2$, $y(0) = 6$, is an initial-value problem.

9. The first-order linear system $x' = e^t x - t^3 y$, $y' = 4x - 5t^2 y$, with auxiliary condition $x(2) = 5$, is an initial-value problem.

10. The first-order linear system $x' = t^2x + 3y$ $y' = 3x - ty$, with auxillary conditions $x(3) = 7$, $y(-3) = -7$, is an initial-value problem.

Problems

For Problems 1–7, solve the given system of differential equations.

1. $x_1' = 2x_1 - 3x_2$, $x_2' = x_1 - 2x_2$.

2. $x_1' = 4x_1 + 2x_2$, $x_2' = -x_1 + x_2$.

3. $x_1' = 2x_1 + 4x_2$, $x_2' = -4x_1 - 6x_2$.

4. $x_1' = 2x_2$, $x_2' = -2x_1$.

5. $x_1' = x_1 - 3x_2$, $x_2' = 3x_1 + x_2$.

6. $x_1' = 2x_1$, $x_2' = x_2 - x_3$, $x_3' = x_2 + x_3$.

7. $x_1' = -2x_1 + x_2 + x_3$, $x_2' = x_1 - x_2 + 3x_3$, $x_3' = -x_2 - 3x_3$.

For Problems 8–10, solve the given initial-value problem.

8. $x_1' = 2x_2$, $x_2' = x_1 + x_2$, $x_1(0) = 3$, $x_2(0) = 0$.

9. $x_1' = 2x_1 + 5x_2$, $x_2' = -x_1 - 2x_2$, $x_1(0) = 0$, $x_2(0) = 1$.

10. $x_1' = 2x_1 + x_2$, $x_2' = -x_1 + 4x_2$, $x_1(0) = 1$, $x_2(0) = 3$.

For Problems 11–13, solve the given nonhomogeneous system.

11. $x_1' = x_1 + 2x_2 + 5e^{4t}$, $x_2' = 2x_1 + x_2$.

12. $x_1' = -2x_1 + x_2 + t$, $x_2' = -2x_1 + x_2 + 1$.

13. $x_1' = x_1 + x_2 + e^{2t}$, $x_2' = 3x_1 - x_2 + 5e^{2t}$.

For Problems 14–15, convert the given system of differential equations to a first-order linear system.

14. $\dfrac{dx}{dt} - ty = \cos t, \quad \dfrac{d^2y}{dt^2} - \dfrac{dx}{dt} + x = e^t.$

15. $\dfrac{d^2x}{dt^2} - 3\dfrac{dy}{dt} + x = \sin t, \quad \dfrac{d^2y}{dt^2} - t\dfrac{dx}{dt} - e^t y = t^2.$

For Problems 16–18, convert the given linear differential equation to a first-order linear system.

16. $y'' + 2ty' + y = \cos t.$

17. $y'' + ay' + by = F(t), \quad a, b$ constants.

18. $y''' + t^2 y' - e^t y = t.$

19. The initial-value problem that governs the behavior of a coupled spring–mass system is (see the introduction to Chapter 7)

$$m_1\dfrac{d^2x}{dt^2} = -k_1 x + k_2(y - x),$$
$$m_2\dfrac{d^2y}{dt^2} = -k_2(y - x),$$
$$x(0) = \alpha_1, \quad x'(0) = \alpha_2, \quad y(0) = \alpha_3, \quad y'(0) = \alpha_4,$$

where $\alpha_1, \alpha_2, \alpha_3,$ and α_4 are constants. Convert this problem into an initial-value problem for an equivalent first-order linear system. (You must give the appropriate initial conditions in the new variables.)

20. Solve the initial-value problem:

$$x_1' = -(\tan t)x_1 + 3\cos^2 t,$$
$$x_2' = x_1 + (\tan t)x_2 + 2\sin t,$$
$$x_1(0) = 4, \quad x_2(0) = 0.$$

7.2　Vector Formulation

The first step in developing the general theory for first-order linear systems is to formulate the problem of solving such a system as an appropriate vector-space problem. The key to this formulation is the realization that the scalar system of equations

$$
\begin{aligned}
x_1' &= a_{11}(t)x_1(t) + a_{12}(t)x_2(t) + \cdots + a_{1n}(t)x_n(t), \\
x_2' &= a_{21}(t)x_1(t) + a_{22}(t)x_2(t) + \cdots + a_{2n}(t)x_n(t), \\
&\vdots \\
x_n' &= a_{n1}(t)x_1(t) + a_{n2}(t)x_2(t) + \cdots + a_{nn}(t)x_n(t),
\end{aligned}
\tag{7.2.1}
$$

can be written as the equivalent vector equation

$$\boxed{\mathbf{x}'(t) = A(t)\mathbf{x}(t) + \mathbf{b}(t),} \tag{7.2.2}$$

where

$$
\mathbf{x}(t) = \begin{bmatrix} x_1(t) \\ x_2(t) \\ \vdots \\ x_n(t) \end{bmatrix}, \qquad
\mathbf{x}'(t) = \begin{bmatrix} x_1'(t) \\ x_2'(t) \\ \vdots \\ x_n'(t) \end{bmatrix}
$$

and

$$
A(t) = \begin{bmatrix}
a_{11}(t) & a_{12}(t) & \cdots & a_{1n}(t) \\
a_{21}(t) & a_{22}(t) & \cdots & a_{2n}(t) \\
\vdots & \vdots & & \vdots \\
a_{n1}(t) & a_{n2}(t) & \cdots & a_{nn}(t)
\end{bmatrix}, \qquad
\mathbf{b}(t) = \begin{bmatrix} b_1(t) \\ b_2(t) \\ \vdots \\ b_n(t) \end{bmatrix}.
$$

Notice that $\mathbf{x}, \mathbf{x}',$ and $\mathbf{b}$ in (7.2.2) are column n-vector functions. We let $V_n(I)$ denote the set of all column n-vector functions defined on an interval I, and define addition and scalar multiplication within this set in the same manner as for column vectors. The following result concerning $V_n(I)$ will be needed in the remaining sections:

Theorem 7.2.1　The set $V_n(I)$ is a vector space.

Proof Verifying that $V_n(I)$ together with the operations of addition and scalar multiplication just defined satisfies Definition 4.2.1 is left as an exercise (Problem 10). ∎

Since $V_n(I)$ is a vector space, we can discuss linear dependence and linear independence of column vector functions. We first need a definition.

DEFINITION   7.2.2

Let $x_1(t), x_2(t), \ldots, x_n(t)$ be vectors in $V_n(I)$. Then the **Wronskian** of these vector functions, denoted $W[x_1, x_2, \ldots, x_n](t)$, is defined by

$$W[x_1, x_2, \ldots, x_n](t) = \det([x_1(t), x_2(t), \ldots, x_n(t)]).$$

Remark   Notice that the Wronskian introduced in this definition refers to column vector functions in the vector space $V_n(I)$, whereas the Wronskian defined previously in the text refers to functions in $C^n(I)$. The relationship between these two Wronskians is investigated in Problem 13.

Example 7.2.3

Determine the Wronskian of the column vector functions

$$x_1(t) = \begin{bmatrix} e^t \\ 2e^t \end{bmatrix}, \qquad x_2(t) = \begin{bmatrix} 3\sin t \\ \cos t \end{bmatrix}.$$

Solution: From Definition 7.2.2, we have

$$W[x_1, x_2](t) = \begin{vmatrix} e^t & 3\sin t \\ 2e^t & \cos t \end{vmatrix} = e^t(\cos t - 6\sin t). \qquad \square$$

Our next theorem indicates that the Wronskian plays a familiar role in determining the linear independence of a set of vectors in $V_n(I)$.

Theorem 7.2.4

Let $x_1(t), x_2(t), \ldots, x_n(t)$ be vectors in $V_n(I)$. If $W[x_1, x_2, \ldots, x_n](t_0)$ is *nonzero* at some point t_0 in I, then $\{x_1(t), x_2(t), \ldots, x_n(t)\}$ is linearly independent on I.

Proof Consider

$$c_1 x_1(t) + c_2 x_2(t) + \cdots + c_n x_n(t) = 0,$$

where $c_1, c_2, \ldots, c_n$ are scalars. Using Theorem 2.2.9, we can write this as the vector equation

$$X(t)c = 0,$$

where $c = [c_1 \ c_2 \ \cdots \ c_n]^T$ and $X(t) = [x_1(t), x_2(t), \ldots, x_n(t)]$. Let t_0 be in I. If we assume that $\det([X(t_0)]) = W[x_1, x_2, \ldots, x_n](t_0) \neq 0$, Corollary 3.2.5 implies that the only solution to this $n \times n$ system of linear equations is $c = 0$. Consequently, $\{x_1(t), x_2(t), \ldots, x_n(t)\}$ is linearly independent on I, as required. ∎

Example 7.2.5   The vector functions

$$\mathbf{x}_1(t) = \begin{bmatrix} e^t \\ 2e^t \end{bmatrix}, \qquad \mathbf{x}_2 = \begin{bmatrix} 3\sin t \\ \cos t \end{bmatrix}$$

are linearly independent on $(-\infty, \infty)$, since $W[\mathbf{x}_1, \mathbf{x}_2](t) = e^t(\cos t - 6\sin t)$ is nonzero, for example, when $t = 0$. $\square$

Example 7.2.6   Given an $n \times n$ matrix function $A(t)$, the function $T : V_n(I) \to V_n(I)$ defined by

$$T(\mathbf{x}(t)) = A(t)\mathbf{x}(t)$$

is a linear transformation. To see this, let $\mathbf{x}(t)$ and $\mathbf{y}(t)$ be column n-vector functions, and let c be a scalar. For clarity, we suppress the variable t from the functions A, $\mathbf{x}$, and $\mathbf{y}$ in the calculations below. We have

$$T(\mathbf{x} + \mathbf{y}) = A(\mathbf{x} + \mathbf{y}) = A\mathbf{x} + A\mathbf{y} = T(\mathbf{x}) + T(\mathbf{y})$$

and

$$T(c\,\mathbf{x}) = A(c\,\mathbf{x}) = c(A\mathbf{x}) = c\,T(\mathbf{x}).$$

Likewise, the reader can verify that the function $D : V_n(I) \to V_n(I)$ defined by

$$D(\mathbf{x}(t)) = \mathbf{x}'(t)$$

is a linear transformation. $\square$

Vector Differential Equations

A system of linear differential equations written in the vector form

$$\mathbf{x}'(t) = A(t)\mathbf{x}(t) + \mathbf{b}(t)$$

will be called a **vector differential equation**. We emphasize that within this formulation the primary unknown is the column vector function $\mathbf{x}(t)$, whose components, $x_1(t), x_2(t), \ldots, x_n(t)$, are the unknowns in the corresponding linear system (7.2.1). The problem of determining all solutions to the general first-order linear system of differential equations (7.2.1) can now be formulated as the vector-space problem

> Find all column vector functions $\mathbf{x}(t) \in V_n(I)$
> satisfying the vector differential equation
> $$\mathbf{x}'(t) = A(t)\mathbf{x}(t) + \mathbf{b}(t).$$

The vector space $V_n(I)$ is not finite-dimensional, since there is no finite set of linearly independent vectors that span $V_n(I)$. The key to solving linear differential systems comes from the realization that, if $A(t)$ is an $n \times n$ matrix function, then the set of all solutions to the homogeneous vector differential equation

$$\mathbf{x}'(t) = A(t)\mathbf{x}(t)$$

is an n-dimensional subspace of $V_n(I)$. This is illustrated in the next example and established in general in the next section.

Example 7.2.7   Consider the homogeneous linear system of differential equations

$$x_1' = x_1 + 2x_2$$
$$x_2' = 2x_1 - 2x_2.$$

In the previous section, we derived the following solution to this scalar system:

$$x_1(t) = c_1 e^{-3t} + c_2 e^{2t}, \qquad x_2(t) = \tfrac{1}{2}(-4c_1 e^{-3t} + c_2 e^{2t}).$$

We can reformulate this problem as an equivalent vector differential equation:

$$\mathbf{x}' = A\mathbf{x}, \qquad A = \begin{bmatrix} 1 & 2 \\ 2 & -2 \end{bmatrix}. \tag{7.2.3}$$

The solution vector to the vector differential equation (7.2.3) is

$$\mathbf{x}(t) = \begin{bmatrix} c_1 e^{-3t} + c_2 e^{2t} \\ \tfrac{1}{2}(-4c_1 e^{-3t} + c_2 e^{2t}) \end{bmatrix},$$

which can be written in the equivalent form

$$\mathbf{x}(t) = c_1 \begin{bmatrix} e^{-3t} \\ -2e^{-3t} \end{bmatrix} + c_2 \begin{bmatrix} e^{2t} \\ \tfrac{1}{2}e^{2t} \end{bmatrix}.$$

Consequently, in this particular example, the set of all solutions to the vector differential equation is the two-dimensional subspace of $V_2(I)$ spanned by the linearly independent column vector functions[2]

$$\mathbf{x}_1(t) = \begin{bmatrix} e^{-3t} \\ -2e^{-3t} \end{bmatrix}, \qquad \mathbf{x}_2(t) = \begin{bmatrix} e^{2t} \\ \tfrac{1}{2}e^{2t} \end{bmatrix}. \qquad \square$$

Exercises for 7.2

Key Terms

Vector formulation of a linear system, Wronskian of vector functions, Vector differential equation.

Skills

- Be able to write a system of first-order linear differential equations as an equivalent vector differential equation.

- Be able to determine the Wronskian of a collection of column vector functions.

- Understand how the Wronskian of a collection of column vector functions relates to the linear independence of those functions.

True-False Review

For Questions 1–7, decide if the given statement is **true** or **false**, and give a brief justification for your answer. If true, you can quote a relevant definition or theorem from the text. If false, provide an example, illustration, or brief explanation of why the statement is false.

1. In the vector equation $\mathbf{x}'(t) = A(t)\mathbf{x}(t) + \mathbf{b}(t)$, the matrix function $A(t)$ must always be a square matrix.

2. The Wronskians $W[\mathbf{x}_1, \mathbf{x}_2](t)$ and $W[\mathbf{x}_2, \mathbf{x}_1](t)$ are the same.

3. Three column vector functions in $V_2(I)$ must be linearly dependent.

4. In order for a set of n column vector functions in $V_n(I)$ to be linearly independent, the Wronskian of these vector functions must be nonzero for all x in I.

---

[2]The linear independence follows since $\mathbf{x}_1$ and $\mathbf{x}_2$ are not proportional.

5. If A is a 2×2 matrix of constants whose determinant is zero, then the vector differential equation $\mathbf{x}'(t) = A\mathbf{x}(t)$ cannot have two linearly independent solutions.

6. A single fourth-order linear differential equation can be rewritten as a 4×4 linear system of differential equations.

7. If $\mathbf{x}_0(t)$ is a solution to the homogeneous vector differential equation $\mathbf{x}'(t) = A(t)\mathbf{x}(t)$, then $\mathbf{x}_0(t) + \mathbf{b}(t)$ is a solution to the nonhomogeneous vector differential equation $\mathbf{x}'(t) = A(t)\mathbf{x}(t) + \mathbf{b}(t)$.

Problems

For Problems 1–5, show that the given vector functions are linearly independent on $(-\infty, \infty)$.

1. $\mathbf{x}_1(t) = \begin{bmatrix} e^t \\ -e^t \end{bmatrix}$, $\mathbf{x}_2(t) = \begin{bmatrix} e^t \\ e^t \end{bmatrix}$.

2. $\mathbf{x}_1(t) = \begin{bmatrix} t \\ t \end{bmatrix}$, $\mathbf{x}_2(t) = \begin{bmatrix} t \\ t^2 \end{bmatrix}$.

3. $\mathbf{x}_1(t) = \begin{bmatrix} t+1 \\ t-1 \\ 2t \end{bmatrix}$, $\mathbf{x}_2(t) = \begin{bmatrix} e^t \\ e^{2t} \\ e^{3t} \end{bmatrix}$,

   $\mathbf{x}_3(t) = \begin{bmatrix} 1 \\ \sin t \\ \cos t \end{bmatrix}$.

4. $\mathbf{x}_1(t) = \begin{bmatrix} t \\ t \end{bmatrix}$, $\mathbf{x}_2(t) = \begin{bmatrix} |t| \\ t \end{bmatrix}$.

   Is there an interval on which $\mathbf{x}_1(t)$ and $\mathbf{x}_2(t)$ in this exercise are not linearly independent?

5. $\mathbf{x}_1(t) = \begin{bmatrix} \sin t \\ \cos t \\ 1 \end{bmatrix}$, $\mathbf{x}_2(t) = \begin{bmatrix} t \\ 1-t \\ 1 \end{bmatrix}$,

   $\mathbf{x}_3(t) = \begin{bmatrix} \sinh t \\ \cosh t \\ 1 \end{bmatrix}$.

For Problems 6–9, show that the given vector functions are linearly dependent on $(-\infty, \infty)$.

6. $\mathbf{x}_1(t) = \begin{bmatrix} t^2 \\ 6-t+t^3 \end{bmatrix}$, $\mathbf{x}_2(t) = \begin{bmatrix} -3t^2 \\ -18+3t-3t^2 \end{bmatrix}$.

7. $\mathbf{x}_1(t) = \begin{bmatrix} e^t \\ 2e^{2t} \end{bmatrix}$, $\mathbf{x}_2(t) = \begin{bmatrix} 4e^t \\ 8e^{2t} \end{bmatrix}$.

8. $\mathbf{x}_1(t) = \begin{bmatrix} \sin^2 t \\ \cos^2 t \\ 2 \end{bmatrix}$, $\mathbf{x}_2(t) = \begin{bmatrix} 2\cos^2 t \\ 2\sin^2 t \\ 1 \end{bmatrix}$,

   $\mathbf{x}_3(t) = \begin{bmatrix} 2 \\ 2 \\ 5 \end{bmatrix}$.

9. $\mathbf{x}_1(t) = \begin{bmatrix} t \\ t^2 \\ -t^3 \end{bmatrix}$, $\mathbf{x}_2(t) = \begin{bmatrix} 2t \\ 3t^2 \\ 0 \end{bmatrix}$, $\mathbf{x}_3(t) = \begin{bmatrix} -t \\ 0 \\ 3t^3 \end{bmatrix}$.

10. Prove that $V_n(I)$ is a vector space.

11. Let $A(t)$ be an $n \times n$ matrix function. Prove that the set of all solutions $\mathbf{x}$ to the system $\mathbf{x}'(t) = A(t)\mathbf{x}(t)$ is a subspace of $V_n(I)$.

12. If

$$A = \begin{bmatrix} 2 & -4 \\ 1 & -3 \end{bmatrix},$$

determine two linearly independent solutions to $\mathbf{x}' = A\mathbf{x}$ on $(-\infty, \infty)$.

Problem 13 investigates the relationship between the Wronskian defined in this section for column vector functions in $V_n(I)$ and the Wronskian defined previously for functions in $C^n(I)$.

13. Consider the differential equation

$$\frac{d^2 y}{dt^2} + a\frac{dy}{dt} + by = 0, \qquad (7.2.4)$$

where a and b are arbitrary functions of t.

(a) Show that Equation (7.2.4) can be replaced by the equivalent linear system

$$\mathbf{x}' = A\mathbf{x}, \qquad (7.2.5)$$

where

$$A = \begin{bmatrix} 0 & 1 \\ -b & -a \end{bmatrix} \qquad \text{and} \qquad x_1 = y, \ x_2 = y'.$$

(b) If $y_1 = f_1(t)$ and $y_2 = f_2(t)$ are solutions to Equation (7.2.4) on an interval I, show that the corresponding solutions to the system (7.2.5) are

$$\mathbf{x}_1(t) = \begin{bmatrix} f_1(t) \\ f_1'(t) \end{bmatrix}, \qquad \mathbf{x}_2(t) = \begin{bmatrix} f_2(t) \\ f_2'(t) \end{bmatrix}.$$

(c) Show that

$$W[\mathbf{x}_1, \mathbf{x}_2](t) = W[y_1, y_2](t).$$

7.3   General Results for First-Order Linear Differential Systems

We now show how the formulation of a linear system of differential equations as a single vector differential equation enables us to derive the underlying theory for linear differential systems as an application of the vector space results from Chapter 4. We emphasize that, although the derivation of the results is based on the vector differential equation formulation, the results themselves apply to any first-order linear system, since such a system can always be formulated as a vector differential equation.

The fundamental theoretical result that will be used in deriving the underlying theory for the solution of vector differential equations is the following existence and uniqueness theorem:

Theorem 7.3.1   The initial-value problem

$$\mathbf{x}'(t) = A(t)\mathbf{x}(t) + \mathbf{b}(t), \qquad \mathbf{x}(t_0) = \mathbf{x}_0,$$

where $A(t)$ and $\mathbf{b}(t)$ are continuous on an interval I, has a unique solution on I.

Proof   The proof is omitted. [See, for example, F. J. Murray and K. S. Miller, *Existence Theorems* (New York: University Press, 1954).]   ∎

Homogeneous Vector Differential Equations

Just as for a single nth-order linear differential equation, the solution to a nonhomogeneous linear differential system can, in theory, be obtained once we have solved the associated homogeneous differential system. Consequently, we begin by developing the theory for homogeneous vector differential equations:

$$\mathbf{x}'(t) = A(t)\mathbf{x}(t), \tag{7.3.1}$$

where A is an $n \times n$ matrix function. This is where the vector space techniques are required. We first show that the set of all solutions to (7.3.1) is an n-dimensional subspace of the vector space of all column n-vector functions.

Theorem 7.3.2   The set of all solutions to $\mathbf{x}'(t) = A(t)\mathbf{x}(t)$, where $A(t)$ is an $n \times n$ matrix function that is continuous on an interval I, is a vector space of dimension n.

Proof   Let S denote the set of all solutions to $\mathbf{x}' = A(t)\mathbf{x}(t)$. By Example 7.2.6, the functions $T(\mathbf{x}) = A\mathbf{x}$ and $D(\mathbf{x}) = \mathbf{x}'$ are linear transformations, hence so is

$$(D - T)(\mathbf{x}) = \mathbf{x}' - A\mathbf{x}.$$

Therefore, since S is simply the kernel of the linear transformation $D - T$, it is a subspace of $V_n(I)$ by Theorem 5.3.5.

We now prove that the dimension of S is n by constructing a basis for S containing n vectors. We first show that there exist n linearly independent solutions to $\mathbf{x}' = A\mathbf{x}$. Let $\mathbf{e}_i$ denote the ith column vector of the identity matrix I_n. Then, from Theorem 7.3.1, for every t_0 in I and every i, the initial-value problem

$$\begin{cases} \mathbf{x}_i'(t) = A(t)\mathbf{x}_i(t), \\ \\ \mathbf{x}_i(t_0) = \mathbf{e}_i, \end{cases} \qquad i = 1, 2, \ldots, n,$$

has a unique solution $\mathbf{x}_i(t)$. Further, $W[\mathbf{x}_1, \mathbf{x}_2, \ldots, \mathbf{x}_n](t_0) = \det(I_n) = 1 \neq 0$ for any t_0 in I, so that $\{\mathbf{x}_1(t), \mathbf{x}_2(t), \ldots, \mathbf{x}_n(t)\}$ is linearly independent on I. Next we establish that these solutions span the solution space. Let $\mathbf{x}(t)$ be any real solution to $\mathbf{x}' = A\mathbf{x}$ on I. Then, since $\{\mathbf{x}_1(t_0), \mathbf{x}_2(t_0), \ldots, \mathbf{x}_n(t_0)\}$ is the standard basis for $\mathbb{R}^n$, we can write

$$\mathbf{x}(t_0) = c_1\mathbf{x}_1(t_0) + c_2\mathbf{x}_2(t_0) + \cdots + c_n\mathbf{x}_n(t_0)$$

for some scalars $c_1, c_2, \ldots, c_n$. It follows that $\mathbf{x}(t)$ is the *unique* solution to the initial-value problem

$$\begin{cases} \mathbf{x}'(t) = A(t)\mathbf{x}(t), \\ \mathbf{x}(t_0) = c_1\mathbf{x}_1(t_0) + c_2\mathbf{x}_2(t_0) + \cdots + c_n\mathbf{x}_n(t_0), \end{cases} \tag{7.3.2}$$

by Theorem 7.3.1. But

$$\mathbf{u}(t) = c_1\mathbf{x}_1(t) + c_2\mathbf{x}_2(t) + \cdots + c_n\mathbf{x}_n(t)$$

also satisfies the initial-value problem (7.3.2), and so, by uniqueness, we must have

$$\mathbf{x}(t) = \mathbf{u}(t) = c_1\mathbf{x}_1(t) + c_2\mathbf{x}_2(t) + \cdots + c_n\mathbf{x}_n(t).$$

We have therefore shown that any solution to $\mathbf{x}' = A\mathbf{x}$ on I can be written as a linear combination of the n linearly independent solutions $\mathbf{x}_1(t), \mathbf{x}_2(t), \ldots, \mathbf{x}_n(t)$, hence $\{\mathbf{x}_1(t), \mathbf{x}_2(t), \ldots, \mathbf{x}_n(t)\}$ is a basis for the solution space. Consequently, the dimension of the solution space is n. ∎

It follows from Theorem 7.3.2 that if $\{\mathbf{x}_1, \mathbf{x}_2, \ldots, \mathbf{x}_n\}$ is *any* set of n linearly independent solutions to (7.3.1), then every solution to the system can be written as

$$\mathbf{x}(t) = c_1\mathbf{x}_1(t) + c_2\mathbf{x}_2(t) + \cdots + c_n\mathbf{x}_n(t), \tag{7.3.3}$$

for appropriate constants $c_1, c_2, \ldots, c_n$. In keeping with the terminology that we have used throughout the text, we will refer to (7.3.3) as the **general solution** to the vector differential equation (7.3.1).

The following definition introduces some important terminology for homogeneous vector differential equations.

DEFINITION 7.3.3

Let $A(t)$ be an $n \times n$ matrix function that is continuous on an interval I. Any set of n solutions, $\{\mathbf{x}_1, \mathbf{x}_2, \ldots, \mathbf{x}_n\}$, to $\mathbf{x}' = A\mathbf{x}$ that is linearly independent on I is called a **fundamental solution set** on I. The corresponding matrix $X(t)$ defined by

$$X(t) = [\mathbf{x}_1, \mathbf{x}_2, \ldots, \mathbf{x}_n]$$

is called a **fundamental matrix** for the vector differential equation $\mathbf{x}' = A\mathbf{x}$.

Remarks

1. Since the vector space of all solutions to $\mathbf{x}' = A\mathbf{x}$ is n-dimensional by Theorem 7.3.2, the fundamental solution set for $\mathbf{x}' = A\mathbf{x}$ is a *basis* for the space of solutions to the system.

2. If $X(t)$ is a fundamental matrix for (7.3.1) then, applying Theorem 2.2.9, the general solution (7.3.3) can be written in vector form as $\mathbf{x}(t) = X(t)\mathbf{c}$, where $\mathbf{c} = [c_1 \ c_2 \ \ldots \ c_n]^T$. This will be our starting point when the variation-of-parameters method is derived in Section 7.6.

Now suppose that $\mathbf{x}_1, \mathbf{x}_2, \ldots, \mathbf{x}_n$ are solutions to $\mathbf{x}' = A\mathbf{x}$ on an interval I. We have shown in the previous section that if $W[\mathbf{x}_1, \mathbf{x}_2, \ldots, \mathbf{x}_n](t) \neq 0$ at some point in I, then the solutions are linearly independent on I. We now prove a converse statement.

Theorem 7.3.4   Let $A(t)$ be an $n \times n$ matrix function that is continuous on an interval I. If $\{\mathbf{x}_1, \mathbf{x}_2, \ldots, \mathbf{x}_n\}$ is a linearly independent set of *solutions* to $\mathbf{x}' = A\mathbf{x}$ on I, then

$$W[\mathbf{x}_1, \mathbf{x}_2, \ldots, \mathbf{x}_n](t) \neq 0$$

at every point t in I.

Proof We prove the equivalent statement that if $W[\mathbf{x}_1, \mathbf{x}_2, \ldots, \mathbf{x}_n](t_0) = 0$ at some point t_0 in I, then $\{\mathbf{x}_1, \mathbf{x}_2, \ldots, \mathbf{x}_n\}$ is linearly dependent on I. We proceed as follows. If $W[\mathbf{x}_1, \mathbf{x}_2, \ldots, \mathbf{x}_n](t_0) = 0$, then from Corollary 4.5.15, the set of vectors $\{\mathbf{x}_1(t_0), \mathbf{x}_2(t_0), \ldots, \mathbf{x}_n(t_0)\}$ is linearly dependent in $\mathbb{R}^n$. Thus, there exist scalars $c_1, c_2, \ldots, c_n$, *not all zero*, such that

$$c_1\mathbf{x}_1(t_0) + c_2\mathbf{x}_2(t_0) + \cdots + c_n\mathbf{x}_n(t_0) = \mathbf{0}. \tag{7.3.4}$$

Now let

$$\mathbf{x}(t) = c_1\mathbf{x}_1(t) + c_2\mathbf{x}_2(t) + \cdots + c_n\mathbf{x}_n(t). \tag{7.3.5}$$

It follows from Equations (7.3.4) and (7.3.5) and Theorem 7.3.1 that $\mathbf{x}(t)$ is the unique solution to the initial-value problem

$$\mathbf{x}' = A(t)\mathbf{x}(t), \qquad \mathbf{x}(t_0) = \mathbf{0}.$$

However, this initial-value problem has the solution $\mathbf{x}(t) = \mathbf{0}$, and so, by uniqueness, we must have

$$c_1\mathbf{x}_1(t) + c_2\mathbf{x}_2(t) + \cdots + c_n\mathbf{x}_n(t) = \mathbf{0}.$$

Since not all of the c_i are zero, it follows that the set of vector functions $\{\mathbf{x}_1, \mathbf{x}_2, \ldots, \mathbf{x}_n\}$ is indeed linearly dependent on I.   ∎

Thus, to determine whether $\{\mathbf{x}_1, \mathbf{x}_2, \ldots, \mathbf{x}_n\}$ is a fundamental solution set for $\mathbf{x}' = A\mathbf{x}$ on an interval I, we can compute the Wronskian of $\mathbf{x}_1, \mathbf{x}_2, \ldots, \mathbf{x}_n$ at *any* convenient point t_0 in I. If $W[\mathbf{x}_1, \mathbf{x}_2, \ldots, \mathbf{x}_n](t_0) \neq 0$, then the solutions are linearly independent on I, whereas if $W[\mathbf{x}_1, \mathbf{x}_2, \ldots, \mathbf{x}_n](t_0) = 0$, then the solutions are linearly dependent on I.

Example 7.3.5   Consider the vector differential equation $\mathbf{x}' = A\mathbf{x}$, where

$$A = \begin{bmatrix} 1 & 2 \\ -2 & 1 \end{bmatrix},$$

and let

$$\mathbf{x}_1(t) = \begin{bmatrix} -e^t \cos 2t \\ e^t \sin 2t \end{bmatrix}, \qquad \mathbf{x}_2(t) = \begin{bmatrix} e^t \sin 2t \\ e^t \cos 2t \end{bmatrix}.$$

(a) Verify that $\{\mathbf{x}_1, \mathbf{x}_2\}$ is a fundamental set of solutions for the vector differential equation on any interval, and write the general solution to the vector differential equation.

(b) Solve the initial-value problem

$$\mathbf{x}' = A\mathbf{x}, \qquad \mathbf{x}(0) = \begin{bmatrix} 3 \\ 2 \end{bmatrix},$$

and write the corresponding scalar solutions.

Solution:

(a) Differentiating the given vector functions with respect to t yields, respectively,

$$\mathbf{x}_1' = \begin{bmatrix} e^t(-\cos 2t + 2\sin 2t) \\ e^t(\sin 2t + 2\cos 2t) \end{bmatrix}, \qquad \mathbf{x}_2' = \begin{bmatrix} e^t(\sin 2t + 2\cos 2t) \\ e^t(\cos 2t - 2\sin 2t) \end{bmatrix},$$

whereas

$$A\mathbf{x}_1 = \begin{bmatrix} 1 & 2 \\ -2 & 1 \end{bmatrix} \begin{bmatrix} -e^t\cos 2t \\ e^t\sin 2t \end{bmatrix} = \begin{bmatrix} e^t(-\cos 2t + 2\sin 2t) \\ e^t(\sin 2t + 2\cos 2t) \end{bmatrix},$$

and

$$A\mathbf{x}_2 = \begin{bmatrix} 1 & 2 \\ -2 & 1 \end{bmatrix} \begin{bmatrix} e^t\sin 2t \\ e^t\cos 2t \end{bmatrix} = \begin{bmatrix} e^t(\sin 2t + 2\cos 2t) \\ e^t(\cos 2t - 2\sin 2t) \end{bmatrix}.$$

Hence,

$$\mathbf{x}_1' = A\mathbf{x}_1 \text{ and } \mathbf{x}_2' = A\mathbf{x}_2,$$

so that $\mathbf{x}_1$ and $\mathbf{x}_2$ are indeed solutions to the given vector differential equation. Furthermore, the Wronskian of these solutions is

$$W[\mathbf{x}_1, \mathbf{x}_2](t) = \begin{vmatrix} -e^t\cos 2t & e^t\sin 2t \\ e^t\sin 2t & e^t\cos 2t \end{vmatrix} = -e^{2t}.$$

Since the Wronskian is never zero, it follows that $\{\mathbf{x}_1, \mathbf{x}_2\}$ is linearly independent on any interval so it is a fundamental set of solutions for the given vector differential equation. Therefore, the general solution to the system is

$$\mathbf{x}(t) = c_1\mathbf{x}_1(t) + c_2\mathbf{x}_2(t) = c_1 \begin{bmatrix} -e^t\cos 2t \\ e^t\sin 2t \end{bmatrix} + c_2 \begin{bmatrix} e^t\sin 2t \\ e^t\cos 2t \end{bmatrix}.$$

Combining the two column vector functions on the right-hand side yields

$$\mathbf{x}(t) = \begin{bmatrix} e^t(-c_1\cos 2t + c_2\sin 2t) \\ e^t(c_1\sin 2t + c_2\cos 2t) \end{bmatrix}.$$

(b) Imposing the given initial condition $\mathbf{x}(0) = \begin{bmatrix} 3 \\ 2 \end{bmatrix}$ on the general solution above requires that

$$\begin{bmatrix} -c_1 \\ c_2 \end{bmatrix} = \begin{bmatrix} 3 \\ 2 \end{bmatrix},$$

so that $c_1 = -3$ and $c_2 = 2$. Hence,

$$\mathbf{x}(t) = \begin{bmatrix} e^t(3\cos 2t + 2\sin 2t) \\ e^t(-3\sin 2t + 2\cos 2t) \end{bmatrix}.$$

The corresponding scalar solutions are

$$x_1(t) = e^t(3\cos 2t + 2\sin 2t), \qquad x_2(t) = e^t(-3\sin 2t + 2\cos 2t). \qquad \square$$

Nonhomogeneous Vector Differential Equations

The preceding results have dealt with the case of a homogeneous vector differential equation. We end this section with the main theoretical result that will be needed for nonhomogeneous vector differential equations. In view of our previous experience with nonhomogeneous linear problems, the following theorem should not be too surprising:

Theorem 7.3.6   Let $A(t)$ be a matrix function that is continuous on an interval I, and let $\{\mathbf{x}_1, \mathbf{x}_2, \ldots, \mathbf{x}_n\}$ be a fundamental solution set on I for the vector differential equation $\mathbf{x}'(t) = A(t)\mathbf{x}(t)$. If $\mathbf{x} = \mathbf{x}_p(t)$ is any particular solution to the nonhomogeneous vector differential equation

$$\mathbf{x}'(t) = A(t)\mathbf{x}(t) + \mathbf{b}(t) \tag{7.3.6}$$

on I, then every solution to (7.3.6) on I is of the form

$$\mathbf{x}(t) = c_1\mathbf{x}_1 + c_2\mathbf{x}_2 + \cdots + c_n\mathbf{x}_n + \mathbf{x}_p.$$

Proof Since $\mathbf{x} = \mathbf{x}_p(t)$ is a solution to $\mathbf{x}'(t) = A(t)\mathbf{x}(t) + \mathbf{b}(t)$ on I, we have

$$\mathbf{x}'_p(t) = A(t)\mathbf{x}_p(t) + \mathbf{b}(t). \tag{7.3.7}$$

Now let $\mathbf{u}(t)$ be any solution to $\mathbf{x}'(t) = A(t)\mathbf{x}(t) + \mathbf{b}(t)$ on I. We then also have

$$\mathbf{u}'(t) = A(t)\mathbf{u}(t) + \mathbf{b}(t). \tag{7.3.8}$$

Subtracting (7.3.7) from (7.3.8) yields

$$(\mathbf{u} - \mathbf{x}_p)' = A(\mathbf{u} - \mathbf{x}_p).$$

Thus, the vector function $\mathbf{x} = \mathbf{u} - \mathbf{x}_p$ is a solution to the associated homogeneous system $\mathbf{x}' = A\mathbf{x}$ on I. Since $\{\mathbf{x}_1, \mathbf{x}_2, \ldots, \mathbf{x}_n\}$ spans the solution space of this system, it follows that

$$\mathbf{u} - \mathbf{x}_p = c_1\mathbf{x}_1 + c_2\mathbf{x}_2 + \cdots + c_n\mathbf{x}_n,$$

for some scalars $c_1, c_2, \ldots, c_n$. Consequently,

$$\mathbf{u} = c_1\mathbf{x}_1 + c_2\mathbf{x}_2 + \cdots + c_n\mathbf{x}_n + \mathbf{x}_p,$$

and the result is proved.   ∎

Theorem 7.3.6 implies that in order to solve a nonhomogeneous vector differential equation, we must first find the general solution to the associated homogeneous system. In the next two sections we will concentrate on homogeneous vector differential equations, and then, in Section 7.6, we will see how the variation-of-parameters technique can be used to determine a particular solution to a nonhomogeneous vector differential equation.

Key Terms

Homogeneous vector differential equation, General solution to vector differential equation, Fundamental solution set, Fundamental matrix, Nonhomogeneous vector differential equation, Particular solution.

Skills

- Know the existence and uniqueness theorem (Theorem 7.3.1) for a vector differential equation together with an initial condition.

- Understand the theory underlying fundamental solution sets and fundamental matrices for a given vector differential equation.

- Be able to use a fundamental solution set to write down the general solution to a homogeneous vector differential equation.

- Know the relationship between the Wronskian of n column n-vector functions that solve $\mathbf{x}' = A\mathbf{x}$ and their linear dependence/independence.

- Be able to write down the general solution to a nonhomogeneous vector differential equation by using a fundamental solution set for the corresponding homogeneous vector differential equation and a particular solution to the vector differential equation.

True-False Review

For Questions 1–4, decide if the given statement is **true** or **false**, and give a brief justification for your answer. If true, you can quote a relevant definition or theorem from the text. If false, provide an example, illustration, or brief explanation of why the statement is false.

1. The set of all solutions to the vector differential equation $\mathbf{x}'(t) = A(t)\mathbf{x}(t) + \mathbf{b}(t)$ (where $A(t)$ is an $n \times n$ matrix function and $\mathbf{b}(t) \neq \mathbf{0}$) is a vector space of dimension n.

2. Any fundamental matrix for the vector differential equation $\mathbf{x}' = A\mathbf{x}$ is invertible.

3. A fundamental solution set to a vector differential equation $\mathbf{x}' = A\mathbf{x}$ forms a spanning set for the vector space of all solutions to $\mathbf{x}' = A\mathbf{x}$.

4. If the general solution to the homogeneous vector differential equation $\mathbf{x}'(t) = A(t)\mathbf{x}(t)$ is $\mathbf{x}_c(t)$, and $\mathbf{x}_p(t)$ is a particular solution to the nonhomogeneous vector differential equation $\mathbf{x}'(t) = A(t)\mathbf{x}(t) + \mathbf{b}(t)$, then the general solution to the nonhomogeneous vector differential equation is $\mathbf{x}_c(t) + \mathbf{x}_p(t)$.

Problems

For Problems 1–5, show that the given functions are solutions of the system $\mathbf{x}'(t) = A(x)\mathbf{x}(t)$ for the given matrix A, and hence, find the general solution to the system (remember to check linear independence). If auxiliary conditions are given, find the particular solution that satisfies these conditions.

1. $\mathbf{x}_1(t) = \begin{bmatrix} e^{4t} \\ 2e^{4t} \end{bmatrix}$, $\mathbf{x}_2(t) = \begin{bmatrix} 3e^{-t} \\ e^{-t} \end{bmatrix}$,

$A = \begin{bmatrix} -2 & 3 \\ -2 & 5 \end{bmatrix}$, $\mathbf{x}(0) = \begin{bmatrix} -2 \\ 1 \end{bmatrix}$.

2. $\mathbf{x}_1(t) = \begin{bmatrix} e^{2t} \\ -e^{2t} \end{bmatrix}$, $\mathbf{x}_2(t) = \begin{bmatrix} e^{2t}(1+t) \\ -te^{2t} \end{bmatrix}$,

$A = \begin{bmatrix} 3 & 1 \\ -1 & 1 \end{bmatrix}$.

3. $\mathbf{x}_1(t) = \begin{bmatrix} -3 \\ 9 \\ 5 \end{bmatrix}$, $\mathbf{x}_2(t) = \begin{bmatrix} e^{2t} \\ 3e^{2t} \\ e^{2t} \end{bmatrix}$,

$\mathbf{x}_3(t) = \begin{bmatrix} e^{4t} \\ e^{4t} \\ e^{4t} \end{bmatrix}$, $A = \begin{bmatrix} 2 & -1 & 3 \\ 3 & 1 & 0 \\ 2 & -1 & 3 \end{bmatrix}$.

4. $\mathbf{x}_1(t) = \begin{bmatrix} 2t \\ e^t \end{bmatrix}$, $\mathbf{x}_2(t) = \begin{bmatrix} 0 \\ 3e^t \end{bmatrix}$,

$A = \begin{bmatrix} 1/t & 0 \\ 0 & 1 \end{bmatrix}$.

5. $\mathbf{x}_1(t) = \begin{bmatrix} t\sin t \\ \cos t \end{bmatrix}$, $\mathbf{x}_2(t) = \begin{bmatrix} -t\cos t \\ \sin t \end{bmatrix}$,

$A = \begin{bmatrix} 1/t & t \\ -1/t & 0 \end{bmatrix}$.

6. If $x_1, x_2, \ldots, x_n$ are solutions to $\mathbf{x}' = A(t)\mathbf{x}$ and $X = [\mathbf{x}_1, \mathbf{x}_2, \ldots, \mathbf{x}_n]$, prove that

$$X' = A(t)X.$$

7. Let $X(t)$ be a fundamental matrix for $\mathbf{x}' = A(t)\mathbf{x}$ on the interval I.

   (a) Show that the general solution to the linear system can be written as

$$\mathbf{x} = X(t)\mathbf{c},$$

   where $\mathbf{c}$ is a vector of constants.

(b) If $t_0 \in I$, show that the solution to the initial-value problem

$$\mathbf{x}' = A\mathbf{x}, \quad \mathbf{x}(t_0) = \mathbf{x}_0,$$

can be written as

$$\mathbf{x} = X(t)X^{-1}(t_0)\mathbf{x}_0.$$

7.4   Vector Differential Equations: Nondefective Coefficient Matrix

The theory that we have developed in the previous section is valid for any first-order linear system. However, in practice, obtaining a fundamental solution set for a given vector differential equation $\mathbf{x}' = A\mathbf{x}$ on an interval I can be a difficult task, so we must make a simplifying assumption in order to develop solution techniques applicable to a broad class of linear systems. The assumption we will make is that the coefficient matrix is a constant matrix.[3] In the next two sections we will consider only homogeneous linear systems

$$\mathbf{x}' = A\mathbf{x},$$

where A is an $n \times n$ matrix of real *constants*.

To motivate the new solution technique to be developed, we recall from Example 7.2.7 that two linearly independent solutions to the vector differential equation

$$\mathbf{x}' = A\mathbf{x}, \quad A = \begin{bmatrix} 1 & 2 \\ 2 & -2 \end{bmatrix}$$

are

$$\mathbf{x}_1(t) = \begin{bmatrix} e^{-3t} \\ -2e^{-3t} \end{bmatrix}, \qquad \mathbf{x}_2(t) = \begin{bmatrix} e^{2t} \\ \frac{1}{2}e^{2t} \end{bmatrix},$$

which we write as

$$\mathbf{x}_1(t) = e^{-3t}\begin{bmatrix} 1 \\ -2 \end{bmatrix}, \qquad \mathbf{x}_2(t) = e^{2t}\begin{bmatrix} 1 \\ 1/2 \end{bmatrix}.$$

The key point to notice is that both of these solutions are of the form

$$\mathbf{x}(t) = e^{\lambda t}\mathbf{v}, \tag{7.4.1}$$

where λ is a scalar and $\mathbf{v}$ is a constant vector. This suggests that the general vector differential equation

$$\mathbf{x}' = A\mathbf{x} \tag{7.4.2}$$

may also have solutions of the form (7.4.1). We now investigate this possibility. Differentiating (7.4.1) with respect to t yields

$$\mathbf{x}' = \lambda e^{\lambda t}\mathbf{v}.$$

---

[3]This assumption should not be too surprising in view of the discussion of linear nth-order differential equations in Chapter 6.

Thus, $\mathbf{x}(t) = e^{\lambda t}\mathbf{v}$ is a solution to (7.4.2) if and only if

$$\lambda e^{\lambda t}\mathbf{v} = e^{\lambda t}A\mathbf{v};$$

that is, if and only if λ and $\mathbf{v}$ satisfy

$$A\mathbf{v} = \lambda\mathbf{v}.$$

But this is the statement that λ and $\mathbf{v}$ must be an eigenvalue/eigenvector pair for A. Consequently, we have established the following fundamental result:

Theorem 7.4.1   Let A be an $n \times n$ matrix of real constants, and let λ be an eigenvalue of A with corresponding eigenvector $\mathbf{v}$. Then

$$\mathbf{x}(t) = e^{\lambda t}\mathbf{v}$$

is a solution to the constant-coefficient vector differential equation $\mathbf{x}' = A\mathbf{x}$ on any interval.

Remark   Notice that we have not assumed that the eigenvalues and eigenvectors of A are real; the preceding result holds in the complex case also.

We now illustrate how Theorem 7.4.1 can be used to find the general solution to constant-coefficient vector differential equations.

Example 7.4.2   Find the general solution to

$$\begin{aligned} x_1' &= 2x_1 + x_2, \\ x_2' &= -3x_1 - 2x_2. \end{aligned} \qquad (7.4.3)$$

Solution: The corresponding vector differential equation is

$$\mathbf{x}' = A\mathbf{x}, \qquad \text{where } A = \begin{bmatrix} 2 & 1 \\ -3 & -2 \end{bmatrix}. \qquad (7.4.4)$$

A straightforward calculation yields

$$\det(A - \lambda I) = \begin{vmatrix} 2 - \lambda & 1 \\ -3 & -2 - \lambda \end{vmatrix} = \lambda^2 - 1,$$

so that A has eigenvalues $\lambda = \pm 1$.
Eigenvalue $\lambda_1 = 1$: In this case, the system $(A - \lambda_1 I)\mathbf{v} = \mathbf{0}$ is

$$\begin{aligned} v_1 + v_2 &= 0, \\ -3v_1 - 3v_2 &= 0, \end{aligned}$$

with solution $\mathbf{v} = r(1, -1)$. Therefore,

$$\mathbf{x}_1(t) = e^t \begin{bmatrix} 1 \\ -1 \end{bmatrix}$$

is a solution to the vector differential equation (7.4.4).
Eigenvalue $\lambda_2 = -1$: In this case, the system $(A - \lambda_2 I)\mathbf{v} = \mathbf{0}$ is

$$\begin{aligned} 3v_1 + v_2 &= 0, \\ -3v_1 - v_2 &= 0, \end{aligned}$$

with solution $\mathbf{v} = s(1, -3)$. Consequently,

$$\mathbf{x}_2(t) = e^{-t}\begin{bmatrix} 1 \\ -3 \end{bmatrix}$$

is also a solution to the vector differential equation (7.4.4).

The Wronskian of the solutions $\mathbf{x}_1$ and $\mathbf{x}_2$ obtained is

$$W[\mathbf{x}_1, \mathbf{x}_2](t) = \begin{vmatrix} e^t & e^{-t} \\ -e^t & -3e^{-t} \end{vmatrix} = -2 \neq 0,$$

so that $\{\mathbf{x}_1, \mathbf{x}_2\}$ is linearly independent on any interval by Theorem 7.2.4. Hence, the general solution to (7.4.4) is

$$\mathbf{x}(t) = c_1\mathbf{x}_1 + c_2\mathbf{x}_2 = c_1 e^t\begin{bmatrix} 1 \\ -1 \end{bmatrix} + c_2 e^{-t}\begin{bmatrix} 1 \\ -3 \end{bmatrix}.$$

Combining the column vectors on the right-hand side yields the solution vector

$$\mathbf{x}(t) = \begin{bmatrix} c_1 e^t + c_2 e^{-t} \\ -(c_1 e^t + 3c_2 e^{-t}) \end{bmatrix}.$$

Therefore, the solution to the linear system of differential equations (7.4.3) is

$$x_1(t) = c_1 e^t + c_2 e^{-t}, \qquad x_2(t) = -(c_1 e^t + 3c_2 e^{-t}). \qquad \square$$

To find the general solution to an $n \times n$ constant-coefficient vector differential equation, we need to find n linearly independent solutions (see Theorem 7.3.2). The preceding example, together with our experience with eigenvalues and eigenvectors, suggests that we will be able to find n such linearly independent solutions, provided the matrix A has n linearly independent eigenvectors—that is, provided that A is nondefective. This is indeed the case, although if the eigenvalues and eigenvectors are complex, we must do some work to obtain real-valued solutions to the system. We first give the result for the case of real eigenvalues.

Theorem 7.4.3   Let A be an $n \times n$ matrix of real constants. If A has n real linearly independent eigenvectors $\mathbf{v}_1, \mathbf{v}_2, \ldots, \mathbf{v}_n$, with corresponding real eigenvalues $\lambda_1, \lambda_2, \ldots, \lambda_n$ (not necessarily distinct), then the vector functions $\{\mathbf{x}_1, \mathbf{x}_2, \ldots, \mathbf{x}_n\}$ defined by

$$\mathbf{x}_k(t) = e^{\lambda_k t}\mathbf{v}_k, \qquad k = 1, 2, \ldots, n,$$

for all t, are linearly independent solutions to $\mathbf{x}' = A\mathbf{x}$ on any interval. The general solution to this vector differential equation is

$$\mathbf{x}(t) = c_1\mathbf{x}_1 + c_2\mathbf{x}_2 + \cdots + c_n\mathbf{x}_n.$$

Proof We have already shown in Theorem 7.4.1 that each $\mathbf{x}_k(t)$ satisfies $\mathbf{x}' = A\mathbf{x}$ for all t. Further, using properties of determinants,

$$W[\mathbf{x}_1, \mathbf{x}_2, \ldots, \mathbf{x}_n] = \det([e^{\lambda_1 t}\mathbf{v}_1, e^{\lambda_2 t}\mathbf{v}_2, \ldots, e^{\lambda_k t}\mathbf{v}_n])$$
$$= e^{(\lambda_1 + \lambda_2 + \cdots + \lambda_n)t}\det([\mathbf{v}_1, \mathbf{v}_2, \ldots, \mathbf{v}_n])$$
$$\neq 0,$$

since the eigenvectors are linearly independent by assumption, and hence the solutions are linearly independent on any interval. Thus, $\{x_1, x_2, \ldots, x_n\}$ is a fundamental solution set to the vector differential equation, from which we immediately deduce the last statement. ∎

Example 7.4.4   Find the general solution to $x' = Ax$ if

$$A = \begin{bmatrix} 0 & 2 & -3 \\ -2 & 4 & -3 \\ -2 & 2 & -1 \end{bmatrix}.$$

Solution: We first determine the eigenvalues and eigenvectors of A. For the given matrix, we have

$$\det(A - \lambda I) = \begin{vmatrix} -\lambda & 2 & -3 \\ -2 & 4-\lambda & -3 \\ -2 & 2 & -1-\lambda \end{vmatrix} = -(\lambda + 1)(\lambda - 2)^2,$$

so that the eigenvalues are

$$\lambda_1 = -1 \text{ (with multiplicity 1)}, \qquad \lambda_2 = 2 \text{ (with multiplicity 2)}.$$

Eigenvalue $\lambda_1 = -1$: It is easily shown that all eigenvectors corresponding to this eigenvalue are of the form $v = r(1, 1, 1)$, so that we can take

$$v_1 = (1, 1, 1).$$

Eigenvalue $\lambda_2 = 2$: In this case, the system for the eigenvectors reduces to the single equation

$$2v_1 - 2v_2 + 3v_3 = 0,$$

which has solution $v = r(1, 1, 0) + s(-3, 0, 2)$. Therefore, two linearly independent eigenvectors corresponding to $\lambda = 2$ are

$$v_2 = (1, 1, 0), \qquad v_3 = (-3, 0, 2).$$

It follows from Theorem 7.4.3 that three linearly independent solutions to the given vector differential equation are

$$x_1(t) = e^{-t} \begin{bmatrix} 1 \\ 1 \\ 1 \end{bmatrix}, \qquad x_2(t) = e^{2t} \begin{bmatrix} 1 \\ 1 \\ 0 \end{bmatrix}, \qquad x_3(t) = e^{2t} \begin{bmatrix} -3 \\ 0 \\ 2 \end{bmatrix}.$$

Consequently, the general solution to the given system is

$$x(t) = c_1 e^{-t} \begin{bmatrix} 1 \\ 1 \\ 1 \end{bmatrix} + c_2 e^{2t} \begin{bmatrix} 1 \\ 1 \\ 0 \end{bmatrix} + c_3 e^{2t} \begin{bmatrix} -3 \\ 0 \\ 2 \end{bmatrix},$$

which can be combined to obtain the solution vector

$$x(t) = \begin{bmatrix} c_1 e^{-t} + c_2 e^{2t} - 3c_3 e^{2t} \\ c_1 e^{-t} + c_2 e^{2t} \\ c_1 e^{-t} + 2c_3 e^{2t} \end{bmatrix}. \qquad \square$$

Theorem 7.4.3 constructs the general solution to the vector differential equation $\mathbf{x}' = A\mathbf{x}$ in the case of a nondefective matrix A with real eigenvalues. For such a matrix, there is another way to arrive at the general solution

$$\mathbf{x}(t) = c_1\mathbf{x}_1 + c_2\mathbf{x}_2 + \cdots + c_n\mathbf{x}_n$$

to $\mathbf{x}' = A\mathbf{x}$ that we briefly introduced in Section 5.8. Namely, we can write

$$A = SDS^{-1},$$

where the columns of S consist of n linearly independent eigenvectors of A,

$$S = [\mathbf{v}_1, \mathbf{v}_2, \ldots, \mathbf{v}_n],$$

and D is a diagonal matrix containing the corresponding eigenvalues:

$$D = \text{diag}(\lambda_1, \lambda_2, \ldots, \lambda_n).$$

We can use the linear change of variables $\mathbf{x} = S\mathbf{y}$ to replace $\mathbf{x}' = A\mathbf{x}$ by

$$S\mathbf{y}' = (SDS^{-1})S\mathbf{y} = SD\mathbf{y},$$

or, using the invertibility of S,

$$\mathbf{y}' = D\mathbf{y}.$$

This is an uncoupled system of differential equations that is easy to solve:

$$\mathbf{y} = [c_1 e^{\lambda_1 t} \quad c_2 e^{\lambda_2 t} \quad \cdots \quad c_n e^{\lambda_n t}]^T.$$

Hence, we find that

$$\mathbf{x} = S\mathbf{y} = c_1 e^{\lambda_1 t}\mathbf{v}_1 + c_2 e^{\lambda_2 t}\mathbf{v}_2 + \cdots + c_n e^{\lambda_n t}\mathbf{v}_n,$$

precisely the general solution guaranteed by Theorem 7.4.3.

We now consider the case when some (or all) of the eigenvalues are complex. Since we are restricting attention to systems of equations with *real* constant coefficients, it follows that the matrix of the system will have real entries, and hence, from Theorem 5.6.8, the eigenvalues *and* eigenvectors will occur in conjugate pairs. The corresponding solutions to $\mathbf{x}' = A\mathbf{x}$ guaranteed by Theorem 7.4.1 will also be complex conjugate. However, as we now show, each conjugate pair gives rise to two real-valued solutions.

Theorem 7.4.5   Let $\mathbf{u}(t)$ and $\mathbf{v}(t)$ be real-valued vector functions. If

$$\mathbf{w}_1(t) = \mathbf{u}(t) + i\mathbf{v}(t) \qquad \text{and} \qquad \mathbf{w}_2(t) = \mathbf{u}(t) - i\mathbf{v}(t)$$

are complex conjugate solutions to $\mathbf{x}' = A\mathbf{x}$, then

$$\mathbf{x}_1(t) = \mathbf{u}(t) \text{ and } \mathbf{x}_2(t) = \mathbf{v}(t)$$

are themselves *real-valued* solutions of $\mathbf{x}' = A\mathbf{x}$.

Proof Since $\mathbf{w}_1$ and $\mathbf{w}_2$ are solutions to the vector differential equation, so is any linear combination of them. In particular,

$$\mathbf{x}_1(t) = \frac{1}{2}[\mathbf{w}_1(t) + \mathbf{w}_2(t)] = \mathbf{u}(t)$$

and

$$\mathbf{x}_2(t) = \frac{1}{2i}[\mathbf{w}_1(t) - \mathbf{w}_2(t)] = \mathbf{v}(t)$$

are solutions to the vector differential equation.   ∎

We now explicitly derive two appropriate real-valued solutions corresponding to a complex conjugate pair of eigenvalues. Suppose that $\lambda = a + ib$ ($b \neq 0$) is an eigenvalue of A with corresponding eigenvector $\mathbf{v} = \mathbf{r} + i\mathbf{s}$. Then, applying Theorem 7.4.1, a complex-valued solution to $\mathbf{x}' = A\mathbf{x}$ is

$$\mathbf{w}(t) = e^{(a+ib)t}(\mathbf{r} + i\mathbf{s}) = e^{at}(\cos bt + i \sin bt)(\mathbf{r} + i\mathbf{s}),$$

which can be written as

$$\mathbf{w}(t) = e^{at}(\cos bt\ \mathbf{r} - \sin bt\ \mathbf{s}) + ie^{at}(\sin bt\ \mathbf{r} + \cos bt\ \mathbf{s}).$$

Theorem 7.4.5 implies that two real-valued solutions to $\mathbf{x}' = A\mathbf{x}$ are

$$\mathbf{x}_1(t) = e^{at}(\cos bt\ \mathbf{r} - \sin bt\ \mathbf{s}), \qquad \mathbf{x}_2(t) = e^{at}(\sin bt\ \mathbf{r} + \cos bt\ \mathbf{s}).$$

It can further be shown that the set of all real-valued solutions obtained in this manner is linearly independent on any interval.

Remark   Notice that we do not have to derive the solution corresponding to the conjugate eigenvalue $\bar{\lambda} = a - ib$, since it does not yield any new linearly independent solutions to $\mathbf{x}' = A\mathbf{x}$.

Example 7.4.6   Find the general solution to the vector differential equation $\mathbf{x}' = A\mathbf{x}$, if

$$A = \begin{bmatrix} 0 & 2 \\ -2 & 0 \end{bmatrix}.$$

Solution: The characteristic polynomial of A is

$$p(\lambda) = \begin{vmatrix} -\lambda & 2 \\ -2 & -\lambda \end{vmatrix} = \lambda^2 + 4.$$

Consequently, A has complex conjugate eigenvalues $\lambda = \pm 2i$. The eigenvectors corresponding to the eigenvalues $\lambda = 2i$ are obtained by solving

$$-2iv_1 + 2v_2 = 0$$

and are therefore of the form $\mathbf{v} = r(-i, 1)$. Hence, a complex-valued solution to the given differential equation is

$$\mathbf{w}(t) = e^{2it}\begin{bmatrix} -i \\ 1 \end{bmatrix}.$$

We must now do some algebra to obtain the corresponding real-valued solutions. Using Euler's formula, we can write

$$\mathbf{w}(t) = (\cos 2t + i \sin 2t)\begin{bmatrix} -i \\ 1 \end{bmatrix} = \begin{bmatrix} \sin 2t - i \cos 2t \\ \cos 2t + i \sin 2t \end{bmatrix}$$

$$= \begin{bmatrix} \sin 2t \\ \cos 2t \end{bmatrix} + i\begin{bmatrix} -\cos 2t \\ \sin 2t \end{bmatrix}.$$

Applying Theorem 7.4.5, we directly obtain the two real-valued functions

$$\mathbf{x}_1(t) = \begin{bmatrix} \sin 2t \\ \cos 2t \end{bmatrix} \quad \text{and} \quad \mathbf{x}_2(t) = \begin{bmatrix} -\cos 2t \\ \sin 2t \end{bmatrix}.$$

Consequently, the general solution to the given vector differential equation is

$$\mathbf{x}(t) = c_1 \begin{bmatrix} \sin 2t \\ \cos 2t \end{bmatrix} + c_2 \begin{bmatrix} -\cos 2t \\ \sin 2t \end{bmatrix}$$

$$= \begin{bmatrix} c_1 \sin 2t - c_2 \cos 2t \\ c_1 \cos 2t + c_2 \sin 2t \end{bmatrix}.$$

We note that, in this case, the eigenvalues of A were pure imaginary and the components of the corresponding solution vector are oscillatory. Once more, this illustrates the importance of complex scalars when modeling oscillatory physical behavior.   ☐

As illustrated in the next example, when the real part of a complex eigenvalue is nonzero, the algebra can become a little bit more tedious.

Example 7.4.7   Find the general solution to the vector differential equation $\mathbf{x}' = A\mathbf{x}$ if

$$A = \begin{bmatrix} 2 & -1 \\ 2 & 4 \end{bmatrix}.$$

Solution: The characteristic polynomial of A is

$$p(\lambda) = \begin{vmatrix} 2 - \lambda & -1 \\ 2 & 4 - \lambda \end{vmatrix} = \lambda^2 - 6\lambda + 10,$$

so that the eigenvalues are $\lambda = 3 \pm i$. We need only find the eigenvectors corresponding to one of these conjugate eigenvalues. When $\lambda = 3 + i$, the eigenvectors are obtained by solving

$$-(1 + i)v_1 - v_2 = 0,$$
$$2v_1 + (1 - i)v_2 = 0,$$

which yield the complex eigenvectors $\mathbf{v} = r(1, -(1 + i))$. Hence a complex-valued solution to the given system is

$$\mathbf{w}(t) = e^{3t}(\cos t + i \sin t) \begin{bmatrix} 1 \\ -(1 + i) \end{bmatrix} = e^{3t} \begin{bmatrix} \cos t + i \sin t \\ -(1 + i)(\cos t + i \sin t) \end{bmatrix}$$

$$= e^{3t} \begin{bmatrix} \cos t + i \sin t \\ (\sin t - \cos t) - i(\sin t + \cos t) \end{bmatrix}$$

$$= e^{3t} \left\{ \begin{bmatrix} \cos t \\ \sin t - \cos t \end{bmatrix} + i \begin{bmatrix} \sin t \\ -(\sin t + \cos t) \end{bmatrix} \right\}.$$

From Theorem 7.4.5, the real and imaginary parts of this complex-valued solution yield the following two *real-valued* linearly independent solutions:

$$\mathbf{x}_1(t) = e^{3t} \begin{bmatrix} \cos t \\ \sin t - \cos t \end{bmatrix} \quad \text{and} \quad \mathbf{x}_2(t) = \begin{bmatrix} \sin t \\ -(\sin t + \cos t) \end{bmatrix}.$$

Hence, the general solution to the given system is

$$\mathbf{x}(t) = c_1 e^{3t} \begin{bmatrix} \cos t \\ \sin t - \cos t \end{bmatrix} + c_2 e^{3t} \begin{bmatrix} \sin t \\ -(\sin t + \cos t) \end{bmatrix}$$

$$= e^{3t} \left\{ c_1 \begin{bmatrix} \cos t \\ \sin t - \cos t \end{bmatrix} + c_2 \begin{bmatrix} \sin t \\ -(\sin t + \cos t) \end{bmatrix} \right\}$$

$$= \begin{bmatrix} e^{3t}(c_1 \cos t + c_2 \sin t) \\ e^{3t}[c_1(\sin t - \cos t) - c_2(\sin t + \cos t)] \end{bmatrix}.$$ ☐

The results of this section are summarized in the next theorem.

Theorem 7.4.8   Let A be an $n \times n$ matrix of real constants.

1. Suppose λ is a real eigenvalue of A with corresponding linearly independent eigenvectors $\mathbf{v}_1, \mathbf{v}_2, \ldots, \mathbf{v}_k$. Then k linearly independent solutions to $\mathbf{x}' = A\mathbf{x}$ are

$$\mathbf{x}_j(t) = e^{\lambda t}\mathbf{v}_j, \qquad j = 1, 2, \ldots, k.$$

2. Suppose $\lambda = a + ib$ is a complex eigenvalue of A with corresponding linearly independent eigenvectors $\mathbf{v}_1, \mathbf{v}_2, \ldots, \mathbf{v}_k$, where $\mathbf{v}_j = \mathbf{r}_j + i\mathbf{s}_j$. Then k complex-valued solutions to $\mathbf{x}' = A\mathbf{x}$ are

$$\mathbf{u}_j(t) = e^{\lambda t}\mathbf{v}_j, \qquad j = 1, 2, \ldots, k$$

and $2k$ *real-valued* linearly independent solutions to $\mathbf{x}' = A\mathbf{x}$ are

$$\begin{array}{ll}
\mathbf{x}_{11}(t) = e^{at}(\cos bt\ \mathbf{r}_1 - \sin bt\ \mathbf{s}_1), & \mathbf{x}_{12}(t) = e^{at}(\sin bt\ \mathbf{r}_1 + \cos bt\ \mathbf{s}_1), \\
\mathbf{x}_{21}(t) = e^{at}(\cos bt\ \mathbf{r}_2 - \sin bt\ \mathbf{s}_2), & \mathbf{x}_{22}(t) = e^{at}(\sin bt\ \mathbf{r}_2 + \cos bt\ \mathbf{s}_2), \\
\quad\vdots & \quad\vdots \\
\mathbf{x}_{k1}(t) = e^{at}(\cos bt\ \mathbf{r}_k - \sin bt\ \mathbf{s}_k), & \mathbf{x}_{k2}(t) = e^{at}(\sin bt\ \mathbf{r}_k + \cos bt\ \mathbf{s}_k).
\end{array}$$

Further, the set of all solutions to $\mathbf{x}' = A\mathbf{x}$ obtained in this manner is linearly independent on any interval.

Corollary 7.4.9   If A is a nondefective $n \times n$ matrix, then the solutions obtained from parts 1 and 2 of Theorem 7.4.8 yield a fundamental set of solutions $\{\mathbf{x}_1, \mathbf{x}_2, \ldots, \mathbf{x}_n\}$ to $\mathbf{x}' = A\mathbf{x}$, and the general solution to this vector differential equation is

$$\mathbf{x}(t) = c_1\mathbf{x}_1 + c_2\mathbf{x}_2 + \cdots + c_n\mathbf{x}_n.$$

Exercises for 7.4

Skills

- Be able to find the general solution to the vector differential equation $\mathbf{x}' = A\mathbf{x}$ in the case where A is a constant, nondefective matrix (Theorem 7.4.8 and Corollary 7.4.9).

- In the case of a complex eigenvalue/eigenvector pair $(\lambda, \mathbf{v})$, be able to obtain two real-valued solutions for the solution $\mathbf{x}(t) = e^{\lambda t}\mathbf{v}$ to $\mathbf{x}' = A\mathbf{x}$.

True-False Review

For Questions 1–6, decide if the given statement is **true** or **false**, and give a brief justification for your answer. If true, you can quote a relevant definition or theorem from the text. If false, provide an example, illustration, or brief explanation of why the statement is false.

1. If $(\lambda, \mathbf{v})$ is an eigenvalue/eigenvector pair for A, then $\mathbf{x}(t) = e^{\lambda t}\mathbf{v}$ is a solution to the vector differential equation $\mathbf{x}' = A\mathbf{x}$.

2. Each pair of complex eigenvalues $\lambda = a \pm ib$ ($b \neq 0$) of a matrix A gives rise to a pair of real-valued solutions to the vector differential equation $\mathbf{x}' = A\mathbf{x}$.

3. If A and B are $n \times n$ matrices with the same characteristic equation, then the solution sets to the vector differential equation $\mathbf{x}' = A\mathbf{x}$ and $\mathbf{x}' = B\mathbf{x}$ are the same.

4. If A and B are $n \times n$ matrices with the same collection of eigenvalue/eigenvector pairs, then the solutions to the vector differential equation $\mathbf{x}' = A\mathbf{x}$ and $\mathbf{x}' = B\mathbf{x}$ are the same.

5. If A is a 2×2 nondefective matrix with eigenvalues $\lambda = a \pm ib$ with $a, b > 0$, then all solutions of the vector differential equation $\mathbf{x}' = A\mathbf{x}$ satisfy $||\mathbf{x}(t)|| \to \infty$ as $t \to \infty$.

6. If the eigenvalues λ_1 and λ_2 of a 2×2 matrix A are real with $\lambda_1 < 0 < \lambda_2$, then a solution $\mathbf{x}(t)$ to the vector differential equation $\mathbf{x}' = A\mathbf{x}$ tends toward the origin as $t \to \infty$ if and only if $\mathbf{x}(0)$ is a vector parallel to the eigenvector $\mathbf{v}_2$ that which corresponds to λ_2.

Problems

For Problems 1–15, determine the general solution to the system $\mathbf{x}' = A\mathbf{x}$ for the given matrix A.

1. $\begin{bmatrix} -2 & -7 \\ -1 & 4 \end{bmatrix}$.

2. $\begin{bmatrix} 0 & -4 \\ 4 & 0 \end{bmatrix}$.

3. $\begin{bmatrix} 1 & -2 \\ 5 & -5 \end{bmatrix}$.

4. $\begin{bmatrix} -1 & 2 \\ -2 & -1 \end{bmatrix}$.

5. $\begin{bmatrix} 2 & 0 & 0 \\ 0 & 5 & -7 \\ 0 & 2 & -4 \end{bmatrix}$.

6. $\begin{bmatrix} -1 & 0 & 0 \\ 1 & 5 & -1 \\ 1 & 6 & -2 \end{bmatrix}$.

7. $\begin{bmatrix} 0 & 1 & 0 \\ -1 & 0 & 0 \\ 0 & 0 & 5 \end{bmatrix}$.

8. $\begin{bmatrix} 2 & 0 & 3 \\ 0 & -4 & 0 \\ -3 & 0 & 2 \end{bmatrix}$.

9. $\begin{bmatrix} 3 & 2 & 6 \\ -2 & 1 & -2 \\ -1 & -2 & -4 \end{bmatrix}$.

10. $\begin{bmatrix} 0 & -3 & 1 \\ -2 & -1 & 1 \\ 0 & 0 & 2 \end{bmatrix}$.

11. $\begin{bmatrix} 3 & 0 & -1 \\ 0 & -3 & -1 \\ 0 & 2 & -1 \end{bmatrix}$.

12. $\begin{bmatrix} 1 & 1 & -1 \\ 1 & 1 & 1 \\ -1 & 1 & 1 \end{bmatrix}$.

13. $\begin{bmatrix} 2 & -1 & 3 \\ 2 & -1 & 3 \\ 2 & -1 & 3 \end{bmatrix}$.

14. $\begin{bmatrix} 1 & 2 & 3 & 4 \\ 4 & 3 & 2 & 1 \\ 4 & 5 & 6 & 7 \\ 7 & 6 & 5 & 4 \end{bmatrix}$.

15. $\begin{bmatrix} 0 & 1 & 0 & 0 \\ -1 & 0 & 0 & 0 \\ 0 & 0 & 0 & -1 \\ 0 & 0 & 1 & 0 \end{bmatrix}$.

For Problems 16–18, solve the initial-value problem $\mathbf{x}' = A\mathbf{x}$, $\mathbf{x}(0) = \mathbf{x}_0$.

16. $A = \begin{bmatrix} -1 & 4 \\ 2 & -3 \end{bmatrix}$, $\mathbf{x}_0 = \begin{bmatrix} 3 \\ 0 \end{bmatrix}$.

17. $A = \begin{bmatrix} -1 & -6 \\ 3 & 5 \end{bmatrix}$, $\mathbf{x}_0 = \begin{bmatrix} 2 \\ 2 \end{bmatrix}$.

18. $A = \begin{bmatrix} 2 & -1 & 3 \\ 3 & 1 & 0 \\ 2 & -1 & 3 \end{bmatrix}$, $\mathbf{x}_0 = \begin{bmatrix} -4 \\ 4 \\ 4 \end{bmatrix}$.

19. Solve the initial-value problem

$$\mathbf{x}' = A\mathbf{x}, \qquad \mathbf{x}(0) = \begin{bmatrix} 1 \\ 1 \end{bmatrix}, \qquad \text{when } A = \begin{bmatrix} 0 & 4 \\ -4 & 0 \end{bmatrix}.$$

Sketch the solution in the $x_1 x_2$-plane.

20. Consider the differential equation

$$\frac{d^2 x}{dt^2} + b\frac{dx}{dt} + cx = 0, \qquad (7.4.5)$$

where b and c are constants.

(a) Show that Equation (7.4.5) can be replaced by the equivalent first-order linear system $\mathbf{x}' = A\mathbf{x}$, where $A = \begin{bmatrix} 0 & 1 \\ -c & -b \end{bmatrix}$.

(b) Show that the characteristic polynomial of A coincides with the auxiliary polynomial of Equation (7.4.5).

21. Let $\lambda = a + ib$, $b \neq 0$, be an eigenvalue of the $n \times n$ (real) matrix A with corresponding eigenvector $\mathbf{v} = \mathbf{r} + i\mathbf{s}$. Then we have shown in the text that two real-valued solutions to $\mathbf{x}' = A\mathbf{x}$ are

$$\mathbf{x}_1(t) = e^{at}[\cos bt\mathbf{r} - \sin bt\mathbf{s}],$$
$$\mathbf{x}_2(t) = e^{at}[\sin bt\mathbf{r} + \cos bt\mathbf{s}].$$

Prove that $\mathbf{x}_1$ and $\mathbf{x}_2$ are linearly independent on any interval. (You may assume that $\mathbf{r}$ and $\mathbf{s}$ are linearly independent in $\mathbb{R}^n$.)

The remaining problems in this section investigate general properties of solutions to $\mathbf{x}' = A\mathbf{x}$, where A is a nondefective matrix.

22. Let A be a 2×2 nondefective matrix. If all eigenvalues of A have negative real part, prove that every solution to $\mathbf{x}' = A\mathbf{x}$ satisfies

$$\lim_{t \to \infty} \mathbf{x}(t) = \mathbf{0}. \qquad (7.4.6)$$

23. Let A be a 2×2 nondefective matrix. If *every solution* to $\mathbf{x}' = A\mathbf{x}$ satisfies (7.4.6), prove that all eigenvalues of A have negative real part.

24. Determine the general solution to $\mathbf{x}' = A\mathbf{x}$ if

$$A = \begin{bmatrix} 0 & b \\ -b & 0 \end{bmatrix}, \qquad \text{where } b > 0.$$

Describe the behavior of the solutions.

25. Describe the behavior of the solutions to $\mathbf{x}' = A\mathbf{x}$, if

$$A = \begin{bmatrix} a & b \\ -b & a \end{bmatrix},$$

where $a < 0$ and $b > 0$.

26. What conditions on the eigenvalues of an $n \times n$ matrix A would guarantee that the system $\mathbf{x}' = A\mathbf{x}$ has at least one solution satisfying

$$\mathbf{x}(t) = \mathbf{x}_0,$$

for all t, where $\mathbf{x}_0$ is a constant vector?

27. The motion of a certain physical system is described by the system of differential equations

$$x_1' = x_2, \qquad x_2' = -bx_1 - ax_2,$$

where a and b are positive constants and $a \neq 2b$. Show that the motion of the system dies out as $t \to +\infty$.

7.5   Vector Differential Equations: Defective Coefficient Matrix

The results of the previous section enable us to solve any constant-coefficient linear system of differential equations $\mathbf{x}' = A\mathbf{x}$, provided that A is nondefective. We recall from Chapter 5 that if m denotes the multiplicity of an eigenvalue of A, then the dimension r of the corresponding eigenspace satisfies the inequality

$$1 \le r \le m,$$

and the condition for A to be nondefective is that $r = m$ for each eigenvalue; that is, the dimension of each eigenspace equals the multiplicity of the corresponding eigenvalue. (See Theorem 5.7.11.) We now turn our attention to the case when A is defective. Then, for at least one eigenvalue λ, the dimension r of the corresponding eigenspace is strictly less than the multiplicity m of the eigenvalue. In this case, there are only r linearly independent eigenvectors corresponding to λ, and so Theorem 7.4.8 will yield only r linearly independent solutions to the vector differential equation $\mathbf{x}' = A\mathbf{x}$. We must therefore find an additional $m - r$ linearly independent solutions. In order to motivate the results of this section, we consider a particular example.

Example 7.5.1   Find the general solution to

$$\mathbf{x}' = A\mathbf{x}, \qquad A = \begin{bmatrix} 0 & 1 \\ -9 & 6 \end{bmatrix}. \qquad (7.5.1)$$

Solution: We try using the technique from the previous section. The coefficient matrix A has the single eigenvalue $\lambda = 3$ of multiplicity 2, and it is straightforward to show that there is just one corresponding linearly independent eigenvector, which we may take to be $\mathbf{v}_0 = (1, 3)$. Consequently, A is defective, and we obtain only one linearly independent solution to the vector differential equation, namely,

$$\mathbf{x}_0(t) = e^{3t} \begin{bmatrix} 1 \\ 3 \end{bmatrix}. \tag{7.5.2}$$

In order to obtain a second linearly independent solution $\mathbf{x}_1(t)$ to the vector differential equation (7.5.1), we explicitly consider the equations of the linear system:

$$x_1' = x_2, \qquad x_2' = -9x_1 + 6x_2.$$

Applying the solution technique introduced in Section 7.1 yields

$$x_1(t) = c_1 e^{3t} + c_2 t e^{3t}, \qquad x_2(t) = 3c_1 e^{3t} + c_3 e^{3t}(3t + 1).$$

Consequently, the general solution to (7.5.1) is

$$\mathbf{x}(t) = \begin{bmatrix} c_1 e^{3t} + c_2 t e^{3t} \\ 3c_1 e^{3t} + c_2 e^{3t}(3t + 1) \end{bmatrix},$$

which can be written in the equivalent form

$$\mathbf{x}(t) = c_1 e^{3t} \begin{bmatrix} 1 \\ 3 \end{bmatrix} + c_2 e^{3t} \left\{ \begin{bmatrix} 0 \\ 1 \end{bmatrix} + t \begin{bmatrix} 1 \\ 3 \end{bmatrix} \right\}.$$

We see that two linearly independent solutions to the given system are

$$\mathbf{x}_0(t) = e^{3t} \begin{bmatrix} 1 \\ 3 \end{bmatrix} \qquad \text{and} \qquad \mathbf{x}_1(t) = e^{3t} \left\{ \begin{bmatrix} 0 \\ 1 \end{bmatrix} + t \begin{bmatrix} 1 \\ 3 \end{bmatrix} \right\}.$$

The solution $\mathbf{x}_0(t)$ coincides with the solution (7.5.2), which was derived using the eigenvalue/eigenvector technique of the previous section. The key point to notice is that, in this particular example, there is a second linearly independent solution of the form

$$\mathbf{x}_1(t) = e^{\lambda t}(\mathbf{v}_1 + t\mathbf{v}_0). \qquad \square$$

As a first step toward generalizing the result of the preceding example, suppose that the $n \times n$ matrix A has precisely one eigenvalue, λ, of multiplicity n, and that the dimension of the corresponding eigenspace is $n - 1$. If $\mathbf{v}_0^{(1)}, \mathbf{v}_0^{(2)}, \ldots, \mathbf{v}_0^{(n-1)}$ denote linearly independent eigenvectors corresponding to λ, then $n - 1$ linearly independent solutions to $\mathbf{x}' = A\mathbf{x}$ are

$$\mathbf{x}_0^{(i)}(t) = e^{\lambda t}\mathbf{v}_0^{(i)}, \qquad i = 1, 2, \ldots, n - 1.$$

Based on the previous example, we look for a further linearly independent solution to $\mathbf{x}' = A\mathbf{x}$ of the form

$$\mathbf{x}_1(t) = e^{\lambda t}(\mathbf{v}_1 + t\mathbf{v}_0),$$

for some eigenvector $\mathbf{v}_0$. Differentiating this proposed solution with respect to t yields

$$\mathbf{x}_1' = e^{\lambda t}[(\lambda\mathbf{v}_1 + \mathbf{v}_0) + \lambda t\mathbf{v}_0].$$

Consequently, $\mathbf{x}_1' = A\mathbf{x}_1$ provided $\mathbf{v}_1$ and $\mathbf{v}_0$ satisfy

$$e^{\lambda t}[(\lambda\mathbf{v}_1 + \mathbf{v}_0) + \lambda t\mathbf{v}_0] = e^{\lambda t}(A\mathbf{v}_1 + tA\mathbf{v}_0);$$

that is,

$$(\lambda \mathbf{v}_1 + \mathbf{v}_0) + \lambda t \mathbf{v}_0 = A\mathbf{v}_1 + tA\mathbf{v}_0.$$

Since this equation must hold for all values of t in the interval of interest, we can equate the corresponding coefficients of powers of t to obtain

$$A\mathbf{v}_0 = \lambda \mathbf{v}_0,$$
$$A\mathbf{v}_1 = \lambda \mathbf{v}_1 + \mathbf{v}_0,$$

which can be rearranged to read

$$(A - \lambda I)\mathbf{v}_0 = \mathbf{0}, \tag{7.5.3}$$
$$(A - \lambda I)\mathbf{v}_1 = \mathbf{v}_0. \tag{7.5.4}$$

Equation (7.5.3) requires that $\mathbf{v}_0$ be an eigenvector of A corresponding to the eigenvalue λ as expected. Consequently, one approach to determining $\mathbf{v}_0$ and $\mathbf{v}_1$ would be to substitute the general expression for the eigenvectors $\mathbf{v}_0$ into the right-hand side of Equation (7.5.4) and solve the resulting system of equations for $\mathbf{v}_1$. Since Equation (7.5.4) is a non-homogeneous system of equations there is no a priori reason why the system should have a solution. However, it is possible to establish that, by an appropriate choice of $\mathbf{v}_0$, consistency can be obtained and therefore $\mathbf{v}_1$ can indeed be determined. Whereas this method for obtaining $\mathbf{v}_0$ and $\mathbf{v}_1$ works well in the present situation, we prefer to introduce an equivalent method that is computationally more efficient and transparent. We proceed as follows. Left multiplying (7.5.4) by $A - \lambda I$ and using (7.5.3) yields

$$(A - \lambda I)^2 \mathbf{v}_1 = \mathbf{0}. \tag{7.5.5}$$

Any vector $\mathbf{v}_1$ satisfying (7.5.5) is called a **generalized eigenvector** of A (see Definition 5.11.1). The key point is that, in the case under consideration, it is always possible to choose a vector $\mathbf{v}_1$ satisfying $(A - \lambda I)^2 \mathbf{v}_1 = \mathbf{0}$ and $(A - \lambda I)\mathbf{v}_1 \neq \mathbf{0}$. For any such $\mathbf{v}_1$, we can use Equation (7.5.4) to *define* the corresponding $\mathbf{v}_0$ by

$$\mathbf{v}_0 = (A - \lambda I)\mathbf{v}_1.$$

Then $\mathbf{v}_0$ is an eigenvector, since

$$(A - \lambda I)\mathbf{v}_0 = (A - \lambda I)^2 \mathbf{v}_1 = \mathbf{0}.$$

Consequently, Equation (7.5.3) is also satisfied. To summarize,

$$\mathbf{x}_1(t) = e^{\lambda t}(\mathbf{v}_1 + t\mathbf{v}_0)$$

is a solution to the system $\mathbf{x}' = A\mathbf{x}$ whenever $\mathbf{v}_0$ and $\mathbf{v}_1$ satisfy

$$(A - \lambda I)^2 \mathbf{v}_1 = \mathbf{0}, \quad (A - \lambda I)\mathbf{v}_1 \neq \mathbf{0}$$
$$\mathbf{v}_0 = (A - \lambda I)\mathbf{v}_1.$$

Furthermore, the resulting n solutions $\mathbf{x}_0^{(1)}, \mathbf{x}_0^{(2)}, \ldots, \mathbf{x}_0^{(n-1)}, \mathbf{x}_1$ are linearly independent on any interval. We illustrate this solution technique with some examples.

Example 7.5.2 | Solve the vector differential equation $\mathbf{x}' = A\mathbf{x}$ if $A = \begin{bmatrix} 6 & -8 \\ 2 & -2 \end{bmatrix}$.

Solution: We first determine the eigenvalues and eigenvectors of A. We have

$$\det(A - \lambda I) = \begin{vmatrix} 6 - \lambda & -8 \\ 2 & -2 - \lambda \end{vmatrix} = (\lambda - 2)^2,$$

so that there is only one eigenvalue $\lambda = 2$, with multiplicity 2. The corresponding eigenvectors are of the form $\mathbf{v} = r(2, 1)$. Therefore, the associated eigenspace is one-dimensional and we have only a single eigenvalue/eigenvector solution to the vector differential equation. It follows from the preceding discussion that there is a second linearly independent solution to $\mathbf{x}' = A\mathbf{x}$ of the form

$$\mathbf{x}_1(t) = e^{2t}(\mathbf{v}_1 + t\mathbf{v}_0)$$

where $\mathbf{v}_1$ and $\mathbf{v}_0$ are determined from

$$(A - 2I)^2\mathbf{v}_1 = \mathbf{0}, \quad (A - 2I)\mathbf{v}_1 \neq \mathbf{0} \tag{7.5.6}$$
$$\mathbf{v}_0 = (A - 2I)\mathbf{v}_1. \tag{7.5.7}$$

In this case, we have

$$A - 2I = \begin{bmatrix} 4 & -8 \\ 2 & -4 \end{bmatrix} \quad \text{and} \quad (A - 2I)^2 = 0_2.$$

Consequently, the equations in (7.5.6) will be satisfied by any vector $\mathbf{v}_1$ such that $(A - 2I)\mathbf{v}_1 \neq \mathbf{0}$. For simplicity, we take

$$\mathbf{v}_1 = \begin{bmatrix} 1 \\ 0 \end{bmatrix}.$$

Then, from Equation (7.5.7),

$$\mathbf{v}_0 = \begin{bmatrix} 4 & -8 \\ 2 & -4 \end{bmatrix} \begin{bmatrix} 1 \\ 0 \end{bmatrix} = \begin{bmatrix} 4 \\ 2 \end{bmatrix}.$$

From the expressions here for $\mathbf{v}_0$ and $\mathbf{v}_1$, we can write down two linearly independent solutions to the vector differential equation:

$$\mathbf{x}_0(t) = e^{2t}\begin{bmatrix} 4 \\ 2 \end{bmatrix} \quad \text{and} \quad \mathbf{x}_1(t) = e^{2t}\left\{\begin{bmatrix} 1 \\ 0 \end{bmatrix} + t\begin{bmatrix} 4 \\ 2 \end{bmatrix}\right\} = e^{2t}\begin{bmatrix} 1 + 4t \\ 2t \end{bmatrix}.$$

Consequently, the general solution to the vector differential equation is

$$\mathbf{x}(t) = c_1 e^{2t}\begin{bmatrix} 4 \\ 2 \end{bmatrix} + c_2 e^{2t}\begin{bmatrix} 1 + 4t \\ 2t \end{bmatrix} = \begin{bmatrix} e^{2t}[4c_1 + c_2(1 + 4t)] \\ e^{2t}(2c_1 + 2c_2 t) \end{bmatrix}. \qquad \square$$

Example 7.5.3 | Determine the general solution to $\mathbf{x}' = A\mathbf{x}$ if $A = \begin{bmatrix} 5 & 2 & -1 \\ 1 & 6 & -1 \\ 3 & 6 & 1 \end{bmatrix}$.

Solution: A short calculation shows that

$$\det(A - \lambda I) = \begin{vmatrix} 5 - \lambda & 2 & -1 \\ 1 & 6 - \lambda & -1 \\ 3 & 6 & 1 - \lambda \end{vmatrix} = -(\lambda - 4)^3.$$

We see that A has a single eigenvalue, $\lambda = 4$, of multiplicity 3. It is easily shown that the associated eigenvectors are of the form

$$\mathbf{v}_0 = (-2a + b, a, b) = a(-2, 1, 0) + b(1, 0, 1), \tag{7.5.8}$$

which means that only two linearly independent solutions $\mathbf{x}_0^{(1)}$ and $\mathbf{x}_0^{(2)}$ can be constructed solely from the eigenvectors of A. Consequently, we seek a third linearly independent solution $\mathbf{x}_1(t)$ to the vector differential equation of the form

$$\mathbf{x}_1(t) = e^{4t}(\mathbf{v}_1 + t\mathbf{v}_0),$$

where $\mathbf{v}_1$ and $\mathbf{v}_0$ are determined from

$$(A - 4I)^2\mathbf{v}_1 = \mathbf{0}, \quad (A - 4I)\mathbf{v}_1 \neq \mathbf{0}. \tag{7.5.9}$$
$$\mathbf{v}_0 = (A - 4I)\mathbf{v}_1. \tag{7.5.10}$$

In this example we have

$$A - 4I = \begin{bmatrix} 1 & 2 & -1 \\ 1 & 2 & -1 \\ 3 & 6 & -3 \end{bmatrix} \quad \text{and} \quad (A - 4I)^2 = 0_3.$$

By (7.5.9), we can therefore choose $\mathbf{v}_1$ to be any vector such that $(A - 4I)\mathbf{v}_1 \neq \mathbf{0}$. For simplicity, we take

$$\mathbf{v}_1 = \begin{bmatrix} 1 \\ 0 \\ 0 \end{bmatrix}.$$

Then, from Equation (7.5.10),

$$\mathbf{v}_0 = \begin{bmatrix} 1 & 2 & -1 \\ 1 & 2 & -1 \\ 3 & 6 & -3 \end{bmatrix} \begin{bmatrix} 1 \\ 0 \\ 0 \end{bmatrix} = \begin{bmatrix} 1 \\ 1 \\ 3 \end{bmatrix}.$$

Therefore, we obtain two linearly independent solutions to the vector differential equation, namely,

$$\mathbf{x}_0^{(1)}(t) = e^{4t} \begin{bmatrix} 1 \\ 1 \\ 3 \end{bmatrix} \quad \text{and} \quad \mathbf{x}_1^{(1)}(t) = e^{4t} \left\{ \begin{bmatrix} 1 \\ 0 \\ 0 \end{bmatrix} + t \begin{bmatrix} 1 \\ 1 \\ 3 \end{bmatrix} \right\} = e^{4t} \begin{bmatrix} 1+t \\ t \\ 3t \end{bmatrix}.$$

Finally, we obtain a third linearly independent solution to $\mathbf{x}' = A\mathbf{x}$ by choosing a second eigenvector $\mathbf{v}_0^{(2)}$ that is non-proportional to $\mathbf{v}_0$. From (7.5.8), we may choose $\mathbf{v}_0^{(2)} = (-2, 1, 0)$. Thus, $\mathbf{x}_0^{(2)}(t) = e^{4t} \begin{bmatrix} -2 \\ 1 \\ 0 \end{bmatrix}$. Consequently, the general solution to the given vector differential equation is

$$\mathbf{x}(t) = c_1 e^{4t} \begin{bmatrix} 1 \\ 1 \\ 3 \end{bmatrix} + c_2 e^{4t} \begin{bmatrix} 1+t \\ t \\ 3t \end{bmatrix} + c_3 e^{4t} \begin{bmatrix} -2 \\ 1 \\ 0 \end{bmatrix},$$

or equivalently,

$$\mathbf{x}(t) = e^{4t} \begin{bmatrix} c_1 + c_2(1+t) - 2c_3 \\ c_1 + c_2 t + c_3 \\ 3c_1 + 3c_2 t \end{bmatrix}. \qquad \square$$

Our next theorem tells us how the preceding technique generalizes to the case of an arbitrary defective coefficient matrix.

Theorem 7.5.4   Let A be an $n \times n$ matrix.

1. Suppose the eigenvalue λ has multiplicity m and that the dimension of the corresponding eigenspace is $r \leq m$. Then there exist m linearly independent solutions to $\mathbf{x}' = A\mathbf{x}$ that can be constructed in r cycles (one for each i with $1 \leq i \leq r$) of the form

$$\mathbf{x}_0^{(i)}(t) = e^{\lambda t} \mathbf{v}_0^{(i)}, \tag{7.5.11}$$

$$\mathbf{x}_1^{(i)}(t) = e^{\lambda t} \left(\mathbf{v}_1^{(i)} + t\mathbf{v}_0^{(i)} \right), \tag{7.5.12}$$

$$\mathbf{x}_2^{(i)}(t) = e^{\lambda t} \left(\mathbf{v}_2^{(i)} + t\mathbf{v}_1^{(i)} + \frac{1}{2!}t^2\mathbf{v}_0^{(i)} \right), \tag{7.5.13}$$

$$\vdots$$

$$\mathbf{x}_{k_i}^{(i)}(t) = e^{\lambda t} \left(\mathbf{v}_{k_i}^{(i)} + t\mathbf{v}_{k_i-1}^{(i)} + \cdots + \frac{1}{k_i!}t^{k_i}\mathbf{v}_0^{(i)} \right), \tag{7.5.14}$$

where each $k_i \geq 0$, and $k_1 + k_2 + \cdots + k_r = m - r$. For each i, the vectors $\mathbf{v}_0^{(i)}, \mathbf{v}_1^{(i)}, \ldots, \mathbf{v}_{k_i}^{(i)}$ can be determined as follows. Choose $\mathbf{v}_{k_i}$ to be any vector satisfying

$$(A - \lambda I)^{k_i+1} \mathbf{v}_{k_i}^{(i)} = \mathbf{0}, \quad (A - \lambda I)^{k_i} \mathbf{v}_{k_i}^{(i)} \neq \mathbf{0}. \tag{7.5.15}$$

Then:

$$\mathbf{v}_{k_i-1}^{(i)} = (A - \lambda I)\mathbf{v}_{k_i}^{(i)}, \tag{7.5.16}$$

$$\mathbf{v}_{k_i-2}^{(i)} = (A - \lambda I)^2\mathbf{v}_{k_i}^{(i)}, \tag{7.5.17}$$

$$\vdots$$

$$\mathbf{v}_1^{(i)} = (A - \lambda I)^{k_i-1}\mathbf{v}_{k_i}^{(i)}, \tag{7.5.18}$$

$$\mathbf{v}_0^{(i)} = (A - \lambda I)^{k_i}\mathbf{v}_{k_i}^{(i)}. \tag{7.5.19}$$

The collection of all solutions $\mathbf{x}_j^{(i)}(t)$ (for $1 \leq i \leq r$ and $0 \leq j \leq k_i$) is a linearly independent set of m solutions for $\mathbf{x}' = A\mathbf{x}$.

2. Applying the results of (1) to each eigenvalue of A generates a set of n solutions to $\mathbf{x}' = A\mathbf{x}$ that are linearly independent on any interval.

Remarks

1. Note that if $k_i = 0$ for each i and for each eigenvalue λ, then $m = r$ for each eigenvalue λ, which is the case in which the matrix A is nondefective, and the result here is compatible with the results in the previous section.

2. A proof of Theorem 7.5.4 based on the Jordan canonical form of an $n \times n$ matrix A is given in Problem 19. For readers who have studied this concept (Section 5.11), observe that r is the number of Jordan blocks corresponding to λ in the Jordan canonical form of A, the numbers $k_i + 1$ are the sizes of the Jordan

blocks corresponding to the eigenvalue λ, and each ordered collection of vectors $\{\mathbf{v}_0^{(i)}, \mathbf{v}_1^{(i)}, \mathbf{v}_2^{(i)}, \ldots, \mathbf{v}_{k_i}^{(i)}\}$ is a cycle of generalized eigenvectors of A corresponding to λ.

We conclude this section with examples to illustrate Theorem 7.5.4.

Example 7.5.5   Find the general solution to $\mathbf{x}' = A\mathbf{x}$ if $A = \begin{bmatrix} 1 & 2 & 0 \\ 1 & 1 & 2 \\ 0 & -1 & 1 \end{bmatrix}$.

Solution: The given matrix has characteristic polynomial

$$\det(A - \lambda I) = \begin{vmatrix} 1 - \lambda & 2 & 0 \\ 1 & 1 - \lambda & 2 \\ 0 & -1 & 1 - \lambda \end{vmatrix} = -(\lambda - 1)^3.$$

Hence, $\lambda = 1$ is the only eigenvalue of A, of multiplicity 3. The corresponding eigenvectors are of the form

$$\mathbf{v}_0 = a(-2, 0, 1),$$

so that the eigenspace has dimension 1. Therefore, $r = 1$ in Theorem 7.5.4, and the only value of i we must consider is $i = 1$. Therefore, we will omit the superscript (i) from the notation used in the theorem. We obtain a solution to $\mathbf{x}' = A\mathbf{x}$ of the form

$$\mathbf{x}_0(t) = e^t \mathbf{v}_0. \tag{7.5.20}$$

According to Theorem 7.5.4, there exist two further linearly independent solutions to $\mathbf{x}' = A\mathbf{x}$ of the form

$$\mathbf{x}_1(t) = e^t (\mathbf{v}_1 + t\mathbf{v}_0), \tag{7.5.21}$$

$$\mathbf{x}_2(t) = e^t (\mathbf{v}_2 + t\mathbf{v}_1 + \frac{1}{2!}t^2\mathbf{v}_0), \tag{7.5.22}$$

where

$$(A - I)^3 \mathbf{v}_2 = \mathbf{0}, \quad (A - I)^2 \mathbf{v}_2 \neq \mathbf{0}$$

and

$$\mathbf{v}_1 = (A - I)\mathbf{v}_2$$
$$\mathbf{v}_0 = (A - I)^2 \mathbf{v}_2.$$

A short calculation shows that

$$A - I = \begin{bmatrix} 0 & 2 & 0 \\ 1 & 0 & 2 \\ 0 & -1 & 0 \end{bmatrix}, \quad (A - I)^2 = \begin{bmatrix} 2 & 0 & 4 \\ 0 & 0 & 0 \\ -1 & 0 & -2 \end{bmatrix}, \quad (A - I)^3 = 0_3,$$

so we may take

$$\mathbf{v}_2 = \begin{bmatrix} 1 \\ 0 \\ 0 \end{bmatrix}, \quad \mathbf{v}_1 = (A - I)\mathbf{v}_2 = \begin{bmatrix} 0 \\ 1 \\ 0 \end{bmatrix}, \quad \mathbf{v}_0 = (A - I)^2 \mathbf{v}_2 = \begin{bmatrix} 2 \\ 0 \\ -1 \end{bmatrix}.$$

Substituting these vectors into the expressions (7.5.20), (7.5.21), and (7.5.22) for $\mathbf{x}_0(t)$, $\mathbf{x}_1(t)$, and $\mathbf{x}_2(t)$ gives us

$$\mathbf{x}_0(t) = e^t \begin{bmatrix} 2 \\ 0 \\ -1 \end{bmatrix},$$

$$\mathbf{x}_1(t) = e^t \left\{ \begin{bmatrix} 0 \\ 1 \\ 0 \end{bmatrix} + t \begin{bmatrix} 2 \\ 0 \\ -1 \end{bmatrix} \right\} = \begin{bmatrix} 2te^t \\ e^t \\ -te^t \end{bmatrix},$$

and

$$\mathbf{x}_2(t) = e^t \left\{ \begin{bmatrix} 1 \\ 0 \\ 0 \end{bmatrix} + t \begin{bmatrix} 0 \\ 1 \\ 0 \end{bmatrix} + \frac{1}{2!} t^2 \begin{bmatrix} 2 \\ 0 \\ -1 \end{bmatrix} \right\} = \begin{bmatrix} e^t(1 + t^2) \\ te^t \\ -\frac{1}{2} t^2 e^t \end{bmatrix}.$$

Hence, the general solution to the vector differential equation is

$$\mathbf{x}(t) = c_1 e^t \begin{bmatrix} 2 \\ 0 \\ -1 \end{bmatrix} + c_2 \begin{bmatrix} 2te^t \\ e^t \\ -te^t \end{bmatrix} + c_3 \begin{bmatrix} e^t(1 + t^2) \\ te^t \\ -\frac{1}{2} t^2 e^t \end{bmatrix}$$

$$= \begin{bmatrix} e^t[2c_1 + 2c_2 t + c_3(1 + t^2)] \\ e^t(c_2 + c_3 t) \\ e^t(-c_1 - c_2 t - \frac{1}{2} c_3 t^2) \end{bmatrix}. \qquad \square$$

Example 7.5.6   Find the general solution to $\mathbf{x}' = A\mathbf{x}$ if $A = \begin{bmatrix} 2 & 1 & 0 & 0 \\ 0 & 2 & 0 & 0 \\ 0 & 0 & 2 & 1 \\ 0 & 0 & 0 & 2 \end{bmatrix}$.

Solution: The matrix A has characteristic polynomial

$$\det(A - \lambda I) = \begin{vmatrix} 2 - \lambda & 1 & 0 & 0 \\ 0 & 2 - \lambda & 0 & 0 \\ 0 & 0 & 2 - \lambda & 1 \\ 0 & 0 & 0 & 2 - \lambda \end{vmatrix} = (2 - \lambda)^4.$$

Hence, $\lambda = 2$ is the only eigenvalue, with multiplicity 4. The corresponding eigenvectors are of the form

$$\mathbf{v}_0 = a(1, 0, 0, 0) + b(0, 0, 1, 0).$$

Consequently, since the eigenspace corresponding to $\lambda = 2$ is two-dimensional, Theorem 7.5.4 guarantees that we have two cycles of linearly independent solutions to $\mathbf{x}' = A\mathbf{x}$. In general, two cycles that produce four generalized eigenvectors must consist either of one cycle of length one and one cycle of length three, or two cycles of length two. In this case, however, we find that

$$A - 2I = \begin{bmatrix} 0 & 1 & 0 & 0 \\ 0 & 0 & 0 & 0 \\ 0 & 0 & 0 & 1 \\ 0 & 0 & 0 & 0 \end{bmatrix}, \quad (A - 2I)^2 = 0_4, \qquad (7.5.23)$$

so no cycles of length 3 are possible (a cycle of length 3 would have to contain a nonzero vector of the form $(A - 2I)^2 \mathbf{v}_2$, which cannot exist in this case). Therefore, we must seek two cycles of length two. From (7.5.23) we see that the two linearly independent vectors

$$\mathbf{v}_1^{(1)} = \begin{bmatrix} 0 \\ 1 \\ 0 \\ 0 \end{bmatrix} \qquad \text{and} \qquad \mathbf{v}_1^{(2)} = \begin{bmatrix} 0 \\ 0 \\ 0 \\ 1 \end{bmatrix}$$

satisfy

$$(A-2I)^2\mathbf{v}_1^{(1)} = \mathbf{0}, \quad (A-2I)\mathbf{v}_1^{(1)} \neq \mathbf{0} \quad \text{and} \quad (A-2I)^2\mathbf{v}_1^{(2)} = \mathbf{0}, \quad (A-2I)\mathbf{v}_1^{(2)} \neq \mathbf{0}.$$

Consequently, we have

$$\mathbf{v}_0^{(1)} = (A-2I)\mathbf{v}_1^{(1)} = \begin{bmatrix} 1 \\ 0 \\ 0 \\ 0 \end{bmatrix} \quad \text{and} \quad \mathbf{v}_0^{(2)} = (A-2I)\mathbf{v}_1^{(2)} = \begin{bmatrix} 0 \\ 0 \\ 1 \\ 0 \end{bmatrix}.$$

Hence, using Theorem 7.5.4, we obtain the solutions

$$\mathbf{x}_0^{(1)} = e^{2t}\begin{bmatrix} 1 \\ 0 \\ 0 \\ 0 \end{bmatrix}, \quad \mathbf{x}_1^{(1)} = e^{2t}\left\{ \begin{bmatrix} 0 \\ 1 \\ 0 \\ 0 \end{bmatrix} + t\begin{bmatrix} 1 \\ 0 \\ 0 \\ 0 \end{bmatrix} \right\} = e^{2t}\begin{bmatrix} t \\ 1 \\ 0 \\ 0 \end{bmatrix} \tag{7.5.24}$$

and

$$\mathbf{x}_0^{(2)} = e^{2t}\begin{bmatrix} 0 \\ 0 \\ 1 \\ 0 \end{bmatrix}, \quad \mathbf{x}_1^{(2)} = e^{2t}\left\{ \begin{bmatrix} 0 \\ 0 \\ 0 \\ 1 \end{bmatrix} + t\begin{bmatrix} 0 \\ 0 \\ 1 \\ 0 \end{bmatrix} \right\} = e^{2t}\begin{bmatrix} 0 \\ 0 \\ t \\ 1 \end{bmatrix} \tag{7.5.25}$$

to $\mathbf{x}' = A\mathbf{x}$. Equations (7.5.24) and (7.5.25) provide four linearly independent solutions to $\mathbf{x}' = A\mathbf{x}$. Therefore, the general solution to this system of differential equations is

$$\mathbf{x}(t) = c_1 e^{2t}\begin{bmatrix} 1 \\ 0 \\ 0 \\ 0 \end{bmatrix} + c_2 e^{2t}\begin{bmatrix} 0 \\ 0 \\ 1 \\ 0 \end{bmatrix} + c_3 e^{2t}\begin{bmatrix} t \\ 1 \\ 0 \\ 0 \end{bmatrix} + c_4 e^{2t}\begin{bmatrix} 0 \\ 0 \\ t \\ 1 \end{bmatrix} = e^{2t}\begin{bmatrix} c_1 + c_3 t \\ c_3 \\ c_2 + t c_4 \\ c_4 \end{bmatrix}. \quad \square$$

Example 7.5.7   Find the general solution to $\mathbf{x}' = A\mathbf{x}$ if $A = \begin{bmatrix} 2 & 1 & 0 & 0 \\ 0 & 2 & 1 & 0 \\ 0 & 0 & 2 & 0 \\ 0 & 0 & 0 & 2 \end{bmatrix}$.

Solution: Here, A has the same characteristic polynomial as in the previous example, with $\lambda = 2$ occurring as an eigenvalue of multiplicity 4. The corresponding eigenvectors in this case are of the form

$$\mathbf{v}_0 = a(1, 0, 0, 0) + b(0, 0, 0, 1), \tag{7.5.26}$$

and so once more we have two cycles of linearly independent solutions to $\mathbf{x}' = A\mathbf{x}$. In this case, we observe that

$$(A-2I)^2 = \begin{bmatrix} 0 & 0 & 1 & 0 \\ 0 & 0 & 0 & 0 \\ 0 & 0 & 0 & 0 \\ 0 & 0 & 0 & 0 \end{bmatrix} \neq 0_4, \quad (A-2I)^3 = 0_4.$$

Therefore, in contrast to the preceding example, the matrix $(A-2I)^2$ here is nonzero. Consequently, it is possible to find a vector $\mathbf{v}_2^{(1)}$ such that

$$(A-2I)^3\mathbf{v}_2^{(1)} = \mathbf{0} \quad \text{and} \quad (A-2I)^2\mathbf{v}_2^{(1)} \neq \mathbf{0}.$$

An appropriate choice is

$$\mathbf{v}_2^{(1)} = \begin{bmatrix} 0 \\ 0 \\ 1 \\ 0 \end{bmatrix}.$$

Therefore, we have

$$\mathbf{v}_2^{(1)} = \begin{bmatrix} 0 \\ 0 \\ 1 \\ 0 \end{bmatrix}, \qquad \mathbf{v}_1^{(1)} = (A - 2I)\mathbf{v}_2^{(1)} = \begin{bmatrix} 0 \\ 1 \\ 0 \\ 0 \end{bmatrix}, \qquad \mathbf{v}_0^{(1)} = (A - 2I)^2\mathbf{v}_2^{(1)} = \begin{bmatrix} 1 \\ 0 \\ 0 \\ 0 \end{bmatrix}.$$

Using (7.5.26), we can select an eigenvector that is non-proportional to $\mathbf{v}_0^{(1)}$, say

$$\mathbf{v}_0^{(2)} = \begin{bmatrix} 0 \\ 0 \\ 0 \\ 1 \end{bmatrix}.$$

Theorem 7.5.4 supplies four linearly independent solutions to $\mathbf{x}' = A\mathbf{x}$, namely,

$$\mathbf{x}_0^{(1)}(t) = e^{2t} \begin{bmatrix} 1 \\ 0 \\ 0 \\ 0 \end{bmatrix}, \qquad \mathbf{x}_1^{(1)}(t) = e^{2t} \left\{ \begin{bmatrix} 0 \\ 1 \\ 0 \\ 0 \end{bmatrix} + t \begin{bmatrix} 1 \\ 0 \\ 0 \\ 0 \end{bmatrix} \right\},$$

$$\mathbf{x}_2^{(1)}(t) = e^{2t} \left\{ \begin{bmatrix} 0 \\ 0 \\ 1 \\ 0 \end{bmatrix} + t \begin{bmatrix} 0 \\ 1 \\ 0 \\ 0 \end{bmatrix} + \frac{t^2}{2!} \begin{bmatrix} 1 \\ 0 \\ 0 \\ 0 \end{bmatrix} \right\},$$

and

$$\mathbf{x}_0^{(2)}(t) = e^{2t} \begin{bmatrix} 0 \\ 0 \\ 0 \\ 1 \end{bmatrix}.$$

We can now give the general solution:

$$\mathbf{x}(t) = c_1 e^{2t} \begin{bmatrix} 0 \\ 0 \\ 0 \\ 1 \end{bmatrix} + c_2 e^{2t} \begin{bmatrix} 1 \\ 0 \\ 0 \\ 0 \end{bmatrix} + c_3 e^{2t} \left\{ \begin{bmatrix} 0 \\ 1 \\ 0 \\ 0 \end{bmatrix} + t \begin{bmatrix} 1 \\ 0 \\ 0 \\ 0 \end{bmatrix} \right\}$$

$$+ c_4 e^{2t} \left\{ \begin{bmatrix} 0 \\ 0 \\ 1 \\ 0 \end{bmatrix} + t \begin{bmatrix} 0 \\ 1 \\ 0 \\ 0 \end{bmatrix} + \frac{t^2}{2!} \begin{bmatrix} 1 \\ 0 \\ 0 \\ 0 \end{bmatrix} \right\} = e^{2t} \begin{bmatrix} c_2 + c_3 t + \frac{c_4}{2} t^2 \\ c_3 + c_4 t \\ c_4 \\ c_1 \end{bmatrix}.$$

□

Remark   The difference between the last two examples is somewhat subtle. In both cases, the eigenspace corresponding to $\lambda = 2$ is two-dimensional, so according to Theorem 7.5.4, two cycles of (linearly independent) generalized eigenvectors can be constructed. In Example 7.5.6, however, both cycles have length two, and in Example 7.5.7, one cycle has length three while the other has length one. From a computational point of view, the reason for this difference is that $(A - \lambda I)^2$ is the zero matrix in the former example (so that no cycles of length greater than 2 are possible), but nonzero in the latter example (so that a cycle of length 3 is possible). Readers who have studied the Jordan canonical form (Section 5.11) should observe that $\text{JCF}(A)$ contains the information about what the lengths of the cycles of generalized eigenvectors of A are. These lengths are precisely the sizes of the Jordan blocks comprising $\text{JCF}(A)$. Therefore, if we are given what $\text{JCF}(A)$ is in advance, then the task of determining the requisite cycles of generalized eigenvectors of A is substantially simplified.

Exercises for 7.5

Key Terms

Cycle of generalized eigenvectors.

Skills

- Be able to solve the vector differential equation $\mathbf{x}' = A\mathbf{x}$ in the case where A is a defective matrix by constructing cycles of generalized eigenvectors and forming corresponding solutions $\mathbf{x}_k(t)$ as in Theorem 7.5.4.

True-False Review

For Questions 1–4, decide if the given statement is **true** or **false**, and give a brief justification for your answer. If true, you can quote a relevant definition or theorem from the text. If false, provide an example, illustration, or brief explanation of why the statement is false.

1. If A is an $n \times n$ defective matrix, then the vector differential equation $\mathbf{x}' = A\mathbf{x}$ cannot have n linearly independent solutions.

2. A cycle of generalized eigenvectors of A corresponding to an eigenvalue λ consisting of k vectors yields k linearly independent solutions to the vector differential equation $\mathbf{x}' = A\mathbf{x}$.

3. The number of linearly independent solutions to $\mathbf{x}' = A\mathbf{x}$ corresponding to λ is equal to the dimension of the eigenspace E_λ.

4. If $\mathbf{v}_0$ is an eigenvector of A corresponding to λ and $\mathbf{v}_1$ is a vector satisfying $(A - \lambda I)\mathbf{v}_1 = \mathbf{v}_0$, then $\mathbf{x}(t) = e^{\lambda t}\mathbf{v}_1$ is a solution to the vector differential equation $\mathbf{x}' = A\mathbf{x}$.

Problems

For Problems 1–13, determine the general solution to the system $\mathbf{x}' = A\mathbf{x}$ for the given matrix A.

1. $\begin{bmatrix} 0 & -2 \\ 2 & 4 \end{bmatrix}$.

2. $\begin{bmatrix} -3 & -2 \\ 2 & 1 \end{bmatrix}$.

3. $\begin{bmatrix} 0 & 1 & 0 \\ 0 & 0 & 1 \\ 1 & 1 & -1 \end{bmatrix}$.

4. $\begin{bmatrix} 2 & 2 & -1 \\ 2 & 1 & -1 \\ 2 & 3 & -1 \end{bmatrix}$.

5. $\begin{bmatrix} -2 & 0 & 0 \\ 1 & -3 & -1 \\ -1 & 1 & -1 \end{bmatrix}$.

6. $\begin{bmatrix} 15 & -32 & 12 \\ 8 & -17 & 6 \\ 0 & 0 & -1 \end{bmatrix}$.

7. $\begin{bmatrix} 4 & 0 & 0 \\ 1 & 4 & 0 \\ 0 & 1 & 4 \end{bmatrix}$.

8. $\begin{bmatrix} 1 & 0 & 0 \\ 0 & 3 & 2 \\ 2 & -2 & -1 \end{bmatrix}$.

9. $\begin{bmatrix} 3 & 1 & 0 \\ -1 & 5 & 0 \\ 0 & 0 & 4 \end{bmatrix}$.

10. $\begin{bmatrix} -1 & 1 & 0 \\ -2 & -3 & 1 \\ 1 & 1 & -2 \end{bmatrix}$.

11. $\begin{bmatrix} 0 & -1 & 0 & 0 \\ 1 & 0 & 0 & 0 \\ 1 & 0 & 2 & 1 \\ 0 & 1 & 0 & 2 \end{bmatrix}$.

12.
$$\begin{bmatrix} -2 & 3 & 0 & 0 \\ 3 & -2 & 0 & 0 \\ 1 & 0 & 1 & -1 \\ 0 & 1 & 0 & 1 \end{bmatrix}.$$

13.
$$\begin{bmatrix} 0 & -1 & 0 & 0 \\ 1 & 0 & 0 & 0 \\ 1 & 0 & 0 & -1 \\ 0 & 1 & 1 & 0 \end{bmatrix}.$$

For Problems 14–15, solve the initial-value problem.

14. $\mathbf{x}' = A\mathbf{x}, \mathbf{x}(0) = \mathbf{x}_0$, where

$$A = \begin{bmatrix} -2 & -1 \\ 1 & -4 \end{bmatrix}, \quad \mathbf{x}_0 = \begin{bmatrix} 0 \\ -1 \end{bmatrix}.$$

15. $\mathbf{x}' = A\mathbf{x}, \mathbf{x}(0) = \mathbf{x}_0$, where

$$A = \begin{bmatrix} -2 & -1 & 4 \\ 0 & -1 & 0 \\ -1 & -3 & 2 \end{bmatrix}, \quad \mathbf{x}_0 = \begin{bmatrix} -2 \\ 1 \\ 1 \end{bmatrix}.$$

16. Show that if the vector differential equation $\mathbf{x}' = A\mathbf{x}$ has a solution of the form

$$\mathbf{x}(t) = e^{\lambda t}\left[\mathbf{v}_2 + t\mathbf{v}_1 + \frac{t^2}{2!}\mathbf{v}_2\right],$$

then

$$(A - \lambda I)\mathbf{v}_0 = \mathbf{0}, (A - \lambda I)\mathbf{v}_1 = \mathbf{v}_0, \text{ and}$$

$$(A - \lambda I)\mathbf{v}_2 = \mathbf{v}_1.$$

17. Let A be a 2×2 real matrix. Prove that all solutions to $\mathbf{x}' = A\mathbf{x}$ satisfy

$$\lim_{t \to \infty} \mathbf{x}(t) = \mathbf{0}$$

if and only if all eigenvalues of A have negative real part.

18. Extend the result of the previous exercise to the system $\mathbf{x}' = A\mathbf{x}$, where A is an arbitrary (real) $n \times n$ matrix.

19. This problem outlines a proof of Theorem 7.5.4 using results from the optional section on Jordan canonical forms, Section 5.11.

(a) Conclude from the summary preceding Example 5.11.11 that there are r cycles of generalized eigenvectors of A corresponding to λ. Let the lengths of these cycles be $l_1, l_2, \ldots, l_r$, respectively.

(b) How are k_i and l_i related for each i? Show that $k_i \geq 0$ for each i and that $k_1 + k_2 + \cdots + k_r = m - r$.

(c) Conclude that for each i we have a cycle of generalized eigenvectors $\{\mathbf{v}_0^{(i)}, \mathbf{v}_1^{(i)}, \ldots, \mathbf{v}_{k_i}^{(i)}\}$ satisfying (7.5.15)–(7.5.19).

(d) By Theorem 5.11.10, the vectors in the cycle in part (c) are linearly independent. Conclude that the corresponding vector functions in (7.5.11)–(7.5.19) are linearly independent.

(e) Show that the functions defined in (7.5.11)–(7.5.14) whose terms satisfy (7.5.15)–(7.5.17) are solutions to the vector differential equation $\mathbf{x}' = A\mathbf{x}$. This proves part 1 of Theorem 7.5.4.

(f) Deduce part 2 of Theorem 7.5.4.

7.6 Variation-of-Parameters for Linear Systems

We now consider solving the nonhomogeneous vector differential equation

$$\mathbf{x}'(t) = A(t)\mathbf{x}(t) + \mathbf{b}(t), \tag{7.6.1}$$

where A is an $n \times n$ matrix function and $\mathbf{b}$ is a column n-vector function. The homogeneous equation associated with Equation (7.6.1) is

$$\mathbf{x}'(t) = A(t)\mathbf{x}(t). \tag{7.6.2}$$

According to Theorem 7.3.6, every solution to the system (7.6.1) is of the form

$$\mathbf{x}(t) = c_1\mathbf{x}_1(t) + c_2\mathbf{x}_2(t) + \cdots + c_n\mathbf{x}_n(t) + \mathbf{x}_p(t),$$

where $\mathbf{x}_1, \mathbf{x}_2, \ldots, \mathbf{x}_n$ are n linearly independent solutions to the associated homogeneous system (7.6.2), and $\mathbf{x}_p$ is a *particular* solution to (7.6.1). In this section, we derive the variation-of-parameters method for determining $\mathbf{x}_p$, assuming that we know n linearly independent solutions to (7.6.2).

| Theorem 7.6.1 | **(Variation-of-Parameters Method)** |

Let $A(t)$ be an $n \times n$ matrix function and let $\mathbf{b}(t)$ be a column n-vector, both of which are continuous on an interval I. If $X(t) = [\mathbf{x}_1(t), \mathbf{x}_2(t), \ldots, \mathbf{x}_n(t)]$ is a fundamental matrix for $\mathbf{x}'(t) = A(t)\mathbf{x}(t)$, then a particular solution to

$$\mathbf{x}'(t) = A(t)\mathbf{x}(t) + \mathbf{b}(t) \tag{7.6.3}$$

is

$$\mathbf{x}_p(t) = X(t)\mathbf{u}(t),$$

where $\mathbf{u}(t)$ satisfies

$$X(t)\mathbf{u}'(t) = \mathbf{b}(t).$$

Explicitly,

$$\mathbf{x}_p(t) = X(t) \int^t X^{-1}(s)\mathbf{b}(s)\, ds.$$

Proof The general solution to $\mathbf{x}'(t) = A(t)\mathbf{x}(t)$ is

$$\mathbf{x}_c(t) = c_1\mathbf{x}_1(t) + c_2\mathbf{x}_2(t) + \cdots + c_n\mathbf{x}_n(t),$$

which can be written in the form

$$\mathbf{x}_c(t) = X(t)\mathbf{c},$$

where $X(t) = [\mathbf{x}_1(t), \mathbf{x}_2(t), \ldots, \mathbf{x}_n(t)]$ and $\mathbf{c} = [c_1\ c_2\ \ldots\ c_n]^T$. We try for a particular solution to Equation (7.6.3) of the form

$$\mathbf{x}_p(t) = X(t)\mathbf{u}(t), \tag{7.6.4}$$

where[4] $\mathbf{u}(t) = [u_1(t)\ u_2(t)\ \ldots\ u_n(t)]^T$. Substituting (7.6.4) into (7.6.3), it follows that $\mathbf{x}_p$ is a solution to (7.6.3), provided that $\mathbf{u}$ satisfies

$$(X\mathbf{u})' = A(X\mathbf{u}) + \mathbf{b}. \tag{7.6.5}$$

Applying the product rule for differentiation to the left-hand side of Equation (7.6.5), we obtain

$$X'\mathbf{u} + X\mathbf{u}' = A(X\mathbf{u}) + \mathbf{b}. \tag{7.6.6}$$

By definition, we have

$$X = [\mathbf{x}_1, \mathbf{x}_2, \ldots, \mathbf{x}_n],$$

so that

$$X' = [\mathbf{x}_1', \mathbf{x}_2', \ldots, \mathbf{x}_n']. \tag{7.6.7}$$

Since each of the vector functions $\mathbf{x} = \mathbf{x}_i$ is a solution to the associated homogeneous equation $\mathbf{x}' = A\mathbf{x}$, we can write (7.6.7) in the form

$$X' = [A\mathbf{x}_1, A\mathbf{x}_2, \ldots, A\mathbf{x}_n].$$

That is,

$$X' = AX.$$

---

[4]That is, we replace the constants in $\mathbf{x}_c$ by arbitrary functions.

Substituting this expression for X' into (7.6.6) yields

$$(AX)\mathbf{u} + X\mathbf{u}' = A(X\mathbf{u}) + \mathbf{b},$$

so that

$$X\mathbf{u}' = \mathbf{b}.$$

This implies that[5]

$$\mathbf{u}' = X^{-1}\mathbf{b}.$$

Consequently,

$$\mathbf{u}(t) = \int^t X^{-1}(s)\mathbf{b}(s)\, ds$$

(we have set the integration constant to zero without loss of generality) and hence, from (7.6.4), a particular solution to (7.6.3) is

$$\mathbf{x}_p(t) = X(t) \int^t X^{-1}(s)\mathbf{b}(s)\, ds. \qquad \blacksquare$$

Remarks

1. There is no need to memorize the formula for $\mathbf{x}_p$ given in the previous theorem. Rather, it should be remembered that a particular solution to $\mathbf{x}' = A\mathbf{x} + \mathbf{b}$ is

$$\mathbf{x}_p(t) = X\mathbf{u},$$

where X is a fundamental matrix for the associated homogeneous vector differential equation and $\mathbf{u}'$ is determined by solving the linear system

$$X\mathbf{u}' = \mathbf{b}. \qquad (7.6.8)$$

2. Whereas the proof of the previous theorem used X^{-1} to obtain a simple formula for the solution to (7.6.8), in practice any of the methods for solving systems of linear equations that we have derived in the text can be applied. For 2×2 systems, Cramer's rule is quite effective. Alternatively, the inverse of X can be determined very quickly using the adjoint method. For systems bigger than 2×2, it is computationally more efficient to use Gaussian elimination to solve (7.6.8) for $\mathbf{u}'$ and then integrate the resulting vector to determine $\mathbf{u}$.

Example 7.6.2   Solve the initial-value problem

$$\mathbf{x}'(t) = A(t)\mathbf{x}(t) + \mathbf{b}(t), \qquad \mathbf{x}(0) = \begin{bmatrix} 3 \\ 0 \end{bmatrix},$$

if

$$A = \begin{bmatrix} 1 & 2 \\ 4 & 3 \end{bmatrix} \qquad \text{and} \qquad \mathbf{b}(t) = \begin{bmatrix} 12e^{3t} \\ 18e^{2t} \end{bmatrix}.$$

Solution: We first solve the associated homogeneous equation $\mathbf{x}'(t) = A(t)\mathbf{x}(t)$. For the given matrix A, we find that

$$\det(A - \lambda I) = (\lambda - 5)(\lambda + 1),$$

---

[5]Note that X^{-1} exists, since $\det(X) \neq 0$. (Why?)

so that the eigenvalues of A are $\lambda_1 = -1$ and $\lambda_2 = 5$. A quick calculation shows that the corresponding eigenvectors are, respectively,

$$\mathbf{v}_1 = \begin{bmatrix} -1 \\ 1 \end{bmatrix} \quad \text{and} \quad \mathbf{v}_2 = \begin{bmatrix} 1 \\ 2 \end{bmatrix}.$$

Consequently, two linearly independent solutions to $\mathbf{x}' = A\mathbf{x}$ are

$$\mathbf{x}_1(t) = e^{-t} \begin{bmatrix} -1 \\ 1 \end{bmatrix}, \qquad \mathbf{x}_2(t) = e^{5t} \begin{bmatrix} 1 \\ 2 \end{bmatrix},$$

and therefore a fundamental matrix for $\mathbf{x}' = A\mathbf{x}$ is

$$X(t) = \begin{bmatrix} -e^{-t} & e^{5t} \\ e^{-t} & 2e^{5t} \end{bmatrix}.$$

It follows from Theorem 7.6.1 that a particular solution to $\mathbf{x}' = A\mathbf{x} + \mathbf{b}$ is

$$\mathbf{x}_p = X\mathbf{u},$$

where

$$X\mathbf{u}' = \mathbf{b}.$$

Since this is a 2×2 system, we will solve for the components of $\mathbf{u}'$ using Cramer's rule. We have

$$\det(X(t)) = -3e^{4t},$$

so that

$$u_1'(t) = \frac{\begin{vmatrix} 12e^{3t} & e^{5t} \\ 18e^{2t} & 2e^{5t} \end{vmatrix}}{-3e^{4t}} = -8e^{4t} + 6e^{3t}$$

and

$$u_2'(t) = \frac{\begin{vmatrix} -e^{-t} & 12e^{3t} \\ e^{-t} & 18e^{2t} \end{vmatrix}}{-3e^{4t}} = 6e^{-3t} + 4e^{-2t}.$$

Integrating these two expressions yields

$$u_1(t) = -2e^{4t} + 2e^{3t} \qquad \text{and} \qquad u_2(t) = -2e^{-3t} - 2e^{-2t},$$

where we have set the integration constants to zero. Hence,

$$\mathbf{u}(t) = \begin{bmatrix} -2e^{4t} + 2e^{3t} \\ -2e^{-2t} - 2e^{-3t} \end{bmatrix}.$$

It follows that a particular solution to the given vector differential equation is

$$\mathbf{x}_p(t) = X(t)\mathbf{u}(t) = \begin{bmatrix} -e^{-t} & e^{5t} \\ e^{-t} & 2e^{5t} \end{bmatrix} \begin{bmatrix} -2e^{4t} + 2e^{3t} \\ -2e^{-2t} - 2e^{-3t} \end{bmatrix} = \begin{bmatrix} -4e^{2t} \\ -2e^{2t} - 6e^{3t} \end{bmatrix}.$$

Consequently, the general solution to the given nonhomogeneous vector differential equation is

$$\mathbf{x}(t) = c_1 e^{-t} \begin{bmatrix} -1 \\ 1 \end{bmatrix} + c_2 e^{5t} \begin{bmatrix} 1 \\ 2 \end{bmatrix} - \begin{bmatrix} 4e^{2t} \\ 2e^{2t} + 6e^{3t} \end{bmatrix},$$

or equivalently,

$$\mathbf{x}(t) = \begin{bmatrix} -c_1 e^{-t} + c_2 e^{5t} - 4e^{2t} \\ c_1 e^{-t} + 2c_2 e^{5t} - 2(e^{2t} + 3e^{3t}) \end{bmatrix}. \tag{7.6.9}$$

The initial condition $\mathbf{x}(0) = \begin{bmatrix} 3 \\ 0 \end{bmatrix}$ requires that

$$c_1 \begin{bmatrix} -1 \\ 1 \end{bmatrix} + c_2 \begin{bmatrix} 1 \\ 2 \end{bmatrix} - \begin{bmatrix} 4 \\ 8 \end{bmatrix} = \begin{bmatrix} 3 \\ 0 \end{bmatrix},$$

which yields the equations

$$-c_1 + c_2 = 7 \qquad \text{and} \qquad c_1 + 2c_2 = 8.$$

Thus, $c_1 = -2$ and $c_2 = 5$. Substituting these values into (7.6.9) yields

$$\mathbf{x}(t) = \begin{bmatrix} 2e^{-t} + 5e^{5t} - 4e^{2t} \\ -2e^{-t} + 10e^{5t} - 2(e^{2t} + 3e^{3t}) \end{bmatrix}. \qquad \square$$

Exercises for 7.6

Key Terms

Variation-of-parameters, Particular solution to a nonhomogeneous vector differential equation.

Skills

- Be able to use the variation-of-parameters technique to find a particular solution $\mathbf{x}_p$ to the vector differential equation $\mathbf{x}'(t) = A(t)\mathbf{x}(t) + \mathbf{b}(t)$ for a given matrix function A and column vector function $\mathbf{b}(t)$.

True-False Review

For Questions 1–6, decide if the given statement is **true** or **false**, and give a brief justification for your answer. If true, you can quote a relevant definition or theorem from the text. If false, provide an example, illustration, or brief explanation of why the statement is false.

1. To apply the variation-of-parameters technique for determining a particular solution to the nonhomogeneous vector differential equation $\mathbf{x}' = A\mathbf{x} + \mathbf{b}$, it is not necessary to determine any solutions to the corresponding homogeneous vector differential equation $\mathbf{x}' = A\mathbf{x}$.

2. If $X(t)$ is a fundamental matrix for $\mathbf{x}' = A\mathbf{x}$, then a particular solution to the nonhomogeneous vector differential equation $\mathbf{x}' = A\mathbf{x} + \mathbf{b}$ is $\mathbf{x}_p(t) = X(t)\mathbf{u}(t)$, where $\mathbf{u}(t)$ is any arbitrary vector function.

3. If $X(t)$ is a fundamental matrix for $\mathbf{x}' = A\mathbf{x}$, then $X'(t) = A(t)X(t)$.

4. If $\mathbf{x}_p$ is a particular solution to the vector differential equation $\mathbf{x}' = A\mathbf{x} + \mathbf{b}$, then so is $c \cdot \mathbf{x}_p$ for any nonzero constant c.

5. Only one particular solution $\mathbf{x}_p$ satisfies the nonhomogeneous vector differential equation $\mathbf{x}' = A\mathbf{x} + \mathbf{b}$.

6. The parameters $\mathbf{u}(t)$ in the equation $\mathbf{x}_p(t) = X(t)\mathbf{u}(t)$, where $X(t)$ is a fundamental matrix for $\mathbf{x}' = A\mathbf{x}$, are determined by solving the system $X(t)\mathbf{u}'(t) = \mathbf{b}(t)$ for $\mathbf{u}'(t)$ and integrating the resulting vector function.

Problems

For Problems 1–8, use the variation-of-parameters technique to find a particular solution $\mathbf{x}_p$ to $\mathbf{x}' = A\mathbf{x} + \mathbf{b}$, for the given A and $\mathbf{b}$. Also obtain the general solution to the system of differential equations.

1. $A = \begin{bmatrix} 2 & -1 \\ -1 & 2 \end{bmatrix}$, $\mathbf{b} = \begin{bmatrix} 0 \\ 4e^t \end{bmatrix}$.

2. $A = \begin{bmatrix} 4 & -3 \\ 2 & -1 \end{bmatrix}$, $\mathbf{b} = \begin{bmatrix} e^{2t} \\ e^t \end{bmatrix}$.

3. $A = \begin{bmatrix} -1 & 1 \\ 3 & 1 \end{bmatrix}$, $\mathbf{b} = \begin{bmatrix} 20e^{3t} \\ 12e^t \end{bmatrix}$.

4. $A = \begin{bmatrix} -1 & 2 \\ -2 & 4 \end{bmatrix}$, $\mathbf{b} = \begin{bmatrix} 54te^{3t} \\ 9e^{3t} \end{bmatrix}$.

5. $A = \begin{bmatrix} 2 & 4 \\ -2 & -2 \end{bmatrix}$, $\mathbf{b} = \begin{bmatrix} 8\sin 2t \\ 8\cos 2t \end{bmatrix}$.

6. $A = \begin{bmatrix} 3 & 2 \\ -2 & -1 \end{bmatrix}$, $\mathbf{b} = \begin{bmatrix} -3e^t \\ 6te^t \end{bmatrix}$.

7. $A = \begin{bmatrix} 1 & 0 & 0 \\ 2 & -3 & 2 \\ 1 & -2 & 2 \end{bmatrix}$, $\mathbf{b} = \begin{bmatrix} -e^t \\ 6e^{-t} \\ e^t \end{bmatrix}$.

8. $A = \begin{bmatrix} -1 & -2 & 2 \\ 2 & 4 & -1 \\ 0 & 0 & 3 \end{bmatrix}$, $\mathbf{b} = \begin{bmatrix} -e^{3t} \\ 4e^{3t} \\ 3e^{3t} \end{bmatrix}$.

9. Let $X(t)$ be a fundamental matrix for the system $\mathbf{x}' = A(t)\mathbf{x}(t)$, where $A(t)$ is an $n \times n$ matrix function. Show that the solution to the initial-value problem

$$\mathbf{x}'(t) = A(t)\mathbf{x}(t) + \mathbf{b}(t), \qquad \mathbf{x}(t_0) = \mathbf{x}_0$$

can be written as

$$\mathbf{x}(t) = X(t)X^{-1}(t_0)\mathbf{x}_0 + X(t)\int_{t_0}^t X^{-1}(s)\mathbf{b}(s)\,ds.$$

10. Consider the nonhomogeneous system

$$x_1' = 2x_1 - 3x_2 + 34\sin t,$$
$$x_2' = -4x_1 - 2x_2 + 17\cos t.$$

Find the general solution to this system by first solving the associated homogeneous system and then using the *method of undetermined coefficients* to obtain a particular solution. [**Hint:** The form of the nonhomogeneous term suggests a trial solution of the form

$$\mathbf{x}_p(t) = \begin{bmatrix} A_1 \cos t + B_1 \sin t \\ A_2 \cos t + B_2 \sin t \end{bmatrix},$$

where the constants A_1, A_2, B_1, and B_2 can be determined by substituting into the given system.]

7.7   Some Applications of Linear Systems of Differential Equations

In this section we analyze the two problems briefly introduced at the beginning of this chapter. We begin with the coupled spring–mass system that consists of two masses m_1 and m_2 connected by two springs whose spring constants are k_1 and k_2, respectively. (See Figure 7.7.1.) Let $x(t)$ and $y(t)$ denote the displacement of m_1 and m_2 from their equilibrium positions, respectively. When the system is in motion, the extension of spring 1 is

$$L_1(t) = x(t),$$

whereas the *net* extension of spring 2 is

$$L_2(t) = y(t) - x(t).$$

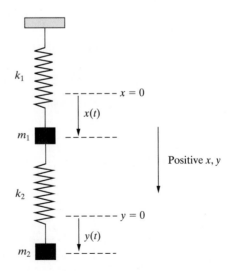

Figure 7.7.1: A coupled spring–mass system.

Consequently, using Hooke's law, the net forces acting on masses m_1 and m_2 at time t are

$$F_1(t) = -k_1 x(t) + k_2[y(t) - x(t)] \quad \text{and} \quad F_2(t) = -k_2[y(t) - x(t)],$$

respectively. Thus, applying Newton's second law to each mass yields the system of differential equations

$$m_1 \frac{d^2 x}{dt^2} = -k_1 x + k_2(y - x),$$ (7.7.1)

$$m_2 \frac{d^2 y}{dt^2} = -k_2(y - x).$$ (7.7.2)

The motion of the spring–mass system will be fully determined, once we have specified appropriate initial conditions of the form

$$x(t_0) = \alpha_1, \quad \frac{dx}{dt}(t_0) = \alpha_2, \quad y(t_0) = \alpha_3, \quad \frac{dy}{dt}(t_0) = \alpha_4,$$ (7.7.3)

where $\alpha_1, \alpha_2, \alpha_3$, and α_4 are constants.

To apply the techniques that we have developed in this chapter for solving systems of differential equations, we must convert Equations (7.7.1) and (7.7.2) into a first-order system. We introduce new variables x_1, x_2, x_3, and x_4 defined by

$$x_1 = x, \quad x_2 = x', \quad x_3 = y, \quad x_4 = y'.$$ (7.7.4)

Using (7.7.4), we can replace Equations (7.7.1) and (7.7.2) by the equivalent system

$$x_1' = x_2, \quad x_2' = -\frac{k_1}{m_1} x_1 + \frac{k_2}{m_1}(x_3 - x_1), \quad x_3' = x_4, \quad x_4' = -\frac{k_2}{m_2}(x_3 - x_1).$$

Rearranging terms yields the first-order linear system

$$x_1' = x_2,$$ (7.7.5)

$$x_2' = -\left(\frac{k_1}{m_1} + \frac{k_2}{m_1}\right) x_1 + \frac{k_2}{m_1} x_3,$$ (7.7.6)

$$x_3' = x_4,$$ (7.7.7)

$$x_4' = \frac{k_2}{m_2} x_1 - \frac{k_2}{m_2} x_3.$$ (7.7.8)

In the new variables, the initial conditions (7.7.3) are

$$x_1(t_0) = \alpha_1, \quad x_2(t_0) = \alpha_2, \quad x_3(t_0) = \alpha_3, \quad x_4(t_0) = \alpha_4.$$

This initial-value problem for x_1, x_2, x_3, x_4 can be written in vector form as

$$\mathbf{x}' = A\mathbf{x}, \quad \mathbf{x}(t_0) = \mathbf{x}_0,$$

where

$$\mathbf{x} = \begin{bmatrix} x_1 \\ x_2 \\ x_3 \\ x_4 \end{bmatrix}, \quad A = \begin{bmatrix} 0 & 1 & 0 & 0 \\ -\frac{1}{m_1}(k_1 + k_2) & 0 & \frac{k_2}{m_1} & 0 \\ 0 & 0 & 0 & 1 \\ \frac{k_2}{m_2} & 0 & -\frac{k_2}{m_2} & 0 \end{bmatrix}, \quad \mathbf{x}_0 = \begin{bmatrix} \alpha_1 \\ \alpha_2 \\ \alpha_3 \\ \alpha_4 \end{bmatrix}.$$

We leave the analysis of the general system for the exercises and consider a particular example.

Example 7.7.1   Consider the spring–mass system with

$$k_1 = 4 \text{ Nm}^{-1}, \qquad k_2 = 2 \text{ Nm}^{-1}, \qquad m_1 = 2 \text{ kg}, \qquad m_2 = 1 \text{ kg}.$$

At $t = 0$, both masses are pulled down a distance 1 m from equilibrium and released from rest. Determine the subsequent motion of the system.

Solution: The motion of the system is governed by the initial-value problem

$$2\frac{d^2x}{dt^2} = -4x + 2(y - x),$$

$$\frac{d^2y}{dt^2} = -2(y - x),$$

$$x(0) = 1, \qquad \frac{dx}{dt}(0) = 0, \qquad y(0) = 1, \qquad \frac{dy}{dt}(0) = 0.$$

Introducing new variables $x_1 = x$, $x_2 = x'$, $x_3 = y$, and $x_4 = y'$ yields the equivalent initial-value problem

$$x_1' = x_2, \quad x_2' = -3x_1 + x_3, \quad x_3' = x_4, \quad x_4' = 2x_1 - 2x_3,$$
$$x_1(0) = 1, \quad x_2(0) = 0, \quad x_3(0) = 1, \quad x_4(0) = 0.$$

In vector form, we have

$$\mathbf{x}' = A\mathbf{x}, \qquad \mathbf{x}(0) = \mathbf{x}_0, \tag{7.7.9}$$

where

$$A = \begin{bmatrix} 0 & 1 & 0 & 0 \\ -3 & 0 & 1 & 0 \\ 0 & 0 & 0 & 1 \\ 2 & 0 & -2 & 0 \end{bmatrix} \qquad \text{and} \qquad \mathbf{x}_0 = \begin{bmatrix} 1 \\ 0 \\ 1 \\ 0 \end{bmatrix}.$$

The characteristic polynomial of A is[6]

$$\det(A - \lambda I) = \begin{vmatrix} -\lambda & 1 & 0 & 0 \\ -3 & -\lambda & 1 & 0 \\ 0 & 0 & -\lambda & 1 \\ 2 & 0 & -2 & -\lambda \end{vmatrix}$$
$$= \lambda^4 + 5\lambda^2 + 4$$
$$= (\lambda^2 + 1)(\lambda^2 + 4).$$

Thus the eigenvalues of A are

$$\lambda = \pm i, \pm 2i.$$

We now determine the eigenvectors.
Eigenvalue $\lambda = i$: The system $(A - \lambda I)\mathbf{v}_1 = \mathbf{0}$ has augmented matrix

$$\begin{bmatrix} -i & 1 & 0 & 0 & 0 \\ -3 & -i & 1 & 0 & 0 \\ 0 & 0 & -i & 1 & 0 \\ 2 & 0 & -2 & -i & 0 \end{bmatrix}$$

---

[6]In Problem 1, the reader is asked to fill in the missing details of this computation.

with reduced row-echelon form

$$\begin{bmatrix} 1 & 0 & 0 & i/2 & 0 \\ 0 & 1 & 0 & -1/2 & 0 \\ 0 & 0 & 1 & i & 0 \\ 0 & 0 & 0 & 0 & 0 \end{bmatrix}.$$

Consequently, the eigenvectors are

$$\mathbf{v}_1 = r(-i, 1, -2i, 2),$$

so that a complex-valued solution to $\mathbf{x}' = A\mathbf{x}$ is

$$\mathbf{u}_1(t) = e^{it} \begin{bmatrix} -i \\ 1 \\ -2i \\ 2 \end{bmatrix} = (\cos t + i \sin t) \begin{bmatrix} -i \\ 1 \\ -2i \\ 2 \end{bmatrix}.$$

Taking the real and imaginary parts of this complex-valued solution yields the two linearly independent real-valued solutions

$$\mathbf{x}_1(t) = \begin{bmatrix} \sin t \\ \cos t \\ 2 \sin t \\ 2 \cos t \end{bmatrix} \quad \text{and} \quad \mathbf{x}_2(t) = \begin{bmatrix} -\cos t \\ \sin t \\ -2 \cos t \\ 2 \sin t \end{bmatrix}.$$

Eigenvalue $\lambda = 2i$: The augmented matrix of the system $(A - \lambda I)\mathbf{v}_2 = \mathbf{0}$ in this case is

$$\begin{bmatrix} -2i & 1 & 0 & 0 & 0 \\ -3 & -2i & 1 & 0 & 0 \\ 0 & 0 & -2i & 1 & 0 \\ 2 & 0 & -2 & -2i & 0 \end{bmatrix}$$

with reduced row-echelon form

$$\begin{bmatrix} 1 & 0 & 0 & -i/2 & 0 \\ 0 & 1 & 0 & 1 & 0 \\ 0 & 0 & 1 & i/2 & 0 \\ 0 & 0 & 0 & 0 & 0 \end{bmatrix}.$$

The corresponding eigenvectors are therefore of the form

$$\mathbf{v}_2 = s(i, -2, -i, 2),$$

so that a complex-valued solution to the system $\mathbf{x}' = A\mathbf{x}$ is

$$\mathbf{u}_2(t) = e^{2it} \begin{bmatrix} i \\ -2 \\ -i \\ 2 \end{bmatrix} = (\cos 2t + i \sin 2t) \begin{bmatrix} i \\ -2 \\ -i \\ 2 \end{bmatrix}.$$

Taking the real and imaginary parts of this complex-valued solution yields the additional real-valued linearly independent solutions

$$\mathbf{x}_3(t) = \begin{bmatrix} -\sin 2t \\ -2\cos 2t \\ \sin 2t \\ 2\cos 2t \end{bmatrix} \quad \text{and} \quad \mathbf{x}_4(t) = \begin{bmatrix} \cos 2t \\ -2\sin 2t \\ -\cos 2t \\ 2\sin 2t \end{bmatrix}.$$

Consequently, the vector differential equation in (7.7.9) has general solution

$$\mathbf{x}(t) = c_1 \begin{bmatrix} \sin t \\ \cos t \\ 2\sin t \\ 2\cos t \end{bmatrix} + c_2 \begin{bmatrix} -\cos t \\ \sin t \\ -2\cos t \\ 2\sin t \end{bmatrix} + c_3 \begin{bmatrix} -\sin 2t \\ -2\cos 2t \\ \sin 2t \\ 2\cos 2t \end{bmatrix} + c_4 \begin{bmatrix} \cos 2t \\ -2\sin 2t \\ -\cos 2t \\ 2\sin 2t \end{bmatrix}.$$

Combining the vector functions yields the solution vector

$$\mathbf{x}(t) = \begin{bmatrix} c_1 \sin t - c_2 \cos t - c_3 \sin 2t + c_4 \cos 2t \\ c_1 \cos t + c_2 \sin t - 2c_3 \cos 2t - 2c_4 \sin 2t \\ 2c_1 \sin t - 2c_2 \cos t + c_3 \sin 2t - c_4 \cos 2t \\ 2(c_1 \cos t + c_2 \sin t + c_3 \cos 2t + c_4 \sin 2t) \end{bmatrix}.$$

Imposing the initial condition

$$\mathbf{x}(0) = \begin{bmatrix} 1 \\ 0 \\ 1 \\ 0 \end{bmatrix},$$

we find that $c_1 = 0$, $c_2 = -\frac{2}{3}$, $c_3 = 0$, $c_4 = \frac{1}{3}$. Thus,

$$\mathbf{x}(t) = \begin{bmatrix} \frac{1}{3}(2\cos t + \cos 2t) \\[2mm] -\frac{2}{3}(\sin t + \sin 2t) \\[2mm] \frac{1}{3}(4\cos t - \cos 2t) \\[2mm] \frac{2}{3}(-2\sin t + \sin 2t) \end{bmatrix}.$$

Since $x = x_1$ and $y = x_3$, it follows that the motion of the spring–mass system is given by

$$x(t) = \tfrac{1}{3}(2\cos t + \cos 2t),$$
$$y(t) = \tfrac{1}{3}(4\cos t - \cos 2t).$$

The motion of both masses is periodic, with period 2π. Differentiating the previous expressions for x and y, we see that

$$x'(t) = -\tfrac{2}{3}(\sin t + \sin 2t) = -\tfrac{2}{3}(1 + 2\cos t)\sin t,$$
$$y'(t) = \tfrac{2}{3}(-2\sin t + \sin 2t) = \tfrac{4}{3}(\cos t - 1)\sin t.$$

Consequently, on the interval $[0, 2\pi]$, x' has zeros when $t = 0$, $2\pi/3$, π, $4\pi/3$, and 2π, whereas the only zeros of y' are 0, π, and 2π. Notice that both y'' and y''' vanish at $t = 0$ and $t = 2\pi$. Hence, the graph of y is very flat in the neighborhood of these points. This motion is depicted in Figure 7.7.2.

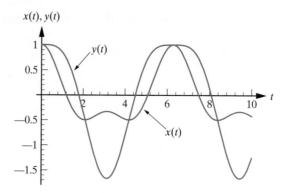

Figure 7.7.2: The solutions for the spring–mass system in Example 7.7.1.

□

Next consider the mixing problem depicted in Figure 7.7.3. Two tanks contain a solution consisting of chemical dissolved in water. A solution containing c_{in} g/L of chemical flows into tank 1 at a rate of r_{in} L/min and solution of concentration c_{out} g/L flows out of tank 2 at a rate of r_{out} L/min. In addition, solution of concentration c_{12} g/L flows into tank 1 from tank 2 at a rate of r_{12} L/min and solution of concentration c_{21} g/L flows into tank 2 from tank 1 at a rate of r_{21} L/min. We wish to determine $A_1(t)$ and $A_2(t)$, the amounts of chemical in tank 1 and tank 2, respectively. The analysis is similar to that used in Section 1.7.

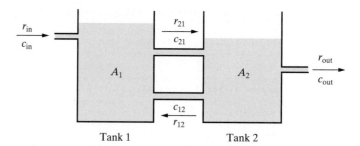

Figure 7.7.3: A general mixing problem.

Assuming that the solution in each tank is well mixed, it follows immediately that

$$c_{12} = c_{out} = \frac{A_2}{V_2}, \qquad c_{21} = \frac{A_1}{V_1},$$

where V_i denotes the volume of solution in tank i at time t. Consider a short time interval Δt. The total amount of chemical entering tank 1 in this time interval is approximately

$$(c_{in}r_{in} + c_{12}r_{12}) \, \Delta t \text{ grams,}$$

whereas approximately

$$c_{21}r_{21} \, \Delta t \text{ grams}$$

of chemical leave tank 1 in the same time interval. Consequently, the change in the amount of chemical in tank 1 in the time interval Δt, denoted ΔA_1, is approximately

$$\Delta A_1 \approx [(c_{in}r_{in} + c_{12}r_{12}) - c_{21}r_{21}] \, \Delta t;$$

that is,

$$\Delta A_1 \approx \left[c_{in}r_{in} + r_{12}\frac{A_2}{V_2} - r_{21}\frac{A_1}{V_1} \right] \Delta t. \qquad (7.7.10)$$

Similarly, the change in the amount of chemical in tank 2 in the time interval Δt, denoted ΔA_2, is approximately

$$\Delta A_2 \approx [r_{21}c_{21} - (r_{12}c_{12} + r_{\text{out}}c_{\text{out}})] \, \Delta t$$

or, equivalently,

$$\Delta A_2 \approx \left[r_{21}\frac{A_1}{V_1} - (r_{12} + r_{\text{out}})\frac{A_2}{V_2} \right] \Delta t. \tag{7.7.11}$$

Dividing Equations (7.7.10) and (7.7.11) by Δt and taking the limit as $\Delta t \to 0^+$ yields the following system of differential equations for A_1 and A_2:

$$\frac{dA_1}{dt} = -r_{21}\frac{A_1}{V_1} + r_{12}\frac{A_2}{V_2} + c_{\text{in}}r_{\text{in}},$$

$$\frac{dA_2}{dt} = r_{21}\frac{A_1}{V_1} - (r_{12} + r_{\text{out}})\frac{A_2}{V_2}.$$

We will now assume that V_1 and V_2 are constant. This imposes the conditions

$$r_{\text{in}} + r_{12} - r_{21} = 0,$$
$$r_{21} - r_{12} - r_{\text{out}} = 0.$$

(See Problem 7.) Consequently, the foregoing system of differential equations reduces to

$$\frac{dA_1}{dt} = -\frac{r_{21}}{V_1}A_1 + \frac{r_{12}}{V_2}A_2 + c_{\text{in}}r_{\text{in}},$$

$$\frac{dA_2}{dt} = \frac{r_{21}}{V_1}A_1 - \frac{r_{21}}{V_2}A_2.$$

This is a constant-coefficient system for A_1 and A_2, and therefore it can be solved using the techniques that we have developed in this chapter.

Example 7.7.2   Two tanks each contain 20 L of a solution consisting of salt dissolved in water. (See Figure 7.7.4.) A solution containing 4 g/L of salt flows into tank 1 at a rate of 3 L/min and the solution in tank 2 flows out at the same rate. In addition, solution flows into tank 1 from tank 2 at a rate of 1 L/min and into tank 2 from tank 1 at a rate of 4 L/min. Initially tank 1 contained 40 g of salt and tank 2 contained 20 g of salt. Find the amount of salt in each tank at time t.

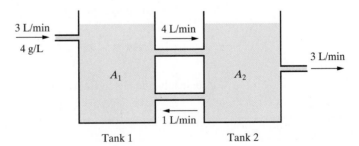

Figure 7.7.4: The mixing problem considered in Example 7.7.2.

Solution: The inflow and outflow rates from each tank are indicated in Figure 7.7.4. We notice that the total amount of solution flowing into tank 1 is 4 L/min, and the same volume of solution flows out of tank 1 per minute. Consequently, the volume of solution in tank 1 remains constant at 20 L. The same is true for tank 2. Let $A_1(t)$ and $A_2(t)$ denote the amounts of salt in tanks 1 and 2, respectively, and let c_{ij} denote the concentration of salt in the solution flowing into tank i from tank j. Now consider a short time interval Δt. The overall change in the amount of salt in tank 1 in this time interval Δt is approximately

$$\Delta A_1 \approx (12 + 1 \cdot c_{12}) \, \Delta t - 4c_{21} \, \Delta t;$$

that is,

$$\Delta A_1 \approx \left(12 + \tfrac{1}{20}A_2 - \tfrac{1}{5}A_1\right) \Delta t. \tag{7.7.12}$$

A similar analysis of the change in the amount of salt in tank 2 in the time interval Δt yields

$$\Delta A_2 \approx \left(\tfrac{1}{5}A_1 - \tfrac{1}{20}A_2 - \tfrac{3}{20}A_2\right) \Delta t,$$

that is

$$\Delta A_2 \approx \left(\tfrac{1}{5}A_1 - \tfrac{1}{5}A_2\right) \Delta t. \tag{7.7.13}$$

Dividing Equations (7.7.12) and (7.7.13) by Δt and taking the limit as $\Delta t \to 0^+$ yields the system of differential equations

$$\frac{dA_1}{dt} = -\tfrac{1}{5}A_1 + \tfrac{1}{20}A_2 + 12,$$
$$\frac{dA_2}{dt} = \tfrac{1}{5}A_1 - \tfrac{1}{5}A_2.$$

We are also given the initial conditions

$$A_1(0) = 40 \quad \text{and} \quad A_2(0) = 20.$$

In vector form, we must therefore solve the initial-value problem

$$\mathbf{x}' = A\mathbf{x} + \mathbf{b}, \qquad \mathbf{x}(0) = \mathbf{x}_0,$$

where

$$\mathbf{x} = \begin{bmatrix} A_1 \\ A_2 \end{bmatrix}, \quad A = \begin{bmatrix} -\tfrac{1}{5} & \tfrac{1}{20} \\ \tfrac{1}{5} & -\tfrac{1}{5} \end{bmatrix}, \quad \mathbf{b} = \begin{bmatrix} 12 \\ 0 \end{bmatrix}, \quad \mathbf{x}_0 = \begin{bmatrix} 40 \\ 20 \end{bmatrix}.$$

The characteristic polynomial of A is

$$\det(A - \lambda I) = \begin{vmatrix} -\tfrac{1}{5} - \lambda & \tfrac{1}{20} \\ \tfrac{1}{5} & -\tfrac{1}{5} - \lambda \end{vmatrix} = (\lambda + \tfrac{1}{5})^2 - \tfrac{1}{100}.$$

Consequently, the eigenvalues of A are

$$\lambda = -\frac{1}{5} \pm \frac{1}{10}.$$

That is,

$$\lambda_1 = -\frac{1}{10} \quad \text{and} \quad \lambda_2 = -\frac{3}{10}.$$

The corresponding eigenvectors are

$$\mathbf{v}_1 = (1, 2) \quad \text{and} \quad \mathbf{v}_2 = (1, -2),$$

respectively, so that two linearly independent solutions to $\mathbf{x}' = A\mathbf{x}$ are

$$\mathbf{x}_1(t) = e^{-t/10} \begin{bmatrix} 1 \\ 2 \end{bmatrix} \qquad \text{and} \qquad \mathbf{x}_2(t) = e^{-3t/10} \begin{bmatrix} 1 \\ -2 \end{bmatrix}.$$

Thus,

$$\mathbf{x}_c(t) = c_1 e^{-t/10} \begin{bmatrix} 1 \\ 2 \end{bmatrix} + c_2 e^{-3t/10} \begin{bmatrix} 1 \\ -2 \end{bmatrix}.$$

We now need a particular solution to $\mathbf{x}' = A\mathbf{x} + \mathbf{b}$. According to the variation-of-parameters technique, a particular solution is

$$\mathbf{x}_p = X\mathbf{u},$$

where

$$X\mathbf{u}' = \mathbf{b} \tag{7.7.14}$$

and

$$X(t) = \begin{bmatrix} e^{-t/10} & e^{-3t/10} \\ 2e^{-t/10} & -2e^{-3t/10} \end{bmatrix}.$$

The system (7.7.14) is

$$e^{-t/10}u_1' + e^{-3t/10}u_2' = 12,$$
$$e^{-t/10}u_1' - e^{-3t/10}u_2' = 0,$$

which has solution

$$u_1' = 6e^{t/10} \qquad \text{and} \qquad u_2' = 6e^{3t/10}.$$

By integrating, we obtain

$$u_1(t) = 60e^{t/10} \qquad \text{and} \qquad u_2(t) = 20e^{3t/10},$$

where we have set the integration constants to zero without loss of generality. Consequently,

$$\mathbf{x}_p(t) = \begin{bmatrix} e^{-t/10} & e^{-3t/10} \\ 2e^{-t/10} & -2e^{-3t/10} \end{bmatrix} \begin{bmatrix} 60e^{t/10} \\ 20e^{3t/10} \end{bmatrix} = \begin{bmatrix} 80 \\ 80 \end{bmatrix}.$$

Hence, the general solution to the system $\mathbf{x}' = A\mathbf{x} + \mathbf{b}$ is

$$\mathbf{x}(t) = c_1 e^{-t/10} \begin{bmatrix} 1 \\ 2 \end{bmatrix} + c_2 e^{-3t/10} \begin{bmatrix} 1 \\ -2 \end{bmatrix} + \begin{bmatrix} 80 \\ 80 \end{bmatrix}.$$

That is,

$$\mathbf{x}(t) = \begin{bmatrix} c_1 e^{-t/10} + c_2 e^{-3t/10} + 80 \\ 2c_1 e^{-t/10} - 2c_2 e^{-3t/10} + 80 \end{bmatrix}.$$

Imposing the initial condition $\mathbf{x}(0) = \begin{bmatrix} 40 \\ 20 \end{bmatrix}$ requires

$$\begin{bmatrix} c_1 + c_2 + 80 \\ 2c_1 - 2c_2 + 80 \end{bmatrix} = \begin{bmatrix} 40 \\ 20 \end{bmatrix}.$$

We quickly solve this for c_1 and c_2: $c_1 = -35$ and $c_2 = -5$. Thus, the solution to the initial-value problem is

$$\mathbf{x}(t) = -35e^{-t/10} \begin{bmatrix} 1 \\ 2 \end{bmatrix} - 5e^{-3t/10} \begin{bmatrix} 1 \\ -2 \end{bmatrix} + \begin{bmatrix} 80 \\ 80 \end{bmatrix}.$$

Consequently, the amounts of salt in tanks 1 and 2 at time t are, respectively,

$$A_1(t) = 80 - 35e^{-t/10} - 5e^{-3t/10},$$
$$A_2(t) = 80 - 70e^{-t/10} + 10e^{-3t/10}.$$

We see that both A_1 and A_2 approach the constant value of 80 g as $t \to \infty$. Why is this a reasonable result? ☐

Exercises for 7.7

Key Terms

Coupled spring–mass system, Mixing problem.

Skills

- Be able to determine the motion of a coupled spring–mass system as a function of time.

- Be able to solve mixing problems to determine the amounts (or concentrations) of chemical present in a system of tanks as a function of time.

True-False Review

For Questions 1–5, decide if the given statement is **true** or **false**, and give a brief justification for your answer. If true, you can quote a relevant definition or theorem from the text. If false, provide an example, illustration, or brief explanation of why the statement is false.

1. A coupled spring–mass system consisting of two masses and two springs can be solved as a first-order linear system $\mathbf{x}' = A\mathbf{x}$, where A is a 2×2 matrix whose entries are determined by the masses and spring constants.

2. The units of the spring constant in the metric system are Newtons per meter.

3. In the metric system, the concentration of chemical in a tank of solution is measured in grams.

4. In a chemical mixing problem involving two tanks, the rate at which fluid moves from tank 1 to tank 2 must be the same as the rate at which fluid moves from tank 2 to tank 1.

5. In a chemical mixing problem with two tanks in which no solution flows in or out of the system from the outside and the rates of flow between the two tanks r_{12} and r_{21} are equal, the amount of chemical in each tank remains constant over time, regardless of the initial conditions.

Problems

1. Derive the eigenvalues and eigenvectors given in Example 7.7.1.

2. Determine the motion of the coupled spring–mass system which has

$$k_1 = 3 \text{ Nm}^{-1}, \quad k_2 = \tfrac{1}{2} \text{ Nm}^{-1},$$
$$m_1 = \tfrac{1}{2} \text{ kg}, \quad m_2 = \tfrac{1}{12} \text{ kg}$$

given that at $t = 0$ both masses are set in motion from their equilibrium positions with a velocity of 1 m/s.

3. Determine the general motion of the coupled spring–mass system that has

$$k_1 = 3 \text{ Nm}^{-1}, \quad k_2 = 4 \text{ Nm}^{-1},$$
$$m_1 = 1 \text{ kg}, \quad m_2 = \tfrac{4}{3} \text{ kg}.$$

4. Determine the general motion of the coupled spring–mass system which has

$$k_1 = 2k_2, \quad m_1 = 2m_2.$$

[**Hint:** Let $\omega^2 = k_2/(2m_2)$.]

5. Consider the general coupled spring–mass system whose motion is governed by the system (7.7.5)–(7.7.8). Show that the coefficient matrix of the system has characteristic equation

$$\lambda^4 + \left[\frac{k_2}{m_2} + \frac{(k_1 + k_2)}{m_1} \right] \lambda^2 + \frac{k_1 k_2}{m_1 m_2} = 0$$

and that the corresponding eigenvalues are of the form

$$\lambda = \pm i\omega_1, \pm i\omega_2,$$

where ω_1 and ω_2 are positive real numbers.

6. Two masses m_1 and m_2 rest on a horizontal friction-less plane. The masses are attached to fixed walls by springs whose spring constants are k_1 and k_3. (See Figure 7.7.5.) The masses are connected by a spring whose spring constant is k_2. Determine a first-order system of differential equations that governs the motion of the system.

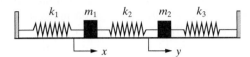

Figure 7.7.5: Three spring system.

7. Show that the assumption that V_1 and V_2 are constant in the general mixing problem considered in the text imposes the conditions

$$r_{\text{in}} + r_{12} = r_{21}, \qquad r_{21} - r_{12} = r_{\text{out}}.$$

8. Solve the initial-value problem arising in Example 7.7.2 using the technique derived in Section 7.1.

9. Solve the mixing problem depicted in Figure 7.7.6, given that at $t = 0$, the volume of solution in both tanks is 60 L, and tank 1 contains 60 g of chemical whereas tank 2 contains 200 g of chemical.

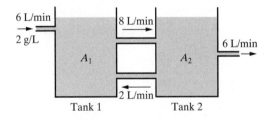

Figure 7.7.6: Mixing problem in Problem 9.

10. In the mixing problem shown in Figure 7.7.7, there is no inflow from or outflow to the outside. For this reason, the system is said to be *closed*. If tank 1 contains 6 L of solution and tank 2 contains 12 L of solution, determine the amount of chemical in each tank at time t, given that initially tank 1 contains 5 g of chemical and tank 2 contains 25 g of chemical.

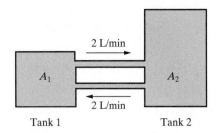

Figure 7.7.7: Mixing problem in Problem 10.

11. Consider the general closed system depicted in Figure 7.7.8.

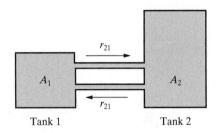

Figure 7.7.8: Mixing problem in Problem 11.

(a) Derive the system of differential equations that governs the behavior of A_1 and A_2.

(b) Define the constant β by $V_2 = \beta V_1$, where V_1 and V_2 denote the volume of solution in tank 1 and tank 2, respectively. Show that the eigenvalues of the coefficient matrix of the system derived in (a) are

$$\lambda_1 = 0 \quad \text{and} \quad \lambda_2 = -\frac{1+\beta}{\beta V_1} r_{21}.$$

(c) Determine A_1 and A_2, given that $A_1(0) = \alpha_1$ and $A_2(0) = \alpha_2$, where α_1 and α_2 are positive constants.

(d) Show that

$$\lim_{t \to +\infty} \frac{A_1}{V_1} = \lim_{t \to +\infty} \frac{A_2}{V_2} = \frac{\alpha_1 + \alpha_2}{(1+\beta)V_1}.$$

Is this result reasonable?

7.8   Matrix Exponential Function and Systems of Differential Equations

The matrix exponential function was first introduced in Section 5.9. Recall that for an $n \times n$ matrix A, the matrix exponential function is defined by

$$e^{At} = I_n + At + \frac{1}{2!}(At)^2 + \frac{1}{3!}(At)^3 + \cdots + \frac{1}{k!}(At)^k + \cdots, \tag{7.8.1}$$

where the infinite series here can be shown to converge for all real numbers t. In this section we investigate the relationship between the matrix exponential function e^{At} and the solutions to the corresponding vector differential equation

$$\mathbf{x}' = A\mathbf{x}.$$

We begin by defining the derivative of e^{At}. It can be shown that the infinite series (7.8.1) defining e^{At} can be differentiated term by term and the resulting series converges for all $t \in (-\infty, \infty)$. Thus, through differentiating (7.8.1), we have

$$\frac{d}{dt}(e^{At}) = A + A^2 t + \frac{1}{2!}A^3 t^2 + \cdots + \frac{1}{(k-1)!}A^k t^{k-1} + \cdots.$$

That is,

$$\frac{d}{dt}(e^{At}) = A\left[I + At + \frac{1}{2!}(At)^2 + \frac{1}{3!}(At)^3 + \cdots + \frac{1}{k!}(At)^k + \cdots\right].$$

Hence,

$$\boxed{\frac{d}{dt}(e^{At}) = Ae^{At}.} \tag{7.8.2}$$

Now recall from Chapter 1 that for all values of the constants a and x_0, the unique solution to the initial-value problem

$$\frac{dx}{dt} = ax, \qquad x(0) = x_0$$

is

$$x(t) = x_0 e^{at}.$$

Our next theorem shows that the same formula holds for the vector differential equation

$$\mathbf{x}' = A\mathbf{x},$$

provided we replace e^{at} by e^{At}. This very elegant result has far-reaching consequences in both the computation of e^{At} and the analysis of vector differential equations.

Theorem 7.8.1   Let $\mathbf{x}_0$ be an arbitrary vector. Then the unique solution to the *initial-value problem*

$$\mathbf{x}' = A\mathbf{x}, \qquad \mathbf{x}(0) = \mathbf{x}_0$$

is

$$\mathbf{x}(t) = e^{At}\mathbf{x}_0.$$

Proof If $\mathbf{x}(t) = e^{At}\mathbf{x}_0$, then by (7.8.2), we have $\mathbf{x}'(t) = Ae^{At}\mathbf{x}_0$. That is,

$$\mathbf{x}' = A\mathbf{x}.$$

Further, setting $t = 0$,

$$\mathbf{x}(0) = e^{0 \cdot A}\mathbf{x}_0 = I\mathbf{x}_0 = \mathbf{x}_0.$$

Consequently, $\mathbf{x}(t) = e^{At}\mathbf{x}_0$ is a solution to the given initial-value problem. The uniqueness of the solution follows from Theorem 7.3.1. ∎

We now investigate how the result of Theorem 7.8.1, combined with our previous techniques for solving $\mathbf{x}' = A\mathbf{x}$, can be used to determine e^{At}. To this end, let $\mathbf{x}_1, \mathbf{x}_2, \ldots, \mathbf{x}_n$ be linearly independent solutions to the vector differential equation

$$\mathbf{x}' = A\mathbf{x}, \tag{7.8.3}$$

where A is an $n \times n$ matrix of constants. We recall from Section 7.3 that the corresponding matrix function

$$X(t) = [\mathbf{x}_1, \mathbf{x}_2, \ldots, \mathbf{x}_n]$$

is called a fundamental matrix for (7.8.3) and that the general solution to (7.8.3) can be written as

$$\mathbf{x}(t) = X(t)\mathbf{c},$$

where $\mathbf{c}$ is a column vector of arbitrary constants. If $X(t)$ is any fundamental matrix for (7.8.3) and B is any *invertible* matrix of constants, then the matrix function

$$Y(t) = X(t)B$$

is also a fundamental matrix for (7.8.3), since its columns are linear combinations of the column vectors of X and hence are linearly independent solutions of (7.8.3). (The linear independence follows since $Y(t)$ is invertible.) We focus our attention on a particular fundamental matrix.

DEFINITION   7.8.2

The unique fundamental matrix for $\mathbf{x}' = A\mathbf{x}$ that satisfies

$$X(0) = I_n$$

is called the **transition matrix** for $\mathbf{x}' = A\mathbf{x}$ based at $t = 0$, and it is denoted by $X_0(t)$.

In terms of the transition matrix, the solution to the initial-value problem

$$\mathbf{x}' = A\mathbf{x}, \qquad \mathbf{x}(0) = \mathbf{x}_0$$

is just

$$\mathbf{x}(t) = X_0(t)\mathbf{x}_0,$$

so that the transition matrix does indeed describe the transition of the system from its state at time $t = 0$ to its state at time t. Further, if $X(t)$ is any fundamental matrix for $\mathbf{x}' = A\mathbf{x}$, then the transition matrix can be determined from (see Problem 1)

$$X_0(t) = X(t)X^{-1}(0). \tag{7.8.4}$$

We now prove that $X_0(t)$ is in fact e^{At}. From Equation (7.8.2), we have

$$\frac{d}{dt}(e^{At}) = Ae^{At},$$

so that the column vectors of e^{At} are solutions to $\mathbf{x}' = A\mathbf{x}$. Further, setting $t = 0$ yields

$$e^{0 \cdot A} = I_n, \tag{7.8.5}$$

which implies that

$$\det(e^{0 \cdot A}) = 1 \neq 0.$$

Hence, the column vectors of e^{At} are linearly independent on any interval. Consequently, e^{At} is a fundamental matrix for $\mathbf{x}' = A\mathbf{x}$. Finally, combining (7.8.5) with the uniqueness of the transition matrix leads to the required conclusion, namely

$$e^{At} = X_0(t). \tag{7.8.6}$$

Thus, if A is an $n \times n$ matrix and $X(t)$ is *any* fundamental matrix for the corresponding vector differential equation $\mathbf{x}' = A\mathbf{x}$, then Equations (7.8.4) and (7.8.6) imply that

$$\boxed{e^{At} = X(t)X^{-1}(0).} \tag{7.8.7}$$

Consequently, to determine e^{At}, we can use the techniques from Sections 7.3 and 7.4 to find a fundamental matrix for $\mathbf{x}' = A\mathbf{x}$, and then e^{At} can be obtained directly from Equation (7.8.7).

Example 7.8.3 | Determine e^{At} if $A = \begin{bmatrix} 6 & -8 \\ 2 & -2 \end{bmatrix}$.

Solution: We first find a fundamental matrix for

$$\mathbf{x}' = A\mathbf{x}.$$

This system has been solved in Example 7.5.2, where it was found that two linearly independent solutions are[7]

$$\mathbf{x}_1(t) = e^{2t} \begin{bmatrix} 2 \\ 1 \end{bmatrix} \quad \text{and} \quad \mathbf{x}_2(t) = e^{2t} \begin{bmatrix} 1 + 4t \\ 2t \end{bmatrix}.$$

Thus, a fundamental matrix for $\mathbf{x}' = A\mathbf{x}$ is

$$X(t) = \begin{bmatrix} 2e^{2t} & (1 + 4t)e^{2t} \\ e^{2t} & 2te^{2t} \end{bmatrix},$$

whose inverse at $t = 0$ is

$$X^{-1}(0) = \begin{bmatrix} 0 & 1 \\ 1 & -2 \end{bmatrix}.$$

Consequently,

$$e^{At} = X(t)X^{-1}(0) = \begin{bmatrix} 2e^{2t} & (1 + 4t)e^{2t} \\ e^{2t} & 2te^{2t} \end{bmatrix} \begin{bmatrix} 0 & 1 \\ 1 & -2 \end{bmatrix}.$$

---

[7] Notice that in this example A is defective.

That is,

$$e^{At} = \begin{bmatrix} (1 + 4t)e^{2t} & -8te^{2t} \\ 2te^{2t} & (1 - 4t)e^{2t} \end{bmatrix},$$

which can be written as

$$e^{At} = e^{2t} \begin{bmatrix} 1 + 4t & -8t \\ 2t & 1 - 4t \end{bmatrix}. \qquad \square$$

We have seen how to take a set of n linearly independent solutions to $\mathbf{x}' = A\mathbf{x}$, where A is an $n \times n$ matrix, and compute the matrix exponential function e^{At}. We now indicate how this process can be reversed. That is, given the matrix exponential function e^{At}, we derive n linearly independent solutions to $\mathbf{x}' = A\mathbf{x}$ as follows.

We begin by writing

$$A = \lambda I + (A - \lambda I),$$

where λ is a (possibly complex) scalar. By writing $B = \lambda I$ and $C = A - \lambda I$ and noting that $BC = CB$, it follows from property 1 of the matrix exponential function (see Section 5.9) that for every $\mathbf{v}$ in $\mathbb{C}^n$

$$e^{At}\mathbf{v} = e^{[\lambda It + (A - \lambda I)t]}\mathbf{v} = e^{\lambda It}e^{(A - \lambda I)t}\mathbf{v}.$$

Further, by Problem 1 in Section 5.9,

$$e^{\lambda It} = \operatorname{diag}(e^{\lambda t}, e^{\lambda t}, \ldots, e^{\lambda t}) = e^{\lambda t}I,$$

so that

$$e^{At}\mathbf{v} = e^{\lambda t}\left[\mathbf{v} + t(A - \lambda I)\mathbf{v} + \frac{t^2}{2!}(A - \lambda I)^2\mathbf{v} + \cdots\right]. \qquad (7.8.8)$$

Theorem 7.8.1 guarantees that $e^{At}\mathbf{v}$ is a solution to $\mathbf{x}' = A\mathbf{x}$, but in general, the preceding series for $e^{At}\mathbf{v}$ contains an infinite number of terms and hence is intractable. However, if we can find vectors $\mathbf{v}$ such that

$$(A - \lambda I)^k\mathbf{v} = \mathbf{0} \qquad (7.8.9)$$

for some positive integer k, then the series will terminate after a finite number of terms. Nonzero vectors $\mathbf{v}$ satisfying (7.8.9) for some positive k were introduced in Section 5.11 as **generalized eigenvectors** of A, and, as indicated there, if λ occurs with multiplicity m in the characteristic polynomial of A, then A has m linearly independent generalized eigenvectors corresponding to λ. Proceeding with each of these generalized eigenvectors $\mathbf{v}$, we get a solution to $\mathbf{x}' = A\mathbf{x}$ in the form

$$e^{At}\mathbf{v} = e^{\lambda t}\left[\mathbf{v} + t(A - \lambda I)\mathbf{v} + \frac{t^2}{2!}(A - \lambda I)^2\mathbf{v} + \cdots + \frac{t^{k-1}}{(k-1)!}(A - \lambda I)^{k-1}\mathbf{v}\right],$$

each of which is a solution with a finite number of terms. Therefore, we have obtained m linearly independent solutions to $\mathbf{x}' = A\mathbf{x}$ corresponding to λ. Proceeding in this manner for each eigenvalue, we can obtain n linearly independent solutions to $\mathbf{x}' = A\mathbf{x}$. This will determine a fundamental matrix for $\mathbf{x}' = A\mathbf{x}$ from which e^{At} can be determined in the usual manner.

Remark   Note, too, that for each $\mathbf{v} \neq \mathbf{0}$ such that (7.8.9) holds with $k = 1$, $\mathbf{v}$ is an eigenvector of A and the series (7.8.8) has only one term. In that case, we obtain the result of Theorem 7.4.1, namely, that $e^{At}\mathbf{v} = e^{\lambda t}\mathbf{v}$ is a solution to $\mathbf{x}' = A\mathbf{x}$ whenever λ and $\mathbf{v}$ are an eigenvalue/eigenvector pair for A. Hence, if A is nondefective, Equation (7.8.8) yields n linearly independent solutions to $\mathbf{x}' = A\mathbf{x}$ in the usual manner.

Example 7.8.4   Let $A = \begin{bmatrix} 6 & 8 & 1 \\ -1 & -3 & 3 \\ -1 & -1 & 1 \end{bmatrix}$.

(a) Determine a fundamental matrix for $\mathbf{x}' = A\mathbf{x}$, and thereby determine the general solution to the vector differential equation.

(b) Determine e^{At}.

Solution:

(a) The characteristic polynomial of A is

$$p(\lambda) = -(\lambda - 3)^2(\lambda + 2).$$

Hence, A has eigenvalues $\lambda_1 = 3$ (multiplicity 2), and $\lambda_2 = -2$.

Eigenvalue $\lambda_1 = 3$: In this case, we determine two linearly independent solutions to

$$(A - 3I)^2\mathbf{v} = \mathbf{0}. \tag{7.8.10}$$

The coefficient matrix of this system is

$$(A - 3I)^2 = \begin{bmatrix} 3 & 8 & 1 \\ -1 & -6 & 3 \\ -1 & -1 & -2 \end{bmatrix}\begin{bmatrix} 3 & 8 & 1 \\ -1 & -6 & 3 \\ -1 & -1 & -2 \end{bmatrix} = \begin{bmatrix} 0 & -25 & 25 \\ 0 & 25 & -25 \\ 0 & 0 & 0 \end{bmatrix},$$

so that the system (7.8.10) reduces to the single equation

$$v_2 - v_3 = 0,$$

which has two free variables. We set $v_1 = r$ and $v_3 = s$, in which case $v_2 = s$. Hence, Equation (7.8.10) has solution

$$\mathbf{v} = r(1, 0, 0) + s(0, 1, 1).$$

Consequently, two linearly independent solutions to Equation (7.8.10) are

$$\mathbf{v}_1 = (1, 0, 0) \qquad \text{and} \qquad \mathbf{v}_2 = (0, 1, 1).$$

Since $(A - 3I)^2\mathbf{v}_1 = \mathbf{0}$, $e^{At}\mathbf{v}_1$ reduces to

$$e^{At}\mathbf{v}_1 = e^{3t}[\mathbf{v}_1 + t(A - 3I)\mathbf{v}_1]$$

$$= e^{3t}\left\{\begin{bmatrix} 1 \\ 0 \\ 0 \end{bmatrix} + t\begin{bmatrix} 3 & 8 & 1 \\ -1 & -6 & 3 \\ -1 & -1 & -2 \end{bmatrix}\begin{bmatrix} 1 \\ 0 \\ 0 \end{bmatrix}\right\}$$

$$= e^{3t}\left\{\begin{bmatrix} 1 \\ 0 \\ 0 \end{bmatrix} + t\begin{bmatrix} 3 \\ -1 \\ -1 \end{bmatrix}\right\}.$$

Exercises for 7.8

Key Terms

Transition matrix.

Skills

- Be able to compute the derivative of the matrix exponential function.

- Be able to use a fundamental matrix for the vector differential equation $\mathbf{x}' = A\mathbf{x}$ to compute the matrix exponential function.

- Be able to use the matrix exponential function to find a fundamental matrix and the solution to the vector differential equation $\mathbf{x}' = A\mathbf{x}$.

True-False Review

For Questions 1–6, decide if the given statement is **true** or **false**, and give a brief justification for your answer. If true, you can quote a relevant definition or theorem from the text. If false, provide an example, illustration, or brief explanation of why the statement is false.

1. The derivative of the matrix exponential function e^{At} with respect to the variable t is Ae^{At}.

2. The transition matrix $X_0(t)$ for the vector differential equation $\mathbf{x}' = A\mathbf{x}$ is precisely the same as the matrix exponential function e^{At}.

3. If the matrix exponential function e^{At} is known, then one can explicitly solve the initial-value problem

$$\mathbf{x}' = A\mathbf{x}, \qquad \mathbf{x}(0) = \mathbf{x}_0.$$

4. The transition matrix for a linear system is always invertible.

5. The matrix exponential function e^{At} can be written as $e^{At} = X(t)X^{-1}(0)$ for any fundamental matrix $X(t)$ for the vector differential equation $\mathbf{x}' = A\mathbf{x}$.

6. If $\{\mathbf{v}_1, \mathbf{v}_2, \ldots, \mathbf{v}_n\}$ is linearly independent in $\mathbb{R}^n$, then so is $\{e^{At}\mathbf{v}_1, e^{At}\mathbf{v}_2, \ldots, e^{At}\mathbf{v}_n\}$ for all $n \times n$ matrices A.

Problems

1. If $X(t)$ is any fundamental matrix for $\mathbf{x}' = A\mathbf{x}$, show that the transition matrix based at $t = 0$ is given by

$$X_0(t) = X(t)X^{-1}(0).$$

For Problems 2–4, use the techniques from Section 7.4 and Section 7.5 to determine a fundamental matrix for $\mathbf{x}' = A\mathbf{x}$, and hence find e^{At}.

2. $A = \begin{bmatrix} 1 & 2 \\ 0 & -1 \end{bmatrix}$.

3. $A = \begin{bmatrix} 2 & 1 \\ 0 & 2 \end{bmatrix}$.

4. $A = \begin{bmatrix} 3 & 0 & 0 \\ 0 & 3 & -1 \\ 0 & 1 & 1 \end{bmatrix}$.

For Problems 5–7, find n linearly independent solutions to $\mathbf{x}' = A\mathbf{x}$ of the form $e^{At}\mathbf{v}$, and hence find e^{At}.

5. $A = \begin{bmatrix} 3 & -1 \\ 4 & -1 \end{bmatrix}$.

6. $A = \begin{bmatrix} -3 & -2 \\ 2 & 1 \end{bmatrix}$.

7. $A = \begin{bmatrix} 2 & 0 & 0 \\ 0 & 1 & -8 \\ 0 & 2 & -7 \end{bmatrix}$.

For Problems 8–10, solve $\mathbf{x}' = A\mathbf{x}$ by determining n linearly independent solutions of the form $\mathbf{x}(t) = e^{At}\mathbf{v}$.

8. $A = \begin{bmatrix} -8 & 6 & -3 \\ -12 & 10 & -3 \\ -12 & 12 & -2 \end{bmatrix}$. You may assume that $p(\lambda) = -(\lambda + 2)^2(\lambda - 4)$.

9. $A = \begin{bmatrix} 0 & 1 & 3 \\ 2 & 3 & -2 \\ 1 & 1 & 2 \end{bmatrix}$. You may assume that $p(\lambda) = -(\lambda + 1)(\lambda - 3)^2$.

10. $A = \begin{bmatrix} 1 & 0 & 0 & 0 \\ 0 & 6 & -7 & 3 \\ 0 & 0 & 3 & -1 \\ 0 & -4 & 9 & -3 \end{bmatrix}$. You may assume that $p(\lambda) = (\lambda - 1)(\lambda - 2)^3$.

11. The matrix

$$A = \begin{bmatrix} 0 & -1 & 0 & 0 \\ 1 & 0 & 0 & 0 \\ 1 & 0 & 0 & -1 \\ 0 & 1 & 1 & 0 \end{bmatrix}$$

has characteristic polynomial $p(\lambda) = (\lambda^2 + 1)^2$. Determine two *complex-valued* solutions to $\mathbf{x}' = A\mathbf{x}$ of the form $\mathbf{x} = e^{At}\mathbf{v}$, and hence find four linearly independent *real-valued* solutions to the differential system.

7.9   The Phase Plane for Linear Autonomous Systems

So far in this chapter, we have developed the general theory for linear systems of differential equations, and we have derived particular solution techniques for solving such systems in the case of constant coefficients. If we drop either the constant-coefficient assumption or the linearity assumption, in general, it is not possible to explicitly solve the resulting systems. Consequently, we need to resort either to a qualitative analysis of the system or to numerical techniques. In the final two sections of this chapter we give a brief introduction to the qualitative approach in the case of systems of the form

$$\frac{dx}{dt} = F(x, y), \tag{7.9.1}$$

$$\frac{dy}{dt} = G(x, y), \tag{7.9.2}$$

where F and G depend only on x and y. Such a system, in which t does not explicitly occur in F and G is called an **autonomous system**. We can interpret the two equations in the system as determining the components of the velocity of a particle that is moving in the xy-plane. As t increases, the particle moves along a curve in the xy-plane called a **trajectory**.[8] The xy-plane itself is referred to as the **phase plane**, and the totality of all trajectories gives the **phase portrait**. Note that each trajectory has a natural direction associated with it—namely, the direction that the particle moves along a trajectory as t increases. From Equations (7.9.1) and (7.9.2), we see that the differential equation determining the trajectories is

$$\frac{dy}{dx} = \frac{G(x, y)}{F(x, y)}.$$

Even if we cannot solve this differential equation, we can obtain much qualitative information about the behavior of the trajectories by constructing, either by hand or using technology, the slope field associated with it.

For the system of equations (7.9.1) and (7.9.2), any values of x and y for which *both* F and G vanish are called **equilibrium points**. If (x_0, y_0) is an equilibrium point, then $x(t) = x_0$, $y(t) = y_0$ is a solution to the system (7.9.1) and (7.9.2) and is called an **equilibrium solution**. We will see that equilibrium points play a key role in the analysis of the phase plane.

Example 7.9.1   Determine all equilibrium points for the system

$$x' = x + y, \quad y' = 2x - 3y.$$

Solution: To determine any equilibrium points, we must solve

$$x + y = 0, \quad 2x - 3y = 0.$$

Since the determinant of the matrix of coefficients of this homogeneous linear system is nonzero, the only solution is $(0, 0)$. Hence, $(0, 0)$ is the only equilibrium point.   □

Example 7.9.2   Determine all equilibrium points for the system

$$x' = 2x + y, \quad y' = 4x + 2y.$$

---

[8]Other terms used for trajectories are *phase paths* or *orbits*.

Solution: Here, we must solve

$$2x + y = 0, \quad 4x + 2y = 0.$$

In this case, any point (x, y) in the plane that lies along the line $y = -2x$ will be an equilibrium point, since the second equation is simply a multiple of the first one.   □

Remark   As the preceding example illustrates, it is possible for a system of differential equations to have an infinite number of equilibrium points. However, we will restrict our attention from now on to the case when there is a unique equilibrium point.

Before analyzing the general autonomous system (7.9.1) and (7.9.2), we need to look at the simpler case when F and G are linear functions of x and y. The system then reduces to the general homogeneous constant coefficient system

$$\frac{dx}{dt} = ax + by, \qquad \frac{dy}{dt} = cx + dy, \qquad (7.9.3)$$

where $a, b, c,$ and d are constants. Consequently, the differential equation for determining the trajectories (or slope field) is

$$\frac{dy}{dx} = \frac{cx + dy}{ax + by},$$

which falls into the first-order homogeneous type that we studied in Chapter 1. Whereas in many cases it is possible to solve this differential equation using the change of variables $y = xV$, we can more easily determine the general behavior in the phase plane by working with the equivalent vector differential equation and using results already obtained in this chapter. Consequently, we write (7.9.3) as the vector differential equation

$$\mathbf{x}' = A\mathbf{x}, \qquad A = \begin{bmatrix} a & b \\ c & d \end{bmatrix}. \qquad (7.9.4)$$

The goal is to determine the general qualitative behavior of the solution curves $\mathbf{x}(t) = (x(t), y(t))$ to (7.9.4).

The equilibrium points of the system (7.9.4) are solutions to the 2×2 homogeneous linear system $A\mathbf{x} = \mathbf{0}$. The Invertible Matrix Theorem therefore guarantees that there will be a unique equilibrium point if and only if A is invertible, in which case the corresponding equilibrium point is $\mathbf{x} = (0, 0)$. Therefore,

> *we will assume for the remainder of this section that the matrix A in question is invertible.*

As we now show, the eigenvalues and eigenvectors of A play a basic role in the structure of the phase plane. Note that since A is invertible, all eigenvalues of A are nonzero. Suppose that $\lambda \neq 0$ is a real eigenvalue of A, with corresponding eigenvector $\mathbf{v}$. Then a solution to (7.9.4) is

$$\mathbf{x}(t) = e^{\lambda t}\mathbf{v}.$$

Since $\mathbf{v}$ is a constant vector, the corresponding trajectories are two half-lines that emanate from the equilibrium point $(0, 0)$ and are parallel to the eigenvector $\mathbf{v}$. The initial conditions would determine which half-line corresponded to a particular motion. If $\lambda > 0$, then, owing to the $e^{\lambda t}$ term, the direction along the trajectory is away from the origin. (See Figure 7.9.1.) Interpreting $(x(t), y(t))$ as the coordinates of a particle at time t, these trajectories correspond to a particle emitted from the equilibrium point at $t = -\infty$ and moving outward along the appropriate half-line. If $\lambda < 0$, then the direction along the trajectory is toward the origin. (See Figure 7.9.2.) In this case, we can interpret the trajectory as corresponding to a point particle that moves along the half-line toward the origin but does not reach the origin in a finite time.

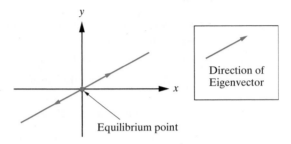

Figure 7.9.1: Trajectories corresponding to a positive eigenvalue and real eigenvector solution to the system $\mathbf{x}' = A\mathbf{x}$.

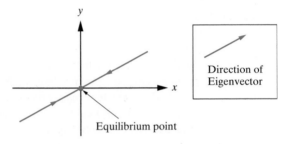

Figure 7.9.2: Trajectories corresponding to a negative eigenvalue and real eigenvector solution to the system $\mathbf{x}' = A\mathbf{x}$.

As the above discussion suggests, the eigenvalue/eigenvector pairs for A play a fundamental role in the general analysis of the phase plane. The eigenvalues of A are the solutions to the equation

$$0 = \det(A - \lambda I) = (a - \lambda)(d - \lambda) - bc = \lambda^2 - (a + d)\lambda + (ad - bc),$$

which can be written as

$$\lambda^2 - \text{tr}(A)\lambda + \det(A) = 0,$$

where we recall that $\text{tr}(A)$ is the trace of A, the sum of the elements on the main diagonal of A. This quadratic equation has roots λ_1, λ_2, where

$$\lambda_1 = \frac{\text{tr}(A) + \sqrt{[\text{tr}(A)]^2 - 4\det(A)}}{2}, \qquad \lambda_2 = \frac{\text{tr}(A) - \sqrt{[\text{tr}(A)]^2 - 4\det(A)}}{2}.$$

The following three cases arise:

1. λ_1, λ_2 are real and distinct. This occurs if and only if $[\text{tr}(A)]^2 > 4\det(A)$.

2. $\lambda_1 = \lambda_2$. This occurs if and only if $[\text{tr}(A)]^2 = 4\det(A)$.

3. λ_1 and λ_2 are complex conjugates. This occurs if and only if $[\text{tr}(A)]^2 < 4\det(A)$.

We now analyze the phase plane in each case.

Case 1: λ_1 and λ_2 are real and distinct.

Let $\mathbf{v}_1$ and $\mathbf{v}_2$ denote corresponding eigenvectors.[9] Then we have the basic solutions

$$\mathbf{x}_1 = e^{\lambda_1 t}\mathbf{v}_1 \quad \text{and} \quad \mathbf{x}_2 = e^{\lambda_2 t}\mathbf{v}_2, \tag{7.9.5}$$

and the general solution to the system is

$$\mathbf{x}(t) = c_1 e^{\lambda_1 t}\mathbf{v}_1 + c_2 e^{\lambda_2 t}\mathbf{v}_2 \tag{7.9.6}$$

Proof   This follows at once from Theorem 7.3.2.   ∎

The two solutions in (7.9.5) give rise to four half-line trajectories, as previously discussed. Since trajectories cannot intersect, the phase plane is divided into four regions. The specific behavior of the remaining trajectories depends on the relationship between λ_1 and λ_2.

(a) $\lambda_2 < \lambda_1 < 0$: The general properties of the trajectories are summarized as follows:

1. Since λ_1 and λ_2 are both negative,

$$\lim_{t \to \infty} \mathbf{x}(t) = \mathbf{0},$$

so that as $t \to \infty$, all trajectories approach the equilibrium point $(0, 0)$.

2. Writing (7.9.6) as

$$\mathbf{x}(t) = e^{\lambda_1 t}[c_1\mathbf{v}_1 + c_2\mathbf{v}_2 e^{(\lambda_2 - \lambda_1)t}]$$

and using the fact that $\lambda_2 - \lambda_1 < 0$ in this case, we see that for large t, and $c_1 \neq 0$, the second term in the brackets is negligible compared to the first term. Consequently, apart from the trajectories corresponding to the eigenvector solution

$$\mathbf{x}(t) = e^{\lambda_2 t}\mathbf{v}_2,$$

all trajectories are parallel to $\mathbf{v}_1$ as $t \to \infty$.

3. Writing (7.9.6) as

$$\mathbf{x}(t) = e^{\lambda_2 t}[c_1\mathbf{v}_1 e^{(\lambda_1 - \lambda_2)t} + c_2\mathbf{v}_2]$$

and using the fact that $\lambda_1 - \lambda_2 > 0$, we see that apart from the trajectories corresponding to the eigenvector solution

$$\mathbf{x}(t) = e^{\lambda_1 t}\mathbf{v}_1,$$

all trajectories are parallel to $\mathbf{v}_2$ as $t \to -\infty$.

A generic sketch of the phase plane in this case is given in Figure 7.9.3. The equilibrium point $(0, 0)$ is called a **node**. It is **stable**, since all solutions *approach* the node as $t \to \infty$.

---

[9]By Theorem 5.7.5 or Problem 32 in Section 5.7, $\mathbf{v}_1$ and $\mathbf{v}_2$ must be linearly independent.

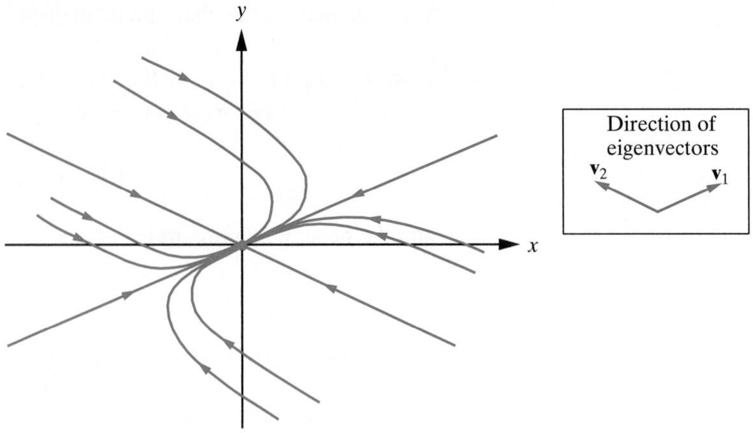

Figure 7.9.3: Typical phase portrait in the case $\lambda_2 < \lambda_1 < 0$.

(b) $0 < \lambda_1 < \lambda_2$: The general behavior in this case is the same as that in Case 1(a), except the arrows are reversed on each trajectory. The equilibrium point is still called a node, but in this case it is **unstable**.

(c) $\lambda_2 < 0 < \lambda_1$: The following general behavior can be identified.

1. The only trajectories that approach the equilibrium point as $t \to \infty$ are those corresponding to the eigenvector solution

$$\mathbf{x}_2(t) = e^{\lambda_2 t}\mathbf{v}_2.$$

2. Writing (7.9.6) as

$$\mathbf{x}(t) = e^{\lambda_1 t}[c_1\mathbf{v}_1 + c_2\mathbf{v}_2 e^{(\lambda_2 - \lambda_1)t}]$$

and using the fact that $\lambda_2 - \lambda_1 < 0$, it follows that, apart from the trajectories corresponding to the eigenvector solution

$$\mathbf{x}_2(t) = e^{\lambda_2 t}\mathbf{v}_2,$$

all trajectories are parallel to $\mathbf{v}_1$ as $t \to \infty$.

3. Writing (7.9.6) as

$$\mathbf{x}(t) = e^{\lambda_2 t}[c_1\mathbf{v}_1 e^{(\lambda_1 - \lambda_2)t} + c_2\mathbf{v}_2]$$

and using the fact that $\lambda_1 - \lambda_2 > 0$, we see that, apart from the trajectories corresponding to the eigenvector solution

$$\mathbf{x}_1(t) = e^{\lambda_1 t}\mathbf{v}_1,$$

all trajectories are parallel to $\mathbf{v}_2$ as $t \to -\infty$.

A typical phase portrait is sketched in Figure 7.9.4. In this case, the equilibrium point is called a **saddle point**.

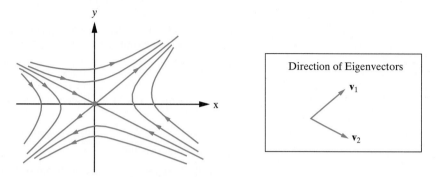

Figure 7.9.4: Typical phase portrait in the case $\lambda_2 < 0 < \lambda_1$. The equilibrium point is a saddle point and is unstable.

Case 2: $\lambda_1 = \lambda_2 = \lambda \neq 0$.

Two main subcases can be distinguished, depending on the structure of the matrix A.

(a) If A is a scalar multiple of the identity matrix—that is, $A = aI$ where a is a nonzero constant—then

$$\det(A - \lambda I) = (a - \lambda)^2$$

so that $\lambda = a$ is a repeated eigenvalue. Furthermore,

$$A - \lambda I = aI - aI = 0,$$

so that every nonzero vector is an eigenvector. Consequently, if we choose two nonproportional eigenvectors $\mathbf{v}_1$ and $\mathbf{v}_2$ of A, it follows that every solution to the system is

$$\mathbf{x}(t) = e^{\lambda t}(c_1\mathbf{v}_1 + c_2\mathbf{v}_2),$$

which, for each pair of values for c_1 and c_2, is the equation of a line through the origin. Moreover, since $\mathbf{v}_1$ and $\mathbf{v}_2$ are nonproportional, all directions in the xy-plane are obtained as c_1 and c_2 assume all possible values. Consequently, the trajectories consist of all half-lines through the origin. The equilibrium point is called a **proper node** in this case. If $\lambda < 0$, then all trajectories approach the equilibrium point as $t \to \infty$, whereas if $\lambda > 0$, the direction along the trajectories is away from the equilibrium point. A representative sketch of the phase portraits is given in Figure 7.9.5.

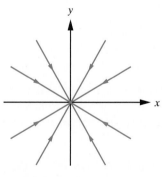

Figure 7.9.5: Typical phase portrait when $A = aI$, $a < 0$. The equilibrium point is a proper node and is stable.

(b) If A is not a scalar multiple of the identity matrix, then, as shown in Problem 27, A is defective, and so all eigenvectors are proportional to one another. If $\mathbf{v}_0$ denotes any such eigenvector, then, from our preceding discussion,

$$\mathbf{x}_0 = e^{\lambda t}\mathbf{v}_0$$

is a solution to the system (7.9.4). Using the material in Section 7.5, we know that a second linearly independent solution to the system in this case is

$$\mathbf{x}_1 = e^{\lambda t}(\mathbf{v}_1 + t\mathbf{v}_0),$$

where $\mathbf{v}_1$ is a vector satisfying the conditions $(A - \lambda I)\mathbf{v}_1 \neq \mathbf{0}$ and $(A - \lambda I)^2 \mathbf{v}_1 = \mathbf{0}$. Consequently, all solutions to the system are of the form

$$\mathbf{x}(t) = e^{\lambda t}[c_0 \mathbf{v}_0 + c_1(\mathbf{v}_1 + t\mathbf{v}_0)].$$

For $c_1 = 0$, we have the trajectories corresponding to the eigenvector solution. If $c_1 \neq 0$, then the dominant term in the general solution as $t \to \pm\infty$ is

$$\mathbf{x}(t) \approx c_1 t e^{\lambda t} \mathbf{v}_0.$$

Consequently, if $\lambda < 0$, we have the following results.

1. As $t \to \infty$, all trajectories approach the equilibrium point $(0, 0)$ tangent to the eigenvector $\mathbf{v}_0$.

2. As $t \to -\infty$, all trajectories are parallel to $\mathbf{v}_0$.

See Figure 7.9.6 for a typical phase portrait. If $\lambda > 0$, then the direction along the trajectories is reversed. In this case, the equilibrium point $(0, 0)$ is called a **degenerate node** and is said to be stable or unstable depending on whether λ is negative or positive, respectively.

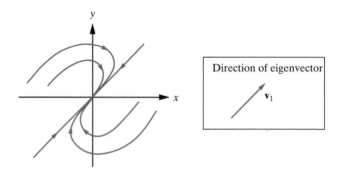

Figure 7.9.6: Typical phase portrait when A has only one linearly independent eigenvector and $\lambda < 0$. The equilibrium point is called a degenerate node.

Case 3:   Complex conjugate eigenvalues $\lambda = a \pm ib$.

If we let $\mathbf{v} = \mathbf{r} + i\mathbf{s}$ denote a complex eigenvector (with $\mathbf{r}$ and $\mathbf{s}$ real valued) corresponding to the eigenvalue $\lambda = a + ib$, then, according to our results from Section 7.4, two linearly independent solutions to the system of differential equations are of the form

$$\mathbf{x}_1(t) = e^{at}(\cos bt\,\mathbf{r} - \sin bt\,\mathbf{s}) \quad \text{and} \quad \mathbf{x}_2(t) = e^{at}(\sin bt\,\mathbf{r} + \cos bt\,\mathbf{s}). \tag{7.9.7}$$

Consequently, the general solution in this case is

$$\mathbf{x}(t) = e^{at}[c_1(\cos bt\ \mathbf{r} - \sin bt\ \mathbf{s}) + c_2(\sin bt\ \mathbf{r} + \cos bt\ \mathbf{s})],$$

or equivalently,

$$\mathbf{x}(t) = e^{at}\mathbf{v}(t), \tag{7.9.8}$$

where

$$\mathbf{v}(t) = c_1(\cos bt\ \mathbf{r} - \sin bt\ \mathbf{s}) + c_2(\sin bt\ \mathbf{r} + \cos bt\ \mathbf{s}).$$

The key point to notice is that

$$\mathbf{v}(t + 2\pi/b) = \mathbf{v}(t).$$

Consequently, $\mathbf{v}(t)$ has period $T = 2\pi/b$, and we can therefore draw the following conclusions.

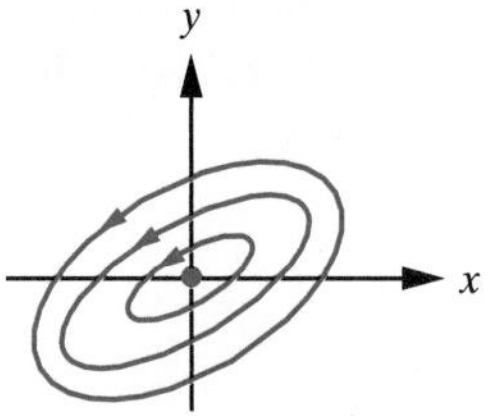

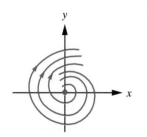

Figure 7.9.7: Typical phase portrait for the case of pure imaginary eigenvalues. The equilibrium point is called a center and is stable.

(a) If $a = 0$, (7.9.8) implies that $\mathbf{x}(t + 2\pi/b) = \mathbf{x}(t)$. All trajectories are therefore closed curves (see Figure 7.9.7) and so the corresponding solutions are periodic. In this case, the equilibrium point $(0, 0)$ is called a **center** and is stable.

(b) If $a \neq 0$, then the trajectories spiral around the origin. (See Figure 7.9.8.) The equilibrium point $(0, 0)$ is called a **spiral point**. Furthermore,

    1. If $a > 0$, the trajectories spiral away from the equilibrium point, and therefore it is called an **unstable spiral point**.

    2. If $a < 0$, the trajectories spiral toward the origin, and therefore it is called a **stable spiral point**.

This completes the classification of the equilibrium point $(0, 0)$ associated with the system (7.9.4) in the case where $\det(A) \neq 0$. The results when $\lambda_1 \neq \lambda_2$ are summarized in Table 7.9.1.

Figure 7.9.8: Typical phase portrait for the case of complex conjugate eigenvalues $\lambda = a \pm ib$, with $a > 0$. The equilibrium point is called a spiral point.

Eigenvalues	Type of Equilibrium Point
Real and negative	Stable node
Real and positive	Unstable node
Opposite sign	Saddle
Pure imaginary	Stable center
Complex with positive real part	Unstable spiral
Complex with negative real part	Stable Spiral

Table 7.9.1: Classification of equilibrium points in terms of eigenvalues.

Based on the preceding analysis, it is not too difficult to obtain a general sketch of the phase plane, once we have determined the eigenvalues and eigenvectors of A. However, as in the case of slope fields considered in Chapter 1, this is an area in which technology is a definite benefit. In the examples that follow we have provided Maple plots of the phase planes.

Example 7.9.3   Characterize the equilibrium point for the linear systems $\mathbf{x}' = A\mathbf{x}$ and sketch the phase portrait.

    (a) $A = \begin{bmatrix} -1 & -2 \\ -2 & -1 \end{bmatrix}$ **(b)** $A = \begin{bmatrix} -1 & -2 \\ 2 & -1 \end{bmatrix}$ **(c)** $A = \begin{bmatrix} 1 & 3 \\ -2 & -4 \end{bmatrix}$.

Solution:

(a) The matrix A has eigenvalues $\lambda_1 = 1$ and $\lambda_2 = -3$ with corresponding non-proportional eigenvectors $\mathbf{v}_1 = (1, -1)$ and $\mathbf{v}_2 = (1, 1)$. Since the eigenvalues have different signs, the equilibrium point $(0, 0)$ is a saddle point. A Maple sketch including the slope field is given in Figure 7.9.9. Notice that we have appended arrowheads to each line segment in the slope field to indicate the direction in which trajectories are traversed. The resulting slope field is usually called a **direction field**.

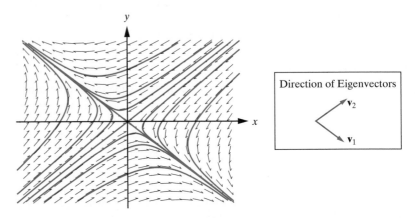

Figure 7.9.9: Phase portrait for the system in part (a) of Example 7.9.3.

(b) In this case, the matrix has complex conjugate eigenvalues $\lambda = -1 \pm 2i$. Since the real part of the eigenvalues is negative, the equilibrium point is a stable spiral. To determine whether the trajectories spiral clockwise or counterclockwise toward the origin, we check the sign of dx/dt at points where the trajectories intersect the positive y-axis. From the given system, when $x = 0$ and $y > 0$, we see that $dx/dt < 0$, so that the trajectories spiral counterclockwise around the origin. A Maple plot of the phase plane is given in Figure 7.9.10.

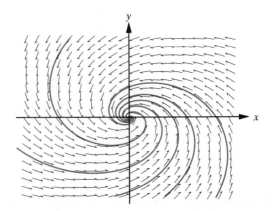

Figure 7.9.10: Phase portrait for the system in part (b) of Example 7.9.3.

(c) In this case, the matrix A has eigenvalues $\lambda_1 = -1$ and $\lambda_2 = -2$ with corresponding nonproportional eigenvectors $\mathbf{v}_1 = (3, -2)$ and $\mathbf{v}_2 = (1, -1)$. We see that the equilibrium point is a stable node. All trajectories approach $(0, 0)$ tangent to the eigenvector $\mathbf{v}_1$ (see Case 1(a)). The phase plane for this system is given in Figure 7.9.11.

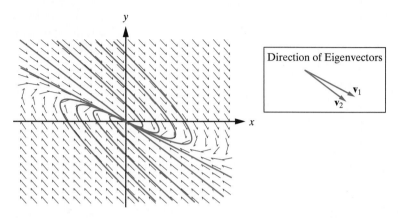

Figure 7.9.11: Phase portrait for the system in part (c) of Example 7.9.3. □

Exercises for 7.9

Key Terms

Autonomous system, Trajectory, Phase plane, Phase portrait, Equilibrium points, Equilibrium solution, Stable node, Unstable node, Saddle point, Proper node, Degenerate node, Stable spiral point, Unstable spiral point, Direction field.

Skills

- Be able to determine the equilibrium point(s) for a linear system of two differential equations.

- Be able to classify the equilibrium point(s) as stable or unstable nodes, saddle points, proper nodes, degenerate nodes, or stable or unstable spiral points.

- Be able to use the eigenvalue/eigenvector pairs for a linear system of two differential equations to predict the qualitative behavior of the trajectories in the phase plane.

True-False Review

For Questions 1–6, decide if the given statement is **true** or **false**, and give a brief justification for your answer. If true, you can quote a relevant definition or theorem from the text. If false, provide an example, illustration, or brief explanation of why the statement is false.

1. An equilibrium point is a point toward which all trajectories of a linear system of differential equations approach as $t \to \infty$.

2. A linear system $\mathbf{x}' = A\mathbf{x}$ for which A has complex eigenvalues $\lambda = \pm ib$ (for some $b \in \mathbb{R}$) gives rise to elliptical trajectories in the phase plane.

3. The equilibrium point of a linear system $\mathbf{x}' = A\mathbf{x}$ that has two positive (real) eigenvalues is called an unstable node.

4. The type of equilibrium point for the linear system $\mathbf{x}' = A\mathbf{x}$ is the same as the type of equilibrium point for the linear system $\mathbf{x}' = (2A)\mathbf{x}$.

5. If all solutions to the linear system $\mathbf{x}' = A\mathbf{x}$ approach $(0, 0)$, then $(0, 0)$ is stable.

6. For a saddle point, the only trajectories that approach $(0, 0)$ as $t \to \infty$ are those pointing along the eigenvector $\mathbf{v}$ corresponding to the positive eigenvalue λ.

Problems

For Problems 1–3, determine all equilibrium points of the given system.

1. $x' = x(x - y + 1), \quad y' = y(y + 2x)$.

2. $x' = x(2x + y), \quad y' = y(x - 2y + 4)$.

3. $x' = x(x^2 + y^2 - 1)$, $y' = 2y(xy - 1)$.

For Problems 4–19, characterize the equilibrium point for the system $\mathbf{x}' = A\mathbf{x}$ and sketch the phase portrait.

4. $A = \begin{bmatrix} 1 & 3 \\ 1 & -1 \end{bmatrix}$.

5. $A = \begin{bmatrix} 0 & 2 \\ -2 & 0 \end{bmatrix}$.

6. $A = \begin{bmatrix} 1 & 0 \\ 3 & 1 \end{bmatrix}$.

7. $A = \begin{bmatrix} 2 & 3 \\ -1 & -2 \end{bmatrix}$.

8. $A = \begin{bmatrix} -2 & 3 \\ -3 & -2 \end{bmatrix}$.

9. $A = \begin{bmatrix} -2 & 1 \\ 1 & -2 \end{bmatrix}$.

10. $A = \begin{bmatrix} 5 & 4 \\ 4 & 5 \end{bmatrix}$.

11. $A = \begin{bmatrix} 0 & -1 \\ 1 & 0 \end{bmatrix}$.

12. $A = \begin{bmatrix} 3 & -2 \\ 2 & -1 \end{bmatrix}$.

13. $A = \begin{bmatrix} 2 & -1 \\ 1 & 2 \end{bmatrix}$.

14. $A = \begin{bmatrix} 2 & -5 \\ 4 & -7 \end{bmatrix}$.

15. $A = \begin{bmatrix} 2 & 1 \\ 3 & 4 \end{bmatrix}$.

16. $A = \begin{bmatrix} 3 & 4 \\ 4 & -3 \end{bmatrix}$.

17. $A = \begin{bmatrix} 1 & 1 \\ -9 & -5 \end{bmatrix}$.

18. $A = \begin{bmatrix} 1 & -1 \\ 1 & 2 \end{bmatrix}$.

19. $A = \begin{bmatrix} 3 & 0 \\ 0 & 3 \end{bmatrix}$.

20. Characterize the equilibrium point $(0, 0)$ for the system $\mathbf{x}' = A\mathbf{x}$ if

$$A = \begin{bmatrix} -1 & 2 \\ -2 & -1 \end{bmatrix}.$$

Solve the system of differential equations, and show that the components of the solution vector satisfy

$$x^2 + y^2 = e^{-4t}(c_1^2 + c_2^2), \qquad (7.9.9)$$

where c_1 and c_2 are constants. As t varies in Equation (7.9.9), describe the curve that is generated in the phase plane.

For Problems 21–24, convert the given differential equation to a first-order system using the substitution $u = y$, $v = dy/dt$, and determine the phase portrait for the resulting system.

21. $\dfrac{d^2y}{dt^2} + 6\dfrac{dy}{dt} + 9y = 0$.

22. $\dfrac{d^2y}{dt^2} + 16y = 0$.

23. $\dfrac{d^2y}{dt^2} + 4\dfrac{dy}{dt} + 5y = 0$.

24. $\dfrac{d^2y}{dt^2} - 25y = 0$.

25. Consider the differential equation

$$\frac{d^2y}{dt^2} + 2c\frac{dy}{dt} + ky = 0,$$

where c and k are positive constants, that governs the behavior of a spring–mass system. Convert the differential equation to a first-order linear system and sketch the corresponding phase portraits. (You will need to distinguish the three cases $c^2 > k$, $c^2 < k$, and $c^2 = k$.) In each case, use your phase portrait to describe the behavior of y for various initial conditions.

26. Verify that the solutions to $\mathbf{x}' = A\mathbf{x}$ appearing in (7.9.7) are not proportional. [**Hint:** Evaluate each solution at $t = 0$ and at $t = 2\pi/b$, and use the fact that both $\mathbf{r}$ and $\mathbf{s}$ are real-valued vectors.]

27. Let A be a 2×2 matrix of real constants with a repeated eigenvalue: $\lambda_1 = \lambda_2 = \lambda$.

(a) Show that $\lambda = (a + d)/2$.

(b) Show that if A has two linearly independent eigenvectors corresponding to λ, then $A = aI$ for some scalar a. [**Hint:** Under this assumption, both $\mathbf{e}_1$ and $\mathbf{e}_2$ must be eigenvectors. Now consider $(A - \lambda I)\mathbf{e}_1 = \mathbf{0}$ and $(A - \lambda I)\mathbf{e}_2 = \mathbf{0}$.]

(c) Conclude from part (b) that if A is a 2×2 matrix of real constants with a repeated eigenvalue that is not a scalar multiple of the identity matrix, then A is defective.

7.10   Nonlinear Systems

We now briefly discuss the qualitative analysis of general autonomous systems of the form

$$\frac{dx}{dt} = F(x, y), \qquad \frac{dy}{dt} = G(x, y), \tag{7.10.1}$$

where, throughout the remainder of the discussion, we will assume that F and G have continuous partial derivatives up to order at least two. We are interested in classifying the equilibrium points for this system as we did in the linear case. The approach that we will take is to approximate (7.10.1) with a corresponding linear system. To see how the approximation arises, we recall from elementary calculus that if we are given a function $f(x, y)$ defined in some region of the xy-plane, then the equation $z = f(x, y)$ defines a surface in space. Further, the tangent plane to this surface at any point (x_0, y_0) has equation

$$z = f(x_0, y_0) + (x - x_0)\frac{\partial f}{\partial x}(x_0, y_0) + (y - y_0)\frac{\partial f}{\partial y}(x_0, y_0),$$

and this plane gives the best linear approximation to $f(x, y)$ at (x_0, y_0). Returning to the system (7.10.1), we define the linear approximation to this system at (x_0, y_0) by

$$\frac{dx}{dt} = F(x_0, y_0) + (x - x_0)\frac{\partial F}{\partial x}(x_0, y_0) + (y - y_0)\frac{\partial F}{\partial y}(x_0, y_0),$$

$$\frac{dy}{dt} = G(x_0, y_0) + (x - x_0)\frac{\partial G}{\partial x}(x_0, y_0) + (y - y_0)\frac{\partial G}{\partial y}(x_0, y_0).$$

In the case when (x_0, y_0) is an equilibrium point of (7.10.1), the approximate system reduces to

$$\frac{dx}{dt} = (x - x_0)\frac{\partial F}{\partial x}(x_0, y_0) + (y - y_0)\frac{\partial F}{\partial y}(x_0, y_0),$$

$$\frac{dy}{dt} = (x - x_0)\frac{\partial G}{\partial x}(x_0, y_0) + (y - y_0)\frac{\partial G}{\partial y}(x_0, y_0).$$

The **Jacobian matrix**, $J(x, y)$, is defined by

$$J(x, y) = \begin{bmatrix} \dfrac{\partial F}{\partial x} & \dfrac{\partial F}{\partial y} \\[2mm] \dfrac{\partial G}{\partial x} & \dfrac{\partial G}{\partial y} \end{bmatrix}.$$

Using this matrix, the linear approximation to (7.10.1) at an equilibrium point (x_0, y_0) can be written as

$$\begin{bmatrix} x'(t) \\ y'(t) \end{bmatrix} = J(x_0, y_0)\begin{bmatrix} x - x_0 \\ y - y_0 \end{bmatrix},$$

or, equivalently, as

$$\mathbf{u}' = J(x_0, y_0)\mathbf{u},$$

where $\mathbf{u} = \begin{bmatrix} x - x_0 \\ y - y_0 \end{bmatrix}$.

Example 7.10.1   Determine the linear approximation to the system

$$x' = \cos x + 3y - 1, \quad y' = 2x + \sin y$$

at the equilibrium point $(0, 0)$.

Solution: For the given system, we have

$$F(x, y) = \cos x + 3y - 1 \quad \text{and} \quad G(x, y) = 2x + \sin y,$$

so that

$$J(x, y) = \begin{bmatrix} -\sin x & 3 \\ 2 & \cos y \end{bmatrix}.$$

Hence,

$$J(0, 0) = \begin{bmatrix} 0 & 3 \\ 2 & 1 \end{bmatrix},$$

and the linear approximation to the given system at $(0, 0)$ is

$$\mathbf{x}' = \begin{bmatrix} 0 & 3 \\ 2 & 1 \end{bmatrix} \begin{bmatrix} x \\ y \end{bmatrix},$$

or, equivalently,

$$\frac{dx}{dt} = 3y, \quad \frac{dy}{dt} = 2x + y. \qquad \square$$

Example 7.10.2   Determine all equilibrium points for the system

$$x' = x(1 - y) \quad \text{and} \quad y' = y(2 - x),$$

and determine the linear approximation to the system at each equilibrium point.

Solution: The system has the two equilibrium points $(0, 0)$ and $(2, 1)$. In this case, the Jacobian matrix is

$$J(x, y) = \begin{bmatrix} 1 - y & -x \\ -y & 2 - x \end{bmatrix}.$$

Thus,

$$J(0, 0) = \begin{bmatrix} 1 & 0 \\ 0 & 2 \end{bmatrix},$$

so that the linear approximation at the equilibrium point $(0, 0)$ is

$$\mathbf{x}' = J(0, 0)\mathbf{x}.$$

That is,

$$x' = x \quad \text{and} \quad y' = 2y.$$

Similarly, the linear approximation at the equilibrium point $(2, 1)$ is

$$\mathbf{u}' = J(2, 1)\mathbf{u} = \begin{bmatrix} 0 & -2 \\ -1 & 0 \end{bmatrix} \mathbf{u},$$

where $\mathbf{u} = \begin{bmatrix} x - 2 \\ y - 1 \end{bmatrix}$. In scalar form, we have

$$x' = -2(y - 1) \quad \text{and} \quad y' = -(x - 2). \qquad \square$$

It may seem reasonable to expect that the behavior of the trajectories to the nonlinear system (7.10.1) is closely approximated by the trajectories of the corresponding linear system, provided that we do not move too far away from (x_0, y_0). This is indeed true in most cases. In Table 7.10.1 we summarize the relationship between the behavior at an equilibrium point of a nonlinear system and the behavior at the corresponding equilibrium point of the linear approximation in the case of distinct eigenvalues. This indicates that, apart from the case of pure imaginary eigenvalues (or a repeated eigenvalue), the phase portrait for a nonlinear system looks similar to the linear approximation in the neighborhood of an equilibrium point.

Eigenvalues	Linear Approximation	Nonlinear System
Real and negative	Stable node	Stable node
Real and positive	Unstable node	Unstable node
Opposite signs	Saddle	Saddle
Complex with negative real part	Stable spiral	Stable spiral
Complex with positive real part	Unstable spiral	Unstable spiral
Pure imaginary	Stable center	Center or spiral point, stability indeterminate

Table 7.10.1: Nonlinear system and linear approximation behavior in terms of eigenvalues.

Example 7.10.3 Determine and classify all equilibrium points for the given system

(a) $x' = x(x - 1), \quad y' = y(2 + xy^2)$.

(b) $x' = x - y, \quad y' = y(2x + y - 3)$.

Solution:

(a) The equilibrium points are obtained by solving

$$x(x - 1) = 0, \quad y(2 + xy^2) = 0.$$

The first of these equations implies that $x = 0$ or $x = 1$. In both cases, substitution into the second equation yields $y = 0$. Consequently, the only equilibrium points are $(0, 0)$ and $(1, 0)$. The Jacobian for the given system is

$$J(x, y) = \begin{bmatrix} 2x - 1 & 0 \\ y^3 & 2 + 3xy^2 \end{bmatrix}.$$

Hence,

$$J(0, 0) = \begin{bmatrix} -1 & 0 \\ 0 & 2 \end{bmatrix},$$

which has eigenvalues $\lambda_1 = 2$ and $\lambda_2 = -1$. Since the eigenvalues have different signs, the equilibrium point $(0, 0)$ is a saddle point in both the linear approximation to the given system and the given system itself. At the equilibrium point $(1, 0)$, we have

$$J(1, 0) = \begin{bmatrix} 1 & 0 \\ 0 & 2 \end{bmatrix}.$$

This matrix has eigenvalues $\lambda_1 = 2$ and $\lambda_2 = 1$, which implies that the equilibrium point is an unstable node. Figure 7.10.1 gives a Maple plot of the phase plane for the given nonlinear system.

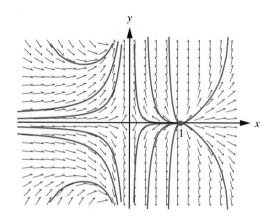

Figure 7.10.1: Phase portrait for the system in part (a) of Example 7.10.3.

(b) To determine the equilibrium points, we must solve

$$x - y = 0, \quad y(2x + y - 3) = 0.$$

Substituting $y = x$ from the first equation into the second yields the condition

$$3x(x - 1) = 0.$$

Hence, the equilibrium points are $(0, 0)$ and $(1, 1)$. The Jacobian for the given system is

$$J(x, y) = \begin{bmatrix} 1 & -1 \\ 2y & 2x + 2y - 3 \end{bmatrix}.$$

At the equilibrium point $(0, 0)$, we have

$$J(0, 0) = \begin{bmatrix} 1 & -1 \\ 0 & -3 \end{bmatrix},$$

which has eigenvalues $\lambda_1 = 1$ and $\lambda_2 = -3$, with corresponding eigenvectors $\mathbf{v}_1 = (1, 0)$ and $\mathbf{v}_2 = (1, 4)$. Since the eigenvalues have different signs, the equilibrium point $(0, 0)$ is a saddle point. At the equilibrium point $(1, 1)$, we have

$$J(1, 1) = \begin{bmatrix} 1 & -1 \\ 2 & 1 \end{bmatrix},$$

which has eigenvalues $\lambda = 1 \pm i\sqrt{2}$. Hence, the equilibrium point is an unstable spiral point. A Maple plot of the phase plane is given in Figure 7.10.2.   □

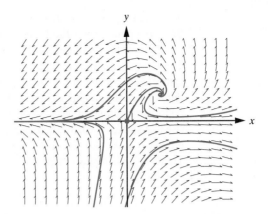

Figure 7.10.2: Phase portrait for the system in part (b) of Example 7.10.3.

A Predator/Prey Model

As an example of an applied problem that is modeled by a nonlinear system of differential equations, we consider the interaction of two species. One species is a predator, and the other is the prey. Let $x(t)$ denote the prey population at time t, and let $y(t)$ denote the predator population. Then the model equations that we discuss are

$$\frac{dx}{dt} = x(a - by), \tag{7.10.2}$$

$$\frac{dy}{dt} = y(cx - d), \tag{7.10.3}$$

where $a, b, c,$ and d are positive constants. To interpret these equations, we see that in the absence of any predators (i.e. $y = 0$), Equation (7.10.2) reduces to the simple Malthusian exponential growth law. We take account of the inclusion of predators by subtracting a term proportional to the number of predators present from the growth rate of the prey. Similarly, from Equation (7.10.3), in the absence of prey (i.e., $x = 0$), a predator population would decay exponentially. To account for the inclusion of prey, a term proportional to the number of prey has been added to the growth rate of the predator. This model is called the Lotka-Volterra system. We see that the system is nonlinear with equilibrium points at $(0, 0)$ and $(d/c, a/b)$. Computing the Jacobian of the system yields

$$J(x, y) = \begin{bmatrix} a - by & -bx \\ cy & cx - d \end{bmatrix},$$

so that

$$J(0, 0) = \begin{bmatrix} a & 0 \\ 0 & -d \end{bmatrix}$$

with eigenvalues $\lambda_1 = a$ and $\lambda_2 = -d$. Consequently, the equilibrium point $(0, 0)$ is a saddle point. We note that nonproportional eigenvectors in this case are $\mathbf{v}_1 = (1, 0)$ and $\mathbf{v}_2 = (0, 1)$. At the equilibrium point $(d/c, a/b)$, we have

$$J(\frac{d}{c}, \frac{a}{b}) = \begin{bmatrix} 0 & -\frac{bd}{c} \\ \frac{ca}{b} & 0 \end{bmatrix},$$

with eigenvalues $\lambda = \pm i \sqrt{ad}$. Consequently, in the linear approximation the equilibrium point at $(d/c, a/b)$ is a center. Therefore, according to the results given in Table 7.10.1, the nonlinear system has either a center or a spiral point. In Figure 7.10.3 we give a Maple plot of the phase plane using typical values for the constants $a, b, c,$ and d. This indicates that the equilibrium point in the nonlinear model is also a center. Consequently, the corresponding solutions for both x and y are periodic in time. The model therefore predicts that the population of both species is periodic, hence both species would survive. The general qualitative behavior starting at small values for both the predator and prey can be seen from the trajectories. The prey initially increases, and the predator population remains approximately constant. Then, since there is plenty of food (prey), the predator population increases, with a corresponding decrease in the prey population. This gives rise to a situation where there are too many predators for the prey population, and so the predator population decreases while the prey population remains approximately constant. Then the cycle repeats itself. We see from the three different trajectories in Figure 7.10.3 that the specific behavior varies quite significantly, depending on the initial conditions.

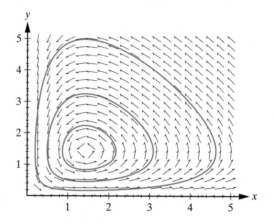

Figure 7.10.3: Representative phase portrait for a predator/prey system.

The Van Der Pol Equation

In concluding this section, we use the nonlinear differential equation

$$\frac{d^2y}{dt^2} + \mu(y^2 - 1)\frac{dy}{dt} + y = 0, \quad \mu > 0, \tag{7.10.4}$$

to illustrate a new type of behavior that does not arise in linear systems. The differential equation (7.10.4) is called the **Van der Pol equation** and arises in the study of nonlinear circuits. If the parameter μ is zero, then Equation (7.10.4) reduces to that of the simple harmonic oscillator, which has periodic solutions and circular trajectories. For μ small and positive and $|y| > 1$, we can presumably interpret the dy/dt term as a damping term and expect the system to behave somewhat like a damped harmonic oscillator. However, for $-1 < y < 1$, the coefficient of dy/dt is negative, and therefore, this term would tend to amplify, rather than dampen, any oscillations. This suggests that there may be an isolated periodic solution (closed trajectory) with the property that all trajectories that start within it approach the closed path as t increases, and all trajectories that start outside the closed path spiral toward it as $t \rightarrow \infty$. Such a closed path, if it exists, is called a **limit cycle**. To analyze the Van der Pol equation, we introduce the phase-plane variables

$$u = y, \qquad v = \frac{dy}{dt},$$

thereby obtaining the equivalent first-order system

$$\frac{du}{dt} = v, \qquad \frac{dv}{dt} = -u - \mu(u^2 - 1)v. \tag{7.10.5}$$

The only equilibrium point of the system (7.10.5) is $(0, 0)$, and the Jacobian of this system is

$$J(u, v) = \begin{bmatrix} 0 & 1 \\ -1 - 2\mu v & -\mu(u^2 - 1) \end{bmatrix}.$$

Hence,

$$J(0, 0) = \begin{bmatrix} 0 & 1 \\ -1 & \mu \end{bmatrix}.$$

This matrix has characteristic polynomial

$$p(\lambda) = \lambda^2 - \mu\lambda + 1,$$

so that the eigenvalues are

$$\lambda = \tfrac{1}{2}(\mu \pm \sqrt{\mu^2 - 4}).$$

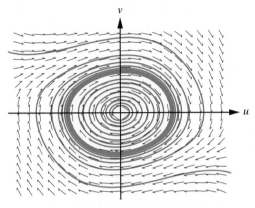

Figure 7.10.4: Maple plot of the phase plane for the Van der Pol equation with $\mu = 0.1$.

For $\mu > 2$, there are two positive eigenvalues, and the equilibrium point is an unstable node. For $\mu < 2$, however, the equilibrium point is an unstable spiral. Hence, the trajectories close to the equilibrium point do indeed spiral outward. However, this local analysis does not give us information about the global behavior of the trajectories. Although we do not have the tools to prove that the Van der Pol equation does indeed have a limit cycle, further convincing evidence for its existence can be obtained by studying the phase portraits associated with the differential equation. Figure 7.10.4 contains a Maple plot in the case when $\mu = 0.1$. The limit cycle is clearly visible and is almost circular, as we would expect with a small μ value. Figure 7.10.5 contains a similar plot with $\mu = 1$. The limit cycle is still visible, but it no longer resembles a circle.

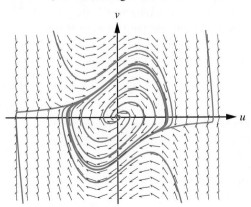

Figure 7.10.5: The phase plane for the Van der Pol equation with $\mu = 1$.

Exercises for 7.10

Key Terms

Jacobian matrix, Lotka-Volterra system, Van der Pol equation, Limit cycle.

Skills

- Be able to determine the linear approximation to a nonlinear system by using the Jacobian matrix.

- Be able to find and classify the equilibrium points for a nonlinear system, as well as for a linear system.

- Be familiar with the Lotka-Volterra system for modeling predator/prey interactions.

- Be familiar with the Van der Pol equation, how to convert it into a first-order nonlinear system, and the qualitative behavior of the solutions to the system, depending on the parameter μ.

True-False Review

For Questions 1–5, decide if the given statement is **true** or **false**, and give a brief justification for your answer. If true, you can quote a relevant definition or theorem from the text. If false, provide an example, illustration, or brief explanation of why the statement is false.

1. The Jacobian matrix for a linear system of the form (7.10.1) is

$$J(x, y) = \begin{bmatrix} \dfrac{\partial F}{\partial x} & \dfrac{\partial F}{\partial y} \\ \dfrac{\partial G}{\partial x} & \dfrac{\partial G}{\partial y} \end{bmatrix}.$$

2. If we replace a nonlinear system of differential equations with a linear approximation to the system at an equilibrium point, then the behavior of the trajectories of the nonlinear system and the behavior of the trajectories of the corresponding linear system are approximately the same throughout the xy-plane.

3. In the Lotka-Volterra predator/prey model, the origin is a saddle point.

4. The only equilibrium point of the linear system of differential equations arising from the Van der Pol equation is the origin.

5. The equilibrium point of the linear system arising from the Van der Pol equation

$$\frac{d^2 y}{dt^2} + 3(y^2 - 1)\frac{dy}{dt} + y = 0$$

is an unstable spiral.

Problems

For Problems 1–9, determine all equilibrium points of the given system and, if possible, characterize them as centers, spirals, saddles, or nodes.

1. $x' = y(3x - 2), \quad y' = 2x + 9y^2$.

2. $x' = y(3x - 2), \quad y' = 2x - 9y^2$.

3. $x' = x - y^2, \quad y' = y(9x - 4)$.

4. $x' = x + 3y^2, \quad y' = y(x - 2)$.

5. $x' = 2x + 5y^2, \quad y' = y(3 - 4x)$.

6. $x' = 2y + \sin x, \quad y' = x(\cos y - 2)$.

7. $x' = x - 2y + 5xy, \quad y' = 2x + y$.

8. $x' = x(1 - y), \quad y' = y(x + 1)$.

9. $x' = 4x - y - y \sin x, \quad y' = x + 2y$.

The remaining problems require the use of some form of technology to generate the phase plane for the system of differential equations.

10. ◇ Sketch the phase portrait of the system in Problem 1 for $-1 \le x \le 1, -1 \le y \le 1$, and thereby determine whether the equilibrium point $(0, 0)$ is a center or a spiral.

11. ◇ Sketch the phase portrait of the system in Problem 6 for $-2 \le x \le 2, -2 \le y \le 2$, and thereby determine whether the equilibrium point $(0, 0)$ is a center or a spiral.

12. ◇ Sketch the phase portrait of the system in Problem 8 for $-2 \le x \le 2, -2 \le y \le 2$. By inspection, guess the equation of one particular trajectory.

For Problems 13–18, sketch the phase portrait of the given system for $-2 \le x \le 2, -2 \le y \le 2$. Comment on the types of equilibrium points.

13. ◇ The system in Problem 2.

14. ◇ The system in Problem 3.

15. ◊ The system in Problem 4.

16. ◊ The system in Problem 5.

17. ◊ The system in Problem 7.

18. ◊ The system in Problem 9.

19. ◊ Consider the predator/prey model

$$\frac{dx}{dt} = x(2 - y), \qquad \frac{dy}{dt} = y(x - 2).$$

Sketch the phase plane for $0 \le x \le 10, 0 \le y \le 10$. Compare the behavior of the two specific cases corresponding to the initial conditions $x(0) = 1$, $y(0) = 0.1$, and $x(0) = 1$, $y(0) = 1$.

20. ◊ Consider the predator/prey model

$$\frac{dx}{dt} = x(3 - x - y), \qquad \frac{dy}{dt} = y(x - 1).$$

Sketch the phase plane for $0 \le x \le 4, 0 \le y \le 4$. What happens to the populations of both species as $t \to +\infty$?

21. ◊ Consider the differential equation

$$\frac{d^2 y}{dt^2} + 0.1(y - 4)(y + 1)\frac{dy}{dt} + y = 0.$$

(a) Convert the differential equation to a first-order system using the substitution $u = y$, $v = dy/dt$, and characterize the equilibrium point $(0, 0)$.

(b) Sketch the phase plane for the system on the square $-2 \le u \le 2, -2 \le v \le 2$. Based on the resulting sketch, do you think the differential equation has a limit cycle?

(c) Repeat (b) using the square $-8 \le u \le 8, -8 \le v \le 8$, and include the trajectories corresponding to the initial conditions $u(0) = 1, v(0) = 0$, and $u(0) = 6, v(0) = 0$.

7.11   Chapter Review

Many problems in applied mathematics involve two or more unknown functions, as well as their derivatives, and therefore require the solution of a system of differential equations. Two such problems that we have considered in this chapter (Section 7.7) are (a) coupled spring–mass systems and (b) mixing problems involving chemicals in a system of two connected tanks.

A first-order linear system of differential equations for n unknown functions $x_1(t)$, $x_2(t), \ldots, x_n(t)$ may be written in the form

$$\mathbf{x}'(t) = A(t)\mathbf{x}(t) + \mathbf{b}(t), \tag{7.11.1}$$

where

$$\mathbf{x}(t) = \begin{bmatrix} x_1(t) \\ x_2(t) \\ \vdots \\ x_n(t) \end{bmatrix}, \quad \mathbf{x}'(t) = \begin{bmatrix} x_1'(t) \\ x_2'(t) \\ \vdots \\ x_n'(t) \end{bmatrix},$$

and

$$A(t) = \begin{bmatrix} a_{11}(t) & a_{12}(t) & \ldots & a_{1n}(t) \\ a_{21}(t) & a_{22}(t) & \ldots & a_{2n}(t) \\ \vdots & \vdots & & \vdots \\ a_{n1}(t) & a_{n2}(t) & \ldots & a_{nn}(t) \end{bmatrix}, \quad \mathbf{b}(t) = \begin{bmatrix} b_1(t) \\ b_2(t) \\ \vdots \\ b_n(t) \end{bmatrix}.$$

Equation (7.11.1) is called a *vector differential equation*. A primary goal of this chapter has been to develop techniques for solving Equation (7.11.1) for the unknown vector function $\mathbf{x}(t)$. To do this, we have assumed throughout much of the chapter that the matrix $A(t)$ is constant.

Homogeneous First-Order Linear Systems

In Sections 7.4 and 7.5, we have developed solution techniques for (7.11.1) in the case when this system is *homogeneous* (i.e., $\mathbf{b}(t) = \mathbf{0}$). In this case, we often abbreviate Equation (7.11.1) as

$$\mathbf{x}' = A\mathbf{x}, \tag{7.11.2}$$

where A is an $n \times n$ matrix of constants. It turns out that the solution set to the homogeneous system (7.11.2) is a vector space of dimension n, and consequently, the goal is to determine n linearly independent solutions $\mathbf{x}_1(t), \mathbf{x}_2(t), \ldots, \mathbf{x}_n(t)$, called a *fundamental solution set*, for the linear system (7.11.2). The fundamental solution set is therefore a basis for the space of all solutions to (7.11.2), and thus allows us to write the *general solution* to (7.11.2) in the form

$$\mathbf{x}(t) = c_1\mathbf{x}_1(t) + c_2\mathbf{x}_2(t) + \cdots + c_n\mathbf{x}_n(t). \tag{7.11.3}$$

This general solution is sometimes written in the form

$$\mathbf{x}(t) = X(t)\mathbf{c},$$

where

$$X(t) = [\mathbf{x}_1(t)\ \mathbf{x}_2(t)\ \ldots\ \mathbf{x}_n(t)] \tag{7.11.4}$$

is called a *fundamental matrix* for the linear system (7.11.2) and $\mathbf{c} = (c_1\ c_2\ \ldots\ c_n)^T$. If values

$$x_1(t_0),\quad x_2(t_0),\quad \ldots,\quad x_n(t_0) \tag{7.11.5}$$

are specified, then we can solve for the values of $c_1, c_2, \ldots, c_n$, thereby determining the solution to the *initial-value problem* (7.11.2) with initial conditions given by (7.11.5).

The eigenvalues and eigenvectors of A play a crucial role in finding a fundamental solution set for (7.11.2). In particular, if λ is an eigenvalue of A with corresponding eigenvector $\mathbf{v}$, then

$$\mathbf{x}(t) = e^{\lambda t}\mathbf{v}$$

is a solution to the system (7.11.2). Moreover, if $\mathbf{v}_1, \mathbf{v}_2, \ldots, \mathbf{v}_n$ are linearly independent eigenvectors, corresponding to (not necessarily distinct) eigenvalues $\lambda_1, \lambda_2, \ldots, \lambda_n$ of A, then the vector functions

$$\mathbf{x}_i(t) = e^{\lambda_i t}\mathbf{v}_i, \qquad i = 1, 2, \ldots, n \tag{7.11.6}$$

are linearly independent. In the case where A is nondefective (Section 7.4), we therefore already have a natural way, by using the vector functions in (7.11.6), to obtain a fundamental solution set to $\mathbf{x}' = A\mathbf{x}$.

The case in which A is a defective matrix is discussed in Section 7.5. In this case, we do not have n linearly independent eigenvectors of A at our disposal, so the strategy becomes to manufacture so-called *generalized eigenvectors* that enable us to generate additional solutions to the system (7.11.2). Although these solutions take a more complicated form, it can be shown that *any* system of the form (7.11.2) has n linearly independent solutions $\mathbf{x}_1(t), \mathbf{x}_2(t), \ldots, \mathbf{x}_n(t)$ that can be built using generalized eigenvectors.

Matrix Exponential Function

Alternatively, we can use the matrix exponential function e^{At}, first introduced in Chapter 5, to directly derive n linearly independent solutions to $\mathbf{x}' = A\mathbf{x}$. To do this, we observed in Section 7.8 that if $\mathbf{v}_1, \mathbf{v}_2, \ldots, \mathbf{v}_n$ are *any* n linearly independent vectors in $\mathbb{R}^n$ (or $\mathbb{C}^n$), then each of the vector functions

$$\mathbf{x}_1(t) = e^{At}\mathbf{v}_1, \quad \mathbf{x}_2(t) = e^{At}\mathbf{v}_2, \quad \ldots, \quad \mathbf{x}_n(t) = e^{At}\mathbf{v}_n \tag{7.11.7}$$

is a solution to $\mathbf{x}' = A\mathbf{x}$. Moreover, the n solutions in (7.11.7) are linearly independent.

Nonhomogeneous First-Order Linear Systems

In Section 7.6 we considered the nonhomogeneous linear system (7.11.1). In this case, the general solution takes the form

$$\mathbf{x}(t) = c_1\mathbf{x}_1 + c_2\mathbf{x}_2 + \cdots + c_n\mathbf{x}_n + \mathbf{x}_p,$$

where $\{\mathbf{x}_1, \mathbf{x}_2, \ldots, \mathbf{x}_n\}$ is a fundamental solution set for the corresponding homogeneous system (7.11.2) and $\mathbf{x}_p$ is one particular solution to (7.11.1). In the text, we used a variation-of-parameters technique for linear systems to derive a particular solution. Explicitly,

$$\mathbf{x}_p(t) = X(t) \int^t X^{-1}(s)\mathbf{b}(s)\,ds,$$

where $X(t)$ is the fundamental matrix given in Equation (7.11.4).

Qualitative Analysis

If the matrix A in the system $\mathbf{x}' = A\mathbf{x}$ is not constant, or if the system of differential equations is nonlinear, we must resort to either a qualitative analysis or numerical techniques to study the system. In Sections 7.9 and 7.10, we considered the qualitative aspects of linear systems in the case of two differential equations and two unknown functions. If

$$A = \begin{bmatrix} a & b \\ c & d \end{bmatrix}$$

is a 2×2 invertible matrix, the system

$$\mathbf{x}'(t) = A\mathbf{x}(t)$$

can be solved for the *trajectory* $\mathbf{x}(t)$ in the xy-plane (called the *phase plane*). The collection of all valid trajectories in the phase plane is known as the *phase portrait*. Once more, the analysis hinges on the eigenvalues λ and corresponding eigenvectors $\mathbf{v}$ of A. Various cases arise according to whether the values of λ are real or complex, repeated or distinct, positive or negative, and so on. The equilibrium point may be a stable node, unstable node, proper node, degenerate node, saddle point, stable spiral point, or unstable spiral point.

Nonlinear Systems of Differential Equations

A system of two differential equations that is not linear can in most cases be approximated at an equilibrium point by a linear system whose behavior is similar near the equilibrium point to the original system.

Additional Problems

1. Verify that

$$\mathbf{x}_1(t) = \begin{bmatrix} e^{t^2 - t} \\ -1 \end{bmatrix}, \qquad \mathbf{x}_2(t) = \begin{bmatrix} 0 \\ 2e^t \end{bmatrix}$$

are linearly independent solutions to $\mathbf{x}' = A\mathbf{x}$, where

$$A = \begin{bmatrix} 2t - 1 & 0 \\ e^{t - t^2} & 1 \end{bmatrix}.$$

Write the general solution to the system $\mathbf{x}' = A\mathbf{x}$.

2. Consider the linear system $\mathbf{x}' = A\mathbf{x} + \mathbf{b}$, where

$$A = \begin{bmatrix} t\cot(t^2) & 0 & t\cos(t^2)/2 \\ 0 & 1/t & -1 \\ \csc(t^2) & 1 & -1 \end{bmatrix}, \mathbf{b} = \begin{bmatrix} 0 \\ 2 - t\sin t \\ 1 - t\cos t \end{bmatrix}.$$

(a) Verify that

$$\mathbf{x}(t) = \begin{bmatrix} \sin(t^2) \\ t\cos t \\ 2 \end{bmatrix}$$

is a solution to this system.

(b) Is it possible for a constant vector $\mathbf{x}_0$ to solve the system? Justify your answer.

For Problems 3–24, determine the general solution to the linear system $\mathbf{x}' = A\mathbf{x}$ for the given matrix A.

3. $\begin{bmatrix} -6 & 1 \\ 6 & -5 \end{bmatrix}.$

4. $\begin{bmatrix} 9 & -2 \\ 5 & -2 \end{bmatrix}.$

5. $\begin{bmatrix} 10 & -4 \\ 4 & 2 \end{bmatrix}.$

6. $\begin{bmatrix} -8 & 5 \\ -5 & 2 \end{bmatrix}.$

7. $\begin{bmatrix} 3 & 0 & 4 \\ 0 & 2 & 0 \\ -4 & 0 & -5 \end{bmatrix}.$

8. $\begin{bmatrix} -3 & -1 & 0 \\ 4 & -7 & 0 \\ 6 & 6 & 4 \end{bmatrix}.$

9. $\begin{bmatrix} 3 & 13 \\ -1 & -3 \end{bmatrix}.$

10. $\begin{bmatrix} -3 & -10 \\ 5 & 11 \end{bmatrix}.$

11. $\begin{bmatrix} -1 & -5 & 1 \\ 4 & -9 & -1 \\ 0 & 0 & 3 \end{bmatrix}.$

12. $\begin{bmatrix} -4 & 0 & 0 \\ 2 & 5 & -9 \\ 0 & 5 & -1 \end{bmatrix}.$

13. $\begin{bmatrix} 2 & -2 & 1 \\ 1 & -4 & 1 \\ 2 & 2 & -3 \end{bmatrix}.$
[**Hint:** The eigenvalues of A are $\lambda = 2, -2, -5$.]

14. $\begin{bmatrix} 2 & -4 & 3 \\ -9 & -3 & -9 \\ 4 & 4 & 3 \end{bmatrix}.$
[**Hint:** The eigenvalues of A are $\lambda = 6, -3, -1$.]

15. $\begin{bmatrix} -17 & 0 & -42 \\ -7 & 4 & -14 \\ 7 & 0 & 18 \end{bmatrix}.$
[**Hint:** The eigenvalues of A are $\lambda = 4, -3$.]

16. $\begin{bmatrix} -16 & 30 & -18 \\ -8 & 8 & 16 \\ 8 & -15 & 9 \end{bmatrix}.$
[**Hint:** The eigenvalues of A are $\lambda = 8, -7, 0$.]

17. $\begin{bmatrix} -7 & -6 & -7 \\ -3 & -3 & -3 \\ 7 & 6 & 7 \end{bmatrix}.$
[**Hint:** The eigenvalues of A are $\lambda = 0, -3$.]

18. $\begin{bmatrix} 3 & -1 & -2 \\ 1 & 6 & 1 \\ 1 & 0 & 6 \end{bmatrix}.$
[**Hint:** The only eigenvalue of A is $\lambda = 5$.]

19. $\begin{bmatrix} -1 & -4 & -2 \\ -4 & -5 & -6 \\ 4 & 8 & 7 \end{bmatrix}.$
[**Hint:** The eigenvalues of A are $\lambda = -1, 1 \pm 2i$.]

20. $\begin{bmatrix} 7 & -2 & 2 \\ 0 & 4 & -1 \\ -1 & 1 & 4 \end{bmatrix}.$
[**Hint:** The only eigenvalue of A is $\lambda = 5$.]

21. $\begin{bmatrix} -3 & -1 & -2 \\ 1 & 0 & 1 \\ 1 & 0 & 0 \end{bmatrix}.$

22. $\begin{bmatrix} -2 & 0 & -1 \\ 0 & -1 & 0 \\ 1 & 0 & 0 \end{bmatrix}$.

23. $\begin{bmatrix} 2 & 13 & 0 & 0 \\ -1 & -2 & 0 & 0 \\ 0 & 0 & 2 & 4 \\ 0 & 0 & 0 & 2 \end{bmatrix}$.

24. $\begin{bmatrix} 7 & 0 & 0 & -1 \\ 0 & 6 & 0 & 0 \\ 0 & 0 & -1 & 0 \\ 2 & 0 & 0 & 5 \end{bmatrix}$.

For Problems 25–29, use the variation-of-parameters method to determine a particular solution to the nonhomogeneous linear system $\mathbf{x}' = A\mathbf{x} + \mathbf{b}$. Also find the general solution to the system.

25. $A = \begin{bmatrix} -6 & 1 \\ 6 & -5 \end{bmatrix}$, $\mathbf{b} = \begin{bmatrix} 1 \\ e^{-t} \end{bmatrix}$.

26. $A = \begin{bmatrix} 9 & -2 \\ 5 & -2 \end{bmatrix}$, $\mathbf{b} = \begin{bmatrix} 9t \\ 0 \end{bmatrix}$.

27. $A = \begin{bmatrix} 10 & -4 \\ 4 & 2 \end{bmatrix}$, $\mathbf{b} = \begin{bmatrix} 0 \\ \dfrac{1}{t}e^{6t} \end{bmatrix}$.

28. $A = \begin{bmatrix} 2 & -4 & 3 \\ -9 & -3 & -9 \\ 4 & 4 & 3 \end{bmatrix}$, $\mathbf{b} = \begin{bmatrix} e^{6t} \\ 1 \\ 0 \end{bmatrix}$.

29. $A = \begin{bmatrix} 2 & -2 & 1 \\ 1 & -4 & 1 \\ 2 & 2 & -3 \end{bmatrix}$, $\mathbf{b} = \begin{bmatrix} t \\ 0 \\ 1 \end{bmatrix}$.

30. True or False: If $X(t)$ is a fundamental matrix for the linear system $\mathbf{x}' = A\mathbf{x}$, then $X(t)^T$ is a fundamental matrix for the linear system $\mathbf{x}' = A^T\mathbf{x}$.

31. True or False: If $\mathbf{x}_0$ is a solution to the linear system $\mathbf{x}' = A\mathbf{x}$, then $\mathbf{x}_0$ is also a solution to the linear system $\mathbf{x}'' = A^2\mathbf{x}$.

32. Show that the function $D : V_n(I) \to V_n(I)$ defined by

$$D(\mathbf{x}(t)) = \mathbf{x}'(t)$$

is a linear transformation.

33. Consider the differential equation

$$\frac{d^3y}{dt^3} + a\frac{d^2y}{dt^2} + b\frac{dy}{dt} + cy = 0, \qquad (7.11.8)$$

where a, b, and c are arbitrary functions of t.

(a) Replace Equation (7.11.8) by an equivalent linear system $\mathbf{x}' = A\mathbf{x}$ for an appropriate 3×3 matrix A.

(b) If $y_1 = f_1(t)$, $y_2 = f_2(t)$, and $y_3 = f_3(t)$ are solutions to Equation (7.11.8) on an interval I, show that the corresponding solutions to the system you derived in part (a) are

$$\mathbf{x}_1(t) = \begin{bmatrix} f_1(t) \\ f_1'(t) \\ f_1''(t) \end{bmatrix}, \quad \mathbf{x}_2(t) = \begin{bmatrix} f_2(t) \\ f_2'(t) \\ f_2''(t) \end{bmatrix},$$

$$\mathbf{x}_3(t) = \begin{bmatrix} f_3(t) \\ f_3'(t) \\ f_3''(t) \end{bmatrix}.$$

(c) Show that

$$W[\mathbf{x}_1, \mathbf{x}_2, \mathbf{x}_3](t) = W[y_1, y_2, y_3](t).$$

For Problems 34–41, characterize the equilibrium point for the system $\mathbf{x}' = A\mathbf{x}$ and sketch the phase portrait.

34. $A = \begin{bmatrix} -3 & 4 \\ 8 & 1 \end{bmatrix}$.

35. $A = \begin{bmatrix} 0 & -6 \\ 1 & -5 \end{bmatrix}$.

36. $A = \begin{bmatrix} 5 & 9 \\ -2 & -1 \end{bmatrix}$.

37. $A = \begin{bmatrix} -4 & 0 \\ 0 & -4 \end{bmatrix}$.

38. $A = \begin{bmatrix} 7 & -2 \\ 1 & 4 \end{bmatrix}$.

39. $A = \begin{bmatrix} -3 & -5 \\ 1 & -7 \end{bmatrix}$.

40. $A = \begin{bmatrix} -2 & -1 \\ 1 & -4 \end{bmatrix}$.

41. $A = \begin{bmatrix} 10 & -8 \\ 2 & 2 \end{bmatrix}$.

For Problems 42–43, convert the given differential equation to a first-order system using the substitution $u = y$, $v = dy/dt$, and determine the phase portrait for the resulting system.

42. $\dfrac{d^2y}{dt^2} + \dfrac{dy}{dt} - 12y = 0$.

43. $\dfrac{d^2y}{dt^2} + 25y = 0$.

CHAPTER

8

The Laplace Transform and Some Elementary Applications

Do not worry too much about your difficulties in mathematics. I can assure you that mine are still greater. — Albert Einstein

8.1 Definition of the Laplace Transform

In this chapter, we introduce another technique for solving linear, constant-coefficient ordinary differential equations. Actually, the technique is used more broadly—for example, in the solution of linear systems of differential equations, partial differential equations, and also integral equations (see Section 8.9). The reader may question the need to introduce a new method for solving linear, constant-coefficient, ordinary differential equations, since our results from Chapter 6 can be applied to any such equation. To answer this question, consider the differential equation

$$y'' + ay' + by = F,$$

where a and b are constants. We have seen how to solve this equation when F is a continuous function on some interval I. However, in many problems that arise in engineering, physics, and applied mathematics, F represents an external force that is acting on the system under investigation, and often this force acts intermittently or even instantaneously.[1] Whereas the techniques from Chapter 6 can be extended to cover these cases, the computations involved are tedious. In contrast, the approach introduced here can handle such problems quite easily.

---

[1] For example, the switch in an RLC circuit may be turned on and off several times, or the mass in a spring–mass system may be dealt an instantaneous blow at $t = t_0$.

First we need a definition.

DEFINITION 8.1.1

Let f be a function defined on the interval $[0, \infty)$. The **Laplace transform** of f is the function $F(s)$ defined by

$$F(s) = \int_0^\infty e^{-st} f(t)\, dt, \tag{8.1.1}$$

provided that the improper integral converges. We will usually denote the Laplace transform of f by $L[f]$.

Recall that the improper integral appearing in (8.1.1) is defined by

$$\int_0^\infty e^{-st} f(t)\, dt = \lim_{N \to \infty} \int_0^N e^{-st} f(t)\, dt,$$

and that this improper integral converges if and only if the limit on the right-hand side exists and is finite. It follows that *not* all functions defined on $[0, \infty)$ have a Laplace transform. In the next section, we will address some of the theoretical aspects associated with determining the types of functions for which (8.1.1) converges. For the remainder of this section, we focus our attention on gaining familiarity with Definition 8.1.1, and we derive some basic Laplace transforms.

Example 8.1.2   Determine the Laplace transform of the following functions:

(a) $f(t) = 1$.

(b) $f(t) = t$.

(c) $f(t) = e^{at}$, where a is constant.

(d) $f(t) = \cos bt$, where b is constant.

Solution:

(a) From the foregoing definition, we have

$$L[1] = \int_0^\infty e^{-st}\, dt = \lim_{N \to \infty} \left[-\frac{1}{s} e^{-st} \right]_0^N$$

$$= \lim_{N \to \infty} \left[\frac{1}{s} - \frac{1}{s} e^{-sN} \right] = \frac{1}{s}, \qquad s > 0.$$

Notice that the restriction $s > 0$ is required for the improper integral to converge, hence the Laplace transform F of $f(t) = 1$ is defined only on the set of positive real numbers.

(b) In this case, we use integration by parts to obtain

$$L[t] = \int_0^\infty e^{-st} t\, dt = \lim_{N \to \infty} \left[-\frac{t e^{-st}}{s} \right]_0^N + \int_0^\infty \frac{1}{s} e^{-st}\, dt.$$

But,

$$\lim_{N \to \infty} N e^{-sN} = 0, \qquad s > 0,$$

so that

$$L[t] = \int_0^\infty \frac{1}{s} e^{-st} \, dt = \lim_{N \to \infty} \left[-\frac{e^{-st}}{s^2} \right]_0^N = \frac{1}{s^2}, \qquad s > 0.$$

It is left as an exercise to show that more generally, for all positive integers n,

$$L[t^n] = \frac{n!}{s^{n+1}}, \qquad s > 0. \tag{8.1.2}$$

(c) In this case, we have

$$L[e^{at}] = \int_0^\infty e^{-st} e^{at} \, dt = \int_0^\infty e^{(a-s)t} \, dt$$

$$= \lim_{N \to \infty} \left[\frac{1}{a-s} e^{(a-s)t} \right]_0^N = \frac{1}{s-a},$$

provided that $s > a$. Thus,

$$L[e^{at}] = \frac{1}{s-a}, \qquad s > a. \tag{8.1.3}$$

(d) From the definition of the Laplace transform,

$$L[\cos bt] = \int_0^\infty e^{-st} \cos bt \, dt.$$

Using the standard integral

$$\int e^{at} \cos bt \, dt = \frac{e^{at}}{a^2 + b^2} (a \cos bt + b \sin bt) + c,$$

it follows that

$$L[\cos bt] = \lim_{N \to \infty} \left[\frac{e^{-st}}{s^2 + b^2} (b \sin bt - s \cos bt) \right]_0^N = \frac{s}{s^2 + b^2}$$

provided that $s > 0$. Thus,

$$L[\cos bt] = \frac{s}{s^2 + b^2}, \qquad s > 0. \tag{8.1.4}$$

Similarly, it can be shown that

$$L[\sin bt] = \frac{b}{s^2 + b^2}, \qquad s > 0. \tag{8.1.5}$$

$\square$

As illustrated by the preceding examples, the range of values s can assume must often be restricted to ensure the convergence of the improper integral (8.1.1). For the remainder of the chapter, we will often take for granted without mention that the Laplace transforms we compute have a restricted domain.

Linearity of the Laplace Transform

Suppose that the Laplace transforms of both f and g exist for $s > \alpha$, where α is a constant. Then, using properties of convergent improper integrals, it follows that for $s > \alpha$,

$$
\begin{aligned}
L[f + g] &= \int_0^\infty e^{-st}[f(t) + g(t)]\, dt \\
&= \int_0^\infty e^{-st} f(t)\, dt + \int_0^\infty e^{-st} g(t)\, dt \\
&= L[f] + L[g].
\end{aligned}
$$

Further, if c is any real number, then

$$
L[cf] = \int_0^\infty e^{-st} cf(t)\, dt = c \int_0^\infty e^{-st} f(t)\, dt = cL[f].
$$

Consequently,

$$
\boxed{\;\textbf{1.}\; L[f + g] = L[f] + L[g] \qquad \textbf{2.}\; L[cf] = cL[f]\;}
$$

so that the Laplace transform satisfies the basic properties of a linear transformation. This linearity of L enables us to determine the Laplace transform of complicated functions from a knowledge of the Laplace transform of some basic functions. This will be used continuously throughout this chapter.

Example 8.1.3   Determine the Laplace transform of

$$
f(t) = 4e^{3t} + 2\sin 5t - 7t^3.
$$

Solution: Since the Laplace transform is linear, it follows that

$$
L[4e^{3t} + 2\sin 5t - 7t^3] = 4L[e^{3t}] + 2L[\sin 5t] - 7L[t^3].
$$

Using the results of the previous example, we therefore obtain

$$
L[4e^{3t} + 2\sin 5t - 7t^3] = \frac{4}{s - 3} + \frac{10}{s^2 + 25} - \frac{42}{s^4}, \qquad s > 3. \qquad \square
$$

Piecewise Continuous Functions

The functions that we have considered in the foregoing examples have all been continuous on $[0, \infty)$. As we will see in later sections, the real power of the Laplace transform comes from the fact that piecewise continuous functions can be transformed. Before illustrating this point, we recall the definition of a piecewise continuous function.

DEFINITION   8.1.4

A function f is called **piecewise continuous** on the interval $[a, b]$ if we can divide $[a, b]$ into a finite number of subintervals in such a manner that

1. f is continuous on each subinterval, and

2. f approaches a finite limit as the endpoints of each subinterval are approached from within.

If f is piecewise continuous on every interval of the form $[0, b]$, where b is a constant, then we say that f is piecewise continuous on $[0, \infty)$.

Example 8.1.5   The function f defined by

$$f(t) = \begin{cases} t^2 + 1, & 0 \leq t \leq 1, \\ 2 - t, & 1 < t \leq 2, \\ 1, & 2 < t \leq 3, \end{cases}$$

is piecewise continuous on $[0, 3]$, whereas

$$f(t) = \begin{cases} \frac{1}{1-t}, & 0 \leq t < 1, \\ t, & 1 \leq t \leq 3, \end{cases}$$

is not piecewise continuous on $[0, 3]$. The graphs of these functions are shown in Figure 8.1.1.

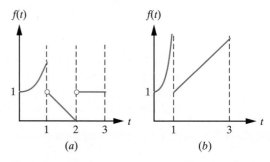

Figure 8.1.1: (a) An example of a piecewise continuous function. (b) An example of a function that is not piecewise continuous.

□

Example 8.1.6   Determine the Laplace transform of the piecewise continuous function

$$f(t) = \begin{cases} t, & 0 \leq t < 1, \\ -1, & t \geq 1. \end{cases}$$

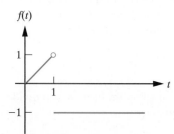

Figure 8.1.2: The piecewise continuous function in Example 8.1.6

Solution: The function is sketched in Figure 8.1.2. To determine the Laplace transform of f, we use Definition 8.1.1.

$$L[f] = \int_0^\infty e^{-st} f(t) \, dt = \int_0^1 e^{-st} f(t) \, dt + \int_1^\infty e^{-st} f(t) \, dt$$

$$= \int_0^1 t e^{-st} \, dt - \int_1^\infty e^{-st} \, dt = \left[-\frac{1}{s} t e^{-st} - \frac{1}{s^2} e^{-st} \right]_0^1 + \lim_{N \to \infty} \left[\frac{1}{s} e^{-st} \right]_1^N$$

$$= -\frac{1}{s} e^{-s} - \frac{1}{s^2} e^{-s} + \frac{1}{s^2} - \frac{1}{s} e^{-s},$$

provided that $s > 0$. Thus,

$$L[f] = \frac{1}{s^2} \left[1 - e^{-s} (2s + 1) \right], \qquad s > 0.$$

□

Exercises for 8.1

Key Terms

Laplace transform, Piecewise continuous function.

Skills

- Be able to determine the Laplace transform of a given function f.

- Be able to use the linearity of the Laplace transform to assist in computing Laplace transforms.

- Be able to determine whether or not a given function is piecewise continuous.

True-False Review

For Questions 1–9, decide if the given statement is **true** or **false**, and give a brief justification for your answer. If true, you can quote a relevant definition or theorem from the text. If false, provide an example, illustration, or brief explanation of why the statement is false.

1. The function

$$f(t) = \begin{cases} t, & \text{if } t \text{ is an integer,} \\ t^2, & \text{if } t \text{ is not an integer,} \end{cases}$$

is piecewise continuous on $[0, \infty)$.

2. The function

$$g(t) = \begin{cases} \frac{1}{\sin t}, & \text{if } t \text{ is not a multiple of } \pi, \\ 0, & \text{if } t \text{ is a multiple of } \pi, \end{cases}$$

is piecewise continuous on $[0, \infty)$.

3. If f and g are piecewise continuous functions on an interval I, then so is $f + g$.

4. If f is piecewise continuous on the interval $[a, b]$ and on the interval $[b, c]$, then f is piecewise continuous on the interval $[a, c]$.

5. The Laplace transform of a function f is a function $F(s)$ defined by

$$F(s) = \int_1^\infty e^{-st} f(t)\, dt,$$

provided that this improper integral converges.

6. The Laplace transform $F(s)$ of $f(t) = e^t$ is only defined for $s > 1$.

7. The Laplace transform $F(s)$ of $f(t) = 2\cos 3t$ is only defined for $s > 3$.

8. For any function f such that $f(x) > 0$ for all x in $[0, \infty)$, $L[1/f] = 1/(L[f])$.

9. For any function f such that $f(x) > 0$ for all x in $[0, \infty)$, $L[f^2] = L[f]^2$.

Problems

For Problems 1–12, use (8.1.1) to determine $L[f]$.

1. $f(t) = e^{2t}$.

2. $f(t) = t - 1$.

3. $f(t) = \sin bt$, where b is constant.

4. $f(t) = te^t$.

5. $f(t) = \cosh bt$, where b is constant.

6. $f(t) = \sinh bt$, where b is constant.

7. $f(t) = 2t$.

8. $f(t) = 3e^{2t}$.

9. $f(t) = \begin{cases} 1, & 0 \le t < 2, \\ -1, & t \ge 2. \end{cases}$

10. $f(t) = \begin{cases} t^2, & 0 \le t \le 1, \\ 1, & t > 1. \end{cases}$

11. $f(t) = e^t \sin t$.

12. $f(t) = e^{2t} \cos 3t$.

For Problems 13–22, use the linearity of L and the formulas derived in this section to determine $L[f]$.

13. $f(t) = 2t - e^{2t}$.

14. $f(t) = 2\sin 3t + 4t^3$.

15. $f(t) = \cosh bt$, where b is constant.

16. $f(t) = \sinh bt$, where b is constant.

17. $f(t) = 3t^2 - 5\cos 2t + \sin 3t$.

18. $f(t) = 7e^{-2t} + 1$.

19. $f(t) = 2e^{-3t} + 4e^t - 5\sin t$.

20. $f(t) = 4\cos(t - \pi/4)$.

21. $f(t) = 4\cos^2 bt$, where b is constant.

22. $f(t) = 2\sin^2 4t - 3$.

For Problems 23–30, sketch the given function and determine whether it is piecewise continuous on $[0, \infty)$.

23. $f(t) = \begin{cases} 3, & 0 \le t \le 1, \\ 0, & 1 \le t < 3, \\ -1, & t \ge 3. \end{cases}$

24. $f(t) = \begin{cases} 1, & 0 \le t \le 1, \\ 1 - t, & t > 2, \\ 1, & t > 2. \end{cases}$

25. $f(t) = \begin{cases} 1, & 0 \le t \le 1, \\ 1/(t - 1), & t > 1. \end{cases}$

26. $f(t) = \begin{cases} t, & 0 \le t \le 1, \\ 1/t^2, & t > 1. \end{cases}$

27. $f(t) = n$, $n \le t < n + 1$, $n = 0, 1, 2, \ldots$.

28. $f(t) = t$, $0 \le t < 1$, $f(t + 1) = f(t)$.

29. $f(t) = \dfrac{1}{t - 2}$.

30. $f(t) = \dfrac{2}{t + 1}$.

For Problems 31–34, sketch the given function and determine its Laplace transform.

31. $f(t) = \begin{cases} t, & 0 \le t \le 1, \\ 0, & t \ge 1. \end{cases}$

32. $f(t) = \begin{cases} 1, & 0 \le t \le 2, \\ -1, & t > 2. \end{cases}$

33. $f(t) = \begin{cases} 0, & 0 \le t \le 1, \\ t, & 1 < t \le 2, \\ 0, & t > 2. \end{cases}$

34. $f(t) = \begin{cases} t, & 0 \le t < 1, \\ 1, & 1 \le t < 3, \\ e^{t-3}, & t > 3. \end{cases}$

35. Recall that according to Euler's formula

$$e^{ibt} = \cos bt + i \sin bt.$$

Since the Laplace transform is linear, it follows that

$$L[\cos bt] = \text{Re}(L[e^{ibt}]),$$
$$L[\sin bt] = \text{Im}(L[e^{ibt}]).$$

Find $L[e^{ibt}]$, and hence, derive Equations (8.1.4) and (8.1.5).

36. Use the technique introduced in the previous problem to determine

$$L[e^{at} \cos bt] \text{ and } L[e^{at} \sin bt],$$

where a and b are arbitrary constants.

37. Use mathematical induction to prove that for every positive integer n,

$$L[t^n] = \frac{n!}{s^{n+1}}.$$

38. **(a)** By making the change of variables $t = x^2/s$ ($s > 0$) in the integral that defines the Laplace transform, show that

$$L[t^{-1/2}] = 2s^{-1/2} \int_0^\infty e^{-x^2} \, dx.$$

(b) Use your result in (a) to show that

$$(L[t^{-1/2}])^2 = 4s^{-1} \int_0^\infty \int_0^\infty e^{-(x^2+y^2)} \, dx \, dy.$$

(c) By changing to polar coordinates, evaluate the double integral in (b), and hence, show that

$$L[t^{-1/2}] = \sqrt{\frac{\pi}{s}}, \quad s > 0.$$

8.2   The Existence of the Laplace Transform and the Inverse Transform

In the previous section, we derived the Laplace transform of several elementary functions. In this section, we address some of the more theoretical aspects of the Laplace transform. The first question we wish to answer is:

What types of functions have a Laplace transform?

We will not be able to answer this question completely, since it requires a deeper mathematical background than we assume of the reader. However, we can identify a very large class of functions that are Laplace transformable.

By definition, the Laplace transform of a function f is

$$L[f] = \int_0^\infty e^{-st} f(t)\, dt, \tag{8.2.1}$$

provided that the integral converges. If f is piecewise continuous on an interval $[a, b]$, then it is a standard result from calculus that f is also integrable over $[a, b]$. Thus, if we restrict attention to functions that are piecewise continuous on $[0, \infty)$, it follows that the integral

$$\int_0^b e^{-st} f(t)\, dt$$

exists for all positive (and finite) b. However, it does not follow that the Laplace transform of f exists, since the improper integral in (8.2.1) may still diverge. To guarantee convergence of the integral, we must ensure that the integrand in (8.2.1) approaches zero rapidly enough as $t \to \infty$. As we show next, this will be the case provided that, in addition to being piecewise continuous, f also satisfies the following definition:

DEFINITION   8.2.1

A function f is said to be of **exponential order** if there exist constants M and α such that

$$|f(t)| \le M e^{\alpha t},$$

for all $t > 0$.

Example 8.2.2   The function $f(t) = 10 e^{7t} \cos 5t$ is of exponential order, since

$$|f(t)| = 10 e^{7t} |\cos 5t| \le 10 e^{7t}. \qquad \square$$

Now let $E(0, \infty)$ denote the set of all functions that are both piecewise continuous on $[0, \infty)$ and of exponential order. If we add two functions that are in $E(0, \infty)$, the result is a new function that is also in $E(0, \infty)$. Similarly, if we multiply a function in $E(0, \infty)$ by a constant, the result is once more a function in $E(0, \infty)$. It follows from Theorem 4.3.2 that $E(0, \infty)$ is a subspace of the vector space of all functions defined on $[0, \infty)$. We will show next that the functions in the vector space $E(0, \infty)$ have a Laplace transform. Before doing so, we need to state a basic result about the convergence of improper integrals.

Lemma 8.2.3   **(The Comparison Test for Improper Integrals)**

Suppose that $0 \leq G(t) \leq H(t)$ for $0 \leq t < \infty$. If $\int_0^\infty H(t)\, dt$ converges, then $\int_0^\infty G(t)\, dt$ converges.

Proof This can be found in any textbook on advanced calculus.   ∎

We also recall that if $\int_0^\infty |F(t)|\, dt$ converges, then so does $\int_0^\infty F(t)\, dt$. We can now establish a key existence theorem for the Laplace transform.

Theorem 8.2.4   If f is in $E(0, \infty)$, then there exists a constant α such that

$$L[f] = \int_0^\infty e^{-st} f(t)\, dt$$

exists for all $s > \alpha$.

Proof Since f is piecewise continuous on $[0, \infty)$, $e^{-st} f(t)$ is integrable over any finite interval. Further, since f is in $E(0, \infty)$, there exist constants M and α such that

$$|f(t)| \leq Me^{\alpha t},$$

for all $t > 0$. We now use the comparison test for integrals to establish that the improper integral defining the Laplace transform converges on the domain (α, ∞). Let

$$F(t) = |e^{-st} f(t)|.$$

Then,

$$F(t) = e^{-st}|f(t)| \leq Me^{(\alpha-s)t}.$$

But, for $s > \alpha$,

$$\int_0^\infty Me^{(\alpha-s)t}\, dt = \lim_{N \to \infty} \int_0^N Me^{(\alpha-s)t}\, dt = \frac{M}{s - \alpha}.$$

Applying the comparison test for improper integrals with $F(t)$ as just defined and $G(t) = Me^{(\alpha-s)t}$, it follows that

$$\int_0^\infty |e^{-st} f(t)|\, dt$$

converges for $s > \alpha$, and hence, so also does

$$\int_0^\infty e^{-st} f(t)\, dt.$$

Thus, we have shown that $L[f]$ exists for $s > \alpha$, as required.   ∎

Remark   The preceding theorem gives only sufficient conditions that guarantee the existence of the Laplace transform. There are functions that are not in $E(0, \infty)$ but that do have a Laplace transform. For example, $f(t) = t^{-1/2}$ is certainly not in $E(0, \infty)$, but $L[t^{-1/2}] = \sqrt{\pi/s}$. (See Problem 38 in the preceding section.)

The Inverse Laplace Transform

Let V denote the subspace of $E(0, \infty)$ consisting of all *continuous* functions of exponential order. We have seen in the previous section that the Laplace transform satisfies

$$L[f + g] = L[f] + L[g], \qquad L[cf] = cL[f].$$

Consequently, L defines a linear transformation of V onto $\text{Rng}(L)$. Further, it can be shown (Problem 22) that L is also one-to-one, and therefore, from the results of Section 5.4, the inverse transformation, L^{-1}, exists and is defined as follows:

DEFINITION 8.2.5

The linear transformation $L^{-1} : \text{Rng}(L) \to V$ defined by

$$L^{-1}[F](t) = f(t) \qquad \text{if and only if} \qquad L[f](s) = F(s) \qquad (8.2.2)$$

is called the **inverse Laplace transform**.

Remark We emphasize the fact that L^{-1} is a *linear transformation*, so that

$$L^{-1}[F + G] = L^{-1}[F] + L^{-1}[G],$$

and

$$L^{-1}[cF] = cL^{-1}[F],$$

for all F and G in $\text{Rng}(L)$ and all real numbers c. This can be seen either directly from (8.2.2) or from the general theory of inverse linear transformations.

In Section 8.1, we derived the transforms

$$L[t^n] = \frac{n!}{s^{n+1}}, \quad L[e^{at}] = \frac{1}{s - a}, \quad L[\cos bt] = \frac{s}{s^2 + b^2}, \quad L[\sin bt] = \frac{b}{s^2 + b^2},$$

from which we directly obtain the inverse transforms

$$L^{-1}\left[\frac{1}{s^{n+1}}\right] = \frac{1}{n!}t^n, \qquad L^{-1}\left[\frac{1}{s - a}\right] = e^{at},$$

$$L^{-1}\left[\frac{s}{s^2 + b^2}\right] = \cos bt, \qquad L^{-1}\left[\frac{b}{s^2 + b^2}\right] = \sin bt.$$

Example 8.2.6 Find $L^{-1}[F](t)$ if

(a) $F(s) = \dfrac{2}{s^2}$.

(b) $F(s) = \dfrac{3s}{s^2 + 4}$.

(c) $F(s) = \dfrac{3s + 2}{(s - 1)(s - 2)}$.

Solution:

(a) $L^{-1}\left[\dfrac{2}{s^2}\right] = 2L^{-1}\left[\dfrac{1}{s^2}\right] = 2t.$

(b) $L^{-1}\left[\dfrac{3s}{s^2+4}\right] = 3L^{-1}\left[\dfrac{s}{s^2+4}\right] = 3\cos 2t.$

(c) In this case, it is not obvious at first sight what the appropriate inverse transform is. However, decomposing $F(s)$ into partial fractions yields[2]

$$F(s) = \frac{3s+2}{(s-1)(s-2)} = \frac{8}{s-2} - \frac{5}{s-1}.$$

Consequently, using linearity of L^{-1},

$$L^{-1}\left[\frac{3s+2}{(s-1)(s-2)}\right] = L^{-1}\left[\frac{8}{s-2}\right] - L^{-1}\left[\frac{5}{s-1}\right]$$

$$= 8L^{-1}\left[\frac{1}{s-2}\right] - 5L^{-1}\left[\frac{1}{s-1}\right]$$

$$= 8e^{2t} - 5e^t. \qquad \square$$

If we relax the assumption that V contain only *continuous* functions of exponential order, then it is no longer true that L is one-to-one, and so, for a given $F(s)$, there will be (infinitely) many piecewise continuous functions f with the property that

$$L[f] = F(s).$$

Thus, we lose the uniqueness of $L^{-1}[F]$. However, it can be shown [see, for example, R. V. Churchill, *Modern Operational Mathematics in Engineering* (New York: McGraw-Hill, 1944)] that if two functions have the same Laplace transform, then they can differ only in their values at points of discontinuity. This does not affect the solution to our problems, and therefore we will use (8.2.2) to determine the inverse Laplace transform, even if f is piecewise continuous.

Example 8.2.7   In the preceding section we have shown that the Laplace transform of the piecewise continuous function

$$f(t) = \begin{cases} t, & 0 \le t < 1, \\ -1, & t \ge 1. \end{cases}$$

is

$$L[f] = \frac{1}{s^2}[1 - e^{-s}(2s+1)], \qquad s > 0.$$

Consequently,

$$L^{-1}\left\{\frac{1}{s^2}[1 - e^{-s}(2s+1)]\right\} = f(t). \qquad \square$$

It is possible to give a general formula for determining the inverse Laplace transform of $F(s)$ in terms of a contour integral in the complex plane. However, this is beyond our

---

[2]It is very important in this chapter to be able to perform partial fractions decompositions. A review of this technique is given in Appendix B.

present scope. In practice, as in the previous examples, we determine inverse Laplace transforms by recognizing $F(s)$ as being the Laplace transform of an appropriate function $f(t)$. In order for this approach to work, we need to memorize a few basic transforms and then be able to use them to determine the inverse Laplace transform of more complicated functions. This is similar to the way we learn how to integrate. The transform pairs that will be needed for the remainder of the text are listed in Table 8.2.1. Several transforms given in this table will be derived in the sections that follow. More generally, very large tables of Laplace transforms have been compiled for use in applications, and most current computer algebra systems (Maple, Mathematica, and so on.) have the built-in capability to determine Laplace transforms.

Table 8.2.1: Transform pairs.

Function $f(t)$	Laplace Transform $F(s)$
$f(t) = t^n$, n a nonnegative integer	$F(s) = \dfrac{n!}{s^{n+1}}, s > 0.$
$f(t) = e^{at}$, a constant	$F(s) = \dfrac{1}{s-a}, s > a.$
$f(t) = \sin bt$, b constant	$F(s) = \dfrac{b}{s^2 + b^2}, s > 0.$
$f(t) = \cos bt$, b constant	$F(s) = \dfrac{s}{s^2 + b^2}, s > 0.$
$f(t) = t^{-1/2}$	$F(s) = (\pi/s)^{1/2}, s > 0.$
$f(t) = u_a(t)$ (see Section 8.7)	$F(s) = \dfrac{1}{s}e^{-as}.$
$f(t) = \delta(t-a)$ (see Section 8.8)	$F(s) = e^{-as}.$
Transform of Derivatives (see Section 8.4)	
f'	$L[f'] = sL[f] - f(0).$
f''	$L[f''] = s^2 L[f] - sf(0) - f'(0).$
Shifting Theorems (see Sections 8.5 and 8.7)	
$e^{at} f(t)$	$F(s-a)$
$u_a(t) f(t-a)$	$e^{-as} F(s).$

Exercises for 8.2

Key Terms

Exponential order, Comparison test for improper integrals, Inverse Laplace transform.

Skills

- Be able to decide whether or not a given function is of exponential order.

- Be able to determine the inverse Laplace transform of a given function.

True-False Review

For Questions 1–6, decide if the given statement is **true** or **false**, and give a brief justification for your answer. If true, you can quote a relevant definition or theorem from the text. If false, provide an example, illustration, or brief explanation of why the statement is false.

1. If f is a function of exponential order, and $|g(x)| < f(x)$ for all x, then g is of exponential order.

2. If f and g are functions of exponential order, then so is $f + g$.

3. If $0 \le G(t) \le H(t)$ for $0 \le t < \infty$ and $\int_0^\infty G(t)\,dt$ converges, then so does $\int_0^\infty H(t)\,dt$.

4. The inverse Laplace transform operator is a linear transformation.

5. The inverse Laplace transform of the function

$$F(s) = \frac{s}{s^2 + 9}$$

is $f(t) = \sin 3t$.

6. The inverse Laplace transform of the function

$$F(s) = \frac{1}{s + 3}$$

is the function $f(t) = e^{3t}$.

Problems

For Problems 1–5, show that the given function is of exponential order.

1. $f(t) = \cos 2t$.

2. $f(t) = e^{2t}$.

3. $f(t) = e^{3t} \sin 4t$.

4. $f(t) = te^{-2t}$.

5. $f(t) = t^n e^{at}$, where a and n are positive integers.

6. Show that if f and g are in $E(0, \infty)$, then so are $f + g$ and cf for any scalar c.

For Problems 7–21, determine the inverse Laplace transform of the given function.

7. $F(s) = \dfrac{2}{s}$.

8. $F(s) = \dfrac{3}{s - 2}$.

9. $F(s) = \dfrac{5}{s + 3}$.

10. $F(s) = \dfrac{1}{s^2 + 4}$.

11. $F(s) = \dfrac{2s}{s^2 + 9}$.

12. $F(s) = \dfrac{4}{s^3}$.

13. $F(s) = \dfrac{s + 6}{s^2 + 1}$.

14. $F(s) = \dfrac{2s + 1}{s^2 + 16}$.

15. $F(s) = \dfrac{2}{s} - \dfrac{3}{s + 1}$.

16. $F(s) = \dfrac{4}{s^2} - \dfrac{s + 2}{s^2 + 9}$.

17. $F(s) = \dfrac{1}{s(s + 1)}$.

18. $F(s) = \dfrac{s - 2}{(s + 1)(s^2 + 4)}$.

19. $F(s) = \dfrac{2s + 3}{(s - 2)(s^2 + 1)}$.

20. $F(s) = \dfrac{s + 4}{(s - 1)(s + 2)(s - 3)}$.

21. $F(s) = \dfrac{2s + 3}{(s^2 + 4)(s^2 + 1)}$.

22. This exercise verifies the claim in the text that the Laplace transform defines a *one-to-one* linear transformation from V to Rng(L). Let f be a continuous function of exponential order. It suffices to prove that if f is not identically zero, then $L[f] \ne 0$. [**Hint:** Show that $L[|f|] > 0$ if f is not identically zero.]

8.3   Periodic Functions and the Laplace Transform

Many of the functions that arise in engineering applications are periodic on some interval. Owing to the symmetry associated with a periodic function, we might suspect that the evaluation of the Laplace transform of such a function can be reduced to an integration over one period of the function. Before establishing this result, we first recall the definition of a periodic function.

DEFINITION 8.3.1

A function f defined on an interval $[0, \infty)$ is said to be **periodic with period** T if T is the smallest positive real number that satisfies the equation

$$f(t + T) = f(t),$$

for all $t \geq 0$.

The most familiar examples of periodic functions are the trigonometric functions sine and cosine, which have period 2π.

Example 8.3.2 The function f defined by

$$f(t) = \begin{cases} 2, & 0 \leq t \leq 1, \\ 1, & 1 < t < 2, \end{cases} \quad f(t + 2) = f(t),$$

is periodic on $[0, \infty)$ with period 2. (See Figure 8.3.1.) $\square$

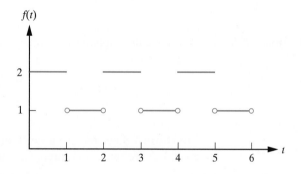

Figure 8.3.1: A function that is periodic on $[0, \infty)$ with period 2.

The following theorem can be used to simplify the evaluation of the Laplace transform of a periodic function.

Theorem 8.3.3 Let f be in $E(0, \infty)$. If f is periodic on $[0, \infty)$ with period T, then

$$L[f] = \frac{1}{1 - e^{-sT}} \int_0^T e^{-st} f(t)\, dt.$$

Proof By definition of the Laplace transform, we have

$$L[f] = \int_0^\infty e^{-st} f(t)\, dt$$
$$= \int_0^T e^{-st} f(t)\, dt + \int_T^{2T} e^{-st} f(t)\, dt + \cdots + \int_{nT}^{(n+1)T} e^{-st} f(t)\, dt + \cdots.$$

Now consider the general integral

$$I = \int_{nT}^{(n+1)T} e^{-st} f(t)\, dt.$$

If we let $x = t - nT$, then $dx = dt$. Further, $t = nT$ corresponds to $x = 0$, whereas $t = (n + 1)T$ corresponds to $x = T$. Hence, I can be written in the equivalent form

$$I = \int_0^T e^{-s(x+nT)} f(x + nT) \, dx = e^{-snT} \int_0^T e^{-sx} f(x) \, dx,$$

where we have used the fact that f is periodic of period T to replace $f(x + nT)$ by $f(x)$. All of the integrals that arise in the expression for $L[f]$ are of the preceding form for an appropriate value of n. It follows, therefore, that we can write

$$L[f] = (1 + e^{-sT} + e^{-2sT} + \cdots + e^{-nsT} + \cdots) \int_0^T e^{-sx} f(x) \, dx. \qquad (8.3.1)$$

However, the series multiplying the integral is just a geometric series with common ratio[3] e^{-sT}. Consequently, the sum of the geometric series is $1/(1 - e^{-sT})$, so that

$$L[f] = \frac{1}{1 - e^{-sT}} \int_0^T e^{-st} f(t) \, dt,$$

where we have replaced the dummy variable x in (8.3.1) by t without loss of generality. ∎

Example 8.3.4   Determine the Laplace transform of

$$f(t) = \begin{cases} \sin t, & 0 \le t \le \pi, \\ 0, & \pi \le t < 2\pi, \end{cases} \qquad f(t + 2\pi) = f(t).$$

Solution: Since the given function is periodic on $[0, \infty)$ with period 2π (see Figure 8.3.2), we can use Theorem 8.3.3 to determine $L[f]$. We have

$$L[f] = \frac{1}{1 - e^{-2\pi s}} \int_0^{2\pi} e^{-st} f(t) \, dt = \frac{1}{1 - e^{-2\pi s}} \int_0^{\pi} e^{-st} \sin t \, dt.$$

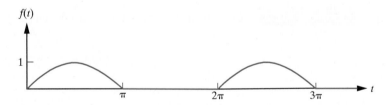

Figure 8.3.2: The periodic function defined in Example 8.3.4.

Using the standard integral

$$\int e^{at} \sin bt \, dt = \frac{1}{a^2 + b^2} e^{at} (a \sin bt - b \cos bt) + c,$$

---

[3]Recall that an infinite series of the form $a + ar + ar^2 + ar^3 + \cdots$ is called a geometric series with common ratio r. If $|r| < 1$, then the sum of such a series is $a/(1 - r)$.

it follows that

$$L[f] = \frac{1}{1 - e^{-2\pi s}} \left\{ -\frac{1}{s^2 + 1} \left[e^{-st}(\cos t + s \sin t) \right]_0^\pi \right\}$$

$$= \frac{1}{1 - e^{-2\pi s}} \left[\frac{e^{-s\pi} + 1}{s^2 + 1} \right].$$

Substituting for

$$1 - e^{-2\pi s} = (1 - e^{-\pi s})(1 + e^{-\pi s})$$

yields

$$L[f] = \frac{1}{(s^2 + 1)(1 - e^{-\pi s})}. \qquad \square$$

Exercises for 8.3

Key Terms

Periodic function, Period.

Skills

- Be able to determine the Laplace transform of a given periodic function.

True-False Review

For Questions 1–8, decide if the given statement is **true** or **false**, and give a brief justification for your answer. If true, you can quote a relevant definition or theorem from the text. If false, provide an example, illustration, or brief explanation of why the statement is false.

1. Any periodic function f has more than one period.

2. If f is a periodic function and g is a function such that $g(t) = f(t + c)$ for some constant c, then g is a periodic function with the same period as f.

3. The function $f(t) = \cos 2t$ is periodic with period $\pi/2$.

4. The function $f(t) = \sin(t^2)$ is periodic.

5. Every piecewise continuous function f is periodic.

6. If m and n are positive integers and f is a periodic function with period m and g is a periodic function with period n, then $f + g$ is a periodic function with period $m + n$.

7. If m, n, f, and g are as in the previous question, then fg is a periodic function with period mn.

8. The Laplace transform of a periodic function is a periodic function.

Problems

For Problems 1–9, determine the Laplace transform of the given function.

1. $f(t) = t$, $0 \leq t < 1$, $f(t + 1) = f(t)$.

2. $f(t) = t^2$, $0 \leq t < 2$, $f(t + 2) = f(t)$.

3. $f(t) = \sin t$, $0 \leq t < \pi$, $f(t + \pi) = f(t)$.

4. $f(t) = \cos t$, $0 \leq t < \pi$, $f(t + \pi) = f(t)$.

5. $f(t) = e^t$, $0 \leq t < 1$, $f(t + 1) = f(t)$.

6. $f(t) = \begin{cases} 1, & 0 \leq t < 1, \\ -1, & 1 \leq t < 2, \end{cases}$
   where $f(t + 2) = f(t)$.

7. $f(t) = \begin{cases} 2t/\pi, & 0 \leq t < \pi/2, \\ \sin t, & \pi/2 \leq t < \pi, \end{cases}$
   where $f(t + \pi) = f(t)$.

8. $f(t) = |\cos t|$, $0 \leq t < \pi$, $f(t + \pi) = f(t)$.

9. The triangular wave function (see Figure 8.3.3)

$$f(t) = \begin{cases} t/a, & 0 \leq t < a, \\ (2a - t)/a, & a \leq t < 2a, \end{cases}$$

where $f(t + 2a) = f(t)$, for a positive constant a.

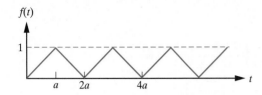

f(t)

1

a $2a$ $4a$ t

Figure 8.3.3: A triangular wave function.

10. Use Theorem 8.3.3, together with the fact that $f(t) = \sin at$ is periodic on the interval $[0, 2\pi/a]$, to determine $L[f]$.

11. Repeat the previous problem for the function $f(t) = \cos at$.

8.4   The Transform of Derivatives and Solution of Initial-Value Problems

The reason we have introduced the Laplace transform is that it provides an alternative technique for solving differential equations. To see how this technique arises, we must first consider how the derivative of a function transforms.

Theorem 8.4.1   Suppose that f is of exponential order on $[0, \infty)$ and that f' exists and is piecewise continuous on $[0, \infty)$. Then $L[f']$ exists and is given by

$$L[f'] = sL[f] - f(0).$$

Proof   For simplicity we consider the case when f' is continuous on $[0, \infty)$. The extension to the case of piecewise continuity is straightforward. Since f is differentiable and of exponential order on $[0, \infty)$, it follows that it belongs to $E(0, \infty)$, and hence its Laplace transform exists. By definition of the Laplace transform, we have

$$L[f'] = \int_0^\infty e^{-st} f'(t)\, dt = \left[e^{-st} f(t)\right]_0^\infty + s \int_0^\infty e^{-st} f(t)\, dt.$$

That is, since f is of exponential order on $[0, \infty)$,

$$L[f'] = sL[f] - f(0). \qquad \blacksquare$$

Example 8.4.2   Solve the initial-value problem

$$\frac{dy}{dt} = t, \qquad y(0) = 1.$$

Solution: This problem can be solved by a direct integration. However, we will use the Laplace transform. Taking the Laplace transform of both sides of the given differential equation and using the result of the previous theorem, we obtain

$$sY(s) - y(0) = \frac{1}{s^2}.$$

This is an *algebraic* equation for $Y(s)$. Substituting in the initial condition and solving algebraically for $Y(s)$ yields

$$Y(s) = \frac{1}{s^3} + \frac{1}{s}.$$

To determine the solution of the original problem, we now take the inverse Laplace transform of both sides of this equation. The result is

$$y(t) = L^{-1}\left[\frac{1}{s^3} + \frac{1}{s}\right].$$

That is, since $L^{-1}\left[\dfrac{1}{s^{n+1}}\right] = \dfrac{1}{n!}t^n$,

$$y(t) = \frac{1}{2}t^2 + 1. \qquad \square$$

The foregoing example illustrates the basic steps in solving an initial-value problem using the Laplace transform. We proceed as follows:

---

1. Take the Laplace transform of the given differential equation, and substitute in the given initial conditions.
2. Solve the resulting equation algebraically for $Y(s)$.
3. Take the inverse Laplace transform of $Y(s)$ to determine the solution $y(t)$ of the given initial-value problem.

These steps are illustrated in Figure 8.4.1.

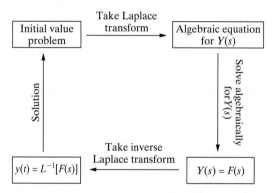

Figure 8.4.1: A schematic representation of the Laplace transform method for solving initial-value problems.

To extend the technique introduced in the previous example to higher-order differential equations, we need to determine how the higher-order derivatives transform. This can be derived quite easily from Theorem 8.4.1. We illustrate for the case of second-order derivatives and leave the derivation of the general case as an exercise.

Assuming that f'' is sufficiently smooth, it follows from Theorem 8.4.1 that

$$L[f''] = sL[f'] - f'(0).$$

Thus, applying Theorem 8.4.1 once more yields

$$\boxed{L[f''] = s^2 L[f] - sf(0) - f'(0).}$$

We leave it as an exercise to establish that, more generally, we have

$$\boxed{L[f^{(n)}] = s^n L[f] - s^{n-1} f(0) - s^{n-2} f'(0) - \cdots - sf^{(n-2)}(0) - f^{(n-1)}(0).}$$

Example 8.4.3   Use the Laplace transform to solve the initial-value problem

$$y'' - y' - 6y = 0, \qquad y(0) = 1, \qquad y'(0) = 2.$$

Solution: We take the Laplace transform of both sides of the differential equation to obtain

$$[s^2 Y(s) - sy(0) - y'(0)] - [sY(s) - y(0)] - 6Y(s) = 0.$$

Substituting in the given initial values and rearranging terms yields

$$(s^2 - s - 6)Y(s) = s + 1.$$

That is,

$$Y(s) = \frac{s+1}{(s-3)(s+2)}.$$

Thus, we have solved for the Laplace transform of $y(t)$. To find y itself, we must take the inverse Laplace transform. We first decompose the right-hand side into partial fractions to obtain

$$Y(s) = \frac{4}{5(s-3)} + \frac{1}{5(s+2)}.$$

We recognize the terms on the right-hand side as being the Laplace transform of appropriate exponential functions. Taking the inverse Laplace transform yields

$$y(t) = \tfrac{4}{5}e^{3t} + \tfrac{1}{5}e^{-2t},$$

and the initial-value problem is solved.   ☐

Example 8.4.4   Solve the initial-value problem

$$y'' + y = e^{2t}, \qquad y(0) = 0, \qquad y'(0) = 1.$$

Solution: Once more we take the Laplace transform of both sides of the differential equation to obtain

$$[s^2 Y(s) - sy(0) - y'(0)] + Y(s) = \frac{1}{s-2}.$$

That is, upon substituting for the given initial conditions and simplifying,

$$Y(s) = \frac{s-1}{(s-2)(s^2+1)}.$$

We must now determine the partial fractions decomposition of the right-hand side. We have

$$\frac{s-1}{(s-2)(s^2+1)} = \frac{A}{s-2} + \frac{Bs+C}{s^2+1},$$

for appropriate constants A, B, and C. Multiplying both sides of this equality by $(s-2)(s^2+1)$ yields

$$s - 1 = A(s^2+1) + (Bs+C)(s-2).$$

Equating coefficients of s^0, s^1, and s^2 results in the three conditions

$$A - 2C = -1, \qquad -2B + C = 1, \qquad A + B = 0.$$

Solving for A, B and C, we obtain

$$A = \tfrac{1}{5}, \qquad B = -\tfrac{1}{5}, \qquad C = \tfrac{3}{5}.$$

Thus,

$$Y(s) = \frac{1}{5(s-2)} - \frac{s-3}{5(s^2+1)}.$$

That is,

$$Y(s) = \frac{1}{5(s-2)} - \frac{s}{5(s^2+1)} + \frac{3}{5(s^2+1)}.$$

Taking the inverse Laplace transform of both sides of this equation yields

$$y(t) = \tfrac{1}{5}e^{2t} - \tfrac{1}{5}\cos t + \tfrac{3}{5}\sin t. \qquad \square$$

The structure of the solution obtained in the previous example has a familiar form. The first term represents a particular solution to the differential equation that could have been obtained by the method of undetermined coefficients, whereas the last two terms come from the complementary function. There are no arbitrary constants in the solution, since we have solved an initial-value problem. Notice the difference between solving an initial-value problem using the Laplace transform and our previous techniques. In the Laplace transform technique, we impose the initial values at the beginning of the problem and just solve the initial-value problem. In our previous techniques, we first found the general solution to the differential equation and *then* imposed the initial values to solve the initial-value problem. We note, however, that the Laplace transform can also be used to determine the general solution of a differential equation (see Problem 27).

It should be apparent from the results of the previous two sections that the main difficulties in applying the Laplace transform technique to the solution of initial-value problems are in steps 1 and 3. In order for the technique to be useful, we need to know the transform and the inverse transform for a large number of functions. So far, we have determined the Laplace transform only of some very basic functions—namely, t^n, e^{at}, $\sin bt$, $\cos bt$. We will show in the remaining sections how these basic transforms can be used to determine the Laplace transform of almost any function that is likely to arise in the applications. The reader once more is strongly advised to memorize the basic transforms.

Exercises for 8.4

Skills

- For functions f of exponential order on $[0, \infty)$ whose derivatives f' exist and are piecewise continuous on $[0, \infty)$, be able to compute $L[f']$.

- Be able to use the Laplace transform to solve initial-value problems.

- Where applicable, be able to compute the Laplace transform of the nth derivative of f by repeated application of Theorem 8.4.1.

True-False Review

For Questions 1–4, decide if the given statement is **true** or **false**, and give a brief justification for your answer. If true, you can quote a relevant definition or theorem from the text.

If false, provide an example, illustration, or brief explanation of why the statement is false.

1. For every function f with a continuous derivative on $[0, \infty)$, the Laplace transform of the derivative is given by $L[f'] = sL[f] - f(0)$.

2. In solving an initial-value problem using the Laplace transform method, the general solution of the differential equation is not explicitly found. Rather, we impose the initial conditions immediately in the procedure.

3. The Laplace transform method for solving an initial-value problem can also be used to find the general solution of the differential equation.

4. The initial conditions of an initial-value problem do not affect the expression $Y(s)$ for the Laplace transform of the solution $y(t)$.

Problems

For Problems 1–26, use the Laplace transform to solve the given initial-value problem.

1. $y' + y = 8e^{3t}$, $y(0) = 2$.

2. $y' + 3y = 2e^{-t}$, $y(0) = 3$.

3. $y' + 2y = 4t$, $y(0) = 1$.

4. $y' - y = 6\cos t$, $y(0) = 2$.

5. $y' - y = 5\sin 2t$, $y(0) = -1$.

6. $y' + y = 5e^t \sin t$, $y(0) = 1$.

7. $y'' + y' - 2y = 0$, $y(0) = 1$, $y'(0) = 4$.

8. $y'' + 4y = 0$, $y(0) = 5$, $y'(0) = 1$.

9. $y'' - 3y' + 2y = 4$, $y(0) = 0$, $y'(0) = 1$.

10. $y'' - y' - 12y = 36$, $y(0) = 0$, $y'(0) = 12$.

11. $y'' + y' - 2y = 10e^{-t}$, $y(0) = 0$, $y'(0) = 1$.

12. $y'' - 3y' + 2y = 4e^{3t}$, $y(0) = 0$, $y'(0) = 0$.

13. $y'' - 2y' = 30e^{-3t}$, $y(0) = 1$, $y'(0) = 0$.

14. $y'' - y = 12e^{2t}$, $y(0) = 1$, $y'(0) = 1$.

15. $y'' + 4y = 10e^{-t}$, $y(0) = 4$, $y'(0) = 0$.

16. $y'' - y' - 6y = 6(2 - e^t)$, $y(0) = 5$, $y'(0) = -3$.

17. $y'' - y = 6\cos t$, $y(0) = 0$, $y'(0) = 4$.

18. $y'' - 9y = 13\sin 2t$, $y(0) = 3$, $y'(0) = 1$.

19. $y'' - y = 8\sin t - 6\cos t$, $y(0) = 2$, $y'(0) = -1$.

20. $y'' - y' - 2y = 10\cos t$, $y(0) = 0$, $y'(0) = -1$.

21. $y'' + 5y' + 4y = 20\sin 2t$, $y(0) = -1$, $y'(0) = 2$.

22. $y'' + 5y' + 4y = 20\sin 2t$, $y(0) = 1$, $y'(0) = -2$.

23. $y'' - 3y' + 2y = 3\cos t + \sin t$, $y(0) = 1$, $y'(0) = 1$.

24. $y'' + 4y = 9\sin t$, $y(0) = 1$, $y'(0) = -1$.

25. $y'' + y = 6\cos 2t$, $y(0) = 0$, $y'(0) = 2$.

26. $y'' + 9y = 7\sin 4t + 14\cos 4t$, $y(0) = 1$, $y'(0) = 2$.

27. Use the Laplace transform to find the general solution to $y'' - y = 0$.

28. Use the Laplace transform to solve the initial-value problem

$$y'' + \omega^2 y = A\sin \omega_0 t + B\cos \omega_0 t$$
$$y(0) = y_0, \qquad y'(0) = y_1,$$

where A, B, ω, and ω_0 are positive constants and $\omega \neq \omega_0$.

29. The current $i(t)$ in an RL circuit is governed by the differential equation

$$\frac{di}{dt} + \frac{R}{L}i = \frac{1}{L}E(t),$$

where R and L are constants.

(a) Use the Laplace transform to determine $i(t)$ if $E(t) = E_0$, a constant. There is no current flowing initially.

(b) Repeat part (a) in the case when $E(t) = E_0 \sin \omega t$, where ω is a constant.

The Laplace transform can also be used to solve initial-value problems for systems of linear differential equations. The remaining problems deal with this.

30. Consider the initial-value problem

$$x_1' = a_{11}x_1 + a_{12}x_2 + b_1(t),$$
$$x_2' = a_{21}x_1 + a_{22}x_2 + b_2(t),$$
$$x_1(0) = \alpha_1, \qquad x_2(0) = \alpha_2,$$

where the a_{ij}, α_1, and α_2 are constants. Show that the Laplace transforms of $x_1(t)$ and $x_2(t)$ must satisfy the linear system

$$(s - a_{11})X_1(s) - \qquad a_{12}X_2(s) = \alpha_1 + B_1(s),$$
$$- a_{21}X_1(s) + (s - a_{22})X_2(s) = \alpha_2 + B_2(s).$$

This system can be solved quite easily (for example, by Cramer's rule) to determine $X_1(s)$ and $X_2(s)$, and then $x_1(t)$ and $x_2(t)$ can be obtained by taking the inverse Laplace transform.

For Problems 31–32, solve the given initial-value problem.

31. $x_1' = -4x_1 - 2x_2$, $x_2' = x_1 - x_2$, $x_1(0) = 0$, $x_2(0) = 1$.

32. $x_1' = -3x_1 + 4x_2$, $x_2' = -x_1 + 2x_2$, $x_1(0) = 2$, $x_2(0) = 1$.

33. Establish the formula for $L[f^{(n)}]$, the Laplace transform of the nth derivative of f, given in the text. [**Hint:** Use induction on n.]

8.5   The First Shifting Theorem

In order for the Laplace transform to be a useful tool for solving differential equations, we need to be able to find $L[f]$ for a large class of functions f. Trying to apply the definition of the Laplace transform to determine $L[f]$ for every function we encounter is not an appropriate way to proceed. Instead, we derive some general theorems that will enable us to obtain the Laplace transform of most elementary functions from a knowledge of the transforms of the functions given in Table 8.2.1. For the remainder of this section, we will be assuming that all of the functions that we encounter do have a Laplace transform.

Theorem 8.5.1   **(First Shifting Theorem)**

If $L[f] = F(s)$, then

$$L[e^{at} f(t)] = F(s - a).$$

Conversely, if $L^{-1}[F(s)] = f(t)$, then

$$L^{-1}[F(s - a)] = e^{at} f(t).$$

Proof From the definition of the Laplace transform, we have

$$L[e^{at} f(t)] = \int_0^\infty e^{-st} e^{at} f(t) \, dt = \int_0^\infty e^{-(s-a)t} f(t) \, dt. \qquad (8.5.1)$$

But,

$$F(s) = \int_0^\infty e^{-st} f(t) \, dt,$$

so that

$$F(s - a) = \int_0^\infty e^{-(s-a)t} f(t) \, dt. \qquad (8.5.2)$$

Comparing (8.5.1) and (8.5.2), we obtain

$$L[e^{at} f(t)] = F(s - a), \qquad (8.5.3)$$

as required. Taking the inverse Laplace transform of both sides of (8.5.3) yields

$$L^{-1}[F(s - a)] = e^{at} f(t). \qquad \blacksquare$$

We illustrate the use of the preceding theorem with several examples. (See also Figure 8.5.1.)

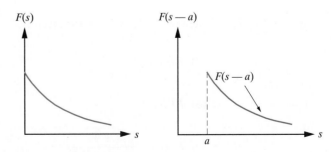

Figure 8.5.1: An illustration of the first shifting theorem. Multiplying $f(t)$ by e^{at} has the effect of shifting $F(s)$ by a units in s-space.

Example 8.5.2   Find $L[f]$ for each $f(t)$.

(a) $f(t) = e^{5t} \cos 4t$.

(b) $f(t) = e^{at} \sin bt$, where a, b are constants.

(c) $f(t) = e^{at} t^n$, where a is a constant and n is a positive integer.

Solution:

(a) From Table 8.2.1, we have

$$L[\cos 4t] = \frac{s}{s^2 + 16},$$

so that applying the first shifting theorem with $a = 5$ yields

$$L[e^{5t} \cos 4t] = \frac{s - 5}{(s - 5)^2 + 16}.$$

(b) Since $L[\sin bt] = \frac{b}{s^2 + b^2}$,

$$L[e^{at} \sin bt] = \frac{b}{(s - a)^2 + b^2}.$$

Similarly, it follows from Table 8.2 and the first shifting theorem that

$$L[e^{at} \cos bt] = \frac{s - a}{(s - a)^2 + b^2}.$$

(c) From Table 8.2.1, we have

$$L[t^n] = \frac{n!}{s^{n+1}},$$

so that

$$L[e^{at} t^n] = \frac{n!}{(s - a)^{n+1}}. \qquad \square$$

The preceding example dealt with the direct use of the first shifting theorem to obtain the Laplace transform of a function. Of equal importance is its use in determining inverse transforms. Once more, we illustrate with several examples.

Example 8.5.3   Determine $L^{-1}[F(s)]$ for the given F.

(a) $F(s) = \dfrac{3}{(s - 2)^2 + 9}$.

(b) $F(s) = \dfrac{6}{(s - 4)^3}$.

(c) $F(s) = \dfrac{s + 4}{s^2 + 6s + 13}$.

(d) $F(s) = \dfrac{s - 2}{s^2 + 2s + 3}$.

Solution:

(a) From Table 8.2.1,

$$L^{-1}\left[\frac{3}{s^2+9}\right] = \sin 3t,$$

so that, by the first shifting theorem,

$$L^{-1}\left[\frac{3}{(s-2)^2+9}\right] = e^{2t}\sin 3t.$$

(b) From Table 8.2.1,

$$L^{-1}\left[\frac{6}{s^3}\right] = 3t^2.$$

Thus, applying the first shifting theorem yields

$$L^{-1}\left[\frac{6}{(s-4)^3}\right] = 3t^2 e^{4t}.$$

(c) In this case,

$$F(s) = \frac{s+4}{s^2+6s+13},$$

which we do not recognize as being a shift of the transform of any of the functions given in Table 8.2.1. However, completing the square in the denominator of $F(s)$ yields[4]

$$F(s) = \frac{s+4}{(s+3)^2+4}. \tag{8.5.4}$$

We still cannot write down the inverse transform directly, but, by the first shifting theorem, we have

$$L^{-1}\left[\frac{s+3}{(s+3)^2+4}\right] = e^{-3t}\cos 2t, \tag{8.5.5}$$

$$L^{-1}\left[\frac{2}{(s+3)^2+4}\right] = e^{-3t}\sin 2t. \tag{8.5.6}$$

This suggests that we rewrite (8.5.4) in the equivalent form

$$F(s) = \frac{s+3}{(s+3)^2+4} + \frac{1}{(s+3)^2+4},$$

so that, using the linearity of L^{-1} and Equations (8.5.5) and (8.5.6),

$$L^{-1}[F(s)] = L^{-1}\left[\frac{s+3}{(s+3)^2+4}\right] + L^{-1}\left[\frac{1}{(s+3)^2+4}\right]$$

$$= e^{-3t}\cos 2t + \frac{1}{2}e^{-3t}\sin 2t.$$

---

[4]Recall that we can always write $x^2 + ax + b = (x + a/2)^2 + b - a^2/4$. This procedure is known as completing the square.

(d) We proceed as in the previous example. In this case, we have

$$F(s) = \frac{s - 2}{(s + 1)^2 + 2},$$

which can be written as

$$F(s) = \frac{s + 1}{(s + 1)^2 + 2} - \frac{3}{(s + 1)^2 + 2}.$$

Then, using Table 8.2.1 and the first shifting theorem, it follows that

$$L^{-1}[F(s)] = e^{-t} \cos \sqrt{2}t - \frac{3}{\sqrt{2}} e^{-t} \sin \sqrt{2}t. \qquad \square$$

Exercises for 8.5

Key Terms

First shifting theorem.

Skills

- Be able to use the first shifting theorem to compute the Laplace transform and inverse Laplace transform of the applicable "shifted" functions in this section.

True-False Review

For Questions 1–8, decide if the given statement is **true** or **false**, and give a brief justification for your answer. If true, you can quote a relevant definition or theorem from the text. If false, provide an example, illustration, or brief explanation of why the statement is false.

1. If $L[f] = F(s)$, then we have $L[e^{-at} f(t)] = F(s + a)$.

2. For every function f, we have $f(t - 1) = f(t) - 1$ for every t.

3. If $f(t + 2) = \dfrac{e^t}{\sqrt{t + 3}}$, then $f(t) = \dfrac{e^{t-2}}{\sqrt{t + 1}}$.

4. If f and g are integrable functions such that $f(x-3) = g(x)$, then

$$\int_0^1 f(t) \, dt = \int_0^1 g(t) \, dt - 3.$$

5. We have $L[e^{-t} \sin 2t] = \dfrac{2}{(s - 1)^2 + 4}$.

6. We have $L[e^{2t} t^3] = \dfrac{6}{(s - 2)^4}$.

7. We have $L^{-1} \left[\dfrac{s + 4}{(s + 4)^2 + 9} \right] = e^{-4t} \cos 3t$.

8. We have $L^{-1} \left[\dfrac{3}{(s + 1)^2 + 36} \right] = 2e^{-t} \sin 6t$.

Problems

For Problems 1–10, determine $f(t - a)$ for the given function f and the given constant a.

1. $f(t) = t, \quad a = 1$.

2. $f(t) = 1, \quad a = 3$.

3. $f(t) = t^2 - 2t, \quad a = -2$.

4. $f(t) = e^{3t}, \quad a = 2$.

5. $f(t) = e^{2t} \cos t, \quad a = \pi$.

6. $f(t) = te^{2t}, \quad a = -1$.

7. $f(t) = e^{-t} \sin 2t, \quad a = \pi/6$.

8. $f(t) = \dfrac{t}{t^2 + 4}, \quad a = 1$.

9. $f(t) = \dfrac{t+1}{t^2 - 2t + 2}$, $a = 2$.

10. $f(t) = e^{-t}(\sin 2t + \cos 2t)$, $a = \pi/4$.

For Problems 11–16, determine $f(t)$.

11. $f(t-1) = (t-1)^2$.

12. $f(t-1) = (t-2)^2$.

13. $f(t-2) = (t-2)e^{3(t-2)}$.

14. $f(t-1) = t\sin[3(t-1)]$.

15. $f(t-3) = te^{-(t-3)}$.

16. $f(t-4) = \dfrac{t+1}{(t-1)^2 + 4}$.

For Problems 17–26, determine the Laplace transform of f.

17. $f(t) = e^{3t}\cos 4t$.

18. $f(t) = e^{-4t}\sin 5t$.

19. $f(t) = te^{2t}$.

20. $f(t) = 3te^{-t}$.

21. $f(t) = t^3 e^{-4t}$.

22. $f(t) = e^t - te^{-2t}$.

23. $f(t) = 2e^{3t}\sin t + 4e^{-t}\cos 3t$.

24. $f(t) = e^{2t}(1 - \sin^2 t)$.

25. $f(t) = t^2(e^t - 3)$.

26. $f(t) = e^{-2t}\sin(t - \pi/4)$.

For Problems 27–41, determine $L^{-1}[F]$.

27. $F(s) = \dfrac{1}{(s-3)^2}$.

28. $F(s) = \dfrac{4}{(s+2)^3}$.

29. $F(s) = \dfrac{2}{\sqrt{s+3}}$.

30. $F(s) = \dfrac{2}{(s-1)^2 + 4}$.

31. $F(s) = \dfrac{s+2}{(s+2)^2 + 9}$.

32. $F(s) = \dfrac{s}{(s-3)^2 + 4}$.

33. $F(s) = \dfrac{5}{(s-2)^2 + 16}$.

34. $F(s) = \dfrac{6}{s^2 + 2s + 2}$.

35. $F(s) = \dfrac{s-2}{s^2 + 2s + 26}$.

36. $F(s) = \dfrac{2s}{s^2 - 4s + 13}$.

37. $F(s) = \dfrac{s}{(s+1)^2 + 4}$.

38. $F(s) = \dfrac{2s+3}{(s+5)^2 + 49}$.

39. $F(s) = \dfrac{4}{s(s+2)^2}$.

40. $F(s) = \dfrac{2s+1}{(s-1)^2(s+2)}$.

41. $F(s) = \dfrac{2s+3}{s(s^2 - 2s + 5)}$.

For Problems 42–52, solve the given initial-value problem.

42. $y'' - y = 8e^t$, $y(0) = 0$, $y'(0) = 0$.

43. $y'' - 4y = 12e^{2t}$, $y(0) = 2$, $y'(0) = 3$.

44. $y'' - y' - 2y = 6e^{-t}$, $y(0) = 0$, $y'(0) = 1$.

45. $y'' + y' - 2y = 3e^{-2t}$, $y(0) = 3$, $y'(0) = -1$.

46. $y'' - 4y' + 4y = 6e^{2t}$, $y(0) = 1$, $y'(0) = 0$.

47. $y'' + 2y' + y = 2e^{-t}$, $y(0) = 2$, $y'(0) = 1$.

48. $y'' - 4y = 2te^t$, $y(0) = 0$, $y'(0) = 0$.

49. $y'' + 3y' + 2y = 12te^{2t}$, $y(0) = 0$, $y'(0) = 1$.

50. $y'' + y = 5te^{-3t}$, $y(0) = 2$, $y'(0) = 0$.

51. $y'' - y = 8e^t \sin 2t$, $y(0) = 2$, $y'(0) = -2$.

52. $y'' + 2y' - 3y = 26e^{2t}\cos t$, $y(0) = 1$, $y'(0) = 0$.

53. Solve the initial-value problem

$$x_1' = 2x_1 - x_2, \qquad x_2' = x_1 + 2x_2,$$
$$x_1(0) = 1, \qquad x_2(0) = 0.$$

54. Solve the initial-value problem

$$x_1' = 3x_1 + 2x_2, \qquad x_2' = -x_1 + 4x_2,$$
$$x_1(0) = -1, \qquad x_2(0) = 1.$$

8.6   The Unit Step Function

In applications of differential equations such as

$$y'' + by' + cy = F(t)$$

to engineering problems, it often arises that the forcing term $F(t)$ is either piecewise continuous or even discontinuous. In such a situation, the Laplace transform is ideally suited for determining the solution of the differential equation as compared with the techniques developed in Chapter 6. To specify piecewise continuous functions in an appropriate manner, it is useful to introduce the unit step function, defined as follows:

DEFINITION   8.6.1

The **unit step function** or **Heaviside step function**, $u_a(t)$, is defined by

$$u_a(t) = \begin{cases} 0, & 0 \le t < a, \\ 1, & t \ge a, \end{cases}$$

where a is any positive number. (See Figure 8.6.1.)

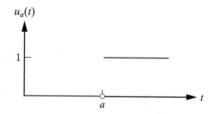

Figure 8.6.1: The unit step function $u_a(t)$.

Example 8.6.2   Sketch the function $f(t) = u_a(t) - u_b(t)$, where $b > a$.

Solution: By definition of the unit step function, we have

$$f(t) = \begin{cases} 0, & 0 \le t < a, \\ 1, & a \le t < b, \\ 0, & t \ge b, \end{cases}$$

so that the graph of f is as given in Figure 8.6.2.

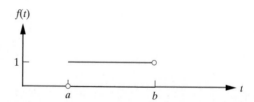

Figure 8.6.2: A sketch of the function given in Example 8.6.2.

$\square$

The real power of the unit step function is that it enables us to model the situation when a force acts intermittently or in a nonsmooth manner. For example, the function f

in Figure 8.6.2 can be interpreted as representing a force of unit magnitude that begins to act at $t = a$ and that stops acting at $t = b$. More generally, it is useful to regard the unit step function $u_a(t)$ as giving a mathematical description of a switch that is turned on at $t = a$.

The remaining examples in this section indicate how $u_a(t)$ can be useful for representing functions that are piecewise continuous.

Example 8.6.3   Express the following function in terms of the unit step function:

$$f(t) = \begin{cases} 0, & 0 \leq t < 1, \\ t - 1, & 1 \leq t < 2, \\ 1, & t \geq 2. \end{cases}$$

Solution: We view the given function in the following way: The contribution $f_1(t) = t - 1$ is "switched on" at $t = 1$ and is "switched off" again at $t = 2$. Mathematically this can be described by

$$f_1(t) = \underbrace{u_1(t)(t - 1)}_{\text{switch on at } t = 1} - \underbrace{u_2(t)(t - 1)}_{\text{switch off at } t = 2}.$$

At $t = 2$, the contribution $f_2(t) = 1$ switches on and remains on for all $t \geq 2$. Mathematically this is described by

$$f_2(t) = u_2(t).$$

The function f is then given by

$$f(t) = f_1(t) + f_2(t) = (t - 1)u_1(t) - (t - 1)u_2(t) + u_2(t),$$

which can be written in the equivalent form

$$f(t) = (t - 1)u_1(t) - (t - 2)u_2(t). \tag{8.6.1}$$

A sketch of $f(t)$ is given in Figure 8.6.3. Notice that this sketch is more easily determined from the original definition of f, rather than from (8.6.1).   □

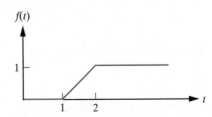

Figure 8.6.3: A sketch of the function given in Example 8.6.3.

Example 8.6.4   Make a sketch of the function $f(t)$ defined by

$$f(t) = \begin{cases} t, & 0 \leq t < 2, \\ -1, & 2 \leq t < 4, \\ t - 4, & 4 \leq t < 5, \\ e^{(5-t)}, & t \geq 5, \end{cases}$$

and express f in terms of the unit step function.

Solution: The function is sketched in Figure 8.6.4. Using the unit step function, we see that f consists of the following different parts:

$$f_1(t) = t[1 - u_2(t)],$$
$$f_2(t) = -[u_2(t) - u_4(t)],$$
$$f_3(t) = (t - 4)[u_4(t) - u_5(t)],$$
$$f_4(t) = e^{(5-t)}u_5(t).$$

Thus,

$$f(t) = t[1 - u_2(t)] - [u_2(t) - u_4(t)] + (t - 4)[u_4(t) - u_5(t)] + e^{(5-t)}u_5(t). \quad \square$$

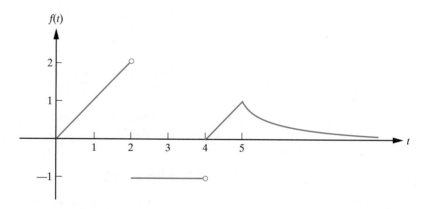

Figure 8.6.4: A sketch of the function defined in Example 8.6.4.

Exercises for 8.6

Key Terms

Unit (Heaviside) step function.

Skills

- Be able to sketch functions that involve the unit step function.

- Be able to express appropriate functions in terms of unit step functions.

True-False Review

For Questions 1–4, decide if the given statement is **true** or **false**, and give a brief justification for your answer. If true, you can quote a relevant definition or theorem from the text. If false, provide an example, illustration, or brief explanation of why the statement is false.

1. The unit step function is defined by

$$\begin{cases} 0, & 0 \leq t \leq a, \\ 1, & t \geq a, \end{cases}$$

where a is any positive number.

2. If $a < b$, the function $f(t) = u_a(t) - u_b(t)$ has value 1 on the interval $[a, b]$ and value 0 elsewhere.

3. If a and b are positive integers with $a < b$, then $u_a(t) \leq u_b(t)$ for all $t \geq 0$.

4. The function

$$\begin{cases} 0, & 0 \leq t < a, \\ 1, & a \leq t < b, \\ 0, & t \geq b \end{cases}$$

can be expressed as $f(t) = u_b(t) - u_a(t)$.

Problems

For Problems 1–6, make a sketch of the given function on the interval $[0, \infty)$.

1. $f(t) = 2u_1(t) - 4u_3(t)$.

2. $f(t) = 1 + (t - 1)u_1(t)$.

3. $f(t) = t\,[1 - u_1(t)]$.

4. $f(t) = u_1(t) + u_2(t) + u_3(t) + u_4(t)$.

5. $f(t) = u_1(t) + u_2(t) + \cdots = \displaystyle\sum_{i=1}^{\infty} u_i(t)$.

6. $f(t) = u_1(t) - u_2(t) + u_3(t) - \cdots$

$$= \sum_{i=1}^{\infty} (-1)^{i+1} u_i(t).$$

For Problems 7–14, make a sketch of the given function on $[0, \infty)$ and express it in terms of the unit step function.

7. $f(t) = \begin{cases} 3, & 0 \le t < 1, \\ -1, & t \ge 1. \end{cases}$

8. $f(t) = \begin{cases} t^2, & 0 \le t < 1, \\ 1, & t \ge 1. \end{cases}$

9. $f(t) = \begin{cases} 2, & 0 \le t < 2, \\ 1, & 2 \le t < 4, \\ -1, & t \ge 4. \end{cases}$

10. $f(t) = \begin{cases} 2, & 0 \le t \le 1, \\ 2e^{(t-1)}, & t > 1. \end{cases}$

11. $f(t) = \begin{cases} t, & 0 \le t < 3, \\ 6 - t, & 3 \le t < 6, \\ 0, & t \ge 6. \end{cases}$

12. $f(t) = \begin{cases} 0, & 0 \le t < 2, \\ 3 - t, & 2 \le t < 4, \\ -1, & t \ge 4. \end{cases}$

13. $f(t) = \begin{cases} 1, & 0 \le t < \pi/2, \\ \sin t, & \pi/2 \le t < 3\pi/2, \\ -1, & t \ge 3\pi/2. \end{cases}$

14. $f(t) = \begin{cases} \sin t, & 2n\pi \le t < (2n+1)\pi, \\ & \quad (n = 0, 1, 2, 3, \ldots), \\ 0, & \text{otherwise.} \end{cases}$

8.7 The Second Shifting Theorem

In the preceding section we saw how the unit step function can be used to represent functions that are piecewise continuous. In this section we show that the Laplace transform provides a straightforward method for solving constant-coefficient linear differential equations that have such functions as driving terms. We first need to determine how the unit step function transforms.

Theorem 8.7.1 **(Second Shifting Theorem)**

Let $L[f(t)] = F(s)$. Then

$$L[u_a(t)f(t - a)] = e^{-as} F(s). \tag{8.7.1}$$

Conversely,

$$L^{-1}[e^{-as} F(s)] = u_a(t)f(t - a). \tag{8.7.2}$$

Proof Once more we must return to the definition of the Laplace transform. We have

$$L[u_a(t)f(t - a)] = \int_0^\infty e^{-st} u_a(t) f(t - a)\, dt = \int_a^\infty e^{-st} f(t - a)\, dt,$$

where we have used the definition of the unit step function. We now make a change of variable in the integral. Let $x = t - a$. Then $dx = dt$, and the lower limit of integration $t = a$ corresponds to $x = 0$, whereas the upper limit of integration is unchanged. Thus,

$$L[u_a(t) f(t - a)] = \int_0^\infty e^{-s(x+a)} f(x) \, dx = e^{-as} \int_0^\infty e^{-sx} f(x) \, dx = e^{-as} L[f],$$

as required. Taking the inverse Laplace transform of both sides of (8.7.1) yields (8.7.2).
∎

This theorem is illustrated in Figure 8.7.1.

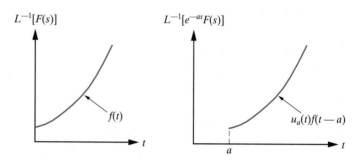

Figure 8.7.1: An illustration of the second shifting theorem. Multiplying $F(s)$ by e^{-as} has the effect of shifting $f(t)$ by a units to the right in t-space.

Corollary 8.7.2   If $L[f(t)] = F(s)$, then

$$L[u_a(t) f(t)] = e^{-as} L[f(t + a)].$$

Proof This is a direct consequence of the previous theorem.   ∎

We illustrate the use of the preceding theorem and corollary with several examples.

Example 8.7.3   Determine $L[f]$ if

$$f(t) = \begin{cases} 0, & 0 \le t < 1, \\ t - 1, & 1 \le t < 2, \\ 1, & t \ge 2. \end{cases}$$

Solution: We have already shown in Example 8.6.3 that the given function can be expressed in terms of the unit step function as

$$f(t) = (t - 1)u_1(t) - (t - 2)u_2(t).$$

If we let $g(t) = t$, then

$$f(t) = g(t - 1)u_1(t) - g(t - 2)u_2(t).$$

Using Theorem 8.7.1, it follows that

$$L[f] = e^{-s} L[g] - e^{-2s} L[g] = \frac{1}{s^2}(e^{-s} - e^{-2s}).$$
    □

Example 8.7.4   Find $L[f]$ if

$$f(t) = \begin{cases} 1, & 0 \le t < 2, \\ e^{-(t-2)}, & t \ge 2. \end{cases}$$

Solution: To determine $L[f]$, we first express f in terms of the unit step function. In this case, we have

$$f(t) = [1 - u_2(t)] + e^{-(t-2)}u_2(t).$$

That is,

$$f(t) = 1 + u_2(t)[e^{-(t-2)} - 1].$$

If we let $g(t) = e^{-t} - 1$, then

$$f(t) = 1 + u_2(t)g(t - 2),$$

so that, from Theorem 8.7.1,

$$L[f] = \frac{1 - e^{-2s}}{s} + \frac{e^{-2s}}{s + 1}. \qquad \square$$

Example 8.7.5   Determine $L^{-1}\left[\dfrac{2e^{-s}}{s^2 + 4}\right]$.

Solution: From Table 8.2, we have

$$L[\sin 2t] = \frac{2}{s^2 + 4}.$$

Consequently,

$$L^{-1}\left[\frac{2e^{-s}}{s^2 + 4}\right] = L^{-1}\{e^{-s}L[\sin 2t]\}$$

$$= u_1(t)\sin[2(t - 1)],$$

using Theorem 8.7.1. $\qquad \square$

Example 8.7.6   Determine $L^{-1}\left[\dfrac{(s - 4)e^{-3s}}{s^2 - 4s + 5}\right]$.

Solution: Let

$$G(s) = \frac{(s - 4)e^{-3s}}{s^2 - 4s + 5}.$$

We first rewrite G in a form more suitable for determining $L^{-1}[G]$. Completing the square in the denominator yields

$$G(s) = \frac{(s - 4)e^{-3s}}{(s - 2)^2 + 1},$$

which can be written in the equivalent form

$$G(s) = e^{-3s}\left[\frac{s - 2}{(s - 2)^2 + 1} - \frac{2}{(s - 2)^2 + 1}\right].$$

Thus,

$$L^{-1}[G] = L^{-1}\left\{e^{-3s}L\left[e^{2t}\cos t - 2e^{2t}\sin t\right]\right\}$$

$$= u_3(t)\left[e^{2(t-3)}\cos(t-3) - 2e^{2(t-3)}\sin(t-3)\right]$$

$$= e^{2(t-3)}u_3(t)\left[\cos(t-3) - 2\sin(t-3)\right]. \qquad \square$$

We now illustrate how the unit step function can be useful in the solution of initial-value problems. For simplicity, we will start with a first-order differential equation.

Example 8.7.7   Solve the initial-value problem

$$y' - y = 1 - (t-1)u_1(t), \qquad y(0) = 0.$$

Solution: In this case, the forcing term on the right-hand side of the differential equation is sketched in Figure 8.7.2. Taking the Laplace transform of both sides of the differential equation yields

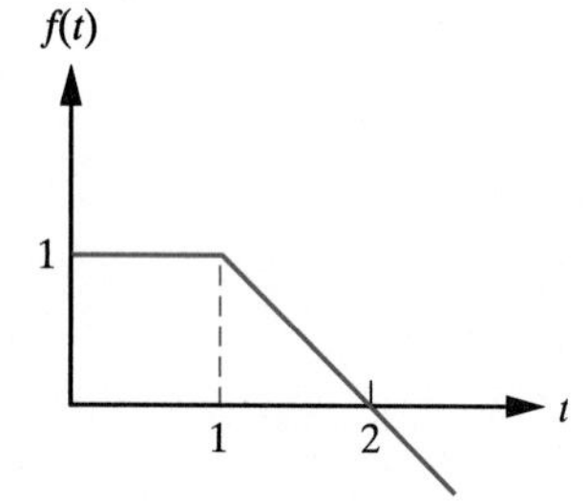

$$sY(s) - Y(s) - y(0) = \frac{1}{s} - \frac{e^{-s}}{s^2}.$$

By imposing the given initial conditions and simplifying, we obtain

$$Y(s) = \frac{1}{s(s-1)} - e^{-s}\left[\frac{1}{s^2(s-1)}\right].$$

Decomposing the terms on the right-hand side into partial fractions yields

$$Y(s) = \frac{1}{s-1} - \frac{1}{s} - e^{-s}\left(\frac{1}{s-1} - \frac{1}{s} - \frac{1}{s^2}\right).$$

Taking the inverse Laplace transform of both sides of this equation, we obtain

$$y(t) = e^t - 1 - u_1(t)[e^{(t-1)} - 1 - (t-1)].$$

That is,

$$y(t) = e^t - 1 - u_1(t)[e^{(t-1)} - t].$$

This problem was solved in Chapter 1 (Example 1.6.5). A comparison of the two solution methods indicates the power of the Laplace transform. $\qquad \square$

Figure 8.7.2: The forcing function in Example 8.7.7.

Example 8.7.8   Solve the initial-value problem

$$y'' + 2y' + 5y = f(t), \qquad y(0) = 0, \qquad y'(0) = 0$$

if

$$f(t) = \begin{cases} 10, & 0 \le t < 4, \\ -10, & 4 \le t < 8, \\ 0, & t \ge 8. \end{cases}$$

Solution: The forcing function $f(t)$ is sketched in Figure 8.7.3. We first express f in terms of the unit step function. In this case, we have

$$f(t) = 10\left[1 - 2u_4(t) + u_8(t)\right],$$

so that the differential equation can be written as

$$y'' + 2y' + 5y = 10\left[1 - 2u_4(t) + u_8(t)\right],$$

and we can now proceed in the usual manner. Taking the Laplace transform of both sides of the differential equation yields

$$[s^2 Y(s) - sy(0) - y'(0)] + 2[sY(s) - y(0)] + 5Y(s) = \frac{10}{s}(1 - 2e^{-4s} + e^{-8s}).$$

That is, by imposing the given initial conditions and simplifying,

$$Y(s) = \frac{10(1 - 2e^{-4s} + e^{-8s})}{s(s^2 + 2s + 5)}.$$

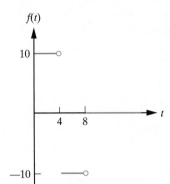

Figure 8.7.3: The forcing function in Example 8.7.8.

The right-hand side of this equation has the following partial fractions decomposition:

$$\frac{1}{s(s^2 + 2s + 5)} = \frac{1}{5s} - \frac{s+2}{5(s^2 + 2s + 5)},$$

so that

$$Y(s) = 2(1 - 2e^{-4s} + e^{-8s})\left(\frac{1}{s} - \frac{s+2}{s^2 + 2s + 5}\right). \tag{8.7.3}$$

Now,

$$L^{-1}\left[\frac{s+2}{s^2 + 2s + 5}\right] = L^{-1}\left[\frac{s+1}{(s+1)^2 + 4} + \frac{1}{(s+1)^2 + 4}\right]$$

$$= e^{-t}\cos 2t + \tfrac{1}{2}e^{-t}\sin 2t.$$

Taking the inverse Laplace transform of both sides of (8.7.3) and using Theorem 8.7.1 yields

$$y(t) = 2\Bigg\{1 - e^{-t}\cos 2t - \frac{1}{2}e^{-t}\sin 2t$$

$$- 2u_4(t)\left[1 - e^{-(t-4)}\cos 2(t-4) - \frac{1}{2}e^{-(t-4)}\sin 2(t-4)\right]$$

$$+ u_8(t)\left[1 - e^{-(t-8)}\cos 2(t-8) - \frac{1}{2}e^{-(t-8)}\sin 2(t-8)\right]\Bigg\}.$$

We can express this solution in the simpler form

$$y(t) = g(t) - 2u_4(t)g(t-4) + u_8(t)g(t-8),$$

where

$$g(t) = 2\left(1 - e^{-t}\cos 2t - \tfrac{1}{2}e^{-t}\sin 2t\right).$$

The given initial-value problem can be interpreted as governing the motion of a damped spring–mass system. Owing to the form of the initial conditions, if the forcing function were the same constant, F_0, for all time, the oscillations would quickly be damped out, and $y(t)$ would approach $F_0/5$. In this problem, the forcing term is constant over different time intervals. Consequently, the mass first performs damped oscillations approaching

$y = 2$. Then the second part of the driving term comes into effect, and the subsequent oscillations are about $y = -2$. After $t = 8$, the forcing function vanishes, and the physical system performs damped oscillations about the equilibrium. These features can be seen in Figure 8.7.4. ☐

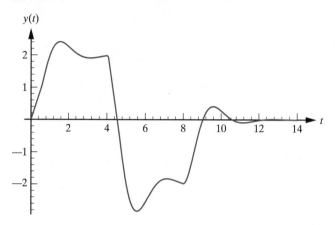

Figure 8.7.4: The response of a damped spring–mass system to the driving term given in Example 8.7.8.

Exercises for 8.7

Key Terms

Second shifting theorem.

Skills

- Be able to apply the second shifting theorem to compute Laplace transforms of functions involving the unit step function.

- Be able to use the second shifting theorem to compute inverse Laplace transforms that result in unit step functions.

- Be able to solve initial-value problems that involve unit step functions.

True-False Review

For Questions 1–7, decide if the given statement is **true** or **false**, and give a brief justification for your answer. If true, you can quote a relevant definition or theorem from the text. If false, provide an example, illustration, or brief explanation of why the statement is false.

1. If $L[f(t)] = F(s)$, then the inverse Laplace transform of the function $e^{as}F(s)$ is $u_a(t)f(t-a)$.

2. If $L[f(t)] = F(s)$, then the Laplace transform of $u_a(t)f(t)$ is $e^{-as}L[f(t+a)]$.

3. If $L[f(t)] = F(s)$, then multiplying $F(s)$ by e^{-as} has the effect of shifting $f(t)$ by a units to the right in t-space.

4. We have $L[u_2(t)\cos 4t] = \dfrac{se^{-2s}}{s^2 + 16}$.

5. We have $L[u_3(t)e^t] = \dfrac{1}{e^{3s}(s-1)}$.

6. We have $L^{-1}\left[\dfrac{e^s}{s^2 + 9}\right] = \dfrac{1}{3}u_1(t)\sin[3(t-1)]$.

7. We have $L^{-1}\left[\dfrac{1}{se^{2s}}\right] = u_2(t)(t-2)$.

Problems

For Problems 1–10, determine the Laplace transform of the given function f.

1. $f(t) = (t-1)u_1(t)$.

2. $f(t) = e^{3(t-2)}u_2(t)$.

3. $f(t) = \sin(t - \pi/4)u_{\pi/4}(t)$.

4. $f(t) = \cos t u_\pi(t)$.

5. $f(t) = (t-2)^2 u_2(t)$.

6. $f(t) = tu_3(t)$.

7. $f(t) = (t-1)^2 u_2(t)$.

8. $f(t) = e^{(t-4)}(t-4)^3 u_4(t)$.

9. $f(t) = e^{-2(t-1)} \sin 3(t-1)u_1(t)$.

10. $f(t) = e^{a(t-c)} \cos b(t-c)u_c(t)$, where a, b, and c are positive constants.

For Problems 11–25, determine the inverse Laplace transform of F.

11. $F(s) = \dfrac{e^{-2s}}{s^2}$.

12. $F(s) = \dfrac{e^{-s}}{s+1}$.

13. $F(s) = \dfrac{e^{-3s}}{s+4}$.

14. $F(s) = \dfrac{se^{-s}}{s^2+4}$.

15. $F(s) = \dfrac{e^{-3s}}{s^2+1}$.

16. $F(s) = \dfrac{e^{-2s}}{s+2}$.

17. $F(s) = \dfrac{e^{-s}}{(s+1)(s-4)}$.

18. $F(s) = \dfrac{e^{-2s}}{s^2+2s+2}$.

19. $F(s) = \dfrac{e^{-s}(s+6)}{s^2+9}$.

20. $F(s) = \dfrac{e^{-5s}}{s^2+16}$.

21. $F(s) = \dfrac{e^{-2s}}{(s-3)^3}$.

22. $F(s) = \dfrac{e^{-4s}(s+3)}{s^2-6s+13}$.

23. $F(s) = \dfrac{e^{-s}(2s-1)}{s^2+4s+5}$.

24. $F(s) = \dfrac{2e^{-2s}}{(s-1)(s^2+1)}$.

25. $F(s) = \dfrac{50e^{-3s}}{(s+1)^2(s^2+4)}$.

For Problems 26–40, solve the given initial-value problem.

26. $y' + 2y = 2u_1(t)$, $y(0) = 1$.

27. $y' - 2y = u_2(t)e^{t-2}$, $y(0) = 2$.

28. $y' - y = 4u_{\pi/4}(t)\cos(t - \pi/4)$, $y(0) = 1$.

29. $y' + 2y = u_\pi(t)\sin 2t$, $y(0) = 3$.

30. $y' + 3y = f(t)$, $y(0) = 1$, where
$$f(t) = \begin{cases} 1, & 0 \le t < 1, \\ 0, & t \ge 1. \end{cases}$$

31. $y' - 3y = f(t)$, $y(0) = 2$, where
$$f(t) = \begin{cases} \sin t, & 0 \le t < \pi/2, \\ 1, & t \ge \pi/2. \end{cases}$$

32. $y' - 3y = 10e^{-(t-a)}\sin[2(t-a)]u_a(t)$, $y(0) = 5$, where a is a positive constant.

33. $y'' - y = u_1(t)$, $y(0) = 2$, $y'(0) = 0$.

34. $y'' - y' - 2y = 1 - 3u_2(t)$, $y(0) = 1$, $y'(0) = -2$.

35. $y'' - 4y = u_1(t) - u_2(t)$, $y(0) = 0$, $y'(0) = 4$.

36. $y'' + y = t - u_1(t)(t-1)$, $y(0) = 2$, $y'(0) = 1$.

37. $y'' + 3y' + 2y = 10u_{\pi/4}(t)\sin(t - \pi/4)$, $y(0) = 1$, $y'(0) = 0$.

38. $y'' + y' - 6y = 30u_1(t)e^{-(t-1)}$, $y(0) = 3$, $y'(0) = -4$.

39. $y'' + 4y' + 5y = 5u_3(t)$, $y(0) = 2$, $y'(0) = 1$.

40. $y'' - 2y' + 5y = 2\sin t + u_{\pi/2}(t)[1 - \sin(t - \pi/2)]$, $y(0) = 0$, $y'(0) = 0$.

For Problems 41–44, solve the given initial-value problem.

41. $y' + y = f(t)$, $y(0) = 2$, where $f(t)$ is given in Figure 8.7.5.

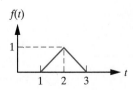

Figure 8.7.5: Forcing term for Problem 41.

42. $y' + 2y = f(t)$, $y(0) = 0$, where $f(t)$ is given in Figure 8.7.6.

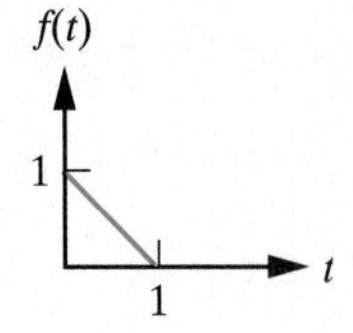

Figure 8.7.6: Forcing term for Problem 42.

43. $y' - y = f(t)$, $y(0) = 2$, where $f(t)$ is given in Figure 8.7.7.

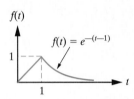

Figure 8.7.7: Forcing term for Problem 43.

44. $y' - 2y = f(t)$, $y(0) = 0$, where $f(t)$ is given in Figure 8.7.8.

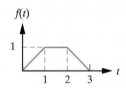

Figure 8.7.8: Forcing term for Problem 44.

45. Solve the initial-value problem

$$y' - y = f(t), \quad y(0) = 1,$$

where

$$f(t) = \begin{cases} 2, & 0 \le t < 1, \\ -1, & t \ge 1, \end{cases}$$

in the following two ways:

(a) Directly using the Laplace transform.

(b) Using the technique for solving first-order linear equations developed in Section 1.6.

46. The current $i(t)$ in an RL circuit is governed by the differential equation

$$\frac{di}{dt} + \frac{R}{L}i = \frac{1}{L}E(t),$$

where R an L are constants and $E(t)$ represents the applied EMF. At $t = 0$, the switch in the circuit is closed, and the applied EMF increases linearly from 0 V to 10 V in a time interval of 5 seconds. The EMF then remains constant for $t \ge 5$. Determine the current in the circuit for $t \ge 0$.

47. The differential equation governing the charge $q(t)$ on the capacitor in an RC circuit is

$$\frac{dq}{dt} + \frac{1}{RC}q = \frac{1}{R}E(t),$$

where R and C are constants and $E(t)$ represents the applied EMF. Over a time interval of 10 seconds, the applied EMF has the constant value 20 V. Thereafter, the EMF decays exponentially according to $E(t) = 20e^{-(t-10)}$. If the capacitor is initially uncharged and $RC \ne 1$, determine the current in the circuit for $t > 0$. [Recall that the current $i(t)$ is related to the charge on the capacitor by $i(t) = dq/dt$.]

8.8 Impulsive Driving Terms: The Dirac Delta Function

Consider the differential equation

$$y'' + by' + cy = f(t).$$

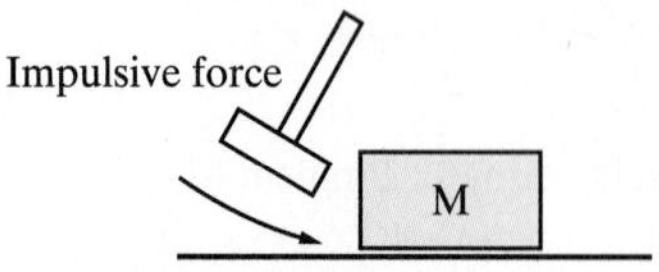

Impulsive force

Figure 8.8.1: An example of an impulsive force.

We have seen in the previous sections that the Laplace transform is useful in the case when the forcing term $f(t)$ is piecewise continuous. We now consider another type of forcing term; namely, that describing an impulsive force. Such a force arises when an object is dealt an instantaneous blow—for example, when an object is hit by a hammer. (See Figure 8.8.1.) The aim of this section is to develop a way of representing impulsive forces mathematically and then to show how the Laplace transform can be used to solve differential equations when the driving term is due to an impulsive force.

Suppose that a force of magnitude F acts on an object over the time interval $[t_1, t_2]$. The **impulse** of this force, I, is defined by[5]

$$I = \int_{t_1}^{t_2} F(t)\, dt.$$

Since $F(t)$ is zero for t outside the interval $[t_1, t_2]$, we can write

$$I = \int_{-\infty}^{\infty} F(t)\, dt.$$

Mathematically, I gives the area under the curve $y = F(t)$ lying over the t-axis. (See Figure 8.8.2.)

We now introduce a mathematical description of a force that instantaneously imparts an impulse of unit magnitude to an object at $t = a$. Thus, the two properties that we wish to characterize are the following:

1. The force acts instantaneously.

2. The force has unit impulse.

We proceed in the following manner. Define the function $d_\epsilon(t - a)$ by

$$d_\epsilon(t - a) = \frac{u_a(t) - u_{a+\epsilon}(t)}{\epsilon}, \qquad (8.8.1)$$

where u_a is the unit step function. (See Figure 8.8.3.) We can interpret $d_\epsilon(t - a)$ as

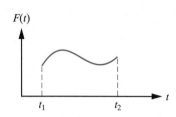

$F(t)$

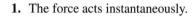

Figure 8.8.2: When a force of magnitude F Newtons acts on an object over a time interval $[t_1, t_2]$ seconds, the impulse of the force is given by the area under the curve.

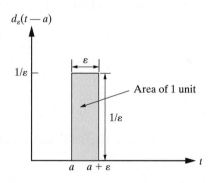

Figure 8.8.3: The function $d_\epsilon(t - a)$.

representing a force of magnitude $1/\epsilon$ that acts for a time interval of ϵ starting at $t = a$. Notice that this force does have unit impulse, since

$$I = \int_{-\infty}^{\infty} d_\epsilon(t - a)\, dt = \int_{-\infty}^{\infty} \frac{u_a(t) - u_{a+\epsilon}(t)}{\epsilon}\, dt = \int_a^{a+\epsilon} \frac{1}{\epsilon}\, dt = 1.$$

To capture the idea of an instantaneous force we take the limit as $\epsilon \to 0^+$. It follows from (8.8.1) that

$$\lim_{\epsilon \to 0^+} d_\epsilon(t - a) = 0 \quad \text{whenever } t \neq a.$$

Also, since $I = 1$ for all t,

$$\lim_{\epsilon \to 0^+} I = 1.$$

---

[5]This represents the change in momentum of the object due to the applied force.

These properties characterize mathematically the idea of a force of unit impulse acting instantaneously at $t = a$. We use them to define the *unit impulse function*.

DEFINITION 8.8.1

The **unit impulse function**, or **Dirac delta function**, $\delta(t - a)$, is the (generalized) function that satisfies

1. $\delta(t - a) = 0$, $t \neq a$.

2. $\displaystyle\int_{-\infty}^{\infty} \delta(t - a)\, dt = 1$.

Remark   The unit impulse function is not a function in the usual sense. It is an example of what is called a *generalized function*. The detailed study of such functions is beyond the scope of the present text. However, all that we will require are the properties 1 and 2 of Definition 8.8.1.

Thus, to summarize,

$\delta(t - a)$ describes a force that instantaneously
imparts a unit impulse to a system at $t = a$.

We now consider the possibility of determining the Laplace transform of $\delta(t - a)$. The natural way to do this is to return to the function $d_\epsilon(t - a)$ and define the Laplace transform of $\delta(t - a)$ in the following manner.

$$L[\delta(t - a)] = \lim_{\epsilon \to 0^+} L[d_\epsilon(t - a)] = \lim_{\epsilon \to 0^+} \int_0^\infty e^{-st}\left[\frac{u_a(t) - u_{a+\epsilon}(t)}{\epsilon}\right] dt$$

$$= \lim_{\epsilon \to 0^+} \frac{1}{\epsilon} \int_a^{a+\epsilon} e^{-st}\, dt = \lim_{\epsilon \to 0^+} \frac{1}{\epsilon}\left\{-\frac{1}{s}[e^{-s(a+\epsilon)} - e^{-sa}]\right\}$$

$$= \frac{1}{s} e^{-sa} \lim_{\epsilon \to 0^+}\left(\frac{1 - e^{-\epsilon s}}{\epsilon}\right).$$

Using L'Hopital's rule to evaluate the preceding limit yields

$$L[\delta(t - a)] = e^{-sa}.$$

In particular,

$$L[\delta(t)] = 1.$$

It can be shown more generally that if g is a continuous function on $(-\infty, \infty)$, then

$$\int_{-\infty}^{\infty} g(t)\delta(t - a)\, dt = g(a).$$

Example 8.8.2   Solve the initial-value problem

$$y'' + 4y' + 13y = \delta(t - \pi), \qquad y(0) = 2, \quad y'(0) = 1.$$

Solution: Taking the Laplace transform of both sides of the given differential equation and imposing the initial conditions yields

$$s^2 Y - 2s - 1 + 4(sY - 2) + 13Y = e^{-\pi s},$$

which implies that

$$Y(s) = \frac{e^{-\pi s} + 2s + 9}{s^2 + 4s + 13} = \frac{e^{-\pi s} + 2s + 9}{(s+2)^2 + 9} = \frac{e^{-\pi s}}{(s+2)^2 + 9} + \frac{2(s+2)}{(s+2)^2 + 9} + \frac{5}{(s+2)^2 + 9}.$$

Taking the inverse Laplace transform of both sides gives

$$y(t) = L^{-1}\left\{\frac{1}{3}e^{-\pi t}L[e^{-2t}\sin 3t]\right\} + 2e^{-2t}\cos 3t + \frac{5}{3}e^{-2t}\sin 3t$$

$$= \frac{1}{3}u_\pi(t)e^{-2(t-\pi)}\sin[3(t-\pi)] + 2e^{-2t}\cos 3t + \frac{5}{3}e^{-2t}\sin 3t.$$

Since $\sin[3(t-\pi)] = -\sin 3t$, we finally obtain

$$y(t) = -\frac{1}{3}u_\pi(t)e^{-2(t-\pi)}\sin 3t + e^{-2t}\left(2\cos 3t + \frac{5}{3}\sin 3t\right). \qquad \square$$

Example 8.8.3 Consider the spring–mass system depicted in Figure 8.8.4. At $t = 0$, the mass is pulled down a distance 1 unit from equilibrium and released from rest. After 3 seconds, the mass is dealt an instantaneous blow that imparts 5 units of impulse in the upward direction. The initial-value problem governing the motion of the mass is

$$\frac{d^2y}{dt^2} + 4y = -5\delta(t-3), \quad y(0) = 1, \quad \frac{dy}{dt}(0) = 0,$$

where $5\,\delta(t-3)$ describes the impulsive force that acts on the mass at $t = 3$, and the positive y-direction is downward. Determine the motion of the mass for all $t > 0$.

Solution: Taking the Laplace transform of the differential equation and imposing the initial conditions yields

$$s^2Y(s) - s + 4Y(s) = -5e^{-3s},$$

so that

$$Y(s) = \frac{-5e^{-3s} + s}{s^2 + 4} = -\frac{5e^{-3s}}{s^2 + 4} + \frac{s}{s^2 + 4}.$$

We can now take the inverse Laplace transform of both sides of this equation to obtain

$$y(t) = L^{-1}\left\{-\frac{5}{2}e^{-3s}L[\sin 2t]\right\} + \cos 2t.$$

Thus,

$$y(t) = -\frac{5}{2}u_3(t)\sin[2(t-3)] + \cos 2t.$$

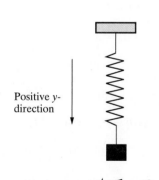

Positive y-direction

Impulsive force acts after 3 s

Figure 8.8.4: A spring–mass system in which friction is neglected and the only external force acting on the system is an impulsive force which imparts 5 units of impulse at $t = 3$.

The first term on the right-hand side represents the contribution from the impulsive force. Obviously this does not affect the motion of the mass until $t = 3$, but then it contributes for all $t \geq 3$. More explicitly, we can write the solution as

$$y(t) = \begin{cases} \cos 2t, & 0 \leq t < 3, \\ \cos 2t - \dfrac{5}{2}\sin 2(t-3), & t \geq 3. \end{cases}$$

The resulting motion is depicted in Figure 8.8.5, where the effect of the impulsive force is apparent. We see that $y(t)$ is continuous at $t = 3$, but that $y'(t)$ is discontinuous at $t = 3$.   □

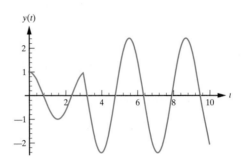

Figure 8.8.5: The motion of the spring–mass system in Example 8.8.3.

Exercises for 8.8

Key Terms

Impulsive force, Unit impulse (Dirac delta) function.

Skills

- Be able to solve initial-value problems involving the Dirac delta function.

True-False Review

For Questions 1–5, decide if the given statement is **true** or **false**, and give a brief justification for your answer. If true, you can quote a relevant definition or theorem from the text. If false, provide an example, illustration, or brief explanation of why the statement is false.

1. The impulse of a force $F(t)$ acting on an object is obtained by integrating $F(t)$ over all values of t.

2. A unit impulse force is a force that acts instantaneously and has unit impulse.

3. The Laplace transform of the unit impulse function $\delta(t - a)$ is e^{as}.

4. The initial-value problem governing the motion of a spring–mass system that is dealt an instantaneous blow at $t = a$ is a second-order nonhomogeneous differential equation involving the unit impulse function.

5. The instantaneous blow delivered to a spring–mass system determines the initial conditions for the nonho-

mogeneous differential equation governing the motion of the mass as a function of time.

Problems

For Problems 1–12, solve the given initial-value problem.

1. $y' - 2y = \delta(t - 2), \quad y(0) = 1.$

2. $y' + 4y = 3\,\delta(t - 1), \quad y(0) = 2.$

3. $y' - 5y = 2e^{-t} + \delta(t - 3), \quad y(0) = 0.$

4. $y'' - 3y' + 2y = \delta(t - 1), \quad y(0) = 1, \quad y'(0) = 0.$

5. $y'' - 4y = \delta(t - 3), \quad y(0) = 0, \quad y'(0) = 1.$

6. $y'' + 2y' + 5y = \delta(t - \pi/2), \quad y(0) = 0, \quad y'(0) = 2.$

7. $y'' - 4y' + 13y = \delta(t - \pi/4), \quad y(0) = 3, \quad y'(0) = 0.$

8. $y'' + 4y' + 3y = \delta(t - 2), \quad y(0) = 1, \quad y'(0) = -1.$

9. $y'' + 6y' + 13y = \delta(t - \pi/4), \quad y(0) = 5, \quad y'(0) = 5.$

10. $y'' + 9y = 15\sin 2t + \delta(t - \pi/6), \quad y(0) = 0, \quad y'(0) = 0.$

11. $y'' + 16y = 4\cos 3t + \delta(t - \pi/3), \quad y(0) = 0, \quad y'(0) = 0.$

12. $y'' + 2y' + 5y = 4\sin t + \delta(t - \pi/6), \quad y(0) = 0, \quad y'(0) = 1.$

13. The motion of a spring–mass system is governed by the initial-value problem

$$\frac{d^2y}{dt^2} + 4y = F_0 \cos 3t, \quad y(0) = 0, \quad \frac{dy}{dt}(0) = 0,$$

where F_0 is a constant. At $t = 5$ seconds, the mass is dealt a blow in the upward (negative) direction that instantaneously imparts 4 units of impulse to the system. Determine the resulting motion of the mass.

14. The motion of a spring–mass system is governed by

$$\frac{d^2y}{dt^2} + 4\frac{dy}{dt} + 13y = 10 \sin 5t,$$
$$y(0) = 0, \quad \frac{dy}{dt}(0) = 0.$$

At $t = 10$ seconds, the mass is dealt a blow in the downward (positive) direction that instantaneously imparts 2 units of impulse to the system. Determine the resulting motion of the mass.

15. Consider the spring–mass system whose motion is governed by the initial-value problem

$$\frac{d^2y}{dt^2} + \omega_0^2 y = F_0 \sin \omega t + A\delta(t - t_0),$$
$$y(0) = 0, \quad \frac{dy}{dt}(0) = 0,$$

where ω_0, ω, F_0, A and t_0 are positive constants and $\omega \neq \omega_0$. Solve the initial-value problem to determine the position of the mass at time t.

8.9   The Convolution Integral

Very often in solving a differential equation using the Laplace transform method we require the inverse Laplace transform of an expression of the form

$$H(s) = F(s)G(s),$$

where $F(s)$ and $G(s)$ are functions whose inverse Laplace transform is known. It is important to note that

$$L^{-1}[H(s)] \neq L^{-1}[F(s)]L^{-1}[G(s)].$$

For example,

$$L^{-1}\left[\frac{1}{(s-1)(s^2-1)}\right] = L^{-1}\left[\frac{1}{2(s-1)} - \frac{s+1}{2(s^2+1)}\right]$$
$$= \frac{1}{2}e^t - \frac{1}{2}\cos t - \frac{1}{2}\sin t,$$

whereas

$$L^{-1}\left[\frac{1}{s-1}\right]L^{-1}\left[\frac{1}{s^2+1}\right] = e^t \sin t.$$

Consequently,

$$L^{-1}\left[\frac{1}{(s-1)(s^2+1)}\right] \neq L^{-1}\left[\frac{1}{s-1}\right]L^{-1}\left[\frac{1}{s^2+1}\right].$$

However, it is possible, at least in theory, to determine $L^{-1}[F(s)G(s)]$ directly in terms of an integral involving $f(t)$ and $g(t)$. Before showing this, we require a definition.

DEFINITION   8.9.1

Suppose that f and g are continuous on the interval $[0, b]$. Then for t in $(0, b]$, the **convolution product**, $f * g$, of f and g is defined by

$$(f * g)(t) = \int_0^t f(t - \tau)g(\tau)\, d\tau.$$

Notice that $f * g$ is indeed a function of t. The integral

$$\int_0^t f(t - \tau)g(\tau)\, d\tau$$

is called a **convolution integral**.

Example 8.9.2   If $f(t) = t$ and $g(t) = \sin t$, determine $f * g$.

Solution: From Definition 8.9.1, we have

$$(f * g)(t) = \int_0^t (t - \tau) \sin \tau\, d\tau = t \int_0^t \sin \tau\, d\tau - \int_0^t \tau \sin \tau\, d\tau$$
$$= t(1 - \cos t) - (\sin t - t \cos t)$$
$$= t - \sin t. \qquad \square$$

The convolution product satisfies the three basic properties of the ordinary multiplicative product:

> **1.** $f * g = g * f.$ (commutative)
> **2.** $f * (g * h) = (f * g) * h.$ (associative)
> **3.** $f * (g + h) = f * g + f * h.$ (distributive over addition)

The proofs of these properties are left as exercises.

We now show how the convolution product can be useful in evaluating inverse Laplace transforms.

Theorem 8.9.3   **(The Convolution Theorem)**

If f and g are in $E(0, \infty)$, then

$$L[f * g] = L[f]L[g]. \tag{8.9.1}$$

Conversely,

$$L^{-1}[F(s)G(s)] = (f * g)(t). \tag{8.9.2}$$

Proof We must use the definition of the Laplace transform and the convolution product:

$$L[f * g] = \int_0^\infty e^{-st} \left\{ \int_0^t f(t - \tau)g(\tau)d\tau \right\} dt$$
$$= \int_0^\infty \int_0^t e^{-st} f(t - \tau)g(\tau)\, d\tau\, dt.$$

It is not clear how to proceed at this point. However, when dealing with an iterated double integral, it is often worth changing the order of integration to see if any simplification arises. In this case, the limits of integration are $0 \leq \tau \leq t, 0 \leq t < \infty$, so that the region of integration is that part of the $t\tau$-plane that lies above the t-axis and below the line $\tau = t$. This region is shown in Figure 8.9.1.

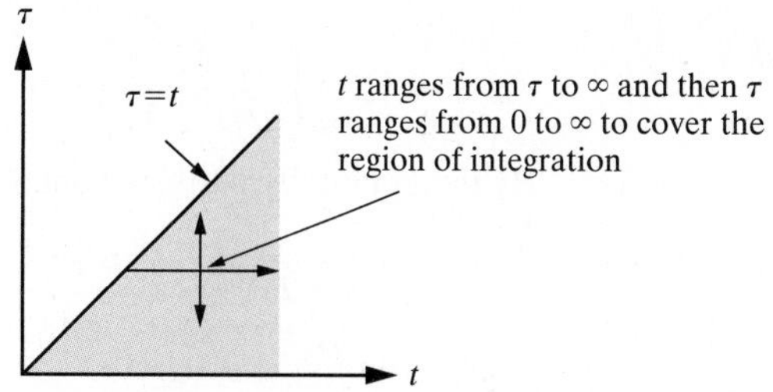

Figure 8.9.1: Changing the order of integration in Theorem 8.9.3.

Reversing the order of integration, the new limits are $\tau \leq t < \infty, 0 \leq \tau < \infty$. Thus, we can write

$$L[(f * g)(t)] = \int_0^\infty \int_\tau^\infty e^{-st} f(t - \tau) g(\tau) \, dt \, d\tau.$$

We now make the change of variable $u = t - \tau$ in the first iterated integral. Then $du = dt$ (remember that τ is treated as a constant when performing the inside integration) and the new u-limits are $0 \leq u < \infty$. Consequently,

$$
\begin{aligned}
L[(f * g)(t)] &= \int_0^\infty \int_0^\infty e^{-s(u+\tau)} g(\tau) f(u) \, du \, d\tau \\
&= \int_0^\infty e^{-s\tau} g(\tau) \left[\int_0^\infty e^{-su} f(u) \, du \right] d\tau. \\
&= \left[\int_0^\infty e^{-su} f(u) \, du \right] \left[\int_0^\infty e^{-s\tau} g(\tau) d\tau \right] = L[f] L[g],
\end{aligned}
$$

as required. The converse, (8.9.2), is obtained in the usual manner by taking the inverse Laplace transform of (8.9.1). ∎

Remark   It can be shown more generally that

$$L^{-1}[F_1(s) F_2(s) \ldots F_n(s)] = (f_1 * f_2 * \cdots * f_n)(t).$$

Example 8.9.4   Determine $L[f]$ if

$$f(t) = \int_0^t \sin(t - \tau) e^{-\tau} \, d\tau.$$

Solution: In this case, we recognize that

$$f(t) = \sin t * e^{-t},$$

so that, by the convolution theorem,

$$L[f] = L[\sin t] L[e^{-t}] = \frac{1}{(s^2 + 1)(s + 1)}. \qquad \square$$

Example 8.9.5   Find $L^{-1}\left[\dfrac{1}{s^2(s - 1)} \right]$.

Solution: We could determine the inverse Laplace transform in the usual manner by first using a partial fractions decomposition. However, we will use the convolution theorem:

$$L^{-1}\left[\frac{1}{s^2(s-1)}\right] = L^{-1}\left[\frac{1}{s^2}\right] * L^{-1}\left[\frac{1}{s-1}\right] = \int_0^t (t-\tau)e^\tau \, d\tau.$$

By integrating by parts, we obtain

$$L^{-1}\left[\frac{1}{s^2(s-1)}\right] = \left[te^\tau - (\tau e^\tau - e^\tau)\right]\Big|_0^t = e^t - t - 1. \qquad \square$$

Example 8.9.6   Find $L^{-1}\left[\dfrac{G(s)}{(s-1)^2+1}\right]$.

Solution: Using the convolution theorem, we see that

$$L^{-1}\left[\frac{G(s)}{(s-1)^2+1}\right] = L^{-1}\left[\frac{1}{(s-1)^2+1}\right] * L^{-1}[G(s)]$$
$$= e^t \sin t * g(t).$$

That is,

$$L^{-1}\left[\frac{G(s)}{(s-1)^2+1}\right] = \int_0^t e^{t-\tau} \sin(t-\tau)g(\tau) \, d\tau. \qquad \square$$

Example 8.9.7   Solve the initial-value problem

$$y'' + \omega^2 y = f(t), \qquad y(0) = \alpha, \quad y'(0) = \beta,$$

where α, β, and ω are constants with $\omega \neq 0$, and $f(t)$ is an arbitrary function in $E(0, \infty)$.

Solution: Taking the Laplace transform of the differential equation and imposing the given initial conditions yields

$$s^2 Y(s) - \alpha s - \beta + \omega^2 Y(s) = F(s),$$

where $F(s)$ denotes the Laplace transform of f. Simplifying, we obtain

$$Y(s) = \frac{F(s)}{s^2+\omega^2} + \frac{\alpha s}{s^2+\omega^2} + \frac{\beta}{s^2+\omega^2}.$$

Taking the inverse Laplace transform of both sides of this equation and using the convolution theorem yields

$$y(t) = \frac{1}{\omega}\int_0^t \sin\left[\omega(t-\tau)\right] f(\tau) \, d\tau + \alpha\cos\omega t + \frac{\beta}{\omega}\sin\omega t. \qquad \square$$

Volterra Integral Equations

The applications of the Laplace transform that we have so far considered have been for solving differential equations. We now briefly discuss another type of equation whose solution can often be obtained by using the Laplace transform.

An equation of the form

$$x(t) = f(t) + \int_0^t k(t-\tau)x(\tau) \, d\tau \qquad (8.9.3)$$

is called a **Volterra integral equation**. In this equation, the unknown function is $x(t)$. The functions f and k are specified, and k is called the **kernel** of the equation. For example,

$$x(t) = 2 \sin t + \int_0^t \cos(t - \tau)x(\tau) \, d\tau$$

is a Volterra integral equation. We now show how the convolution theorem for Laplace transforms can be used to determine, up to the evaluation of an inverse transform, the function $x(t)$ that satisfies Equation (8.9.3). The key to solving Equation (8.9.3) is to notice that the integral that appears in this equation is, in fact, a convolution integral. Thus, taking the Laplace transform of both sides of Equation (8.9.3), we obtain

$$X(s) = F(s) + L[k(t) * x(t)].$$

That is, using the convolution theorem,

$$X(s) = F(s) + K(s)X(s).$$

Solving algebraically for $X(s)$ yields

$$X(s) = \frac{F(s)}{1 - K(s)},$$

so that

$$x(t) = L^{-1}\left[\frac{F(s)}{1 - K(s)}\right].$$

This technique can be used to solve a wide variety of Volterra integral equations.

Example 8.9.8   Solve the Volterra integral equation

$$x(t) = 3 \cos t + 5 \int_0^t \sin(t - \tau)x(\tau) \, d\tau.$$

Solution: Taking the Laplace transform of the given integral equation and using the convolution theorem yields

$$X(s) = \frac{3s}{s^2 + 1} + \frac{5}{s^2 + 1}X(s).$$

That is,

$$X(s)\left(\frac{s^2 - 4}{s^2 + 1}\right) = \frac{3s}{s^2 + 1},$$

so that

$$X(s) = \frac{3s}{s^2 - 4}.$$

Decomposing the right-hand side into partial fractions, we obtain

$$X(s) = \frac{3}{2}\left(\frac{1}{s - 2} + \frac{1}{s + 2}\right).$$

Taking the inverse Laplace transform yields

$$x(t) = \frac{3}{2}(e^{2t} + e^{-2t}) = 3 \cosh 2t. \qquad \square$$

Exercises for 8.9

Key Terms

Convolution product, Convolution integral, Convolution Theorem, Volterra integral equation, Kernel of the Volterra integral equation.

Skills

- Be able to compute the convolution product of two functions f and g.

- Be able to prove the basic properties of the convolution product.

- Be able to use the Convolution Theorem to compute the Laplace transform of a convolution product.

- Be able to use the Convolution Theorem to compute the inverse Laplace transform of a product of functions.

- Be able to solve initial-value problems up to the evaluation of a convolution integral.

- Be able to solve Volterra integral equations.

True-False Review

For Questions 1–7, decide if the given statement is **true** or **false**, and give a brief justification for your answer. If true, you can quote a relevant definition or theorem from the text. If false, provide an example, illustration, or brief explanation of why the statement is false.

1. For all continuous functions f and g, $f * g = g * f$.

2. If the functions f and g are continuous and positive on $[0, b]$, then the convolution product $f * g$ is an increasing function of t.

3. If f and g are in $E(0, \infty)$, then $L[fg] = L[f] * L[g]$.

4. Any equation of the form

$$x(t) = f(t) + \int_0^t k(t - \tau)x(\tau)\, d\tau$$

is called a Volterra integral equation.

5. If f, g, and h are continuous functions on $[0, b]$ and if $f * g = f * h$, then $g = h$.

6. The Convolution Theorem states (in part) that the inverse Laplace transform of the product $F(s)G(s)$ is the convolution product $(f * g)(t)$, where f and g are in $E(0, \infty)$.

7. If a is a constant and f and g are continuous on the interval $[0, b]$, then

$$a(f * g) = (af) * g = f * (ag).$$

Problems

For Problems 1–5, determine $f * g$.

1. $f(t) = t, \quad g(t) = 1.$

2. $f(t) = \cos t, \quad g(t) = t.$

3. $f(t) = e^t, \quad g(t) = t.$

4. $f(t) = t^2, \quad g(t) = e^t.$

5. $f(t) = e^t, \quad g(t) = e^t \sin t.$

6. Prove that $f * g = g * f$.

7. Prove that $f * (g * h) = (f * g) * h$.

8. Prove that $f * (g + h) = f * g + f * h$.

For Problems 9–13, determine $L[f * g]$.

9. $f(t) = t, \quad g(t) = \sin t.$

10. $f(t) = e^{2t}, \quad g(t) = 1.$

11. $f(t) = \sin t, \quad g(t) = \cos 2t.$

12. $f(t) = e^t, \quad g(t) = te^{2t}.$

13. $f(t) = t^2, \quad g(t) = e^{2t} \sin 2t.$

For Problems 14–19, determine $L^{-1}[F(s)G(s)]$ in the following two ways:

(a) using the Convolution Theorem,

(b) using partial fractions.

14. $F(s) = \dfrac{1}{s}, \quad G(s) = \dfrac{1}{s - 2}.$

15. $F(s) = \dfrac{1}{s + 1}, \quad G(s) = \dfrac{1}{s}.$

16. $F(s) = \dfrac{s}{s^2 + 4}, \quad G(s) = \dfrac{2}{s}.$

17. $F(s) = \dfrac{1}{s + 2}, \quad G(s) = \dfrac{s + 2}{s^2 + 4s + 13}.$

18. $F(s) = \dfrac{1}{s^2 + 9}, \quad G(s) = \dfrac{2}{s^3}.$

19. $F(s) = \dfrac{1}{s^2}, \quad G(s) = \dfrac{e^{-\pi s}}{s^2 + 1}.$

For Problems 20–24, express $L^{-1}[F(s)G(s)]$ in terms of a convolution integral.

20. $F(s) = \dfrac{4}{s^3}, \quad G(s) = \dfrac{s - 1}{s^2 - 2s + 5}.$

21. $F(s) = \dfrac{s + 1}{s^2 + 2s + 2}, \quad G(s) = \dfrac{1}{(s + 3)^2}.$

22. $F(s) = \dfrac{2}{s^2 + 6s + 10}, \quad G(s) = \dfrac{2}{s - 4}.$

23. $F(s) = \dfrac{s + 4}{s^2 + 8s + 25}, \quad G(s) = \dfrac{se^{-\pi/2}}{s^2 + 16}.$

24. $F(s) = \dfrac{1}{s - 4}, \quad G(s)$ arbitrary.

For Problems 25–31, solve the given initial-value problem up to the evaluation of a convolution integral.

25. $y'' + y = e^{-t}, \quad y(0) = 0, \quad y'(0) = 1.$

26. $y'' - 2y' + 10y = \cos 2t, \quad y(0) = 0, \quad y'(0) = 1.$

27. $y'' + 16y = f(t), \quad y(0) = \alpha, \quad y'(0) = \beta,$ where α and β are constants.

28. $y' - ay = f(t), \quad y(0) = \alpha,$ where a and α are constants.

29. $y'' - a^2 y = f(t), \quad y(0) = \alpha, \quad y'(0) = \beta,$ where $a, \alpha,$ and β are constants and $a \neq 0.$

30. $y'' - (a+b)y' + aby = f(t), \quad y(0) = \alpha, \quad y'(0) = \beta,$ where $a, b, \alpha,$ and β are constants and $a \neq b.$

31. $y'' - 2ay' + (a^2 + b^2)y = f(t), \quad y(0) = \alpha, \quad y'(0) = \beta,$ where $a, b, \alpha,$ and β are constants, and $b \neq 0.$

For Problems 32–37, solve the given Volterra integral equation.

32. $x(t) = e^{-t} + 4\int_0^t (t - \tau)x(\tau)\, d\tau.$

33. $x(t) = 2e^{3t} - \int_0^t e^{2(t-\tau)}x(\tau)\, d\tau.$

34. $x(t) = 4e^t + 3\int_0^t e^{-(t-\tau)}x(\tau)\, d\tau.$

35. $x(t) = 1 + 2\int_0^t \sin(t - \tau)x(\tau)\, d\tau.$

36. $x(t) = e^{2t} + 5\int_0^t \cos[2(t - \tau)]x(\tau)\, d\tau.$

37. $x(t) = 2\left\{1 + \int_0^t \cos[2(t - \tau)]x(\tau)\, d\tau\right\}.$

38. Show that the initial-value problem

$$y'' + y = f(t), \qquad y(0) = 0, \quad y'(0) = 0$$

can be reformulated as the integral equation

$$x(t) = f(t) - \int_0^t (t - \tau)x(\tau)\, d\tau,$$

where $y''(t) = x(t).$

39. Verify that for functions $f, g,$ and $h,$ we have
$f * (g + h) = f * g + f * h.$

8.10 Chapter Review

Laplace transforms are a powerful tool in solving differential equations. In particular, the real power of the Laplace transform comes from the simplification of differential equations with a forcing term that is piecewise continuous or periodic in nature. The Laplace transform of a function f defined on an interval $[0, \infty)$ is the function $F(s)$ defined by

$$F(s) = \int_0^\infty e^{-st} f(t)\, dt, \tag{8.10.1}$$

provided that the improper integral converges. Throughout the chapter, we have developed formulas for the Laplace transforms of a number of basic functions. These results are summarized in the accompanying table.

The Laplace transform satisfies the linearity properties

$$L[f + g] = L[f] + L[g] \qquad \text{and} \qquad L[cf] = cL[f]$$

Table 8.10.1: Summary of Laplace Transforms

Function $f(t)$	**Laplace Transform** $F(s)$
$f(t) = t^n$, n a nonnegative integer	$F(s) = \dfrac{n!}{s^{n+1}}$, $s > 0$.
$f(t) = e^{at}$, a constant	$F(s) = \dfrac{1}{s-a}$, $s > a$.
$f(t) = \sin bt$, b constant	$F(s) = \dfrac{b}{s^2 + b^2}$, $s > 0$.
$f(t) = \cos bt$, b constant	$F(s) = \dfrac{s}{s^2 + b^2}$, $s > 0$.
$f(t) = t^{-1/2}$	$F(s) = (\pi/s)^{1/2}$, $s > 0$.
$f(t) = u_a(t)$ (see Section 8.7)	$F(s) = \dfrac{1}{s} e^{-as}$.
$f(t) = \delta(t - a)$ (see Section 8.8)	$F(s) = e^{-as}$.

Transform of Derivatives (see Section 8.4)

f'	$L[f'] = sL[f] - f(0)$.
f''	$L[f''] = s^2 L[f] - sf(0) - f'(0)$.

Shifting Theorems (see Sections 8.5 and 8.7)

$e^{at} f(t)$	$F(s - a)$
$u_a(t) f(t - a)$	$e^{-as} F(s)$.

for all transformable functions f and g and constants c. Therefore, we can take linear combinations of the functions from the above table and compute their Laplace transforms as well.

In order to transform a first or second-order differential equation with unknown function $y(t)$, we use the formulas given in the accompanying table for $L[f']$ or $L[f'']$, respectively, and obtain the resulting algebraic equation for the Laplace transform $Y(s)$ of $y(t)$. After solving this equation for $Y(s)$, we find the solution $y(t)$ of the initial-value problem by taking the inverse Laplace transform: $y(t) = L^{-1}[Y(s)]$. Note that $y(t) = L^{-1}[Y(s)]$ if and only if $L[y](s) = Y(s)$. Higher-order differential equations can also be handled by this technique, using a generalization of the formulas for Laplace transforms of derivatives in the table. Figure 8.10.1 summarizes this technique.

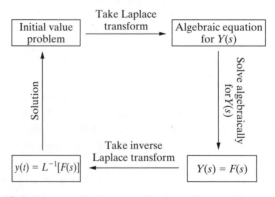

Figure 8.10.1: Using the Laplace transform to solve an initial-value problem.

Additional Problems

For Problems 1–10, use (8.10.1) to determine $L[f]$.

1. $f(t) = 3t - 4$.

2. $f(t) = \sin 2t$.

3. $f(t) = 4t^2$.

4. $f(t) = 5e^{-3t}$.

5. $f(t) = 7te^{-t}$.

6. $f(t) = \sin at \cos bt$, where a, b are positive constants.

7. $f(t) = \sin^2 at$, where a is a positive constant.

8. $f(t) = \begin{cases} 2, & 0 \le t \le 1, \\ t, & t > 1. \end{cases}$

9. $f(t) = \begin{cases} t+1, & 0 \le t \le 3, \\ t^2 - 1, & t < 3. \end{cases}$

10. $f(t) = \begin{cases} 2, & 0 \le t \le 1, \\ 1 - t, & 1 < t \le 2, \\ 0, & t > 2. \end{cases}$

For Problems 11–19, use properties of the Laplace transform and the table of Laplace transforms to determine $L[f]$.

11. $f(t) = 5\cos 2t - 7e^{-t} - 3t^6$.

12. $f(t) = e^{-5t}/\sqrt{t}$.

13. $f(t) = e^{3t} \cos 5t - e^{-t} \sin 2t$.

14. $f(t) = 6t^4 e^{-2t} - 2te^{t+1} + \sqrt{10t}$.

15. $f(t) = e^{-5t}/\sqrt{t}$.

16. $f(t) = 2(t - 5)u_5(t)$.

17. $f(t) = 2 + 2(e^{-t} - 1)u_1(t)$.

18. $f(t) = \int_0^t (t - w) \cos 2w \, dw$.

19. $f(t) = \int_0^t (t - w)^2 e^w \, dw$.

For Problems 20–25, determine a function $f(t)$ that has the given Laplace transform $F(s)$.

20. $F(s) = \dfrac{3}{s^2}$.

21. $F(s) = \dfrac{4s + 5}{s^2 + 9}$.

22. $F(s) = \dfrac{s - 2}{s^2 + 2s + 2}$.

23. $F(s) = \dfrac{2}{s(s^2 + 16)}$.

24. $F(s) = \dfrac{2s + 5}{s(s^2 + 4s + 20)}$.

25. $F(s) = \dfrac{2s + 5}{s(s^2 + 4s + 20)}$.

For Problems 26–28, sketch $f(t)$, express $f(t)$ in terms of $u_a(t)$, and determine $L\{f(t)\}$.

26. $f(t) = \begin{cases} 2, & 0 \le t < 1 \\ 3 - t, & t \ge 1 \end{cases}$.

27. $f(t) = \begin{cases} 1, & 0 \le t < \ln 2 \\ 2e^{-t}, & t \ge \ln 2 \end{cases}$.

28. $f(t) = \begin{cases} t, & 0 \le t \le 1 \\ 1, & 1 < t \le 2 \\ 3 - t, & 2 < t \le 3 \\ 0, & t > 3. \end{cases}$

29. Let $f \in E(0, \infty)$ and let a be a positive real number. Define the function f_a as follows

$$f_a(t) = \begin{cases} f(t - a), & \text{if } t \ge a, \\ 0, & \text{if } 0 \le t < a. \end{cases}$$

Show that $L\{f_a(t)\} = f(s + a)$.

30. Use the Convolution Theorem and the table of Laplace transforms to show that

$$\int_0^x (x - w)^a w^b \, dw = \frac{a! \, b!}{(a + b + 1)!} x^{a+b+1},$$

$a > -1, b > -1, x > 0$.

31. Let $f(x) = xe^{ax}$, where a is a constant.

(a) Show that

$$L\{f'(x)\} = aL\{f(x)\} + \frac{1}{s - a}.$$

(b) Use the result from (a), together with the expression for the Laplace transform of the derivative of a function, to determine $L\{xe^{ax}\}$ without integrating.

(c) Use mathematical induction to establish that

$$L\{x^n e^{ax}\} = \frac{n!}{(s-a)^{n+1}}, \quad s > a, \quad n = 1, 2, \ldots.$$

32. Let $y(t)$ be the solution to the initial-value problem $y' + ay = f(t)$, $y(0) = y_0$, where a and y_0 are constants. Verify that

$$L[y] = \frac{L[f]}{s+a} + \frac{y_0}{s+a},$$

and show that

$$y(t) = y_0 e^{-at} + \int_0^t e^{-a(t-w)} f(w) \, dw.$$

33. Show that the general solution to the initial-value problem

$$y^{(n)} + a_1 y^{(n-1)} + \cdots + a_n y = f(t),$$
$$y(0) = 0, \ y'(0) = 0, \ldots, y^{(n-1)}(0) = 0,$$

is

$$y(t) = \int_0^t K(t-w) f(w) \, dw,$$

for an appropriate function $K(t)$ that should be determined.

For Problems 34–40, use the Laplace transform to solve the given initial-value problem.

34. $y'' - 3y' - 4y = 4e^{-t}$, $\quad y(0) = 1$, $\quad y'(0) = 1$.

35. $y'' - 2y' - 8y = 5$, $\quad y(0) = 1$, $\quad y'(0) = 0$.

36. $y'' + 9y = 8\cos 3t$, $\quad y(0) = 1$, $\quad y'(0) = 0$.

37. $y'' + y = f(t)$, $\quad y(0) = 0$, $\quad y'(0) = 1$, where

$$f(t) = \begin{cases} 1, & 0 \le t < \pi/2, \\ 0, & t \ge \pi/2. \end{cases}$$

38. $y'' + 4y = 4\sin t + 3\delta(t-2)$, $\quad y(0) = 2$, $\quad y'(0) = 1$.

39. $y'' + 2y' + y = \delta(t-4)$, $\quad y(0) = 0$, $\quad y'(0) = 0$.

40. $y'' + 4y' + 4y = \delta(t-4)$, $\quad y(0) = 1$, $\quad y'(0) = 2$.

For Problems 41–44, use the Laplace transform to solve the given system of differential equations subject to the given initial conditions.

41. $\dfrac{dx_1}{dt} = x_1 + 2x_2$, $\quad \dfrac{dx_2}{dt} = 2x_1 + x_2$,

$x_1(0) = 1$, $\quad x_2(0) = 0$.

42. $\dfrac{dx_1}{dt} = 2x_2$, $\quad \dfrac{dx_2}{dt} = -2x_1$,

$x_1(0) = 0$, $\quad x_2(0) = 1$.

43. $\dfrac{dx_1}{dt} = -2x_2$, $\quad \dfrac{dx_2}{dt} = 2x_1 + 4x_2$,

$x_1(0) = 1$, $\quad x_2(0) = 1$.

44. $\dfrac{dx_1}{dt} = 2x_1 + 4x_2 + 16\sin 2t$,

$\dfrac{dx_2}{dt} = -2x_1 - 2x_2 + 16\cos 2t$,

$x_1(0) = 0$, $\quad x_2(0) = 1$.

For Problems 45–48, use the Laplace transform to solve the given integral equation.

45. $x(t) = 2t + \displaystyle\int_0^t \sin(t-\tau) x(\tau) \, d\tau$.

46. $x(t) = 2t^2 + \displaystyle\int_0^t (t-\tau) x(\tau) \, d\tau$.

47. $x(t) = 2t^2 + \displaystyle\int_0^t \sin[2(t-\tau)] x(\tau) \, d\tau$.

48. $x(t) = 3 + 4 \displaystyle\int_0^t x(t-\tau) \cos \tau \, d\tau$.

Series Solutions to Linear Differential Equations

It is clear that the chief end of mathematical study must be to make the students think. — John Wesley Young

So far, the techniques that we have developed for solving differential equations have involved determining a closed-form solution for a given equation (or system) in terms of familiar elementary functions. Essentially, the only differential equations of order two or more that we can derive such solutions for are as follows:

1. Constant-coefficient equations.

2. Cauchy-Euler equations.

For example, we cannot yet determine the solution to the seemingly simple differential equation

$$y'' + e^x y = 0.$$

In this chapter, we consider the possibility of representing solutions to linear differential equations in the form of some type of infinite series. We begin in Section 9.2 with the simplest case, namely, differential equations whose solutions can be represented as a convergent power series,

$$y(x) = \sum_{n=0}^{\infty} a_n x^n,$$

where a_n are constants. This can be considered as a generalization of the method of undetermined coefficients to the case when we have an infinite number of constants. We will determine the appropriate values of these constants by substitution into the differential equation.

Not all differential equations have solutions that can be represented by a convergent power series. We will find that the next-simplest type of series solution that is applicable to a broad class of linear differential equations is one of the form

$$y(x) = x^r \sum_{n=0}^{\infty} a_n x^n,$$

called a Frobenius series. Here, in addition to the coefficients a_n, we must also determine the value of the constant r (which in general will *not* be a positive integer). The analysis of this problem is quite involved, and the computations can be extremely tedious. However, the technique is an important and useful addition to the applied mathematician's tools for solving differential equations.

Before beginning the development of the theory, we note that for simplicity we will restrict our attention in this chapter to second-order linear homogeneous differential equations whose standard form is

$$y'' + p(x)y' + q(x)y = 0,$$

where p and q are functions that are specified on some interval I. The techniques can be extended easily to higher order, and also to systems of linear differential equations.

9.1   A Review of Power Series

We begin with a very brief review of the main facts about power series, which should be familiar from a previous calculus course. They will be required throughout the remainder of the chapter.

DEFINITION   9.1.1

An infinite series of the form

$$\sum_{n=0}^{\infty} a_n (x - x_0)^n, \tag{9.1.1}$$

where a_n and x_0 are constants, is called a **power series centered at** $x = x_0$.

The substitution $u = x - x_0$ has the effect of transforming (9.1.1) to

$$\sum_{n=0}^{\infty} a_n u^n,$$

so that we can, without loss of generality, restrict attention to power series of the form

$$\sum_{n=0}^{\infty} a_n x^n, \tag{9.1.2}$$

whose center is $x = 0$. The series (9.1.2) is said to **converge** at $x = x_1$ if

$$\lim_{k \to \infty} \sum_{n=0}^{k} a_n x_1^n$$

exists and is finite. The set of all x for which (9.1.2) converges is called the **interval of convergence**.

Theorem 9.1.2   **(Basic Convergence Theorem)**

For the power series (9.1.2), precisely one of the following is true:

1. $\displaystyle\sum_{n=0}^{\infty} a_n x^n$ converges only at $x = 0$.

2. $\displaystyle\sum_{n=0}^{\infty} a_n x^n$ converges for all real x.

3. There is a positive number R such that $\displaystyle\sum_{n=0}^{\infty} a_n x^n$ converges (absolutely) for $|x| < R$ and diverges for $|x| > R$.

Remark   The number R occurring in possibility 3 is called the **radius of convergence**. (See Figure 9.1.1.) The convergence or divergence of the series at the endpoints $x = \pm R$ must be treated separately. In possibility 2, we define the radius of convergence to be $R = \infty$.

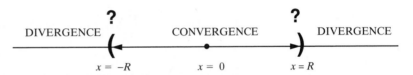

DIVERGENCE   CONVERGENCE   DIVERGENCE

$x = -R$ $x = 0$ $x = R$

Figure 9.1.1: The radius of convergence of a power series.

Ratio Test

For the power series $\displaystyle\sum_{n=0}^{\infty} a_n x^n$, if

$$\lim_{n \to \infty} \left| \frac{a_{n+1}}{a_n} \right| = L,$$

then the radius of convergence of the power series is $R = 1/L$. If $L = 0$, the series converges for all x, whereas if $L = \infty$, series converges only at $x = 0$.

Example 9.1.3   Determine the radius of convergence of $\displaystyle\sum_{n=0}^{\infty} \frac{n^2 x^n}{3^n}$.

Solution: In this case, we have

$$\lim_{n \to \infty} \left| \frac{a_{n+1}}{a_n} \right| = \lim_{n \to \infty} \frac{(n+1)^2}{3^{n+1}} \cdot \frac{3^n}{n^2} = \lim_{n \to \infty} \frac{(n+1)^2}{3n^2} = \frac{1}{3}.$$

Thus, $L = \frac{1}{3}$, so that the radius of convergence is $R = 3$. It is easy to see that the series diverges at the endpoints[1] $x = \pm 3$, so that the interval of convergence is $(-3, 3)$. $\square$

---

[1] Recall that a necessary (but not sufficient) condition for the convergence of the infinite series $\displaystyle\sum_{n=0}^{\infty} a_n x^n$ is

that $\displaystyle\lim_{n \to \infty} a_n = 0$.

The Algebra of Power Series

Two power series $\sum_{n=0}^{\infty} a_n x^n$ and $\sum_{n=0}^{\infty} b_n x^n$ are equal if and only if each corresponding coefficient is equal; that is, $a_n = b_n$ for all n. In particular,

$$\sum_{n=0}^{\infty} a_n x^n = 0 \qquad \text{if and only if } a_n = 0 \text{ for every } n.$$

We will use this result repeatedly throughout the chapter.

Now let $\sum_{n=0}^{\infty} a_n x^n$ and $\sum_{n=0}^{\infty} b_n x^n$ be power series with radii of convergence R_1 and R_2, respectively, and let $R = \min\{R_1, R_2\}$. For $|x| < R$, define the functions f and g by

$$f(x) = \sum_{n=0}^{\infty} a_n x^n, \qquad g(x) = \sum_{n=0}^{\infty} b_n x^n.$$

Then, for $|x| < R$,

1. $f(x) + g(x) = \sum_{n=0}^{\infty} (a_n + b_n) x^n$ (addition of power series).

2. $cf(x) = \sum_{n=0}^{\infty} (c a_n) x^n$ (multiplication of a power series by a real number c).

3. $f(x)g(x) = \sum_{n=0}^{\infty} c_n x^n$, where $c_n = \sum_{k=0}^{n} a_{n-k} b_k$ (multiplication of power series).

The coefficients c_n appearing in this formula can be written in the equivalent form

$$c_n = \sum_{k=0}^{n} a_k b_{n-k}.$$

Example 9.1.4　Assume that the coefficients in the expansion

$$f(x) = \sum_{n=0}^{\infty} a_n x^n$$

satisfy

$$\sum_{n=1}^{\infty} n a_n x^n - \sum_{n=0}^{\infty} a_n x^{n+1} = 0. \qquad (9.1.3)$$

Express all a_n in terms of a_0.

Solution:　We replace n by $n - 1$ in the second summation in (9.1.3) (and alter the range of n appropriately) to obtain a common power of x^n in both sums. The result is

$$\sum_{n=1}^{\infty} n a_n x^n - \sum_{n=1}^{\infty} a_{n-1} x^n = 0,$$

which can be written as

$$\sum_{n=1}^{\infty}(na_n - a_{n-1})x^n = 0.$$

It follows that the a_n must satisfy the recurrence relation[2]

$$na_n - a_{n-1} = 0, \qquad n = 1, 2, 3, \ldots;$$

that is,

$$a_n = \frac{1}{n}a_{n-1}, \quad n = 1, 2, 3, \ldots.$$

Substituting for successive values of n, we obtain

$n = 1$: $a_1 = a_0$,

$n = 2$: $a_2 = \dfrac{1}{2}a_1 = \dfrac{1}{2}a_0$,

$n = 3$: $a_3 = \dfrac{1}{3}a_2 = \dfrac{1}{3 \cdot 2}a_0$,

$n = 4$: $a_4 = \dfrac{1}{4}a_3 = \dfrac{1}{4 \cdot 3 \cdot 2}a_0$.

Continuing in this manner, we see that

$$a_n = \frac{1}{n!}a_0.$$

Consequently, we can write[3]

$$f(x) = a_0 \sum_{n=0}^{\infty}\frac{1}{n!}x^n. \qquad \square$$

Differentiation of Power Series

Suppose that $\sum_{n=0}^{\infty} a_n x^n$ has radius of convergence R, and let

$$f(x) = \sum_{n=0}^{\infty} a_n x^n, \qquad |x| < R.$$

Then $f(x)$ can be differentiated an arbitrary number of times on the interval $|x| < R$. Furthermore, the derivatives can be obtained by termwise differentiation. Thus,

$$f'(x) = \sum_{n=0}^{\infty} na_n x^{n-1} = \sum_{n=1}^{\infty} na_n x^{n-1},$$

$$f''(x) = \sum_{n=1}^{\infty} n(n-1)a_n x^n = \sum_{n=2}^{\infty} n(n-1)a_n x^n,$$

and so on for higher-order derivatives. Similar statements can be made for integration of power series, but these will not be needed in this text.

---

[2] A recurrence relation is an equation that expresses an element a_n in a sequence of numbers in terms of the previous term(s) $a_{n-1}, a_{n-2}, \ldots,$ of the sequence.

[3] This power series is the Maclaurin expansion of e^x, so that we can write $f(x) = a_0 e^x$.

Analytic Functions and Taylor Series

We now introduce one of the main definitions of the section.

DEFINITION 9.1.5

A function is said to be **analytic at** $x = x_0$ if it can be represented by a convergent power series centered at $x = x_0$ with nonzero radius of convergence.

In a previous calculus course the reader should have seen that if a function is analytic at $x = x_0$, then the power series representation of that function is unique and is given by

$$f(x) = \sum_{n=0}^{\infty} \frac{f^{(n)}(x_0)}{n!} (x - x_0)^n. \tag{9.1.4}$$

This is the **Taylor series** expansion of $f(x)$ about $x = x_0$. If $x_0 = 0$, then (9.1.4) reduces to

$$f(x) = \sum_{n=0}^{\infty} \frac{f^{(n)}(0)}{n!} x^n,$$

which is called the **Maclaurin series** expansion of $f(x)$. Many of the familiar elementary functions are analytic at all points. In particular, the Maclaurin expansions of e^x, $\sin x$, and $\cos x$ are, respectively,

$$e^x = 1 + x + \frac{1}{2!}x^2 + \frac{1}{3!}x^3 + \cdots + \frac{1}{n!}x^n + \cdots = \sum_{n=0}^{\infty} \frac{1}{n!}x^n,$$

$$\sin x = x - \frac{1}{3!}x^3 + \frac{1}{5!}x^5 - \cdots + \frac{(-1)^n}{(2n+1)!}x^{2n+1} + \cdots = \sum_{n=0}^{\infty} \frac{(-1)^n}{(2n+1)!}x^{2n+1},$$

$$\cos x = 1 - \frac{1}{2!}x^2 + \frac{1}{4!}x^4 - \cdots + \frac{(-1)^n}{(2n)!}x^{2n} + \cdots = \sum_{n=0}^{\infty} \frac{(-1)^n}{(2n)!}x^{2n}.$$

We can determine many other analytic functions using the next theorem, whose proof is omitted.

Theorem 9.1.6   If $f(x)$ and $g(x)$ are analytic at $x = x_0$, then so also are $f(x) \pm g(x)$, $f(x)g(x)$, and $f(x)/g(x)$ (provided that $g(x_0) \neq 0$).

Of particular importance to us throughout this chapter will be polynomial functions—that is, functions of the form

$$p(x) = a_0 + a_1 x + a_2 x^2 + \cdots + a_n x^n, \tag{9.1.5}$$

where $a_0, a_1, \ldots, a_n$ are real numbers. Such a function is analytic at all points. In particular, (9.1.5) can be considered as the Maclaurin series expansion of p about $x = 0$. Since the series has only a finite number of terms, it converges for all real x. Now suppose that $p(x)$ and $q(x)$ are polynomials, and hence are analytic at all points. According to Theorem 9.1.6, the rational function defined by $r(x) = p(x)/q(x)$ is analytic at all points $x = x_0$ such that $q(x_0) \neq 0$. However, Theorem 9.1.6 gives no indication of the radius of convergence of the series representation of $r(x)$. The next theorem deals with this issue.

Theorem 9.1.7   If $p(x)$ and $q(x)$ are polynomials and $q(x_0) \neq 0$, then the power series representation of p/q has radius of convergence R, where R is the distance, in the complex plane, from x_0 to the nearest root of q.

Once more, we omit the proof of this result; it can be found in texts on advanced calculus. If $z = a + ib$ is a root of q, then the distance from the center, $x = x_0$, of the power series to z is (see Figure 9.1.2)

$$|z - x_0| = \sqrt{(a - x_0)^2 + b^2}.$$

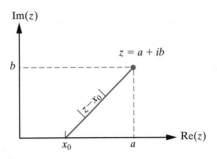

Figure 9.1.2: Determining the radius of convergence of the power series representation of a rational function centered at $x = x_0$.

Example 9.1.8   Determine the radius of convergence of the power series representation of the function

$$f(x) = \frac{1 - x}{x^2 - 4}$$

centered at **(a)** $x = 0$, **(b)** $x = 1$.

Solution: Taking $p(x) = 1 - x$ and $q(x) = x^2 - 4$, we have

$$f(x) = \frac{p(x)}{q(x)},$$

and the roots of q are $x = \pm 2$.

(a) In this case, the center of the power series is $x = 0$, so that the distance to the nearest root of q is 2. (See Figure 9.1.3.) Consequently, the radius of convergence of the power series representation of f centered at $x = 0$ is $R = 2$.

(b) If the center of the power series is $x = 1$, then the nearest root of q is at $x = 2$ (see Figure 9.1.3), hence the radius of convergence of the power series representation of f is $R = 1$.   □

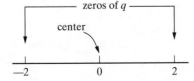 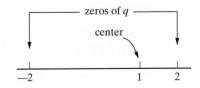

Figure 9.1.3: Determining the radius of convergence of the power series representation of the function given in Example 9.1.8.

Example 9.1.9 Determine the radius of convergence of the power series expansion of

$$f(x) = \frac{1 - x}{(x^2 + 2x + 2)(x - 2)}$$

centered at $x = 0$.

Solution: We take $p(x) = 1 - x$ and $q(x) = (x^2 + 2x + 2)(x - 2)$. According to Theorem 9.1.7, the radius of convergence of the required power series will be given by the distance from $x = 0$ to the nearest root of q. It is easily seen that the roots of q are $x_1 = -1 + i$, $x_2 = -1 - i$, and $x_3 = 2$. The corresponding distances from $x = 0$ are $d_1 = \sqrt{2}$, $d_2 = \sqrt{2}$, and $d_3 = 2$, so the radius of convergence is $R = \sqrt{2}$. (See Figure 9.1.4.)   □

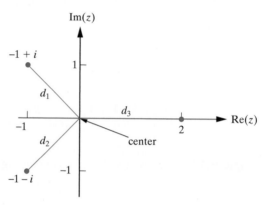

Figure 9.1.4: Determination of the radius of convergence of the power series representation of the rational function given in Example 9.1.9.

Exercises for 9.1

Key Terms

Power series, Converge, Interval of convergence, Radius of convergence, Ratio test, Analytic function, Analytic at $x = x_0$, Taylor series, Maclaurin series.

Skills

- Be able to determine the radius and interval of convergence of a power series, using, for example, the ratio test or Theorem 9.1.2.

- Be able to determine all points at which a function is analytic.

- Be familiar with the basic algebra and calculus of power series.

- Know the relationship between the roots of a function $g(x)$ and the points of analyticity of a function of the form $f(x)/g(x)$.

True-False Review

For Questions 1–10, decide if the given statement is **true** or **false**, and give a brief justification for your answer. If true, you can quote a relevant definition or theorem from the text. If false, provide an example, illustration, or brief explanation of why the statement is false.

1. The radius of convergence of the power series representation of a function $f(x)$ depends on the point x_0 about which the power series is centered.

2. If $\displaystyle\sum_{n=0}^{\infty} a_n x^n$ and $\displaystyle\sum_{n=0}^{\infty} b_n x^n$ both converge at $x = x_1$, then so does $\displaystyle\sum_{n=0}^{\infty} (a_n + b_n)x^n$.

3. If $\displaystyle\sum_{n=0}^{\infty} a_n x^n$ and $\displaystyle\sum_{n=0}^{\infty} b_n x^n$ both fail to converge at $x = x_1$, then so does $\displaystyle\sum_{n=0}^{\infty} (a_n + b_n) x^n$.

4. If the radius of convergence of the power series $\displaystyle\sum_{n=0}^{\infty} a_n x^n$ is R, then the power series converges at $x = \pm R$, as well as $|x| < R$.

5. The product of two analytic functions at $x = x_0$ remains analytic at $x = x_0$.

6. Every infinitely differentiable function f can be represented by the formula

$$f(x) = \sum_{n=0}^{\infty} \frac{f^{(n)}(x_0)}{n!}(x - x_0)^n,$$

which holds for all real values of x in the domain of f.

7. Every polynomial has a power series representation about any point x_0 with an infinite radius of convergence.

8. If f (resp. g) has a power series representation centered at $x = 0$ of radius R_1 (resp. R_2), then the product fg has a power series representation centered at $x = 0$ of radius R, where $R = \min\{R_1, R_2\}$.

9. The coefficient of x^n in the product $\displaystyle\sum_{n=0}^{\infty} a_n x^n \sum_{n=0}^{\infty} b_n x^n$ is $a_n b_n$.

10. A Maclaurin series is a Taylor series which is centered at $x = 0$.

Problems

For Problems 1–5, determine the radius of convergence of the given power series.

1. $\displaystyle\sum_{n=0}^{\infty} \frac{x^n}{2^{2n}}.$

2. $\displaystyle\sum_{n=0}^{\infty} \frac{x^n}{n^2}.$

3. $\displaystyle\sum_{n=0}^{\infty} \frac{2^n x^n}{n}.$

4. $\displaystyle\sum_{n=0}^{\infty} n! x^n.$

5. $\displaystyle\sum_{n=0}^{\infty} \frac{5^n x^n}{n!}.$

For Problems 6–10, determine the radius of convergence of the power series representation of the given function with center x_0.

6. $f(x) = \dfrac{x^2 - 1}{x + 2}, \quad x_0 = 0.$

7. $f(x) = \dfrac{x}{x^2 + 1}, \quad x_0 = 0.$

8. $f(x) = \dfrac{2x}{x^2 + 16}, \quad x_0 = 1.$

9. $f(x) = \dfrac{x^2 - 3}{x^2 - 2x + 5}, \quad x_0 = 0.$

10. $f(x) = \dfrac{x}{(x^2 + 4x + 13)(x - 3)}, \quad x_0 = -1.$

11. **(a)** Determine all values of x at which the function

$$f(x) = \frac{1}{x^2 - 1} \qquad (9.1.6)$$

is analytic.

(b) Determine the radius of convergence of a power series representation of the function (9.1.6) centered at $x = x_0$. (You will need to consider the cases $-1 < x_0 < 1$ and $|x_0| > 1$ separately.)

12. By redefining the ranges of the summations appearing on the left-hand side, show that

$$\sum_{n=2}^{\infty} n(n-1)a_{n-1} x^{n-2} + \sum_{n=1}^{\infty} n a_n x^{n-1}$$

$$= \sum_{n=0}^{\infty} (n+1)(n+3) a_{n+1} x^n.$$

13. If $f(x) = \displaystyle\sum_{n=0}^{\infty} a_n x^n$, where the coefficients in the expansion satisfy

$$\sum_{n=0}^{\infty} n(n+2) a_n x^n + \sum_{n=1}^{\infty} (n-3) a_{n-1} x^n = 0,$$

determine $f(x)$.

14. Suppose it is known that the coefficients in the expansion

$$f(x) = \sum_{n=0}^{\infty} a_n x^n$$

satisfy

$$\sum_{n=0}^{\infty} (n+2)a_{n+1}x^n - \sum_{n=0}^{\infty} a_n x^n = 0.$$

Show that

$$f(x) = \frac{a_0}{x} \sum_{n=0}^{\infty} \frac{1}{(n+1)!} x^{n+1},$$

and express this in terms of familiar elementary functions.

15. If

$$\sum_{n=1}^{\infty} (n+1)(n+2)a_{n+1}x^n - \sum_{n=1}^{\infty} na_{n-1}x^n = 0,$$

show that for $k = 1, 2, 3, \ldots$, we have

$$a_{2k} = \frac{1 \cdot 3 \cdot 5 \cdots (2k-1)}{(2k+1)!}a_0, \quad a_{2k+1} = \frac{2^{k+1}k!}{(2k+2)!}a_1.$$

9.2   Series Solutions about an Ordinary Point

We now consider the second-order linear homogeneous differential equation written in standard form:

$$y'' + p(x)y' + q(x)y = 0.$$

Our aim is to determine a series representation of the general solution to this differential equation centered at $x = x_0$. We will see that the existence and form of the solution are dependent on the behavior of the functions p and q at $x = x_0$.

DEFINITION   9.2.1

The point $x = x_0$ is called an **ordinary point** of the differential equation

$$y'' + p(x)y' + q(x)y = 0 \qquad (9.2.1)$$

if p and q are *both* analytic at $x = x_0$. Any point that is not an ordinary point of (9.2.1) is called a **singular point** of the differential equation.

Example 9.2.2   The differential equation

$$y'' + \frac{1}{x^2 - 4}y' + \frac{1}{x+1}y = 0$$

has

$$p(x) = \frac{1}{x^2 - 4} \qquad \text{and} \qquad q(x) = \frac{1}{x+1}.$$

We see by inspection that the only points at which p fails to be analytic are $x = \pm 2$, whereas q is analytic at all points except $x = -1$. Consequently, the only singular points of the differential equation are $x = -1, \pm 2$. All other points are ordinary points.   □

In this section we restrict our attention to ordinary points. Since the functions p and q are analytic at an ordinary point, we might suspect that any solution to (9.2.1) valid at $x = x_0$ is also analytic there and hence can be represented as a convergent power series

$$y(x) = \sum_{n=0}^{\infty} a_n (x - x_0)^n,$$

for appropriate constants a_n. This is indeed the case, but before stating the general result, we consider a familiar example.

Example 9.2.3 | Determine two linearly independent power series solutions to the differential equation

$$y'' + y = 0 \qquad (9.2.2)$$

centered at $x = 0$. Identify the solutions in terms of familiar elementary functions.

Solution: Since $x = 0$ is an ordinary point of the differential equation, we try for a power series solution of the form

$$y(x) = \sum_{n=0}^{\infty} a_n x^n. \qquad (9.2.3)$$

We proceed in a similar manner to the method of undetermined coefficients by substituting (9.2.3) into (9.2.2) and determining the values of the a_n such that (9.2.3) is indeed a solution. Differentiating (9.2.3) twice with respect to x yields

$$y'(x) = \sum_{n=1}^{\infty} n a_n x^{n-1}, \qquad y''(x) = \sum_{n=2}^{\infty} n(n-1) a_n x^{n-2},$$

where we have shifted the starting point on the summations without loss of generality. Substituting into (9.2.2), it follows that (9.2.3) does define a solution, provided that

$$\sum_{n=2}^{\infty} n(n-1) a_n x^{n-2} + \sum_{n=0}^{\infty} a_n x^n = 0.$$

If we replace n by $k + 2$ in the first summation, and replace n by k in the second summation, the result is

$$\sum_{k=0}^{\infty} (k+2)(k+1) a_{k+2} x^k + \sum_{k=0}^{\infty} a_k x^k = 0.$$

Combining the summations yields

$$\sum_{k=0}^{\infty} \left[(k+2)(k+1) a_{k+2} + a_k \right] x^k = 0.$$

This implies that the coefficients of x^k must vanish for $k = 0, 1, 2, \ldots$. Consequently, we obtain the recurrence relation

$$(k+2)(k+1) a_{k+2} + a_k = 0, \qquad k = 0, 1, 2, \ldots.$$

Since $(k+2)(k+1)$ is never zero, we can write this recurrence relation in the equivalent form

$$a_{k+2} = -\frac{1}{(k+2)(k+1)} a_k, \qquad k = 0, 1, 2, \ldots. \qquad (9.2.4)$$

We now use this relation to determine the appropriate values of the coefficients. It is convenient to consider separately the two cases (1) k is even and (2) k is odd.

Case 1: k is even.   Substituting successively into (9.2.4), we obtain the coefficients as follows:

When $k = 0$, we have

$$a_2 = -\tfrac{1}{2}a_0.$$

When $k = 2$, we have

$$a_4 = -\frac{1}{4 \cdot 3}a_2 = \frac{1}{4 \cdot 3 \cdot 2}a_0.$$

That is,

$$a_4 = \frac{1}{4!}a_0.$$

When $k = 4$, we have

$$a_6 = -\frac{1}{6 \cdot 5}a_4,$$

so that

$$a_6 = -\frac{1}{6!}a_0.$$

Continuing in this manner, we soon recognize the pattern that is emerging—namely,

$$a_{2n} = \frac{(-1)^n}{(2n)!}a_0. \tag{9.2.5}$$

Thus, all of the even coefficients are determined in terms of a_0, but a_0 itself is arbitrary.

Case 2: k is odd.   Now consider the recurrence relation (9.2.4) when k is an odd positive integer:

When $k = 1$, we have

$$a_3 = -\frac{1}{2 \cdot 3}a_1 = -\frac{1}{3!}a_1.$$

When $k = 3$, we have

$$a_5 = -\frac{1}{4 \cdot 5}a_3 = \frac{1}{5!}a_1.$$

When $k = 5$, we have

$$a_7 = -\frac{1}{6 \cdot 7}a_5 = -\frac{1}{7!}a_1.$$

Continuing in this manner, we see that the general odd coefficient is

$$a_{2n+1} = \frac{(-1)^n}{(2n+1)!}a_1. \tag{9.2.6}$$

Thus, using Equations (9.2.5) and (9.2.6), we have shown that for all values of a_0, a_1, a solution to the given differential equation is

$$y(x) = a_0 \left(1 - \frac{1}{2!}x^2 + \frac{1}{4!}x^4 - \cdots \right) + a_1 \left(x - \frac{1}{3!}x^3 + \frac{1}{5!}x^5 - \cdots \right).$$

That is,

$$y(x) = a_0 \sum_{n=0}^{\infty} \frac{(-1)^{2n}}{(2n)!} x^{2n} + a_1 \sum_{n=0}^{\infty} \frac{(-1)^n}{(2n+1)!} x^{2n+1}. \qquad (9.2.7)$$

Setting $a_0 = 1$ and $a_1 = 0$ yields the solution

$$y_1(x) = \sum_{n=0}^{\infty} \frac{(-1)^{2n}}{(2n)!} x^{2n},$$

whereas setting $a_0 = 0$ and $a_1 = 1$ yields the solution

$$y_2(x) = \sum_{n=0}^{\infty} \frac{(-1)^n}{(2n+1)!} x^{2n+1}.$$

Applying the ratio test, it is straightforward to show that both of the foregoing series converge for all real x. Finally, since y_1 and y_2 are not proportional, they are linearly independent on any interval. It follows that (9.2.7) is the general solution of the given differential equation. Indeed, the power series representing y_1 is just the Maclaurin series expansion of $\cos x$, whereas the series defining y_2 is the Maclaurin series expansion of $\sin x$. Thus, we can write (9.2.7) in the more familiar form

$$y(x) = a_0 \cos x + a_1 \sin x. \qquad \square$$

The solution of the previous example consisted of the following four steps.

1. Assume that a power series solution of the form $y(x) = \sum_{n=0}^{\infty} a_n x^n$ exists.

2. Determine the values of the coefficients, a_n, such that y is a formal solution of the differential equation. This led to two distinct solutions, one determined in terms of the constant a_0, and the other in terms of the constant a_1.

3. Use the ratio test to determine the radius of convergence of the solutions and hence the interval over which the solutions are valid.

4. Check that the solutions are linearly independent on the interval of existence.

The next theorem justifies the preceding steps and shows that the technique can be applied about any ordinary point of a differential equation.

Theorem 9.2.4

Let p and q be analytic at $x = x_0$, and suppose that their power series expansions are valid for $|x - x_0| < R$. Then the general solution to the differential equation

$$y'' + p(x)y' + q(x)y = 0 \qquad (9.2.8)$$

can be represented as a power series centered at $x = x_0$, with radius of convergence *at least* R. The coefficients in this series solution can be determined in terms of a_0 and a_1 by directly substituting

$$y(x) = \sum_{n=0}^{\infty} a_n (x - x_0)^n$$

into (9.2.8). The resulting solution is of the form

$$y(x) = a_0 y_1(x) + a_1 y_2(x),$$

where y_1 and y_2 are linearly independent solutions to (9.2.8) on the interval of existence. If the initial conditions $y(x_0) = \alpha$, $y'(x_0) = \beta$ are imposed, then $a_0 = \alpha$, $a_1 = \beta$.

Idea Behind Proof We outline the steps required to prove this theorem but do not give details. The first step is to expand p and q in a power series centered at $x = x_0$. Then we assume a solution exists of the form

$$y(x) = \sum_{n=0}^{\infty} a_n(x - x_0)^n$$

and substitute this into the differential equation. Upon collecting the coefficients of like powers of $x - x_0$, we obtain a recurrence relation, and it can be shown that this relation determines all of the coefficients in terms of a_0 and a_1. These steps are computationally tedious, but quite straightforward. The hard part is to show that the power series solution that has been obtained has a radius of convergence at least equal to R. This requires some ideas from advanced calculus. A power series solution having been determined, y_1 and y_2 arise as the special cases $a_0 = 1, a_1 = 0$ and $a_0 = 0, a_1 = 1$, respectively. The Wronskian of these functions satisfies $W[y_1, y_2](x_0) = 1$, so that they are linearly independent on their interval of existence. Finally, it is easy to show that $y(x_0) = a_0$ and that $y'(x_0) = a_1$.   ∎

We now illustrate the use of the preceding theorem with some examples.

Example 9.2.5   Show that

$$(1 + x^2)y'' + 3xy' + y = 0 \qquad (9.2.9)$$

has two linearly independent series solutions centered at $x = 0$, and determine a lower bound on the radius of convergence of these solutions.

Solution: We first rewrite (9.2.9) in the standard form

$$y'' + \frac{3x}{1 + x^2}y' + \frac{1}{1 + x^2}y = 0,$$

from which we conclude that $x = 0$ is an ordinary point; hence (9.2.9) does indeed have two linearly independent series solutions centered at $x = 0$. In this case,

$$p(x) = \frac{3x}{1 + x^2} \qquad \text{and} \qquad q(x) = \frac{1}{1 + x^2}.$$

According to Theorem 9.2.4, the radius of convergence of the power series solutions will be *at least* equal to the smaller of the radii of convergence of the power series representations of p and q. Using Theorem 9.1.7, we see directly that the series expansions of both p and q about $x = 0$ have radius of convergence $R = 1$, so that a lower bound on the radius of convergence of the power series solutions to (9.2.9) is also $R = 1$.   □

Example 9.2.6   Determine two linearly independent series solutions in powers of x to

$$y'' - 2xy' - 4y = 0, \qquad (9.2.10)$$

and find the radius of convergence of these solutions.

Solution: The point $x = 0$ is an ordinary point of the differential equation, and therefore Theorem 9.2.4 can be applied with $x_0 = 0$. In this case, we have

$$p(x) = -2x, \qquad q(x) = -4.$$

Since these are both polynomials, their power series expansions about $x = 0$ are valid for all x, and hence, from Theorem 9.2.4, the general solution to (9.2.10) can be expressed in the form

$$y(x) = \sum_{n=0}^{\infty} a_n x^n, \tag{9.2.11}$$

and this power series solution will converge for all real x. Differentiating (9.2.11), we obtain

$$y'(x) = \sum_{n=1}^{\infty} n a_n x^{n-1}, \qquad y''(x) = \sum_{n=2}^{\infty} n(n-1) a_n x^{n-2}.$$

Substitution into (9.2.10) yields

$$\sum_{n=2}^{\infty} n(n-1) a_n x^{n-2} - 2 \sum_{n=1}^{\infty} n a_n x^n - 4 \sum_{n=0}^{\infty} a_n x^n = 0.$$

We now redefine the ranges in the summation in order to obtain a common x^k in all terms. This is accomplished by replacing n with $k + 2$ in the first summation, and, for consistency in notation, we replace n with k in the other summations. The result is

$$\sum_{k=0}^{\infty} \left[(k+2)(k+1) a_{k+2} - 2k a_k - 4 a_k \right] x^k = 0.$$

This equation requires that the coefficient of x^k vanish, and hence, we obtain the recurrence relation

$$(k+2)(k+1) a_{k+2} - 2k a_k - 4 a_k = 0, \qquad k = 0, 1, 2, \ldots,$$

which can be written in the equivalent form

$$a_{k+2} = \frac{2(k+2)}{(k+1)(k+2)} a_k, \qquad k = 0, 1, 2, \ldots;$$

that is,

$$a_{k+2} = \frac{2}{k+1} a_k, \qquad k = 0, 1, 2, \ldots. \tag{9.2.12}$$

We see from this relation that, as in Example 9.2.3, all of the even coefficients can be expressed in terms of a_0, whereas all of the odd coefficients can be expressed in terms of a_1. We now determine the exact form of these coefficients.

Case 1: k is even.   From (9.2.12), we have the following:
When $k = 0$,

$$a_2 = 2 a_0.$$

When $k = 2$,

$$a_4 = \frac{2}{3} a_2 = \frac{2^2}{3} a_0.$$

When $k = 4$,

$$a_6 = \frac{2}{5} a_4 = \frac{2^3}{1 \cdot 3 \cdot 5} a_0.$$

The general even term is thus

$$a_{2n} = \frac{2^n}{1 \cdot 3 \cdot 5 \cdot (2n-1)} a_0.$$

Case 2: k is odd.   Substituting successively into (9.2.12) yields the following:
When $k = 1$,

$$a_3 = a_1.$$

When $k = 3$,

$$a_5 = \frac{2}{4} a_3 = \frac{1}{1 \cdot 2} a_1.$$

When $k = 5$,

$$a_7 = \frac{2}{6} a_5 = \frac{1}{1 \cdot 2 \cdot 3} a_1.$$

When $k = 7$,

$$a_9 = \frac{2}{8} a_7 = \frac{1}{1 \cdot 2 \cdot 3 \cdot 4} a_1.$$

The general odd term can therefore be written as

$$a_{2n+1} = \frac{1}{n!} a_1.$$

Substituting back into (9.2.11), we obtain the solution

$$y(x) = a_0 \left[1 + 2x^2 + \frac{2^2}{1 \cdot 3} x^4 + \frac{2^3}{1 \cdot 3 \cdot 5} x^6 + \cdots + \frac{2^n}{1 \cdot 3 \cdot 5 \cdots (2n-1)} x^{2n} + \cdots \right]$$
$$+ a_1 \left(x + x^3 + \frac{1}{2!} x^5 + \frac{1}{3!} x^7 + \cdots + \frac{1}{n!} x^{2n+1} + \cdots \right).$$

That is,

$$y(x) = a_0 \left[1 + \sum_{n=1}^{\infty} \frac{2^n}{1 \cdot 3 \cdot 5 \cdots (2n-1)} x^{2n} \right] + a_1 \sum_{n=0}^{\infty} \frac{1}{n!} x^{2n+1}.$$

Consequently, from Theorem 9.2.4, two linearly independent solutions to (9.2.10) on $(-\infty, \infty)$ are

$$y_1(x) = 1 + \sum_{n=1}^{\infty} \frac{2^n}{1 \cdot 3 \cdot \cdots \cdot (2n-1)} x^{2n}, \qquad y_2(x) = \sum_{n=0}^{\infty} \frac{1}{n!} x^{2n+1}. \qquad \square$$

In Examples 9.2.3 and 9.2.6, we were able to solve the recurrence relation that arose from the power series technique. In general, this will not be possible, hence we must be satisfied with obtaining just a finite number of terms of each power series solution.

Example 9.2.7   Determine the terms up to x^5 in each of the two linearly independent power series solutions to

$$y'' + (2 - 4x^2) y' - 8xy = 0$$

centered at $x = 0$. Also find the radius of convergence of these solutions.

Solution: The functions $p(x) = 2 - 4x^2$ and $q(x) = -8x$ are polynomials, and hence, from Theorem 9.2.4, the power series solutions will converge for all real x. We now determine the solutions. Substituting

$$y(x) = \sum_{n=0}^{\infty} a_n x^n \tag{9.2.13}$$

into the given differential equation yields

$$\sum_{n=2}^{\infty} n(n-1)a_n x^{n-2} + 2\sum_{n=1}^{\infty} na_n x^{n-1} - 4\sum_{n=1}^{\infty} na_n x^{n+1} - 8\sum_{n=0}^{\infty} a_n x^{n+1} = 0.$$

Replacing n by $k+2$ in the first summation, $k+1$ in the second summation, and $k-1$ in the third and fourth summations, we obtain

$$\sum_{k=0}^{\infty}(k+2)(k+1)a_{k+2}x^k + 2\sum_{k=0}^{\infty}(k+1)a_{k+1}x^k - 4\sum_{k=2}^{\infty}(k-1)a_{k-1}x^k - 8\sum_{k=1}^{\infty}a_{k-1}x^k,$$

which equals zero. Separating out the terms corresponding to $k = 0$ and $k = 1$, it follows that this can be written as

$$(2a_2 + 2a_1) + (6a_3 + 4a_2 - 8a_0)x +$$

$$\sum_{k=2}^{\infty}\left\{(k+2)(k+1)a_{k+2} + 2(k+1)a_{k+1} - [4(k-1)+8]a_{k-1}\right\}x^k = 0.$$

Setting the coefficients of all powers x^k to zero yields the following:
For $k = 0$ and $k = 1$, we have

$$2a_2 + 2a_1 = 0 \quad \text{and} \quad 6a_3 + 4a_2 - 8a_0 = 0, \tag{9.2.14}$$

and for $k \geq 2$, we have

$$(k+2)(k+1)a_{k+2} + 2(k+1)a_{k+1} - [4(k-1)+8]a_{k-1} = 0. \tag{9.2.15}$$

It follows from (9.2.14) that

$$a_2 = -a_1, \qquad a_3 = \tfrac{2}{3}(2a_0 + a_1), \tag{9.2.16}$$

and (9.2.15) yields the general recurrence relation

$$a_{k+2} = \frac{4a_{k-1} - 2a_{k+1}}{k+2}, \qquad k = 2, 3, 4, \ldots \tag{9.2.17}$$

In this case, the recurrence relation is quite difficult to solve. However, we were only asked to determine terms up to x^5 in the series solution, and so we proceed to do so. We already have a_2 and a_3 expressed in terms of a_0 and a_1. Setting $k = 2$ in (9.2.17) yields

$$a_4 = \tfrac{1}{4}(4a_1 - 2a_3) = \tfrac{1}{4}\left[4a_1 - \tfrac{4}{3}(2a_0 + a_1)\right],$$

where we have substituted from (9.2.16) for a_3. Simplifying this expression, we obtain

$$a_4 = \tfrac{2}{3}(a_1 - a_0).$$

We still require one more term. Setting $k = 3$ in (9.2.17) yields

$$a_5 = \tfrac{1}{5}(4a_2 - 2a_4) = \tfrac{1}{5}\left[-4a_1 - \tfrac{4}{3}(a_1 - a_0)\right],$$

so that

$$a_5 = \tfrac{4}{15}(a_0 - 4a_1).$$

Substituting for the coefficients a_2, a_3, a_4, and a_5 into (9.2.13), we obtain

$$y(x) = a_0 + a_1 x - a_1 x^2 + \tfrac{2}{3}(2a_0 + a_1)x^3$$
$$+ \tfrac{2}{3}(a_1 - a_0)x^4 + \tfrac{4}{15}(a_0 - 4a_1)x^5 + \cdots .$$

That is,

$$y(x) = a_0\left(1 + \tfrac{4}{3}x^3 - \tfrac{2}{3}x^4 + \tfrac{4}{15}x^5 + \cdots\right)$$
$$+ a_1\left(x - x^2 + \tfrac{2}{3}x^3 + \tfrac{2}{3}x^4 - \tfrac{16}{15}x^5 + \cdots\right).$$

Thus, two linearly independent solutions to the given differential equation on $(-\infty, \infty)$ are

$$y_1(x) = 1 + \tfrac{4}{3}x^3 - \tfrac{2}{3}x^4 + \tfrac{4}{15}x^5 + \cdots$$

and

$$y_2(x) = x - x^2 + \tfrac{2}{3}x^3 + \tfrac{2}{3}x^4 - \tfrac{16}{15}x^5 + \cdots . \qquad \square$$

Exercises for 9.2

Key Terms

Ordinary point, Singular point.

Skills

- Be able to decide if a given point $x = x_0$ is an ordinary point or a singular point of a differential equation of the form (9.2.1).

- Be able to determine two linearly independent power series solutions to a differential equation of the form (9.2.1) centered at an ordinary point $x = x_0$.

- In cases where the recurrence relation for the coefficients of a power series solution to a differential equation cannot be explicitly solved, be able to obtain a specified finite number of terms of two linearly independent power series solutions.

- Be able to determine a lower bound on the radius of convergence of a power series solution to a differential equation.

- Be able to determine the general solution to a differential equation of the form (9.2.1) as a linear combination of two linearly independent power series solutions, and be able to find its radius of convergence.

- Be able to solve initial-value problems via the technique of power series solutions.

True-False Review

For Questions 1–10, decide if the given statement is **true** or **false**, and give a brief justification for your answer. If true, you can quote a relevant definition or theorem from the text. If false, provide an example, illustration, or brief explanation of why the statement is false.

1. If $p(x)$ and $q(x)$ are polynomials, then the differential equation $y'' + p(x)y' + q(x)y = 0$ has no singular points.

2. The radius of convergence of a power series solution to the differential equation

$$y'' + \frac{1}{x}y' + \frac{1}{x+1}y = 0$$

centered at $x = -3$ is at most 2.

3. The radius of convergence of a power series solution to the differential equation

$$y'' + \frac{1}{x^2 - 1}y' + \frac{1}{x+2}y = 0$$

centered at $x = 2$ is at least 2.

4. The coefficients a_0 and a_1 in the power series solution to an initial-value problem $y'' + p(x)y' + q(x)y = 0$, where $p(x)$ and $q(x)$ are analytic at $x = x_0$, are the values $y(x_0)$ and $y'(x_0)$, respectively.

5. A power series solution to $y'' + p(x)y' + q(x) = 0$ centered at an ordinary point $x = x_0$ always exists and has a positive radius of convergence.

6. Two linearly independent power series solutions to the differential equation $y'' + p(x)y' + q(x)y = 0$ cannot contain any of the same powers of x.

7. If $\displaystyle\sum_{n=0}^{\infty} a_n x^n$ is a power series solution to the differential equation $y'' + p(x)y' + q(x)y = 0$, then so is $\displaystyle\sum_{n=0}^{\infty} a_n x^{n+1}$.

8. The recurrence relation $a_k = a_{k-2}$ has a unique solution $a_0, a_1, a_2, a_3, \ldots$, provided that the value of a_0 is specified.

9. The recurrence relation $a_k = 3a_{k-1} - 2a_{k-3}$ has a unique solution $a_0, a_1, a_2, a_3, \ldots$, provided that the values of a_0, a_1, and a_2 are specified.

10. If the recurrence relation arising in the power series method of solution of a differential equation cannot be solved, then the differential equation has no solution.

Problems

For Problems 1–8, determine two linearly independent power series solutions to the given differential equation centered at $x = 0$. Also determine the radius of convergence of the series solutions.

1. $y'' - y = 0$.

2. $y'' - 2xy' - 2y = 0$.

3. $y'' + 2xy' + 4y = 0$.

4. $y'' + xy = 0$.

5. $y'' - x^2 y' - 2xy = 0$.

6. $y'' - x^2 y' - 3xy = 0$.

7. $y'' + xy' + 3y = 0$.

8. $y'' + 2x^2 y' + 2xy = 0$.

For Problems 9–12, determine two linearly independent power series solutions to the given differential equation centered at $x = 0$. Give a lower bound on the radius of convergence of the series solutions obtained.

9. $(1 + x^2)y'' + 4xy' + 2y = 0$.

10. $(x^2 - 3)y'' - 3xy' - 5y = 0$.

11. $(x^2 - 1)y'' - 6xy' + 12y = 0$.

12. $(1 - 4x^2)y'' - 20xy' - 16y = 0$.

For Problems 13–16, determine terms up to and including x^5 in two linearly independent power series solutions of the given differential equation. State the radius of convergence of the series solutions.

13. $y'' + xy' + (2 + x)y = 0$.

14. $y'' + 2y' + 4xy = 0$.

15. $y'' - e^x y = 0$. [**Hint:** $e^x = 1 + x + \frac{1}{2!}x^2 + \frac{1}{3!}x^3 + \cdots$.]

16. $y'' + (\sin x)y' + y = 0$.

17. Consider the differential equation

$$xy'' - (x - 1)y' - xy = 0. \qquad (9.2.18)$$

(a) Is $x = 0$ an ordinary point?

(b) Determine the first three nonzero terms in each of two linearly independent series solutions to Equation (9.2.18) centered at $x = 1$. [**Hint:** Make the change of variables $z = x - 1$ and obtain a series solution in powers of z.] Give a lower bound on the radius of convergence of each of your solutions.

18. Determine a series solution to the initial-value problem

$$(1 + 2x^2)y'' + 7xy' + 2y = 0,$$
$$y(0) = 0, \qquad y'(0) = 1.$$

19. (a) Determine a series solution to the initial-value problem

$$4y'' + xy' + 4y = 0, \quad y(0) = 1, \quad y'(0) = 0. \tag{9.2.19}$$

(b) Find a polynomial that approximates the solution to Equation (9.2.19) with an error less than 10^{-5} on the interval $[-1, 1]$. [**Hint:** The series obtained is a convergent alternating series.]

The power series technique can also be used to solve nonhomogeneous differential equations of the form

$$y'' + p(x)y' + q(x)y = r(x),$$

provided that p, q, and r are analytic at the point about which we are expanding. For Problems 20-21, determine terms up to x^6 in the power series representation of the general solution to the given differential equation centered at $x = 0$. Identify those terms in your solution that correspond to the complementary function and those that correspond to a particular solution to the differential equation.

20. $y'' + 2x^2y' + xy = 2\cos x.$

21. $y'' + xy' - 4y = 6e^x.$

9.3   The Legendre Equation

There are several linear differential equations that arise frequently in applied mathematics and whose solutions can be obtained only by using a power series technique. Among the most important of these are the following:

$$(1 - x^2)y'' - 2xy' + \alpha(\alpha + 1)y = 0, \qquad \text{(Legendre equation)}$$
$$y'' - 2xy' + 2\alpha y = 0, \qquad \text{(Hermite equation)}$$
$$(1 - x^2)y'' - xy' + \alpha^2 y = 0, \qquad \text{(Chebyshev equation)}$$

where α is an arbitrary constant. Since $x = 0$ is an ordinary point of these equations, we can obtain a series solution in powers of x. We will consider only the Legendre equation and leave the analysis of the others for the exercises.

The Legendre equation is

$$(1 - x^2)y'' - 2xy' + \alpha(\alpha + 1)y = 0, \tag{9.3.1}$$

where α is an arbitrary constant. To determine a lower bound on the radius of convergence of the series solution about $x = 0$ to this equation, we divide by $1 - x^2$ to obtain

$$y'' - \frac{2x}{1 - x^2}y' + \frac{\alpha(\alpha + 1)}{1 - x^2}y = 0.$$

Since the power series expansion of $1/(1 - x^2)$ about $x = 0$ is valid for $|x| < 1$, it follows that a lower bound on the radius of convergence of the power series solutions to Equation (9.3.1) about $x = 0$ is 1. We now determine the series solutions. Substituting

$$y(x) = \sum_{n=0}^{\infty} a_n x^n$$

into Equation (9.3.1) yields

$$\sum_{n=2}^{\infty} n(n-1)a_n x^{n-2} - \sum_{n=2}^{\infty} n(n-1)a_n x^n - 2\sum_{n=1}^{\infty} na_n x^n + \sum_{n=0}^{\infty} \alpha(\alpha+1)a_n x^n = 0.$$

That is, upon redefining the ranges of the summations,

$$\sum_{n=0}^{\infty} \left[(n+2)(n+1)a_{n+2} - n(n-1)a_n - 2na_n + \alpha(\alpha+1)a_n \right] x^n = 0.$$

Thus, we obtain the recurrence relation

$$a_{n+2} = \frac{n(n+1) - \alpha(\alpha+1)}{(n+1)(n+2)} a_n, \qquad n = 0, 1, 2, \ldots, \qquad (9.3.2)$$

which can be written as

$$a_{n+2} = -\frac{(\alpha-n)(\alpha+n+1)}{(n+1)(n+2)} a_n, \qquad n = 0, 1, 2, \ldots.$$

Even values of n:

$$n = 0 \implies a_2 = -\frac{\alpha(\alpha+1)}{2} a_0,$$

$$n = 2 \implies a_4 = -\frac{(\alpha-2)(\alpha+3)}{3 \cdot 4} a_2 = \frac{(\alpha-2)\alpha(\alpha+1)(\alpha+3)}{4!} a_0,$$

$$n = 4 \implies a_6 = -\frac{(\alpha-4)(\alpha+5)}{5 \cdot 6} a_4$$

$$= -\frac{(\alpha-4)(\alpha-2)\alpha(\alpha+1)(\alpha+3)(\alpha+5)}{6!} a_0.$$

In general, for $k = 1, 2, 3, \ldots$, we have

$$a_{2k} = (-1)^k \frac{(\alpha-2k+2)(\alpha-2k+4)\cdots(\alpha-2)\alpha(\alpha+1)\cdots(\alpha+2k-1)}{(2k)!} a_0.$$

Odd values of n:

$$n = 1 \implies a_3 = -\frac{(\alpha-1)(\alpha+2)}{2 \cdot 3} a_1,$$

$$n = 3 \implies a_5 = -\frac{(\alpha-3)(\alpha+4)}{4 \cdot 5} a_3 = \frac{(\alpha-3)(\alpha-1)(\alpha+2)(\alpha+4)}{5!} a_1,$$

$$n = 5 \implies a_7 = -\frac{(\alpha-5)(\alpha+6)}{6 \cdot 7} a_5$$

$$= -\frac{(\alpha-5)(\alpha-3)(\alpha-1)(\alpha+2)(\alpha+4)(\alpha+6)}{7!} a_1.$$

In general, for $k = 1, 2, 3, \ldots$, we have

$$a_{2k+1} = (-1)^k \frac{(\alpha-2k+1)\cdots(\alpha-3)(\alpha-1)(\alpha+2)(\alpha+4)\cdots(\alpha+2k)}{(2k+1)!} a_1.$$

Consequently, for $a_0 \neq 0$ and $a_1 \neq 0$, two linearly independent solutions to the Legendre equation are

$$y_1(x) = a_0 \left[1 - \frac{\alpha(\alpha+1)}{2} x^2 + \frac{(\alpha-2)\alpha(\alpha+1)(\alpha+3)}{4!} x^4 \right.$$

$$\left. - \frac{(\alpha-4)(\alpha-2)\alpha(\alpha+1)(\alpha+3)(\alpha+5)}{6!} x^6 + \cdots \right] \qquad (9.3.3)$$

and

$$y_2(x) = a_1 \left[x - \frac{(\alpha-1)(\alpha+2)}{3!} x^3 + \frac{(\alpha-3)(\alpha-1)(\alpha+2)(\alpha+4)}{5!} x^5 + \cdots \right], \qquad (9.3.4)$$

and both of these solutions are valid for $|x| < 1$.

The Legendre Polynomials

Of particular importance in applications is the special case of the Legendre equation

$$(1 - x^2)y'' - 2xy' + N(N + 1)y = 0$$

in which N is a nonnegative *integer*. In this case, the recurrence relation (9.3.2) is

$$a_{n+2} = \frac{n(n + 1) - N(N + 1)}{(n + 1)(n + 2)} a_n, \qquad n = 0, 1, 2, \ldots,$$

which implies that

$$a_{N+2} = a_{N+4} = \cdots = 0.$$

Consequently, one of the solutions (9.3.3) or (9.3.4) to Legendre's equation, depending on whether N is even or odd, is a polynomial of degree N. (Notice that such a solution converges for all x, hence we have a radius of convergence greater than that guaranteed by Theorem 9.2.4.)

DEFINITION   9.3.1

Let N be a nonnegative integer. The **Legendre polynomial of degree** N, denoted $P_N(x)$, is defined to be the polynomial solution to

$$(1 - x^2)y'' - 2xy' + N(N + 1)y = 0,$$

which has been normalized so that $P_N(1) = 1$.

Example 9.3.2   Determine P_0, P_1, and P_2.

Solution: Substituting $\alpha = N$ into (9.3.3) and (9.3.4) yields

$N = 0:\ y_1(x) = a_0$, which implies that $P_0(x) = 1$;
$N = 1:\ y_2(x) = a_1 x$, which implies that $P_1(x) = x$;
$N = 2:\ y_1(x) = a_0(1 - 3x^2)$, and imposing the normalizing condition $y_1(1) = 1$, we
     require that $a_0 = -\frac{1}{2}$, so that $P_2(x) = \frac{1}{2}(3x^2 - 1)$.   □

In general, it is tedious to determine $P_N(x)$ directly from (9.3.3) and (9.3.4), and various other methods have been derived. Among the most useful are the following:

Rodrigues' Formula

$$P_N(x) = \frac{1}{2^N N!} \frac{d^N}{dx^N}(x^2 - 1)^N, \qquad N = 0, 1, 2, \ldots.$$

Recurrence Relation

$$P_{N+1}(x) = \frac{(2N + 1)x P_N(x) - N P_{N-1}(x)}{N + 1}, \qquad N = 1, 2, 3, \ldots.$$

We can use Rodrigues' formula to obtain P_N directly. Alternatively, starting with P_0 and P_1, we can use the recurrence relation to generate all P_N.

Example 9.3.3   According to Rodrigues' formula,

$$P_2(x) = \frac{1}{8}\frac{d^2}{dx^2}(x^2 - 1)^2 = \frac{1}{8}\frac{d}{dx}[4x(x^2 - 1)] = \frac{1}{2}(3x^2 - 1),$$

which does indeed coincide with that given in the previous example.   □

The first five Legendre polynomials are given in Table 9.3.1.

N	**Legendre Polynomial of Degree N**
0	$P_0(x) = 1$
1	$P_1(x) = x$
2	$P_2(x) = \frac{1}{2}(3x^2 - 1)$
3	$P_3(x) = \frac{1}{2}x(5x^2 - 3)$
4	$P_4(x) = \frac{1}{8}(35x^4 - 30x^2 + 3)$

Table 9.3.1: The First Five Legendre Polynomials

Orthogonality of the Legendre Polynomials

In Section 4.11, we defined an inner product on the vector space $C^0[a, b]$ by

$$\langle f, g \rangle = \int_a^b f(x)g(x)\, dx,$$

for all f and g in $C^0[a, b]$. We now show that the Legendre polynomials are orthogonal relative to the above inner product on the interval $[-1, 1]$.

Theorem 9.3.4   The set of Legendre polynomials $\{P_0, P_1, P_2, \ldots\}$ is an orthogonal set of functions on the interval $[-1, 1]$. That is,

$$\int_{-1}^1 P_M(x)P_N(x)\, dx = 0 \qquad \text{whenever } M \neq N.$$

Proof Using the product rule for differentiation, we see easily that Legendre's equation

$$(1 - x^2)y'' - 2xy' + \alpha(\alpha + 1)y = 0$$

can be written in the form

$$[(1 - x^2)y']' + \alpha(\alpha + 1)y = 0.$$

Consequently, the Legendre polynomials $P_N(x)$ and $P_M(x)$ satisfy

$$[(1 - x^2)P_N']' + N(N + 1)P_N = 0, \tag{9.3.5}$$

$$[(1 - x^2)P_M']' + M(M + 1)P_M = 0, \tag{9.3.6}$$

respectively. Multiplying Equation (9.3.5) by P_M and Equation (9.3.6) by P_N and subtracting yields

$$\left[(1 - x^2)P_N'\right]' P_M - \left[(1 - x^2)P_M'\right]' P_N + [N(N + 1) - M(M + 1)]\, P_M P_N = 0,$$

which can be written as

$$\left\{[(1 - x^2)P_N' P_M]' - (1 - x^2)P_N' P_M'\right\} - \left\{[(1 - x^2)P_M' P_N]' - (1 - x^2)P_N' P_M'\right\}$$
$$+ \; [N(N + 1) - M(M + 1)] \, P_M P_N = 0.$$

That is,

$$\left[(1 - x^2)(P_N' P_M - P_M' P_N)\right]' + [N(N + 1) - M(M + 1)] \, P_M P_N = 0.$$

Integrating over the interval $[-1, 1]$, we obtain

$$\left[(1 - x^2)(P_N' P_M - P_M' P_N)\right]_{-1}^{1} + [N(N + 1) - M(M + 1)] \int_{-1}^{1} P_M(x) P_N(x) \, dx = 0.$$

The first term vanishes at $x = \pm 1$, and the term multiplying the integral can be factorized to yield

$$(N - M)(N + M + 1) \int_{-1}^{1} P_M(x) P_N(x) \, dx = 0.$$

Since M and N are nonnegative integers, the previous formula implies that

$$\int_{-1}^{1} P_M(x) P_N(x) \, dx = 0 \qquad \text{whenever } M \neq N. \qquad\blacksquare$$

It can also be shown, although it is more difficult [see N. N. Lebedev, *Special Functions and Their Applications* (New York: Dover, 1972)], that

$$\int_{-1}^{1} P_N^2(x) \, dx = \frac{2}{2N + 1}. \tag{9.3.7}$$

Consequently, $\left\{\sqrt{\frac{2N+1}{2}} \, P_N(x)\right\}$ is an orthonormal set of polynomials on $[-1, 1]$.

Since the set of Legendre polynomials $\{P_0, P_1, \ldots, P_N\}$ is linearly independent on any interval,[4] it is a basis for the vector space of all polynomials of degree less than or equal to N. Thus, if $p(x)$ is any polynomial, there exist scalars $a_0, a_1, \ldots, a_N$ such that

$$p(x) = \sum_{k=0}^{N} a_k P_k(x). \tag{9.3.8}$$

Because the Legendre polynomials form an orthogonal basis, we can use Theorem 4.12.7 to determine the coefficients a_k in this expansion. More explicitly, multiplying (9.3.8) by $P_j(x)$ for $0 \leq j \leq N$ and integrating over the interval $[-1, 1]$ yields

$$\int_{-1}^{1} p(x) P_j(x) \, dx = \int_{-1}^{1} \sum_{k=0}^{N} a_k P_k(x) P_j(x) \, dx = \sum_{k=0}^{N} a_k \int_{-1}^{1} P_k(x) P_j(x) \, dx.$$

However, owing to the orthogonality of the Legendre polynomials, all of the terms in the summation with $k \neq j$ vanish, so that

$$\int_{-1}^{1} p(x) P_j(x) \, dx = a_j \int_{-1}^{1} P_j(x) P_j(x) \, dx.$$

---

[4]This is true, for example, by Theorems 9.3.4 and 4.12.5, or from the fact that the polynomials in the set each have a different degree.

Consequently, using (9.3.7), we obtain

$$\int_{-1}^{1} p(x) P_j(x)\, dx = \frac{2}{2j+1} a_j,$$

which implies that

$$a_j = \frac{2j+1}{2} \int_{-1}^{1} p(x) P_j(x)\, dx. \tag{9.3.9}$$

Example 9.3.5   Expand $f(x) = x^2 - x + 2$ as a series of Legendre polynomials.

Solution: Since $f(x)$ has degree 2, we can write

$$x^2 - x + 2 = a_0 P_0 + a_1 P_1 + a_2 P_2,$$

where, from (9.3.7), the coefficients are given by

$$a_j = \frac{2j+1}{2} \int_{-1}^{1} (x^2 - x + 2) P_j(x)\, dx.$$

From Table 9.3.1,

$$P_0(x) = 1, \qquad P_1(x) = x, \qquad P_2(x) = \tfrac{1}{2}(3x^2 - 1),$$

so that

$$a_0 = \tfrac{1}{2} \int_{-1}^{1} (x^2 - x + 2)\, dx = \tfrac{7}{3},$$
$$a_1 = \tfrac{3}{2} \int_{-1}^{1} (x^2 - x + 2) x\, dx = -1,$$
$$a_2 = \tfrac{5}{2} \int_{-1}^{1} \tfrac{1}{2}(x^2 - x + 2)(3x^2 - 1)\, dx = \tfrac{2}{3}.$$

Consequently,

$$x^2 - x + 2 = \tfrac{7}{3} P_0 - P_1 + \tfrac{2}{3} P_2. \qquad \square$$

More generally, the following expansion theorem plays a fundamental role in many applications of mathematics to physics, chemistry, engineering, and so on.

Theorem 9.3.6   Let f and f' be continuous on the interval $(-1, 1)$. Then, for $-1 < x < 1$,

$$f(x) = a_0 P_0(x) + a_1 P_1(x) + \cdots + a_n P_n(x) + \cdots = \sum_{n=0}^{\infty} a_n P_n(x), \tag{9.3.10}$$

where

$$a_n = \frac{2n+1}{2} \int_{-1}^{1} f(x) P_n(x)\, dx. \tag{9.3.11}$$

Proof Establishing the existence of a convergent series of the form (9.3.10) is best left for a course on Fourier analysis or partial differential equations. The derivation that the coefficients in such an expansion must be given by (9.3.11) follows steps similar to those leading to Equation (9.3.9) and is left as an exercise. ∎

Example 9.3.7   Determine the terms leading up to and including $P_3(x)$ in the Legendre series expansion of $f(x) = \sin \pi x$, $-1 < x < 1$.

Solution: According to the previous theorem, the given function does have a Legendre series expansion with coefficients given by

$$a_n = \frac{2n + 1}{2} \int_{-1}^{1} \sin(\pi x) P_n(x) \, dx.$$

Thus,

$$a_0 = \frac{1}{2} \int_{-1}^{1} \sin(\pi x) \, dx = 0;$$

$$a_1 = \frac{3}{2} \int_{-1}^{1} x \sin(\pi x) \, dx = \frac{3}{\pi},$$

$$a_2 = \frac{5}{2} \int_{-1}^{1} \frac{1}{2}(3x^2 - 1) \sin(\pi x) \, dx = 0,$$

$$a_3 = \frac{7}{2} \int_{-1}^{1} \frac{1}{2}x(5x^2 - 3) \sin(\pi x) \, dx = \frac{7}{\pi^3}(\pi^2 - 15).$$

Consequently, Theorem 9.3.6 implies that, for $-1 < x < 1$,

$$\sin \pi x = \frac{3}{\pi} P_1(x) + \frac{7}{\pi^3}(\pi^2 - 15) P_3(x) + \cdots . \qquad (9.3.12)$$

To illustrate how good this approximation is, in Figure 9.3.1 we have sketched the functions

$$f(x) = \sin \pi x \quad \text{and} \quad g(x) = \frac{3}{\pi} P_1(x) + \frac{7}{\pi^3}(\pi^2 - 15) P_3(x).$$

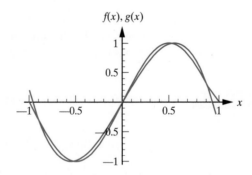

Figure 9.3.1: Comparison of $f(x) = \sin \pi x$ and its Legendre polynomial expansion of degree 3.

For comparison, in Figure 9.3.2 we sketch $f(x)$ and the fifth-order Taylor approximation

$$h(x) = \pi x - \frac{1}{3!}(\pi x)^3 + \frac{1}{5!}(\pi x)^5.$$

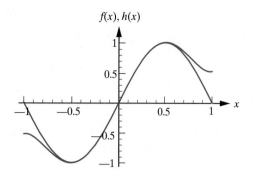

Figure 9.3.2: Comparison of $f(x) = \sin \pi x$ and its Taylor polynomial expansion of degree 5.

In Figure 9.3.3, we sketch $f(x)$ together with the Legendre polynomial approximation

$$k(x) = \frac{3}{\pi}P_1(x) + \frac{7}{\pi^3}(\pi^2 - 15)P_3(x) + \frac{11}{\pi^5}(945 - 105\pi^2 + \pi^4)P_5(x)$$

that arises when we include the next nonzero term in (9.3.12). We see that $k(x)$ gives an excellent approximation to $f(x) = \sin \pi x$ at all points in $[-1, 1]$ (including the endpoints).   □

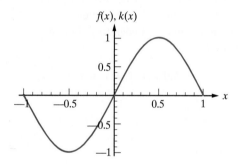

Figure 9.3.3: Comparison of $f(x) = \sin \pi x$ and its Legendre polynomial expansion of degree 5.

Remark   Computation by hand of the coefficients in a Legendre polynomial expansion can be very tedious. However, computer algebra systems, such as Maple or Mathematica, have the Legendre polynomials as built-in functions and therefore can be useful in computing the coefficients. For example, in Maple the command $P(n, x)$ generates $P_N(x)$.

Exercises for 9.3

Key Terms

Legendre equation, Legendre polynomial of degree N, Rodrigues' formula.

Skills

- Be able to compute the Legendre polynomials $P_N(x)$ directly for small n, either from (9.3.3) and (9.3.4), or from Rodrigues' formula, or from the recurrence relation for the Legendre polynomials.

- Provided that a function f has a continuous derivative on $(-1, 1)$, be able to use the orthogonality of the Legendre polynomials to expand f as a linear combination of Legendre polynomials.

True-False Review

For Questions 1–4, decide if the given statement is **true** or **false**, and give a brief justification for your answer. If true, you can quote a relevant definition or theorem from the text. If false, provide an example, illustration, or brief explanation of why the statement is false.

1. The radius of convergence of the power series solutions to Legendre's equation about $x = 0$ is 1.

2. The Legendre polynomials define an orthogonal basis for the vector space P_n.

3. For an integer value of α, one (but not both) of the formulas (9.3.3) and (9.3.4) is a polynomial.

4. Each of the Legendre polynomials contains terms with all odd powers of x or with all even powers of x.

Problems

1. Use Equations (9.3.3) and (9.3.4) to determine polynomial solutions to Legendre's equation when $\alpha = 3$ and $\alpha = 4$. Hence, determine the Legendre polynomials $P_3(x)$ and $P_4(x)$.

2. Starting with $P_0(x) = 1$ and $P_1(x) = x$, use the recurrence relation

$$(n+1)P_{n+1} + nP_{n-1} = (2n+1)xP_n, \quad n = 1, 2, 3, \ldots$$

to determine P_2, P_3, and P_4.

3. Use Rodrigues' formula to determine the Legendre polynomial of degree 3.

4. Determine all values of the constants a_0, a_1, a_2, and a_3 such that

$$x^3 + 2x = a_0 P_0 + a_1 P_1 + a_2 P_2 + a_3 P_3.$$

5. Express $p(x) = 2x^3 + x^2 + 5$ as a linear combination of Legendre polynomials.

6. Let $Q(x)$ be a polynomial of degree less than N. Prove that $\int_{-1}^{1} Q(x) P_N(x) \, dx = 0$.

7. Show that

$$\frac{d^2Y}{d\phi^2} + \cot \phi \frac{dY}{d\phi} + \alpha(\alpha + 1)Y = 0, \quad 0 < \phi < \pi,$$

is transformed into Legendre's equation by the change of variables $x = \cos \phi$.

Problems 8–10 deal with Hermite's equation:

$$y'' - 2xy' + 2\alpha y = 0, \quad -\infty < x < \infty. \tag{9.3.13}$$

8. Determine two linearly independent series solutions to Hermite's equation centered at $x = 0$.

9. Show that if $\alpha = N$, a positive integer, then Equation (9.3.13) has a polynomial solution. Determine the polynomial solutions when $\alpha = 0, 1, 2,$ and 3.

10. When suitably normalized, the polynomial solutions to Equation (9.3.13) are called the **Hermite polynomials** and are denoted by $H_N(x)$.

(a) Use Equation (9.3.13) to show that $H_N(x)$ satisfies

$$(e^{-x^2} H_N')' + 2Ne^{-x^2} H_N = 0. \tag{9.3.14}$$

[**Hint:** Replace α with N in Equation (9.3.13) and multiply the resulting equation by e^{-x^2}.]

(b) Use Equation (9.3.14) to prove that the Hermite polynomials satisfy

$$\int_{-\infty}^{\infty} e^{-x^2} H_N(x) H_M(x) \, dx = 0, \quad M \neq N. \tag{9.3.15}$$

[**Hint:** Follow the steps taken in proving orthogonality of the Legendre polynomials. You will need to recall that

$$\lim_{x \to \pm\infty} e^{-x^2} p(x) = 0,$$

for any polynomial p.]

(c) Let $p(x)$ be a polynomial of degree N. Then we can write

$$p(x) = \sum_{k=1}^{N} a_k H_k(x). \tag{9.3.16}$$

Given that

$$\int_{-\infty}^{\infty} e^{-x^2} H_N^2(x) \, dx = 2^N N! \sqrt{\pi},$$

use (9.3.15) to prove that the constants in (9.3.16) are given by

$$a_j = \frac{1}{2^j j! \sqrt{\pi}} \int_{-\infty}^{\infty} e^{-x^2} H_j(x) p(x) \, dx.$$

11. ◇   Use some form of technology to determine the coefficients in the Legendre expansion of the polynomial $p(x) = 3x^3 - 1$.

For Problems 12–13, use some form of technology to determine the first five terms in the Legendre series expansion of the given function on the interval $(-1, 1)$. Plot the given function and the approximations

$$S_1 = a_0 P_0, \qquad S_3 = a_0 P_0 + a_1 P_1 + a_2 P_2,$$
$$S_5 = a_0 P_0 + a_1 P_1 + a_2 P_2 + a_3 P_3 + a_4 P_4$$

on the interval $(-1, 1)$. Comment on the convergence of the Legendre series to the given function for $-1 < x < 1$.

12. ◇ $f(x) = \cos \pi x$.

13. ◇ $f(x) = x(1 - x^2)e^x$.

9.4   Series Solutions about a Regular Singular Point

The power series technique for solving

$$y'' + P(x)y' + Q(x)y = 0 \tag{9.4.1}$$

developed in the previous section is directly applicable only at ordinary points—that is, points where P and Q are both analytic. According to Definition 9.2.1, any points at which P or Q fail to be analytic are called *singular points* of Equation (9.4.1), and the general analysis of the behavior of solutions to Equation (9.4.1) in the neighborhood of a singular point is quite complicated. However, singular points often turn out to be the points of major interest in an applied problem, and so it is of some importance that we pursue this analysis. In the next two sections we will show that, provided that the functions P and Q are not too badly behaved at a singular point, the power series technique can be extended to obtain solutions of the corresponding differential equations that are valid in the neighborhood of the singular point. We will restrict our attention to differential equations whose singular points satisfy the following definition.

DEFINITION   9.4.1

The point $x = x_0$ is called a **regular singular point** of the differential equation (9.4.1) if and only if the following two conditions are satisfied:

1. x_0 is a singular point of Equation (9.4.1).

2. *Both* of the functions

$$p(x) = (x - x_0)P(x) \qquad \text{and} \qquad q(x) = (x - x_0)^2 Q(x)$$

are analytic at $x = x_0$.

A singular point of Equation (9.4.1) that does not satisfy condition (2) is called an **irregular singular point**.

Example 9.4.2   Determine the ordinary points, regular singular points, and irregular singular points of the differential equation

$$y'' + \frac{1}{x(x-1)^2}y' + \frac{x+1}{x(x-1)^3}y = 0. \tag{9.4.2}$$

Solution: In this case, we have

$$P(x) = \frac{1}{x(x-1)^2} \qquad \text{and} \qquad Q(x) = \frac{x+1}{x(x-1)^3},$$

and, by inspection, P and Q are analytic at all points except $x = 0$ and $x = 1$. Hence the only singular points of Equation (9.4.2) are $x = 0$ and $x = 1$. Consequently, every x with $x \neq 0$ and $x \neq 1$ is an ordinary point of the differential equation. We now determine whether the singular points $x = 0$ and $x = 1$ are regular or irregular.

Consider the singular point $x = 0$. The functions

$$p(x) = xP(x) = \frac{1}{(x-1)^2} \quad \text{and} \quad q(x) = x^2 Q(x) = \frac{x(x+1)}{(x-1)^3}$$

are both analytic at $x = 0$, so that $x = 0$ is a *regular singular point* of Equation (9.4.2). Now consider the singular point $x = 1$. Since

$$p(x) = (x-1)P(x) = \frac{1}{x(x-1)}$$

is nonanalytic at $x = 1$, it follows that $x = 1$ is an *irregular singular point* of Equation (9.4.2).   □

Now suppose that $x = x_0$ is a regular singular point of the differential equation

$$y'' + P(x)y' + Q(x)y = 0.$$

Multiplying this equation by $(x - x_0)^2$ yields

$$(x - x_0)^2 y'' + (x - x_0)[(x - x_0)P(x)]y' + (x - x_0)^2 Q(x)y = 0,$$

which we can write as

$$(x - x_0)^2 y'' + (x - x_0)p(x)y' + q(x)y = 0,$$

where

$$p(x) = (x - x_0)P(x) \quad \text{and} \quad q(x) = (x - x_0)^2 Q(x).$$

Since, by assumption, $x = x_0$ is a regular singular point, it follows that the functions p and q are analytic at $x = x_0$. By the change of variables $z = x - x_0$, we can always transform a regular singular point to $x = 0$, and so we will restrict our attention to differential equations that can be written in the form

$$\boxed{x^2 y'' + xp(x)y' + q(x)y = 0,} \tag{9.4.3}$$

where p and q are analytic at $x = 0$. *This is the standard form of any differential equation that has a regular singular point at $x = 0$.* The simplest type of equation that falls into this category is the second-order Cauchy-Euler equation

$$x^2 y'' + p_0 xy' + q_0 y = 0, \tag{9.4.4}$$

where p_0 and q_0 are constants. The solution techniques that we will develop for solving Equation (9.4.3) will be motivated by the solutions to Equation (9.4.4). Recall from Section 6.8 that (9.4.4) has solutions on the interval $(0, \infty)$ of the form $y(x) = x^r$, where r is a root of the indicial equation

$$r(r-1) + p_0 r + q_0 = 0.$$

Now consider Equation (9.4.3). Since, by assumption, p and q are analytic at $x = 0$, we can write

$$p(x) = p_0 + p_1 x + p_2 x^2 + \cdots, \qquad q(x) = q_0 + q_1 x + q_2 x^2 + \cdots, \tag{9.4.5}$$

for x in some interval of the form $(-R, R)$. It follows that Equation (9.4.3) can be written as

$$x^2 y'' + x(p_0 + p_1 x + p_2 x^2 + \cdots)y' + (q_0 + q_1 x + q_2 x^2 + \cdots)y = 0.$$

If $|x| << 1$, this is approximately the Cauchy-Euler equation (9.4.4), and so it is reasonable to *expect* that for x in the interval $(0, R)$, Equation (9.4.3) has solutions of the form

$$y(x) = x^r \sum_{n=0}^{\infty} a_n x^n, \qquad a_0 \neq 0, \tag{9.4.6}$$

where r is a root of the **indicial equation**

$$r(r - 1) + p_0 r + q_0 = 0. \tag{9.4.7}$$

A series of the form (9.4.6) is called a **Frobenius series**. We can assume without loss of generality that $a_0 \neq 0$, since if this were not the case, we could always factor the leading power of x out of the series and combine it into x^r.

The following theorem confirms our expectations:

Theorem 9.4.3   Consider the differential equation

$$x^2 y'' + xp(x)y' + q(x)y = 0, \qquad x > 0, \tag{9.4.8}$$

where p and q are analytic at $x = 0$. Suppose that

$$p(x) = \sum_{n=0}^{\infty} p_n x^n, \qquad q(x) = \sum_{n=0}^{\infty} q_n x^n,$$

for $|x| < R$. Let r_1 and r_2 denote the roots of the indicial equation

$$r(r - 1) + p_0 r + q_0 = 0,$$

and assume that $r_1 \geq r_2$ if these roots are real. Then Equation (9.4.8) has a solution of the form

$$y_1(x) = x^{r_1} \sum_{n=0}^{\infty} a_n x^n, \qquad a_0 \neq 0.$$

This solution is valid (at least) for $0 < x < R$. Further, provided that r_1 and r_2 are distinct and do not differ by an integer, then there exists a second solution to (9.4.8) that is valid (at least) for $0 < x < R$ of the form

$$y_2(x) = x^{r_2} \sum_{n=0}^{\infty} b_n x^n, \qquad b_0 \neq 0.$$

The solutions y_1 and y_2 are linearly independent on their intervals of existence.

Proof The proof of this theorem, as well as its extension to the case when the roots of the indicial equation do differ by an integer, will be discussed fully in the next section. ∎

Remark   Using the formula for the Maclaurin expansion of p and q, it follows that the constants p_0 and q_0 appearing in (9.4.5) are given by

$$p_0 = p(0) \qquad \text{and} \qquad q_0 = q(0).$$

Consequently, the indicial equation (9.4.7) for

$$x^2 y'' + x p(x) y' + q(x) y = 0$$

can be written directly as

$$r(r-1) + p(0)r + q(0) = 0.$$

We conclude this section with some examples that illustrate the implementation of the above theorem.

Example 9.4.4   Show that the differential equation

$$x^2 y'' + x e^{2x} y' - 2(\cos x) y = 0, \qquad x > 0$$

has two linearly independent Frobenius series solutions, and determine the interval on which these solutions are valid.

Solution: Comparing the given differential equation with the standard form (9.4.3), we see that

$$p(x) = e^{2x} \qquad \text{and} \qquad q(x) = -2 \cos x.$$

Consequently,

$$p(0) = 1 \qquad \text{and} \qquad q(0) = -2,$$

and so the indicial equation is

$$r(r-1) + r - 2 = 0.$$

That is,

$$r^2 - 2 = 0.$$

Thus, the roots of the indicial equation are $r_1 = \sqrt{2}$ and $r_2 = -\sqrt{2}$. Since r_1 and r_2 are distinct and do not differ by an integer, it follows from Theorem 9.4.3 that the given differential equation has two Frobenius series solutions of the form

$$y_1(x) = x^{\sqrt{2}} \sum_{n=0}^{\infty} a_n x^n, \qquad y_2(x) = x^{-\sqrt{2}} \sum_{n=0}^{\infty} b_n x^n.$$

Further, since the power series expansions of p and q about $x = 0$ are valid for all x, the preceding solutions will be defined and linearly independent on $(0, \infty)$. $\quad\square$

Example 9.4.5   Find the general solution to

$$2x^2 y'' + x y' - (1+x) y = 0, \qquad x > 0. \tag{9.4.9}$$

Solution: In this case, $p(x) = \frac{1}{2}$ and $q(x) = -\frac{1}{2}(1+x)$, both of which are analytic at $x = 0$. Thus, $x = 0$ is a regular singular point of Equation (9.4.9). The indicial equation $r(r - 1) + \frac{1}{2}r - \frac{1}{2} = 0$ can be written as $2r^2 - r - 1 = 0$, which factors as $(2r - 1)(r + 1) = 0$. Hence, the roots of the indicial equation are

$$r_1 = 1 \qquad \text{and} \qquad r_2 = -\tfrac{1}{2}. \tag{9.4.10}$$

Since these roots are distinct and do not differ by an integer, it follows from Theorem 9.4.3 that the given differential equation has two Frobenius series solutions of the form

$$y_1(x) = x \sum_{n=0}^{\infty} a_n x^n, \qquad y_2(x) = x^{-1/2} \sum_{n=0}^{\infty} b_n x^n.$$

Now we wish to determine the coefficients a_i and b_i. To do this, we let

$$y(x) = x^r \sum_{n=0}^{\infty} a_n x^n = \sum_{n=0}^{\infty} a_n x^{r+n}, \qquad a_0 \neq 0,$$

so that

$$y'(x) = \sum_{n=0}^{\infty} (r + n) a_n x^{r+n-1}, \qquad y''(x) = \sum_{n=0}^{\infty} (r + n)(r + n - 1) a_n x^{r+n-2}.$$

Substituting into (9.4.9) yields

$$\sum_{n=0}^{\infty} 2(r+n)(r+n-1) a_n x^{r+n} + \sum_{n=0}^{\infty} (r+n) a_n x^{r+n} - \sum_{n=0}^{\infty} a_n x^{r+n} - \sum_{n=0}^{\infty} a_n x^{r+n+1} = 0.$$

That is, combining the first three terms and replacing n with $n - 1$ in the fourth sum,

$$\sum_{n=0}^{\infty} \left[2(r+n)(r+n-1) + (r+n) - 1 \right] a_n x^n - \sum_{n=1}^{\infty} a_{n-1} x^n = 0. \tag{9.4.11}$$

Thus, the coefficients of x^n must vanish for $n = 0, 1, 2, \ldots$. The constant term in (9.4.11), corresponding to $n = 0$, leads directly to the indicial equation and the roots given in (9.4.10). On the other hand, when $n = 1, 2, \ldots$, in (9.4.11), we obtain the recurrence relation

$$(r + n - 1)(2r + 2n + 1) a_n - a_{n-1} = 0. \tag{9.4.12}$$

We now substitute the values of r obtained in (9.4.10) into (9.4.12) to determine the corresponding Frobenius series solutions.

Let $r = 1$: Substituting into the recurrence relation (9.4.12) yields

$$a_n = \frac{1}{n(2n + 3)} a_{n-1}, \qquad n = 1, 2, 3, \ldots.$$

Thus,

$$\begin{aligned}
n = 1: \qquad & a_1 = \frac{1}{1 \cdot 5} a_0, \\
n = 2: \qquad & a_2 = \frac{1}{2 \cdot 7} a_1 = \frac{1}{(2!)(5 \cdot 7)} a_0, \\
n = 3: \qquad & a_3 = \frac{1}{3 \cdot 9} a_2 = \frac{1}{(3!)(5 \cdot 7 \cdot 9)} a_0, \\
n = 4: \qquad & a_4 = \frac{1}{4 \cdot 11} a_3 = \frac{1}{(4!)(5 \cdot 7 \cdot 9 \cdot 11)} a_0.
\end{aligned}$$

It follows that, in general,

$$a_n = \frac{1}{(n!)[5 \cdot 7 \cdot 9 \cdots (2n+3)]} a_0, \qquad n = 1, 2, 3, \ldots,$$

so that the corresponding Frobenius series solution is

$$y_1(x) = x \left\{ 1 + \frac{1}{5}x + \frac{1}{(2!)(5 \cdot 7)}x^2 + \cdots + \frac{1}{(n!)[5 \cdot 7 \cdot 9 \cdots (2n+3)]}x^n + \cdots \right\},$$

where we have set $a_0 = 1$. We can write this solution as

$$y_1(x) = x \left\{ 1 + \sum_{n=1}^{\infty} \frac{1}{(n!)[5 \cdot 7 \cdot 9 \cdots (2n+3)]}x^n \right\}, \qquad x > 0.$$

Let $r = -\frac{1}{2}$: For the second Frobenius series solution, we replace the coefficients a_i in the preceding work by b_i. In this case, the recurrence relation (9.4.12) reduces to

$$b_n = \frac{1}{n(2n-3)} b_{n-1}, \qquad n = 1, 2, \ldots.$$

We therefore obtain

$$
\begin{aligned}
n = 1: &\qquad b_1 = -b_0, \\
n = 2: &\qquad b_2 = \frac{1}{2 \cdot 1} b_1 = -\frac{1}{2!} b_0, \\
n = 3: &\qquad b_3 = \frac{1}{3 \cdot 3} b_2 = -\frac{1}{(3!)(1 \cdot 3)} b_0, \\
n = 4: &\qquad b_4 = \frac{1}{4 \cdot 5} b_3 = -\frac{1}{(4!)(1 \cdot 3 \cdot 5)} b_0.
\end{aligned}
$$

In general, we have

$$b_n = -\frac{1}{(n!)[1 \cdot 3 \cdot 5 \cdots (2n-3)]} b_0, \qquad n = 1, 2, 3, \ldots.$$

It follows that a second linearly independent Frobenius series solution to the differential equation (9.4.9) on $(0, \infty)$ is

$$y_2(x) = x^{-1/2} \left\{ 1 - x - \frac{1}{2!}x^2 - \frac{1}{(3!)(1 \cdot 3)}x^3 - \cdots \right.$$
$$\left. - \frac{1}{(n!)[1 \cdot 3 \cdots (2n-3)]}x^n - \cdots \right\},$$

where we have set $b_0 = 1$. This can be written as

$$y_2(x) = x^{-1/2} \left\{ 1 - \sum_{n=1}^{\infty} \frac{1}{(n!)[1 \cdot 3 \cdot 5 \cdots (2n-3)]}x^n \right\}, \qquad x > 0.$$

Consequently, the general solution to the differential equation (9.4.9) on $(0, \infty)$ is

$$y(x) = c_1 y_1(x) + c_2 y_2(x). \qquad \qquad \square$$

The differential equation in the previous example had two linearly independent Frobenius series solutions, and we were therefore able to determine its general solution. In the following example, there is only one linearly independent *Frobenius series* solution.

Example 9.4.6   Determine a Frobenius series solution to

$$x^2 y'' + x(3+x)y' + (1+3x)y = 0, \quad x > 0. \tag{9.4.13}$$

Solution: By inspection, we see that $x = 0$ is a regular singular point of the differential equation (9.4.13), and so from Theorem 9.4.3 the differential equation has at least one Frobenius series solution. Further, since $p(x) = 3 + x$ and $q(x) = 1 + 3x$ are both polynomials, their power series expansions about $x = 0$ are valid for all x. It follows that any Frobenius series solution will be valid for $0 < x < \infty$. To determine a solution, we let

$$y(x) = x^r \sum_{n=0}^{\infty} a_n x^n = \sum_{n=0}^{\infty} a_n x^{r+n}.$$

Differentiating twice with respect to x yields

$$y'(x) = \sum_{n=0}^{\infty} (r+n)a_n x^{r+n-1}, \quad y''(x) = \sum_{n=0}^{\infty} (r+n)(r+n-1)a_n x^{r+n-2}$$

so that y is a solution to (9.4.13) provided that a_n and r satisfy

$$\sum_{n=0}^{\infty} (r+n)(r+n-1)a_n x^{r+n} + 3 \sum_{n=0}^{\infty} (r+n)a_n x^{r+n} + \sum_{n=0}^{\infty} (r+n)a_n x^{r+n+1}$$

$$+ \sum_{n=0}^{\infty} a_n x^{r+n} + 3 \sum_{n=0}^{\infty} a_n x^{r+n+1} = 0.$$

Dividing by x^r and replacing n by $n-1$ in the third and fifth sums yields

$$\sum_{n=0}^{\infty} \left[(r+n)(r+n-1) + 3(r+n) + 1 \right] a_n x^n + \sum_{n=1}^{\infty} (r+n+2)a_{n-1} x^n = 0.$$

$$\tag{9.4.14}$$

This implies that the coefficients of x^n must vanish for $n = 0, 1, 2, \ldots$. When $n = 0$, we obtain the indicial equation

$$r(r-1) + 3r + 1 = 0.$$

That is,

$$(r+1)^2 = 0.$$

Thus, the only value of r for which a Frobenius series solution exists is

$$r = -1.$$

For $n \geq 1$, (9.4.14) yields the recurrence relation

$$\left[(r+n)(r+n-1) + 3(r+n) + 1 \right] a_n + (r+n+2)a_{n-1} = 0.$$

Setting $r = -1$, we obtain

$$\left[(n-1)(n-2) + 3(n-1) + 1 \right] a_n + (n+1)a_{n-1} = 0,$$

which can be written as

$$a_n = -\frac{n+1}{n^2} a_{n-1}, \quad n = 1, 2, \ldots.$$

Solving this recurrence relation, we have

$$n = 1: \qquad a_1 = -2a_0,$$

$$n = 2: \qquad a_2 = -\frac{3}{4}a_1 = \frac{3 \cdot 2}{4}a_0,$$

$$n = 3: \qquad a_3 = -\frac{4}{9}a_2 = -\frac{4!}{4 \cdot 9}a_0.$$

The general term is

$$a_n = (-1)^n \frac{(n+1)!}{2^2 \cdot 3^2 \cdots n^2} a_0, \qquad n = 1, 2, 3, \ldots,$$

which can be written as

$$a_n = (-1)^n \frac{(n+1)!}{(n!)^2} a_0.$$

That is,

$$a_n = (-1)^n \frac{n+1}{n!} a_0, \qquad n = 1, 2, 3, \ldots.$$

Consequently, the corresponding Frobenius series solution is

$$y(x) = x^{-1}\left[1 + \sum_{n=1}^{\infty}(-1)^n \frac{n+1}{n!}x^n\right], \qquad x > 0,$$

where we set $a_0 = 1$.   □

Notice that in this problem, there is only *one* linearly independent Frobenius series solution to the given differential equation. To determine a second linearly independent solution, we could, for example, use the reduction-of-order method introduced in Section 6.9. We will have more to say about this in the next section.

Exercises for 9.4

Key Terms

Regular singular point, Irregular singular point, Frobenius series.

Skills

- Be able to classify singular points as regular or irregular.

- Be familiar with the indicial equation and be able to find its roots.

- Be able to use the roots of the indicial equation to determine a Frobenius series solution to Equation (9.4.3).

- In the case where the indicial equation has two distinct roots that do not differ by an integer, be able to find the general solution to Equation (9.4.3).

True-False Review

For Questions 1–5, decide if the given statement is **true** or **false**, and give a brief justification for your answer. If true, you can quote a relevant definition or theorem from the text. If false, provide an example, illustration, or brief explanation of why the statement is false.

1. A point $x = x_0$ is a regular singular point of the differential equation $y'' + P(x)y' + Q(x)y = 0$, provided that $P(x)$ and $Q(x)$ are not analytic at $x = x_0$, but $(x - x_0)P(x)$ and $(x - x_0)Q(x)$ are analytic at $x = x_0$.

2. If r_1 and r_2 are the roots of the indicial equation associated with the differential equation $x^2 y'' + xp(x)y' + q(x)y = 0$, then $x^{r_1} \sum_{n=0}^{\infty} a_n x^n$ and $x^{r_2} \sum_{n=0}^{\infty} b_n x^n$ (where $a_0, b_0 \neq 0$) are two linearly independent solutions to the differential equation.

3. The coefficients in a Frobenius series solution to the differential equation $x^2 y'' + xp(x)y' + q(x)y = 0$ are obtained by substituting the series solution and its derivatives into the differential equation and matching coefficients of the powers of x on each side of the equation.

4. It is possible for a given differential equation of the form $y'' + P(x)y' + Q(x)y = 0$ to have ordinary points, regular singular points, and irregular singular points.

5. It is possible for all of the singular points of a given differential equation of the form $y'' + P(x)y' + Q(x)y = 0$ to be irregular.

Problems

For Problems 1–4, determine all singular points of the given differential equation and classify them as regular or irregular singular points.

1. $y'' + \dfrac{1}{1-x}y' + xy = 0$.

2. $x^2 y'' + \dfrac{x}{(1-x^2)^2}y' + y = 0$.

3. $(x-2)^2 y'' + (x-2)e^x y' + \dfrac{4}{x}y = 0$.

4. $y'' + \dfrac{2}{x(x-3)}y' - \dfrac{1}{x^3(x+3)}y = 0$.

For Problems 5–8, determine the roots of the indicial equation of the given differential equation.

5. $x^2 y'' + x(1-x)y' - 7y = 0$.

6. $4x^2 y'' + xe^x y' - y = 0$.

7. $4xy'' - xy' + 2y = 0$.

8. $x^2 y'' - x(\cos x)y' + 5e^{2x}y = 0$.

For Problems 9–16, show that the indicial equation of the given differential equation has distinct roots that do *not* differ by an integer and find two linearly independent Frobenius series solutions on $(0, \infty)$.

9. $4x^2 y'' + 3xy' + xy = 0$.

10. $6x^2 y'' + x(1+18x)y' + (1+12x)y = 0$.

11. $x^2 y'' + xy' - (2+x)y = 0$.

12. $2xy'' + y' - 2xy = 0$.

13. $3x^2 y'' - x(x+8)y' + 6y = 0$.

14. $2x^2 y'' - x(1+2x)y' + 2(4x-1)y = 0$.

15. $x^2 y'' + x(1-x)y' - (5+x)y = 0$.

16. $3x^2 y'' + x(7+3x)y' + (1+6x)y = 0$.

17. Consider the differential equation

$$x^2 y'' + xy' + (1-x)y = 0, \quad x > 0. \quad (9.4.15)$$

(a) Find the indicial equation, and show that the roots are $r = \pm i$.

(b) Determine the first three terms in a complex-valued Frobenius series solution to Equation (9.4.15).

(c) Use the solution in (b) to determine two linearly independent real-valued solutions to Equation (9.4.15).

18. Determine the first five nonzero terms in each of two linearly independent Frobenius series solutions to

$$3x^2 y'' + x(1+3x^2)y' - 2xy = 0, \quad x > 0.$$

19. Consider the differential equation

$$4x^2 y'' - 4x^2 y' + (1+2x)y = 0.$$

(a) Show that the indicial equation has only one root, and find the corresponding Frobenius series solution.

(b) Use the *reduction-of-order* technique to find a second linearly independent solution on $(0, \infty)$. [**Hint:** To evaluate $\displaystyle\int \dfrac{e^x}{x}\, dx$, expand e^x in a Maclaurin series.]

20. Find two linearly independent solutions to

$$x^2 y'' + x(3-2x)y' + (1-2x)y = 0$$

on $(0, \infty)$.

9.5   Frobenius Theory

In the preceding section we saw how Frobenius series solutions can be obtained to the differential equation

$$x^2 y'' + xp(x)y' + q(x)y = 0, \qquad x > 0. \qquad (9.5.1)$$

In this section we give some justification for Theorem 9.4.3 and extend this theorem to the case when the roots of the indicial equation for (9.5.1) differ by an integer. We will first assume that $x > 0$, since our results can easily be extended to $x < 0$. We begin by establishing the existence of at least one Frobenius series solution.

Assuming that $x = 0$ is a regular singular point of (9.5.1), it follows that p and q are analytic at $x = 0$, hence we can write

$$p(x) = \sum_{n=0}^{\infty} p_n x^n, \qquad q(x) = \sum_{n=0}^{\infty} q_n x^n,$$

for $|x| < R$. Consequently, (9.5.1) can be written as

$$x^2 y'' + x \sum_{n=0}^{\infty} p_n x^n y' + \sum_{n=0}^{\infty} q_n x^n y = 0. \qquad (9.5.2)$$

We try for a Frobenius series solution and therefore let

$$y(x) = x^r \sum_{n=0}^{\infty} a_n x^n, \qquad a_0 \neq 0,$$

where r and a_n are constants to be determined. Differentiating y twice yields

$$y' = \sum_{n=0}^{\infty} (r+n)a_n x^{r+n-1}, \qquad y'' = \sum_{n=0}^{\infty} (r+n)(r+n-1)a_n x^{r+n-2}.$$

We now substitute into Equation (9.5.2) to obtain

$$\sum_{n=0}^{\infty} (r+n)(r+n-1)a_n x^n + \left(\sum_{n=0}^{\infty} p_n x^n \right) \left(\sum_{n=0}^{\infty} (r+n)a_n x^n \right)$$
$$+ \left(\sum_{n=0}^{\infty} q_n x^n \right) \left(\sum_{n=0}^{\infty} a_n x^n \right) = 0.$$

Using the formula given in Section 9.1 for the product of two infinite series gives

$$\sum_{n=0}^{\infty} (r+n)(r+n-1)a_n x^n + \sum_{n=0}^{\infty} \sum_{k=0}^{n} p_{n-k} x^{n-k} (k+r)a_k x^k$$
$$+ \sum_{n=0}^{\infty} \sum_{k=0}^{n} q_{n-k} x^{n-k} a_k x^k = 0,$$

which can be written as

$$\sum_{n=0}^{\infty} \left\{ (r+n)(r+n-1)a_n + \sum_{k=0}^{n} \left[p_{n-k}(k+r) + q_{n-k} \right] a_k \right\} x^n = 0.$$

Thus, a_n must satisfy the recurrence relation

$$(r+n)(r+n-1)a_n + \sum_{k=0}^{n} \left[p_{n-k}(k+r) + q_{n-k} \right] a_k = 0, \qquad n = 1, 2, \dots.$$

$$(9.5.3)$$

Evaluating Equation (9.5.3) when $n = 0$ yields

$$[r(r-1) + p_0 r + q_0]a_0 = 0,$$

so that, since $a_0 \neq 0$ (by assumption), r must satisfy

$$r(r-1) + p_0 r + q_0 = 0, \qquad (9.5.4)$$

which we recognize as being the indicial equation for (9.5.1). When $n \geq 1$, we combine the coefficients of a_n in (9.5.3) to obtain

$$\left[(r+n)(r+n-1) + p_0(r+n) + q_0 \right]a_n + \sum_{k=0}^{n-1} \left[p_{n-k}(k+r) + q_{n-k} \right] a_k = 0.$$

That is,

$$\left[(r+n)(r+n-1) + p_0(r+n) + q_0 \right]a_n$$
$$= -\sum_{k=0}^{n-1} \left[p_{n-k}(k+r) + q_{n-k} \right] a_k, \qquad n = 1, 2, \dots \qquad (9.5.5)$$

If we define $F(r)$ by

$$F(r) = r(r-1) + p_0 r + q_0,$$

then

$$F(r+n) = (r+n)(r+n-1) + p_0(r+n) + q_0,$$

so that the indicial equation (9.5.4) is

$$F(r) = 0,$$

whereas the recurrence relation (9.5.5) can be written as

$$F(r+n)a_n = -\sum_{k=0}^{n-1} \left[p_{n-k}(k+r) + q_{n-k} \right] a_k, \qquad n = 1, 2, 3, \dots. \qquad (9.5.6)$$

It is tempting to divide (9.5.6) by $F(r+n)$, thereby determining a_n in terms of $a_0, a_1, \dots,$ a_{n-1}. However, we can do this only if $F(r+n) \neq 0$. Let r_1 and r_2 denote the roots of Equation (9.5.4). The following three familiar cases arise:

1. r_1 and r_2 are real and distinct.
2. r_1 and r_2 are real and coincident.
3. r_1 and r_2 are complex conjugates.

If r_1 and r_2 are real, we assume without loss of generality that $r_1 \geq r_2$. Consider (9.5.6) when $r = r_1$. We have

$$F(r_1+n)a_n = -\sum_{k=0}^{n-1} \left[p_{n-k}(k+r_1) + q_{n-k} \right] a_k, \qquad n = 1, 2, 3, \dots. \qquad (9.5.7)$$

In cases 1 and 2 it follows, since $r = r_1$ is the largest root of $F(r) = 0$, that $F(r_1+n) \neq 0$, for any n. Also, in case 3, $F(r_1 + n) \neq 0$, and so in all three cases we can write (9.5.7) as

$$a_n = -\frac{1}{F(r_1 + n)} \sum_{k=0}^{n-1} \left[p_{n-k}(k + r_1) + q_{n-k} \right] a_k, \qquad n = 1, 2, 3, \ldots. \qquad (9.5.8)$$

Starting from $n = 1$, we can therefore determine all of the a_n in terms of a_0, and so we formally obtain the Frobenius series solution

$$y_1(x) = x^{r_1} \left[1 + \sum_{n=1}^{\infty} a_n(r_1) x^n \right],$$

where $a_n(r_1)$ denotes the coefficients obtained from (9.5.8) upon setting $a_0 = 1$. A fairly delicate analysis shows that this series solution converges for (at least) $0 < x < R$. This justifies the steps in the preceding derivation and establishes the first part of Theorem 9.4.3 stated in the previous section.

We now consider the problem of determining a second linearly independent solution to Equation (9.5.1). We must consider the three cases separately.

Case 1: r_1 and r_2 are real and distinct.   Setting $r = r_2$ in (9.5.6) yields

$$F(r_2 + n)a_n = -\sum_{k=0}^{n-1} \left[p_{n-k}(k + r_2) + q_{n-k} \right] a_k. \qquad (9.5.9)$$

Thus, provided that $F(r_2 + n) \neq 0$ for any positive integer n, the same procedure that we used when $r = r_1$ will yield a second Frobenius series solution. But, since r_1 and r_2 are the only zeros of F, it follows that $F(r_2 + n) = 0$ if and only if there exists a positive integer n such that $r_2 + n = r_1$. Consequently, provided that $r_1 - r_2$ is not a positive integer, there exists a second Frobenius series solution of the form

$$y_2(x) = x^{r_2} \left[1 + \sum_{n=1}^{\infty} a_n(r_2) x^n \right],$$

where $a_n(r_2)$ denotes the values of the coefficients obtained from (9.5.6) when $r = r_2$, and once more we have set $a_0 = 1$. Since $r_1 \neq r_2$, it follows that the Frobenius series solutions y_1 and y_2 are linearly independent on (at least) $0 < x < R$.

Now suppose that $r_1 - r_2 = N$, where N is a positive integer. Then, substituting for $r_2 = r_1 - N$ in (9.5.9), we obtain

$$F(r_1 + (n - N))a_n = -\sum_{k=0}^{n-1} \left[p_{n-k}(k + r_2) + q_{n-k} \right] a_k, \qquad n = 1, 2, \ldots,$$

which, when $n = N$, leads to the consistency condition

$$0 \cdot a_N = -\sum_{k=0}^{N-1} \left[p_{N-k}(k + r_2) + q_{N+k} \right] a_k. \qquad (9.5.10)$$

Since all of the coefficients $a_1, a_2, \ldots, a_{N-1}$ will already have been determined in terms of a_0, (9.5.10) will be of the form

$$0 \cdot a_N = \alpha a_0, \qquad (9.5.11)$$

where α is a constant. By assumption, a_0 is nonzero, and therefore two possibilities arise. (Notice that (9.5.11) is *not* an equation for determining α; rather, it is a consistency condition for the validity of the recurrence relation.)

(a) First it may happen that $\alpha = 0$. If this occurs, then from (9.5.11), a_N can be specified arbitrarily, and (9.5.9) determines all of the remaining Frobenius coefficients. We therefore do obtain a second linearly independent Frobenius series solution.

(b) In the more general case, α will be nonzero. Then, since $a_0 \neq 0$ (by assumption), (9.5.11) *cannot* be satisfied, and so we cannot compute the Frobenius coefficients. Hence, there *does not exist* a Frobenius series solution corresponding to $r = r_2$. The reduction-of-order technique can be used, however, to prove that there exists a second linearly independent solution to (9.5.1) on $(0, R)$, of the form

$$y_2(x) = Ay_1(x)\ln x + x^{r_2}\sum_{n=0}^{\infty}b_n x^n,$$

where the constants A and b_n can be determined by direct substitution into the differential equation (9.5.1). The derivation is straightforward, but quite longwinded, and so the details have been relegated to Appendix D. Notice that the foregoing form includes case (a), which arises when $A = 0$.

Case 2: $r_1 = r_2$.   In this case, there certainly cannot exist a second linearly independent *Frobenius series* solution. However, once more the reduction-of-order technique can be used to establish the existence of a second linearly independent solution of the form

$$y_2(x) = y_1(x)\ln x + x^{r_1}\sum_{n=1}^{\infty}b_n x^n$$

valid for (at least) $0 < x < R$. (See Appendix D.) The coefficients b_n can be obtained by substituting this expression for y_2 into the differential equation (9.5.1).

Case 3: r_1 **and** r_2 **are complex conjugates.**   In this case, the solution just obtained when $r = r_1$ will be a complex-valued solution. Owing to the linearity of the differential equation, it follows that the real and imaginary parts of this complex-valued solution will themselves be real-valued solutions. It can be shown that these real-valued solutions are linearly independent on their intervals of existence. Thus, we can always, in theory, obtain the general solution in this case.

Finally, we mention that the validity of the above solutions can be extended to $-R < x < 0$ by the replacement

$$x^{r_1} \to |x|^{r_1}, \qquad x^{r_2} \to |x|^{r_2}.$$

The preceding discussion is summarized in the next theorem.

Theorem 9.5.1   Consider the differential equation

$$x^2 y'' + xp(x)y' + q(x)y = 0, \tag{9.5.12}$$

where p and q are analytic at $x = 0$. Suppose that

$$p(x) = \sum_{n=0}^{\infty}p_n x^n, \qquad q(x) = \sum_{n=0}^{\infty}q_n x^n,$$

for $|x| < R$. Let r_1 and r_2 denote the roots of the indicial equation and assume that $r_1 \geq r_2$ if these roots are real. Then (9.5.12) has two linearly independent solutions valid (at least) on the interval $(0, R)$. The form of the solution is determined as follows:

1. $r_1 - r_2$ not an integer:

$$y_1(x) = x^{r_1} \sum_{n=0}^{\infty} a_n x^n, \qquad a_0 \neq 0, \qquad (9.5.13)$$

$$y_2(x) = x^{r_2} \sum_{n=0}^{\infty} b_n x^n, \qquad b_0 \neq 0. \qquad (9.5.14)$$

2. $r_1 = r_2 = r$:

$$y_1(x) = x^{r} \sum_{n=0}^{\infty} a_n x^n, \qquad a_0 \neq 0, \qquad (9.5.15)$$

$$y_2(x) = y_1(x) \ln x + x^{r} \sum_{n=1}^{\infty} b_n x^n, \qquad a_0 \neq 0. \qquad (9.5.16)$$

3. $r_1 - r_2$ a positive integer:

$$y_1(x) = x^{r_1} \sum_{n=0}^{\infty} a_n x^n, \qquad a_0 \neq 0, \qquad (9.5.17)$$

$$y_2(x) = A y_1(x) \ln x + x^{r_2} \sum_{n=0}^{\infty} b_n x^n, \qquad b_0 \neq 0. \qquad (9.5.18)$$

The coefficients in each of these solutions can be determined by direct substitution into Equation (9.5.12). Finally, if x^{r_1} and x^{r_2} are replaced by $|x|^{r_1}$ and $|x|^{r_2}$, respectively, we obtain linearly independent solutions that are valid for (at least) $0 < |x| < R$.

Remark   Since a solution of a homogeneous linear differential equation is defined only up to a multiplicative constant, we can use this freedom to set $a_0 = 1$ in (9.5.13), (9.5.15), and (9.5.17) and to set $b_0 = 1$ in (9.5.14) and (9.5.18). It is often convenient to make these choices in solving our problems.

We now consider several examples to illustrate the use of the preceding theorem.

Example 9.5.2   Consider the differential equation

$$x^2 y'' - x(3 + x) y' + (4 - x) y = 0. \qquad (9.5.19)$$

Determine the general form of two linearly independent series solutions in the neighborhood of the regular singular point $x = 0$.

Solution: By inspection we see that $x = 0$ is a regular singular point of Equation (9.5.19) and that the indicial equation is

$$r(r - 1) - 3r + 4 = r^2 - 4r + 4 = (r - 2)^2 = 0.$$

Since $r = 2$ is a repeated root, it follows from Theorem 9.5.1 that there exist two linearly independent solutions to Equation (9.5.19) of the form

$$y_1(x) = x^2 \sum_{n=0}^{\infty} a_n x^n, \qquad y_2(x) = y_1(x) \ln |x| + x^2 \sum_{n=1}^{\infty} b_n x^n.$$

The coefficients in each of these series solutions could be obtained by direct substitution into Equation (9.5.19). In this problem, we have

$$p(x) = -(3 + x), \qquad q(x) = 4 - x.$$

Since these are polynomials, their power series expansions about $x = 0$ converge for all real x. It follows from Theorem 9.5.1 that the series solutions to Equation (9.5.19) will be valid for all $x \neq 0$.   □

Example 9.5.3   Consider the differential equation

$$x^2 y'' + x(1 + 2x)y' - \tfrac{1}{4}(1 - \gamma x)y = 0, \qquad x > 0, \tag{9.5.20}$$

where γ is a constant. Determine the form of two linearly independent series solutions to this differential equation about the regular singular point $x = 0$.

Solution: In this case, we have $p_0 = 1$ and $q_0 = -\tfrac{1}{4}$, so that

$$F(r) = r(r - 1) + r - \tfrac{1}{4} = r^2 - \tfrac{1}{4}.$$

Thus, the roots of the indicial equation are $r = \pm\tfrac{1}{2}$. Setting $r_1 = \tfrac{1}{2}$ and $r_2 = -\tfrac{1}{2}$, it follows that $r_1 - r_2 = 1$, so that we are in Case 3 of Theorem 9.5.1. Thus, there exist two linearly independent solutions to Equation (9.5.20) on $(0, \infty)$ of the form

$$y_1(x) = x^{1/2} \sum_{n=0}^{\infty} a_n x^n, \qquad y_2(x) = A y_1(x) \ln x + x^{-1/2} \sum_{n=0}^{\infty} b_n x^n. \tag{9.5.21}$$

To determine whether the constant A is zero or nonzero, we need the general recurrence relation. Substituting

$$y_1(x) = x^r \sum_{n=0}^{\infty} a_n x^n, \qquad a_0 \neq 0,$$

into (9.5.20) yields

$$\left[\frac{4(r + n)^2 - 1}{4} \right] a_n = -\left[2(r + n - 1) + \gamma \right] a_{n-1}, \qquad n = 1, 2, \ldots .$$

When $r = -\tfrac{1}{2}$, this reduces to

$$n(n - 1)a_n = -(2n - 3 + \gamma)a_{n-1}. \tag{9.5.22}$$

As predicted from our general theory, the coefficient of a_n is zero when $n = r_1 - r_2 = 1$. Thus, a second Frobenius series solution exists [$A = 0$ in (9.5.21)] if and only if the term on the right-hand side also vanishes when $n = 1$. Setting $n = 1$ in (9.5.22) yields the consistency condition

$$0 \cdot a_1 = (1 - \gamma)a_0. \tag{9.5.23}$$

Since $a_0 \neq 0$, it follows from (9.5.23) that when $\gamma \neq 1$, there does not exist a second linearly independent Frobenius series solution, and so the constant A in (9.5.21) is necessarily nonzero. If $\gamma = 1$, however, then (9.5.23) is identically satisfied independently of the value of a_1. In this case, we can specify a_1 arbitrarily and then the recurrence relation (9.5.22) will determine the remaining coefficients in a second linearly independent Frobenius series solution. To summarize:

1. If $\gamma \neq 1$, there exist two linearly independent solutions of the form (9.5.21) with A necessarily nonzero.

2. If $\gamma = 1$, then there exist two linearly independent Frobenius series solutions

$$y_1(x) = x^{1/2} \sum_{n=0}^{\infty} a_n x^n, \qquad y_2(x) = x^{-1/2} \sum_{n=0}^{\infty} b_n x^n.$$

In both cases, the series solutions will be valid for $0 < x < \infty$.   □

Example 9.5.4   Determine two linearly independent series solutions to

$$x^2 y'' + x(3 - x)y' + y = 0, \qquad x > 0. \tag{9.5.24}$$

Solution: We see by inspection that $x = 0$ is a regular singular point. Furthermore, since the functions

$$p(x) = 3 - x \qquad \text{and} \qquad q(x) = 1$$

are polynomials, the linearly independent series solutions obtained in Theorem 9.5.1 will be valid on $(0, \infty)$. We begin by obtaining a Frobenius series solution. Substituting

$$y(x) = x^r \sum_{n=0}^{\infty} a_n x^n, \qquad a_0 \neq 0,$$

into (9.5.24) yields

$$\sum_{n=0}^{\infty} (r + n)(r + n - 1)a_n x^{r+n} + 3 \sum_{n=0}^{\infty} (r + n)a_n x^{r+n}$$
$$- \sum_{n=0}^{\infty} (r + n)a_n x^{r+n+1} + \sum_{n=0}^{\infty} a_n x^{r+n} = 0,$$

which, upon collecting coefficients of x^{r+n} and replacing n by $n - 1$ in the third sum, can be written as

$$\sum_{n=0}^{\infty} [(r + n)(r + n + 2) + 1] a_n x^{r+n} - \sum_{n=1}^{\infty} (r + n - 1)a_{n-1} x^{r+n} = 0.$$

Dividing by x^r yields

$$\sum_{n=0}^{\infty} [(r + n)(r + n + 2) + 1] a_n x^n - \sum_{n=1}^{\infty} (r + n - 1)a_{n-1} x^n = 0. \tag{9.5.25}$$

We determine the a_n in the usual manner. When $n = 0$, we obtain

$$[r(r + 2) + 1]a_0 = 0,$$

so that the indicial equation is

$$r^2 + 2r + 1 = 0;$$

that is,

$$(r + 1)^2 = 0.$$

Hence there is only one root:
$$r = -1.$$

From (9.5.25), the remaining coefficients must satisfy the recurrence relation

$$[(r+n)(r+n+2)+1]a_n - (r+n-1)a_{n-1} = 0, \qquad n = 1, 2, \ldots.$$

Setting $r = -1$, this simplifies to

$$a_n = \frac{n-2}{n^2}a_{n-1}, \qquad n = 1, 2, 3, \ldots.$$

Consequently,

$$a_1 = -a_0, \qquad a_2 = a_3 = a_4 = \cdots = 0.$$

Thus, a Frobenius series solution to (9.5.24) is

$$y_1(x) = x^{-1}(1-x), \tag{9.5.26}$$

where we set $a_0 = 1$.

Since the indicial equation for (9.5.24) has only one root, there does not exist a second linearly independent Frobenius series solution. However, according to Theorem 9.5.1, there is a second linearly independent solution of the form

$$y_2(x) = y_1(x)\ln x + x^{-1}\sum_{n=1}^{\infty} b_n x^n, \tag{9.5.27}$$

where the coefficients b_n can be determined by substitution into (9.5.24). We now determine such a solution. The computations are quite straightforward, but they are long and tedious. The reader is encouraged to pay full attention and not to be overwhelmed by the formidable look of the equations. Differentiating (9.5.27) with respect to x yields

$$y_2' = y_1'\ln x + x^{-1}y_1 + \sum_{n=1}^{\infty}(n-1)b_n x^{n-2},$$

$$y_2'' = y_1''\ln x + 2x^{-1}y_1' - x^{-2}y_1 + \sum_{n=1}^{\infty}(n-2)(n-1)b_n x^{n-3}.$$

Substituting into (9.5.24), we obtain the following equation for the b_n:

$$x^2\left[y_1''\ln x + 2x^{-1}y_1' - x^{-2}y_1 + \sum_{n=1}^{\infty}(n-2)(n-1)b_n x^{n-3}\right]$$

$$+ x(3-x)\left[y_1'\ln x + x^{-1}y_1 + \sum_{n=1}^{\infty}(n-1)b_n x^{n-2}\right]$$

$$+ y_1\ln x + x^{-1}\sum_{n=1}^{\infty}b_n x^n = 0,$$

or, equivalently,

$$\left[(x^2 y_1'' + x(3-x)y_1' + y_1)\right]\ln x + 2xy_1' + 2y_1 - xy_1 + \sum_{n=1}^{\infty}(n-1)(n-2)b_n x^{n-1}$$

$$+ 3\sum_{n=1}^{\infty}(n-1)b_n x^{n-1} - \sum_{n=1}^{\infty}(n-1)b_n x^n + \sum_{n=1}^{\infty}b_n x^{n-1} = 0.$$

Since y_1 is a solution to Equation (9.5.24), the combination of terms multiplying $\ln x$ will vanish.[5] Combining the coefficients of x^{n-1}, we obtain

$$2xy_1' + 2y_1 - xy_1 + \sum_{n=1}^{\infty}[(n-1)(n-2) + 3(n-1) + 1]b_n x^{n-1} - \sum_{n=1}^{\infty}(n-1)b_n x^n = 0.$$

Simplifying the terms in the first sum and replacing n by $n-1$ in the second sum yields

$$2xy_1' + 2y_1 - xy_1 + \sum_{n=1}^{\infty} n^2 b_n x^{n-1} - \sum_{n=2}^{\infty}(n-2)b_{n-1} x^{n-1} = 0. \qquad (9.5.28)$$

From (9.5.26), we have

$$y_1(x) = x^{-1}(1-x), \qquad y_1'(x) = -x^{-2}.$$

Substituting these expressions into (9.5.28) gives

$$-2x^{-1} + 2x^{-1}(1-x) - (1-x) + \sum_{n=1}^{\infty} n^2 b_n x^{n-1} - \sum_{n=2}^{\infty}(n-2)b_{n-1} x^{n-1} = 0.$$

That is,

$$-3 + x + \sum_{n=1}^{\infty} n^2 b_n x^{n-1} - \sum_{n=2}^{\infty}(n-2)b_{n-1} x^{n-1} = 0.$$

Equating the corresponding coefficients of x^{n-1} to zero for $n = 1, 2, \ldots$, we obtain the b_n in a familiar manner.

$n = 1$: $-3 + b_1 = 0$, which implies that $b_1 = 3$.

$n = 2$: $1 + 4b_2 = 0$, which implies that $b_2 = -\frac{1}{4}$.

$n \geq 3$: $n^2 b_n - (n-2)b_{n-1} = 0$.

That is,

$$b_n = \frac{n-2}{n^2} b_{n-1}, \qquad n = 3, 4, \ldots. \qquad (9.5.29)$$

When $n = 3$, we have

$$b_3 = \frac{1}{3^2} b_2.$$

Substituting for $b_2 = -\frac{1}{4} = -\frac{1}{2^2}$ yields

$$b_3 = -\frac{1}{2^2 \cdot 3^2}.$$

When $n = 4$, (9.5.29) implies that

$$b_4 = \frac{2}{4^2} b_3 = -\frac{1 \cdot 2}{2^2 \cdot 3^2 \cdot 4^2}.$$

We see that in general, for $n \geq 3$,

$$b_n = -\frac{(n-2)!}{2^2 \cdot 3^2 \cdots n^2},$$

---

[5]Note that this is not just a fortuitous result for this particular example. In general, the combination of terms multiplying $\ln x$ will vanish at this stage of the computation.

which can be written as

$$b_n = -\frac{(n-2)!}{(n!)^2}, \qquad n = 3, 4, \ldots.$$

Finally, substituting for b_n into (9.5.27) yields the following solution to Equation (9.5.24):

$$y_2(x) = x^{-1}(1-x)\ln x + x^{-1}\left[3x - \frac{1}{4}x^2 - \sum_{n=3}^{\infty}\frac{(n-2)!}{(n!)^2}x^n\right].$$

Theorem 9.5.1 guarantees that y_1 and y_2 are linearly independent on $(0, \infty)$.   □

We give one final example to illustrate the case when the roots of the indicial equation differ by an integer.

Example 9.5.5   Determine two linearly independent solutions to

$$x^2 y'' + xy' - (4+x)y = 0, \qquad x > 0. \tag{9.5.30}$$

Solution: Since $x = 0$ is a regular singular point, we try for Frobenius series solutions. Substituting

$$y(x) = \sum_{n=0}^{\infty} a_n x^{r+n}$$

into (9.5.30) and simplifying yields the indicial equation

$$r^2 - 4 = 0$$

and the recurrence relation

$$[(r+n)^2 - 4]a_n = a_{n-1}, \qquad n = 1, 2, \ldots. \tag{9.5.31}$$

It follows that the roots of the indicial equation are

$$r_1 = 2 \qquad \text{and} \qquad r_2 = -2,$$

which differ by an integer. Substituting $r = 2$ into (9.5.31) yields

$$a_n = \frac{1}{n(n+4)}a_{n-1}, \qquad n = 1, 2, \ldots.$$

This is easily solved to obtain

$$a_n = \frac{4!}{n!(n+4)!}a_0.$$

Consequently, one Frobenius series solution to (9.5.30) is

$$y_1(x) = a_0 \sum_{n=0}^{\infty}\frac{4!}{n!(n+4)!}x^{n+2}.$$

Choosing $a_0 = 1/4!$, this solution reduces to

$$y_1(x) = \sum_{n=0}^{\infty}\frac{1}{n!(n+4)!}x^{n+2}. \tag{9.5.32}$$

We now determine whether there exists a second linearly independent *Frobenius series* solution. Substituting $r = -2$ into (9.5.31) yields

$$n(n-4)a_n = a_{n-1}, \qquad n = 1, 2, \ldots. \tag{9.5.33}$$

Thus, when $n = 1$, $n = 2$, or $n = 3$, we obtain

$$a_1 = -\tfrac{1}{3}a_0, \qquad a_2 = \tfrac{1}{12}a_0, \qquad a_3 = -\tfrac{1}{36}a_0.$$

However, when $n = 4$, (9.5.33) requires

$$0 \cdot a_4 = a_3 = -\tfrac{1}{36}a_0,$$

which is clearly impossible, since $a_0 \neq 0$. It follows that a second linearly independent *Frobenius series* solution does not exist. However, according to Theorem 9.5.1, there is a second linearly independent solution of the form

$$y_2(x) = Ay_1(x)\ln x + x^{-2}\sum_{n=0}^{\infty} b_n x^n, \tag{9.5.34}$$

where the constants A and b_n can be determined by substitution into (9.5.30). Differentiating y_2 yields

$$y_2' = Ay_1'\ln x + Ax^{-1}y_1 + \sum_{n=0}^{\infty}(n-2)b_n x^{n-3},$$

$$y_2'' = Ay_1''\ln x + 2Ax^{-1}y_1' - Ax^{-2}y_1 + \sum_{n=0}^{\infty}(n-3)(n-2)b_n x^{n-4}.$$

By substituting into (9.5.30) and simplifying, we obtain

$$A\left[x^2 y_1'' + xy_1' - (4+x)y\right]\ln x + 2Axy_1' + \sum_{n=0}^{\infty}(n-3)(n-2)b_n x^{n-2}$$

$$+ \sum_{n=0}^{\infty}(n-2)b_n x^{n-2} - 4\sum_{n=0}^{\infty}b_n x^{n-2} - \sum_{n=0}^{\infty}b_n x^{n-1} = 0.$$

The combination of terms multiplying $\ln x$ will vanish,[6] since y_1 is a solution of (9.5.30). Combining the coefficients of x^{n-2} and simplifying yields

$$2Axy_1' + \sum_{n=1}^{\infty}n(n-4)b_n x^{n-2} - \sum_{n=0}^{\infty}b_n x^{n-1} = 0. \tag{9.5.35}$$

We must now determine y_1'. Differentiating (9.5.32), we obtain

$$y_1'(x) = \sum_{n=0}^{\infty}\frac{n+2}{n!(n+4)!}x^{n+1}.$$

Substituting into (9.5.35) yields

$$2A\sum_{n=0}^{\infty}\frac{n+2}{n!(n+4)!}x^{n+2} + \sum_{n=1}^{\infty}n(n-4)b_n x^{n-2} - \sum_{n=0}^{\infty}b_n x^{n-1} = 0.$$

---

[6]Once more, we note that this will always happen at this stage of the computation.

We now replace n with $n - 4$ in the first sum and replace n with $n - 1$ in the third sum to obtain a common power of x in all sums. The result is

$$2A \sum_{n=4}^{\infty} \frac{n - 2}{(n - 4)!n!} x^{n-2} + \sum_{n=1}^{\infty} n(n - 4)b_n x^{n-2} - \sum_{n=1}^{\infty} b_{n-1} x^{n-2} = 0.$$

Upon multiplying through by x^2, this becomes

$$2A \sum_{n=4}^{\infty} \frac{n - 2}{(n - 4)!n!} x^n + \sum_{n=1}^{\infty} n(n - 4)b_n x^n - \sum_{n=1}^{\infty} b_{n-1} x^n = 0.$$

We can now determine the appropriate values of the constants by setting successive coefficients of x^n to zero in the usual manner.

$n = 1$: $-3b_1 - b_0 = 0$; hence

$$b_1 = -\tfrac{1}{3}b_0.$$

$n = 2$: $-4b_2 - b_1 = 0$; hence

$$b_2 = -\tfrac{1}{4}b_1 = \tfrac{1}{12}b_0.$$

$n = 3$: $-3b_3 - b_2 = 0$; hence

$$b_3 = -\tfrac{1}{3}b_2 = -\tfrac{1}{36}b_0.$$

$n = 4$: $\tfrac{1}{6}A - b_3 = 0$; hence

$$A = 6b_3 = -\tfrac{1}{6}b_0.$$

$n \geq 5$:

$$b_n = \frac{1}{n(n - 4)}\left[b_{n-1} - \frac{2A(n - 2)}{(n - 4)!n!}\right].$$

Using this recurrence relation, we could continue to determine values of the b_n. Notice that b_4 is unconstrained in this problem and so can be set equal to any convenient value. For example, we could set $b_4 = 0$. Substituting the values of the coefficients so far obtained into (9.5.34) gives

$$y_2(x) = b_0\left[-\tfrac{1}{6}y_1(x)\ln x + x^{-2}\left(1 - \tfrac{1}{3}x + \tfrac{1}{12}x^2 - \tfrac{1}{36}x^3 + \cdots\right)\right]. \qquad (9.5.36)$$

Thus, two linearly independent solutions to the given differential equation on $(0, \infty)$ are

$$y_1(x) = \sum_{n=0}^{\infty} \frac{1}{n!(n + 4)!} x^{n+2}$$

and

$$y_2(x) = y_1(x)\ln x - 6x^{-2}\left(1 - \tfrac{1}{3}x + \tfrac{1}{12}x^2 - \tfrac{1}{36}x^3 + \cdots\right),$$

where we set $b_0 = -6$ in (9.5.36). $\square$

Remark   In the previous example, we had to be content with determining only a few terms in the second linearly independent solution, since we could not solve the recurrence relation that arose for the b_n. This is usually the case in these types of problems.

Exercises for 9.5

Skills

- Understand the three cases arising from the roots of the indicial equation of the differential equation $x^2y'' + xp(x)y' + q(x)y = 0$, where p and q are analytic at $x = 0$. The results are summarized in Theorem 9.5.1.

- Be able to determine the general form for two linearly independent series solutions to the differential equation $x^2y'' + xp(x)y' + q(x)y = 0$

- Be able to solve for the coefficients in the general form of a series solution to the differential equation $x^2y'' + xp(x)y' + q(x)y = 0$ by substituting the series into the differential equation.

True-False Review

For Questions 1–6, decide if the given statement is **true** or **false**, and give a brief justification for your answer. If true, you can quote a relevant definition or theorem from the text. If false, provide an example, illustration, or brief explanation of why the statement is false.

1. The roots of the indicial equation for the differential equation

$$x^2y'' - (x - x^2)y' + (1 + x^3)y = 0$$

are distinct and differ by an integer.

2. The roots of the indicial equation for the differential equation

$$x^2y'' - (2\sqrt{5} - 1)xy' + (\tfrac{19}{4} - 3x^2)y = 0$$

are distinct and differ by an integer.

3. The roots of the indicial equation for the differential equation

$$x^2y'' + (9x - 2x^5)y' + (25 + 5x^2 + 10x^4)y = 0$$

are distinct and differ by an integer.

4. The roots of the indicial equation for the differential equation

$$x^2y'' + (4x + \tfrac{1}{2}x^2 - \tfrac{1}{3}x^3)y' - \tfrac{7}{4}y = 0$$

are distinct and differ by an integer.

5. The roots of the indicial equation for the differential equation

$$x^2y'' + x^2y' + xy = 0$$

are distinct and differ by an integer.

6. Two linearly independent Frobenius series solutions exist for the differential equation $x^2y'' + xp(x)y' + q(x)y = 0$ (where p and q are analytic at $x = 0$) only if the roots r_1 and r_2 of the indicial equation differ by an integer.

Problems

For Problems 1–7, determine the roots of the indicial equation of the given differential equation. Also obtain the general *form* of two linearly independent solutions to the differential equation on an interval $(0, R)$. Finally, if $r_1 - r_2$ equals a positive integer, obtain the recurrence relation and determine whether the constant A in

$$y_2(x) = Ay_1(x)\ln x + x^{r_2}\sum_{n=0}^{\infty} b_n x^n$$

is zero or nonzero.

1. $4x^2y'' + 2x^2y' + y = 0$.

2. $x^2y'' + x(\cos x)y' - 2e^x y = 0$.

3. $x^2y'' + x^2y' - (2 + x)y = 0$.

4. $x^2y'' + 2x^2y' + (x - \tfrac{3}{4})y = 0$.

5. $x^2y'' + xy' + (2x - 1)y = 0$.

6. $x^2y'' + x^3y' - (2 + x)y = 0$.

7. $x^2(x^2 + 1)y'' + 7xe^x y' + 9(1 + \tan x)y = 0$.

8. Determine all values of the constant α for which

$$x^2y'' + x(1 - 2x)y' + [2(\alpha - 1)x - \alpha^2]y = 0$$

has two linearly independent *Frobenius series* solutions on $(0, \infty)$.

9. The indicial equation and recurrence relation for the differential equation

$$x^2y'' + x[(2 - b) + x]y' - (b - \gamma x)y = 0$$
$$(9.5.37)$$

are, respectively,

$$(r + 1)(r - b) = 0,$$
$$(r + n + 1)(r + n - b)a_n = -[(r + n - 1) + \gamma]a_{n-1},$$
$$n = 1, 2, 3, \ldots,$$

in the usual notation, where b and γ are constants. Determine the *form* of two linearly independent series solutions to Equation (9.5.37) on $(0, \infty)$ in the following cases:

(a) b not an integer.

(b) $b = -1$.

(c) $b = N$, a nonnegative integer. [For solutions containing a term of the form $Ay_1(x) \ln x$, you must determine whether A is zero or nonzero.]

10. Show that

$$x^2(1 + x)y'' + x^2 y' - 2y = 0$$

has two linearly independent Frobenius series solutions on $(-1, 1)$ and find them.

11. Consider the differential equation

$$x^2 y'' + 3xy' + (1 - x)y = 0, \qquad x > 0. \quad (9.5.38)$$

(a) Determine the indicial equation and show that it has a repeated root $r = -1$.

(b) Obtain the corresponding Frobenius series solution.

(c) It follows from Theorem 9.5.1 that Equation (9.5.38) has a second linearly independent solution of the form

$$y_2(x) = y_1(x) \ln x + \frac{1}{x} \sum_{n=1}^{\infty} b_n x^n.$$

Show that $b_1 = -2$ and that in general,

$$b_n = \frac{1}{n^2}\left[b_{n-1} - \frac{2n}{n!(n-1)!}\right],$$

$$n = 2, 3, 4, \dots$$

Use this to find the first three terms of y_2.

12. Consider the differential equation

$$xy'' - y = 0, \qquad x > 0. \quad (9.5.39)$$

(a) Show that the roots of the indicial equation are $r_1 = 1$ and $r_2 = 0$, and determine the Frobenius series solutions corresponding to $r_1 = 1$.

(b) Show that there does not exist a second linearly independent Frobenius series solution.

(c) According to Theorem 9.5.1, Equation (9.5.39) has a second linearly independent solution of the form

$$y_2(x) = Ay_1(x) \ln x + \sum_{n=0}^{\infty} b_n x^n.$$

Show that $A = b_0$, and determine the first three terms in y_2.

For Problems 13–26, determine two linearly independent solutions to the given differential equation on $(0, \infty)$.

13. $x^2 y'' + x(1 - x)y' - y = 0$.

14. $x^2 y'' + x(6 + x^2)y' + 6y = 0$.

15. $xy'' + y' - 2y = 0$.

16. $4x^2 y'' + (1 - 4x)y = 0$.

17. $x^2 y'' - x(x + 3)y' + 4y = 0$.

18. $x^2 y'' + xy' - (1 + x)y = 0$.

19. $x^2 y'' - x^2 y' - (3x + 2)y = 0$.

20. $x^2 y'' - x^2 y' - 2y = 0$.

21. $4x^2 y'' + 4x(1 - x)y' + (2x - 9)y = 0$.

22. $x^2 y'' + x(5 - x)y' + 4y = 0$.

23. $x^2 y'' - x(1 - x)y' + (1 - x)y = 0$.

24. $x^2 y'' + 2x(2 + x)y' + 2(1 + x)y = 0$.

25. $4x^2 y'' - (3 + 4x)y = 0$.

26. $4x^2 y'' + 4x(1 + 2x)y' + (4x - 1)y = 0$.

For Problems 27–28, determine a Frobenius series solution to the given differential equation and use the *reduction-of-order* technique to find a second linearly independent solution on $(0, \infty)$.

27. $xy'' - xy' + y = 0$.

28. $x^2 y'' + x(4 + x)y' + (2 + x)y = 0$.

29. Consider the Laguerre differential equation

$$x^2 y'' + x(1 - x)y' + Nxy = 0, \quad (9.5.40)$$

where N is a constant. Show that in the case when N is a positive integer, Equation (9.5.40) has a solution that is a polynomial of degree N, and find it. When properly normalized, these solutions are called the **Laguerre polynomials**.

30. Consider the differential equation

$$x^2 y'' + x(1 + 2N + x)y' + N^2 y = 0, \quad (9.5.41)$$

where N is a positive integer.

(a) Show that there is only one Frobenius series solution and that it terminates after $N + 1$ terms. Find this solution.

(b) Show that the change of variables $Y = x^N y$ transforms Equation (9.5.41) into the Laguerre differential equation (9.5.40).

9.6   Bessel's Equation of Order p

One of the most important differential equations in applied mathematics and mathematical physics is **Bessel's equation of order** p, defined by

$$x^2 y'' + xy' + (x^2 - p^2)y = 0, \quad (9.6.1)$$

where p is a *nonnegative* constant. In general, it is not possible to obtain closed-form solutions to this equation. However, since $x = 0$ is a regular singular point, we can apply the Frobenius series technique to obtain series solutions. We will assume that $x > 0$. The indicial equation for (9.6.1) is

$$r(r - 1) + r - p^2 = 0$$

with roots

$$r = \pm p.$$

Consequently, provided that $2p$ is not an integer, there will exist two linearly independent Frobenius series solutions. In order to obtain these solutions, we let

$$y(x) = x^r \sum_{n=0}^{\infty} a_n x^n \quad (9.6.2)$$

so that

$$y'(x) = \sum_{n=0}^{\infty} (r + n)a_n x^{r+n-1}, \qquad y''(x) = \sum_{n=0}^{\infty} (r + n)(r + n - 1)a_n x^{r+n-2}.$$

Substituting into (9.6.1) and rearranging yields

$$\sum_{n=0}^{\infty} [(r + n)^2 - p^2]a_n x^{r+n} + \sum_{n=2}^{\infty} a_{n-2} x^{r+n} = 0. \quad (9.6.3)$$

When $n = 0$, we obtain the indicial equation whose roots are, as we have seen above,

$$r = \pm p.$$

When $n = 1$, (9.6.3) implies that

$$[(r + 1)^2 - p^2]a_1 = 0 \quad (9.6.4)$$

and for $n \geq 2$, we obtain the general recurrence relation

$$[(r + n)^2 - p^2]a_n = -a_{n-2}, \qquad n = 2, 3, \ldots. \quad (9.6.5)$$

Consider the root $r = p$. In this case, (9.6.4) reduces to

$$(2p + 1)a_1 = 0$$

so that, since $p \geq 0$,

$$a_1 = 0. \tag{9.6.6}$$

Setting $r = p$ in (9.6.5) yields

$$a_n = -\frac{1}{n(2p+n)} a_{n-2}, \qquad n = 2, 3, \dots. \tag{9.6.7}$$

It follows from (9.6.6) and (9.6.7) that all of the odd coefficients are zero; that is,

$$a_{2k+1} = 0, \qquad k = 0, 1, 2, \dots. \tag{9.6.8}$$

Now consider the even coefficients. From (9.6.7), we obtain

$$a_2 = -\frac{1}{2(2p+2)} a_0, \qquad a_4 = -\frac{1}{4(2p+4)} a_2 = \frac{1}{2 \cdot 4(2p+4)(2p+2)} a_0,$$

and so on. The general even coefficient is

$$a_{2k} = \frac{(-1)^k}{2 \cdot 4 \cdots (2k)(2p+2)(2p+4) \cdots (2p+2k)} a_0, \qquad k = 1, 2, \dots,$$

which can be written as

$$a_{2k} = \frac{(-1)^k}{2^{2k} k! (p+1)(p+2) \cdots (p+k)} a_0, \qquad k = 1, 2, 3, \dots.$$

Consequently, the corresponding Frobenius series solution to Bessel's equation is

$$y_1(x) = a_0 x^p \left[1 + \sum_{k=1}^{\infty} \frac{(-1)^k}{2^{2k} k! (p+1)(p+2) \cdots (p+k)} x^{2k} \right]. \tag{9.6.9}$$

This solution is valid for all $x > 0$.

Bessel Functions of the First Kind[7]

In order to study the solutions of Bessel's equation obtained above, it is convenient to first rewrite (9.6.9) in a different, but equivalent form. The analysis splits into two cases:

Case 1: $p = N$, a nonnegative integer: In this case, the solution (9.6.9) can be written as

$$y_1(x) = a_0 x^N \left[1 + \sum_{k=1}^{\infty} \frac{(-1)^k}{2^{2k} k! (N+1)(N+2) \cdots (N+k)} x^{2k} \right], \tag{9.6.10}$$

where the constant a_0 can be chosen arbitrarily. It is convenient to make the choice

$$a_0 = \frac{1}{N! 2^N}.$$

The corresponding solution of Bessel's equation is denoted $J_N(x)$ and is called the **Bessel function of the first kind of integer order** N. Substituting for a_0 into (9.6.10), we obtain

$$J_N(x) = \sum_{k=0}^{\infty} \frac{(-1)^k}{k! (N+k)!} \left(\frac{x}{2} \right)^{2k+N}. \tag{9.6.11}$$

---

[7]The remainder of this section includes only a brief introduction to Bessel functions. For more details and the proofs of the results stated in this section, the reader is referred to N. N. Lebedev, *Special Functions and Their Applications* (New York: Dover, 1972).

The most important Bessel functions of integer order are $J_0(x)$ and $J_1(x)$, since, as we shall see shortly, all other integer-order Bessel functions of the first kind can be expressed in terms of these two. Writing out the first few terms in (9.6.11) when $N = 0, 1$ yields, respectively,

$$J_0(x) = 1 - \tfrac{1}{4}x^2 + \tfrac{1}{64}x^4 - \cdots$$

$$J_1(x) = \tfrac{1}{2}x \left(1 - \tfrac{1}{8}x^2 + \tfrac{1}{192}x^4 - \cdots \right).$$

An analysis of these functions shows that they both oscillate with decaying amplitude. Further, each has an infinite number of nonnegative zeros. A sketch of J_0 and J_1 on the interval $(0, 10]$ is given in Figure 9.6.1.

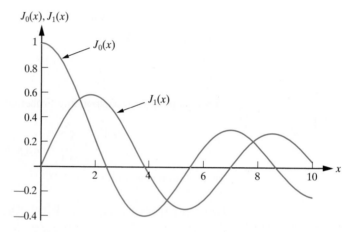

Figure 9.6.1: The Bessel functions of the first kind $J_0(x)$, $J_1(x)$.

Case 2: p a noninteger: In order to obtain a formula analogous to (9.6.11) for the Frobenius series solution (9.6.9) when p is not an integer, we need to introduce the *gamma function*. This function can be considered as the generalization of the factorial function to the case of noninteger real numbers.

DEFINITION   9.6.1

The **gamma function**, Γ, is defined by

$$\Gamma(p) = \int_0^\infty t^{p-1}e^{-t}\, dt, \qquad p > 0.$$

It can be shown that the above improper integral converges for all $p > 0$, so that the gamma function is well defined for all such p. To show that the gamma function is a generalization of the factorial function, we first require the following result:

Lemma 9.6.2   For all $p > 0$,

$$\Gamma(p + 1) = p\Gamma(p). \tag{9.6.12}$$

Proof The proof consists of integrating the expression for $\Gamma(p + 1)$ by parts:

$$\Gamma(p + 1) = \int_0^\infty t^p e^{-t}\, dt = \left[-t^p e^{-t}\right]_0^\infty + p \int_0^\infty t^{p-1} e^{-t}\, dt$$
$$= p\Gamma(p).$$ ∎

We also require

$$\Gamma(1) = \int_0^\infty e^{-t}\, dt = 1. \tag{9.6.13}$$

Equations (9.6.12) and (9.6.13) imply that

$$\Gamma(2) = 1\Gamma(1) = 1, \qquad \Gamma(3) = 2\Gamma(2) = 2!, \qquad \Gamma(4) = 3\Gamma(3) = 3!,$$

and in general, for all nonnegative integers N,

$$\Gamma(N + 1) = N!.$$

This justifies the claim that the gamma function generalizes the factorial function. We now extend the definition of the gamma function to $p < 0$. From (9.6.12),

$$\Gamma(p) = \frac{\Gamma(p + 1)}{p} \tag{9.6.14}$$

for $p > 0$. We use this expression to *define* $\Gamma(p)$ for $p < 0$ as follows. If p is in the interval $(-1, 0)$, then $p + 1$ is in the interval $(0, 1)$ and so $\Gamma(p)$ given in (9.6.14) is well defined. We continue in this manner to successively define $\Gamma(p)$ in the intervals $(-2, -1), (-3, -2), \ldots$. From (9.6.13) and (9.6.14), it follows that

$$\lim_{p \to 0^+} \Gamma(p) = +\infty, \qquad \lim_{p \to 0^-} \Gamma(p) = -\infty$$

so that the graph of the gamma function has the general form given in Figure 9.6.2

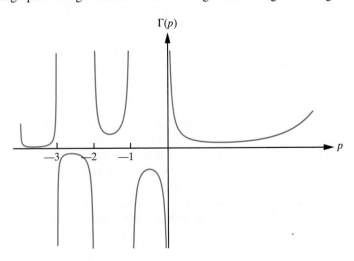

Figure 9.6.2: The gamma function.

We note that the gamma function is continuous and in fact infinitely differentiable at all points of its domain, which consists of all real numbers with the exception of the collection of integers ≤ 0. Finally, before returning to our discussion of Bessel's equation, we require the following formula:

$$\Gamma(p+1)[(p+1)(p+2)\cdots(p+k)] = \Gamma(p+k+1). \tag{9.6.15}$$

The proof of this follows by repeated application of (9.6.14), in the form

$$p\Gamma(p) = \Gamma(p+1),$$

to the left-hand side of the above equality and is left as an exercise.

Now let us return to the solution (9.6.9) of Bessel's equation. Once more we make a specific choice for a_0. We set

$$a_0 = \frac{1}{2^p \Gamma(p+1)}.$$

Substituting this value for a_0 into (9.6.9) and using (9.6.15) yields the **Bessel function of the first kind of order** p, $J_p(x)$, defined by

$$J_p(x) = \sum_{k=0}^{\infty} \frac{(-1)^k}{\Gamma(k+1)\Gamma(p+k+1)} \left(\frac{x}{2}\right)^{2k+p}. \tag{9.6.16}$$

Notice that this does reduce to $J_N(x)$ when N is a nonnegative integer.

Bessel Functions of the Second Kind

Now consider determining the general solution of Bessel's equation. For all $p \geq 0$, we have shown above that one solution to Bessel's equation on $(0, \infty)$ is given in (9.6.16). We therefore require a second linearly independent solution. Since the roots of the indicial equation are $r = \pm p$, it follows from our general Frobenius theory that, provided $2p$ is not equal to an integer, there will exist a second linearly independent Frobenius series solution corresponding to the root $r = -p$. It is not too difficult to see that this solution can be obtained by replacing p by $-p$ in (9.6.16). Thus, we obtain

$$J_{-p}(x) = \sum_{k=0}^{\infty} \frac{(-1)^k}{\Gamma(k+1)\Gamma(k-p+1)} \left(\frac{x}{2}\right)^{2k-p}. \tag{9.6.17}$$

Consequently, the general solution to Bessel's equation of order p when $2p$ is not equal to an integer is

$$y(x) = c_1 J_p(x) + c_2 J_{-p}(x). \tag{9.6.18}$$

When $2p$ is equal to an integer, two subcases arise depending on whether p is itself an integer or a half-integer (that is, $\frac{1}{2}, \frac{3}{2}, \ldots$). In the latter case, with p of the form $(2j+1)/2$ (j a nonnegative integer), a straightforward analysis of the recurrence relation (9.6.5) with $r = -p$ shows that a second Frobenius series also exists in this case, and it is, in fact, given by (9.6.17). Thus, (9.6.18) represents the general solution of Bessel's equation, provided p is not equal to an integer. In practice, rather than using J_{-p} as the second linearly independent solution of Bessel's equation, it is usual to use the following linear combination of these two solutions:

$$Y_p(x) = \frac{J_p(x)\cos p\pi - J_{-p}(x)}{\sin p\pi}. \tag{9.6.19}$$

The function defined in (9.6.19) is called the **Bessel function of the second kind of order** p. Using Y_p, we can therefore write the general solution of Bessel's equation, when p is not equal to a positive integer, in the form

$$y(x) = c_1 J_p(x) + c_2 Y_p(x). \tag{9.6.20}$$

The determination of a second linearly independent solution to Bessel's equation when p is a positive integer, n, is quite a bit more complicated. From our general Frobenius theory, we certainly know the form of the second linearly independent solution—namely,

$$y_2(x) = A J_n(x) \ln x + x^{-n} \sum_{k=0}^{\infty} b_k x^k$$

and the coefficients A, b_n could be determined by direct substitution into (9.6.1). However, if we extend the definition of the Bessel function of the second kind to the case of positive integers by

$$Y_n(x) = \lim_{p \to n} \left[\frac{J_p(x) \cos p\pi - J_{-p}(x)}{\sin p\pi} \right], \tag{9.6.21}$$

it can be shown that the above limit exists and that the resulting function is indeed a second linearly independent solution of Bessel's equation. We could derive the series representation of (9.6.21) by evaluating the limit explicitly. The calculations are lengthy and tedious and so we omit them. The result of these calculations is

$$Y_n(x) = \frac{2}{\pi} J_n(x) \left[\ln \left(\frac{x}{2} \right) + \gamma \right] - \frac{1}{\pi} \left[\sum_{k=0}^{n-1} \frac{(n-k-1)!}{k!} \left(\frac{x}{2} \right)^{2k-n} + \frac{s_n}{n!} \left(\frac{x}{2} \right)^n \right.$$
$$\left. + \sum_{k=1}^{\infty} (-1)^k \frac{s_k + s_{n+k}}{k!(n+k)!} \left(\frac{x}{2} \right)^{2k+n} \right],$$

where

$$s_k = 1 + \frac{1}{2} + \frac{1}{3} + \cdots + \frac{1}{k}$$

and

$$\gamma = \lim_{k \to \infty} (s_k - \ln k) \approx 0.577215664\ldots$$

The value γ is called *Euler's constant*.

Thus (9.6.20) represents the general solution to Bessel's equation of arbitrary order p.

Properties of Bessel Functions of the First Kind

In practice, we are usually interested in solutions of Bessel's equation that are bounded at $x = 0$. However, the Bessel functions of the second kind are always unbounded at $x = 0$. (When p is not an integer, this follows from the fact that x^{-p} has a negative exponent, and when p is an integer, it is due to the second term in the expansion of Y_n given previously.) Thus we usually require only the Bessel functions of the first kind. In this section, we list various properties of the Bessel functions of the first kind that help in either tabulating values of the functions or in working with the functions themselves.

We first recall from (9.6.16) the definition of the Bessel functions of the first kind of order p, namely

$$J_p(x) = \sum_{k=0}^{\infty} \frac{(-1)^k}{\Gamma(k+1)\Gamma(p+k+1)} \left(\frac{x}{2} \right)^{2k+p}. \tag{9.6.22}$$

Property 1:

$$\frac{d}{dx}[x^p J_p(x)] = x^p J_{p-1}(x).\tag{9.6.23}$$

Property 2:

$$\frac{d}{dx}[x^{-p} J_p(x)] = -x^{-p} J_{p+1}(x).\tag{9.6.24}$$

Proof of Property 1: Multiplying (9.6.22) by x^p and differentiating the result with respect to x yields

$$\frac{d}{dx}[x^p J_p(x)] = \sum_{k=0}^{\infty} \frac{(-1)^k}{\Gamma(k+1)\Gamma(p+k+1)}(2k+2p)2^{p-1}\left(\frac{x}{2}\right)^{2k+2p-1}$$

$$= x^p \sum_{k=0}^{\infty} \frac{(-1)^k}{\Gamma(k+1)\Gamma(p+k+1)}(k+p)2^{p-1}\left(\frac{x}{2}\right)^{2k+p-1}.$$

But, from (9.6.14), $\Gamma(p+k+1) = (p+k)\Gamma(p+k)$, so that

$$\frac{d}{dx}[x^p J_p(x)] = x^p \sum_{k=0}^{\infty} \frac{(-1)^k}{\Gamma(k+1)\Gamma(p+k)}\left(\frac{x}{2}\right)^{2k+p-1}$$

$$= x^p J_{p-1}(x).$$

Property 2 is proved similarly.   ∎

We now derive two identities satisfied by the derivatives of J_p. Expanding the derivatives on the right-hand sides of (9.6.23) and (9.6.24) and dividing the resulting equations by x^p and x^{-p} respectively yields

$$J_p'(x) + x^{-1}p J_p(x) = J_{p-1}(x)\tag{9.6.25}$$

$$J_p'(x) - x^{-1}p J_p(x) = -J_{p+1}(x).\tag{9.6.26}$$

Subtracting (9.6.26) from (9.6.25) and rearranging terms, we obtain:

Property 3:

$$J_{p+1}(x) = 2x^{-1}p J_p(x) - J_{p-1}(x).\tag{9.6.27}$$

Similarly, adding (9.6.25) and (9.6.26) and rearranging yields:

Property 4:

$$J_p'(x) = \tfrac{1}{2}[J_{p-1}(x) - J_{p+1}(x)].\tag{9.6.28}$$

These formulas allow us to express high-order Bessel functions and their derivatives in terms of lower-order functions. For example, all integer-order Bessel functions can be expressed in terms of $J_0(x)$ and $J_1(x)$.

Example 9.6.3   Express $J_2(x)$ and $J_3(x)$ in terms of $J_0(x)$ and $J_1(x)$.

Solution: Applying (9.6.27) with $p = 1$, we obtain

$$J_2(x) = 2x^{-1}J_1(x) - J_0(x).$$

Similarly, when $p = 2$,

$$J_3(x) = 4x^{-1}J_2(x) - J_1(x) = x^{-2}(8 - x^2)J_1(x) - 4x^{-1}J_0(x). \qquad \square$$

A Bessel Function Expansion Theorem

It can be shown that every Bessel function of the first kind has an infinite number of positive zeros. We now show how this can be used to obtain a Bessel function expansion of an arbitrary function.

Let $\lambda_1, \lambda_2, \ldots$ denote the positive zeros of the Bessel function $J_p(x)$, where $p \geq 0$ is fixed, and consider the corresponding functions

$$u_m(x) = J_p(\lambda_m x), \qquad u_n(x) = J_p(\lambda_n x)$$

for x in the interval $[0, 1]$. We begin by deriving the orthogonality relation for the functions u_m and u_n. Since $J_p(x)$ solves Bessel's equation

$$y'' + \frac{1}{x} y' + \left(1 - \frac{p^2}{x^2} \right) y = 0,$$

it is not too difficult to show (see Problem 15) that u_m and u_n satisfy

$$u_m'' + \frac{1}{x} u_m' + \left(\lambda_m^2 - \frac{p^2}{x^2} \right) u_m = 0, \tag{9.6.29}$$

$$u_n'' + \frac{1}{x} u_n' + \left(\lambda_n^2 - \frac{p^2}{x^2} \right) u_n = 0. \tag{9.6.30}$$

Multiplying (9.6.29) by u_n and (9.6.30) by u_m, and subtracting, we get

$$(u_m'' u_n - u_n'' u_m) + \frac{1}{x}(u_m' u_n - u_n' u_m) + (\lambda_m^2 - \lambda_n^2) u_m u_n = 0,$$

which can be written as

$$(u_m' u_n - u_n' u_m)' + \frac{1}{x}(u_m' u_n - u_n' u_m) + (\lambda_m^2 - \lambda_n^2) u_m u_n = 0.$$

Multiplying this equation by x and combining the first two terms of the resulting equation, we obtain

$$\frac{d}{dx}\left[x(u_m' u_n - u_n' u_m) \right] + x(\lambda_m^2 - \lambda_n^2) u_m u_n = 0.$$

Integrating from 0 to 1 and using the fact that $u_m(1) = u_n(1) = 0$ (since λ_m and λ_n are zeros of $J_p(x)$) yields

$$(\lambda_m^2 - \lambda_n^2) \int_0^1 x u_m u_n \, dx = 0;$$

that is, since λ_m and λ_n are distinct and positive,

$$\int_0^1 x u_m u_n \, dx = 0.$$

Substituting for u_m and u_n, we finally obtain

$$\int_0^1 x J_p(\lambda_m x) J_p(\lambda_n x) \, dx = 0, \qquad \text{whenever } m \neq n. \tag{9.6.31}$$

In this case, we say that the set of functions $\{J_p(\lambda_n x)\}_{n=1}^{\infty}$ is *orthogonal on* (0, 1) *relative to the weight function* $w(x) = x$. It can be further shown (see Problem 16) that when $m = n$,

$$\int_0^1 x [J_p(\lambda_n x)]^2 \, dx = \tfrac{1}{2}[J_{p+1}(\lambda_n)]^2. \tag{9.6.32}$$

We can now state a Bessel function expansion theorem.

Theorem 9.6.4   If f and f' are continuous on the interval $[0, 1]$, then for $0 < x < 1$,

$$f(x) = a_1 J_p(\lambda_1 x) + a_2 J_p(\lambda_2 x) + \cdots + a_n J_p(\lambda_n x) + \cdots = \sum_{n=1}^{\infty} a_n J_p(\lambda_n x),$$

$$(9.6.33)$$

where the coefficients can be determined from

$$a_n = \frac{2}{[J_{p+1}(\lambda_n)]^2} \int_0^1 x f(x) J_p(\lambda_n x) \, dx. \qquad (9.6.34)$$

Proof   We show only that if a series of the form (9.6.33) exists, then the coefficients are given by (9.6.34). The proof of convergence is omitted. Multiplying (9.6.33) by $x J_p(\lambda_m x)$ and integrating the resulting equation with respect to x from 0 to 1, we obtain

$$\int_0^1 x f(x) J_p(\lambda_m x) \, dx = a_m \int_0^1 x [J_p(\lambda_m x)]^2 \, dx,$$

where we have used (9.6.31). Substituting from (9.6.32) for the integral on the right-hand side of the preceding equation and rearranging terms yields

$$a_m = \frac{2}{[J_{p+1}(\lambda_m)]^2} \int_0^1 x f(x) J_p(\lambda_m x) \, dx,$$

which is what we wished to show.   ∎

Remark   An expansion of the form (9.6.33) is called a **Fourier-Bessel expansion of f.**

Example 9.6.5   Determine the Fourier-Bessel expansion of $f(x) = 1$ in terms of the functions $J_0(\lambda_n x)$.

Solution: According to Theorem 9.6.4, for $0 < x < 1$ we can write

$$1 = \sum_{n=1}^{\infty} a_n J_0(\lambda_n x),$$

where

$$a_n = \frac{2}{[J_1(\lambda_n)]^2} \int_0^1 x J_0(\lambda_n x) \, dx. \qquad (9.6.35)$$

But,

$$\int_0^1 x J_0(\lambda_n x) \, dx = \frac{1}{\lambda_n^2} \int_0^{\lambda_n} u J_0(u) \, du,$$

where $u = \lambda_n x$. Applying (the integrated form of) (9.6.23) with $p = 1$ yields

$$\int_0^1 x J_0(\lambda_n x) \, dx = \frac{1}{\lambda_n^2} [u J_1(u)]_0^{\lambda_n} = \frac{1}{\lambda_n} J_1(\lambda_n).$$

Substitution into (9.6.35) gives

$$a_n = \frac{2}{\lambda_n J_1(\lambda_n)},$$

so that the appropriate Fourier-Bessel expansion is

$$1 = 2 \sum_{n=1}^{\infty} \frac{1}{\lambda_n J_1(\lambda_n)} J_0(\lambda_n x), \qquad 0 < x < 1. \qquad \square$$

Exercises for 9.6

Key Terms

Bessel's equation of order p, Bessel functions of the first kind, Gamma function, Bessel functions of the second kind, Fourier-Bessel expansion of a function.

Skills

- Be able to determine two linearly independent solutions to Bessel's equation.

- Be able to evaluate the gamma function for all p at which the gamma function is defined.

- Be familiar with Properties 1–4 of the Bessel functions of the first kind.

- If f is a continuous function with a continuous derivative on [0, 1], be able to compute a Fourier-Bessel expansion of f on the interval (0, 1).

True-False Review

For Questions 1–7, decide if the given statement is **true** or **false**, and give a brief justification for your answer. If true, you can quote a relevant definition or theorem from the text. If false, provide an example, illustration, or brief explanation of why the statement is false.

1. Provided that p is a positive noninteger, two linearly independent Frobenius series solutions can be obtained to Bessel's equation of order p.

2. Equations (9.6.18) and (9.6.20) are both valid general solutions to Bessel's equation of order p in the case when $2p$ is not an integer.

3. The gamma function is defined for all real numbers p.

4. $\Gamma(p) < 0$ if and only if the greatest integer less than or equal to p is odd and negative.

5. The Bessel function $J_p(x)$ can be written as a linear combination of the Bessel functions $J_{p-1}(x)$ and $J_{p-2}(x)$.

6. The Bessel functions are differentiable with $J_p'(x) = p J_{p-1}(x)$.

7. If $\lambda_1, \lambda_2, \ldots$, denote the positive zeros of the Bessel function $J_p(x)$, then the functions $J_p(\lambda_n x)$ and $J_p(\lambda_m x)$ are orthogonal on (0, 1) relative to the inner product $\langle f, g \rangle = \int_0^1 f(x)g(x)\, dx$.

Problems

1. Use the relations (9.6.4) and (9.6.5) to show that if p is a half-integer, then Bessel's equation of order p has two linearly independent Frobenius series solutions.

2. Find two linearly independent solutions to

$$x^2 y'' + xy' + \left(x^2 - \tfrac{9}{4}\right) y = 0$$

on the interval $(0, \infty)$.

3. Let $\Gamma(p)$ denote the gamma function. Show that

$$\Gamma(p+1)[(p+1)(p+2)\cdots(p+k)] = \Gamma(p+k+1).$$

4. **(a)** By making the change of variables $t = x^2$ in the integral that defines the gamma function, show that

$$\Gamma(1/2) = 2 \int_0^{\infty} e^{-x^2}\, dx.$$

(b) Use your result from (a) to show that

$$[\Gamma(1/2)]^2 = 4 \int_0^{\infty} \int_0^{\infty} e^{-(x^2+y^2)}\, dx\, dy.$$

(c) By changing to polar coordinates, evaluate the double integral in (b) and hence show that

$$\Gamma(\tfrac{1}{2}) = \sqrt{\pi}.$$

(d) Use your result from (c) to find $\Gamma(\tfrac{3}{2})$ and $\Gamma(-\tfrac{1}{2})$.

5. Let $J_p(x)$ denote the Bessel function of the first kind of order p. Show that

$$\frac{d}{dx}(x^{-p}J_p(x)) = -x^{-p}J_{p+1}(x).$$

6. By manipulating the general expression for $J_{1/2}(x)$, show that it can be written in closed form as

$$J_{1/2}(x) = \sqrt{\frac{2}{\pi x}}\sin x.$$

7. Given that

$$J_{1/2}(x) = \sqrt{\frac{2}{\pi x}}\sin x, \qquad J_{-1/2}(x) = \sqrt{\frac{2}{\pi x}}\cos x,$$

express $J_{3/2}(x)$ and $J_{-3/2}(x)$ in closed form. Convince yourself that all half-integer order Bessel functions of the first kind can be expressed as a finite sum of terms involving products of $\sin x$, $\cos x$, and powers of x.

8. By integrating the recurrence relation for derivatives of the Bessel functions of the first kind, show that

(a) $\int x^p J_{p-1}(x)\,dx = x^p J_p(x) + C.$

(b) $\int x^{-p} J_{p+1}(x)\,dx = -x^{-p}J_p(x) + C.$

9. Show that

(a) $J_0''(x) = -J_0(x) - x^{-1}J_0'(x).$

(b) $J_0'''(x) = x^{-1}J_0(x) + x^{-2}(2 - x^2)J_0'(x).$

10. Show that

$$J_4(x) = 8x^{-3}(6 - x^2)J_1(x) - x^{-2}(24 - x^2)J_0(x).$$

11. Show that

$$J_2'(x) = 2x^{-1}J_0(x) + x^{-2}(4 - x^2)J_0'(x).$$

12. Show that

(a) $J_2(x) = J_0(x) + 2J_0''(x).$

(b) $J_3(x) = 3J_1(x) + 4J_1''(x).$

13. Determine the Fourier-Bessel expansion in the functions $J_p(\lambda_n x)$ for $f(x) = x^p$, on the interval $(0, 1)$. [Here the λ_n denote the positive zeros of $J_p(x)$.] [**Hint:** You will need to use one of the results from Problem 8.]

14. Determine the Fourier-Bessel expansion in the functions $J_0(\lambda_k x)$ of $f(x) = x^2$ on the interval $(0, 1)$. [Here the λ_k denote the positive zeros of $J_0(x)$.]

15. Let $J_p(x)$ denote the Bessel function of the first kind of order p, and let λ be a positive real number. If $u(x) = J_p(\lambda x)$, show that u satisfies the differential equation

$$\frac{d^2u}{dx^2} + \frac{1}{x}\frac{du}{dx} + \left(\lambda^2 - \frac{p^2}{x^2}\right)u = 0.$$

16. Let λ and μ be positive real numbers. Then $J_p(\lambda x)$ and $J_p(\mu x)$ satisfy

$$\frac{d}{dx}\left\{x\frac{d}{dx}[J_p(\lambda x)]\right\} + \left(\lambda^2 x - \frac{p^2}{x}\right)J_p(\lambda x) = 0,$$

(9.6.36)

$$\frac{d}{dx}\left\{x\frac{d}{dx}[J_p(\mu x)]\right\} + \left(\mu^2 x - \frac{p^2}{x}\right)J_p(\mu x) = 0,$$

(9.6.37)

respectively.

(a) Show that for $\lambda \neq \mu$,

$$\int_0^1 x J_p(\lambda x) J_p(\mu x)\,dx$$
$$= \frac{\mu J_p(\lambda)J_p'(\mu) - \lambda J_p(\mu)J_p'(\lambda)}{\lambda^2 - \mu^2}.$$

(9.6.38)

[**Hint:** Multiply (9.6.36) by $J_p(\mu x)$, (9.6.37) by $J_p(\lambda x)$, subtract the resulting equations and integrate over $(0, 1)$.] If λ and μ are distinct zeros of $J_p(x)$, what does your result imply?

(b) In order to compute $\int_0^1 x[J_p(\mu x)]^2\,dx$, we take the limit as $\lambda \to \mu$ in (9.6.38). Use L'Hopital's rule to compute this limit and thereby show that

$$\int_0^1 x[J_p(\mu x)]^2\,dx$$
$$= \frac{\mu[J_p'(\mu)]^2 - J_p(\mu)J_p'(\mu) - \mu J_p(\mu)J_p''(\mu)}{2\mu}.$$

(9.6.39)

Substituting from Bessel's equation for $J_p''(\mu)$, show that (9.6.39) can be written as

$$\int_0^1 x[J_p(\mu x)]^2 \, dx = \frac{1}{2}\left\{[J_p'(\mu)]^2 + \left(1 - \frac{p^2}{\mu^2}\right)[J_p(\mu)]^2\right\}.$$

(c) In the case when μ is a zero of $J_p(x)$, use (9.6.26) to show that your result in (b) can be written as

$$\int_0^1 x[J_p(\mu x)]^2 \, dx = \frac{1}{2}[J_{p+1}(\mu)]^2.$$

9.7   Chapter Review

In this chapter our focus has been on second-order homogeneous nonconstant-coefficient differential equations of the form

$$y'' + p(x)y' + q(x)y = 0, \tag{9.7.1}$$

for $x \in I$. In general it is not possible to determine two linearly independent solutions to such a differential equation in closed form, so our approach has been to try to represent solutions in the form of some type of convergent infinite series.

Series Solution about an Ordinary Point

The simplest situation occurs when both of the functions p and q in (9.7.1) are analytic at $x_0 \in I$, which means that each can be represented as power series centered at $x = x_0$ with a radius of convergence denoted by R. We say that x_0 is an *ordinary point* of the differential equation (9.7.1). In such a case it is possible to expand the solutions to (9.7.1) in terms of a convergent power series of the form

$$y(x) = \sum_{n=0}^{\infty} a_n(x - x_0)^n \tag{9.7.2}$$

Substitution of this expression for y into (9.7.1) and matching coefficients on each side of the differential equation leads to a recurrence relation on the coefficients. Therefore, all coefficients a_n can be expressed in terms of the initial coefficients a_0 and a_1. In fact, assuming that a_0 and a_1 are nonzero, we obtain the general solution in the form

$$y(x) = a_0 y_1(x) + a_1 y_2(x),$$

where y_1 and y_2 are linearly independent solutions to (9.7.1). This solution is valid for (at least) all x with $|x - x_0| < R$.

Legendre's Equation

A particularly important example of a differential equation of the form (9.7.1) is the Legendre equation

$$(1 - x^2)y'' - 2xy' + \alpha(\alpha + 1)y = 0, \quad -1 < x < 1. \tag{9.7.3}$$

Since the functions

$$p(x) = -\frac{2x}{1 - x^2}, \quad q(x) = \frac{\alpha(\alpha + 1)}{1 - x^2}$$

both have power series expansions about $x = 0$ with radius of convergence $R = 1$, it follows that we can obtain two linearly independent power series solutions to the Legendre equation which are also valid for (at least) $|x| < 1$. The key results that we have established about the solutions to the Legendre equation are as follows:

1. If $\alpha = N$, a nonnegative integer, then the Legendre equation has solutions (unique up to a multiplicative constant) that are polynomials of degree N. Imposing the condition that $y(1) = 1$ yields the degree-N Legendre polynomial denoted $P_N(x)$.

2. The set of all Legendre polynomials is an orthogonal set of functions on the interval $[-1, 1]$; that is,

$$\int_{-1}^{1} P_M(x) P_N(x) \, dx = 0, \qquad \text{whenever } M \neq N.$$

3. If f and f' are continuous on the interval $(-1, 1)$, then for $-1 < x < 1$,

$$f(x) = \sum_{n=0}^{\infty} a_n P_n(x),$$

where

$$a_n = \frac{2n+1}{2} \int_{-1}^{1} f(x) P_n(x) \, dx.$$

Series Solution about a Regular Singular Point

We now rewrite (9.7.1) in the form

$$y'' + P(x)y' + Q(x)y = 0. \tag{9.7.4}$$

If $P(x)$ and/or $Q(x)$ fail to be analytic at $x_0 \in I$, then x_0 is called a *singular point* of the differential equation (9.7.4). In this chapter we have considered the special type of singular point that arises when both of the functions

$$p(x) = (x - x_0)P(x) \quad \text{and} \quad q(x) = (x - x_0)^2 Q(x)$$

are analytic at $x = x_0$. In such a case the point x_0 is called a *regular singular point*. Without loss of generality we can restrict our attention to the case when $x_0 = 0$, in which case (9.7.4) can be written as

$$x^2 y'' + x p(x) y' + q(x) y = 0, \tag{9.7.5}$$

where p and q are assumed to be analytic at $x = 0$. We let R denote the smaller of the radii of convergence of the power series expansions of $p(x)$ and $q(x)$ about $x = 0$. If r_1 and r_2 denote the roots of the indicial equation

$$r(r - 1) + p(0)r + q(0) = 0, \tag{9.7.6}$$

with $r_1 \geq r_2$ if these roots are real, then we have seen in this chapter that two linearly independent solutions to (9.7.5) on the interval $(0, R)$ can be determined as follows.

1. If $r_1 - r_2 \neq$ integer, then

$$y_1(x) = x^{r_1} \sum_{n=0}^{\infty} a_n x^n, \quad a_0 \neq 0, \qquad y_2(x) = x^{r_2} \sum_{n=0}^{\infty} b_n x^n, \quad b_0 \neq 0.$$

2. If $r_1 = r_2 = r$, then

$$y_1(x) = x^r \sum_{n=0}^{\infty} a_n x^n, \quad a_0 \neq 0, \qquad y_2(x) = y_1(x) \ln x + x^r \sum_{n=1}^{\infty} b_n x^n.$$

3. If $r_1 - r_2 =$ positive integer, then

$$y_1(x) = x^r \sum_{n=0}^{\infty} a_n x^n, \quad a_0 \neq 0, \qquad y_2(x) = A y_1(x) \ln x + x^{r_2} \sum_{n=0}^{\infty} b_n x^n, \quad b_0 \neq 0.$$

The coefficients in each of these solutions can be determined by direct substitution into (9.7.5).

Bessel's Equation of Order p

An important differential equation that has a regular singular point at $x = 0$ is *Bessel's equation of order p*:

$$x^2 y'' + x y' + (x^2 - p^2) y = 0, \quad x > 0, \tag{9.7.7}$$

where p is a nonnegative constant. This differential equation has general solution

$$y(x) = c_1 J_p(x) + c_2 Y_p(x),$$

where J_p and Y_p denote the Bessel functions of the first and second kinds of order p, respectively. Our major interest is in the Bessel functions of the first kind. We let $\lambda_1, \lambda_2, \ldots,$ denote the positive zeros of $J_p(x)$, $p \geq 0$, and consider the corresponding functions $J_p(\lambda_n x)$. The key results about these functions are as follows:

1. The set of functions $\{J_p(\lambda_n x)\}_{n=1}^{\infty}$ is orthogonal on the interval $(0, 1)$ relative to the weight function $w(x) = x$; that is,

$$\int_0^1 x J_p(\lambda_m x) J_p(\lambda_n x) \, dx = 0, \quad \text{whenever} \quad m \neq n.$$

2. If f and f' are continuous on the interval $[0, 1]$, then for $0 < x < 1$,

$$f(x) = \sum_{n=1}^{\infty} a_n J_p(\lambda_n x),$$

where

$$a_n = \frac{2}{\left[J_{p+1}(\lambda_n x) \right]^2} \int_0^1 x f(x) J_p(\lambda_n x) \, dx.$$

Additional Problems

For Problems 1–13, determine whether $x = 0$ is an ordinary point or a regular singular point of the given differential equation. Then obtain two linearly independent solutions to the differential equation and state the maximum interval on which your solutions are valid.

1. $y'' + xy = 0.$

2. $y'' - x^2 y = 0.$

3. $(1 - x^2) y'' - 6xy' - 4y = 0.$

4. $xy'' + y' + 2y = 0.$

5. $xy'' + 2y' + xy = 0.$

6. $2xy'' + 5(1 - 2x) y' - 5y = 0.$

7. $xy'' + y' + xy = 0.$

8. $(1 + 4x^2) y'' - 8y = 0.$

9. $x^2 y'' + xy' + \left(x^2 - \frac{1}{4} \right) y = 0.$

10. $4xy'' + 3y' + 3y = 0.$

11. $x^2 y'' + \frac{3}{2} xy' - \frac{1}{2}(1 + x) y = 0.$

12. $x^2 y'' - x(2-x)y' + (2+x^2)y = 0.$

13. $x^2 y'' - 3xy' + 4(x+1)y = 0.$

14. Consider the **hypergeometric equation**

$$x(1-x)y'' + [c - (a+b+1)x]y' - aby = 0, \tag{9.7.8}$$

where a, b, c are constants.

(a) Verify that $x = 0$ is a regular singular point of (9.7.8).

(b) Verify that the roots of the indicial equation are $r_1 = 0, r_2 = 1 - c.$

(c) Show that the coefficients in a Frobenius series solution to Equation (9.7.8) centered at $x = 0$ must satisfy the recurrence relation:

$$a_{n+1} = \frac{(n+r+a)(n+r+b)}{(n+r+1)(n+r+c)}a_n,$$
$$n = 0, 1, \ldots .$$

(d) Assuming that c is not an integer, derive the following solutions to (9.7.8):

$$y_1(x) = F(a, b, c; x) = 1 + \frac{ab}{c}x + \frac{a(a+1)b(b+1)}{c(c+1)}\frac{x^2}{2!}$$
$$+ \cdots +$$
$$\frac{a(a+1)\cdots(a+n-1)b(b+1)\cdots(b+n-1)}{c(c+1)\cdots(c+n-1)}\frac{x^n}{n!},$$
$$y_2(x) = x^{1-c}F(a+1-c, b+1-c, 2-c; x).$$

15. Consider the differential equation

$$(x^2 - 1)y'' + [1 - (a+b)]xy' + aby = 0, \tag{9.7.9}$$

where a and b are constants.

(a) Show that the coefficients in a Frobenius series solution to Equation (9.7.9) centered at $x = 0$ must satisfy the recurrence relation

$$a_{n+2} = \frac{(n-a)(n-b)}{(n+2)(n+1)}a_n, \quad n = 0, 1, \ldots,$$

and determine two linearly independent series solutions.

(b) Show that if either a or b is a nonnegative integer, then one of the solutions obtained in (a) is a polynomial.

(c) Show that if a is an odd positive integer and b is an even positive integer, then *both* of the solutions defined in (a) are polynomials.

(d) If $a = 5$ and $b = 4$, determine two linearly independent polynomial solutions to Equation (9.7.9). Notice that in this case the radius of convergence of the solutions obtained is $R = \infty$, whereas Theorem 9.2.4 only guarantees a radius of convergence $R \geq 1$.

16. Consider the Chebyshev equation

$$(1 - x^2)y'' - xy' + a^2 y = 0, \tag{9.7.10}$$

where a is a constant.

(a) Show that if $a = N$, a nonnegative integer, then Equation (9.7.10) has a polynomial solution of degree N. When suitably normalized, these polynomials are called the **Chebyshev polynomials** and are denoted by $T_N(x)$.

(b) Use Equation (9.7.10) to show that $T_N(x)$ satisfies

$$[\sqrt{1-x^2}T_N']' + \frac{N}{\sqrt{1-x^2}}T_N = 0.$$

(c) Use the result from (b) to prove that

$$\int_{-1}^{1} \frac{T_N(x)T_M(x)}{\sqrt{1-x^2}}dx = 0, \quad M \neq N.$$

17. Consider the differential equation

$$x^2 y'' + x(1 + 2N - x)y' + N^2 y = 0, \quad x > 0. \tag{9.7.11}$$

(a) Find the indicial equation, and show that it has only one root, $r = -N.$

(b) If N is a nonnegative integer, show that the Frobenius series solution to Equation (9.7.11) terminates after N terms.

(c) Determine the Frobenius series solutions when $N = 0, 1, 2, 3.$

(d) Show that if N is a positive integer, then the Frobenius series solution to Equation (9.7.11) can be written as

$$y(x) =$$
$$x^{-N}\left[1 + \sum_{k=1}^{N}(-1)^k \frac{N(N-1)\cdots(N+1-k)}{1^2 \cdot 2^2 \cdots k^2}x^k\right]$$
$$= x^{-N}\left[1 + \sum_{k=1}^{N}(-1)^k \prod_{i=1}^{k} \frac{(N+1-i)}{i^2}x^k\right].$$

18. Consider the general "perturbed" Cauchy-Euler equation

$$x^2 y'' + x[1 - (a + b) + cx]y' + (ab + dx)y = 0,$$
$$x > 0,$$

(9.7.12)

where a, b, c, d are constants. Assuming that a and b are distinct and do not differ by an integer, determine two linearly independent Frobenius series solutions to Equation (9.7.12). [**Hint:** Use symmetry to get the second solution.]

19. Consider the differential equation

$$x^2 y'' + x(1 + bx)y' + [b(1 - N)x - N^2]y = 0,$$
$$x > 0,$$

(9.7.13)

where N is a positive integer and b is a constant.

(a) Show that the roots of the indicial equation are $r = \pm N$.

(b) Show that the Frobenius series solution corresponding to $r = N$ is

$$y_1(x) = a_0 x^N \sum_{n=0}^{\infty} \frac{(2N)!(-b)^n}{(2N + n)!} x^n$$

and that by an appropriate choice of a_0, one solution to (9.7.13) is

$$y_1(x) = x^{-N} \left[e^{-bx} - \sum_{n=0}^{2N-1} \frac{(-bx)^n}{n!} \right].$$

(c) Show that Equation (9.7.13) has a second linearly independent Frobenius series solution that can be taken as

$$y_2(x) = x^{-N} \sum_{n=0}^{2N-1} \frac{(-bx)^n}{n!}.$$

Hence, conclude that Equation (9.7.13) has linearly independent solutions

$$y_1(x) = x^{-N} e^{-bx}, \quad y_2(x) = x^{-N} \sum_{n=0}^{2N-1} \frac{(-bx)^n}{n!}.$$

20. Show that the change of variables $y = x^{1/2}u$ transforms the differential equation

$$y'' + \left(1 - \frac{3}{4x^2} \right) y = 0$$

into the Bessel equation

$$x^2 u'' + xu' + (x^2 - 1)u = 0$$

and thereby write the general solution to the given differential equation.

Review of Complex Numbers

Any number z of the form $z = a + ib$, where a and b are real numbers and $i = \sqrt{-1}$, is called a **complex number**. If $z = a + ib$, then we refer to a as the **real part** of z, denoted $\text{Re}(z)$, and we refer to b as the **imaginary part** of z, denoted $\text{Im}(z)$. Thus

$$\text{If } z = a + ib, \qquad \text{then } \text{Re}(z) = a \text{ and } \text{Im}(z) = b.$$

Example A.1 If $z = 2 - 3i$, then $\text{Re}(z) = 2$ and $\text{Im}(z) = -3$. □

Complex numbers can be added, subtracted, and multiplied in the usual manner, and the result is once more a complex number. Further, these operations satisfy all of the basic properties satisfied by the real numbers. All we need to remember is that whenever we encounter the term i^2, it must be replaced by -1.

Example A.2 If $z_1 = 3 + 4i$ and $z_2 = -1 + 2i$, find $z_1 - 3z_2$ and $z_1 z_2$.

Solution: We have

$$z_1 - 3z_2 = (3 + 4i) - 3(-1 + 2i) = 6 - 2i = 2(3 - i)$$

and

$$z_1 z_2 = (3 + 4i)(-1 + 2i) = -3 + 6i - 4i + 8i^2 = -11 + 2i.$$ □

Example A.3 If $z_1 = 4 + 3i$ and $z_2 = 4 - 3i$, determine $z_1 z_2$.

Solution: In this case, $z_1 z_2 = (4+3i)(4-3i) = 16 - 12i + 12i - 9i^2 = 16 + 9 = 25$. ☐

Notice that in the previous example the product $z_1 z_2$ turned out to be a real number. This was not an accident. If we look at the definition of z_2, we see that it can be obtained from z_1 by replacing the imaginary part of z_1 by its negative. Complex numbers that are related in this manner are called conjugates of one another.

DEFINITION   A.4

If $z = a + ib$, then the complex number $\bar{z}$ defined by

$$\bar{z} = a - ib$$

is called the **conjugate** of z.

Example A.5 If $z = 2 + 5i$, then $\bar{z} = 2 - 5i$, whereas if $z = 3 - 4i$, then $\bar{z} = 3 + 4i$. ☐

Properties of the Conjugate
1. $\bar{\bar{z}} = z$.
2. $z\bar{z} = \bar{z}z = a^2 + b^2$.

Proof

1. If $z = a + ib$, then $\bar{z} = a - ib$, so that $\bar{\bar{z}} = a + ib = z$.
2. We have $z\bar{z} = (a + ib)(a - ib) = a^2 - iab + iab - (ib)^2 = a^2 + b^2$. ∎

If $z = a + ib$, then the real number $\sqrt{a^2 + b^2}$ is often called the **modulus** of z or the **absolute value** of z and is denoted $|z|$. It follows from Property 2 that

$$|z|^2 = z\bar{z}.$$

Example A.6 Determine $|z|$ if $z = 2 - 3i$.

Solution: By definition,

$$|z| = \sqrt{2^2 + (-3)^2} = \sqrt{13}.$$ ☐

We now recall from elementary algebra that an expression of the form $1/(a+\sqrt{b})$ can always be written with the radical in the numerator. To accomplish this, we multiply by

$$\frac{a - \sqrt{b}}{a - \sqrt{b}},$$

and the result is

$$\frac{a - \sqrt{b}}{a^2 - b}.$$

The reason that this works is because

$$(a + \sqrt{b})(a - \sqrt{b}) = a^2 - b.$$

This is similar to $(a + ib)(a - ib) = a^2 + b^2$. Now consider an expression of the form

$$\frac{1}{a + ib}.$$

As this is written, we cannot say that it is a complex number, since it is not of the form $a + ib$. However, if we multiply by

$$\frac{a - ib}{a - ib}$$

and use Property 2 of the conjugate, we obtain

$$\frac{1}{a + ib} = \frac{1}{(a + ib)} \frac{(a - ib)}{(a - ib)} = \frac{a - ib}{a^2 + b^2},$$

which is a complex number.

Example A.7   Express $z = 1/(2 + 5i)$ in the form $a + ib$.

Solution: We have

$$z = \frac{1}{2 + 5i} = \frac{1}{(2 + 5i)} \frac{(2 - 5i)}{(2 - 5i)} = \frac{2}{29} - \frac{5}{29}i. \qquad \square$$

More generally, if $z_1 = a + ib$ and $z_2 = x + iy$, then

$$\frac{z_1}{z_2} = \frac{a + ib}{x + iy} = \frac{(a + ib)}{(x + iy)} \frac{(x - iy)}{(x - iy)} = \frac{1}{x^2 + y^2} \left[(ax + by) + i(ay + bx) \right].$$

This illustrates that we can divide two complex numbers, and the result is once more a complex number.

Example A.8   If $z_1 = 2 + 3i$ and $z_2 = 3 + 4i$, determine z_1/z_2.

Solution: In this case, we have

$$\frac{z_1}{z_2} = \frac{2 + 3i}{3 + 4i} = \frac{(2 + 3i)}{(3 + 4i)} \frac{(3 - 4i)}{(3 - 4i)} = \frac{1}{25}(18 + i). \qquad \square$$

Complex-Valued Functions

A function $w(x)$ of the form

$$w(x) = u(x) + iv(x),$$

where u and v are real-valued functions of a real variable x (and $i^2 = -1$), is called a **complex-valued function** of a real variable. An example of such a function is

$$w(x) = 3 \cos 2x + 4i \sin 3x.$$

The Complex Exponential Function: Recall that for all real x, the function e^x has the Maclaurin expansion

$$e^x = \sum_{n=0}^{\infty} \frac{1}{n!} x^n.$$

We can also discuss convergence of infinite series of complex numbers. We define e^{ib}, where b is a real number, by

$$e^{ib} = \sum_{n=0}^{\infty} \frac{1}{n!}(ib)^n = 1 + ib + \frac{1}{2!}(ib)^2 + \frac{1}{3!}(ib)^3 + \cdots + \frac{1}{n!}(ib)^n + \cdots .$$

Factoring the even and odd powers of b and using the formulas

$$i^{2k} = (-1)^k \qquad \text{and} \qquad i^{2k+1} = (-1)^k i$$

yields

$$e^{ib} = \left[1 - \frac{1}{2!}b^2 + \frac{1}{4!}b^4 + \cdots + \frac{(-1)^k}{(2k)!}b^{2k} + \cdots \right]$$
$$+ i \left[b - \frac{1}{3!}b^3 + \frac{1}{5!}b^5 - \cdots + \frac{(-1)^k}{(2k+1)!}b^{2k+1} + \cdots \right].$$

That is,

$$e^{ib} = \sum_{n=0}^{\infty}(-1)^n \frac{b^{2n}}{(2n)!} + i \sum_{n=0}^{\infty}(-1)^n \frac{b^{2n+1}}{(2n+1)!}.$$

The two series appearing the foregoing equation are, respectively, the Maclaurin series expansions of $\cos b$ and $\sin b$, both of which converge for all real b. Thus, we have shown that

$$\boxed{e^{ib} = \cos b + i \sin b,} \qquad (A.1)$$

which is called **Euler's formula**. It is now natural to *define* e^{a+ib} by

$$e^{a+ib} = e^a \cdot e^{ib} = e^a(\cos b + i \sin b), \qquad (A.2)$$

where a and b are any real numbers.

A function of the form $f(x) = e^{rx}$, where $r = a + ib$ and x is a real variable, is called a **complex exponential function**. Replacing ib with ibx in (A.1) and $a + ib$ with $(a + ib)x$ in (A.2) yields the following important formulas:

$$\boxed{e^{ibx} = \cos bx + i \sin bx, \qquad e^{(a+ib)x} = e^{ax}(\cos bx + i \sin bx).}$$

Replacing i with $-i$, we obtain

$$e^{-ibx} = \cos bx - i \sin bx, \qquad e^{(a-ib)x} = e^{ax}(\cos bx - i \sin bx).$$

Example A.9   Express $e^{(3-5i)x}$ in terms of trigonometric functions.

Solution: We have

$$e^{(3-5i)x} = e^{3x}(\cos 5x - i \sin 5x). \qquad \square$$

The preceding definition of $e^{(a+ib)x}$ also enables us to attach a meaning to $x^{(a+ib)}$. We recall that for a nonrational number r and $x > 0$, x^r is defined by $x^r = e^{r \ln x}$. We now extend this definition to the case when r is complex and therefore define

$$\boxed{x^{a+ib} = e^{(a+ib)\ln x}.}$$

Using Euler's formula, this can be written as

$$x^{a+ib} = x^a e^{ib \ln x} = x^a [\cos(b \ln x) + i \sin(b \ln x)].$$

For example,

$$x^{2+3i} = x^2 [\cos(3 \ln x) + i \sin(3 \ln x)].$$

Differentiation of Complex-Valued Functions

We now return to the general complex-valued function $w(x) = u(x) + iv(x)$. If $u'(x)$ and $v'(x)$ exist, then we define the derivative of w by

$$w'(x) = u'(x) + iv'(x).$$

Higher-order derivatives are defined similarly. In particular, we have the following important result:

$$\boxed{\frac{d}{dx}(e^{rx}) = re^{rx} \qquad \text{when } r \text{ is complex.}}$$

This coincides with the usual formula for the derivative of e^{rx} when r is a real number. To establish the above formula, we proceed as follows. If $r = a + ib$, then

$$e^{rx} = e^{(a+ib)x} = e^{ax}(\cos bx + i \sin bx).$$

Differentiating with respect to x using the product rule yields

$$\begin{aligned}
\frac{d}{dx}(e^{rx}) &= ae^{ax}(\cos bx + i \sin bx) + be^{ax}(-\sin bx + i \cos bx) \\
&= ae^{ax}(\cos bx + i \sin bx) + ibe^{ax}(\cos bx + i \sin bx) \\
&= (a + ib)e^{ax}(\cos bx + i \sin bx) = re^{rx},
\end{aligned}$$

as required.

Similarly, it can be shown that

$$\boxed{\frac{d}{dx}(x^r) = rx^{r-1} \qquad \text{when } r \text{ is complex.}}$$

Exercises for A

Problems

For Problems 1–5, determine $\bar{z}$ and $|z|$ for the given complex number.

1. $z = 2 + 5i$.

2. $z = 3 - 4i$.

3. $z = 5 - 2i$.

4. $z = 7 + i$.

5. $z = 1 + 2i$.

For Problems 6–10, express $z_1 z_2$ and z_1/z_2 in the form $a + ib$.

6. $z_1 = 1 + i$, $z_2 = 3 + 2i$.

7. $z_1 = -1 + 3i$, $z_2 = 2 - i$.

8. $z_1 = 2 + 3i$, $z_2 = 1 - i$.

9. $z_1 = 4 - i$, $z_2 = 1 + 3i$.

10. $z_1 = 1 - 2i$, $z_2 = 3 + 4i$.

11. Show that if z_1 and z_2 are complex numbers, then

$$\overline{z_1 + z_2} = \overline{z_1} + \overline{z_2}.$$

12. Generalize the previous example to the case when $z_1, z_2, \ldots, z_n$ are complex numbers.

13. Show that if z_1 and z_2 are complex numbers, then

$$\overline{z_1 z_2} = \overline{z_1}\,\overline{z_2}.$$

14. Show that if z_1 and z_2 are complex numbers then

$$\overline{(z_1/z_2)} = \overline{z_1}/\overline{z_2}.$$

For Problems 15–22, express the given complex-valued function in the form $u(x) + iv(x)$ for appropriate real-valued functions u and v.

15. e^{2ix}.

16. $e^{(3+4i)x}$.

17. e^{-5ix}.

18. $e^{-(2+i)x}$.

19. x^{2-i}.

20. x^{3i}.

21. x^{-1+2i}.

22. $x^{2i} e^{(3+4i)x}$.

23. Derive the famous mathematical formula

$$e^{i\pi} + 1 = 0.$$

24. Show that

$$\cos bx = \tfrac{1}{2}(e^{ibx} + e^{-ibx})$$

and

$$\sin bx = \frac{1}{2i}(e^{ibx} - e^{-ibx}).$$

[A comparison of these formulas with the corresponding formulas

$$\cosh bx = \tfrac{1}{2}(e^{bx} + e^{-bx})$$

and

$$\sinh bx = \tfrac{1}{2}(e^{bx} - e^{-bx})$$

indicates why the trigonometric and hyperbolic functions satisfy similar identities.]

For Problems 25–27, use the result of Problem 24 to express the given functions in terms of complex exponential functions.

25. $\sin 4x$.

26. $\cos 8x$.

27. $\tan x$.

28. Use the result of Problem 24 to verify the identity $\sin^2 x + \cos^2 x = 1$.

Review of Partial Fractions

In this appendix we review the partial fraction decomposition of rational functions. Recall that a function of the form

$$p(x) = a_n x^n + a_{n-1} x^{n-1} + \cdots + a_1 x + a_0 \tag{B.1}$$

with $a_n \neq 0$ is called a **polynomial of degree** n. According to the fundamental theorem of algebra, the equation $p(x) = 0$ has precisely n roots (not all necessarily distinct). If we let $x_1, x_2, \ldots, x_n$ denote these roots, then $p(x)$ can be factored as

$$p(x) = K(x - x_1)(x - x_2) \cdots (x - x_n), \tag{B.2}$$

where K is a constant. Some of the roots may be complex. We will assume that the coefficients in (B.1) are real numbers in which case any complex roots must occur in conjugate pairs.

A quadratic factor of the form $ax^2 + bx + c$ (where $a \neq 0$) that has no *real* linear factors is said to be **irreducible**.

Theorem B.1   Any real polynomial[1] can be factored into linear and irreducible quadratic terms with real coefficients.

Proof Let $p(x)$ be a real polynomial and suppose that $x = \alpha$ is a complex root of $p(x) = 0$. Then $x = \overline{\alpha}$ is also a root. Thus, (B.2) will contain the terms $(x - \alpha)(x - \overline{\alpha})$. These linear terms have complex coefficients. However, if we expand the product, the result is

$$(x - \alpha)(x - \overline{\alpha}) = x^2 - (\alpha + \overline{\alpha})x + \alpha\overline{\alpha}.$$

---

[1] By a "real polynomial" we mean a polynomial with real coefficients.

But,

$$\alpha + \overline{\alpha} = 2\text{Re}(\alpha) \qquad \text{and} \qquad \alpha\overline{\alpha} = |\alpha|^2$$

are both real, so that the irreducible quadratic term does indeed have real coefficients. ∎

If $p(x)$ and $q(x)$ are two polynomials (not necessarily of the same degree) then a function of the form

$$R(x) = \frac{p(x)}{q(x)}$$

is called a rational function. Suppose that $q(x)$ has been factored into linear and irreducible quadratic terms. Then $q(x)$ will consist of a product of terms of the form

$$(ax - b)^k \qquad \text{or} \qquad (ax^2 + bx + c)^k, \tag{B.3}$$

where a, b, c, and k are constants. For example,

$$\frac{x^2 - 1}{(x + 2)(x^2 + 3)}$$

is a factorized rational function. The idea behind a partial fraction decomposition is to express a rational function as a sum of terms whose denominators are of the form (B.3). The following rules tell us the form that such a decomposition must take.

1. Each factor of the form $(ax - b)^k$ in $q(x)$ contributes the following terms to the partial fraction decomposition of $p(x)/q(x)$:

$$\frac{A_1}{(ax - b)} + \frac{A_2}{(ax - b)^2} + \cdots + \frac{A_k}{(ax - b)^k},$$

where $A_1, A_2, \ldots, A_k$ are constants.

2. Each irreducible quadratic factor of the form $(ax^2 + bx + c)^k$ contributes the following terms to the partial fraction decomposition of $p(x)/q(x)$:

$$\frac{A_1x + B_1}{ax^2 + bx + c} + \frac{A_2x + B_2}{(ax^2 + bx + c)^2} + \cdots + \frac{A_kx + B_k}{(ax^2 + bx + c)^k}.$$

Thus, for example,

$$\frac{x^2 + 1}{x(x - 1)(x^2 + 4)} = \frac{A}{x} + \frac{B}{x - 1} + \frac{Cx + D}{x^2 + 4}$$

for appropriate values of the constants A, B, C, D. Similarly,

$$\frac{x - 2}{(x + 2)^2(x^2 + 2x + 2)} = \frac{A}{x + 2} + \frac{B}{(x + 2)^2} + \frac{Cx + D}{(x^2 + 2x + 2)}$$

for appropriate values of the constants A, B, C, D.

The preceding rules give only the form of a partial fraction decomposition. The next question that needs answering is: How do we determine the constants that arise in the partial fraction decomposition? A standard way to proceed is as follows:

1. Determine the general form of the partial fraction decomposition of $p(x)/q(x)$.
2. Multiply both sides of the resulting decomposition by $q(x)$.
3. Equate the coefficients of like powers of x on both sides of the resulting equation in order to determine the constants in the partial fraction decomposition.

We illustrate the procedure with several examples.

Example B.2   Determine the partial fraction decomposition of

$$\frac{2x}{(x-1)(x+3)}.$$

Solution: The general form of the partial fraction decomposition is

$$\frac{2x}{(x-1)(x+3)} = \frac{A}{x-1} + \frac{B}{x+3}.$$

Multiplying both sides of this equation by $(x-1)(x+3)$ yields

$$2x = A(x+3) + B(x-1).$$

We now equate coefficients of like powers of x on both sides of this equation to obtain

$$A + B = 2 \quad \text{and} \quad 3A - B = 0.$$

Consequently,

$$A = \frac{1}{2} \quad \text{and} \quad B = \frac{3}{2}$$

so that

$$\frac{2x}{(x-1)(x+3)} = \frac{1}{2(x-1)} + \frac{3}{2(x+3)}. \qquad \square$$

Example B.3   Determine the partial fraction decomposition of

$$\frac{x^2+1}{(x+1)(x^2+4)}.$$

Solution: In this case the general form of the partial fraction decomposition is

$$\frac{x^2+1}{(x+1)(x^2+4)} = \frac{A}{x+1} + \frac{Bx+C}{x^2+4}.$$

Multiplying both sides by $(x+1)(x^2+4)$, we obtain

$$x^2 + 1 = A(x^2+4) + (Bx+C)(x+1).$$

Equating coefficients of like powers of x on both sides of this equality yields

$$A + B = 1, \quad B + C = 0, \quad 4A + c = 1.$$

Solving this system of equations, we obtain

$$A = \frac{2}{5}, \quad B = \frac{3}{5}, \quad C = -\frac{3}{5},$$

so that

$$\frac{x^2+1}{(x+1)(x^2+4)} = \frac{2}{5(x+1)} + \frac{3(x-1)}{5(x^2+4)}. \qquad \square$$

Example B.4   Determine the partial fraction decomposition of

$$\frac{2x-1}{(x+2)^2(x^2+2x+2)}.$$

Solution: The term x^2+2x+2 is irreducible. Thus, the partial fraction decomposition has the general form:

$$\frac{2x-1}{(x+2)^2(x^2+2x+2)} = \frac{A}{x+2} + \frac{B}{(x+2)^2} + \frac{Cx+D}{x^2+2x+2}.$$

Clearing the fractions yields

$$2x-1 = A(x+2)(x^2+2x+2) + B(x^2+2x+2) + (Cx+D)(x+2)^2.$$

Equating the coefficients of like powers of x, we obtain

$$
\begin{array}{rcrcrcrcl}
A & + & & & C & & & = & 0, \\
4A & + & B & + & 4C & + & D & = & 0, \\
6A & + & 2B & + & 4C & + & 4D & = & 2, \\
4A & + & 2B & & & + & 4D & = & -1.
\end{array}
$$

Solving this system yields

$$A = -\frac{3}{2}, \qquad B = -\frac{5}{2}, \qquad C = \frac{3}{2}, \qquad D = \frac{5}{2}.$$

Consequently,

$$\frac{2x-1}{(x+2)^2(x^2+2x+2)} = \frac{3x+5}{2(x^2+2x+2)} - \frac{3}{2(x+2)} - \frac{5}{2(x+2)^2}. \qquad \square$$

Some Shortcuts

The preceding technique for determining the constants that arise in the partial fraction decomposition of a rational function will always work. However, in practice it is often tedious to apply. We now present, without justification, some shortcuts that can circumvent many of the computations.

1. *Linear Factors:* The Cover-up Rule

    If $q(x)$ contains a linear factor $x-a$, then this factor contributes a term of the form

    $$\frac{A}{x-a}$$

    to the partial fraction decomposition of $p(x)/q(x)$. Let $P(x)$ denote the expression obtained by omitting the $x-a$ term in $p(x)/q(x)$. Then the constant A is given by

    $$\boxed{A = P(a).}$$

Example B.5   Determine the partial fraction decomposition of

$$\frac{3x-1}{(x-3)(x+2)}.$$

Solution: The general form of the decomposition is

$$\frac{3x - 1}{(x - 3)(x + 2)} = \frac{A}{x - 3} + \frac{B}{x + 2}.$$

To determine A, we neglect the $x - 3$ term in the given rational function and set

$$P(x) = \frac{3x - 1}{x + 2}.$$

Then, according to the preceding rule,

$$A = P(3) = \frac{8}{5}.$$

Similarly, to determine B we neglect the $x + 2$ term in the given function, and set

$$P(x) = \frac{3x - 1}{x - 3}.$$

Using the cover-up rule, it then follows that

$$B = P(-2) = \frac{-7}{-5} = \frac{7}{5}.$$

Consequently,

$$\frac{3x - 1}{(x - 3)(x + 2)} = \frac{8}{5(x - 3)} + \frac{7}{5(x + 2)}.$$

The idea behind the technique is to cover up the linear factor $x - a$ in the given rational function and set $x = a$ in the remaining part of the function. The result will be the constant A in the contribution $A/(x-a)$ to the partial fraction decomposition of the rational function. ☐

2. *Repeated Linear Factors:* The cover-up rule can be extended to the case of repeated linear factors also. Suppose that $q(x)$ contains a factor of the form $(x - a)^k$. Then this contributes the terms

$$\frac{A_1}{(ax - b)} + \frac{A_2}{(ax - b)^2} + \cdots + \frac{A_k}{(ax - b)^k}$$

to the partial fraction decomposition of $p(x)/q(x)$. Let $P(x)$ be the expression obtained when the $(x - a)^k$ term is neglected in $p(x)/q(x)$. Then the constants $A_1, A_2, \ldots, A_k$ are given by

$$A_k = P(a), \quad A_{k-1} = P'(a), \quad A_{k-2} = \frac{1}{2!} P''(a), \quad \ldots,$$

$$A_1 = \frac{1}{(k - 1)!} P^{(k-1)}(a)$$

where a prime symbol denotes differentiation with respect to x.

Remarks

1. The above formulae look rather formidable to begin with. However, they are easy to apply in practice.
2. Notice that in the case $k = 1$ we are back to the cover-up rule.

Example B.6   Determine the partial fraction decomposition of

$$\frac{x}{(x-1)(x+2)^2}.$$

Solution: The general form of the partial fraction decomposition is

$$\frac{x}{(x-1)(x+2)^2} = \frac{A_1}{x+2} + \frac{A_2}{(x+2)^2} + \frac{A_3}{x-1}.$$

To determine A_1 and A_2 we omit the term $(x+2)^2$ in the given function to obtain

$$P(x) = \frac{x}{x-1}.$$

Applying the preceding rule with $k = 2$ yields

$$A_2 = P(-2) = \frac{2}{3}, \qquad A_1 = P'(-2) = -\frac{1}{(x-1)^2}\bigg|_{x=-2} = -\frac{1}{9}.$$

We now use the cover-up rule to determine A_3. Neglecting the $x-1$ term in the given function and setting $x = 1$ in the resulting expression yields $A_3 = \frac{1}{9}$. Thus,

$$\frac{x}{(x-1)(x+2)^2} = -\frac{1}{9(x+2)} + \frac{2}{3(x+2)^2} + \frac{1}{9(x-1)}.$$

3. *Irreducible Quadratic Factors of the Form $x^2 + a^2$:* The final case that we will consider is when $q(x)$ contains a factor of the form $x^2 + a^2$. This will contribute a term

$$\frac{Ax + B}{x^2 + a^2}$$

to the partial fraction decomposition of $p(x)/q(x)$. Let $P(x)$ be the expression obtained by deleting the term $x^2 + a^2$ in $p(x)/q(x)$. Then the constants A and B are given by

$$\boxed{A = \frac{1}{a}\, \mathrm{Im}[P(ia)], \qquad B = \mathrm{Re}[P(ia)].}$$ □

Example B.7   Determine the partial fraction decomposition of

$$\frac{x-1}{(x+2)(x^2+4)}.$$

Solution: In this case the general form of the partial fraction decomposition is

$$\frac{x-1}{(x+2)(x^2+4)} = \frac{Ax+B}{x^2+4} + \frac{C}{x+2}.$$

In order to determine A and B we delete the $x^2 + 4$ term from the given function to obtain

$$P(x) = \frac{x-1}{x+2}.$$

Since in this case $a = 2i$, we first compute

$$P(2i) = \frac{2i - 1}{2i + 2} = \frac{1}{4}(1 + 3i).$$

Thus,

$$A = \frac{1}{2} \text{Im} \left[\frac{1}{4}(1 + 3i) \right] = \frac{3}{8}, \qquad B = \text{Re} \left[\frac{1}{4}(1 + 3i) \right] = \frac{1}{4}.$$

In order to determine C we use the cover-up rule. Neglecting the $x + 2$ factor in the given function and setting $x = -2$ in the result yields $C = -\frac{3}{8}$. Thus,

$$\frac{x - 1}{(x + 2)(x^2 + 4)} = \frac{3x + 2}{8(x^2 + 4)} - \frac{3}{8(x + 2)}. \qquad \square$$

Remark   These techniques can be extended to the case of irreducible factors of the form $(ax^2 + bx + c)^k$.

Exercises for B

Problems

For Problems 1–18, determine the partial fraction decomposition of the given rational function.

1. $\dfrac{2x - 1}{(x + 1)(x + 2)}$.

2. $\dfrac{x - 2}{(x - 1)(x + 4)}$.

3. $\dfrac{x + 1}{(x - 3)(x + 2)}$.

4. $\dfrac{x^2 - x + 4}{(x + 3)(x - 1)(x + 2)}$.

5. $\dfrac{2x - 1}{(x + 4)(x - 2)(x + 1)}$.

6. $\dfrac{3x^2 - 2x + 14}{(2x - 1)(x + 5)(x + 2)}$.

7. $\dfrac{2x + 1}{(x + 2)(x + 1)^2}$.

8. $\dfrac{5x^2 + 3}{(x + 1)(x - 1)^2}$.

9. $\dfrac{3x + 4}{x^2(x^2 + 4)}$.

10. $\dfrac{3x - 2}{(x - 5)(x^2 + 1)}$.

11. $\dfrac{x^2 + 6}{(x - 2)(x^2 + 16)}$.

12. $\dfrac{10}{(x - 1)(x^2 + 9)}$.

13. $\dfrac{7x + 2}{(x - 2)(x + 2)^2}$.

14. $\dfrac{7x^2 - 20}{(x - 2)(x^2 + 4)}$.

15. $\dfrac{7x + 4}{(x + 1)^3(x - 2)}$.

16. $\dfrac{x(2x + 3)}{(x + 1)(x^2 + 2x + 2)}$.

17. $\dfrac{3x + 4}{(x - 3)(x^2 + 4x + 5)}$.

18. $\dfrac{7 - 2x^2}{(x - 1)(x^2 + 4)}$.

Review of Integration Techniques

In this appendix, we review some of the basic integration techniques that are required throughout the text. This is a very brief refresher and should *not* be considered as a substitute for a calculus text.

1. **Integration by Parts.** The basic formula for integration by parts can be written in the form

$$\int u \, dv = uv - \int v \, du.$$

To derive this, we start with the product rule for differentiation, namely

$$\frac{d}{dx}[u(x)v(x)] = u\frac{dv}{dx} + v\frac{du}{dx}.$$

Integrating both sides of this equation with respect to x yields

$$u(x)v(x) = \int \left(u\frac{dv}{dx} + v\frac{du}{dx} \right) dx$$

or, upon rearranging terms,

$$\int u\frac{dv}{dx} \, dx = uv - \int v\frac{du}{dx} \, dx.$$

Consequently,

$$\int u \, dv = uv - \int v \, du.$$

Then the given integral can be written in the form

$$\int \frac{1}{x} (\ln x)^2 \, dx = \int u^2 \, du = \frac{1}{3} u^3 + c.$$

Substituting back for $u = \ln x$ yields

$$\int \frac{1}{x} (\ln x)^2 \, dx = \frac{1}{3} (\ln x)^3 + c. \qquad \Box$$

Now consider the general integral

$$\int \frac{f'(x)}{f(x)} \, dx.$$

If we let $u = f(x)$, then $du = f'(x) \, dx$, so that

$$\int \frac{f'(x)}{f(x)} \, dx = \frac{1}{u} \, du = \ln |u| + c.$$

Substituting back for $u = f(x)$ yields the important formula

$$\boxed{\int \frac{f'(x)}{f(x)} \, dx = \ln |f(x)| + c.}$$

Example C.5   Evaluate $\displaystyle\int \frac{x - 2}{x^2 - 4x + 3} \, dx.$

Solution: If we rewrite the integral in the equivalent form

$$\int \frac{x - 2}{x^2 - 4x + 3} \, dx = \frac{1}{2} \int \frac{2(x - 2)}{x^2 - 4x + 3} \, dx,$$

then we see that the numerator in the second integral is the derivative of the denominator. Thus

$$\int \frac{x - 2}{x^2 - 4x + 3} \, dx = \frac{1}{2} \ln |x^2 - 4x + 3| + c. \qquad \Box$$

3. **Integration by Partial Fractions.** We can always evaluate an integral of the form

$$\int \frac{p(x)}{q(x)} \, dx \qquad\qquad (C.1)$$

when $p(x)$ and $q(x)$ are polynomials in x. Consider first the case when the degree of $p(x)$ is *less than* the degree of $q(x)$. To evaluate (C.1) we first determine the partial fraction decomposition of the integrand (a review of partial fractions is given in Appendix B). The result will always be integrable, although we might need a substitution to carry out this integration.

Example C.6   Evaluate $\displaystyle\int \frac{3x+2}{(x+1)(x+2)}\,dx$.

Solution: In this case we require the partial fraction decomposition of the integrand. Using the rules for partial fractions, it follows that

$$\frac{3x+2}{(x+1)(x+2)} = \frac{A}{x+1} + \frac{B}{x+2}.$$

Multiplying both sides of this equality by $(x+1)(x+2)$ yields

$$3x+2 = A(x+2) + B(x+1).$$

Equating coefficients of like powers of x on both sides of this equality, we obtain

$$A+B = 3 \qquad \text{and} \qquad 2A+B = 2.$$

Solving for A and B yields

$$A = -1 \qquad \text{and} \qquad B = 4.$$

Thus,

$$\frac{3x+2}{(x+1)(x+2)} = -\frac{1}{x+1} + \frac{4}{x+2}$$

so that

$$\int \frac{3x+2}{(x+1)(x+2)}\,dx = -\ln|x+1| + 4\ln|x+2| + c. \qquad \square$$

Example C.7   Evaluate $\displaystyle\int \frac{2x-3}{(x-2)(x^2+1)}\,dx$.

Solution: We first determine the partial fraction decomposition of the integrand. From the general rules of partial fractions it follows that there are constants A, B, and C such that

$$\frac{2x-3}{(x-2)(x^2+1)} = \frac{A}{x-2} + \frac{Bx+C}{x^2+1}. \qquad \text{(C.2)}$$

In order to determine the values of A, B, and C, we multiply both sides of (C.2) by $(x-2)(x^2+1)$. This yields

$$2x-3 = A(x^2+1) + (Bx+C)(x-2).$$

Equating coefficients of like powers of x on both sides of this equality we obtain

$$A+B = 0 \qquad \text{and} \qquad -2B+C = 2 \qquad \text{and} \qquad A-2C = -3.$$

Solving for A, B, and C yields

$$A = \frac{1}{5} \qquad \text{and} \qquad B = -\frac{1}{5} \qquad \text{and} \qquad C = \frac{8}{5}$$

so that

$$\frac{2x - 3}{(x - 2)(x^2 + 1)} = \frac{1}{5(x - 2)} + \frac{8 - x}{5(x^2 + 1)}.$$

Thus,

$$\int \frac{2x - 3}{(x - 2)(x^2 + 1)} \, dx = \frac{1}{5} \int \frac{1}{x - 2} \, dx + \frac{8}{5} \int \frac{1}{x^2 + 1} \, dx - \frac{1}{5} \int \frac{x}{x^2 + 1} \, dx$$

$$= \frac{1}{5} \ln |x - 2| + \frac{8}{5} \tan^{-1} x - \frac{1}{10} \ln(x^2 + 1) + c. \qquad \square$$

Now return to the integral (C.1). If the degree of $p(x)$ is greater than or equal to the degree of $q(x)$, then we first divide $q(x)$ into $p(x)$. The resulting expression will be integrable, although in general we will need to perform a partial fraction decomposition.

Example C.8   Evaluate $\int \frac{2x - 1}{x + 3} \, dx.$

Solution: We first divide the denominator into the numerator to obtain

$$\frac{2x - 1}{x + 3} = 2 - \frac{7}{x + 3}.$$

Thus,

$$\int \frac{2x - 1}{x + 3} \, dx = \int 2 \, dx - 7 \int \frac{1}{x + 3} \, dx$$

$$= 2x - 7 \ln |x + 3| + c. \qquad \square$$

Example C.9   Evaluate $\int \frac{3x^2 + 5}{x^2 - 3x + 2} \, dx.$

Solution: Once more we must first divide the denominator into the numerator. It is easily shown that

$$\frac{3x^2 + 5}{x^2 - 3x + 2} = 3 + \frac{9x - 1}{x^2 - 3x + 2}. \qquad (C.3)$$

The next step is to determine the partial fraction decomposition of the second term on the right-hand side. We first notice that the denominator can be factored as $(x - 2)(x - 1)$. Using the rules for partial fraction decomposition, it follows that

$$\frac{9x - 1}{(x - 2)(x - 1)} = \frac{A}{x - 2} + \frac{B}{x - 1}.$$

Clearing the fractions yields

$$9x - 1 = A(x - 1) + B(x - 2).$$

Equating coefficients of like powers of x on both sides of this equation, we obtain

$$A + B = 9 \qquad \text{and} \qquad A + 2B = 1.$$

Solving for A and B yields

$$A = 17 \qquad \text{and} \qquad B = -8.$$

Substitution into (C.3) gives

$$\frac{3x^2 + 5}{x^2 - 3x + 2} = 3 + \frac{17}{x - 2} - \frac{8}{x - 1},$$

so that

$$\int \frac{3x^2 + 5}{x^2 - 3x + 2}\, dx = 3x + 17 \ln |x - 2| - 8 \ln |x - 1| + c. \qquad \square$$

Table C.1 lists some of the more important integrals. Notice that we have omitted the integration constant.

Function $F(x)$	$\int F(x)\, dx$
$x^n, n \neq -1$	$\dfrac{1}{n + 1} x^{n+1}$
x^{-1}	$\ln |x|$
$e^{ax}, a \neq 0$	$\dfrac{1}{a} e^{ax}$
$\sin x$	$-\cos x$
$\cos x$	$\sin x$
$\tan x$	$\ln |\sec x|$
$\sec x$	$\ln |\sec x + \tan x|$
$\csc x$	$\ln |\csc x - \cot x|$
$e^{ax} \sin bx$	$\dfrac{1}{a^2 + b^2} e^{ax} (a \sin bx - b \cos bx)$
$e^{ax} \cos bx$	$\dfrac{1}{a^2 + b^2} e^{ax} (a \cos bx + b \sin bx)$
$\ln x$	$x \ln x - x$
$\dfrac{1}{a^2 + x^2}$	$\dfrac{1}{a} \tan^{-1}\left(\dfrac{x}{a}\right)$
$\dfrac{1}{\sqrt{a^2 - x^2}}, a > 0$	$\sin^{-1}\left(\dfrac{x}{a}\right)$
$\dfrac{1}{\sqrt{x^2 + a^2}}$	$\ln |x + \sqrt{a^2 + x^2}|$
$\dfrac{f'(x)}{f(x)}$	$\ln |f(x)|$
$e^{u(x)} \dfrac{du}{dx}$	$e^{u(x)}$

Table C.1: Some Basic Integrals

Exercises for C

Problems

For Problems 1–22, evaluate the given integral.

1. $\int x \cos x \, dx.$

2. $\int x^2 e^{-x} \, dx.$

3. $\int \ln x \, dx.$

4. $\int \tan^{-1} x \, dx.$

5. $\int x^3 e^{x^2} \, dx.$

6. $\int \dfrac{x}{x^2 + 1} \, dx.$

7. $\int \dfrac{x - 1}{x + 2} \, dx.$

8. $\int \dfrac{x + 2}{(x - 1)(x + 3)} \, dx.$

9. $\int \dfrac{2x + 1}{x(x^2 + 4)} \, dx.$

10. $\int \dfrac{x^2 + 5}{(x - 1)(x + 4)} \, dx.$

11. $\int \dfrac{x + 3}{2x - 1} \, dx.$

12. $\int \dfrac{2x + 3}{x^2 + 3x + 4} \, dx.$

13. $\int \dfrac{3x + 2}{x(x + 1)^2} \, dx.$

14. $\int \dfrac{1}{\sqrt{4 - x^2}} \, dx.$

15. $\int \dfrac{1}{x^2 + 2x + 2} \, dx.$

16. $\int \dfrac{1}{x \ln x} \, dx.$

17. $\int \tan x \, dx.$

18. $\int \dfrac{x + 1}{x^2 - x - 6} \, dx.$

19. $\int \cos^2 x \, dx.$

20. $\int \sqrt{1 - x^2} \, dx.$

21. $\int e^{3x} \sin 2x \, dx.$

22. $\int e^x \sin^2 x \, dx.$

Linearly Independent Solutions to $x^2 y'' + xp(x)y' + q(x)y = 0$

Consider the differential equation

$$x^2 y'' + xp(x)y' + q(x)y = 0, \tag{D.1}$$

where p and q are analytic at $x = 0$. Writing

$$p(x) = \sum_{n=0}^{\infty} p_n x^n \qquad \text{and} \qquad q(x) = \sum_{n=0}^{\infty} q_n x^n \tag{D.2}$$

on $(0, R)$, it follows that the indicial equation for (D.1) is

$$r(r-1) + p_0 r + q_0. \tag{D.3}$$

If the roots of the indicial equation are distinct and do not differ by an integer, then there exist two linearly independant Frobenius series solutions in the neighborhood of $x = 0$. In this appendix, we derive the general form for two linearly independant solutions to (D.1) in the case that the roots of the indicial equation differ by an integer. The analysis of the case when the roots of the indicial equation coincide is left as an exercise. Let r_1 and r_2 denote the roots of the indicial equation and suppose that

$$r_1 - r_2 = N, \tag{D.4}$$

where N is a positive integer. Then, we know that one solution to (D.1) on $(0, R)$ is given by the Frobenius series

$$y_1(x) = x^{r_1}(1 + a_1 x + a_2 x^2 + \cdots). \tag{D.5}$$

According to the reduction-of-order technique, a second linearly independant solution to (D.1) on $(0, R)$ is given by

$$y_2(x) = u(x)y_1(x), \tag{D.6}$$

where the function u can be determined by substitution into (D.1). Differentiating (D.6) twice and substituting into (D.1) yields

$$x^2(y_1 u'' + 2y_1' u' + y_1'' u) + xp(x)(y_1 u' + y_1' u) + q(x)uy_1 = 0;$$

that is, since y_1 is a solution to (D.1),

$$x^2(y_1u'' + 2y_1'u') + xp(x)(y_1u') = 0.$$

Consequently, u can be determined by solving

$$\frac{u''}{u'} = -\left(\frac{p(x)}{x} + 2\frac{y_1'}{y_1}\right)$$

which, upon integrating, yields

$$u' = y^{-2}e^{-\int [p(x)/x]\,dx}. \tag{D.7}$$

We now determine the two terms that appear on the right-hand side. From (D.2) we can write

$$\frac{p(x)}{x} = \frac{p_0}{x} + p_1 + p_2x + \cdots$$

so that

$$\int \frac{p(x)}{x}\,dx = p_0 \ln x + P(x),$$

where

$$P(x) = p_1x + \tfrac{1}{2}p_2x^2 + \tfrac{1}{3}p_3x^3 + \cdots.$$

Consequently,

$$e^{-\int [p(x)/x]\,dx} = x^{-p_0}e^{P(x)}. \tag{D.8}$$

Since $P(x)$ is analytic at $x = 0$, it follows that

$$e^{P(x)} = \alpha_0 + \alpha_1x + \alpha_2x^2 + \cdots$$

for appropriate constants $\alpha_0, \alpha_1, \alpha_2, \ldots$. Hence, from (D.8),

$$e^{-\int [p(x)/x]dx} = x^{-p_0}(\alpha_0 + \alpha_1x + \alpha_2x^2 + \cdots). \tag{D.9}$$

Now consider y_1^{-2}. From (D.5),

$$y_1^{-2} = \frac{x^{-2r_1}}{(1 + a_1x + a_2x^2 + \cdots)^2}. \tag{D.10}$$

Further, since the series

$$1 + a_1x + a_2x^2 + \cdots$$

converges for $0 \le x < R$ and is nonzero at $x = 0$, it follows that $(1 + a_1x + a_2x^2 + \cdots)^{-2}$ is analytic at $x = 0$, hence there exist constants $\beta_1, \beta_2, \ldots$, such that

$$(1 + a_1x + a_2x^2 + \cdots)^{-2} = 1 + \beta_1x + \beta_2x^2 + \cdots.$$

We can therefore write (D.10) in the form

$$y_1^{-2} = x^{-2r_1}(1 + \beta_1x + \beta_2x^2 + \cdots). \tag{D.11}$$

Substituting from (D.9) and (D.11) into (D.7) yields

$$u' = x^{-(p_0+2r_1)}(\alpha_0 + \alpha_1x + \alpha_2x^2 + \cdots)(1 + \beta_1x + \beta_2x^2 + \cdots),$$

which can be written as

$$u' = x^{-(p_0 + 2r_1)}(A_0 + A_1 x + A_2 x^2 + \cdots) \tag{D.12}$$

for appropriate constants $A_0, A_1, A_2, \ldots$. Now, since the roots of the indicial equation are r_1 and $r_2 = r_1 - N$, it follows that the indicial equation has factored form

$$(r - r_1)(r - r_1 + N) = 0,$$

which upon expansion gives

$$r^2 - (2r_1 - N)r + r_1(r_1 - N) = 0.$$

Comparison with (D.3) reveals that

$$p_0 - 1 = -(2r_1 - N),$$

so that

$$p_0 + 2r_1 = N + 1.$$

Consequently, (D.12) can be written as

$$u' = x^{-(N+1)}(A_0 + A_1 x + A_2 x^2 + \cdots);$$

that is,

$$u' = A_0 x^{-(N+1)} + A_1 x^{-N} + \cdots + A_N x^{-1} + A_{N+1} + A_{N+2} + \cdots,$$

which can be integrated directly to yield

$$u(x) = -\frac{A_0}{N}x^{-N} + \frac{A_1}{1-N}x^{1-N} + \cdots - A_{N-1}x^{-1} + A_N \ln x + A_{N+1}x + \cdots.$$

Rearranging terms, we can write this as

$$u(x) = A \ln x + x^{-N}(B_0 + B_1 x + B_2 x^2 + \cdots),$$

where we have redefined the coefficients. Substituting this expression for u into (D.6) gives

$$y_2(x) = [A \ln x + x^{-N}(B_0 + B_1 x + B_2 x^2 + \cdots)]y_1(x).$$

Hence,

$$y_2(x) = Ay_1(x) \ln x + x^{-N}y_1(x)(B_0 + B_1 x + B_2 x^2 + \cdots).$$

Substituting for y_1 from (D.5) into the second term on the right-hand side yields

$$y_2(x) = Ay_1(x) \ln x + x^{r_1 - N}(1 + a_1 x + a_2 x^2 + \cdots)(B_0 + B_1 x + B_2 x^2 + \cdots).$$

Finally, multiplying the two power series together and substituting $r_2 = r_1 - N$ from (D.4), we obtain

$$y_2(x) = Ay_1(x) \ln x + x^{r_2}(b_0 + b_1 x + b_2 x^2 + \cdots)$$

for appropriate constants $b_0, b_1, b_2, \ldots$, and so we have

$$y_2(x) = Ay_1(x) \ln x + x^{r_2} \sum_{n=0}^{\infty} b_n x^n.$$

The derivation of a second solution to the differential equation (D.1) in the case when the indicial equation has two equal roots follows exactly the same lines as the derivative above and is left as an exercise.

Answers to Odd-Numbered Exercises

Chapter 1

Section 1.1

True-False Review

1. False
3. True
5. False
7. False
9. False
11. False

Problems

1. $y(t) = gt^2/2; \quad t = \sqrt{\frac{200}{g}} \approx 4.52$ sec.
3. **(a)** $\sqrt{180g} \approx 42.0$ m/sec.

   (b) $t = \sqrt{\frac{180}{g}} \approx 4.28$ sec.
5. $y(t) = gt^2/2 - 2t; \quad h \approx 470$ m.
7. $A = \pm a; \quad \phi = n\pi$ (n an integer).
13. After 4 p.m.
15. $y = kx^4$.
17. $x^2 + 2y^2 = k$.
19. $y = ke^{-x}$.

21. $y = -\frac{1}{m}x + k$.
23. $y^m = kx$.
25. $y = kx^2$.

Section 1.2

True-False Review

1. False
3. True
5. False

Problems

1. 2, nonlinear.
3. 2, nonlinear.
5. 4, linear.
7. $(-\infty, \infty)$.
9. $(-\infty, -4)$ or $(-4, \infty)$.
11. $(-\infty, \infty)$.
13. $(-\infty, 0)$ or $(0, \infty)$.
15. $(-\infty, \infty)$.
17. $(-\infty, \infty)$.
19. $r = -3, 1$.
21. $r = \pm 1$.

27. $y(x) = \dfrac{\ln x}{x}, \quad x > 0.$

29. $y(x) = \dfrac{\sqrt{1 + \sin x}}{x}, \quad x > 0.$

31. $y(x) = 2\sqrt{x} + c$ for all $x > 0$.

33. If $n \neq -1, -2$,

$$y(x) = \frac{x^{n+2}}{(n+1)(n+2)} + c_1 x + c_2.$$

If $n = -1$, $y(x) = x \ln|x| + c_1 x + c_2$. If $n = -2$, $y(x) = c_1 x + c_2 - \ln|x|$. Solution is valid on $(-\infty, \infty)$ for $n \geq 0$ and on $(-\infty, 0)$ or $(0, \infty)$ for $n < 0$.

35. $y(x) = 3 + x - \cos x.$

37. $y(x) = xe^x - 2e^x + 5x + 5.$

39. $y(x) = e^{-x} - \frac{1}{e}x.$

49(a) $y(x) = 1 - \frac{1}{4}x^2 + \frac{1}{64}x^4 + \cdots .$

Section 1.3

True-False Review

1. True

3. False

5. True

7. True.

Problems

1. $y' = -\dfrac{y}{x}.$

3. $y' = \dfrac{y^2 - x^2}{2xy}.$

5. $y' = \dfrac{x}{-y \pm \sqrt{x^2 + y^2}}.$

7. $y' = -\dfrac{x^2 + 2xy - y^2}{y^2 + 2xy - x^2}.$

15. (c)   (i) $y(x) = \dfrac{1}{x^2 + 1}$ on $(-\infty, \infty)$,

   (ii) $y(x) = \dfrac{1}{x^2}$ on $(0, \infty)$,

   (iii) $y(x) = \dfrac{1}{x^2 - 1}$ on $(-1, 1)$.

   (d) $y(x) = 0.$

Section 1.4

True-False Review

1. True

3. True

5. False

7. True

9. True

Problems

1. $y(x) = ce^{x^2}.$

3. $y(x) = \ln(c - e^{-x}).$

5. $y(x) = c(x - 2).$

7. $y(x) = \dfrac{cx - 3}{2x - 1}.$

9. $y(x) = \dfrac{(x-1) + c(x-2)^2}{(x-1) - c(x-2)^2}$ and $y(x) = -1.$

11. $y(x) = c + c_1 \left(\dfrac{x-a}{x-b}\right)^{1/(a-b)}.$

13. $y(x) = a(1 + \sqrt{1 - x^2}).$

15. $y(x) = 0.$

17.   (a) $v(t) = a\left(\dfrac{e^{gt/a} - e^{-gt/a}}{e^{gt/a} + e^{-gt/a}}\right) = a\tanh(gt/a)$, where $a = \sqrt{mg/k}.$

   (b) No.

   (c) $y(t) = \dfrac{a^2}{g}\ln\left[\cosh(gt/a)\right].$

19. $y(x) = \ln(e^x - e^3 + e).$

21.   (a) No.

   (b) $\frac{1}{2}\ln(1 + v_0^2).$

25. $T(40) = 225°$ F; $t \approx 96.4$ minutes.

27.   (a) $500°$ F.

   (b) $t \approx 6 : 07$ p.m.

Section 1.5

True-False Review

1. True

3. True

5. True

7. False

9. False

Problems

1. 2560.

3. $t \approx 35.86$ hours.

5. 1091.

7.   (a) $P_1 > \dfrac{2P_0 P_2}{P_0 + P_2}$

   (b) No.

9.   (a) Equilibrium solutions: $P(t) = 0$, $P(t) = T$.
   Isoclines: $P = \frac{1}{2}\left(T \pm \sqrt{\frac{rT^2 + 4k}{r}}\right).$
   Slope: positive for $P > T$, negative for $0 < T < P$. Concave up for $P > T/2$, concave down for $0 < P < T/2$.

   (c) For $0 < P_0 < T$, population dies out; For $P_0 > T$, population grows.

11. Equilibrium solutions: $P(t) = 0$, $P(t) = T$, $P(t) = C$.
Slope: positive for $T < P < C$, negative for $0 < P < T$ and for $P > C$. Changes in concavity occur when $P = \frac{1}{3}\left(C + T \pm \sqrt{C^2 - CT + T^2}\right)$. For $0 < P_0 < T$, population dies out; for $T < P_0 < C$, population grows asymptotically toward C; for $P_0 > C$, population declines asymptotically toward C.

13. $P(t) = Ce^{\ln(P_0/k)e^{-rt}}$.

15. (a) More.

   (b) 31.06 min.

   (c) 85.02 min.

19. Approx. 52 years.

Section 1.6

True-False Review

1. False

3. True

5. True

Problems

1. $y(x) = e^x(e^x + c)$.

3. $y(x) = x^2 - 1 + ce^{-x^2}$.

5. $y(x) = \dfrac{1}{1+x^2}(4\tan^{-1}x + c)$.

7. $y(x) = \dfrac{x^3(3\ln x - 1) + c}{\ln x}$.

9. $x(t) = \dfrac{4e^t(t-1) + c}{t^2}$.

11. $y(x) = (\tan x + c)\cos x$.

13. $y(x) = \begin{cases} \dfrac{1}{\alpha + \beta}e^{\beta x} + ce^{-\alpha x}, & \alpha + \beta \neq 0, \\ e^{-\alpha x}(x + c), & \alpha + \beta = 0. \end{cases}$

15. $y(x) = x^{-2}(x^4 + 1)$.

17. $x(t) = (4 - t)(1 + t)$.

19. $y(x) = \begin{cases} 2e^{-x} + 1, & \text{if } x \leq 1, \\ e^{-x}(e + 2), & \text{if } x > 1. \end{cases}$

29. $y(x) = x^{-1}(\cos x + x\sin x + c)$.

31. $y(x) = \sin x + c \cdot \csc x$.

Section 1.7

True-False Review

1. True

3. True

5. False

7. True

Problems

1. 196 g.

3. 300 g.

5. (a) 6.75 g.

   (b) $15\sqrt[3]{2}$ L.

9. $i(t) = 5(1 - e^{-40t})$.

11. $i(t) = \frac{3}{5}(3\sin 4t - 4\cos 4t + 4e^{-3t})$.

13. $q(t) = 5e^{-t/RC}$.

15. $i(t) = \dfrac{E_0}{R^2 + L^2\omega^2}(R\sin\omega t - \omega L\cos\omega t) + Ae^{-Rt/L}$,

   $i_S(t) = \dfrac{E_0}{R^2 + L^2\omega^2}(R\sin\omega t - \omega L\cos\omega t)$,

   $i_T(t) = Ae^{-Rt/L}$.

17. $i(t) = \dfrac{E_0 C}{1 - aRC}\left(\dfrac{1}{RC}e^{-t/RC} - ae^{-at}\right)$.

Section 1.8

True-False Review

1. True

3. False

5. True

7. True

9. True

Problems

1. $F(V) = (1 - V^2)/V$.

3. $F(V) = \dfrac{\sin(1/V) - V\cos V}{V}$.

5. Not homogeneous.

7. $F(V) = -\sqrt{1 + V^2}$.

9. $y^2 = ce^{-3x/y}$.

11. $y(x) = x\cos^{-1}(c/x)$.

13. $y + \sqrt{9x^2 + y^2} = cx^2$.

15. $y(x) = xe^{(cx+1)}$.

17. $y^2 = x^2\ln(\ln cx)$.

19. $y^2 = c^2 - 2cx$.

21. $y(x) = x\sin^{-1}cx$.

25. $2y^2 + xy - x^2 = 2$.

27. $\sin^{-1}(y/2x) = \ln x + c$, and $y(x) = \pm 2x$.

29. $x^2 + y^2 = 2kx$.

31. (b) $(x + cm)^2 + (y - c)^2 = c^2(1 + m^2)$.

33. $(3y - x)^3 = k(2y - x)^4$.

35. (a) $(y^2 - x^2)\tan(\alpha_0) - 2xy = k$.

37. $y^2 = x^2(8\sin x + c)$.

39. $y^{2/3} = x[x^2(2\ln x - 1) + c]$.

41. $y(x) = \dfrac{1}{x^2(c - 2x^3)}$.

43. $y(x) = \dfrac{1}{4}\left(\dfrac{x-b}{x-a}\right)^2[x + (b-a)\ln|x-b| + c]^2$.

45. $y(x) = [(x^2 - 1) + ce^{-x^2}]^2$.

47. $y(x) = \left(\dfrac{x^3 + c}{x}\right)^{1/(1-\pi)}$.

49. $y(x) = \left(1 + \dfrac{c}{\sec x + \tan x}\right)^{1/(1-\sqrt{3})}$.

51. $y^2 = \dfrac{1}{\sin^2 x(2\cos x + 1)}$.

53. $y(x) = 3(3x - \tanh 3x)$.

55. $y(x) = \frac{1}{3}[3x - \tan^{-1}(3x + c) + 1]$.

57. $y(x) = x^{-1}e^{cx}$.

59. $y(x) = x^{-1}\left[\dfrac{1}{c - \ln x} - 1\right]$.

61.   **(b)** $y(x) = x^{-1}\left[1 + \dfrac{1}{c - 3\ln x}\right]$.

63. $y(x) = xe^{x^2}$.

65. $y(x) = \tan^{-1}(1 + ce^{-\sqrt{1+x}})$.

Section 1.9

True-False Review

1. False

3. False

5. False

7. False

9. False

Problems

1. Exact.

3. Not exact.

5. $xy^2 - \cos y + \sin x = c$.

7. $6e^{2x} + 3x^2y - 3xy^2 + y^3 = c$.

9. $\tan^{-1}(y/x) + \ln|x| = c$.

11. $\sin xy + \cos x = c$.

13. $y(x) = x^{-1}(x^3 \ln x + 5)$.

15. $y(x) = x^{-1}\ln(2 - \sin x)$.

17. Yes.

19. Yes.

21. $2x - y^4 = cy^2$.

23. $y(x) = \dfrac{c + 2x^{5/2}}{10\sqrt{x}}$.

25. $y(x) = \dfrac{c + \tan^{-1}x}{1 + x^2}$.

27. $r = 2, s = 4$.

29. $r = 1, s = 2$.

Section 1.10

True-False Review

1. True

3. False

Problems

1. $y(0.5) \approx 4.8938$.

3. $y(0.5) \approx 1.0477$.

5. $y(1) \approx 0.8564$.

7. $y(1) \approx 0.5012$.

9. $y(1) \approx 0.7115$.

11. $y(0.5) \approx 5.79167$.

13. $y(0.5) \approx 1.08785$.

15. $y(1) \approx 0.999996$.

Section 1.11

Problems

1. $y(x) = c_1x^3 + x^4 + c_2$.

3. $y(x) = \sqrt[3]{c_1 + c_2e^x}$.

5. $y(x) = c_2 - \ln|c_1 - \sin x|$.

7. $y(x) = \frac{1}{3}x^6 + c_1x^3 + c_2$.

9. $y(x) = \dfrac{1}{\alpha}\ln|c_1 + c_2e^{\beta x}|$.

11. $y(x) = c_1 \tan^{-1}x + c_2$.

13. $y(x) = \ln(\sec x) + c_1 \ln(\sec x + \tan x) + c_2$.

15. $y(x) = a\cosh(\omega x)$.

19.   **(a)** $\theta = \theta_0 \cos(\sqrt{gt/L})$.

Section 1.12

Additional Problems

1. Yes.

3. $x^2 + 3y^2 = k$.

5. $x^2 + 2y = k$.

7.   **(b)** $2y^2 = kx^6 + x^2$.

13. $y(x) = x^2[(\ln x)^2 + c]$.

15. $\coth^{-1}\left(\dfrac{y + x}{\sqrt{2x}}\right) = -\sqrt{2}\ln|x| + c$.

17. $y(x) = \dfrac{1}{e^{2x} + 1}(-x + \ln|e^{2x} - 1| + c)$.

19. $y\sin x + x\sin y - y + x = c$.

21. $y(x) = \frac{1}{2}\ln\,(c - 2e^{-x})$.

23. $\sqrt{y} = \dfrac{x - e^{-x} + c}{1 + e^x}$.

25. $e^{-y}x + x^2 - y = c$.

27. $y^2 = cx^3 - x^2$.

29. $y(x) = c(x + 1)e^x$.

31. $y(x) = x^{-1}\tan^{-1}(c - x^2)$.

33. $\dfrac{y}{x} + \dfrac{y^3}{3x^3} = \ln|x| + c$.

35. $y^2 = \dfrac{25(\ln x)^2 + c}{2x^2}$.

37. $y(x) = \sin x(c - \ln(\sin x))$.

39. $e^{-6y} = -\frac{3}{2}e^{-4x} + \frac{5}{2}$.

41. $ye^{\cos x} = x + 1$.

43. ≈ 12.32 minutes.

45. $T(t) = T_m - \dfrac{1}{kt + (T_m - T_0)^{-1}}$.

47.   **(a)** $20,000$.

    (b) $11,696$.

    (c) 57.9 days after April 1.

49. $i(t) = \frac{3}{178}[64\cos(2t) - 40\sin(2t)] + \frac{1185}{356}e^{5t/4}$.

51. 157.5 g.

53. $y(1.5) \approx 3.67185$.

55. $y(1.5) \approx 3.66576$.

57. $y(1.5) \approx 3.66565$.

Chapter 2
Section 2.1

True-False Review

1. True
3. True
5. True
7. True
9. True

Problems

1. $a_{31} = 0, a_{24} = -1, a_{14} = 2, a_{32} = 2, a_{21} = 7, a_{34} = 4.$

3. $\begin{bmatrix} 2 & 1 & -1 \\ 0 & 4 & -2 \end{bmatrix}, 2 \times 3.$

5. $\begin{bmatrix} 1 & -3 & -2 \\ 3 & 6 & 0 \\ 2 & 7 & 4 \\ -4 & -1 & 5 \end{bmatrix}, 4 \times 3.$

7. 4.

9. $-1.$

11. Column vectors: $\begin{bmatrix} 1 \\ -1 \\ 2 \end{bmatrix}, \begin{bmatrix} 3 \\ -2 \\ 6 \end{bmatrix}, \begin{bmatrix} -4 \\ 5 \\ 7 \end{bmatrix}.$

   Row vectors: $[1 \ 3 \ -4], [-1 \ -2 \ 5], [2 \ 6 \ 7].$

13. $A = \begin{bmatrix} 1 & 2 \\ 3 & 4 \\ 5 & 1 \end{bmatrix}.$ Column vectors: $\begin{bmatrix} 1 \\ 3 \\ 5 \end{bmatrix}, \begin{bmatrix} 2 \\ 4 \\ 1 \end{bmatrix}.$

15. $q \times p.$

21. For example, $A(t) = \begin{bmatrix} \sqrt{t+2} & 0 & 0 \\ 1/\sqrt{3-t} & 0 & 0 \end{bmatrix}.$

25. For example, $A(t) = [t]$ and $B(t) = [t^2].$

27. $a_{12} = -1, a_{13} = -3, a_{32} = 1, a_{11} = a_{22} = a_{33} = 0.$

Section 2.2

True-False Review

1. False
3. True
5. False
7. False
9. True
11. False

Problems

1. $2A = \begin{bmatrix} 2 & 4 & -2 \\ 6 & 10 & 4 \end{bmatrix}, -3B = \begin{bmatrix} -6 & 3 & -9 \\ -3 & -12 & -15 \end{bmatrix},$

   $A - 2B = \begin{bmatrix} -3 & 4 & -7 \\ 1 & -3 & -8 \end{bmatrix}.$

3. $AB = \begin{bmatrix} 5 & 10 & -3 \\ 27 & 22 & 3 \end{bmatrix}, CA$ not possible,

   $DB = [6 \ 14 \ -4], CD = \begin{bmatrix} 2 & -2 & 3 \\ -2 & 2 & -3 \\ 4 & -4 & 6 \end{bmatrix}.$

5. $AB = \begin{bmatrix} -9 - 21i & 11 + 10i \\ -1 + 19i & 9 + 15i \end{bmatrix}.$

7. $ABC = \begin{bmatrix} -12 & -22 \\ 14 & -126 \end{bmatrix}.$

9. $\begin{bmatrix} 0 \\ -38 \end{bmatrix}.$

11. $\begin{bmatrix} -7 \\ 13 \\ 29 \end{bmatrix}.$

13. (a) $A^2 = \begin{bmatrix} -1 & -4 \\ 8 & 7 \end{bmatrix}, A^3 = \begin{bmatrix} -9 & -11 \\ 22 & 13 \end{bmatrix},$

   $A^4 = \begin{bmatrix} -31 & -24 \\ 48 & 17 \end{bmatrix}.$

   (b) $A^2 = \begin{bmatrix} -2 & 0 & 1 \\ 4 & -3 & 0 \\ 2 & 4 & -1 \end{bmatrix}, A^3 = \begin{bmatrix} 4 & -3 & 0 \\ 6 & 4 & -3 \\ -12 & 3 & 4 \end{bmatrix},$

   $A^4 = \begin{bmatrix} 6 & 4 & -3 \\ -20 & 9 & 4 \\ 10 & -16 & 3 \end{bmatrix}.$

17. $x = 2, y = -1$ or $x = -1, y = 2.$

19. $\begin{bmatrix} -6 & 1 \\ 2 & 6 \end{bmatrix}.$

27. $A^T = \begin{bmatrix} 1 & 2 & 3 \\ -1 & 0 & 4 \\ 1 & 2 & -1 \\ 4 & -3 & 0 \end{bmatrix}, AA^T = \begin{bmatrix} 19 & -8 & -2 \\ -8 & 17 & 4 \\ -2 & 4 & 26 \end{bmatrix},$

   $B^T A^T = \begin{bmatrix} 10 & -4 & -5 \\ 4 & 1 & 10 \end{bmatrix}.$

29. (a) $2I_4.$

   (b) $7I_3.$

33. $S = 0.$

37. $A'(t) = \begin{bmatrix} -2e^{-2t} \\ \cos t \end{bmatrix}.$

39. $A'(t) = \begin{bmatrix} e^t & 2e^{2t} & 2t \\ 2e^t & 8e^{2t} & 10t \end{bmatrix}.$

43. $\int_0^1 A(t)\, dt = \begin{bmatrix} e - 1 & 1 - \dfrac{1}{e} \\ 2(e-1) & 5(1 - \dfrac{1}{e}) \end{bmatrix}.$

45. $\int_0^1 A(t)\, dt = \begin{bmatrix} e - 1 & \frac{1}{2}(e^2 - 1) & \frac{1}{3} \\ 2(e-1) & 2(e^2 - 1) & \frac{5}{3} \end{bmatrix}.$

47. $\int A(t)\, dt = \begin{bmatrix} -\cos t & \sin t & 0 \\ -\sin t & -\cos t & \frac{1}{2}t^2 \\ 0 & \frac{3}{2}t^2 & t \end{bmatrix}.$

49. $\int A(t)\, dt = \begin{bmatrix} \frac{1}{2}e^{2t} & -\frac{1}{2}\cos 2t \\ \frac{1}{3}t^3 - 5t & te^t - e^t \\ \tan t & \frac{3}{2}t^2 + \cos t \end{bmatrix}.$

Section 2.3

True-False Review

1. False
3. False
5. True

Problems

7. $A = \begin{bmatrix} 1 & 1 & 1 & -1 \\ 2 & 4 & -3 & 7 \end{bmatrix}$, $\mathbf{b} = \begin{bmatrix} 3 \\ 2 \end{bmatrix}$, $A^\# = \begin{bmatrix} 1 & 1 & 1 & -1 & 3 \\ 2 & 4 & -3 & 7 & 2 \end{bmatrix}$.

9. $\begin{aligned} x_1 - x_2 + 2x_3 + 3x_4 &= 1 \\ x_1 + x_2 - 2x_3 + 6x_4 &= -1 \\ 3x_1 + x_2 + 4x_3 + 2x_4 &= 2. \end{aligned}$

11.   **(b)** No.

13. $\begin{bmatrix} x_1' \\ x_2' \end{bmatrix} = \begin{bmatrix} t^2 & -t \\ -\sin t & 1 \end{bmatrix} \begin{bmatrix} x_1 \\ x_2 \end{bmatrix}$.

15. $\begin{bmatrix} x_1' \\ x_2' \\ x_3' \end{bmatrix} = \begin{bmatrix} 0 & -\sin t & 1 \\ -e^t & 0 & t^2 \\ -t & t^2 & 0 \end{bmatrix} \begin{bmatrix} x_1 \\ x_2 \\ x_3 \end{bmatrix} + \begin{bmatrix} t \\ t^3 \\ 1 \end{bmatrix}$.

Section 2.4

True-False Review

1. True

3. True

5. False

7. True

9. True

Problems

1. Row-echelon form.

3. Reduced row-echelon form.

5. Reduced row-echelon form.

7. Reduced row-echelon form.

9. $\begin{bmatrix} 1 & -3 \\ 0 & 1 \end{bmatrix}$, rank$(A) = 2$.

11. $\begin{bmatrix} 1 & 1 & 2 \\ 0 & 1 & 0 \\ 0 & 0 & 0 \end{bmatrix}$, rank$(A) = 2$.

13. $\begin{bmatrix} 1 & 3 \\ 0 & 1 \\ 0 & 0 \end{bmatrix}$, rank$(A) = 2$.

15. $\begin{bmatrix} 1 & -2 & 1 & 3 \\ 0 & 1 & \frac{1}{3} & -\frac{2}{3} \\ 0 & 0 & 0 & 0 \end{bmatrix}$, rank$(A) = 2$.

17. $\begin{bmatrix} 1 & 2 & 1 & 2 \\ 0 & 1 & 0 & 1 \\ 0 & 0 & 0 & 0 \\ 0 & 0 & 0 & 0 \end{bmatrix}$, rank$(A) = 2$.

19. I_2, rank$(A) = 2$.

21. $\begin{bmatrix} 1 & -1 & 2 \\ 0 & 0 & 0 \\ 0 & 0 & 0 \end{bmatrix}$, rank$(A) = 1$.

23. I_4, rank$(A) = 4$.

25. $\begin{bmatrix} 0 & 1 & 0 & 0 \\ 0 & 0 & 1 & 0 \\ 0 & 0 & 0 & 1 \end{bmatrix}$, rank$(A) = 3$.

Section 2.5

True-False Review

1. False

3. True

5. False

Problems

1. $(-2, 2, -1)$.

3. No solution.

5. $(2, -1, 3)$.

7. $\{(1 - 2r + s - t, r, s, t) : r, s, t \in \mathbb{R}\}$.

9. $\{(1 - 2r + 3s - t, r, 2 - 4s + 3t, s, t) : r, s, t \in \mathbb{R}\}$.

11. No solution.

13. $\{(-s + t, -3 + s + t, s, t) : s, t \in \mathbb{R}\}$.

15. $(1, -3, 4, -4, 2)$.

17. $\{(-5t, -2t - 1, t) : t \in \mathbb{R}\}$.

19. No solution.

21.   **(a)** $k = 2$.

   (b) $k = -2$.

   (c) $k \neq \pm 2$.

23.   **(a)** $a = 3$ and $b \neq 10$.

   (b) $a = 3$ and $b = 10$.

   (c) $a \neq 3$ and $b \in \mathbb{R}$.

31.   **(b)** $(2, 1, -1)$.

33. Trivial solution.

35. $\{(-t, -3t, t) : t \in \mathbb{R}\}$.

37. $\{(t, 0, -3t) : t \in \mathbb{R}\}$.

39. $\{(2t(-1 + i), t, (2 - i)t) : t \in \mathbb{C}\}$.

41. $\{(2r - 3s, r, s) : r, s \in \mathbb{R}\}$.

43. Trivial solution.

45. $\{((1 - i)t, t) : t \in \mathbb{C}\}$.

47. Trivial solution.

49. $\{((1 - 5i)r + (5 + i)s, 13r, 13s) : r, s \in \mathbb{C}\}$.

51. $\{(-3t, -2t, t) : t \in \mathbb{R}\}$.

53. $\{(3s, r, s, t) : r, s, t \in \mathbb{R}\}$.

Section 2.6

True-False Review

1. False

3. True

5. False

7. True

9. True

Problems

5. $A^{-1} = \begin{bmatrix} -1 & 1 + i \\ 1 - i & -1 \end{bmatrix}$.

7. A is not invertible.

9. $A^{-1} = \begin{bmatrix} 8 & -29 & 3 \\ -5 & 19 & -2 \\ 2 & -8 & 1 \end{bmatrix}$.

11. $A^{-1} = \begin{bmatrix} 18 & -34 & -1 \\ -29 & 55 & 2 \\ 1 & -2 & 0 \end{bmatrix}$.

13. $A^{-1} = \begin{bmatrix} -i & 1 & 0 \\ 1 - 5i & i & 2i \\ -2 & 0 & 1 \end{bmatrix}$.

15. $A^{-1} = \begin{bmatrix} 27 & 10 & -27 & 35 \\ 7 & 3 & -8 & 11 \\ -14 & -5 & 14 & -18 \\ 3 & 1 & -3 & 4 \end{bmatrix}.$

17. $\begin{bmatrix} 1 \\ -2 \\ -1 \end{bmatrix}.$

19. $(-2, 2, 1).$

21. $(-6, 1, 3).$

33. No.

37. **(b)** The set of right inverses is $\left\{ \begin{bmatrix} 7+5r & -3+5s \\ -2-2r & 1-2s \\ r & s \end{bmatrix} \right\}.$

Section 2.7

True-False Review

1. True

3. False

5. False

7. False

9. False

Problems

3. $M_2(-\frac{1}{7})A_{12}(-5)P_{12}.$

5. $A_{23}(2)M_2(-1)A_{13}(-3)A_{12}(-2).$

7. $A_{12}(-2)P_{12}A_{12}(-2)A_{21}(-1)M_2(-1).$

9. $P_{12}A_{12}(4)M_2(-21)A_{21}(4).$

11. $P_{12}A_{13}(-2)A_{21}(\frac{1}{4})M_2(-4)M_3(8)A_{32}(\frac{1}{2})A_{31}(\frac{7}{2}).$

13. $P_{12}, A_{12}(-2), M_2(-\frac{1}{7}), A_{21}(-3).$

15. $L = \begin{bmatrix} 1 & 0 \\ \frac{5}{2} & 1 \end{bmatrix}, U = \begin{bmatrix} 2 & 3 \\ 0 & -\frac{13}{2} \end{bmatrix}.$

17. $L = \begin{bmatrix} 1 & 0 & 0 \\ 2 & 1 & 0 \\ -1 & 4 & 1 \end{bmatrix}, U = \begin{bmatrix} 3 & -1 & 2 \\ 0 & 1 & -3 \\ 0 & 0 & 16 \end{bmatrix}.$

19. $L = \begin{bmatrix} 1 & 0 & 0 & 0 \\ 2 & 1 & 0 & 0 \\ 3 & 1 & 1 & 0 \\ 1 & 2 & 2 & 1 \end{bmatrix}, U = \begin{bmatrix} 1 & -1 & 2 & 3 \\ 0 & 2 & -1 & -10 \\ 0 & 0 & 2 & 9 \\ 0 & 0 & 0 & 4 \end{bmatrix}.$

21. $\mathbf{x} = (-11, 7).$

23. $\mathbf{x} = (-\frac{1}{12}, \frac{1}{3}, \frac{1}{2}).$

25. $\mathbf{x}_1 = (-4, -11), \mathbf{x}_2 = (-\frac{13}{2}, -15), \mathbf{x}_3 = (-3, -11).$

Section 2.8

True-False Review

1. False

3. False

Section 2.9

Additional Problems

1. $\begin{bmatrix} 11 & -18 & -9 & -24 \\ 4 & 2 & -17 & -1 \end{bmatrix}.$

3. $\begin{bmatrix} 4 & -52 \\ -52 & 676 \end{bmatrix}.$

5. $\frac{1}{320}\begin{bmatrix} 17 & 8 \\ 6 & -16 \end{bmatrix}.$

7. **(a)** $\begin{bmatrix} -5+3a & 2a+4b \\ -14+7a & 5a+9b \end{bmatrix}; a = 2, b = -1.$

   (b) $\begin{bmatrix} 1 & 1 & 2 \\ 0 & 2 & 2 \\ 0 & -1 & -1 \end{bmatrix}.$

13. $\begin{bmatrix} -3e^{-3t} & -2\sec^2 t \tan t \\ 6t^2 & -\sin t \\ 6/t & -5 \end{bmatrix}.$

15. Not possible.

17. $(-\frac{17}{2}, \frac{1}{2}, 0).$

19. $\{(5 - t, t - 2, t) : t \in \mathbb{R}\}.$

21. $\{(-\frac{2}{7}, -\frac{2}{7}, -\frac{41}{7}, -2, 0) + t(\frac{16}{7}, -\frac{33}{7}, -\frac{15}{7}, -4, 1) : t \in \mathbb{R}\}.$

23. $\{(0, -\frac{1}{3}, 0) + t(\frac{1}{2}(1 - 3i), \frac{1}{6}(1 + i), 1) : t \in \mathbb{R}\}.$

25. Infinitely many solutions for all k.

27. **(a)** $k = 2.$

   (b) $k \neq 0, 2.$

   (c) $k = 0.$

29. **(a)** $\begin{bmatrix} 1 & \frac{7}{4} \\ 0 & 1 \end{bmatrix}.$

   (b) $\text{rank}(A) = 2.$

   (c) $\begin{bmatrix} \frac{5}{34} & -\frac{7}{34} \\ \frac{1}{17} & \frac{2}{17} \end{bmatrix}.$

31. **(a)** $\begin{bmatrix} 1 & -\frac{1}{3} & 2 \\ 0 & 1 & \frac{3}{2} \\ 0 & 0 & 0 \end{bmatrix}.$

   (b) $\text{rank}(A) = 2.$

   (c) No inverse.

33. **(a)** $\begin{bmatrix} 1 & 0 & 0 \\ 0 & 1 & -2 \\ 0 & 0 & 1 \end{bmatrix}.$

   (b) $\text{rank}(A) = 3.$

   (c) $\begin{bmatrix} \frac{1}{3} & 0 & 0 \\ -\frac{1}{9} & \frac{2}{3} & \frac{1}{3} \\ -\frac{2}{9} & \frac{1}{3} & \frac{2}{3} \end{bmatrix}.$

35. For $\mathbf{e}_1$: $(25, 3, -7)$. For $\mathbf{e}_2$: $(-7, -1, 2)$. For $\mathbf{e}_3$: $(4, 1, -1)$.

39. **(a)** $A = M_1(4)A_{12}(-2)M_2(17/2)A_{21}(7/4).$

   (b) $L = \begin{bmatrix} 1 & 0 \\ -\frac{1}{2} & 1 \end{bmatrix}; U = \begin{bmatrix} 4 & 7 \\ 0 & \frac{17}{2} \end{bmatrix}.$

41. **(a)** $A = M_1(3)A_{13}(1)P_{23}A_{23}(-2)M_3(3)A_{32}(2)M_2(-1).$

   (b) $L = \begin{bmatrix} 1 & 0 & 0 \\ 0 & 1 & 0 \\ \frac{1}{3} & -\frac{1}{2} & 1 \end{bmatrix}; U = \begin{bmatrix} 3 & 0 & 0 \\ 0 & 2 & -1 \\ 0 & 0 & \frac{3}{2} \end{bmatrix}.$

43. **(b)** All terms containing two factors of A and one factor of B become subtracted.

   (c) $2^k.$

45. **(a)** 6.

   (b) 4.

   (c) 15.

   (d) $\binom{n}{m} = \dfrac{n!}{m!(n-m)!}.$

Chapter 3

Section 3.1

True-False Review

1. True
3. False
5. False
7. True

Problems

1. Odd
3. Even
5. Even
9. 0.
11. -17.
13. -8.
15. -60.
17. 48.
19. -20.
21. $42e^t$.
23. **(b)** $(d_1, d_2, d_3) = (-1, 1, 1)$.
25. 70.
27. -315.
31. No.
33. Yes, Even, Plus.
35. $p = 1, q = 4$, 2 inversions, Plus.
37. $p = 2, q = 1$, 3 inversions, Minus.
39. $(-1)^{n(n-1)/2}$.

Section 3.2

True-False Review

1. False
3. True
5. False

Problems

1. -36.
3. 45.
5. -103.
7. 624.
9. 21.
11. 84.
13. Invertible
15. Invertible
17. Invertible
19. Not invertible
21. $k = -4, \frac{1}{3}$.
23. All $k \neq 0, 1$.
25. 40.
31. -2.

33. 24.
35. -12.
37. 6075.
39. $\frac{5}{16}$.
41. **(a)** $|8 + 8k|$.
   (b) No change.
   (c) $k \neq -1$.
47. **(b)** -26.

Section 3.3

True-False Review

1. False
3. True
5. False
7. True

Problems

1. $M_{11} = 4, M_{12} = 2, M_{21} = -3, M_{22} = 1,$
   $C_{11} = 4, C_{12} = -2, C_{21} = 3, C_{22} = 1.$
3. $M_{11} = -5, M_{12} = 0, M_{13} = 4, M_{21} = 47, M_{22} = -2,$
   $M_{23} = -38, M_{31} = 3, M_{32} = 0, M_{33} = -2,$
   $C_{11} = -5, C_{12} = 0, C_{13} = 4, C_{21} = -47, C_{22} = -2,$
   $C_{23} = 38, C_{31} = 3, C_{32} = 0, C_{33} = -2.$
5. 5.
7. -153.
9. 0.
11. 9.
13. 3.
15. -4.
17. 11,997.
19. -170.
23. **(a)** 7.

   (b) $M_C = \begin{bmatrix} 1 & -4 \\ 2 & -1 \end{bmatrix}$.

   (c) $\text{adj}(A) = \begin{bmatrix} 1 & 2 \\ -4 & -1 \end{bmatrix}$.

   (d) $A^{-1} = \begin{bmatrix} \frac{1}{7} & \frac{2}{7} \\ -\frac{4}{7} & -\frac{1}{7} \end{bmatrix}$.

25. **(a)** 26.

   (b) $M_C = \begin{bmatrix} 7 & -4 & -2 \\ 6 & 4 & 2 \\ -15 & -10 & 8 \end{bmatrix}$.

   (c) $\text{adj}(A) = \begin{bmatrix} 7 & 6 & -15 \\ -4 & 4 & -10 \\ -2 & 2 & 8 \end{bmatrix}$.

   (d) $A^{-1} = \begin{bmatrix} \frac{7}{26} & \frac{3}{13} & -\frac{15}{26} \\ -\frac{2}{13} & \frac{2}{13} & -\frac{5}{13} \\ -\frac{1}{13} & \frac{1}{13} & \frac{4}{13} \end{bmatrix}$.

27. (a) 6.

(b) $M_C = \begin{bmatrix} -11 & -1 & 8 \\ 9 & -3 & -6 \\ -2 & 2 & 2 \end{bmatrix}$.

(c) $\text{adj}(A) = \begin{bmatrix} -11 & 9 & -2 \\ -1 & -3 & 2 \\ 8 & -6 & 2 \end{bmatrix}$.

(d) $A^{-1} = \begin{bmatrix} -\frac{11}{6} & \frac{3}{2} & -\frac{1}{3} \\ -\frac{1}{6} & -\frac{1}{2} & \frac{1}{3} \\ \frac{4}{3} & -1 & \frac{1}{3} \end{bmatrix}$.

29. (a) 14.

(b) $M_C = \begin{bmatrix} -9 & 1 & 7 \\ 32 & -2 & -14 \\ -13 & 3 & 7 \end{bmatrix}$.

(c) $\text{adj}(A) = \begin{bmatrix} -9 & 32 & -13 \\ 1 & -2 & 3 \\ 7 & -14 & 7 \end{bmatrix}$.

(d) $A^{-1} = \frac{1}{14} \begin{bmatrix} -9 & 32 & -13 \\ 1 & -2 & 3 \\ 7 & -14 & 7 \end{bmatrix}$.

31. (a) 402.

(b) $M_C = \begin{bmatrix} 84 & -46 & -29 & 81 \\ -162 & 60 & 99 & -27 \\ 18 & 38 & -11 & 3 \\ -30 & 26 & 130 & -72 \end{bmatrix}$.

(c) $\text{adj}(A) = \begin{bmatrix} 84 & -162 & 18 & -30 \\ -46 & 60 & 38 & 26 \\ -29 & 99 & -11 & 130 \\ 81 & -27 & 3 & -72 \end{bmatrix}$.

(d) $A^{-1} = \frac{1}{402} \begin{bmatrix} 84 & -162 & 18 & -30 \\ -46 & 60 & 38 & 26 \\ -29 & 99 & -11 & 130 \\ 81 & -27 & 3 & -72 \end{bmatrix}$.

33. -1.

35. $\frac{9}{16}$.

37. $\begin{bmatrix} e^{-t}\sin 2t & e^{-t}\cos 2t \\ -e^t\cos 2t & e^t\sin 2t \end{bmatrix}$.

39. 0_3; $\det(A) = 0$.

41. $(\frac{3}{2}, 0, -\frac{1}{2})$.

43. $(\frac{11}{3}, -\frac{17}{3}, -\frac{16}{3}, -2)$.

45. $\frac{9}{4}$.

Section 3.4

Problems

1. 38.

3. -3.

5. $3abc - a^3 - b^3 - c^3$.

7. -2196.

9. $\det A = -18$, $A^{-1} = \frac{1}{18}\begin{bmatrix} -5 & 1 & 7 \\ 1 & 7 & -5 \\ 7 & -5 & 1 \end{bmatrix}$.

11. $\det A = 116$, $A^{-1} = \frac{1}{116}\begin{bmatrix} -51 & 8 & 31 \\ -32 & -20 & 24 \\ 54 & 12 & -26 \end{bmatrix}$.

13. $(\frac{37}{24}, -\frac{1}{8})$.

15. $(\frac{1}{4}, \frac{1}{16}, \frac{21}{16})$.

17. $\approx (3.77, 0.66, -1.46)$.

19. $\det(2A) = 24$, $\det(A^{-1}) = \frac{1}{3}$, $\det(A^T B) = -12$, $\det(B^5) = -1024$, $\det(B^{-1}AB) = 3$.

Section 3.5

Additional Problems

1. 37.

3. 22.

5. 24.

7. -32.

9. -12.

11. -6.

13. -8.

15. $\det(A)$ not defined, $\det(C) = -18$, $\det(AB) = 474$, $\det(ACB) = 4104$.

17. $A^{-1} = \begin{bmatrix} \frac{4}{7} & \frac{1}{7} & -\frac{1}{7} \\ -\frac{1}{28} & \frac{5}{28} & \frac{1}{14} \\ -\frac{5}{28} & -\frac{3}{28} & \frac{5}{14} \end{bmatrix}$.

19. $A^{-1} = \begin{bmatrix} -\frac{11}{6} & -\frac{2}{3} & -\frac{1}{3} & \frac{1}{12} \\ \frac{1}{2} & -\frac{1}{4} & -\frac{1}{4} & -\frac{1}{8} \\ \frac{5}{6} & \frac{5}{12} & \frac{1}{12} & \frac{1}{24} \\ 1 & 0 & 0 & 0 \end{bmatrix}$.

21. $A^{-1} = \begin{bmatrix} \frac{7}{2} & 0 & -3 \\ -1 & 1 & 0 \\ 0 & -1 & 1 \end{bmatrix}$.

23. False.

25. (a) $k = 0, 4$.

(b) $|4k - k^2|$, Yes.

29. $(-\frac{5}{7}, \frac{6}{7})$.

31. $(\frac{1}{2}, \frac{1}{2}, -\frac{3}{2})$.

Chapter 4
Section 4.1

True-False Review

1. False.

3. True.

5. False.

7. True.

9. False.

11. False.

Problems

1. $\mathbf{v}_1 = (6, 2)$, $\mathbf{v}_2 = (-3, 6)$, $\mathbf{v}_3 = (3, 8)$.

3. $\mathbf{v} = (22, -19, -53, 39)$, $-\mathbf{v} = (-22, 19, 53, -39)$.

Section 4.2

True-False Review

1. True

3. False

5. True

7. False

Problems

1. (A1) holds, (A2) fails.

3. (A1), (A2) both fail.

5. (A1), (A2) both hold.

9. $\mathbf{0} = 0_{2\times3} = \begin{bmatrix} 0 & 0 & 0 \\ 0 & 0 & 0 \end{bmatrix}$, $-\begin{bmatrix} a & b & c \\ d & e & f \end{bmatrix} = \begin{bmatrix} -a & -b & -c \\ -d & -e & -f \end{bmatrix}$.

11. No.

13. Only (A1), (A2), (A5), (A6) hold.

15. Only (A1), (A2), (A4), (A5), (A7), (A8) hold.

19. Yes.

Section 4.3

True-False Review

1. False

3. True

5. True

7. False

Problems

3. $S = \{(x, y) \in \mathbb{R}^2 : 3x + 2y = 0\}$, Yes.

5. $S = \{(x, y, z) \in \mathbb{R}^3 : x + y + z = 1\}$, No.

7. $S = \{(x, y) \in \mathbb{R}^2 : x^2 - y^2 = 0\}$, No.

9. $S = \{A \in M_n(\mathbb{R}) : A \text{ is lower triangular}\}$, Yes.

11. $S = \{A \in M_2(\mathbb{R}) : A \text{ is invertible}\}$, No.

13. $S = \{f \in V : f(a) = f(b)\}$, Yes.

15. $S = \{f \in V : f(-x) = f(x) \text{ for all } x \in \mathbb{R}\}$, Yes.

17. $S = \{p \in V : p(x) = ax^2 + 1 \text{ for some } a \in \mathbb{R}\}$, No.

19. $S = \{y \in V : y'' + 2y' - y = 1 \text{ on } I\}$, No.

21. $\text{nullspace}(A) = \{(8r + 8s, -2r - 3s, r, s) : r, s \in \mathbb{R}\}$.

Section 4.4

True-False Review

1. True

3. True

5. True

7. False

9. True

11. False

Problems

1. Yes.

3. No.

5. Yes.

9. $(x_1, x_2, x_3) = \frac{1}{16}(-3x_1 + x_2 + 5x_3)\mathbf{v}_1 + \frac{1}{16}(-x_1 - 5x_2 + 7x_3)\mathbf{v}_2 + \frac{1}{16}(7x_1 + 3x_2 - x_3)\mathbf{v}_3$.

13. One possible answer: $\{(2, 1, 0), (1, 0, 1)\}$.

15. One possible answer: $\{(1, 1, -1, 0)\}$.

17. One possible answer: $\begin{bmatrix} 0 & -1 \\ 1 & 0 \end{bmatrix}$.

19. $\{\mathbf{v} \in \mathbb{R}^3 : \mathbf{v} = (a + 2b, -a - b, 2a + 3b), \ a, b \in \mathbb{R}\}$. Geometrically, $\text{span}\{\mathbf{v}_1, \mathbf{v}_2\}$ is the plane through $(0, 0, 0)$ determined by the two given vectors.

23. Yes.

25. Yes.

27. $\text{span}\{A_1, A_2\} = \left\{ A \in M_n(\mathbb{R}) : A = \begin{bmatrix} a - 2b & 2a + b \\ -a + b & 3a - b \end{bmatrix} \right\}$, $B \in \text{span}\{A_1, A_2\}$.

29. $\{(0, 0, 0)\}$.

31. Plane spanned by $\{\mathbf{v}_1, \mathbf{v}_2\}$.

Section 4.5

True-False Review

1. False

3. False

5. True

7. True

9. False

Problems

1. Linearly independent.

3. Linearly independent.

5. Linearly dependent: $3(-2, 4, -6) + 2(3, -6, 9) = (0, 0, 0)$.

7. Linearly dependent: $-2(-1, 1, 2) + 3(0, 2, -1) + (3, 1, 2) + 5(-1, -1, 1) = (0, 0, 0)$.

9. Linearly independent.

11. **(b)** No.

13. $k \neq -1, 2$.

15. Linearly dependent.

17. Linearly dependent.

19. Linearly dependent.

21. Linearly dependent.

23. One possible answer: $\{(3, 1, 5), (1, 2, -1), (-1, 2, 3)\}$.

25. One possible answer: $\{(1, 1, -1, 1), (2, -1, 3, 1), (1, 1, 2, 1)\}$.

27. One possible answer: $\{2 - 5x, 3 + 7x\}$.

35. Linearly dependent.

Section 4.6

True-False Review

1. False
3. True
5. False
7. False
9. False
11. False

Problems

1. Yes.
3. No.
5. Yes.
7. One possible basis: $\{1, x, x^2, x^3\}$.
9. $\dim[\text{nullspace}(A)] = 1$.
11. $\dim[\text{nullspace}(A)] = 1$.
13. One possible basis: $\{(3, 1, 0), (-1, 0, 1)\}$; $\dim[S] = 2$.
15. One possible basis: $\left\{\begin{bmatrix} 1 & 0 \\ 0 & 0 \end{bmatrix}, \begin{bmatrix} 0 & 1 \\ 0 & 0 \end{bmatrix}, \begin{bmatrix} 0 & 0 \\ 0 & 1 \end{bmatrix}\right\}$; $\dim[S] = 3$.
17. One possible basis: $\{\mathbf{v}_1, \mathbf{v}_2\}$; $\dim[S] = 2$.
19. One possible basis: $\left\{\begin{bmatrix} 1 & 3 \\ -1 & 2 \end{bmatrix}, \begin{bmatrix} -1 & 4 \\ 1 & 1 \end{bmatrix}\right\}$.
23. $\alpha \neq -1$.
27. **(b)** $(x, y, z, w) = (-2a + 3b, 7a - 8b, 5a, 5b)$, where $a, b \in \mathbb{R}$.
29. **(a)** $\begin{bmatrix} 1 & 0 & 0 \\ 0 & 0 & 1 \\ 0 & 1 & 0 \end{bmatrix}, \begin{bmatrix} 0 & 1 & 0 \\ 0 & 0 & 1 \\ 1 & 0 & 0 \end{bmatrix}, \begin{bmatrix} 0 & 0 & 1 \\ 0 & 0 & 1 \\ 1 & 1 & -1 \end{bmatrix}, \begin{bmatrix} 0 & 0 & 0 \\ 1 & 0 & -1 \\ -1 & 0 & 1 \end{bmatrix},$
$\begin{bmatrix} 0 & 0 & 0 \\ 0 & 1 & -1 \\ 0 & -1 & 1 \end{bmatrix}, \dim[S] = 5$.
    (b) To the basis in (a), add $E_{11}, E_{12}, E_{23}, E_{31}$.
31. $\dim[Sym_n(\mathbb{R})] = n(n + 1)/2$, $\dim[Skew_n(\mathbb{R})] = n(n - 1)/2$.
33. Basis: $\left\{\begin{bmatrix} 1 & 0 \\ 0 & 1 \end{bmatrix}, \begin{bmatrix} 0 & 1 \\ 1 & 0 \end{bmatrix}\right\}$; Extension: add E_{11}, E_{12}.
37. **(a)** $2n$; One possible basis:
    $\{\mathbf{e}_j : j = 1, 2, \ldots, n\} \cup \{i \cdot \mathbf{e}_j : j = 1, 2, \ldots, n\}$.

    (b) n; One possible basis: $\{\mathbf{e}_j : j = 1, 2, \ldots, n\}$.

Section 4.7

True-False Review

1. True
3. True
5. True
7. False

Problems

1. $[\mathbf{v}]_B = (3, -1)$.
3. $[\mathbf{v}]_B = (-4, 1, -3)$.
5. $[\mathbf{v}]_B = (-1, 4, 6)$.
7. $[p(x)]_B = (-4, 3, 0)$.
9. $[p(x)]_B = (5, -2, 3, -2)$.
11. $[A]_B = (2, -3, -1, -1)$.
13. $[A]_B = (-\frac{34}{3}, 12, -\frac{55}{3}, \frac{56}{3})$.

15. $(2a_0 - a_1 - a_2, 2a_0 - 2a_1 - a_2, -a_0 + a_1 + a_2)$.
17. $P_{C \leftarrow B} = \begin{bmatrix} -1 & 4 \\ 1 & -20 \end{bmatrix}$.
19. $P_{C \leftarrow B} = \begin{bmatrix} 0 & 3 & 5 \\ \frac{5}{3} & -\frac{2}{3} & 4 \\ -\frac{7}{3} & \frac{1}{3} & -4 \end{bmatrix}$.
21. $P_{C \leftarrow B} = \begin{bmatrix} -1 & 0 & 2 \\ 3 & 0 & -1 \\ -3 & 2 & -2 \end{bmatrix}$.
23. $P_{C \leftarrow B} = \begin{bmatrix} 3 & -4 & -2 \\ 3 & 2 & 1 \\ -1 & 3 & -5 \end{bmatrix}$.
25. $P_{C \leftarrow B} = \begin{bmatrix} 0 & 1 & 0 & 0 \\ 0 & 0 & 0 & 1 \\ 0 & 0 & 1 & 0 \\ 1 & 0 & 0 & 0 \end{bmatrix}$.
27. $P_{B \leftarrow C} = -\frac{1}{16}\begin{bmatrix} 20 & 4 \\ 1 & 1 \end{bmatrix}$.
29. $P_{B \leftarrow C} = \frac{1}{7}\begin{bmatrix} 1 & 2 \\ -2 & 3 \end{bmatrix}$.
31. $P_{B \leftarrow C} = \begin{bmatrix} 0 & 0 & 0 & 1 \\ 1 & 0 & 0 & 0 \\ 0 & 0 & 1 & 0 \\ 0 & 1 & 0 & 0 \end{bmatrix}$.

Section 4.8

True-False Review

1. True
3. False
5. True

Problems

1. Basis for rowspace(A): $\{(1, -2)\}$.
   Basis for colspace(A): $\{(1, -3)\}$.
3. Basis for rowspace(A): $\{(1, 2, 3), (0, 1, 2)\}$.
   Basis for colspace(A): $\{(1, 5, 9), (2, 6, 10)\}$.
5. Basis for rowspace(A): $\{(1, 2, -1, 3), (0, 0, 0, 1)\}$.
   Basis for colspace(A): $\{(1, 3, 1, 5), (3, 5, -1, 7)\}$.
7. Basis: $\{(1, -1, 2), (0, 1, -9)\}$.
9. Basis: $\{(1, 1, -1, 2), (0, 1, -5, 8)\}$.
11. Basis for rowspace(A): $\{(1, -3)\}$.
    Basis for colspace(A): $\{(-3, 1)\}$.

Section 4.9

True-False Review

1. False
3. True
5. True
7. False
9. True

Problems

1. $\text{nullspace}(A) = \{(x, y, z, w) : x - 6z - w = 0\}$; $\text{nullity}(A) = 3$; $\text{rank}(A) = 1$.

3. $\text{nullspace}(A) = \{\mathbf{0}\}$; $\text{nullity}(A) = 0$; $\text{rank}(A) = 3$.

5. 1.

7. 1.

9. $\mathbf{x} = c\mathbf{x}_1 + \mathbf{x}_p$, where $\mathbf{x}_1 = (34, -11, 1)$ and $\mathbf{x}_p = (-5, 3, 0)$.

11. $\mathbf{x} = \mathbf{x}_p = (2, -3, 1)$.

13. No.

Section 4.10

True-False Review

1. True
3. False
5. True
7. False
9. False

Section 4.11

True-False Review

1. False
3. True
5. False
7. False

Problems

1. $\theta = 0.95$ rad.

3. $\langle \mathbf{u}, \mathbf{v} \rangle = 19 + 11i$, $\|\mathbf{u}\| = \sqrt{35}$, $\|\mathbf{v}\| = \sqrt{22}$.

7. $\langle A, B \rangle = 13$, $\|A\| = \sqrt{13}$, $\|B\| = \sqrt{7}$.

11. (a) 9.

   (b) 0.

Section 4.12

True-False Review

1. True
3. True
5. True
7. True

Problems

1. Orthonormal set: $\left\{ \dfrac{1}{\sqrt{6}}(2, -1, 1), \dfrac{1}{\sqrt{3}}(1, 1, -1), \dfrac{1}{\sqrt{2}}(0, 1, 1) \right\}$.

3. Not orthogonal.

5. Orthonormal set: $\left\{ \dfrac{1}{\sqrt{14}}(1, 2, 3), \dfrac{1}{\sqrt{3}}(1, 1, -1), \dfrac{1}{\sqrt{42}}(5, -4, 1) \right\}$.

7. Orthonormal set:
$\left\{ \dfrac{1}{\sqrt{5}}(1 - i, 1 + i, i), \dfrac{1}{\sqrt{3}}(0, i, 1 - i), \dfrac{1}{\sqrt{30}}(-3 + 3i, 2 + 2i, 2i) \right\}$.

9. Orthonormal set: $\left\{ \dfrac{1}{\sqrt{2}}, \sin \pi x, \cos \pi x \right\}$.

13. $A_4 = k \begin{bmatrix} 3 & -1 \\ 2 & 0 \end{bmatrix}$.

15. Orthonormal basis: $\left\{ \dfrac{1}{3}(2, 1, -2), \dfrac{1}{3\sqrt{2}}(-1, 4, 1) \right\}$.

17. Orthonormal basis: $\{ \dfrac{1}{\sqrt{2}}(1, 0, -1, 0), (0, 1, 0, 0),$

$\dfrac{1}{\sqrt{6}}(-1, 0, -1, 2) \}$.

19. Orthonormal basis:
$\left\{ \dfrac{1}{\sqrt{3}}(1, 1, -1, 0), \dfrac{1}{\sqrt{15}}(-1, 2, 1, 3), \dfrac{1}{\sqrt{15}}(3, -1, 2, 1) \right\}$.

21. Orthonormal basis: $\left\{ \dfrac{1}{\sqrt{3}}(1 - i, 0, i), \dfrac{1}{\sqrt{21}}(1, 3 + 3i, 1 - i) \right\}$.

23. Orthogonal basis: $\left\{ 1, \dfrac{1}{2}(2x - 1), \dfrac{1}{6}(6x^2 - 6x + 1) \right\}$.

25. Orthogonal basis: $\left\{ 1, \sin x, \dfrac{1}{\pi}(\pi \cos x - 2) \right\}$.

27. Orthogonal basis: $\left\{ \begin{bmatrix} 0 & 1 \\ 1 & 0 \end{bmatrix}, \begin{bmatrix} 1 & 0 \\ 0 & 0 \end{bmatrix}, \begin{bmatrix} 0 & 0 \\ 0 & 1 \end{bmatrix} \right\}$, the subspace of all symmetric matrices in $M_2(\mathbb{R})$.

29. Orthogonal basis: $\left\{ 1 + x^2, 1 - x - x^2 + x^3, -3 - 5x + 3x^2 + x^3 \right\}$.

35. $W^\perp = \text{span}\{(0, 1, 1, 0), (-3, -3, 0, 1)\}$.

Section 4.13

Additional Problems

3. No.

5. No.

7. No.

9. Yes.

11. No.

13. No.

19. No.

21. No.

23. Yes.

25. (b) only.

27. (b) only.

29. (a) and (b).

31. (b) only.

35. (b) One possible basis:
$\left\{ \begin{bmatrix} 0 & -1 & 0 \\ 1 & 0 & 0 \\ 0 & 0 & 0 \end{bmatrix}, \begin{bmatrix} 0 & 0 & -1 \\ 0 & 0 & 0 \\ 1 & 0 & 0 \end{bmatrix}, \begin{bmatrix} 0 & 0 & 0 \\ 0 & 0 & -1 \\ 0 & 1 & 0 \end{bmatrix} \right\}$.

   (c) Add $E_{11}, E_{22}, E_{33}, E_{12}, E_{13},$ and E_{23}.

41. Basis for $\text{rowspace}(A)$: $\{(-3, -6)\}$. Basis for $\text{colspace}(A)$: $\{(-3, -6)\}$. Basis for $\text{nullspace}(A)$: $\{(-2, 1)\}$.

43. Basis for $\text{rowspace}(A)$: $\{(1, 0, 0), (0, 1, 0), (0, 0, 1)\}$. Basis for $\text{colspace}(A)$: $\{(-4, 0, 6, -2), (0, 10, 5, 5), (3, 13, 2, 10)\}$. Basis for $\text{nullspace}(A)$: $\{(0, 0, 0)\}$.

45. Orthonormal basis for rowspace(A):

$$\left\{ \frac{1}{\sqrt{41}}(1, 2, 6), \frac{1}{\sqrt{369}}(-14, 13, -2) \right\}.$$

Orthonormal basis for colspace(A):

$$\left\{ \frac{1}{\sqrt{6}}(1, 2, 0, 1), \sqrt{\frac{3}{10}}\left(\frac{4}{3}, -\frac{1}{3}, 1, -\frac{2}{3}\right) \right\}.$$

Orthonormal basis for nullspace(A): $\{(-\frac{2}{3}, -\frac{2}{3}, \frac{1}{3})\}$.

47. Orthogonal basis = $\{(5, -1, 2), (1, \frac{11}{5}, -\frac{7}{5})\}$.

49. Orthogonalizing standard basis on P_3 gives:

$$\{1, x - \tfrac{1}{2}, x^2 - x + \tfrac{1}{6}, x^3 - \tfrac{3}{2}x^2 + \tfrac{3}{5}x - \tfrac{1}{20}\}.$$

51. $\theta = \cos^{-1}(5/\sqrt{221}) \approx 1.23$ radians.

53. $\theta = \cos^{-1}(13/\sqrt{561}) \approx 0.99$ radians.

55. ≈ 0.100.

Chapter 5

Section 5.1

True-False Review

1. False

3. False

5. True

Problems

11. $A = \begin{bmatrix} 3 & -2 \\ 1 & 5 \end{bmatrix}$.

13. $A = \begin{bmatrix} 1 & -1 & 1 \\ -1 & 0 & 1 \end{bmatrix}$.

15. $A = \begin{bmatrix} -1 & 0 & 1 \\ -1 & 0 & 0 \\ 3 & 0 & 2 \\ 0 & 0 & 0 \end{bmatrix}$.

17. $T(x_1, x_2, x_3) = (2x_1 - x_2 + 5x_3, 3x_1 + x_2 - 2x_3)$.

19. $T(x) = (-3x, -2x, 0, x)$.

23. **(b)** $T(x_1, x_2) = (\frac{1}{2}x_1 + \frac{3}{2}x_2, 2x_1 + x_2)$; $T(4, -2) = (-1, 6)$.

25. $A = \begin{bmatrix} 3 & 2 & -6 & 3 \\ -2 & 3 & -1 & 2 \end{bmatrix}$.

27. $A = \begin{bmatrix} 0 & -\frac{2}{3} & \frac{1}{3} \\ \frac{3}{2} & -1 & 1 \\ -\frac{1}{4} & \frac{2}{5} & -\frac{2}{5} \\ -\frac{1}{4} & \frac{17}{15} & \frac{8}{15} \end{bmatrix}$.

29. $T(\mathbf{v}_1) = 8\mathbf{v}_1 - 4\mathbf{v}_2$, $T(\mathbf{v}_2) = -5\mathbf{v}_1 + 3\mathbf{v}_2$.

31. Writing $\mathbf{v} = a\mathbf{v}_1 + b\mathbf{v}_2$, we have $T(\mathbf{v}) = (3a + b)\mathbf{v}_1 + (2b - a)\mathbf{v}_2$.

35. Matrix of $T_1 + T_2$ is $\begin{bmatrix} 5 & 6 \\ 2 & -2 \end{bmatrix}$. Matrix of cT_1 is $\begin{bmatrix} 3c & c \\ -c & 2c \end{bmatrix}$.

Section 5.2

True-False Review

1. False

3. False

5. False

Problems

5. Shear parallel to the x-axis.

7. Shear parallel to the y-axis.

9. Shear parallel to the x-axis, followed by a linear stretch in the y-direction, followed by a shear parallel to the y-axis.

11. Reflection in the x-axis followed by a linear stretch in the y-direction.

Section 5.3

True-False Review

1. False

3. False

5. True

Problems

1. **(a)** $T(\mathbf{x}) = \mathbf{0}$.

   (b) $T(\mathbf{x}) = (-2, 3)$.

   (c) $T(\mathbf{x}) = \mathbf{0}$.

3. $\text{Ker}(T) = \mathbf{0}$, $\dim[\text{Ker}(T)] = 0$; $\text{Rng}(T) = \mathbb{R}^3$, $\dim[\text{Rng}(T)] = 3$.

5. $\text{Ker}(T) = \{\mathbf{x} \in \mathbb{R}^3 : \mathbf{x} = r(-2, 0, 1) + s(1, 1, 0), \; r, s \in \mathbb{R}\}$, $\dim[\text{Ker}(T)] = 2$; $\text{Rng}(T) = \{\mathbf{y} \in \mathbb{R}^2 : \mathbf{y} = t(1, -3), \; t \in \mathbb{R}\}$, $\dim[\text{Rng}(T)] = 1$.

7. $\text{Ker}(T) = \{\mathbf{0}\}$, $\text{Rng}(T) = \text{Span}\{(-\frac{5}{3}, \frac{1}{3}, \frac{5}{3}, -1), (-\frac{2}{3}, \frac{1}{3}, -\frac{1}{3}, 1)\}$.

9. $\text{Ker}(T) = \{\mathbf{0}\}$, $\text{Rng}(T) = \mathbb{R}^3$.

11. **(a)** $\text{Ker}(T) = \{\mathbf{v} \in \mathbb{R}^3 : \mathbf{v} \text{ is orthogonal to } \mathbf{u}\}$, the plane of vectors orthogonal to $\mathbf{u}$, $\dim[\text{Ker}(T)] = 2$.

   (b) $\text{Rng}(T) = \mathbb{R}$, $\dim[\text{Rng}(T)] = 1$.

13. $\text{Ker}(T)$ is the set of all matrices that commute with B.

15. $\text{Ker}(T) = \{r(-x^2 + x + 1) : r \in \mathbb{R}\}$, $\dim[\text{Ker}(T)] = 1$; $\text{Rng}(T) = P_2$, $\dim[\text{Rng}(T)] = 2$.

17. $\text{Ker}(T) = \{\mathbf{v} \in V : r(-\mathbf{v}_1 + \mathbf{v}_2 + \mathbf{v}_3), r \in \mathbb{R}\}$, $\dim[\text{Ker}(T)] = 1$. $\text{Rng}(T) = W$, $\dim[\text{Rng}(T)] = 2$.

Section 5.4

True-False Review

1. False

3. True

5. True

7. False

9. True

11. False

Problems

1. $(T_1 T_2)(x, y) = (-5(x + y), x + 15y)$, $(T_2 T_1)(x, y) = (7(2x + y), 2(x - 2y))$, No.

3. $\text{Ker}(T_1) = \{\mathbf{x} \in \mathbb{R}^2 : \mathbf{x} = r(1, 1), r \in \mathbb{R}\}$, $\text{Ker}(T_2) = \{\mathbf{0}\}$, $\text{Ker}(T_1 T_2) = \{\mathbf{x} \in \mathbb{R}^2 : \mathbf{x} = s(2, 1)\}$, $\text{Ker}(T_2 T_1) = \text{Ker}(T_1)$.

5.   **(a)** $[T_1(f)](x) = \cos(x - a)$, $[T_2(f)](x) = 1 - \cos(x - a)$.

7. If $\mathbf{v} = a\mathbf{v}_1 + b\mathbf{v}_2$, $(T_2 T_1)(\mathbf{v}) = -a\mathbf{v}_1 - 9a\mathbf{v}_2$.

9. $\text{Ker}(T) = \{r(-2, 1) : r \in \mathbb{R}\}$, $\text{Rng}(T) = \{r(1, -2) : r \in \mathbb{R}^2\}$, T is neither one-to-one nor onto.

11. $\text{Ker}(T) = \{r(1, -2, 1) : r \in \mathbb{R}\}$, $\text{Rng}(T) = \{r(0, 3, 5, 2) + s(1, 4, 4, 1) : r, s \in \mathbb{R}\}$, T is neither one-to-one nor onto.

13. $T^{-1}(\mathbf{x}) = \dfrac{1}{\lambda}\mathbf{x}$.

15. T is onto, but not one-to-one.

17. $(T_1 T_2)(\mathbf{v}) = (T_2 T_1)(\mathbf{v}) = \mathbf{v}$.

19. One example is $T(a, b, c) = \begin{bmatrix} a & b \\ 0 & c \end{bmatrix}$.

21. One example is $T(a, b, c) = \begin{bmatrix} a & b \\ b & c \end{bmatrix}$.

23. $n = 5$: $T(a_0 + a_2 x^2 + a_4 x^4 + a_6 x^6 + a_8 x^8) = (a_0, a_2, a_4, a_6, a_8)$.

25. $T_2^{-1}(\mathbf{x}) = \begin{bmatrix} -\frac{1}{3} & -\frac{1}{6} \\ \frac{1}{3} & \frac{2}{3} \end{bmatrix} \mathbf{x}$.

27. $T_4^{-1}(\mathbf{x}) = \begin{bmatrix} 8 & -29 & 3 \\ -5 & 19 & -2 \\ 2 & -8 & 1 \end{bmatrix}$.

29. Matrix of $T_4 T_3$: $\begin{bmatrix} 6 & 13 & 18 \\ 4 & 8 & 6 \\ 23 & 43 & 11 \end{bmatrix}$; Matrix of $T_3 T_4$: $\begin{bmatrix} 10 & 25 & 23 \\ 5 & 14 & 15 \\ 12 & 19 & 1 \end{bmatrix}$.

Section 5.5

True-False Review

1. False

3. False

5. True

Problems

1.   **(a)** $\begin{bmatrix} 1 & 0 & -3 \\ 2 & 1 & -2 \end{bmatrix}$.

(b) $\begin{bmatrix} -1 & -\frac{5}{3} & -\frac{4}{3} \\ 1 & \frac{4}{3} & -\frac{1}{3} \end{bmatrix}$.

3.   **(a)** $\begin{bmatrix} 1 & 0 & -2 \\ 0 & 3 & 1 \end{bmatrix}$.

(b) $\begin{bmatrix} 4 & -11.5 & -4.5 \\ 0 & 2.5 & 6.5 \end{bmatrix}$.

5.   **(a)** $\begin{bmatrix} 0 & 1 & 0 & 0 \\ 0 & 0 & 2 & 0 \\ 0 & 0 & 0 & 3 \end{bmatrix}$.

(b) $\begin{bmatrix} 0 & 0 & 1 & -2 \\ -3 & -3 & -3 & -1 \\ 3 & 3 & 3 & 3 \end{bmatrix}$.

7.   **(a)** $\begin{bmatrix} 1 & 0 & 0 & 0 \\ 0 & 2 & -1 & 0 \\ 0 & -1 & 2 & 0 \\ 0 & 0 & 0 & 1 \end{bmatrix}$.   **(b)** $\begin{bmatrix} -1 & 1 & 0 & 0 \\ -2 & 0 & -8 & 7 \\ -2 & 3 & 7 & -2 \\ -3 & 2 & -2 & 0 \end{bmatrix}$.

9. $T(\mathbf{v}) = \begin{bmatrix} -5 & 0 \\ -6 & 5 \end{bmatrix}$.

11. $T(\mathbf{v}) = \begin{bmatrix} -19 & -2 \\ 0 & 8 \end{bmatrix}$.

13. $T(\mathbf{v}) = 3$.

15. $T(p(x)) = -8$.

17.   **(a)** 0_2.

19. No.

Section 5.6

True-False Review

1. False

3. True

5. True

7. False

9. True

Problems

5. Vectors along the x-axis are eigenvectors with $\lambda = 1$ and vectors along the y-axis are eigenvectors with $\lambda = -1$.

7. $\theta = 0, \pi$: all nonzero vectors are eigenvectors (corresponding to $\lambda = 1$ or $\lambda = -1$, respectively).

9. $\lambda_1 = 4$, $\mathbf{v}_1 = r(1, -1)$; $\lambda_2 = -2$, $\mathbf{v}_2 = s(1, 5)$.

11. $\lambda = 5$, $\mathbf{v} = r(-2, 1)$.

13. $\lambda_1 = 1 + 2i$, $\mathbf{v}_1 = r(1, 1 - i)$; $\lambda_2 = 1 - 2i$, $\mathbf{v}_2 = s(1, 1 + i)$.

15. $\lambda = 2$, $\mathbf{v} = r(3, 2, 0) + s(-1, 0, 1)$.

17. $\lambda = 1$, $\mathbf{v} = r(0, -1, 1)$.

19. $\lambda = -1$, $\mathbf{v} = r(-3, 0, 4) + s(1, 1, 0)$.

21. $\lambda_1 = 1$, $\mathbf{v}_1 = r(1, 0, 0)$; $\lambda_2 = i$, $\mathbf{v}_2 = s(0, 1, i)$; $\lambda_3 = -i$, $\mathbf{v}_3 = t(0, 1, -i)$.

23. $\lambda_1 = 0$, $\mathbf{v}_1 = r(-3, 9, 5)$; $\lambda_2 = 2$, $\mathbf{v}_2 = s(1, 3, 1)$; $\lambda_3 = 4$, $\mathbf{v} = t(1, 1, 1)$.

25. $\lambda_1 = -2$, $\mathbf{v}_1 = r(-1, 0, 1) + s(-1, 1, 0)$; $\lambda_2 = 4$, $\mathbf{v}_2 = t(1, 1, 1)$.

27. $\lambda_1 = i$, $\mathbf{v}_1 = a(0, 0, i, 1) + b(-i, 1, 0, 0)$; $\lambda_2 = -i$, $\mathbf{v}_2 = r(0, 0, -i, 1) + s(i, 1, 0, 0)$.

29.   **(c)** $A^{-1} = \begin{bmatrix} \frac{2}{3} & \frac{1}{6} \\ -\frac{1}{3} & \frac{1}{6} \end{bmatrix}$.

31. $(12, -3)$.

37.   **(a)** $\lambda\mu$.
       (b) $\lambda + \mu$.

Section 5.7

True-False Review

1. True

3. True

5. True

7. True

Problems

1. $\lambda_1 = 5$, $m_1 = 1$, basis for E_1: $\{(1, 1)\}$, $n_1 = 1$; $\lambda_2 = -1$, $m_2 = 1$, basis for E_2: $\{(-2, 1)\}$, $n_2 = 1$; nondefective.

3. $\lambda_1 = 3, m_1 = 2$, basis for E_1: $\{(1, 1)\}, n_1 = 1$; defective.

5. $\lambda_1 = -2, m_1 = 1$, basis for E_1: $\{(1, 1, 1)\}, n_1 = 1; \lambda_2 = 3, m_2 = 2$, basis for E_2: $\{(1, 0, 0), (0, -1, 4)\}, n_2 = 2$; nondefective.

7. $\lambda_1 = 4, m_1 = 3$, basis for E_1: $\{(0, 0, 1), (1, 1, 0)\}, n_1 = 2$; defective.

9. $\lambda_1 = 2, m_1 = 2$, basis for E_1: $\{(-1, 2, 0)\}, n_1 = 1$;
$\lambda_2 = -3, m_2 = 1$, basis for E_2: $\{(-1, 1, 1)\}, n_2 = 1$; defective.

11. $\lambda_1 = -1, m_1 = 3$, basis for E_1: $\{(-3, 0, 4), (1, 1, 0)\}, n_1 = 2$; defective.

13. $\lambda_1 = 0, m_1 = 2$, basis for E_1: $\{(-2, 0, 1), (1, 1, 0)\}, n_1 = 2$; $\lambda_2 = 2, m_2 = 1$, basis for E_2: $\{(1, 1, 1)\}, n_2 = 1$; nondefective.

15. $\lambda_1 = 1, m_1 = 2$, basis for E_1: $\{(-1, 0, 1), (-1, 1, 0)\}, n_1 = 2$; $\lambda_2 = -2, m_2 = 1$, basis for E_2: $\{(1, 1, 1)\}, n_2 = 1$; nondefective.

17. Defective.

19. Nondefective.

21. $\lambda_1 = 1$, basis for E_1: $\{(-1, 1)\}; \lambda_2 = 5$, basis for E_2: $\{(1, 3)\}$.

23. $\lambda_1 = 5$, basis for E_1: $\{(1, 0), (0, 1)\}$.

25. $\lambda_1 = -2$, basis for E_1: $\{(1, 1, 0)\}$.

27. $\lambda_1 = 2$, orthogonal basis for E_1: $\{(1, 0, 1), (-1, 2, 1)\}; \lambda_2 = -1$, basis for E_2: $\{(-1, -1, 1)\}$. Vectors in E_2 are orthogonal to the vectors in E_1.

Section 5.8

True-False Review

1. True

3. False

5. True

7. True

Problems

1. $S = \begin{bmatrix} 1 & 2 \\ -2 & 1 \end{bmatrix}, S^{-1}AS = \text{diag}(3, -2)$.

3. Not diagonalizable.

5. $S = \begin{bmatrix} 15 & 0 & 0 \\ -7 & 7 & 1 \\ 2 & 1 & -1 \end{bmatrix}, S^{-1}AS = \text{diag}(1, 4, -4)$.

7. $S = \begin{bmatrix} 1 & -1 & -1 \\ 1 & 0 & 1 \\ 1 & 1 & 0 \end{bmatrix}, S^{-1}AS = \text{diag}(-4, 2, 2)$.

9. Not diagonalizable.

11. $S = \begin{bmatrix} -2 & 4+3i & 4-3i \\ 1 & -2+6i & -2-6i \\ 2 & 5 & 5 \end{bmatrix}, \ S^{-1}AS = \text{diag}(0, 3i, -3i)$.

13. $S = \begin{bmatrix} 3 & -3 & 1 \\ 0 & 1 & 2 \\ 1 & 0 & 2 \end{bmatrix}, S^{-1}AS = \text{diag}(2, 2, 1)$.

15. $x_1(t) = c_1 e^{5t} - 2c_2 e^{-t}, x_2(t) = c_1 e^{5t} + c_2 e^{-t}$.

17. $x_1(t) = -c_1 e^{3t} + 3c_2 e^{-t}, x_2(t) = c_1 e^{3t} - 5c_2 e^{-t}$.

19. $x_1(t) = c_1 \sin t - c_2 \cos t, x_2(t) = c_1 \cos t + c_2 \sin t$.

21. $x_1(t) = e^{2t}(c_1 - c_2) + c_3 e^{-t}, x_2(t) = c_1 e^{2t} - c_3 e^{-t}$, $x_3(t) = c_2 e^{2t} + c_3 e^{-t}$.

25. **(c)** One possible answer: $\begin{bmatrix} \frac{12}{5} & -\frac{2}{5} \\ -\frac{3}{5} & \frac{13}{5} \end{bmatrix}$.

31. $S = \begin{bmatrix} -4 & 1 \\ 1 & 1 \end{bmatrix}$;Eigenvectors with $\lambda_1 = -3$: $r(1, -1)$. Eigenvectors with $\lambda_2 = 2$: $s(-1, -4)$.

33. One possible answer: $S = \begin{bmatrix} 1 & 0 \\ 1 & 1 \end{bmatrix}$.

Section 5.9

True-False Review

1. True

3. False

5. False

Problems

7. $e^{At} = \begin{bmatrix} \frac{1}{2}(e^{4t} + e^{2t}) & \frac{1}{2}(e^{4t} - e^{2t}) \\ \frac{1}{2}(e^{4t} - e^{2t}) & \frac{1}{2}(e^{4t} + e^{2t}) \end{bmatrix}$.

9. $e^{At} = \begin{bmatrix} e^{-t}\cos 3t & e^{-t}\sin 3t \\ -e^{-t}\sin 3t & e^{-t}\cos 3t \end{bmatrix}$.

11. $e^{At} = \begin{bmatrix} 2e^{2t} - e^t & 2(e^t - e^{2t}) & e^t - e^{3t} \\ e^{2t} - e^t & 2e^t - e^{2t} & e^t - e^{3t} \\ 0 & 0 & e^{3t} \end{bmatrix}$.

13. $e^{At} = \begin{bmatrix} 1-3t & 9t \\ -t & 1+3t \end{bmatrix}$.

15. $e^{At} = \begin{bmatrix} 1 & 0 & 0 \\ t & 1 & 0 \\ \frac{1}{2}t^2 & t & 1 \end{bmatrix}$.

17. $e^{At} = \begin{bmatrix} 1 & t & \frac{1}{2}t^2 & \frac{1}{6}t^3 \\ 0 & 1 & t & \frac{1}{2}t^2 \\ 0 & 0 & 1 & t \\ 0 & 0 & 0 & 1 \end{bmatrix}$.

Section 5.10

True-False Review

1. True

3. False

5. True

7. True

Problems

1. $S = \begin{bmatrix} -\frac{1}{\sqrt{5}} & \frac{2}{\sqrt{5}} \\ \frac{2}{\sqrt{5}} & \frac{1}{\sqrt{5}} \end{bmatrix}, S^T AS = \text{diag}(-2, 3)$.

3. $S = \begin{bmatrix} \frac{1}{\sqrt{2}} & -\frac{1}{\sqrt{2}} \\ \frac{1}{\sqrt{2}} & \frac{1}{\sqrt{2}} \end{bmatrix}, S^T AS = \text{diag}(3, -1)$.

5. $S = \begin{bmatrix} -\frac{1}{\sqrt{2}} & -\frac{1}{\sqrt{3}} & \frac{1}{\sqrt{6}} \\ 0 & \frac{1}{\sqrt{3}} & \frac{2}{\sqrt{6}} \\ \frac{1}{\sqrt{2}} & -\frac{1}{\sqrt{3}} & \frac{1}{\sqrt{6}} \end{bmatrix}, S^T AS = \text{diag}(0, 0, 6)$.

7. $S = \begin{bmatrix} -\frac{1}{\sqrt{2}} & 0 & \frac{1}{\sqrt{2}} \\ \frac{1}{\sqrt{2}} & 0 & \frac{1}{\sqrt{2}} \\ 0 & 1 & 0 \end{bmatrix}$, $S^T AS = \text{diag}(-1, 1, 1)$.

9. $S = \begin{bmatrix} \frac{1}{\sqrt{6}} & \frac{1}{\sqrt{2}} & -\frac{1}{\sqrt{3}} \\ -\frac{1}{\sqrt{6}} & \frac{1}{\sqrt{2}} & \frac{1}{\sqrt{3}} \\ \frac{2}{\sqrt{6}} & 0 & \frac{1}{\sqrt{3}} \end{bmatrix}$, $S^T AS = \text{diag}(-1, 1, 2)$.

11. $S = \begin{bmatrix} \frac{1}{\sqrt{3}} & -\frac{1}{\sqrt{2}} & -\frac{1}{\sqrt{6}} \\ \frac{1}{\sqrt{3}} & \frac{1}{\sqrt{2}} & -\frac{1}{\sqrt{6}} \\ \frac{1}{\sqrt{3}} & 0 & \frac{2}{\sqrt{6}} \end{bmatrix}$, $S^T AS = \text{diag}(1, -5, -5)$.

13. Principal axes: $\left\{ \left(\frac{1}{\sqrt{2}}, \frac{1}{\sqrt{2}} \right), \left(\frac{1}{\sqrt{2}}, -\frac{1}{\sqrt{2}} \right) \right\}$,
reduced quadratic form: $4y_1^2 - 2y_2^2$.

15. Principal axes:

$$\left\{ \left(\frac{1}{\sqrt{2}}, \frac{1}{\sqrt{2}}, 0 \right), \left(-\frac{1}{\sqrt{6}}, \frac{1}{\sqrt{6}}, \frac{2}{\sqrt{6}} \right), \left(\frac{1}{\sqrt{3}}, -\frac{1}{\sqrt{3}}, \frac{1}{\sqrt{3}} \right) \right\},$$

reduced quadratic form: $2y_1^2 + 2y_2^2 - y_3^2$.

19. $\mathbf{v}_2 = (-2, 1)$.

21. **(a)** Basis for E_2: $\{(1, 1, 0), (-1, 0, 1)\}$;

$$S = \begin{bmatrix} \frac{1}{\sqrt{3}} & \frac{1}{\sqrt{2}} & -\frac{1}{\sqrt{6}} \\ -\frac{1}{\sqrt{3}} & \frac{1}{\sqrt{2}} & \frac{1}{\sqrt{6}} \\ \frac{1}{\sqrt{3}} & 0 & \frac{2}{\sqrt{6}} \end{bmatrix}.$$

(b) $A = \frac{1}{3} \begin{bmatrix} \lambda_1 + 2\lambda_2 & -\lambda_1 + \lambda_2 & \lambda_1 - \lambda_2 \\ -\lambda_1 + \lambda_2 & \lambda_1 + 2\lambda_2 & -\lambda_1 + \lambda_2 \\ \lambda_1 - \lambda_2 & -\lambda_1 + \lambda_2 & \lambda_1 + 2\lambda_2 \end{bmatrix}$.

25. $\lambda_1 = 0, \mathbf{v}_1 = r(-5, -6, 1)$;
$\lambda_2 = \sqrt{26}i, \mathbf{v}_2 = s(-966 - 31\sqrt{26}i, -26 + 30\sqrt{26}i, -156 - 5\sqrt{26}i)$;
$\lambda_3 = -\sqrt{26}i, \mathbf{v}_3 = t(-966 + 31\sqrt{26}i, -26 - 30\sqrt{26}i, -156 - 5\sqrt{26}i)$.

Section 5.11

True-False Review

1. True
3. False
5. True
7. True
9. True
11. True

Problems

1. 3. Block sizes: (3), (2, 1), (1, 1, 1).
3. 7. Block Sizes: (5), (4, 1), (3, 2), (3, 1, 1), (2, 2, 1), (2, 1, 1, 1), (1, 1, 1, 1, 1).
5. 75.
7. 20.
9. 3. Block sizes: (3, 3), (3, 2, 1), (3, 1, 1, 1).
11. Block structures: (3; 2), (2, 1; 2), (1, 1, 1; 2), (3; 1, 1), (2, 1; 1, 1), (1, 1, 1; 1, 1).
13. Block structures: (2; 5), (1, 1; 5), (2; 4, 1), (1, 1; 4, 1), (2; 3, 2), (1, 1; 3, 2), (2; 3, 1, 1), (1, 1; 3, 1, 1), (2; 2, 2, 1), (1, 1; 2, 2, 1), (2; 2, 1, 1, 1), (1, 1; 2, 1, 1, 1), (2; 1, 1, 1, 1, 1), (1, 1; 1, 1, 1, 1, 1).

15. $A = \begin{bmatrix} 0 & 1 \\ 0 & 0 \end{bmatrix}$; $\mathbf{v} = (0, 1)$.

17. $J = \begin{bmatrix} 2 & 1 \\ 0 & 2 \end{bmatrix}$, $S = \begin{bmatrix} 1 & 0 \\ 1 & 1 \end{bmatrix}$.

19. $J = \begin{bmatrix} 4 & 1 & 0 \\ 0 & 4 & 0 \\ 0 & 0 & 4 \end{bmatrix}$, $S = \begin{bmatrix} 1 & 1 & 0 \\ 1 & 0 & 1 \\ 1 & 0 & 0 \end{bmatrix}$.

21. $J = \begin{bmatrix} -5 & 1 & 0 \\ 0 & -5 & 0 \\ 0 & 0 & -6 \end{bmatrix}$, $S = \begin{bmatrix} 1 & 0 & 1 \\ 1 & 1 & 0 \\ 0 & 1 & 1 \end{bmatrix}$.

23. $J = \begin{bmatrix} 5 & 1 & 0 \\ 0 & 5 & 1 \\ 0 & 0 & 5 \end{bmatrix}$, $S = \begin{bmatrix} 2 & 2 & 1 \\ 1 & 0 & 0 \\ -1 & -1 & 0 \end{bmatrix}$.

25. $J = \begin{bmatrix} 2 & 1 & 0 & 0 \\ 0 & 2 & 0 & 0 \\ 0 & 0 & 2 & 0 \\ 0 & 0 & 0 & 3 \end{bmatrix}$, $S = \begin{bmatrix} 1 & 0 & 1 & 1 \\ 1 & 0 & 0 & 0 \\ 1 & -1 & 0 & 0 \\ 1 & 1 & 0 & 1 \end{bmatrix}$.

27. Block sizes: (3, 1, 1).
29. Block sizes: (2, 2, 2, 2).
31. Yes.

33. $\mathbf{x}(t) = c_1 e^t \begin{bmatrix} 1 \\ 1 \\ 1 \end{bmatrix} + c_2 e^{-t} \begin{bmatrix} t + 1 \\ -t \\ t - 1 \end{bmatrix} + c_3 e^{-t} \begin{bmatrix} 1 \\ -1 \\ 1 \end{bmatrix}$.

35. $\mathbf{x}(t) = e^{4t} \left(c_1 \begin{bmatrix} 1 \\ t \\ t^2/2 \end{bmatrix} + c_2 \begin{bmatrix} 0 \\ 1 \\ t \end{bmatrix} + c_3 \begin{bmatrix} 0 \\ 0 \\ 1 \end{bmatrix} \right)$.

Section 5.12

Additional Problems

1. No.
3. Yes, onto only, Basis for $\text{Ker}(T) = \{(1, 2, 0)\}$, $\dim[\text{Ker}(T)] = 1$, Basis for $\text{Rng}(T) = \{(1, 0), (0, 1)\}$, $\dim[\text{Rng}(T)] = 2$.
5. Yes, onto only, Basis for $\text{Ker}(T) = \{(1, -1)\}$, $\dim[\text{Ker}(T)] = 1$, Basis for $\text{Rng}(T) = \{1\}$, $\dim[\text{Rng}(T)] = 1$.
7. Yes, one-to-one only, Basis for $\text{Ker}(T) = \emptyset$, $\dim[\text{Ker}(T)] = 0$, Basis for $\text{Rng}(T) = \left\{ \begin{bmatrix} -1 & 0 \\ -1 & 0 \end{bmatrix}, \begin{bmatrix} -1 & 0 \\ 0 & -2 \end{bmatrix}, \begin{bmatrix} 0 & 0 \\ 3 & 0 \end{bmatrix} \right\}$, $\dim[\text{Rng}(T)] = 3$.
9. Yes, neither, Basis for $\text{Ker}(T) = \{(0, 1, 2)\}$, $\dim[\text{Ker}(T)] = 1$, Basis for $\text{Rng}(T) = \{x^2 + 1, 2x - 2\}$, $\dim[\text{Rng}(T)] = 2$.
11. $T(x, y, z) = (-x + 8y, 2x - 2y - 5z)$.
13. $T(x) = (-\frac{1}{2}x, \frac{5}{2}x, 0, -x)$.
15. $T(a + bx + cx^2) = \begin{bmatrix} 2a - 5b - c & a - 2b + 2c \\ a - 1.5b - 2.5c & 3a - 8.5b - 0.5c \end{bmatrix}$.
17. 6.
19. Not diagonalizable.
21. Not diagonalizable.
23. $S = \begin{bmatrix} 1 & 1 & -1 \\ 0 & 0 & 2 \\ 1 & 2 & 0 \end{bmatrix}$; $D = \text{diag}(4, -1, -1)$.
29. $J = \begin{bmatrix} -1 & 1 & 0 \\ 0 & -1 & 0 \\ 0 & 0 & 3 \end{bmatrix}$.

31. Block sizes: (5), 1 linearly independent eigenvector, maximum cycle length: 5; (4, 1), 2 linearly independent eigenvectors, maximum cycle length: 4; (3, 2), 2 linearly independent eigenvectors, maximum cycle length: 3; (3, 1, 1), 3 linearly independent eigenvectors, maximum cycle length: 3; (2, 2, 1), 3 linearly independent eigenvectors, maximum cycle length: 2; (2, 1, 1, 1), 4 linearly independent eigenvectors, maximum cycle length: 2; (1, 1, 1, 1, 1), 5 linearly independent eigenvectors, maximum cycle length: 1.

35. False.

37. False.

41. $\lambda_1 \lambda_2 \cdots \lambda_k$.

Chapter 6

Section 6.1

True-False Review

1. True

3. False

5. True

7. True

Problems

1. $L(2x - 3e^{2x}) = 2(1 - x^2) + 3e^{2x}(x - 2)$.

3. $L(2x - 3e^{2x}) = 24e^{2x}(x - 1)$.

9. $\text{Ker}(L) = \{ce^{x^2} : c \in \mathbb{R}\}$.

11. $\text{Ker}(L) = \{c_1 e^{-5x} + c_2 e^{3x} : c_1, c_2 \in \mathbb{R}\}$.

13. $L_1 L_2 = D^2 + (1 - 2x^2)D - 2x(x + 2)$,
$L_2 L_1 = D^2 + (1 - 2x^2)D - 2x^2$.

15. $L_2 = D + a_1(x) + c$, where c is a constant.

17. $(D^2 + 4xD - 6x^2)y = x^2 \sin x$, $y'' + 4xy' - 6x^2 y = 0$.

21. $y(x) = c_1 e^{-2x} + c_2 e^{-5x}$.

23. $y(x) = c_1 + c_2 e^{-4x}$.

25. $y(x) = c_1 e^{2x} + c_2 e^{-2x} + c_3 e^{-3x}$.

27. $y(x) = c_1 + c_2 e^{-x} + c_3 e^{2x}$.

29. $y(x) = c_1 e^{-x} + c_2 + c_3 e^x + c_4 e^{2x}$.

31. $y(x) = c_1 x^2 + c_2 x^{-4}$.

33. $y(x) = c_1 x + c_2 x^2 + c_3 x^{-1}$.

35. $y(x) = c_1 e^{-3x} + c_2 e^{2x} + 3e^{3x}$.

37. $y(x) = c_1 e^{-2x} + c_2 e^x + c_3 e^{-x} + \frac{1}{10} e^{3x}$.

39. $y(x) = c_1 + c_2 e^{-2x} + c_3 e^{-3x} - \frac{3}{168} e^{4x}$.

Section 6.2

True-False Review

1. False

3. True

5. False

7. True

Problems

1. $\{e^x, e^{-3x}\}$.

3. $\{e^{3x} \cos 4x, e^{3x} \sin 4x\}$.

5. $y(x) = c_1 e^{-x} + c_2 e^{2x}$.

7. $y(x) = e^{-3x}(c_1 \cos 4x + c_2 \sin 4x)$.

9. $y(x) = e^{-2x}(c_1 + c_2 x)$.

11. $y(x) = e^{-5x}(c_1 + c_2 x)$.

13. $y(x) = e^{-4x}(c_1 \cos 2x + c_2 \sin 2x)$.

15. $y(x) = c_1 e^{4x} + c_2 e^{-2x}$.

17. $y(x) = c_1 e^x + c_2 \cos x + c_3 \sin x$.

19. $y(x) = c_1 e^{2x} + c_2 e^{4x} + c_3 e^{-4x}$.

21. $y(x) = c_1 e^{-x} + c_2 \cos 2x + c_3 \sin 2x + x(c_4 \cos 2x + c_5 \sin 2x)$.

23. $y(x) = c_1 + c_2 x + c_3 e^x$.

25. $y(x) = c_1 e^{2x} + c_2 e^{-2x} + c_3 \cos 2x + c_4 \sin 2x$.

27. $y(x) = c_1 e^x + c_2 e^{-5x} + e^{2x}(c_3 \cos x + c_4 \sin x)$.

29. $y(x) = c_1 e^{-x} + c_2 e^x + e^x(c_3 \cos x + c_4 \sin x) + xe^x(c_5 \cos x + c_6 \sin x)$.

31. $y(x) = c_1 \cos 3x + c_2 \sin 3x + x(c_3 \cos 3x + c_4 \sin 3x) + x^2(c_5 \cos 3x + c_6 \sin 3x)$.

33. $y(x) = e^{2x}(3 \cos x - \sin x)$.

35. $y(x) = 2(e^{2x} - e^{-2x} - xe^{-2x})$.

37. $y(x) = e^{mx}(c_1 e^{kx} + c_2 e^{-kx})$.

39.   **(b)** $u(x, y) = e^{x/\alpha}[e^{-p\xi}(A \sin q\xi + B \cos q\xi)$.

41. All roots must have negative real part.

Section 6.3

True-False Review

1. False

3. True

5. False

7. False

Problems

1. $D^2(D - 1)$.

3. $(D^2 + 16)(D - 7)^4$.

5. $(D^2 - 2D + 5)(D^2 + 4)$.

7. $(D^2 - 10D + 26)^3$.

9. $(D - 4)^2(D^2 - 8D + 41)D^2(D^2 + 4D + 5)^3$.

11. $(D^2 + 9)^2$.

13. $D^5 + 20D^3 + 64D$.

15. $D^3 + 64D$.

17. $y(x) = e^{-2x}(c_1 + c_2 x + \frac{5}{6} x^3)$.

19. $y(x) = c_1 e^{2x} + c_2 e^{-x} + \frac{5}{3} xe^{2x}$.

21. $y(x) = c_1 e^{2x} + c_2 e^{3x} - 7xe^{2x}$.

23. $y(x) = c_1 \cos \sqrt{6}x + c_2 \sin \sqrt{6}x + \frac{1}{80} \cos 4x + \frac{1}{48}$.

25. $y(x) = c_1 e^{2x} + c_2 e^{-x} + c_3 e^{-3x} - \frac{37}{27} + \frac{10}{9} x - \frac{2}{3} x^2$.

27. $y(x) = e^{-x}(c_1 + c_2 x + c_3 x^2 + \frac{1}{3} x^3) + \frac{1}{9} e^{2x}$.

29. $y(x) = c_1 e^x + c_2 e^{-x} + c_3 e^{2x} + \frac{1}{2} e^{3x}$.

31. $y(x) = c_1 e^{-3x} + c_2 e^{2x} + c_3 e^{-2x} - \frac{6}{25} \cos x - \frac{2}{25} \sin x$.

33. $y(x) = e^{-x}(c_1 + c_2 x + c_3 x^2 + \frac{5}{6} x^3)$.

35. $y_p(x) = x^2 e^{-2x}(A_0 \cos 3x + B_0 \sin 3x)$.

37. $y_p(x) = A_0 xe^x + x^2(A_1 \sin 2x + B_1 \cos 2x)$.

39. $y_p(x) = A_0 xe^{3x} + xe^{2x}(A_1 \sin x + B_1 \cos x)$.

Section 6.4

True-False Review

1. True
3. False
5. False

Problems

1. $y_p(x) = -(3\cos 3x + 4\sin 3x)$.
3. $y_p(x) = -(12\cos 3x + 5\sin 3x)$.
5. $y_p(x) = \frac{3}{10}e^x(2\sin 2x - \cos 2x)$.
7. $y_p(x) = e^x[2\sin x - 14\cos x - 10x(2\sin x + \cos x)]$.
9. $y_p(x) = 4xe^x\sin 3x$.
11. $y_p(t) = \dfrac{F_0}{\omega_0^2 - \omega^2}\cos \omega t$, if $\omega_0 \neq \omega$;

   $y_p(t) = \dfrac{F_0}{2\omega_0}t\sin \omega_0 t$, if $\omega = \omega_0$.

Section 6.5

True-False Review

1. True
3. True
5. False
7. True
9. True

Problems

1. $\omega_0 = 2$, $A_0 = 2\sqrt{2}$, $\phi = \pi/4$, $T = \pi$.
3. (a) $k = 3$.
   (b) $\omega_0 = \sqrt{3}/2$, $A_0 = 2\sqrt{3}/3$, $\phi = 5\pi/6$,
      $T = 4\pi\sqrt{3}/3$.
5. Overdamped, $y(t) = e^{-2t}(2e^t - 1)$.
7. Critically damped, $y(t) = e^{-t}(t - 1)$.
9. Underdamped, $y(t) = \frac{2}{3}e^{-2t}(3\cos\sqrt{3}t + 5\sqrt{3}\sin\sqrt{3}t)$.
11. $t = \ln 2$; Max. displacement: $1/27$.
13. (a) $y(t) = -e^{-t} + 2e^{-2t}$.
   (b) $\ln 2$.
17. $\theta(t) = \frac{1}{10}\cos\sqrt{2g}\cdot t$.
19. $L \approx 0.993$ m.
23. (a) Underdamped.
   (b) $y(t) = 2e^{-t}\cos 2t - 4\cos 2t + \sin 2t$; Transient part:
      $2e^{-t}\cos 2t$, Steady-state part: $-4\cos 2t + \sin 2t$.
25. $y_p(t) = \sqrt{10}\sin(t - \tan^{-1}(3))$.
31. (a) Transient part: $e^{-t}(c_1\cos 2t + c_2\sin 2t)$, Steady-state part:

   $$\frac{40 - 8\omega^2}{\omega^4 - 6\omega^2 + 25}\cos \omega t + \frac{16\omega}{\omega^4 - 6\omega^2 + 25}\sin \omega t.$$

   (b) $\omega = \sqrt{3}$, $y_p(t) = 2\cos(\sqrt{3}t - \pi/3)$.
33. $y(t) = A_0\cos(4t - \phi) + e^{-t}(8\cos t - \sin t)$; Transient part:
   $e^{-t}(8\cos t - \sin t)$, Steady-state part: $A_0\cos(4t - \phi)$.

Section 6.6

True-False Review

1. True

3. True
5. False

Problems

1. $i_S(t) = 2(3\cos 3t + 2\sin 3t)$.
5. $i(t) = -A_0e^{-3t}[3\cos(t - \phi) + \sin(t - \phi)] - \dfrac{2\omega}{H}\sin(\omega t - \eta)$, where

   $H = \dfrac{1}{2}\sqrt{(10 - \omega^2)^2 + 36\omega^2}$, $\cos\eta = \dfrac{10 - \omega^2}{2H}$, $\sin\eta = \dfrac{3\omega}{H}$. The
   maximum value of the amplitude occurs when $\omega = \sqrt{10}$.

7. $i(t) = A_0e^{-Rt/(2L)}\cos(\mu t - \phi) - \dfrac{aCE_0}{(a^2LC - aCR + 1)}e^{-at}$,

   where $\mu = \dfrac{\sqrt{(4L/C) - R^2}}{2L}$.

Section 6.7

True-False Review

1. True
3. False

Problems

1. $y(x) = e^{-3x}[c_1 + c_2x + 2x\tan^{-1}x - \ln(x^2 + 1)]$.
3. $y(x) = e^{2x}[\cos x(c_1 - \ln|\sec x + \tan x|) + c_2\sin x]$.
5. $y(x) = e^{-2x}(c_1 + c_2x - \ln x)$.
7. $y(x) = c_1e^x + c_2e^{-x} + 4\tan^{-1}(e^x)\cosh x$.
9. $y(x) = e^x[c_1 + c_2x + x^{-1}(2\ln x + 3)]$.
11. $y(x) = e^{-x}\left[c_1\cos 4x + c_2\sin 4x + \cos 4x\ln\left(\dfrac{\cos 4x + 2}{\cos 4x - 2}\right) + \dfrac{4}{\sqrt{3}}\sin 4x\tan^{-1}\left(\dfrac{\sin 4x}{\sqrt{3}}\right)\right]$.
13. $y(x) = e^{5x}[c_1 + c_2x - \ln(4 + x^2) + x\tan^{-1}(x/2)]$.
15. $y(x) = c_1\cos x + c_2\sin x + 2e^x + \cos x\ln(\cos x) + x\sin x$.
17. $y(x) =$
   $c_1e^{-2x} + c_2xe^{-2x} + 3\cos x + 4\sin x + \frac{15}{4}x^2e^{-2x}(2\ln x - 3)$.
19. $y_p(x) = x^3e^{2x}(6\ln x - 11)$.
21. $y_p(x) = e^{-x}[x + (x^2 - 1)\tan^{-1}x - x\ln(1 + x^2)]$.
23. $y_p(x) = \displaystyle\int_{x_0}^x \sinh(x - t)F(t)\,dt$.
25. $y_p(x) = \dfrac{1}{3}\displaystyle\int_{x_0}^x [e^{t-x} - e^{4(t-x)}]F(t)\,dt$.
27. $y(x) = \sin x + x\sin x + [\ln(\cos x)]\cos x$.
29. (a) $y_p(x) = \alpha e^{ax}\left[\dfrac{x}{\beta}\tan^{-1}(x/\beta) - \dfrac{1}{2}\ln\left(\dfrac{x^2 + \beta^2}{\beta^2}\right)\right]$.
   (b) $y_p(x) = \alpha e^{ax}[x\sin^{-1}(x/\beta) + \sqrt{\beta^2 - x^2} - \beta^2]$.
   (c) $\alpha = -1$: $y_p(x) = \dfrac{xe^{ax}}{2}[(\ln x)^2 - 2\ln x - 2]$;
      $\alpha = -2$: $y_p(x) = -e^{ax}[(\ln x)^2/2 + \ln x + 1]$.
      $\alpha \neq -1, -2$:
      $y_p(x) = \dfrac{e^{ax}x^{\alpha+2}}{(\alpha + 1)(\alpha + 2)}\left[\ln x - \dfrac{2\alpha + 3}{(\alpha + 1)(\alpha + 2)}\right]$.

33. $y_p(x) = \dfrac{1}{48} \displaystyle\int_{x_0}^{x} F(t) \left\{ e^{3(x-t)} + 3e^{-5(x-t)} - 4e^{-3(x-t)} \right\} dt.$

35. $y_p(x) = \dfrac{1}{36} \displaystyle\int_{x_0}^{x} F(t) \left\{ e^{-4(x-t)}[6(t-x) - 1] + e^{2(x-t)} \right\} dt.$

37. $y + p(x) = \frac{1}{30} \int_{x_0}^{x} F(t) \left\{ e^{-4(x-t)} + 5e^{2(x-t)} - 6e^{x-t} \right\} dt$

Section 6.8

True-False Review

  1. False

  3. False

  5. True

Problems

  1. $y(x) = x[c_1 \sin(2 \ln x) + c_2 \cos(2 \ln x)].$

  3. $y(x) = x^2(c_1 + c_2 \ln x).$

  5. $y(x) = x^{-1}(c_1 + c_2 \ln x).$

  7. $y(x) = c_1 x^7 + c_2 x^{-5}.$

  9. $y(x) = c_1 x^m + c_2 x^{-m}.$

  11. $y(x) = x^m[c_1 \cos(k \ln x) + c_2 \sin(k \ln x)].$

  15. $y(x) = c_1 x^3 + x^2(c_2 - \sin x).$

  17. $y(x) = x^2[c_1 + c_2 \ln x + \ln x(\ln |\ln x| - 1)].$

  19. $y(x) = c_1 \cos(3 \ln x) + c_2 \sin(3 \ln x) + \ln x.$

  21. $y(x) = y_c(x) + y_p(x),$ where $y_c(x) = c_1 x^m + c_2 x^m \ln x,$ and

$$y_p(x) = \begin{cases} \dfrac{x^m (\ln x)^{k+2}}{(k+1)(k+2)}, & \text{if } k \neq -1, -2, \\ (\ln |\ln x| - 1)x^m \ln x, & \text{if } k = -1, \\ -x^m(1 + \ln |\ln x|), & \text{if } k = -2. \end{cases}$$

  23.   **(b)** $t = e^{\pi(6n+5)/30}.$

    (d) No.

Section 6.9

Problems

  1. $y_2(x) = x^2 \ln x.$

  3. $y_2(x) = e^x \ln x.$

  5. $y_2(x) = \dfrac{1}{2} x \ln\left(\dfrac{1+x}{1-x} \right) - 1.$

  7.   **(a)** $y_1(x) = x^m.$

    (b) $y_2(x) = x^m \ln x.$

  11. $y(x) = e^{2x}[c_1 + c_2 x + x^2(2 \ln x - 3)].$

  13. $y(x) = c_1 \cos x + c_2 \sin x + \sin x[\ln(\sin x) - x \cot x].$

  15. $y(x) = x^2(c_1 + c_2 \ln x + 2x^2).$

Section 6.10

Additional Problems

  1. $3e^{x^3}[3x^4 + 2x + 1].$

  3. $-\frac{4}{x} \sin x + 4x \cos x - 8 \sin x.$

  5. $\frac{2}{x^3} + \frac{2}{x} + 5x + 480x^2 - 40x^3 + 480x^4 + 200x^6 - \frac{\cos x}{x}.$

  7. $y(x) = c_1 e^x + c_2 e^{-2x} + c_3 x e^{-2x}.$

  9. $y(x) = c_1 \cos 2x + c_2 \sin 2x + c_3 \cos 3x + c_4 \sin 3x.$

  11. $y(x) = c_1 e^{-3x} + c_2 x e^{-3x} + c_3 x^2 e^{-3x} + e^{2x}(c_4 \cos 3x + c_5 \sin 3x).$

13. $y(x) = c_1 e^{3x} + c_2 e^{-2x} + c_3 x e^{-2x}.$

15. $D^2 - 6D + 10.$

17. $(D+2)(D^2+1)^2.$

19. $y_p(x) = A_0 e^{-2x}.$

21. $y_p(x) = A_0 \cos 4x + B_0 \sin 4x.$

23. $y_p(x) = x^3(A_0 \cos x + B_0 \sin x).$

25. $y(x) = c_1 + e^{3x}(c_2 \cos 4x + c_3 \sin 4x) + \frac{22}{15625} x + \frac{6}{625} x^2 + \frac{1}{75} x^3.$

27. $y(x) = c_1 e^{2x} + c_2 e^{-2x} - \frac{5}{3} e^x.$

29. $y(x) = c_1 e^x + c_2 e^{-x} + 2x e^x.$

31. Annihilators cannot be used.

33. Annihilators cannot be used.

35. $y_p(x) = A_0 x^2 e^{4x}.$

37. $y_p(x) = A_0 \cos x + B_0 \sin x + A_1 e^x \cos x + A_2 e^x \sin x.$

39. $y_p(x) = A_1 e^{at} + A_2 t e^{at} + A_0 t e^{at} \cos bt + B_0 t e^{at} \sin bt.$

41. $y(x) = c_1 e^x + c_2 e^{-3x} - \frac{1}{8} x e^{-3x} - \frac{1}{4} x^2 e^{-3x}.$

43. $y(x) = c_1 \cos 2x + c_2 \sin 2x + 2x \sin 2x.$

45. $y(x) = c_1 e^x + c_2 e^{-x} + e^{2x} - \frac{1}{2} \sin x.$

47. $y(x) = c_1 \cos x + c_2 \sin x - x \cos x + (\ln |\sin x|) \sin x.$

49. $y(x) = c_1 e^{mx} + c_2 x e^{mx} - e^{mx}\left[\frac{1}{2}x^2 \ln x - \frac{1}{4}x^2\right] + xe^{mx}[x \ln x - x].$

51. $y(x) = c_1 e^x + c_2 x e^x - e^x\left[\frac{1}{2}x^2 \ln x - \frac{1}{4}x^2\right] + xe^x[x \ln x - x].$

53. $y(x) = c_1 x^{-4} + c_2 x^{-4} \ln x.$

55. $y(x) = c_1 x^6 \cos x + c_2 x^6 \sin x.$

57. $y(x) = c_1 x^6 + c_2 x^{-3}.$

59. $y(x) = c_1 x^{-4} + c_2 x^{-4} \ln x - x^{-4}[x \ln x - x] + x^{-3} \ln x.$

61. $y(x) = c_1 x^3 \cos x + c_2 x^3 \sin x + x^3 \cos x(x \cos x - \sin x) + x^3 \sin x(x \sin x + \cos x).$

63. $y_p(x) = -e^{-2x}(x \ln x - x) + xe^{-2x}(\ln x)^2/2.$

65. $y_p(x) = \frac{x^4 e^{5x}}{4} - e^{5x}(x^3 - 3x^2 + 6x - 6).$

Chapter 7
Section 7.1

True-False Review

  1. True

  3. False

  5. True

  7. False

  9. False

Problems

  1. $x_1(t) = c_1 e^t + c_2 e^{-t}, x_2(t) = \frac{1}{3}(c_1 e^t + 3c_2 e^{-t}).$

  3. $x_1(t) = e^{-2t}(c_1 + c_2 t), x_2(t) = \frac{1}{4} e^{-2t}[-4c_1 + c_2(1 - 4t)].$

  5. $x_1(t) = e^t(c_1 \cos 3t + c_2 \sin 3t), x_2(t) = e^t(c_1 \sin 3t - c_2 \cos 3t).$

7. $x_1(t) = -e^{-2t}(2c_1 + c_2 \cos t + c_3 \sin t)$,

$x_2(t) = -e^{2t}[c_1 + c_2(\cos t - \sin t) + c_3(\sin t + \cos t)]$,

$x_3(t) = e^{-2t}(c_1 + c_2 \cos t + c_3 \sin t)$.

9. $x_1(t) = 5 \sin t, x_2(t) = \cos t - 2 \sin t$.

11. $x_1(t) = c_1 e^{-t} + c_2 e^{3t} + 3e^{4t}, x_2(t) = -c_1 e^{-t} + c_2 e^{3t} + 2e^{4t}$.

13. $x_1(t) = c_1 e^{2t} + c_2 e^{-2t} + 2te^{2t}$,

$x_2(t) = c_1 e^{2t} - 3c_2 e^{-2t} + e^{2t}(1 + 2t)$.

15. $x_1' = x_2, x_2' = -x_1 + 3x_4 + \sin t, x_3' = x_4, x_4' = tx_2 + e^t x_3 + t^2$.

17. $x_1' = x_2, x_2' = -bx_1 - ax_2 + F(t)$.

19. $x_1' = x_2, x_2' = -\dfrac{(k_1 + k_2)}{m_1}x_1 + \dfrac{k_2}{m_1}x_3, x_3' = x_4, x_4' = \dfrac{k_2}{m_2}x_1 -$

$\dfrac{k_2}{m_2}x_3, x_1(0) = \alpha_1, x_2(0) = \alpha_2, x_3(0) = \alpha_3, x_4(0) = \alpha_4$.

Section 7.2

True-False Review

1. True
3. False
5. False
7. False

Section 7.3

True-False Review

1. False
3. True

Problems

1. General solution: $\mathbf{x}(t) = c_1 \begin{bmatrix} e^{4t} \\ 2e^{4t} \end{bmatrix} + c_2 \begin{bmatrix} 3e^{-t} \\ e^{-t} \end{bmatrix}$.

Particular solution: $\mathbf{x}(t) = \begin{bmatrix} e^{4t} & 3e^{-t} \\ 2e^{4t} & e^{-t} \end{bmatrix} \begin{bmatrix} 1 \\ -1 \end{bmatrix}$.

3. General solution: $\mathbf{x}(t) = \begin{bmatrix} -3 & e^{2t} & e^{4t} \\ 9 & 3e^{2t} & e^{4t} \\ 5 & e^{2t} & e^{4t} \end{bmatrix} \begin{bmatrix} c_1 \\ c_2 \\ c_3 \end{bmatrix}$.

5. General solution: $\mathbf{x}(t) = c_1 \begin{bmatrix} t \sin t \\ \cos t \end{bmatrix} + c_2 \begin{bmatrix} -t \cos t \\ \sin t \end{bmatrix}$.

Section 7.4

True-False Review

1. True
3. False
5. True

Problems

1. $\mathbf{x}(t) = c_1 e^{-3t} \begin{bmatrix} 7 \\ 1 \end{bmatrix} + c_2 e^{5t} \begin{bmatrix} -1 \\ 1 \end{bmatrix}$.

3. $\mathbf{x}(t) = e^{-2t} \left\{ c_1 \begin{bmatrix} 3 \cos t - \sin t \\ 5 \cos t \end{bmatrix} + c_2 \begin{bmatrix} \cos t + 3 \sin t \\ 5 \sin t \end{bmatrix} \right\}$.

5. $\mathbf{x}(t) = c_1 e^{-2t} \begin{bmatrix} 0 \\ 1 \\ 1 \end{bmatrix} + c_2 e^{2t} \begin{bmatrix} 1 \\ 0 \\ 0 \end{bmatrix} + c_3 e^{3t} \begin{bmatrix} 0 \\ 7 \\ 2 \end{bmatrix}$.

7. $\mathbf{x}(t) = c_1 e^{5t} \begin{bmatrix} 0 \\ 0 \\ 1 \end{bmatrix} + c_2 \begin{bmatrix} \sin t \\ \cos t \\ 0 \end{bmatrix} + c_3 \begin{bmatrix} -\cos t \\ \sin t \\ 0 \end{bmatrix}$.

9. $\mathbf{x}(t) = c_1 e^{-3t} \begin{bmatrix} -1 \\ 0 \\ 1 \end{bmatrix} + c_2 e^t \begin{bmatrix} -1 \\ -2 \\ 1 \end{bmatrix} + c_3 e^{2t} \begin{bmatrix} 2 \\ -10 \\ 3 \end{bmatrix}$.

11. $\mathbf{x}(t) = c_1 e^{3t} \begin{bmatrix} 1 \\ 0 \\ 0 \end{bmatrix} +$

$e^{-2t} \left\{ c_2 \begin{bmatrix} 5 \cos t - \sin t \\ -13(\cos t + \sin t) \\ 26 \cos t \end{bmatrix} + c_3 \begin{bmatrix} \cos t + 5 \sin t \\ 13(\cos t - \sin t) \\ 26 \sin t \end{bmatrix} \right\}$.

13. $\mathbf{x}(t) = c_1 \begin{bmatrix} -3 \\ 0 \\ 2 \end{bmatrix} + c_2 \begin{bmatrix} 1 \\ 2 \\ 0 \end{bmatrix} + c_3 e^{4t} \begin{bmatrix} 1 \\ 1 \\ 1 \end{bmatrix}$.

15. $\mathbf{x}(t) = c_1 \begin{bmatrix} 0 \\ 0 \\ -\sin t \\ \cos t \end{bmatrix} + c_2 \begin{bmatrix} 0 \\ 0 \\ \cos t \\ \sin t \end{bmatrix} + c_3 \begin{bmatrix} \sin t \\ \cos t \\ 0 \\ 0 \end{bmatrix} +$

$c_4 \begin{bmatrix} -\cos t \\ \sin t \\ 0 \\ 0 \end{bmatrix}$.

17. $\mathbf{x}(t) = e^{2t} \begin{bmatrix} 2 \cos 3t - 6 \sin 3t \\ 2 \cos 3t + 4 \sin 3t \end{bmatrix}$.

19. $\mathbf{x}(t) = \begin{bmatrix} \cos 4t + \sin 4t \\ -\sin 4t + \cos 4t \end{bmatrix}$.

Section 7.5

True-False Review

1. False
3. False

Problems

1. $\mathbf{x}(t) = e^{2t} \left\{ c_1 \begin{bmatrix} -1 \\ 1 \end{bmatrix} + c_2 \begin{bmatrix} 1 - 2t \\ 2t \end{bmatrix} \right\}$.

3. $\mathbf{x}(t) = c_1 e^t \begin{bmatrix} 1 \\ 1 \\ 1 \end{bmatrix} + e^{-t} \left\{ c_2 \begin{bmatrix} 1 \\ -1 \\ 1 \end{bmatrix} + c_3 \begin{bmatrix} 1 + t \\ -t \\ t - 1 \end{bmatrix} \right\}$.

5. $\mathbf{x}(t) = e^{-2t} \left\{ c_1 \begin{bmatrix} 1 \\ 0 \\ 1 \end{bmatrix} + c_2 \begin{bmatrix} 1 \\ 1 \\ 0 \end{bmatrix} + c_3 \begin{bmatrix} 1 \\ t \\ -t \end{bmatrix} \right\}$.

7. $\mathbf{x}(t) = e^{4t} \left\{ c_1 \begin{bmatrix} 0 \\ 0 \\ 1 \end{bmatrix} + c_2 \begin{bmatrix} 0 \\ 1 \\ t \end{bmatrix} + c_3 \begin{bmatrix} 2 \\ 2t \\ t^2 \end{bmatrix} \right\}$.

9. $\mathbf{x}(t) = e^{4t}\left\{c_1\begin{bmatrix}0\\0\\1\end{bmatrix} + c_2\begin{bmatrix}1\\1\\0\end{bmatrix} + c_3\begin{bmatrix}t\\1+t\\0\end{bmatrix}\right\}.$

11. $\mathbf{x}(t) = c_1\begin{bmatrix}5(2\sin t - \cos t)\\-5(2\cos t + \sin t)\\-2(\cos t + 2\sin t)\\5\cos t\end{bmatrix} + c_2\begin{bmatrix}-5(2\cos t + \sin t)\\5(\cos t - 2\sin t)\\2(\cos t - \sin t)\\5\sin t\end{bmatrix}$

$+\, e^{2t}\left\{c_3\begin{bmatrix}0\\0\\1\\0\end{bmatrix} + c_4\begin{bmatrix}0\\0\\t\\1\end{bmatrix}\right\}.$

13. $\mathbf{x}(t) = c_1\begin{bmatrix}0\\0\\-\sin t\\\cos t\end{bmatrix} + c_2\begin{bmatrix}0\\0\\\cos t\\\sin t\end{bmatrix} + c_3\begin{bmatrix}0\\\cos t\\\cos t - t\sin t\\\sin t + t\cos t\end{bmatrix} +$

$c_4\begin{bmatrix}\cos t\\\sin t\\\sin t + t\cos t\\t\sin t - \cos t\end{bmatrix}.$

15. $\mathbf{x}(t) = \begin{bmatrix}7 - 2t - 9e^{-t}\\e^{-t}\\3 - t - 2e^{-t}\end{bmatrix}.$

Section 7.6

True-False Review

1. False
3. True
5. False

Problems

1. $\mathbf{x}_p(t) = \begin{bmatrix}e^t(2t+1)\\e^t(2t-1)\end{bmatrix}.$ General solution:

$\mathbf{x}(t) = c_1 e^t\begin{bmatrix}1\\1\end{bmatrix} + c_2 e^{3t}\begin{bmatrix}-1\\1\end{bmatrix} + \begin{bmatrix}e^t(2t+1)\\e^t(2t-1)\end{bmatrix}.$

3. $\mathbf{x}_p(t) = \begin{bmatrix}4e^t(2e^{2t}-1)\\4e^t(3e^{2t}-2)\end{bmatrix}.$ General solution:

$\mathbf{x}(t) = c_1 e^{-2t}\begin{bmatrix}-1\\1\end{bmatrix} + c_2 e^{2t}\begin{bmatrix}1\\3\end{bmatrix} + \begin{bmatrix}4e^t(2e^{2t}-1)\\4e^t(3e^{2t}-2)\end{bmatrix}.$

5. $\mathbf{x}_p(t) = \begin{bmatrix}(12t+1)\sin 2t + (1-4t)\cos 2t\\(8t-1)\cos 2t - 4t\sin 2t\end{bmatrix}.$ General solution:

$\mathbf{x}(t) = c_1\begin{bmatrix}-\cos 2t + \sin 2t\\\cos 2t\end{bmatrix} + c_2\begin{bmatrix}\cos 2t + \sin 2t\\-\sin 2t\end{bmatrix} +$

$\begin{bmatrix}(12t+1)\sin 2t + (1-4t)\cos 2t\\(8t-1)\cos 2t - 4t\sin 2t\end{bmatrix}.$

7. $\mathbf{x}_p(t) = \begin{bmatrix}-te^t\\9e^{-t}\\te^t + 6e^{-t}\end{bmatrix}.$ General solution:

$\mathbf{x}(t) = c_1 e^{-2t}\begin{bmatrix}0\\2\\1\end{bmatrix} + c_2 e^t\begin{bmatrix}2\\1\\0\end{bmatrix} + c_3 e^t\begin{bmatrix}-1\\0\\1\end{bmatrix} +$

$\begin{bmatrix}-te^t\\9e^{-t}\\te^t + 6e^{-t}\end{bmatrix}.$

Section 7.7

True-False Review

1. False
3. False
5. False

Problems

3. $x(t) = 2(c_1\sin t - c_2\cos t - c_3\sin 3t + c_4\cos 3t),$
$y(t) = 3c_1\sin t - 3c_2\cos t + c_3\sin 3t - c_4\cos 3t.$

9. $A_1(t) = 120 - 50e^{-t/5} - 10e^{-t/15},$
$A_2(t) = 120 + 100e^{-t/5} - 20e^{-t/15}.$

11. **(c)** $A_1 = \dfrac{\alpha_1 + \alpha_2}{1+\beta} + \dfrac{\beta\alpha_1 - \alpha_2}{1+\beta}e^{\lambda_2 t},$

$A_2 = \beta\dfrac{\alpha_1 + \alpha_2}{1+\beta} + \dfrac{\alpha_2 - \beta\alpha_1}{1+\beta}e^{\lambda_2 t}.$

Section 7.8

True-False Review

1. True
3. True
5. True

Problems

3. $e^{At} = e^{2t}\begin{bmatrix}1 & t\\0 & 1\end{bmatrix}.$

5. $\mathbf{x}_1(t) = e^t\begin{bmatrix}1+2t\\4t\end{bmatrix},\ \mathbf{x}_2(t) = e^t\begin{bmatrix}-t\\1-2t\end{bmatrix},\ e^{At} =$

$e^t\begin{bmatrix}1+2t & -t\\4t & 1-2t\end{bmatrix}.$

7. $\mathbf{x}_1(t) = e^{2t}\begin{bmatrix}1\\0\\0\end{bmatrix},\ \mathbf{x}_2(t) = e^{-3t}\begin{bmatrix}0\\1+4t\\2t\end{bmatrix},\ \mathbf{x}_3(t) =$

$e^{-3t}\begin{bmatrix}0\\-8t\\1-4t\end{bmatrix},\ e^{At} = \begin{bmatrix}e^{2t} & 0 & 0\\0 & e^{-3t}(1+4t) & -8te^{-3t}\\0 & 2te^{-3t} & e^{-3t}(1-4t)\end{bmatrix}.$

9. $\mathbf{x}(t) = c_1 e^{-t}\begin{bmatrix}-7\\4\\1\end{bmatrix} + e^{3t}\left\{c_2\begin{bmatrix}1\\0\\1\end{bmatrix} + c_3\begin{bmatrix}t\\1\\t\end{bmatrix}\right\}.$

11. $\mathbf{x}_1(t) = \begin{bmatrix}-\sin t\\\cos t\\-t\sin t\\t\cos t\end{bmatrix},\ \mathbf{x}_2(t) = \begin{bmatrix}\cos t\\\sin t\\t\cos t\\t\sin t\end{bmatrix},$

$\mathbf{x}_3(t) = \begin{bmatrix}0\\0\\-\sin t\\\cos t\end{bmatrix},\ \mathbf{x}_4(t) = \begin{bmatrix}0\\0\\\cos t\\\sin t\end{bmatrix}.$

Section 7.9

True-False Review

1. False
3. True
5. True

Problems

1. $(0,0)$, $(-1,0)$, $(-\frac{1}{3}, \frac{2}{3})$.
3. $(0,0)$, $(1,0)$, $(-1,0)$.
5. Stable center.
7. Saddle.
9. Stable node.
11. Stable center.
13. Unstable spiral.
15. Unstable node.
17. Stable degenerate node.
19. Unstable proper node.

Section 7.10

True-False Review

1. True
3. False
5. False

Problems

1. $(0,0)$, center or spiral.
3. $(0,0)$, saddle; $(\frac{4}{9}, \frac{2}{3})$, unstable spiral; $(\frac{4}{9}, -\frac{2}{3})$, unstable spiral.
5. $(0,0)$, unstable node.
7. $(0,0)$, unstable spiral; $(\frac{1}{2}, -1)$, saddle.
9. $(0,0)$, unstable degenerate node.

Section 7.11

Additional Problems

3. $\mathbf{x}(t) = c_1 e^{-3t} \begin{bmatrix} 1 \\ 3 \end{bmatrix} + c_2 e^{-8t} \begin{bmatrix} -1 \\ 2 \end{bmatrix}$.

5. $\mathbf{x}(t) = c_1 e^{6t} \begin{bmatrix} 1 \\ 1 \end{bmatrix} + c_2 e^{6t} \left\{ \begin{bmatrix} 1 \\ \frac{3}{4} \end{bmatrix} + t \begin{bmatrix} 1 \\ 1 \end{bmatrix} \right\}$.

7. $\mathbf{x}(t) = c_1 e^{2t} \begin{bmatrix} 0 \\ 1 \\ 0 \end{bmatrix} + c_2 e^{-t} \begin{bmatrix} 1 \\ 0 \\ -1 \end{bmatrix} +$
$c_3 e^{-t} \left\{ \begin{bmatrix} 1 \\ 0 \\ -3/4 \end{bmatrix} + t \begin{bmatrix} 1 \\ 0 \\ -1 \end{bmatrix} \right\}$.

9. $\mathbf{x}(t) = c_1 \begin{bmatrix} 3\cos 2t - 2\sin 2t \\ -\cos 2t \end{bmatrix} + c_2 \begin{bmatrix} 2\cos 2t + 3\sin 2t \\ -\sin 2t \end{bmatrix}$.

11. $\mathbf{x}(t) = c_1 e^{3t} \begin{bmatrix} 1 \\ 0 \\ 4 \end{bmatrix} + c_2 e^{-5t} \begin{bmatrix} 5\cos 2t \\ 4\cos 2t + 2\sin 2t \\ 0 \end{bmatrix} +$
$c_3 e^{-5t} \begin{bmatrix} -5\sin 2t \\ 2\cos 2t - 4\sin 2t \\ 0 \end{bmatrix}$.

13. $\mathbf{x}(t) = c_1 e^{2t} \begin{bmatrix} 4 \\ 1 \\ 2 \end{bmatrix} + c_2 e^{-2t} \begin{bmatrix} 0 \\ 1 \\ 2 \end{bmatrix} + c_3 e^{-5t} \begin{bmatrix} 0 \\ -2 \\ 3 \end{bmatrix}$.

15. $\mathbf{x}(t) = c_1 e^{4t} \begin{bmatrix} -2 \\ 0 \\ 1 \end{bmatrix} + c_2 e^{4t} \begin{bmatrix} 0 \\ 1 \\ 0 \end{bmatrix} + c_3 e^{-3t} \begin{bmatrix} -3 \\ -1 \\ 1 \end{bmatrix}$.

17. $\mathbf{x}(t) = c_1 \begin{bmatrix} -1 \\ 0 \\ 1 \end{bmatrix} + c_2 \begin{bmatrix} -t \\ -1 \\ 1+t \end{bmatrix} + c_3 e^{-3t} \begin{bmatrix} -2 \\ -1 \\ 2 \end{bmatrix}$.

19. $\mathbf{x}(t) = c_1 e^{-t} \begin{bmatrix} -2 \\ -1 \\ 2 \end{bmatrix} + c_2 e^{t} \begin{bmatrix} \sin 2t \\ -\cos 2t \\ \cos 2t - \sin 2t \end{bmatrix} +$
$c_3 e^{t} \begin{bmatrix} \cos 2t \\ \sin 2t \\ -\cos 2t - \sin 2t \end{bmatrix}$.

21. $\mathbf{x}(t) = c_1 e^{-t} \begin{bmatrix} -1 \\ 0 \\ 1 \end{bmatrix} + c_2 e^{-t} \begin{bmatrix} -t \\ -1 \\ 1+t \end{bmatrix} + c_3 e^{-t} \begin{bmatrix} -\frac{t^2}{2} \\ -2 - t \\ 1+t+\frac{t^2}{2} \end{bmatrix}$.

23. $\mathbf{x}(t) = c_1 \begin{bmatrix} 3\sin 3t - 2\cos 3t \\ \cos 3t \\ 0 \\ 0 \end{bmatrix} + c_2 \begin{bmatrix} 3\cos 3t - 2\sin 3t \\ -\sin 3t \\ 0 \\ 0 \end{bmatrix}$
$+ c_3 e^{2t} \begin{bmatrix} 0 \\ 0 \\ 4 \\ 0 \end{bmatrix} + c_4 e^{2t} \begin{bmatrix} 0 \\ 0 \\ 4t \\ 1 \end{bmatrix}$.

25. $\mathbf{x}_p(t) = \begin{bmatrix} \frac{5}{24} + \frac{1}{14} e^{-t} \\ \frac{1}{4} + \frac{5}{14} e^{-t} \end{bmatrix}$. General solution: Add $\mathbf{x}_p$ to Problem 3 answer.

27. $\mathbf{x}_p(t) = e^{6t} \begin{bmatrix} 4t(1 - \ln|t|) \\ \ln|t| + 4t(1 - \ln|t|) \end{bmatrix}$. General solution: Add $\mathbf{x}_p$ to Problem 5 answer.

29. $\mathbf{x}_p(t) = - \begin{bmatrix} \frac{t}{2} + \frac{1}{4} \\ \frac{t}{4} - \frac{3}{35} \\ \frac{t}{2} - \frac{13}{35} \end{bmatrix}$. General solution: Add $\mathbf{x}_p$ to Problem 13 answer.

Chapter 8

Section 8.1

True-False Review

1. False
3. True
5. False
7. False
9. False

Problems

1. $\dfrac{1}{s-2}$.

3. $\dfrac{b}{s^2+b^2}$.

5. $\dfrac{s}{s^2-b^2}$.

7. $\dfrac{2}{s^2}$.

9. $\dfrac{1}{s}(1-2e^{-2s})$.

11. $\dfrac{1}{(s-1)^2+1}$.

13. $\dfrac{2}{s^2}-\dfrac{1}{s-3}$.

15. $\dfrac{s}{s^2-b^2}$.

17. $\dfrac{6}{s^3}-\dfrac{5s}{s^2+4}+\dfrac{3}{s^2+9}$.

19. $\dfrac{2}{s+3}+\dfrac{4}{s-1}-\dfrac{5}{s^2+1}$.

21. $\dfrac{4(s^2+2b^2)}{s(s^2+4b^2)}$.

23. Piecewise continuous.

25. Not piecewise continuous.

27. Piecewise continuous.

29. Not piecewise continuous.

31. $\dfrac{1}{s^2}[1-e^{-s}(s+1)]$.

33. $\dfrac{1}{s^2}e^{-2s}[e^s(s+1)-(2s+1)]$.

35. $L[e^{ibt}]=\dfrac{s}{s^2+a^2}+\dfrac{b}{s^2+a^2}i$.

Section 8.2

True-False Review

1. True

3. False

5. False

Problems

7. 2.

9. $5e^{-3t}$.

11. $2\cos 3t$.

13. $\cos t+6\sin t$.

15. $2-3e^{-t}$.

17. $1-e^{-t}$.

19. $\frac{1}{5}(7e^{2t}-7\cos t-4\sin t)$.

21. $\frac{1}{6}(4\cos t+6\sin t-4\cos 2t-3\sin 2t)$.

Section 8.3

True-False Review

1. False

3. False

5. False

7. False

Problems

1. $F(s)=\dfrac{1}{s^2(1-e^{-s})}[1-e^{-s}(s+1)]$.

3. $F(s)=\dfrac{1+e^{-\pi s}}{(1-e^{-\pi s})(s^2+1)}$.

5. $F(s)=\dfrac{e^{1-s}-1}{(1-s)(1-e^{-s})}$.

7. $F(s)=\dfrac{1}{1-e^{-\pi s}}\left[\dfrac{2-e^{-\pi s/2}(\pi s+2)}{\pi s^2}+\dfrac{se^{-\pi s/2}+e^{-\pi s}}{s^2+1}\right]$.

9. $F(s)=\dfrac{1}{as^2}\tanh\left(\dfrac{as}{2}\right)$.

11. $F(s)=\frac{s}{s^2+a^2}$.

Section 8.4

True-False Review

1. False

3. True

Problems

1. $y(t)=2e^{3t}$.

3. $y(t)=2t-1+2e^{-2t}$.

5. $y(t)=e^t-\sin 2t-2\cos 2t$.

7. $y(t)=2e^t-e^{-2t}$.

9. $y(t)=2+3e^{2t}-5e^t$.

11. $y(t)=2e^t+3e^{-2t}-5e^{-t}$.

13. $y(t)=3e^{2t}+2e^{-3t}-4$.

15. $y(t)=2\cos 2t+\sin 2t+2e^{-t}$.

17. $y(t)=\frac{7}{2}e^t-\frac{1}{2}e^{-t}-3\cos t$.

19. $y(t)=e^t-2e^{-t}-4\sin t+3\cos t$.

21. $y(t)=2e^{-t}-e^{-4t}-2\cos 2t$.

23. $y(t)=\frac{1}{5}(7e^{2t}-5e^t+3\cos t-4\sin t)$.

25. $y(t)=2(\cos t+\sin t-\cos 2t)$.

27. $y(t)=y(0)\cosh t+y'(0)\sinh t$.

29. (a) $i(t)=\dfrac{E_0}{R}[1-e^{-(R/L)t\cdot}]$.

   (b) $i(t)=\dfrac{E_0\omega}{\sqrt{R^2+L^2\omega^2}}\sin(\omega t-\theta)+\dfrac{E_0 L\omega}{R^2+L^2\omega^2}e^{-Rt/L}$,

   where $\tan\theta=\omega L/R$.

31. $x_1(t)=2(e^{-3t}-e^{-2t})$, $x_2(t)=2e^{-2t}-e^{-3t}$.

Section 8.5

1. True
3. True
5. False
7. True

Problems

1. $t - 1$.
3. $t(t + 2)$.
5. $-e^{2(t-\pi)} \cos t$.
7. $\frac{1}{2}e^{-(t-\pi/6)}(\sin 2t - \sqrt{3}\cos 2t)$.
9. $\dfrac{t - 1}{t^2 - 6t + 10}$.
11. t^2.
13. te^{3t}.
15. $(t + 3)e^{-t}$.
17. $\dfrac{s - 3}{(s - 3)^2 + 16}$.
19. $\dfrac{1}{(s - 2)^2}$.
21. $\dfrac{6}{(s + 4)^4}$.
23. $\dfrac{2(2s^3 - 9s^2 + 10s + 30)}{[(s - 3)^2 + 1][(s + 1)^2 + 9]}$.
25. $\dfrac{2}{(s - 1)^3} - \dfrac{6}{s^3}$.
27. te^{3t}.
29. $\dfrac{2}{e^{3t}\sqrt{t\pi}}$.
31. $e^{-2t}\cos 3t$.
33. $\frac{5}{4}e^{2t}\sin 4t$.
35. $\frac{1}{5}e^{-t}(5\cos 5t - 3\sin 5t)$.
37. $\frac{1}{2}e^{-t}(2\cos 2t - \sin 2t)$.
39. $1 - e^{-2t}(1 + 2t)$.
41. $\frac{1}{10}[6 + e^t(13\sin 2t - 6\cos 2t)]$.
43. $y(t) = e^{2t}(1 + 3t) + e^{-2t}$.
45. $y(t) = 2e^t + e^{-2t}(1 - t)$.
47. $y(t) = e^{-t}(2 + 3t + t^2)$.
49. $y(t) = \frac{7}{3}e^{-t} - \frac{7}{4}e^{-2t} + \frac{1}{12}e^{2t}(12t - 7)$.
51. $y(t) = 2e^t + e^{-t} - e^t(\sin 2t + \cos 2t)$.
53. $x_1(t) = e^{2t}\cos t$, $x_2(t) = e^{2t}\sin t$.

Section 8.6

1. False
3. True

Problems

7. $f(t) = 3 - 4u_1(t)$.
9. $f(t) = 2 - u_2(t) - 2u_4(t)$.
11. $f(t) = t + 2(3 - t)u_3(t) - (6 - t)u_6(t)$.
13. $f(t) = 1 + (\sin t - 1)u_{\pi/2}(t) - (\sin t + 1)u_{3\pi/2}(t)$.

Section 8.7

1. False
3. True
5. True
7. False

Problems

1. $F(s) = \dfrac{1}{s^2}e^{-s}$.
3. $F(s) = \dfrac{e^{-\pi s/4}}{s^2 + 1}$.
5. $F(s) = \dfrac{2}{s^3}e^{-2s}$.
7. $F(s) = \left(\dfrac{s^2 + 2s + 2}{s^3}\right)e^{-2s}$.
9. $F(s) = \dfrac{3}{(s + 2)^2 + 9}e^{-s}$.
11. $f(t) = (t - 2)u_2(t)$.
13. $f(t) = e^{-4(t-3)}u_3(t)$.
15. $f(t) = u_3(t)\sin(t - 3)$.
17. $f(t) = \frac{1}{5}u_1(t)[e^{4(t-1)} - e^{-(t-1)}]$.
19. $f(t) = [\cos 3(t - 1) + 2\sin 3(t - 1)]u_1(t)$.
21. $f(t) = \frac{1}{2}u_2(t)(t - 2)^2 e^{3(t-2)}$.
23. $f(t) = e^{-2(t-1)}[2\cos(t - 1) - 5\sin(t - 1)]u_1(t)$.
25. $f(t) = [2e^{-(t-3)}(5t - 13) - 4\cos 2(t - 3) - 3\sin 2(t - 3)]u_3(t)$.
27. $y(t) = 2e^{2t} - u_2(t)[e^{(t-2)} - e^{2(t-2)}]$.
29. $y(t) = 3e^{-2t} + \frac{1}{4}u_\pi(t)\left[e^{-2(t-\pi)} - \cos 2t + \sin 2t\right]$.
31. $y(t) = \dfrac{1}{10}[21e^{3t} - \cos t - 3\sin t + \dfrac{u_{\frac{\pi}{2}}(t)}{3}[e^{3(t-\frac{\pi}{2})} - 10$
    $+ 9\sin t + 3\cos t]]$.
33. $y(t) = 2\cosh t + u_1(t)[\cosh(t - 1) - 1]$.
35. $y(t) = 2\sinh 2t + \frac{1}{4}u_1(t)[\cosh 2(t - 1)$
    $- 1] - \frac{1}{4}u_2(t)[\cosh 2(t - 2) - 1]$.
37. $y(t) = 2e^{-t} - e^{-2t} + u_{\pi/4}(t)[5e^{-(t-\pi/4)} - 2e^{-2(t-\pi/4)} - $
    $3\cos(t - \pi/4) + \sin(t - \pi/4)]$.
39. $y(t) = e^{-2t}(2\cos t + 5\sin t)$
    $+ u_3(t)\left\{1 - e^{-2(t-3)}[\cos(t - 3) - 2\sin(t - 3)]\right\}$.
41. $y(t) = 2e^{-t} + u_1(t)[e^{-(t-1)} + t - 2]$
    $- 2u_2(t)\left[e^{-(t-2)} + t - 3\right] + u_3(t)\left[e^{-(t-3)} + t - 4\right]$.
43. $y(t) = 3e^t - 1 - t - \frac{1}{2}u_1(t)[3e^{t-1} + e^{-(t-1)} - 2 - 2t]$.
45. $y(t) = 3e^t - 2 + 3u_1(t)(1 - e^{t-1})$.
47. $i(t) = \dfrac{20}{R}\left\{e^{-at} + u_{10}(t)\left[\dfrac{1}{a - 1}(ae^{-a(t-10)} - e^{-(t-10)})\right.\right.$
    $\left.\left. - e^{-at}\right]\right\}$, where $a = 1/RC$.

Section 8.8

True-False Review

1. True
3. False
5. False

Problems

1. $y(t) = e^{2t} + u_2(t)e^{2(t-2)}$.
3. $y(t) = \frac{1}{3}(e^{5t} - e^{-t}) + u_3(t)e^{5(t-3)}$.
5. $y(t) = \frac{1}{2}[\sinh 2t + u_3(t) \sinh 2(t-3)]$.
7. $y(t) =$

$$e^{2t}(3\cos 3t - 2\sin 3t) - \frac{\sqrt{2}}{6}e^{2(t-\pi/4)}(\sin 3t + \cos 3t)u_{\pi/4}(t).$$

9. $y(t) = 5e^{-3t}(\cos 2t + 2\sin 2t) - \frac{1}{2}e^{-3(t-\pi/4)}u_{\pi/4}(t)\cos 2t$.
11. $y(t) = \frac{4}{7}(\cos 3t - \cos 4t) + \frac{1}{4}u_{\pi/3}(t)\sin[4(t-\pi/3)]$.
13. $y(t) = \frac{F_0}{5}(\cos 2t - \cos 3t) - \frac{5}{2}u_1(t)\sin[2(t-1)]$.
15. $y(t) =$

$$\frac{F_0}{\omega_0(\omega^2 - \omega_0^2)}(\omega\sin\omega_0 t - \omega_0\sin\omega t) + \frac{A}{\omega_0}u_{t_0}(t)\sin[\omega_0(t-t_0)].$$

Section 8.9

True-False Review

1. True
3. False
5. False
7. True

Problems

1. $\frac{1}{2}t^2$.
3. $e^t - 1 - t$.
5. $e^t(1 - \cos t)$.
9. $\dfrac{1}{s^2(s^2+1)}$.
11. $\dfrac{s}{(s^2+1)(s^2+4)}$.
13. $\dfrac{4}{s^3[(s-3)^2+4]}$.
15. $1 - e^{-t}$.
17. $\frac{1}{3}e^{-2t}\sin 3t$.
19. $\begin{cases} 0, & \text{if } t < \pi, \\ t - \pi + \sin t, & \text{if } t \geq \pi. \end{cases}$
21. $\int_0^t e^{-(t+2\tau)}\tau\cos(t-\tau)\,d\tau$.
23. $\begin{cases} 0, & 0 \leq t < \pi/2, \\ \int_{\pi/2}^t e^{-4(t-\tau)}\cos[3(t-\tau)]\cos 4\tau\,d\tau, & t \geq \pi/2. \end{cases}$
25. $y(t) = \int_0^t e^{-\tau}\sin(t-\tau)\,d\tau + \sin t$.
27. $y(t) = \frac{1}{4}\int_0^t f(t-\tau)\sin 4\tau\,d\tau + \frac{1}{4}(4\alpha\cos 4t + \beta\sin 4t)$.

29. $y(t) = \frac{1}{a}\int_0^t f(t-\tau)\sinh a\tau\,d\tau + \frac{1}{a}(\alpha a\cosh at + \beta\sinh at)$.
31. $y(t) = \frac{1}{b}\int_0^t f(t-\tau)e^{a\tau}\sin b\tau\,d\tau + e^{at}\left[\alpha\cos bt + \frac{\beta - \alpha a}{b}\sin bt\right]$.
33. $x(t) = e^{3t} + e^t$.
35. $x(t) = 2\cosh t - 1$.
37. $x(t) = 2 + \frac{4}{\sqrt{3}}e^t\sin\sqrt{3}t$.

Section 8.10

Additional Problems

1. $F(s) = \frac{3}{s^2} - \frac{4}{s}$.
3. $F(s) = \frac{8}{s^3}$.
5. $F(s) = \frac{7}{(s+1)^2}$.
7. $F(s) = \frac{1}{2s} - \frac{s}{2(s^2+4a^2)}$.
9. $F(s) = \frac{(4s^2+5s+2)e^{-3s}}{s^3} + \frac{s+1}{s^2}$.
11. $F(s) = \frac{5s}{s^2+4} - \frac{7}{s+1} - \frac{360}{s^7}$.
13. $F(s) = \frac{s-3}{(s-3)^2+25} - \frac{2}{(s+1)^2+4}$.
15. $\sqrt{\dfrac{\pi}{s+5}}$.
17. $F(s) = \frac{2}{s}(1 - e^{-s}) + \frac{2e^{-(s+1)}}{s+1}$.
19. $F(s) = \frac{2}{s^3(s-1)}$.
21. $f(t) = 4\cos 3t + \frac{5}{3}\sin 3t$.
23. $f(t) = \frac{1}{8}[1 - \cos 4t]$.
25. $f(t) = \frac{1}{8}[2 + 3e^{-2t}\sin 4t - 2e^{-2t}\cos 4t]$.
27. $f(t) = 1 + u_{\ln 2}(t)[2e^{-t} - 1]$; $L[f] = \frac{1}{s} - \frac{1}{s(s+1)2^s}$.
35. $y(t) = \frac{13}{24}e^{4t} + \frac{13}{12}e^{-2t} - \frac{5}{8}$.
37. $y(t) = 1 + \sin t - \cos t - u_{\pi/2}(t)[1 - \cos(t - \pi/2)]$.
39. $y(t) = u_4(t)(t-4)e^{-(t-4)}$.
41. $x_1(t) = \frac{1}{4}e^{3t} + \frac{3}{4}e^{-t}$, $x_2(t) = \frac{1}{4}e^{3t} - \frac{3}{4}e^{-t}$.
43. $x_1(t) = e^{2t} - 4te^{2t}$, $x_2(t) = e^{2t} + 4te^{2t}$.
45. $x(t) = 2t + t^3/3$.
47. $x(t) = -2 + 4t^2 + 2\cos\sqrt{2}t$.

Chapter 9
Section 9.1

True-False Review

1. True
3. False
5. True
7. True
9. False

Problems

1. $R = 4$.
3. $R = \frac{1}{2}$.
5. $R = \infty$.
7. $R = 1$.
9. $R = \sqrt{5}$.
11. **(a)** Analytic for $x \neq \pm 1$.
    (b) $R = \begin{cases} 1 - |x_0|, & \text{if } |x_0| < 1, \\ |x_0| - 1, & \text{if } |x_0| > 1. \end{cases}$
13. $f(x) = a_0(1 + \frac{2}{3}x + \frac{1}{12}x^2)$.

Section 9.2

True-False Review

1. True
3. False
5. True
7. False
9. True

Problems

1. $y_1(x) = \sum_{n=0}^{\infty} \frac{1}{(2n)!} x^{2n}$, $y_2(x) = \sum_{n=0}^{\infty} \frac{1}{(2n+1)!} x^{2n+1}$, $R = \infty$.

3. $y_1(x) = 1 + \sum_{n=1}^{\infty} \frac{(-2)^n}{1 \cdot 3 \cdots (2n-1)} x^{2n}$,

   $y_2(x) = \sum_{n=0}^{\infty} \frac{(-1)^n}{n!} x^{2n+1}$, $R = \infty$.

5. $y_1(x) = \sum_{n=0}^{\infty} \frac{1}{3^n n!} x^{3n}$,

   $y_2(x) = \sum_{n=0}^{\infty} \frac{1}{1 \cdot 4 \cdots (3n+1)} x^{3n+1}$, $R = \infty$.

7. $y_1(x) = \sum_{n=0}^{\infty} (-1)^n \frac{1 \cdot 3 \cdots (2n+1)}{(2n)!} x^{2n}$,

   $y_2(x) = \sum_{n=0}^{\infty} (-2)^n \frac{(n+1)!}{(2n+1)!} x^{2n+1}$, $R = \infty$.

9. $y_1(x) = \sum_{n=0}^{\infty} (-1)^n x^{2n}$, $y_2(x) = \sum_{n=0}^{\infty} (-1)^n x^{2n+1}$, $R = 1$.

11. $y_1(x) = x(1 + x^2)$, $y_2(x) = 1 + 6x^2 + x^4$, $R = \infty$.
13. $y_1(x) = 1 - x^2 - \frac{1}{6}x^3 + \frac{1}{3}x^4 + \frac{11}{120}x^5 + \cdots$, $y_2(x)$
    $= 1 - \frac{1}{2}x^3 - \frac{1}{12}x^4 + \frac{1}{8}x^5 + \cdots$, $R = \infty$.

15. $y_1(x) = 1 + \frac{1}{2}x^2 + \frac{1}{6}x^3 + \frac{1}{12}x^4 + \frac{1}{24}x^5 + \cdots$,
    $y_2(x) = x + \frac{1}{6}x^3 + \frac{1}{12}x^4 + \frac{1}{30}x^5 + \cdots$, $R = \infty$.

17. **(a)** No.
    (b) $y_1(x) = 1 + \frac{1}{2}(x - 1)^2 + \frac{1}{8}(x - 1)^4 + \cdots$,
    $y_2(x) = (x - 1) + \frac{1}{3}(x - 1)^3 - \frac{1}{12}(x - 1)^4 + \cdots$, $R \geq 1$.

19. **(a)** $y(x) = \sum_{n=0}^{\infty} (-1)^n \frac{(n+1)!}{2^n (2n)!} x^{2n}$.

    (b) Polynomial approximation: $y_8(x) = 1 - \frac{1}{2}x^2 + \frac{1}{16}x^4 - \frac{1}{240}x^6 + \frac{1}{5376}x^8$, error $< 6.3 \times 10^{-6}$ on $[-1, 1]$.

21. $y(x) = a_0(1 + 2x^2 + \frac{1}{3}x^4 + \cdots) + a_1(x + \frac{1}{2}x^3 + \frac{1}{40}x^5 + \cdots) + (3x^2 + x^3 + \frac{13}{24}x^4 + \frac{1}{10}x^5 + \frac{1}{120}x^6 + \cdots)$.

Section 9.3

True-False Review

1. False
3. True
5. False

Problems

1. $\alpha = 3 : y_2(x) = a_1 x(1 - \frac{5}{3}x^2)$; $\alpha = 4 : y_1(x) = a_0(1 - 10x^2 + \frac{35}{3}x^4)$; $P_3(x) = -\frac{3}{2}x(1 - \frac{5}{3}x^2)$; $P_4(x) = \frac{3}{8}(1 - 10x^2 + \frac{35}{3}x^4)$.

5. $2x^3 + x^2 + 5 = \frac{16}{3}P_0 + \frac{6}{5}P_1 + \frac{2}{3}P_2 + \frac{4}{5}P_3$.

9. $\alpha = 0 : y_1(x) = 1$;
   $\alpha = 1 : y_2(x) = x$;
   $\alpha = 2 : y_1(x) = 1 - 2x^2$; $\alpha = 3 : y_2(x) = x(1 - \frac{2}{3}x^2)$.

Section 9.4

True-False Review

1. False
3. True
5. True

Problems

1. $x = 1$ is a regular singular point. All other points are ordinary points.
3. $x = 0, 2$ are regular singular points. All other points are ordinary points.
5. $r = \pm\sqrt{7}$.
7. $r = 0, 1$.
9. $y_1(x) = x^{1/4} \left\{ 1 + \sum_{n=1}^{\infty} \frac{(-1)^n}{n![5 \cdot 9 \cdots (4n+1)]} x^n \right\}$,

   $y_2(x) = 1 + \sum_{n=1}^{\infty} \frac{(-1)^n}{n![3 \cdot 7 \cdots (4n-1)]} x^n$.

11. $y_1(x) = x^{\sqrt{2}}\left[1+\displaystyle\sum_{n=1}^{\infty}\frac{1}{n!(1+2\sqrt{2})(2+2\sqrt{2})\cdots(n+2\sqrt{2})}x^n\right]$,

$y_2(x) = x^{-\sqrt{2}}\left[1+\displaystyle\sum_{n=1}^{\infty}\frac{1}{n!(1-2\sqrt{2})(2-2\sqrt{2})\cdots(n-2\sqrt{2})}x^n\right]$.

13. $y_1(x) = x^3\left[1+\displaystyle\sum_{n=1}^{\infty}\frac{(n+1)(n+2)}{10\cdot13\cdots(3n+7)}x^n\right]$,

$y_2(x) = x^{2/3}\left[1+\displaystyle\sum_{n=1}^{\infty}\frac{(3n-4)(3n-1)}{n!3^n}x^n\right]$.

15. $y_1(x) = x^{\sqrt{5}}\left[1+\displaystyle\sum_{n=1}^{\infty}\frac{(1+\sqrt{5})(2+\sqrt{5})\cdots(n+\sqrt{5})}{n!(1+2\sqrt{5})(2+2\sqrt{5})\cdots(n+2\sqrt{5})}x^n\right]$,

$y_2(x) = x^{-\sqrt{5}}\left[1+\displaystyle\sum_{n=1}^{\infty}\frac{(1-\sqrt{5})(2-\sqrt{5})\cdots(n-\sqrt{5})}{n!(1-2\sqrt{5})(2-2\sqrt{5})\cdots(n-2\sqrt{5})}x^n\right]$.

17. $y_1(x) = (1+\frac{1}{5}x-\frac{3}{100}x^2+\cdots)\cos(\ln x)$
$+\frac{1}{25}x(10+x+\cdots)\sin(\ln x)$,
$y_2(x) = (1+\frac{1}{5}x-\frac{3}{100}x^2+\cdots)\sin(\ln x)$
$-\frac{1}{25}x(10+x+\cdots)\cos(\ln x)$.

19. $y_1(x) = \sqrt{x}$, $y_2(x) = \sqrt{x}\left(\ln x+\displaystyle\sum_{n=1}^{\infty}\frac{1}{n\cdot n!}x^n\right)$.

Section 9.5

True-False Review

1. False
3. False
5. True

Problems

1. $y_1(x) = \sqrt{x}\displaystyle\sum_{n=0}^{\infty}a_nx^n$, $y_2(x) = y_1(x)\ln x+\sqrt{x}\displaystyle\sum_{n=1}^{\infty}b_nx^n$.

3. $y_1(x) = x^2\displaystyle\sum_{n=0}^{\infty}a_nx^n$, $y_2(x) = x^{-1}\displaystyle\sum_{n=0}^{\infty}b_nx^n$.

5. $y_1(x) = x\displaystyle\sum_{n=0}^{\infty}a_nx^n$, $y_2(x) = Ay_1(x)\ln x+x^{-1}\displaystyle\sum_{n=0}^{\infty}b_nx^n$, $A\neq$
0.

7. $y_1(x) = x^{-3}\displaystyle\sum_{n=0}^{\infty}a_nx^n$, $y_2(x) = y_1(x)\ln x+x^{-3}\displaystyle\sum_{n=1}^{\infty}b_nx^n$.

9.   **(b)** $y_1(x) = x^{-1}\displaystyle\sum_{n=0}^{\infty}a_nx^n$,

$y_2(x) = y_1(x)\ln x+x^{-1}\displaystyle\sum_{n=0}^{\infty}b_nx^n$.

(c) $y_1(x) = x^N\displaystyle\sum_{n=0}^{\infty}a_nx^n$,

$y_2(x) = Ay_1(x)\ln x+x^{-1}\displaystyle\sum_{n=0}^{\infty}b_nx^n$, where $A=0$ if and
only if $\gamma = 1, 0, -1, \ldots, 1-N$.

11. $y_1(x) = x^{-1}\displaystyle\sum_{n=0}^{\infty}\frac{1}{(n!)^2}x^n$,

$y_2(x) = y_1(x)\ln x-(2+\frac{3}{4}x+\frac{11}{108}x^2+\cdots)$.

13. $y_1(x) = x\displaystyle\sum_{n=0}^{\infty}\frac{1}{(n+2)!}x^n = x^{-1}(e^x-x-1)$,

$y_2(x) = x^{-1}(1+x)$.

15. $y_1(x) = \displaystyle\sum_{n=0}^{\infty}\frac{(2x)^n}{(n!)^2}$,

$y_2(x) = y_1(x)\ln x-(4x+3x^2+\frac{22}{27}x^3+\cdots)$.

17. $y_1(x) = x^2\displaystyle\sum_{n=0}^{\infty}\frac{(n+1)}{n!}x^n$,

$y_2(x) = y_1(x)\ln x-x^2(3x+\frac{13}{4}x^2+\frac{31}{18}x^3+\cdots)$.

19. $y_1(x) = x^2\left[1+\displaystyle\sum_{n=1}^{\infty}\frac{(n+4)}{n!}x^n\right]$,

$y_2(x) = 2y_1(x)\ln x+x^{-1}(1-x+\frac{3}{2}x^2-12x^4+\cdots)$.

21. $y_1(x) = x^{3/2}\displaystyle\sum_{n=0}^{\infty}\frac{1}{(n+3)!}x^n = x^{-3/2}(e^x-1-x-\frac{1}{2}x^2)$,

$y_2(x) = x^{-3/2}(1+x+\frac{1}{2}x^2)$.

23. $y_1(x) = x$, $y_2(x) = y_1(x)\ln x+\displaystyle\sum_{n=1}^{\infty}\frac{(-1)^n}{n\cdot n!}x^{n+1}$.

25. $y_1(x) = x^{3/2}\displaystyle\sum_{n=0}^{\infty}\frac{1}{n!(n+2)!}x^n$,

$y_2(x) = 2y_1(x)\ln x-5x^{-1/2}\left(1-x+\frac{1}{5}x^3+\frac{37}{960}x^4+\cdots\right)$.

27. $y_1(x) = x$, $y_2(x) = x\ln x-1+\displaystyle\sum_{n=2}^{\infty}\frac{1}{n!(n-1)}x^n$.

29. $y(x) = 1+\displaystyle\sum_{k=1}^{N}\frac{(-1)^kN(N-1)\cdots(N-k+1)}{(k!)^2}x^k$.

Section 9.6

True-False Review

1. False
3. False
5. False
7. False

Problems

7. $J_{3/2}(x) = \sqrt{\dfrac{2}{\pi}}x^{-3/2}(\sin x-x\cos x)$,

$J_{-3/2}(x) = -\sqrt{\dfrac{2}{\pi}}x^{-3/2}(x\sin x+\cos x)$.

13. $x^p = \displaystyle\sum_{n=1}^{\infty}\frac{2}{\lambda_nJ_{p+1}(\lambda_n)}J_p(\lambda_nx)$.

Section 9.7

Additional Problems

1. Ordinary point.

$$y_1(x) = 1 + \sum_{k=1}^{\infty} \frac{(-1)^k (3k-2)(3k-5)\cdots 7 \cdot 4 \cdot 1}{(3k)!} x^{3k},$$

$$y_2(x) = x + \sum_{k=1}^{\infty} \frac{(-1)^k (3k-1)(3k-4)\cdots 5 \cdot 2}{(3k+1)!} x^{3k+1}.$$

Solutions valid on $(-\infty, \infty)$.

3. Ordinary point.

$$y_1(x) = 1 + 2x^2 + 3x^4 + 4x^5 + \cdots,$$

$$y_2(x) = x + \frac{5}{3}x^3 + \frac{7}{3}x^5 + \frac{9}{3}x^7 + \cdots.$$

Solutions valid on $(-1, 1)$.

5. Regular singular point.

$$y_1(x) = 1 - \frac{1}{3!}x^2 + \frac{1}{5!}x^4 - \frac{1}{7!}x^6 + \cdots,$$

$$y_2(x) = x^{-1}\left[1 - \frac{1}{2}x^2 + \frac{1}{4!}x^4 + \cdots\right].$$

Solutions valid on $(0, \infty)$.

7. $y_1(x) = \sum_{n=0}^{\infty} \frac{(-1)^n}{2^2 \cdot 4^2 \cdots (2n)^2} x^{2n}, \quad y_2(x) = y_1(x)\ln x +$
$\left[\frac{1}{4}x^2 - \frac{3}{128}x^4 + \cdots\right]$

9. $y_1(x) = J_{\frac{1}{2}}(x) = \sqrt{\dfrac{2}{\pi x}} \sin x, \quad y_2(x) = J_{-\frac{1}{2}}(x) = \sqrt{\dfrac{2}{\pi x}} \cos x.$

11. $y_1(x) = x^{-1}\left[1 - x - \frac{1}{2}x^2 - \frac{1}{18}x^3 - \frac{1}{360}x^4 - \frac{1}{12600}x^5 + \cdots\right]$
$y_2(x) = x^{1/2}\left[1 + \frac{1}{5}x + \frac{1}{70}x^2 + \frac{1}{1890}x^3 + \frac{1}{83160}x^4 \right.$
$\left. + \frac{1}{5405400}x^5 + \cdots\right]$

13. $y_1(x) = x^2\left[1 - 4x + 4x^2 - \frac{16}{9}x^3 + \frac{4}{9}x^4 - \frac{16}{225}x^5 + \cdots\right]$
$y_2(x) = y_1(x)\ln x + x^2\left[8x - 12x^2 + \frac{176}{27}x^3 - \frac{50}{27}x^4\right.$
$\left. + \frac{1096}{3375}x^5 + \cdots\right]$

15. **(a)** $y_1(x) = 1 + \dfrac{ab}{2!}x^2 + \dfrac{ab(2-a)(2-b)}{4!}x^4 + \cdots,$
$y_2(x) = x + \dfrac{(1-a)(1-b)}{3!}x^3$
$+ \dfrac{(1-a)(1-b)(3-a)(3-b)}{5!}x^5 + \cdots.$
(d) $y_1(x) = 1 + 21x^2 + 35x^4, \ y_2(x) = x + 5x^3 + 15x^5.$

17. **(c)** $N = 0: \ y(x) = 1;$
$N = 1: \ y(x) = x^{-1}(1-x);$
$N = 2: \ y(x) = x^{-1}[1 - 2x + \frac{1}{2}x^2];$
$N = 3: \ y(x) = x^{-1}[1 - 3x + \frac{3}{2}x^2 - \frac{1}{6}x^3].$

Appendix A

1. $\bar{z} = 2 - 5i, \ |z| = \sqrt{29}.$

3. $\bar{z} = 5 + 2i, \ |z| = \sqrt{29}.$

5. $\bar{z} = 1 - 2i, \ |z| = \sqrt{5}.$

7. $z_1 z_2 = 1 + 7i, \ \dfrac{z_1}{z_2} = -1 + i.$

9. $z_1 z_2 = 7 + 11i, \ \dfrac{z_1}{z_2} = \dfrac{1}{10}(1 - 13i).$

Appendix B

1. $\dfrac{5}{x+2} - \dfrac{3}{x+1}.$

3. $\dfrac{4}{5(x-3)} + \dfrac{1}{5(x+2)}.$

5. $\dfrac{1}{3(x+1)} + \dfrac{1}{6(x-2)} - \dfrac{1}{2(x+4)}.$

7. $\dfrac{3}{x+1} - \dfrac{1}{(x+1)^2} - \dfrac{3}{x+2}.$

9. $\dfrac{3}{4x} + \dfrac{1}{x^2} - \dfrac{3x+4}{4(x^2+4)}.$

11. $\dfrac{1}{2(x-2)} + \dfrac{x+2}{2(x^2+16)}.$

13. $\dfrac{1}{x-2} - \dfrac{1}{x+2} + \dfrac{3}{(x+2)^2}.$

15. $\dfrac{2}{3(x-2)} - \dfrac{2}{3(x+1)} - \dfrac{2}{(x+1)^2} + \dfrac{1}{(x+1)^3}.$

17. $\dfrac{1}{2(x-3)} - \dfrac{x+1}{2(x^2+4x+5)}.$

Appendix C

Note that we have omitted the integration constants.

1. $\cos x + x \sin x.$

3. $x \ln x - x.$

5. $\frac{1}{2}e^{x^2}(x^2 - 1).$

7. $x - 3 \ln |x + 2|.$

9. $\frac{1}{4} \ln |x| - \frac{1}{8} \ln(x^2 + 4) + \tan^{-1}\dfrac{x}{2}.$

11. $\frac{1}{2}x + \frac{7}{4} \ln |2x - 1|.$

13. $2 \ln |x| - \dfrac{1}{x+1} - 2 \ln |x + 1|.$

15. $\tan^{-1}(x + 1).$

17. $-\ln |\cos x|.$

19. $\frac{1}{4}(2x + \sin 2x).$

21. $\frac{1}{13}e^{3x}(3 \sin 2x - 2 \cos 2x).$

Index

A

Absolute value
  around a matrix, 193–194
  of a complex number, 740

Addition
  associativity of, 236, 241
  closure under, 241, 251
  commutativity of, 236, 241
  in $\mathbb{R}^n$, 237–238
  of geometric vectors, 235–236
  of linear transformations, 353
  of matrices, 119

Additive inverse
  existence of, 241
  in $\mathbb{R}^n$, 236–238
  properties of, 246

Adjoint method, 217–219, 228, 231
  for computing matrix inverse, 218

Algebra of power series, 674–675

Alternating current (AC), 62
  phase-amplitude form of a solution, 63
  steady-state, 63
  transient, 63

Amplitude, 488
  of a steady-state current, 501

Analytic functions, 676–678

Angle between vectors
  in $\mathbb{R}^3$, 314
  in a real inner product space, 318

Annihilators, 470–480, 530

Anti-symmetric matrix, 116

Applied Linear Algebra (Noble/Daniel), 179

Archimedes' principle, 496*fn*

Area of a parallelogram, 195–196

Associated homogeneous differential equation, 456, 460, 470, 503–504, 508, 510, 525

Augmented matrix, 135–136, 184

Autonomous system, 596–605

Auxiliary equation, 460, 464

Auxiliary polynomial, 460

B

Back substitution, 142

Basis, 281–290, 336
  change of, 293–299
  defined, 282
  of eigenvectors, 404
  ordered, 294
  orthogonal, 325–327
  orthonormal, 325–327
  standard basis, 282–284

Bernoulli equations, 73–74

Bessel
  equation of order p, 722–730, 735
  Fourier-Bessel expansion, 731–732
  function expansion theorem, 729–731
  function of the first kind, 723–726
    properties of, 727–728
  function of the second kind, 726–727

Bound parameters, 153

Bound variables, 153

Boundary-value problem, 20

C

Capacitance, 60

Capacitor, 60

Carrying capacity, 43

CAS, *See* Computer algebra systems (CAS)

Catenary, 104

Cauchy-Euler equation, 513–522, 531, 700–701

Cauchy-Schwarz inequality, 317

Center, 603, 609

E

THE METHOD OF UNDETERMINED COEFFICIENTS

The following table lists trial solutions for the differential equation $P(D)y = F(x)$, where $P(D)$ is a polynomial differential operator (see Section 6.3).

$F(x)$	Usual trial solution	Modified trial solution
$cx^k e^{ax}$	If $P(a) \neq 0$: $$y_p(x) = e^{ax}(A_0 + A_1 x + \cdots + A_k x^k)$$	If a is a root of $P(r) = 0$ of multiplicity m: $$y_p(x) = x^m e^{ax}(A_0 + A_1 x + \cdots + A_k x^k)$$
$cx^k e^{ax} \cos bx$ or $cx^k e^{ax} \sin bx$	If $P(a + ib) \neq 0$: $$y_p(x) = e^{ax}\left[\sum_{i=0}^{k} x^i (A_i \cos bx + B_i \sin bx)\right]$$	If $a + ib$ is a root of $P(r) = 0$ of multiplicity m: $$y_p(x) = x^m e^{ax}\left[\sum_{i=0}^{k} x^i (A_0 \cos bx + B_0 \sin bx)\right]$$

VARIATION-OF-PARAMETERS

Consider $y'' + a_1(x)y' + a_2(x)y = F(x)$, where a_1, a_2, F are continuous. If y_1 and y_2 are linearly independent solutions to $y'' + a_1(x)y' + a_2(x)y = 0$, then a particular solution to the nonhomogeneous differential equation is $y_p = u_1 y_1 + u_2 y_2$, where $u_1(x) = -\int \dfrac{y_2 F}{y_1 y_2' - y_1' y_2}\,dx$ and $u_2(x) = \int \dfrac{y_1 F}{y_1 y_2' - y_1' y_2}\,dx$. Section 6.7 contains a generalization to differential equations of order > 2.

A SHORT TABLE OF LAPLACE TRANSFORMS

Function $f(t)$	Laplace Transform $F(s)$
$f(t) = t^n$ (n = nonnegative integer)	$F(s) = \dfrac{n!}{s^{n+1}}, \quad s > 0$
$f(t) = e^{at}$ (a = constant)	$F(s) = \dfrac{1}{s - a}, \quad s > a$
$f(t) = \sin bt$ (b = constant)	$F(s) = \dfrac{b}{s^2 + b^2}, \quad s > 0$
$f(t) = \cos bt$ (b = constant)	$F(s) = \dfrac{s}{s^2 + b^2}, \quad s > 0$
$f(t) = t^{-1/2}$	$F(s) = \sqrt{\pi/s}, \quad s > 0$
$f(t) = u_a(t)$	$F(s) = \dfrac{1}{s}e^{-as}$
$f(t) = \delta(t - a)$	$F(s) = e^{-as}$

Transform of Derivatives (Section 8.4)

f' $\qquad L[f'] = sL[f] - f(0)$

f'' $\qquad L[f''] = s^2 L[f] - sf(0) - f'(0)$

Shifting Theorems (Sections 8.5 and 8.7)

$e^{at} f(t)$ $\qquad F(s - a)$

$u_a(t) f(t - a)$ $\qquad e^{-as} F(s)$

MATRICES AND SYSTEMS OF LINEAR EQUATIONS

A real linear system of m equations in n unknowns can be written $A\mathbf{x} = \mathbf{b}$, where

$$A \text{ is an } m \times n \text{ matrix}, \quad \mathbf{x} \text{ is in } \mathbb{R}^n, \quad \mathbf{b} \text{ is in } \mathbb{R}^m.$$

$$\text{rank}(A) = \text{number of nonzero rows in any row-echelon form of } A.$$

If $r = \text{rank}(A)$ and $r^{\#} = \text{rank}(A^{\#})$, where $A^{\#}$ is the augmented matrix $[A \ \mathbf{b}]$, then (see Theorem 2.5.9)

1. If $r < r^{\#}$, the system $A\mathbf{x} = \mathbf{b}$ is inconsistent.

2. If $r = r^{\#}$, the system is consistent and:

   (a) There exists a unique solution if and only if $r^{\#} = n$.
   (b) There exists an infinite number of solutions if and only if $r^{\#} < n$.

INVERTIBLE MATRIX THEOREM (Theorems 2.8.1, 3.2.4, 4.10.1, and 5.4.20)

Let A be an $n \times n$ matrix with real elements. The following conditions are equivalent:

   (a) A is invertible (i.e., there exists a matrix A^{-1} with $AA^{-1} = A^{-1}A = I_n$).

   (b) The equation $A\mathbf{x} = \mathbf{b}$ has a unique solution for every $\mathbf{b}$ in $\mathbb{R}^n$.

   (c) The equation $A\mathbf{x} = \mathbf{0}$ has only the trivial solution $\mathbf{x} = \mathbf{0}$.

   (d) $\text{rank}(A) = n$.

   (e) A can be expressed as a product of elementary matrices.

   (f) A is row-equivalent to I_n.

   (g) $\det(A) \neq 0$.

   (h) $\text{nullity}(A) = 0$.

   (i) $\text{nullspace}(A) = \{\mathbf{0}\}$.

   (j) The columns of A form a linearly independent set of vectors in $\mathbb{R}^n$.

   (k) $\text{colspace}(A) = \mathbb{R}^n$ (that is, the columns of A span $\mathbb{R}^n$).

   (l) The columns of A form a basis for $\mathbb{R}^n$.

   (m) The rows of A form a linearly independent set of vectors in $\mathbb{R}^n$.

   (n) $\text{rowspace}(A) = \mathbb{R}^n$ (that is, the rows of A span $\mathbb{R}^n$).

   (o) The rows of A form a basis for $\mathbb{R}^n$.

   (p) A^T is invertible.

   (q) The transformation $T : \mathbb{R}^n \to \mathbb{R}^n$ given by $T(\mathbf{x}) = A\mathbf{x}$ is one-to-one (i.e. $\text{Ker}(T) = \{\mathbf{0}\}$).

   (r) The transformation in (q) is onto (i.e. $\text{Rng}(T) = \mathbb{R}^n$).

   (s) The transformation in (q) is an isomorphism.

- If these conditions hold, the inverse of A is given by the formula $A^{-1} = \dfrac{1}{\det(A)} \text{adj}(A)$, where $\text{adj}(A)$ denotes the adjoint of A. (Theorem 3.3.16)

- **(Cramer's Rule)** If these conditions hold, the unique solution $\mathbf{x}$ in (b) is $(x_1, x_2, \ldots, x_n)$, where $x_k = \dfrac{\det(B_k)}{\det(A)}$ $(k = 1, 2, \ldots, n)$, and B_k denotes the matrix obtained by replacing the kth column vector of A by $\mathbf{b}$. (Theorem 3.3.19)